# ONE WEEK LOAN

23 JAN 06

# LIST OF EXAMPLES IN TEXT

## CHAPTER 2

2.1 Construction of a true-stress–true-strain curve    34
2.2 Calculation of ultimate tensile strength    36
2.3 Instability in torsion of a thin-walled tube    47
2.4 Calculation of the modulus of resilience    53
2.5 Elimination of residual stresses by tension    59
2.6 Yielding of a thin-walled shell    62
2.7 Correction factor for true-stress–true-strain curves    63
2.8 Expansion of a thin-walled spherical shell    70
2.9 Temperature rise in compression    72

## CHAPTER 4

4.1 Determining the coefficient of friction    139
4.2 Adhesive wear in sliding    142

## CHAPTER 5

5.1 Determining the number of phases in carbon steel    194
5.2 Solidification times for various solid shapes    204
5.3 Investment-cast superalloy components for gas turbines    226

## CHAPTER 6

6.1 Upsetting force    268
6.2 Transition from sliding to sticking friction    268
6.3 Back tension required to cause roll slip    288
6.4 Power required in rolling    290
6.5 Strain rate in extrusion    304
6.6 Force in hot extrusion    306
6.7 Power and die pressure in rod drawing    315
6.8 Maximum reduction per pass for a strain-hardening material    317

## CHAPTER 7

7.1 Calculation of maximum punch force    344
7.2 Use of tailor-welded blanks in the automotive industry    348
7.3 Estimating springback    355
7.4 Work done in stretch forming    361
7.5 Peak pressure in explosive forming    370
7.6 Estimating the limiting drawing ratio    381
7.7 Theoretical limiting drawing ratio    382
7.8 Estimating cup diameter and earing    382
7.9 Estimating diameter of expansion    389
7.10 Strains in the sheet-metal body of an automobile    390

## CHAPTER 8

8.1 Relative energies in cutting    422
8.2 Comparison of forming and machining energies    422
8.3 Increasing tool life by reducing the cutting speed    431
8.4 Effect of cutting fluids on machining    453
8.5 Material removal rate and cutting force in turning    459
8.6 Typical parts made on CNC turning machine tools    462
8.7 Calculation of material removal rate, power required, and cutting time in face milling    473
8.8 Machining outer bearing races on a turning center    485

## CHAPTER 9

9.1 Chip dimensions in grinding    513
9.2 Forces in surface grinding    515
9.3 Action of a grinding wheel    520
9.4 Grinding vs. hard turning    527

9.5   Machining time in electrochemical machining vs. drilling   540
9.6   Machining time in electrical-discharge machining vs. drilling   543

**CHAPTER 10**
10.1   Degree of polymerization in polyvinyl chloride (PVC)   564
10.2   Lowering the viscosity of a polymer   574
10.3   Stress relaxation in a thermoplastic member under tension   575
10.4   Properties of a graphite–epoxy-reinforced plastic   589
10.5   Blown film   594
10.6   Injection molding of gears   598
10.7   Tennis rackets made of composite materials   607
10.8   Rapid prototyping of an injection manifold   618

**CHAPTER 11**
11.1   Particle shape-factor determination   634
11.2   Density of metal-powder–lubricant mix   638
11.3   Pressure decay in compaction   640
11.4   Shrinkage in sintering   646
11.5   Production of automotive-engine main bearing caps by powder metallurgy   648
11.6   Production of tungsten carbide for tools and dies   650
11.7   Effect of porosity on properties   663
11.8   Dimensional changes during shaping of ceramic components   668
11.9   Metal-matrix-composite brake rotors and cylinder liners   677

**CHAPTER 12**
12.1   Laser-beam welding of razor blades   700
12.2   Roll bonding of the U.S. quarter   713
12.3   Heat generated in resistance spot welding   718
12.4   Weld design selection   741

**CHAPTER 13**
13.1   Processing of a $p$-type region in $n$-type silicon   772
13.2   Comparison of wet and dry etching   775
13.3   Surface micromachining of a hinge for a mirror actuation system   790
13.4   Operation and fabrication sequence for a thermal ink-jet printer   792
13.5   Production of rare-earth magnets   797

**CHAPTER 15**
15.1   Simulation of plant-scale manufacturing   865
15.2   Manufacturing cells in small shops   874

**CHAPTER 16**
16.1   An example of DFMA application   897
16.2   Sustainable manufacturing in the production of Nike athletic shoes   900
16.3   Changes made in materials between C-5A and C-5B military cargo aircraft   906
16.5   Manufacturing a sheet-metal part by various methods   914

**CASE STUDIES**
Permanent-mold casting of aluminum automotive pistons   250
Manufacturing solid rocket-motor case segments for the Space Shuttle   325
Can manufacturing   394
High-Speed Dry Machining of Cast-Iron Engine Blocks   493
Manufacture of Stents   551
The EPOCH hip   622
Production of High-Temperature Superconducting Tapes   679
Light-Curing Acrylic Adhesives for Medical Products   744
Accelerometer for Automotive Air Bags   800
Development of a Modular Fixture   845
CAD Model Development for Automotive Components   886
Concurrent Engineering for Intravenous Solution Containers   920

# Manufacturing Processes
# for Engineering Materials

To grandson
**Bennett Matthew Kalpakjian**
and
**Shelly Petronis**

# Manufacturing Processes for Engineering Materials

FOURTH EDITION

## Serope Kalpakjian
*Illinois Institute of Technology, Chicago, Illinois*

## Steven R. Schmid
*University of Notre Dame, Notre Dame, Indiana*

Prentice Hall

Pearson Education International

Vice President and Editorial Director, ECS: *Marcia J. Horton*
Acquisitions Editor: *Dorothy Marrero*
Editorial Assistant: *Brian Hoehl*
Vice President and Director of Production and Manufacturing, ESM: *David W. Riccardi*
Executive Managing Editor: *Vince O'Brien*
Managing Editor: *David A. George*
Production Editor: *Tamar Savir*
Director of Creative Services: *Paul Belfanti*
Creative Director: *Carole Anson*
Art Editor: *Xiaohong Zhu*
Cover Designer: *Geoffrey Cassar*
Manufacturing Manager: *Trudy Pisciotti*
Manufacturing Buyer: *Lisa McDowell*
Marketing Manager: *Holly Stark*

© 2003 by Pearson Education, Inc.
Pearson Education, Inc.
Upper Saddle River, NJ 07458

Printed in the United States of America.

10   9   8   7   6   5   4   3

ISBN 0-13-045373-0

Pearson Education Ltd., *London*
Pearson Education Australia Pty. Ltd., *Sydney*
Pearson Education Singapore, Pte. Ltd.
Pearson Education North Asia Ltd., *Hong Kong*
Pearson Education Canada, Inc., *Toronto*
Pearson Educación de Mexico, S.A. de C.V.
Pearson Education—Japan, *Tokyo*
Pearson Education Malaysia, Pte. Ltd.
Pearson Education, Inc., *Upper Saddle River*, *New Jersey*

# CONTENTS

Preface    xiii

## 1  Introduction    1

1.1   What Is Manufacturing?    1
1.2   Product Design and Concurrent Engineering    6
1.3   Design for Manufacture, Assembly, Disassembly, and Service    8
1.4   Environmentally Conscious Design and Manufacturing    10
1.5   Selecting Materials    11
1.6   Selecting Manufacturing Processes    14
1.7   Computer-Integrated Manufacturing    17
1.8   Quality Assurance and Total Quality Management    20
1.9   Product Liability    21
1.10  Manufacturing Costs, Lean Production, and Agile Manufacturing    22
1.11  General Trends in Manufacturing    23
1.12  Responsibilities of Manufacturing Engineers    24
      Summary    24

## 2  Fundamentals of the Mechanical Behavior of Materials    26

2.1   Introduction    26
2.2   Tension    27
2.3   Compression    43
2.4   Torsion    46
2.5   Bending    48
2.6   Hardness    49
2.7   Fatigue    54
2.8   Creep    55
2.9   Impact    56
2.10  Residual Stresses    57
2.11  Triaxial Stresses and Yield Criteria    60
2.12  Work of Deformation    68
      Summary    72
      Summary of Equations    74
      Bibliography    75
      Questions    75
      Problems    77

# 3    Structure and Manufacturing Properties of Metals    81

3.1    Introduction    81
3.2    The Crystal Structure of Metals    82
3.3    Deformation and Strength of Single Crystals    83
3.4    Grains and Grain Boundaries    89
3.5    Plastic Deformation of Polycrystalline Metals    92
3.6    Recovery, Recrystallization, and Grain Growth    93
3.7    Cold, Warm, and Hot Working    95
3.8    Failure and Fracture    96
3.9    Physical Properties    103
3.10    Properties and Applications of Ferrous Alloys    108
3.11    Properties and Applications of Nonferrous Metals and Alloys    112
        Summary    124
        Summary of Equations    125
        Bibliography    125
        Questions    126
        Problems    127

# 4    Surfaces, Tribology, Dimensional Characteristics, Inspection, and Product Quality Assurance    128

4.1    Introduction    128
4.2    Surface Structure and Properties    129
4.3    Surface Texture    130
4.4    Tribology: Friction, Wear, and Lubrication    134
4.5    Surface Treatments, Coatings, and Cleaning    151
4.6    Engineering Metrology and Instrumentation    160
4.7    Dimensional Tolerances    165
4.8    Testing and Inspection    167
4.9    Quality Assurance    170
        Summary    180
        Summary of Equations    181
        Bibliography    182
        Questions    183
        Problems    184
        Design    185

# 5    Metal-Casting Processes and Equipment; Heat Treatment    186

5.1    Introduction    186
5.2    Solidification of Metals    187
5.3    Cast Structures    195
5.4    Fluid Flow and Heat Transfer    199

5.5  Melting Practice and Furnaces  207
5.6  Casting Alloys  209
5.7  Ingot Casting and Continuous Casting  215
5.8  Casting Processes: Expendable Mold, Permanent Pattern  218
5.9  Casting Processes: Expendable Mold, Expendable Pattern  224
5.10  Casting Processes: Permanent Mold  227
5.11  Heat Treatment  235
5.12  Design Considerations  245
5.13  Economics of Casting  249
CASE STUDY: Permanent-Mold Casting of Aluminum
                Automotive Pistons  250
Summary  252
Summary of Equations  253
Bibliography  254
Questions  254
Problems  256
Design  258

## 6  Bulk Deformation Processes  260

6.1  Introduction  260
6.2  Forging  260
6.3  Rolling  282
6.4  Extrusion  300
6.5  Rod, Wire, and Tube Drawing  312
6.6  Swaging  321
6.7  Die Manufacturing Methods  322
6.8  Die Failures  324
CASE STUDY: Manufacturing Solid Rocket-Motor Case Segments
                for the Space Shuttle  325
Summary  327
Summary of Equations  328
Bibliography  329
Questions  330
Problems  332
Design  335

## 7  Sheet-Metal Forming Processes  336

7.1  Introduction  336
7.2  Sheet-Metal Characteristics  336
7.3  Shearing  341
7.4  Bending of Flat Sheet and Plate  349
7.5  Stretch Forming  360
7.6  Bulging  362
7.7  Rubber Forming and Hydroforming  363
7.8  Spinning  364
7.9  High-Energy-Rate Forming  368
7.10  Superplastic Forming  371
7.11  Various Forming Methods  372

7.12   Deep Drawing   374
7.13   Formability of Sheet Metals   385
7.14   Equipment for Sheet-Metal Forming   391
7.15   Design Considerations   392
7.16   Economics of Sheet-Metal Forming   394
       CASE STUDY: Can Manufacturing   394
       Summary   397
       Summary of Equations   399
       Bibliography   399
       Questions   400
       Problems   401
       Design   403

## 8    Material-Removal Processes: Cutting   404

8.1    Introduction   404
8.2    Mechanics of Chip Formation   405
8.3    Tool Wear and Failure   426
8.4    Surface Finish and Integrity   434
8.5    Machinability   437
8.6    Cutting-Tool Materials   440
8.7    Cutting Fluids   451
8.8    Cutting Processes and Machine Tools for Producing Round Shapes   454
8.9    Cutting Processes and Machine Tools for Producing Various Shapes   468
8.10   Machining and Turning Centers   482
8.11   Vibration and Chatter   486
8.12   Machine-Tool Structures   488
8.13   Design Considerations   489
8.14   Economics of Machining   490
       CASE STUDY: High-Speed Dry Machining of Cast-Iron Engine Blocks   493
       Summary   494
       Summary of Equations   496
       Bibliography   497
       Questions   498
       Problems   501
       Design   502

## 9    Material Removal Processes: Abrasive, Chemical, Electrical, and High-Energy Beams   504

9.1    Introduction   504
9.2    Abrasives   505
9.3    Bonded Abrasives   506
9.4    Mechanics of Grinding   510

9.5    Grinding-Wheel Wear    517
9.6    Grinding Operations and Machines    521
9.7    Ultrasonic Machining    527
9.8    Finishing Operations    528
9.9    Deburring    532
9.10   Chemical Machining    533
9.11   Electrochemical Machining    537
9.12   Electrochemical Grinding    539
9.13   Electrical-Discharge Machining    540
9.14   High-Energy-Beam Machining    544
9.15   Water-Jet, Abrasive Water-Jet, and Abrasive-Jet
       Machining    547
9.16   Design Considerations    549
9.17   Process Economics    549
       CASE STUDY: Manufacture of Stents    551
       Summary    553
       Summary of Equations    554
       Bibliography    554
       Questions    555
       Problems    557
       Design    558

# 10 Properties and Processing of Polymers and Reinforced Plastics; Rapid Prototyping and Rapid Tooling    559

10.1   Introduction    559
10.2   The Structure of Polymers    560
10.3   Thermoplastics: Behavior and Properties    568
10.4   Thermosets: Behavior and Properties    575
10.5   Thermoplastics: General Characteristics and Applications    576
10.6   Thermosets: General Characteristics and Applications    578
10.7   High-Temperature Polymers, Electrically Conducting Polymers,
       and Biodegradable Plastics    579
10.8   Elastomers (Rubbers): General Characteristics
       and Applications    581
10.9   Reinforced Plastics    582
10.10  Processing of Plastics    590
10.11  Processing of Polymer-Matrix-Reinforced Plastics    605
10.12  Rapid Prototyping and Rapid Tooling    610
10.13  Design Considerations    619
10.14  Economics of Processing Plastics    621
       CASE STUDY: The EPOCH Hip    622
       Summary    623
       Summary of Equations    624
       Bibliography    625
       Questions    626
       Problems    628
       Design    629

# 11 Properties and Processing of Metal Powders, Ceramics, Glasses, Composites, and Superconductors    630

11.1   Introduction    630
11.2   Production of Metal Powders    631
11.3   Compaction of Metal Powders    636
11.4   Sintering    644
11.5   Secondary and Finishing Operations    649
11.6   Design Considerations for Powder Metallurgy    651
11.7   Economics of Powder Metallurgy    653
11.8   Ceramics: Structure, Properties, and Applications    655
11.9   Shaping Ceramics    663
11.10  Glasses: Structure, Properties, and Applications    668
11.11  Forming and Shaping Glass    671
11.12  Design Considerations for Ceramics and Glasses    675
11.13  Graphite and Diamond    675
11.14  Processing Metal-Matrix and Ceramic-Matrix Composites    676
11.15  Processing Superconductors    679
       CASE STUDY: Production of High-Temperature Superconducting
                   Tapes    679
       Summary    682
       Summary of Equations    683
       Bibliography    683
       Questions    685
       Problems    686
       Design    687

# 12 Joining and Fastening Processes    688

12.1   Introduction    688
12.2   Arc-Welding Processes: Consumable Electrode    690
12.3   Arc-Welding Processes: Nonconsumable Electrode    697
12.4   High-Energy-Beam Welding    699
12.5   The Fusion-Welded Joint    701
12.6   Cold Welding    712
12.7   Ultrasonic Welding    713
12.8   Friction Welding    714
12.9   Resistance Welding    716
12.10  Explosion Welding    721
12.11  Diffusion Bonding    722
12.12  Brazing and Soldering    723
12.13  Adhesive Bonding    730
12.14  Mechanical Fastening    734
12.15  Joining Plastics, Ceramics, and Glasses    737
12.16  Design Considerations in Joining    740
12.17  Economic Considerations    743
       CASE STUDY: Light-Curing Acrylic Adhesives for Medical
                   Products    744

Summary   746
Bibliography   747
Questions   749
Problems   751
Design   752

## 13 Fabrication of Microelectronic and Micromechanical Devices    753

13.1    Introduction   753
13.2    Semiconductors and Silicon   756
13.3    Crystal Growing and Wafer Preparation   757
13.4    Films and Film Deposition   759
13.5    Oxidation   761
13.6    Lithography   762
13.7    Diffusion and Ion Implantation   765
13.8    Etching   766
13.9    Metallization and Testing   776
13.10   Wire Bonding and Packaging   778
13.11   Yield and Reliability of Chips   781
13.12   Printed Circuit Boards   782
13.13   Micromachining of MEMS Devices   785
13.14   The LIGA Microfabrication Process   794
13.15   Solid Free-Form Fabrication of Devices   799
        CASE STUDY: Accelerometer for Automotive Air Bags   800
        Summary   803
        Bibliography   804
        Questions   805
        Problems   806
        Design   806

## 14 Automation of Manufacturing Processes and Systems    808

14.1    Introduction   808
14.2    Automation   809
14.3    Numerical Control   815
14.4    Programming for Numerical Control   821
14.5    Adaptive Control   822
14.6    Material Handling and Movement   825
14.7    Industrial Robots   827
14.8    Sensor Technology   833
14.9    Flexible Fixturing   837
14.10   Assembly, Disassembly, and Service   839
14.11   Design Considerations   842
14.12   Economic Considerations   845
        CASE STUDY: Development of a Modular Fixture   845
        Summary   847
        Bibliography   848
        Questions   849
        Design   850

# 15 Computer-Integrated Manufacturing Systems 852

15.1 Introduction 852
15.2 Manufacturing Systems 853
15.3 Computer-Integrated Manufacturing 854
15.4 Computer-Aided Design and Engineering 856
15.5 Computer-Aided Manufacturing 861
15.6 Computer-Aided Process Planning 861
15.7 Computer Simulation of Manufacturing Processes and Systems 864
15.8 Group Technology 866
15.9 Cellular Manufacturing 872
15.10 Flexible Manufacturing Systems 874
15.11 Just-in-Time Production 876
15.12 Communications Networks in Manufacturing 878
15.13 Artificial Intelligence 880
15.14 The Factory of the Future 883
15.15 Economic Considerations 885
CASE STUDY: CAD Model Development for Automotive Components 886
Summary 889
Bibliography 890
Questions 891
Design 892

# 16 Product Design and Competitive Aspects of Manufacturing 893

16.1 Introduction 893
16.2 Product Design and Robust Design 894
16.3 Product Quality and Life Expectancy 898
16.4 Life-Cycle Assessment, Life-Cycle Engineering, and Sustainable Manufacturing 899
16.5 Selection of Materials for Products 901
16.6 Substitution of Materials in Products 905
16.7 Capabilities of Manufacturing Processes 907
16.8 Selection of Manufacturing Processes 911
16.9 Manufacturing Costs and Cost Reduction 915
16.10 Value Analysis 919
CASE STUDY: Concurrent Engineering for Intravenous Solution Containers 920
Summary 922
Bibliography 923
Questions 924
Design 925

Index 928

# PREFACE

The fourth edition of this text, like the previous editions, continues to make an attempt at presenting a comprehensive, balanced, and up-to-date coverage of the relevant fundamentals and real-world applications of engineering materials and manufacturing processes and systems and the analytical approaches involved. The topics covered throughout the 16 chapters reflect the rapid and significant advances that have taken place in various areas in manufacturing, and they are organized and described in such a manner to draw the interest of students. The case studies, mostly from industry, make the subject of manufacturing science and engineering less abstract by showing students the practical aspects of process applications.

Integration of product design and manufacturing has justly resulted in greater recognition and prestige for these disciplines, and this edition therefore puts greater emphasis and better focus on this increasingly important subject. While studying this text, students should be able to assess the capabilities, limitations, and potential of production processes, including, particularly, the processes' economic and competitive aspects. The discussions throughout the chapters are aimed at motivating and challenging students to explore technically and economically viable solutions to a wide variety of important questions and problems in product design and manufacturing.

This book has been written mainly for undergraduate students in mechanical, industrial, and metallurgical and materials engineering programs; it is also useful for graduate courses in manufacturing science and engineering. The text, as well as the numerous examples and case studies in each chapter, clearly shows that manufacturing engineering is an interdisciplinary and complex subject and that it is as important, exciting, and challenging as any other engineering discipline.

## What's new in this edition

- A new chapter has been added that covers the fabrication of microelectronic and micromechanical devices.
- The discussion of design considerations in each chapter have been expanded.
- Several new examples and case studies have been added throughout all chapters.
- A total of 1230 questions and problems has been added—30% more than in the third edition.
- Questions and problems now include a total of 140 design problems.
- Figures have been improved for better graphical impact.
- More cross-references to sections and chapters have been added throughout the text.
- All chapter bibliographies have been thoroughly updated.

## New or expanded topics for this edition include the following:

- Automated guided vehicles
- Biodegradable plastics
- Cryogenic machining and grinding
- Cryogenic treatment of cutting tools
- Design considerations for casting
- Design considerations for powder metallurgy

- Design considerations for sheet-metal forming
- Electrically conducting adhesives
- Enterprise resource planning
- Flexible fixturing
- ISO 9000 and 14000 standards
- Life-cycle assessment
- LIGA process
- Metal foams
- Microelectromechanical systems fabrication
- Microelectronics device fabrication
- Micromachining
- Modeling of casting
- Nanomaterials
- Plasma and wet etching
- Polymer processing
- Printed circuit boards
- Rapid protoyping and rapid tooling
- Rotary ultrasonic machining
- Silicon microstructure
- Six-sigma quality
- Solid free-form fabrication
- Stick–slip in extrusion
- Superconductor processing
- Surface-mount technology
- Value assessment
- Vibration and chatter in rolling
- Water-jet peening

## Acknowledgments

We gratefully acknowledge the following colleagues and associates for their contributions to the various sections of this book:

K. Anderson (Baxter Healthcare Corp.), P. J. Courtney (Loctite Corp.), D. Furrer (Ladish Corp.), K. L. Graham (Guidant Corp.), M. Hawkins (Zimmer, Inc.), K. M. Kalpakjian (Micron Technology, Inc., Mr. Kalpakjian is principal author of the sections on fabrication of microelectronic devices in Chapter 13), R. Kassing (University of Kassel), K. M. Kulkarni (Advanced Manufacturing Practices, Inc.), M. Madou (Nanogen, Inc.), A. Marsan (Ford Motor Co.), S. Paolucci (University of Notre Dame), C. Petronis (ST Microelectronics, Inc.), S. Petronis (Zimmer Holdings, Inc.), M. Pradheeradhi (Concurrent Technologies Corp.), Y. Rong (Worcester Polytechnic Institute), P. Saha (Boeing Co.), S. Shepel (University of Notre Dame), M. T. Siniawski (Northwestern University), P. Stewart (Ford Motor Co.), S. Vaze (Concurrent Technologies Corp.), J. E. Wang (Texas A&M University), K. J. Weinmann (Michigan Technological University), and K. R. Williams (Agilent, Inc.).

Many thanks to our colleagues at various institutions for their help, their detailed reviews, comments contributions, and many constructive suggestions for this edition:

D. D. Arola (University of Maryland), S. Mantell (University of Minnesota), M. H. Miller (Michigan Technological University), J. Moller (Miami University), D. J. Morrison (Clarkson University), U. Pal (Boston University), B. S. Thakkar (Lucent Technologies), and A. Tseng (Arizona State University), Dr. Rajiv Shivpuri (Ohio State University), Dr. Shaochen Chen (Iowa State University), Dr. Nicholas Zabaras (Cornell University), Dr. Mark Tuttle (University of Washington), Dr. Donald W. Radford (Colorado State University), Dr. Mica Grujicic (Clemson University).

We would also like to acknowledge the dedication and continued help of our editor, Laura Fischer, and the editorial staff at Prentice Hall, including Tamar Savir, David George, and Xiaohong Zhu. We appreciate the help of the many organizations that supplied us with various case studies and numerous illustrations. And, finally, many thanks to Jean Kalpakjian for her help in the final preparation of this book.

SEROPE KALPAKJIAN
*Chicago, Illinois*

STEVEN R. SCHMID
*Notre Dame, Indiana*

# ABOUT THE AUTHORS

**Serope Kalpakjian** taught and conducted research at the Illinois Institute of Technology for 38 years prior to his retirement in 2001 as professor emeritus of mechanical and materials engineering. After graduating from Robert College (high honors), Harvard University, and the Massachusetts Institute of Technology, he joined Cincinnati Milacron, where he was a research supervisor in advanced metal-forming processes. He is the author of numerous technical papers and several articles in handbooks and encyclopedias and has edited various conference proceedings. In addition, Professor Kalpakjian has served on the editorial boards of several journals and the *Encyclopedia Americana* and is the coauthor of *Lubricants and Lubrication in Metalworking Operations*. Both first editions of his textbooks, *Manufacturing Processes for Engineering Materials* and *Manufacturing Engineering and Technology*, have received the M. Eugene Merchant Manufacturing Textbook Award. He is a Life Fellow of the ASME, Fellow of the SME, Fellow and Life Member of ASM International, emeritus member of CIRP (International Institution for Production Engineering Research), and a founding member and past president of NAMRI/SME.

Among the awards Professor Kalpakjian has received are the Forging Industry Educational and Research Foundation Best Paper Award (1966), an Excellence in Teaching Award from IIT (1970), the Centennial Medallion from the ASME (1980), the International Education Award from the SME (1989), a Person of the Millennium Award from IIT (1999), and the Albert Easton White Distinguished Teacher Award from ASM International (2000). SME named the Outstanding Young Manufacturing Engineer Award after Professor Kalpakjian for the year 2002.

**Steven R. Schmid** is an associate professor with the Department of Aerospace and Mechanical Engineering, University of Notre Dame, where he teaches and performs research in the general areas of manufacturing, machine design, and tribology. As the director of the Manufacturing Tribology Laboratory at the university, he oversees industry- and government-funded research on a variety of manufacturing topics, including tribological issues in rolling, forging and sheet-metal forming, polymer processing, medical-device design and manufacture, and nanomechanics.

He received the B.S. degree in mechanical engineering from the Illinois Institute of Technology (with honors) and M.S. and Ph.D. degrees, both in mechanical engineering, from Northwestern University. He has received numerous awards, including the John T. Parsons Award from the SME (2000), the Newkirk Award from the ASME (2000), and the Kaneb Center Teaching Award (2000). He is the recipient of a National Science Foundation CAREERS Award (1996) and an ALCOA Foundation Award (1994). Dr. Schmid is the author of over 50 technical papers; has edited three conference proceedings; has coauthored two books, *Fundamentals of Machine Elements* and *Manufacturing Engineering and Technology*; and has contributed two chapters to the *CRC Handbook of Modern Tribology*. He serves on the Tribology Division Executive Committee of the ASME, is an associate editor of the *Journal of Manufacturing Science and Engineering*, and is a registered professional engineer and certified manufacturing engineer.

# Introduction

## 1.1 | What Is Manufacturing?

As you read this Introduction, take a few moments to inspect the different objects around you: your watch, chair, stapler, pencil, calculator, telephone, and light fixtures. You will soon realize that all these objects have been transformed from various raw materials into individual parts and assembled into specific products. Some objects, such as nails, bolts, and paper clips, are made of one material; however, most objects, such as light bulbs, toasters, bicycles, computers, all types of instruments and machinery, and automotive engines (Fig. 1.1), and countless other products, are made of numerous parts from a wide variety of materials. A ballpoint pen, for example, consists of about a dozen parts, a lawnmower about 300 parts, a grand piano

1.1 What Is Manufacturing?
1.2 Product Design and Concurrent Engineering
1.3 Design for Manufacture, Assembly, Disassembly, and Service
1.4 Environmentally Conscious Design and Manufacturing
1.5 Selecting Materials
1.6 Selecting Manufacturing Processes
1.7 Computer-Integrated Manufacturing
1.8 Quality Assurance and Total Quality Management
1.9 Product Liability
1.10 Manufacturing Costs, Lean Production, and Agile Manufacturing
1.11 General Trends in Manufacturing
1.12 Responsibilities of Manufacturing Engineers

Polymer manifolds removed for clarity

Copper hydraulic tubing

Cast aluminum cylinder head

Brass and steel bolts

Ceramic housing, platinum electrode spark plugs

Blanked steel spur gear

Micro-honed cast iron cylinder bore liners

Forged steel crankshaft

Powder metal connecting rods

Copper brushes (inside alternator)

Graphite coated, cast aluminum pistons

**FIGURE 1.1** Section of an automotive engine—the Duravec V-6—showing various components and the materials used in making them. *Source*: Courtesy of Ford Motor Company. Illustration by David Kimball.

12,000 parts, a typical automobile 15,000 parts, a C-5A transport plane more than 4 million parts, and a Boeing 747-400 about 6 million parts; all are produced by a combination of various processes called *manufacturing*.

**Manufacturing**, in its broadest sense, is the process of converting raw materials into products; it encompasses the design and fabrication of goods by means of various production methods and techniques. Manufacturing began around 5000 to 4000 BC with the production of various articles of wood, ceramic, stone, and metal (Table 1.1). The word *manufacturing* is derived from the Latin *manu factus*, meaning made by hand; the word *manufacture* first appeared in AD 1567 and the word *manufacturing* in 1683. The word **production** is also used interchangeably with the word *manufacturing*. Manufacturing represents approximately 20% to 30% of the value of all goods and services produced in industrialized countries. Generally, the higher the level of manufacturing in a country, the higher the standard of living of its people. Manufactured products are also used to make other products; examples include large presses to form sheet metal for car bodies, metalworking machinery used to make parts for other products, and sewing machines for making clothing.

Manufacturing may produce *discrete products*, meaning individual parts or pieces of parts, or it may produce continuous products. Nails, gears, steel balls, beverage cans, and engine blocks are examples of discrete products. Wire, metal or plastic sheet, hose, and pipe are *continuous products* that may be cut into individual pieces and thereby become discrete products.

Because a manufactured item has undergone a number of changes during which raw material has become a useful product, it has **added value**, defined as monetary worth in terms of price. For example, clay has a certain value when mined. When the clay is used to make a ceramic dinner plate, cutting tool, or electrical insulator, value is added to the clay; similarly, a wire coat hanger or a nail has added value over and above the cost of a piece of wire.

Manufacturing is generally a complex activity involving people who have a broad range of disciplines and skills, together with a wide variety of machinery, equipment, and tools with various levels of automation, including computers, robots, and material-handling equipment. Manufacturing activities must be responsive to several demands and trends:

1. A product must fully meet **design requirements** and **specifications** and **standards**.
2. A product must be manufactured by the most **economical** and **environmentally friendly** methods.
3. **Quality** must be built into the product at each stage, from design to assembly, rather than relying on quality testing after the product is made.
4. In a highly competitive environment, production methods must be sufficiently **flexible** to respond to changing market demands, types of products, production rates, and production quantities and to provide on-time delivery to the customer.
5. New developments in **materials, production methods**, and **computer integration** of both technological and managerial activities in a manufacturing organization must constantly be evaluated with respect to their timely and economic implementation.
6. Manufacturing activities must be viewed as a large **system** in which all individual components are interrelated. Such systems can now be modeled in order to study the effects of various factors, such as changes in market demands, product design, materials, costs, and production methods, on product quality and cost.

**TABLE 1.1**

**Historical Development of Materials and Manufacturing Processes**

| Period | Dates | Metals and casting | Various materials and composites | Forming and shaping | Joining | Tools, machining, and manufacturing systems |
|---|---|---|---|---|---|---|
| | Before 4000 BC | Gold, copper, meteoric iron | Earthenware, glazing, natural fibers | Hammering | | Tools of stone, flint, wood, bone, ivory, composite tools |
| | 4000–3000 BC | Copper casting, stone and metal molds, lost-wax process, silver, lead, tin, bronze | | Stamping, jewelry | Soldering (Cu–Au, Cu–Pb, Pb–Sn) | Corundum (alumina, emery) |
| | 3000–2000 BC | Bronze casting and drawing, gold leaf | Glass beads, potter's wheel, glass vessels | Wire by slitting sheet metal | Riveting, brazing | Hoe making, hammered axes, tools for ironmaking and carpentry |
| | 2000–1000 BC | Wrought iron, brass | | | | |
| | 1000–1 BC | Cast iron, cast steel | Glass pressing and blowing | Stamping of coins | Forge welding of iron and steel, gluing | Improved chisels, saws, files, woodworking lathes |
| | AD 1–1000 | Zinc, steel | Venetian glass | Armor, coining, forging, steel swords | | Etching of armor |
| | 1000–1500 | Blast furnace, type metals, casting of bells, pewter | Crystal glass | Wire drawing, gold- and silversmith work | | Sandpaper, windmill-driven saw |
| | 1500–1600 | Cast-iron cannon, tinplate | Cast plate glass, flint glass | Water power for metalworking, rolling mill for coinage strips | | Hand lathe for wood |
| | 1600–1700 | Permanent-mold casting, brass from copper and metallic zinc | Porcelain | Rolling (lead, gold, silver), shape rolling (lead) | | Boring, turning, screw-cutting lathe, drill press |

Period (bracket spanning rows):
Egypt: ~3100 BC to ~300 BC
Greece: ~1100 BC to ~146 BC
Roman Empire: ~500 BC to AD 476
Middle Ages: ~476 to 1492
Renaissance: 14th to 16th centuries

*(continues on next page)*

3

# Historical Development of Materials and Manufacturing Processes (cont.)

| Period | Dates | Metals and casting | Various materials and composites | Forming and shaping | Joining | Tools, machining, and manufacturing systems |
|---|---|---|---|---|---|---|
| Industrial Revolution: ~1750 to 1850 | 1700–1800 | Malleable cast iron, crucible steel (iron bars and rods) | | Extrusion (lead pipe), deep drawing, rolling | | |
| Industrial Revolution: ~1750 to 1850 | 1800–1900 | Centrigufal casting, Bessemer process, electrolytic aluminum, nickel steels, babbitt, galvanized steel, powder metallurgy, open-hearth steel | Window glass from slit cylinder, light bulb, vulcanization, rubber processing, polyester, styrene, celluloid, rubber extrusion, molding | Steam hammer, steel rolling, seamless tube, steel-rail rolling, continuous rolling, electroplating | | Shaping, milling, copying lathe for gunstocks, turret lathe, universal milling machine, vitrified grinding wheel |
| | 1900–1920 | | Automatic bottle making, bakelite, borosilicate glass | Tube rolling, hot extrusion | Oxyacetylene; arc, electrical-resistance, and thermit welding | Geared lathe, automatic screw machine, hobbing, high-speed-steel tools, aluminum oxide and silicon carbide (synthetic) |
| WW I | 1920–1940 | Die casting | Development of plastics, casting, molding, polyvinyl chloride, cellulose acetate, polyethylene, glass fibers | Tungsten wire from metal powder | Coated electrodes | Tungsten carbide, mass production, transfer machines |
| WW II | 1940–1950 | Lost-wax process for engineering parts | Acrylics, synthetic rubber, epoxies, photosensitive glass | Extrusion (steel), swaging, powder metals for engineering parts | Submerged arc welding | Phosphate conversion coatings, total quality control |
| | 1950–1960 | Ceramic mold, nodular iron, semiconductors, continuous casting | Acrylonitrile-butadiene-styrene, silicones, fluorocarbons, polyurethane, float glass, tempered glass, glass ceramics | Cold extrusion (steel), explosive forming, thermomechanical processing | Gas metal arc, gas tungsten arc, and electroslag welding; explosive welding | Electrical and chemical machining, automatic control |

4

**Historical Development of Materials and Manufacturing Processes (cont.)**

| Period | Dates | Metals and casting | Various materials and composites | Forming and shaping | Joining | Tools, machining, and manufacturing systems |
|---|---|---|---|---|---|---|
| Space Age | 1960–1970 | Squeeze casting, single-crystal turbine blades | Acetals, polycarbonate, cold forming of plastics, reinforced plastics, filament winding | Hydroforming, hydrostatic extrusion, electroforming | Plasma-arc and electron-beam welding, adhesive bonding | Titanium carbide, synthetic diamond, numerical control, integrated circuit chip |
| | 1970–1990 | Compacted graphite, vacuum casting, organically bonded sand, automation of molding and pouring, rapid solidification, metal-matrix composites, semisolid metalworking, amorphous metals, shape-memory alloys (smart materials), computer simulation | Adhesives, composite materials, semiconductors, optical fibers, structural ceramics, ceramic-matrix composites, biodegradable plastics, electrically conducting polymers | Precision forging, isothermal forging, superplastic forming, dies made by computer-aided design and manufacturing, net-shape forging and forming, computer simulation | Laser beam, diffusion bonding (also combined with superplastic forming), surface-mount soldering | Cubic boron nitride, coated tools, diamond turning, ultraprecision machining, computer-integrated manufacturing, industrial robots, machining and turning centers, flexible-manufacturing systems, sensor technology, automated inspection, expert systems, artificial intelligence, computer simulation and optimization |
| Information Age | 1990–2000s | Rheocasting, computer-aided design of molds and dies, rapid tooling | Nanophase materials, metal foams, advanced coatings, high-temperature superconductors. machinable ceramics, diamondlike carbon | Rapid prototyping, rapid tooling, environmentally friendly metalworking fluids | Friction stir welding, lead-free solders, laser butt-welded (tailored) sheet-metal blanks, electrically conducting adhesives | Micro- and nano-fabrication, LIGA (a German acronym for a process involving lithography, electroplating, and molding), dry etching, linear motor drives, artificial neural networks, six sigma |

*Source*: J.A. Schey, C.S. Smith, R.F. Tylecote, T.K. Derry, T.I Williams, S.R. Schmid, and S. Kalpakjian.

5

7. The manufacturer must work with the customer to get timely feedback for **continuous product improvement**.

8. The manufacturing organization must constantly strive for higher **productivity**, defined as the optimum use of all its resources: materials, machines, energy, capital, labor, and technology. Output per employee per hour in all phases must be maximized.

## 1.2 | Product Design and Concurrent Engineering

*Product design* is a critical activity: It has been estimated that 70% to 80% of the cost of product development and manufacture is determined at the initial design stages. The design process for a product first requires a clear understanding of the functions and the performance expected of the product. The product may be new, or it may be a new and improved model of an existing product. The market for a product and the product's anticipated uses must be defined clearly, with the assistance of sales personnel, market analysts, and others in the organization.

Traditionally, design and manufacturing activities have taken place sequentially rather than concurrently or simultaneously (Fig. 1.2a). Designers would spend considerable effort and time in analyzing components and preparing detailed drawings of parts; these drawings would then be forwarded to other departments in the organization, such as materials departments, where, for example, particular alloys and vendor sources would be identified. The specifications would then be sent to the manufacturing department, where the detailed drawings would be reviewed and processes selected for efficient production. While this approach seems logical and straightforward in theory, it has been found in practice to be extremely wasteful of resources.

In theory, a product can flow from one department in an organization to another and then directly to the marketplace, but in practice difficulties are usually encountered. For example, a manufacturing engineer may wish to taper the flange on a part to improve its castability or may decide that a different alloy is desirable; such changes necessitate a repeat of the design analysis stage in order to ensure that the product will still function satisfactorily. These iterations, also shown in Fig. 1.2a, certainly waste resources, but, more importantly, they waste time.

A more advanced approach to product development is shown in Fig. 1.2b. While it still has a general product flow from market analysis to design to manufacturing, it contains deliberate iterations. The main difference between it and the older approach is that all disciplines are now involved in the early design stages and the stages progress concurrently, so that the iterations (which occur by nature) result in less wasted effort and less lost time. A key to the new approach is the recognition of the importance of communication among and within disciplines. That is, while there must be communication between engineering, marketing, and service functions, so, too, must there be avenues of interaction between engineering subdisciplines—for example, design for manufacture, design for recyclability, and design for safety.

**Concurrent engineering** (CE), also called *simultaneous engineering*, is a systematic approach that integrates the design and manufacture of products with the view of optimizing all elements involved in the life cycle of the product. **Life cycle** means that all aspects of a product, such as design, development, production, distribution, use, its ultimate recycling and disposal and its environmental impact are

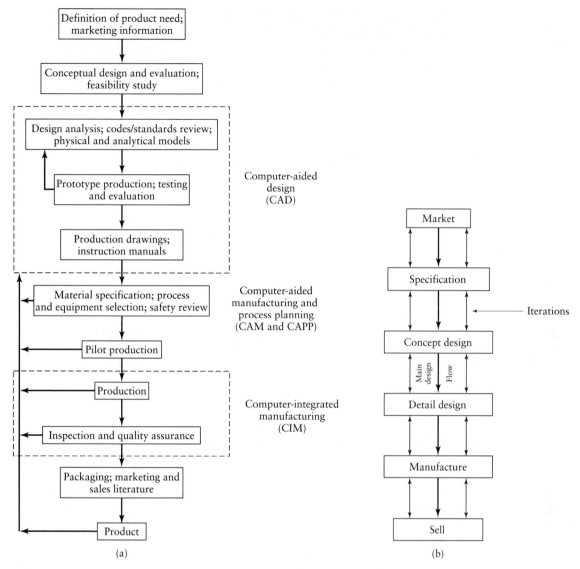

**FIGURE  1.2**    (a) Chart showing the various steps involved in designing and manufacturing a product. Depending on the complexity of the product and the type of materials used, the time span between the original concept and the marketing of a product may range from a few months to many years. (b) Chart showing general product flow, from market analysis to selling the product, and depicting concurrent engineering. *Source*: After S. Pugh, *Total Design*, Addison-Wesley, 1991.

considered simultaneously. The basic goal of concurrent engineering is to minimize product design and engineering changes and the time and costs involved in taking the product from design concept to production to introduction into the market-place. An extension of concurrent engineering, called **direct engineering**, has also been proposed. It uses a database representing the engineering logic applied during the design of each component of a product. If a design modification is made on a part, direct engineering will determine the manufacturing consequences of that change.

Although the concept of concurrent engineering appears to be logical and efficient, its implementation can take considerable time and effort when those using it either do not work as a team or fail to appreciate its real benefits. It is apparent that for concurrent engineering to succeed, it must (1) have the full support of top management, (2) have a multifunctional and interactive work team, including support groups, and (3) use all available technologies.

For both large and small companies, product design often involves preparing analytical and physical models of the product as an aid for analyzing factors such as forces, stresses, deflections, and optimal part shape. The necessity for such models depends on product complexity; today, however, the process of constructing and studying analytical models is simplified through the use of **computer-aided design**, **engineering**, and **manufacturing** techniques. On the basis of these models, the product designer selects and specifies the final shape and dimensions of the product, its dimensional tolerances and surface finish, and the materials to be used. The selection of materials is generally made with the advice and cooperation of materials engineers, unless the design engineer is sufficiently experienced and qualified in this area. An important design consideration is how a particular component is to be assembled into the final product. To understand the concept of assembly, take apart a ballpoint pen or a toaster, or lift the hood of a car and observe how hundreds of components are put together in a confined space.

A powerful and effective tool, particularly for complex production systems, is **computer simulation**, which can be used to evaluate the performance of the product and plan the manufacturing system to produce it. Computer simulation also helps in early detection of design flaws, identification of possible problems in a particular production system, and optimization of manufacturing lines for minimum product cost. Several computer simulation languages that use animated graphics and have various capabilities have been developed.

The next step in the production process is to make and test a **prototype**, that is, an original working model of the product. An important development is **rapid prototyping** (Section 10.12), which relies on computer-aided design/computer-aided manufacturing (CAD/CAM) and various manufacturing techniques (typically using polymers or metal powders) to rapidly produce prototypes in the form of a solid physical model of a part. Rapid prototyping can significantly cut costs as well as development times. These techniques are currently being advanced further to such an extent that they can now be used for low-volume economical production of parts.

Tests on prototypes must be designed to simulate as closely as possible the conditions under which the product is to be used. Such factors include environmental conditions such as temperature and humidity, as well as the effects of vibration and repeated use and misuse of the product. During this stage, modifications of the original design, materials selected, or production methods may be necessary. After this phase has been completed, appropriate process plans, manufacturing methods (Table 1.2), equipment, and tooling are selected with the cooperation of manufacturing engineers, process planners, and all other involved in production.

## 1.3 | Design for Manufacture, Assembly, Disassembly, and Service

It is apparent that design and manufacturing must be closely interrelated; they should never be viewed as separate disciplines or activities. Each part or component of a product must be designed so that it not only meets design requirements and specifications, but also can be manufactured economically and with relative ease.

**TABLE 1.2**

## Shapes and Some Common Methods of Production

| Shape or feature | Production method |
| --- | --- |
| Flat surfaces | Rolling, planing, broaching, milling, shaping, grinding |
| Parts with cavities | End milling, electrical-discharge machining, electrochemical machining, ultrasonic machining, casting |
| Parts with sharp features | Permanent-mold casting, machining, grinding, fabricating, powder metallurgy |
| Thin hollow shapes | Slush casting, electroforming, fabricating |
| Tubular shapes | Extrusion, drawing, roll forming, spinning, centrifugal casting |
| Tubular parts | Rubber forming, expanding with hydraulic pressure, explosive forming, spinning |
| Curvature on thin sheets | Stretch forming, peen forming, fabricating |
| Openings in thin sheets | Blanking, chemical blanking, photochemical blanking, laser |
| Square edges | Fine blanking, machining, shaving, grinding |
| Small holes | Laser, electrical-discharge machining, electrochemical machining, micromachining |
| Surface textures | Knurling, wire brushing, grinding, belt grinding, shot blasting, etching, deposition, laser texturing |
| Detailed surface features | Coining, investment casting, permanent-mold casting, machining |
| Threaded parts | Thread cutting, thread rolling, thread grinding, chasing |
| Very large parts | Casting, forging, fabricating |
| Very small parts | Investment casting, machining, etching, powder metallurgy, nanofabrication, LIGA, micromachining |

This approach improves productivity and allows a manufacturer to remain competitive.

This broad concept is now recognized as the area of **design for manufacture** (DFM): a comprehensive approach to the production of goods that integrates the product design process with selection of materials, consideration of manufacturing methods, process planning, assembly, testing, and quality assurance. Effective implementation of design for manufacture requires that designers acquire a fundamental understanding of the characteristics, capabilities, and limitations of materials, production methods, and related operations, machinery, and equipment. This knowledge includes an understanding of characteristics such as variability in machine performance, dimensional accuracy and surface finish of the workpiece, processing time, and the effect of processing method on part quality.

Designers and product engineers must be able to assess the impact of design modifications on manufacturing-process selection, tools and dies, assembly, inspection, and product cost. Establishment of quantitative relationships is essential in order to optimize the design for ease of manufacturing and assembly at *minimum cost* (also called *producibility*). Computer-aided design, engineering, manufacturing, and process-planning techniques, using powerful computer programs, have become indispensable to those conducting such analysis; such computerized tools include **expert systems**, which are computer programs with optimization capabilities, that expedite the traditional iterative process in design optimization.

After individual parts have been manufactured, they are assembled into a product. **Assembly** is an important phase of the overall manufacturing operation and requires

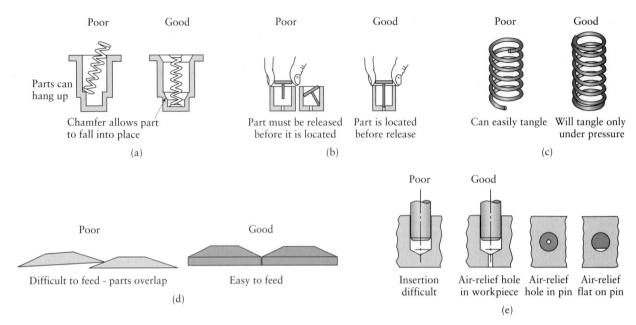

**FIGURE 1.3**   Redesign of parts to facilitate automated assembly. *Source*: Reprinted from G. Boothroyd and P. Dewhurst, *Product Design for Assembly*, 1989, by courtesy of Marcel Dekker, Inc.

consideration of the ease, speed, and cost of putting parts together (Fig. 1.3). Also, products must be designed so that **disassembly** is possible with relative ease and less time involved, enabling the products to be taken apart for maintenance, servicing, or recycling of their components. Because assembly operations can contribute significantly to product cost, **design for assembly** (DFA) and **design for disassembly** are now recognized as important aspects of manufacturing; typically, a product that is easy to assemble is also easy to disassemble. The trend now also includes **design for service**, ensuring that individual parts in a product are easy to reach and service. In addition, current methods combine design for manufacture and design for assembly into the more comprehensive **design for manufacture and assembly** (DFMA), which recognizes the inherent and important interrelationships among design, manufacturing, and assembly.

Design principles for economic production may be summarized as follows:

- The design should be as simple as possible to manufacture, assemble, disassemble, service, and recycle.
- Materials should be chosen for their appropriate manufacturing characteristics as well as for their service life.
- Specified dimensional accuracy and surface finish should be as broad as permissible.
- Because they can add significantly to cost, secondary and finishing operations should be avoided or minimized.

## 1.4   Environmentally Conscious Design and Manufacturing

In the United States alone, more than five billion kilograms of plastic products are discarded each year. Every three months, U.S. industries and consumers discard enough aluminum to rebuild the country's commercial air fleet. Globally, countless tons of automobiles, television sets, appliances, and computer equipment are discarded each year. Furthermore, metalworking fluids such as lubricants and coolants are often used in machining, grinding, and forming operations, and various fluids and solvents are used in cleaning manufactured products; some of these fluids can pollute the air and waters if discarded in an untreated form. Likewise, many byproducts from manufacturing plants have, for years, been discarded: sand with additives used in metal-casting processes; water, oil, and other fluids from heat-treating facilities and plating operations; slag from foundries and welding operations; and a wide variety of metallic and nonmetallic scrap produced in operations such as sheet forming, casting, and molding. Consider also the effects of water and air pollution, acid rain, ozone depletion, the greenhouse effect, hazardous wastes, landfill seepage, and global warming.

The present and potential future adverse effects of these activities, their damage to our environment and to the earth's ecosystem, and, ultimately, their effect on the quality of human life are now well recognized by the public as well as by local and federal governments. In response, a wide range of laws and regulations have been promulgated by local, state, and federal governments as well as professional organizations in the United States and other industrial countries. These regulations are generally stringent, and their implementation can have a major impact on the economic operation and financial health of industrial organizations. Regardless, and notwithstanding arguments by some that environmental damage from these activities is overestimated, manufacturing engineers and the management of companies have a major responsibility to plan and implement environmentally safe materials and manufacturing operations.

Major developments also have been taking place regarding **design for recycling** (DFR), indicating universal awareness of the aforementioned problems, and waste has become unacceptable. A major emphasis is on **design for the environment** (DFE), or **green design** ("green" meaning environmentally safe and friendly). This approach anticipates the possible negative environmental impact of materials, products, and processes so that it can be considered at the earliest stages of design and production.

It is apparent that it is essential to conduct a thorough analysis of the product, materials used, and the manufacturing processes and practices employed in making the product. Certain guidelines can be followed in this regard:

- Reduce waste of materials at their source by implementing refinements in product design and reducing the amount of materials used.

- Reduce the use of hazardous materials in products and processes.

- Ensure proper handling and disposal of all waste.

- Make improvements in recycling, waste treatment, and reuse of materials.

# 1.5 | Selecting Materials

An ever-increasing variety of materials is now available, each having its own characteristics, applications, advantages, and limitations (Fig. 1.4). The following are the general types of materials used in manufacturing today, either individually or in combination:

1. **Ferrous metals:** carbon, alloy, stainless, and tool and die steels (Chapter 3);
2. **Nonferrous metals and alloys:** aluminum, magnesium, copper, nickel, titanium, superalloys, refractory metals, beryllium, zirconium, low-melting alloys, and precious metals (Chapter 3);

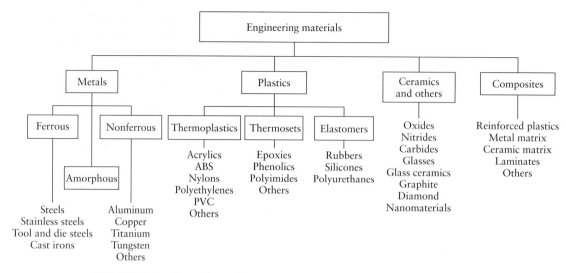

**FIGURE 1.4**   An outline of engineering materials.

3. **Plastics**: thermoplastics, thermosets, and elastomers (Chapter 10);
4. **Ceramics**: glass ceramics, glasses, graphite, and diamond (Chapter 11);
5. **Composite materials**: reinforced plastics, metal-matrix and ceramic-matrix composites, and honeycomb structures—these materials are also known as *engineered materials* (Chapters 10 and 11);
6. **New materials**: nanomaterials, shape-memory alloys, amorphous alloys, superconductors, and various other materials with unique properties (Chapter 3).

As new materials are developed, important trends arise with respect to their selection and application. Aerospace structures as well as products such as sporting goods especially have been at the forefront of new-material usage. Because of the vested interest of the producers of different types of metallic and nonmetallic materials, there are constantly shifting trends in the usage of these materials, driven principally by economics. For example, by demonstrating steel's technical and economic feasibility, steel producers are countering the increased use of plastics in cars and aluminum in beverage cans. Likewise, aluminum producers are countering the use of various materials in cars (Fig. 1.5). Obviously, making materials-selection choices that facilitate recycling should be a fundamental part of these considerations.

Some examples of the use or substitution of materials in common products are the following: steel vs. plastic paper clips, plastic vs. sheet-metal light-switch plates, wood vs. metal handles for hammers, glass vs. metal water pitchers, plastic vs. leather car seats, sheet-metal vs. reinforced-plastic chairs, galvanized-steel vs. copper nails, and aluminum vs. cast-iron frying pans.

**Properties of materials.**   When selecting materials for products, the first consideration generally involves **mechanical properties** (Chapter 2), which typically are strength, toughness, ductility, hardness, elasticity, and fatigue and creep resistance. The mechanical properties of materials can be significantly modified by various heat treatment methods (Chapter 5). The strength-to-weight and stiffness-to-weight ratios of materials are also important, particularly for aerospace and automotive applications.

Die-cast nodes are thin walled to maximize weight reduction, yet provide high performance.

Robotically applied, advanced arc-welding processes provide consistent, high-quality assembly of castings, extrusions, and sheet components.

Strong, thin-walled extrusions exhibit high ductility, energy absorption, and toughness.

Advanced extrusion bending processes support complex shapes and tight radii.

(a)

(b)

**FIGURE 1.5** (a) The Audi A8 automobile, an example of advanced materials construction. (b) The aluminum body structure, showing various components made by extrusion, sheet forming, and casting processes. *Source:* Courtesy of ALCOA, Inc.

Aluminum, titanium, and reinforced plastics, for example, have higher strength-to-weight ratios than those of steels and cast irons. The mechanical properties specified for a product and its components should, of course, be appropriate for the conditions under which the product is expected to function.

Further aspects to consider are the **physical properties** (Chapter 3) of the materials, such as density, specific heat, thermal expansion and conductivity, melting point, and electrical and magnetic properties. **Chemical properties** can play a significant role in hostile as well as normal environments. Oxidation, corrosion, general degradation of properties, and flammability of materials are among the important factors to consider, as is toxicity. (Note, for example, the development of lead-free solders; see Section 12.12.3). Both the physical and the chemical properties of materials are important in advanced machining processes (Chapter 9). Finally, the **manufacturing properties** of materials determine whether they can be cast, formed, shaped, machined, welded, or heat treated with relative ease. An important consideration is that the methods used to process materials to the desired shapes should not adversely affect the product's final properties, service life, and cost.

**Cost and availability.** The economic aspects of material selection are as important as the technological considerations of properties and characteristics of materials. The cost and availability of raw and processed materials and manufactured components are major concerns in manufacturing. If raw or processed materials or manufactured components are not commercially available in the desired shapes, dimensions, and quantities, substitutes or additional processing will be required, which can contribute significantly to product cost (Chapter 16). For example, if we need a round bar of a certain diameter and it is not available from the supplier, then we will have to purchase a larger rod and reduce its diameter by a process such as machining, drawing though a die, or grinding.

Reliability of both supply and demand affects cost. Most countries import numerous raw materials that are essential for production. The United States, for example, imports the majority of such raw materials as natural rubber, diamond, cobalt, titanium, chromium, aluminum, and nickel from other countries. The broad political implications of such reliance on other countries are self-evident. In addition, various costs are involved in processing materials by different methods. Some methods require expensive machinery, others require extensive labor, and still others require personnel with special skills, a high level of formal education, or specialized training.

**Service life and recycling.** Time- and service-dependent phenomena such as wear, fatigue, creep, and dimensional stability are important considerations, as they can significantly affect a product's performance and, if not controlled, can lead to failure of the product. Likewise, compatibility of materials used in a product is important; for example, galvanic action between mating parts made of dissimilar metals causes them to corrode. Recycling or proper disposal of components in a product at the end of its useful life has become increasingly important as we become more and more aware of the importance of conserving materials and energy and living in a clean and healthy environment. The proper treatment and disposal of toxic wastes is also a crucial consideration.

# 1.6 | Selecting Manufacturing Processes

As can be seen in Table 1.2, a wide range of manufacturing processes is used to produce a variety of parts, shapes, and sizes. Also note that there is usually more than one method for manufacturing a part from a given material, and, as expected, each of these processes has its own advantages, limitations, production rates, and cost (Fig. 1.6). The broad categories of processing methods for materials are as follows:

- **Casting:** expendable mold and permanent mold (Chapter 5);
- **Forming and shaping:** rolling, forging, extrusion, drawing, sheet forming, powder metallurgy, and molding (Chapters 6, 7, 10, 11);
- **Machining:** turning; boring; drilling; milling; planing; shaping; broaching; grinding; ultrasonic machining; chemical, electrical, and electrochemical machining; and high-energy beam machining (Chapters 8 and 9);
- **Joining:** welding, brazing, soldering, diffusion bonding, adhesive bonding, and mechanical joining (Chapter 12);
- **Finishing:** honing, lapping, polishing, burnishing, deburring, surface treating, coating, and plating (Chapter 9).

Selection of a particular manufacturing process or a series of processes depends not only on the component or part shape to be produced, but also on many other factors. Brittle and hard materials, for example, cannot be shaped easily, whereas they

**FIGURE 1.6** Various methods of making a simple part: (a) casting or powder metallurgy, (b) forging or upsetting, (c) extrusion, (d) machining, (e) joining two pieces.

(a)    (b)    (c)    (d)    (e)

can be cast or machined by several methods. The manufacturing process usually alters the properties of materials; metals that are formed at room temperature, for example, become stronger, harder, and less ductile than they were before processing. Thus, it is important to study characteristics such as *castability*, *formability*, *machinability*, and *weldability* of materials. Some examples of manufacturing processes used or substituted for common products are the following: forging vs. casting of crankshafts, sheet-metal vs. cast hubcaps, casting vs. stamping sheet metal for frying pans, machined vs. powder-metallurgy gears, thread rolling vs. machining of bolts, and casting vs. welding of machine structures.

Manufacturing engineers are constantly challenged to find new solutions to production problems as well as to find means for significant cost reduction. For a long time, for example, sheet-metal parts were cut and made by traditional tools such as punches and dies. Although they are still widely used, these operations are now being replaced by laser cutting techniques. With advances in computer controls, path of the laser can be automatically controlled, thus producing a wide variety of shapes accurately, repeatedly, and economically without the use of expensive tooling.

**Size and dimensional accuracy.** Obviously, size, shape complexity, and thickness of a part have a major influence on the process selected to produce the part. Complex parts, for example, cannot be formed easily and economically, whereas they may be cast, injection molded, or fabricated from individual pieces or by powder-metallurgy techniques; likewise, flat parts with thin cross-sections cannot be cast properly. Dimensional tolerances and surface finish (Chapter 4) obtained in hot-working operations cannot be as fine as those obtained in cold-working operations, because dimensional changes, warping, and surface oxidation occur during processing at elevated temperatures. Also, some casting processes produce a better surface finish than others, because of the different types of mold materials used. The appearance of materials after they have been manufactured into products influences their appeal to the consumer; color, feel, and surface texture are characteristics that we all consider when making a purchasing decision.

The size and shape of manufactured products vary widely. For example, the main landing gear for the twin-engine, 400-passenger Boeing 777 jetliner is 4.3 m (14 ft) high, with three axles and six wheels, made by forging and machining processes (Chapters 6, 8, and 9). At the other extreme is the manufacturing of microscopic parts and mechanisms (Fig. 1.7). Such components are produced through surface micromachining operations, referred to as **nanotechnology** and **nanofabrication**, typically using electron-beam, laser-beam, and etching techniques on materials such as silicon. Such mechanisms, called **microelectromechanical systems** (MEMS; see Chapter 13), can potentially power microrobots to repair human cells, and produce microknives for surgery and camera shutters for precise photography, and some are now widely used in sensors, ink-jet printing mechanisms, and magnetic storage devices. The most recent trend is the development of **nanoelectromechanical systems** (NEMS), which operate on the same scale as biological molecules.

**Ultraprecision manufacturing** techniques and machinery are now being developed and are coming into more common use. For machining mirrorlike surfaces, for example, the cutting tool is a very sharp diamond tip, and the equipment has very high stiffness and must be operated in a room where the temperature is controlled within less than one degree. Highly sophisticated techniques such as molecular-beam epitaxy and scanning-tunneling engineering are being implemented to reach dimensional accuracies on the order of the atomic lattice (nanometer).

**FIGURE 1.7** An example of nanofabrication: gear assembly driven by resonant combdrives. (a) A view of the entire assembly. (b) Details of the rotor-driver portion of the gear assembly. *Source*: R. Muller, University of California at Berkeley.

(a)

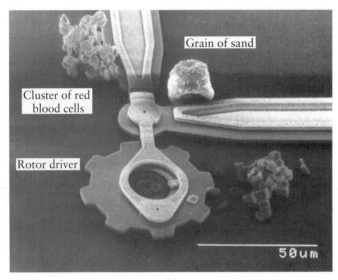

(b)

**Manufacturing and operational cost considerations.** The design and cost of tooling, the lead time required to begin production, and the effect of workpiece materials on tool and die life are major considerations in manufacturing (Chapter 16). The cost of tooling can be substantial, depending on tool size, design, and expected life; for example, a set of steel dies for stamping sheet-metal fenders for automobiles may cost about $2 million. For parts made from expensive materials, the lower the scrap rate, the more economical the production process will be. Because machining takes longer time and produces chips, it may not be as economical as forming operations, all other factors being equal.

The number of parts or products required and the desired production rate (in pieces per hour) help determine the processes to be used and the economics of production. Beverage cans and transistors, for example, are consumed in numbers and at rates much higher than telescopes and propellers for ships. The availability of machines and equipment, operating experience, and economic considerations within the manufacturing facility are also important cost factors. If certain parts are not produced internally, they must be manufactured by outside firms; automobile manufacturers, for example, purchase numerous parts from outside vendors or have the parts made by outside firms, according to their specifications.

The operation of machinery has significant environmental and safety implications. Depending on the type of operation and the machinery involved, some processes adversely affect the environment. Unless properly controlled, such processes may cause air, water, and noise pollution. The safe use of machinery is another important consideration, requiring precautions to eliminate hazards in the workplace.

**Net-shape manufacturing.** Because not all manufacturing operations produce finished parts or products, additional processes may be necessary. For example, a forged part may not have the desired dimensional accuracy or surface finish; thus, additional operations such as machining or grinding may be necessary. Likewise, it may be difficult, impossible, or economically undesirable to produce a part with holes in it by using just one manufacturing process. The holes produced by a particular process may not have the proper roundness, dimensional accuracy, or surface finish, thus necessitating additional operations such as boring, drilling, and reaming.

These additional operations can contribute significantly to the cost of a product. Consequently, there is currently an important trend, called *net-shape* or *near-net-shape manufacturing*, in which the part is made as close to the final desired dimensions, tolerances, and specifications as possible. Typical examples of such manufacturing methods are near-net-shape forging and casting of parts, powder-metallurgy techniques, metal injection molding of metal powders, and injection molding of plastics (Chapters 5, 6, 10, and 11).

# 1.7 | Computer-Integrated Manufacturing

The major goals of automation in manufacturing facilities (Chapters 14 and 15) are to integrate various operations so as to improve productivity, increase product quality and uniformity, minimize cycle times, and reduce labor costs. Beginning in the 1940s, automation has accelerated because of rapid advances in control systems for machines and in computer technology.

Few developments in the history of manufacturing have had a more significant impact than computers. Computers are now used in a very broad range of applications, including control and optimization of manufacturing processes, material handling, assembly, automated inspection and testing of products (Section 4.8), inventory control, and numerous management activities. Beginning with computer graphics and computer-aided design and manufacturing, the use of computers has been extended to *computer-integrated manufacturing* (CIM).

Computer-integrated manufacturing is particularly effective because of its capability for making the following possible: (1) responsiveness to rapid changes in market demand and product modification; (2) better use of materials, machinery, and personnel, and reduction in inventory; (3) better control of production and management of the total manufacturing operation; and (4) manufacture of high-quality products at low cost.

The following is an outline of the major applications of computers in manufacturing (Chapters 14 and 15):

a) **Computer numerical control** (CNC). This is a method of controlling the movements of machine components by direct insertion of coded instructions in the

(a)

(b)

(c)

**FIGURE 1.8**   Machining a mold cavity for making sunglasses. (a) Computer model of the sunglasses as designed and viewed on the monitor. (b) Machining the die cavity using a computerized numerical-control milling machine. (c) Final product produced from the mold. *Source*: Courtesy of Mastercam/CNC Software, Inc.

form of numerical data (Fig. 1.8). Numerical control was first implemented in the early 1950s and was a major advance in the automation of machines.

b) **Adaptive control** (AC). The parameters in a manufacturing process are adjusted automatically to optimize production rate and product quality and to minimize cost. Parameters such as forces, temperatures, surface finish, and dimensions of the part are monitored constantly. If they move outside the acceptable range, the system adjusts the process variables until the parameters again fall within the acceptable range.

c) **Industrial robots.** Introduced in the early 1960s, industrial robots have been replacing humans in operations that are repetitive, boring, and dangerous, thus reducing the possibility of human error, decreasing variability in product quality, and improving productivity. Robots with sensory-perception capabilities (intelligent robots) are being developed, with movements that simulate those of humans (Section 14.7).

d) **Automated handling.** Computers have made possible highly efficient handling of materials and products in various stages of completion (i.e., work in progress, WIP), such as when the materials and products are being moved from storage to machines or from machine to machine and at the points of inspection, inventory, and shipment.

e) **Automated and robotic assembly systems.** These systems are replacing costly assembly by human operators. Products are being designed or redesigned so that they can be assembled more easily by machine.

f) **Computer-aided process planning** (CAPP). This approach is capable of improving productivity in a plant by optimizing process plans, reducing planning costs, and improving the consistency of product quality and reliability. Functions such as estimation of cost and monitoring of work standards (i.e., the time required to perform a certain operation) can also be incorporated into the system.

g) **Group technology** (GT). The concept of group technology is that parts can be grouped and produced by classifying them into families, according to similarities in design and similarities in the manufacturing processes employed to produce each part. In this way, part designs and process plans can be standardized, and families of like parts can be produced efficiently and economically.

h) **Just-in-time production** (JIT). The principal of JIT is that supplies are delivered just in time to be used, parts are produced just in time to be made into subassemblies and assemblies, and products are finished just in time to be delivered to the customer. In this way, inventory-carrying costs are low, part defects are detected right away, productivity is increased, and high-quality products are made at low cost.

i) **Cellular manufacturing.** Cellular manufacturing involves workstations, called *manufacturing cells*, usually containing several machines controlled by a central robot, each machine performing a different operation on the part.

j) **Flexible manufacturing systems** (FMS). This methodology integrates manufacturing cells into a large unit, all interfaced with a central computer. Flexible manufacturing systems have the highest level of efficiency, sophistication, and productivity of all manufacturing systems (Fig. 1.9). Although very costly, they are capable of efficiently producing parts in small runs and of changing manufacturing sequences on different parts quickly. This flexibility enables them to meet rapid changes in market demand for various types of products.

k) **Expert systems.** These systems are basically complex computer programs. They are rapidly developing the capability to perform tasks and solve difficult real-life problems much as human experts would.

l) **Artificial intelligence** (AI). This important field involves the use of machines and computers to replace human intelligence. Computer-controlled systems are becoming capable of learning from experience and of making decisions that optimize operations and minimize costs. Artificial neural networks, which are designed to simulate the thought processes of the human brain, have the capability of modeling and simulating production facilities, monitoring and controlling manufacturing processes, diagnosing problems in machine performance, conducting financial planning, and managing a company's manufacturing strategy.

Although large corporations can implement modern technology and afford to take risks, smaller companies, with their limited resources, personnel, and capital, generally have difficulty in doing so. More recently, the concept of *shared manufacturing* has been proposed. This system would consist of a regional or nationwide network of manufacturing facilities with state-of-the-art equipment for training, prototype

**FIGURE 1.9** General view of a flexible manufacturing system, showing several machines (machining centers) and an automated guided vehicle (AGV) moving along the aisle. *Source*: Courtesy of Cincinnati Machine, a UNOVA Company.

development, and small-scale production runs, and it would be available to help small companies develop products that compete in the global marketplace.

In view of the aforementioned advances, some experts have envisioned *the factory of the future*. Although highly controversial, and viewed as unrealistic by some, this is a system in which production will take place with little or no direct human intervention. The human role is expected to be confined to the supervision, maintenance, and upgrading of machines, computers, and software.

The implementation of some of the modern technologies requires significant technical and economic expertise, time, and capital investment. Some of the advanced technologies can be applied improperly, or they can be implemented on too large or ambitious a scale, involving major expenditures with questionable return on investment. Consequently, it is essential to perform a comprehensive analysis and assessment of the real and specific needs of a company and of the market for its products, as well as of whether there is good communication among the parties involved, such as the vendors.

## 1.8 | Quality Assurance and Total Quality Management

Product quality has always been one of the most important concerns in manufacturing, as it directly influences the marketability of a product and customer satisfaction. Traditionally, quality assurance has been obtained by inspecting parts after they have been manufactured; parts are inspected to ensure that they conform to a detailed set of specifications and standards, such as dimensional tolerances, surface finish, and mechanical and physical properties.

However, quality cannot be inspected into a product after it is made. The practice of inspecting products after they are made has now been replaced by the broader view that *quality must be built into a product*, from the design stage through all subsequent stages of manufacture and assembly. Because products typically are made by using several manufacturing processes that can have significant variations in their performance, even within a brief period of time, the control of processes is a critical factor in product quality.

Thus, *we control processes, and not products.* Producing defective products can be very costly to the manufacturer, creating difficulties in assembly operations, necessitating repairs in the field, and resulting in customer dissatisfaction. Contrary to general public opinion, low-quality products do not necessarily cost less than high-quality products. Although it can be described in various ways, **product integrity** is a term that can be used to define the degree to which a product (1) is suitable for its intended purpose, (2) fills a real market need, (3) functions reliably during its life expectancy, and (4) can be maintained with relative ease.

**Total quality management** (TQM) and *quality assurance* are the responsibility of everyone involved in designing and manufacturing a product. Awareness of the technological and economic importance of built-in product quality has been heightened by pioneers in quality control, primarily Deming, Taguchi, and Juran. They have pointed out the importance of management's commitment to product quality, pride of workmanship at all levels of production, and the use of powerful techniques such as **statistical process control** (SPC) and *control charts* for on-line monitoring of part production and rapid identification of the sources of quality problems (Section 4.9). Ultimately, the major goal is to prevent defects from occurring rather than to identify defective products; for example, computer chips can now be produced in such a way that only a few parts out of a million may be defective. The importance of this trend to the reliability of products and customer satisfaction is self-evident.

Important developments in quality assurance include the implementation of **design of experiments**, a technique in which the factors involved in a manufacturing process and their interactions are studied simultaneously. Thus, for example, variables affecting dimensional accuracy or surface finish in a machining operation can readily be identified, allowing appropriate actions to be taken. The use of computers has greatly enhanced our capability to rapidly utilize such techniques.

The major trend toward global competitiveness has created a need for international conformity in the use of (and for consensus regarding the establishment of) quality control methods. This need has resulted in the International Organization for Standardization ISO 9000 series (Quality Management and Quality Assurance Standards), as well as QS 9000. A company's registration for this standard, which is a quality process certification, not a product certification, means that the company conforms to consistent practices as specified by its own quality system. These two standards have permanently influenced the manner in which companies conduct business in world trade, and they have become the world standard for quality.

# 1.9 | Product Liability

We are all familiar with the consequences of using a product that has malfunctioned, causing bodily injury, or even death, and the financial loss to an employee as well as the organization manufacturing that product. This important topic is referred to as **product liability**. Because of the related technical and legal aspects, for which laws can vary from state to state and from country to country, this subject is complex and can have a major economic impact on all the parties involved.

Designing and manufacturing safe products is an important and integral part of a manufacturer's responsibilities. All those involved with product design, manufacture, and marketing must be fully cognizant of the consequences of product failure, including failures occurring during possible misuse of the product. Numerous examples of products that could involve liability can be cited: (1) a grinding wheel that shatters and blinds a worker; (2) a cable that snaps, allowing a platform to drop; (3) brakes that become inoperative due to the failure of one component; (4) machines with no guards or inappropriate guards; and (5) electric or pneumatic tools without proper warning labels.

**Human-factors engineering** and **ergonomics** (human–machine interactions) are important aspects of the design and manufacture of safe products. Examples of products for which these factors are faulty include (1) an uncomfortable or unstable workbench or chair; (2) a mechanism that is difficult to operate manually, causing back injury; and (3) a poorly designed computer keyboard, causing pain to the user's hands and arms as a result of repetitive use.

## 1.10  | Manufacturing Costs, Lean Production, and Agile Manufacturing

The cost of a product is often the overriding consideration in its marketability and general customer satisfaction. Typically, manufacturing costs represent about 40% of a product's selling price. The total cost of manufacturing a product consists of costs of materials, tooling, and labor, as well as fixed and capital costs, and several factors are involved in each cost category. Manufacturing costs can be minimized by analyzing the product's design to determine whether part size and shape are optimal and the materials selected are the least costly, while still possessing the desired properties and characteristics. The possibility of substituting materials is an important consideration in minimizing costs (Chapter 16).

While economic considerations in manufacturing have always been a major factor, they have become even more so as international competition (**global competitiveness**) for high-quality products and low prices becomes entrenched in worldwide markets. The markets have become multinational and dynamic, product lines have become extensive and technically complex, and demands by customers for quality have become commonplace (**world-class manufacturing**). To respond to these needs while keeping costs low is a constant challenge to manufacturing companies and is crucial to their very survival. These approaches require that manufacturers **benchmark** their operations; this concept means understanding the competitive position of other manufacturers with respect to one's own position and setting realistic goals for the future. Benchmarking is thus a reference from which various measurements can be made and compared.

The trends just briefly outlined have led to the concept of **lean production**, also called *lean manufacturing*. Not a novel concept, it basically involves a major assessment of each activity of a company: the efficiency and effectiveness of its various operations, the necessity of retaining some of its operations and managers, the efficiency of the machinery and equipment used in the operation while maintaining and improving quality, the number of personnel involved in a particular operation, and a thorough analysis to reduce the cost of each activity, including both productive and nonproductive labor. This concept requires a fundamental change in corporate culture as well as necessitates cooperation and teamwork between management and the workforce. It does not necessarily mean cutting back resources, but aims at *continuously improving* efficiency and profitability of the company by removing all kinds of waste from the operations (*zero-base waste*) and dealing with problems right away.

**Agile manufacturing** is a term that has been coined to indicate the use of the principles of lean production on a broader scale. The principle behind agile manufacturing is ensuring flexibility (agility) in the manufacturing enterprise so that it can quickly respond to changes in product variety and demand and customer needs. This agility is to be achieved through machines and equipment with built-in flexibility (**reconfigurable machines**), using modular components that can be arranged and rearranged in different ways, advanced computer hardware and software, reduced changeover time, and advanced communications systems. It has been predicted, for example, that in the future, the automotive industry will be able to configure and build a car in three days and that, eventually, the traditional assembly line will be replaced by a system in which a nearly custom-made car will be produced by connecting modules.

# 1.11  | General Trends in Manufacturing

With advances in all aspects of materials, processes, and production control, there have been certain important trends in manufacturing, briefly outlined as follows:

**Materials.**  The trend is for better control of material compositions, purity, and defects (e.g. impurities, inclusions, flaws) in order to enhance the material's overall properties, manufacturing characteristics, reliability, and service life while keeping costs low. Developments are continuing on superconductors, semiconductors, nanomaterials and nanopowders, amorphous alloys, shape-memory alloys (*smart materials*), coatings, and various other engineered metallic and nonmetallic materials. Testing methods and equipment are being improved, through methods including the use of advanced computers and software, particularly for materials such as ceramics, carbides, and various composites.

Concerns over energy and material savings are leading to better recyclability and higher strength- and stiffness-to-weight ratios. Thermal treatment of materials is being conducted under better control of relevant variables for more predictable and reliable results, and surface treatment methods are being advanced rapidly. Included in these developments are advances in tool, die, and mold materials, with better resistance to a wide variety of process variables, thus improving the efficiency and economics of manufacturing processes. As a result of these developments, production of goods has become more efficient, with high-quality products and at low cost.

**Processes, equipment, and systems.**  Continuing developments in computers, controls, industrial robots, automated inspection, handling and assembly, and sensor technology are having a major impact on the efficiency and reliability of all manufacturing processes and equipment. Advances in computer hardware and software, communications systems, adaptive control, expert systems, and artificial intelligence and neural networks have all helped enable the effective implementation of concepts such as group technology, cellular manufacturing, and flexible manufacturing systems, as well as modern practices in the efficient administration of manufacturing organizations.

Computer simulation and modeling are becoming widely used in design and manufacturing, resulting in the optimization of processes and production systems, and better prediction of the effects of relevant variables on product integrity. As a result of such efforts, the speed and efficiency of product design and manufacturing are improving greatly, which also affects the overall economics of production and reduces product costs in an increasingly competitive marketplace.

## 1.12 | Responsibilities of Manufacturing Engineers

The various manufacturing activities and functions described in this chapter must be organized and managed efficiently and effectively in order to maximize productivity and minimize costs while maintaining high quality standards. Because of the complex interactions among the various factors involved in manufacturing (materials, machines, people, information, power, and capital), proper coordination and administration of diverse functions and responsibilities are essential.

**Manufacturing engineers** traditionally have had several major responsibilities:

1. Plan the manufacture of a product, and select the processes to be used. This function requires a thorough knowledge of the product, its expected performance, and standards and specifications.

2. Identify the machines, equipment, tooling, and personnel needed to carry out the plan. This function requires evaluation of the capabilities of machines, tools, and workers so that proper functions and responsibilities can be assigned.

3. Interact with design and materials engineers to optimize productivity and minimize production costs.

4. Cooperate with **industrial engineers** when planning plant-floor activities such as plant layout, machine arrangement, material-handling equipment, time-and-motion study, production methods analysis, production planning and scheduling, and maintenance. Some of these activities are carried out under the name **plant engineering,** and some are interchangeably performed by both manufacturing and industrial engineers.

Manufacturing engineers, in cooperation with industrial engineers, are also responsible for evaluating new technologies, their applications, and how they can be implemented. In view of the rapidly growing amount of global technical information, this task in itself can present a major challenge. Gaining a broad perspective of computer capabilities, applications, and integration in all phases of manufacturing operations is crucial. This knowledge is particularly essential for long-range production-facility planning, because of the wide variety of products available, competition, and constantly changing global markets.

## SUMMARY

- Manufacturing is the process of converting raw materials into products by means of a variety of processes and methods (Section 1.1).
- Product design is an integral part of manufacturing, as evidenced by trends in concurrent engineering, design for manufacture, design for assembly, disassembly, and service (Sections 1.2, 1.3).
- Designing and manufacturing safe and environmentally friendly products are important and integral parts of a manufacturer's responsibilities (Section 1.4).
- A key task is to select appropriate materials and an optimal manufacturing method from several possible alternatives, given product design goals, process capabilities, and cost considerations (Sections 1.5, 1.6).
- Computer-integrated manufacturing technologies use computers to automate a wide range of design, analysis, manufacturing, and quality-control tasks (Section 1.7).
- Ensuring product quality is now a concurrent engineering process rather than a last step in the manufacture of a product. Total quality management and statistical

process control techniques have increased our ability to build quality into a product at every step of the design and manufacturing process (Section 1.8).

- Human-factors engineering, ergonomics, and product liability are important considerations in design and manufacturing (Section 1.9).

- Lean production and agile manufacturing are approaches that focus on the efficiency and flexibility of the entire organization in order to help manufacturers respond to global competitiveness and economic challenges (Section 1.10).

- There are certain general trends that have an important bearing on the future of manufacturing processes and materials, and systems, including better control of materials, development of new materials improved recyclability, and increased availability and application of computer hardware and software (Section 1.11).

- Manufacturing engineers, in cooperation with other professionals in a variety of disciplines, have major responsibilities regarding all aspects outlined previously (Section 1.12).

# 2

# Fundamentals of the Mechanical Behavior of Materials

2.1 Introduction

2.2 Tension

2.3 Compression

2.4 Torsion

2.5 Bending

2.6 Hardness

2.7 Fatigue

2.8 Creep

2.9 Impact

2.10 Residual Stresses

2.11 Triaxial Stresses and Yield Criteria

2.12 Work of Deformation

## 2.1 | Introduction

Chapter 1 outlined the manufacturing methods and systems by which materials can be shaped into useful products. In manufacturing, one of the most important groups of processes is **plastic deformation** (also known as **deformation processing**), namely, shaping materials by applying forces in various ways. It includes *bulk deformation* (forging, rolling, extrusion, and rod and wire drawing) and *sheet-forming processes* (bending, drawing, spinning, and general pressworking). This chapter describes the fundamental aspects of the mechanical behavior of materials during deformation. Individual topics described are deformation modes, stresses, forces, work of deformation, effects of rate of deformation and temperature, hardness, residual stresses, and yield criteria.

In stretching a piece of metal to make an object such as an automobile fender or a length of wire, the material is subjected to *tension*. A solid cylindrical piece of metal is forged in the making of a turbine disk, thus subjecting the material to *compression*. Sheet metal undergoes shearing stresses when, for example, a hole is punched through its cross-section. A piece of plastic tubing is expanded by internal pressure to make a bottle, subjecting the material to tension in various directions.

In all these processes, the material is subjected to one or more of the basic modes of deformation, namely, tension, compression, and shear, as shown in Fig. 2.1. The degree of deformation to which the material is subjected is defined as **strain**. For tension or compression, the **engineering strain**, or **nominal strain**, is defined as

$$e = \frac{\ell - \ell_o}{\ell_o}. \tag{2.1}$$

**FIGURE 2.1** Types of strain. (a) Tensile, (b) compressive, and (c) shear. All deformation processes in manufacturing involve strains of these types. Tensile strains are involved in stretching sheet metal to make car bodies, compressive strains in forging metals to make turbine disks, and shear strains in making holes by punching.

(a)  (b)  (c)

Note that in tension the strain is positive, and in compression it is negative. The **shear strain** in Fig. 2.1c is defined as

$$\gamma = \frac{a}{b}. \tag{2.2}$$

In order to change the geometry of the elements or bodies shown in Fig. 2.1, *forces* must be applied to them, shown by the arrows. The determination of these forces as a function of strain is very important in manufacturing processes. We have to know the forces in order to design the proper equipment, select the tool and die materials for proper strength, and determine whether a specific metalworking operation can be accomplished on certain equipment. Thus, the relation between a force and the deformation it produces is an essential parameter in manufacturing.

## 2.2 | Tension

Because of its relative simplicity, the *tension test* is the most common test for determining the *strength-deformation characteristics* of materials. It involves the preparation of a test specimen (according to ASTM specifications) and testing it under tension on any of a variety of available testing equipment.

The specimen has an original length $\ell_o$ and an original cross-sectional area $A_o$ (Fig. 2.2a). Although most specimens are solid and round, flat-sheet and tubular specimens are also tested under tension. The original length is the distance between **gage marks** on the specimen and is generally 50 mm (2 in). Longer lengths may be used for larger specimens, such as structural members, and shorter lengths can be used as well.

Figure 2.2b shows typical results from a tension test. The **engineering stress**, or **nominal stress**, is defined as the ratio of the applied load to the original area,

$$\sigma = \frac{P}{A_o}, \tag{2.3}$$

and the engineering strain is given by Eq. (2.1).

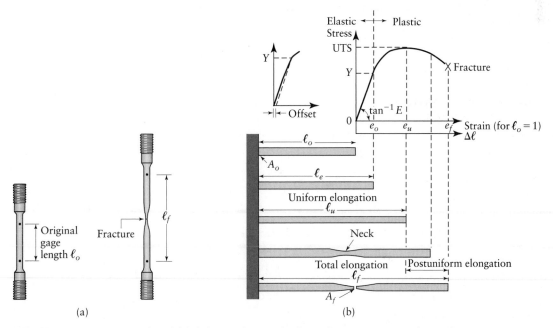

(a)                                      (b)

**FIGURE 2.2**    (a) Original and final shape of a standard tensile-test specimen. (b) Outline of a tensile-test sequence showing different stages in the elongation of the specimen.

When the load is first applied, the specimen elongates proportionately to the load up to the **proportional limit**; this is known as **linear elastic behavior**. The material will continue to deform elastically, although not strictly linearly, up to the **yield point** Y. If the load is removed before the yield point is reached, the specimen will return to its original length. The **modulus of elasticity**, or **Young's modulus**, E, is defined as follows:

$$E = \frac{\sigma}{e}.$$  (2.4)

This linear relationship between stress and strain is known as **Hooke's law**, the more generalized forms of which are given in Section 2.11. The elongation of the specimen is accompanied by a contraction of its lateral dimensions. The absolute value of the ratio of the lateral strain to the longitudinal strain is known as **Poisson's ratio, ν**. Table 2.1 gives typical values for E and ν for various materials.

The area under the stress–strain curve up to the yield point Y of a material is known as the **modulus of resilience**:

$$\text{Modulus of resilience} = \frac{Ye_o}{2} = \frac{Y^2}{2E}.$$  (2.5)

**TABLE 2.1**

**Typical Mechanical Properties of Various Materials at Room Temperature**

|  | E (GPa) | Y (MPa) | UTS (MPa) | Elongation in 50 mm (%) | Poisson's Ratio (ν) |
|---|---|---|---|---|---|
| **METALS (WROUGHT)** | | | | | |
| Aluminum and its alloys | 69–79 | 35–550 | 90–600 | 45–5 | 0.31–0.34 |
| Copper and its alloys | 105–150 | 76–1100 | 140–1310 | 65–3 | 0.33–0.35 |
| Lead and its alloys | 14 | 14 | 20–55 | 50–9 | 0.43 |
| Magnesium and its alloys | 41–45 | 130–305 | 240–380 | 21–5 | 0.29–0.35 |
| Molybdenum and its alloys | 330–360 | 80–2070 | 90–2340 | 40–30 | 0.32 |
| Nickel and its alloys | 180–214 | 105–1200 | 345–1450 | 60–5 | 0.31 |
| Steels | 190–200 | 205–1725 | 415–1750 | 65–2 | 0.28–0.33 |
| Stainless steels | 190–200 | 240–480 | 480–760 | 60–20 | 0.28–0.30 |
| Titanium and its alloys | 80–130 | 344–1380 | 415–1450 | 25–7 | 0.31–0.34 |
| Tungsten and its alloys | 350–400 | 550–690 | 620–760 | 0 | 0.27 |
| **NONMETALLIC MATERIALS** | | | | | |
| Ceramics | 70–1000 | – | 140–2600 | 0 | 0.2 |
| Diamond | 820–1050 | – | – | – | – |
| Glass and porcelain | 70–80 | – | 140 | 0 | 0.24 |
| Rubbers | 0.01–0.1 | – | – | – | 0.5 |
| Thermoplastics | 1.4–3.4 | – | 7–80 | 1000–5 | 0.32–0.40 |
| Thermoplastics, reinforced | 2–50 | – | 20–120 | 10–1 | – |
| Thermosets | 3.5–17 | – | 35–170 | 0 | 0.34 |
| Boron fibers | 380 | – | 3500 | 0 | – |
| Carbon fibers | 275–415 | – | 2000–5300 | 1–2 | – |
| Glass fibers (S, E) | 73–85 | – | 3500–4600 | 5 | – |
| Kevlar fibers (29, 49, 129) | 70–113 | – | 3000–3400 | 3–4 | – |
| Spectra fibers (900, 1000) | 73–100 | – | 2400–2800 | 3 | – |

*Note:* In the upper table, the lowest values for E, Y, and UTS and the highest values for elongation are for the pure metals. Multiply GPa by 145,000 to obtain psi, and MPa by 145 to obtain psi. For example, 100 GPa = 14,500 ksi, and 100 MPa = 14,500 psi.

This area has the units of **energy per unit volume** and indicates the **specific energy** that the material can store elastically. Typical values for the modulus of resilience are $2.1 \times 10^4$ N-m/m$^3$ (3 in-lb/in$^3$) for annealed copper, $1.9 \times 10^5$ (28) for annealed medium-carbon steel, and $2.7 \times 10^6$ (385) for spring steel.

With increasing load, the specimen begins to yield; that is, it begins to undergo **plastic (permanent) deformation**, and the relationship between stress and strain is no longer linear. The rate of change in the slope of the stress–strain curve beyond the yield point is very small for most materials, so the determination of $Y$ can be difficult. The usual practice is to define the yield stress as the point on the curve that is *offset* by a strain of (usually) 0.2%, or 0.002 (Fig 2.2b); other offset strains may also be used and should be specified when reporting the yield stress of a material.

It is important to note that yielding does not necessarily mean failure. In the design of structures and load-bearing members, yielding is not acceptable, since it leads to permanent deformation. However, yielding is essential in metalworking processes, such as bending, forging, rolling, and sheet-metal forming, where materials have to be subjected to permanent deformation.

As the specimen continues to elongate under increasing load beyond $Y$, its cross-sectional area decreases permanently and uniformly throughout its gage length. If the specimen is unloaded from a stress level higher than $Y$, the curve follows a straight line downward and parallel to the original elastic slope (Fig. 2.3). As the load, and hence the engineering stress, is further increased, it eventually reaches a maximum and then begins to decrease. The maximum stress is known as the **tensile strength**, or **ultimate tensile strength** (UTS), of the material (Table 2.1). Ultimate tensile strength is a practical measure of the overall strength of a material.

When the specimen is loaded beyond its UTS, it begins to *neck* (Fig. 2.2a), and the elongation is no longer uniform. That is, the change in the cross-sectional area of the specimen is no longer uniform between the gage marks of the specimen, but is concentrated locally in a "neck" formed in the specimen (known as **necking**, or *necking down*). As the test progresses, the engineering stress drops further, and the specimen finally fractures in the necked region. The final stress level (marked by an $\times$ in Fig. 2.2b) is known as the **breaking**, or **fracture, stress**.

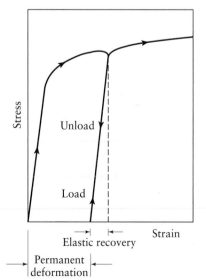

**FIGURE 2.3**   Schematic illustration of loading and unloading of a tensile-test specimen. Note that during unloading, the curve follows a path parallel to the original elastic slope.

### 2.2.1 Ductility

The strain at fracture is a measure of *ductility*, that is, how large a strain a material withstands before failure. Note from Fig. 2.2b that until the UTS is reached, elongation is *uniform*; the strain up to the UTS is known as **uniform strain**. The elongation at fracture is known as the **total elongation**. It is measured between the original gage marks after the two pieces of the broken specimen are placed together.

Two quantities commonly used to define ductility in a tension test are *elongation* and *reduction of area*. **Elongation** is defined as

$$\text{Elongation} = \frac{\ell_f - \ell_o}{\ell_o} \times 100 \qquad (2.6)$$

and is based on the total elongation (Table 2.1).

Necking is a *local* phenomenon. If we put a series of gage marks at different points on the specimen, pull and break the specimen under tension, and then calculate the percent elongation for each pair of gage marks, we will find that with decreasing gage length, the percent elongation increases (Fig. 2.4). The closest pair of gage marks undergoes the largest elongation, because these gage marks are closest to the necked region. However, the curves do not approach zero elongation with increasing gage length, because the specimens have all undergone a finite permanent elongation before fracture. It is therefore important to report gage length in conjunction with elongation data; other tensile properties, however, are generally independent of gage length.

A second measure of ductility is **reduction of area**, defined as

$$\text{Reduction of area} = \frac{A_o - A_f}{A_o} \times 100. \qquad (2.7)$$

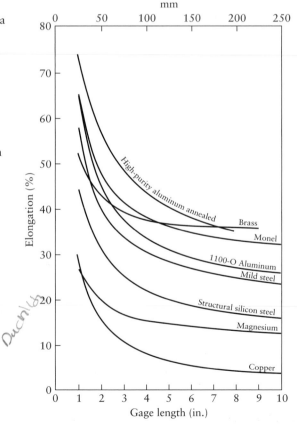

**FIGURE 2.4**  Total elongation of a specimen in a tension test as a function of original gage length for various metals. Because necking is a local phenomenon, elongation decreases with gage length. Standard gage length is usually 50 mm (2 in.), although shorter lengths can be used if larger specimens are not available.

Thus, a material that necks down to a point at fracture, such as a glass rod at elevated temperature, has 100% reduction of area.

The elongation and reduction of area are generally related to each other for many engineering metals and alloys. Elongation ranges approximately between 10% and 60%, and values between 20% and 90% are typical for reduction of area for most materials. *Thermoplastics* (Chapter 10) and *superplastic* materials (Section 2.2.7) exhibit much higher ductility. Brittle materials, by definition, have little or no ductility. Examples of brittle materials include glass at room temperature, a piece of chalk, gray cast iron, and ceramics.

## 2.2.2  True stress and true strain

Because stress is defined as the ratio of force to area, *true stress* is defined as

$$\sigma = \frac{P}{A}, \tag{2.8}$$

where $A$ is the actual (hence true) or instantaneous area supporting the load.

Likewise, the complete tension test may be regarded as a series of incremental tension tests where, for each succeeding increment, the original specimen is a little longer than the previous one. Thus, *true strain* (or *natural or logarithmic strain*), $\varepsilon$, can be defined as

$$\varepsilon = \int_{\ell_o}^{\ell} \frac{d\ell}{\ell} = \ln\left(\frac{\ell}{\ell_o}\right). \tag{2.9}$$

Note that, for small values of engineering strain, $e = \varepsilon$, since $\ln(1 + e) = \varepsilon$. For larger strains, however, the values diverge rapidly, as shown in Table 2.2.

The volume of a metal specimen in the plastic region of the test remains constant (*volume constancy*, Section 2.11.5). Hence, the true strain within the uniform elongation range can be expressed as

$$\varepsilon = \ln\left(\frac{\ell}{\ell_o}\right) = \ln\left(\frac{A_o}{A}\right) = \ln\left(\frac{D_o}{D}\right)^2 = 2\ln\left(\frac{D_o}{D}\right). \tag{2.10}$$

Once necking begins, the true strain at any point in the specimen can be calculated from the change in cross-sectional area at that point; thus, by definition, the largest strain is at the narrowest region of the neck.

We have seen that, for small strains, the engineering and true strains are very close, and therefore either one can be used in calculations. However, for the large strains encountered in metalworking, the true strain should be used. This value is the true measure of the strain and can be illustrated by the following two examples: (a) Assume that a tension specimen is elongated to twice its original length. This deformation is equivalent to compressing a specimen to one half its original height. Using the subscripts $t$ and $c$ for tension and compression, respectively, we find that $\varepsilon_t = 0.69$ and $\varepsilon_c = -0.69$, whereas $e_t = 1$ and $e_c = -0.5$. Hence, true strain is a correct measure of strain. (b) As another example, assume that a specimen 10 mm in

**TABLE 2.2**

## Comparison of Engineering and True Strains in Tension

| $e$ | 0.01 | 0.05 | 0.1 | 0.2 | 0.5 | 1 | 2 | 5 | 10 |
|---|---|---|---|---|---|---|---|---|---|
| $\varepsilon$ | 0.01 | 0.049 | 0.095 | 0.18 | 0.4 | 0.69 | 1.1 | 1.8 | 2.4 |

height is compressed to a final thickness of zero; thus, $\varepsilon_c = -\infty$, whereas $e_c = -1$. We have deformed the specimen infinitely, which is exactly what the value of the true strain indicates. From these examples, it is obvious that true strain is consistent with the actual physical phenomenon, which is not the case for engineering strain.

### 2.2.3 True-stress–true-strain curves

The relation between engineering and true values for stress and strain can now be used to construct *true-stress–true-strain curves* from a curve such as that in Fig. 2.2b. The procedure for this construction is given in Example 2.1. Figure 2.5a shows typical true-stress–true-strain curve. For convenience, such a curve is typically approximated by the equation

$$\sigma = K\varepsilon^n. \tag{2.11}$$

Note that Eq. 2.11 indicates neither the elastic region nor the yield point $Y$ of the material. These quantities, however, are readily available from the engineering stress–strain curve. (Since the strains at the yield point are very small, the difference between the true and engineering yield stress is negligible for metals. The reason is that the difference in the cross-sectional areas $A_o$ and $A$ at yielding is very small.)

We can rewrite Eq. (2.11) as

$$\log \sigma = \log K + n \log \varepsilon.$$

If we now plot the true-stress–true-strain curve on a log–log scale, we obtain the graph in Fig. 2.5b. The slope $n$ is known as the **strain-hardening exponent**, and $K$ is known

**FIGURE 2.5**    (a) True-stress–true-strain curve in tension. Note that, unlike in an engineering stress–strain curve, the slope is always positive, and the slope decreases with increasing strain. Although stress and strain are proportional in the elastic range, the total curve can be approximated by the power expression shown. On this curve, $Y$ is the yield stress and $Y_f$ is the flow stress. (b) True-stress–true-strain curve plotted on a log–log scale. (c) True-stress–true-strain curve in tension for 1100-O aluminum plotted on a log–log scale. Note the large difference in the slopes in the *elastic* and *plastic* ranges. *Source*: After R. M. Caddell and R. Sowerby.

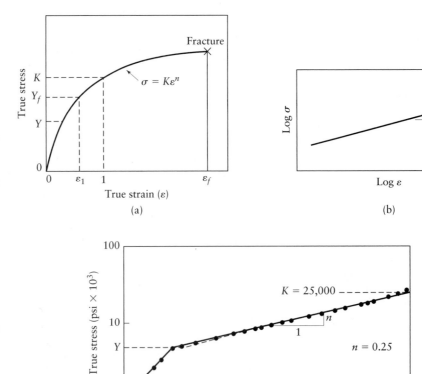

**TABLE 2.3**

| Typical Values for K and n in Eq. (2.11) at Room Temperature | | |
|---|---|---|
| Material | $K$ (MPa) | $n$ |
| Aluminum, 1100-O | 180 | 0.20 |
| 2024-T4 | 690 | 0.16 |
| 5052-O | 210 | 0.13 |
| 6061-O | 205 | 0.20 |
| 6061-T6 | 410 | 0.05 |
| 7075-O | 400 | 0.17 |
| Brass, 70-30, annealed | 895 | 0.49 |
| 85-15, cold rolled | 580 | 0.34 |
| Bronze (phosphor), annealed | 720 | 0.46 |
| Cobalt-base alloy, heat treated | 2070 | 0.50 |
| Copper, annealed | 315 | 0.54 |
| Molybdenum, annealed | 725 | 0.13 |
| Steel, low carbon, annealed | 530 | 0.26 |
| 1045 hot rolled | 965 | 0.14 |
| 1112 annealed | 760 | 0.19 |
| 1112 cold rolled | 760 | 0.08 |
| 4135 annealed | 1015 | 0.17 |
| 4135 cold rolled | 1100 | 0.14 |
| 4340 annealed | 640 | 0.15 |
| 17–4 P–H, annealed | 1200 | 0.05 |
| 52100 annealed | 1450 | 0.07 |
| 304 stainless, annealed | 1275 | 0.45 |
| 410 stainless, annealed | 960 | 0.10 |

*Note:* 100 MPa = 14,500 psi.

as the **strength coefficient**. Note that $K$ is the true stress at a true strain of unity. Table 2.3 lists values of $K$ and $n$ for a variety of engineering materials. The true-stress–true-strain curves for some materials are given in Fig. 2.6. Some differences between the values listed in Table 2.3 and these curves exist because of different sources of data and test conditions.

**FIGURE 2.6**   True-stress–true-strain curves in tension at room temperature for various metals. The point of intersection of each curve at the ordinate is the yield stress Y; thus, the elastic portions of the curves are not indicated. When the K and n values are determined from these curves, they may not agree with those given in Table 2.3, because of the different sources from which they were collected. *Source*: S. Kalpakjian.

In Fig. 2.5a, $Y_f$ is the flow stress, defined as the true stress required to continue plastic deformation at a particular true strain $\varepsilon_1$. Thus, for strain-hardening materials, the flow stress increases with increasing strain. Note from Fig. 2.5c that the elastic strains are much smaller than the plastic strains. Consequently, and although both effects exist, we will ignore elastic strains in our calculations for forming processes throughout the rest of this text; the plastic strain will be the total strain.

The area under the true-stress–true-strain curve is known as the **toughness** and can be expressed as

$$\text{Toughness} = \int_0^{\varepsilon_f} \sigma \, d\varepsilon, \qquad (2.12)$$

where $\varepsilon_f$ is the true strain at fracture. Toughness is hence defined as the energy per unit volume (specific energy) that has been dissipated up to the point of fracture. It is important to remember that this specific energy pertains only to that volume of material at the narrowest region of the neck; any volume of material away from the neck has undergone less strain and thus has dissipated less energy.

As used here, toughness is different than the concept of fracture toughness as treated in textbooks on fracture mechanics. Fracture mechanics (the study of the initiation and propagation of cracks in a solid medium) is beyond the scope of this text. It is of limited relevance to metalworking processes, except in die design and die life.

---

**EXAMPLE 2.1  Construction of a true-stress–true-strain curve**

The following data are taken from a stainless-steel tension-test specimen:

| Load $P$ (lb) | Extension $\Delta\ell$ (in) |
|---|---|
| 1600 | 0 |
| 2500 | 0.02 |
| 3000 | 0.08 |
| 3600 | 0.20 |
| 4200 | 0.40 |
| 4500 | 0.60 |
| 4600 (max.) | 0.86 |
| 3300 (fracture) | 0.98 |

$A_o = 0.056 \text{ in}^2$
$A_f = 0.016 \text{ in}^2$
$\ell_o = 2 \text{ in.}$

Draw the true-stress–true-strain curve for the material.

**Solution.** In order to determine the true stress and true strain, we use the following relationships:

$$\text{True stress } \sigma = \frac{P}{A};$$

$$\text{True strain } \varepsilon = \ln\left(\frac{\ell}{\ell_o}\right) \text{ up to necking;}$$

$$\varepsilon_f = \ln\left(\frac{A_o}{A_f}\right) \text{ at fracture;}$$

$$\ell = \ell_o + \Delta\ell \text{ up to necking.}$$

Assuming that the volume of the specimen remains constant, we have

$$A_o\ell_o = A\ell,$$

or

$$A = \frac{A_o \ell_o}{\ell} = \frac{(0.056)(2)}{\ell} = \frac{0.112}{\ell} \text{ in}^2.$$

Using these relationships, we obtain the following data:

| $\Delta\ell$ (in) | $\ell$ (in) | $\varepsilon$ | $A$ (in$^2$) | True stress (psi) |
|---|---|---|---|---|
| 0 | 2.00 | 0 | 0.056 | 28,600 |
| 0.02 | 2.02 | 0.01 | 0.055 | 45,000 |
| 0.08 | 2.08 | 0.039 | 0.054 | 55,800 |
| 0.20 | 2.20 | 0.095 | 0.051 | 70,800 |
| 0.40 | 2.40 | 0.182 | 0.047 | 90,000 |
| 0.60 | 2.60 | 0.262 | 0.043 | 104,500 |
| 0.86 | 2.86 | 0.357 | 0.039 | 117,300 |
| 0.98 | 2.98 | 1.253 | 0.016 | 206,000 |

The true stress and true strain are plotted (solid line) in Fig. 2.7. The point at necking is connected to the point at fracture by a straight line, because there are no data on the instantaneous areas after necking begins. The reason for the correction on this curve (broken line) is explained in Example 2.7.

## 2.2.4 Instability in simple tension

As noted previously, the onset of necking in a tension test corresponds to the ultimate tensile strength, UTS, of the material. The slope of the load–elongation curve at this point is zero (or $dP = 0$), and it is here that *instability* begins; that is, the specimen begins to neck and cannot support the load, because the neck is becoming smaller in cross-sectional area.

Using the relationships

$$\varepsilon = \ln\left(\frac{A_o}{A}\right), \quad A = A_o e^{-\varepsilon}, \quad \text{and} \quad P = \sigma A = \sigma A_o e^{-\varepsilon},$$

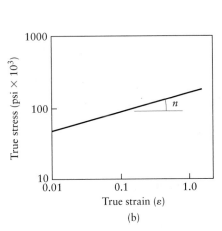

FIGURE 2.7 (a) True-stress–true-strain curve developed from the given data. Note that this curve has a positive slope, indicating that the material is becoming stronger as it is strained. (b) True-stress–true-strain curve plotted on a log–log scale and based on the corrected curve. The steeper the slope $n$, the higher the capacity of the material to become stronger and harder as it is worked.

we can determine $dP/d\varepsilon$:

$$\frac{dP}{d\varepsilon} = A_o\left(\frac{d\sigma}{d\varepsilon}e^{-\varepsilon} - \sigma e^{-\varepsilon}\right).$$

Because $dP = 0$ at the UTS where necking begins, we set this expression equal to zero, obtaining

$$\frac{d\sigma}{d\varepsilon} = \sigma.$$

However,

$$\sigma = K\varepsilon^n,$$

and consequently,

$$nK\varepsilon^{n-1} = K\varepsilon^n,$$

or

$$\varepsilon = n. \tag{2.13}$$

Thus, the true strain at the onset of necking (i.e., the termination of uniform elongation) is numerically equal to the strain-hardening exponent $n$. Hence, the higher the value of $n$, the greater the strain to which a piece of material can be stretched before necking begins. Table 2.3 shows that metals such as annealed copper, brass, and stainless steel can be stretched uniformly to a greater extent than the other materials listed. These observations are covered in greater detail in Chapters 7 and 10, because of their relevance to sheet-forming processes.

Instability in a tension test can be viewed as a phenomenon in which two competing processes are taking place simultaneously. As the load on the specimen is increased, its cross-sectional area becomes smaller; this behavior is more pronounced in the region where necking begins. With increasing strain, however, the material becomes stronger, due to strain hardening. Because the load on the specimen is the product of area and strength, instability sets in when the rate of decrease of cross-sectional area is greater than the rate of increase of strength; this condition is known as **geometric softening**.

---

**EXAMPLE 2.2  Calculation of ultimate tensile strength**

A material has a true-stress–true-strain curve given by

$$\sigma = 100{,}000\varepsilon^{0.5} \text{ psi}.$$

Calculate the true UTS and the engineering UTS of this material.

**Solution.** Since the necking strain corresponds to the maximum load, and the necking strain for this material is given as

$$\varepsilon = n = 0.5,$$

then from

$$\sigma = Kn^n,$$

we have

$$\text{UTS}_{\text{true}} = 100{,}000(0.5)^{0.5} = 70{,}710 \text{ psi}.$$

The cross-sectional area at the onset of necking is obtained from

$$\ln\left(\frac{A_o}{A_{neck}}\right) = n = 0.5.$$

Consequently,

$$A_{neck} = A_o e^{-0.5},$$

and the maximum load $P$ is expressed as

$$P = \sigma A = \sigma A_o e^{-0.5},$$

where $\sigma$ is the true UTS. Hence,

$$P = (70,710)(0.606)(A_o) = 42,850\, A_o \text{ lb}.$$

since the engineering UTS $= P/A_o$, we have

$$UTS = 42,850 \text{ psi}.$$

### 2.2.5  Types of stress–strain curves

Each material has a differently shaped stress–strain curve; its shape depends on its composition and many other factors, to be treated in detail later. Some of the major types of curves are shown in Fig. 2.8 with their associated stress–strain equations and have the following characteristics:

1. A **perfectly elastic** material behaves like a spring with stiffness $E$. The behavior of brittle materials, such as glass, ceramics, and some cast irons, may be represented by such a curve (Fig. 2.8a). There is a limit to the stress the material can sustain, after which it breaks. Permanent deformation, if any, is negligible.

2. A **rigid, perfectly plastic** material has, by definition, an infinite value of $E$. Once the stress reaches the yield stress $Y$, it continues to undergo deformation at the same stress level. When the load is released, the material has undergone permanent deformation, with no elastic recovery (Fig. 2.8b).

3. An **elastic, perfectly plastic** material is a combination of the first two types of material: it has a finite elastic modulus, and it undergoes elastic recovery when the load is released (Fig. 2.8c).

4. A **rigid, linearly strain-hardening** material requires an increasing stress level to undergo further strain; thus, its **flow stress** (the magnitude of the stress required to maintain plastic deformation at a given strain; see Fig. 2.5a) increases with increasing strain. It has no elastic recovery upon unloading (Fig. 2.8d).

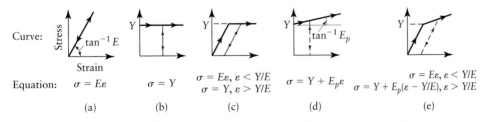

Curve:

Equation:  (a) $\sigma = E\varepsilon$  (b) $\sigma = Y$  (c) $\sigma = E\varepsilon,\ \varepsilon < Y/E$ ; $\sigma = Y,\ \varepsilon > Y/E$  (d) $\sigma = Y + E_p\varepsilon$  (e) $\sigma = E\varepsilon,\ \varepsilon < Y/E$ ; $\sigma = Y + E_p(\varepsilon - Y/E),\ \varepsilon > Y/E$

**FIGURE 2.8**  Schematic illustration of various types of idealized stress–strain curves. (a) Perfectly elastic. (b) Rigid, perfectly plastic. (c) Elastic, perfectly plastic. (d) Rigid, linearly strain hardening. $Ep$ is referred to as the *plastic modulus* and is the slope of the stress–strain curve after yiedling. (e) Elastic, linearly strain hardening. The broken lines and arrows indicate unloading and reloading, respectively, during the test. Most engineering metals exhibit a behavior similar to that shown in curve (e).

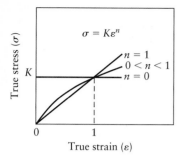

**FIGURE 2.9** The effect of the strain-hardening exponent, $n$, on the shape of true-stress–true-strain curves. When $n = 1$, the material is elastic, and when $n = 0$, it is rigid and perfectly plastic.

5. An **elastic, linearly strain-hardening** curve (Fig. 2.8e) is an approximation of the behavior of most engineering materials, with the modification that the plastic portion of the curve has a decreasing slope with increasing strain (see Fig. 2.5a).

Note that some of these curves can be expressed by Eq. 2.11 by changing the value of $n$ (Fig. 2.9), or by other equations of a similar nature.

### 2.2.6  Effects of temperature

The various factors that have an influence on the shape of stress–strain curves will be discussed in this and subsequent sections. The first factor is temperature. Although it is somewhat difficult to generalize, increasing temperature usually increases ductility and toughness and lowers the modulus of elasticity, yield stress, and ultimate tensile strength (Fig. 2.10). Temperature also affects the strain-hardening exponent, $n$, of most metals, in that $n$ decreases with increasing temperature. Depending on the type of material and its composition and level of impurities, elevated temperatures can have other significant effects as well, as detailed in Chapter 3. However, the influence of temperature is best discussed in conjunction with strain rate, for the reasons explained in the next section.

### 2.2.7  Effects of strain rate

Depending on the particular manufacturing operation and equipment, a piece of material may be formed at low or high speeds; thus, in performing a tension test, the specimen can be strained at different rates to simulate the actual deformation process. Whereas the **deformation rate** may be defined as the *speed* (in m/s, for instance) at which a tension test is being carried out, the strain rate is a function of the geometry of the specimen.

The **engineering strain rate**, $\dot{e}$, is defined as

$$\dot{e} = \frac{de}{dt} = \frac{d\left(\dfrac{\ell - \ell_o}{\ell_o}\right)}{dt} = \frac{1}{\ell_o}\frac{d\ell}{dt} = \frac{v}{\ell_o} \tag{2.14}$$

and the **true strain rate**, $\dot{\varepsilon}$ as

$$\dot{\varepsilon} = \frac{d\varepsilon}{dt} = \frac{d\left[\ln\left(\dfrac{\ell}{\ell_o}\right)\right]}{dt} = \frac{1}{\ell}\frac{d\ell}{dt} = \frac{v}{\ell}, \tag{2.15}$$

**FIGURE 2.10** Typical effects of temperature on engineering stress–strain curves. Temperature affects the modulus of elasticity, yield stress, ultimate tensile strength, and toughness of materials.

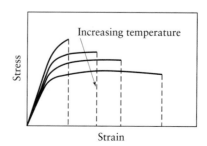

TABLE 2.4

**Typical Ranges of Strain, Deformation Speed, and Strain Rates in Metalworking Processes**

| Process | True Strain | Deformation Speed (m/s) | Strain Rate (s$^{-1}$) |
|---|---|---|---|
| Cold working | | | |
|     Forging, rolling | 0.1–0.5 | 0.1–100 | $1-10^3$ |
|     Wire and tube drawing | 0.05–0.5 | 0.1–100 | $1-10^4$ |
|     Explosive forming | 0.05–0.2 | 10–100 | $10-10^5$ |
| Hot working and warm working | | | |
|     Forging, rolling | 0.1–0.5 | 0.1–30 | $1-10^3$ |
|     Extrusion | 2–5 | 0.1–1 | $10^{-1}-10^2$ |
| Machining | 1–10 | 0.1–100 | $10^3-10^6$ |
| Sheet-metal forming | 0.1–0.5 | 0.05–2 | $1-10^2$ |
| Superplastic forming | 0.2–3 | $10^{-4}-10^{-2}$ | $10^{-4}-10^{-2}$ |

where $v$ is the speed or rate of deformation (e.g., the speed of the jaws of the testing machine in which the specimen is clamped).

Although the deformation rate $v$ and the engineering strain rate, $\dot{e}$, are proportional, the true strain rate, $\dot{\varepsilon}$, is not proportional to $v$. Thus, in a tension test with constant $v$, the true strain rate decreases as the specimen becomes longer. Therefore, in order to maintain a constant $\dot{\varepsilon}$, the speed must be increased accordingly; however, for small changes in length of the specimen during a test, this difference is not significant.

Table 2.4 lists typical deformation speeds employed in various metalworking processes, as well as the strain rates involved. Note that there are considerable differences in the magnitudes. Because of this wide range, strain rates are quoted in orders of magnitudes, such as $10^2\,\text{s}^{-1}$, $10^4\,\text{s}^{-1}$, etc.

The general effects of temperature and strain rate on the strength of metals are shown in Fig. 2.11. The figure clearly indicates that increasing strain rate increases strength and that the sensitivity of strength to the strain rate increases with

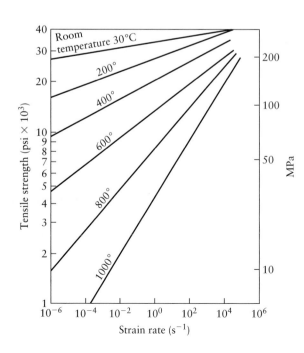

**FIGURE 2.11** The effect of strain rate on the ultimate tensile strength of aluminum. Note that as temperature increases, the slope increases. Thus, tensile strength becomes more sensitive to strain rate as temperature increases. *Source*: After J. H. Hollomon.

**FIGURE 2.12**

Dependence of the strain-rate sensitivity exponent $m$ on the homologous temperature $T/T_m$ for various materials. $T$ is the testing temperature, and $T_m$ is the melting point of the metal, both on the absolute scale. The transition in the slopes of the curve occurs at about the recrystallization temperature of the metals. *Source*: After F. W. Boulger.

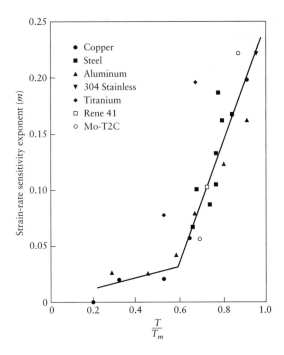

temperature, as shown in Fig. 2.12. Note, however, that this effect is relatively small at room temperature. Figure 2.11 shows that the same strength can be obtained either at low temperature and low strain rate or at high temperature and high strain rate. These relationships are important in estimating the resistance of materials to deformation when processing them at various strain rates and temperatures.

The effect of strain rate on strength also depends on the particular level of strain: it increases with strain. The strain rate also affects the strain-hardening exponent, $n$, because $n$ decreases as the strain rate increases. The effect of strain rate on the strength of materials is generally expressed as

$$\sigma = C\dot{\varepsilon}^m, \qquad (2.16)$$

where $C$ is the **strength coefficient**, similar to $K$ in Eq. (2.11), and $m$ is the **strain-rate sensitivity exponent** of the material. A general range of values for $m$ is as follows: up to 0.05 for cold working, 0.05 to 0.4 for hot working, and 0.3 to 0.85 for superplastic materials. For a *Newtonian fluid*, for which the shear stress increases linearly with the rate of shear, the value of $m$ is unity. Table 2.5 lists some specific values for $C$ and $m$. (See also Section 10.3.)

The term **superplastic** refers to the capability of some materials to undergo large uniform elongation prior to failure. The elongation may be on the order of a few hundred percent to as much as 2000%. An extreme example is hot glass; other materials exhibiting superplastic behavior include polymers at elevated temperatures, very fine-grain alloys of zinc–aluminum, and titanium alloys.

The magnitude of $m$ has a significant effect on necking in a tension test. Experimental observations show that, with high $m$ values, the material stretches to a greater length before it fails; this behavior is an indication that necking is delayed with increasing $m$. When necking is about to begin, the region's strength with respect to the rest of the specimen increases, due to strain hardening. However, the strain rate in the neck region is also higher than in the rest of the specimen, because the material is elongating faster there. Since the material in the necked region becomes stronger as it is strained at a higher rate, the region exhibits a greater resistance to necking. This increased resistance to necking thus depends on the magnitude of $m$. As the test progresses, necking becomes more *diffuse*, and the specimen becomes longer before fracture; hence, total elongation increases with increasing values of *m* (Fig. 2.13). As expected, the elongation after necking (*postuniform elongation*) also increases with increasing *m*. It has been observed that the value of *m* decreases with metals of increasing strength.

**TABLE 2.5**

**Approximate Range of Values for C and *m* in Eq. (2.16) for Various Annealed Metals at True Strains Ranging from 0.2 to 1.0**

| Material | Temperature, °C | C psi × 10³ | C MPa | *m* |
|---|---|---|---|---|
| Aluminum | 200–500 | 12–2 | 82–14 | 0.07–0.23 |
| Aluminum alloys | 200–500 | 45–5 | 310–35 | 0–0.20 |
| Copper | 300–900 | 35–3 | 240–20 | 0.06–0.17 |
| Copper alloys (brasses) | 200–800 | 60–2 | 415–14 | 0.02–0.3 |
| Lead | 100–300 | 1.6–0.3 | 11–2 | 0.1–0.2 |
| Magnesium | 200–400 | 20–2 | 140–14 | 0.07–0.43 |
| Steel | | | | |
|     Low carbon | 900–1200 | 24–7 | 165–48 | 0.08–0.22 |
|     Medium carbon | 900–1200 | 23–7 | 160–48 | 0.07–0.24 |
|     Stainless | 600–1200 | 60–5 | 415–35 | 0.02–0.4 |
| Titanium | 200–1000 | 135–2 | 930–14 | 0.04–0.3 |
| Titanium alloys | 200–1000 | 130–5 | 900–35 | 0.02–0.3 |
| Ti-6Al-4V* | 815–930 | 9.5–1.6 | 65–11 | 0.50–0.80 |
| Zirconium | 200–1000 | 120–4 | 830–27 | 0.04–0.4 |

* at a strain rate of $2 \times 10^{-4}\,\mathrm{s}^{-1}$.

*Note*: As temperature increases, C decreases and *m* increases. As strain increases, C increases and *m* may increase or decrease, or it may become negative within certain ranges of temperature and strain.

*Source*: After T. Altan and F.W. Boulger.

**FIGURE 2.13**   (a) The effect of the strain-rate sensitivity exponent *m* on the total elongation for various metals. Note that elongation at high values of *m* approaches 1000%. *Source*: After D. Lee and W. A. Backofen. (b) The effect of the strain-rate sensitivity exponent on the postuniform (after necking) elongation for various metals. *Source*: After A. K. Ghosh.

Because the formability of materials depends largely on their ductility, it is important to recognize the effect of temperature and strain rate on ductility. Generally, higher strain rates have an adverse effect on the ductility of materials. (The increase in ductility due to the strain-rate sensitivity of materials has been exploited in superplastic forming of metals, as described in Section 7.10.)

### 2.2.8 Effects of hydrostatic pressure

Although most tests are generally carried out at ambient pressure, tests also have been performed under hydrostatic conditions, with pressures ranging up to around $10^3$ MPa ($10^5$ psi). Three important observations have been made concerning the effects of high hydrostatic pressure on the behavior of materials: (1) It substantially increases the strain at fracture (Fig. 2.14); (2) it has little or no effect on the shape of the true-stress–true-strain curve, but only extends it; and (3) it has no effect on the strain or the maximum load at which necking begins. Experiments have shown that, generally, the mechanical properties of metals are not altered after the material has been subjected to hydrostatic pressure.

The increase in ductility due to hydrostatic pressure has also been observed in other tests, such as compression and torsion tests. This increase in ductility has been observed not only with ductile metals, but also with brittle metals and nonmetallic materials. Materials such as cast iron, marble, and various rocks acquire some ductility (or an increase in ductility) and thus deform plastically when subjected to hydrostatic pressure. The level of pressure required to enhance ductility depends on the material.

### 2.2.9 Effects of radiation

In view of the various nuclear applications of many metals and alloys, studies have been conducted on the effects of radiation on material properties. Typical changes in the mechanical properties of steels and other metals exposed to high-energy radiation are (1) increased yield stress, (2) increased tensile strength and hardness, and (3) decreased ductility and toughness. The magnitudes of these changes depend on the material and its condition, temperature, and level of radiation.

**FIGURE 2.14** The effect of hydrostatic pressure on true strain at fracture in tension for various metals. Note that even cast iron becomes ductile under high pressure. *Source*: After H. L. D. Pugh and D. Green.

## 2.3 | Compression

Many operations in metalworking, such as forging, rolling, and extrusion, are performed with the workpieces under compressive forces. The compression test, in which the specimen is subjected to a compressive load, as shown in Fig. 2.1b, can give useful information for these processes, such as stresses required and the behavior of the material under compression. The deformation shown in Fig. 2.1b is ideal. The compression test is usually carried out by compressing a solid cylindrical specimen between two flat platens; the friction between the specimen and the dies is an important factor. Friction causes **barreling** (Fig. 2.15). In other words, friction prevents the top and bottom surfaces from expanding freely.

This situation makes it difficult to obtain relevant data and to construct properly the compressive stress–strain curve, because (1) the cross-sectional area of the specimen changes along its height; and (2) friction dissipates energy, and this energy is supplied through an increased compressive force. Therefore, it is difficult to obtain results that are truly indicative of the properties of the material. With effective lubrication or other means, it is possible, of course, to minimize friction (and hence barreling) to obtain a reasonably constant cross-sectional area during this test.

The *engineering strain rate*, $\dot{e}$, in compression is given by

$$\dot{e} = -\frac{v}{h_o},\tag{2.17}$$

where $v$ is the speed of the die and $h_o$ is the original height of the specimen. The *true strain rate*, $\dot{\varepsilon}$, is given by

$$\dot{\varepsilon} = -\frac{v}{h},\tag{2.18}$$

where $h$ is the instantaneous height of the specimen. If $v$ is constant, the true strain rate increases as the test progresses. In order to conduct this test at a constant true strain rate, a **cam plastometer** has been designed that, through a cam action, reduces $v$ proportionately as $h$ decreases during the test.

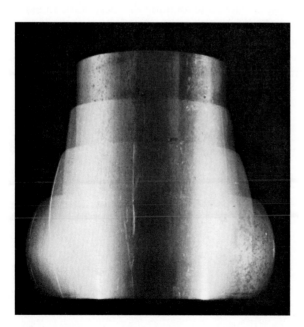

**FIGURE 2.15** Barreling in compression of a round solid cylindrical specimen (7075-O aluminum) between flat dies. Barreling is caused by friction at the die–specimen interfaces, which retards the free flow of the material. See also Figs. 6.1 and 6.2. *Source*: K. M. Kulkarni and S. Kalpakjian.

**FIGURE 2.16**   Schematic illustration of the plane-strain compression test. The dimensional relationships shown should be satisfied for this test to be useful and reproducible. This test gives the yield stress of the material in plane strain, Y'. *Source*: After A. Nadai and H. Ford.

$w > 5h$
$w > 5b$
$b > 2h$ to $4h$

The compression test can also be used to determine the ductility of a metal by observing the cracks that form on the barreled cylindrical surfaces of the specimen. Hydrostatic pressure has a beneficial effect in delaying the formation of these cracks. With a sufficiently ductile material and effective lubrication, compression tests can be carried out uniformly to large strains. This behavior is unlike that in the tension test, where, even for very ductile materials, necking sets in after relatively little elongation of the specimen.

### 2.3.1 Plane-strain compression test

Another useful test is the plane-strain compression test (Fig. 2.16), which simulates processes such as forging and rolling (Chapter 6). In this test, the die and workpiece geometries are such that the width of the specimen does not undergo any significant change during compression; that is, the material under the dies is in the condition of *plane strain*. (See Section 2.11.3.) The yield stress of a material in plane strain, Y', is given by

$$Y' = \frac{2}{\sqrt{3}} Y = 1.15 Y,$$   (2.19)

according to the distortion-energy yield criterion. (See Section 2.11.2.) As the geometric relationships in Fig. 2.16 show, the test parameters must be chosen properly to make the results meaningful. Furthermore, caution should be exercised in test procedures, such as preparing the die surfaces and aligning the dies, lubricating the surfaces, and accurately measuring the load.

For *ductile* metals, when the results of tension and compression tests on the same material are compared, the true-stress–true-strain curves for both tests coincide (Fig. 2.17); however, this is not the case for brittle materials, particularly in regard to ductility. (See also Section 3.8.)

**FIGURE 2.17**   True-stress–true-strain curve in tension and compression for aluminum. For ductile metals, the curves for tension and compression are identical. *Source*: Courtesy of Alan Cottrell.

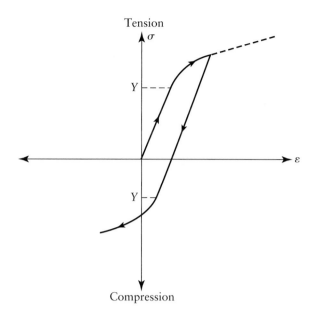

**FIGURE 2.18** Schematic illustration of the Bauschinger effect. The arrows show loading and unloading paths. Note the decrease in the yield stress in compression after the specimen has been subjected to tension. The same result is obtained if compression is applied first, followed by tension, whereby the yield stress in tension decreases.

### 2.3.2 Bauschinger effect

In deformation processing of materials, a workpiece is sometimes first subjected to tension and then to compression, or vice versa. Examples of such situations are bending and unbending, roller leveling (see Fig. 6.41), and reverse drawing (see Fig. 7.62). When a metal with a tensile yield stress, $Y$, is subjected to tension into the plastic range and then the load is released and applied in compression, the yield stress in compression is lower than that in tension (Fig. 2.18). This phenomenon is known as the *Bauschinger effect* and is exhibited in varying degrees by all metals and alloys. This effect is also observed when the loading path is reversed, i.e., compression followed by tension. Because of the lowered yield stress in the reverse direction of load application, this phenomenon is also called **strain softening**, or **work softening**. This phenomenon is also observed in torsion (Section 2.4).

### 2.3.3 The disk test

For brittle materials, such as ceramics and glasses, a *disk test* has been developed in which the disk is subjected to compression between two hardened flat platens (Fig. 2.19). When the disk is loaded as shown in Fig. 2.19, tensile stresses develop perpendicular to the vertical centerline along the disk, fracture begins, and the disk splits vertically in half. (See also *rotary tube piercing* in Section 6.3.5.)

The tensile stress, $\sigma$, in the disk is uniform along the centerline and can be calculated from the formula

$$\sigma = \frac{2P}{\pi dt}, \qquad (2.20)$$

where $P$ is the load at fracture, $d$ is the diameter of the disk and $t$ is its thickness. In order to avoid premature failure at the two contact points, thin strips of soft metal are placed between the disk and the platens. These strips also protect the platens from being damaged during the test.

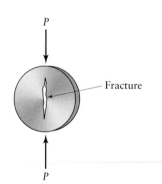

**FIGURE 2.19** Disk test on a brittle material, showing the direction of loading and the fracture path. This test is useful for brittle materials, such as ceramics, carbides, and round specimens from grinding wheels.

## 2.4 | Torsion

Another method of determining properties of materials is the *torsion test*. In order to obtain an approximately uniform stress and strain distribution along the cross-section, this test is generally carried out on a tubular specimen (Fig. 2.20). The *shear stress,* $\tau$, can be determined from the equation

$$\tau = \frac{T}{2\pi r^2 t},\tag{2.21}$$

where $T$ is the torque applied, $r$ is the mean radius, and $t$ is its thickness of the reduced section of the tube. The *shear strain,* $\gamma$, is determined from the equation

$$\gamma = \frac{r\phi}{\ell},\tag{2.22}$$

where $\ell$ is the length of the reduced section and $\phi$ is the angle of twist, in radians. With the shear stress and shear strain thus obtained from torsion tests, we can construct the shear-stress–shear-strain curve of the material. (See also Section 2.11.7.)

In the elastic range, the ratio of the shear stress to shear strain is known as the **shear modulus,** or the **modulus of rigidity,** $G$:

$$G = \frac{\tau}{\gamma}.\tag{2.23}$$

The shear modulus and the modulus of elasticity are related by the formula

$$G = \frac{E}{2(1 + \nu)},\tag{2.24}$$

which is based on a comparison of **simple-shear** and **pure-shear** strains (Fig. 2.21). Note from Fig. 2.21 that the difference between the two strains is that simple shear is equivalent to pure shear plus a rotation of $\gamma/2$ degrees. (See also Section 2.11.7 for the state of stress shown in Fig. 2.21 for pure shear.)

**FIGURE 2.20**    A typical torsion-test specimen. It is mounted between the two heads of a machine and is twisted. Note the shear deformation of an element in the reduced section of the tubular specimen.

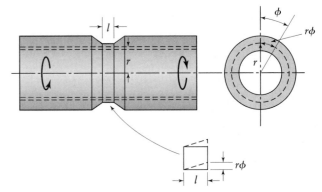

**FIGURE 2.21**
Comparison of (a) simple shear and (b) pure shear. Note that simple shear is equivalent to pure shear plus a rotation.

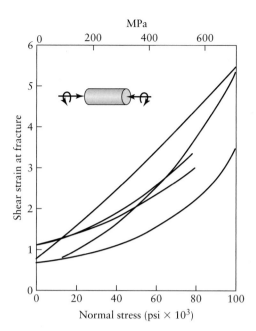

**FIGURE 2.22**   The effect of axial compressive stress on the shear strain at fracture in torsion for various steels. Note that the effect on ductility is similar to that of hydrostatic pressure, shown in Fig. 2.14. *Source:* Based on data in P. W. Bridgman, *Large Plastic Flow and Fracture*, McGraw-Hill, 1952.

In Example 2.3, we show that a thin-walled tube does not neck in torsion, unlike the tension-test specimens described in Section 2.2.4; therefore, we do not have to be concerned with changes in the cross-sectional area of the specimen in torsion testing. The shear-stress–shear-strain curves obtained from torsion tests increase monotonically just as do in true-stress–true-strain curves.

Torsion tests are performed on solid round bars at elevated temperatures in order to estimate the *forgeability* of metals. (See Section 6.2.6.) The greater the number of twists prior to failure, the greater the forgeability of the material. Torsion tests have also been conducted on round bars that are compressed axially (Fig. 2.22). The maximum shear strain at fracture is measured as a function of the compressive stress. The shear strain at fracture increases substantially as the compressive stress is increased. This observation again indicates the beneficial effect of a compressive environment on the ductility of materials. Other experiments show that, with increasing normal *tensile* stresses, the curves in Fig. 2.22 follow a trend downward and to the left, signifying reduced ductility.

The effect of compressive stresses in increasing the maximum shear strain at fracture has also been observed in metal cutting. (See Section 8.2.) The normal compressive stress has no effect on the magnitude of shear stresses required to cause yielding or to continue the deformation, just as hydrostatic pressure has no effect on the general shape of the stress–strain curve.

---

## EXAMPLE 2.3   Instability in torsion of a thin-walled tube

Show that necking cannot take place in the torsion of a thin-walled tube made of a material whose true-stress–true-strain curve is given by $\sigma = K\varepsilon^n$.

**Solution.**   According to Eq. (2.21), the expression for torque, $T$, is

$$T = 2\pi r^2 t\tau,$$

where for this case the *shear stress*, $\tau$, can be related to the *normal stress*, $\sigma$, using Eq. (2.56) for the distortion-energy criterion, as $\tau = \sigma/\sqrt{3}$. (See Section 2.11.7 for details.) The criterion for instability in torsion is

$$\frac{dT}{d\varepsilon} = 0.$$

Because $r$ and $t$ are constant, we have

$$\frac{dT}{d\varepsilon} = \frac{2}{\sqrt{3}} \frac{\pi r^2 \tau \, d\sigma}{d\varepsilon}.$$

For a material represented by $\sigma = K\varepsilon^n$,

$$\frac{d\sigma}{d\varepsilon} = nK\varepsilon^{n-1}.$$

Therefore,

$$\frac{dT}{d\varepsilon} = \frac{2}{\sqrt{3}} \pi r^2 \tau n K \varepsilon^{n-1}.$$

Since none of the quantities in this expression are zero, $dT/d\varepsilon$ cannot be zero. Consequently, a tube in torsion does not undergo necking.

## 2.5 | Bending

Preparing specimens from brittle materials, such as ceramics and carbides, is difficult, because of the problems involved in shaping and machining the materials to proper dimensions. Furthermore, because of their sensitivity to surface defects and notches, clamping brittle specimens for testing is difficult. Also, improper alignment of the test specimen may result in nonuniform stress distribution along the cross-section of the specimen.

A commonly used test method for brittle materials is the *bend*, or *flexure*, *test*, usually involving a specimen that has a rectangular cross-section and is supported at its ends (Fig. 2.23). The load is applied vertically, either at one or two points; hence, these tests are referred to as **three-point** and **four-point bending**, respectively. The stresses in these specimens are tensile at their lower surfaces and compressive at their upper surfaces, and they can be calculated by using simple beam equations, described in texts on the mechanics of solids. The stress at fracture in bending is known as the **modulus of rupture**, or **transverse rupture strength**, and is obtained from the formula

$$\sigma = \frac{Mc}{I}, \tag{2.25}$$

where $M$ is the bending moment, $c$ is one half of the specimen depth, and $I$ is the moment of inertia of the cross-section.

Note that there is a basic difference between the two loading conditions in Fig. 2.23. In three-point bending, the maximum stress is at the center of the beam,

**FIGURE 2.23** Two bend-test methods for brittle materials: (a) three-point bending; (b) four-point bending. The lower sketches represent the bending-moment diagrams, described in texts on the mechanics of solids. Note the region of constant maximum bending moment in diagram (b), whereas the maximum bending moment occurs only at the center of the specimen in diagram (a).

whereas in the four-point test, the maximum stress is constant between the two loading points. The stress is the same in both cases when all other parameters are maintained. However, there is a greater probability for defects and imperfections to be present in the volume of material between the loading points in the four-point test than in the much smaller volume under the single load in the three-point test. This difference means that the four-point test is likely to result in a lower modulus of rupture than that of the three-point test, as also verified by experiments. Furthermore, the results of the four-point test show less scatter than those of the three-point test. The reason is that the strength of a small volume of brittle material (such as under the single load in the three-point test) is subject to greater variations than is the strength of the weakest point in a larger volume (such as in the four-point test).

# 2.6 | Hardness

One of the most common tests for assessing the mechanical properties of materials is the hardness test. The *hardness* of a material is generally defined as its resistance to permanent indentation. Less commonly, hardness may also be defined as resistance to scratching or to wear. Various techniques have been developed to measure the hardness of materials, using different indenter geometries and materials. Because resistance to indentation depends on the shape of the indenter and the load applied, hardness is not a fundamental property. Among the most common standardized hardness tests are the Brinell, Rockwell, Vickers, Knoop (Fig. 2.24), and scleroscope tests.

| Test | Indenter | Shape of indentation Side view | Top view | Load $P$ | Hardness number | |
|---|---|---|---|---|---|---|
| Brinell | 10-mm steel or tungsten carbide ball | $\leftarrow D \rightarrow$ $\rightarrow d \leftarrow$ | $\rightarrow d \leftarrow$ | 500 kg 1500 kg 3000 kg | $HB = \dfrac{2P}{(\pi D)(D \sqrt{D^2 - d^2})}$ | |
| Vickers | Diamond pyramid | 136° | $L$ | 1–120 kg | $HV = \dfrac{1.854P}{L^2}$ | |
| Knoop | Diamond pyramid | $t$ $L/b = 7.11$ $b/t = 4.00$ | $b$ $\leftarrow L \rightarrow$ | 25g–5kg | $HK = \dfrac{14.2P}{L^2}$ | |
| Rockwell A C D | Diamond cone | 120° $t =$ mm | | kg 60 150 100 | HRA HRC HRD | = 100–500t |
| B F G | $\frac{1}{16}$-in-diameter steel ball | $t =$ mm | | 100 60 150 | HRB HRF HRG | = 130–500t |
| E | $\frac{1}{8}$-in-diameter steel ball | | | 100 | HRE | |

**FIGURE 2.24** General characteristics of hardness testing methods. The Knoop test is known as a *microhardness* test, because of the light load and small impressions. *Source:* Courtesy of Dr. William G. Moffatt.

### 2.6.1 Brinell Test

In the *Brinell test*, a steel or tungsten carbide ball 10 mm in diameter is pressed against a surface with a load of 500, 1500, or 3000 kg. The *Brinell hardness number* (HB) is defined as the ratio of the load $P$ to the curved area of indentation, or

$$\text{HB} = \frac{2P}{(\pi D)\left(D - \sqrt{D^2 - d^2}\right)} \text{ kg/mm}^2, \qquad (2.26)$$

where $D$ is the diameter of the ball and $d$ is the diameter of the impression, in mm.

Depending on the condition of the material tested, different types of impressions are obtained on the surface after a Brinell hardness test has been performed. Annealed materials generally have a rounded profile, whereas cold-worked (strain-hardened) materials have a sharp profile (Fig. 2.25). The correct method of measuring the indentation diameter $d$ for both cases is shown in Fig. 2.25.

Because the indenter (with a finite elastic modulus) also undergoes elastic deformation under the applied load $P$, measurements of hardness may not be as accurate as expected. One method of minimizing this effect is to use tungsten carbide balls, which, because of their high modulus of elasticity, deform less than steel balls. Also, since harder workpiece materials produce very small impressions, a 1500-kg or 3000-kg load is recommended in order to obtain impressions that are sufficiently large for accurate measurement. Tungsten carbide balls are generally recommended for Brinell hardness numbers higher than 500. In reporting the test results for these high hardnesses, the type of ball used should be cited.

Because the impressions made by the same indenter at different loads are not geometrically similar, the Brinell hardness number depends on the load used. Consequently, the load employed should also be cited with the test results. The Brinell test is generally suitable for materials of low to medium hardness.

**Brinelling** is a term used to describe permanent indentations on a surface between contacting bodies—for example, a component with a hemispherical protrusion or a ball bearing resting on a flat surface. Under fluctuating loads or vibrations, such as during transportation or vibrating foundations, a permanent indentation may be produced on the flat surface by dynamic loading.

### 2.6.2 Rockwell test

In the *Rockwell test*, the depth of penetration is measured. The indenter is pressed on the surface, first with a minor load and then with a major one. The difference in the depth of penetration under the loads is a measure of the hardness. There are several Rockwell hardness test scales that use different loads, indenter materials, and indenter geometries. Some of the more common hardness scales and the indenters used are listed in Fig. 2.24. The Rockwell hardness number, which is read directly from a dial on the testing machine, is expressed as follows: if the hardness number is 55 using the C scale, then it is written as 55 HRC. *Rockwell*

**FIGURE 2.25** Indentation geometry for Brinell hardness testing: (a) annealed metal; (b) work-hardened metal. Note the difference in metal flow at the periphery of the impressions.

(a)

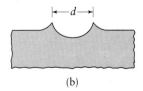

(b)

*superficial hardness tests*, that use lighter loads and the same type of indenters, are also available.

### 2.6.3  Vickers test

The *Vickers test*, also known as the *diamond pyramid hardness test*, uses a pyramid-shaped diamond indenter (see Fig. 2.24) with loads ranging from 1 to 120 kg. The Vickers hardness number (HV) is given by the formula

$$\text{HV} = \frac{1.854P}{L^2} \qquad (2.27)$$

The impressions are typically less than 0.5 mm on the diagonal. The Vickers test gives essentially the same hardness number regardless of the load. It is suitable for testing materials with a wide range of hardnesses, including very hard steels.

### 2.6.4  Knoop test

The *Knoop test* uses a diamond indenter in the shape of an elongated pyramid (see Fig. 2.24) with loads ranging generally from 25 g to 5 kg. The Knoop hardness number (HK) is given by the formula

$$\text{HK} = \frac{14.2P}{L^2}. \qquad (2.28)$$

The Knoop test is referred to as a *microhardness test* because of the light loads used; hence, it is suitable for very small or thin specimens and for hard and brittle materials, such as gemstones, carbides, and glass. This test is also used for measuring the hardness of individual grains in a metal. The size of the indentation is generally in the range of 0.01 to 0.10 mm; therefore, surface preparation is very important. Because the hardness number obtained depends on the applied load, test results should always cite the load applied.

### 2.6.5  Scleroscope

The *scleroscope* is an instrument in which a diamond-tipped indenter (hammer), enclosed in a glass tube, is dropped on the specimen from a certain height. The hardness is determined by the *rebound* of the indenter; the higher the rebound, the harder the specimen. Indentation is slight, and because the instrument is portable, it is useful for measuring the hardness of large objects.

### 2.6.6  Mohs test

The *Mohs test* is based on the capability of one material to *scratch* another. Mohs hardness is on a scale of 1 to 10, with 1 for talc and 10 for diamond (the hardest substance known); thus, a material with a higher Mohs hardness can scratch materials with a lower Mohs hardness. The Mohs scale generally is used by mineralogists and geologists. However, some of the materials tested are of interest in manufacturing, as discussed in Chapters 8, 9, and 11. Although the Mohs scale is qualitative, good correlation is obtained with Knoop hardness. Soft metals have a Mohs hardness of 2 to 3, hardened steels about 6, and aluminum oxide (used in grinding wheels) 9.

### 2.6.7 Durometer

The hardness of elastomers, rubbers, plastics, and similar soft and elastic materials is generally measured with an instrument called a *durometer*. The durometer provides an empirical test in which an indenter is pressed against the surface, with a constant load applied rapidly. The depth of penetration is measured after one second; the hardness is inversely related to the penetration. There are two commonly used scales for this test. Type A has a blunt indenter, a load of 1 kg, and is used for softer materials. Type D has a sharper indenter, a load of 5 kg, and is used for harder materials. The hardness numbers in these tests range up to 100; there are other scales, the lowest of which is 000 for measuring the hardness of soft foams.

### 2.6.8 Hardness and strength

Since hardness is the resistance to permanent indentation, testing for it is equivalent to performing a compression test on a small portion of a material's surface. Therefore, we would expect some correlation between hardness and yield stress, $Y$, in the form of

$$\text{Hardness} = cY, \tag{2.29}$$

where $c$ is a proportionality constant.

Theoretical studies, based on plane-strain slip-line analysis with a smooth flat punch indenting the surface of a semi-infinite body (see Fig. 6.12), show that for a perfectly plastic material of yield stress, $Y$, the value of $c$ is about 3. This value is in reasonably good agreement with experimental data (Fig. 2.26). Note that cold-worked materials (which are close to being perfectly plastic in their behavior) show better agreement than annealed ones. The higher $c$ value for the annealed materials is explained by the fact that, due to strain hardening, the average yield stress they exhibit during indentation is higher than their initial yield stress.

The reason that hardness, as a compression test, gives higher values than the uniaxial yield stress, $Y$, of the material can be seen in the following analysis: if the volume under the indenter were a column of material, it would exhibit a uniaxial compressive yield stress, $Y$. However, the volume being deformed under the indenter

**FIGURE 2.26** Relation between Brinell hardness and yield stress for aluminum and steels. For comparison, the Brinell hardness (which is always measured in kg/mm$^2$) is converted to psi units in the scale on the left.

**FIGURE 2.27** Bulk deformation in mild steel under a spherical indenter. Note that the depth of the deformed zone is about one order of magnitude larger than the depth of indentation. For a hardness test to be valid, the material should be allowed to fully develop this zone. This difference in depth is why thinner specimens require smaller indentations. *Source:* Courtesy of M. C. Shaw and C. T. Yang.

is, in reality, surrounded by a rigid mass (Fig. 2.27). The surrounding mass prevents this volume of material from deforming freely. In fact, this volume is under *triaxial compression*. As shown in Section 2.11, on yield criteria, this material requires a normal compressive yield stress that is higher than the uniaxial yield stress of the material.

More practically, a relationship also has been observed between the ultimate tensile strength (UTS) and Brinell hardness number (HB) for steels,

$$UTS = 500(HB), \qquad (2.30)$$

where UTS is in psi and HB is in kg/mm$^2$ as measured with a load of 3000 kg. In SI units, the relationship is given by

$$UTS = 3.5(HB), \qquad (2.31)$$

where UTS is in MPa.

**Hot hardness tests** can also be carried out using conventional testers with certain modifications, such as enclosing the specimen and indenter in a small electric furnace. The hot hardness of materials is important in applications where the materials are subjected to elevated temperatures, such as in cutting tools in machining and dies for metalworking (See Fig. 8.31).

---

**EXAMPLE 2.4 Calculation of the modulus of resilience**

A piece of steel is highly deformed at room temperature. Its hardness is found to be 300 HB. Estimate the modulus of resilience for this material.

**Solution.** Since the steel has been subjected to large strains at room temperature, it may be assumed that its stress–strain curve has flattened considerably, thus approaching the shape of a perfectly plastic curve. According to Eq. (2.29) and using a value of $c = 3$, we obtain

$$Y = \frac{300}{3} = 100 \, \text{kg/mm}^2 = 142,250 \, \text{psi}.$$

The modulus of resilience is defined by Eq. (2.5):

$$\text{Modulus of resilience} = \frac{Y^2}{2E}.$$

For steel, $E = 30 \times 10^6$ psi. Hence,

$$\text{Modulus of resilience} = \frac{(142,250)^2}{(2 \times 30 \times 10^6)} = 337 \text{ in} \cdot \text{lb/in}^3.$$

## 2.7 | Fatigue

Various structures and components in manufacturing operations, such as tools, dies, gears, cams, shafts, and springs, are subjected to rapidly fluctuating (cyclic or periodic) loads, as well as static loads. Cyclic stresses may be caused by fluctuating mechanical loads, such as gear teeth, or by thermal stresses, such as a cool die coming into repeated contact with hot workpieces. Under these conditions, the part fails at a stress level below which failure would occur under static loading. This phenomenon is known as **fatigue failure** and is responsible for the majority of failures in mechanical components.

Fatigue test methods involve testing specimens under various states of stress, usually in a combination of tension and compression, or torsion. The test is carried out at various stress amplitudes, $S$, and the number of cycles, $N$, to cause total failure of the specimen or part is recorded. Stress amplitude is the maximum stress, in tension and compression, to which the specimen is subjected.

A typical plot, known as an *S–N curve*, is shown in Fig. 2.28. S–N curves are based on complete reversal of the stress–that is, maximum tension, maximum compression, maximum tension, and so on—such as that obtained by bending an eraser or piece of wire alternately in one direction, then the other. The test can also be performed on a rotating shaft with a constant downward load. The maximum stress to which the material can be subjected without fatigue failure, regardless of the number of cycles, is known as the **endurance limit**, or **fatigue limit**.

The endurance limit for metals is related to their ultimate tensile strength (Fig. 2.29). For steels, the endurance limit is about one half their tensile strength. Although many metals, especially steels, have a definite endurance limit, aluminum alloys do not, and their S–N curve continues its downward trend. For metals exhibiting such behavior (most face–centered cubic metals; see Fig. 3.2b), the fatigue strength is specified at a certain number of cycles, such as $10^7$. In this way, the useful service life of the component can be specified.

**FIGURE 2.28**  Typical S–N curves for two metals. Note that, unlike steel, aluminum does not have an endurance limit.

**FIGURE 2.29**   Ratio of endurance limit to tensile strength for various metals, as a function of tensile strength. The fatigue strength of a metal can thus be estimated from its tensile strength.

# 2.8 | Creep

*Creep* is the permanent elongation of a component under a static load maintained for a period of time. It is a phenomenon of metals and certain nonmetallic materials, such as thermoplastics and rubbers, and can occur at any temperature. Lead, for example, creeps under a constant tensile load at room temperature, and the thickness of window glass in old houses has been found to be greater at the bottom than at the top of windows, the glass having undergone creep by its own weight over many years.

For metals and their alloys, creep of any significance occurs at elevated temperatures, beginning at about 200°C (400°F) for aluminum alloys, and up to about 1500°C (2800°F) for refractory alloys. The mechanism of creep at elevated temperatures in metals is generally attributed to *grain-boundary sliding*. Creep is especially important in high-temperature applications, such as gas-turbine blades and similar components in jet engines and rocket motors. High-pressure steam lines and nuclear-fuel elements also are subject to creep. Creep deformation also can occur in tools and dies that are subjected to high stresses at elevated temperatures during hot-working operations such as forging and extrusion.

A creep test typically consists of subjecting a specimen to a constant tensile load, and hence a constant engineering stress, at a certain temperature and measuring the change in length over a period of time. A typical creep curve (Fig. 2.30)

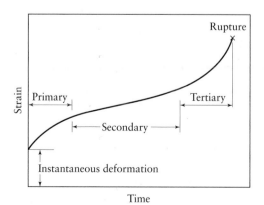

**FIGURE 2.30**   Schematic illustration of a typical creep curve. The linear segment of the curve (constant slope) is useful in designing components for a specific creep life.

usually consists of primary, secondary, and tertiary stages. The specimen eventually fails by necking and fracture, as in a tension test; this failure is called **rupture** or **creep rupture**. As expected, the creep rate increases with temperature and the applied load.

Design against creep usually involves a knowledge of the secondary (linear) range and its slope, because the creep rate can be determined reliably when the curve has a constant slope. Generally, resistance to creep increases with the melting temperature of a material; this trend serves as a general guideline for design purposes. Stainless steels, superalloys, and refractory metals and alloys are commonly used in applications where creep resistance is required.

**Stress relaxation** is closely related to creep. In stress relaxation, the stresses resulting from external loading of a structural component decrease in magnitude over a period of time, even though the dimensions of the component remain constant. Examples include rivets, bolts, guy wires, and similar parts under tension, compression, or bending. Stress relaxation is particularly common and important in thermoplastics. (See Section 10.3.)

## 2.9 | Impact

In many manufacturing operations, as well as during the service life of components, materials are subjected to *impact* (or *dynamic*) *loading*, as in high-speed metalworking operations. Tests have been developed to determine the impact toughness of metallic and nonmetallic materials under impact loading. A typical impact test consists of placing a *notched* specimen in an impact tester and breaking it with a swinging pendulum. In the **Charpy test**, the specimen is supported at both ends (Fig. 2.31a), whereas in the **Izod test**, it is supported at one end, like a cantilever beam (Fig. 2.31b). From the amount of swing of the pendulum, the *energy* dissipated in breaking the specimen is obtained; this energy is the **impact toughness** of the material.

**FIGURE 2.31** Impact tests: (a) Charpy; (b) Izod.

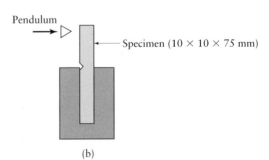

Impact tests are particularly useful in determining the ductile–brittle *transition temperature* of materials. (See Fig. 3.26.) Materials that have high impact resistance are generally those that have high strength and high ductility, and hence high toughness. Sensitivity to surface defects (**notch sensitivity**) is important, as it lowers impact toughness.

## 2.10 | Residual Stresses

In this section, we show that inhomogeneous deformation leads to *residual stresses*, which are stresses that remain within a part after it has been deformed and all external forces have been removed. A typical example of inhomogeneous deformation is the bending of a beam (Fig. 2.32). The bending moment first produces a linear elastic stress distribution. As the moment is increased, the outer fibers begin to yield, and, for a typical strain-hardening material, the stress distribution shown in Fig. 2.32b is eventually obtained. After the part is bent (permanently, since it has undergone plastic deformation), the moment is removed by unloading. This unloading is equivalent to applying an equal and opposite moment to the beam.

As previously shown in Fig. 2.3, all recovery is elastic; thus, in Fig. 2.32c, the moments of the areas *oab* and *oac* about the neutral axis must be equal. (For purposes of this treatment, let's assume that the neutral axis does not shift.) The difference between the two stress distributions produces the residual-stress pattern shown within the beam in Fig. 2.32d. Note that there are compressive residual stresses in layers *ad* and *oe*, and tensile residual stresses in layers *do* and *ef*. With no external forces, the residual stresses in the beam must be in static equilibrium. Although this example involves stresses in one direction only, in most situations in deformation processing, the residual stresses are three dimensional.

The equilibrium of residual stresses may be disturbed by altering the shape of the beam, such as by removing a layer of material by machining. The beam will then acquire a new radius of curvature in order to balance the internal forces. Another example of this effect is the drilling of round holes on surfaces with residual stresses. It

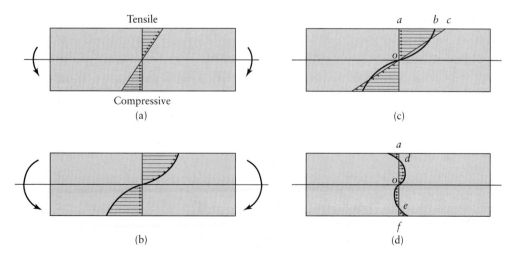

**FIGURE 2.32**   Residual stresses developed in bending a beam made of an elastic, strain-hardening material. Note that unloading is equivalent to applying an equal and opposite moment to the bent part, as shown in diagram (c). Because of nonuniform deformation, most parts made by plastic-deformation processes contain residual stresses. Note that for equilibrium, the forces and moments due to residual stresses must be internally balanced.

**FIGURE 2.33** Distortion of parts with residual stresses after cutting or slitting: (a) rolled sheet or plate; (b) drawn rod; (c) thin-walled tubing. Due to the presence of residual stresses on the surfaces of parts, a round drill may produce an oval-shaped hole, because of relaxation of stresses when a portion is removed.

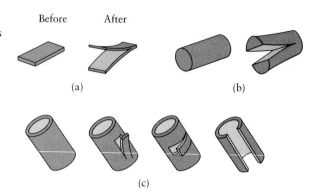

may be found that, as a result of removing the material, the equilibrium of the residual stresses is disturbed, and the hole becomes oval. Such disturbances of residual stresses lead to *warping*, some simple examples of which are shown in Fig. 2.33.

The equilibrium of residual stresses may also be disturbed by *relaxation* of these stresses over a period of time, which results in instability of the dimensions and shape of the component. These dimensional changes can be an important consideration for precision machinery and measuring equipment.

Residual stresses may also be caused by *phase changes* in metals during or after processing, because of density differences between phases (such as between ferrite and martensite in steels). Phase changes cause microscopic volumetric changes and result in residual stresses. This phenomenon is important in warm and hot working and in heat treatment following cold working.

Residual stresses can also be caused by *temperature gradients* within a body, as during the cooling cycle of a casting (Chapter 5), when applying brakes to a railroad wheel, or in a grinding operation (Section 9.4). The expansion and contraction due to temperature gradients are analogous to nonuniform plastic deformation.

### 2.10.1 Effects of residual stresses

Tensile residual stresses on the surface of a part are generally undesirable, because they lower the fatigue life and fracture strength of the part. A surface with tensile residual stresses will sustain lower additional tensile stresses due to external loading than can a surface that is free from residual stresses. This is particularly true for relatively brittle materials, where fracture takes place with little or no plastic deformation. Tensile residual stresses in manufactured products can also lead to **stress cracking**, or **stress-corrosion cracking**, (see Section 3.8.2) over a period of time.

Conversely, compressive residual stresses on a surface are generally desirable. In fact, in order to increase the fatigue life of components, compressive residual stresses are imparted on surfaces by techniques such as **shot peening** and **surface rolling**. (See Section 4.5.1.)

### 2.10.2 Reduction of residual stresses

Residual stresses may be reduced or eliminated either by **stress-relief annealing** or by further *plastic deformation*. Given sufficient time, residual stresses may also be diminished at room temperature by *relaxation*; the time required can be greatly reduced by increasing the temperature of the component. Relaxation of residual stresses by stress-relief annealing is generally accompanied by warpage of the part; hence, a "machining allowance" is commonly provided to compensate for dimensional changes during stress relieving.

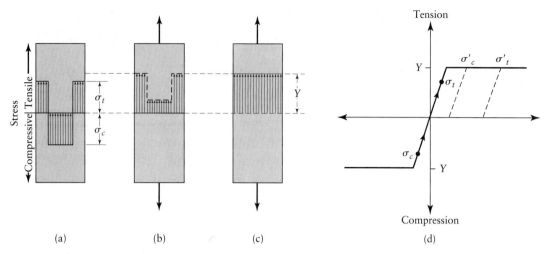

**FIGURE 2.34**   Elimination of residual stresses by stretching. Residual stresses can be also reduced or eliminated by thermal treatments, such as stress relieving or annealing.

The mechanism of residual-stress reduction or elimination by plastic deformation is as follows: assume that a piece of metal has the residual stresses shown in Fig. 2.34a, namely, tensile on the outside and compressive on the inside; these stresses are in equilibrium. Also assume that the material is elastic and perfectly plastic, as shown in Fig. 2.34d. The levels of the residual stresses are shown on the stress–strain diagram; both are below the yield stress, $Y$. If a uniformly distributed tension is now applied to this specimen, points $\sigma_c$ and $\sigma_t$ in the diagram move up on the stress–strain curve, as shown by the arrows. The maximum level that these stresses can reach is the tensile yield stress, $Y$. With sufficiently high loading, the stress distribution eventually becomes uniform throughout the part, as shown in Fig. 2.34c. If the load is then removed, the stresses recover elastically, and the part has no residual stresses. Note that very little stretching is required to relieve these residual stresses. The reason is that the elastic portions of the stress–strain curves for metals are very steep; hence the elastic stresses can be raised to the yield stress with very little strain.

The technique for reducing or relieving residual stresses by plastic deformation, such as by stretching as described, requires sufficient straining to establish a uniformly distributed stress in the part. A material such as the elastic, linearly strain-hardening type (Fig. 2.8e) therefore can never reach this condition, since the compressive stress, $\sigma_c'$, will always lag behind $\sigma_t'$. If the slope of the stress–strain curve in the plastic region is small, the difference between $\sigma_c'$ and $\sigma_t'$ will be rather small, and little residual stress will be left in the part after unloading.

**EXAMPLE 2.5   Elimination of residual stresses by tension**

Refer to Fig. 2.34a, and assume that $\sigma_t = 140$ MPa and $\sigma_c = -140$ MPa. The material is aluminum, and the length of the specimen is 0.25 m. Calculate the length to which this specimen should be stretched so that, when unloaded, it will be free from residual stresses. Assume that the yield stress of the material is 150 MPa.

**Solution.**   Stretching should be to the extent that $\sigma_c$ reaches the yield stress in tension, $Y$. Therefore, the total strain should be equal to the sum of the strain required to bring the compressive residual stress to zero and the strain required to bring it to the tensile yield stress. Hence,

$$\varepsilon_{\text{total}} = \frac{\sigma_c}{E} + \frac{Y}{E}. \tag{2.32}$$

For aluminum, let $E = 70\,\text{GPa}$. Thus,

$$\varepsilon_{\text{total}} = \frac{140}{70 \times 10^3} + \frac{150}{70 \times 10^3} = 0.00414.$$

Hence, the stretched length should be

$$\ln\left(\frac{\ell_f}{0.25}\right) = 0.00414, \quad \text{or} \quad \ell_f = 0.2510\,\text{m}.$$

As the strains are very small, we may use engineering strains in these calculations. Hence,

$$\text{strain} = \frac{\ell_f - 0.25}{0.25} = 0.00414, \quad \text{or} \quad \ell_f = 0.2510\,\text{m}.$$

## 2.11 | Triaxial Stresses and Yield Criteria

In most manufacturing operations involving deformation processing, the material, unlike that in a simple tension or compression test, is generally subjected to triaxial stresses. For example, in the expansion of a thin-walled spherical shell under internal pressure, an element in the shell is subjected to equal biaxial tensile stresses (Fig. 2.35a). In drawing a rod or wire through a conical die, an element in the deformation zone is subjected to tension in its length direction and to compression on its conical surface (Fig. 2.35b). An element in the flange in deep drawing of sheet metal is subjected to a tensile radial stress and compressive stresses on its surface and in the circumferential direction (Fig. 2.35c). As shown in subsequent chapters, many

**FIGURE 2.35**   The state of stress in various metalworking processes. (a) Expansion of a thin-walled spherical shell under internal pressure. (b) Drawing of round rod or wire through a conical die to reduce its diameter; see Section 6.5 (c) Deep drawing of sheet metal with a punch and die to make a metal cup; see Section 7.12.

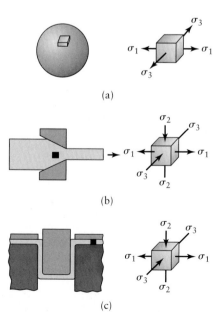

similar examples can be given in which the material is subjected to various normal and shear stresses during processing. The three-dimensional stresses shown in Fig. 2.35 cause distortion of the elements.

In the *elastic range*, the strains are represented by the *generalized Hooke's law* equations:

$$\varepsilon_1 = \frac{1}{E}[\sigma_1 - \nu(\sigma_2 + \sigma_3)]; \tag{2.33a}$$

$$\varepsilon_2 = \frac{1}{E}[\sigma_2 - \nu(\sigma_1 + \sigma_3)]; \tag{2.33b}$$

$$\varepsilon_3 = \frac{1}{E}[\sigma_3 - \nu(\sigma_1 + \sigma_2)]. \tag{2.33c}$$

Therefore, for simple tension where $\sigma_2 = \sigma_3 = 0$,

$$\varepsilon_1 = \frac{\sigma_1}{E},$$

and

$$\varepsilon_2 = \varepsilon_3 = -\frac{\nu\sigma_1}{E}.$$

The negative sign indicates a contraction of the element in the 2 and 3 directions, respectively.

In a simple tension or simple compression test, when the applied stress reaches the uniaxial yield stress $Y$, the material will deform *plastically*. However, if the material is subjected to a more complex state of stress, relationships between the stresses will predict yielding. These relationships are known as *yield criteria*, the most common ones being the maximum-shear-stress criterion and the distortion-energy criterion.

## 2.11.1 Maximum-shear-stress criterion

The maximum-shear-stress criterion, also known as the **Tresca criterion**, states that yielding occurs when the *maximum shear stress* within an element is equal to or exceeds a critical value. As shown in Section 3.3, this critical value of the shear stress is a *material property* and is called **shear yield stress**, $k$; hence, for yielding to occur, the following condition must be met:

$$\tau_{max} \geq k. \tag{2.34}$$

The most convenient way of determining the stresses acting on an element is by use of **Mohr's circles for stresses**, described extensively in textbooks on mechanics of solids. **Principal stresses** and their directions can be determined easily from Mohr's circles or from stress-transformation equations.

If the maximum shear stress, as determined from the Mohr's circles or from appropriate equations, is equal to or exceeds $k$, then yielding will occur. It can be seen that there are many combinations of stresses (or *states of stress*) that can produce the same maximum shear stress. From the simple tension test, we note that

$$k = \frac{Y}{2}, \tag{2.35}$$

where $Y$ is the uniaxial yield stress of the material. If, for some reason, we are unable to increase the stresses on the element in order to cause yielding, the solution is simply to lower $Y$ by raising the temperature of the material; this solution is the basis of and one major reason for hot working of materials. We can now write the maximum-shear-stress criterion as

$$\sigma_{max} - \sigma_{min} = Y. \tag{2.36}$$

This relationship means that the maximum and minimum normal stresses produce the largest circle and hence the *largest* shear stress. Consequently, *the intermediate stress has no effect on yielding*. It should be emphasized that the left-hand side of Eq. (2.36) represents the *applied stresses* and that the right-hand side is a *material property*.

We have assumed here that the material is *continuous, homogeneous,* and *isotropic* (i.e., it has the same properties in all directions) and that tensile stresses are positive, compressive stresses negative, and the yield stress in tension and that in compression are equal (see the discussion of the Bauschinger effect in Section 2.3.2). All of these assumptions are important.

## 2.11.2 Distortion-energy criterion

The distortion-energy criterion, also called the **von Mises criterion**, states that yielding occurs when the relationship between the principal applied stresses and uniaxial yield stress $Y$ of the material is

$$(\sigma_1 - \sigma_2)^2 + (\sigma_2 - \sigma_3)^2 + (\sigma_3 - \sigma_1)^2 = 2Y^2. \tag{2.37}$$

Note that, unlike the case for the maximum-shear-stress criterion, the intermediate principal stress is included in this equation. Here again, the left-hand side of the equation represents the applied stresses and the right-hand side a material property.

---

**EXAMPLE 2.6**  **Yielding of a thin-walled shell**

A thin-walled spherical shell is under internal pressure, $p$. The shell has a diameter of 20 in. and is 0.1 in. thick. It is made of a perfectly plastic material with a yield stress of 20,000 psi. Calculate the pressure required to cause yielding of the shell according to both yield criteria.

**Solution.**  For this shell under internal pressure, the membrane stresses are given by

$$\sigma_1 = \sigma_2 = \frac{pr}{2t}, \tag{2.38}$$

where $r = 10$ in. and $t = 0.1$ in. The stress in the thickness direction, $\sigma_3$, is negligible, because of the high $r/t$ ratio of the shell. Thus, according to the maximum-shear-stress criterion,

$$\sigma_{max} - \sigma_{min} = Y,$$

or

$$\sigma_1 - 0 = Y$$

and

$$\sigma_2 - 0 = Y$$

Hence, $\sigma_1 = \sigma_2 = 20,000$ psi. The pressure required is then

$$p = \frac{2tY}{r} = \frac{(2)(0.1)(20,000)}{10} = 400 \text{ psi.}$$

According to the distortion-energy criterion,

$$(\sigma_1 - \sigma_2)^2 + (\sigma_2 - \sigma_3)^2 + (\sigma_3 - \sigma_1)^2 = 2Y^2,$$

or

$$0 + \sigma_2^2 + \sigma_1^2 = 2Y^2.$$

Hence, $\sigma_1 = \sigma_2 = Y$. Therefore, the answer is the same: $p = 400$ psi.

---

**EXAMPLE 2.7**  **Correction factor for true-stress–true-strain curves**

Explain why a correction factor has to be applied in the construction of a true-stress–true-strain curve, shown in Fig. 2.7a based on tensile-test data.

**Solution.** The neck of a specimen is subjected to a triaxial state of stress, as shown in Fig. 2.36. The reason for this stress state is that each element in the region has a different cross-sectional area; the smaller the area, the greater the tensile stress on the element. Hence, element 1 will contract laterally more than element 2, and so on; however, element 1 is restrained from contracting freely by element 2, and element 2 is restrained by element 3, and so on. This restraint causes radial and circumferential tensile stresses in the necked region, resulting in an axial tensile stress distribution, as shown in Fig. 2.36.

The *true* uniaxial stress in tension is $\sigma$, whereas the *calculated* value of true stress at fracture is the *average* stress; hence, a correction has to be made.

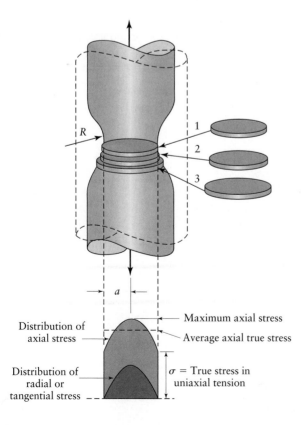

**FIGURE 2.36**  Stress distribution in the necked region of a tensile-test specimen.

A mathematical analysis by P. W. Bridgman gives the ratio of true stress to average stress as

$$\frac{\sigma}{\sigma_{av}} = \frac{1}{\left(1 + \dfrac{2R}{a}\right)\left[\ln\left(1 + \dfrac{a}{2R}\right)\right]}, \tag{2.39}$$

where $R$ is the radius of curvature of the neck and $a$ is the radius of the specimen at the neck. However, because $R$ is difficult to measure during a test, an empirical relation has been established between $a/R$ and the true strain at the neck. The corrected true-stress–true-strain curve is shown in Fig. 2.7a in Example 2.1.

### 2.11.3 Plane stress and plane strain

Plane stress and plane strain are important in the application of yield criteria. *Plane stress* is the state of stress in which one or two of the pairs of faces on an elemental cube are free from stress. An example is the torsion of a thin-walled tube (See Fig. 2.20). There are no stresses normal to the inside or outside surface of the tube; hence the state of the stress of the tube is one of plane stress. Other examples are shown in Fig. 2.37.

The state of stress where one of the pairs of faces on an element undergoes zero strain is known as *plane strain* (e.g., Figs. 2.37c and d). Another example is the **plane-strain compression** test shown in Fig. 2.16. There, by proper choice of specimen dimensions, the width of the specimen essentially is kept constant. (Note that an element does not have to be physically constrained on the pair of faces for plane-strain conditions to exist.) A third example is the torsion of a thin-walled tube in which the wall thickness remains constant. (See Section 2.11.7.)

**FIGURE 2.37** Examples of states of stress: (a) Plane stress in sheet stretching; there are no stresses acting on the surfaces of the sheet. (b) Plane stress in compression; there are no stresses acting on the sides of the specimen being compressed. (c) Plane strain in tension; the width of the sheet remains constant while being stretched. (d) Plane strain in compression (see also Fig. 2.16); the width of the specimen remains constant, due to the restraint by the groove. (e) Triaxial tensile stresses acting on an element. (f) Hydrostatic compression of an element. Note also that an element on the cylindrical portion of a thin-walled tube in torsion is in the condition of both plane stress and plane strain. (See also Section 2.11.7.)

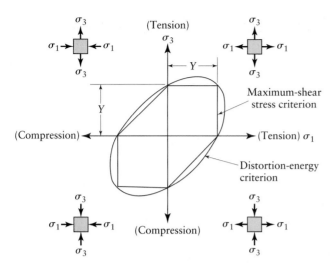

**FIGURE 2.38** Plane-stress diagrams for maximum-shear-stress and distortion-energy criteria; note that $\sigma_2 = 0$.

A review of the two yield criteria previously outlined indicates that the *plane-stress* condition (where $\sigma_2 = 0$) can be represented by the diagram in Fig. 2.38. The maximum-shear-stress criterion gives an envelope of straight lines. In the first quadrant, where $\sigma_1 > 0$ and $\sigma_3 > 0$, and $\sigma_2$ is always zero for plane stress, Eq. (2.36) reduces to $\sigma_{max} = Y$. Therefore, the maximum value that either $\sigma_1$ or $\sigma_3$ can acquire is $Y$—hence the straight lines in the diagram. In the third quadrant, the same situation exists, because $\sigma_1$ and $\sigma_3$ are both compressive. In the second and fourth quadrants, $\sigma_2$ (which is zero for the plane-stress condition) is the intermediate stress. Thus, for the second quadrant, Eq. (2.36) reduces to

$$\sigma_3 - \sigma_1 = Y, \tag{2.40}$$

and for the fourth quadrant, it reduces to

$$\sigma_1 - \sigma_3 = Y. \tag{2.41}$$

Equations (2.40) and (2.41) represent the 45° lines in Fig. 2.38.

The distortion-energy criterion for plane stress reduces to

$$\sigma_1^2 + \sigma_3^2 - \sigma_1\sigma_3 = Y^2 \tag{2.42}$$

and is shown graphically in Fig. 2.38. Whenever a point (with its coordinates representing the two principal stresses) falls on these boundaries, the element will yield.

The three-dimensional *elastic* stress–strain relationships are shown in Eqs. (2.33a)–(2.33c). When the stresses are sufficiently high to cause *plastic deformation*, the stress–strain relationships are obtained from **flow rules** (*Lévy–Mises equations*), described in detail in texts on plasticity. These relationships are in strain increments in the following equations:

$$d\varepsilon_1 = \frac{d\bar{\varepsilon}}{\bar{\sigma}}\left[\sigma_1 - \frac{1}{2}(\sigma_2 + \sigma_3)\right]; \tag{2.43a}$$

$$d\varepsilon_2 = \frac{d\bar{\varepsilon}}{\bar{\sigma}}\left[\sigma_2 - \frac{1}{2}(\sigma_1 + \sigma_3)\right]; \tag{2.43b}$$

$$d\varepsilon_3 = \frac{d\bar{\varepsilon}}{\bar{\sigma}}\left[\sigma_3 - \frac{1}{2}(\sigma_1 + \sigma_2)\right]. \tag{2.43c}$$

Note that these expressions are similar to those given by the elastic stress–strain relationships of the generalized Hooke's law.

For the plane-strain condition shown in Figs. 2.37c and d, we have $\varepsilon_2 = 0$. Therefore,

$$\sigma_2 = \frac{\sigma_1 + \sigma_3}{2}. \tag{2.44}$$

Note that $\sigma_2$ is now an intermediate stress.

For the *plane-strain compression* of Figs. 2.16 and 2.37d, the distortion-energy criterion (which includes the intermediate stress) reduces to

$$\sigma_1 - \sigma_3 = \frac{2}{\sqrt{3}} Y = 1.15Y = Y'. \tag{2.45}$$

Note that, whereas for the maximum-shear-stress criterion $k = Y/2$, we have $k = Y/\sqrt{3}$ for the distortion-energy criterion for the plane-strain condition.

### 2.11.4  Experimental verification of yield criteria

The yield criteria described thus far have been tested experimentally. A suitable specimen commonly used is a thin-walled tube under internal pressure and/or torsion. Under such loading, it is possible to generate different states of plane stress. Various experiments, with a variety of *ductile* materials, have shown that the distortion-energy criterion agrees better with the experimental data than does the maximum-shear-stress criterion.

These findings suggest use of the distortion-energy criterion for the analysis of metalworking processes (Chapters 6 and 7), which generally make use of ductile materials. The simpler maximum-shear-stress criterion, however, can also be used, particularly by designers. The difference between the two criteria is negligible for most practical applications.

### 2.11.5  Volume strain

By summing the three equations of the generalized Hooke's law [Eq. (2.33)], we obtain

$$\varepsilon_1 + \varepsilon_2 + \varepsilon_3 = \frac{1 - 2\nu}{E}(\sigma_1 + \sigma_2 + \sigma_3), \tag{2.46}$$

where it can be shown that the left-hand side of the equation is the *volume strain*, or **dilatation**, $\Delta$. Thus,

$$\Delta = \frac{\text{volume change}}{\text{original volume}} = \frac{1 - 2\nu}{E}(\sigma_1 + \sigma_2 + \sigma_3) \tag{2.47}$$

In the plastic range, where $\nu = 0.5$, the volume change is zero. Hence, in plastic working of metals,

$$\varepsilon_1 + \varepsilon_2 + \varepsilon_3 = 0, \tag{2.48}$$

which is a convenient means of determining a third strain if two strains are known.

The **bulk modulus** is defined as

$$\text{Bulk modulus} = \frac{\sigma_m}{\Delta} = \frac{E}{3(1 - 2\nu)}, \tag{2.49}$$

where $\sigma_m$ is the *mean stress*, defined as

$$\sigma_m = \frac{1}{3}(\sigma_1 + \sigma_2 + \sigma_3). \tag{2.50}$$

From Eq. (2.47), in the *elastic* range, where $0 < v < 0.5$, it can be seen that the volume of a tension-test specimen increases, and that of a compression-test specimen decreases during the test.

## 2.11.6  Effective stress and effective strain

A convenient means of expressing the state of stress on an element is by the *effective* (*equivalent* or *representative*) *stress, $\bar{\sigma}$,* and the *effective strain, $\bar{\varepsilon}$.* For the maximum-shear-stress criterion, the effective stress is

$$\bar{\sigma} = \sigma_1 - \sigma_3, \tag{2.51}$$

and for the distortion-energy criterion, it is

$$\bar{\sigma} = \frac{1}{\sqrt{2}}[(\sigma_1 - \sigma_2)^2 + (\sigma_2 - \sigma_3)^2 + (\sigma_3 - \sigma_1)^2]^{1/2}. \tag{2.52}$$

The factor $1/\sqrt{2}$ is chosen so that, for simple tension, the effective stress is equal to the uniaxial yield stress, $Y$.

The strains are likewise related to the *effective strain.* For the maximum-shear-stress criterion, the effective strain is

$$\bar{\varepsilon} = \frac{2}{3}(\varepsilon_1 - \varepsilon_3), \tag{2.53}$$

and for the distortion-energy criterion, it is

$$\bar{\varepsilon} = \frac{\sqrt{2}}{3}[(\varepsilon_1 - \varepsilon_2)^2 + (\varepsilon_2 - \varepsilon_3)^2 + (\varepsilon_3 - \varepsilon_1)^2]^{1/2}. \tag{2.54}$$

Again, the factors $2/3$ and $\sqrt{2}/3$ are chosen so that for simple tension, the effective strain is equal to the uniaxial tensile strain. It is apparent that stress–strain curves may also be called effective-stress–effective-strain curves. (See also Example 2.8.)

## 2.11.7  Comparison of normal stress/normal strain and shear stress/shear strain

Stress–strain curves in tension and torsion for the same material are, of course, comparable. Also, it is possible to construct one curve from the other, since the material is the same; the procedure for doing so is outlined next.

The following observations are made with regard to tension and torsional states of stress:

- In the tension test, the uniaxial stress $\sigma_1$ is also the effective stress and the principal stress;
- in the torsion test, the principal stresses occur on planes whose normals are at 45° to the longitudinal axis; the principal stresses $\sigma_1$ and $\sigma_3$ are equal in magnitude, but opposite in sign;
- the magnitude of the principal stress in torsion is the same as that of the maximum shear stress.

We now have the following relationships:

$$\sigma_1 = -\sigma_3, \quad \sigma_2 = 0, \quad \text{and} \quad \sigma_1 = \tau_1.$$

Substituting these stresses into Eqs. (2.51) and (2.52) for effective stress, the following relationships are obtained: for the maximum-shear-stress criterion,

$$\bar{\sigma} = \sigma_1 - \sigma_3 = \sigma_1 + \sigma_1 = 2\sigma_1 = 2\tau_1, \tag{2.55}$$

and for the distortion-energy criterion,

$$\bar{\sigma} = \frac{1}{\sqrt{2}}[(\sigma_1 - 0)^2 + (0 + \sigma_1)^2 + (-\sigma_1 - \sigma_1)^2]^{1/2} = \sqrt{3}\sigma_1 = \sqrt{3}\tau_1. \quad (2.56)$$

With regard to strains, the following observations can be made:

- In the tension test, $\varepsilon_2 = \varepsilon_3 = -\varepsilon_1/2$;
- in the torsion test, $\varepsilon_1 = -\varepsilon_3 = \gamma/2$;
- the strain in the thickness direction of the tube is zero, i.e., $\varepsilon_2 = 0$.

The third observation is true because the thinning caused by the principal tensile stress is countered by the thickening effect under the principal compressive stress of the same magnitude. Hence, $\varepsilon_2 = 0$. Since $\sigma_2$ is also zero, a thin-walled tube under torsion is both a plane-stress and a plane-strain situation.

Substituting these strains into Eqs. (2.53) and (2.54) for effective strain, the following relationships can be obtained: for the maximum-shear-stress criterion,

$$\bar{\varepsilon} = \frac{2}{3}(\varepsilon_1 - \varepsilon_3) = \frac{2}{3}(\varepsilon_1 + \varepsilon_1) = \frac{4}{3}\varepsilon_1 = \frac{2}{3}\gamma, \quad (2.57)$$

and for the distortion-energy criterion,

$$\bar{\varepsilon} = \frac{\sqrt{2}}{3}[(\varepsilon_1 - 0)^2 + (0 + \varepsilon_1)^2 + (\varepsilon_1 - \varepsilon_1)^2]^{1/2} = \frac{2}{\sqrt{3}}\varepsilon_1 = \frac{1}{\sqrt{3}}\gamma. \quad (2.58)$$

This set of equations provides a means by which tensile-test data can be converted to torsion-test data, and vice versa.

### 2.11.8 Instability in biaxial tension

In Section 2.2.4, we showed that in simple tension and for a material whose behavior can be represented by $\sigma = K\varepsilon^n$, instability occurs at a true strain of $\varepsilon = n$. Using the equations obtained from flow rules, one can derive the strain at instability for various conditions.

## 2.12 | Work of Deformation

Work is defined as the product of collinear force and distance, so a quantity equivalent to work per unit volume is the product of stress and strain. Because the relation between stress and strain in the plastic range depends on the particular stress–strain curve, this work is best calculated by referring to Fig. 2.39.

Note that the area under the true-stress–true-strain curve for any strain $\varepsilon$ is the **energy per unit volume $u$ (specific energy)** *of the material deformed.* It is expressed as

$$u = \int_0^{\varepsilon_1} \sigma \, d\varepsilon. \quad (2.59)$$

As seen in Section 2.2.3, true-stress–true-strain curves can be represented by the expression

$$\sigma = K\varepsilon^n.$$

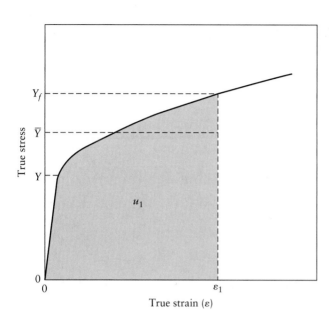

**FIGURE 2.39**  Schematic illustration of a true-stress–true-strain curve showing the yield stress $Y$, the average flow stress $\bar{Y}$, the specific energy $u_1$, and the flow stress $Y_f$.

Hence, Eq. (2.59) can be written as

$$u = K \int_0^{\varepsilon_1} \varepsilon^n \, d\varepsilon,$$

or

$$u = \frac{K\varepsilon_1^{n+1}}{n+1} = \bar{Y}\varepsilon_1, \qquad (2.60)$$

where $\bar{Y}$ is the **average flow stress** of the material.

This energy represents the work dissipated in uniaxial deformation. For triaxial states of stress, a general expression is given by

$$du = \sigma_1 \, d\varepsilon_1 + \sigma_2 \, d\varepsilon_2 + \sigma_3 \, d\varepsilon_3.$$

For an application of this equation, see Example 2.8. For a more general condition, the effective stress and effective strain can be used. The energy per unit volume is then

$$u = \int_0^{\bar{\varepsilon}} \bar{\sigma} \, d\bar{\varepsilon}. \qquad (2.61)$$

To obtain the work expended, we multiply $u$ by the volume of the material deformed:

$$\text{Work} = (u)(\text{volume}). \qquad (2.62)$$

The energy represented by Eq. (2.62) is the *minimum energy*, or the *ideal energy*, required for uniform (homogeneous) deformation. The energy required for actual deformation involves two additional factors. One factor is the energy required to overcome *friction* at the die–workpiece interfaces; the other is the **redundant work** of deformation, which is described as follows.

In Fig. 2.40a, a block of material is being deformed into shape by forging, extrusion, or drawing through a die, as described in Chapter 6. As shown in Fig. 2.40b, this deformation is uniform, or homogeneous. In reality, however, the material more often than not deforms as in Fig. 2.40c, from the effects of friction and die geometry. The difference between $b$ and $c$ is that $c$ has undergone additional shearing along horizontal planes.

**FIGURE 2.40**
Deformation of grid patterns in a workpiece: (a) the original pattern; (b) after ideal deformation; (c) after inhomogeneous deformation, requiring redundant work of deformation. Note that grid pattern (c) is basically the same as grid pattern (b), but with additional shearing, especially at the outer layers. Thus, grid pattern (c) requires higher work of deformation than grid pattern (b). See also Figs. 6.2 and 6.51.

(a)　　　　　　　　(b)　　　　　　　　(c)

This shearing requires expenditure of energy, because additional plastic work has to be done in subjecting the various layers to shear strains. This additional work is known as *redundant work*; the word "redundant" reflects the fact that this work does not contribute to the shape change of the material. (Note that grid patterns (b) and (c) in Fig. 2.40 have the same overall shape and dimensions.)

The *total specific energy* required can now be written as

$$u_{total} = u_{ideal} + u_{friction} + u_{redundant}. \tag{2.63}$$

The *efficiency*, $\eta$, of a process is defined as

$$\eta = \frac{u_{ideal}}{u_{total}}. \tag{2.64}$$

The magnitude of this efficiency varies widely, depending on the particular process, frictional conditions, die geometry, and other process parameters. Typical values are estimated to be 30% to 60% for extrusion and 75% to 95% for rolling.

---

**EXAMPLE 2.8  Expansion of a thin-walled spherical shell**

A thin-walled spherical shell made of a perfectly plastic material of yield stress, Y, original radius, $r_o$, and thickness, $t_o$, is being expanded by internal pressure. (a) Calculate the work done in expanding this shell to a radius of $r_f$. (b) If the diameter expands at a constant rate, what changes take place in the power consumed as the radius increases?

**Solution.**  The membrane stresses are given by

$$\sigma_1 = \sigma_2 = Y$$

(from Example 2.7), where $r$ and $t$ are instantaneous dimensions. The true strains in the membrane are given by

$$\varepsilon_1 = \varepsilon_2 = \ln\left(\frac{2\pi r_f}{2\pi r_o}\right) = \ln\left(\frac{r_f}{r_o}\right).$$

Because an element in this shell is subjected to equal biaxial stretching, the specific energy is

$$u = \int_0^{\varepsilon_1} \sigma_1 \, d\varepsilon_1 + \int_0^{\varepsilon_2} \sigma_2 \, d\varepsilon_2 = 2\sigma_1 \varepsilon_1 = 2Y \ln\left(\frac{r_f}{r_o}\right)$$

Since the volume of the shell's material is $4\pi r_o^2 t_o$, the work $W$ done is

$$W = (u)(volume) = 8\pi Y r_o^2 t_o \ln\left(\frac{r_f}{r_o}\right).$$

The specific energy can also be calculated from the effective stresses and strains. Thus, according to the distortion-energy criterion,

$$\bar{\sigma} = \frac{1}{\sqrt{2}}[(0)^2 + (\sigma_2)^2 + (-\sigma_1)^2]^{1/2} = \sigma_1 = \sigma_2$$

and

$$\bar{\varepsilon} = \frac{\sqrt{2}}{3}[(0)^2 + (\varepsilon_2 + 2\varepsilon_2)^2 + (-2\varepsilon_2 - \varepsilon_2)^2]^{1/2} = 2\varepsilon_2 = 2\varepsilon_1.$$

(The thickness strain $\varepsilon_3 = -2\varepsilon_2 = -2\varepsilon_1$, because of volume constancy in plastic deformation, where $\varepsilon_1 + \varepsilon_2 + \varepsilon_3 = 0$.) Hence,

$$u = \int_0^{\bar{\varepsilon}} \bar{\sigma}\,d\bar{\varepsilon} = \int_0^{2\varepsilon_1} \sigma_1\,d\varepsilon_1 = 2\sigma_1\varepsilon_1.$$

Thus, the answer is the same.

Power is defined as work per unit time; thus,

$$\text{Power} = \frac{dW}{dt}.$$

Since all other factors in the expression are constant, the expression for work can be written as being proportional to strain:

$$W \propto \ln\left(\frac{r}{r_o}\right) \propto (\ln r - \ln r_o).$$

Therefore,

$$\text{Power} \propto \frac{1}{r}\frac{dr}{dt}.$$

Because the shell is expanding at a constant rate, $dr/dt = $ constant. Hence, the power is related to the instantaneous radius $r$ by the following expression:

$$\text{Power} \propto \frac{1}{r}.$$

## 2.12.1  Work, heat, and temperature rise

Almost all the mechanical work of deformation in plastic working is converted into heat. This conversion is not 100% because a small portion of energy is stored within the deformed material as elastic energy. This is known as *stored energy* (see Section 3.6); it is generally 5% to 10% of the total energy input, but it may be as high as 30% in some alloys.

In a simple frictionless process, and assuming that work is completely converted into heat, the *temperature rise*, $\Delta T$, is given by

$$\Delta T = \frac{u_{\text{total}}}{\rho c}, \tag{2.65}$$

where $u_{\text{total}}$ is the specific energy from Eq. (2.63), $\rho$ is the density, and $c$ is the specific heat of the material. Thus, higher temperatures are associated with large areas under the stress–strain curve and smaller values of specific heat. The theoretical temperature rise for a true strain of 1 (such as a 27-mm-high specimen compressed down to 10 mm) has been calculated to be as follows: aluminum, 75°C (165°F); copper, 140°C (285°F); low-carbon steel, 280°C (535°F); and titanium, 570°C (1060°F).

The temperature rise given by Eq. (2.65) is for an ideal situation, where there is no heat loss. In actual operations, heat is lost to the environment, to tools and dies, and to any lubricants or coolants used. If the process is performed very rapidly, these losses are relatively small. Under extreme conditions, an *adiabatic* state is approached, with a very high temperature rise, leading to **incipient melting**. The rise in temperature can be calculated if the stress–strain curve used is at the appropriate strain-rate level (see Section 2.2.7); however, if the process is carried out slowly, the actual temperature rise will be a small portion of the calculated value. It should also be noted that properties such as specific heat and thermal conductivity also depend on temperature, and they should be taken into account in the calculations.

---

**EXAMPLE 2.9 Temperature rise in compression**

A cylindrical specimen 1 in. diameter and 1 in. high is being compressed by dropping a weight of 100 lb on it from a certain height. The material has the following properties: $K = 15,000$ psi; $n = 0.5$; density $= 0.1$ lb/in$^3$ and specific heat $= 0.3$ Btu/lb-°F; Assuming no heat loss and no friction, calculate the final height of the specimen if the temperature rise is 100°F.

**Solution.** The expression for heat is given by

$$\text{Heat} = (c_p)(\rho)(\text{volume})(\Delta T),$$

or

$$\text{Heat} = (0.3)(0.10)[(\pi)(1^2)/4](100)(778),$$

where 778 ft-lb = 1BTU. Thus,

$$\text{Heat} = 1830 \text{ ft-lb} = 22,000 \text{ in.-lb.}$$

Also, ideally,

$$\text{Heat} = \text{Work} = (\text{volume})(u) = \frac{\pi}{4} \frac{15,000 \varepsilon^{1.5}}{1.5}$$

and

$$\varepsilon^{1.5} = \frac{(22,000)(1.5)(4)}{(\pi)(15,000)} = 2.8;$$

hence,

$$\varepsilon = 1.99.$$

Using absolute values, we have

$$\ln\left(\frac{h_o}{h_f}\right) = \ln\left(\frac{1}{h_f}\right) = 1.99,$$

and therefore, $h_f = 0.137$ in.

---

# SUMMARY

- Many manufacturing processes involve shaping materials by plastic deformation; consequently, mechanical properties such as strength, elasticity, ductility, hardness, toughness, and the energy required for plastic deformation are important factors. The behavior of materials, in turn, depends on the particular material and its condition, as well as other variables, particularly temperature, strain rate, and the state of stress. (Section 2.1)

- Mechanical properties measured by tension tests include the modulus of elasticity ($E$), yield stress ($Y$), ultimate tensile strength (UTS), and Poisson's ratio ($\nu$). Ductility, as measured by elongation and reduction of area, can also be determined by tension tests. True-stress–true-strain curves are important in determining such mechanical properties as strength coefficient ($K$), strain-hardening exponent ($n$), strain-rate sensitivity exponent ($m$), and toughness. (Section 2.2)

- Compression tests model manufacturing processes such as forging, rolling, and extrusion. Properties measured by compression tests are subject to inaccuracy, due to the presence of friction and barreling. (Section 2.3)

- Torsion tests are typically conducted on tubular specimens that are subjected to twisting. These tests model such manufacturing processes as shearing, cutting, and various machining processes. (Section 2.4)

- Bend, or flexure, tests are commonly used for brittle materials; the outer fiber stress at fracture in bending is known as the modulus of rupture, or transverse rupture strength. The forces applied in bend tests model those incurred in such manufacturing processes as forming of sheet and plate, as well as testing of tool and die materials. (Section 2.5)

- A variety of hardness tests is available to test the resistance of a material to permanent indentation. Hardness is related to strength and wear resistance, but is itself not a fundamental property of a material. (Section 2.6)

- Fatigue tests model manufacturing processes whereby components are subjected to rapidly fluctuating loads. The quantity measured is the endurance limit, or fatigue limit—that is, the maximum stress to which a material can be subjected without fatigue failure, regardless of the number of cycles. (Section 2.7)

- Creep is the permanent elongation of a component under a static load maintained for a period of time. A creep test typically consists of subjecting a specimen to a constant tensile load at a certain temperature and measuring its change in length over a period of time. The specimen eventually fails by necking and rupture. (Section 2.8)

- Impact tests model some high-rate manufacturing operations as well as the service conditions in which materials are subjected to impact (or dynamic) loading, such as drop forging or the behavior of tool and die materials in interrupted cutting or high-rate deformation. Impact tests determine the energy required to fracture the specimen, known as impact toughness. Impact tests are also useful in determining the ductile–brittle transition temperature of materials. (Section 2.9)

- Residual stresses are those stresses that remain in a part after it has been deformed and all external forces have been removed. The nature and level of residual stresses depend on the manner in which the part has been plastically deformed. Residual stresses may be reduced or eliminated by stress-relief annealing, by further plastic deformation, or by relaxation. (Section 2.10)

- In metalworking operations, the workpiece material is generally subjected to three-dimensional stresses through various tools and dies. Yield criteria establish relationships between the uniaxial yield stress of the material and the stresses applied; the two most widely used yield criteria are the maximum-shear-stress (Tresca) and the distortion-energy (von Mises) criteria. (Section 2.11)

- Because energy is required to deform materials, the work of deformation per unit volume of material, $u$, is an important parameter and is composed of ideal, frictional, and redundant work components. In addition to supplying information on force and energy requirements, the work of deformation also indicates the amount of heat developed and, hence, the temperature rise in the workpiece during plastic deformation. (Section 2.12)

# SUMMARY OF EQUATIONS

Engineering strain, $e = \dfrac{\ell - \ell_o}{\ell_o}$

Engineering strain rate, $\dot{e} = \dfrac{v}{\ell_o}$

Engineering stress, $\sigma = \dfrac{P}{A_o}$

True strain, $\varepsilon = \ln\left(\dfrac{\ell}{\ell_o}\right)$

True strain rate, $\dot{\varepsilon} = \dfrac{v}{\ell}$

True stress, $\sigma = \dfrac{P}{A}$

Modulus of elasticity, $E = \dfrac{\sigma}{e}$

Shear modulus, $G = \dfrac{E}{2(1 + v)}$

Modulus of resilience $= \dfrac{Y^2}{2E}$

Elongation $= \dfrac{\ell_f - \ell_o}{\ell_o} \times 100$

Reduction of area $= \dfrac{A_o - A_f}{A_o} \times 100$

Shear strain in torsion, $\gamma = \dfrac{r\phi}{\ell}$

Hooke's law, $\varepsilon_1 = \dfrac{1}{E}[\sigma_1 - v(\sigma_2 + \sigma_3)]$, etc.

Effective strain (Tresca), $\bar{\varepsilon} = \dfrac{2}{3}(\varepsilon_1 - \varepsilon_3)$

Effective strain (von Mises), $\bar{\varepsilon} = \dfrac{\sqrt{2}}{3}[(\varepsilon_1 - \varepsilon_2)^2 + (\varepsilon_2 - \varepsilon_3)^2 + (\varepsilon_3 - \varepsilon_1)^2]^{1/2}$

Effective stress (Tresca), $\bar{\sigma} = \sigma_1 - \sigma_3$

Effective stress (von Mises), $\bar{\sigma} = \dfrac{1}{\sqrt{2}}[(\sigma_1 - \sigma_2)^2 + (\sigma_2 - \sigma_3)^2 + (\sigma_3 - \sigma_1)^2]^{1/2}$

True-stress–true-strain relationship (power law), $\sigma = K\varepsilon^n$

True-stress–true-strain rate relationship, $\sigma = C\dot{\varepsilon}^m$

Flow rules, $d\varepsilon_1 = \dfrac{d\bar{\varepsilon}}{\bar{\sigma}}\left[\sigma_1 - \dfrac{1}{2}(\sigma_2 + \sigma_3)\right]$, etc.

Maximum-shear-stress criterion (Tresca), $\sigma_{max} - \sigma_{min} = Y$

Distortion-energy criterion (von Mises) , $(\sigma_1 - \sigma_2)^2 + (\sigma_2 - \sigma_3)^2 + (\sigma_3 - \sigma_1)^2 = 2Y^2$

Shear yield stress, $k = Y/2$ for Tresca and $k = Y/\sqrt{3}$ for von Mises (plane strain)

Volume strain (dilatation), $\Delta = \dfrac{1 - 2v}{E}(\sigma_1 + \sigma_2 + \sigma_3)$

Bulk modulus $= \dfrac{E}{3(1 - 2v)}$

# BIBLIOGRAPHY

*ASM Handbook*, Vol. 8: *Mechanical Testing*, ASM International, 2000.

*ASM Handbook*, Vol. 10: *Materials Characterization*, ASM International, 1986.

Boyer, H. E. (ed.), *Atlas of Creep and Stress-Rupture Curves*, ASM International, 1986.

————, *Atlas of Fatigue Curves*, ASM International, 1986.

————, *Atlas of Stress–Strain Curves*, ASM International, 1986.

Budinski, K. G., *Engineering Materials: Properties and Selection*, 6th ed., Prentice Hall, 1998.

Cheremisinoff, N. P., and P. N. Cheremisinoff, *Handbook of Advanced Materials Testing*, Dekker, 1994.

Courtney, T. H., *Mechanical Behavior of Materials*, 2d ed., McGraw-Hill, 1999.

Davis, J. R. (ed.), *Metals Handbook: Desk Edition*, ASM International, 1998.

Dieter, G. E., *Mechanical Metallurgy*, 3d ed., McGraw-Hill, 1986.

Dowling, N. E., *Mechanical Behavior of Materials: Engineering Methods for Deformation, Fracture, and Fatigue*, 2d ed., Prentice Hall, 1998.

Han, P. (ed.), *Tensile Testing*, ASM International, 1992.

Hsu, T. H., *Stress and Strain Data Handbook*, Gulf Publishing Co., 1986.

Pohlandt, K., *Material Testing for the Metal Forming Industry*, Springer, 1989.

# QUESTIONS

**2.1**  Can you calculate the percent elongation of materials based only on the information given in Fig. 2.6? Explain.

**2.2**  Describe your observations regarding the contents of Table 2.1.

**2.3**  Explain if it is possible for the curves in Fig. 2.4 to reach 0% elongation as the gage length is increased further.

**2.4**  Explain why the difference between engineering strain and true strain becomes larger as strain increases. Is this phenomenon true for both tensile and compressive strains? Explain.

**2.5**  Using the same scale for stress, we note that the tensile true-stress–true-strain curve is higher than the engineering stress–strain curve. Explain whether this condition also holds for a compression test.

**2.6**  Which of the two tests, tension or compression, requires a higher capacity testing machine than the other? Explain.

**2.7**  Explain how the modulus of resilience of a material changes, if at all, as it is strained: (1) for an elastic, perfectly plastic material, and (2) for an elastic, linearly strain-hardening material.

**2.8**  Take an element under the anvils in the deformation zone in the plane-strain compression test shown in Fig. 2.16. Is the element subjected to simple shear or pure shear during deformation? Explain. (*Hint*: View this element from the sides, in the width direction, and compare it with Fig. 2.21.)

**2.9**  If you pull and break a tensile-test specimen rapidly, where would the temperature be highest? Explain why.

**2.10**  Comment on the temperature distribution if the specimen in Question 2.9 is pulled very slowly.

**2.11**  In a tension test, the area under the true-stress–true-strain curve is the work done per unit volume (the specific work). We also know that the area under the load–elongation curve represents the work done on the specimen. If you divide this latter work by the volume of the specimen between the gage marks, you will determine the work done per unit volume (assuming that all deformation is confined between the gage marks). Will this specific work be the same as the area under the true-stress–true-strain curve? Explain. Will your answer be the same for any value of strain? Explain.

**2.12**  The note at the bottom of Table 2.5 states that as temperature increases, $C$ decreases and $m$ increases. Explain why.

**2.13**  You are given the $K$ and $n$ values of two different materials, respectively. Is this information sufficient to determine which material is tougher? If not, what additional information do you need, and why?

**2.14**  Modify the curves in Fig. 2.8 to indicate the effects of temperature. Explain your changes.

**2.15**  Using a specific example, show why the deformation rate, say in m/s, and the true strain rate are not the same.

**2.16**  Referring to Fig. 2.3, explain how the magnitude of elastic recovery will change as an elastic, linearly strain-hardening material is strained.

**2.17**  It has been stated that the higher the value of $m$, the more diffuse the neck is, and likewise, the lower the value of $m$, the more localized the neck is. Explain the reason for this behavior.

**2.18**  Explain why materials with high $m$ values, such as hot glass and silly putty, when stretched slowly, undergo large elongations before failure. Consider events taking place in the necked region of the specimen.

**2.19** Assume that you are running four-point bending tests on a number of identical specimens of the same length and cross-section, but with increasing distance between the upper points of loading. (See Fig. 2.23b.) What changes, if any, would you expect in the test results? Explain.

**2.20** Would Eq. (2.10) hold true in the elastic range? Explain.

**2.21** Why have different types of hardness tests been developed? How would you measure the hardness of a very large object?

**2.22** Which hardness tests and scales would you use for very thin strips of material, such as aluminum foil? Why?

**2.23** Would the answer to Example 2.2 double if the strength coefficient of the material is doubled? Explain.

**2.24** List and explain the factors that you would consider in selecting an appropriate hardness test and scale for a particular application.

**2.25** Describe the difference between creep and stress relaxation, giving two examples for each as they relate to engineering applications.

**2.26** Referring to the two impact tests shown in Fig. 2.31, explain how different the results would be if the specimens were impacted from the opposite directions.

**2.27** If you remove the layer *ad* from the part shown in Fig. 2.32d, such as by machining or grinding, which way will the specimen curve? (*Hint*: Assume that the part in diagram (d) is composed of four horizontal springs held at the ends. Thus, from the top down, we have compression, tension, compression, and tension springs.)

**2.28** Is it possible to completely remove residual stresses in a piece of material by the technique described in Fig. 2.34 if the material is elastic, linearly strain hardening? Explain.

**2.29** Referring to Fig. 2.34, would it be possible to eliminate residual stresses by compression? Assume that the piece of material will not buckle under the uniaxial compressive force.

**2.30** List and explain the desirable mechanical properties for (1) an elevator cable, (2) a paper clip, (3) a leaf spring, (4) a bracket for a bookshelf, (5) a piano wire, (6) a coat hanger, (7) a gas-turbine blade, and (8) a staple.

**2.31** Using the generalized Hooke's law [Eq. (2.33)], show whether or not a thin-walled tube undergoes any change in thickness when subjected to torsion in the elastic range.

**2.32** Make a sketch showing the nature and distribution of the residual stresses in Figs. 2.33a and b before the parts were cut. Assume that the split parts are free from any stresses. (*Hint*: Force these parts back to the shape they were in before they were cut.)

**2.33** It is possible to calculate the work of plastic deformation by measuring the temperature rise in a workpiece, assuming that there is no heat loss and that the temperature distribution is uniform throughout. If the specific heat of the material decreases with increasing temperature, will the work of deformation calculated using the specific heat at room temperature be higher or lower than the actual work done? Explain.

**2.34** Explain whether or not the volume of a homogeneous, isotropic specimen changes when the specimen is subjected to a state of (a) uniaxial compressive stress and (b) uniaxial tensile stress, all in the elastic range.

**2.35** How would you explain the general behavior of a metal as shown in Fig. 2.11 to someone over the telephone?

**2.36** We know that it is relatively easy to subject a specimen to hydrostatic compression, such as by using a chamber filled with a liquid. Devise a means whereby the specimen (say, in the shape of a cube or a round disk) can be subjected to hydrostatic tension, or one approaching this state of stress. (Note that a thin-walled, internally pressurized spherical shell is not a correct answer, because it is subjected only to a state of plane stress.)

**2.37** Referring to Fig. 2.20, make sketches of the state of stress for an element in the reduced section of the tube when it is subjected to (1) torsion only, (2) torsion while the tube is internally pressurized, and (3) torsion while the tube is externally pressurized. Assume that the tube is a closed-end tube.

**2.38** How would you conduct the test shown in Fig. 2.22? Make a sketch of the setup you would use.

**2.39** A penny-shaped piece of soft metal of a certain diameter is brazed to the ends of two flat, round steel rods of the same diameter as the piece. The assembly is then subjected to uniaxial tension. What is the state of stress to which the soft metal is subjected? Explain.

**2.40** A circular disk of soft metal is being compressed between two flat, hardened circular steel punches of having the same diameter as the disk. Assume that the disk material is perfectly plastic and that there is no friction or any temperature effects. Explain the change, if any, in the magnitude of the punch force as the disk is being compressed plastically to, say, a fraction of its original thickness.

**2.41** A perfectly plastic metal is yielding under the stress state $\sigma_1, \sigma_2, \sigma_3$, where $\sigma_1 > \sigma_2 > \sigma_3$. Explain what happens if $\sigma_1$ is increased.

**2.42** What is the dilatation of a material with a Poisson's ratio of 0.5? Is it possible for a material to have a Poisson's ratio of 0.7? Give a rationale for your answer.

**2.43** Can a material have a negative Poisson's ratio? Give a rationale for your answer.

**2.44** As clearly as possible, define plane stress and plane strain.

**2.45** What test would you use to evaluate the hardness of a coating on a metal surface? Would it matter if the coating was harder or softer than the substrate? Explain.

**2.46** Explain clearly the difference between resilience and toughness. Give an application for which each is important.

**2.47** List the advantages and limitations of the stress–strain relationships given in Fig. 2.8.

**2.48** Plot the data in Table 2.1 on a bar chart, showing the range of values, and comment on the results.

**2.49** A hardness test is conducted on as-received metal as a quality check. The results show indicate that the hardness is too high, indicating that the material may not have sufficient ductility for the intended application. The supplier is reluctant to accept the return of the material, instead claiming that the diamond cone used in the Rockwell testing was worn and blunt, and hence the test needed to be recalibrated. Is this explanation plausible? Explain.

**2.50** Explain why a 0.2% offset is used to obtain the yield strength in a tension test.

**2.51** Referring to Question 2.50, would the offset method be necessary for a highly strained hardened material? Explain.

# PROBLEMS

**2.52** A strip of metal is originally 1.5 m long. It is stretched in three steps: first to a length of 2.0 mm, then to 2.5 m, and finally to 3.0 m. Show that the total true strain is the sum of the true strains in each step, that is, that the strains are additive. Show that, using engineering strains, the strain for each step cannot be added to obtain the total strain.

**2.53** A paper clip is made of wire 1.25 mm in diameter. If the original material from which the wire is made is a rod 15 mm in diameter, calculate the longitudinal and diametrical engineering and true strains that the wire has undergone during processing.

**2.54** A material has the following properties: UTS = 50,000 psi and $n = 0.3$. Calculate its strength coefficient $K$.

**2.55** Based on the information given in Fig. 2.6, calculate the ultimate tensile strength of annealed copper.

**2.56** Identify the two materials in Fig. 2.6 that have the lowest and highest uniform elongations, respectively. Calculate these quantities as percentages of the original gage lengths.

**2.57** Calculate the ultimate tensile strength (engineering) of a material whose strength coefficient is 800 MPa and of a tensile-test specimen that necks at a true strain of 0.30.

**2.58** A cable is made of four parallel strands of different materials, all behaving according to the equation $\sigma = K\varepsilon^n$, where $n = 0.5$. The materials, strength coefficients and cross-sections are as follows:

Material $A$: $K = 500$ MPa, $A_o = 7$ mm$^2$;
Material $B$: $K = 800$ MPa, $A_o = 2$ mm$^2$;
Material $C$: $K = 400$ MPa, $A_o = 3$ mm$^2$;
Material $D$: $K = 600$ MPa, $A_o = 2$ mm$^2$.

(a) Calculate the maximum tensile force that this cable can withstand prior to necking.

(b) Explain how you would arrive at an answer if the $n$ values of the three strands were different from each other.

**2.59** Using only Fig. 2.6, calculate the maximum load in tension testing of a 304 stainless-steel specimen with an original diameter of 0.5 in.

**2.60** Using the data given in Table 2.1, calculate the values of the shear modulus $G$ for the metals listed in the table.

**2.61** Derive an expression for the toughness of a material represented by the equation $\sigma = K(\varepsilon + 0.2)^n$ and whose fracture strain is denoted as $\varepsilon_f$.

**2.62** A cylindrical specimen made of a brittle material 1 in. high and with a diameter of 1 in. is subjected to a compressive force along its axis. It is found that fracture takes place at an angle of 45° under a load of 40,000 lb. Calculate the shear stress and the normal stress, respectively, acting on the fracture surface.

**2.63** What is the modulus of resilience of a highly cold-worked piece of steel with a hardness of 350 HB? Of a piece of highly cold-worked copper with a hardness of 125 HB?

**2.64** Calculate the work done in frictionless compression of a solid cylinder 40 mm high and 20 mm in diameter to a reduction in height of 50% for the following materials: (1) 1100-O aluminum, (2) annealed copper, (3) annealed 302 stainless steel, and (4) 70–30 brass, annealed.

**2.65** Calculate the toughness of the material in Example 2.1 by using the area under the stress–strain curve. Then determine the $K$ and $n$ values for this material and obtain the toughness by integration. Compare the two answers.

**2.66** A material has a strength coefficient $K = 125,000$ psi and $n = 0.2$. Assuming that a tensile-test specimen made from this material begins to neck at a true strain of 0.2, show that the ultimate tensile strength of this material is 74,200 psi.

**2.67** A tensile-test specimen is made of a material represented by the equation $\sigma = K(\varepsilon + n)^n$. (a) Determine the true strain at which necking will begin. (b) Show that it is possible for an engineering material to exhibit this behavior.

**2.68** Take two solid cylindrical specimens of equal diameter, but different heights. Assume that both specimens are compressed (frictionless) by the same percent reduction, say 50%. Prove that the final diameters will be the same.

**2.69** A horizontal rigid bar $c$–$c$ is subjecting specimen $a$ to tension and specimen $b$ to frictionless compression such that the bar remains horizontal. (See the accompanying figure.) The force $F$ is located at a distance ratio of $2:1$. Both specimens have an original cross-sectional area of $1\ \text{in}^2$, and the original lengths are $a = 10$ in and $b = 5$ in The material for specimen $a$ has a true-stress–true-strain curve of $\sigma = 100,000\varepsilon^{0.5}$ psi. Plot the true-stress–true-strain curve that the material for specimen $b$ should have for the bar to remain horizontal.

**2.70** Inspect the curve that you obtained in Problem 2.69. Does a typical strain-hardening material behave in that manner? Explain.

**2.71** In a disk test performed on a specimen 30 mm in diameter and 5 mm thick, the specimen fractures at a stress of 500 MPa. What was the load on it at fracture?

**2.72** In Fig. 2.34a, let the tensile and compressive residual stresses both be 30,000 psi, and the modulus of elasticity of the material be $30 \times 10^6$ psi, with a modulus of resilience of 30 in.-lb/in³. If the original length in diagram (a) is 40 in., what should be the stretched length in diagram (b) so that, when unloaded, the strip will be free of residual stresses?

**2.73** Show that you can take a bent bar made of an elastic, perfectly plastic material and straighten it by stretching it into the plastic range. (*Hint*: Observe the events shown in Fig. 2.34.)

**2.74** A bar 0.5 m long is bent and then stress relieved. The radius of curvature to the neutral axis is 0.30 m. The bar is 30 mm thick and is made of an elastic, perfectly plastic material with $Y = 600$ MPa and $E = 200$ GPa. Calculate the length to which this bar should be stretched so that, after unloading, it will become and remain straight.

**2.75** Derive the three generalized Hooke's law equations given in Eq. (2.33).

**2.76** A thin cylindrical sleeve of thickness $t$, length $L$, and inside radius $r$ is placed on a rigid shaft with no clearance. (See the accompanying figure.) The sleeve is then stretched elastically by tension $\sigma_a$ in the axial direction. Let the coefficient of friction at the sleeve–shaft interface be denoted as $\mu$. Derive an expression for the force necessary to strip the sleeve from the shaft. Assume that $\sigma_a$ does not affect the stripping force.

**2.77** Assume that a material with a uniaxial yield stress $Y$ yields under a stress state of principal stresses $\sigma_1, \sigma_2, \sigma_3$, where $\sigma_1 > \sigma_2 > \sigma_3$. Show that the superposition of a hydrostatic stress $p$ on this system (such as placing the specimen in a chamber pressurized with a liquid) does not affect yielding. In other words, the material will still yield according to yield criteria.

**2.78** Give two different and specific examples in which the maximum-shear-stress and the distortion-energy criteria give the same answer.

**2.79** A thin-walled spherical shell with a yield stress $Y$ is subjected to an internal pressure $p$. With appropriate equations, show whether or not the pressure required to yield this shell depends on the particular yield criterion used.

**2.80** Show that, according to the distortion-energy criterion, the yield stress in plane strain is $1.15Y$, where $Y$ is the uniaxial yield stress of the material.

**2.81** What would be the answer to Problem 2.80 if the maximum shear stress criterion were used?

**2.82** A closed-end, thin-walled cylinder of original length $\ell$, thickness $t$, and internal radius $r$ is subjected to an internal pressure $p$. Using the generalized Hooke's law quations, show the change, if any, that occurs in the length of this cylinder when it is pressurized. Let $\nu = 0.33$.

**2.83** A round, thin-walled tube is subjected to tension in the elastic range. Show that both the thickness and the diameter decrease as tension increases.

**2.84** In Problem 2.83, assume that the tube is subjected to compression. Explain what changes, if any, take place in the thickness and diameter of the tube. Assume that the tube does not buckle.

**2.85** Take a long cylindrical balloon and, with a thin felt-tip pen, mark a small square on it. What will be the shape of this square after you blow up the balloon, (1) a larger square, (2) a rectangle with its long axis in the circumferential directions, (3) a rectangle with its long axis in the longitudinal direction, or (4) an ellipse? Perform this experiment, and, based on your observations, explain the results, using appropriate equations. Assume that the material the balloon is made up of is perfectly elastic and isotropic and that this situation represents a thin-walled closed-end cylinder under internal pressure.

**2.86** Take a cubic piece of metal a side length $\ell_o$ and deform it plastically to the shape of a rectangular parallelepiped of dimensions $\ell_1, \ell_2$, and $\ell_3$. Assuming that the material is rigid and perfectly plastic, show that volume constancy requires that the following expression be satisfied: $\varepsilon_1 + \varepsilon_2 + \varepsilon_3 = 0$.

**2.87** What is the diameter of an originally 20-mm-diameter solid steel ball when the ball is subjected to a hydrostatic pressure of 10 GPa?

**2.88** Determine the effective stress and effective strain in plane-strain compression according to the distortion-energy criterion.

**2.89** It can be shown that a thin-walled tube under longitudinal tension (in the elastic range) decreases in diameter. It is, of course, possible to internally pressurize the tube so that its diameter remains constant while being stretched. Assuming that the tube is open ended, derive an expression for the pressure required to keep the diameter constant as the tube is stretched.

**2.90** What is the magnitude of compressive stress for the plane-strain compression test shown in Fig. 2.16, according to the maximum-shear-stress criterion?

**2.91** (a) Calculate the work done in expanding a 1-mm-thick spherical shell from a diameter of 100 mm to 140 mm, where the shell is made of a material for which $\sigma = 200 + 50\varepsilon^{0.5}$ MPa. (b) Does your answer depend on the particular yield criterion used? Explain.

**2.92** A cylindrical slug that has a diameter of 1 in. and is 1 in. high is placed at the center of a 2-in.-diameter cavity in a rigid die. (See the accompanying figure.) The slug is surrounded by a compressible matrix, the pressure of which is given by the relation

$$p_m = 40,000\ \Delta V/V_{om}\ \text{psi},$$

where $m$ denotes matrix and $V_{om}$ is the original volume of the compressible matrix. Both the slug and the matrix are being compressed by a piston and without any friction. The initial pressure of the matrix is zero, and the slug material has the true-stress–true-strain curve of $\sigma = 25,000\varepsilon^{0.5}$. Obtain an expression for the force $F$ versus piston travel $d$ up to $d = 0.5$ in.

Compressible matrix

**2.93** A specimen in the shape of a cube 10 mm on each side is being compressed without friction in a die cavity, as shown in Fig. 2.37d, where the width of the groove is 15 mm. Assume that the linearly strain-hardening material has the true-stress–true-strain curve given by $\sigma = 75 + 20\varepsilon$ MPa. Calculate the compressive force required when the height of the specimen is 3 mm, according to both yield criteria.

**2.94** Assume that you are asked to give a quiz to students on the contents of this chapter. Prepare five quantitative problems, and supply the answers.

**2.95** Obtain expressions for the specific energy for a material for each of the stress–strain curves shown in Fig. 2.8, similar to those shown in Section 2.12.

**2.96** A material with a yield stress of 100 MPa is subjected to principal (normal) stresses of $\sigma_1, \sigma_2 = 0$, and $\sigma_3 = -\sigma_1/2$. What is the value of $\sigma_1$ when the metal yields according to the von Mises criterion? What if $\sigma_2 = \sigma_1/3$?

**2.97** A steel plate is 100 mm $\times$ 100 mm $\times$ 5 mm thick. It is subjected to biaxial tension $\sigma_1 = \sigma_2$, with the stress in the thickness direction of $\sigma_3 = 0$. What is the largest possible change in volume at yield, using the von Mises criterion? What would this change in volume be if the plate were made of copper?

**2.98**   A 25-mm-wide, 1-mm-thick strip is rolled to a final thickness of 0.6 mm. It is noted that the strip has increased in width to 26 mm. What is the strain in the rolling direction?

**2.99**   An aluminum alloy yields at a stress of 50 MPa in uniaxial tension. If this material is subjected to the stresses $\sigma_1 = 25$ MPa, $\sigma_2 = 15$ MPa, and $\sigma_3 = -26$ MPa, will it yield? Explain.

**2.100**   A cylindrical specimen 1 in. in diameter and 1 in. high is being compressed by dropping a weight of 200 lb on it from a certain height. After deformation, it is found that the temperature rise in the specimen is 300°F. Assuming no heat loss and no friction, calculate the final height of the specimen, using the following data for the material: $K = 30{,}000$ psi, $n = 0.5$, density $= 0.1$ lb/in$^3$, and specific heat $= 0.3$ BTU/lb · °F.

# Structure and Manufacturing Properties of Metals

## 3.1 | Introduction

The structure of metals greatly influences their *behavior* and *properties*. A knowledge of structures guides us in controlling and predicting the behavior and performance of metals in various manufacturing processes. Understanding the structure of metals also allows us to predict and evaluate their properties. This then helps us make appropriate selections for specific applications under particular force, temperature, and environmental conditions. For example, we will come to appreciate the reasons for the development of single-crystal turbine blades (Fig. 3.1) for use in jet engines; these blades have better properties than those made conventionally by casting or forging.

This chapter begins with a review of the crystal structure of metals and the role of various imperfections as they affect plastic deformation and work-hardening characteristics. We will also describe the importance of grain size, grain boundaries, and environment. We will then review the failure and fracture of metals, both ductile and brittle, together with factors that influence fracture, such as state of stress, temperature, strain rate, and external and internal defects in metals. The rest of the

3.1 Introduction
3.2 The Crystal Structure of Metals
3.3 Deformation and Strength of Single Crystals
3.4 Grains and Grain Boundaries
3.5 Plastic Deformation of Polycrystalline Metals
3.6 Recovery, Recrystallization, and Grain Growth
3.7 Cold, Warm, and Hot Working
3.8 Failure and Fracture
3.9 Physical Properties
3.10 Properties and Applications of Ferrous Alloys
3.11 Properties and Applications of Nonferrous Metals and Alloys

(a)          (b)          (c)

**FIGURE 3.1** Turbine blades for jet engines, manufactured by three different methods: (a) conventionally cast; (b) directionally solidified, with columnar grains, as can be seen from the vertical streaks; and (c) single crystal. Although more expensive, single-crystal blades have properties at high temperatures that are superior to those of other blades. *Source*: Courtesy of United Technologies Pratt and Whitney.

chapter is devoted to a brief discussion of physical and other properties of ferrous and nonferrous metals and alloys relevant to their use as engineering materials. Some data on properties are also given as a guide in design considerations and material selection.

## 3.2 | The Crystal Structure of Metals

When metals solidify from a molten state, the atoms arrange themselves into various orderly configurations, called **crystals**. The arrangement of the atoms in the crystal is called **crystalline structure**. A crystalline structure is like metal scaffolding in front of a building under construction, with uniformly repetitive horizontal and vertical metal pipes and braces.

The smallest group of atoms showing the characteristic **lattice structure** of a particular metal is known as a **unit cell**. It is the building block of a crystal, and a single crystal can have many unit cells. Think of each brick in a wall as a unit cell. The wall has a crystalline structure, that is, it consists of an orderly arrangement of bricks; thus, the wall is like a single crystal, consisting of many unit cells.

The three basic patterns of atomic arrangement found in most metals are as follows (Figs. 3.2a–c):

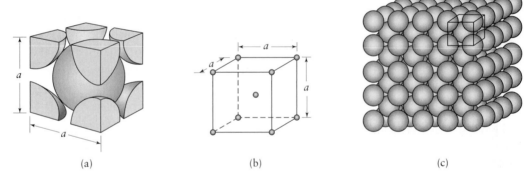

     (a)          (b)          (c)

**FIGURE 3.2a**    The body-centered cubic (bcc) crystal structure: (a) hard-ball model; (b) unit cell; and (c) single crystal with many unit cells. *Source*: Courtesy of Dr. William G. Moffatt.

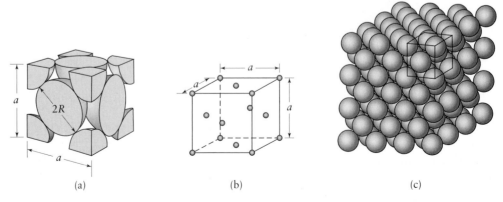

     (a)          (b)          (c)

**FIGURE 3.2b**    The face-centered cubic (fcc) crystal structure: (a) hard-ball model; (b) unit cell; and (c) single crystal with many unit cells. *Source*: Courtesy of Dr. William G. Moffatt.

**FIGURE 3.2c** The hexagonal close-packed (hcp) crystal structure: (a) unit cell; and (b) single crystal with many unit cells. *Source*: Courtesy of Dr. William G. Moffatt.

(a)                    (b)

a) body-centered cubic (bcc),

b) face-centered cubic (fcc), and

c) hexagonal close packed (hcp).

The order of magnitude of the distance between the atoms in these crystal structures is 0.1 nm ($10^{-8}$ in.). The manner in which the atoms are arranged has a significant influence on the properties of a particular metal. These arrangements can be modified by adding atoms of some other metal or metals, known as **alloying**, and often improves the properties of the metal. (See Section 5.2.)

In the three structures illustrated in Figs. 3.2a–c, the fcc and hcp crystals have the most densely packed configurations. In the hcp structure, the top and bottom planes are called **basal planes**. The different crystal structures form due to differences in the energy required to form these structures. Thus, for example, tungsten forms a bcc structure because it requires less energy than other structures; similarly, aluminum forms an fcc structure. However, at different temperatures, the same metal may form different structures, because of differences in the amount of energy required at each temperature. For example, iron forms a bcc structure below 912°C (1674°F)) and above 1394°C (2541°F), but it forms an fcc structure between 912°C and 1394°C. The appearance of more than one type of crystal structure is known as **allotropism**, or **polymorphism**; as described in Section 5.2.5, these structural changes are an important aspect of heat treatment of metals and alloys.

# 3.3 | Deformation and Strength of Single Crystals

When crystals are subjected to an external force, they first undergo **elastic deformation**; that is, they return to their original shape when the force is removed. (See Section 2.2.) An analogy to this type of behavior is a helical spring that stretches when loaded and returns to its original shape when the load is removed. However, if the force on the crystal structure is increased sufficiently, the crystal undergoes **plastic deformation**, or *permanent deformation*; that is, it does not return to its original shape when the force is removed.

There are two basic mechanisms by which plastic deformation may take place in crystal structures:

**FIGURE 3.3** Permanent deformation, also called plastic deformation, of a single crystal subjected to a shear stress: (a) structure before deformation; and (b) deformation by slip. The $b/a$ ratio influences the magnitude of the shear stress required to cause slip.

(a)    (b)

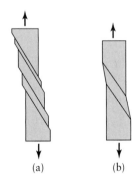

(a)    (b)

**FIGURE 3.4**

(a) Permanent deformation of a single crystal under a tensile load. Note that the slip planes tend to align themselves in the direction of pulling. This behavior can be simulated using a deck of cards with a rubber band around them. (b) Twinning in tension.

1. **Slip.** The first deformation mechanism is the slipping of one plane of atoms over an adjacent plane (the **slip plane**) under shear stress (Fig. 3.3). The deformation of a single-crystal specimen by slip is shown schematically in Fig. 3.4; this situation is much like sliding playing cards against each other. Just as it takes a certain amount of force to slide playing cards against each other, so a crystal requires a certain amount of shear stress (**critical shear stress**) to undergo permanent deformation. Thus, there must be a shear stress of sufficient magnitude within a crystal for plastic deformation to take place.

   The **maximum theoretical shear stress**, $\tau_{max}$, to cause permanent deformation in a perfect crystal is obtained as follows: when there is no stress, the atoms in the crystal are in equilibrium (Fig. 3.5). Under shear stress, the upper row of atoms is moved to the right, where the position of an atom is denoted as $x$. Thus, when $x = 0$ or $x = b$, the shear stress is zero. Each atom of the upper row is attracted to the nearest atom of the lower row, resulting in nonequilibrium at positions 2 and 4, where the stresses are maximum, but opposite in sign. At position 3, the shear stress is again zero, since this position is symmetric. In Fig. 3.5, we assume that, as a first approximation, the shear stress varies sinusoidally. Hence, the shear stress at a displacement $x$ is

$$\tau = \tau_{max} \sin \frac{2\pi x}{b}, \tag{3.1}$$

which, for small values of $x/b$, can be written as

$$\tau = \tau_{max} \frac{2\pi x}{b}.$$

From Hooke's law, we have

$$\tau = G\gamma = G\left(\frac{x}{a}\right).$$

**FIGURE 3.5** Variation of shear stress in moving a plane of atoms over another plane.

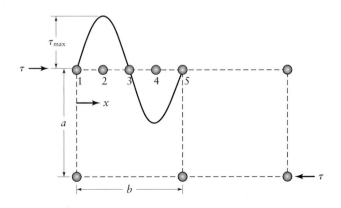

Hence,

$$\tau_{\text{max}} = \frac{Gb}{2\pi a}. \tag{3.2}$$

If we assume that $b$ is approximately equal to $a$, then

$$\tau_{\text{max}} = \frac{G}{2\pi}. \tag{3.3}$$

More refined calculations have set the value of the maximum theoretical shear stress as between $G/10$ and $G/30$.

The shear stress required to cause slip in single crystals is directly proportional to the ratio $b/a$ in Fig. 3.5. It can therefore be stated that slip in a crystal takes place along planes of maximum atomic density, or that slip takes place in closely packed planes and in closely packed directions. Because the $b/a$ ratio is different for different directions within the crystal, a single crystal has different properties when tested in different directions; thus, a crystal is anisotropic. A common example of **anisotropy** is woven cloth, which stretches differently when we pull it in different directions, or plywood, which is much stronger in the planar direction than along its thickness direction (where it splits easily).

2.  **Twinning.**  The second mechanism of plastic deformation is *twinning*, in which a portion of the crystal forms structurally a mirror image of itself across the *plane of twinning* (Fig. 3.4b). Twins form abruptly and are the cause of the creaking sound (*tin cry*) when a tin or zinc rod is bent at room temperature. Twinning usually occurs in hcp and bcc metals by plastic deformation and in fcc metals by annealing. (See Section 5.11.4.)

## 3.3.1  Slip systems

The combination of a slip plane and its direction of slip is known as a *slip system*. In general, metals with five or more slip systems are ductile, whereas those with less than five are not ductile. Each pattern of atomic arrangement has a different number of potential slip systems:

1.  In body-centered cubic crystals, there are 48 possible slip systems; thus, the probability is high that an externally applied shear stress will operate on one of the systems and cause slip. However, because of the relatively high $b/a$ ratio, the required shear stress is high. Metals with bcc structures (such as titanium, molybdenum, and tungsten) have good strength and moderate ductility.

2.  In face-centered cubic crystals, there are 12 slip systems. The probability of slip is moderate, and the required shear stress is low. Metals with fcc structures (such as aluminum, copper, gold, and silver) have moderate strength and good ductility.

3.  The hexagonal close-packed crystal has three slip systems and thus has a low probability of slip. However, more systems become active at elevated temperatures. Metals with hcp structures (such as beryllium, magnesium, and zinc) are generally brittle at room temperature.

Note in Fig. 3.4a that the portions of the single crystal that have slipped have rotated from their original angular position toward the direction of the tensile force. Note also that slip has taken place along certain planes only. With the use of electron microscopy, it has been shown that what appears to be a single slip plane is actually a **slip band**, consisting of a number of slip planes (Fig. 3.6).

### 3.3.2 Theoretical tensile strength of metals

In addition to obtaining the maximum theoretical shear strength, we can also calculate the theoretical tensile strength of metals. In the bar shown in Fig. 3.7, the interatomic distance is $a$ when no stress is applied. In order to increase this distance, we have to apply a force to overcome the cohesive force between the atoms, the cohesive force being zero in the unstrained equilibrium condition. When the tensile stress reaches $\sigma_{max}$, the atomic bonds between two neighboring atomic planes break; this $\sigma_{max}$ is known as the *ideal tensile strength*.

**FIGURE 3.6**   Schematic illustration of slip lines and slip bands in a single crystal subjected to a shear stress. A slip band consists of a number of slip planes. The crystal at the center of the upper drawing is an individual grain surrounded by other grains.

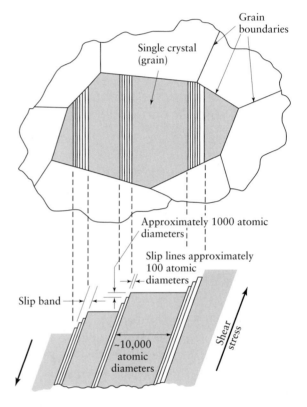

**FIGURE 3.7**   Variation of cohesive stress as a function of distance between a row of atoms.

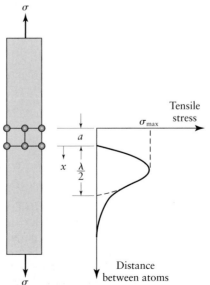

The ideal tensile strength can be derived by the following approach: Referring to Fig. 3.7, it can be shown that

$$\sigma_{max} = \frac{E\lambda}{2\pi a} \qquad (3.4a)$$

and that the work done per unit area in breaking the bond between the two atomic planes (i.e., the area under the cohesive-force curve) is given by

$$\text{Work} = \frac{\sigma_{max}\lambda}{\pi}. \qquad (3.4b)$$

This work is dissipated in creating two new fracture surfaces, thus involving surface energy of the material, $\gamma$. The total surface energy then is $2\gamma$. Combining these equations, we find that

$$\sigma_{max} = \sqrt{\frac{E\gamma}{a}}. \qquad (3.4c)$$

When appropriate values are substituted into this equation, we have

$$\sigma_{max} \simeq \frac{E}{10}. \qquad (3.5)$$

Thus the theoretical strength of steel, for example, would be on the order of 20 GPa ($3 \times 10^6$ psi).

### 3.3.3  Imperfections

The actual strength of metals is approximately one to two orders of magnitude lower than the strength levels obtained from the foregoing theoretical calculations. This discrepancy has been explained in terms of *imperfections* in the crystal structure. Unlike the idealized models described previously, actual metal crystals contain a large number of imperfections and defects, which are categorized as follows:

1. *Point defects*, such as a **vacancy** (missing atom), an **interstitial atom** (extra atom in the lattice), or an **impurity atom** (foreign atom that has replaced an atom of the pure metal) (Fig. 3.8);

2. *Linear*, or *one-dimensional, defects*, called **dislocations** (Fig. 3.9);

3. *Planar*, or *two-dimensional, imperfections*, such as **grain boundaries** and **phase boundaries**;

4. *Volume*, or *bulk, imperfections*, such as **voids, inclusions** (nonmetallic elements such as oxides, sulfides, and silicates), other **phases**, or **cracks**.

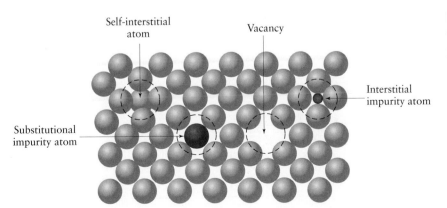

**FIGURE 3.8**  Various defects in a single-crystal lattice. *Source*: Courtesy of Dr. William G. Moffatt.

**FIGURE 3.9** (a) Edge dislocation, a linear defect at the edge of an extra plane of atoms. (b) Screw dislocation, a helical defect in a three–dimentional lattice of atoms.

Screw dislocation

(a)

(b)

A slip plane containing a dislocation requires lower shear stress to cause slip than does a plane in a perfect lattice (Fig. 3.10). One analogy used to describe the movement of an edge dislocation is the earthworm, which moves forward through a hump that starts at the tail and moves toward the head. A second analogy is that of moving a large carpet by forming a hump at one end and moving the hump forward toward the other end. The force required to move a carpet in this way is much less than that required to slide the whole carpet along the floor. *Screw dislocations* are so named because the atomic planes form a spiral ramp.

The **density of dislocations** (i.e., the total length of dislocation line per unit volume, $mm/mm^3 = mm^{-2}$) increases with increasing plastic deformation, by as much as $10^6 \ mm^{-2}$ at room temperature. The dislocation densities for some conditions are as follows:

a) very pure single crystals: 0 to $10^3 \ mm^{-2}$;
b) annealed single crystals: $10^5$ to $10^6$;
c) annealed polycrystals: $10^7$ to $10^8$;
d) highly cold-worked metals: $10^{11}$ to $10^{12}$.

The mechanical properties of metals, such as yield and fracture strength, as well as electrical conductivity, are affected by lattice defects; these properties are known as **structure-sensitive properties**. On the other hand, physical properties, such as melting point, specific heat, coefficient of thermal expansion, and elastic constants are not sensitive to these defects and are thus known as **structure-insensitive properties**.

### 3.3.4 Strain hardening (work hardening)

Although the presence of a dislocation lowers the shear stress required to cause slip, dislocations can (1) become entangled and interfere with each other, and (2) be impeded by barriers, such as grain boundaries and impurities and inclusions in the material. Entanglement and impediments increase the shear stress required for slip.

**FIGURE 3.10** Movement of an edge dislocation across the crystal lattice under a shear stress. Dislocations help explain why the actual strength of metals is much lower than that predicted by theory.

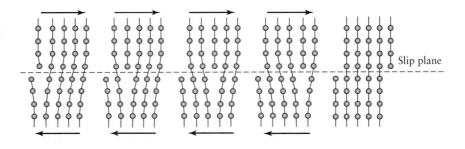

Slip plane

The increase in the shear stress required, and hence the increase in the overall strength of the metal, is known as *strain hardening*, or *work hardening*. (See Section 2.2.3.) The greater the deformation, the more the entanglements, which increase the metal's strength. Work hardening is used extensively to strengthen metals in metal-working processes at ambient temperature. Typical examples are strengthening wire by drawing it through a die to reduce its cross-section (Section 6.5), producing the head on a bolt by forging it (Section 6.2.4), and producing sheet metal for automobile bodies and aircraft fuselages by rolling (Section 6.3). As shown by Eq. (2.11), the degree of strain hardening is indicated by the strain-hardening exponent, $n$. Among the three crystal structures, hcp has the lowest $n$ value, followed by bcc and then fcc, which has the highest.

# 3.4 | Grains and Grain Boundaries

Metals commonly used for manufacturing various products are composed of many individual, randomly oriented crystals (*grains*). We are thus dealing with metal structures that are not single crystals, but **polycrystals**, that is, many crystals.

When a mass of molten metal begins to solidify, crystals begin to form independently of each other, with random orientations and at various locations within the liquid mass (Fig. 3.11). Each of the crystals grows into a crystalline structure, or grain. The number and size of the grains developed in a unit volume of the metal depend on the rate at which **nucleation** (the initial stage of formation of crystals) takes place. The number of different sites in which individual crystals begin to form (seven in Fig. 3.11a) and the rate at which these crystals grow affect the size of grains developed. Generally, rapid cooling produces smaller grains, whereas slow cooling produces larger grains. (See also Section 5.3.)

If the crystal nucleation rate is high, the number of grains in a unit volume of metal will be greater, and consequently grain size will be small. Conversely, if the rate of growth of the crystals is high compared with their nucleation rate, there will be fewer, but larger, grains per unit volume.

Note in Fig. 3.11 how the grains eventually interfere with and impinge upon one another. The surfaces that separate the individual grains are called *grain boundaries*.

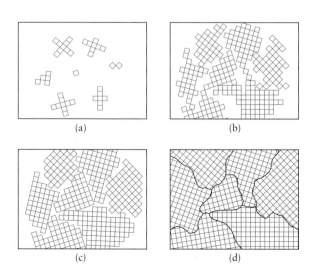

(a)        (b)

(c)        (d)

**FIGURE 3.11** Schematic illustration of the various stages during solidification of molten metal. Each small square represents a unit cell. (a) Nucleation of crystals at random sites in the molten metal. Note that the crystallographic orientation of each site is different. (b) and (c) Growth of crystals as solidification continues. (d) Solidified metal, showing individual grains and grain boundaries. Note the different angles at which neighboring grains meet each other. *Source*: W. Rosenhain.

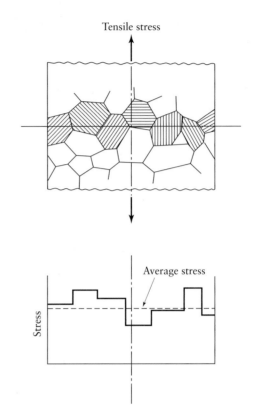

**FIGURE 3.12** Variation of tensile stress across a plane of polycrystalline metal specimen subjected to tension. Note that the strength exhibited by each grain depends on its orientation.

Each grain consists of either a single crystal (for pure metals) or a polycrystalline aggregate (for alloys). Note that the crystallographic orientation changes abruptly from one grain to the next across the grain boundaries. Recall from Section 3.3 that the behavior of a single crystal or single grain is *anisotropic*. The ideal behavior of a piece of polycrystalline metal (Fig. 3.12) is *isotropic*, because the grains have random crystallographic orientations; thus, the metal's properties do not vary with the direction of testing. In practice, however, this situation rarely exists.

### 3.4.1 Grain size

*Grain size* significantly influences the mechanical properties of metals; large grain size is generally associated with low strength, low hardness, and high ductility. Large grains, particularly in sheet metals, also have a rough surface appearance after being stretched. (See Section 3.6.) The yield strength, $Y$, is the most sensitive property and is related to grain size by the empirical formula (*Hall-Petch equation*):

$$Y = Y_i + kd^{-1/2}, \tag{3.6}$$

where $Y_i$ is a basic yield stress that can be regarded as the stress opposing the motion of dislocations, $k$ is a constant indicating the extent to which dislocations are piled up at barriers (such as grain boundaries), and $d$ is the grain diameter. Equation (3.6) is valid below the recrystallization temperature of the material.

Grain size is usually measured by counting the number of grains in a given area or the number of grains that intersect a given length of a line randomly drawn on an enlarged photograph of the grains on a polished and etched specimen taken under a microscope. Grain size may also be determined by referring to a standard chart. The American Society for Testing and Materials (ASTM) grain-size number, $n$, is related to the number of grains, $N$, per square inch at a magnification of $100\times$ (equal to $0.0645$ mm$^2$ of actual area) by the expression

$$N = 2^{n-1}. \tag{3.7}$$

Because grains generally are extremely small, many grains occupy a unit volume of metal. Grains of sizes between 5 and 8 are generally considered fine grains. A grain size of 7 generally is acceptable for sheet metals used to make car bodies, appliances, and kitchen utensils. Grains can also be large enough to be visible to the naked eye, such as grains of zinc on the surface of galvanized sheet steel.

### 3.4.2  Influence of grain boundaries

Grain boundaries have an important influence on the strength and ductility of metals. Also, because they interfere with the movement of dislocations, grain boundaries also influence strain hardening. These effects depend on temperature, rate of deformation, and the type and amount of impurities present along the grain boundaries. Grain boundaries are more reactive than the grains themselves, because the atoms along the grain boundaries are packed less efficiently and are more disordered and hence have a higher energy than the atoms in the orderly arrangement within the grains.

At elevated temperatures and in materials whose properties depend on the rate of deformation, plastic deformation also takes place by means of **grain-boundary sliding**. The **creep mechanism** (elongation under stress over a period of time, usually at elevated temperatures; see Section 2.8) results from grain-boundary sliding.

When brought into close atomic contact with certain low-melting-point metals, a normally ductile and strong metal can crack under very low stresses, a phenomenon referred to as **grain boundary embrittlement** (Fig. 3.13). Examples include aluminum wetted with a mercury–zinc amalgam, and liquid gallium and copper at elevated temperature wetted with lead or bismuth; these elements weaken the grain boundaries of the metal by *embrittlement*. The term **liquid-metal embrittlement** is used to describe such phenomena, because the embrittling element is in a liquid state. However, embrittlement can also occur at temperatures well below the melting point of the embrittling element, a phenomenon known as **solid-metal embrittlement**.

**Hot shortness** is caused by local melting of a constituent or an impurity in the grain boundary at a temperature below the melting point of the metal itself. When such a metal is subjected to plastic deformation at elevated temperatures (*hot working*; see Section 3.7), the piece of metal crumbles and disintegrates along its grain

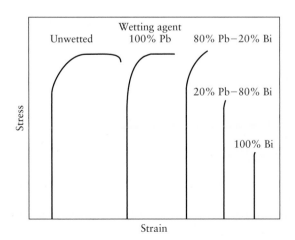

**FIGURE  3.13**
Embrittlement of copper by lead and bismuth at 350°C (660°F). Embrittlement has important effects on the strength, ductility, and toughness of materials. *Source*: After W. Rostoker.

boundaries; examples include antimony in copper, and leaded steels and brass. To avoid hot shortness, the metal is usually worked at a lower temperature, which prevents softening and melting along the grain boundaries. Another form of embrittlement is **temper embrittlement** in alloy steels, which is caused by segregation (movement) of impurities to the grain boundaries.

## 3.5 | Plastic Deformation of Polycrystalline Metals

If a piece of polycrystalline metal with uniform equiaxed grains (i.e., having equal dimensions in all directions, as shown in the model in Fig. 3.14) is subjected to plastic deformation at room temperature (cold working), the grains become permanently deformed and elongated. The deformation process may be carried out either by compressing the metal (as in forging) or by subjecting it to tension (as in stretching sheet metal). The deformation within each grain takes place by the mechanisms described in Section 3.3 for a single crystal.

During plastic deformation, the grain boundaries remain intact, and mass continuity is maintained. The deformed metal exhibits higher strength, because of the entanglement of dislocations with grain boundaries. The increase in strength depends on the amount of deformation (strain) to which the metal is subjected; the greater the deformation, the stronger the metal becomes. Furthermore, the increase in strength is greater for metals with smaller grains, because they have a larger grain-boundary surface area per unit volume of metal.

**Anisotropy (texture).** Figure 3.14 shows that, as a result of plastic deformation, the grains in a piece of metal become elongated in one direction and contracted in the other. Consequently, the metal has become *anisotropic*, and thus its properties in the vertical direction are different from those in the horizontal direction. Many products develop anisotropy of mechanical properties after they have been processed by metalworking techniques. The degree of anisotropy depends on how uniformly the metal is deformed. Note from the direction of the crack in Fig. 3.15, for example, that the ductility of the cold-rolled sheet in the vertical (transverse) direction is lower than in its longitudinal direction.

Anisotropy influences both mechanical and physical properties of metals. For example, sheet steel for electrical transformers is rolled in such a way that the resulting deformation imparts anisotropic magnetic properties to the sheet, thus reducing magnetic-hysteresis losses and improving the efficiency of transformers. (See

**FIGURE 3.14** Plastic deformation of idealized (equiaxed) grains in a specimen subjected to compression, such as is done in rolling or forging of metals: (a) before deformation; and (b) after deformation. Note the alignment of grain boundaries along a horizontal direction.

(a)

(b)

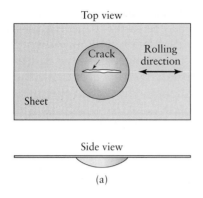

Top view

Side view

(a)

(b)

**FIGURE 3.15**
(a) Illustration of a crack in sheet metal subjected to bulging, such as by pushing a steel ball against the sheet. Note the orientation of the crack with respect to the rolling direction of the sheet. This material is anisotropic. (b) Aluminum sheet with a crack (vertical dark line at the center) developed in a bulge test. *Source*: Courtesy of J. S. Kallend, Illinois Institute of Technology.

also the discussion of amorphous alloys in Section 3.11.9.) There are two general types of anisotropy in metals: *preferred orientation* and *mechanical fibering*.

**Preferred orientation.** Also called **crystallographic anisotropy**, preferred orientation can be best described by reference to Fig. 3.4. When a metal crystal is subjected to tension, the sliding blocks rotate toward the direction of pulling; thus, slip planes and slip bands tend to align themselves with the direction of deformation. Similarly, for a polycrystalline aggregate, with grains in various orientations (Fig. 3.12), all slip directions tend to align themselves with the direction of pulling. Conversely, under compression, the slip planes tend to align themselves in a direction perpendicular to the direction of compression.

**Mechanical fibering.** Mechanical fibering results from the alignment of impurities, inclusions (stringers), and voids in the metal during deformation. Note that if the spherical grains in Fig. 3.14 were coated with impurities, these impurities would align themselves generally in a horizontal direction after deformation. Since impurities weaken the grain boundaries, this piece of metal would be weak and less ductile when tested in the vertical direction than in the horizontal direction. An analogy to this case would be plywood, which is strong in tension along its planar directions, but peels off easily when tested in tension in its thickness direction.

# 3.6 | Recovery, Recrystallization, and Grain Growth

We have shown that plastic deformation at room temperature results in the deformation of grains and grain boundaries, a general increase in strength, and a decrease in ductility, as well as causing anisotropic behavior. These effects can be reversed and the properties of the metal brought back to their original levels by heating the metal to within a specific temperature range for a period of time. The temperature range and amount of time depend on the material and several other factors. Three events take place consecutively during the heating process:

1. **Recovery.** During recovery, which occurs at a certain temperature range below the **recrystallization temperature** of the metal, the stresses in the highly deformed regions are relieved and the number of mobile dislocations is reduced. Subgrain boundaries begin to form (**polygonization**) with no appreciable change in mechanical properties, such as hardness and strength, but with some increase in ductility (Fig. 3.16).

2. **Recrystallization.** The process in which, at a certain temperature range, new equiaxed and strain-free grains are formed, replacing the older grains, is called **recrystallization**. The temperature for recrystallization ranges approximately between $0.3T_m$ and $0.5T_m$, where $T_m$ is the melting point of the metal on the absolute scale. The recrystallization temperature is generally defined as the temperature at which complete recrystallization occurs within approximately one hour. Recrystallization decreases the density of dislocations and lowers the strength, but raises the ductility, of the metal (Fig. 3.16). Metals such as lead, tin, cadmium, and zinc recrystallize at about room temperature.

Recrystallization depends on the degree of prior cold work (work hardening); the higher the amount of cold work, the lower the temperature required for recrystallization to occur. The reason for this relationship is that as the amount of cold work increases, the number of dislocations and the amount of energy stored in the dislocations (*stored energy*) also increase. The stored energy supplies the work required for recrystallization. Recrystallization is a function of time, because it involves *diffusion*, that is, movement and exchange of atoms across grain boundaries.

The effects on recrystallization of temperature, time, and reduction in the thickness or height of the workpiece by cold working are as follows (Fig. 3.17):

**FIGURE 3.16** Schematic illustration of the effects of recovery, recrystallization, and grain growth on mechanical properties and shape and size of grains. Note the formation of small new grains during recrystallization. *Source*: G. Sachs.

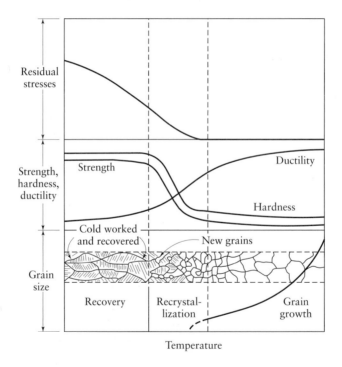

**FIGURE 3.17** Variation of strength and hardness with recrystallization temperature, time, and prior cold work. Note that the more a metal is cold worked, the less time it takes to recrystallize, because of the higher stored energy from cold working due to increased dislocation density.

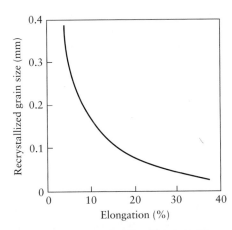

**FIGURE 3.18** The effect of prior cold work on the recrystallized grain size of alpha brass. Below a critical elongation (strain), typically 5%, no recrystallization occurs.

**FIGURE 3.19** Surface roughness on the cylindrical surface of an aluminum specimen subjected to compression. *Source*: A. Mulc and S. Kalpakjian.

- For a constant amount of deformation by cold working, the time required for recrystallization decreases with increasing temperature.
- The more prior cold work done, the lower the temperature required for recrystallization.
- The higher the amount of deformation, the smaller the grain size during recrystallization (Fig. 3.18). Subjecting a metal to deformation is a common method of converting a coarse-grained structure to one of fine grain, with improved properties.
- Anisotropy due to preferred orientation usually persists after recrystallization. To restore isotropy, a temperature higher than that required for recrystallization may be necessary.

3. **Grain growth.** If we continue to raise the temperature of the metal, the grains begin to grow, and their size may eventually exceed the original grain size. This phenomenon is known as *grain growth*, and it has a slightly adverse effect on mechanical properties (Fig. 3.16). However, large grains produce a rough surface appearance on sheet metals (**orange-peel effect**) when they are stretched to form a part or when a piece of metal is subjected to compression (Fig. 3.19), such as in forging operations.

# 3.7 | Cold, Warm, and Hot Working

When plastic deformation is carried out above or below the recrystallization temperature, it is called *hot working* or *cold working*, respectively. As the name implies, *warm working* is carried out at an intermediate temperature; thus, warm working is a compromise between cold and hot working. The temperature ranges for these three

**TABLE 3.1**

| Homologous Temperature Ranges for Various Processes | |
| --- | --- |
| Process | $T/T_m$ |
| Cold working | <0.3 |
| Warm working | 0.3 to 0.5 |
| Hot working | >0.6 |

categories of plastic deformation are given in Table 3.1 in terms of a ratio, where $T$ is the working temperature and $T_m$ is the melting point of the metal, both on the absolute scale. Although it is a dimensionless quantity, this ratio is known as the **homologous temperature**.

There are important technological differences in products that are processed by cold, warm, or hot working. For example, compared with cold-worked products, hot-worked products generally have less dimensional accuracy, because of uneven thermal expansion and contraction during processing, and a rougher surface appearance and finish, because of the oxide layer that usually develops during heating. Other important manufacturing characteristics, such as formability, machinability, and weldability, are also affected by cold, warm, and hot working to different degrees.

## 3.8  Failure and Fracture

*Failure* is one of the most important aspects of a material's behavior, because it directly influences the selection of a material for a certain application, the methods of manufacturing, and the service life of the component. Because of the many factors involved, failure and fracture of materials is a complex area of study. In this section, we consider only those aspects of failure that are of particular significance to selecting and processing materials.

There are two general types of failure: (1) **fracture** and separation of the material, through either internal or external cracking; and (2) **buckling** (Fig. 3.20). Fracture is further divided into two general categories: ductile and brittle (Fig. 3.21). Although failure of materials is generally regarded as undesirable, certain products are designed to fail. Typical examples include (a) food and beverage containers with tabs or entire tops that are removed by tearing the sheet metal along a prescribed path, and (b) screw caps for bottles.

### 3.8.1  Ductile fracture

*Ductile fracture* is characterized by plastic deformation, which precedes failure of the part. In a tension test, highly ductile materials, such as gold and lead, may neck down to a point and then fail (Fig. 3.21d). Most metals and alloys, however, neck

**FIGURE 3.20**  Schematic illustration of types of failure in materials: (a) necking and fracture of ductile materials; (b) buckling of ductile materials under a compressive load; (c) fracture of brittle materials in compression; (d) cracking on the barreled surface of ductile materials in compression. (See also Fig. 6.1b)

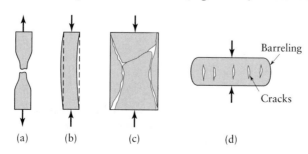

(a)    (b)    (c)    (d)

down to a finite area and then fail. Ductile fracture generally takes place along planes on which the *shear stress is a maximum*. In torsion, for example, a ductile metal fractures along a plane perpendicular to the axis of twist, that is, the plane on which the shear stress is a maximum. Fracture in shear is a result of extensive slip along slip planes within the grains.

Upon close examination of the surface of a ductile fracture (Fig. 3.22), we see a *fibrous* pattern with *dimples*, as if a number of very small tension tests have been carried out over the fracture surface. Failure is initiated with the formation of tiny voids, usually around small inclusions or preexisting voids, which then grow and coalesce, developing cracks that grow in size and lead to fracture. In a tension-test specimen, fracture begins at the center of the necked region from the growth and coalescence of cavities (Fig. 3.23). The central region becomes one large crack and propagates to the periphery of this necked region. Because of its appearance, the fracture surface of a tension-test specimen is called a **cup-and-cone fracture**.

**Effects of inclusions.** Because they are nucleation sites for voids, *inclusions* have an important influence on ductile fracture and thus on the formability of materials (Fig. 3.24). Inclusions may consist of impurities of various kinds and second-phase particles such as oxides, carbides, and sulfides. The extent of their influence depends on factors such as their shape, hardness, distribution, and volume fraction. The greater the volume fraction of inclusions, the lower will be the ductility of the material. Voids and porosity developed during processing, such as from casting, reduce the ductility of a material. (See Section 5.4.6.)

Two factors affect void formation. The first is the strength of the bond at the interface of an inclusion and the matrix; if the bond is strong, there is less tendency for void formation during plastic deformation. The second factor is the hardness of

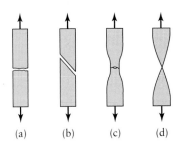

**FIGURE 3.21** Schematic illustration of the types of fracture in tension: (a) brittle fracture in polycrystalline metals; (b) shear fracture in ductile single crystals (see also Fig. 3.4a); (c) ductile cup-and-cone fracture in polycrystalline metals (see also Fig. 2.2); (d) complete ductile fracture in polycrystalline metals, with 100% reduction of area.

**FIGURE 3.22** Surface of ductile fracture in low-carbon steel, showing dimples. Fracture is usually initiated at impurities, inclusions, or preexisting voids in the metal. *Source*: K.-H. Habig and D. Klaffke. Photo by BAM, Berlin, Germany.

**FIGURE 3.23** Sequence of events in necking and fracture of a tensile-test specimen: (a) early stage of necking; (b) small voids begin to form within the necked region; (c) voids coalesce, producing an internal crack; (d) rest of cross-section begins to fail at the periphery by shearing; (e) final fracture surfaces, known as cup-(top fracture surface) and-cone (bottom surface) fracture.

the inclusion. If the inclusion is soft, such as manganese sulfide, it will conform to the overall change in shape of the specimen or workpiece during plastic deformation. If it is hard, such as a carbide or oxide, it could lead to void formation (Fig. 3.25). Hard inclusions may also break up into smaller particles during deformation, because of their brittle nature. The alignment of inclusions during plastic deformation leads to **mechanical fibering**. Subsequent processing of such a material must therefore involve considerations of the proper direction of working for maximum ductility and strength.

**Transition temperature.** Many metals undergo a sharp change in ductility and toughness across a narrow temperature range called the *transition temperature* (Fig. 3.26). This phenomenon occurs in body-centered cubic and some hexagonal close-packed metals; it is rarely exhibited by face-centered cubic metals. The transi-

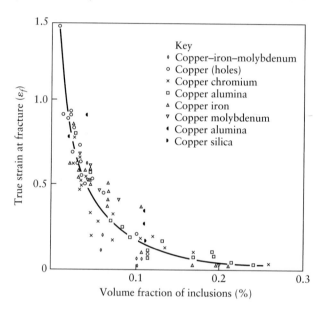

**FIGURE 3.24** The effect of volume fraction of various second-phase particles on the true strain at fracture in a tensile test for copper. *Source*: After B. I. Edelson and W. M. Baldwin.

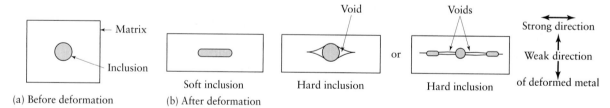

(a) Before deformation  (b) After deformation

**FIGURE 3.25** Schematic illustration of the deformation of soft and hard inclusions and their effect on void formation in plastic deformation. Note that hard inclusions, because they do not comply with the overall deformation of the ductile matrix, can cause voids.

**FIGURE 3.26** Schematic illustration of transition temperature. Note the narrow temperature range across which the behavior of the metal undergoes a major transition.

**FIGURE 3.27** Strain aging and its effect on the shape of the true-stress–true-strain curve for 0.03% C rimmed steel at 60°C (140°F). *Source*: A. S. Keh and W. C. Leslie.

tion temperature depends on factors such as composition, microstructure, grain size, surface finish and shape of the specimen, and rate of deformation. High rates, abrupt changes in shape, and surface notches raise the transition temperature.

**Strain aging.**  *Strain aging* is a phenomenon in which carbon atoms in steels segregate to dislocations, thereby pinning them and thus increasing the steel's resistance to dislocation movement. The result is increased strength and reduced ductility. Figure 3.27 shows the effects of strain aging on the shape of the stress–strain curve for low-carbon steel at room temperature. Curve *abc* is the original curve, with upper and lower yield points typical of these steels. If the tension test is stopped at point *e* and the specimen is unloaded and tested again, curve *dec* is obtained. Note that the upper and lower yield points have disappeared. However, if four hours pass before the specimen is strained again, curve *dfg* is obtained; if 126 hours pass, curve *dhi* is obtained. Instead of taking place over several days at room temperature, this same phenomenon can occur in just a few hours at a higher temperature; it is called **accelerated strain aging**. For steels, the phenomenon is called **blue brittleness**.

### 3.8.2  Brittle fracture

*Brittle fracture* occurs with little or no gross plastic deformation preceding the separation of the material into two or more pieces. In tension, fracture takes place along a crystallographic plane, called a **cleavage plane**, on which the normal tensile stress is a maximum. Face-centered cubic metals usually do not fail in brittle fracture, whereas body-centered cubic and some hexagonal close-packed metals fail by cleavage. In general, low temperature and high rates of deformation promote brittle fracture.

In a polycrystalline metal under tension, the fracture surface has a bright granular appearance, because of the changes in the direction of the cleavage planes as the crack propagates from one grain to another. Figure 3.28 shows an example of the surface of brittle fracture. Brittle fracture of a specimen in compression is more complex and may follow a path that is theoretically 45° to the direction of the applied force.

Examples of fracture along a cleavage plane include the splitting of rock salt and the peeling of layers of mica. Tensile stresses normal to the cleavage plane, caused by

**FIGURE 3.28** Typical fracture surface of steel that has failed in a brittle manner. The fracture path is transgranular (through the grains). Compare this surface with the ductile fracture surface shown in Fig. 3.22. Magnification: 200×. *Source*: Courtesy of Packer Engineering.

pulling, initiate and control the propagation of fracture. Another example is the behavior of brittle materials, such as chalk, gray cast iron, and concrete. In tension, they typically fail in the manner shown in Fig. 3.21a. In torsion, however, they fail along a plane at 45° to the axis of twist, that is, along a plane on which the tensile stress is a maximum.

**Defects.** An important factor in fracture is the presence of *defects* such as scratches, flaws, and external or internal cracks. Under tension, the tip of a crack is subjected to high tensile stresses, which propagate the crack rapidly, because the material has little capacity to dissipate energy. The tensile strength of a specimen with a crack perpendicular to the direction of pulling is related to the length of the crack as follows:

$$\sigma \propto \frac{1}{\sqrt{\text{crack length}}}. \tag{3.8}$$

The presence of defects is essential in explaining why brittle materials are so weak in tension compared with their strength in compression. Under tensile stresses, cracks propagate rapidly, causing what is known as *catastrophic failure*. With polycrystalline metals, the fracture paths most commonly observed are **transgranular** (*transcrystalline*, or *intragranular*), meaning that the crack propagates *through* the grain. **Intergranular** fracture, where the crack propagates *along* the grain boundaries (Fig. 3.29), generally occurs when the grain boundaries are soft, contain a brittle phase, or have been weakened by liquid-or solid-metal embrittlement (Section 3.4.2).

The maximum crack velocity in a brittle material is about 62% of the elastic wave-propagation (or acoustic) velocity of the material; this velocity is given by the formula $\sqrt{E/\rho}$, where $E$ is the elastic modulus and $\rho$ is the mass density. Thus, for steel, the maximum crack velocity is 2000 m/s (6600 ft/s).

**FIGURE 3.29** Intergranular fracture, at two different magnifications. Grains and grain boundaries are clearly visible in this micrograph. The fracture path is along the grain boundaries. Magnification: left, 100×; right, 500×. *Source*: Courtesy of Packer Engineering.

As Fig. 3.30 shows, cracks may be subjected to stresses in different directions. Mode I is tensile stress applied perpendicular to the crack. Modes II and III are shear stresses applied in two different directions. Tearing paper, cutting sheet metal with shears, and opening pop-top cans are examples of Mode III fracture.

**Fatigue fracture.** *Fatigue fracture* is basically of a brittle nature. Minute external or internal cracks develop at flaws or defects in the material. The cracks then propagate and eventually lead to total failure of the part. The fracture surface in fatigue is generally characterized by the term **beach marks**, because of its appearance (Fig. 3.31). With large magnification, such as higher than 1000×, a series of **striations** can be seen on fracture surfaces, where each beach mark consists of a number of striations.

Fatigue life is greatly influenced by the method of preparation of the surfaces of the part or specimen (Fig. 3.32). The fatigue strength of manufactured products can generally be improved by the following methods:

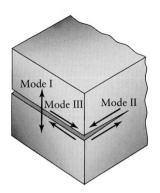

**FIGURE 3.30** Three modes of fracture. Mode I has been studied extensively, because it is the most commonly observed in engineering structures and components. Mode II is rare. Mode III is the tearing process; examples include opening a pop-top can, tearing a piece of paper, and cutting materials with a pair of scissors.

**FIGURE 3.31** Typical fatigue fracture surface on metals, showing beach marks. Most components in machines and engines fail by fatigue and not by excessive static loading. Magnification: left, 500×; right, 1000×. *Source*: Courtesy of Packer Engineering.

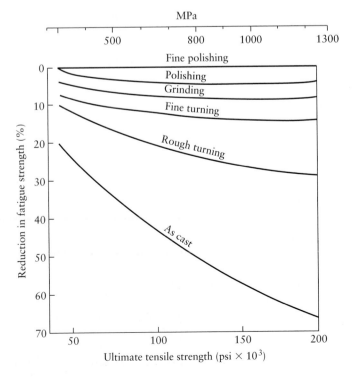

**FIGURE 3.32** Reduction in fatigue strength of cast steels subjected to various surface-finishing operations. Note that the reduction is greater as the surface roughness and strength of the steel increase. *Source*: Shigley, J.E., and Mitchell, L.D., Mechanical Engineering Design, New York, McGraw-Hill, 1983.

1. Inducing compressive residual stresses on surfaces, such as by shot peening or roller burnishing (Section 4.5.1);

2. Surface (case) hardening by various means (Section 5.11.3);

3. Providing fine surface finish, thereby reducing the effects of notches and other surface imperfections;

4. Selecting appropriate materials and ensuring that they are free from significant amounts of inclusions, voids, and impurities.

Conversely, the following factors and processes can reduce fatigue strength: decarburization, surface pits due to corrosion that act as stress raisers, hydrogen embrittlement (see below), galvanizing, and electroplating. Two of these processes are described as follows:

**Stress-corrosion cracking.**   An otherwise ductile metal can fail in a brittle manner by *stress-corrosion cracking* (also called *stress cracking* or *season cracking*). After forming, parts may, either over a period of time or soon after being made into a product, develop cracks. Crack propagation may be intergranular or transgranular.

The susceptibility of metals to stress-corrosion cracking depends mainly on the material, the presence and magnitude of tensile residual stresses, and the environment. Brass and austenitic stainless steels are among the metals that are highly susceptible to stress cracking. Environmental factors such as salt water or other chemicals could be corrosive to metals. The usual procedure to avoid stress-corrosion cracking is to stress relieve the part just after it is formed. Full annealing may also be done, but this treatment reduces the strength of cold-worked parts.

**Hydrogen embrittlement.**   The presence of hydrogen can reduce ductility and cause severe embrittlement in many metals, alloys, and nonmetallic materials, leading to premature failure. This phenomenon is known as *hydrogen embrittlement* and is especially severe in high-strength steels. Possible sources of hydrogen are during melting of the metal, during pickling (removal of surface oxides by chemical or electrochemical reaction), through electrolysis in electroplating, from water vapor in the atmosphere, or from moist electrodes and fluxes used during welding. Oxygen can also cause embrittlement in metals, especially in copper alloys.

### 3.8.3 Size effect

The dependence of the properties of a material on its size is knows as *size effect*. Note from the foregoing discussions and from Eq. (3.8) that as the size decreases, defects, cracks, imperfections, and the like are less likely to be present; hence, the strength and ductility of a specimen increase with decreasing size. Strength is related to cross-sectional area, but the length of the specimen is also important: The greater the length, the greater the probability of defects. As an analogy, a long chain is more likely to be weaker than a shorter chain, because the probability of one of the links being weak increases with the number of links (i.e., the length of chain). Size effect is also demonstrated by **whiskers**, which, because of their small size, are either free from imperfections or do not contain the types of imperfections that affect their strength. The actual strength in such cases approaches the theoretical strength. (See also Sections 8.6.10 and 10.9.2 for some applications of whiskers.)

---

**EXAMPLE 3.1**  Brittle fracture of steel plates of the hull of the R.M.S. Titanic

A detailed analysis of the Titanic disaster in 1912 has indicated that the ship sank not so much because it hit an iceberg as because of structural weaknesses. The plates composing its hull were made of low-grade steel with a high sulfur content; they had low toughness (as determined by the Charpy test; see Section 2.9) when chilled, as was the case in the Atlantic Ocean, and subjected to an external impact loading. With such a material, a crack that starts in one part of a welded steel structure can propagate rapidly and completely around the hull and cause a large ship to split in two. Although Titanic was built with brittle plates, as we know from the physical and photographic observations of the sunken ship, not all ships of that time were built with such low-grade steel. Furthermore, better construction techniques could have been employed, among them better welding techniques (Chapter 12) to improve the structural strength of the hull.

---

## 3.9 | Physical Properties

In addition to their mechanical properties, the *physical properties* of materials should also be considered in selecting and processing materials. Properties of particular interest are density, melting point, specific heat, thermal conductivity and expansion, electrical and magnetic properties, and resistance to oxidation and corrosion, as described next.

### 3.9.1 Density

The *density* of a material is its weight per unit volume; if expressed relative to the density of water, the density of a material is known as **specific gravity** and thus has no units. The density of materials depends on atomic weight, atomic radius, and the packing of the atoms. Alloying elements generally have a minor effect on the density of metals, where the strength of the effect depends on the density of the alloying elements. Table 3.2 lists the range of densities for a variety of materials at room temperature.

The most significant role that density plays is in the **specific strength** (strength-to-weight ratio) and **specific stiffness** (stiffness-to-weight ratio) of materials and structures. (See also Section 10.9.1.) Weight reduction is particularly important for aircraft and aerospace structures, automotive bodies and components, and other products for which energy consumption and power limitations are major concerns. Substitution of materials for weight saving and economy is a major factor in advanced equipment and machinery as well as in consumer products such as automobiles.

Density also is an important factor in the selection of materials for high-speed equipment, such as the use of magnesium in printing and textile machinery, many components of which usually operate at very high speeds. To obtain an exposure time of 1/4000 s in cameras without sacrificing accuracy, the shutters of some high-quality 35-mm cameras are made of titanium. The resulting light weight of the components in these high-speed operations reduces inertial forces that otherwise could lead to vibrations, inaccuracies, and even part failure over time. However, there are also applications for which higher levels of weight are desirable. Examples include counterweights for various mechanisms (using lead and steel), flywheels, and components for self-winding watches (using high-density materials such as tungsten).

### 3.9.2 Melting point

The *melting point* of a material depends on the energy required to separate its atoms. As shown in Table 3.2, the melting point of an alloy is a wide range of temperatures, unlike the case for pure metals, which have a definite melting point. The melting points of alloys depend on the alloys' particular composition. (See Section 5.2.) The melting point of a material has a number of indirect effects on manufacturing operations. The choice of a material for high-temperature applications is the most obvious consideration, such as in jet engines and furnaces. The choice of a material for applications where high frictional heat is generated, such as with high sliding speeds of machine components, is also affected by melting point. Sliding down a rope or rubbing your hands together fast is a simple demonstration of how high temperatures can become with increasing sliding speed. (See also Section 4.4.1.)

Since the recrystallization temperature of a metal is related to its melting point (Section 3.6), operations such as annealing, heat treating, and hot working require a

**TABLE 3.2**

**Physical Properties of Various Materials at Room Temperature**

| | Density (kg/m$^3$) | Melting Point (°C) | Specific Heat (J/kg K) | Thermal Conductivity (W/m K) | Coefficient of Thermal Expansion ($\mu$m/m°C) |
|---|---|---|---|---|---|
| **METAL** | | | | | |
| Aluminum | 2700 | 660 | 900 | 222 | 23.6 |
| Aluminum alloys | 2630–2820 | 476–654 | 880–920 | 121–239 | 23.0–23.6 |
| Beryllium | 1854 | 1278 | 1884 | 146 | 8.5 |
| Columbium (niobium) | 8580 | 2468 | 272 | 52 | 7.1 |
| Copper | 8970 | 1082 | 385 | 393 | 16.5 |
| Copper alloys | 7470–8940 | 885–1260 | 337–435 | 29–234 | 16.5–20 |
| Gold | 19300 | 1063 | 129 | 317 | 19.3 |
| Iron | 7860 | 1537 | 460 | 74 | 11.5 |
| Steels | 6920–9130 | 1371–1532 | 448–502 | 15–52 | 11.7–17.3 |
| Lead | 11350 | 327 | 130 | 35 | 29.4 |
| Lead alloys | 8850–11,350 | 182–326 | 126–188 | 24–46 | 27.1–31.1 |
| Magnesium | 1745 | 650 | 1025 | 154 | 26.0 |
| Magnesium alloys | 1770–1780 | 610–621 | 1046 | 75–138 | 26.0 |
| Molybdenum alloys | 10210 | 2610 | 276 | 142 | 5.1 |
| Nickel | 8910 | 1453 | 440 | 92 | 13.3 |
| Nickel alloys | 7750–8850 | 1110–1454 | 381–544 | 12–63 | 12.7–18.4 |
| Silicon | 2330 | 1423 | 712 | 148 | 7.63 |
| Silver | 10500 | 961 | 235 | 429 | 19.3 |
| Tantalum alloys | 16600 | 2996 | 142 | 54 | 6.5 |
| Titanium | 4510 | 1668 | 519 | 17 | 8.35 |
| Titanium alloys | 4430–4700 | 1549–1649 | 502–544 | 8–12 | 8.1–9.5 |
| Tungsten | 19290 | 3410 | 138 | 166 | 4.5 |
| **NONMETALLIC** | | | | | |
| Ceramics | 2300–5500 | – | 750–950 | 10–17 | 5.5–13.5 |
| Glasses | 2400–2700 | 580–1540 | 500–850 | 0.6–1.7 | 4.6–70 |
| Graphite | 1900–2200 | – | 840 | 5–10 | 7.86 |
| Plastics | 900–2000 | 110–330 | 1000–2000 | 0.1–0.4 | 72–200 |
| Wood | 400–700 | – | 2400–2800 | 0.1–0.4 | 2–60 |

knowledge of the melting points of the metals involved. These considerations, in turn, influence the selection of tool and die materials in manufacturing operations. Another major influence of the melting point is in the selection of the equipment and melting practice in casting operations; the higher the melting point of the material, the more difficult the operation becomes (Section 5.5). Also, the selection of die materials, as in die casting, depends on the melting point of the workpiece material. In the electrical-discharge machining process (Section 9.13), the melting points of metals are related to the rate of material removal and tool wear.

### 3.9.3  Specific heat

*Specific heat* is the energy required to raise the temperature of a unit mass of material by one degree. Alloying elements have a relatively minor effect on the specific heat of metals. The temperature rise in a workpiece, resulting from forming or machining operations, is a function of the work done and the specific heat of the workpiece material (see Section 2.12.1); thus, the lower the specific heat, the higher the temperature will rise in the material. If excessively high, temperature can have detrimental effects on product quality by adversely affecting surface finish and dimensional accuracy, can cause excessive tool and die wear, and can result in adverse metallurgical changes in the material.

### 3.9.4  Thermal conductivity

Whereas specific heat indicates the temperature rise in a material as a result of energy input, *thermal conductivity* indicates the rate with which heat flows within and through the material. Metallically bonded materials (metals) generally have high thermal conductivity, whereas ionically or covalently bonded materials (ceramics and plastics) have poor conductivity. Because of the large difference in their thermal conductivities, alloying elements can have a significant effect on the thermal conductivity of alloys, as can be seen in Table 3.2 by comparing the metals with their alloys.

When heat is generated by plastic deformation or friction, the heat should be conducted away at a high enough rate to prevent a severe rise in temperature. The main difficulty in machining titanium, for example, is caused by its very low thermal conductivity. Low thermal conductivity can also result in high thermal gradients and thus cause inhomogeneous deformation in metalworking processes.

### 3.9.5  Thermal expansion

*Thermal expansion* of materials can have several significant effects. Generally, the coefficient of thermal expansion is inversely proportional to the melting point of the material; alloying elements have a relatively minor effect on the thermal expansion of metals. Shrink fits use thermal expansion; other examples where relative expansion or contraction is important include electronic and computer components, glass-to-metal seals, metal-to-ceramic subassemblies (see also Section 11.8.2), struts on jet engines, and moving parts in machinery that require certain clearances for proper functioning.

Improper selection of materials and assembly can cause **thermal stresses**, *cracking, warping*, or loosening of components in the structure during their service life. Thermal conductivity, in conjunction with thermal expansion, plays the most significant role in causing thermal stresses, both in manufactured components and in tools and dies. This phenomenon is particularly important in a forging operation, for example, when hot workpieces are placed over relatively cool dies, making the die surfaces undergo thermal cycling. To reduce thermal stresses, a combination of high thermal conductivity and low thermal expansion is desirable.

Thermal stresses can lead to cracks in ceramic parts and in tools and dies made of relatively brittle materials. **Thermal fatigue** results from thermal cycling and causes a number of surface cracks; **thermal shock** is the term generally used to describe development of cracks after a single thermal cycle. Thermal stresses may be caused both by temperature gradients and by **anisotropy of thermal expansion**, which is generally observed in hexagonal close-packed metals and in ceramics.

The influence of temperature on dimensions and the modulus of elasticity can be a significant problem in precision instruments and equipment. A spring, for example, will have a lower stiffness as its temperature increases, because of reduced elastic modulus; similarly, a tuning fork or a pendulum will have different frequencies at different temperatures. Dimensional changes resulting from thermal expansion can be significant in measurements for quality control and in equipment such as measuring tapes for geodetic applications. To alleviate some of the problems of thermal expansion, a family of iron–nickel alloys that have very low thermal-expansion coefficients has been developed, called **low-expansion alloys**. Typical compositions are 64% Fe-36% Ni (Invar) and 54% Fe-28% Ni-18 Co (Kovar). Consequently, these alloys also have good thermal-fatigue resistance.

### 3.9.6 Electrical and magnetic properties

Electrical conductivity and dielectric properties of materials are of great importance, not only in electrical equipment and machinery, but also in manufacturing processes such as magnetic-pulse forming of sheet metals (Section 7.9.3) and electrical-discharge machining and electrochemical grinding of hard and brittle materials (Chapter 9).

The **electrical conductivity** of a material can be defined as a measure of the ease with which the material conducts electric current. Materials with high conductivity, such as metals, are generally referred to as **conductors**. The units of electrical conductivity are mho/m or mho/ft, where mho is the inverse of ohm (the unit for electrical resistance). **Electrical resistivity** is the inverse of conductivity. Materials with high resistivity are referred to as **dielectrics**, or **insulators**. The influence of the type of atomic bonding on the electrical conductivity of materials is the same as that for thermal conductivity. Alloying elements have a major effect on the electrical conductivity of metals: the higher the conductivity of the alloying element, the higher the conductivity of the alloy. **Dielectric strength** of materials is the resistivity to direct electric current and is defined as the voltage required per unit distance for electrical breakdown; the unit is V/m or V/ft.

**Superconductivity** is the phenomenon of almost zero electrical resistivity that occurs in some metals and alloys below a critical temperature. The highest temperature at which superconductivity has to date been exhibited, at about $-123°C$ ($-190°F$), is with an alloy of lanthanum, strontium, copper, and oxygen, although other material compositions are also being studied. Developments in superconductivity indicate that the efficiency of electrical components such as large high-power magnets, high-voltage power lines, and various other electronic and computer components can be markedly improved.

The electrical properties of materials such as single-crystal silicon, germanium, and gallium arsenide make them extremely sensitive to the presence and type of minute impurities as well as to temperature. Thus, by controlling the concentration and type of impurities (**dopants**), such as phosphorus and boron in silicon, electrical conductivity can be controlled as well. This property is exploited in **semiconductor** (solid state) devices used extensively in miniaturized electronic circuitry as described in Chapter 13. Such devices are very compact, efficient, and relatively inexpensive; consume little power; and require no warm-up time for operation.

**Ferromagnetism** is the large and permanent magnetization resulting from an exchange interaction between neighboring atoms (such as iron, nickel, and cobalt) that aligns their magnetic moments in parallel. It has important applications in electric motors, generators, transformers, and microwave devices. **Ferrimagnetism** is the permanent and large magnetization exhibited by some ceramic materials, such as cubic ferrites.

The **piezoelectric** effect (*piezo* from Greek, meaning *to press*), in which there is a reversible interaction between an elastic strain and an electric field, is exhibited by some materials, such as certain ceramics and quartz crystals. This property is utilized in making *transducers*, which are devices that convert the strain from an external force to electrical energy. Typical applications of the piezoelectric effect include force or pressure transducers, strain gages, sonar detectors, and microphones.

**Magnetostriction** is the phenomenon of expansion or contraction of a material when subjected to a magnetic field. Some materials, such as pure nickel and some iron–nickel alloys, exhibit this behavior. Magnetostriction is the principle behind ultrasonic machining equipment. (See Section 9.7.)

### 3.9.7  Resistance to corrosion

**Corrosion** is the deterioration of metals and ceramics, while **degradation** is a similar phenomenon in plastics. Corrosion resistance is an important aspect of selection, especially for applications in the chemical, food, and petroleum industries. In addition to various possible chemical reactions from the elements and compounds present, environmental oxidation and corrosion of components and structures are a major concern, particularly at elevated temperatures and in automobiles and other transportation equipment.

Resistance to corrosion depends on the particular environment, as well as the composition of the material. Chemicals (acids, alkali, and salts), the environment (oxygen, pollution, and acid rain), and water (fresh or salt) may all act as corrosive media. Nonferrous metals, stainless steels, and nonmetallic materials generally have high corrosion resistance. Steels and cast irons generally have poor resistance and must be protected by various coatings and surface treatments.

Corrosion can occur over an entire surface, or it can be localized (such as in **pitting**). It can occur along grain boundaries of metals as **intergranular corrosion** and at the interface of bolted or riveted joints as **crevice corrosion**. Two dissimilar metals may form a galvanic cell (two electrodes in an electrolyte in a corrosive environment, including moisture) and cause **galvanic corrosion**. Two-phase alloys (Section 5.2.3) are more susceptible to galvanic corrosion, because of the two different metals involved, than are single-phase alloys or pure metals; thus, heat treatment can have an influence on corrosion resistance.

Corrosion may also act in indirect ways. **Stress-corrosion cracking** is an example of the effect of a corrosive environment on the integrity of a product that, as manufactured, had residual stresses in it. Likewise, metals that are cold worked are likely to contain residual stresses and hence are more susceptible to corrosion as compared with hot-worked or annealed metals.

Tool and die materials also can be susceptible to chemical attack by lubricants and coolants (Section 4.4.4). The chemical reaction alters their surface finish and adversely influences the metalworking operation. An example is carbide tools and dies that have cobalt as a binder (Section 8.6.4), where the cobalt is attacked by elements in the cutting fluid (**selective leaching**). Thus, compatibility of the tool, die, and workpiece materials and the metalworking fluid under actual operating conditions is an important consideration in the selection of materials. Note that chemical reactions should not be regarded as having adverse effects only. Nontraditional machining processes such as chemical machining and electrochemical machining are indeed based on controlled reactions (Chapter 9). These processes remove material by chemical reactions, similar to the etchings of metallurgical specimens.

The usefulness of some level of *oxidation* is exhibited by the corrosion resistance of aluminum, titanium, and stainless steel. Aluminum develops a thin (a few atomic layers), strong, and adherent hard oxide film ($Al_2O_3$) that protects the surface from further environmental corrosion. Titanium develops a film of titanium oxide ($TiO_2$). A similar phenomenon occurs in stainless steels, which, because of the chromium present in the alloy, develop a similar protective film on their surfaces. These processes are known as **passivation** (Section 4.2). When the protective film is scratched, thus exposing the metal underneath, a new oxide film forms in due time.

# 3.10 | Properties and Applications of Ferrous Alloys

By virtue of their very wide range of mechanical, physical, and chemical properties, **ferrous alloys** are among the most useful of all metals. Ferrous metals and alloys contain iron as their base metal and are variously categorized as carbon and alloy steels, stainless steels, tool and die steels, cast irons, and cast steels. Ferrous alloys are produced as sheet steel for automobiles, appliances, and containers; as plates for ships, boilers, and bridges; as structural members (such as I-beams), bar products for leaf springs, gears, axles, crankshafts, and railroad rails; as stock for tools and dies; as music wire; and as fasteners such as bolts, rivets, and nuts.

A typical U.S. passenger car now contains about 816 kg (1800 lb) of steel, accounting for about 55% of its weight. As an example of their widespread use, ferrous materials comprise 70% to 85% by weight of virtually all structural members and mechanical components. Carbon steels are the least expensive of all metals, but stainless steels can be costly.

## 3.10.1 Carbon and alloy steels

Carbon and alloy steels are among the most commonly used metals and have a wide variety of applications. The composition and processing of these steels are controlled in a manner that makes them suitable for numerous applications. They are available in various basic product shapes: plate, sheet, strip, bar, wire, tube, castings, and forgings.

Several elements are added to steels to impart various properties such as hardenability, strength, hardness, toughness, wear resistance, workability, weldability, and machinability. Generally, the higher the percentages of these elements in steels, the higher the particular properties that they impart. Thus, for example, the higher the carbon content, the higher the hardenability of the steel and the higher its strength, hardness, and wear resistance. Conversely, ductility, weldability, and toughness are reduced with increasing carbon content.

**Carbon steels.** *Carbon steels* are generally classified as low, medium, and high:

1. **Low-carbon steel**, also called **mild steel**, has less than 0.30% carbon. It is generally used for common industrial products, such as bolts, nuts, sheet plates, tubes, and machine components that do not require high strength.

2. **Medium-carbon steel** has 0.30% to 0.60% carbon. It is generally used in applications requiring higher strength than those using low-carbon steels, such as machinery, automotive and agricultural equipment parts (e.g., gears, axles, connecting rods, crankshafts), railroad equipment, and parts for metalworking machinery.

3. **High-carbon steel** has more than 0.60% carbon. It is generally used for parts requiring strength, hardness, and wear resistance, such as cutting tools, cable, music wire, springs, cutlery, and rails. After being manufactured into shapes, the parts are usually heat treated and tempered. The higher the carbon content of the steel, the higher its hardness, strength, and wear resistance after heat treatment.

4. Carbon steels containing sulfur or phosphorus, known as **resulfurized** and **rephosphorized and resulfurized carbon steels,** are available. These steels have improved machinability, as described in Section 8.5.1.

**Alloy steels.**   Steels containing significant amounts of alloying elements are called *alloy steels*, and they are usually made with more care than are carbon steels. **Structural-grade** alloy steels, as identified by ASTM specifications, are used mainly in the construction and transportation industries, because of their high strength. Other alloy steels are used in applications where strength, hardness, resistance to creep, fatigue, and toughness are required. These steels may also be heat treated to obtain the desired properties.

**High-strength low-alloy steels.**   In order to improve the strength-to-weight ratio of steels, a number of *high-strength low-alloy* (HSLA) steels have been developed. These steels have a low carbon content (usually less than 0.30%), and are characterized by a microstructure consisting of fine-grain ferrite and a hard second phase of carbides, carbonitrides, or nitrides. First developed in the 1930s, HSLA steels are usually produced in sheet form by microalloying and controlled hot rolling. Plates, bars, and structural shapes are made from these steels. However, the ductility, formability, and weldability of HSLA steels are generally inferior to those of conventional low-alloy steels. To improve these properties, dual-phase steels have been developed.

Sheet products of HSLA steels typically are used in certain parts of automobile bodies to reduce weight (and hence fuel consumption) and in transportation equipment, mining equipments agricultural equipment, and various other industrial applications. Plates made of HSLA steel are used in ships, bridges, and building construction, and shapes such as I-beams, channels, and angles are used in buildings and various other structures.

**Dual-phase steels.**   Designated by the letter D, these steels are processed specially and have a mixed ferrite and martensite structure. Developed in the late 1960s, dual-phase steels have high work-hardening characteristics (i.e., a high $n$ value; see Section 2.2.3), which improve their ductility and formability.

## 3.10.2  Stainless steels

*Stainless steels* are characterized primarily by their corrosion resistance, high strength and ductility, and high chromium content. They are called *stainless* because, in the presence of oxygen (air), they develop a thin, hard adherent film of *chromium oxide* that protects the metal from corrosion (passivation). This protective film builds up again if the surface is scratched. For passivation to occur, the minimum chromium content of the steel should be 10% to 12% by weight.

In addition to chromium, other alloying elements in stainless steels typically are nickel, molybdenum, copper, titanium, silicon, manganese, columbium, aluminum, nitrogen, and sulfur. The higher the carbon content, the lower the corrosion resistance of stainless steels. The reason is that the carbon combines with chromium in the steel and forms chromium carbide, which lowers the passivity of the steel; the chromium carbides introduce a second phase in the metal, which promotes galvanic corrosion. The letter *L* is used to identify low-carbon stainless steels.

Developed in the early 1900s, stainless steels are made by techniques similar to those used in other types of steelmaking, using electric furnaces or the basic-oxygen process. The level of purity is controlled by various refining techniques. Stainless steels are available in a wide variety of shapes. Typical applications are for cutlery, kitchen equipment, health care and surgical equipment, and automotive trim and in the chemical, food-processing, and petroleum industries.

Stainless steels are generally divided into five types (Table 3.3): austenitic, ferritic, martensitic, precipitation-hardening, and duplex-structure steels. These types are described as follows:

1. The **austenitic steels** (200 and 300 series) are generally composed of chromium, nickel, and manganese in iron. They are nonmagnetic and have excellent corrosion resistance, but are susceptible to stress-corrosion cracking. Austenitic stainless steels are hardened by cold working; they are the most ductile of all stainless steels and hence can be formed easily. However, with increasing cold work, their formability is reduced. These steels are used in a wide variety of applications, such as kitchenware, fittings, welded construction, lightweight transportation equipment, furnace and heat-exchanger parts, and components for severe chemical environments.

2. The **ferritic steels** (400 series) have a high chromium content: up to 27%. They are magnetic and have good corrosion resistance, but have lower ductility than austenitic stainless steels. Ferritic stainless steels are hardened by cold working and are not heat treatable. They are generally used for nonstructural applications, such as kitchen equipment and automotive trim.

**TABLE 3.3**

**Room-Temperature Mechanical Properties and Typical Applications of Annealed Stainless Steels**

| AISI (UNS) | Ultimate Tensile Strength (MPa) | Yield Strength (MPa) | Elongation (%) | Characteristics and Typical Applications |
|---|---|---|---|---|
| 303 (S30300) | 550–620 | 240–260 | 50–53 | Screw-machine products, shafts, valves, bolts, bushings, and nuts; aircraft fittings; rivets; screws; studs. |
| 304 (S30400) | 565–620 | 240–290 | 55–60 | Chemical and food-processing equipment, brewing equipment, cryogenic vessels, gutters, downspouts, and flashings. |
| 316 (S31600) | 550–590 | 210–290 | 55–60 | High corrosion resistance and high creep strength. Chemical and pulp-handling equipment, photographic equipment, brandy vats, fertilizer parts, ketchup-cooking kettles, and yeast tubs. |
| 410 (S41000) | 480–520 | 240–310 | 25–35 | Machine parts, pump shafts, bolts, bushings, coal chutes, cutlery, fishing tackle, hardware, jet engine parts, mining machinery, rifle barrels, screws, and valves. |
| 416 (S41600) | 480–520 | 275 | 20–30 | Aircraft fittings, bolts, nuts, fire extinguisher inserts, rivets, and screws. |

3. Most of the **martensitic steels** (400 and 500 series) do not contain nickel and are hardenable by heat treatment; their chromium content may be as high as 18%. These steels are magnetic and have high strength, hardness, and fatigue resistance, and good ductility, but moderate corrosion resistance. Martensitic stainless steels are used for cutlery, surgical tools, instruments, valves, and springs.

4. The **precipitation-hardening (PH) steels** contain chromium and nickel, along with copper, aluminum, titanium, or molybdenum. They have good corrosion resistance, good ductility, and high strength at elevated temperatures. Their main application is in aircraft and aerospace structural components.

5. The **duplex-structure steels** have a mixture of austenite and ferrite. They have good strength and higher resistance to corrosion in most environments and to stress-corrosion cracking than does the 300 series of austenitic steels. Typical applications of duplex-structure steels are in water-treatment plants and heat-exchanger components.

### 3.10.3  Tool and die steels

*Tool and die steels* are specially alloyed steels (Table 3.4) designed for high strength, impact toughness, and wear resistance at room and elevated temperatures. They are commonly used in forming and machining of metals. Table 3.5 lists various tool and die materials for a variety of manufacturing applications. The main categories of these materials are described as follows:

1. **High-speed steels** (HSS), first developed in the early 1900s, are the most highly alloyed tool and die steels and maintain their hardness and strength at elevated operating temperatures. There are two basic types of high-speed steels: the *molybdenum type* (M series) and the *tungsten type* (T series). The M-series steels contain up to about 10% molybdenum, with chromium, vanadium, tungsten, and cobalt as other alloying elements. The T-series steels contain 12% to 18% tungsten, with chromium, vanadium, and cobalt as other alloying elements. As compared with the T-series steels, the M-series steels generally have higher abrasion resistance, have less distortion in heat treatment, and are less expensive. The M-series steels constitute about 95% of all high-speed steels produced in the United States. High-speed steel tools can be coated with titanium nitride and titanium carbide for better resistance to wear.

**TABLE 3.4**

| Basic Types of Tool and Die Steels | |
|---|---|
| Type | AISI |
| High speed | M (molybdenum base) |
| | T (tungsten base) |
| Hot work | H1 to H19 (chromium base) |
| | H20 to H39 (tungsten base) |
| | H40 to H59 (molybdenum base) |
| Cold work | D (high carbon, high chromium) |
| | A (medium alloy, air hardening) |
| | O (oil hardening) |
| Shock resisting | S |
| Mold steels | P1 to P19 (low carbon) |
| | P20 to P39 (others) |
| Special purpose | L (low alloy) |
| | F (carbon–tungsten) |
| Water hardening | W |

**TABLE 3.5**

## Typical Tool and Die Materials for Various Processes

| Process | Material |
|---|---|
| Die casting | H13, P20 |
| Powder metallurgy | |
|     Punches | A2, S7, D2, D3, M2 |
|     Dies | WC, D2, M2 |
| Molds for plastic and rubber | S1, O1, A2, D2, 6F5, 6F6, P6, P20, P21, H13 |
| Hot forging | 6F2, 6G, H11, H12 |
| Hot extrusion | H11, H12, H13 |
| Cold heading | W1, W2, M1, M2, D2, WC |
| Cold extrusion | |
|     Punches | A2, D2, M2, M4 |
|     Dies | O1, W1, A2, D2 |
| Coining | 52100, W1, O1, A2, D2, D3, D4, H11, H12, H13 |
| Drawing | |
|     Wire | WC, diamond |
|     Shapes | WC, D2, M2 |
|     Bar and tubing | WC, W1, D2 |
| Rolls | |
|     Rolling | Cast iron, cast steel, forged steel, WC |
|     Thread rolling | A2, D2, M2 |
|     Shear spinning | A2, D2, D3 |
| Sheet metals | |
|     Shearing | |
|         Cold | D2, A2, A9, S2, S5, S7 |
|         Hot | H11, H12, H13 |
|     Pressworking | Zinc alloys, 4140 steel, cast iron, epoxy composites, A2, D2, O1 |
|     Deep drawing | W1, O1, cast iron, A2, D2 |
| Machining | Carbides, high-speed steels, ceramics, diamond, cubic boron nitride |

2. **Hot-work steels** (H series) are designed for use at elevated temperatures and have high toughness and high resistance to wear and cracking. The alloying elements are generally tungsten, molybdenum, chromium, and vanadium.

3. **Cold-work steels** (A, D, and O series) are used for cold-working operations. They generally have high resistance to wear and cracking. These steels are available as oil-hardening or air-hardening types.

4. **Shock-resisting steels** (S series) are designed for impact toughness and are used in applications such as header dies, punches, and chisels. Other properties of these steels depend on the particular composition.

## 3.11 | Properties and Applications of Nonferrous Metals and Alloys

*Nonferrous metals and alloys* cover a wide range of materials, from the more common metals, such as aluminum, copper, and magnesium, to high-strength, high-temperature alloys, such as tungsten, tantalum, and molybdenum. Although more expensive than ferrous metals, nonferrous metals and alloys have important applications, because of their numerous mechanical, physical, and chemical properties.

**FIGURE  3.33**   Cross-section of a jet engine (PW2037) showing various components and the alloys used in making them. *Source*:  Courtesy of United Aircraft Pratt & Whitney.

A turbofan jet engine for the Boeing 757 aircraft typically contains the following nonferrous metals and alloys: 38% titanium, 37% nickel, 12% chromium, 6% cobalt, 5% aluminum, 1% niobium (columbium), and 0.02% tantalum. Without these materials, a jet engine (Fig. 3.33) could not be designed, manufactured, and operated at the required energy and efficiency levels.

Typical examples of the applications of nonferrous metals and alloys include aluminum for cooking utensils and aircraft bodies, copper wire for electricity and copper tubing for water in residences, titanium for jet-engine turbine blades and prosthetic devices (such as artificial joints), and tantalum for rocket engines. In this section, we discuss the general properties, production methods, and important engineering applications of nonferrous metals and alloys. We describe the manufacturing properties of these materials, such as formability, machinability, and weldability, in various chapters throughout this text.

## 3.11.1  Aluminum and aluminum alloys

The important factors in selecting *aluminum* (Al) and its alloys are their high strength-to-weight ratio, resistance to corrosion by many chemicals, high thermal and electrical conductivity, nontoxicity, reflectivity, appearance, and ease of formability and machinability; they are also nonmagnetic.

Principal uses of aluminum and its alloys, in decreasing order of consumption, include containers and packaging (aluminum cans and foil), buildings and other types of construction, transportation (aircraft and aerospace applications, buses, automobiles (see Fig. 1.5), railroad cars, and marine craft), electrical products (nonmagnetic

**TABLE 3.6**

### Properties of Various Aluminum Alloys at Room Temperature

| Alloy (UNS) | Temper | Ultimate Tensile Strength (MPa) | Yield Strength (MPa) | Elongation in 50 mm (%) |
|---|---|---|---|---|
| 1100 (A91100) | O | 90 | 35 | 35–45 |
| 1100 | H14 | 125 | 120 | 9–20 |
| 2024 (A92024) | O | 190 | 75 | 20–22 |
| 2024 | T4 | 470 | 325 | 19–20 |
| 3003 (A93003) | O | 110 | 40 | 30–40 |
| 3003 | H14 | 150 | 145 | 8–16 |
| 5052 (A95052) | O | 190 | 90 | 25–30 |
| 5052 | H34 | 260 | 215 | 10–14 |
| 6061 (A96061) | O | 125 | 55 | 25–30 |
| 6061 | T6 | 310 | 275 | 12–17 |
| 7075 (A97075) | O | 230 | 105 | 16–17 |
| 7075 | T6 | 570 | 500 | 11 |

and economical electrical conductors), consumer durables (appliances, cooking uten-sils, and furniture), and portable tools (Tables 3.6 and 3.7). Nearly all high-voltage transmission wiring is made of aluminum. With respect to their structural (load-bearing) components, 82% of a Boeing 747 aircraft and 79% of a Boeing 757 air-craft is made of aluminum.

As in all cases of material selection, because each aluminum alloy has its par-ticular properties, its selection can be critical with respect to manufacturing as well as economic considerations. For example, a typical aluminum beverage can consists of the following alloys, where all sheet is in the H19 condition, which is the highest cold-worked state: 3004 or 3104 for the can body, 5182 for the lid, and 5042 for the tab.

Aluminum alloys are available as mill products, that is, wrought product made into various shapes by rolling, extrusion, drawing, and forging. Aluminum ingots are available for casting, as are powder metals for powder metallurgy applications. There are two types of wrought alloys of aluminum: (1) alloys that can be hardened by cold

**TABLE 3.7**

### Manufacturing Properties and Typical Applications of Wrought Aluminum Alloys

| Alloy | Characteristics* | | | Typical Applications |
|---|---|---|---|---|
| | Corrosion Resistance | Machinability | Weldability | |
| 1100 | A | D–C | A | Sheet-metal work, spun hollow parts, tin-stock. |
| 2014 | C | C–B | C–B | Heavy-duty forgings, plate and extrusions for aircraft structural components, wheels. |
| 3003 | A | D–C | A | Cooking utensils, chemical equipment, pressure vessels, sheet-metal work, builders' hardware, storage tanks. |
| 5054 | A | D–C | A | Welded structures, pressure vessels, tube for marine uses. |
| 6061 | B | D–C | A | Trucks, canoes, furniture, structural applications. |
| 7005 | D | B–D | B | Extruded structural members, large heat exchangers, tennis racquets, softball bats. |

* From A (excellent) to D (poor).

working (designated by the letter H) and are not heat treatable, and (2) alloys that are hardenable by heat treatment (designated by the letter T). The letter O indicates the annealed condition. Techniques have been developed whereby most aluminum alloys can be machined, formed, and welded with relative ease.

## 3.11.2  Magnesium and magnesium alloys

*Magnesium* (Mg) is the lightest engineering metal available and also has good vibration-damping characteristics. Its alloys are used in structural and nonstructural applications where weight is of primary importance. Magnesium is also an alloying element in various nonferrous metals.

Typical uses of magnesium alloys include aircraft and missile components, material-handling equipment, portable power tools (such as drills and sanders), ladders, luggage, bicycles, sporting goods, and general lightweight components. These alloys are available as either castings or wrought products, such as extruded bars and shapes, forgings, and rolled plate and sheet. Magnesium alloys are also used in printing and textile machinery to minimize inertial forces in high-speed components.

Because it is not sufficiently strong in its pure form, magnesium is alloyed with various elements (Table 3.8) to impart certain specific properties, particularly a high strength-to-weight ratio. A variety of magnesium alloys has good casting, forming, and machining characteristics. Because magnesium alloys oxidize rapidly (that is, they are *pyrophoric*) they are a potential fire hazard, and precautions must be taken when machining, grinding, or sand casting magnesium alloys. However, products made of magnesium and its alloys are not a fire hazard.

## 3.11.3  Copper and copper alloys

First produced in about 4000 BC, *copper* (Cu) and its alloys have properties somewhat similar to those of aluminum alloys. In addition, they are among the best conductors of electricity and heat and have good resistance to corrosion. Because of these properties, copper and its alloys are among the most important metals. They can be processed easily by various forming, machining, casting, and joining techniques.

Copper alloys often are attractive for applications where combined properties, such as electrical and mechanical properties, corrosion resistance, thermal conductivity, and wear resistance, are required. Such applications include electrical and electronic components, springs, cartridges for small arms, plumbing, heat exchangers, and marine hardware, as well as some consumer goods, such as cooking utensils, jewelry, and other decorative objects.

### TABLE 3.8

**Properties and Typical Forms of Various Wrought Magnesium Alloys**

| Alloy | Composition (%) | | | | Condition | Ultimate Tensile Strength (MPa) | Yield Strength (MPa) | Elongation in 50 mm (%) | Typical Forms |
|---|---|---|---|---|---|---|---|---|---|
| | Al | Zn | Mn | Zr | | | | | |
| AZ31B | 3.0 | 1.0 | 0.2 | | F | 260 | 200 | 15 | Extrusions |
| | | | | | H24 | 290 | 220 | 15 | Sheet and plates |
| AZ80A | 8.5 | 0.5 | 0.2 | | T5 | 380 | 380 | 7 | Extrusions and forgings |
| HK31A* | | | | 0.7 | H24 | 255 | 255 | 8 | Sheet and plates |
| ZK60A | | 5.7 | | 0.55 | T5 | 365 | 365 | 11 | Extrusions and forgings |

\*  HK31A also contains 3%Th.

**TABLE 3.9**

## Properties and Typical Applications of Various Wrought Copper and Brasses

| Type and UNS Number | Nominal Composition (%) | Ultimate Tensile Strength (MPa) | Yield Strength (MPa) | Elongation in 50 mm (%) | Typical Applications |
|---|---|---|---|---|---|
| Oxygen-free electronic (C10100) | 99.99 Cu | 220–450 | 70–365 | 55–4 | Bus bars, waveguides, hollow conductors, lead in wires, coaxial cables and tubes, microwave tubes, rectifiers. |
| Red brass, 85% (C23000) | 85.0 Cu 15.0 Zn | 270–72 | 70–435 | 55–3 | Weather stripping, conduit, sockets, fasteners, fire extinguishers, condenser and heat-exchanger tubing. |
| Low Brass, 80% (C24000) | 80.0 Cu 20.0 Zn | 300–850 | 80–450 | 55–3 | Battery caps, bellows, musical instruments, clock dials, flexible hose. |
| Free-cutting brass (C36000) | 61.5 Cu, 3.0 Pb, 35.5 Zn | 340–470 | 125–310 | 53–18 | Gears, pinions, automatic high-speed screw-machine parts. |
| Naval brass (C46400 to C46700) | 60.0 Cu, 39.25 Zn, 0.75 Sn | 380–610 | 170–455 | 50–17 | Aircraft turnbuckle barrels, balls, bolts, marine hardware, valve stems, condensor plates. |

Copper alloys can acquire a wide variety of properties by the addition of alloying elements and by heat treatment to improve their manufacturing characteristics. The most common copper alloys are brasses and bronzes. **Brass**, which is an alloy of copper and zinc, was one of the earliest alloys developed and has numerous applications, including decorative objects (Table 3.9). **Bronze** is an alloy of copper and tin (Table 3.10). There are also other bronzes, such as aluminum bronze, an alloy of copper and aluminum, and tin bronzes. Beryllium–copper, or **beryllium bronze**, and **phosphor bronze** have good strength and hardness for applications such as springs and bearings. Other major copper alloys are copper nickels and nickel silvers.

### 3.11.4 Nickel and nickel alloys

*Nickel* (Ni), a silver-white metal discovered in 1751, is a major alloying element that imparts strength, toughness, and corrosion resistance to metals. It is used extensively in stainless steels and nickel-base alloys. These alloys are used for high-temperature applications, such as jet-engine components, rockets, and nuclear power plants, as well as in food-handling and chemical-processing equipment, coins, and marine applications. Because nickel is magnetic, nickel alloys are also used in electromagnetic applications, such as solenoids. The principal use of nickel as a metal is in electroplating for appearance and for improving resistance to corrosion and wear.

Nickel alloys have high strength and corrosion resistance at elevated temperatures. Alloying elements in nickel are chromium, cobalt, and molybdenum. The behavior of nickel alloys in machining, forming, casting, and welding can be modified by various other alloying elements. A variety of nickel alloys that have a range of strengths at different temperatures have been developed (Table 3.11). *Monel* is a nickel–copper alloy, and *Inconel* is a nickel–chromium alloy. A nickel–molybdenum–chromium alloy (*Hastelloy*) has good corrosion resistance and high strength at elevated temperatures. *Nichrome*, an alloy of nickel, chromium, and iron, has high oxidation and electrical resistance and is used for electrical-heating elements. *Invar*, an alloy of iron and nickel, has relatively low sensitivity to temperature. (See Section 3.9.5.)

**TABLE 3.10**

**Properties and Typical Applications of Various Wrought Bronzes**

| Type and UNS Number | Nominal Composition (%) | Ultimate Tensile Strength (MPa) | Yield Strength (MPa) | Elongation in 50 mm (%) | Typical Applications |
|---|---|---|---|---|---|
| Architectural bronze (C38500) | 57.0 Cu, 3.0 Pb, 40.0 Zn | 415 | 140 (as extruded) | 30 | Architectural extrusions, storefronts, thresholds, trim, butts, hinges. |
| Phosphor bronze, 5% A (C51000) | 95.0 Cu, 5.0 Sn, trace P | 325–960 | 130–550 | 64–2 | Bellows, clutch disks, cotter pins, diaphragms, fasteners, wire brushes, chemical hardware, textile machinery. |
| Free-cutting phosphor bronze (C54400) | 88.0 Cu, 4.0 Pb, 4.0 Zn, 4.0 Sn | 300–520 | 130–435 | 50–15 | Bearings, bushings, gears, pinions, shafts, thrust washers, valve parts. |
| Low-silicon bronze, B (C65100) | 98.5 Cu, 1.5 Si | 275–655 | 100–475 | 55–11 | Hydraulic pressure lines, bolts, marine hardware, electrical conduits, heat-exchanger tubing. |
| Nickel–silver, 65–18 (C74500) | 65.0 Cu, 17.0 Zn, 18.0 Ni | 390–710 | 170–620 | 45–3 | Rivets, screws, zippers, camera parts, base for silver plate, nameplates, etching stock. |

**TABLE 3.11**

**Properties and Typical Applications of Various Nickel Alloys (All Alloy Names are Trade Names)**

| Alloy (Condition) | Principal Alloying Elements (%) | Ultimate Tensile Strength (MPa) | Yield Strength (MPa) | Elongation in 50 mm (%) | Typical Applications |
|---|---|---|---|---|---|
| Nickel 200 (annealed) | None | 380–550 | 100–275 | 60–40 | Chemical- and food-processing industry, aerospace equipment, electronic parts. |
| Duranickel 301 (age hardened) | 4.4 Al, 0.6 Ti | 1300 | 900 | 28 | Springs, plastics-extrusion equipment, molds for glass, diaphragms. |
| Monel R-405 (hot rolled) | 30 Cu | 525 | 230 | 35 | Screw-machine products, water-meter parts. |
| Monel K-500 (age hardened) | 29 Cu, 3 Al | 1050 | 750 | 20 | Pump shafts, valve stems, springs. |
| Inconel 600 (annealed) | 15 Cr, 8 Fe | 640 | 210 | 48 | Gas-turbine parts, heat-treating equipment, electronic parts, nuclear reactors. |
| Hastelloy C-4 (solution treated and quenched) | 16 Cr, 15 Mo | 785 | 400 | 54 | High-temperature stability, resistance to stress-corrosion cracking. |

## 3.11.5 Superalloys

*Superalloys* are important in high-temperature applications; hence, they are also known as **heat-resistant**, or **high-temperature, alloys**. Major applications of superalloys are in jet engines and gas turbines, with other applications in reciprocating engines; rocket engines; tools and dies for hot working of metals; and in the nuclear, chemical, and petrochemical industries.

**TABLE 3.12**

**Properties and Typical Applications of Various Nickel-Base Superalloys at 870°C (1600°F) (All Alloy Names Are Trade Names)**

| Alloy | Condition | Ultimate Tensile Strength (MPa) | Yield Strength (MPa) | Elongation in 50 mm (%) | Typical Applications |
|---|---|---|---|---|---|
| Astroloy | Wrought | 770 | 690 | 25 | Forgings for high-temperature applications. |
| Hastelloy X | Wrought | 255 | 180 | 50 | Jet-engine sheet parts. |
| IN-100 | Cast | 885 | 695 | 6 | Jet-engine blades and wheels. |
| IN-102 | Wrought | 215 | 200 | 110 | Superheater and jet-engine parts. |
| Inconel 625 | Wrought | 285 | 275 | 125 | Aircraft engines and structures, chemical-processing equipment. |
| Inconel 718 | Wrought | 340 | 330 | 88 | Jet-engine and rocket parts. |
| MAR-M 200 | Cast | 840 | 760 | 4 | Jet-engine blades. |
| MAR-M 432 | Cast | 730 | 605 | 8 | Integrally cast turbine wheels. |
| René 41 | Wrought | 620 | 550 | 19 | Jet-engine parts. |
| Udimet 700 | Wrought | 690 | 635 | 27 | Jet-engine parts. |
| Waspaloy | Wrought | 525 | 515 | 35 | Jet-engine parts. |

These alloys are also referred to as **iron-base, cobalt-base,** or **nickel-base superalloys.** They contain nickel, chromium, cobalt, and molybdenum as major alloying elements. Other alloying elements are aluminum, tungsten, and titanium. Superalloys are generally identified by trade names or by special numbering systems and are available in a variety of shapes. Most superalloys have a maximum service temperature of about 1000°C (1800°F) for structural applications. The temperatures can be as high as 1200°C (2200°F) for non-load-bearing components. Superalloys generally have good resistance to corrosion, mechanical and thermal fatigue, mechanical and thermal shock, creep, and erosion at elevated temperatures.

Iron-base superalloys generally contain 32% to 67% iron, 15% to 22% chromium, and 9% to 38% nickel. Common alloys in this group are the *Incoloy* series. Cobalt-base superalloys generally contain 35% to 65% cobalt, 19% to 30% chromium, and up to 35% nickel. Cobalt (Co) is a white-colored metal that resembles nickel. These superalloys are not as strong as nickel-base superalloys, but they retain their strength at higher temperatures than do nickel-base superalloys. Nickel-base superalloys are the most common of the superalloys and are available in a wide variety of compositions (Table 3.12). The range of nickel is from 38% to 76%; they also contain up to 27% chromium and 20% cobalt. Common alloys in this group include the *Hastelloy, Inconel, Nimonic, René, Udimet, Astroloy,* and *Waspaloy* series.

### 3.11.6 Titanium and titanium alloys

*Titanium* (Ti) was discovered in 1791, but was not commercially produced until the 1950s. Although titanium is expensive, its high strength-to-weight ratio and its corrosion resistance at room and elevated temperatures make it attractive for applications such as aircraft, jet-engine, racing-car, chemical, petrochemical, and marine components; submarine hulls; and biomaterials such as orthopedic implants (Table 3.13). Titanium alloys have been developed for service at 550°C (1000°F) for long periods of time and at up to 750°C (1400°F) for shorter periods. Unalloyed titanium, known as commercially pure titanium, has excellent corrosion resistance for applications where strength considerations are secondary. Aluminum, vanadium, molybdenum, manganese, and other alloying elements are added to titanium alloys to impart properties such as improved workability, strength, and hardenability.

TABLE 3.13

**Properties and Typical Applications of Wrought Titanium Alloys**

| Nominal Composition (%) | UNS | Condition | Room Temperature | | | | Various Temperatures | | | | | Typical Applications |
|---|---|---|---|---|---|---|---|---|---|---|---|---|
| | | | Ultimate Tensile Strength (MPa) | Yield Strength (MPa) | Elongation (%) | Reduction of Area (%) | Temp (°C) | Ultimate Tensile Strength (MPa) | Yield Strength (MPa) | Elongation (%) | Reduction of Area (%) | |
| 99.5 Ti | R50250 | Annealed | 330 | 240 | 30 | 55 | 300 | 150 | 95 | 32 | 80 | Airframes; chemical, desalination, and marine parts; plate-type heat exchangers. |
| 5 Al, 2.5 Sn | R54520 | Annealed | 860 | 810 | 16 | 40 | 300 | 565 | 450 | 18 | 45 | Aircraft-engine compressor blades and ducting; steam-turbine blades. |
| 6 Al, 4 V | R56400 | Annealed | 1000 | 925 | 14 | 30 | 300 | 725 | 650 | 14 | 35 | Rocket motor cases; blades and disks for aircraft turbines and compressors; orthopedic implants; structural forgings; fasteners. |
| | | | | | | | 425 | 670 | 570 | 18 | 40 | |
| | | | | | | | 550 | 530 | 430 | 35 | 50 | |
| | | Solution + age | 1175 | 1100 | 10 | 20 | 300 | 980 | 900 | 10 | 28 | |
| 13 V, 11 Cr, 3 Al | R58010 | Solution + age | 1275 | 1210 | 8 | – | 425 | 1100 | 830 | 12 | – | High-strength fasteners; aerospace components; honeycomb panels. |

The properties and manufacturing characteristics of titanium alloys are extremely sensitive to small variations in both alloying and residual elements. Thus, control of composition and processing is important, including prevention of surface contamination by hydrogen, oxygen, or nitrogen during processing. These elements cause embrittlement of titanium, resulting in reduced toughness and ductility.

The body-centered cubic structure of titanium (beta-titanium, above 880°C[1600°F]) is ductile, whereas its hexagonal close-packed structure (alpha-titanium) is somewhat brittle and is very sensitive to stress corrosion. A variety of other structures (alpha, near alpha, alpha–beta, and beta) can be obtained by alloying and heat treating, such that the properties can be optimized for specific applications. **Titanium aluminide intermetallics** (TiAl and $Ti_3Al$) have higher stiffness and lower density and can withstand temperatures higher than conventional titanium alloys.

### 3.11.7 Refractory metals

Molybdenum, columbium, tungsten, and tantalum are *refractory metals*; they are called *refractory* because of their high melting point. Although refractory-metal elements were discovered about 200 years ago—and have been used as important alloying elements in steels and superalloys—their use as engineering metals and alloys did not begin until about the 1940s. More than most other metals and alloys, these metals maintain their strength at elevated temperatures; thus, they are of great importance and use in rocket engines, gas turbines, and various other aerospace applications; in the electronics, nuclear power, and chemical industries; and as tool and die materials. The temperature range for some of these applications is on the order of 1100°C to 2200°C (2000°F to 4000°F), where strength and oxidation are of major concern.

1. **Molybdenum.** *Molybdenum* (Mo), a silvery white metal, was discovered in the 18th century. It has a high melting point, a high modulus of elasticity, good resistance to thermal shock, and good electrical and thermal conductivity. Typical applications of molybdenum are in solid-propellent rockets, jet engines, honeycomb structures, electronic components, heating elements, and dies for die casting. Principal alloying elements for molybdenum are titanium and zirconium. Molybdenum is used in greater amounts than any other refractory metal. It is an important alloying element in cast and wrought alloys, such as steels and heat-resistant alloys, and imparts strength, toughness, and corrosion resistance. A major disadvantage of molybdenum alloys is their low resistance to oxidation at temperatures above 500°C (950°F), thus necessitating the use of protective coatings.

2. **Niobium.** *Niobium* (Nb), or *columbium* (after the mineral *columbite*), possesses good ductility and formability and has greater resistance to oxidation than do other refractory metals. With various alloying elements, niobium alloys can be produced with moderate strength and good fabrication characteristics. These alloys are used in rockets; missiles; and nuclear, chemical, and superconductor applications. Niobium, first identified in 1801, is also an alloying element in various alloys and superalloys.

3. **Tungsten.** *Tungsten* (W, from *wolframite*) was first identified in 1781 and is the most plentiful of all refractory metals. Tungsten has the highest melting point of any metal (3410°C [6170°F]), and thus it is characterized by high strength at elevated temperatures. On the other hand, it has high density, brittleness at low temperatures, and poor resistance to oxidation. Tungsten and its alloys are used for applications involving temperatures above 1650°C (3000°F), such as nozzle throat liners in missiles and in the hottest parts of jet and rocket

engines, circuit breakers, welding electrodes, and spark-plug electrodes. The filament wire in incandescent light bulbs is made of pure tungsten, using powder metallurgy and wire-drawing techniques. Because of its high density, tungsten is used in balancing weights and counterbalances in mechanical systems, including self-winding watches. Tungsten is an important element in tool and die steels, imparting strength and hardness at elevated temperatures. Tungsten carbide, with cobalt as a binder for the carbide particles, is one of the most important tool and die materials.

4. **Tantalum.** Identified in 1802, *tantalum* (Ta) is characterized by a high melting point (3000°C [5425°F]), good ductility, and good resistance to corrosion. However, it has high density and poor resistance to chemicals at temperatures above 150°C (300°F). Tantalum is also used as an alloying element. Tantalum is used extensively in electrolytic capacitors and various components in the electrical, electronic, and chemical industries, as well as for thermal applications, such as in furnaces and acid-resistant heat exchangers. A variety of tantalum-base alloys is available in many forms for use in missiles and aircraft.

## 3.11.8  Various other nonferrous metals

1. **Beryllium.** Steel gray in color, *beryllium* (Be) has a high strength-to-weight ratio. Unalloyed beryllium is used in nuclear and X-ray applications, because of its low neutron absorption characteristics, and in rocket nozzles, space and missile structures, aircraft disc brakes, and precision instruments and mirrors. Beryllium is also an alloying element, and its alloys of copper and nickel are used in applications such as springs (beryllium–copper), electrical contacts, and non-sparking tools for use in explosive environments, such as mines and in metal-powder production. Beryllium and its oxide are toxic and should not be inhaled.

2. **Zirconium.** *Zirconium* (Zr) is silvery in appearance, has good strength and ductility at elevated temperatures, and has good corrosion resistance because of an adherent oxide film. The element is used in electronic components and nuclear power reactor applications, because of its low neutron absorption characteristics.

3. **Low melting point metals.** The major metals in this category are lead, zinc, and tin and are described as follows:

   a. **Lead**  *Lead* (Pb, after *plumbum*) has properties of high density, good resistance to corrosion (by virtue of the stable lead-oxide layer that forms and protects the surface), softness, low strength, high ductility, and good workability. Alloying with various elements, such as antimony and tin, enhances lead's properties, making it suitable for piping, collapsible tubing, bearing alloys, cable sheathing, roofing, and lead-acid storage batteries. Lead is also used for damping sound and vibrations, for radiation shielding against X rays, in printing (*type metals*), in weights, and in the chemical and paint industries. The oldest lead artifacts were made around 3000 BC. Lead pipes made by the Romans and installed in the Roman baths in Bath, England, two millennia ago are still in use. Lead is also an alloying element in solders, steels, and copper alloys and promotes corrosion resistance and machinability. Because of its toxicity, environmental contamination by lead is a significant concern.

   b. **Zinc**  Industrially, *zinc* (Zn), which has a bluish–white color, is the fourth most utilized metal, after iron, aluminum, and copper. Although known about for many centuries, zinc was not studied and developed until the 18th century. It has two major uses: for galvanizing iron, steel sheet, and wire; and as an alloy base for casting. In **galvanizing**, zinc serves as the anode and protects the steel (cathode) from corrosive attack should the coating be scratched or punctured. Zinc is also used as an alloying element; brass, for example, is an alloy of copper and zinc.

Major alloying elements in zinc are aluminum, copper, and magnesium. They impart strength and provide dimensional control during casting of the metal. Zinc-base alloys are used extensively in die casting for making products such as fuel pumps and grills for automobiles, components for household appliances (such as vacuum cleaners, washing machines, and kitchen equipment), various other machine parts, and photoengraving plates. Another use for zinc is in superplastic alloys, which have good formability characteristics by virtue of their capacity to undergo large deformation without failure. Very fine-grained 78% Zn–22% Al sheet is a common example of a superplastic zinc alloy that can be formed by methods used for forming plastics or metals.

c. **Tin** Although used in small amounts, *tin* (Sn, after *stannum*) is an important metal. The most extensive use of tin, a silvery white, lustrous metal, is as a protective coating on steel sheet (**tin plate**), which is used in making containers (tin cans) for food and various other products. Inside a sealed can, the steel is cathodic and is protected by the tin (anode), so that the steel does not corrode. The low shear strength of the tin coatings on steel sheet also improves its performance in deep drawing and general pressworking operations.

Unalloyed tin is used in applications such as lining material for water distillation plants and as a molten layer of metal over which plate glass is made. Tin-base alloys (also called **white metals**) generally contain copper, antimony, and lead. The alloying elements impart hardness, strength, and corrosion resistance. Because of their low friction coefficients, which result from low shear strength and low adhesion, tin alloys are used as journal-bearing materials. These alloys are known as **babbitts** and contain tin, copper, and antimony. **Pewter** is an alloy of tin, copper, and antimony. It was developed in the 15th century and is used for tableware, hollowware, and decorative artifacts. Organ pipes are made of tin alloys. Tin is an alloying element for type metals; dental alloys; and bronze (copper–tin alloy), titanium, and zirconium alloys. Tin–lead alloys are common soldering materials, with a wide range of compositions and melting points. (See Section 12.12.3.)

4. **Precious metals.** Gold, silver, and platinum are the most important *precious* (costly) *metals* and are also called *noble metals.*

a. **Gold** (Au, after *aurum*) is soft and ductile and has good corrosion resistance at any temperature. Typical applications include jewelry, coinage, reflectors, gold leaf for decorative purposes, dental work, and electroplating, as well as such important applications as electric contact and terminals.

b. **Silver** (Ag, after *argentum*) is a ductile metal and has the highest electrical and thermal conductivity of any metal. It does, however, develop an oxide film that affects its surface properties and appearance. Typical applications for silver include tableware, jewelry, coinage, electroplating, and photographic film, as well as electrical contacts, solders, bearings, and food and chemical equipment. *Sterling silver* is an alloy of silver and 7.5% copper.

c. **Platinum** (Pt) is a soft, ductile, grayish–white metal that has good corrosion resistance, even at elevated temperatures. Platinum alloys are used as electrical contacts, spark-plug electrodes, catalysts for automobile pollution-control devices, filaments, nozzles, dies for extruding glass fibers, and thermocouples, and in the electrochemical industry; other applications include jewelry and dental work.

## 3.11.9  Special metals and alloys

1. **Shape-memory alloys.** Shape-memory alloys, after being plastically deformed at room temperature into various shapes, return to their original shapes upon heating. For example, a piece of straight wire made of this material can be wound into a helical spring; when heated with a match, the spring uncoils and returns to the original straight shape. A typical shape-memory alloy is 55% Ni and 45% Ti; other alloys under development are copper–aluminum–nickel, copper–zinc–aluminum, iron–manganese–silicon, and nickel–titanium alloys.

   The behavior of shape-memory alloys can be *reversible*, that is, the shape can switch back and forth repeatedly upon application and removal of heat. These alloys generally have good ductility and corrosion resistance and high electrical conductivity. Typical applications include temperature sensors, clamps, connectors, fasteners, and seals that are easy to install.

2. **Amorphous alloys.** *Amorphous alloys* are a class of metal alloys that, unlike ordinary metals, do not have a long-range crystalline structure. (See also Section 5.10.8.) These alloys have no grain boundaries, and the atoms are randomly and tightly packed. Because their structure resembles that of glasses (Section 11.10), these alloys are also called **metallic glasses**. Amorphous alloys typically contain iron, nickel, and chromium, alloyed with carbon, phosphorus, boron, aluminum, and silicon. They are available in the form of wire, ribbon, strip, and powder. These alloys exhibit excellent corrosion resistance, good ductility, high strength, and very low loss from magnetic hysteresis. The latter property is utilized in making magnetic steel cores for transformers, generators, motors, lamp ballasts, magnetic amplifiers, and linear accelerators, with greatly improved efficiency.

   The amorphous structure was first obtained in the late 1960s by extremely rapid cooling of the molten alloy. One such method is called *splat cooling* or *melt spinning* (see Fig. 5.35), in which the alloy is propelled at a very high speed against a rotating metal surface; since the rate of cooling is on the order of $10^6$ K/s to $10^8$ K/s the molten alloy does not have sufficient time to crystallize. If an amorphous alloy's temperature is raised and then cooled, the alloy develops a crystalline structure.

3. **Nanomaterials.** Important recent developments involve the production of materials with grains, fibers, films, and composites which have particles that are on the order of 1 nm to 100 nm in size. First investigated in the early 1980s and generally called *nanomaterials*, these materials have some properties that are often superior to those of traditional and commercially available materials. These characteristics include strength, hardness, ductility, wear resistance and corrosion resistance suitable for structural (load-bearing) and nonstructural applications, in combination with unique electrical, magnetic and optical properties. Current and potential applications for nanomaterials include cutting tools, metal powders, computer chips, flat-panel displays for laptops, sensors, and various electrical and magnetic components. (See Sections 8.6.10 and 11.8.1.)

   The composition of nanomaterials can be any combination of chemical elements; among the more important compositions are carbides, oxides, nitrides, metals and alloys, organic polymers, and various composites. Synthesis methods include inert-gas condensation, plasma synthesis, electrodeposition, sol-gel synthesis, and mechanical alloying or ball milling.

4. **Metal foams.** A recent development is the production of *metal foams*, usually of aluminum alloys, but also of titanium or tantalum, where the metal consists of only 5% to 20% of the structure's volume. Metal foams can be produced by blowing air into a molten metal and tapping the froth that forms at the surface, which solidifies into a foam. Alternative approaches to creating metal foams include chemical vapor deposition onto a polymer or carbon foam lattice, and doping molten or powder metals with titanium hydride ($TiH_2$), which then releases hydrogen gas at the elevated casting or sintering temperatures. Metal foams present unique combinations of strength- and stiffness-to-density ratios, but these ratios are not as high as that of the base metal itself. However, the foams are very lightweight, making them an attractive material for aerospace applications; other applications include filters, lightweight beams, and orthopedic implants.

# SUMMARY

- Manufacturing properties of metals and alloys depend largely on their mechanical and physical properties. These properties, in turn, are governed mainly by their crystal structure, grain boundaries, grain size, texture, and various imperfections. (Sections 3.1, 3.2)

- Dislocations are responsible for the lower shear stress required to cause slip for plastic deformation. However, when dislocations become entangled with one another or are impeded by barriers such as grain boundaries, impurities, and inclusions, the shear stress required to cause slip increases; this phenomenon is called work hardening, or strain hardening. (Section 3.3)

- Grain size and the nature of grain boundaries significantly affect the properties of metals, including strength, ductility, and hardness, as well as increase the tendency for embrittlement with reduced toughness. Grain boundaries enhance strain hardening by interfering with the movement of dislocations. (Section 3.4)

- Plastic deformation at room temperature (cold working) of polycrystalline materials results in higher strength of the material as dislocations become entangled. The deformation also generally causes anisotropy, whereby the mechanical properties are different in different directions. (Section 3.5)

- The effects of cold working can be reversed by heating the metal within a specific temperature range and for a specific period of time, through the consecutive processes of recovery, recrystallization, and grain growth. (Section 3.6)

- Metals and alloys can be worked at room, warm, or high temperatures. Their overall behavior, force and energy requirements, and workability depend largely on whether the working temperature is below or above their recrystallization temperature. (Section 3.7)

- Failure and fracture of a workpiece when subjected to deformation in metalworking operations is an important consideration. Two types of fracture are ductile fracture and brittle fracture. Ductile fracture is characterized by plastic deformation preceding fracture and requires a considerable amount of energy. Because it is not preceded by plastic deformation, brittle fracture can be catastrophic and requires much less energy than does ductile fracture. Impurities and inclusions, as well as factors such as the environment, strain rate, and state of stress, can play a major role in the fracture behavior of metals and alloys. (Section 3.8)

- Physical and chemical properties of metals and alloys can significantly affect design considerations; service requirements; compatibility with other materials, including tools and dies; and the behavior of the metals and alloys during processing. (Section 3.9)

- A very wide variety of metals and alloys is now available, with a wide range of properties, such as strength, toughness, hardness, ductility, and resistance to high temperatures and oxidation. These materials can be classified generally as ferrous metals and alloys; nonferrous metals and alloys; superalloys; refractory metals and their alloys; and various other types, such as precious metals. More recently developed materials include amorphous alloys, shape-memory alloys, nanomaterials, and metal foams. (Sections 3.10, 3.11)

## SUMMARY OF EQUATIONS

Theoretical shear strength of metals, $\tau_{max} = \dfrac{G}{2}\pi$

Theoretical tensile strength of metals, $\sigma_{max} = \sqrt{\dfrac{E\gamma}{a}} \simeq \dfrac{E}{10}$

Hall–Petch equation, $Y = Y_i + kd^{-1/2}$

ASTM grain-size number, $N = 2^{n-1}$

Tensile strength vs. crack length, $\sigma \propto \dfrac{1}{\sqrt{\text{crack length}}}$

## BIBLIOGRAPHY

*Aerospace Structural Metals Handbook*, 5 vols., Metals and Ceramics Information Center, Battelle, 1987.

Ashby, M. F., and D. R. H. Jones, *Engineering Materials*, Vol. 1, "*An Introduction to Their Properties and Applications*," 2d. ed., Pergamon, 1996; Vol. 2, "*An Introduction to Microstructures, Processing and Design*," Pergamon, 1986; Vol. 3, "*Materials Failure Analysis: Case Studies and Design Implications*," Pergamon, 1993.

Ashby, M. F., *Materials Selection in Mechanical Design*, 2d. ed., Pergamon, 1999.

Askeland, D. R., *The Science and Engineering of Materials*, 5th ed., PWS-KENT, 1994.

*ASM Handbook*, various volumes, ASM International.

*ASM Specialty Handbooks*, various volumes. ASM International.

Boyer, H. E. (ed.), *Selection of Materials for Component Design: Source Book*, American Society for Metals, 1986.

Budinski, K. G., *Engineering Materials: Properties and Selection*, 5th ed. Prentice Hall, 1996.

Callister, W. D., Jr., *Materials Science and Engineering*, 5th ed., Wiley, 2000.

Dieter, G. E., *Engineering Design: A Materials and Processing Approach*, 3d. ed., McGraw-Hill, 1999.

_____, *Mechanical Metallurgy*, 3d. ed., McGraw-Hill, 1986.

Farag, M. M., *Materials Selection for Engineering Design*. Prentice Hall, 1997.

Flinn, R. A., and P. K. Trojan, *Engineering Materials and Their Applications*, 4th ed., Houghton Mifflin, 1990.

Harper, C. (ed.), *Handbook of Materials for Product Design*, 3d. ed., McGraw-Hill, 2001.

Hertzberg, R. W., *Deformation and Fracture Mechanics of Engineering Materials*, 4th ed. Wiley, 1996.

Mangonon, P. C., *The Principles of Material Selection for Engineering Design*. Prentice Hall, 1999.

*Material Selector*, annual publication of *Materials Engineering Magazine*, Penton/IPC.

Meyers, M. A., and K. K. Chawla, *Mechanical Metallurgy: Principles and Applications*. Prentice-Hall, 1984.

Pollock, D. D., *Physical Properties of Materials for Engineers*, 2d. ed., CRC Press, 1993.

Roberge, P. R., and M. Tullmin, *Handbook of Corrosion Engineering*, McGraw-Hill, 1999.

Roberts, G. A., G. Krauss, R. Kennedy, and R. A. Cary, *Tool Steels*, 5th ed., ASM International, 1998.

Schaffer, J., A. Saxena, S. Antalovich, T. Sanders, and S. Warner, *The Science & Design of Engineering Materials*, 2d. ed., McGraw-Hill, 1999.

Shackelford, J. F., *Introduction to Materials Science for Engineers*, 5th ed., Macmillan, 2000.

Smith, W. F., *Principles of Materials Science and Engineering*, 3d. ed., McGraw-Hill, 1995.

*Tool and Manufacturing Engineers Handbook*, 4th ed., Vol. 3, "*Materials, Finishing and Coating*," Society of Manufacturing Engineers, 1985.

*Woldman's Engineering Alloys*, 9th ed., ASM International, 2000.

Wroblewski, A. J., and S. Vanka, *MaterialTool: A Selection Guide of Materials and Processes for Designers*. Prentice Hall, 1997.

Wulpi, D. J., *Understanding How Components Fail*, 2d. ed., ASM International, 1999.

# QUESTIONS

**3.1**   What is the difference between a unit cell and a single crystal?

**3.2**   Explain why we should study the crystal structure of metals.

**3.3**   What effects does recrystallization have on the properties of metals?

**3.4**   What is the significance of a slip system?

**3.5**   Explain what is meant by structure-sensitive and structure-insensitive properties of metals.

**3.6**   What is the relationship between nucleation rate and the number of grains per unit volume of a metal?

**3.7**   Explain the difference between recovery and recrystallization.

**3.8**   (a) Is it possible for two pieces of the same metal to have different recrystallization temperatures? Explain. (b) Is it possible for recrystallization to take place in some regions of a workpiece before other regions in the same workpiece? Explain.

**3.9**   Make a list of various phenomena that result from the presence of grain boundaries in metals.

**3.10**   Describe why different crystal structures exhibit different strengths and ductilities.

**3.11**   Explain the difference between preferred orientation and mechanical fibering.

**3.12**   Name some analogies to mechanical fibering (e.g., layers of thin dough sprinkled with flour).

**3.13**   A cold-worked piece of metal has been recrystallized. When tested, it is found to be anisotropic. Explain the probable reason for its anisotropy.

**3.14**   Does recrystallization eliminate mechanical fibering in a workpiece? Explain.

**3.15**   Explain why we may have to be concerned with the orange-peel effect on metal surfaces.

**3.16**   Do you think it might be important to know whether a raw material for a manufacturing process has anisotropic properties? What about anisotropy in the finished product? Explain, using appropriate examples.

**3.17**   How can you tell the difference between two parts made of the same metal, one shaped by cold working and the other by hot working? Explain the differences you might observe. Note that there are several methods that can be used to determine the differences between the two parts.

**3.18**   Explain why the strength of a polycrystalline metal at room temperature decreases as its grain size increases.

**3.19**   Describe the technique by which you would reduce the orange-peel effect.

**3.20**   What is the significance of metals such as lead and tin having recrystallization temperatures at about room temperature?

**3.21**   You are given a deck of playing cards held together a rubber band. Which of the material-behavior phenomena described in this chapter could you demonstrate? What are the effects of increasing the number of rubber bands holding the cards together? Explain. (*Hint*: Inspect Figs. 3.4 and 3.6.)

**3.22**   Using the information given in Chapters 2 and 3, describe the conditions that induce brittle fracture in an otherwise ductile piece of metal.

**3.23**   A part fractures while being subjected to plastic deformation at an elevated temperature. List the factors that could have played a role in the fracture.

**3.24**   Make a list of metals that would be suitable for a (1) paper clip, (2) buckle for a safety belt, (3) nail, (4) battery cable, and (5) gas-turbine blade.

**3.25**   Explain the advantages and limitations of cold, warm, and hot working of metals, respectively.

**3.26**   Describe the significance of structures and machine components made of two materials with different coefficients of thermal expansion.

**3.27**   Explain why parts may crack when suddenly subjected to extremes of temperature.

**3.28**   From your own experience and observations, list three applications each for the following metals and their alloys: (1) steel, (2) aluminum, (3) copper, (4) lead, and (5) gold.

**3.29**   Name products that would not have been developed to their advanced stages as we find them today if alloys with high strength and corrosion and creep resistance at elevated temperatures had not been developed.

**3.30**   Inspect several metal products and components and make an educated guess as to what materials they are made from. Give reasons for your guess. If you list two or more possibilities, explain your reasoning.

**3.31**   List three engineering applications for which the following physical properties would be desirable: (1) high density, (2) low melting point, and (3) high thermal conductivity.

**3.32**   Two physical properties that have a major influence on the cracking of workpieces or dies during thermal cycling are thermal conductivity and thermal expansion. Explain why.

**3.33**   What might be the adverse effects of using a material with low specific heat in manufacturing?

**3.34**  Would a solid-metal shell with a metal-foam interior be useful in applications such as a piston in an internal combustion engine? Explain your answer.

**3.35**  Describe the advantages of nanomaterials over traditional materials.

**3.36**  Aluminum has been cited as a possible substitute material for steel in automobiles. What concerns, if any, would you have before purchasing an aluminum automobile?

**3.37**  Lead shot is popular among sportsmen for hunting, but birds commonly ingest the pellets (along with gravel) to help digest food. What substitute materials would you recommend for lead, and why?

# PROBLEMS

**3.38**  The unit cells shown in Figs. 3.2a–c can be represented by hard-ball or hard-sphere models, similar to tennis balls arranged in various configurations in a box. In such an arrangement, the *atomic packing factor* (APF) is defined as the ratio of the volume of atoms to the volume of the unit cell. Show that the APF is 0.68 for the bcc structure and 0.74 for the fcc structure.

**3.39**  Show that the lattice constant $a$ in Fig. 3.2b is related to the atomic radius by expression

$$a = 2\sqrt{2}R,$$

where $R$ is the radius of the atom as depicted by the hard-ball model.

**3.40**  Show that for the fcc unit cell, the radius $r$ of the largest hole is given by the expression

$$r = 0.414R.$$

Determine the size of the largest hole for the iron atom in the fcc structure.

**3.41**  Calculate the theoretical (1) shear strength and (2) tensile strength for aluminum, plain-carbon steel, and tungsten. Estimate the ratios of their theoretical strength to actual strength.

**3.42**  A technician determines that the grain size of a certain etched specimen is 4. Upon further checking, it is found that the magnification used was 150, instead of 100, the latter of which is required by ASTM standards. What is the correct grain size?

**3.43**  Estimate the number of grains in a regular paper clip if its ASTM grain size is 10.

**3.44**  Plot the following for the materials described in this chapter: (a) yield strength vs. density, and (b) elastic modulus vs. density. Explain your observations.

**3.45**  Determine the maximum crack velocity in aluminum, plain–carbon steel, and tungsten. (See Section 3.8.2 and Tables 2.1 and 3.3.)

**3.46**  Calculate the maximum theoretical shear stress $\tau_{max}$, assuming that in Fig. 3.5 the shear stress $\tau$ varies linearly with $x$.

**3.47**  The natural frequency $f$ of a cantilever beam is given by the expression

$$f = 0.56\sqrt{\frac{EIg}{wL^4}},$$

where $E$ is the modulus of elasticity, $I$ is the moment of inertia, $g$ is the gravitational constant, $w$ is the weight of the beam per unit length, and $L$ is the length of the beam. How does the natural frequency of the beam change, if at all, as its temperature is increased?

**3.48**  A strip of metal is reduced in thickness by cold working from 25 mm to 15 mm. A similar strip is reduced from 25 mm to 10 mm. Which one of these strips will recrystallize at a lower temperature? Why?

**3.49**  Assume that you are asked to give a quiz to students on the contents of this chapter. Prepare three quantitative problems and three qualitative questions, and supply the answers.

**3.50**  A 1-m long, simply supported beam with a round cross-section is subjected to a load of 25 kg at its center. (a) If the shaft is made from AISI 303 steel and has a diameter of 20 mm, what is the deflection under the load? (b) For shafts made from 2024-T4 aluminum, architectural bronze, and 99.5% titanium, respectively, what must the diameter of the shaft be for the shaft to have the same deflection as in part (a)?

# 4

# Surfaces, Tribology, Dimensional Characteristics, Inspection, and Product Quality Assurance

## 4.1 | Introduction

Because of the various mechanical, physical, thermal, and chemical effects induced by its processing history, the **surface** of a manufactured part generally has properties and behavior that are considerably different from those of its bulk. Although the **bulk material** generally determines a component's overall mechanical properties, the component's surface directly influences several important properties and characteristics of the manufactured part:

* Friction and wear properties of the part during subsequent processing when it comes in direct contact with tools and dies, or when it is placed in service;

* Effectiveness of lubricants during manufacturing processes, as well as throughout the part's service life;

* Appearance and geometric features of the part and their role in subsequent operations such as painting, coating, welding, soldering, and adhesive bonding, as well as corrosion resistance of the part;

* Initiation of cracks due to surface defects, such as roughness, scratches, seams, and heat-affected zones, which could lead to weakening and premature failure of the part by fatigue or other fracture mechanisms;

* Thermal and electrical conductivity of contacting bodies; for example, rough surfaces have higher thermal and electrical resistance than smooth surfaces.

Friction, wear, and lubrication (**tribology**) are surface phenomena. Friction influences forces, power requirements, and surface quality of parts. Wear alters the surface geometry of tools and dies, which, in turn, adversely affects the quality of manufactured products and the economics of production. Lubrication, with few exceptions, is an integral aspect of all manufacturing operations, as well as in the proper functioning of machinery and equipment. We describe those aspects of lubrication that are relevant to manufacturing operations, such as the types of wear encountered, how to reduce wear, and the proper selection and application of lubricants and coolants.

4.1 Introduction
4.2 Surface Structure and Properties
4.3 Surface Texture
4.4 Tribology: Friction, Wear, and Lubrication
4.5 Surface Treatments, Coatings, and Cleaning
4.6 Engineering Metrology and Instrumentation
4.7 Dimensional Tolerances
4.8 Testing and Inspection
4.9 Quality Assurance

Another important aspect in surface technology is **surface treatments**, which modify the properties and characteristics of surfaces. Several mechanical, thermal, electrical, and chemical methods can be used to modify surfaces of parts to improve frictional behavior, effectiveness of lubricants, resistance to wear and corrosion, and surface finish and appearance.

**Measurement** of the relevant dimensions and features of parts is an integral aspect of interchangeable manufacture, the basic concept of standardization and mass production. In this chapter, we describe the principles involved and the various instruments used in measurement. Another important aspect is **testing and inspection** of manufactured products, using both destructive and nondestructive testing methods. One of the most critical aspects of manufacturing is **product quality**. This chapter emphasizes the technological and economic importance of *building quality into a product*, rather than inspecting the product after it is made, and it describes the statistical techniques and control charts in achieving these goals.

## 4.2 | Surface Structure and Properties

Upon close examination of the surface of a piece of metal, we find that it generally consists of several layers (Fig. 4.1). The part of the metal starting at the interior and moving outward to the surface is called the *bulk metal*, or the **metal substrate**; its structure depends on the composition and processing history of the metal. Above this bulk metal is a layer that usually has been plastically deformed and work hardened during the manufacturing process.

The depth and properties of the work-hardened layer (the **surface structure**) depend on factors such as the processing method used and the extent of frictional sliding to which the surface was subjected. For example, sharp tools and selection of proper process parameters produce surfaces with little or no disturbance. If the surface is produced by machining with a dull tool or under poor cutting conditions, or is ground with a dull grinding wheel, this layer will be relatively thick. Also, nonuniform surface deformation or severe temperature gradients during manufacturing operations usually causes *residual stresses* in this work-hardened layer. A **Beilby**, or **amorphous**, **layer** may or may not be present on top of the work-hardened layer; it consists of a structure that is microcrystalline or amorphous. This layer develops in some machining and finishing operations where melting and surface flow, followed by rapid quenching, are encountered. Unless the metal is processed and kept in an inert (oxygen-free) environment, or it is a noble metal, such as gold or platinum, an **oxide layer** usually lies on top of the work-hardened or Beilby layer. Some examples of metals with an oxide layer include the following:

1. *Iron* has an oxide structure, with FeO adjacent to the bulk metal, followed by a layer of $Fe_3O_4$, and then a layer of $Fe_2O_3$, which is exposed to the environment.

1–100 nm — Contaminant
1 nm — Adsorbed gas
10–100 nm — Oxide
1–100 nm — Beilby (amorphous) layer
1–100 nm — Work-hardened layer
1–100 μm — Metal substrate

**FIGURE 4.1** Schematic illustration of the cross-section of the surface structure of metals. The thickness of the individual layers depends on processing conditions and the environment. *Source*: After E. Rabinowicz and B. Bhushan.

2. *Aluminum* has a dense *amorphous* (i.e., without crystalline structure) layer of $Al_2O_3$, with a thick, porous hydrated aluminum-oxide layer over it.

3. *Copper* has a bright shiny surface when freshly scratched or machined. Soon after, however, it develops a $Cu_2O$ layer, which is then covered with a layer of CuO; these layers give copper its somewhat dull color, such as seen in kitchen utensils.

4. *Stainless steels* are "stainless" because they develop a protective layer of chromium oxide, CrO (**passivation**).

Under normal environmental conditions, surface oxide layers are generally covered with *adsorbed* layers of gas and moisture. Finally, the outermost surface of the metal may be covered with contaminants, such as dirt, dust, grease, lubricant residues, cleaning-compound residues, and pollutants from the environment. Thus, surfaces generally have properties that are very different from those of the substrate. The oxide on a metal's surface, for example, is generally much harder than the base metal; hence, oxides tend to be brittle and abrasive. This surface characteristic has, in turn, several important effects on friction, wear, and lubrication in materials processing and subsequent coatings on products. The factors involved in the surface structure of metals are also relevant, to a great extent, to the surface structure of plastics and ceramics. The surface texture of these materials depends, as with metals, on the method of production; environmental conditions also influence the surface characteristics of these materials.

**Surface integrity.** *Surface integrity* describes not only the topological (geometric) aspects of surfaces, but also their mechanical and metallurgical properties and characteristics. Surface integrity is an important consideration in manufacturing operations, because it influences the properties of the product, such as its fatigue strength, resistance to corrosion, and service life.

Several **defects** caused by and produced during manufacturing can be responsible for lack of surface integrity. These defects are usually caused by a combination of factors, such as defects in the original material; the method by which the surface is produced; and lack of proper control of process parameters, which can result in excessive stresses and temperatures. The major surface defects found in practice usually consist of one or more of the following: *cracks, craters, folds, laps, seams, splatter, inclusions, intergranular attack, heat-affected zones, metallurgical transformations, plastic deformation*, and *residual stresses*.

## 4.3 | Surface Texture

Regardless of the method of production, all surfaces have their own set of characteristics, which is referred to as **surface texture**. The description of surface texture as a geometrical property is complex; however, certain guidelines have been established for identifying surface texture in terms of well-defined and measurable quantities (Fig. 4.2):

1. **Flaws,** or **defects**, are random irregularities, such as scratches, cracks, holes, depressions, seams, tears, and inclusions.

2. **Lay,** or **directionality**, is the direction of the predominant surface pattern and is usually visible to the naked eye.

3. **Waviness** is a recurrent deviation from a flat surface, much like waves on the surface of water. It is measured and described in terms of (a) the space between

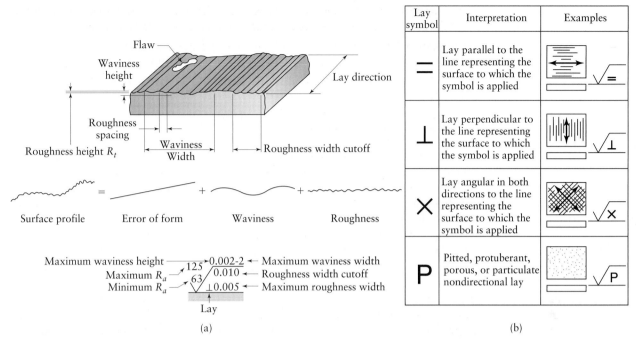

**FIGURE 4.2**   (a) Standard terminology and symbols used to describe surface finish. The quantities are given in $\mu$ in. (b) Common surface-lay symbols.

adjacent crests of the waves (*waviness width*) and (b) the height between the crests and valleys of the waves (*waviness height*). Waviness may be caused by deflections of tools, dies, and the workpiece; warping from forces or temperature; uneven lubrication; and vibration or any periodic mechanical or thermal variations in the system during the manufacturing process.

4. **Roughness** consists of closely spaced, irregular deviations on a scale smaller than that for waviness; roughness may be superimposed on waviness. Roughness is expressed in terms of its height, its width, and the distance on the surface along which it is measured.

**Surface roughness.**   Surface roughness is generally described by two methods: arithmetic mean value and root-mean-square average. The **arithmetic mean value**, $R_a$, formerly identified as AA (for *arithmetic average*) or CLA (for *center-line average*), is based on the schematic illustration of a rough surface, as shown in Fig. 4.3. The arithmetic mean value is defined as

$$R_a = \frac{y_a + y_b + y_c + \cdots + y_n}{n} = \frac{1}{n}\sum_{i=1}^{n} y_i = \frac{1}{\ell}\int_0^l |y|\, dx, \qquad (4.1)$$

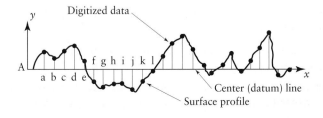

**FIGURE 4.3**   Coordinates used for measurement of surface roughness, using Eqs. (4.1) and (4.2).

where all ordinates $y_a, y_b, y_c, \ldots$, are absolute values. The last term in Eq. (4.1) refers to the $R_a$ of a continuous surface or wave, as is commonly encountered in signal processing. $\ell$ refers to the total profile length measured.

The **root-mean-square average**, $R_q$, formerly identified as RMS, is defined as

$$R_q = \sqrt{\frac{y_a^2 + y_b^2 + y_c^2 + \cdots + y_n^2}{n}} = \sqrt{\frac{1}{n}\sum_1^n y_i^2} = \left[\frac{1}{\ell}\int_0^l y^2\, dx\right]^{1/2}. \quad (4.2)$$

The center (*datum*) line in Fig. 4.3 is located so that the sum of the areas above the line is equal to the sum of the areas below the line. The units generally used for surface roughness are $\mu$m (micrometer, or micron) or $\mu$in. (microinch), where $1\ \mu$m $= 40\ \mu$in. and $1\ \mu$in. $= 0.025\ \mu$m. In addition, the **maximum roughness height**, $R_t$, may also be used as a measure of roughness. It is defined as the height from the deepest trough to the highest peak. It indicates the amount of material that has to be removed in order to obtain a smooth surface by polishing or other means.

Because of its simplicity, the arithmetic mean value was adopted internationally in the mid-1950s and is widely used in engineering practice. Equations (4.1) and (4.2) show that there is a relationship between $R_q$ and $R_a$. For a surface roughness in the shape of a sine curve, $R_q$ is larger than $R_a$ by a factor of 1.11. This factor is 1.1 for most machining processes by cutting, 1.2 for grinding, and 1.4 for lapping and honing. $R_q$ is more sensitive to the largest asperity peaks and deepest valleys, and since these peaks and valleys are important for friction and lubrication, $R_q$ is often used even though it is slightly more complicated to calculate than $R_a$.

In general, a surface cannot be adequately described by its $R_a$ or $R_q$ value alone, since these values are averages. Two surfaces may have the same roughness value, but their actual topography may be quite different. A few deep troughs, for example, affect the roughness values insignificantly. However, such differences in the surface profile can be significant in terms of fatigue, friction, and the wear characteristics of a manufactured product.

**Symbols for surface roughness.** Acceptable limits for surface roughness are specified on technical drawings by the symbols shown around the check mark in the lower portion of Fig. 4.2, and their values are placed to the left of the check mark. Symbols used to describe a surface specify only the roughness, waviness, and lay; they do not include flaws. Whenever flaws are important, a special note is included in technical drawings to describe the method to be used to inspect for surface flaws.

**Measuring surface roughness.** Various commercially available instruments, called **surface profilometers**, are used to measure and record surface roughness. The most commonly used instruments feature a diamond stylus traveling along a straight line over the surface (Figs. 4.4a and b). The distance that the stylus travels, which can be varied, is called the **cutoff**. (See Fig. 4.2.) To highlight the roughness, profilometer traces are recorded on an exaggerated vertical scale (a few orders of magnitude greater than the horizontal scale; Figs. 4.4c–f), called **gain** on the recording instrument. Thus, the recorded profile is significantly distorted, and the surface appears to be much rougher than it actually is. The recording instrument compensates for any surface waviness and indicates only roughness. A record of the surface profile is made by mechanical and electronic means.

Surface roughness can be observed directly through interferometry, optical or scanning-electron microscopy, or atomic-force microscopy. Scanning-electron and atomic-force microscopy are especially useful for imaging very smooth surfaces for which features cannot be captured by the less sensitive instruments. Stereoscopic photographs are particularly useful for three-dimensional views of surfaces and can also be used to measure surface roughness.

**FIGURE  4.4**    (a) Measuring surface roughness with a stylus. The rider supports the stylus and guards against damage. (b) Path of the stylus in measurements of surface roughness (broken line) compared with the actual roughness profile. Note that the profile of the stylus' path is smoother than the actual surface profile. *Source*: D. H. Buckley. Typical surface profiles produced by (c) lapping, (d) finish grinding, (e) rough grinding, and (f) turning processes. Note the difference between the vertical and horizontal scales. *Source*:  from D. B. Dallas (ed.), *Tool and Manufacturing Engineers Handbook*, 3d ed.

**Surface roughness in practice.**    Surface-roughness design requirements for typical engineering applications can vary by as much as two orders of magnitude for different parts. The reasons and considerations for this wide range include the following:

1. *Precision required on mating surfaces*, such as seals, fittings, gaskets, tools, and dies (for example, ball bearings and gages require very smooth surfaces, whereas surfaces for gaskets and brake drums can be quite rough);

2. *Tribological considerations*, that is, the effect of roughness on friction, wear, and lubrication;

3. *Fatigue and notch sensitivity*, because rougher surfaces often have shorter fatigue lives;

4. *Electrical and thermal contact resistance*, because the rougher the surface is, the higher the resistance will be;

5. *Corrosion resistance*, because the rougher the surface is, the greater will be the possibility that there are entrapped corrosive media;

6. *Subsequent processing*, such as painting and coating, in which a certain degree of roughness can result in better bonding;

7. *Appearance*, because, depending on the application, a rough or smooth surface may be preferred;

8. *Cost considerations*, because the finer the finish, the higher is the cost.

These factors should be carefully considered before a decision is made as to the surface-roughness recommendation for a given product. As in all manufacturing processes, the cost involved in the selection should also be a major consideration.

# 4.4 | Tribology: Friction, Wear, and Lubrication

*Tribology* is the science and technology of interacting surfaces; thus, it involves friction, wear, and lubrication. All these factors can significantly influence the technology and economics of manufacturing processes and operations.

## 4.4.1 Friction

*Friction* is defined as the resistance to relative sliding between two bodies in contact under a normal load. Metalworking and manufacturing processes are significantly affected by friction, because of the relative motion and the forces present between tools, dies, and workpieces. Friction is an energy-dissipating process that results in the generation of heat. The subsequent rise in temperature can have a major effect on the overall operation. Furthermore, because it impedes movement at the tool, die, and workpiece interfaces, friction significantly affects the flow and deformation of materials in metalworking processes.

However, friction is not always undesirable; the presence of friction is necessary for the success or optimization of many operations, the best example of which is the rolling of metals to make sheet and plate. Without friction, it would be impossible to roll materials, just as it would be impossible to drive a car on the road. Various theories have been proposed to explain the phenomenon of friction. An early Coulomb model states that friction results from the mechanical interlocking of rough surface asperities, thereby requiring some force to slide the two bodies along each other.

For a particular theory of friction to be valid, it must explain the frictional behavior of two bodies under different conditions, including load, relative sliding speed, temperature, surface condition, environment, etc. Many models have been proposed, with varying degrees of success. The most commonly used theory of friction is based on adhesion, because of its reasonable agreement with experimental observations.

**Adhesion theory of friction.** The *adhesion theory of friction* is based on the observation that two clean, dry (unlubricated) metal surfaces, regardless of how smooth they are, contact each other at only a fraction of their apparent area of contact (Fig. 4.5). The static load at the interface is thus supported by the contacting asperities. The sum of the contacting areas is known as the **real area of contact**, $A_r$. For most engi-

**FIGURE 4.5** (a) Schematic illustration of the interface of two contacting surfaces, showing the real areas of contact. (b) Sketch illustrating the proportion of the apparent area to the real area of contact. The ratio of the areas can be as high as four to five orders of magnitude.

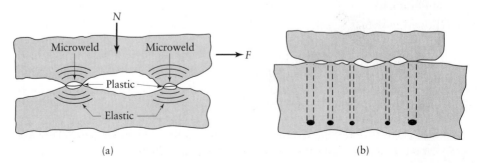

neering surfaces, the average angle that the asperity hill makes with a horizontal line is found to be between 5° and 15°. (Illustrations of surface roughness are highly exaggerated, because of the scales involved.)

Under light loads and with a large real area of contact, the normal stress at the asperity contacts is *elastic*. As the load increases, the stresses increase, and eventually, *plastic* deformation takes place at the **junctions**. Furthermore, with increasing load, the asperities in contact increase in area, and new junctions are formed with other asperities between the two surfaces in contact with each other. Because of the random asperity height distribution, some junctions are in elastic contact, while others are in plastic contact. The intimate contact of asperities creates an **adhesive bond**.

The nature of the adhesive bond is complex, involving atomic interactions, and may include mutual solubility and diffusion. The strength of the bond depends on the physical and mechanical properties of the metals in contact, the temperature, and the nature and thickness of any oxide film or other contaminants present on the surfaces. In many manufacturing processes (see Chapters 6 through 9), the load at the interface is generally high so that the normal stress on the asperity reaches the yield stress. Plastic deformation of the asperities then takes place, causing adhesion of the junctions. In other words, the asperities form **microwelds**. The cleaner the interface, the stronger are the adhesive bonds.

**Coefficient of friction.** Sliding between two bodies under a normal load, N, is possible only by the application of a tangential force F. According to the adhesion theory, F is the force required to *shear* the junctions (friction force). The *coefficient of friction*, $\mu$, at the interface is defined as

$$\mu = \frac{F}{N} = \frac{\tau A_r}{\sigma A_r} = \frac{\tau}{\sigma}, \tag{4.3}$$

where $\tau$ is the shear strength of the junction and $\sigma$ is the normal stress, which, for a plastically deformed asperity, is its yield stress. Thus, the numerator in Eq. (4.3) is a *surface* property, whereas the denominator is a *bulk* property. Because an asperity is surrounded by a large mass of material, the normal stress on an asperity is equivalent to the hardness of the material. (See Section 2.6.) The coefficient of friction can now be defined as

$$\mu = \frac{\tau}{\text{Hardness}}. \tag{4.4}$$

The nature and strength of the interface is the most significant factor in friction. A strong interface requires a high friction force for relative sliding, and a weak interface requires a low friction force. Equation (4.4) indicates that the coefficient of friction can be reduced not only by decreasing the numerator, but also by increasing the denominator. This observation suggests that placing thin films of low shear strength over a substrate with high hardness is the ideal method for reducing abrasive friction; in fact, this result is exactly what is achieved with a lubricant layer.

Two additional phenomena take place in asperity interactions under plastic stresses. For a strain-hardening material, the peak of the asperity is stronger than the bulk material, because of plastic deformation at the junction. Thus, under ideal conditions, the breaking of the bonds under a tensile load will likely follow a path below or above the geometric interface. Secondly, a tangential motion at the junction under load will cause *junction growth*; that is, the contact area increases. The reason is that, according to the yield criteria described in Section 2.11, the effective yield stress of the material decreases when the material is subjected to a shear stress. Hence, the junction has to grow in area to support the normal load. In other words, the two surfaces tend to approach each other under a tangential force.

**FIGURE 4.6** Schematic illustration of the relation between friction force $F$ and normal force $N$. Note that as the real area of contact approaches the apparent area, the friction force reaches a maximum and stabilizes. Most machine components operate in the first region. The second and third regions are encountered in metalworking operations, because of the high contact pressures involved between sliding surfaces, i.e., die and workpiece.

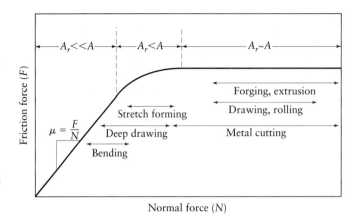

If the normal load is increased further, the real area of contact, $A_r$, ideally continues to increase, especially with soft and ductile metals. In the absence of contaminants or fluids trapped at the interface, the real area of contact eventually reaches the apparent area of contact. This area is the maximum contact area that can be obtained. Because the shear strength at the interface is constant, the friction force (shearing force) reaches a maximum and levels off (Fig. 4.6).

This condition, known as **sticking**, creates difficulty in defining friction. Sticking in a sliding interface does not necessarily mean complete adhesion at the interface, as in welding. Rather, it means that the frictional stress at the surface has reached the shear yield stress, $k$, of the material. When two clean surfaces are pressed together with sufficiently high forces, *cold pressure welding* can take place, but usually the frictional stress is limited by contamination or by the presence of oxide layers. As the normal load, $N$, is increased further, the friction force, $F$, remains constant, and hence, by definition, the coefficient of friction decreases. This situation, which is an anomaly, indicates that there might be a different, and more realistic, means of expressing the frictional condition at an interface.

A more recent trend is to define a **friction factor**, or **shear factor**, $m$, as

$$m = \frac{\tau_i}{k}, \tag{4.5}$$

where $\tau_i$ is the shear strength of the interface and $k$ is the *shear yield stress* of the softer material in a sliding pair. (The quantity $k$ is equal to $Y/2$ according to Eq. (2.35) and to $Y/\sqrt{3}$ according to the distortion-energy criterion, where $Y$ is the uniaxial yield stress of the material.) Equation (4.5) is often referred to as a **Tresca** friction model.

The value of $\tau_i$ is difficult to measure, because we have to know the real area of contact; average values will not be meaningful. However, note that when $m = 0$ in this definition, there is no friction, and when $m = 1$, complete sticking takes place at the interface. The magnitude of $m$ is independent of the normal force or stress; the reason is that the shear yield stress of a thin layer of material is unaffected by the magnitude of the normal stress.

The sensitivity of friction to different factors can be illustrated by the following examples: A typical value for the coefficient of friction of steel sliding on lead is 1.0; for steel on copper, it is 0.9. However, for steel sliding on copper coated with a thin layer of lead, it is 0.2. The coefficient of friction of pure nickel sliding on nickel in a hydrogen or nitrogen atmosphere is 5; in air or oxygen, it is 3; and in the presence of water vapor, it is 1.6. Coefficients of friction in sliding contact, as measured experimentally, vary from as low as 0.02 to as high as 100, or even higher. This range is not surprising in view of the many variables involved in the friction process. In metalworking opera-

TABLE 4.1

## Coefficient of Friction in Metalworking Processes

| | Coefficient of Friction ($\mu$) | |
| --- | --- | --- |
| Process | Cold | Hot |
| Rolling | 0.05–0.1 | 0.2–0.7 |
| Forging | 0.05–0.1 | 0.1–0.2 |
| Drawing | 0.03–0.1 | – |
| Sheet-metal forming | 0.05–0.1 | 0.1–0.2 |
| Machining | 0.5–2 | – |

tions that use various lubricants, the range for $\mu$ is much narrower, as shown in Table 4.1. The effects of load, temperature, speed, and environment on the coefficient of friction are difficult to generalize, as each situation has to be studied individually.

**Abrasion theory of friction.** If the upper body in the model shown in Fig. 4.5(a) is harder than the lower one, or if its surface contains protruding hard particles, then as it slides over the softer body, it can scratch and produce grooves on the lower surface. This phenomenon is known as **plowing** and is an important aspect in abrasive frictional behavior; in fact, it can be a dominant mechanism for situations where adhesion is not strong.

Plowing may involve two different mechanisms. One is the generation of a *groove*, whereby the surface is deformed plastically. The other is the formation of a groove because the cutting action of the upper body generates a *chip* or *sliver*. Both of these processes involve work supplied by a force that manifests itself as friction force. The plowing force can contribute significantly to friction and to the measured coefficient of friction at the interface.

**Measuring friction.** The coefficient of friction is usually determined experimentally, either during actual manufacturing processes or in simulated tests using small-scale specimens of various shapes. The techniques used generally involve measurements of either forces or dimensional changes in the specimen. One test that has gained wide acceptance, particularly for bulk deformation processes such as forging, is the **ring compression test**. In this test, a flat ring is compressed plastically between two flat platens (Fig. 4.7). As its height is reduced, the ring expands radially outward.

Good lubrication          Poor lubrication

(a)

(b)

**FIGURE 4.7** (a) The effects of lubrication on barreling in the ring compression test: (a) With good lubrication, both the inner and outer diameters increase as the specimen is compressed; and with poor or no lubrication, friction is high, and the inner diameter decreases. The direction of barreling depends on the relative motion of the cylindrical surfaces with respect to the flat dies. (b) Test results: (1) original specimen, and (2–4) the specimen under increasing friction. *Source*: A. T. Male and M. G. Cockcroft.

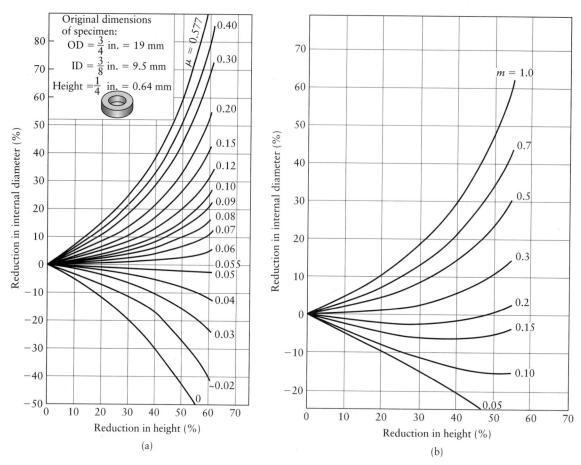

**FIGURE 4.8** Charts to determine friction in ring compression tests: (a) coefficient of friction, $\mu$; (b) friction factor, $m$. Friction is determined from these charts from the percent reduction in height and by measuring the percent change in the internal diameter of the specimen after compression.

If friction at the interfaces is zero, both the inner and outer diameters of the ring expand as if the ring were a solid disk; with increasing friction, the inner diameter of the ring becomes increasingly smaller.

For a particular reduction in height, there is a critical friction value at which the internal diameter increases (from the original) if $\mu$ is low and decreases if $\mu$ is high. By measuring the change in the specimen's internal diameter and using the curves shown in Fig. 4.8, which are obtained through theoretical analyses, we can determine the coefficient of friction. The geometry of each ring has its own specific set of curves; the most common geometry has specimen proportions of outer diameter to inner diameter to height of 6:3:2. The actual size of the specimen usually is not relevant in these tests; thus, once the percentage reductions in internal diameter and height are known, $\mu$ can be determined from the appropriate chart.

**Temperature rise caused by friction.** Almost all of the energy dissipated in overcoming friction is converted into heat; a small portion remains in the bodies as stored energy, and some of it goes into surface energy during generation of new surfaces and wear particles. Frictional heat raises the interface temperature. The magnitude of this temperature rise and its distribution depend not only on the friction force, but also on speed, surface roughness, and the physical properties of the materials, such as thermal conductivity and specific heat. The interface temperature increases with increasing friction and speed and with low thermal conductivity and specific heat of the materials. The interface temperature can be high enough to soften and melt the surface. (The temperature cannot, of course, exceed the melting point of the material.) A number of analytical expressions have been obtained to calculate the temperature rise in sliding; these expressions have been derived by methods including studies on the flash temperatures developed when a pair of asperities are sheared. The flash temperatures are highly localized and are much higher than the overall surface temperature during sliding.

**Reducing friction.** Friction can be reduced by selecting materials that exhibit low adhesion, such as relatively hard materials, like carbides and ceramics, and by using surface films and coatings. Lubricants, such as oils, or solid films, such as graphite, interpose an adherent film between tool, die, and workpiece. The film minimizes adhesion of one surface to the other and interactions between surfaces, thus reducing friction. Abrasive friction can be reduced by lowering the surface roughness of the harder of the two mating surfaces; a smooth surface inherently has less aggressive asperity peaks, and therefore, plowing will be less of a concern. However, it should be noted that adhesive friction is more of a concern with smooth surfaces. Friction can also be reduced significantly by subjecting the die–workpiece interface to **ultrasonic vibrations**, generally at 20 kHz. Depending on the amplitude, these vibrations momentarily separate the die and the workpiece, thus allowing the lubricant to flow more freely into the interface.

**Friction in plastics and ceramics.** Although their strength is low compared with that of metals, plastics (see Chapter 10) generally possess low frictional characteristics. This property makes polymers attractive for applications such as bearings, gears, seals, prosthetic joints, and general low-friction applications. In fact, polymers are sometimes called **self-lubricating** materials. The factors involved in metal friction are also generally applicable to polymers. In sliding, the plowing component of friction in thermoplastics and elastomers is a significant factor, because of their viscoelastic behavior (i.e., they exhibit both viscous and elastic behavior) and subsequent *hysteresis loss*.

An important factor in plastics applications is the effect of temperature rise at the sliding interfaces caused by friction. Thermoplastics lose their strength and become soft as temperature increases. Their low thermal conductivity and low melting points are thus significant in terms of heat generation by friction. If the temperature rise is not controlled, sliding surfaces can undergo plastic deformation and thermal degradation.

In view of the importance of ceramics (Section 11.9), their frictional behavior is now being studied intensively. Initial investigations indicate that the origin of friction in ceramics is similar to that in metals; thus, adhesion and plowing at interfaces contribute to the friction force in ceramics.

---

**EXAMPLE 4.1** Determining the coefficient of friction

In a ring compression test, a specimen 10 mm in height with outside diameter (OD) = 30 mm and inside diameter (ID) = 15 mm is reduced in thickness by 50%. Determine the coefficient of friction, $\mu$, and the friction factor, $m$, if the OD after deformation is 39 mm.

**Solution.** We first have to determine the new ID; this diameter is obtained from volume constancy, as follows:

$$\text{Volume} = \frac{\pi}{4}(30^2 - 15^2)10 = \frac{\pi}{4}(39^2 - \text{ID}^2)5.$$

From this equation, we find that the new ID = 13 mm. Thus,

$$\text{Change in ID} = \frac{15 - 13}{15} \times 100\% = 13\% \text{ (decrease)}.$$

For a 50% reduction in height and a 13% reduction in ID, the following values are interpolated from Fig. 4.8:

$$\mu = 0.09 \quad \text{and} \quad m = 0.4.$$

### 4.4.2 Wear

*Wear* is defined as the progressive loss or unwanted removal of material from a surface. Wear has important technological and economic significance, because it changes the shape of the workpiece, tool, and die interfaces; hence, it affects the process and the size and quality of the parts produced. The magnitude of wear is evident in the countless numbers of parts and components that continually have to be replaced or repaired. Examples of wear in manufacturing processes include dull drills that have to be reground, worn cutting tools that have to be indexed or resharpened, and forming tools and dies that have to be repaired or replaced. **Wear plates**, which are subjected to high loads, are important components in some metalworking machinery. These plates, also known as *wear parts*, because they are expected to wear, can easily be replaced.

Although wear generally alters the surface topography and can result in severe surface damage, it also has a beneficial effect: it can reduce surface roughness by removing the peaks from asperities (Fig. 4.9). Thus, under controlled conditions, wear may be regarded as a kind of smoothening or polishing process; the **running-in period** for various machines and engines produces this type of wear. Wear is usually classified into the following categories:

**FIGURE 4.9**  Changes in originally (a) wire-brushed and (b) ground-surface profiles after wear. *Source*: E. Wild and K. J. Mack.

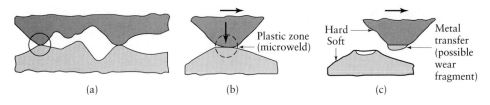

**FIGURE 4.10** Schematic illustration of (a) two asperities contacting, (b) adhesion between two asperities, and (c) the formation of a wear particle.

**Adhesive wear.** If a tangential force is applied to the model shown in Fig. 4.5, shearing can take place either at the original interface or along a path below or above it (Fig. 4.10), causing *adhesive wear*. The fracture path depends on whether or not the strength of the adhesive bond of the asperities is greater than the cohesive strength of either of the two sliding bodies.

Because of factors such as strain hardening at the asperity contact, diffusion, and mutual solid solubility of the two bodies, the adhesive bonds are often stronger than the base metals. Thus, during sliding, fracture at the asperity usually follows a path in the weaker or softer component. A *wear fragment* is then generated. Although this fragment is attached to the harder component (the upper member in Fig. 4.10), it eventually becomes detached during further rubbing at the interface and develops into a loose wear particle. This process is known as *adhesive wear*, or *sliding wear*. In more severe cases, such as with high loads and strongly bonded asperities, adhesive wear is described as *scuffing, smearing, tearing, galling*, or *seizure*. Oxide layers on surfaces greatly influence adhesive wear; they can act as a protective film, resulting in what is known as **mild wear**, consisting of small wear particles.

Based on the probability that a junction between two sliding surfaces will lead to the formation of a wear particle, the **Archard wear law** provides an expression for adhesive wear,

$$V = k\frac{LW}{3p}, \tag{4.6}$$

where $V$ is the volume of material removed by wear from the surface, $k$ is the *wear coefficient* (dimensionless), $L$ is the length of travel, $W$ is the normal load, and $p$ is the indentation hardness of the softer body. In some references, the factor 3 is deleted from the denominator and incorporated into the wear coefficient $k$.

Table 4.2 lists typical values of $k$ for a combination of materials sliding in air. The wear coefficient for the same pair of materials can vary by a factor of 3, depending on whether wear is measured as a loose particle or as a transferred particle (to the other body). Loose particles have a lower $k$ value. Likewise, when mutual solubility is a significant parameter in adhesion, similar metal pairs have higher $k$ values than dissimilar pairs.

**TABLE 4.2**

| Approximate Order of Magnitude for the Wear Coefficient $k$ in Air | | | |
|---|---|---|---|
| Unlubricated | $k$ | Lubricated | $k$ |
| Mild steel on mild steel | $10^{-2}$ to $10^{-3}$ | 52100 steel on 52100 steel | $10^{-7}$ to $10^{-10}$ |
| 60–40 brass on hardened tool steel | $10^{-3}$ | Aluminum bronze on hardened steel | $10^{-8}$ |
| Hardened tool steel on hardened tool steel | $10^{-4}$ | Hardened steel on hardened steel | $10^{-9}$ |
| Polytetrafluoroethylene (PTFE) on tool steel | $10^{-5}$ | | |
| Tungsten carbide on mild steel | $10^{-6}$ | | |

Thus far, we have based the treatment of adhesive wear on the assumption that the surface layers of the two contacting bodies are clean and free from contaminants. Under these conditions, the adhesive-wear rate can be very high, in which case it is known as **severe wear**. As described in Section 4.2, however, metal surfaces are almost always covered with oxide layers, with thicknesses generally ranging between 10 nm and 100 nm. Although this thickness may at first be regarded as insignificant, the oxide layer has a profound effect on wear behavior.

Oxide layers are usually hard and brittle. When such a surface is subjected to rubbing, the layer can have the following effects: if the load is light and the oxide layer is strongly adhering to the bulk metal, the strength of the junctions between the asperities is weak, and wear is low. In this situation, the oxide layer acts as a protective film, and the type of wear is known as **mild wear**.

The oxide layer can be broken up under high normal loads if the layer is brittle and is not strongly adhering to the bulk metal, or if the asperities are rubbing against each other repeatedly, so that the oxide layer breaks up by fatigue. However, if the surfaces are smooth, the respective oxide layers are more difficult to break up. When one of the layers is eventually broken up, a wear particle is formed. An upper asperity can then form a strong junction with the lower asperity, which is now unprotected by the oxide layer; the wear rate will now be higher until a fresh oxide layer is formed.

In addition, a surface exposed to the environment is covered with adsorbed layers of gas and usually with other contaminants. (See Fig. 4.1.) Such films, even if they are very thin, generally weaken the interfacial bond strength of contacting asperities by separating the metal surfaces. The net effect of these films is to provide a protective layer by reducing adhesion and thus reducing wear. The magnitude of the effect of adsorbed gases and contaminants depends on many factors. For instance, it has been shown repeatedly that even small differences in humidity can have a profound influence on the wear rate.

---

**EXAMPLE 4.2  Adhesive wear in sliding**

The end of a rod made of 60–40 brass is sliding over the unlubricated surface of hardened tool steel with a load of 200 lb. The hardness of brass is 120 HB. What is the distance traveled to produce a wear volume of 0.001 in$^3$ by adhesive wear of the brass rod?

**Solution.**  The parameters in Eq. (4.6) for adhesive wear are as follows:

$$V = 0.001 \text{ in}^3;$$
$$k = 10^{-3} \text{ (from Table 4.2)};$$
$$W = 200 \text{ lb};$$
$$p = 120 \text{ kg/mm}^2 = 170{,}700 \text{ lb/in}^2.$$

Therefore, the distance traveled is

$$L = \frac{3Vp}{kW} = \frac{(3)(0.001)(170{,}700)}{(10^{-3})(200)} = 2560 \text{ in.} = 213 \text{ ft.}$$

---

**Abrasive wear.**  *Abrasive wear* is caused by a hard and rough surface, or a surface containing hard protruding particles, sliding along another surface. This type of wear removes particles by forming microchips or *slivers*, thereby producing grooves or

**FIGURE 4.11**   Schematic illustration of abrasive wear in sliding. Longitudinal scratches on a surface usually indicate abrasive wear.

scratches on the softer surface (Fig. 4.11). In fact, the abrasive processes described in Chapter 9, such as grinding, ultrasonic machining, and abrasive-jet machining, act in this manner. The difference is that in those operations, we control the process parameters to produce desired shapes and surfaces, whereas abrasive wear is unintended and unwanted.

The abrasive wear resistance of pure metals and ceramics is directly proportional to their hardness. Abrasive wear can thus be reduced by increasing the hardness of materials (such as by heat treating and microstructural changes) or by reducing the normal load. Other materials that resist abrasive wear are elastomers and rubbers, because they deform elastically and then recover when abrasive particles cross over their surfaces. The best example is automobile tires, which have long lives even though they are operated on abrasive road surfaces; even hardened steels would not last long under such conditions.

There are two basic types of abrasive wear. **Two-body wear** is the basis of **erosive wear**. The abrasive particles (such as sand) are usually carried in a jet of fluid or air, and they remove material from the surface by erosion. In **three-body wear**, the lubricant between the die and the workpiece may carry with it wear particles (generated over a period of time) and cause abrasive wear. Three-body wear is particularly important in metal-forming processes; thus, the necessity for proper inspection and filtering of the metalworking fluid is apparent. There is also the possibility that particles (from machining, grinding, the environment, etc.) may contaminate the lubricating system of the machine components and thus scratch and damage the surfaces.

**Corrosive wear.**   *Corrosive wear*, also known as **oxidation**, or **chemical**, **wear**, is caused by chemical or electrochemical reactions between the surfaces and the environment. The fine corrosive products on the surface constitute the wear particles. When the corrosive layer is destroyed or removed, as by sliding or abrasion, another layer begins to form, and the process of removal and corrosive-layer formation is repeated. Among corrosive media are water (including seawater); oxygen, acids and chemicals, and atmospheric hydrogen sulfide and sulfur dioxide. Corrosive wear can be reduced by selecting materials that will resist environmental attack, controlling the environment, and reducing operating temperatures to lower the rate of chemical reaction.

**Fatigue wear.**   *Fatigue wear*, also called **surface fatigue**, or **surface fracture**, **wear**, is caused when the surface of a material is subjected to cyclic loading, such as in rolling contact in bearings. The wear particles are usually formed by **spalling** or **pitting**. Another type of fatigue wear is **thermal fatigue**. In thermal fatigue, cracks on the surface are generated by thermal stresses from thermal cycling, such as a cool die repeatedly contacting hot workpieces (*heat checking*). These cracks then join, and the surface begins to spall, producing fatigue wear. This type of wear usually occurs in hot-working and die-casting dies. Fatigue wear can be reduced by lowering contact stresses; reducing thermal cycling; and improving the quality of materials by removing impurities, inclusions, and various other flaws that may act as local points for crack initiation.

**FIGURE 4.12** Types of wear observed in a single die used for hot forging. *Source:* T. A. Dean.

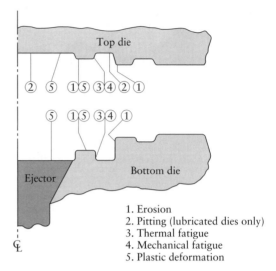

1. Erosion
2. Pitting (lubricated dies only)
3. Thermal fatigue
4. Mechanical fatigue
5. Plastic deformation

**Other types of wear.** Several other types of wear are important in manufacturing processes. **Erosion** is caused by loose abrasive particles abrading a surface. **Fretting** corrosion occurs at interfaces that are subjected to very small movements, such as in machinery. **Impact wear** is the removal of small amounts of materials from a surface by impacting particles. Deburring by vibratory finishing and tumbling (see Section 9.9) is an example of impact wear. In many cases, component wear is the result of a combination of different types of wear. Note in Fig. 4.12, for example, that even in the same forging die, various types of wear take place in different locations. A similar situation exists in cutting tools. (See Fig. 8.24.)

**Wear of plastics and ceramics.** The wear behavior of plastics is similar to that of metals; thus, wear may occur in ways similar to those described earlier. Abrasive-wear behavior depends partly on the ability of the polymer to deform and recover elastically, similar to the behavior of elastomers. There is evidence that the parameter describing this behavior may be the ratio of hardness to elastic modulus; thus, the polymer's resistance to abrasive wear increases as this ratio increases.

Typical polymers with good wear resistance include polyimides, nylons, polycarbonate, polypropylene, acetals, and high-density polyethylene (Chapter 10). These polymers are molded or machined to make gears, pulleys, sprockets, and similar mechanical components. Because plastics can be made with a wide variety of compositions, they can also be blended with internal lubricants, such as PTFE, silicon, graphite, and molybdenum disulfide, and with rubber particles that are interspersed within the polymer matrix.

Wear resistance of reinforced plastics (Section 10.9) depends on the type, amount, and direction of reinforcement in the polymer matrix. Carbon, glass, and aramid fibers all improve wear resistance. Wear takes place when fibers are pulled out of the matrix (*fiber pullout*). Wear is highest when the sliding direction is parallel to the fibers, because the fibers can be pulled out more easily in this direction. Long fibers increase the wear resistance of composites, because such fibers are more difficult to pull out, and cracks in the matrix cannot propagate to the surface as easily.

When ceramics slide against metals, wear is caused by small-scale plastic deformation and brittle surface fracture, surface chemical reactions, plowing, and possibly some fatigue. Metals can be transferred to the oxide-type ceramics surfaces, forming metal oxides; thus, sliding actually takes place between the metal and the metal-oxide surface. Conventional lubricants do not appear to influence ceramics wear significantly.

**Measuring wear.**  Several methods can be used to observe and measure wear. The choice of a particular method depends on the accuracy desired and the physical constraints of the system, such as specimen or part size and difficulty of disassembly. Although not quantitative, the simplest method is visual and tactile (touching) inspection. Measuring dimensional changes, gaging the worn component, profilometry, and weighing are more accurate methods. However, for large workpieces or tools and dies, the weighing method is not accurate, because the amount of wear is usually very small compared with the overall weight of the components involved. Performance and noise level can also be monitored; worn machinery components emit more noise than new parts.

Radiography is a method in which wear particles from an irradiated surface are transferred to the mating surface, whose amount of radiation is then measured. An example is the transfer of wear particles from irradiated cutting tools to the back side of chips. In other situations, the lubricant can be analyzed for wear particles (spectroscopy); this method is precise and is used widely for applications such as checking jet-engine component wear.

**General observations.**  It is apparent from the foregoing discussions that wear is a complex phenomenon and that each type of wear is influenced by several factors. Numerous theories have been advanced to explain and interpret the experimental data and observations made regarding wear. Because of the many factors involved, quantifying wear data is difficult; in fact, it is well known that wear data are difficult to reproduce.

The most significant aspects of wear are the nature and geometry of the contacting asperities. The strength of the bond at the asperity junction relative to the cohesive strength of the weaker body is an important factor. When a polymer such as PTFE (*Teflon*) is rubbed against steel, the surface of the polymer (which is much weaker than the steel) becomes embedded with steel wear particles. Also, when we rub gold with our fingers, wear particles of gold remain on our fingers; the mechanisms for this phenomenon are not clear.

The question also arises as to whether friction and wear for a pair of metals are directly related. Although, at first, one would expect such a relationship, experimental evidence for unlubricated metal pairs indicates that this is not always the case. These results are not surprising, however, in view of the many factors involved in friction and wear. For instance, ball bearings and races have very little friction, yet they undergo wear after a large number of cycles.

### 4.4.3  Lubrication

In manufacturing processes, the interface between tools, dies, and workpieces is subjected to a wide range of variables. Among the major variables are the following:

1. **Contact pressure,** ranging from low values of elastic stresses to multiples of the yield stress of the workpiece material;

2. **Speed,** ranging from very low (such as in some superplastic forming operations) to very high speeds (such as in explosive forming, thin-wire drawing, grinding, and high-speed metal cutting);

3. **Temperature,** ranging from ambient to almost melting, such as in hot extrusion and squeeze casting.

If two surfaces slide against each other under high pressure, speed, and/or temperature, and with no protective layers at the interface, friction and wear will be high. To reduce friction and wear, surfaces should be held as far apart as possible; this is generally done by using metalworking lubricants, which may be solid, semisolid, or liquid in nature. Metalworking lubricants are not only fluids or solids with certain desirable physical properties, such as viscosity, but also are chemicals that can react with the surfaces of the tools, dies, and workpieces and alter their properties.

**Lubrication regimes.** Four regimes of lubrication are relevant to metalworking processes (Fig. 4.13):

1. **Thick film**: The metal surfaces are completely separated by a fluid film that has a thickness of about one order of magnitude greater than that of the surface roughness; thus, there is no metal-to-metal contact. The normal load is light and is supported by the **hydrodynamic fluid film** formed by the wedge effect caused by the relative velocity of the two bodies and the viscosity of the fluid. The coefficient of friction is very low, usually ranging between 0.001 and 0.02, and wear is practically nonexistent.

2. **Thin film**: As the normal load increases or as the speed and viscosity of the fluid decrease (caused by, say, a rise in temperature), the thickness of the film is reduced to between 3 and 10 times that of the surface roughness. There may be some metal-to-metal contact at the higher asperities; this contact increases friction and leads to slightly increased wear.

3. **Mixed**: In this situation, a significant portion of the load is carried by the metal-to-metal contact of the asperities, and the rest of the load is carried by the pressurized fluid film present in hydrodynamic pockets, such as in the valleys of asperities. The average film thickness is less than three times that of the surface roughness. With proper selection of lubricants, a strongly adhering *boundary film* a few molecules thick can be formed on the surfaces; this film prevents direct metal-to-metal contact and thus reduces wear. Depending on the strength of the boundary film and other parameters, the friction coefficient in mixed lubrication may range up to about 0.4.

4. **Boundary lubrication**: The load here is supported by the contacting surfaces, which are covered with a *boundary layer*. A lubricant may or may not be present in the surface valleys, but it is not pressurized enough to support sig-

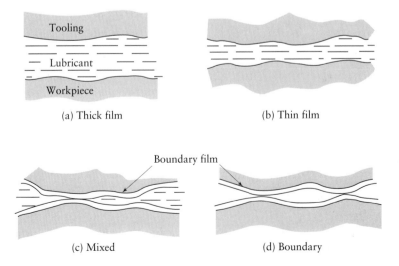

**FIGURE 4.13** Types of lubrication generally occurring in metalworking operations. *Source*: After W. R. D. Wilson.

Tooling

Lubricant

Workpiece

(a) Thick film

(b) Thin film

Boundary film

(c) Mixed

(d) Boundary

nificant load. Depending on the boundary film thickness and its strength, the friction coefficient ranges from about 0.1 to 0.4. Wear can be relatively high if the boundary layer is destroyed.

Typical *boundary lubricants* are natural oils, fats, fatty acids, and soaps. Boundary films form rapidly on metal surfaces. As the film thickness decreases and metal-to-metal contact takes place, the chemical aspects and roughness of surfaces become significant. (In thick-film lubrication, the viscosity of the lubricant is the important parameter in controlling friction and wear; chemical aspects are not particularly significant, except as they affect corrosion and staining of metal surfaces.) A boundary film can *break down*, or it can be removed by being disturbed or rubbed off during sliding, or because of *desorption* due to high temperatures at the interface. The metal surfaces may thus be deprived of this protective layer. The clean metal surfaces then contact each other, and, as a consequence, severe wear and scoring can occur; thus, adherence of boundary films is an important aspect in lubrication.

**Surface roughness and geometric effects.**  Surface roughness, particularly in mixed lubrication, is important; roughness can serve to create local reservoirs or pockets for lubricants. The lubricant can be trapped in the valleys of the surface, and, because these fluids are incompressible, they can support a substantial portion of the normal load. The pockets also supply lubricant to regions where the boundary layer has been destroyed. There is an optimal roughness for lubricant entrapment. In metalworking operations, it is generally desirable for the workpiece, not the die, to have the rougher surface. Otherwise, the workpiece surface will be damaged by the rougher and harder die surface. The recommended surface roughness on most dies is about 0.40 $\mu$m (15 $\mu$in).

In addition to surface roughness, the overall *geometry* of the interacting bodies is an important consideration in lubrication. Figure 4.14 shows the typical metalworking processes of drawing, extrusion, and rolling. The movement of the workpiece into the deformation zone must allow a supply of the lubricant to be carried into the die–workpiece interfaces. Analysis of hydrodynamic lubrication shows that the *inlet angle* is an important parameter. As this angle decreases, more lubricant is entrained; thus, lubrication is improved, and friction is reduced. With proper selection of parameters, a relatively thick film of lubricant can be maintained at the die–workpiece interface.

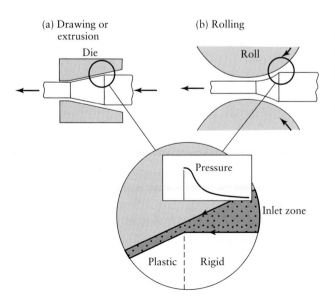

**FIGURE 4.14**  Inlet zones for entrainment of lubricants in drawing, extrusion, and rolling. Entrainment of lubricants is important in reducing friction and wear in metalworking operations. *Source*: After W. R. D. Wilson.

The reduction of friction and wear by *entrainment* or *entrapment of lubricants* may not always be desirable. The reason is that with a thick fluid film, the metal surface cannot come into full contact with the die surfaces and hence does not acquire the shiny appearance that is generally required on the product. (Note, for instance, the shiny appearance of a piece of copper wire.) A thick film of lubricant generates a grainy, dull surface on the workpiece, the degree of roughness depending on the grain size of the material. In operations such as coining and forging (Section 6.2), trapped lubricants are undesirable, because they interfere with the forming process and prevent precise shape generation.

### 4.4.4 Metalworking fluids

On the basis of the foregoing discussions, the functions of metalworking fluids can be summarized as follows:

- Reduce *friction*, thus reducing force and energy requirements and preventing increases in temperature;
- Reduce *wear, seizure,* and *galling*;
- Improve *material flow* in tools, dies, and molds;
- Act as a *thermal barrier* between the workpiece and tool and die surfaces, thus preventing workpiece cooling in hot-working processes;
- Act as a *release* or *parting agent* to help in the removal or ejection of parts from dies and molds.

Several types of metalworking fluids are now available to fulfill these requirements. Because of the lubricants' diverse chemistries, properties, and characteristics, the behavior and performance of lubricants can be complex. The general properties of the most commonly used lubricants are described as follows:

1. **Oils** have high film strength on the surface of a metal, as you well know if you have ever tried to clean an oily surface. Although they are very effective in reducing friction and wear, oils have low thermal conductivity and specific heat; thus, they are not effective in conducting away the heat generated by friction and plastic deformation in metalworking operations. Oils also are difficult to remove from component surfaces that subsequently are to be painted or welded, and they are difficult to dispose of. The sources of oils are **mineral** (petroleum), **animal, vegetable,** and **fish**. Oils may be **compounded** with a variety of additives or with other oils to impart special properties, such as lubricity, viscosity, detergents for cleaning ability, biocides, etc.

2. Metalworking fluids are usually blended with several **additives,** including oxidation inhibitors, rust preventatives, odor control agents, antiseptics, and foam inhibitors. Important additives in oils are sulfur, chlorine, and phosphorus. Known as **extreme-pressure (EP) additives** and used singly or in combination, they react chemically with metal surfaces and form adherent surface films of metallic sulfides and chlorides. These films have low shear strength and good antiweld properties and thus effectively reduce friction and wear. While EP additives are important in boundary lubrication, these lubricants may be undesirable in some cases. For example, they may preferentially attack the cobalt binder in tungsten carbide tools and dies (Section 8.6), causing undesirable changes in their surface roughness and integrity.

3. An **emulsion** is a mixture of two immiscible liquids, usually mixtures of oil and water in various proportions, along with additives. A typical emulsion will be

provided as a mixture of base oil and additives and then mixed with water by the user. Emulsions are of two types: direct and indirect. In a direct emulsion, oil is dispersed in water as very small droplets, with average diameters greater than 0.1 $\mu$m (4 $\mu$in.), but usually less than 30 $\mu$m (1200 $\mu$in.) in practice. In an indirect emulsion, water droplets are dispersed in the oil. Direct emulsions are important fluids, because the presence of water (usually 95% of the emulsion by volume) gives them high cooling capacity, yet they still provide good lubrication. They are particularly effective in high-speed metal-forming and cutting operations, where increases in temperature have detrimental effects on tool life, workpiece surface integrity, and dimensional accuracy.

4. **Synthetic solutions** are fluids that contain inorganic and other chemicals dissolved in water. Various chemical agents are added to impart different properties. **Semisynthetic solutions** are basically synthetic solutions to which small amounts of emulsifiable oils have been added.

5. **Soaps** are generally reaction products of sodium or potassium salts with fatty acids. Alkali soaps are soluble in water, but other metal soaps are generally insoluble. Soaps are effective boundary lubricants and can also form thick film layers at die–workpiece interfaces, particularly when applied on conversion coatings (see later) for cold metalworking applications.

6. **Greases** are solid or semisolid lubricants and generally consist of soaps, mineral oil, and various additives. They are highly viscous and adhere well to metal surfaces. Although used extensively in machinery, greases have limited use in manufacturing processes.

7. **Waxes** may be of animal or plant (*paraffin*) origin and have complex structures. Compared with greases, waxes are less "greasy" and are more brittle. They have limited use in metalworking operations, except for copper and, as chlorinated paraffin, for stainless steels and high-temperature alloys.

**Solid lubricants.** Because of their unique properties and characteristics, several solid materials are used as lubricants in manufacturing operations:

1. **Graphite.** The properties of *graphite* are described in Section 11.14. Graphite is weak in shear along its layers and has a low coefficient of friction in that direction; thus, it can be a good solid lubricant, particularly at elevated temperatures. However, its friction is low only in the presence of air or moisture. In a vacuum or an inert-gas atmosphere, its friction is very high; in fact, graphite can be quite abrasive in these environments. Graphite may be applied either by rubbing it on surfaces or as a *colloidal* (dispersion of small particles) suspension in liquid carriers, such as water, oil, or alcohols.

   A more recent development is the production of soccer-ball-shaped carbon molecules, called **buckyballs** or **fullerenes** (both after Buckminster Fuller). These chemically inert spherical molecules are produced from soot and act much like solid lubricant particles.

2. **Molybdenum disulfide.** *Molybdenum disulfide* ($MoS_2$), another widely used lamellar solid lubricant, is somewhat similar in appearance to graphite. However, unlike graphite, it has a high friction coefficient in an ambient environment. Oils are commonly used as carriers for molybdenum disulfide, which is used as a lubricant at room temperature. Molybdenum disulfide can also be rubbed onto the surfaces of a workpiece.

3. **Soft metals and polymer coatings.** Because of the low strength of *soft metals* and *polymer coatings*, thin layers of them are used as solid lubricants. Suitable metals are lead, indium, cadmium, tin, and silver, and suitable polymers

include PTFE, polyethylene, and methacrylates. However, these coatings have limited applications, because of their lack of strength under high stresses and at elevated temperatures. Soft metals are used to coat high-strength metals such as steels, stainless steels, and high-temperature alloys. Copper or tin, for example, is chemically deposited on the surface before the metal is processed. If the oxide of a particular metal has low friction and is sufficiently thin, the oxide layer can serve as a solid lubricant, particularly at elevated temperatures.

4. **Glass.** Although a solid material, *glass* becomes viscous at high temperatures and hence can serve as a liquid lubricant. Its viscosity is a function of temperature, but not of pressure, and depends on the type of glass. Poor thermal conductivity also makes glass attractive, since it acts as a thermal barrier between hot workpieces and relatively cool dies. Typical glass-lubrication applications include hot extrusion (Section 6.4) and forging (Section 6.2).

**Lubricant selection.** Selecting a lubricant for a particular manufacturing process and workpiece material involves consideration of several factors:

- The particular manufacturing process;
- Compatibility of the lubricant with the workpiece and tool and die materials;
- The surface preparation required;
- The method of the lubricant application;
- Removal of the lubricant after processing;
- Contamination of the lubricant by other lubricants, such as those used to lubricate the machinery;
- Treatment of waste lubricant;
- Storage and maintenance of lubricants;
- Biological and ecological considerations;
- Costs involved in all of the foregoing aspects.

In selecting an oil as a lubricant, the importance of its viscosity-temperature-pressure characteristics should be recognized. Low viscosity can have a significant detrimental effect on friction and wear. The different functions of a metalworking fluid, whether primarily a lubricant or a coolant, must also be taken into account. Water-base fluids are very effective coolants, but as lubricants are not as effective as oils.

Metalworking fluids should not leave any harmful residues that could interfere with machinery operations. The fluids should not stain or corrode workpieces or equipment. The fluids should be checked periodically for deterioration caused by bacterial growth, accumulation of oxides, chips, and wear debris and also for general degradation and breakdown due to temperature and time. A lubricant may carry wear particles within it and cause damage to the system, so proper inspection and filtering of metalworking fluids are important. These precautions are necessary for all types of machinery, such as they are for internal combustion and jet engines.

After completion of manufacturing operations, metal surfaces are usually covered with lubricant residues, which should be removed prior to further workpiece processing such as welding or painting. Various cleaning solutions and techniques can be used for this purpose (section 4.5.2). *Biological and ecological considerations*, with their accompanying health and legal ramifications, are also important. Potential health hazards may be involved in contacting or inhaling some metalworking fluids. Recycling and disposal of waste fluids are additional important factors to be considered.

# 4.5 | Surface Treatments, Coatings, and Cleaning

After a component is manufactured, all or parts of its surfaces may have to be processed further in order to impart certain properties and characteristics. *Surface treatments* may be necessary to

- Improve resistance to wear, erosion, and indentation (e.g., for slideways in machine tools; wear surfaces of machinery; and shafts, rolls, cams, and gears);
- Control friction (sliding surfaces on tools, dies, bearings, and machine ways);
- Reduce adhesion (electrical contacts);
- Improve lubrication (surface modification to retain lubricants);
- Improve resistance to corrosion and oxidation (sheet metals for automotive or other outdoor uses, gas-turbine components, and medical devices);
- Improve fatigue resistance (bearings and shafts with fillets);
- Rebuild surfaces on components (worn tools, dies, and machine components);
- Improve surface roughness (appearance, dimensional accuracy, and frictional characteristics);
- Impart decorative features, color, or special surface texture.

## 4.5.1 Surface treatment processes

Numerous processes are used for surface treatments, based on mechanical, chemical, thermal, and physical methods. Their principles and characteristics are outlined as follows:

1. **Shot peening, water-jet peening, and laser peening.** In *shot peening*, the surface of the workpiece is hit repeatedly with a large number of cast-steel, glass, or ceramic shot (small balls), making overlapping indentations on the surface; this action causes plastic deformation of the surface to depths up to 1.25 mm (0.05 in.), using shot sizes ranging from 0.125 mm to 5 mm (0.005 in. to 0.2 in.) in diameter. Because plastic deformation is not uniform throughout a part's thickness, shot peening imparts compressive residual stresses on the surface, thus improving the fatigue life of the component. This process is used extensively on shafts, gears, springs, oil-well drilling equipment, and jet-engine parts (such as turbines and compressor blades).

   In *water-jet peening*, a water jet at pressures as high as 400 MPa (60 ksi) impinges on the surface of the workpiece, inducing compressive residual stresses to an extent similar to that of shot peening. This method has been used successfully on steels and aluminum alloys.

   In *laser peening*, the surface is subjected to laser shocks from high-powered lasers (up to 1 kW) and at energy levels of 100 J/pulse. This method has been used successfully on jet-engine fan blades and on materials such as titanium and nickel alloys, with compressive surface residual stresses deeper than 1 mm (0.04 in.).

**FIGURE 4.15** Examples of roller burnishing of (a) a conical surface and (b) a flat surface and the burnishing tools used. *Source*: Cogsdill Tool Products.

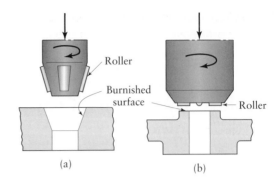

2. **Roller burnishing (surface rolling).** The surface of the component is cold worked by a hard and highly polished roller or series of rollers; this process is used on various flat, cylindrical, or conical surfaces (Fig. 4.15). Roller burnishing improves surface finish by removing scratches, tool marks, and pits. Consequently, corrosion resistance is also improved, since corrosive products and residues cannot be entrapped. Roller burnishing is used to improve the mechanical properties of surfaces, as well as the shape and surface finish of components. It can be used either singly or in combination with other finishing processes, such as grinding, honing, and lapping. Soft and ductile, as well as very hard, metals can be roller burnished. Typical applications of roller burnishing include hydraulic-system components, seals, valves, spindles, and fillets on shafts. Internal cylindrical surfaces are burnished by a similar process, called **ballizing**, or **ball burnishing**, in which a smooth ball slightly larger than the bore diameter is pushed through the length of the hole.

3. **Explosive hardening.** The surface is subjected to high transient pressures by placing a layer of explosive sheet directly on the workpiece surface and detonating it. Large increases in surface hardness can be obtained by this method, with very little change (less than 5%) in the shape of the component. Railroad rail surfaces can be hardened by this method.

4. **Cladding (clad bonding).** Metals are bonded with a thin layer of corrosion-resistant metal by applying pressure with rolls or other means. A typical application is cladding of aluminum (*Alclad*), in which a corrosion-resistant layer of aluminum alloy is clad over pure aluminum; other applications of cladding include steels clad with stainless steel or nickel alloys. The cladding material may also be applied through dies, as in cladding steel wire with copper, or by explosives.

5. **Mechanical plating (mechanical coating, impact plating, peen plating).** Fine particles of metal are compacted over the workpiece surfaces by impacting them with spherical glass, ceramic, or porcelain beads. The beads are propelled by rotary means. This process typically is used for hardened-steel parts for automobiles, with plating thickness usually less than 0.025 mm (0.001 in.).

6. **Case hardening (carburizing, carbonitriding, cyaniding, nitriding, flame hardening, induction hardening).** These processes are described in Section 5.11.3 and summarized in Tables 5.8. In addition to the common heat sources of gas and electricity, laser beams are also used as a heat source in surface hardening of both metals and ceramics. Case hardening, as well as some of the other surface-treatment processes, induces residual stresses on surfaces. The formation of martensite in case hardening of steels causes compressive residual stresses on surfaces. Such stresses are desirable, because they improve the fatigue life of components by delaying the initiation of fatigue cracks.

7. **Hard facing.** A relatively thick layer, edge, or point of wear-resistant hard metal is deposited on the surface by any of the welding techniques described in Chapter 12. A number of layers are usually deposited (*weld overlay*). Hard coatings of tungsten carbide, chromium, and molybdenum carbide can also be deposited using an electric arc (*spark hardening*). Hard-facing alloys are available as electrodes, rods, wire, and powder. Typical applications for hard facing include valve seats, oil-well drilling tools, and dies for hot metalworking. Worn parts are also hard faced for extended use.

8. **Thermal spraying.** In *thermal spraying*, also called **metallizing**, metal in the form of rod, wire, or powder is melted in a stream of oxyacetylene flame, electric arc, or plasma arc, and the droplets are sprayed on the preheated surface at speeds up to 100 m/s (20,000 ft/min) with a compressed-air spray gun. All of the surfaces to be sprayed should be cleaned and roughened to improve bond strength. Typical applications for thermal spraying include steel structures, storage tanks, and tank cars, sprayed with zinc or aluminum up to 0.25 mm (0.010 in.) in thickness.

   There are several types of thermal-spraying processes, listed in decreasing order of performance as follows: (a) **plasma**, either conventional, high energy, or vacuum. It produces temperatures on the order of 8300°C (15,000°F) and has very good bond strength with very low oxide content. (b) **Detonation gun,** in which a controlled explosion takes place using an oxyfuel gas mixture. It has a performance similar to that of plasma. (c) **High-velocity oxyfuel** (HVOF) **gas spraying,** which has a similarly high performance as those of the foregoing processes, but is less expensive. (d) **Wire arc,** in which an arc is formed between two consumable wire electrodes. It has good bond strength and is the least expensive process. (e) **Flame wire spraying,** in which the oxyfuel flame melts the wire and deposits it on the surface. Its bond is of medium strength, but the process is relatively inexpensive.

9. **Surface texturing.** Manufactured surfaces can be further modified by secondary operations for technical, functional, optical, or aesthetic reasons. These additional processes generally consist of (1) etching, using chemicals or sputtering techniques, (2) electric arcs, (3) laser pulses and (4) atomic oxygen, which reacts with surfaces and produces fine, conelike surface textures. The possible adverse effects of any of these processes on the properties and performance of materials should be considered.

10. **Ceramic coating.** Techniques have been developed for spraying ceramic coatings for high-temperature and electrical-resistance applications, such as to withstand repeated arcing. Powders of hard metals and ceramics are also used as spraying materials. Plasma-arc temperatures may reach 15,000°C (27,000°F), which is much higher than that obtained with flames. Typical applications are nozzles for rocket motors and wear-resistant parts.

11. **Vapor deposition.** In this process, the substrate is subjected to chemical reactions by gases that contain chemical compounds of the materials to be deposited; the thickness of the coating is usually a few $\mu$m. The deposited materials may consist of metals, alloys, carbides, nitrides, borides, ceramics, or various oxides. The substrate may be metal, plastic, glass, or paper. Typical applications include coatings for cutting tools, drills, reamers, milling cutters, punches, dies, and wear surfaces.

   There are two major deposition processes: physical vapor deposition and chemical vapor deposition. These techniques allow effective control of coating composition, thickness, and porosity. The three basic types of **physical vapor deposition** (PVD) processes are vacuum or arc evaporation (PV/ARC), sputtering, and ion plating. These processes are carried out in a high vacuum at temperatures in the range of 200°C–500°C (400°F–900°F). In physical vapor deposition, the particles to be deposited are carried physically to the workpiece, rather than by chemical reactions as in chemical vapor deposition.

In **vacuum evaporation**, the metal to be deposited is evaporated at high temperatures in a vacuum and deposited on the substrate, which is usually at room temperature or slightly higher. Uniform coatings can be obtained on complex shapes with this method. In PV/ARC, the coating material (cathode) is evaporated by a number of arc evaporators, using highly localized electric arcs. The arcs produce a highly reactive plasma consisting of ionized vapor of the coating material; the vapor condenses on the substrate (anode) and coats it. Applications for this process may be functional (e.g., oxidation-resistant coatings for high-temperature applications, electronics, and optics) or decorative (e.g., hardware, appliances, and jewelry).

In **sputtering**, an electric field ionizes an inert gas (usually argon). The positive ions bombard the coating material (cathode) and cause sputtering (ejecting) of its atoms; these atoms then condense on the workpiece, which is heated to improve bonding. In **reactive sputtering**, the inert gas is replaced by a reactive gas, such as oxygen, in which case the atoms are oxidized and the oxides are deposited. **Radiofrequency** (RF) **sputtering** is used for nonconductive materials such as electrical insulators and semiconductor devices. **Ion plating** is a generic term describing the combined processes of sputtering and vacuum evaporation. An electric field causes a glow discharge, generating a plasma; the vaporized atoms in this process are only partially ionized. **Dual ion-beam-assisted deposition** is a more recent development in which PVD is combined with simultaneous ion-beam bombardment. It results in good adhesion on metals, ceramics, and polymers; ceramic bearings and dental instruments are two examples of its applications.

**Chemical vapor deposition** (CVD) is a thermochemical process. In a typical application, such as for coating cutting tools with titanium nitride (TiN) (Fig. 4.16), the tools are placed on a graphite tray and heated to 950°C–1050°C (1740°F–1920°F) in an inert atmosphere. Titanium tetrachloride (a vapor), hydrogen, and nitrogen are then introduced into the chamber; the chemical reactions form titanium nitride on the tool's surfaces. For coating with titanium carbide, methane is substituted for the gases. Coatings obtained by chemical vapor deposition are usually thicker than those obtained via PVD. A more recent development is **medium-temperature CVD** (MTCVD), which produces coatings with higher resistance to crack propagation than CVD coatings.

In **ion implantation**, ions are introduced into the surface of the workpiece material. The ions are accelerated in a vacuum to such an extent that they penetrate the substrate to a depth of a few $\mu$m. Ion implantation (not to be confused

**FIGURE 4.16**   Schematic illustration of the chemical vapor deposition process.

with ion plating) modifies surface properties by increasing surface hardness and improving resistance to friction, wear, and corrosion. This process can be controlled accurately, and the surface can be masked to prevent ion implantation in unwanted places. When used in specific applications, such as semiconductors, this process is called **doping** (meaning alloying with small amounts of various elements).

12. **Diffusion coating.** In this process, an alloying element is diffused into the surface of the substrate, thus altering its properties. Such elements can be supplied in solid, liquid, or gaseous states. This process is given different names, depending on the diffused element, that describe diffusion processes, such as carburizing, nitriding, and boronizing. (See Table 5.8.)

13. **Electroplating.** The workpiece (cathode) is plated with a different metal (anode) while both are suspended in a bath containing a water-base electrolyte solution (Fig. 4.17). Although the plating process involves a number of reactions, the basic procedure is that the metal ions from the anode are discharged under the potential from the external source of electricity, combine with the ions in the solution, and are deposited on the cathode. All metals can be electroplated, with thicknesses ranging from a few atomic layers to a maximum of about 0.05 mm (0.002 in.). Complex shapes may have varying plating thicknesses. Common plating materials include chromium, nickel, cadmium, copper, zinc, and tin. **Chromium plating** is carried out by plating the metal first with copper, then with nickel, and finally with chromium. **Hard chromium plating** is done directly on the base metal and has a hardness of up to 70 HRC.

   Typical electroplating applications include copper plating aluminum wire and phenolic boards for printed circuits, chrome plating hardware, tin plating copper electrical terminals for ease of soldering, and plating various components for good appearance and resistance to wear and corrosion. Because they do not develop oxide films, noble metals (such as gold, silver, and platinum; see Section 3.11.8) are important electroplating materials for the electronics and jewelry industries. Plastics such as ABS, polypropylene, polysulfone, polycarbonate, polyester, and nylon also can have metal coatings applied through electroplating. Because they are not electrically conductive, plastics must be preplated by such processes as electroless nickel plating. (See below.) Parts to be coated may be simple or complex, and size is not a limitation.

14. **Electroless plating.** This process is carried out by chemical reactions and without the use of an external source of electricity. The most common application uses nickel, although copper is also used. In electroless nickel plating, nickel

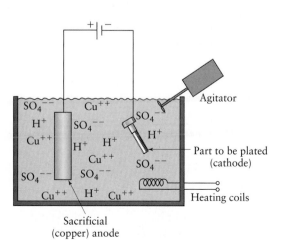

Sacrificial
(copper) anode

**FIGURE 4.17** Schematic illustration of the electroplating process.

chloride (a metallic salt) is reduced (with sodium hypophosphite as the reducing agent) to nickel metal, which is then deposited on the workpiece. The hardness of nickel plating ranges between 425 HV and 575 HV, and the plating can be heat treated to 1000 HV. The coating has excellent wear and corrosion resistance.

15. **Electroforming.** A variation of electroplating, electroforming is actually a metal-fabricating process. Metal is electrodeposited on a *mandrel* (also called a *mold* or *matrix*), which is then removed; thus, the coating itself becomes the product. Simple and complex shapes can be produced by electroforming, with wall thicknesses as small as 0.025 mm (0.001 in.). Parts may weigh from a few grams to as much as 270 kg (600 lb). The electroforming process is particularly suited to low production quantities and intricate parts (such as molds, dies, waveguides, nozzles, and bellows) made of nickel, copper, gold, and silver; it is also suitable for aerospace, electronics, and electro-optics applications. Production rates can be increased with multiple mandrels.

16. **Anodizing.** Anodizing is an oxidation process (*anodic oxidation*) in which the surfaces of the workpiece are converted to a hard and porous oxide layer that provides corrosion resistance and a decorative finish. The workpiece is the anode in an electrolytic cell immersed in an acid bath, resulting in chemical adsorption of oxygen from the bath. Organic dyes of various colors (typically black, red, bronze, gold, and gray) can be used to produce stable, durable surface films. Typical applications for anodizing include aluminum furniture and utensils, architectural shapes, automobile trim, picture frames, keys, and sporting goods. Anodized surfaces also serve as a good base for painting, especially for aluminum, which otherwise is difficult to paint.

17. **Conversion coating.** In this process, also called **chemical-reaction priming**, a coating forms on metal surfaces as a result of chemical or electrochemical reactions. Various metals, particularly steel, aluminum, and zinc, can be conversion coated. Oxides that naturally form on their surfaces are a form of conversion coating; phosphates, chromates, and oxalates are used to produce conversion coatings. These coatings are used for purposes such as prepainting, decorative finishes, and protection against corrosion. The two common methods of coating are immersion and spraying. The equipment involved depends on the method of application, the type of product, and considerations of quality.

    An important application is in conversion coating of workpieces as a lubricant carrier in cold-forming operations (Section 4.4.4). Lubricants may not always adhere properly to workpiece surfaces, particularly when under high normal and shearing stresses; this condition is a problem particularly in forging, extrusion, and wire drawing of steels, stainless steels, and high-temperature alloys. For these applications, acids transform the workpiece surface by chemically reacting with it, leaving a somewhat rough and spongy surface that acts as a carrier for the lubricant. After treatment, borax or lime is used to remove any excess acid from the surfaces. A liquid lubricant, such as a soap, is then applied to the coated surface. The lubricant film adheres to the surface and cannot be scraped off easily. **Zinc phosphate** conversion coatings are often used on carbon and low-alloy steels. **Oxalate** coatings are used for stainless steels and high-temperature alloys.

18. **Coloring.** As the name implies, coloring involves processes that alter the color of metals, alloys, and ceramics. The change is caused by the conversion of surfaces (by chemical, electrochemical, or thermal processes) into chemical compounds, such as oxides, chromates, and phosphates. In **blackening**, iron and steels develop a lustrous black-oxide film, using solutions of hot caustic soda.

19. **Hot dipping.** In this process, the workpiece, usually steel or iron, is dipped into a bath of molten metal, such as zinc (for galvanized-steel sheet and plumbing supplies), tin (for tin plate and tin cans for food containers), aluminum (*aluminizing*), and *terne* (lead alloyed with 10% to 20% tin). Hot-dipped coatings on discrete parts or sheet metal provide galvanized pipe, plumbing supplies, and many other products with long-term resistance to corrosion. The coating thickness is usually given in terms of coating weight per unit surface area of the sheet, typically $150 \, g/m^2 - 900 \, g/m^2$ ($0.5 \, oz/ft^2 - 3 \, oz/ft^2$). The service life depends on the thickness of the zinc coating and the environment to which it is exposed. Various **precoated sheet steels** are used extensively in automobile bodies. Proper draining to remove excess coating materials is an important consideration.

20. **Porcelain enameling.** Metals may be coated with a variety of glassy (vitreous) coatings to provide corrosion and electrical resistance and for service at elevated temperatures. These coatings are usually classified as porcelain enamels and generally include enamels and ceramics (Section 11.8). The word **enamel** is also used for glossy paints, indicating a smooth, hard coating. Porcelain enamels are glassy inorganic coatings consisting of various metal oxides. **Enameling** involves fusing the coating material on the substrate by heating them both to $425°C - 1000°C$ ($800°F - 1800°F$) to liquefy the oxides. Depending on their composition, enamels have varying resistances to alkali, acids, detergents, cleansers, and water and are available in different colors.

    Typical applications for porcelain enameling include household appliances, plumbing fixtures, chemical-processing equipment, signs, cookware, and jewelry; porcelain enamels are also used as protective coatings on jet-engine components. The coating may be applied by dipping, spraying, or electrodeposition, and thicknesses are usually $0.05 \, mm - 0.6 \, mm$ ($0.002 \, in. - 0.025 \, in.$). Metals that are coated with porcelain enamel are typically steels, cast iron, and aluminum. Glasses are used as lining for chemical resistance, and the thickness is much greater than in enameling. **Glazing** is the application of glassy coatings on ceramic and earthenwares to give them decorative finishes and to make them impervious to moisture.

21. **Organic coatings.** Metal surfaces may be coated or *precoated* with a variety of organic coatings, films, and laminates to improve appearance and corrosion resistance. Coatings are applied to the coil stock on continuous lines, with thicknesses generally of $0.0025 \, mm - 0.2 \, mm$ ($0.0001 \, in. - 0.008 \, in.$). Such coatings have a wide range of properties: flexibility, durability, hardness, resistance to abrasion and chemicals, color, texture, and gloss. Coated sheet metal is subsequently formed into various products, such as TV cabinets, appliance housings, paneling, shelving, siding for residential buildings, gutters, and metal furniture.

    More critical applications of organic coatings involve, for example, coatings for naval aircraft that are subjected to high humidity, rain, seawater, pollutants (such as from ship exhaust stacks), aviation fuel, deicing fluids, and battery acid and that are also impacted by particles such as dust, gravel, stones, and deicing salts. For aluminum structures, organic coatings have consisted typically of an epoxy primer and a polyurethane topcoat, with a lifetime of four to six years. Primer performance is very important for coating durability.

22. **Ceramic coatings.** Ceramics such as aluminum oxide and zirconium oxide are applied to a surface at room temperature, usually by thermal-spraying techniques. Such coatings serve as a thermal barrier in applications such as hot extrusion dies, diesel-engine components, and turbine blades.

23. **Painting.** Because of its decorative and functional properties, paint is widely used as a surface coating. Paints are basically classified as enamels, lacquers, and water-base paints, with a wide range of characteristics and applications. They are applied by brushing, dipping, spraying, or electrostatically.

24. **Diamond coating.** Important advances have been made in diamond coating of metals, glass, ceramics, and plastics, using various chemical and plasma-assisted vapor deposition processes and ion-beam enhanced deposition techniques. Techniques have also been developed to produce *free-standing diamond films* on the order of 1 mm (0.040 in.) thick and up to 125 mm (5 in.) in diameter, including smooth and optically clear diamond films. Development of these techniques, combined with important properties of diamonds, such as hardness, wear resistance, high thermal conductivity, and transparency to ultraviolet light and microwave frequencies, has enabled the production of various aerospace and electronic parts and components.

    Examples of diamond-coated products include scratchproof windows (such as for aircraft and missile sensors to protect against sandstorms), sunglasses, cutting tools (such as drills and end mills), calipers, surgical knives, razors, electronic and infrared heat seekers and sensors, light-emitting diodes, speakers for stereo systems, turbine blades, and fuel-injection nozzles. Studies are continuing on growing diamond films on crystalline copper substrate by implantation of carbon ions; an important application for such diamond films is in making computer chips. Diamond can be doped to form *p*- and *n*-type ends on semiconductors to make transistors (Chapter 13), and its high thermal conductivity allows closer packing of chips than with silicon or gallium–arsenide chips, thus significantly increasing the speed of computers.

25. **Diamond-like carbon** (DLC). By using a low-temperature, ion-beam-assisted deposition process, this relatively recently developed material is applied as a coating of a few nanometers in thickness. Less expensive than diamond films, but with similar properties as diamond, DLC has applications in such areas as tools and dies, gears, bearings, microelectromechanical systems, and microscale probes.

### 4.5.2 Cleaning of surfaces

We have stressed the importance of surfaces and the influence of deposited or adsorbed layers of various elements and contaminants on surfaces. A clean surface can have both beneficial and detrimental effects. Although an unclean surface may reduce the tendency for adhesion and galling, cleanliness generally is essential for more effective application of metalworking fluids, coating and painting, adhesive bonding, welding, brazing, soldering, reliable functioning of manufactured parts in machinery, manufacturing of food and beverage containers, effective storage, and in assembly operations.

*Cleaning* involves the removal of solid, semisolid, or liquid contaminants from a surface, and it is an important part of manufacturing operations and the economics of production. The word *clean*, or the degree of cleanliness of a surface, is somewhat difficult to define; how, for example, would you test the cleanliness of a dinner plate? Two simple and common tests are based on the following procedures:

1. Wiping the area with a clean cloth and observing any *residues* on the cloth; or

2. Observing whether water continuously coats the surface (called the *waterbreak test*). If water collects as individual droplets, the surface is not clean. Test this phenomenon yourself by wetting dinner plates that have been cleaned to varying degrees.

The type of cleaning process required depends on the type of contaminants to be removed. *Contaminants*, also called *soils*, may consist of rust, scale, chips and other metallic and nonmetallic debris, metalworking fluids, solid lubricants, pigments, polishing and lapping compounds, and general environmental elements.

Basically, there are two types of cleaning methods: mechanical and chemical. **Mechanical cleaning methods** consist of physically disturbing the contaminants, as with wire or fiber brushing, dry or wet abrasive blasting, tumbling, and steam jets. Many of these processes are particularly effective in removing rust, scale, and other solid contaminants. Ultrasonic cleaning may also be placed in this category.

**Chemical cleaning methods** usually involve the removal of oil and grease from surfaces. They consist of one or more of the following processes:

- *Solution:* The soil dissolves in the cleaning solution.
- *Saponification:* A chemical reaction that converts animal or vegetable oils into a soap that is soluble in water.
- *Emulsification:* The cleaning solution reacts with the soil or lubricant deposits and forms an emulsion; the soil and the emulsifier then become suspended in the emulsion.
- *Dispersion:* The concentration of soil on the surface is decreased by surface-active materials in the cleaning solution.
- *Aggregation:* Lubricants are removed from the surface by various agents in the cleaning fluid and collect as large dirt particles.

Some common **cleaning fluids** are used in conjunction with electrochemical processes for more effective cleaning. These fluids include *alkaline solutions, emulsions, solvents, hot vapors, acids, salts*, and mixtures of *organic compounds*.

In **vapor degreasing**, a heated, evaporated solvent collects on workpieces that are at room temperature. Dirt and lubricant on the workpiece are removed and dispersed, emulsified, or dissolved in the solvent. As the solvent drips off the workpiece into the solvent tank, the contaminants are removed. An advantage of vapor degreasing is that the solvent does not become dilute, since the vapor does not contain previously removed contaminant. A drawback is that the fumes can be irritating or toxic, so that environmental controls are needed.

Mechanical agitation of the surface can aid in removal of contaminants; examples of this method include wire brushing, abrasive blasting, abrasive jets, and ultrasonic vibration of a solvent bath (**ultrasonic cleaning**). In **electrolytic cleaning**, a charge is applied to the workpiece in an aqueous, often alkaline, cleaning solution. This charge results in bubbles of hydrogen or oxygen being released at the workpiece surface, depending on the polarity of the charge; these abrasive bubbles aid in the removal of contaminants.

Cleaning discrete parts that have complex shapes can be difficult. Design engineers should be aware of this difficulty; provide alternative designs, such as avoiding deep blind holes or making several smaller components instead of one large component that may be difficult to clean; and provide appropriate drain holes in the part to be cleaned.

## 4.6 | Engineering Metrology and Instrumentation

*Engineering metrology* is defined as the measurement of dimensions such as length, thickness, diameter, taper, angle, flatness, and profiles. Traditionally, measurements have been made after the part has been produced, a procedure known as *postprocess inspection*. The practice in manufacturing now is to make measurements while the part is being produced on the machine, a procedure known as *in-process, on-line*, or *real-time inspection*. Here, the term *inspection* means to check the dimensions of what is produced or being produced, and to observe whether it complies with the specified dimensional accuracy. Much progress has been made in developing new and automated measuring instruments that are highly precise and sensitive.

### 4.6.1 Measuring instruments

1. **Line-graduated instruments.** Line-graduated instruments are used for measuring length (linear measurements) or angles (angular measurements). *Graduated* means marked to indicate a certain quantity. The simplest and most commonly used instrument for making linear measurements is a *steel rule* (*machinist's rule*), bar, or tape with fractional or decimal graduations. Lengths are measured directly, to an accuracy that is limited to the nearest division, usually 1 mm or $\frac{1}{64}$ in. Rules may be rigid or flexible and may be equipped with a hook at one end for ease of measuring from an edge. Rule-depth gages are similar to rules and slide along a special head.

   **Vernier calipers** have a graduated beam and a sliding jaw with a *vernier*; these instruments are also called **caliper gages**. The two jaws of the caliper contact the part being measured, and the dimension is read at the matching graduated lines. Vernier calipers can be used to measure inside or outside lengths. The vernier improves the sensitivity of a simple rule by indicating fractions of the smallest division on the graduated beam, usually to 25 $\mu$m (0.001 in.). Calipers are also equipped with *digital readouts*, which are easy to read and less subject to human error. Vernier height gages are vernier calipers with setups similar to that of a depth gage and have similar sensitivity as well.

   **Micrometers**, which have a graduated, threaded spindle, are commonly used for measuring the thickness and inside or outside diameters of parts. Circumferential vernier readings to a sensitivity of 2.5 $\mu$m (0.0001 in.) can be obtained. Micrometers are also available for measuring depths (*micrometer depth gage*) and internal diameters (*inside micrometer*) with the same sensitivity. Micrometers are also equipped with digital readouts to reduce errors in reading. The anvils on micrometers can be equipped with conical or ball contacts; they are used to measure inside recesses, threaded rod diameters, and wall thicknesses of tubes and curved sheets.

   **Diffraction gratings** consist of two flat optical glasses with closely spaced parallel lines scribed on their surfaces. The grating on the shorter glass is inclined slightly; as a result, interference fringes develop when it is viewed over the longer glass. The position of these fringes depends on the relative position of the two sets of glasses. With modern equipment, using electronic counters and photoelectric sensors, sensitivities of 2.5 $\mu$m (0.0001 in.) can be obtained with gratings that have 40 lines/mm (1000 lines/in.).

2. **Indirect-reading instruments.** Indirect-reading instruments typically are calipers and dividers without any graduated scales. They are used to transfer the size measured to a direct-reading instrument, such as a rule. After the legs of

the instrument have been adjusted to contact the part at the desired location, the instrument is held against a graduated rule, and the dimension is read. Because of both the experience required to use them and their dependence on graduated scales, the accuracy of this type of indirect measurement is limited. **Telescoping gages** are available for indirect measurement of holes or cavities.

**Angles** are measured in degrees, radians, or minutes and seconds of arc. Because of the geometry involved, angles are usually more difficult to measure than are linear dimensions. A **bevel protractor** is a direct-reading instrument similar to a common protractor, except that it has a movable member. The two blades of the protractor are placed in contact with the part being measured, and the angle is read directly on the vernier scale. The sensitivity of the instrument depends on the graduations of the vernier. Another type of bevel protractor is the **combination square**, which is a steel rule equipped with devices for measuring 45° and 90° angles.

Measuring with a **sine bar** involves placing the part on an inclined bar or plate and adjusting the angle by placing gage blocks on a surface plate. After the part is placed on the sine bar, a dial indicator (see later) is used to scan the top surface of the part. **Gage blocks** are added or removed as necessary until the top surface is parallel to the surface plate. The angle on the part is then calculated from geometric relationships. Angles can also be measured by using **angle gage blocks**. These blocks have different tapers that can be assembled in various combinations and used in a manner similar to that for sine bars. Angles on small parts can also be measured through microscopes, with graduated eyepieces, or with optical projectors.

3. **Comparative length-measuring instruments.** Unlike the instruments described thus far, instruments used for measuring *comparative lengths*, also called *deviation-type* instruments, amplify and measure variations or deviations in distance between two or more surfaces. These instruments compare dimensions, hence the word *comparative*. The common types of instruments used for making comparative measurements are described next.

**Dial indicators** are simple mechanical devices that convert linear displacements of a pointer to rotation of an indicator on a circular dial. The indicator is set to zero at a certain reference surface, and the instrument or the surface to be measured (either external or internal) is brought into contact with the pointer. The movement of the indicator is read directly on the circular dial (as either plus or minus) to accuracies as high as $1 \mu$m ($40 \mu$in.).

Unlike mechanical systems, **electronic gages** sense the movement of the contacting pointer through changes in the electrical resistance of a strain gage or through inductance or capacitance. The electrical signals are then converted and displayed as linear dimensions. A commonly used electronic gage is the **linear variable differential transformer** (LVDT), used extensively for measuring small displacements. Although they are more expensive than other types of gages, electronic gages have advantages such as ease of operation, rapid response, digital readout, less possibility of human error, versatility, flexibility, and the capability to be integrated into automated systems through microprocessors and computers.

4. **Measuring straightness, flatness, roundness, and profile.** The geometric features of straightness, flatness, roundness, and profile are important aspects of engineering design and manufacturing. For example, piston rods, instrument components, and machine-tool slideways should all meet certain requirements with regard to these characteristics in order to function properly. Consequently, their accurate measurement is essential.

**Straightness** can be checked with straight edges or dial indicators. **Autocollimators**, resembling a telescope with a light beam that bounces back from the object, are used for accurately measuring small angular deviations on a flat surface. Optical means such as **transits** and **laser beams** are used for aligning individual machine elements in the assembly of machine components.

**Flatness** can be measured by mechanical means, using a surface plate and a dial indicator. This method can also be used for measuring perpendicularity, which can be measured with the use of precision steel squares as well. Another method for measuring flatness is by **interferometry**, using an **optical flat**. The device, a glass or fused quartz disk with parallel flat surfaces, is placed on the surface of the workpiece. When a monochromatic (i.e., of only one wavelength) light beam is aimed at the surface at an angle, the optical flat splits it into two beams, appearing as light and dark bands to the naked eye. The number of fringes that appear is related to the distance between the surface of the part and the bottom surface of the optical flat. Consequently, a truly flat workpiece surface (that is, when the angle between the two surfaces is zero) will not split the light beam, and fringes will not appear. When surfaces are not flat, fringes are curved. The interferometry method is also used for observing surface textures and scratches through microscopes for better visibility.

**Roundness** is usually described as deviations from true roundness (mathematically, a circle). The term **out of roundness** is actually more descriptive of the shape of the part. Roundness is essential to the proper functioning of rotating shafts, bearing races, pistons, cylinders, and steel balls in bearings. The various methods of measuring roundness fall into two categories. In the first method, the round part is placed on a V-block or between centers and is rotated, with the pointer of a dial indicator in contact with the surface. After a full rotation of the workpiece, the difference between the maximum and minimum readings on the dial is noted; this difference is called the **total indicator reading** (TIR), or **full indicator movement**. This method is also used for measuring the straightness (*squareness*) of shaft end faces. In the second method, called **circular tracing**, the part is placed on a platform, and its roundness is measured by rotating the platform. Conversely, the probe can be rotated around a stationary part to make the measurement.

**Profile** may be measured by several methods. In one method, a surface is compared with a template or profile gage to check shape conformity; radii or fillets can be measured by this method. Profile may also be measured with a number of dial indicators or similar instruments. Profile-tracing instruments are the latest development for the measurement of profile.

**Threads** and **gear teeth** have several features that require specific dimensions and tolerances. These dimensions must be produced accurately for the smooth operation of gears, reduction of wear and noise level, and part interchangeability. These features are measured by means of thread gages of various designs that compare the thread produced with a standard thread. Some of the gages used are threaded plug gages, screw-pitch gages (similar to radius gages), micrometers with cone-shaped points, and snap gages with anvils in the shape of threads. Gear teeth are measured with instruments that are similar to dial indicators, with calipers and with micrometers using pins or balls of various diameters.

Special profile-measuring equipment is also available, including optical projectors. **Optical projectors**, also called **optical comparators**, were first developed in the 1940s to check the geometry of cutting tools for machining screw threads, but are now used for checking all profiles. The part is mounted on a

**FIGURE 4.18**  A coordinate measuring machine, measuring dimensions on an engine block. *Source*:  Courtesy of Sheffield Measurement Division, Giddings & Lewis.

table, or between centers, and the image is projected on a screen at magnifications up to 100× or higher. Linear and angular measurements are made directly on the screen, which is equipped with reference lines and circles. The screen can be rotated to allow angular measurements as small as 1 min, using verniers.

5. **Coordinate measuring machines and layout machines.** These machines consist of a platform on which the workpiece being measured is placed and moved linearly or rotated (Fig. 4.18); a stylus, attached to a head capable of lateral and vertical movements, records all measurements. *Coordinate measuring machines* (CMMs), also called **measuring machines**, are versatile in their capability to record measurements of complex profiles with high sensitivity ($0.25 \ \mu$m [$10 \ \mu$in.]).

   These machines are built rigidly and are very precise. They are equipped with digital readout or can be linked to computers for on-line inspection of parts. They can be placed close to machine tools for efficient inspection and rapid feedback for correction of processing parameters before the next part is made. They are also being made more rugged to resist environmental effects in manufacturing plants, such as temperature variations, vibration, and dirt. Dimensions of large parts are measured by **layout machines**, which are equipped with digital readout. These machines are equipped with scribing tools for marking dimensions on large parts, with an accuracy of 0.04 mm (0.0016 in.).

6. **Gages.** Thus far, we have used the word *gage* to describe some types of measuring instruments, such as a caliper gage, depth gage, telescoping gage, electronic gage, strain gage, and radius gage. The reason for this word choice is that the words *instrument* and *gage* (also spelled *gauge*) have traditionally been used interchangeably. However, *gage* also has a variety of other meanings, such as in pressure gage; gage length of a tension-test specimen; and gages for sheet metal, wire, railroad rail, and the bore of shotguns.

**Gage blocks** are individual square, rectangular, or round metal or ceramic blocks of various sizes, made very precisely from heat-treated and stress-relieved alloy steels or from carbides. Their surfaces are lapped and are flat and parallel within a range of 0.02 $\mu$m–0.12 $\mu$m (1 $\mu$in.–5 $\mu$in.). Gage blocks are available in sets of various sizes, some sets containing almost 100 gage blocks; the blocks can be assembled in many combinations to obtain desired lengths. Dimensional accuracy can be as high as 0.05 $\mu$m (2 $\mu$in.). Environmental-temperature control is important in using gages for high-precision measurements. Although their use requires some skill, gage-block assemblies are commonly utilized in industry as an accurate reference length. Angle blocks are made similarly and are available for angular gaging.

**Fixed gages** are replicas of the shapes of the parts to be measured. **Plug gages** are commonly used for holes. The *GO gage* is smaller than the *NOT GO* (or *NO GO*) *gage* and slides into any hole whose smallest dimension is less than the diameter of the gage. The NOT GO gage must not go into the hole. Two gages, one GO gage and one NOT GO gage, are required for such measurements, although both may be on the same device, either at opposite ends or in two steps at one end (*step-type gage*). Plug gages are also available for measuring internal tapers (in which deviations between the gage and the part are indicated by the looseness of the gage), splines, and threads (in which the GO gage must screw into the threaded hole).

**Ring gages** are used to measure shafts and similar round parts, and **ring thread gages** are used to measure external threads. The GO and NOT GO features on these gages are identified by the type of knurling on the outside diameter of the rings. **Snap gages** are commonly used to measure external dimensions. They are made with adjustable gaging surfaces for use with parts that have different dimensions. One of the gaging surfaces may be set at a different gap than that of the other, thus making a one-unit GO–NOT-GO gage. Although fixed gages are easy to use and inexpensive, they only indicate whether a part is too small or too large compared with an established standard; they do not measure actual dimensions.

There are several types of **pneumatic gages**, also called *air gages*. The gage head has holes through which pressurized air, supplied by a constant-pressure line, escapes. The smaller the gap between the gage and the hole, the more difficult it is for the air to escape, and hence the back pressure is higher. The back pressure, sensed and indicated by a pressure gage, is calibrated to read dimensional variations of holes.

7. **Microscopes.** Microscopes are optical instruments used to view and measure very fine details, shapes, and dimensions on small and medium-sized tools, dies, and workpieces. The most common and versatile microscope used in tool rooms is the **toolmaker's microscope**. It is equipped with a stage that is movable in two principal directions and can be read to 2.5 $\mu$m (0.0001 in.). Several models of microscopes are available with various features for specialized inspection, including models with digital readout. The **light-section microscope** is used to measure small surface details, such as scratches, and the thickness of deposited films and coatings. A thin light band is applied obliquely to the surface, and the reflection is viewed at 90°, showing surface roughness, contours, and other features. Unlike ordinary optical microscopes, the **scanning electron microscope** (SEM) has excellent depth of field. As a result, all regions of a complex part are in focus and can be viewed and photographed to show extremely fine detail. This type of microscope is particularly useful for studying surface textures and fracture patterns. Although expensive, such microscopes are capable of magnifications greater than 100,000×.

## 4.6.2 Automated measurement

With increasing automation in all aspects of manufacturing processes and operations, the need for *automated measurement* (also called **automated inspection**) has become much more apparent. Flexible manufacturing systems and manufacturing cells (see Chapter 15) have led to the adoption of advanced measurement techniques and systems. In fact, installation and use of these systems is now a necessary and not an optional manufacturing technology.

Traditionally, a batch of parts was manufactured and sent for measurement in a separate quality-control room, and if the parts passed measurement inspection, they were put into inventory. Automated inspection, however, is based on various on-line sensor systems that monitor the dimensions of parts being made and use the feedback from these measurements to correct the process, if necessary. To appreciate the importance of on-line monitoring of dimensions, let's find the answer to the following question: if a machine has been producing a certain part with acceptable dimensions, what factors contribute to subsequent deviation in the dimension of the same part produced by the same machine? The major factors are as follows:

1. Static and dynamic deflections of the machine because of vibrations and fluctuating forces, caused by variations such as in the properties and dimensions of the incoming material;

2. Deformation of the machine because of thermal effects, including changes in the temperature of the environment, of metalworking fluids, and of machine bearings and components;

3. Wear of tools and dies.

As a result of these factors, the dimensions of parts produced will vary, necessitating monitoring of dimensions during production. **In-process workpiece control** is accomplished by special gaging and is used in a variety of applications, such as high-quantity machining and grinding.

## 4.7 | Dimensional Tolerances

*Dimensional tolerance* is defined as the permissible or acceptable variation in the dimensions (height, width, depth, diameter, angles) of a part. Tolerances are unavoidable, because it is virtually impossible (and unnecessary) to manufacture two parts that have precisely the same dimensions. Furthermore, because close dimensional tolerances substantially increase the product cost, a narrow tolerance range is undesirable economically. Tolerances become important only when a part is to be assembled or mated with another part. Surfaces that are free and not functional do not need close dimensional tolerance control; thus, for example, the accuracies of the dimensions of the holes and the distance between the holes for a connecting rod are far more critical than the accuracy of the rod's width and thickness at various locations along its length.

To illustrate the importance of dimensional tolerances, imagine that we want to assemble a simple shaft (axle) and a wheel with a hole, assuming that we want the axle's diameter to be 1 in. (Fig. 4.19a). We go to the hardware store and purchase a 1-in.-thick round rod and a wheel with a hole that has a diameter of 1 in. Will the rod fit into the hole without being forced, or will it be loose in the hole? We can't be sure. The 1-in. dimension is the *nominal size* of the shaft. If we purchase such a rod

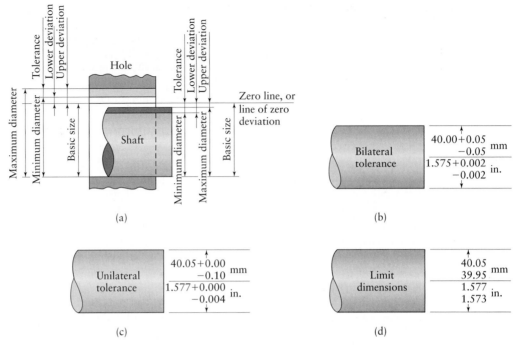

(a)

(b)

(c)

(d)

**FIGURE  4.19**    (a) Basic size, deviation, and tolerance on a shaft, according to the ISO system. (b)–(d) Various methods of assigning tolerances on a shaft. *Source*:  L. E. Doyle.

**FIGURE  4.20**    Tolerances and surface roughness obtained in various manufacturing processes. These tolerances apply to a 25-mm (1-in.) workpiece dimension. *Source*: Schey, J.A., *Introduction to Manufacturing Processes*, 3rd ed. New York, McGraw-Hill, 2000, p. 68.

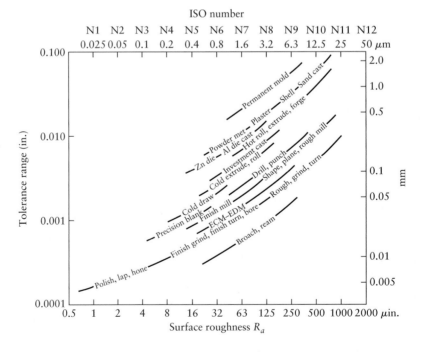

from different stores or at different times, or select one randomly from a lot of, say, 100 shafts, the chances are that each rod will have a slightly different diameter. Machines may, with the same setup, produce rods of slightly different diameters, depending on a number of factors, such as speed of operation, temperature, lubrication, variations in the incoming material, and similar variables.

If we now specify a range of diameters for both the rod and the hole of the wheel, we can predict correctly the type of fit that we will have after assembly. Certain terminology has been established to clearly define these geometric quantities, such as the ISO system shown in Fig. 4.19a. Note that both the shaft and the hole have minimum and maximum diameters, respectively, the difference being the dimensional tolerance for each member. A proper engineering drawing should specify these parameters with numerical values, as shown in Fig. 4.19b.

The range of dimensional tolerances obtained in various manufacturing processes is given in Fig. 4.20. There is a general relationship between tolerances and surface finish of parts manufactured by various processes. Note the wide range of tolerances and surface finishes that can be obtained. Also, the larger the part, the greater the obtainable tolerance range becomes. Experience has shown that dimensional inaccuracies of manufactured parts are approximately proportional to the cube root of the size of the part; thus, doubling the size of a part increases the inaccuracies by $2^{1/3} = 1.26$ times, or 26%.

# 4.8 | Testing and Inspection

Before they are marketed, manufactured parts and products are inspected for several characteristics. This inspection routine is particularly important for products or components whose failure or malfunction has potentially serious implications, such as bodily injury or fatality. Typical examples of such failures include cables breaking, switches malfunctioning, brakes failing, grinding wheels breaking, railroad wheels fracturing, turbine blades failing, pressure vessels bursting, and welded joints failing. In this section, we identify and describe the various methods that are commonly used to inspect manufactured products.

*Product quality* has always been one of the most important elements in manufacturing operations, and with increasing domestic and international competition, it has become even more important. Prevention against defects in products and on-line inspection are now major goals in all manufacturing activities. We again emphasize that quality must be built into a product and not merely checked after the product has been made. Close cooperation and communication between design and manufacturing engineers are thus essential.

## 4.8.1 Nondestructive testing

*Nondestructive testing* (NDT) is carried out in such a way that product integrity and surface texture remain unchanged. These techniques generally require considerable operator skill. Interpreting test results accurately may be difficult, because test results can be quite subjective. However, the use of computer graphics and other enhancement techniques has reduced the likelihood of human error in nondestructive testing. The basic principles of the more commonly used nondestructive testing techniques are described as follows:

1. In the **liquid-penetrants technique**, fluids are applied to the surfaces of the part and allowed to penetrate into surface openings, cracks, seams, and porosity; the penetrant can seep into cracks as small as 0.1 $\mu$m (4 $\mu$in.) in width. Two common types of liquids are (a) fluorescent penetrants with various sensitivities, which fluoresce under ultraviolet light; and (b) visible penetrants, using dyes usually red in color, which appear as bright outlines on the surface.

    The surface to be inspected is first thoroughly cleaned and dried. The liquid is brushed or sprayed on the surface to be inspected and allowed to remain long enough to seep into surface openings. Excess penetrant is then wiped off or washed away with water or solvent. A developing agent is then added to allow the penetrant to seep back to the surface and spread to the edges of openings, thus magnifying the size of defects. The surface is then inspected for defects, either visually in the case of dye penetrants or with fluorescent lighting. This method is capable of detecting a variety of surface defects and is used extensively. The equipment is simple and easy to use, can be portable, and is less costly to operate than other methods. However, this method can detect only defects that are open to the surface, not internal defects.

2. The **magnetic-particle inspection technique** consists of placing fine ferromagnetic particles on the surface of the part. The particles can be applied either dry or in a liquid carrier such as water or oil. When the part is magnetized with a magnetic field, a discontinuity (defect) on the surface causes the particles to gather visibly around it. The collected particles generally take the shape and size of the defect. Subsurface defects can also be detected by this method, provided that they are not deep. The ferromagnetic particles may be colored with pigments for better visibility on metal surfaces. Wet particles are used for detecting fine discontinuities, such as fatigue cracks.

3. In **ultrasonic inspection**, an ultrasonic beam travels through the part. An internal defect, such as a crack, interrupts the beam and reflects back a portion of the ultrasonic energy. The amplitude of the energy reflected and the time required for return indicate the presence and location of any flaws in the workpiece. The ultrasonic waves are generated by transducers, called *search units* or *probes*, of various types and shapes. They operate on the principle of piezoelectricity (Section 3.9.6), using materials such as quartz, lithium sulfate, and various ceramics. Most inspections are carried out at a frequency range of 1–25 MHz. Couplants are used to transmit the ultrasonic waves from the transducer to the test piece; typical couplants are water, oil, glycerin, and grease.

   The ultrasonic inspection method has high penetrating power and sensitivity. It can be used to inspect flaws in large volumes of material, such as railroad wheels, pressure vessels, and die blocks, from various directions. Its accuracy is higher than that of other nondestructive inspection methods. However, this method requires experienced personnel to carry out the inspection and interpret the results correctly.

4. The **acoustic-emission technique** detects signals (high-frequency stress waves) generated by the workpiece itself during plastic deformation, crack initiation and propagation, phase transformation, and sudden reorientation of grain boundaries. Bubble formation during boiling, and friction and wear of sliding interfaces, are other sources of acoustic signals. Acoustic-emission inspection is typically performed by stressing elastically the part or structure, such as bending a beam, applying torque to a shaft, and pressurizing a vessel; the technique is particularly effective for continuous surveillance of load-bearing structures. Acoustic emissions are detected by sensors consisting of piezoelectric ceramic elements.

5. The **acoustic-impact technique** consists of tapping the surface of an object and listening to and analyzing the signals in order to detect discontinuities and flaws. The principle is basically the same as tapping walls, desktops, or countertops in various locations with your fingers or a hammer and listening to the sound emitted. Vitrified grinding wheels are tested in a similar manner (using a ring test) to detect cracks in the wheel that may not be visible to the naked eye. The

acoustic-impact technique can be instrumented and automated and is easy to perform. However, the results depend on the geometry and mass of the part, thus requiring a reference standard to identify flaws.

6. **Radiography** involves X-ray inspection to detect internal flaws or variations in density and thickness in the part. The radiation source is typically an X-ray tube, and a visible permanent image is made on an X-ray film or radiographic paper.

7. The **eddy-current inspection method** is based on the principle of electromagnetic induction. The part is placed in or adjacent to an electric coil through which alternating current (exciting current) flows at frequencies ranging from 6 MHz to 60 MHz. This current induces eddy currents in the part. Defects in the part impede and change the direction of eddy currents, causing changes in the electromagnetic field. These changes affect the exciting coil (inspection coil), whose voltage is monitored to determine the presence of flaws.

8. **Thermal inspection** involves observing temperature changes by contact- or noncontact-type heat-sensing devices. Defects in the workpiece, such as cracks, debonded regions in laminated structures, and poor joints, cause a change in the temperature distribution. In **thermographic inspection**, materials such as heat-sensitive paints and papers, liquid crystals, and other coatings are applied to the surface of the part. Any changes in their color or appearance indicate defects.

9. The **holography technique** creates a three-dimensional image of the workpiece, using an optical system. This technique is generally used on simple shapes and highly polished surfaces, and the image is recorded on a photographic film. Use of this technique has been extended to inspection of parts (**holographic interferometry**) that have various shapes and surface conditions. Through double- and multiple-exposure techniques while the part is being subjected to external forces or time-dependent variations, changes in the images reveal defects in the part. In **acoustic holography**, information on internal defects is obtained directly from the image of the interior of the part. In **liquid–surface acoustical holography**, the workpiece and two ultrasonic transducers (one for the object beam and the other for the reference beam) are immersed in a water-filled tank. The holographic image is then obtained from the ripples in the tank. In **scanning acoustical holography**, only one transducer is used, and the hologram is produced by electronic-phase detection. This system is more sensitive, the equipment is usually portable, and very large workpieces can be accommodated by using a water column instead of a tank.

## 4.8.2 Destructive testing

As the name suggests, the part or product tested using *destructive testing methods* no longer maintains its integrity, original shape, or surface texture. *Mechanical testing methods* (see Chapter 2) are all destructive, in that a sample or specimen has to be removed from the product in order to test it. In addition to mechanical testing, other destructive tests include speed testing of grinding wheels to determine their bursting speed and high-pressure testing of pressure vessels to determine their bursting pressure. Hardness tests that leave large impressions may be regarded as destructive testing methods; however, microhardness tests may be regarded as nondestructive, because only very small permanent indentations are made. This distinction is based on the assumption that the material is not *notch sensitive* (Section 2.9). Most glasses, highly heat-treated metals, and ceramics are notch sensitive; that is, the small indentation produced by the indenter may lower the material's strength and toughness.

### 4.8.3  Automated inspection and sensors

Note that in all the preceding examples, the testing methods involve parts or products that have already been manufactured. Traditionally, individual parts and assemblies of parts have been manufactured in batches, sent to inspection in quality-control rooms, and, if approved, put in inventory. If products do not pass the quality inspection, they are either scrapped or kept on the basis of a certain acceptable deviation from the standard. Obviously, such a system lacks flexibility, requires maintenance of an inventory, and inevitably results in some defective parts going through the system. The traditional method actually tracks the defects after they occur (**postprocess inspection**), and in no way attempts to prevent defects.

In contrast, one of the most important practices in modern manufacturing is **automated inspection**. This method uses a variety of sensor systems that monitor the relevant parameters *during* the manufacturing process (**on-line inspection**). Then, using these measurements, the process automatically corrects itself to produce acceptable parts; thus, further inspection of the part at another location in the plant is unnecessary. Parts may also be inspected immediately after they are produced (**in-process inspection**).

The use of accurate sensors (see later) and computer-control systems (see Chapter 15) has integrated automated inspection into manufacturing operations. Such a system ensures that no part is moved from one manufacturing process to another (for example, a turning operation on a lathe followed by cylindrical grinding) unless the part is made correctly and meets the standards of the first operation. Automated inspection is flexible and responsive to product design changes. Furthermore, because the system uses automated equipment, less operator skill is required; productivity is increased; and parts have higher quality, reliability, and dimensional accuracy.

**Sensors for automated inspection.**  Rapid advances in **sensor technology** are making on-line, or real-time, monitoring of manufacturing processes feasible. (See also Section 14.8.) Directly or indirectly, and with the use of various probes, sensors can detect dimensional accuracy, surface roughness, temperature, force, power, vibration, tool wear, and the presence of external or internal defects. Sensors operate on the same principles as those of strain gages, inductance, capacitance, ultrasonics, acoustics, pneumatics, infrared radiation, optics, lasers, and various electronic gages. Sensors may be tactile (touching) or nontactile.

Sensors, in turn, are linked to microprocessors and computers for graphical display of data. This capability allows rapid on-line adjustment of one or more processing parameters in order to produce parts that are consistently within specified standards of tolerance and quality. Such systems are already implemented as standard equipment on many metal-cutting machine tools and grinding machines.

## 4.9  |  Quality Assurance

*Quality assurance* is the total effort by a manufacturer to ensure that its products conform to a detailed set of specifications and standards. These standards cover several parameters, such as dimensions; surface finish; dimensional tolerances; composition; color; and mechanical, physical, and chemical properties. In addition, standards are usually written to ensure proper assembly using interchangeable, defect-free components and the fabrication of a product that performs as intended by its designers. Quality assurance is the responsibility of everyone involved with design and manufacturing. The often-repeated statement that quality must be built into a product reflects this important concept: Quality cannot be inspected into a finished product. Although product quality has always been important, increased domestic and international competition has caused quality assurance to become even more important. Every aspect of design and manufacturing operations, such as material selection, production, and assembly, is now analyzed in detail to ensure that quality is truly built into the final product.

If you were in charge of product quality for a manufacturing plant, how would you make sure that the final product is of acceptable quality? The best method is to control materials and processes in such a manner that the products are made correctly in the first place. However, 100% inspection is usually too costly to maintain. Therefore, several methods of inspecting smaller, statistically relevant sample lots have been devised. These methods all use **statistics** to determine the probability of defects occurring in the total production batch.

Inspection involves a series of steps:

1. Inspecting incoming materials to make sure that they meet certain property, dimension, and surface finish and integrity requirements;

2. Inspecting individual product components to make sure that they meet specifications;

3. Inspecting the product to make sure that individual parts have been assembled properly;

4. Testing the product to make sure that it functions as designed and intended.

Inspections must be continued during production, because there are always variations in the dimensions and properties of incoming materials; variations in the performance of tools, dies, and machines used in various stages of manufacturing; possibilities of human error; and errors made during assembly of the product. As a result, no two products are ever made exactly alike.

Another important aspect of quality control is the capability to analyze defects and promptly eliminate them or reduce them to acceptable levels. In an even broader sense, quality control involves evaluating the product design and customer satisfaction; the sum total of all these activities is referred to as **total quality management** (TQM). Thus, in order to control quality, we have to be able to (1) measure quantitatively the level of quality, and (2) identify all the material and process variables that can be controlled. The level of quality obtained during production can then be established by inspecting the product to determine whether it meets the specifications for dimensional tolerances, surface finish, defects, and other characteristics. The identification of material and process variables and their effect on product quality is now possible through the extensive knowledge gained from research and development activities in all aspects of manufacturing.

## 4.9.1 Statistical methods of quality control

**Statistics** deals with the collection, analysis, interpretation, and presentation of large amounts of numerical data. The use of statistical techniques in modern manufacturing operations is necessary because of the large number of material and process variables involved. Events that occur randomly, that is, without any particular trend or pattern, are called **chance variations**. Events that can be traced to specific causes are called **assignable variations**. The existence of **variability** in production operations has been recognized for centuries, but Eli Whitney (1765–1825) first grasped its full significance when he found that interchangeable parts were indispensable to the mass production of firearms. Modern statistical concepts relevant to manufacturing engineering were first developed in the early 1900s.

To understand **statistical quality control** (SQC), we first need to define some of the terms that are commonly used in this field.

a) **Sample size:** The number of parts to be inspected in a sample, whose properties are studied to gain information about the whole population.

b) **Random sampling:** Taking a sample from a population or lot in which each item has an equal chance of being included in the sample; thus, when taking samples from a large bin, an inspector does not take only those items that happen to be within reach.

c) **Population** (also called the **universe**): The totality of individual parts of the same design from which samples are taken.

d) **Lot size:** A subset of population. A lot or several lots can be considered subsets of the population and may be treated as representative of the population.

Samples are inspected for certain characteristics and features, such as dimensional tolerances, surface finish, and defects, with the instruments and techniques described earlier in this chapter. These characteristics fall into two categories: those that can be measured quantitatively (method of variables) and those that can be measured qualitatively (method of attributes).

The **method of variables** is the quantitative measurement of characteristics such as dimensions, tolerances, surface finish, and physical or mechanical properties. Such measurements are made for each of the units in the group under consideration, and the results are compared with the specifications for the part. The **method of attributes** involves observing the presence or absence of qualitative characteristics, such as external or internal defects in machined, formed, or welded parts, or dents in sheet-metal products, for each of the units in the group under consideration. Sample size for attributes-type data is generally larger than that for variables-type data.

During the inspection process, measurement results will vary. For example, assume that you are measuring the diameter of turned shafts as they are produced on a lathe, using a micrometer. You soon note that their diameters vary, even though ideally you want all the shafts to be exactly the same size. Let's now turn to consideration of statistical quality-control techniques, which allow us to evaluate these variations and set limits for the acceptance of parts.

If we list the measured diameters of the turned shafts in a given population, we note that one or more parts have the smallest diameter, and one or more have the largest diameter; the majority of the turned shafts have diameters that lie between these extremes. We can then group these diameters and plot them on a bar graph, where each bar represents the number of parts in each diameter group (Fig. 4.21a). The bars show a **distribution**, also called a **spread** or **dispersion**, of the shaft-diameter measurements. The bell-shaped curve in Fig. 4.21a is called **frequency distribution** and represents the frequency with which parts within each diameter group are being produced.

Data from manufacturing processes often fit curves represented by a mathematically derived **normal distribution curve** (Fig. 4.21b); these curves are also called *Gaussian curves* and are developed on the basis of probability. The bell-shaped normal distribution curve fitted to the data in Fig. 4.21b has two important features. First, it shows that most part diameters tend to cluster around an *average* value (**arithmetic mean**). This average is usually designated as $\bar{x}$ and is calculated from the expression

$$\bar{x} = \frac{x_1 + x_2 + x_3 + \cdots + x_n}{n},\tag{4.7}$$

where the numerator is the sum of all measured values (diameters) and $n$ is the number of measurements (the number of shafts in this case).

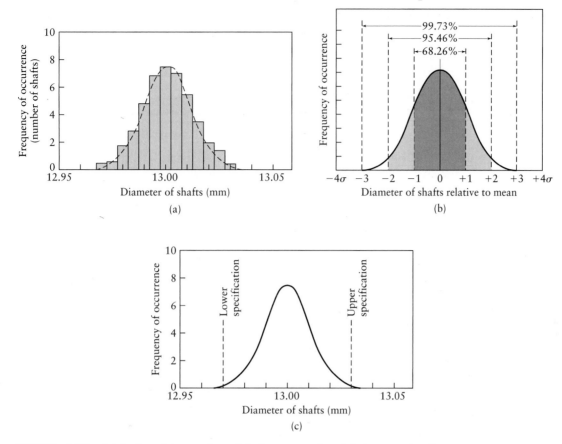

**FIGURE 4.21**   (a) A plot of the number of shafts measured and their respective diameters. This type of curve is called a *frequency distribution*. (b) A normal distribution curve indicating areas within each range of standard deviation. *Note*: The greater the range, the higher the percentage of parts that fall within it. (c) Frequency distribution curve, showing lower and upper specification limits.

The second feature of this curve is its width, indicating the *dispersion* of the diameters measured; the wider the curve, the greater is the dispersion. The difference between the largest value and smallest value is called the *range R*:

$$R = x_{max} - x_{min}. \tag{4.8}$$

Dispersion is estimated by the **standard deviation**, which is generally denoted as $\sigma$ and is obtained from the expression

$$\sigma = \sqrt{\frac{(x_1 - \bar{x})^2 + (x_2 - \bar{x})^2 + (x_3 - \bar{x})^2 + \cdots + (x_n - \bar{x})^2}{n - 1}}, \tag{4.9}$$

where $x$ is the measured value for each part. Note from the numerator in Eq. (4.9) that as the curve widens, the standard deviation becomes greater; also note that $\sigma$ has units of linear dimension. In comparing Eqs. (4.8) and (4.9), we see that the range $R$ is a simpler and more convenient measure of dispersion.

Since we know the number of parts that fall within each group, we can calculate the percentage of the total population represented by each group. Thus, Fig. 4.21b shows that the diameters of 99.73% of the turned shafts fall within the range of $\pm 3\sigma$, 95.46% within $\pm 2\sigma$, and 68.26% within $\pm 1\sigma$; only 0.2% fall outside the $\pm 3\sigma$ range.

**Six sigma.** An important trend in manufacturing operations, as well as in business and service industries, is the concept of *six sigma*. Note from the preceding discussion that maintaining a standard of three sigma in manufacturing would result in 0.27%, or 2700 defective parts per million parts. In modern manufacturing, this is an unacceptable rate; in fact, it has been estimated that at the three-sigma level, virtually no modern computer would function properly and reliably, and in the service industries, 270 million incorrect credit-card transactions would be recorded each year in the United States alone. It has further been estimated that companies operating at three- and four-sigma levels lose about 10% to 15% of their total revenue because of defects. Consequently, led by major companies such as Motorola and General Electric, extensive efforts are being made to eliminate defects in products and processes. The resulting savings are reported in billions of dollars.

Six sigma is a set of statistical tools, based on well-known total-quality-management principles, to continually measure the quality of products and services in selected projects. It includes considerations such as customer satisfaction, delivering defect-free products, and understanding process capabilities (see later). The approach consists of a clear focus on defining the problem; measuring relevant quantities; and analyzing, improving, and controlling processes and activities. Because of its major impact on business, six sigma is now well recognized as a management philosophy.

### 4.9.2 Statistical process control

If the number of parts that do not meet set standards (i.e., defective parts) increases during a production run, we must be able to determine the cause (e.g., incoming materials, machine controls, degradation of metalworking fluids, operator boredom) and take appropriate action. Although this statement at first appears to be self-evident, it was only in the early 1950s that a systematic statistical approach was developed to guide operators in manufacturing plants.

This approach advises the operator of certain measures to take and when to take them, in order to avoid producing further defective parts. Known as *statistical process control* (SPC), this technique consists of several elements: control charts and setting control limits, capabilities of the particular manufacturing process, and characteristics of the machinery involved.

**Control charts.** The frequency distribution curve in Fig. 4.21a shows a range of shaft diameters being produced that may fall beyond the predetermined design tolerance range. Figure 4.21c shows the same bell-shaped curve, which now includes the specified tolerances for the diameters of turned shafts. *Control charts* graphically represent the variations of a process over a period of time. They consist of data plotted during production, and there typically are two plots. The quantity $\bar{x}$ in Fig. 4.22a is the average for each subset of samples taken and inspected, say, a subset consisting of 5 parts. (A sample size of between 2 and 10 parts is sufficiently accurate, provided that sample size is held constant throughout the inspection.)

The frequency of sampling depends on the nature of the process. Some processes may require continuous sampling, whereas others may require only one sample per day. Quality-control analysts are best qualified to determine this frequency for a particular situation. Since the measurements shown in Fig. 4.22a are made consecutively, the abscissa of these control charts also represents time. The solid horizontal line in this figure is the **average of averages (grand average)**, denoted as $\bar{\bar{x}}$, and represents the *population mean*. The upper and lower horizontal broken lines in these control charts indicate the control limits for the process.

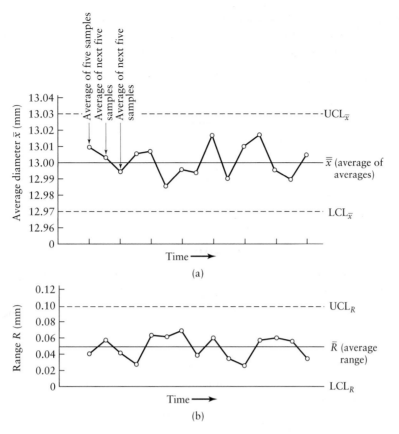

**FIGURE 4.22** Control charts used in statistical quality control. The process shown is in good statistical control, because all points fall within the lower and upper control limits. In this illustration, the sample size is five, and the number of samples is 15.

The **control limits** are set on these charts according to statistical-control formulas designed to keep actual production within the usually acceptable $\pm 3\sigma$ range. Thus, for $\bar{x}$,

$$\text{Upper control limit (UCL}_{\bar{x}}) = \bar{x} + 3\sigma = \bar{\bar{x}} + A_2\bar{R} \qquad (4.10)$$

and

$$\text{Lower control limit (LCL}_{\bar{x}}) = \bar{x} - 3\sigma = \bar{\bar{x}} - A_2\bar{R}, \qquad (4.11)$$

where $A_2$ is read from Table 4.3 and $\bar{R}$ is the average of the $R$ values.

**TABLE 4.3**

**Constants for Control Charts**

| Sample Size | $A_2$ | $D_4$ | $D_3$ | $d_2$ |
|---|---|---|---|---|
| 2 | 1.880 | 3.267 | 0 | 1.128 |
| 3 | 1.023 | 2.575 | 0 | 1.693 |
| 4 | 0.729 | 2.282 | 0 | 2.059 |
| 5 | 0.577 | 2.115 | 0 | 2.326 |
| 6 | 0.483 | 2.004 | 0 | 2.534 |
| 7 | 0.419 | 1.924 | 0.078 | 2.704 |
| 8 | 0.373 | 1.864 | 0.136 | 2.847 |
| 9 | 0.337 | 1.816 | 0.184 | 2.970 |
| 10 | 0.308 | 1.777 | 0.223 | 3.078 |
| 12 | 0.266 | 1.716 | 0.284 | 3.258 |
| 15 | 0.223 | 1.652 | 0.348 | 3.472 |
| 20 | 0.180 | 1.586 | 0.414 | 3.735 |

These limits are calculated based on the historical production capability of the equipment itself; they are generally not associated with design tolerance specifications and dimensions. They indicate the limits within which a certain percentage of measured values is normally expected to fall due to the inherent variations of the process itself, upon which the limits are based. The major goal of statistical process control is to improve the manufacturing process via the aid of control charts to eliminate assignable causes. The control chart continually indicates progress in this area.

The second control chart (Fig. 4.22b) shows the range $R$ in each subset of samples. The solid horizontal line represents the average of $R$ values in the lot, denoted as $\bar{R}$, and is a measure of the variability of the samples. The upper and lower control limits for $R$ are obtained from the equations

$$\mathrm{UCL}_R = D_4\bar{R} \tag{4.12}$$

and

$$\mathrm{LCL}_R = D_3\bar{R}, \tag{4.13}$$

where the constants $D_4$ and $D_3$ are obtained from Table 4.3. This table also includes values for the constant $d_2$, which is used in estimating the standard deviation from the equation

$$\sigma = \frac{\bar{R}}{d_2}. \tag{4.14}$$

When the curve of a control chart is like that shown in Fig. 4.22b, the process is said to be **in good statistical control**. In other words, there is no clear discernible trend in the pattern of the curve, the points (measured values) are random with time, and they do not exceed the control limits. However, you can see that curves such as those shown in Figs. 4.23a–c indicate certain trends. For example, about halfway in Fig. 4.23a, the diameter of the shafts increases with time; the reason for this increase may be a change in one of the process variables, such as wear of the cutting tool. If, as in Fig. 4.23b, the trend is toward consistently larger diameters, hovering around the upper control limit, it could mean that the tool settings on the lathe may not be correct and, as a result, the parts being turned are consistently too large. Figure 4.23c shows two distinct trends that may be caused by factors such as a change in the properties of the incoming material or a change in the performance of the cutting fluid (for example, its degradation). These situations place the process **out of control**.

Analyzing control-chart patterns and trends requires considerable experience in order to identify the specific cause(s) of an out-of-control situation. **Overcontrol** of the manufacturing process, that is, setting upper and lower control limits too close to each other (and hence setting a smaller standard-deviation range), is a further reason for out-of-control situations. Avoidance of overcontrol is the reason that we calculate control limits on process capability rather than on a potentially inapplicable statistic.

It is evident that operator training is critical for successful implementation of SPC on the shop floor. Once process-control guidelines are established, operators also should, in the interest of efficiency of the operation, have some responsibility to make adjustments in processes that begin to become out of control. The capabilities of individual operators should be taken into account so that the operators are not overloaded with data input and interpretation. This task is, however, being made easier by a variety of software that is being continually developed. For example, digital readouts on electronic measuring devices are integrated directly into a computer system for real-time SPC. Figure 4.24 shows such a multifunctional computer system in which the output from a digital caliper or micrometer is analyzed in real time and displayed in several ways, such as frequency distribution curves and control charts.

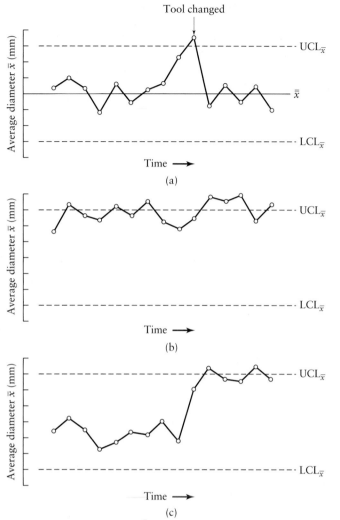

**FIGURE 4.23** Control charts. (a) Process begins to become out of control, because of factors such as tool wear. The tool is changed, and the process is then in good statistical control. (b) Process parameters are not set properly; thus, all parts are around the upper control limit. (c) Process becomes out of control, because of factors such as a sudden change in the properties of the incoming material.

**FIGURE 4.24** Schematic illustration showing integration of digital gages with a miniprocessor for real-time data acquisition and SPC/SQC capabilities. Note the examples on the CRT displays, such as frequency distribution and control charts. *Source*: Mitutoyo Corp.

**Process capability.** *Process capability* is defined as the limits within which individual measurement values resulting from a particular manufacturing process would normally be expected to fall when only random variation is present; it tells us that the process can produce parts within certain limits of precision. Since a manufacturing process involves materials, machinery, and operators, each of these aspects can be analyzed individually to identify a problem when process capabilities do not meet part specifications.

### 4.9.3  Total quality control

The *total quality control* (TQC) concept is a management system that emphasizes the fact that quality must be designed and built into a product; **defect prevention**, rather than defect detection, is the major goal. Total quality control is a systems approach in that both management and workers make an integrated effort to manufacture high-quality products consistently. All tasks concerning quality improvements and responsibilities in the organization are clearly identified. The TQC concept also requires 100% inspection of parts, usually by automated inspection systems; no defective parts are allowed to continue through the production line.

A related concept that has gained some acceptance is **the quality circle**. This activity consists of regular meetings by groups of workers who discuss how to improve and maintain product quality at all stages of the manufacturing process. Worker involvement and responsibility are emphasized. Comprehensive training is provided so that the worker can become capable of analyzing statistical data, identifying causes of poor quality, and taking immediate action to correct the situation. Putting this concept into practice recognizes the importance of quality assurance as a major companywide management policy, affecting all personnel and all aspects of production.

**Quality engineering as a philosophy.**  Many of the quality-control concepts and methods described have been put into larger perspective by certain experts in quality control. Notable among these experts are W. E. Deming, G. Taguchi, and J. M. Juran, whose philosophies of quality and product cost have had a major impact on modern manufacturing.

During World War II, Deming and several others developed new methods of statistical process control in manufacturing plants for wartime industry. The need for statistical control arose from the recognition that there were variations in the performance of machines and people and the quality and dimensions of raw materials. Their efforts, however, involved not only statistical methods of analysis, but also a new way of looking at manufacturing operations to improve quality and lower costs. Recognizing the fact that manufacturing organizations are systems of management, workers, machines, and products, Deming emphasized communication, direct worker involvement, and education in statistics, as well as use of modern manufacturing technology. Fundamental to his philosophy is the dedication of management to quality and a shifting of emphasis from short-term profits to a longer term emphasis on improvement of products and processes.

**Taguchi methods.**  In the *Taguchi methods*, high quality and low cost are achieved by combining engineering and statistical methods to optimize product design and manufacturing processes. *Loss of quality* is defined as the financial loss to society after the product is shipped, with the following results: (1) Poor quality leads to customer dissatisfaction; (2) costs are involved in servicing and repairing defective products, some in the field; (3) the manufacturer's credibility is diminished in the marketplace; and (4) the manufacturer eventually loses its share of the market.

The Taguchi methods of quality engineering emphasize the importance of the following functions:

1. Enhancing cross-functional team interaction. In this interaction, design engineers and process or manufacturing engineers communicate with each other in a common language. They quantify the relationships between design requirements and the manufacturing process.

2. Implementing experimental design, in which the factors involved in a process or operation and their interactions are studied simultaneously.

3. Minimizing variation between parts, so that a consistent, high quality level is achieved.

In **experimental design**, the effects of controllable and uncontrollable variables on the product are identified; this approach minimizes variations in product dimensions and properties, bringing the mean level of product quality to the desired level. The methods used for experimental design are complex, involving the use of fractional factorial design and orthogonal arrays, which reduce the number of experiments required. These methods are also capable of identifying the effect on the product of variables that cannot be controlled (called *noise*), such as changes in environmental conditions.

The use of these methods allows rapid identification of the controlling variables and determination of the best method of process control. These variables are then controlled, without the need for costly new equipment or major modifications to existing equipment. For example, variables affecting dimensional tolerances in machining a particular component can be readily identified, and the correct cutting speed, feed, cutting tool, and cutting fluids can be specified.

**Taguchi loss function.**   Because traditional accounting practices had no real way of calculating losses on parts that met design specifications, the Taguchi loss function was introduced. In the traditional accounting approach, a part is defective and incurs a loss to the company when it exceeds its design tolerances; otherwise, there is no loss to the company. The Taguchi loss function calculates an increasing loss to the company the further the component is from the design objective.

## 4.9.4 The ISO 9000 and 14000 Standards

With increasing international trade, global manufacturing, and price-sensitive competition, a wide range of industrial and consumer products is now becoming available. Also, customers are increasingly demanding high-quality products and services at low prices and are looking for suppliers that can respond to this demand consistently and reliably. This strong trend, in turn, has created the need for international conformity and consensus regarding the establishment of quality-control methods and reliability and safety standards for products made in different countries and traded internationally.

The **ISO 9000** standard (**Quality Management and Quality Assurance Standards**) is a deliberately generic series of quality-system management standards. It has permanently influenced the way in which manufacturing companies conduct business in world trade and has become the world standard for quality. For certification, a company's plants are visited and audited by accredited and independent third-party teams to certify that the standard's 20 key elements are in place and functioning properly. Depending on the extent to which a company does not meet the requirements of the standard, registration may or may not be recommended at that time. The audit team does not advise or consult with the company on how to fix discrepancies, but merely describes the nature of the noncompliance. Periodic audits are required to maintain certification. The certification process can take from six months to a year or more and can cost tens of thousands of dollars, depending on the size, number of plants, and product line of the company.

The ISO 9000 standard is not a product certification, but a *quality-process certification*. Companies establish their own criteria and practices for quality. However, the documented quality system must be in compliance with the ISO 9000 standard; thus, a company cannot write into the system any criterion that opposes the intent of the standard. Registration means conformity to consistent practices, as specified by the company's own quality system (such as quality in design, development, production, installation, and servicing), including proper documentation of such practice. Thus, customers, including government agencies, are assured that specified practices are being followed by the supplier of the product or service (which may or may not be within the same country). In fact, manufacturing companies are themselves assured of such practice regarding their own suppliers who have ISO 9000 registration; they may even demand that their suppliers be registered.

**ISO 14000** is a family of standards pertaining to international **Environmental Management Systems** (EMS) and concerns the way in which an organization's activities affect the environment throughout the life of its products. These activities may be internal or external to the organization; they range from production to ultimate disposal of the product and include effects such as depletion of natural resources, pollution, waste generation and disposal, noise, and efficiency of energy use.

**QS 9000** was jointly developed by the U.S. automotive companies and is often described as an ISO 9000 standard with certain extras and additional requirements. The QS 9000 standards generally consist of production-part approval process, continuous improvement, manufacturing capabilities, and customer-specific requirements.

## SUMMARY

- The surface of a workpiece can significantly affect its properties and characteristics, including friction and wear properties, the effectiveness of lubricants, appearance and geometric features, tendency for cracking, and thermal and electrical conductivity of contacting bodies. (Section 4.1)

- The surface structure of a metal has usually been plastically deformed and work hardened during prior processing. Surface roughness can be quantified by a variety of techniques. Surface integrity can be compromised by a variety of defects. Flaws, directionality, roughness, and waviness are measurable quantities that are used to describe surface texture. (Sections 4.2, 4.3)

- Tribology is the science and technology of interacting surfaces, and it encompasses friction, wear, and lubrication. Friction may be desirable or undesirable, depending on the specific manufacturing circumstances. The adhesion theory of friction is most commonly used to model and predict the frictional behavior between sliding metal surfaces, because it agrees reasonably well with experimental observations. (Section 4.4)

- Wear is defined as the progressive loss or removal of material from a surface. Wear alters the geometry of the workpiece surface and tool and die interfaces and thus affects the manufacturing process, dimensional accuracy, and the quality of parts produced. Friction and wear can be reduced by using various liquid or solid lubricants, as well as by applying ultrasonic vibrations. Four regimes of lubrication are relevant to metalworking processes. (Section 4.4)

- Several surface treatments are used to impart certain physical and mechanical properties to the workpiece, such as resistance to environmental attack, wear, and fatigue. The techniques employed typically include mechanical working, physical and chemical means, heat treatments, and coatings. Cleaning of a manufactured workpiece involves removal of solid, semisolid, and liquid contaminants from a surface by various means. (Section 4.5)

- Many parts are made with a high degree of dimensional accuracy, thus requiring measuring instruments with various features and characteristics. The selection of a particular instrument depends on factors such as the type of measurement, the environment, and the accuracy of measurement. Major advances have been made in automated measurement, linking measuring equipment to microprocessors and computers for accurate in-process control of manufacturing operations. (Section 4.6)
- Dimensional tolerances and their specification are important factors in manufacturing, as they not only affect subsequent assembly of the parts made and the accuracy and operation of all types of machinery and equipment, but also can significantly influence product cost. (Section 4.7)
- Several nondestructive and destructive testing techniques, each having its own applications, advantages, and limitations, are available for inspection of completed parts and products. Unlike the traditional approach, the practice now is towards 100% inspection of all parts being manufactured, using automated and reliable inspection techniques. (Section 4.8)
- Quality assurance is the total effort by a manufacturer to ensure that its products conform to a detailed set of specifications and standards. A variety of statistical quality-control and statistical process-control techniques are now employed in defect detection and prevention. The total-quality-control philosophy focuses on prevention, rather than detection, of defects. (Section 4.9)
- Management philosophies such as those developed by Deming summarize the characteristics of companies that manufacture high-quality products. Chief among these characteristics are dedication to long-term improvement in quality and performance, direct involvement of all employees associated with a product, and continuing education. (Section 4.9)
- The Taguchi methods describe a powerful approach for experimentally characterizing and optimizing a complex system such as a manufacturing process or product design. The Taguchi loss function is a tool that emphasizes minimization of product variation in order to optimize quality. Also, the ISO 9000 and 14000 standards are now universally employed. (Section 4.9)

## SUMMARY OF EQUATIONS

Arithmetic mean value, $R_a = \dfrac{y_a + y_b + y_c + \cdots + y_n}{n} = \dfrac{1}{n}\sum_{i=1}^{n} y_i = \dfrac{1}{l}\int_0^l |y|\,dx$

Root-mean-square average, $R_q = \sqrt{\dfrac{y_a^2 + y_b^2 + y_c^2 + \cdots + y_n^2}{n}} = \sqrt{\dfrac{1}{n}\sum_1^n y_i^2} = \left[\dfrac{1}{l}\int_0^1 y^2\,dx\right]^{1/2}$

Coefficient of friction, $\mu = \dfrac{F}{N} = \dfrac{\tau}{\text{Hardness}}$

Friction (shear) factor, $m = \dfrac{\tau_i}{k}$

Adhesive wear, $V = \dfrac{kLW}{3p}$

Arithmetic mean, $\bar{x} = \dfrac{x_1 + x_2 + x_3 + \cdots + x_n}{n}$

Range, $R = x_{\max} - x_{\min}$

Standard deviation, $\sigma = \sqrt{\dfrac{(x_1 - \bar{x})^2 + (x_2 - \bar{x})^2 + (x_3 - \bar{x})^2 + \cdots + (x_n - \bar{x})^2}{n-1}}$

Upper control limit, $\text{UCL}_{\bar{x}} = \bar{x} + 3\sigma = \bar{\bar{x}} + A_2\bar{R}$

Lower control limit, $\text{LCL}_{\bar{x}} = \bar{x} - 3\sigma = \bar{\bar{x}} - A_2\bar{R}$

# BIBLIOGRAPHY

Aft, L. S., *Fundamentals of Industrial Quality Control*, 3d. ed., Addison-Wesley, 1998.

Arnell, R. D., P. B. Davies, J. Halling, and T. L. Whomes, *Tribology: Principles and Design Applications*, Springer, 1991.

*ASM Handbook*, Vol. 5: *Surface Engineering*, ASM International, 1994.

*ASM Handbook*, Vol. 17: *Nondestructive Evaluation and Quality Control*, ASM International, 1989.

*ASM Handbook*, Vol. 18: *Friction, Lubrication, and Wear Technology*, ASM International, 1992.

Bayer, R. G., *Mechanical Wear Prediction and Prevention*, Dekker, 1995.

Besterfield, D. H., *Quality Control*, Prentice Hall, 1997.

Bhushan, B., *Principles and Applications of Tribology*, Wiley, 1999.

Bhushan, B. (ed.), *Modern Tribology Handbook*, CRC Press, 2001.

Bhushan, B., and B. K. Gupta, *Handbook of Tribology: Materials, Coatings, and Surface Treatments*, McGraw-Hill, 1991.

Booser, E. R. (ed.), *Tribology Data Handbook*, CRC Press, 1998.

Bosch, J. A. (ed.), *Coordinate Measuring Machines and Systems*, Dekker, 1995.

Bothe, D. R., *Measuring Process Capability: Techniques and Calculations for Quality and Manufacturing Engineers*, McGraw-Hill, 1997.

Breyfogle, F., *Implementing Six Sigma: Smarter Solutions Using Statistical Methods*, Wiley, 1999.

Budinski, K. G., *Surface Engineering for Wear Resistance*, Prentice Hall, 1988.

Burakowski, T., and T. Wiershon, *Surface Engineering of Metals: Principles, Equipment, Technologies*, CRC Press, 1998.

Byers, J. P. (ed.), *Metalworking Fluids*, Dekker, 1994.

Cartz, L., *Nondestructive Testing*, ASM International, 1995.

Chattopadhyay, R., *Surface Wear: Analysis, Treatment, and Prevention*, ASM International, 2001.

Clausing, D., *Total Quality Development*, ASME Press, 1994.

Clements, R., *Handbook of Statistical Methods in Manufacturing*, Prentice Hall, 1991.

Deming, W. E., *Out of the Crisis*, MIT Press, 1982.

Dowson, D., *History of Tribology*, 2d. ed. ASME, 1999.

Drake, P. J., *Dimensioning and Tolerancing Handbook*, McGraw-Hill, 1999.

Edwards, J., *Coating and Surface Treatment Systems for Metals: A Comprehensive Guide to Selection*, ASM International, 1997.

Farrago, F. T., and M. A. Curtis, *Handbook of Dimensional Measurement*, 3d. ed., Industrial Press, 1994.

Feigenbaum, A. V., *Total Quality Control*, 3d. ed., McGraw-Hill, 1991.

Glasser, W. A., *Materials for Tribology*, Elsevier, 1992.

Grant, E. L., and R. S. Leavenworth, *Statistical Quality Control*, McGraw-Hill, 1997.

Henzhold, G., *Handbook of Geometric Tolerancing*, Wiley, 1995.

Holmberg, K., and A. Matthews, *Coating Tribology: Properties, Techniques, and Applications*, Elsevier, 1994.

Hutchings, I. M., *Tribology: Friction and Wear of Engineering Materials*, CRC Press, 1992.

Juran, J. M., and A. B. Godfrey (eds.), *Juran's Quality Handbook*, 5th ed., McGraw-Hill, 1999.

Juran, J. M., and F. M. Gryna, Jr., *Quality Planning and Analysis*, 4th ed., McGraw-Hill, 2000.

Kear, F. W., *Statistical Process Control in Manufacturing Practice*, Dekker, 1998.

Krulikowski, A., *Fundamentals of Geometric Dimensioning and Tolerancing*, Delmar, 1997.

Lansdown, A. R., and A. L. Price, *Materials to Resist Wear: A Guide to Their Selection and Use*, Pergamon Press, 1986.

Lindsay, J. H. (ed.), *Coatings and Coating Processes for Metals*, ASM International, 1998.

Ludema, K. C., *Friction, Wear, Lubrication: A Textbook in Tribology*, CRC Press, 1996.

Meadows, J. D., *Geometric Dimensioning and Tolerancing*, Dekker, 1995.

_____, *Measurement of Geometric Tolerances in Manufacturing*, Dekker, 1998.

Menon, H. G., *TQM in New Product Manufacturing*, McGraw-Hill, 1992.

Moore, D. F., *Principles and Applications of Tribology*, Pergamon, 1975.

Murphy, S. D., *In-Process Measurement and Control*, Dekker, 1990.

Nachtman, E. S., and S. Kalpakjian, *Lubricants and Lubrication in Metalworking Operations*, Dekker, 1985.

Neele, M. J. (ed.), *the Tribology Handbook*, 2d. ed., Butterworth-Heinemann, 1995.

Peterson, M. B., and W. O. Winer (eds.), *Wear Control Handbook*, American Society of Mechanical Engineers, 1980.

Puncochar, D. E., *Interpretation of Geometric Dimensioning and Tolerancing*, 2d. ed., Industrial Press, 1997.

Rabinowicz, E., *Friction and Wear of Materials*, 2d. ed., Wiley, 1995.

Robinson, S. L., and R. K. Miller, *Automated Inspection and Quality Assurance*, Dekker, 1989.

Roy, R., *A Primer on the Taguchi Method*, Society of Manufacturing Engineers, 1990.

Rudzki, G. J., *Surface Finishing Systems: Metal and Non-Metal Finishing Handbook-Guide*, Finishing Publications Ltd., 1983.

Schey, J. A., *Tribology in Metalworking: Friction, Lubricating and Wear*, ASM International, 1983.

Stachowiak, G. W., and A. W. Batchelor, *Engineering Tribology*, Butterworth-Heinemann, 2001.

Stern, K. H. (ed.), *Metallurgical and Ceramic Protective Coatings*, Chapman & Hall, 1996.

Sudarshan, T. S. (ed.), *Surface Modification Technologies*, ASM International, 1998.

*Tool and Manufacturing Engineers Handbook*, 4th ed., Vol. 3: *Materials, Finishing and Coating*, Society of Manufacturing Engineers, 1985.

*Tool and Manufacturing Engineers Handbook*, 4th ed., Vol. 4: *Assembly, Testing, and Quality Control*, Society of Manufacturing Engineers, 1986.

Wadsworth, H. M., *Handbook of Statistical Control Methods for Engineers and Scientists*, 2d. ed., McGraw-Hill, 1998.

Whitehouse, D. J., *Handbook of Surface Metrology*, Institute of Physics, 1994.

Winchell, W., *Inspection and Measurement in Manufacturing*, Society of Manufacturing Engineers, 1996.

## QUESTIONS

**4.1**  Explain what is meant by surface integrity. Why should we be interested in it?

**4.2**  Why are surface-roughness design requirements in engineering so broad? Give appropriate examples.

**4.3**  We have seen that a surface has various layers. Describe the factors that influence the thickness of these layers.

**4.4**  What is the consequence of oxides of metals being generally much harder than the base metal? Explain.

**4.5**  What factors would you consider in specifying the lay of a surface?

**4.6**  Describe the effects of various surface defects (see Section 4.3) on the performance of engineering components in service. How would you go about determining whether or not each of these defects is important for a particular application?

**4.7**  Explain why the same surface roughness values do not necessarily represent the same type of surface.

**4.8**  In using a surface roughness measuring instrument, how would you go about determining the cutoff value? Give appropriate examples.

**4.9**  What is the significance of the fact that the stylus path and the actual surface profile are generally not the same?

**4.10**  Give two examples each in which waviness of a surface would be (1) desirable and (2) undesirable.

**4.11**  Explain why surface temperature increases when two bodies are rubbed against each other. What is the significance of a temperature rise due to friction?

**4.12**  To what factors would you attribute the fact that the coefficient of friction in hot working is higher than in cold working, as shown in Table 4.1?

**4.13**  In Section 4.4.1, we note that the values of the coefficient of friction can be much higher than unity. Explain why.

**4.14**  Describe the tribological differences between ordinary machine elements (such as meshing gears, cams in contact with followers, and ball bearings with inner and outer races) and elements of metalworking processes (such as forging, rolling, and extrusion, which involve workpieces in contact with tools and dies).

**4.15**  Give the reasons that an originally round specimen in a ring-compression test may become oval after deformation.

**4.16**  Can the temperature rise at a sliding interface exceed the melting point of the metals? Explain.

**4.17**  Explain why each of the terms in the Archard formula for adhesive wear, Eq. (4.6), should affect the wear volume.

**4.18**  How can adhesive wear be reduced? How can fatigue wear be reduced?

**4.19**  It has been stated that as the normal load decreases, abrasive wear is reduced. Explain why this is so.

**4.20**  Why is the abrasive wear resistance of a material a function of its hardness?

**4.21**  We have seen that wear can have detrimental effects on engineering components, tools, dies, etc. Can you visualize situations in which wear could be beneficial? Give some examples. (*Hint*: Note that writing with a pencil is a wear process.)

**4.22**  On the basis of the topics discussed in this chapter, do you think there is a direct correlation between friction and wear of materials? Explain.

**4.23**  You have undoubtedly replaced parts in various appliances and automobiles because they were worn. Describe the methodology you would follow in determining the type(s) of wear these components have undergone.

**4.24**  Why is the study of lubrication regimes important?

**4.25**  Explain why so many different types of metalworking fluids have been developed.

**4.26**  Differentiate between (1) coolants and lubricants, (2) liquid and solid lubricants, (3) direct and indirect emulsions, and (4) plain and compounded oils.

**4.27** Explain the role of conversion coatings. Based on Fig. 4.13, what lubrication regime is most suitable for application of conversion coatings?

**4.28** Explain why surface treatment of manufactured products may be necessary. Give several examples.

**4.29** Which surface treatments are functional, and which are decorative? Give examples.

**4.30** Give examples of several typical applications of mechanical surface treatment.

**4.31** Explain the difference between case hardening and hard facing.

**4.32** List several applications for coated sheet metal, including galvanized steel.

**4.33** Explain how roller-burnishing processes induce residual stresses on the surface of workpieces.

**4.34** List several products or components that could not be made properly, or function effectively in service, without implementation of the knowledge involved in Sections 4.2 through 4.5.

**4.35** Explain the difference between direct- and indirect-reading linear measurements.

**4.36** Why have coordinate measuring machines become important instruments in modern manufacturing? Give some examples of applications.

**4.37** Give reasons that the control of dimensional tolerances is important.

**4.38** Give examples where it may be preferrable to specify unilateral tolerances as opposed to bilateral tolerances in design.

**4.39** Explain why a measuring instrument may not have sufficient precision.

**4.40** Comment on the differences, if any, between (1) roundness and circularity, (2) roundness and eccentricity, and (3) roundness and cylindricity.

**4.41** It has been stated that dimensional tolerances for nonmetallic stock, such as plastics, are usually wider than for metals. Explain why. Consider physical and mechanical properties.

**4.42** Describe the basic features of nondestructive testing techniques that use electrical energy.

**4.43** Identify the nondestructive techniques that are capable of detecting internal flaws and those that detect external flaws only.

**4.44** Which of the nondestructive inspection techniques are suitable for nonmetallic materials? Why?

**4.45** Why is automated inspection becoming an important part of manufacturing engineering?

**4.46** Describe situations in which the use of destructive testing techniques is unavoidable.

**4.47** Should products be designed and built for a certain expected life? Explain.

**4.48** What are the consequences of setting lower and upper specifications closer to the peak of the curve in Fig. 4.23?

**4.49** Identify factors that can cause a process to become out of control. Give several examples of such factors.

**4.50** In reading this chapter, you will have noted that the specific term *dimensional tolerance* is often used, rather than just the word *tolerance*. Do you think this distinction is important? Explain.

**4.51** Why is it important to minimize variation among manufactured parts?

**4.52** What are the key elements of the Deming philosophy?

**4.53** Does the presence of a lubricant affect abrasive wear? Explain.

**4.54** Give an example of an assignable variation and a chance variation.

**4.55** Explain why GO–NOT-GO gages are incompatible with the Taguchi-loss-function approach.

**4.56** Explain how you would estimate the magnitude of the wear coefficient for a pencil writing on paper.

**4.57** Describe a test method for determining the wear coefficient $k$ in Eq. (4.9). What would be the difficulties in applying the results from this test to a manufacturing application, such as predicting the life of tools and dies?

# PROBLEMS

**4.58** Referring to the surface profile in Fig. 4.3, give some numerical values for the vertical distances from the center line. Calculate the $R_a$ and $R_q$ values. Then give another set of values for the same general profile, and calculate the same quantities. Comment on your results.

**4.59** Using Fig. 4.8a, make a plot of the coefficient of friction versus the change in internal diameter for a reduction in height of (1) 20%, (2) 40%, and (3) 60%.

**4.60** In Example 4.1, assume that the coefficient of friction is 0.30. If all other initial parameters remain the same, what is the new internal diameter of the ring specimen?

**4.61** How would you go about estimating forces required for roller burnishing? (*Hint*: Consider hardness testing.)

**4.62** Assume that a steel rule expands by 1% because of an increase in environmental temperature. What will be the indicated diameter of a shaft whose actual diameter is 75.00 mm?

**4.63** Assume that in Example 4.3, the number of samples was 7 instead of 10. Using the top half of the data given in the table, recalculate the control limits and the standard deviation. Compare your results with the results obtained by using 10 samples.

**4.64** Calculate the control limits for averages and ranges for the following: number of samples $= 5$; $\bar{\bar{x}} = 60$; $\bar{R} = 7$.

**4.65** Calculate the control limits for the following: number of samples $= 5$; $\bar{\bar{x}} = 36.5$; $\mathrm{UCL}_R = 4.75$.

**4.66** In an inspection with a sample size of 7 and a sample number of 40, it was found that the average range was 10 and the average of averages was 75. Calculate the control limits for averages and ranges.

**4.67** Determine the control limits for the data shown in the following table:

| $x_1$ | $x_2$ | $x_3$ | $x_4$ |
|-------|-------|-------|-------|
| 0.65 | 0.70 | 0.67 | 0.65 |
| 0.69 | 0.65 | 0.70 | 0.68 |
| 0.65 | 0.60 | 0.65 | 0.61 |
| 0.64 | 0.67 | 0.60 | 0.60 |
| 0.68 | 0.68 | 0.70 | 0.66 |
| 0.70 | 0.71 | 0.65 | 0.71 |

**4.68** The average of averages of a number of samples of size 5 was determined to be 125. The average range was 17.82, and the standard deviation was 6. The following measurements were taken in a sample: 120, 132, 124, 130, 118, 132, 121, and 127. Is the process in control?

**4.69** Assume that you are asked to give a quiz to students on the contents of this chapter. Prepare three quantitative problems and three qualitative questions, and supply the answers.

# DESIGN

**4.70** Design an experimental fixture that allows a conventional profilometer to be indexed across a surface and thereby measure the surface in three dimensions.

**4.71** List several products or components that could not be designed properly, or function as effectively in service, without implementation of the information covered in this chapter.

**4.72** Survey the available literature, and prepare a report on the environmental considerations regarding the application of the various processes and operations described in this chapter.

**4.73** Comment on the design challenges that have to be faced in installing surface-measuring instruments directly on production machines. Consider the typical factory environment.

**4.74** Are there product designs for which wear is actually desirable? Explain. (See also Question 4.21.)

**4.75** Consider products that you are familiar with and for which wear is a serious problem. What design changes would you recommend to reduce wear in these products? Explain your reasoning.

**4.76** An artificial implant has a porous surface so that bone will grow into the surface of the implant and produce a strong joint. Give recommendations for producing a porous surface.

**4.77** Solar energy has been used as an energy source for posttreatment of coatings and deposited films. Design a suitable system to implement solar energy for this purpose.

**4.78** As you know, vandalism of public places, such as graffiti, scratching, spray painting, etc., is a serious problem. Make a list of possible methods of minimizing such damage, and explain what design changes would be required to make these methods effective.

# CHAPTER

# 5

# Metal-Casting Processes and Equipment; Heat Treatment

## 5.1 | Introduction

As described throughout this text, several methods can be used to shape materials into useful products. Making parts by **casting** molten metal into a mold and letting it solidify is an obvious choice. Indeed, casting is among the oldest methods of net-shape and near-net-shape manufacturing and was first used in about 4000 BC to make ornaments, copper arrowheads, and various other objects. Basically, casting processes involve the introduction of molten metal into a mold cavity; upon solidification, the metal takes the shape of the cavity. The casting process is thus capable of producing intricate shapes in a single piece, including shapes with internal cavities; very large, very small, and hollow parts can be produced economically by casting techniques. Typical cast products are engine blocks, cylinder heads, transmission housings, pistons, turbine disks, railroad and automotive wheels, and ornamental artifacts.

Almost all metals can be cast in (or nearly in) the final shape desired, often with only minor finishing required. With appropriate control of material and process parameters, parts can be cast with uniform properties throughout. As with all other manufacturing processes, a knowledge of certain fundamental relationships is essential to the production of good-quality and economical castings; these relationships help establish proper techniques for mold design and casting practice. The objective is to produce castings that are free from defects and that meet requirements for strength, dimensional accuracy, and surface finish.

Fundamentally, the important factors in casting operations are as follows:

a) **Solidification** of the metal from its molten state (which usually is accompanied by shrinkage);

b) **Flow** of the molten metal into the mold cavity;

c) **Heat transfer** during solidification and cooling of the metal in the mold;

d) **Mold material** and its influence on the casting process.

We first discuss these aspects of casting and then continue with a description of melting practice and the characteristics of casting alloys. The remainder of the chapter covers ingot and continuous casting; various expendable-mold and permanent-mold casting processes, including their advantages and limitations; processing of castings; heat treatment; casting design; and the economics of casting.

5.1 Introduction
5.2 Solidification of Metals
5.3 Cast Structures
5.4 Fluid Flow and Heat Transfer
5.5 Melting Practice and Furnaces
5.6 Casting Alloys
5.7 Ingot Casting and Continuous Casting
5.8 Casting Processes: Expendable Mold, Permanent Pattern
5.9 Casting Processes: Expendable Mold, Expendable Pattern
5.10 Casting Processes: Permanent Mold
5.11 Heat Treatment
5.12 Design Considerations
5.13 Economics of Casting

## 5.2 | Solidification of Metals

This section presents an overview of the solidification of metals and alloys. We cover topics that are particularly relevant to casting and are essential to our understanding of the structures developed in casting and the structure–property relationships obtained in the casting processes described in this chapter.

**Pure metals** have clearly defined melting or freezing points, and solidification takes place at a constant temperature (Fig. 5.1a). When the temperature of a molten metal is reduced to the freezing point, the latent heat of fusion is given off while the temperature remains constant. At the end of this isothermal phase change, solidification is complete, and the solid metal cools to room temperature. Objects contract as they cool; with castings, this cooling-induced contraction is due to (a) contraction from a superheated state to the metal's solidification temperature, and (b) cooling as a solid from the solidification temperature to room temperature (Fig. 5.1b). In addition, a significant density change can occur as a result of phase change from liquid to solid.

Unlike pure metals, **alloys** solidify over a range of temperatures. Solidification begins when the temperature of the molten metal drops below the **liquidus**; it is completed when the temperature reaches the **solidus**. Within this temperature range, the alloy is in a mushy or pasty state; its composition and state are described by the particular alloy's phase diagram (Section 5.2.4).

### 5.2.1 Solid solutions

Two terms are essential in describing alloys: *solute* and *solvent*. **Solute** is the minor element (such as salt or sugar) that is added to the **solvent**, which is the major element (such as water). In terms of the elements involved in a metal's crystal structure, the solute is the element (*solute atoms*) added to the solvent (*host atoms*). When the particular crystal structure of the solvent is maintained during alloying, the alloy is called a *solid solution*.

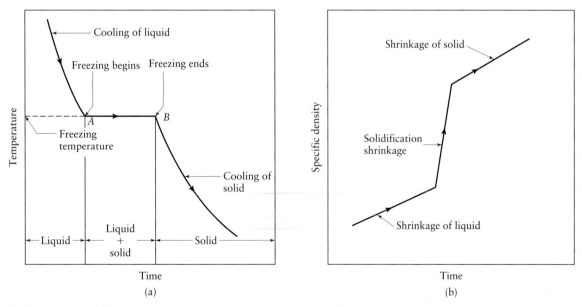

**FIGURE 5.1** (a) Temperature as a function of time for the solidification of pure metals. Note that freezing takes place at a constant temperature. (b) Density as a function of time.

**Substitutional solid solutions.** If the size of the solute atom is similar to that of the solvent atom, solute atoms can replace solvent atoms and form a *substitutional solid solution*. (See Fig. 3.8.) An example of this phenomenon is brass, an alloy of zinc and copper, in which zinc (solute atoms) is introduced into the lattice of copper (solvent atoms). The properties of brasses can thus be altered over a certain range by controlling the amount of zinc in the copper.

**Interstitial solid solutions.** If the size of the solute atom is much smaller than that of the solvent atom, the solute atom occupies an interstitial position (See Fig. 3.8) and forms an *interstitial solid solution*. An important example of an interstitial solid solution is steel, an alloy of iron and carbon, in which carbon atoms are present in an interstitial position between iron atoms. As will be shown in Section 5.11, the properties of steel can be varied over a wide range by controlling the amount of carbon in iron. This is one reason that, in addition to being inexpensive, steel is such a versatile and useful material with a wide variety of properties and applications.

### 5.2.2 Intermetallic compounds

*Intermetallic compounds* are complex structures in which solute atoms are present among solvent atoms in certain specific proportions; thus, some intermetallic compounds have solid solubility. The type of atomic bonds may range from metallic to ionic. Intermetallic compounds are strong, hard, and brittle. An example is copper in aluminum, where an intermetallic compound of $CuAl_2$ can be made to precipitate from an aluminum–copper alloy; this process is an example of *precipitation hardening*. (See Section 5.11.2.)

### 5.2.3 Two-phase alloys

A solid solution is one in which two or more elements are soluble in a solid state, forming a single homogeneous solid phase in which the alloying elements are uniformly distributed throughout the solid mass. However, there is a limit to the concentration of solute atoms in a solvent-atom lattice, just as there is a solubility limit to sugar in water. Most alloys consist of two or more solid phases and may be regarded as mechanical mixtures. Such a system with two solid phases is called a **two-phase system**; each phase is a homogeneous part of the total mass and has its own characteristics and properties.

A typical example of a two-phase system in metals is lead added to copper in the molten state. After the mixture solidifies, the structure consists of two phases: one phase has a small amount of lead in solid solution in copper; in the other phase, lead particles, roughly spherical in shape, are *dispersed* throughout the matrix of the primary phase (Fig. 5.2a). This copper–lead alloy has properties that are different than those of either copper or lead alone. Alloying with finely dispersed particles (*second-phase particles*) is an important method of strengthening alloys and controlling their properties. Generally, in two-phase alloys, the second-phase particles present obstacles to dislocation movement, which increases the alloy's strength.

**FIGURE 5.2** (a) Schematic illustration of grains, grain boundaries, and particles dispersed throughout the structure of a two-phase system, such as lead–copper alloy. The grains represent lead in a solid solution of copper, and the particles are lead as a second phase. (b) Schematic illustration of a two-phase system consisting of two sets of grains: dark and light. Dark and light grains have their own compositions and properties.

(a)

(b)

Another example of a two-phase alloy is the aggregate structure shown in Fig. 5.2b; this alloy system contains two sets of grains, each with its own composition and properties. The darker grains may, for example, have a different structure than the lighter grains and may be brittle, whereas the lighter grains may be ductile.

## 5.2.4  Phase diagrams

A *phase diagram*, also called an **equilibrium diagram** or **constitutional diagram**, graphically illustrates the relationships among temperature, composition, and the phases present in a particular alloy system. *Equilibrium* means that the state of a system remains constant over an indefinite period of time. *Constitutional* indicates the relationships among structure, composition, and physical makeup of the alloy.

An example of a phase diagram is shown in Fig. 5.3 for the nickel–copper alloy; it is called a **binary phase diagram** because of the two elements (nickel and copper) in the system. The left boundary of this phase diagram (100% Ni) indicates the melting point of nickel, and the right boundary (100% Cu) indicates the melting point of copper. (All percentages in this discussion are by weight.) Note that for a composition of, say, 50% Cu-50% Ni, the alloy begins to solidify at a temperature of

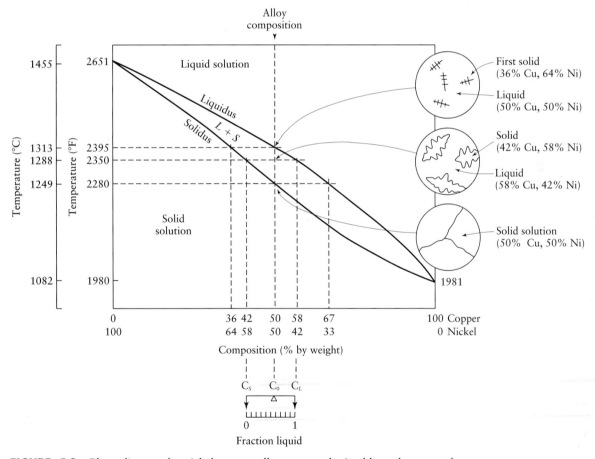

**FIGURE 5.3**   Phase diagram for nickel–copper alloy system obtained by a slow rate of solidification. Note that pure nickel and pure copper each have one freezing or melting temperature. The top circle on the right depicts the nucleation of crystals. The second circle shows the formation of dendrites. The bottom circle shows the solidified alloy with grain boundaries.

1313°C (2395°F), and solidification is complete at 1249°C (2280°F). Above 1313°C, a homogeneous liquid of 50% Cu-50% Ni exists. When cooled slowly to 1249°C, a homogeneous solid solution of 50% Cu-50% Ni results. However, between the liquidus and solidus curves, say, at a temperature of 1288°C (2350°F), is a two-phase region: a *solid phase* composed of 42% Cu-58% Ni, and a *liquid phase* of 58% Cu-42% Ni. To determine the composition of the solid phase, go left horizontally to the solidus curve and read down to obtain 42% Cu-58% Ni. The composition of the liquid phase can be determined similarly by going to the right to the liquidus curve (58% Cu-42% Ni).

The completely solidified alloy in the phase diagram shown in Fig. 5.3 is a **solid solution**, because the alloying element (Cu, the solute atom) is completely dissolved in the host metal (Ni, the solvent atom), and each grain has the same composition. The mechanical properties of solid solutions of Cu–Ni depend on composition. By increasing the nickel content, the properties of pure copper are improved. The improvements in properties result from pinning (blocking) of dislocations at solute atoms of nickel, which may also be regarded as impurity atoms. (See Fig. 3.8.) As a result, dislocations cannot move as freely, and consequently the strength of the alloy increases.

**Lever rule.**  The composition of various phases in a phase diagram can be determined by a procedure called the *lever rule*. As shown in the lower portion of Fig. 5.3, we first construct a lever (*tie line*) between the solidus and liquidus lines; the lever is balanced (on the triangular support) at the nominal weight composition $C_0$ of the alloy. The left end of the lever represents the composition $C_S$ of the solid phase, and the right end represents the composition $C_L$ of the liquid phase. Note from the graduated scale in the figure that we also have indicated the liquid fraction along this tie line, ranging from zero at the left (fully solid) to unity at the right (fully liquid).

The lever rule states that the **weight fraction of solid** is proportional to the distance between $C_0$ and $C_L$:

$$\frac{S}{S + L} = \frac{C_0 - C_L}{C_S - C_L}.$$  (5.1)

Likewise, the **weight fraction of liquid** is proportional to the distance between $C_S$ and $C_0$:

$$\frac{L}{S + L} = \frac{C_S - C_0}{C_S - C_L}.$$  (5.2)

Note that these quantities are fractions and must be multiplied by 100 to obtain percentages.

From inspection of the tie line in Fig. 5.3 (and for a nominal alloy composition of $C_0 = 50\%$ Cu-50% Ni), we note that, because $C_0$ is closer to $C_L$ than it is to $C_S$, the solid phase contains less copper than does the liquid phase. Measuring on the phase diagram and using the lever-rule equations, we find that the composition of the solid phase is 42% Cu and the composition of the liquid phase is 58% Cu, as stated in the middle circle at the right in Fig. 5.3.

Note that these calculations refer to copper. If we now reverse the phase diagram in the figure, so that the left boundary is 0% nickel (whereby nickel becomes the alloying element in copper), these calculations give us the compositions of the solid and liquid phases in terms of nickel. The lever rule is also known as the *inverse lever rule*, because, as Eqs. (5.1) and (5.2) show, the amount of each phase is proportional to the length of the opposite end of the lever.

## 5.2.5 The iron–carbon system

Steel and cast iron are represented by the *iron–carbon binary system*. Commercially pure iron contains up to 0.008% C, steels up to 2.11% C, and cast irons up to 6.67% C, although most cast irons contain less than 4.5% C. In this section, we describe the iron–carbon system in order to evaluate and discuss how to modify the properties of these important materials for specific applications.

Figure 5.4 shows the **iron–iron-carbide phase diagram**. Although this diagram can be extended to the right (to 100% C, pure graphite), the range that is significant to engineering applications is up to 6.67% C, where cementite forms. Pure iron melts at a temperature of 1538°C (2800°F), as shown at the left in Fig. 5.4; as iron cools, it first forms δ-iron, then γ-iron, and finally α-iron, which are described as follows:

1. **Ferrite.** *Delta ferrite* is stable only at very high temperatures and has no significant or practical engineering applications. *Alpha ferrite*, or simply *ferrite*, is a solid solution of bcc iron and has a maximum solid solubility of 0.022% C at a temperature of 727°C (1341°F). Just as there is a solubility limit to salt in water (with any extra amount precipitating as solid salt at the bottom of the container), so there also is a solid solubility limit to carbon in iron.

   Ferrite is relatively soft and ductile and is magnetic from room temperature up to 768°C (1414°F). Although very little carbon can dissolve interstitially in bcc iron, the amount of carbon in the composition can significantly affect the mechanical properties of ferrite. Also, significant amounts of chromium, manganese, nickel, molybdenum, tungsten, and silicon can be contained in iron in solid solution, imparting certain desirable properties.

2. **Austenite.** Between 1394°C (2541°F) and 912°C (1674°F) iron undergoes an *allotropic transformation* from the bcc structure to the fcc structure, becoming what is known as *gamma iron*, or more commonly, *austenite*. This structure has a solid solubility of up to 2.11% C at 1148°C (2098°F); thus, the solid solubility of austenite is about two orders of magnitude higher than that of ferrite, with the carbon occupying interstitial positions in austenite (Fig. 5.5). Note that the atomic radius of Fe is 0.124 nm and of C is 0.071 nm. Austenite is an important phase in the heat treatment of steels (Section 5.11). It is denser than ferrite, and its single-phase fcc structure is ductile at elevated temperatures; thus, it possesses good formability. Large amounts of nickel and manganese can also be dissolved

**FIGURE 5.4** The iron–iron-carbide phase diagram. Because of the importance of steel as an engineering material, this diagram is one of the most important phase diagrams.

in fcc iron to impart various properties. Austenitic steel is nonmagnetic at high temperatures, and austenitic stainless steels are nonmagnetic at room temperature.

3. **Cementite.** The right boundary of Fig. 5.4 represents *cementite*, also called **carbide**, which is 100% iron carbide ($Fe_3C$), with a carbon content of 6.67%. (This carbide should not be confused with carbides used for tool and die materials, described in Section 8.6.4) Cementite is a very hard and brittle intermetallic compound and significantly influences the properties of steels. It can be alloyed with elements such as chromium, molybdenum, and manganese.

### 5.2.6 The iron–iron-carbide phase diagram

The region of the iron–iron-carbide phase diagram that is significant for steels is shown in Fig. 5.6, an enlargement of the lower left portion of Fig. 5.4. Various microstructures

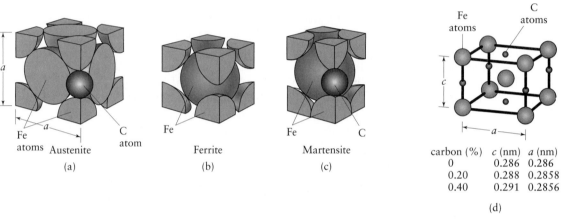

| carbon (%) | $c$ (nm) | $a$ (nm) |
|---|---|---|
| 0 | 0.286 | 0.286 |
| 0.20 | 0.288 | 0.2858 |
| 0.40 | 0.291 | 0.2856 |

(d)

**FIGURE 5.5**   The unit cell for (a) austenite, (b) ferrite, and (c) martensite. The effect of the percentage of carbon (by weight) on the lattice dimensions for martensite is shown in (d). Note the interstitial position of the carbon atoms (see also Fig. 3.8) and the increase in dimension $c$ with increasing carbon content. Thus, the unit cell of martensite is in the shape of a rectangular prism.

**FIGURE 5.6**   Schematic illustration of the microstructures for an iron–carbon alloy of eutectoid composition (0.77% carbon) above and below the eutectoid temperature of 727°C (1341°F).

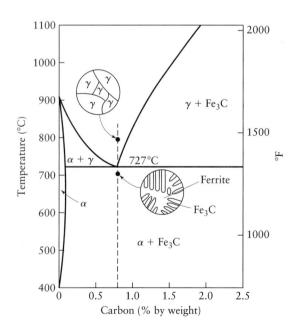

can be developed in steels, depending on the carbon content and the method of heat treatment (Section 5.11). For example, let's consider iron with a 0.77% C content being cooled very slowly from a temperature of, say, 1100°C (2012°F) in the austenite phase. The reason for the slow cooling rate is to maintain equilibrium; higher rates of cooling are used in heat treating. At 727°C (1341°F), a reaction takes place in which austenite is transformed into alpha ferrite (bcc) and cementite. As the solid solubility of carbon in ferrite is only 0.022%, the extra carbon forms cementite.

This reaction is called a **eutectoid** (meaning *eutecticlike*) **reaction**, indicating that at a certain temperature, a single solid phase (austenite) is transformed into two other solid phases (ferrite and cementite). The structure of the eutectoid steel is called **pearlite**, because it resembles mother of pearl at low magnifications. The microstructure of pearlite consists of alternate layers (*lamellae*) of ferrite and cementite (Fig. 5.6). Consequently, the mechanical properties of pearlite are intermediate between ferrite (soft and ductile) and cementite (hard and brittle).

In iron with less than 0.77% C, the microstructure formed consists of a pearlite phase (ferrite and cementite) and a ferrite phase. The ferrite in the pearlite is called *eutectoid ferrite*. The ferrite phase is called *proeutectoid ferrite*, *pro* meaning before, because it forms at a temperature higher than the eutectoid temperature of 727°C (1341°F). If the carbon content is higher than 0.77%, the austenite transforms into pearlite and cementite. The cementite in the pearlite is called *eutectoid cementite*, and the cementite phase is called *proeutectoid cementite*, because it forms at a temperature higher than the eutectoid temperature.

**Effects of alloying elements in iron.**  Although carbon is the basic element that transforms iron into steel, other elements are also added to impart various desirable properties. The effect of these alloying elements on the iron–iron-carbide phase diagram is to shift the eutectoid temperature and eutectoid composition (the percentage of carbon in steel at the eutectoid point). The eutectoid temperature may be raised or lowered from 727°C (1341°F), depending on the particular alloying element. However, alloying elements always lower the eutectoid composition; that is, the carbon content becomes less than 0.77%. Lowering the eutectoid temperature means increasing the austenite range; hence, an alloying element such as nickel is known as an **austenite former**. Because nickel has an fcc structure, it tends to favor the fcc structure of austenite. Conversely, chromium and molybdenum have a bcc structure, causing these elements to favor the bcc structure of ferrite; these elements are known as **ferrite formers**.

## 5.2.7  Cast irons

The term *cast iron* refers to a family of ferrous alloys composed of iron, carbon (ranging from 2.11% to about 4.5%), and silicon (up to about 3.5%). Cast irons (Section 5.6.1) are usually classified according to their solidification morphology:

1. **Gray cast iron**, or gray iron;
2. **Nodular cast iron**, ductile cast iron, or spheroidal graphite cast iron;
3. **White cast iron**;
4. **Malleable iron**;
5. **Compacted graphite iron**.

Cast irons are also classified by their structure: ferrite, pearlite, quenched and tempered, and austempered.

**FIGURE 5.7** Phase diagram for the iron–carbon system with graphite, instead of cementite, as the stable phase. Note that this figure is an extended version of Fig. 5.4.

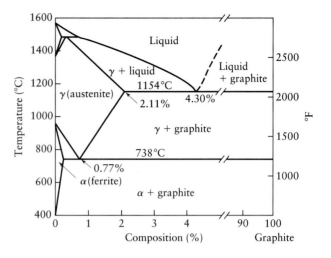

Figure 5.7 shows the equilibrium diagram relevant to cast irons, in which the right boundary is 100% C, that is, pure graphite. The horizontal liquid + graphite line is at 1154°C (2109°F); thus, cast irons are completely liquid at temperatures lower than those required for liquid steels. Consequently, cast irons have lower melting temperatures, which is why the casting process is so suitable for iron with high carbon content. Although cementite exists in steels almost indefinitely, it is not completely stable. That is, it is *metastable*, with an extremely low rate of decomposition. However, cementite can be made to decompose into alpha ferrite and graphite. The formation of graphite (*graphitization*) can be controlled, promoted, and accelerated by modifying the composition and the rate of cooling, and by the addition of silicon.

---

**EXAMPLE 5.1  Determining the amount of phases in carbon steel**

Determine the amount of gamma and alpha phases in a 10-kg, 1040 steel casting as it is being cooled slowly to the following temperatures: (a) 900°C, (b) 728°C, and (c) 726°C.

**Solution.** (a) Referring to Fig. 5.6, we draw a vertical line at 0.40% C at 900°C. We are in the single-phase austenite region, so the percent gamma phase is 100 (10 kg) and the percent alpha phase is zero. (b) At 728°C, the alloy is in the two-phase gamma–alpha field. When we draw the phase diagram in greater detail, we can find the weight percentages of each phase by the lever rule:

$$\text{Percent alpha} = \left(\frac{C_\gamma - C_0}{C_\gamma - C_\alpha}\right)100 = \left(\frac{0.77 - 0.40}{0.77 - 0.022}\right)100 = 50\%, \text{ or 5 kg;}$$

$$\text{Percent gamma} = \left(\frac{C_0 - C_\alpha}{C_\gamma - C_\alpha}\right)100 = \left(\frac{0.40 - 0.022}{0.77 - 0.022}\right)100 = 50\%, \text{ or 5 kg.}$$

(c) At 726°C, the alloy is in the two-phase alpha–$Fe_3C$ region. No gamma phase is present. Again, we use the lever rule to find the weight percentage of alpha phase present:

$$\text{Percent alpha} = \left(\frac{6.67 - 0.40}{6.67 - 0.022}\right)100 = 94\%, \text{ or 9.4 kg.}$$

# 5.3 | Cast Structures

The type of *cast structure* developed during solidification of metals and alloys depends on the composition of the particular alloy, as well as on the rate of heat transfer and the flow of the liquid metal during the casting process. As described in other sections in this chapter, the structures developed, in turn, affect the properties of the finished castings.

## 5.3.1 Pure metals

Figure 5.8a shows the grain structure of a pure metal solidified in a square mold. At the walls of the mold, the metal cools rapidly, because the walls are at ambient temperature. As a result of cooling first at the mold walls, the casting develops a solidified **skin**, or *shell*, or fine **equiaxed grains**. The grains grow in the direction opposite to the direction of heat transfer from the mold. Grains that have a favorable orientation, called **columnar grains**, grow preferentially (Fig. 5.9); grains that have substantially different orientations are blocked from further growth.

(a)              (b)              (c)

**FIGURE 5.8** Schematic illustration of three cast structures of metals solidified in a square mold: (a) pure metals; (b) solid-solution alloys; and (c) the structure obtained by heterogenous nucleation of grains, using nucleating agents. *Source*: G. W. Form, J. F. Wallace, J. L. Walker, and A. Cibula.

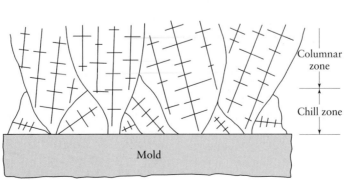

**FIGURE 5.9** Development of a preferred texture at a cool mold wall. Note that only favorably oriented grains grow away from the surface of the mold.

### 5.3.2 Alloys

Pure metals have limited mechanical properties; however, their properties can be enhanced and modified by **alloying**. The vast majority of metals used in engineering applications are some form of an *alloy*, consisting of two or more chemical elements, at least one of which is a metal. Because many consumer and industrial products contain cast alloys, a study of their solidification is essential.

Solidification in alloys begins when the temperature drops below the liquidus, $T_L$, and is complete when it reaches the solidus, $T_S$ (Fig. 5.10). Within this temperature range, the alloy is in a mushy or pasty state and has **columnar dendrites** (from the Greek words *dendron*, meaning *akin to*, and *drys*, meaning *tree*). Note the presence of liquid metal between the dendrites' arms. Dendrites have three-dimensional arms and branches (secondary arms), and the arms eventually interlock, as shown in Fig. 5.11. The width of the mushy zone, where both liquid and solid phases are pre-

**FIGURE 5.10** Schematic illustration of alloy solidification and temperature distribution in a solidifying metal. Note the formation of dendrites in the semisolid (mushy) zone.

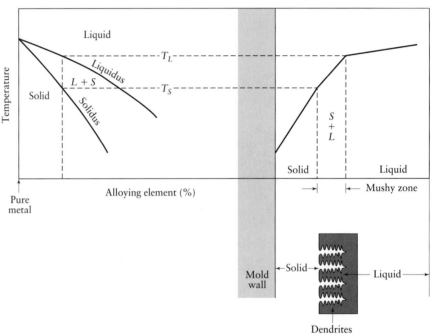

**FIGURE 5.11**
(a) Solidification patterns for gray cast iron in a 180-mm (7-in.) square casting. Note that after 11 minutes of cooling, dendrites reach each other, but the casting is still mushy throughout. It takes about two hours for this casting to solidify completely. (b) Solidification of carbon steels in sand and chill (metal) molds. Note the difference in solidification patterns as the carbon content increases. *Source*: H. F. Bishop and W. S. Pellini.

sent, is an important factor during solidification. This zone is described in terms of a temperature difference, known as the **freezing range**, as follows:

$$\text{Freezing range} = T_L - T_S. \tag{5.3}$$

Figure 5.10 shows that pure metals have a freezing range that approaches zero and that the *solidification front* moves as a plane front, without forming a mushy zone. Eutectics solidify in a similar manner, with an approximately planar front. The type of solidification structure developed depends on the composition of the eutectic. For alloys with a nearly symmetrical phase diagram, the structure is generally lamellar, with two or more solid phases present, depending on the alloy system. When the volume fraction of the minor phase of the alloy is less than about 25%, the structure generally becomes *fibrous*. These conditions are particularly important for cast irons.

For alloys, although it is not precise, a short freezing range generally involves a temperature difference of less than 50°C (90°F), and a long freezing range is generally at temperatures greater than 110°C (200°F). Ferrous castings typically have narrow semisolid (mushy) zones, whereas aluminum and magnesium alloys have wide mushy zones. Consequently, these alloys are in a semisolid state throughout most of the solidification process.

**Effects of cooling rate.**  Slow cooling rates (on the order of $10^2$ K/s) or long local solidification times result in coarse dendritic structures with large spacing between the dendrites' arms. For faster cooling rates (on the order of $10^4$ K/s) or short local solidification times, the structure becomes finer, with smaller dendrite arm spacing. For still higher cooling rates (on the order of $10^6$ to $10^8$ K/s), the structures developed are *amorphous* (without any ordered crystalline structure), as described in Section 5.10.8. The structures developed and the resulting grain size, in turn, influence the properties of the casting. As grain size decreases, (a) the strength and ductility of the cast alloy increase (see the Hall-Petch equation, discussed in Section 3.4.1), (b) microporosity (interdendritic shrinkage voids) in the casting decreases, and (c) the tendency for the casting to crack (*hot tearing*) during solidification decreases. Lack of uniformity in grain size and distribution results in castings with anisotropic properties.

### 5.3.3  Structure–property relationships

Because all castings are expected to possess certain properties in order to meet design and service requirements, the relationships between these properties and the structures developed during solidification are important aspects of casting. In this section, we describe these relationships in terms of dendrite morphology and the concentration of alloying elements in various regions.

The compositions of dendrites and the liquid metal are given by the phase diagram of the particular alloy. When the alloy is cooled very slowly, each dendrite develops a uniform composition. Under normal cooling encountered in practice, however, **cored dendrites** are formed, which have a composition at their surface different from that at their centers (i.e., a *concentration gradient*). The surface has a higher concentration of alloying elements than does the core, because of solute rejection from the core toward the surface during solidification of the dendrite; this phenomenon is known as **microsegregation**. The darker shading in the interdendritic liquid near the dendrite roots shown in Fig. 5.12 indicates that these regions have a higher solute concentration; thus, microsegregation in these regions is much more pronounced than in others.

In contrast to microsegregation, **macrosegregation** involves differences in composition throughout the casting itself. In situations where the solidifying front moves

**FIGURE 5.12** Schematic illustration of three basic types of cast structures: (a) columnar dendritic; (b) equiaxed dendritic; and (c) equiaxed nondendritic. *Source*: After D. Apelian.

(a)                (b)                (c)

away from the surface of a casting as a plane front (Fig. 5.13), constituents with a lower melting point in the solidifying alloy are driven toward the center (**normal segregation**). Consequently, such a casting has a higher concentration of alloying elements at its center than at its surfaces. In dendritic structures such as for solid-solution alloys (Fig. 5.8b), the opposite occurs: the center of the casting has a lower concentration of alloying elements (**inverse segregation**). The reason inverse segregation occurs is that liquid metal (which has a higher concentration of alloying elements) enters the cavities developed from solidification shrinkage in the dendrite arms, which have solidified sooner. Another form of segregation is due to gravity (**gravity segregation**), whereby higher density inclusions or compounds sink, and lighter elements (such as antimony in an antimony–lead alloy) float to the surface.

Figure 5.8b shows a typical cast structure of a solid-solution alloy with an inner zone of equiaxed grains. This inner zone can be extended throughout the casting, as shown in Fig. 5.8c, by adding an **inoculant** (nucleating agent) to the alloy. The inoculant induces nucleation of grains throughout the liquid metal (**heterogeneous nucleation**); an example is $TiB_2$ in AlSi alloys.

Because of the presence of thermal gradients in a solidifying mass of liquid metal, and because of gravity (and hence density differences), *convection* has a strong influence on the structures developed. Convection promotes the formation of an outer chill zone, refines grain size, and accelerates the transition from columnar to equiaxed grains. The structure shown in Fig. 5.12b can also be obtained by increasing convection within the liquid metal, whereby dendrite arms separate (**dendrite multiplication**). Conversely, reducing or eliminating convection results in coarser and longer columnar-dendritic grains. Convection can be enhanced by the use of mechanical or electromagnetic methods. The dendrite arms are not particularly strong, and they can be broken up by agitation or mechanical vibration in the early stages of solidification (**rheocasting**). This process results in finer grain size, with equiaxed-nondendritic grains distributed more uniformly throughout the casting (Fig. 5.12c). Experiments concerning the effects of gravity on the microstructure of castings are currently being conducted during space flights.

**FIGURE 5.13** Schematic illustration of cast structures in (a) plane-front, single-phase alloys; and (b) plane-front, two-phase alloys. *Source*: After D. Apelian.

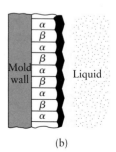

(a)                (b)

## 5.4 | Fluid Flow and Heat Transfer

### 5.4.1 Fluid flow

To emphasize the importance of fluid flow, let's briefly describe a basic gravity casting system, as shown in Fig. 5.14. The molten metal is poured through a **pouring basin**, or *cup*. Next, it flows through the **sprue** to the **well** and then into **runners** and into the mold cavity. **Risers**, also called *feeders*, serve as reservoirs of molten metal to supply the metal necessary to prevent porosity due to shrinkage, as described in Section 5.4.6. Although such a **gating system** appears to be relatively simple, successful casting requires careful design and control of the solidification process to ensure adequate fluid flow during solidification in the system. For example, one of the most important functions of the gating system is to trap contaminants (such as oxides and other inclusions) in the molten metal by having the contaminants adhere to the walls of the gating system, thereby preventing them from reaching the actual mold cavity. Furthermore, proper design of the gating system avoids or minimizes problems such as premature cooling, turbulence, and gas entrapment. Even before it reaches the mold cavity, the molten metal must be handled carefully to avoid the formation of oxides on molten metal surfaces from exposure to the environment or introduction of impurities into the molten metal.

Two basic principles of fluid flow are relevant to gating design: Bernoulli's theorem and the law of mass continuity.

**Bernoulli's theorem.**   Bernoulli's theorem is based on the principle of conservation of energy and relates pressure, velocity, elevation of the fluid at any location in the system, and the frictional losses in a system that is full of liquid as

$$h + \frac{p}{\rho g} + \frac{v^2}{2g} = \text{constant}, \tag{5.4}$$

where $h$ is the elevation above a certain reference plane, $p$ is the pressure at that elevation, $v$ is the velocity of the liquid at that elevation, $\rho$ is the density of the fluid (assuming that it is incompressible), and $g$ is the gravitational constant. Conservation of energy requires that, at any particular location in the system, the following relationship be satisfied:

$$h_1 + \frac{p_1}{\rho g} + \frac{v_1^2}{2g} = h_2 + \frac{p_2}{\rho g} + \frac{v_2^2}{2g} + f. \tag{5.5}$$

In this equation, the subscripts 1 and 2 represent two different elevations, respectively, and $f$ represents the frictional loss in the liquid as it travels downward through the gating system. The frictional loss includes such factors as energy loss at the liquid–mold-wall interfaces and turbulence in the liquid.

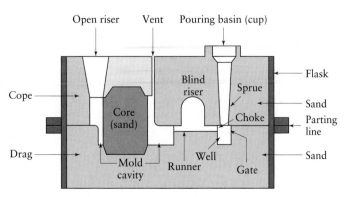

**FIGURE 5.14**   Schematic illustration of a sand mold, showing various features.

**Mass continuity.** The law of mass continuity states that, for incompressible liquids and in a system with impermeable walls, the rate of flow is constant and is expressed as

$$Q = A_1 v_1 = A_2 v_2,  \tag{5.6}$$

where $Q$ is the volumetric rate of flow, such as in $m^3/s$; $A$ is the cross-sectional area of the liquid stream; $v$ is the velocity of the liquid in that cross-sectional location; and the subscripts 1 and 2 pertain to two different locations in the system. Thus, the rate of flow must be maintained everywhere in the system. The permeability of the walls of the system is important, because otherwise some liquid will permeate through (for example, in sand molds) and the rate of flow will decrease as the liquid moves through the system. Coatings are often used to inhibit such behavior in sand molds.

**Sprue design.** An application of the two foregoing principles is the traditional tapered design of sprues (Fig. 5.14); we can determine the shape of the sprue by using Eqs. (5.5) and (5.6). Assuming that the pressure at the top of the sprue is equal to the pressure at the bottom and that there are no frictional losses, the relationship between height and cross-sectional area at any point in the sprue is given by the parabolic relationship

$$\frac{A_1}{A_2} = \sqrt{\frac{h_2}{h_1}},  \tag{5.7}$$

where, let's say, the subscript 1 denotes the top of the sprue and the subscript 2 the bottom. Moving downward from the top, the cross-sectional area of the sprue must decrease. Recall that in a free-falling liquid, such as water from a faucet, the cross-sectional area of the stream decreases as it gains velocity downward. If we design a sprue with a constant cross-sectional area and pour the molten metal, regions may develop where the liquid loses contact with the walls of the sprue. As a result, **aspiration** may take place, whereby air is sucked or entrapped in the liquid. A popular alternative to tapered sprues is the use of straight-sided sprues with a choking mechanism at the bottom, consisting of either a choke core or a runner choke (as shown in Fig. 5.14). The choke slows flow sufficiently to prevent aspiration.

Another application of the foregoing equations is in the *modeling of mold filling*. For example, consider the situation shown in Fig. 5.14, where molten metal is poured into a pouring basin; it flows through a sprue to a gate and runner and fills the mold cavity. If the pouring basin has a much larger cross-sectional area than that of the bottom of the sprue, then the velocity of the molten metal at the top of the pouring basin is very low. If frictional losses are due to viscous dissipation of energy, then $f$ in Eq. (5.5) can be taken as a function of vertical distance, often approximated as a linear function, Therefore, the velocity of the molten metal leaving the gate is obtained from Eq. (5.5) as

$$v = c\sqrt{2gh},  \tag{5.8}$$

where $h$ is the distance from the base of the sprue to the height of the liquid metal and $c$ is a friction factor. For frictionless flow, $c$ equals unity; it is always between 0 and 1. The magnitude of $c$ varies with mold material, runner layout, and channel size and can include energy losses due to turbulence as well as viscous effects.

If the level of the liquid metal has reached a height $x$, then the gate velocity is

$$v = c\sqrt{(2g(h - x))}.  \tag{5.9}$$

The rate of flow through the gate will be the product of this velocity and the gate area, according to Eq. (5.6). The shape of the casting will determine the height as a function of time. Integrating Eq. (5.9) gives the mean fill time and flow rate, and dividing the casting volume by this mean flow rate gives the fill time of the mold.

Simulation of mold filling assists designers in the specification of runner diameter, as well as the size and number of sprues and pouring basins. To ensure that the runners stay open, the fill time must be a small fraction of the solidification time (see Section 5.4.4), but the velocity should not be so high as to erode the mold material (*mold wash*) or to result in too high of a Reynolds number (see below), or else turbulence and associated air entrainment result. Fortunately, many computational tools are now available to evaluate gating designs and to assist in the sizing of components. (See Section 5.12.3.)

**Flow characteristics.** An important consideration in fluid flow in gating systems is the presence of **turbulence**, as opposed to **laminar flow** of fluids; the **Reynolds number** is used to characterize this aspect of fluid flow. The Reynolds number, Re, represents the ratio of the inertia to the viscous forces in fluid flow and is defined as

$$\text{Re} = \frac{vD\rho}{\eta},\tag{5.10}$$

where $v$ is the velocity of the liquid, $D$ is the diameter of the channel, and $\rho$ and $\eta$ are the density and viscosity, respectively, of the liquid. The higher the Reynolds number, the greater is the tendency for turbulent flow. In ordinary gating systems, Re ranges from 2000 to 20,000. An Re value of up to 2000 represents laminar flow; between 2000 and 20,000, there is a mixture of laminar and turbulent flow, generally regarded as harmless in gating systems. However, Re values in excess of 20,000 represent severe turbulence, resulting in air entrainment and *dross* formation (the scum that forms on the surface of the molten metal) from the reaction of the liquid metal with air and other gases. Techniques for minimizing turbulence generally involve avoidance of sudden changes in flow direction and in the geometry of channel cross-sections in the design of the gating system.

The elimination of dross and *slag* (nonmetallic products from mutual dissolution of flux and nonmetallic impurities) is another important consideration in fluid flow. It can be achieved by skimming, using *dross traps*; properly designing pouring basins and gating systems; or using filters. Filters are usually made of ceramics, mica, or fiberglass, and their proper location and placement is important for effective filtering of dross and slag.

## 5.4.2  Fluidity of molten metal

*Fluidity* is a term commonly used to describe the capability of molten metal to fill mold cavities. This term consists of two basic factors: (1) characteristics of the molten metal and (2) casting parameters. The following characteristics of molten metal influence fluidity:

1. **Viscosity.** As viscosity and its sensitivity to temperature (*viscosity index*) increase, fluidity decreases.
2. **Surface tension.** A high surface tension of the liquid metal reduces fluidity. Oxide films developed on the surface of a molten metal thus have a significant adverse effect on fluidity; for example, the oxide film on the surface of pure molten aluminum triples the surface tension.

3. **Inclusions.** As insoluble particles, inclusions can have a significant adverse effect on fluidity. This effect can be verified by observing the viscosity of a liquid such as oil with and without sand particles in it; the latter has higher viscosity.

4. **Solidification pattern of the alloy.** The manner in which solidification occurs, as described in Section 5.3, can influence fluidity. Moreover, fluidity is inversely proportional to the freezing range [see Eq. (5.3)]; hence, the shorter the freezing range (as with pure metals and eutectics), the higher the fluidity becomes; conversely, alloys with long freezing ranges (such as solid-solution alloys) have lower fluidity.

The following casting parameters influence fluidity and also influence the fluid flow and thermal characteristics of the system:

1. **Mold design.** The design and dimensions of components such as the sprue, runners, and risers both influence fluidity.

2. **Mold material and its surface characteristics.** The higher the thermal conductivity of the mold and the rougher its surfaces, the lower the fluidity of the molten metal becomes. Heating the mold improves fluidity, even though it slows down solidification of the metal; also, the casting develops coarse grains, hence, it has less strength.

3. **Degree of superheat.** Defined as the increment of temperature above the melting point of an alloy, superheat improves fluidity by delaying solidification.

4. **Rate of pouring.** The slower the rate at which the molten metal is poured into the mold, the lower the fluidity becomes, because of the higher local rate of cooling.

5. **Heat transfer.** Heat transfer directly affects the viscosity of the liquid metal.

Although the interrelationships among these factors are complex, the term **castability** is generally used to describe the ease with which a metal can be cast to obtain a part with good quality. Obviously, this term includes not only fluidity, but casting practices as well.

**Tests for fluidity.** Although none is accepted universally, several tests have been developed to quantify fluidity. In one such test, the molten metal is made to flow along a channel at room temperature; the distance of metal flow before it solidifies and stops is a measure of the metal's fluidity.

### 5.4.3 Heat transfer

An important consideration in casting is the heat transfer during the complete cycle from pouring to solidification and cooling to room temperature. *Heat flow* at different locations in the system is a complex phenomenon; it depends on many factors relating to the casting material and the mold and process parameters. For instance, in casting thin sections, the metal flow rates must be high enough to avoid premature chilling and solidification. However, the flow rate must not be so fast as to cause excessive turbulence, with its detrimental effects on the properties of the casting.

Figure 5.15 shows a typical temperature distribution in the mold–liquid-metal interface. Heat from the liquid metal is given off through the mold wall and the surrounding air. The temperature drop at the air–mold and mold–metal interfaces is caused by the presence of boundary layers and imperfect contact at these interfaces. The shape of the curve depends on the thermal properties of the molten metal and the mold.

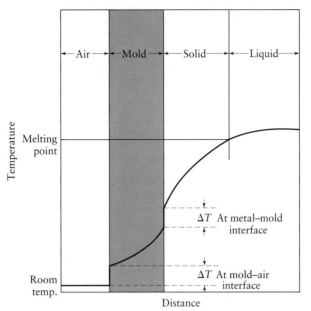

**FIGURE 5.15**
Temperature distribution at the mold wall and liquid–metal interface during solidification of metals in casting.

### 5.4.4  Solidification time

During the early stages of solidification, a thin solidified skin begins to form at the cool mold walls, and, as time passes, the skin thickens. With flat mold walls, this thickness is proportional to the square root of time; thus, doubling the amount of time will make the skin $\sqrt{2} = 1.41$ times, or 41%, thicker. The *solidification time* is a function of the volume of a casting and its surface area (**Chvorinov's rule**) and is expressed as

$$\text{Solidification time} = C\left(\frac{\text{volume}}{\text{surface area}}\right)^2, \qquad (5.11)$$

where $C$ is a constant that reflects the mold material, the properties of the metal (including latent heat), and temperature. Thus, a large sphere solidifies and cools to ambient temperature at a much slower rate than does a smaller sphere. The reason is that the volume of a sphere is proportional to the cube of its diameter, and the surface area is proportional to the square of its diameter. Similarly, it can be shown that molten metal in a cube-shaped mold will solidify faster than in a spherical mold of the same volume.

Figure 5.16 depicts the effects of mold geometry and elapsed time on skin thickness and shape. As illustrated, the unsolidified molten metal has been poured from the mold at different time intervals, ranging from five seconds to six minutes. Note that the skin thickness increases with elapsed time, but that the skin is thinner at internal angles (location $A$ in the figure) than at external angles (location $B$). This latter condition is caused by slower cooling at internal angles than at external angles.

5 s      1 min      2 min      6 min

**FIGURE 5.16**  Solidified skin on a steel casting. The remaining molten metal is poured out at the times indicated in the figure. Hollow ornamental and decorative objects are made by a process called *slush casting*, which is based on this principle. *Source*: H. F. Taylor, J. Wulff, and M. C. Flemings.

---

**EXAMPLE 5.2 Solidification times for various solid shapes**

Three pieces of metal being cast have the same volume, but different shapes. One is a sphere, one a cube, and one a cylinder with a height equal to its diameter. Which piece will solidify the fastest, and which one will solidify the slowest?

**Solution.** Since all three pieces have the same volume, we can just set this volume at unity, so from Eq. (5.11), we have

$$\text{Solidification time} \propto \frac{1}{(\text{surface area})^2}.$$

The respective surface areas are calculated as follows:

Sphere:  $V = \left(\frac{4}{3}\right)\pi r^3$, so $r = \left(\frac{3}{4\pi}\right)^{1/3}$; thus, $A = 4\pi r^2 = 4\pi \left(\frac{3}{4\pi}\right)^{2/3} = 4.84.$

Cube:  $V = a^3$, so $a = 1$; thus, $A = 6a^2 = 6.$

Cylinder:  $V = \pi r^2 h = 2\pi r^3$, so $r = \left(\frac{1}{2\pi}\right)^{1/3}$; thus

$$A = 2\pi r^2 + 2\pi rh = 6\pi r^2 = 6\pi \left(\frac{1}{2\pi}\right)^{2/3} = 5.54.$$

Therefore, the respective solidification times $t$ are

$$t_{\text{sphere}} = 0.043C, \qquad t_{\text{cube}} = 0.028C, \quad \text{and} \quad t_{\text{cylinder}} = 0.033C.$$

Therefore, the cube-shaped casting will solidify the fastest, and the sphere-shaped casting will solidify the slowest.

---

### 5.4.5 Shrinkage

Metals shrink (contract) during solidification and cooling. *Shrinkage*, which causes dimensional changes (and, sometimes, cracking) is the result of the following factors:

1. Contraction of the molten metal as it cools prior to solidification;
2. Contraction of the metal during phase change from liquid to solid (latent heat of fusion);
3. Contraction of the solidified metal (the casting) as its temperature drops to ambient temperature.

The largest amount of shrinkage occurs during cooling of the casting. Table 5.1 lists the percentages of contraction for various metals during solidification. Note, however, that some metals expand, including gray cast iron. The reason for the expansion of gray cast iron is that graphite has a relatively high specific volume, and when it precipitates as graphite flakes during solidification, a net expansion of the metal occurs.

### 5.4.6 Defects in castings

Depending on casting design and method, several defects can develop in castings. Because different names have been used to describe the same defect, the International Committee of Foundry Technical Associations has developed standardized nomenclature, consisting of seven basic categories of casting defects:

**TABLE 5.1**

| Volumetric Solidification Contraction or Expansion Percentages for Various Cast Metals | | | |
|---|---|---|---|
| Contraction (%) | | Expansion (%) | |
| Aluminum | 7.1 | Bismuth | 3.3 |
| Zinc | 6.5 | Silicon | 2.9 |
| Al, 4.5% Cu | 6.3 | Gray iron | 2.5 |
| Gold | 5.5 | | |
| White iron | 4–5.5 | | |
| Copper | 4.9 | | |
| Brass (70–30) | 4.5 | | |
| Magnesium | 4.2 | | |
| 90% Cu, 10% Al | 4 | | |
| Carbon steels | 2.5–4 | | |
| Al, 12% Si | 3.8 | | |
| Lead | 3.2 | | |

A) **Metallic projections**, consisting of fins, flash, or massive projections such as swells and rough surfaces.

B) **Cavities**, consisting of rounded or rough internal or exposed cavities, including blowholes, pinholes, and shrinkage cavities.

C) **Discontinuities**, such as cracks, cold or hot tearing, and cold shuts. If the solidifying metal is constrained from shrinking freely, cracking and tearing can occur. Although many factors are involved in tearing, coarse grain size and the presence of low-melting segregates along the grain boundaries (intergranular) increase the tendency for hot tearing. Incomplete castings result from the molten metal being at too low a temperature or from the metal being poured too slowly. *Cold shut* is an interface in a casting that lacks complete fusion because of the meeting of two streams of partially solidified metal.

D) **Defective surface**, such as surface folds, laps, scars, adhering sand layers, and oxide scale.

E) **Incomplete casting**, such as misruns (due to premature solidification), insufficient volume of metal poured, and runout (due to loss of metal from the mold after pouring).

F) **Incorrect dimensions or shape**, owing to factors such as improper shrinkage allowance, pattern-mounting error, irregular contraction, deformed pattern, or warped casting.

G) **Inclusions**, which form during melting, solidification, and molding. Generally nonmetallic, they are regarded as harmful because they act like stress raisers and reduce the strength of the casting. Hard inclusions (spots) also tend to chip or break tools in machining. They can be filtered out during processing of the molten metal. Inclusions may form during melting because of reaction of the molten metal with the environment (usually with oxygen) or the crucible material. Chemical reactions among components in the molten metal may produce inclusions; slags and other foreign material entrapped in the molten metal also become inclusions. Reactions between the metal and the mold material may produce inclusions as well. In addition, spalling of the mold and core surfaces produces inclusions, indicating the importance of maintaining melt quality and monitoring the conditions of the molds.

**FIGURE 5.17** Various types of (a) internal and (b) external chills (dark areas at corners), used in castings to eliminate porosity caused by shrinkage. Chills are placed in regions where there is a large volume of metal, as shown in (c).

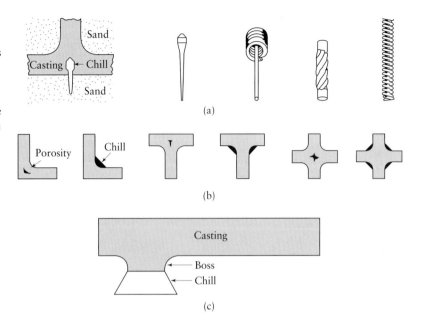

**Porosity.** *Porosity* in a casting may be caused by *shrinkage, trapped gases,* or both. Porosity is detrimental to the ductility of a casting and its surface finish, making it permeable and thus affecting pressure tightness of a cast pressure vessel. Porous regions can develop in castings because of shrinkage of the solidified metal. Thin sections in a casting solidify sooner than thicker regions. As a result, molten metal cannot be fed into the thicker regions that have not yet solidified. Because of contraction, as the surfaces of the thicker regions begin to solidify, porous regions develop at their centers. **Microporosity** can also develop when the liquid metal solidifies and shrinks between dendrites and between dendrite branches.

Porosity caused by shrinkage can be reduced or eliminated by various means. Basically, adequate liquid-metal feeding should be provided to all parts of the solidifying casting to avoid porosity. External or internal **chills**, used in sand casting (Fig. 5.17), also are an effective means of reducing shrinkage porosity. The function of chills is to increase the rate of solidification in critical regions. Internal chills are usually made of the same material as the castings. External chills may be made of the same material or may be made of iron, copper, or graphite. With alloys, porosity can be reduced or eliminated by making the temperature gradient steep, using, for example, mold materials that have high thermal conductivity. Subjecting the casting to **hot isostatic pressing** (see Section 11.3.3) is another method of reducing porosity; however, this process is costly and is used mainly for critical aircraft components.

Gases have much greater **solubility** in liquid metals than in solids (Fig. 5.18). When a metal begins to solidify, the dissolved gases are expelled from the solution. Gases may also result from reactions of the molten metal with the mold materials. Gases either accumulate in regions of existing porosity, such as in interdendritic areas, or they cause microporosity in the casting, particularly in cast iron, aluminum, and copper. Dissolved gases may be removed from the molten metal by flushing or purging with an inert gas or by melting and pouring the metal in a vacuum. If the dissolved gas is oxygen, the molten metal can be *deoxidized*. Steel is usually deoxidized with aluminum or silicon and copper-base alloys, where the copper alloy contains 15% phosphorus.

Whether microporosity is a result of shrinkage or is caused by gases may be difficult to determine. If the porosity is spherical and has smooth walls, much like the

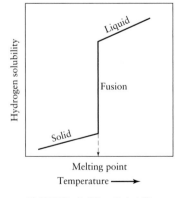

**FIGURE 5.18** Solubility of hydrogen in aluminum. Note the sharp decrease in solubility as the molten metal begins to solidify.

shiny surfaces of holes in Swiss cheese, it is generally from trapped gases. If the walls are rough and angular, the porosity is likely to be from shrinkage between dendrites. Gross porosity, or *macroporosity*, occurs from shrinkage and is usually called **shrinkage cavities**.

# 5.5 | Melting Practice and Furnaces

Melting practice is an important aspect of casting operations because it has a direct bearing on the quality of castings. Furnaces are charged with melting stock consisting of liquid and/or solid metal, alloying elements, and various other materials such as flux and slag-forming constituents. **Fluxes** are inorganic compounds that refine the molten metal by removing dissolved gases and various impurities. Fluxes have several functions, depending on the metal. For example, for aluminum alloys, there are cover fluxes (to form a barrier to oxidation), cleaning fluxes, drossing fluxes, refining fluxes, and wall-cleaning fluxes (because of the detrimental effect that some fluxes have on furnace linings, particularly in induction furnaces). Fluxes may be added manually or can be injected automatically into the molten metal.

To protect the surface of the molten metal against atmospheric reaction and contamination (and to refine the melt), the pour must be insulated against heat loss. Insulation is usually provided by covering the surface or mixing the melt with compounds that form a **slag**. In casting steels, the composition of the slag includes CaO, $SiO_2$, MnO, and FeO. A small quantity of liquid metal is usually tapped and its composition analyzed; necessary additions or inoculations are then made prior to pouring the metal into the molds.

The metal charge may be composed of commercially pure **primary metals**, which are remelted scrap; clean scrapped castings, gates, and risers may also be included in the charge. If the melting points of the alloying elements are sufficiently low, pure alloying elements are added to obtain the desired composition in the melt. If the melting points are too high, the alloying elements do not mix readily with the low-melting-point metals. In this case, **master alloys**, or **hardeners**, are often used. These alloys usually consist of lower-melting-point alloys with higher concentrations of one or two of the needed alloying elements. Differences in specific gravity of master alloys should not be great enough to cause segregation in the melt.

**Melting furnaces.** The melting furnaces commonly used in foundries are electric-arc furnaces, induction furnaces, crucible furnaces, and cupolas.

**Electric-arc furnaces** are used extensively in foundries and have advantages such as a high rate of melting (and hence high production rates), much less pollution output than that of other types of furnaces, and the ability to hold the molten metal for any length of time for alloying purposes.

**Induction furnaces** are especially useful in smaller foundries and produce composition-controlled smaller melts. There are two basic types of induction furnaces. The *coreless induction furnace* consists of a crucible completely surrounded with a water-cooled copper coil through which high-frequency current passes. Because there is a strong electromagnetic stirring action during induction heating, this type of furnace has excellent mixing characteristics for alloying and adding new charge of metal. The other type is called a *core furnace*, or *channel furnace*; it uses low frequency (as low as 60 Hz) and has a coil that surrounds only a small portion of the unit.

**Crucible furnaces,** which have been used extensively throughout history, are heated with various fuels, such as commercial gases, fuel oil, fossil fuel, and electricity. They may be stationary, tilting, or movable. Many ferrous and nonferrous metals are melted in these furnaces.

**Cupolas** are basically refractory-lined vertical steel vessels that are charged with alternating layers of metal, coke, and flux. Although they require major investments and are being replaced by induction furnaces, cupolas operate continuously, have high melting rates, and produce large amounts of molten metal.

**Levitation melting** involves magnetic suspension of the molten metal. An induction coil simultaneously heats a solid billet and stirs and confines the metal, thus eliminating the need for a crucible, which could be a source of contamination with oxide inclusions. The molten metal then flows downward into an investment-casting mold placed directly below the coil. Experiments indicate that investment castings made by this method are free of refractory inclusions and gas porosity and have uniform fine-grained structure.

**Furnace selection** requires careful consideration of several factors that can significantly influence the quality of castings, as well as the economics of casting operations. A variety of furnaces meets requirements for melting and casting metals and alloys in foundries.

**Foundries and foundry automation.** The casting operations described in this chapter are usually carried out in *foundries*. Although casting operations have traditionally involved much manual labor, modern foundries have automated and computer-integrated facilities for all aspects of their operations. They produce a wide variety and sizes of castings at high production rates, at low cost, and with excellent quality control.

Foundry operations initially involve two separate activities: (a) The first activity is pattern and mold making, which now regularly use computer-aided design and manufacturing and **rapid prototyping** techniques (see Section 10.12), minimizing the number of trial-and-error operations and improving efficiency. A variety of automated machinery is used to minimize labor costs, which can be significant in the production of castings. (b) The second activity is melting the metals while controlling their composition and impurities. The rest of the operations, such as pouring metals into molds carried along conveyors, shakeout, cleaning, heat treatment, and inspection, are also automated. Automation minimizes labor, reduces the possibility of human error, increases the production rate, and can be used to attain higher levels of quality. Industrial robots (Section 14.7) are being used extensively in various foundry operations such as removing die castings; cleaning; cutting risers; venting molds; spraying molds; pouring; sorting; inspecting; and automatically storing and retrieving systems for cores and patterns, using automated guided vehicles (Section 14.6).

The level of *automation* in foundries is an important economic consideration, particularly in light of the fact that many foundries are small businesses. The degree of automation depends on the type of products made. A die-casting facility or a foundry making parts for the automotive industry may involve production runs in the hundreds of thousands; thus, a high level of automation is desirable, and such facilities can afford automation. On the other hand, a jobbing foundry that has short production runs may not be as automated. Furthermore, foundries generally still tend to be somewhat dirty, hot, and labor-intensive operations. It is difficult to find qualified personnel to work in such an environment; consequently, automation has become increasingly necessary to compensate for the decline in worker availability. As in all manufacturing operations, safety in foundries is a major consideration.

## 5.6 | Casting Alloys

The general properties of various casting alloys and processes are summarized in Fig. 5.19 and Tables 5.2 through 5.5.

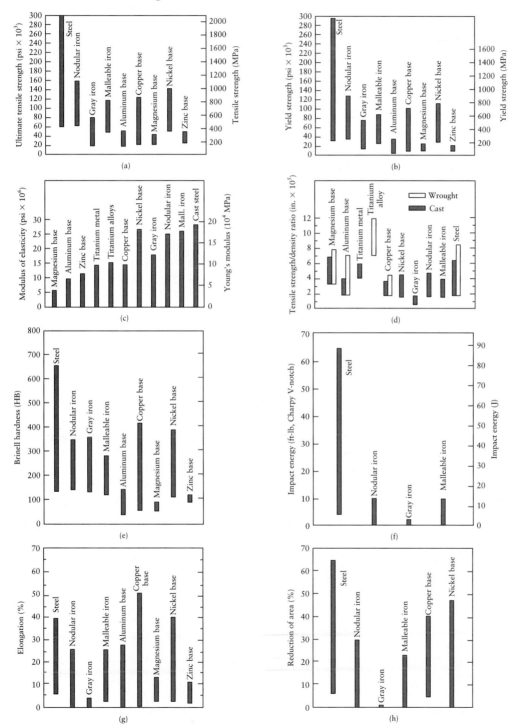

**FIGURE 5.19** Mechanical properties of various groups of cast alloys. Compare this figure with the various tables of properties in Chapter 3. *Source*: Courtesy of Steel Founders' Society of America.

## TABLE 5.2
## General Characteristics of Casting Processes

| | Sand | Shell | Evaporative pattern | Plaster | Investment | Permanent mold | Die | Centrifugal |
|---|---|---|---|---|---|---|---|---|
| Typical materials cast | All | All | All | Nonferrous (Al, Mg, Zn, Cu) | All | All | Nonferrous (Al, Mg, Zn, Cu) | All |
| Weight (kg): | | | | | | | | |
| minimum | 0.01 | 0.01 | 0.01 | 0.01 | 0.001 | 0.1 | <0.01 | 0.01 |
| maximum | No limit | 100+ | 100+ | 50+ | 100+ | 300 | 50 | 5000+ |
| Typ. surface finish ($R_a$, in $\mu$m) | 5–25 | 1–3 | 5–25 | 1–2 | 0.3–2 | 2–6 | 1–2 | 2–10 |
| Porosity[1] | 3–5 | 4–5 | 3–5 | 4–5 | 5 | 2–3 | 1–3 | 1–2 |
| Shape complexity[1] | 1–2 | 2–3 | 1–2 | 1–2 | 1 | 2–3 | 3–4 | 3–4 |
| Dimensional accuracy[1] | 3 | 2 | 3 | 2 | 1 | 1 | 1 | 3 |
| Section thickness (mm): | | | | | | | | |
| minimum | 3 | 2 | 2 | 1 | 1 | 2 | 0.5 | 2 |
| maximum | No limit | — | — | — | 75 | 50 | 12 | 100 |
| Typ. dimensional tolerance (mm/mm) | 1.6–4 mm (0.25 mm for small parts) | ±0.003 | — | ±0.005–0.010 | ±0.005 | ±0.015 | ±0.001–0.005 | 0.015 |
| Cost[1,2] | | | | | | | | |
| Equipment | 3–5 | 3 | 2–3 | 3–5 | 3–5 | 2 | 1 | 1 |
| Pattern/die | 3–5 | 2–3 | 2–3 | 3–5 | 2–3 | 2 | 1 | 1 |
| Labor | 1–3 | 3 | 3 | 1–2 | 1–2 | 3 | 5 | 5 |
| Typical lead time[2,3] | Days | Weeks | Weeks | Days | Weeks | Weeks | Weeks–months | Months |
| Typical production rate[2,3] (parts/mold-hour) | 1–20 | 5–50 | 1–20 | 1–10 | 1–1000 | 5–50 | 2–200 | 1–1000 |
| Minimum quantity[2,3] | 1 | 100 | 500 | 10 | 10 | 1000 | 10,000 | 10–10,000 |

*Notes:* 1. Relative rating, from 1 (best) to 5 (worst). For example, die casting has relatively low porosity, mid to low shape complexity, high dimensional accuracy, high equipment and die costs, and low labor costs. These ratings are only general; significant variations can occur, depending on the manufacturing methods used.
2. Approximate values without the use of rapid prototyping technologies.
*Source:* Data taken from J. A. Schey, *Introduction to Manufacturing Processes*, 3d. ed., 2000.

**TABLE 5.3**

## Typical Applications and Characteristics for Castings

| Type of Alloy | Application | Castability* | Weldability* | Machin-ability* |
|---|---|---|---|---|
| Aluminum | Pistons, clutch housings, intake manifolds, engine blocks, heads, cross members, valve bodies, oil pans, suspension components. | G–E | F | G–E |
| Copper | Pumps, valves, gear blanks, marine propellers. | F–G | F | G–E |
| Gray iron | Engine blocks, gears, brake disks and drums, machine bases. | E | D | G |
| Magnesium | Crankcases, transmission housings, portable computer housings, toys. | G–E | G | E |
| Malleable iron | Farm and construction machinery, heavy-duty bearings, railroad rolling stock. | G | D | G |
| Nickel | Gas-turbine blades, pump and valve components for chemical plants | F | F | F |
| Nodular iron | Crankshafts, heavy-duty gears. | G | D | G |
| Steel (carbon and low alloy) | Die blocks, heavy-duty gear blanks, aircraft undercarriage members, railroad wheels. | F | E | F–G |
| Steel (high alloy) | Gas-turbine housings, pump and valve components, rock-crusher jaws. | F | E | F |
| White iron (Fe$_3$C) | Mill liners, shot-blasting nozzles, railroad brake shoes, crushers, pulverizers. | G | VP | VP |
| Zinc | Door handles, radiator grills. | E | D | E |

* E, excellent; G, good; F, fair; VP, very poor; D, difficult.

**TABLE 5.4**

## Properties and Typical Applications of Cast Irons

| Cast Iron | Type | Ultimate Tensile Strength (MPa) | Yield Strength (MPa) | Elongation in 50 mm (%) | Typical Applications |
|---|---|---|---|---|---|
| Gray | Ferritic | 170 | 140 | 0.4 | Pipe, sanitary ware. |
| | Pearlitic | 275 | 240 | 0.4 | Engine blocks, machine tools. |
| | Martensitic | 550 | 550 | 0 | Wearing surfaces |
| Ductile (Nodular) | Ferritic | 415 | 275 | 18 | Pipe, general service. |
| | Pearlitic | 550 | 380 | 6 | Crankshafts, highly stressed parts. |
| | Tempered martensite | 825 | 620 | 2 | High-strength machine parts, wear resistance. |
| Malleable | Ferritic | 365 | 240 | 18 | Hardware, pipe fittings. |
| | Pearlitic | 450 | 310 | 10 | Couplings. |
| | Tempered | 700 | 550 | 2 | Gears, connecting rods. |
| White | Pearlitic | 275 | 275 | 0 | Wear resistance, mill rolls. |

**TABLE 5.5**

**Typical Properties of Nonferrous Casting Alloys**

| Alloy | Condition | Casting Method* | UTS (MPa) | Yield Stress (MPa) | Elongation in 50 mm (%) | Hardness (HB) |
|---|---|---|---|---|---|---|
| Aluminum | | | | | | |
| 357 | T6 | S | 345 | 296 | 2.0 | 90 |
| 380 | F | D | 331 | 165 | 3.0 | 80 |
| 390 | F | D | 279 | 241 | 1.0 | 120 |
| Magnesium | | | | | | |
| AZ63A | T4 | S, P | 275 | 95 | 12 | — |
| AZ91A | F | D | 230 | 150 | 3 | — |
| QE22A | T6 | S | 275 | 205 | 4 | — |
| Copper | | | | | | |
| Brass C83600 | — | S | 255 | 177 | 30 | 60 |
| Bronze C86500 | — | S | 490 | 193 | 30 | 98 |
| Bronze C93700 | — | P | 240 | 124 | 20 | 60 |
| Zinc | | | | | | |
| No. 3 | — | D | 283 | — | 10 | 82 |
| No. 5 | — | D | 331 | — | 7 | 91 |
| ZA27 | — | P | 425 | 365 | 1 | 115 |

* S, sand; D, die; P, permanent mold.

(a)    (b)    (c)

**FIGURE 5.20** Microstructure for cast irons. Magnification: 100×. (a) Ferritic gray iron with graphite flakes. (b) Ferritic nodular iron (ductile iron), with graphite in nodular form. (c) Ferritic malleable iron. This cast iron solidified as white cast iron, with the carbon present as cementite ($Fe_3C$), and was heat treated to graphitize the carbon. *Source*: ASM International, Materials Park, OH.

### 5.6.1 Ferrous casting alloys

Cast irons represent the largest amount (by weight) of all metals cast. They generally possess several desirable properties, such as hardness, resistance to wear, and good machinability (Chapter 8). These alloys can easily be cast into complex shapes. The term **cast iron** refers to a family of alloys: gray cast iron (gray iron), nodular (ductile or spheroidal) iron, white cast iron, malleable iron, and compacted-graphite iron. These alloys are described as follows:

a) **Gray cast iron.** In this structure, graphite exists largely in the form of flakes (Fig. 5.20a). It is called *gray cast iron*, or *gray iron*, because when broken, the fracture path is along the graphite flakes, and hence it has a gray, sooty appearance. These flakes act as stress raisers; consequently, gray iron has negligible ductility and is weak in tension, although strong in compression, as are other brittle materials. The graphite flakes give this material the capacity to dampen vibrations by the internal friction (and hence energy dissipation) caused by the flakes. This capacity makes gray cast iron a suitable and commonly used material for constructing machine-tool bases and structures (Section 8.12).

The various types of gray cast iron are called *ferritic*, *pearlitic*, and *martensitic*. (See Section 5.11.) Because their structures are different, each has its own respective properties and applications. The structure of **ferritic gray iron**, also known as *fully gray iron*, consists of graphite flakes in an alpha-ferrite matrix. **Pearlitic gray iron** has a structure of graphite in a matrix of pearlite; although still brittle, it is stronger than gray iron. **Martensitic gray iron** is obtained by austenitizing a pearlitic gray iron, followed by rapid quenching to produce a structure of graphite in a martensite matrix; as a result, it is very hard.

Castings of gray iron have relatively few shrinkage cavities and little porosity. Typical uses of gray cast iron are for engine blocks, machine bases, electric-motor housings, pipes, and machine wear surfaces. Gray cast irons are specified by a two-digit ASTM designation. Class 20, for example, specifies that the material must have a minimum tensile strength of 140 MPa (20 ksi).

b) **Ductile iron (nodular iron).** In the structure of ductile iron, graphite is in *nodular* or *spheroid* form (Fig. 5.20b); this shape permits the material to be somewhat ductile and shock resistant. The shape of graphite flakes can be changed into nodules (spheres) by small additions of magnesium and/or cerium to the molten metal prior to pouring. Ductile iron can be made ferritic or pearlitic by heat treatment; it can also be heat treated to obtain a structure of tempered martensite. Typically used for machine parts, pipe, and crankshafts, ductile cast irons are specified by a set of two-digit numbers. Thus, class or grade 80-55-06, for example, indicates that the material has a minimum tensile strength of 550 MPa (80 ksi), a minimum yield strength of 380 MPa (55 ksi), and a 6% elongation in 50 mm (2 in.).

c) **White cast iron.** The structure of white cast iron is very hard, wear resistant, and brittle, because it contains large amounts of iron carbide instead of graphite. White cast iron is obtained either by rapid cooling of gray iron or by adjusting the material's composition by keeping the carbon and silicon content low. This type of cast iron is also called *white iron*, because the lack of graphite gives the fracture surface a white crystalline appearance. Because of its extreme hardness and wear resistance, white cast iron is used mainly for liners for machinery that processes abrasive materials, rolls for rolling mills, and railroad-car brake shoes.

d) **Malleable iron.** Malleable iron is obtained by annealing white cast iron in an atmosphere of carbon monoxide and carbon dioxide between 800°C and 900°C (1470°F and 1650°F) for up to several hours, depending on the size of the part. During this process, the cementite decomposes (*dissociates*) into iron and graphite. The graphite exists as *clusters* (Fig. 5.20c) in a ferrite or pearlite matrix and thus has a structure similar to that of nodular iron; this structure promotes ductility, strength, and shock resistance, hence the term *malleable*. Typical uses are for railroad equipment and various hardware. Malleable irons are specified by a five-digit designation; for example, 35018 indicates that the yield strength of the material is 35 ksi (240 MPa) and the elongation is 18% in 50 mm (2 in.).

e) **Compacted-graphite iron.** The compacted-graphite structure has short, thick, and interconnected flakes with undulating surfaces and rounded extremities. The mechanical and physical properties of this type of cast iron are intermediate between those of flake-graphite and nodular-graphite cast irons. Its machinability is better than that of nodular iron; it is easy to cast and has consistent properties throughout the casting. Typical applications include automotive engine blocks, crankcases, ingot molds, cylinder heads, and brake disks. It is particularly suitable for components at elevated temperatures and resists thermal fatigue.

f) **Cast steels.** Because of the high temperatures required to melt cast steels, up to about 1650°C (3000°F), their casting requires considerable knowledge and experience. The high temperatures involved present difficulties in the selection of mold materials (particularly in light of the high reactivity of steels with oxygen) in melting and pouring the metal. Steel castings possess properties that are more uniform (isotropic) than those made by mechanical working processes. Cast steels can be welded; however, welding alters the cast microstructure in the heat-affected zone (see Section 12.5), influencing the strength, ductility, and toughness of the base metal. Subsequent heat treatment must be performed to restore the mechanical properties of the casting. Cast weldments have gained importance where complex configurations or the size of the casting may prevent the part from being cast economically in one piece.

g) **Cast stainless steels.** Casting of stainless steels involves considerations similar to those for casting steels in general. Stainless steels generally have a long freezing range and high melting temperatures; they develop various structures, depending on their composition and the process parameters. Cast stainless steels are available in various compositions, and they can be heat treated and welded. These cast products have high resistance to heat and corrosion. Nickel-base casting alloys are used for severely corrosive environments and very high-temperature service.

## 5.6.2 Nonferrous casting alloys

1. **Aluminum-base alloys.** Alloys with an aluminum base have a wide range of mechanical properties, mainly because of various hardening mechanisms and heat treatments that can be used with them. Their fluidity depends on oxides and alloying elements in the metal. These alloys have high electrical and thermal conductivity and generally good corrosion resistance to most elements (except alkali). They are nontoxic and lightweight (and hence are called *light-metal castings*) and have good machinability; however, except for alloys with silicon, they generally have low resistance to wear and abrasion. Aluminum-base alloys have many applications, including architectural, decorative, aerospace, and electrical applications. Increasingly, engine blocks of automobiles are made of aluminum-alloy castings.

2. **Magnesium-base alloys.** Alloys with a magnesium base have good corrosion resistance and moderate strength. These characteristics depend on the particular heat treatment and the proper use of protective coating systems.

3. **Copper-base alloys.** Alloys with a copper base have the advantages of good electrical and thermal conductivity, corrosion resistance, nontoxicity (unless they contain lead), and wear properties suitable for bearing materials. Their mechanical properties and fluidity are influenced by the alloying elements. These alloys also possess good machinability characteristics.

4. **Zinc-base alloys.** Alloys with a zinc base have good fluidity and sufficient strength for structural applications. These alloys are commonly used in die casting.

5. **High-temperature alloys.** High-temperature alloys, which have a wide range of properties, typically require temperatures of up to 1650°C (3000°F) for casting titanium and superalloys; refractory alloys require higher temperatures. Special techniques are used to cast these alloys into parts for jet- and rocket-engine components. Some of these alloys are more suitable and economical for casting than for shaping by other manufacturing methods such as forging.

# 5.7 | Ingot Casting and Continuous Casting

Traditionally, the first step in metal processing is the shaping of the molten metal into a solid form (an **ingot**) for further processing by rolling into shapes, casting into semi-finished forms, or forging. The shaping process is being rapidly replaced by *continuous casting*, which improves efficiency and eliminates the need for ingots. (See Section 5.7.2.) In ingot casting, the molten metal is poured (*teemed*) from the ladle into ingot molds in which the metal solidifies; this process is basically equivalent to sand casting or, more commonly, to permanent-mold casting.

Gas entrapment can be reduced by bottom pouring, using an insulated collar on top of the mold, or by using an exothermic compound that produces heat when it contacts the molten metal (hot top); these processes slow the cooling and result in a higher yield of high-quality metal. The cooled ingots are removed (*stripped*) from the molds and lowered into soaking pits, where they are reheated to a uniform temperature of about 1200°C (2200°F) for subsequent processing by rolling. Ingots may be square, rectangular, or round in cross-section, and their weight ranges from a few hundred pounds to 40 tons.

## 5.7.1 Ferrous-alloy ingots

Several reactions take place during solidification of an ingot; each reaction has an important influence on the quality of the steel produced. For example, significant amounts of oxygen and other gases can dissolve in the molten metal during steelmaking. However, much of these gases is rejected during solidification of the metal, because the solubility limit of gases in the metal decreases sharply as its temperature decreases. The rejected oxygen combines with carbon, forming carbon monoxide, which causes porosity in the solidified ingot. Depending on the amount of gas evolved during solidification, three types of steel ingots can be produced: killed, semikilled, and rimmed.

1. **Killed steel** is a fully deoxidized steel; that is, oxygen is removed, and thus porosity is eliminated. In the deoxidation process, the dissolved oxygen in the molten metal is made to react with elements such as aluminum (the most common such element) that are added to the melt. Other elements, such as vanadium, titanium, and zirconium, are also used. These elements have an affinity for oxygen and form metallic oxides. If aluminum is used, the product is called aluminum-killed steel. The term *killed* comes from the fact that the steel lies quietly after being produced into the mold.

   If sufficiently large, the oxide inclusions in the molten bath float out and adhere to, or are dissolved in, the slag. A fully killed steel is thus free from porosity caused by gases; it is also free from *blowholes* (large spherical holes near the surfaces of the ingot). Consequently, the chemical and mechanical properties of killed steels are relatively uniform throughout the ingot. However, because of metal shrinkage during solidification, an ingot of this type develops a *pipe* at its top. Also called a *shrinkage cavity*, the pipe has a funnellike appearance;

this pipe can constitute a substantial portion of the ingot and has to be cut off and scrapped.

2. **Semikilled steel** is a partially deoxidized steel. It contains some porosity, generally in the upper central section of the ingot, but has little or no pipe; thus, scrap is reduced. Piping in semikilled steel is less because it is compensated for by the presence of porosity in that region. Semikilled steels are economical to produce.

3. In a **rimmed steel**, which generally has a low carbon content (less than 0.15%), the evolved gases are only partially killed or controlled by the addition of elements such as aluminum. The gases form blowholes along the outer rim of the ingot, hence the term *rimmed*. Blowholes are generally not objectionable unless they break through the outer skin of the ingot. Rimmed steels have little or no piping and have a ductile skin with good surface finish.

## 5.7.2 Continuous casting

The traditional method of casting ingots is a batch process; that is, each ingot has to be stripped from its mold after solidification and processed individually. Piping and microstructural and chemical variations also are present throughout the ingot. These problems are alleviated by continuous casting processes, which produce higher quality steels for less cost. Conceived in the 1860s, *continuous casting*, or **strand casting**, was first developed for casting nonferrous metal strip. Now used for steel production, the process yields major improvements in efficiency and productivity and significant reductions in cost. Figure 5.21a shows schematically one system for continuous casting. The molten metal in the ladle is cleaned and equalized in temperature by blowing nitrogen gas through it for 5 to 10 min. The metal is then poured into a refractory-lined intermediate pouring vessel (**tundish**), where impurities are skimmed off. The tundish holds as much as three tons of metal. The molten metal travels through water-cooled copper molds and begins to solidify as it travels downward along a path supported by rollers (*pinch rolls*).

Before the casting process is started, a solid *starter*, or *dummy bar*, is inserted into the bottom of the mold. The molten metal is then poured and solidifies on the starter bar. (See the bottom of Fig. 5.21a.) The bar is withdrawn at the same rate the metal is poured. The cooling rate is such that the metal develops a solidified skin (shell) to support itself during its travel downward, at speeds typically of 25 mm/s (1 in./s). The shell thickness at the exit end of the mold is about 12–18 mm (0.5–0.75 in.). Additional cooling is provided by water sprays along the travel path of the solidifying metal. The molds generally are coated with graphite or similar solid lubricants to reduce friction and adhesion at the mold–metal interfaces. The molds are vibrated to further reduce friction and sticking.

The continuously cast metal may be cut into desired lengths by shearing or torch cutting, or it may be fed directly into a rolling mill for further reductions in thickness and for shape rolling of products such as channels and I-beams. In addition to costing less, continuously cast metals have more uniform compositions and properties than metals obtained by ingot casting. Although the thickness of steel strand is usually about 250 mm (10 in.), recent developments have reduced this thickness to about 15 mm (0.6 in.) or less. The thinner strand reduces the number of rolling operations required and improves the economy of the overall operation. After they are hot rolled, steel plates or shapes undergo one or more further processes, such as cleaning and pickling by chemicals to remove surface oxides, cold rolling to improve strength and surface finish, annealing, and coating (galvanizing or aluminizing) for resistance to corrosion.

**FIGURE 5.21**    (a) The continuous-casting process for steel. Typically, the solidified metal decends at a speed of 25 mm/s (1 in/s). Note that the platform is about 20 m (65 ft) above ground level. *Source: Metalcaster's Reference and Guide*, American Foundryman's Society. (b) Continuous strip casting of nonferrous metal strip. *Source*: Courtesy of Hazelett Corporation.

## 5.7.3  Strip casting

In strip casting, thin slabs or strips are produced from molten metal; the metal solidifies in similar fashion to strand casting, but the hot solid is then rolled to form the final shape (Fig. 5.21b). The compressive stresses in rolling (see Section 6.3) serve to reduce porosity and provide better material properties. In effect, strip casting eliminates a hot-rolling operation in the production of metal strips or slabs. In modern facilities, final thicknesses on the order of 2–6 mm (0.08–0.25 in.) can be obtained for carbon, stainless and electrical steels, and other metals.

# 5.8 | Casting Processes: Expendable Mold, Permanent Pattern

Casting processes are generally classified according to mold materials, molding processes, and methods of feeding the mold with the molten metal. The two major categories are *expendable-mold* and *permanent-mold* casting (Section 5.10). Expendable-mold processes are further categorized as permanent-pattern or expendable-pattern processes. Expendable molds are made of sand, plaster, ceramics, and similar materials, which generally are mixed with various **binders** or bonding agents. These materials are refractories; that is, they have the capability to withstand the high temperatures of molten metals. After the casting has solidified, the molds in these processes are broken up to remove the casting. Permanent molds and patterns are reused many times to make large production runs more economical.

## 5.8.1 Sand casting

The traditional method of expendable-mold casting is in sand molds. Simply stated, *sand casting* consists of placing a pattern (having the shape of the desired casting) in sand to make an imprint, incorporating a gating system, filling the resulting cavity with molten metal, allowing the metal to cool until it solidifies, breaking away the sand mold, and removing the casting. While the origins of sand casting date to ancient times, it is still the most prevalent form of casting. In the United States alone, about 15 million tons of metal are cast by this method each year. Typical parts made by sand casting include machine-tool bases, engine blocks, cylinder heads, and pump housings.

**Sands.** Most sand-casting operations use silica sands ($SiO_2$). Sand is the product of the disintegration of rocks over extremely long periods of time. It is inexpensive and is suitable as mold material because of its resistance to high temperatures. There are two general types of sand: *naturally bonded* (*bank sands*) and *synthetic* (*lake sands*). Because its composition can be controlled more accurately, synthetic sand is preferred by most foundries. Several factors are important in the selection of sand for molds. Sand that has fine, round grains can be closely packed and forms a smooth mold surface. Good permeability of molds and cores allows gases and steam evolved during casting to escape easily. The mold should have good **collapsibility** in order to avoid defects in the casting, such as hot tearing and cracking (because the casting shrinks while cooling). The selection of sand involves certain trade-offs with respect to properties. For example, fine-grained sand enhances mold strength, but the fine grains also lower mold permeability. Sand is typically conditioned before use. *Mulling* machines are used to uniformly mull (mix thoroughly) sand with additives; clay is used as a cohesive agent to bond sand particles, giving the sand better strength.

**Types of sand molds.** Sand molds are characterized by the types of sand and by the methods used to produce them; the major components of a sand mold are shown in Fig. 5.14. There are three basic types of sand molds: green-sand, cold-box, and no-bake molds. The most common mold material is *green molding sand*. The term *green* refers to the fact that the sand in the mold is moist or damp while the metal is being poured into it. Green molding sand is a mixture of sand, clay, and water. Green-sand molding is the least expensive method of making molds.

In the **skin-dried** method, the surfaces of the mold are dried, by either storing the mold in air or drying it with torches. Skin-dried molds are generally used for large castings, because of the higher strength. Sand molds are also oven dried (*baked*) prior to receiving the molten metal. They are stronger than green-sand molds and impart better dimensional accuracy and surface finish to the casting. However, distortion of the mold is greater; the castings are more susceptible to hot tearing, because of the lower collapsibility of the mold; and the production rate is slower, because of the drying time required.

In the **cold-box mold** process, various organic and inorganic binders are blended into the sand to bond the grains chemically for greater strength. These molds are dimensionally more accurate than green-sand molds, but are more expensive. In the **no-bake mold** process, a synthetic liquid resin is mixed with the sand, and the mixture hardens at room temperature. Because bonding of the mold in this and the cold-box process takes place without heat, they are called *cold-setting processes*.

**Patterns.** *Patterns* are used to mold the sand mixture into the shape of the casting. They may be made of wood, plastic, or metal. Patterns can also be made directly by **rapid prototyping** techniques. (See Section 10.12.) The selection of a pattern material depends on the size and shape of the casting, the dimensional accuracy and quantity of castings required, and the molding process to be used. Because patterns are used repeatedly to make molds, the strength and durability of the material selected for patterns must reflect the number of castings the mold will produce. Patterns may be made of a combination of materials in order to reduce wear in critical regions. Patterns are usually coated with a **parting agent** to facilitate their removal from the molds.

Patterns can be designed with a variety of features to fit application and economic requirements. **One-piece patterns** are generally used for simpler shapes and low-quantity production. They are generally made of wood and are inexpensive. **Split patterns** are two-piece patterns made so that each part forms a portion of the cavity for the casting. In this way, castings that have complicated shapes can be produced. **Match-plate patterns** are a popular type of mounted pattern in which two-piece patterns are constructed by securing each half of one or more split patterns to the opposite sides of a single plate. In such constructions, the gating system can be mounted on the drag side of the pattern. This type of pattern most often is used in conjunction with molding machines and large production runs.

**Cores.** Cores are used for castings with internal cavities or passages, such as in an automotive engine block or a valve body. Cores are placed in the mold cavity before casting to form the interior surfaces of the casting and are removed from the finished part during shakeout and further processing. Like molds, cores must possess strength, permeability, ability to withstand heat, and collapsibility. Therefore, cores are typically made of sand aggregates. The core is anchored by **core prints**, which are recesses that are added to the pattern to support the core and to provide vents for the escape of gases. A common problem with cores is that for certain casting requirements, as in the case for which a recess is required, they may lack sufficient structural support in the cavity. To keep the core from shifting, metal supports, known as **chaplets,** may be used to anchor the core in place.

Cores are generally made in a manner similar to that used for making molds, and most are made with shell, no-bake, or cold-box processes. Cores are formed in *core boxes*, which are used much as patterns are used to form sand molds. The sand can be packed into the boxes with sweeps or blown into the box by compressed air from *core blowers*; core boxes have the advantages of producing uniform cores and operating at very high production rates.

**Sand-molding machines.** The oldest known method of molding, which is still used for simple castings, is to compact the sand by hand hammering or ramming it around the pattern. For most operations, however, the sand mixture is compacted around the pattern by molding machines. These machines eliminate arduous labor, offer higher quality casting by improving the application and distribution of forces, manipulate the mold in a carefully controlled fashion, and increase the rate of production. Mechanization of the molding process can be further assisted by **jolting** the assembly. The flask, molding sand, and pattern are placed on a pattern plate mounted on an anvil and are jolted upward by air pressure at rapid intervals; the inertial forces compact the sand around the pattern.

In **vertical flaskless molding**, the halves of the pattern form a vertical chamber wall against which sand is blown and compacted. Then the mold halves are packed horizontally, with the parting line oriented vertically and moved along a pouring conveyor. This operation is simple and eliminates the need to handle flasks, making potential production rates very high, particularly when other aspects of the operation, such as coring and pouring, are automated.

**Sandslingers** fill the flask uniformly with sand under a stream of high pressure. Sandthrowers are used to fill large flasks and are typically operated by machine. An impeller in the machine throws sand from its blades or cups at such high speeds that the machine not only places the sand, but also rams it appropriately.

In **impact molding**, the sand is compacted by controlled explosion or instantaneous release of compressed gases. This method produces molds with uniform strength and good permeability. In **vacuum molding**, also known as the "V" *process*, the pattern is covered tightly by a thin sheet of plastic. A flask is placed over the coated pattern and filled with sand. A second sheet of plastic is placed on top of the sand, and a vacuum action hardens the sand so that the pattern can be withdrawn. Both halves of the mold are made this way and then assembled. During pouring, the mold remains under a vacuum, but the casting cavity does not. When the metal has solidified, the vacuum is turned off, and the sand falls away, releasing the casting. Vacuum molding produces castings that have very good detail and accuracy; however, it is quite expensive. It is especially well suited for large, relatively flat castings.

**The sand-casting operation.** After the mold has been shaped and the cores have been placed in position, the two halves (*cope* and *drag*) are closed, clamped, and weighted down. They are weighted to prevent the separation of the mold sections under the pressure exerted when the molten metal is poured into the mold cavity. The design of the gating system is important for proper delivery of the molten metal into the mold cavity. Turbulence must be minimized, air and gases must be allowed to escape by vents or other means, and proper temperature gradients must be established and maintained to eliminate shrinkage and porosity. The design of risers is also important for supplying the necessary molten metal during solidification of the casting; the pouring basin may also serve as a riser. After solidification, the casting is shaken out of its mold, and the sand and oxide layers adhering to the casting are removed by vibration (using a shaker) or by sand blasting. The risers and gates are cut off by oxyfuel-gas cutting, sawing, shearing, and abrasive wheels, or they are trimmed in dies.

Almost all commercially used alloys can be sand cast. The surface finish obtained is largely a function of the materials used in making the mold. The dimensional accuracy of sand casting is not as good as that of other casting processes. However, intricate shapes, such as cast-iron engine blocks and very large propellers for ocean liners and impellers, can be cast by this process. Sand casting can be economical for relatively small production runs as well as large production runs; equipment costs are generally low.

## 5.8.2 Shell-mold casting

The use of *shell-mold casting* has grown significantly, because it can produce many types of castings with close dimensional tolerances and good surface finishes at a low cost. In this process, a mounted pattern, made of a ferrous metal or aluminum, is heated to 175–370°C (350–700°F), coated with a parting agent such as silicone, and clamped to a box or chamber holding a fine sand containing a 2.5% to 4.0% thermosetting resin binder, such as phenol-formaldehyde, which coats the sand particles. The sand mixture is blown over the heated pattern, coating it evenly. The assembly is often, but not always, placed in an oven for a short period of time to complete the curing of the resin. The shell hardens around the pattern and is removed from the pattern, using built-in ejector pins. Two half-shells are made in this manner and are bonded or clamped together in preparation for pouring.

The shells are light and thin, usually 5–10 mm (0.2–0.4 in.) thick, and consequently their thermal characteristics are different than those for thicker molds. The thin shells allow gases to escape during solidification of the metal. The mold is generally used vertically and is supported by surrounding it with steel shot in a cart. The walls of the mold are relatively smooth, offering low resistance to flow of the molten metal and producing castings with sharper corners, thinner sections, and smaller projections than are possible in green-sand molds. With the use of multiple gating systems, several castings can be made in a single mold.

Shell-mold casting may be more economical than other casting processes, depending on various production factors, particularly energy cost. The relatively high cost of metal patterns becomes a smaller factor as the size of the production run increases. The high quality of the finished casting can significantly reduce cleaning, machining, and other finishing costs. Complex shapes can be produced with less labor, and the process can be automated fairly easily. Shell-mold casting applications include small mechanical parts that require high levels of precision, such as gear housings, cylinder heads, and connecting rods. Shell molding is also widely used in producing high-precision molding cores, such as engine-block water jackets.

**Composite molds.** *Composite molds* are made of two or more different materials and are used in shell molding and other casting processes. They are generally used for casting complex shapes, such as impellers for turbines. Figure 5.22 shows an example of composite molds. Commonly used molding materials include shells (made as previously described), plaster, sand with binder, metal, and graphite. These molds may also include cores and chills to control the rate of solidification in critical areas of castings. Composite molds have increased strength, improve the dimensional accuracy and surface finish of castings, and may help reduce overall costs and processing time.

**Sodium silicate process.** The mold material in the *sodium silicate process* is a mixture of sand and 1.5% to 6% sodium silicate (*waterglass*) as the binder for sand. This mixture is then packed around the pattern and hardened by blowing carbon dioxide ($CO_2$) gas through it. This process, also known as *silicate-bonded sand* or the *carbon-dioxide process*, was first used in the 1950s and has been developed further, for example, by using various other chemicals for binders. Cores made by this process reduce the tendency to tear, because of their compliance at elevated temperatures.

**Rammed-graphite molding.** Rammed graphite, instead of sand, is used to make molds for casting reactive metals, such as titanium and zirconium. Sand cannot be used because these metals react vigorously with silica. In this specialty method, the molds

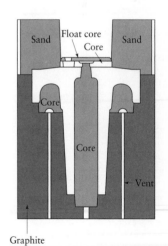

**FIGURE 5.22** Schematic illustration of a semipermanent composite mold. *Source: Steel Castings Handbook*, 5th ed., Steel Founders' Society of America, 1980.

are packed rather like sand molds and are then air dried, baked at 175°C (350°F), fired at 870°C (1600°F), and stored under controlled humidity and temperature. The casting procedures are similar to those for sand molds.

### 5.8.3  Plaster-mold casting

The *plaster-mold casting* process and the ceramic-mold and investment-casting processes are known as **precision casting**, because of the high dimensional accuracy and good surface finish obtained. Typical parts made by these methods include lock components, gears, valves, fittings, tooling, and ornaments, weighing as little as 1 g (0.035 oz). In the plaster-mold casting process, the mold is made of *plaster of paris* (gypsum, or calcium sulfate), with the addition of talc and silica flour to improve strength and to control the time required for the plaster to set. These components are mixed with water, and the resulting slurry is poured over the pattern. After the plaster sets, usually within 15 minutes, the pattern is removed, and the mold is dried to remove the moisture. The mold halves are then assembled to form the mold cavity and preheated to about 120°C (250°F) for 16 hours. Next, the molten metal is poured into the mold. Because plaster molds have very low permeability, gases evolved during solidification of the metal cannot escape. Consequently, the molten metal is poured either in a vacuum or under pressure. Plaster-mold permeability can be increased substantially by the *Antioch process*. In this process, the molds are dehydrated in an autoclave (pressurized oven) for 6 to 12 hours and are then rehydrated in air for 14 hours. Another method of increasing permeability is to use foamed plaster, which contains trapped air bubbles.

Patterns for plaster molding are generally made of aluminum alloys, thermosetting plastics, brass, or zinc alloys. Wood patterns are not suitable for making a large number of molds, because the patterns are repeatedly subjected to the water-based plaster slurry, which causes the wood to swell. Because there is a limit to the maximum temperature that the plaster mold can withstand, generally about 1200°C (2200°F), plaster-mold casting is used only for aluminum, magnesium, zinc, and some copper-base alloys. The castings have fine detail and good surface finish. Since plaster molds have lower thermal conductivity than other types of molds, the castings cool slowly, yielding more uniform grain structure with less warpage and better mechanical properties.

### 5.8.4  Ceramic-mold casting

The *ceramic-mold casting* is another precision-casting process and is similar to the plaster-mold process, with the exception that it uses refractory mold materials suitable for high-temperature applications; the process is also called **cope-and-drag investment casting**. The slurry is a mixture of fine-grained zircon ($ZrSiO_4$), aluminum oxide, and fused silica, which are mixed with bonding agents and poured over the pattern (Fig. 5.23), which has been placed in a flask. The pattern may be made of wood or metal. After setting, the molds (ceramic facings) are removed, dried, burned off to remove volatile matter, and baked. The molds are clamped firmly and used as all-ceramic molds. In the *Shaw process*, the ceramic facings are backed by fireclay (clay used in making firebricks that resist high temperatures) to give the molds strength. The facings are then assembled into a complete mold, ready to be poured.

The high-temperature resistance of the refractory molding materials allows these molds to be used in casting ferrous and other high-temperature alloys, stainless steels, and tool steels. The castings have good dimensional accuracy and surface finish over a wide range of sizes and intricate shapes, but the process is somewhat expensive. Typical parts made are impellers, cutters for machining, dies for metalworking, and molds for making plastic or rubber components, with some parts weighing as much as 700 kg (1500 lb).

**FIGURE 5.23** Sequence of operations in making a ceramic mold. *Source: Metals Handbook*, 8th ed., Vol. 5: *Forging and Casting*, ASM International, 1970.

### 5.8.5 Vacuum casting

Figure 5.24 shows a schematic illustration of the *vacuum-casting process*, or *counter-gravity low-pressure (CL) process* (not to be confused with the vacuum-molding process described in Section 5.8.1). A mixture of fine sand and urethane is molded over metal dies and cured with amine vapor. Then the mold is held with a robot arm and partially immersed into molten metal held in an induction furnace. The metal may be melted in air (*CLA process*) or in a vacuum (*CLV process*). The vacuum reduces the air pressure inside the mold to about two thirds of atmospheric pressure, drawing the molten metal into the mold cavities through a gate in the bottom of the mold. The molten metal in the furnace is usually at a temperature 55°C (100°F) above the liquidus temperature; consequently, the molten metal begins to solidify within a fraction of a second. After the mold is filled, it is withdrawn from the molten metal.

This process is an alternative to investment, shell-mold, and green-sand casting and is particularly suitable for thin-walled (0.75 mm [0.03 in.]) complex shapes with uniform properties. Carbon and low-and high-alloy steel and stainless-steel parts, weighing as much as 70 kg (155 lb), have been vacuum cast by this method. CLA parts are easily made at high volume and relatively low cost. CLV parts usually contain reactive metals, such as aluminum, titanium, zirconium, and hafnium. These parts, which are often in the form of superalloys for gas turbines, may have walls as thin as 0.5 mm (0.02 in.). The process can be automated, and production costs are similar to those for green-sand casting.

Induction furnace

(a)             (b)

**FIGURE 5.24** Schematic illustration of the vacuum-casting process. Note that the mold has a bottom gate. (a) Before and (b) after immersion of the mold into the molten metal. *Source: After R. Blackburn.*

## 5.9 | Casting Processes: Expendable Mold, Expendable Pattern

### 5.9.1 Evaporative-pattern casting (lost foam)

The *evaporative-pattern casting process* uses a polystyrene pattern, which evaporates upon contact with molten metal to form a cavity for the casting. The process is also known as *lost-pattern casting* and under the trade name *Full-Mold process*. It was formerly known as the *expanded polystyrene process* and has become one of the more important casting processes for ferrous and nonferrous metals, particularly for the automotive industry. First, raw expandable polystyrene (EPS) beads, containing 5% to 8% pentane (a volatile hydrocarbon), are placed in a preheated die, usually made of aluminum. The polystyrene expands and takes the shape of the die cavity. Additional heat is applied to fuse and bond the beads together. The die is then cooled and opened, and the polystyrene pattern is removed. Complex patterns may also be made by bonding various individual sections of the pattern, using hot-melt adhesive. In such compound foam-mold assembly, precision of the bonding operation is critically important for maintaining overall dimensional accuracy.

The pattern is then coated with a water-base refractory slurry, dried, and placed in a flask. The flask is filled with loose fine sand, which surrounds and supports the pattern. The sand may be dried or mixed with bonding agents to give it additional strength. The sand is periodically compacted by various means. Then, without removing the polystyrene pattern, the molten metal is poured into the mold. This action immediately vaporizes the pattern (an ablation process) and fills the mold cavity, completely replacing the space previously occupied by the polystyrene pattern. The heat degrades (depolymerizes) the polystyrene, and the degradation products are vented into the surrounding sand.

The flow velocity in the mold depends on the rate of degradation of the polymer. Studies have shown that the flow of the molten metal is basically laminar, with Reynolds numbers in the range of 400 to 3000. (See Section 5.4.1.) The velocity of the molten metal at the metal–polymer pattern front is estimated to be in the range of 0.1–1.0 m/s (4–40 in./s). The velocity can be controlled by producing patterns with cavities or hollow sections; thus, the velocity will increase as the molten metal crosses these hollow regions, similar to pouring in an empty cavity, as is done in sand casting. Furthermore, because the polymer requires considerable energy to degrade, large thermal gradients are present at the metal–polymer interface; in other words, the molten metal cools faster than it would if it were poured into a cavity. Consequently, fluidity (see Section 5.4.2) is less than in sand casting. This characteristic has important effects on the microstructure throughout the casting and also leads to directional solidification of the metal.

Typical applications for this process include cylinder heads, crankshafts, brake components and manifolds for automobiles, and machine bases. The aluminum engine blocks and other components of the General Motors Saturn automobiles are made by this process. Recent developments include the use of polymethylmethacrylate and polyalkylene carbonate as pattern materials for ferrous castings. In a modification of the evaporative-pattern process, a polystyrene pattern is surrounded by a ceramic shell (*Replicast C-S process*). The pattern is burned out prior to pouring the molten metal into the mold. The principal advantage of this process over investment casting (with its wax patterns) is that carbon pickup into the metal is entirely avoided.

More recent developments in evaporative-pattern casting include production of metal-matrix composites (Section 11.14). During the process of molding the

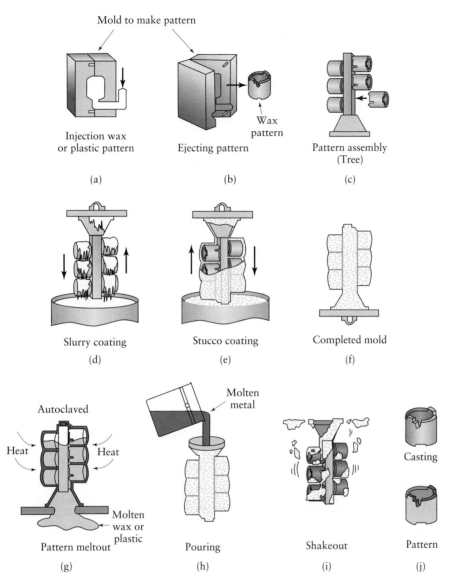

Mold to make pattern

Injection wax
or plastic pattern

(a)

Ejecting pattern

Wax
pattern

(b)

Pattern assembly
(Tree)

(c)

Slurry coating

(d)

Stucco coating

(e)

Completed mold

(f)

Autoclaved

Heat          Heat

Molten
wax or
plastic

Pattern meltout

(g)

Molten
metal

Pouring

(h)

Shakeout

(i)

Casting

Pattern

(j)

**FIGURE 5.25** Schematic illustration of investment casting (lost-wax process). Castings by this method can be made with very fine detail and from a variety of metals. *Source*: Steel Founders' Society of America.

polymer pattern, the pattern is embedded throughout with fibers or particles, which then become an integral part of the casting. Further studies are being conducted on modification and grain refinement of the casting by the use of grain refiners and modifier master alloys (Section 5.5) within the pattern while it is being molded.

## 5.9.2 Investment casting (lost wax)

The *investment-casting process*, also called the *lost-wax process*, was first used during the period 4000–3000 BC. In this process, the pattern is made of wax or a plastic such as polystyrene. The sequences involved in investment casting are shown in Fig. 5.25.

**Solid mold.** The pattern is made by injecting semisolid or liquid wax or plastic into a metal die in the shape of the pattern. The pattern is then removed and dipped into a slurry of refractory material, such as very fine silica and binders, ethyl silicate, and acids. After this initial coating has dried, the pattern is coated repeatedly to increase its thickness. The term *investment* comes from investing the pattern with the refractory material. Wax patterns require careful handling, because they are not strong enough to withstand the forces involved during mold making. The one-piece mold is dried in air and heated to a temperature of 90–175°C (200–350°F) for about four hours, depending on the metal to be cast, to drive off the water of crystallization (chemically combined water). After the metal has been poured and has solidified, the mold is broken up and the casting is removed. A number of patterns can be joined to make one mold called a **tree**, thus increasing process productivity.

Although the labor and materials involved make the lost-wax process costly, it is suitable for casting high-melting-point alloys with a good surface finish and close dimensional tolerances; thus, little or no finishing is required, which otherwise would add significantly to the total cost of the casting. This process is capable of producing intricate shapes from a wide variety of ferrous and nonferrous metals and alloys, with parts generally weighing from 1 g to 100 kg (0.035 oz to 75 lb) and as much as 1140 kg (2500 lb). Typical parts made include components for office equipment; mechanical components such as gears, cams, valves, and ratchets; and jewelry.

**Ceramic-shell investment casting.**   A variation of the investment-casting process is *ceramic-shell investment casting*. It uses the same type of wax or plastic pattern, which is dipped first in a slurry with colloidal silica or ethyl silicate binder and then into a fluidized bed of fine-grained fused silica or zircon flour. Next, the pattern is dipped into coarse-grain silica to build up additional coatings and thickness to withstand the thermal shock of pouring. This process is economical and is used extensively for precision casting of steels, aluminum, and high-temperature alloys.

If ceramic cores are used in the casting, they are removed by leaching with caustic solutions under high pressure and temperature. The molten metal may be poured in a vacuum to extract evolved gases and reduce oxidation, thus improving the quality of the casting. To further reduce microporosity, the castings made by this and other processes are subjected to hot isostatic pressing. Aluminum castings, for example, are subjected to a gas pressure of up to about 100 MPa (15 ksi) at 500°C (900°F).

---

### EXAMPLE 5.3   Investment-cast superalloy components for gas turbines

Since the 1960s, investment-cast superalloys have been replacing wrought counterparts in high-performance gas turbines. Much development has been taking place in producing cleaner superalloys (nickel base and cobalt base). Improvements have been made in melting and casting techniques, such as by vacuum-induction melting, using microprocessor controls. Impurity and inclusion levels have continually been reduced, thus improving the strength, ductility, and overall reliability of these components. Such control is essential, because these parts operate at a temperature only about 50°C (90°F) below the solidus.

---

**FIGURE 5.26**   The top rotor was investment cast; the lower rotor was cast conventionally. *Source: Advanced Materials and Processes*, ASM International, October 1990, p. 25.

The microstructure of an integrally investment-cast, gas-turbine rotor is shown in the upper portion of Fig. 5.26. Note the fine, uniform, equiaxed grain size throughout the cross-section. Recent techniques to obtain this result include the use of a nucleant addition to the molten metal, as well as close control of its superheat, pouring techniques, and control of cooling rate of the casting. In contrast, the lower portion of Fig. 5.26 shows the same type of rotor cast conventionally; note the coarse grain structure. This rotor will have inferior properties compared with those of the fine-grained rotor. Due to developments in these processes, the proportion of cast parts to other parts in aircraft engines has increased from 20% to about 45% by weight.

# 5.10  |  Casting Processes: Permanent Mold

*Permanent molds*, as the name implies, are used repeatedly and are designed so that the casting can be easily removed and the mold reused. These molds are made of metals that maintain their strength at high temperatures and therefore can be used repeatedly. Because metal molds are better heat conductors than expendable molds, the solidifying casting is subjected to a higher rate of cooling, which in turn affects the microstructure and grain size within the casting, as described in Section 5.3.

In permanent-mold casting, two halves of a mold are made from materials such as steel, bronze, refractory metal alloys, or graphite (**semipermanent mold**). The mold cavity and gating system are machined into the mold and thus become an integral part of it. To produce castings with internal cavities, cores made of metal or sand aggregate are placed in the mold prior to casting. Typical core materials are shell or no-bake cores, gray iron, low-carbon steel, and hot-work die steel. Gray iron is the most commonly used core material, particularly for large molds for aluminum and magnesium castings. Inserts are also used for various parts of the mold. In order to increase the life of permanent molds, the surfaces of the mold cavity are sometimes coated with a refractory slurry or sprayed with graphite every few castings. These coatings also serve as parting agents and thermal barriers, controlling the rate of cooling of the casting. Mechanical ejectors, such as pins located in various parts of the mold, may be needed for removal of complex castings. Ejectors usually leave small round impressions on castings and gating systems.

The molds are clamped together by mechanical means and heated to facilitate metal flow and reduce thermal damage to the dies. The molten metal is then poured through the gating system. After solidification, the molds are opened and the casting is removed. Special methods of cooling the mold include water or the use of air-cooled fins, similar to those found on motorcycle and lawnmower engines that cool the engine block and cylinder head. Although the permanent-mold casting operation can be performed manually, the process can be automated for large production runs. This process is used mostly for aluminum, magnesium, and copper alloys, because of their generally lower melting points; steels can also be cast in graphite or heat-resistant metal molds.

The permanent-mold process produces castings with good surface finish, close dimensional tolerances, and uniform and good mechanical properties at high production rates. Typical parts made include automobile pistons, cylinder heads, connecting rods, gear blanks for appliances, and kitchenware. Although equipment costs can be high because of die costs, the process can be mechanized, thus keeping labor costs low. Permanent-mold casting is not economical for small production runs. Furthermore, because of the difficulty in removing the casting from the mold, intricate shapes cannot be cast by this process. However, easily collapsed sand cores can be used and removed from castings to leave intricate internal cavities.

### 5.10.1  Slush casting

It was noted in Fig. 5.16 that a solidified skin develops first in a casting and that this skin becomes thicker with time. Using this principle, hollow castings with thin walls can be made by permanent-mold casting, a process called *slush casting*. The molten metal is poured into the metal mold, and after the desired thickness of solidified skin is obtained, the mold is inverted or slung, and the remaining liquid metal is poured out. The mold halves are then opened, and the casting is removed. The process is suitable for small production runs and is generally used for making ornamental and decorative objects and toys from low-melting-point metals, such as zinc, tin, and lead alloys.

### 5.10.2  Pressure casting

In the two permanent-mold processes just described, the molten metal flows into the mold cavity by gravity. In the *pressure-casting process*, also called *pressure pouring* or *low-pressure casting* (Fig. 5.27), the molten metal is forced upward by gas pressure into a graphite or metal mold. The pressure is maintained until the metal has completely solidified in the mold. The molten metal may also be forced upward by a vacuum, which removes dissolved gases and produces a casting with lower porosity as well.

### 5.10.3  Die casting

The *die-casting* process, developed in the early 1900s, is a further example of permanent-mold casting. The molten metal is forced into the die cavity at pressures ranging from 0.7 to 700 MPa (0.1 to 100 ksi). The European term *pressure die casting*, or simply *die casting*, described in this section, is not to be confused with the term *pressure casting* described in Section 5.10.2. Typical parts made by die casting include transmission housings, valve bodies, carburetors, motors, business machine and appliance components, hand tools, and toys. The weight of most castings ranges from less than 90 g (3 oz) to about 25 kg (55 lb).

a) **Hot-chamber process.** The *hot-chamber process* (Fig. 5.28) involves the use of a piston, which traps a certain volume of molten metal and forces it into the die cavity through a gooseneck and nozzle. The pressures range up to 35 MPa (5000 psi), with an average of about 15 MPa (2000 psi). The metal is held under

**FIGURE 5.27**  The pressure-casting process uses graphite molds for the production of steel railroad wheels. *Source*: Griffin Wheel Division of Amsted Industries Incorporated.

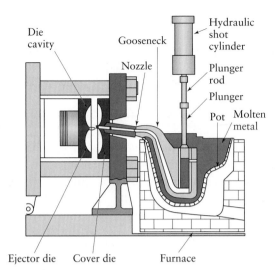

**FIGURE 5.28** Sequence of steps in die casting of a part in the hot-chamber process. *Source*: Courtesy of Foundry Management and Technology.

pressure until it solidifies in the die. To improve die life and to aid in rapid metal cooling, thus reducing cycle time, dies are usually cooled by circulating water or oil through various passageways in the die block. Cycle times usually range up to 900 shots (individual injections) per hour for zinc, although very small components, such as zipper teeth, can be cast at 18,000 shots per hour. Low-melting-point alloys such as zinc, tin, and lead are commonly cast by this process.

b) **Cold-chamber process.** In the *cold-chamber* process (Fig. 5.29), molten metal is introduced into the injection cylinder (*shot chamber*). The shot chamber is not heated, hence the term *cold chamber*. The metal is forced into the die cavity at pressures usually ranging from 20 to 70 MPa (3 to 10 ksi), although they may be as high as 150 MPa (20 ksi). The machines may be horizontal or vertical; in the latter, the shot chamber is vertical, and the machine is similar to a vertical press.

High-melting-point alloys of aluminum, magnesium, and copper are normally cast by this method, although other metals (including ferrous metals) can also be cast in this manner. Temperatures of molten metals start at about 600°C (1150°F) for aluminum and magnesium alloys and increase considerably for copper-base and iron-base alloys.

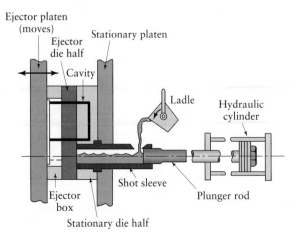

**FIGURE 5.29** Sequence of operations in die casting of a part in the cold-chamber process. *Source*: Courtesy of Foundry Management and Technology.

**Process capabilities and machine selection.** Because of the high pressures involved, the dies have a tendency to part unless clamped together tightly. Die-casting machines are rated according to the clamping force that can be exerted to keep the dies closed. The capacities of commercially available machines range from about 25 to 3000 tons. Other factors involved in the selection of die-casting machines are die size, piston stroke, shot pressure, and cost.

Die-casting dies may be made as single-cavity, multiple-cavity (several identical cavities), combination-cavity (several different cavities), or unit dies (simple small dies that can be combined in two or more units in a master holding die). Dies are usually made of hot-work die steels or mold steels. Die wear increases with the temperature of the molten metal. **Heat-checking** of the dies (for surface cracking from repeated heating and cooling of the die) can be a problem. When die materials are selected and maintained properly, dies may last more than half a million shots before die wear becomes significant.

Die design includes taper (draft) to allow the removal of the casting. The sprues and runners may be removed either manually or by using trim dies in a press. The entire die-casting and finishing process can be highly automated. Lubricants (parting agents) are usually applied as thin coatings on die surfaces. Alloys, except magnesium alloys, generally require lubricants, which usually have a water base, with graphite or other compounds in suspension. Because of the high cooling capacity of water, these lubricants also are effective in keeping die temperatures low.

Die casting has the capability for high production rates of parts with good strength, high-quality parts with complex shapes, and good dimensional accuracy and surface detail, thus requiring little or no subsequent machining or finishing operations (*net-shape forming*). Components such as pins, shafts, and fasteners can be cast integrally (**insert molding**), a process similar to putting wooden sticks (the pin) in popsicles (the casting). Ejector marks remain, as do small amounts of *flash* (thin material squeezed out between the dies; see also Section 6.2.3) at the die parting line.

In the fabrication of certain parts, die casting can compete favorably with other manufacturing methods, such as sheet-metal stamping and forging, or other casting processes. In addition, because the molten metal chills rapidly at the die walls, the casting has a fine-grain, hard skin with higher strength than in the center. Consequently, the strength-to-weight ratio of die-cast parts increases with decreasing wall thickness. With good surface finish and dimensional accuracy, die casting can produce bearing surfaces that would normally be machined. Equipment costs, particularly the cost of dies, are somewhat high, but labor costs are generally low, because the process is semi-or fully automated. Die casting is economical for large production runs. Table 5.6 lists the properties and typical applications of common die-casting alloys.

## 5.10.4  Centrifugal casting

As its name implies, the *centrifugal casting process* uses the inertial forces caused by rotation to distribute the molten metal into the mold cavities. There are three types of centrifugal casting: true centrifugal casting, semicentrifugal casting, and centrifuging.

**True centrifugal casting.** In *true centrifugal casting*, hollow cylindrical parts, such as pipes, gun barrels, and lampposts, are produced by the technique shown in Fig. 5.30, in which molten metal is poured into a rotating mold. The axis of rotation is usually horizontal, but can be vertical for short workpieces. Molds are made of steel, iron, or graphite and may be coated with a refractory lining to increase mold life. Mold

**TABLE 5.6**

## Properties and Typical Applications of Common Die-Casting Alloys

| Alloy | Ultimate Tensile Strength (MPa) | Yield Strength (MPa) | Elongation in 50 mm (%) | Applications |
|---|---|---|---|---|
| Aluminum 380 (3.5 Cu, 8.5 Si) | 320 | 160 | 2.5 | Appliances, automotive components, electrical motor frames and housings, engine blocks. |
| 13 (12Si) | 300 | 150 | 2.5 | Complex shapes with thin walls, parts requiring strength at elevated temperatures. |
| Brass 858 (60 Cu) | 380 | 200 | 15 | Plumbing fixtures, lock hardware, bushings, ornamental castings. |
| Magnesium AZ91B (9 Al, 0.7 Zn) | 230 | 160 | 3 | Power tools, automotive parts, sporting goods. |
| Zinc No. 3 (4 Al) | 280 | — | 10 | Automotive parts, office equipment, household utensils, building hardware, toys. |
| 5 (4 Al, 1 Cu) | 320 | — | 7 | Appliances, automotive parts, building hardware, business equipment. |

*Source:* Data taken from The North American Die Casting Association.

surfaces can be shaped so that pipes with various outer shapes, including square or polygonal, can be cast. The inner surface of the casting remains cylindrical, because the molten metal is uniformly distributed by centrifugal forces. However, because of differences in density, lighter elements such as dross, impurities, and pieces of the refractory lining tend to collect on the inner surface of the casting.

Cylindrical parts ranging from 13 mm (0.5 in.) to 3 m (10 ft) in diameter and 16 m (50 ft) long can be cast centrifugally, with wall thicknesses ranging from 6 to 125 mm (0.25 to 5 in.). The pressure generated by the centrifugal force is high, as much as 150 g's, and is necessary for casting thick-walled parts. Castings of good quality, dimensional accuracy, and external surface detail are obtained by this process. In addition to pipes, typical parts made include bushings, engine cylinder liners, and bearing rings with or without flanges.

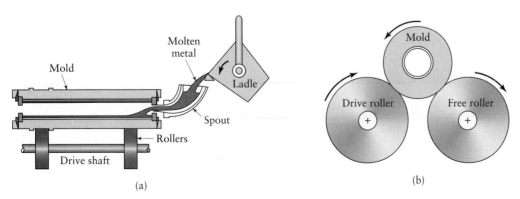

(a)                                    (b)

**FIGURE 5.30**   Schematic illustration of the centrifugal casting process. Pipes, cylinder liners, and similarly shaped parts can be cast by this process.

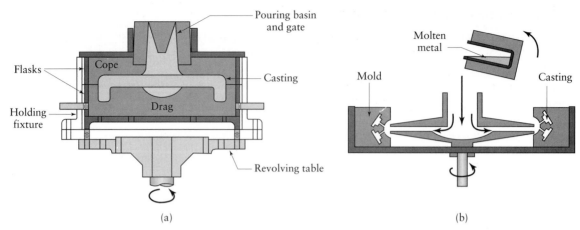

(a) (b)

**FIGURE 5.31** (a) Schematic illustration of the semicentrifugal casting process. Wheels with spokes can be cast by this process. (b) Schematic illustration of casting by centrifuging. The molds are placed at the periphery of the machine, and the molten metal is forced into the molds by centrifugal forces.

**Semicentrifugal casting.** Figure 5.31a shows an example of *semicentrifugal casting*. This method is used to cast parts with rotational symmetry, such as a wheel with spokes.

**Centrifuging.** In *centrifuging*, also called *centrifuge casting*, mold cavities of any shape are placed at a certain distance from the axis of rotation. The molten metal is poured at the center and is forced into the mold by centrifugal forces (Fig. 5.31b). The properties within the castings vary by the distance from the axis of rotation.

### 5.10.5 Squeeze casting

The *squeeze-casting process*, developed in the 1960s, involves solidification of the molten metal under high pressure; hence, the process is a combination of casting and forging (Fig. 5.32). The machinery includes a die, punch, and ejector pin. The pressure applied by the punch keeps the entrapped gases in solution (especially hydrogen in aluminum alloys), and the high-pressure contact at the die–metal interface promotes heat transfer, resulting in a fine microstructure with good mechanical properties and limited microporosity. Parts can be made to *near-net shape*, with complex shapes and fine surface detail, from both nonferrous and ferrous alloys. Typical products made include automotive wheels and mortar bodies (a short-barreled cannon). The pressures required in squeeze casting are typically higher than those used in pressure die casting, but lower than those for hot or cold forging.

**FIGURE 5.32** Sequence of operations in the squeeze-casting process. This process combines the advantages of casting and forging.

Melt metal | Pour molten metal into die | Close die and apply pressure | Eject squeeze casting, charge melt stock, and repeat cycle

(a) (b) (c) (d)

## 5.10.6  Semisolid metal forming

Also called *semisolid metalworking*, semisolid metal forming was developed in the 1970s. The metal or alloy has a nondentric, roughly spherical, and fine-grained structure when it enters the die or mold. The alloy exhibits *thixotropic* behavior, that is, its viscosity decreases when agitated; hence, this process is also known as **thixoforming** or **thixocasting**. For example, at rest and above its solidus temperature, the alloy has the consistency of table butter, but when agitated vigorously, its consistency is more like that of machine oils. This behavior has been used in developing technologies that combine casting and forging of parts, with cast billets that are forged when 30% to 40% is liquid. Semisolid-metal-forming technology was in commercial production by 1981 and is also used in making cast metal-matrix composites. Another technique for forming in a semisolid state is **rheocasting**, in which a slurry is produced in a mixer and delivered to the mold or die. However, this process has not yet been commercially successful.

## 5.10.7  Casting techniques for single-crystal components

These techniques can be illustrated by describing the casting of gas-turbine blades, which are generally made of nickel-base superalloys. The procedures involved can also be used for casting other alloys and components.

**Conventional casting of turbine blades.**  The conventional casting process involves investment casting using a ceramic mold. (See Fig. 5.23.) The molten metal is poured into the mold and begins to solidify at the ceramic walls. The grain structure developed is polycrystalline, and the presence of grain boundaries makes this structure susceptible to creep and cracking along those boundaries under the centrifugal forces at elevated temperatures.

**Directionally solidified blades.**  In the *directional solidification process* (Fig. 5.33a), first developed in 1960, the ceramic mold is prepared basically by investment-casting techniques and is preheated by radiant heating. The mold is supported by a water-cooled chill plate. After the metal is poured into the mold, the assembly is lowered slowly. Crystals begin to grow at the chill-plate surface and upward. The blade is thus directionally solidified, with longitudinal, but no transverse, grain boundaries. Consequently, the blade is stronger in the direction of centrifugal forces developed in the gas turbine.

(a)          (b)          (c)

**FIGURE 5.33**  Methods of casting turbine blades: (a) directional solidification; (b) method to produce a single-crystal blade; and (c) a single-crystal blade with a constriction portion still attached. *Source:* (a) and (b) adapted from "Advanced Metal" by B. H. Kear, copyright © 1986 by Scientific American, Inc. All Rights Reserved. (c) *Advanced Materials and Processes*, ASM International, October 1990, p. 29.

**Single-crystal blades.**  In the growing process for single-crystal blades, developed in 1967, the mold is prepared basically by investment-casting techniques and has a construction in the shape of a corkscrew (Figs. 5.33b and c), the cross-section of which allows only one crystal through. As the assembly is lowered slowly, a single crystal grows upward through the constriction and begins to grow in the mold. Strict control of the rate of movement is necessary. The solidified mass in the mold is a single-crystal blade. Although these blades are more expensive than other blades, the lack of grain boundaries makes these blades resistant to creep and thermal shock; thus, they have a longer and more reliable service life.

**Single-crystal growing.**  With the advent of the semiconductor industry, *single-crystal growing* has become a major activity in the manufacture of microelectronic devices (Chapter 13). There are two methods of crystal growing. In the **crystal-pulling method,** known as the **Czochralski, or CZ, process,** (Fig. 5.34a), a seed crystal is dipped into the molten metal and then pulled out slowly, at a rate of about $10\ \mu m/s$ ($400\ \mu in./s$), while being rotated about 1 rev/s. The liquid metal begins to

(c)

**FIGURE 5.34**    Two methods of crystal growing: (a) crystal-pulling method (Czochralski process) and (b) floating-zone method. Crystal growing is especially important in the semiconductor industry. *Source*:  L. H. Van Vlack, *Materials for Engineering* (Fig. 18.5.1, p. 513), © 1982 Addison-Wesley Publishing Company, Inc. Reprinted by permission of Addison-Wesley Longman Publishing Co., Inc. (c) A single-crystal silicon ingot produced by the Czochralski process. *Source*:  Intel Corp.

solidify on the seed, and the crystal structure of the seed is continued throughout. *Dopants* (alloying elements) may be added to the liquid metal to impart special electrical properties. Single crystals of silicon, germanium, and various other elements are grown by this process. Single-crystal ingots up to 225 mm (9 in.) in diameter and over 1 m (40 in.) in length have been produced by this technique, although 300-mm (12-in.) cylinders are under development (Fig. 5.34c).

The second technique for crystal growing is the **floating-zone method** (Fig. 5.34b). The process starts with a rod of polycrystalline silicon resting on a single crystal; an induction coil heats these two pieces while moving slowly upward. The single crystal grows upward while maintaining its orientation. Thin wafers are then cut from the rod, cleaned, and polished for use in microelectronic-device fabrication. Because of the limited diameters that can be produced through this process, it is being displaced by the Czochralski process.

### 5.10.8  Rapid solidification (amorphous alloys)

First developed in the 1960s, *rapid solidification* involves cooling of molten metal at rates as high as $10^6$ K/s, whereby the metal does not have sufficient time to crystallize. These alloys are called **amorphous alloys** or **metallic glasses**, because they do not have a long-range crystalline structure. They typically contain iron, nickel, and chromium, which are alloyed with carbon, phosphorus, boron, aluminum, and silicon. Among other effects, rapid solidification results in a significant extension of solid solubility, grain refinement, and reduced microsegregation.

Amorphous alloys exhibit excellent corrosion resistance, good ductility, and high strength. Furthermore, they exhibit very little loss from magnetic hysteresis, high resistance to eddy currents, and high permeability. The latter properties are used in making magnetic steel cores for transformers, generators, motors, lamp ballasts, magnetic amplifiers, and linear accelerators, with greatly improved efficiency. Another major application is superalloys of rapidly solidified powders consolidated into near-net shapes for use in aerospace engines. These alloys are produced in the form of wire, ribbon, strip, and powder. In one method, called **melt spinning** (Fig. 5.35), the alloy is melted by induction in a ceramic crucible and propelled under high gas pressure at very high speed against a rotating copper disk (chill block) and chills rapidly (splat cooling).

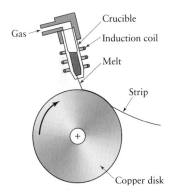

**FIGURE 5.35** Schematic illustration of the melt-spinning process to produce thin strips of amorphous metal.

## 5.11 | Heat Treatment

The various microstructures developed during metal processing can be modified by *heat treatment* techniques, that is, by controlled heating and cooling of the alloys at various rates (also known as **thermal treatment**). These treatments induce phase transformations that greatly influence mechanical properties such as strength, hardness, ductility, toughness, and wear resistance of the alloys. The effects of thermal treatment depend primarily on the alloy, its composition and microstructure, the degree of prior cold work, and the rates of heating and cooling during heat treatment. The processes of recovery, recrystallization, and grain growth (see Section 3.6) are examples of thermal treatment, involving changes in the grain structure of the alloy.

### 5.11.1  Heat treating ferrous alloys

The microstructural changes that occur in the iron–carbon system, discussed in Section 5.2.5, are described as follows:

**Pearlite.** If the ferrite and cementite lamellae in the pearlite structure (see Section 5.2.6) of the eutectoid steel are thin and closely packed, the microstructure is called **fine pearlite**. If the lamellae are thick and widely spaced, the microstructure is called **coarse pearlite**. The difference between the two depends on the rate of cooling through the eutectoid temperature, a reaction in which austenite is transformed into pearlite. If the rate of cooling is relatively high, as in air, fine pearlite is produced; if it is slow, as in a furnace, coarse pearlite is produced.

The transformation from austenite to pearlite (and for other structures) is best illustrated by Fig. 5.36b and c. These diagrams are called **isothermal transformation (IT) diagrams**, or **time-temperature-transformation (TTT) diagrams**. They are constructed from the data in Fig. 5.36a, which show the percentage of austenite transformed into pearlite as a function of temperature and time. The higher the temperature

(a)

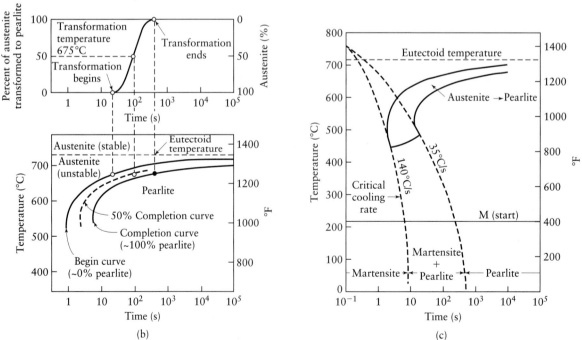

(b)　　　　　　　　　　　　　　　　(c)

**FIGURE 5.36** (a) Austenite-to-pearlite transformation of iron–carbon alloys as a function of time and temperature. (b) Isothermal transformation diagram obtained from (a) for a transformation temperature of 675°C (1247°F). (c) Microstructures obtained for a eutectoid iron–carbon alloy as a function of cooling rate. *Source*: ASM International, Materials Park, OH.

and/or the longer the time, the greater is the percentage of austenite transformed to pearlite. Note that for each temperature, a minimum amount of time is required for the transformation to begin and that sometime later, all the austenite is transformed to pearlite.

**Spheroidite.**  When pearlite is heated to just below the eutectoid temperature and held at that temperature for a period of time, say, for a day at 700°C (1300°F), the cementite lamellae (see Fig. 5.6) transform to *spherical* shapes. Unlike the lamellar shape of cementite, which acts as stress raisers, *spheroidites* (spherical particles) are less conducive to stress concentration, because of their rounded shapes. Consequently, this structure has higher toughness and lower hardness than that of the pearlite structure. In this form, the structure can be cold worked, since the ductile ferrite has high toughness, and the spheroidal carbide particles prevent the propagation of cracks within the material.

**Bainite.**  Visible only using electron microscopy, *bainite* has a very fine microstructure, consisting of ferrite and cementite. It can be produced in steels with alloying elements and at cooling rates that are higher than those required for transformation to pearlite. This structure, called *bainitic steel*, is generally stronger and more ductile than pearlitic steel at the same hardness level.

**Martensite.**  When austenite is cooled rapidly, as by quenching in water, its fcc structure is transformed to a *body-centered tetragonal* (bct) structure. This structure, called *martensite*, can be described as a body-centered rectangular prism that is slightly elongated along one of its principal axes. Because it does not have as many slip systems as does a bcc structure, and the carbon is in interstitial positions, martensite is extremely hard and brittle, lacks toughness, and therefore has limited use. Martensite transformation takes place almost instantaneously (Fig. 5.36c), because it does not involve the diffusion process, a time-dependent phenomenon that is the mechanism in other transformations.

Transformations of martensite involve volume changes, because of the different densities of the various phases in the structure. For example, when austenite transforms to martensite, its volume increases (and hence its density decreases) by as much as 4%. A similar, but smaller, volume expansion also occurs when austenite transforms to pearlite. These expansions, and thus the thermal gradients present in a quenched part, can cause internal stresses within the body. These stresses may cause parts to crack during heat treatment, as in **quench cracking** of steels caused by rapid cooling during quenching.

**Retained austenite.**  If the temperature to which the alloy is quenched is not sufficiently low, only a portion of the structure is transformed to martensite. The rest is *retained austenite*, which is visible as white areas in the structure along with dark needlelike martensite. Retained austenite can cause dimensional instability and cracking and lowers alloy hardness and strength.

**Tempered martensite.**  *Tempering* is a heating process that reduces martensite's hardness and improves its toughness. The body-centered tetragonal martensite is heated to an intermediate temperature, where it transforms to a two-phase microstructure, consisting of body-centered cubic alpha ferrite and small particles of cementite. Longer tempering time and higher temperature decrease martensite's hardness. The reason is that the cementite particles coalesce and grow, and the distance between the particles in the soft ferrite matrix increases as the less stable, smaller carbide particles dissolve.

**Hardenability of ferrous alloys.** The capability of an alloy to be hardened by heat treatment is called its *hardenability*; it is a measure of the depth of hardness that can be obtained by heating and subsequent quenching. The term *hardenability* should not be confused with *hardness* (Section 2.6). Hardenability of ferrous alloys depends on the carbon content, the grain size of the austenite, and the alloying elements present in the material. A test (the **Jominy test**) has been developed in order to determine alloy hardenability.

**Quenching media.** Quenching may be carried out in water, brine (saltwater), oils, molten salts, or air. Caustic solutions, polymer solutions, and gases are also used. Because of the differences in the thermal conductivity, specific heat, and heat of vaporization of these media, the rate of cooling of the alloy (**severity of quench**) will also be different. In relative terms and in decreasing order from 5 to 0, the cooling capacity of several quenching media is as follows: agitated brine, 5; still water, 1; still oil, 0.3; cold gas, 0.1; and still air, 0.02. Agitation is also a significant factor in the rate of cooling. In tool steels, the quenching medium is specified by a letter (see Table 3.4), such as W for water hardening, O for oil hardening, and A for air hardening. The cooling rate also depends on the surface-area-to-volume ratio of the part. (See Eq. 5.11.) The higher this ratio, the higher is the cooling rate; thus, for example, a thick plate cools more slowly than a thin plate with the same surface area.

Water is a common medium for rapid cooling. However, the heated metal may form a **vapor blanket** along its surfaces from water-vapor bubbles that form when water boils at the metal–water interface. This blanket creates a barrier to heat conduction, because of the lower thermal conductivity of the vapor. Agitating the fluid or the part helps to reduce or eliminate the blanket. Also, water may be sprayed on the part under high pressure. Brine is an effective quenching medium, because salt helps to nucleate bubbles at the interfaces, thus improving agitation. However, brine can corrode the part. **Die quenching** is a term used to describe the process of clamping the part to be heat treated to a die, which chills selected regions of the part. In this way, cooling rates and warpage can be controlled.

## 5.11.2 Heat treating nonferrous alloys and stainless steels

Nonferrous alloys and some stainless steels generally cannot be heat treated by the techniques used with ferrous alloys. The reason is that nonferrous alloys do not undergo phase transformations as steels do. The hardening and strengthening mechanisms for these alloys are fundamentally different. Heat-treatable aluminum alloys, copper alloys, and martensitic and precipitation-hardening stainless steels are hardened and strengthened by a process called **precipitation hardening**. This is a technique in which small particles (of a different phase and called *precipitates*) are uniformly dispersed in the matrix of the original phase. (See Fig. 5.2a.) In this process, precipitate forms, because the solid solubility of one element (one component of the alloy) in the other is exceeded.

Three stages are involved in precipitation hardening; they can best be described by referring to Fig. 5.37, the phase diagram for the aluminum–copper system. For an alloy with the composition of 95.5% Al, 4.5% Cu, a single-phase (kappa) substitutional solid solution of copper (solute) in aluminum (solvent) exists between 500°C and 570°C (930°F and 1060°F). The kappa phase is aluminum rich, with an fcc structure, and is ductile. Below the lower temperature, that is, below the lower solubility curve, two phases are present: kappa and theta (a hard intermetallic compound of $CuAl_2$). This alloy can be heat treated and its properties modified by solution treatment or precipitation.

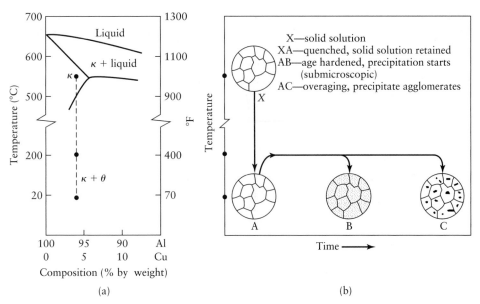

**FIGURE 5.37** (a) Phase diagram for the aluminum–copper alloy system. (b) Various microstructures obtained during the age-hardening process. *Source*: L. H. Van Vlack, *Materials for Engineering* (Fig. 5.6.1, p. 123), © 1982 Addison-Wesley Publishing Company, Inc. Reprinted by permission of Addison-Wesley Longman Publishing Co., Inc.

**Solution treatment.** In *solution treatment*, the alloy is heated to within the solid-solution kappa phase, say 540°C (1000°F), and cooled rapidly, such as by quenching in water. The structure obtained soon after quenching (*A* in Fig. 5.37b), consists only of the single-phase kappa. This alloy has moderate strength and considerable ductility.

**Precipitation hardening.** The structure obtained in A in Fig. 5.37b can be strengthened by precipitation hardening. The alloy is reheated to an intermediate temperature and held there for a period of time, during which precipitation takes place. The copper atoms diffuse to nucleation sites and combine with aluminum atoms, producing the theta phase, which forms as submicroscopic precipitates, shown in B by the small dots within the grains of the kappa phase. This structure is stronger than that in A, although it is less ductile. The increase in strength is attributed to increased resistance to dislocation movement in the region of the precipitates.

**Aging.** Because the precipitation process is one of time and temperature, it is also called *aging*, and the property improvement is known as *age hardening*. If carried out above room temperature, the process is called **artificial aging**. However, several aluminum alloys harden and become stronger over a period of time at room temperature, by a process known as **natural aging**. Such alloys are first quenched and then, if desired, are formed at room temperature into various shapes and allowed to gain strength and hardness by natural aging. Natural aging can be slowed by refrigerating the quenched alloy.

In the precipitation process, if the reheated alloy is held at that elevated temperature for an extended period of time, the precipitates begin to coalesce and grow. They become larger, but fewer, as shown by the larger dots in C in Fig. 5.37b. This process is called **overaging**, which makes the alloy softer and weaker; thus, there is an optimal time–temperature relationship in the aging process for obtaining desired properties. Obviously, an aged alloy can be used only up to a certain maximum temperature in service; otherwise, it will overage and lose its strength and hardness. Although weaker, an overaged part has better dimensional stability.

**Maraging.** Maraging is a precipitation-hardening treatment for a special group of high-strength iron-base alloys. The word *maraging* is derived from the words martensite and aging. In this process, one or more intermetallic compounds are pre-cipitated in a matrix of low-carbon martensite. A typical maraging steel may contain 18% nickel, in addition to other elements, and aging is done at 480°C (900°F). Hard-ening by maraging does not depend on the cooling rate; thus, full uniform hardness can be obtained throughout large parts with minimal distortion. Typical uses of maraging steels are for dies and tooling for casting, molding, forging, and extrusion.

### 5.11.3 Case hardening

The heat treatment processes described so far involve microstructural alterations and property changes in the *bulk* of the material or component by **through hardening**. In many situations, however, alteration of only the *surface* properties of a part (hence the term *case hardening*) is desirable. This method is particularly useful for improv-ing resistance to surface indentation, fatigue, and wear. Typical applications for case hardening include gear teeth, cams, shafts, bearings, fasteners, pins, automotive clutch plates, and tools and dies. Through hardening of these parts would not be desirable, because a hard part lacks the necessary toughness for these applications; a small surface crack can propagate rapidly through the part and cause total failure.

Various surface-hardening processes are available (Table 5.7): **carburizing** (*gas*, *liquid*, and *pack carburizing*), **carbonitriding, cyaniding, nitriding, boronizing,** and **flame** and **induction hardening**. Basically, these processes are heat-treating operations in which the component is heated in an atmosphere containing elements (such as car-bon, nitrogen, or boron) that alter the composition, microstructure, and properties of surfaces.

For steels with sufficiently high carbon content, surface hardening takes place without using any of these additional elements. Only the heat-treatment processes described in Section 5.10.1 are needed to alter the microstructures, usually by flame hardening or induction hardening. Laser beams and electron beams are also used ef-fectively to harden both small and large surfaces and also for through hardening of relatively small parts.

Because case hardening is a localized heat treatment, case-hardened parts have a hardness gradient. Typically, the hardness is greatest at the surface and decreases below the surface, the rate of decrease depending on the composition of the metal and the process variables. Surface-hardening techniques can also be used for tempering, thus modifying the properties of surfaces that have been subjected to heat treatment. Various other processes and techniques for surface hardening, such as shot peening and surface rolling, improve wear resistance and various other characteristics. (See Section 4.5.1.)

**Decarburization** is the phenomenon in which alloys containing carbon lose car-bon from their surfaces as a result of heat treatment or hot working in a medium, usu-ally oxygen, that reacts with the carbon. Decarburization is undesirable, because it affects the hardenability of the surfaces of the part by lowering the carbon content. It also adversely affects the hardness, strength, and fatigue life of steels by signifi-cantly lowering their endurance limits. Decarburization is best avoided by process-ing the alloy in an inert atmosphere or a vacuum or by using neutral salt baths during heat treatment.

### 5.11.4 Annealing

*Annealing* is a general term used to describe the restoration of a cold-worked or heat-treated metal or alloy to its original properties, so as to increase ductility (and hence formability), reduce hardness and strength, or modify the microstructure. Annealing is also used to relieve residual stresses in a manufactured part for improved machinability

TABLE 5.7

## Outline of Heat Treatment Processes for Surface Hardening

| Process | Metals Hardened | Element Added To Surface | Procedure | General Characteristics | Typical Applications |
|---|---|---|---|---|---|
| Carburizing | Low-carbon steel (0.2% C), alloy steels (0.08–0.2% C) | C | Heat steel at 870–950°C (1600–1750°F) in an atmosphere of carbonaceous gasses (gas carburizing) or carbon-containing solids (pack carburizing). Then quench. | A hard, high-carbon surface is produced. Hardness of 55 to 65 HRC. Case depth <0.5–1.5 mm (<0.020 to 0.060 in.) Some distortion of the part during heat treatment. | Gears, cams, shafts, bearings, piston rings, sprockets, clutch plates. |
| Carbonitriding | Low-carbon steel | C and N | Heat steel at 700–800°C (1300–1600°F) in an atmosphere of carbonaceous gas and ammonia. Then quench in oil. | Surface hardness 55 to 62 HRC. Case depth 0.07–0.5 mm (0.003–0.020 in.) Less distortion than in carburizing. | Bolts, nuts, gears. |
| Cyaniding | Low-carbon steel (0.2% C), alloy steels (0.08–0.2% C) | C and N | Heat steel at 760–845°C (1400–1550°F) in a molten bath of solutions of cyanide (e.g., 30% sodium cyanide) and other salts. | Surface hardness up to 65 HRC. Case depth 0.025–0.25 mm (0.001–0.010 in.) Some distortion. | Bolts, nuts, screws, small gears. |
| Nitriding | Steels (1% Al, 1.5% Cr, 0.3% Mo), alloy steels (Cr, Mo), stainless steels, high-speed tool steels | N | Heat steel at 500–600°C (925–1100°F) in an atmosphere of ammonia gas or mixtures of molten cyanide salts. No further treatment. | Surface hardness up to 1100 HV. Case depth 0.1–0.6 mm (0.005–0.030 in.) and 0.02–0.07 mm (0.001–0.003 in.) for high-speed steel. | Gears, shafts, sprockets, valves, cutters, boring bars, fuel-injection pump parts. |
| Boronizing | Steels | B | Part is heated using boron-containing gas or solid in contact with the part. | Extremely hard and wear-resistant surface. Case depth 0.025–0.075 mm (0.001–0.003 in.). | Tool and die steels. |
| Flame hardening | Medium-carbon steels, cast irons | None | Surface is heated with an oxyacetylene torch and then quenched with water spray or other quenching methods. | Surface hardness 50 to 60 HRC. Case depth 0.7–6 mm (0.030–0.25 in.) Little distortion. | Gear and sprocket teeth, axles, crankshafts, piston rods, lathe beds and centers. |
| Induction hardening | Medium-carbon steels, cast irons | None | Metal part is placed in copper induction coils and is heated by high-frequency current and then quenched. | Surface hardness 50 to 60 HRC. Case depth 0.7–6 mm (0.030–0.25 in.) Little distortion. | Gear and sprocket teeth, axles, crankshafts, piston rods, lathe beds and centers. |

**FIGURE 5.38** Heat-treating temperature ranges for plain-carbon steels, as indicated on the iron–iron-carbide phase diagram. *Source*: ASM International, Materials Park, OH.

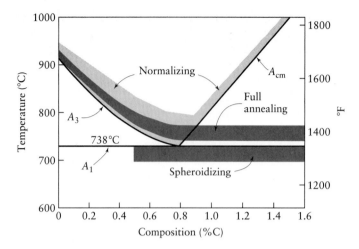

and dimensional stability. The term *annealing* also applies to thermal treatment of glasses and similar products (Section 11.11.2) and weldments (Section 12.5.1).

The annealing process involves (1) heating the workpiece to within a specific range of temperatures, (2) holding it at that temperature for a period of time (soaking), and (3) cooling it slowly. The process may be carried out in an inert or controlled atmosphere or performed at low temperatures to prevent or minimize surface oxidation. Annealing temperatures may be higher than the recrystallization temperature, depending on the degree of cold work (and hence stored energy). For example, the recrystallization temperature for copper ranges between 200°C and 300°C (400°F and 600°F), whereas the annealing temperature needed to fully recover the original properties ranges from 260°C to 650°C (500°F to 1200°F), depending on the degree of prior cold work.

**Full annealing** is a term applied to annealing ferrous alloys, generally low- and medium-carbon steels. The steel is heated to above $A_1$ or $A_3$ in Fig. 5.38, and cooling takes place slowly, at a rate of, say, 10°C (20°F) per hour, in a furnace after it has been turned off. The structure obtained in full annealing is coarse pearlite, which is soft and ductile and has small, uniform grains. Excessive softness in the annealing of steels can be avoided if the entire cooling cycle is carried out in still air; this process is called **normalizing**, in which the part is heated to a temperature of $A_3$ or $A_{cm}$ to transform the structure to austenite. It results in somewhat higher strength and hardness and lower ductility than in full annealing. The structure obtained is fine pearlite with small, uniform grains. Normalizing is generally done to refine the grain structure, obtain uniform structure (*homogenization*), decrease residual stresses, and improve machinability.

**Process annealing.** In *process annealing* (also called *intermediate annealing*, *subcritical annealing*, or *in-process annealing*), the workpiece is annealed to restore its ductility, part or all of which may have been exhausted by work hardening during cold working. In this way, the part can be worked further into the final desired shape. If the temperature is high and/or the time of annealing is long, grain growth may result, with adverse effects on the formability of annealed parts.

**Stress-relief annealing.** To reduce or eliminate residual stresses, a workpiece is generally subjected to *stress-relief annealing*, or simply **stress relieving**. The temperature and amount of time required for this process depend on the material and the magnitude of residual stresses present. The residual stresses may have been induced during forming, machining, or other shaping processes or caused by volume changes during phase transformations. For steels, the part is heated to below $A_1$, thus avoiding phase transformations. Slow cooling rates, such as in still air, are generally employed. Stress relieving promotes dimensional stability in situations where subsequent relaxing of residual stresses present may cause distortion of the part over a period of time when it is in service. It also reduces the tendency for stress-corrosion cracking.

## 5.11.5 Tempering

If steels are hardened by heat treatment, *tempering* is used in order to reduce brittleness, increase ductility and toughness, and reduce residual stresses. The term *tempering* is also used for glasses (Section 11.11.2). In tempering, the steel is heated to a specific temperature, depending on composition, and cooled at a prescribed rate. Alloy steels may undergo **temper embrittlement**, caused by the segregation of impurities along the grain boundaries at temperatures between 480°C and 590°C (900°F and 1100°F).

In **austempering**, the heated steel is quenched from the austenitizing temperature rapidly enough to avoid formation of ferrite or pearlite. It is then held at a certain temperature until isothermal transformation from austenite to bainite is complete. Next, it is cooled to room temperature, usually in still air, at a moderate rate to avoid thermal gradients within the part. The quenching medium most commonly used is molten salt, at temperatures ranging from 160°C to 750°C (320°F to 1380°F). Austempering is often substituted for conventional quenching and tempering, either to reduce the tendency for cracking and distortion during quenching or to improve ductility and toughness while maintaining hardness. Because of the relatively short cycle time, this process is economical for many applications. In *modified austempering*, a mixed structure of pearlite and bainite is obtained. The best example of this practice is **patenting**, which provides high ductility and moderately high strength, such as in the patented wire used in the wire industry (Section 6.5.3).

In **martempering** (**marquenching**), the steel or cast iron is quenched from the austenitizing temperature into a hot fluid medium, such as hot oil or molten salt. It is held at that temperature until the temperature is uniform throughout the part and then cooled at a moderate rate, such as in air, to avoid temperature gradients within the part. The part is then tempered, because the structure thus far obtained is primarily untempered martensite and is not suitable for most applications. Martempered steels have less of a tendency to crack, distort, and develop residual stresses during heat treatment. In modified martempering, the quenching temperature is lower, and hence the cooling rate is higher. The process is suitable for steels with lower hardenability.

In **ausforming**, also called **thermomechanical processing**, the steel is formed into desired shapes within controlled ranges of temperature and amount of time to avoid formation of nonmartensitic transformation products. The part is then cooled at various rates to obtain the desired microstructures. Ausformed parts have superior mechanical properties.

## 5.11.6 Cryogenic treatment

In *cryogenic tempering*, steel is lowered from room temperature to −180°C (−300°F) at a slow rate of as low as 2°C per minute in order to avoid thermal shock. The material is then maintained at this temperature for 24 to 36 hours. At this temperature, the conversion of austenite to martensite occurs slowly, but almost completely, compared with typically only 50% to 90% in conventional quenching. Additional precipitates of carbon (with chromium, tungsten, and other elements) form, relieving residual stresses and leaving a refined grain structure. After the 24-to-36-hour "soak" time, the parts are tempered to stabilize the martensite.

Cryogenically treated steels have higher levels of hardness and wear resistance than those of untreated steels. For example, the wear resistance of D-2 tool steels (see Section 3.10.3) can increase by over 800% after cryogenic treatment to $-180°C$ ($-300°F$), although most tool steels show a 100% to 200% increase in tool life. (See Section 8.3.) Similar property improvements have been observed in nonferrous alloys and polymers, but these improvements are attributable to refinement of grain structure, since no martensitic transformation occurs. Applications of cryogenic treatment include cutting tools and dies, dental instruments, aerospace materials, sporting goods such as golf club heads, and gun barrels.

### 5.11.7  Design for heat treating

In addition to the metallurgical factors described thus far, successful heat treating involves design considerations of avoiding problems such as cracking, warping, and nonuniform properties throughout the heat-treated part. The rate of cooling during quenching may not be uniform, particularly with complex shapes of varying cross-sections and thicknesses, thus producing severe temperature gradients. These gradients lead to variations in contraction, which induce thermal stress and may cause the part to crack. Furthermore, nonuniform cooling causes residual stresses in the part, which can lead to stress-corrosion cracking; hence, the method selected and care taken in quenching, as well as the proper choice of quenching media and temperatures, are important considerations.

As a general guideline for part design for heat treating, internal or external sharp corners should be avoided. Otherwise, stress concentrations at these corners raise stress levels high enough to cause cracking. Parts should have as nearly uniform thicknesses as possible, or the transition between regions of different thicknesses should be smooth. Parts with holes, grooves, keyways, splines, and unsymmetrical shapes may also be difficult to heat treat, because they may crack during quenching. Large surfaces with thin cross-sections are likely to warp. Hot forgings and hot steel-mill products may have a *decarburized skin* and thus may not successfully respond to heat treatment.

### 5.11.8  Cleaning, finishing, and inspection of castings

After solidification and removal from the mold or die, castings may be subjected to a variety of additional processes. In sand casting, the casting is shaken out of its mold, and the sand and oxide layers adhering to the casting are removed by vibration (using a shaker) or by sand blasting. In sand casting and other casting processes, and depending on size, the risers and gates are cut off by oxyfuel-gas cutting, sawing, shearing, and abrasive wheels, or they are trimmed in dies. Castings may be cleaned electrochemically or by pickling with chemicals to remove surface oxides. Maintaining a clean surface of castings is important in subsequent machining operations, because machinability can be adversely affected if the castings are not cleaned properly. If regions of the casting have not formed properly or have formed incompletely, the defects may be repaired by welding, thus filling them with weld metal. Sand castings generally have rough, grainy surfaces, although the extent depends on the quality of the mold and the materials used. *Finishing operations* for castings may include straightening or forging with dies and machining to obtain final dimensions.

Inspection is an important final step and is done to ensure that the casting meets all design and quality-control requirements. Several methods are available for inspection of castings to determine the quality of the castings and whether any defects are present. Castings can be inspected visually or optically for surface defects.

Subsurface and internal defects are investigated using various nondestructive techniques, described in Section 4.8. In destructive testing, test specimens are removed from various sections of a casting and tested for strength, ductility, and other mechanical properties and to determine the location of any defects.

*Pressure tightness* of cast components such as valves, pumps, and pipes is usually determined by sealing the openings in the casting and pressurizing it with water, oil, or air. The casting is then inspected for leaks while the pressure is maintained. Unacceptable or defective castings are melted down for reprocessing. Because of their major economic impact, the types of defects present in castings and their causes must be investigated. Control of all stages during casting, from mold preparation to removal of castings from molds or dies, is important in maintaining good quality.

# 5.12 | Design Considerations

As in all engineering practice and manufacturing operations, certain guidelines and design principles pertaining to casting have been developed over many years. Although these principles were established primarily through practical experience, analytical methods and computer-aided design and manufacturing techniques are now coming into wider use, improving productivity and the quality of castings. Moreover, careful design can result in significant cost savings. Table 5.8 lists some of the advantages and limitations of casting processes that impact design.

**TABLE 5.8**

## Casting Processes, and their Advantages and Limitations

| Process | Advantages | Limitations |
|---|---|---|
| Sand | Almost any metal is cast; no limit to size, shape, or weight; low tooling cost. | Some finishing required; somewhat coarse finish; wide tolerances. |
| Shell mold | Good dimensional accuracy and surface finish; high production rate. | Part size limited; expensive patterns and equipment required. |
| Expandable pattern | Most metals cast with no limit to size; complex shapes may be cast. | Patterns have low strength and can be costly for low quantities. |
| Plaster mold | Intricate shapes may be cast; good dimensional accuracy and finish; low porosity. | Limited to nonferrous metals; limited size and volume of production; mold-making time relatively long. |
| Ceramic mold | Intricate shapes may be cast; close tolerance parts; good surface finish. | Limited size. |
| Investment | Intricate shapes may be cast; excellent surface finish and accuracy; almost any metal may be cast. | Part size limited; expensive patterns, molds, and labor. |
| Permanent mold | Good surface finish and dimensional accuracy; low porosity; high production rate. | High mold cost; limited shapes and intricacy may be cast; not suitable for high-melting-point metals. |
| Die | Excellent dimensional accuracy and surface finish; high production rate. | Die cost is high; part size limited; usually limited to nonferrous metals; long lead time. |
| Centrifugal | Large cylindrical parts with good quality; high production rate. | Equipment is expensive; part shape limited. |

**FIGURE  5.39**
(a) Suggested design modifications to avoid defects in castings. Note that sharp corners are avoided to reduce stress concentrations. (b)–(d) Examples of designs that show the importance of maintaining uniform cross-sections in castings to avoid hot spots and shrinkage cavities.

**FIGURE  5.40**   Examples of casting design modifications. *Source*: Steel Castings Handbook, 5th ed., Steel Founders' Society of America, 1980. Used with permission.

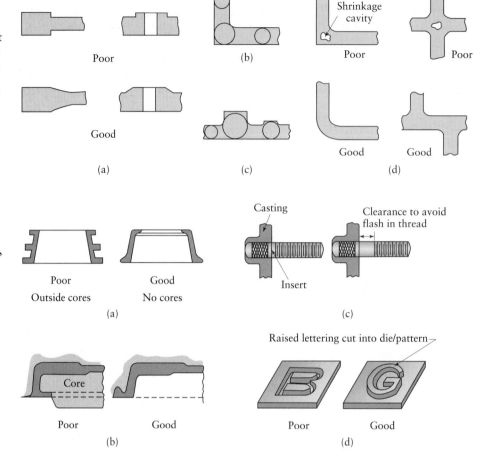

## 5.12.1  Design principles for expendable-mold casting

The guidelines given in this section generally apply to all types of castings. In these guidelines, we identify and describe the most significant design considerations. Figures 5.39 and 5.40 show some typical design modifications for casting.

1.  **Risers.**  A major concern in the design of castings is the size and placement of risers. (See Fig. 5.14.) Traditionally, riser design has been based on experience and on considerations of fluid flow and heat transfer. More recently, commercial computer programs, based on finite-difference algorithms, have become available. Risers are designed according to six basic rules: (1) The riser must not solidify before the casting; (2) The riser volume must be large enough to provide sufficient liquid metal to compensate for shrinkage in the casting; (3) Junctions between the casting and feeder should not create a **hot spot** where shrinkage porosity can occur; (4) Risers must be placed so that liquid metal can be delivered to locations where it is most needed; (5) There must be sufficient pressure to drive liquid metal into locations in the mold where it is needed; (6) The pressure head from the riser should suppress cavity formation and encourage complete filling of cavities.

2.  **Corners, angles, and section thickness.**  Sharp corners, angles, and fillets should be avoided, because they may cause cracking and tearing during solidification of the metal. Fillet radii should be selected to reduce stress concentrations and to ensure proper liquid-metal flow during the pouring process. If the fillet radii are too large, the volume of the material in those regions will also be large, and, consequently, the rate of cooling will be lower. Section changes in

castings should smoothly blend into each other. The location of the largest circle that can be inscribed in a particular region is critical as far as shrinkage cavities are concerned (Fig. 5.39). Because the cooling rate in regions with larger circles is lower than in regions with smaller circles, regions with the larger circles are called *hot spots*. These regions could develop *shrinkage cavities* and *porosity*. Although they increase the cost of production, metal paddings or chills in the mold can eliminate or minimize hot spots.

3. **Flat areas.**  Large flat areas (plain surfaces) should be avoided, as they may warp, because of temperature gradients during cooling, or develop poor surface finish, because of uneven flow of metal during pouring. Flat surfaces can be broken up with ribs and serrations.

4. **Parting lines.**  In general, the parting line should be along a flat plane, rather than a contoured plane. Whenever possible, the parting line should be at the corners or edges of castings, rather than on flat surfaces in the middle of the casting. In this way, the **flash** at the parting line (i.e., the material running out between the two halves of the mold; see also Fig. 6.14) will not be as visible. The location of the parting line is important, because it influences mold design, ease of molding, number and shape of cores, method of support, and the gating system. Preparation of dry-sand cores requires additional time and cost, so these cores should be avoided or minimized; this consideration can usually be made by reviewing and simplifying the design of castings. Figure 5.40 shows two examples of casting design modifications.

5. **Draft.**  A small draft (taper) is provided in sand-mold patterns to enable removal of the pattern without damaging the mold. Depending on the quality of the pattern, draft angles usually range from 0.5° to 2°. The angles on the inside surfaces of molds are typically twice this range; they have to be higher than those for outer surfaces, because the casting shrinks inward toward the core.

6. **Dimensional tolerances.**  Dimensional tolerances depend on the particular casting process, size of the casting, and type of pattern used, with typical values given in Table 5.2. Tolerances are smallest within one part of the mold and, because they are cumulative, increase between different parts of the mold. Tolerances should be as wide as possible, within the limits of good part performance; otherwise, the cost of the casting increases. In commercial practice, tolerances usually are about ±0.8 mm (1/32 in.) for small castings and increase with size of castings, say, to 6 mm (1/4 in.) for large castings.

7. **Shrinkage.**  Allowance for shrinkage during solidification should be provided for, so as to avoid cracking of the casting, particularly in permanent molds. Pattern dimensions should also provide for shrinkage of the metal during solidification and cooling. Allowances for shrinkage, also known as **patternmaker's shrinkage allowances**, usually range from about 10 to 20 mm/m (1/8 to 1/4 in./ft). The normal shrinkage allowance for metals commonly sand cast ranges approximately between 1% and 2%.

8. **Allowances.**  A *shrinkage allowance* should be designed into castings so that the castings achieve the desired shape upon removal from the mold. Because most expendable-mold castings require some additional finishing operations, such as machining, allowances should be made in casting design for these operations. **Machining allowances** are included in pattern dimensions and depend on the type of casting. They increase with the size and section thickness of castings, usually ranging from about 2 to 5 mm (0.1 to 0.2 in.) for small castings to more than 25 mm (1 in.) for large castings. Since blowholes and other defects will occur in larger concentrations near the surface, larger machining allowances are needed for applications where surface finish is critical.

### 5.12.2  Design principles for permanent-mold casting

The design principles for permanent-mold casting are similar to those for expendable-mold casting. Typical design guidelines and examples for permanent-mold casting are shown schematically in Fig. 5.41 for die casting. Note that the cross-sections have been reduced in order to decrease solidification time and save material. Special considerations are involved in designing and tooling for die casting. Designs may be modified to eliminate the draft for better dimensional accuracy.

### 5.12.3  Computer modeling of casting processes

Because casting processes involve complex interactions among material and process variables, a quantitative study of these interactions is essential to the proper design of castings and the production of high-quality castings. In the past, such studies have presented major difficulties. However, rapid advances in computers and modeling techniques have led to important innovations in modeling various aspects of casting, including fluid flow, heat transfer, and microstructures developed during solidification under various casting-process conditions.

Modeling of *fluid flow* is based on Bernoulli's equation and the continuity equations (see Section 5.4). It predicts the behavior of the metal during pouring into the gating system and its travel into the mold cavity, as well as velocity and pressure distributions in the system. Progress has also been made in modeling of *heat transfer* in casting. Studies of this issue include investigation of the coupling of fluid flow and heat transfer and the effects of surface conditions, thermal properties of the materials involved, and natural and forced convection on cooling. Note that the surface conditions vary during solidification, as a layer of air develops between the casting and the mold wall as a result of shrinkage. Similar studies are being conducted on modeling the development of *microstructures* in casting. These studies encompass heat flow, temperature gradients, nucleation and growth of crystals, formation of dendritic and equiaxed structures, impingement of grains on each other, and movement of the liquid–solid interface during solidification.

Such models are now capable of predicting, for example, the width of the mushy zone (see Fig. 5.10) during solidification and grain size in castings. Similarly, the capability to calculate isotherms gives insight into possible hot spots and subsequent development of shrinkage cavities. With the availability of user-friendly computers and advances in computer-aided design and manufacturing (see Chapter 15), modeling techniques are becoming easier to implement. The benefits are increased productivity, improved quality, easier planning and cost estimation, and quicker response to design changes. Several commercial software programs are now available on modeling and casting processes, such as Magmasoft, ProCast, Solidia, and AFSsolid.

**FIGURE 5.41**  Examples of undesirable and desirable design practices for die-cast parts. Note that section-thickness uniformity is maintained throughout the part. *Source*:  Courtesy of The North American Die Casting Association.

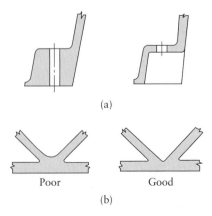

(a)

Poor          Good

(b)

# 5.13 | Economics of Casting

When reviewing various casting processes, we noted that some require more labor than others, some require expensive dies and machinery, and some take a great deal of time to complete. Outlined in Table 5.2, each of these important factors affects to a greater or lesser degree the overall cost of a casting operation. As described in greater detail in Chapter 16, the total cost of a product involves the costs of materials, labor, tooling, and equipment. Preparations for casting a product include making molds and dies, which requires raw materials, time, and effort, all of which contribute to costs. Relatively little cost is involved in making molds for sand casting. At the other extreme, die-casting dies require expensive materials and a great deal of machining and preparation. In addition to molds and dies, facilities are required for melting and pouring the molten metal into the molds or dies. These facilities include furnaces and related machinery, their costs depending on the level of automation desired. Finally, costs are involved in cleaning and inspecting castings.

The amount of labor required for these operations can vary considerably, depending on the particular process and level of automation. Investment casting, for example, requires a great deal of labor, because of the large number of steps involved in the operation. Conversely, operations such as highly automated die casting can maintain high production rates with little labor required.

The cost of equipment per casting (*unit cost*), however, decreases as the number of parts cast increases (Fig. 5.42); thus, sustained high production rates can justify the high cost of dies and machinery. However, if demand is relatively small, the cost per casting increases rapidly. It then becomes more economical to manufacture the parts by sand casting or by other manufacturing processes. Note that the graphical comparison in Fig. 5.42 can include other casting processes suitable for making the same part. The sand and die-casting processes compared in the figure produce castings with significantly different dimensional and surface-finish characteristics; thus, not all manufacturing decisions are based purely on economic considerations. In fact, parts can usually be made by more than one or two processes. The final decision depends on both economic and technical considerations. The competitive aspects of manufacturing processes are discussed further in Chapter 16.

**FIGURE 5.42** Economic comparison of making a part by different casting processes. Note that because of the high cost of equipment, die casting is economical for large production runs. *Source*: The North American Die Casting Association.

**CASE STUDY** | Permanent-Mold Casting of Aluminum Automotive Pistons

Figure 5.43 shows an aluminum piston used in automotive internal-combustion engines. These products must be manufactured at very high rates with very tight dimensional tolerances and strict material requirements in order to achieve proper operation. Economic concerns are obviously paramount, and it is essential that pistons be produced with a minimum of expensive finishing operations and with few rejected parts.

Aluminum pistons are manufactured by the permanent-mold casting process, because of its capability to produce near-net shaped parts at the required production rates. However, with poorly designed molds, underfills or excess porosity can cause parts to be rejected, adding to the cost of the process. These defects were traditionally controlled through the use of large machining allowances coupled with intuitive design of molds, based on experience.

Permanent-mold casting is a high-volume production, where low cycle times are achieved by using metallic molds with integral cooling water channels. Depending on the particular alloy cast and the size of the casting and mold, cycle times can vary from less than a minute to eight minutes. The molds are machined from H13 hot-work tool steel. (See Section 3.10.3.) Depending on the piston geometry, a permanent or expendable core may be required; expendable cores are constructed from sand or plaster, in which case the process is properly referred to as semipermanent-mold casting.

The pistons are produced from high-silicon alloys such as 413.0 aluminum alloy. This alloy has high fluidity and can create high-definition surfaces through permanent-mold casting; it also has high resistance to corrosion, good weldability, and low specific gravity. The universal acceptance of aluminum pistons for internal-combustion- engine applications is due mainly to the pistons' light weight and high thermal conductivity. Their low inertia allows for higher engine speeds and reduced counterweighting in the crankshaft, and their higher thermal conductivity allows more efficient heat transfer from the engine.

**FIGURE 5.43** Aluminum piston for an internal combustion engine. (a) As cast; (b) after machining.

(a)                                    (b)

The mold is preheated to 200–450°C (400–850°F), depending on the cast alloy and part size. Initially, the preheat is achieved with a handheld torch, but after a few castings, the mold reaches a steady-state temperature profile. The molten aluminum is heated to 100–200°C (200–400°F) above its liquidus temperature, and then a shot is placed into the infeed section of the mold. Once the molten-metal shot is in place, a piston drives the metal into the mold. Because of the high thermal conductivity of the mold material, heat extraction from the molten metal is rapid, and the metal can solidify in small channels before filling the mold completely. Solidification usually starts at one end of the casting before the mold is fully filled.

As with most alloys, it is desired to begin solidification at one extreme end of the casting and have the solidification front proceed across the volume, resulting in a directionally solidified microstructure and elimination of gross porosity that arises when two solidification fronts meet inside a casting. Regardless, casting defects such as undercuts, hot spots, porosity, and cracking and entrapped air-zone defects such as blowholes and scabs can occur. With poor mold design, defects can make up 5% of the castings.

In order to improve the reliability and reduce the costs associated with permanent-mold casting, a mathematical modeling capability for the permanent-mold casting process was developed by researchers at the University of Notre Dame working with engineers at A. E. Goetze Corporation. The goal of the simulations is to identify conditions that lead to casting defects and allow evaluation of design alterations before the cost of mold production is incurred. Since molds are extremely expensive, numerical simulation is an essential step in evaluating the designs of new molds.

A typical problem is illustrated in Fig. 5.44, which shows a mold with cooling channels and the desired piston geometry. Initial modeling work was coupled with experiments that implemented thermocouples placed in the cavity and mold. The simulation of permanent-mold casting requires a coupling of the fluid mechanics equations of mold filling with the heat transfer and materials science equations of alloy solidification and is a very computationally intensive process. The goal of the model is to identify combinations of mold design, cooling-channel location and size, and cast-alloy superheat that reduce the numbers of defects.

Figure 5.45 illustrates some temperature profiles and solidification fronts that develop during casting in the part, with the mold removed for clarity. Note that the solidification starts during mold filling, and the developing mushy zone is very large.

**FIGURE 5.44** Schematic illustration of the permanent mold used to produce aluminum pistons, showing the position of four cooling channels.

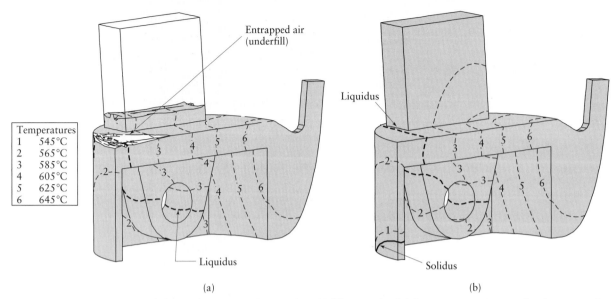

| Temperatures | |
| --- | --- |
| 1 | 545°C |
| 2 | 565°C |
| 3 | 585°C |
| 4 | 605°C |
| 5 | 625°C |
| 6 | 645°C |

(a)

(b)

**FIGURE 5.45**    Simulation of mold filling and solidification. (a) 3.7 seconds after start of pour. Note that the mushy zone has been established before the mold is completely filled. (b) Using a vent in the mold for removal of entrapped air five seconds after pour.

Figure 5.45a shows a piston after 3.7 s from initial injection of metal into the mold, where a region of entrapped air has been identified that would normally lead to a systemic underfill in these castings. Figure 5.45b shows the piston with a venting system incorporated to correct the entrapped-air problem; it also shows the contours of constant temperature, or *isotherms*, that exist five seconds into the casting process. The temperature contours and their evolution can be used to predict locations of macrodefects, the size and shape of grains in the cast part, and the microporosity as a function of location in the part. Therefore, alternative designs can be quickly evaluated for their ability to address any of these issues. Numerical simulation has proven to be an invaluable tool in the evaluation of casting molds and is currently in widespread use.

*Source*:  S. Shepel and S. Paolucci, University of Notre Dame.

## SUMMARY

- Metal casting is among the oldest and most common manufacturing processes. Solidification of pure metals takes place at a clearly defined, constant temperature, whereas solidification of alloys occurs over a range of temperatures, depending on composition. Phase diagrams are useful tools for identifying the solidification point or points for metals and alloys. (Sections 5.1 and 5.2)

- The composition and cooling rate of the melt affect the size and shape of grains and dendrites in the solidifying metal and alloy and thus influence the properties of the solidified casting. (Section 5.3)

- Most metals shrink during solidification, although gray cast iron and a few others actually expand; either way, loss of dimensional control and, sometimes, cracking are difficulties that can arise during solidification and cooling. Seven basic categories of casting defects have been identified. (Section 5.3)

- In casting, the molten metal or alloy may flow through a variety of passages, including pouring basins, sprues, runners, risers, and gating systems, before reaching the final mold cavity. Bernoulli's theorem, the continuity law, and the Reynolds number are the analytical tools used in designing an appropriate system and eliminating defects associated with fluid flow. Heat transfer affects fluid flow and solidification time in casting; solidification time is a function of the volume of a casting and its surface area (Chvorinov's rule). (Section 5.4)

- Melting practices have a direct effect on the quality of castings. Factors that affect melting include the following: (1) fluxes, which are inorganic compounds added to the molten metal to remove dissolved gases and various impurities; (2) the type of furnace used; and (3) foundry operations, that is, pattern and mold making, pouring of the melt, removal of cast items from molds, cleaning, heat treatment, and inspection. (Section 5.5)

- Several ferrous and nonferrous casting alloys are available, with a wide range of properties, casting characteristics, and applications. Because castings are often designed and produced to be assembled with other mechanical components and structures, various other considerations, such as weldability, machinability, and surface conditions, are also important. (Section 5.6)

- The traditional ingot-casting process has been replaced, for many applications, by continuous-casting methods for both ferrous and nonferrous metals. (Section 5.7)

- Casting processes are generally classified as either expendable-mold or permanent-mold casting. The most common expendable-mold methods are sand, shell-mold, plaster, ceramic-mold, and investment casting. Common permanent-mold methods include slush casting, pressure casting, and die casting. Compared with permanent-mold casting, expendable-mold casting usually involves lower mold and equipment costs, but produces parts with lower dimensional accuracy. (Sections 5.8–5.10)

- Castings, as well as wrought products made by other processes, may be subjected to further heat-treatment processes to enhance their properties and service life. As temperature is reduced at different rates, various transformations take place in microstructures that have widely varying characteristics and properties. Important mechanisms of hardening and strengthening involve thermal treatments, including quenching and precipitation hardening. (Section 5.11)

- Several general principles have been established to aid designers in producing castings that are free from defects and meet dimensional tolerance and service requirements; these principles typically concern the shape of the casting and various techniques to minimize hot spots in order to avoid shrinkage cavities. Because of the large number of variables involved, close control of all parameters is essential, particularly those related to the nature of liquid-metal flow into the molds and dies and the rate of cooling in different regions. (Section 5.12)

- Within the limits of good performance, the economic aspects of casting are just as important as the technical considerations. Factors affecting the overall cost of castings include the cost of materials, molds, dies, equipment, and labor, each of which varies with the particular casting process. An important parameter is the cost per casting, which, for large production runs, can justify major expenditures for automated machinery. (Section 5.13)

## SUMMARY OF EQUATIONS

Bernoulli's theorem: $h + \dfrac{p}{\rho g} + \dfrac{v^2}{2g} = \text{constant}$

Continuity equation: $Q = A_1 v_1 = A_2 v_2$

Sprue contour: $\dfrac{A_1}{A_2} = \sqrt{\dfrac{h_2}{h_1}}$

Metal velocity after gate: $v = c\sqrt{2gh}$

Reynolds number: $\text{Re} = \dfrac{vD\rho}{\eta}$

Chvorinov's rule: Solidification time $= C\left(\dfrac{\text{volume}}{\text{surface area}}\right)^2$

# BIBLIOGRAPHY

*Abrasion-Resistant Cast Iron Handbook*, American Foundry Society, 2000.

Alexiades, V., *Mathematical Modeling of Melting and Freezing Processes*, Hemisphere, 1993.

Allsop, D. F., and D. Kennedy, *Pressure Die Casting—Part II: The Technology of the Casting and the Die*, Pergamon, 1983.

*An Introduction to Die Casting*, American Die Casting Institute, 1981.

*Analysis of Casting Defects*, American Foundrymen's Society, 1974.

*ASM Handbook*, Vol. 3: *Alloy Phase Diagrams*, ASM International, 1992.

*ASM Handbook*, Vol. 4: *Heat Treating*, ASM International, 1991.

*ASM Handbook*, Vol. 15: *Casting*, ASM International, 1988.

*ASM Specialty Handbook: Cast Irons*, ASM International, 1996.

Bradley, E. F., *High-Performance Castings: A Technical Guide*, Edison Welding Institute, 1989.

Campbell, J., *Castings*, Butterworth-Heinemann, 1991.

*Case Hardening of Steel*, ASM International, 1987.

"Casting," in *Tool and Manufacturing Engineers Handbook*, Volume II: *Forming*, Society of Manufacturing Engineers, 1984.

*Casting Defects Handbook*, American Foundrymen's Society, 1972.

Clegg, A. J., *Precision Casting Processes*, Pergamon, 1991.

Davis, J. R. (ed.), *Cast Irons.*, ASM International, 1996.

Elliott, R., *Cast Iron Technology*, Butterworths, 1988.

Flemings, M. C., *Solidification Processing*, McGraw-Hill, 1974.

Heine, R. W., C. R. Loper, Jr., and C. Rosenthal, *Principles of Metal Casting*, 2d. ed., McGraw-Hill, 1967.

*Investment Casting Handbook*, Investment Casting Institute, 1997.

Johns, R., *Casting Design*, American Foundrymen's Society, 1987.

Karlsson, L. (ed.), *Modeling in Welding, Hot Powder Forming and Casting*, ASM International, 1997.

Kaye, A., and A. C. Street, *Die Casting Metallurgy*, Butterworth, 1982.

Krauss, G., *Steels: Heat Treatment and Processing Principles*, ASM International, 1990.

Kurz, W., and D. J. Fisher, *Fundamentals of Solidification*, Trans Tech Pub., 1994.

Liebermann, H. H. (ed.), *Rapidly Solidified Alloys*, Dekker, 1993.

Minkoff, I., *Solidification and Cast Structure*, Wiley, 1986.

Powell, G. W., S.-H. Cheng, and C. E. Mobley, Jr., *A Fractography Atlas of Casting Alloys*, Battelle Press, 1992.

*Product Design for Die Casting*, Diecasting Development Council, 1988.

Rowley, M. T. (ed.), *International Atlas of Casting Defects*, American Foundrymen's Society, 1974.

Society of Die Casting Engineers, *The Metallurgy of Die Castings*, 1986.

Street, A. C., *The Diecasting Book*, 2d ed., Portcullis Press, 1986.

Szekely, J., *Fluid Flow Phenomena in Metals Processing*, Academic Press, 1979.

Totten, G. E., and M. A. H. Howes, *Steel Heat Treatment*, Dekker, 1997.

Upton, B., *Pressure Die Casting-Part 1: Metals, Machines, Furnaces*, Pergamon, 1982.

Walton, C. F., and T. J. Opar (eds.), *Iron Castings Handbook*, 3d. ed., Iron Castings Society, 1981.

Wieser, P. P. (ed.), *Steel Castings Handbook*, 6th ed., ASM International, 1995.

Young, K. P., *Semi-solid Processing*, Chapman & Hall, 1997.

Yu, K.–O. (ed.), *Modeling for Casting and Solidification Processing*, Dekker, 2001.

# QUESTIONS

**5.1**   Describe the characteristics of (1) an alloy, (2) pearlite, (3) austenite, (4) martensite, and (5) cementite.

**5.2**   What are the effects of mold materials on fluid flow and heat transfer?

**5.3**   How does the shape of graphite in cast iron affect its properties?

**5.4**   Explain the difference between short and long freezing ranges. How are they determined? Why are they important?

**5.5**   We know that pouring molten metal at a high rate into a mold has certain disadvantages. Are there any disadvantages to pouring it very slowly? Explain.

**5.6**   Why does porosity have detrimental effects on the mechanical properties of castings? Which physical properties are also affected adversely by porosity?

**5.7**   A spoked handwheel is to be cast in gray iron. In order to prevent hot tearing of the spokes, would you insulate the spokes or chill them? Explain.

**5.8** Which of the following considerations is/are important for a riser to function properly? (1) Having a surface area larger than that of the part being cast; (2) being kept open to atmospheric pressure; (3) solidifying sooner. Why?

**5.9** Explain why the constant $C$ in Eq. (5.11) depends on mold material, metal properties, and temperature.

**5.10** Explain why gray iron undergoes expansion, rather than contraction, during solidification.

**5.11** How can you tell whether cavities in a casting are due to porosity or to shrinkage?

**5.12** Explain the reasons for hot tearing in castings.

**5.13** Would you be concerned about the fact that parts of internal chills are left within the casting? What materials do you think chills should be made of, and why?

**5.14** Are external chills as effective as internal chills? Explain.

**5.15** Do you think early formation of dendrites in a mold can impede the free flow of molten metal into the mold? Explain.

**5.16** Is there any difference in the tendency for shrinkage void formation for metals with short freezing ranges and long freezing ranges, respectively? Explain.

**5.17** It has long been observed by foundry-men that low pouring temperatures, i.e., low superheat, promote equiaxed grains over columnar grains. Also, equiaxed grains become finer as the pouring temperature decreases. Explain these phenomena.

**5.18** What are the reasons for the large variety of casting processes that has been developed over the years?

**5.19** Why can blind risers be smaller than open-top risers?

**5.20** Would you recommend preheating molds in permanent-mold casting? Also, would you remove the casting soon after it has solidified? Explain.

**5.21** In a sand-casting operation, what factors determine the time at which you would remove the casting from the mold?

**5.22** Explain why the strength-to-weight ratio of die-cast parts increases with decreasing wall thickness.

**5.23** We note that the ductility of some cast alloys is nil. (See Fig. 5.19.) Do you think that this phenomenon should be a significant concern in engineering applications of castings? Explain.

**5.24** The modulus of elasticity, $E$, of gray iron varies significantly with its type, such as the ASTM class. Explain why.

**5.25** List and explain the considerations for selecting pattern materials.

**5.26** Why is the investment-casting process capable of producing fine surface detail on castings?

**5.27** Explain why a casting may have a slightly different shape than the pattern used to make the mold.

**5.28** Explain why squeeze casting produces parts with better mechanical properties, dimensional accuracy, and surface finish than that of parts produced by expendable-mold processes.

**5.29** Why are steels more difficult to cast than cast irons?

**5.30** What would you do to improve the surface finish in expendable-mold casting processes?

**5.31** You have seen that even though die casting produces thin parts, there is a minimum thickness that can be produced. Why can't even thinner parts be made by this process?

**5.32** What differences, if any, would you expect to find in the properties of castings made by permanent-mold versus sand-casting methods?

**5.33** Which of the casting processes would be suitable for making small toys in large numbers? Why?

**5.34** Why are allowances provided for in making patterns? What do the allowances depend on?

**5.35** Explain the difference in the importance of drafts in green-sand casting versus permanent-mold casting.

**5.36** Make a list of the mold and die materials used in the casting processes described in this chapter. Under each type of material, list the casting processes that are used, and explain why these processes are suitable for that particular mold or die material.

**5.37** Explain why carbon is so effective in imparting strength to iron in the form of steel.

**5.38** Describe the engineering significance of the existence of a eutectic point in phase diagrams.

**5.39** Explain the difference between hardness and hardenability.

**5.40** Explain why it may be desirable for castings to be subjected to various heat treatments.

**5.41** Describe the differences between case hardening and through hardening insofar as engineering applications are concerned.

**5.42** *Type metal* is a bismuth alloy used to cast type for printing. Explain why bismuth is ideal for this process.

**5.43** Do you expect to see larger solidification shrinkage for a material with a bcc crystal structure or a material with an fcc crystal structure? Explain.

**5.44** Describe the drawbacks to having a riser that is (a) too large and (b) too small.

**5.45** If you were to incorporate lettering on a sand casting, would you make the letters protrude from the surface or recess into the surface? What if the part were to be made by investment casting?

**5.46**    List and briefly explain the three mechanisms by which metals shrink during casting.

**5.47**    Explain the significance of the "tree" in investment casting.

**5.48**    Sketch the microstructure you would expect for a slab cast through (a) continuous casting, (b) strip casting, and (c) melt spinning.

**5.49**    The general design recommendations for a well in sand casting are that (a) its diameter should be twice the exit diameter of the sprue, and (b) the depth should be approximately twice the depth of the runner. Explain the consequences of deviating from these rules.

# PROBLEMS

**5.50**    Using Fig. 5.3, estimate the following quantities for a 20% Cu 80% Ni alloy: (1) liquidus temperature, (2) solidus temperature, (3) percentage of nickel in the liquid at 1400°C (2550°F), (4) the major phase at 1400°C, and (5) the ratio of solid to liquid at 1400°C.

**5.51**    Determine the amount of gamma and alpha phases (see Fig. 5.6) in a 5-kg, 1060 steel casting as it is being cooled to the following temperatures: (1) 750°C, (2) 728°C, and (3) 726°C.

**5.52**    A round casting is 0.2 m in diameter and 0.5 m in length. Another casting of the same metal is elliptical in cross-section, with a major-to-minor-axis ratio of 3, and has the same length and cross-sectional area as the round casting. Both pieces are cast under the same conditions. What is the difference in the solidification times of the two castings?

**5.53**    Derive Eq. (5.7).

**5.54**    Two halves of a mold (cope and drag) are weighted down to keep them from separating under the pressure exerted by the molten metal (buoyancy). Consider a solid, spherical steel casting with a diameter of 7 in. that is being produced by sand casting. Each flask (see Fig. 5.14) is 25 in. by 25 in. and 15 in. deep. The parting line is at the middle of the part. Estimate the clamping force required. Assume the metal's density is $500 \text{ lb/ft}^3$ and the sand density is $100 \text{ lb/ft}^3$.

**5.55**    Would changing the position of the parting line for the casting in Problem 5.54 change your answer to the problem? Explain.

**5.56**    Plot the clamping force in Problem 5.54 as a function of increasing diameter of the casting, from 10 in. to 20 in.

**5.57**    Sketch a graph of specific volume versus temperature for a metal that shrinks as it cools from the liquid state to room temperature. On the graph, mark the area where shrinkage is compensated for by risers.

**5.58**    A round casting has the same dimensions as that in Problem 5.52. Another casting of the same metal is rectangular in cross section, with a width-to-thickness ratio of 2, and has the same length and cross-sectional area as the round casting. Both pieces are cast under the same conditions. What is the difference in the solidification times of the two castings?

**5.59**    A 50-mm-thick square plate and a right circular cylinder with a radius of 100 mm and height of 50 mm each have the same volume. If each is to be cast using a cylindrical riser, will the two parts require the same-size riser to ensure proper feeding? Explain.

**5.60**    Assume that the top of a round sprue has a diameter of 4 in. and is at a height of 8 in. from the runner and 3 in. below the metal level in the pouring basin. Based on Eq. (5.7), plot the profile of the sprue's diameter as a function of the height of the sprue.

**5.61**    Estimate the clamping force for a die-casting machine in which the casting is rectangular, with projected dimensions of 100 mm × 150 mm. Would your answer depend on whether or not the process is a hot-chamber or cold-chamber process? Explain.

**5.62**    When designing patterns for casting, pattern makers use special rulers that automatically incorporate solid-shrinkage allowances into their designs. Therefore, a 12-in. pattern maker's ruler is longer than a foot. How long is a pattern maker's ruler designed for the making of patterns for (1) aluminum castings (2), phosphor bronze casting, and (3) high-manganese steel castings?

**5.63**    The blank for the spool shown in the figure accompanying this problem is to be sand cast out of A-319, an aluminum casting alloy. Make a sketch of the wooden pattern for this part. Include all necessary allowances for shrinkage and machining.

**5.64** Repeat Problem 5.63, but assume that the aluminum spool is to be cast using expendable-pattern casting. Explain the important differences between the two patterns.

**5.65** In sand casting, it is important that the cope mold half be held down with enough force to keep it from floating when the molten metal is poured in. For the casting shown in the accompanying figure, calculate the minimum amount of weight necessary to keep the cope from floating up as the molten metal is poured in. (*Hint:* The buoyancy force exerted by the molten metal on the cope is related to the effective height of the metal head above the cope.)

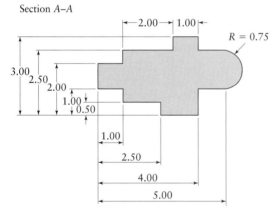

Section *A–A*

Material: Low-carbon steel
Density: 0.26 lb/in$^3$
All dimensions in inches

**5.66** The optimum shape of a riser is spherical to ensure that the riser cools more slowly than the casting it feeds. Spherically shaped risers, however, are difficult to cast. (1) Sketch the shape of a blind riser that is easy to mold, but also has the smallest possible surface-area-to-volume ratio. (2) Compare the solidification time of the riser in part (a) to that of a riser shaped like a right circular cylinder. Assume that the volume of each riser is the same and that for each riser, the height is equal to the diameter. (See Example 5.2.)

**5.67** Assume that you are asked to give a quiz to students on the contents of this chapter. Prepare three quantitative problems and three qualitative questions, and supply the answers.

**5.68** The part shown in the accompanying figure is a semispherical shell used as an acetabular (mushroom shaped) cup in a total hip replacement. Select a casting process for this part, and provide a sketch of all patterns or tooling needed if the part is to be produced from cobalt–chrome alloy.

Dimensions in mm

**5.69** A cylinder with a height-to-diameter ratio of unity solidifies in three minutes in a sand-casting operation. What is the solidification time if the height of the cylinder is doubled? What is the solidification time if the diameter is doubled?

**5.70** Steel piping is to be produced by centrifugal casting. The length of the pipe is to be 10 ft, the diameter 3 ft, and the thickness 0.5 in. Using basic equations from dynamics and statics, determine the rotational speed needed to have the centripetal force be 70 times the weight of the pipe.

**5.71** A sprue is 10 in. long and has a diameter of 5 in. at the top, where the metal is poured. The molten metal level in the pouring basin is taken as 3 in. from the top of the sprue for design purposes. If a flow rate of 40 in$^3$/s is to be achieved, what should be the diameter of the bottom of the sprue? Will the sprue aspirate? Explain.

**5.72** Small amounts of slag often persist after skimming and are introduced into the molten-metal flow in casting. Recognizing that the slag is much less dense than the metal, design mold features that will remove small amounts of slag before the metal reaches the mold cavity.

**5.73** Pure lead is poured into a sand mold. The metal level in the pouring basin is 8 in. above the metal level in the mold, and the runner is circular, with a 0.5-in. diameter. What is the velocity and rate of the flow of the metal into the mold? Is the flow turbulent or laminar?

**5.74** For the sprue described in Problem 5.73, what runner diameter is needed to ensure a Reynolds number of 2000? How long will a 25-in$^3$ casting take to fill with such a runner?

**5.75** How long would it take for the sprue in Problem 5.73 to feed a casting with a square cross-section of 5 in. per side and a height of 4 in.? Assume that the sprue is frictionless.

# DESIGN

**5.76** Design test methods to determine the fluidity of metals in casting. (See Section 5.4.2.) Make appropriate sketches, and explain the important features of each design.

**5.77** The accompanying figures indicate various defects and discontinuities in cast products. Review each one, and offer design solutions to avoid them.

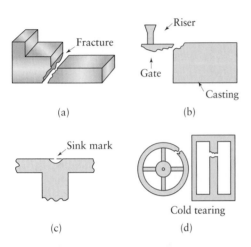

(a)

(b)

(c)

(d)

**5.78** Using the equipment and materials available in a typical kitchen, design an experiment to reproduce results similar to those shown in Fig. 5.16.

**5.79** Design a test method to measure the permeability of sand for sand casting.

**5.80** Describe the procedures that would be involved in making a bronze statue. Which casting process(es) would be suitable? Why?

**5.81** Porosity developed in the boss of a casting is illustrated in the accompanying figure. Show that by simply repositioning the parting line of this casting, this problem can be eliminated.

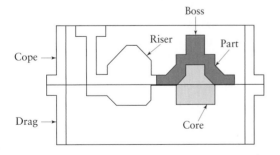

**5.82** For the wheel illustrated in the accompanying figure, show how (a) riser placement, (b) core placement, (c) padding, and (d) chills may be used to help feed molten metal and eliminate porosity in the isolated hob boss.

**5.83** In the accompanying figure, the original casting design shown in (a) was changed to the design shown in (b). The casting is round, with a vertical axis of symmetry. What advantages do you think the new design, as a functional part, has over the old one?

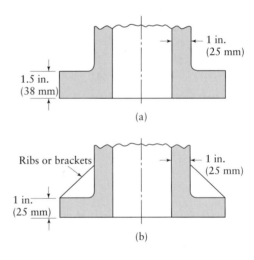

**5.84** An incorrect and a correct design for casting are shown, respectively, in the accompanying figure. Review the changes made to the incorrect design, and comment on their advantages.

**5.85**   Three sets of designs for die casting are shown in the accompanying figure. Note the changes made to original die design (1), and comment on the reasons for these changes.

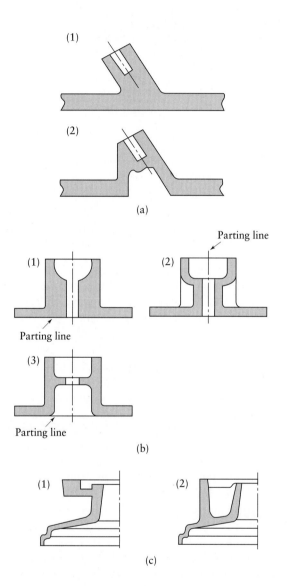

(1)

(2)

(a)

Parting line

(1)

(2)

Parting line

(3)

Parting line

(b)

(1)

(2)

(c)

# 6 Bulk Deformation Processes

## 6.1 | Introduction

Deformation processes in manufacturing are operations that induce shape changes on the workpiece by plastic deformation under forces applied by various tools and dies. Deformation processes can be grouped by temperature, size, and shape of the workpiece and type of operation. For example, using temperature as a criterion, deformation processes may be divided into three basic categories of cold (room temperature), warm, and hot working. Also, deformation processes may be classified by type of operation as primary working or secondary working.

**Primary-working** operations are those that take a solid piece of metal (generally from a cast state, such as an ingot) and break it down successively into shapes such as slabs, plates, and billets. Traditionally, the primary-working processes are forging, rolling, and extrusion. **Secondary working** involves further processing of the products from primary working into final or semifinal products such as bolts, sheet-metal parts, and wire. These classifications are not rigid, because some operations belong to both primary and secondary working. For instance, some forgings result in a product as final as those that result from secondary working; likewise, in extrusion, some products are ready for use without any further processing.

A more recent trend has been to classify deformation processes according to the size and shape of the workpiece. **Bulk deformation** is the processing of workpieces that have a relatively small surface-area-to-volume (or surface-area-to-thickness) ratio, hence the term *bulk*. In all bulk-deformation processing, the thickness or cross-section of the workpiece changes. In **sheet forming**, the surface-area-to-thickness ratio is relatively large. In general, the material is subjected to shape changes by various dies. Changes in thickness are usually undesirable and, in fact, can lead to failure.

In this chapter, we describe the basic bulk-deformation processes for metal: forging, rolling, extrusion, and drawing of rod and wire (Table 6.1). Bulk-deformation processing of plastics and nonmetallic materials is covered in Chapters 10 and 11, respectively.

6.1 Introduction
6.2 Forging
6.3 Rolling
6.4 Extrusion
6.5 Rod, Wire, and Tube Drawing
6.6 Swaging
6.7 Die Manufacturing Methods
6.8 Die Failures

## 6.2 | Forging

*Forging* denotes a family of processes by which plastic deformation of the workpiece is carried out by *compressive forces*. Forging is one of the oldest metalworking operations known, dating back to 5000 BC, and is used in making parts of widely varying sizes and shapes from a variety of metals. Typical parts made by forging today are crankshafts and connecting rods for engines, turbine disks, gears, wheels, bolt heads, hand tools, and many types of structural components for machinery and transportation equipment.

**TABLE 6.1**

| General Characteristics of Bulk Deformation Processes | |
|---|---|
| Process | General Characteristics |
| Forging | Production of discrete parts with a set of dies; some finishing operations usually necessary; similar parts can be made by casting and powder-metallurgy techniques; usually performed at elevated temperatures; dies and equipment costs are high; moderate to high labor costs; moderate to high operator skill. |
| Rolling | |
|     Flat | Production of flat plate, sheet, and foil at high speeds and with good surface finish, especially in cold rolling; requires very high capital investment; low to moderate labor cost. |
|     Shape | Production of various structural shapes, such as I-beams and rails, at high speeds; includes thread and ring rolling; requires shaped rolls and expensive equipment; low to moderate labor costs; moderate operator skill. |
| Extrusion | Production of long lengths of solid or hollow products with constant cross-sections, usually performed at elevated temperatures; product is then cut to desired lengths; can be competitive with roll forming; cold extrusion has similarities to forging and is used to make discrete products; moderate to high die and equipment cost; low to moderate labor cost; low to moderate operator skill. |
| Drawing | Production of long rod, wire, and tubing, with round or various cross-sections; smaller cross-sections than in extrusion; good surface finish; low to moderate die, equipment, and labor costs; low to moderate operator skill. |
| Swaging | Radial forging of discrete or long parts with various internal and external shapes; generally carried out at room temperature; low to moderate operator skill. |

Forging can be carried out at room temperature (*cold working*), or at elevated temperatures, a process called *warm* or *hot forging*, depending on the temperature. The temperature range for these categories is given in Table 3.1 in terms of the homologous temperature, $T/T_m$, where $T_m$ is the melting point of the workpiece material on the absolute scale. Note that the homologous recrystallization temperature for metals is about 0.5. (See also Fig. 3.16.) Simple forgings can be made with a heavy hammer and an anvil by techniques used by blacksmiths for centuries. Usually, though, a set of dies and a press are required. The three basic categories of forging are open die, impression die, and closed die.

## 6.2.1 Open-die forging

In its simplest form, open-die forging generally involves placing a solid cylindrical workpiece between two flat dies (platens) and reducing its height by compressing it (Fig. 6.1a). This operation is also known as **upsetting**. The die surfaces may be shaped, such as a conical or curved cavity, thereby forming the ends of the cylindrical workpiece during upsetting. Under ideal conditions, a solid cylinder deforms as shown in Fig. 6.1a; this is known as **homogeneous deformation**. Because the volume of the cylinder remains constant, any reduction in its height increases its diameter. For a specimen that has been reduced in height from $h_0$ to $h_1$,

$$\text{Reduction in height} = \frac{h_0 - h_1}{h_0} \times 100\%. \qquad (6.1)$$

From Eqs. (2.1) and (2.9) and using absolute values (as is generally the case with bulk-deformation processes), we have

$$e_1 = \frac{h_0 - h_1}{h_0} \qquad (6.2)$$

and

$$\varepsilon_1 = \ln\left(\frac{h_0}{h_1}\right). \qquad (6.3)$$

**FIGURE 6.1**   (a) Ideal deformation of a solid cylindrical specimen compressed between flat frictionless dies. This process is known as *upsetting*. (b) Deformation in upsetting with friction at the die–workpiece interfaces.

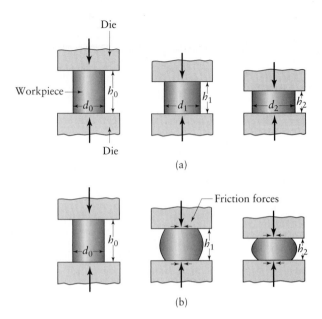

With a relative velocity $v$ between the platens, the specimen is subjected to a strain rate, according to Eqs. (2.17) and (2.18), of

$$\dot{e}_1 = -\frac{v}{h_0} \tag{6.4}$$

and

$$\dot{\varepsilon}_1 = -\frac{v}{h_1}. \tag{6.5}$$

If the specimen is reduced in height from $h_0$ to $h_2$, the subscript 1 in the foregoing equations is replaced by subscript 2. Note that the true strain rate $\dot{\varepsilon}$ increases rapidly as the height of the specimen approaches zero. Positions 1 and 2 in Fig. 6.1a may be regarded as instantaneous positions during a continuous upsetting operation.

Actually, the specimen develops a **barrel** shape, as shown in Fig. 6.1b. Barreling is caused primarily by *frictional forces* at the die–workpiece interfaces that oppose the outward flow of the material at these interfaces. Barreling also occurs in upsetting hot workpieces between cool dies. The material at and near the interfaces cools rapidly, while the rest of the specimen is relatively hot. Because the strength of the material decreases with increasing temperature, the ends of the specimen show a greater resistance to deformation than does the center; thus, the central portion of the cylinder deforms to a greater extent than do its ends.

In barreling, the material flow within the specimen becomes *nonuniform*, or *inhomogeneous*, as can be seen from the polished and etched section of a barreled cylindrical specimen, as shown in Fig. 6.2. Note the roughly triangular stagnant (*dead*) zones at the top and bottom surfaces, which have a wedge effect on the rest of the material, thus causing barreling. This inhomogeneous deformation can also be observed in grid patterns, as shown in Fig. 6.1b, which are obtained experimentally using various techniques (Section 6.4.1).

In addition to the single barreling shown in Fig. 6.3, **double barreling** can also be observed. It occurs at large ratios of height to cross-sectional area and when friction at the die–workpiece interfaces is high. Under these conditions, the stagnant zone under the platen is sufficiently far from the midsection of the workpiece to allow the

**FIGURE 6.2** Grain flow lines in upsetting a solid steel cylinder at elevated temperatures. Note the highly inhomogeneous deformation and barreling. The different shape of the bottom section of the specimen (as compared with the top) results from the hot specimen resting on the lower, cool die before deformation proceeded. The bottom surface was chilled; thus it exhibits greater strength and hence deforms less than the top surface. *Source*: J. A. Schey et al., IIT Research Institute.

midsection to deform uniformly while the top and bottom portions barrel out, hence the term *double barreling*. Barreling caused by friction can be minimized by applying an effective lubricant or **ultrasonic vibration** of the platens. The use of heated platens or a thermal barrier at interfaces will also reduce barreling in hot working.

**Forces and work of deformation under ideal conditions.** If friction at the interfaces is zero and the material is perfectly plastic with a yield stress of $Y$, then the normal compressive stress on the cylindrical specimen is uniform at a level $Y$. The force at any height $h_1$ is then

$$F = YA_1, \tag{6.6}$$

where $A_1$ is the cross-sectional area and is obtained from volume constancy:

$$A_1 = \frac{A_0 h_0}{h_1}.$$

The ideal work of deformation is the product of the volume of the specimen and the specific energy $u$ [in Eq. (2.59)] and is expressed as

$$\text{Work} = \text{volume} \int_0^{\varepsilon_1} \sigma \, d\varepsilon, \tag{6.7}$$

where $\varepsilon_1$ is obtained from Eq. (6.3). If the material is strain hardening, with a true-stress–true-strain curve given by

$$\sigma = K\varepsilon^n,$$

then the expression for force at any stage during deformation becomes

$$F = Y_f A_1, \tag{6.8}$$

(a)   (b)   (c)

**FIGURE 6.3** Schematic illustration of grid deformation in upsetting: (a) original grid pattern; (b) after deformation, without friction; (c) after deformation, with friction. Such deformation patterns can be used to calculate the strains within a deforming body.

where $Y_f$ is the flow stress of the material, corresponding to the true strain given by Eq. (6.3) and Fig. 2.39.

The expression for the work done is

$$\text{Work} = (\text{volume})(\bar{Y})(\varepsilon_1), \qquad (6.9)$$

where $\bar{Y}$ is the average flow stress and is given by

$$\bar{Y} = \frac{K \int_0^{\varepsilon_1} \varepsilon^n \, d\varepsilon}{\varepsilon_1} = \frac{K\varepsilon_1^n}{n+1}. \qquad (6.10)$$

### 6.2.2 Methods of analysis

There are several methods of analysis to theoretically determine various quantities, such as stresses, strains, strain rates, forces, and local temperature rise, in deformation processing, outlined as follows:

**Slab method.** The *slab method* is one of the simpler methods of analyzing the stresses and loads in forging and other bulk-deformation processes. It requires the selection of an element in the workpiece and identification of all the normal and frictional stresses acting on that element.

a) **Forging of a rectangular workpiece in plane strain.** Let's take the case of simple compression with friction (Fig. 6.4), which is the basic deformation process in forging. As the flat dies compress the part, it is reduced in thickness, and, because the volume of the past remains constant, the part expands laterally. This relative movement at the die–workpiece interfaces causes frictional forces acting in opposition to the movement of the piece. These frictional forces are shown by the horizontal arrows in Fig. 6.4. For simplicity, let's also assume that the deformation is in plane strain; that is, the workpiece is not free to flow in the z-direction.

Let's now take an element and indicate all the stresses acting on it (Fig. 6.4b). Note the correct direction of the frictional stresses. Also note the difference between the horizontal stresses acting on the sides of the element; this difference is caused by the presence of frictional stresses on the element. We assume that the lateral stress distribution, $\sigma_x$, is uniform along the height, $h$.

The next step in this analysis is to balance the horizontal forces on this element, because it must be in static equilibrium. Assuming unit width, we have

$$(\sigma_x + d\sigma_x)h + 2\mu\sigma_y \, dx - \sigma_x h = 0,$$

**FIGURE 6.4** Stresses on an element in plane-strain compression (forging) between flat dies. The stress $\sigma_x$ is assumed to be uniformly distributed along the height $h$ of the element. Identifying the stresses on an element (slab) is the first step in the slab method of analysis for metalworking processes.

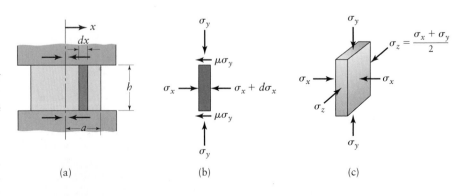

(a)         (b)         (c)

or

$$d\sigma_x + \frac{2\mu\sigma_y}{h} dx = 0.$$

Note that we now have one equation, but two unknowns: $\sigma_x$ and $\sigma_y$. The necessary second equation is obtained from the yield criteria, as follows: As shown in Fig. 6.4c, this element is subjected to triaxial compression. Using the distortion-energy criterion for plane strain, we obtain

$$\sigma_y - \sigma_x = \frac{2}{\sqrt{3}} Y = Y'. \tag{6.11}$$

Hence,

$$d\sigma_y = d\sigma_x.$$

Note that we assumed $\sigma_y$ and $\sigma_x$ to be *principal stresses*. In the strictest sense, $\sigma_y$ cannot be a principal stress, because a shear stress is also acting on the same plane. However, this assumption is acceptable for low values of the coefficient of friction, $\mu$, and is the standard practice in this method of analysis. Note also that $\sigma_z$ in Fig. 6.4c is derived in a manner similar to that in Eq. (2.44). We now have two equations that can be solved by noting that

$$\frac{d\sigma_y}{\sigma_y} = -\frac{2\mu}{h} dx, \quad \text{or} \quad \sigma_y = Ce^{-2\mu x/h}. \tag{6.12}$$

The boundary conditions are such that at $x = a$, $\sigma_x = 0$, and hence $\sigma_y = Y'$ at the edges of the specimen. (All stresses are compressive, so we may ignore negative signs for stresses, which is traditional practice in such analyses.) Hence, the value of $C$ becomes

$$C = Y' e^{2\mu a/h},$$

and therefore,

$$p = \sigma_y = Y' e^{2\mu(a-x)/h}. \tag{6.13}$$

Also,

$$\sigma_x = \sigma_y - Y' = Y'[e^{2\mu(a-x)/h} - 1]. \tag{6.14}$$

Equation (6.13) is plotted qualitatively in Fig. 6.5 in dimensionless form. Note that the pressure increases exponentially toward the center of the part and also that it increases with the $a/h$ ratio and increasing friction. For a strain-hardening material, $Y'$ in Eqs. (6.13) and (6.14) is replaced by $Y'_f$. Because of its shape, the pressure-distribution curve in Fig. 6.5 is referred to as the *friction hill*. Note that the pressure with friction is higher than it is without friction. This result is expected, because the work required to overcome friction must be supplied by the upsetting force.

The area under the pressure curve in Fig. 6.5 is the upsetting **force per unit width** of the specimen. This area can be obtained by integration, but an approximate expression for the *average pressure* $p_{av}$ is

$$p_{av} \simeq Y'\left(1 + \frac{\mu a}{h}\right). \tag{6.15}$$

Again note the significant influence of $a/h$ and friction on the pressure required,

**FIGURE 6.5** Distribution of die pressure, in terms of $p/Y'$, in plane-strain compression with sliding friction. Note that the pressure at the left and right boundaries is equal to the yield stress in plane strain, $Y'$. Sliding friction means that the frictional stress is directly proportional to the normal stress.

**FIGURE 6.6**  Normal stress (pressure) distribution in the compression of a rectangular workpiece with sliding friction under conditions of plane stress, using the *distortion-energy criterion*. Note that the stress at the corners is equal to the uniaxial yield stress, Y, of the material.

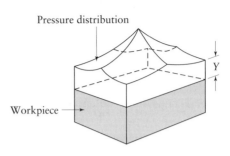

**FIGURE 6.7**  Increase in contact area of a rectangular specimen (viewed from the top) compressed between flat dies with friction. Note that the length of the specimen has increased proportionately less than its width. Likewise, a specimen in the shape of a cube acquires the shape of a pancake after deformation with friction.

especially at high $a/h$ ratios. The *forging force*, $F$, is the product of the average pressure and the contact area; that is,

$$F = (p_{av})(2a)(\text{width}).  \tag{6.16}$$

Note that the expressions for the pressure $p$ are in terms of an instantaneous height $h$; thus, the force at any $h$ during a continuous upsetting operation must be calculated individually.

A *rectangular specimen* can be upset without being constrained on its sides (*plane stress*). According to the distortion-energy criterion, the normal stress distribution can be given qualitatively by the plot in Fig. 6.6. The pressure is Y at the corners, because the elements at the corners are in uniaxial compression. A friction hill remains along the edges of the specimen, because of friction along the edges.

Figure 6.7 shows the lateral expansion (top view) of the edges of a rectangular specimen in actual plane-stress upsetting. Whereas the increase in the length of the specimen is 40%, the increase in width is 230%. The reason for the significantly larger increase in width is that, as expected, *the material flows in the direction of least resistance*. The width has less frictional resistance than the length. Likewise, after upsetting, a specimen in the shape of an equilateral cube acquires a pancake shape. The diagonal direction expands at a slower rate than the other directions.

b) **Forging of a solid cylindrical workpiece.**  Using the slab method of analysis, we can also determine the pressure distribution in forging of a solid cylindrical specimen (Fig. 6.8). We first isolate a segment of angle $d\theta$ in a cylinder of radius $r$ and height $h$, take a small element of radial length $dx$, and place on this element all the normal and frictional stresses acting on it. Following a similar approach as for the plane-strain case, we obtain an expression for the pressure, $p$, at any radius $x$ as

$$p = Ye^{2\mu(r-x)/h}.  \tag{6.17}$$

The *average pressure*, $p_{av}$, can be given approximately as

$$p_{av} \simeq Y\left(1 + \frac{2\mu r}{3h}\right).  \tag{6.18}$$

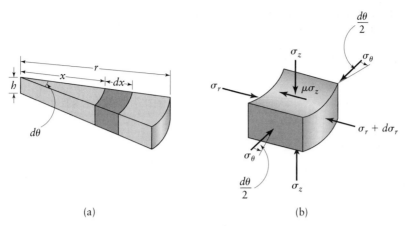

**FIGURE 6.8** Stresses on an element in forging of a solid cylindrical workpiece between flat dies. Compare with Fig. 6.4. *Source:* (a) After J. F. W. Bishop, *J. Mech. Phys. Solids,* Vol. 6, 1958, pp. 132–144. (b) Adapted from W. Schroeder and D. A. Webster, *Trans. ASME,* Vol. 71, 1949, pp. 289–294.

The *forging force* is

$$F = (p_{av})(\pi r^2). \qquad (6.19)$$

For strain-hardening materials, $Y$ in Eqs. (6.17) and (6.18) is replaced by the flow stress, $Y_f$. In forging operations, the value of the coefficient of friction, $\mu$, in Eqs. (6.17) and (6.18) can be estimated to be 0.05 to 0.1 for cold forging and 0.1 to 0.2 for hot forging. The exact value depends on the effectiveness of the lubricant and can be higher than stated, especially in regions where the workpiece surface is devoid of lubricant.

Figure 6.9 illustrates the effects of friction and the aspect ratio of the specimen ($a/h$ or $r/h$) on the average pressure $p_{av}$ in upsetting. The pressure is given in a dimensionless form and can be regarded as a pressure-multiplying factor. These curves are convenient to use and again show the significance of friction and the aspect ratio of the specimen on upsetting pressure.

c) **Forging under sticking condition.** The product of $\mu$ and $p$ is the *frictional stress* (surface shear stress) at the interface at any location $x$ from the center of the specimen. As $p$ increases toward the center, $\mu p$ also increases. However, the value of $\mu p$ cannot be greater than the shear yield stress $k$ of the material. When $\mu p = k$, **sticking** takes place. (In plane strain, the value of $k$ is $Y'/2$; see Section 2.11.) Sticking does not necessarily mean adhesion at the interface; it reflects the fact that, relative to the platen surfaces, the material does not move.

For the sticking condition, the normal stress distribution in plane strain can be shown to be

$$p = Y'\left(1 + \frac{a-x}{h}\right). \qquad (6.20a)$$

**FIGURE 6.9** Ratio of average die pressure to yield stress as a function of friction and aspect ratio of the specimen: (a) plane-strain compression; and (b) compression of a solid cylindrical specimen. Note that the yield stress in (b) is $Y$, not $Y'$, as in the plane-strain compression shown in (a). *Source:* Reprinted from J.F.W. Bishop, J. Mech. Phys. Solids, v. 6, 1958, pp. 132-144 with permission from Elsevier Science.

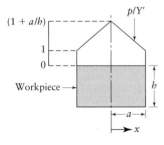

**FIGURE 6.10**
Distribution of die pressure, in terms of $p/Y'$, in the compression of a rectangular specimen in plane strain and under sticking conditions. The pressure at the edges is the uniaxial yield stress of the material in plane strain, $Y'$.

The pressure varies linearly with $x$, as shown in Fig. 6.10. The normal stress distribution for a cylindrical specimen under sticking condition can be shown to be

$$p = Y\left(1 + \frac{r - x}{h}\right). \tag{6.20b}$$

The stress distribution again is linear, as shown in Fig. 6.10 for plane strain.

---

**EXAMPLE 6.1  Upsetting force**

A cylindrical specimen made of annealed 4135 steel has a diameter of 6 in. and is 4 in. high. It is upset by open-die forging with flat dies to a height of 2 in. at room temperature. Assuming that the coefficient of friction is 0.2, calculate the force required at the end of the stroke. Use the average-pressure formula.

**Solution.**  The average-pressure formula, from Eq. (6.18), is

$$p_{av} \simeq Y\left(1 + \frac{2\mu r}{3h}\right),$$

where $Y$ is replaced by $Y_f$, because the workpiece material is strain hardening. From Table 2.3, we have $K = 1015$ MPa (147,000 psi) and $n = 0.17$.
   The absolute value of the true strain is

$$\varepsilon_1 = \ln\left(\frac{4}{2}\right) = 0.693.$$

Therefore,

$$Y_f = K\varepsilon_1^n = (147,000)(0.693)^{0.17} = 138,000 \text{ psi.}$$

The final height $h_1$ is 2 in. The radius $r$ at the end of the stroke is found from volume constancy:

$$\frac{\pi 6^2}{4} \cdot 4 = \pi r_1^2 \cdot 2 \quad \text{and} \quad r_1 = 4.24 \text{ in.}$$

Therefore,

$$p_{av} \simeq 138,000\left[1 + \frac{(2)(0.2)(4.24)}{(3)(2)}\right] = 177,000 \text{ psi.}$$

The upsetting force is then

$$F = (177,000)\pi \cdot (4.24)^2 = 10^7 \text{ lb.}$$

[*Note*: Figure 9.9b can also be used to solve this problem.]

---

**EXAMPLE 6.2  Transition from sliding to sticking friction**

In plane-strain upsetting, the frictional stress cannot be higher than the shear yield stress, $k$, of the workpiece material; thus, there may be a distance $x$ in Fig. 6.4 where a transition occurs from sliding to sticking friction. Derive an expression for $x$ in terms of $a$, $h$, and $\mu$ only.

**Solution.**  The shear stress at the interface due to friction can be expressed as

$$\tau = \mu p.$$

However, the shear stress cannot exceed the yield shear stress, $k$, of the material, which, for plane strain, is $Y'/2$. The pressure curve in Fig. 6.5 is given by Eq. (6.13); thus, in the limit, we have

$$\mu Y' e^{2\mu(a-x)/h} = Y'/2,$$

or

$$2\mu \frac{(a-x)}{h} = \ln\left(\frac{1}{2\mu}\right).$$

Hence,

$$x = a - \left(\frac{h}{2\mu}\right) \ln\left(\frac{1}{2\mu}\right).$$

Note that, as expected, the magnitude of $x$ decreases as $\mu$ decreases. However, the pressures must be sufficiently high to cause sticking, i.e., the $a/h$ ratio must be high. For example, let $a = 10$ mm and $h = 1$ mm. Then, for $\mu = 0.2$, $x = 7.71$ mm, and for $\mu = 0.4$, $x = 9.72$ mm.

**Finite-element method.**   In the *finite-element method*, the deformation zone in an elastic–plastic body is divided into a number of elements interconnected at a finite number of nodal points. The actual velocity distribution is then approximated for each element. Next, a set of simultaneous equations is developed that represents unknown velocity vectors. From the solution of these equations, actual velocity distributions and the stresses are calculated. This technique can incorporate friction conditions at the die–workpiece interfaces and actual properties of the workpiece material. It has been applied to relatively complex geometries in bulk-deformation and sheet-forming problems. Its accuracy is influenced by the number and shape of the finite elements, the deformation increment, and the methods of calculation. To ensure accuracy, complex problems require extensive computations. The finite-element method gives a detailed picture of the actual stresses and strain distributions throughout the workpiece.

Figure 6.11 illustrates the results of applying the finite-element method of analysis on a solid cylindrical workpiece in impression-die forging. Only one quarter of the workpiece is shown, with the vertical and horizontal lines as the axes of symmetry. The grid pattern is known as a *mesh*, and its distortion is predicted from theory, aided by the use of computers, with the results displayed on the video output screen. Note how the workpiece deforms as it is being forged.

The finite-element method is also capable of predicting microstructural changes in the material during hot-working operations, temperature distribution throughout the workpiece, and the onset of defects without using experimental data. This knowledge is then used in modifying die design to avoid making defective parts. Application of this technique requires inputs such as the stress–strain characteristics of the material as a function of strain rate and temperature, and frictional and heat-transfer characteristics of the die and workpiece.

**Other methods of analysis.**   A number of other methods of metal-forming analysis have been used, but are becoming less popular, because of the availability of powerful commercial software. The **slip-line analysis** method is generally applied to plane-strain conditions. The technique consists of the construction of a family of straight or curvilinear lines that intersect each other orthogonally. These lines, known

**FIGURE 6.11** Plastic deformation in forging as predicted by DEFORM using the finite-element method of analysis. *Source*: Scientific Forming Technologies Corporation.

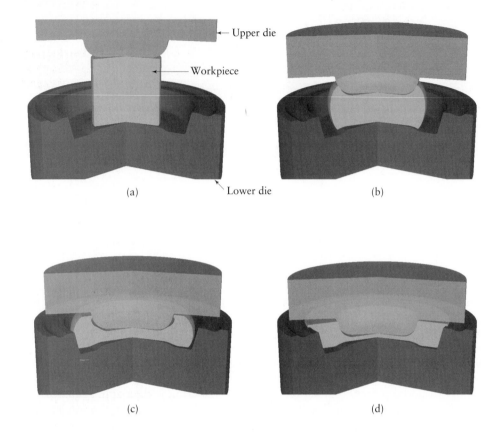

(a)    (b)

(c)    (d)

as a *slip-line field*, correspond to the directions of the maximum shear stress and allow calculation of stresses and strain rates in metal forming.

In the **upper-bound technique**, the overall deformation zone is approximated by a number of smaller zones where the velocity is constant or continuous. However, the velocity in adjacent zones may be different and discontinuous, as long as all discontinuities are tangential to the zone surface. In this approach, a velocity field that minimizes the total calculated power is found and taken as the actual velocity field and subsequently compared with experimental data.

**Deformation-zone geometry.** The observations made concerning Fig. 2.27 are important in estimating and calculating forces in forging and other bulk-deformation processes. As shown previously for hardness testing, the compressive stress required for indentation is, ideally, about three times the yield stress, $Y$, required for uniaxial compression. (See Section 2.6.8.) Moreover, (1) the deformation under the indenter is localized, making the overall deformation highly nonuniform; and (2) the deformation zone is relatively small compared with the size of the specimen.

However, in a simple frictionless compression test with flat dies, the top and bottom surfaces of the specimen are always in contact with the dies, and the specimen deforms uniformly. Various situations occur between these two extreme examples of specimen-deformation geometry. The *deformation zones* and the pressures required are shown in Fig. 6.12 for the frictionless condition as obtained from *slip-line analysis*. Note that the ratio $h/L$ is the important parameter in determining the inhomogeneity of deformation. The frictionless nature of these examples is important, because friction significantly affects forces, particularly at small values of $h/L$. Deformation-zone geometry depends on the particular process and such parameters as die geometry and percent reduction of the material (Fig. 6.13).

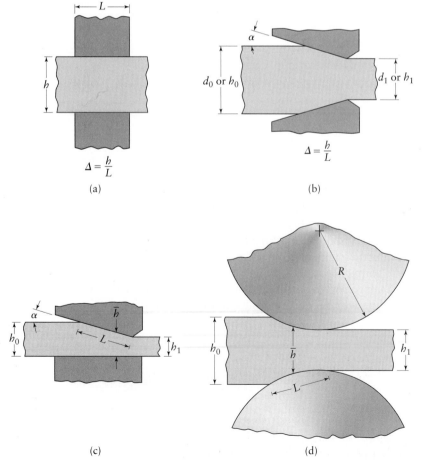

**FIGURE 6.12**  Die pressures required in *frictionless* plane-strain conditions for a variety of metalworking operations. The geometric relationship between contact area of the dies and workpiece dimensions is an important factor in predicting forces in plastic deformation of materials. *Source*: After W. A. Backofen.

**FIGURE 6.13**  Examples of plastic deformation processes in plane strain, showing the $h/L$ ratio. (a) Indenting with flat dies. This operation is similar to cogging, shown in Fig. 6.19. (b) Drawing or extrusion of strip with a wedge-shaped die, described in Sections 6.4 and 6.5. (c) Ironing; see also Fig. 7.54. (d) Rolling, described in Section 6.3. As shown in Fig. 6.12, the larger the $h/L$ ratio, the higher the die pressure becomes. In actual processing, however, the smaller this ratio, the greater is the effect of friction at the die–workpiece interfaces. The reason is that contact area, and hence friction, increases with a decreasing $h/L$ ratio.

**FIGURE 6.14** Schematic illustration of stages in impression-die forging. Note the formation of flash, or excess material that is subsequently trimmed off.

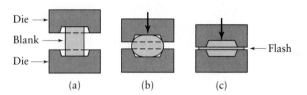

### 6.2.3 Impression-die forging

In *impression-die forging*, the workpiece acquires the shape of the die cavities (impressions) while it is being upset between the closing dies. A typical example is shown in Fig. 6.14. Some of the material flows radially outward and forms a **flash**. Because of its high length-to-thickness ratio (equivalent to a high $a/h$ ratio), the flash is subjected to high pressure, which in turn signifies high frictional resistance to material flow in the radial direction in the flash gap. Since high friction encourages the filling of the die cavities, the flash has a significant role in the flow of material in impression-die forging. Furthermore, if the operation is carried out at elevated temperatures, the flash, because of its high surface-area-to-thickness ratio, cools faster than the bulk of the workpiece; consequently, the flash resists deformation more than the bulk does and helps fill the die cavities.

Because of the more complex shapes involved, accurate calculation of forces in impression-die forging is difficult. Depending on its position, each element within the workpiece is generally subject to different strains and strain rates. Consequently, the level of strength that the material exhibits at each location depends not only on the strain and strain rate, but also on the exponents $n$ and $m$ of the workpiece material in Eqs. (2.11) and (2.16). Because of the many difficulties involved in calculating forces in impression-die forging, certain pressure-multiplying factors $K_p$ have been recommended, as shown in Table 6.2. These factors are to be used with the expression

$$F = (K_p)(Y_f)(A), \tag{6.21}$$

where $F$ is the forging load, $A$ is the projected area of the forging (including the flash), and $Y_f$ is the flow stress of the material at the strain and the strain rate to which the material is subjected.

A typical impression-die forging load as a function of the stroke of the die is shown in Fig. 6.15. For this axisymmetric workpiece, the force increases gradually as the cavity is filled (Fig. 6.15b). The force then increases rapidly as flash forms. In order for us to obtain the final dimensions and details on the forged part, the dies must close further, with an even steeper rise in the forging load. Note that the flash has a finite contact length with the die, called **land**. (See Fig. 6.27.) The land ensures that the flash generates sufficient resistance of the outward flow of the material to aid in die filling without contributing excessively to the forging load.

**TABLE 6.2**

| Range of $K_p$ Values in Eq. (6.21) for Impression-Die Forging | |
|---|---|
| Simple shapes, without flash | 3–5 |
| Simple shapes, with flash | 5–8 |
| Complex shapes, with flash | 8–12 |

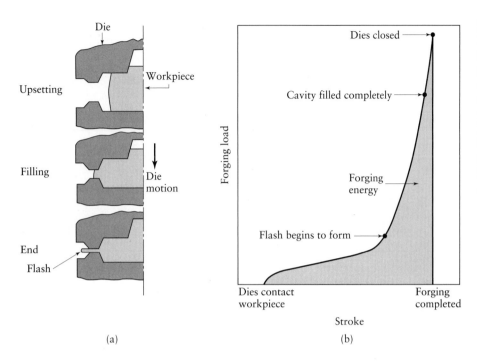

(a)  (b)

**FIGURE 6.15** Typical load–stroke curve for impression-die forging. Note the sharp increase in load after the flash begins to form. In hot-forging operations, the flash requires high levels of stress, because it is thin—that is, it has a small $h$—and cooler than the bulk of the forging. *Source:* After T. Altan.

**Closed-die forging.** Although not quite correct, the example shown in Fig. 6.14 is also referred to as *closed-die forging*. In true closed-die forging, no flash is formed, and the workpiece is completely surrounded by the dies, while in impression-die forging, any excess metal in the die cavity is formed into a flash. Since no flash can be formed in closed-die forging, proper control of the volume of material is essential to obtain a forging of desired dimensions. Undersized blanks in closed-die forging prevent the complete filling of the die; oversized blanks may cause premature die failure or jamming of the dies.

Precision forging, or flashless forging, and similar operations where the part formed is close to the final dimensions of the desired component are also known as **near-net-shape production.** Any excess material is subsequently removed by various machining processes. In precision forging, special dies are made and machined to greater accuracy than in ordinary impression-die forging. Precision forging requires higher capacity forging equipment than do other forging processes. Aluminum and magnesium alloys are particularly suitable for precision forging, because of the low forging loads and temperatures required. Also, they result in little die wear and produce a good surface finish. Steels and other alloys are more difficult to precision forge. The choice between conventional forging and precision forging requires an economic analysis. Precision forging requires special dies. However, much less machining is involved, because the part is closer to the desired final shape.

In the **isothermal-forging** process, also known as **hot-die forging,** the dies are heated to the same temperature as the hot blank. In this way, cooling of the workpiece is eliminated, the low flow stress of the material is maintained, and material flow within the die cavities is improved. The dies are generally made of nickel alloys, and complex parts with good dimensional accuracy can be forged in one stroke in hydraulic presses. Isothermal forging generally is expensive; however, it can be economical for intricate forgings of expensive materials, provided that the quantity required is large enough to justify die costs.

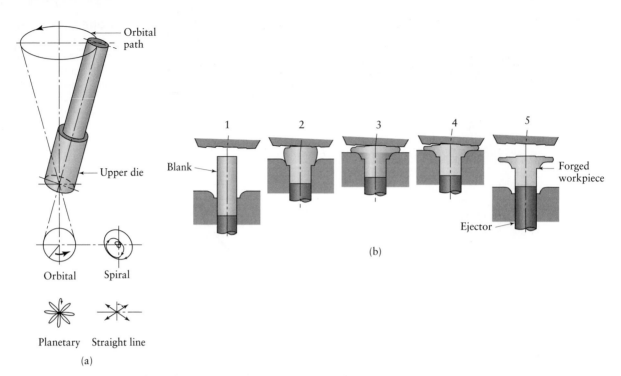

**FIGURE 6.16** Schematic illustration of the orbital-forging process. Note that the die is in contact with only a portion of the workpiece surface. This process is also called *rotary forging*, *swing forging*, and *rocking-die forging* and can be used for forming bevel gears, wheels, and bearing rings.

**Orbital forging** is a relatively new process in which the die moves along an orbital path (Fig. 6.16) and forges or forms the part incrementally. Consequently the forces are lower, the operation is more quiet, and part flexibility can be achieved.

**Incremental forging** is a process in which a blank is forged into a shape with a tool or die that forms it in several small steps, a process somewhat similar to cogging (see Fig. 6.19). Consequently, this operation requires much lower forces as compared with conventional forging.

### 6.2.4 Miscellaneous forging operations

a) **Coining.** Another example of closed-die forging is the *minting of coins*, where the slug is shaped in a completely closed cavity. The pressures required can be as great as five to six times the flow stress of the material in order to produce the fine details of a coin or a medallion. Several coining operations may be necessary in order to obtain full detail on some parts. The coining process is also used with forgings and other products to improve surface finish and impart the desired dimensional accuracy (**sizing**) of the products. The pressures are large, and usually very little change in shape takes place. Lubricants cannot be tolerated in coining, because they can be trapped in die cavities and prevent reproduction of fine die-surface details.

b) **Heading.** Basically an upsetting operation, *heading* is performed at the end of a rod (usually round) to form a shape with a larger cross-section. Typical products formed by heading include the heads of bolts, screws, and nails (Fig. 6.17). An important aspect of the upsetting process during heading is the tendency for buckling if the length-to-diameter ratio is too high. Heading is done on

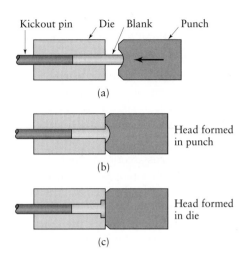

**FIGURE 6.17**   Forming heads on fasteners such as bolts and rivets. These processes are called *heading*.

machines called *headers*, which are usually highly automated horizontal machines with high production rates. The operation can be carried out cold, warm, or hot. Heading forces can be estimated reliably by using the slab method of analysis described in Section 6.2.2.

c) **Piercing.**   In *piercing*, a punch indents the surface of a workpiece to produce a cavity or an impression (Fig. 6.18). The workpiece may be confined in a die cavity, or it may be unconstrained. The piercing force depends on the punch's cross-sectional area and tip geometry, the flow stress of the material, and the friction at the interfaces. Punch pressures may range from three to five times the flow stress of the material. The term *piercing* is also used to describe the process of cutting of holes with a punch and die (Section 7.3).

d) **Hubbing.**   In *hubbing*, a hardened punch with a particular tip geometry is pressed into the surface of a block. This process is used to produce a die cavity, which is then used for subsequent forming operations. The cavity is usually shallow; for deeper cavities, some material may be removed from the surface by machining prior to hubbing. The pressure required to generate a cavity by hubbing is approximately three times the ultimate tensile strength (UTS) of the material of the block; thus, the force required is

$$\text{Hubbing force} = 3(\text{UTS})(A), \qquad (6.22)$$

where $A$ is the projected area of the impression. Note that the factor 3 is in agreement with the observations made with regard to hardness of materials in Section 2.6.8.

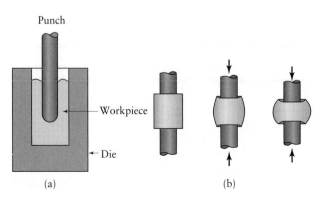

**FIGURE 6.18**   Examples of piercing operations.

e) **Cogging.** In *cogging*, also called *drawing out*, the thickness of a bar is reduced by successive steps at certain intervals (Fig. 6.19). A long section of a bar can be reduced in thickness without large forces, because the contact area is small.

f) **Fullering and edging.** These operations are performed, usually on bar stock, to distribute the material in certain regions of a forging. In *fullering*, the material is distributed away from an area; in *edging* (see Fig. 6.26b) it is gathered into a localized area. The parts are then formed into final shapes by other forging processes.

g) **Roll forging.** In *roll forging*, the cross-sectional area of a bar is reduced and altered in shape by passing it through a pair of rolls with grooves of various shapes (Fig. 6.20). This operation may also be used to produce a part that is basically the final product, such as tapered shafts, tapered leaf springs, table knives, and numerous tools. Roll forging is also used as a preliminary forming operation, followed by other forging processes. Typical products made by this method include crankshafts and other automotive components.

h) **Skew rolling.** A process similar to roll forging is *skew rolling* (Fig. 6.21), which is used for making ball bearings. Round wire or rod stock is fed into the roll gap, and spherical blanks are formed continuously by the rotating rolls. (Another method of forming blanks for ball bearings is by cutting pieces from a round bar and upsetting them, as shown in Fig. 6.22. The balls are then ground and polished in special machinery.)

**FIGURE 6.19** Schematic illustration of a cogging operation on a rectangular bar. With simple tools, the thickness and cross-section of a bar can be reduced by multiple cogging operations. Note the barreling after cogging. Blacksmiths use a similar procedure to reduce the thickness of parts in small increments by heating the workpiece and hammering it numerous times.

**FIGURE 6.20** Schematic illustration of a roll forging (cross-rolling) operation. Tapered leaf springs and knives can be made by this process with specially designed rolls. *Source*: After J. Holub.

**FIGURE 6.21** Production of steel balls for bearings by the skew-rolling process. Balls for bearings can also be made by the forging process shown in Fig. 6.22.

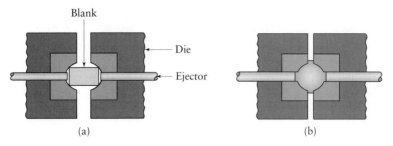

Blank

Die

Ejector

(a)                                    (b)

**FIGURE 6.22** Production of steel balls by upsetting of a cylindrical blank. Note the formation of flash. The balls are subsequently ground and polished for use as ball bearings and in other mechanical components.

## 6.2.5 Defects in forging

In addition to *surface cracking*, other defects in forgings can be caused by the material flow patterns in the die cavity. As shown in Fig. 6.23, excess material in the web of a forging can buckle during forging and develop laps. If the web is thick, the excess material flows past the already forged portions and develops internal cracks (Fig. 6.24). These examples indicate the importance of properly distributing material and controlling the flow in the die cavity.

The various radii in the die cavity can significantly affect formation of defects. In Fig. 6.25, the material follows a large corner radius better than a small radius. With small radii, the material can fold over itself and produce a lap, called **cold shut**. During the service life of the forged component, such defects can lead to fatigue failure and other problems, such as excessive corrosion and wear. The importance of inspecting forgings before they are put into service, particularly for critical applications, is obvious. (For a discussion of inspection techniques, see Section 4.8.)

Although it may not be considered a flaw, another important aspect of quality in a forging is the **grain-flow pattern**. At times, the grain-flow lines reach a surface perpendicularly, exposing the grain boundaries directly to the environment. The exposed boundaries are known as **end grains**. In service, they can be attacked by the environment, develop a rough surface, and act as stress raisers. For critical components, end grains in forgings can be avoided by proper selection of the original workpiece orientation in the die cavity and by control of material flow.

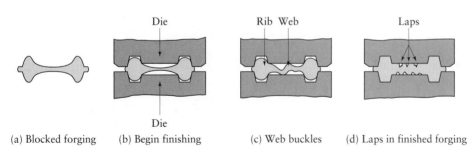

Die

Rib  Web

Laps

Die

(a) Blocked forging    (b) Begin finishing    (c) Web buckles    (d) Laps in finished forging

**FIGURE 6.23** Laps formed by buckling of the web during forging.

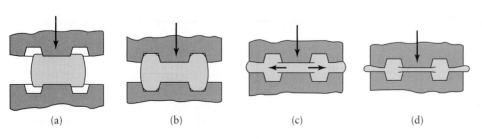

(a)                (b)                (c)                (d)

**FIGURE 6.24** Internal defects produced in a forging because of an oversized billet. The die cavities are filled prematurely, and the material at the center of the part flows past the filled regions as deformation continues.

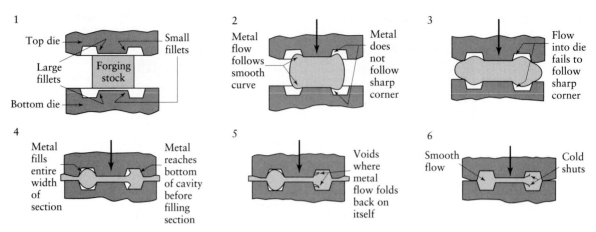

**FIGURE 6.25**    Effect of fillet radius on defect formation in forging. Small fillets (right side of drawings) cause the defects. *Source*: Aluminium Company of America.

Because the metal flows in various directions in a forging, and because of temperature variations, the properties of a forging are generally *anisotropic*. (See Section 3.5.) Strength and ductility vary significantly in test specimens taken from different locations and orientations in a forged part.

### 6.2.6 Forgeability

The forgeability of a metal can be defined as its capability to undergo deformation by forging without cracking. This definition can be expanded to include the flow strength of the metal; thus, a material with good forgeability is one that can be shaped with low forces without cracking. A number of tests have been developed to measure forgeability, although none is universally accepted. The more commonly used tests are described as follows:

**Upsetting test.**    A standard test is to upset a solid cylindrical specimen and observe any cracking on the barreled surfaces (Fig. 3.20d). The greater the reduction in height prior to cracking, the greater is the forgeability of the metal. The cracks are caused by **secondary tensile stresses** on the barreled surfaces. They are referred to as *secondary* because no external tensile stress is applied to the material, and the stresses develop because of changes in the material's shape. Because barreling is a result of friction at the die–workpiece interfaces, friction has a marked effect on cracking in upsetting. As friction increases, the specimen cracks at a lower reduction in height. Also of interest is whether the crack is at a 45° angle or vertical; the angle depends on the sign of the axial stress $\sigma_z$ on the barreled surface. If this stress is negative (compressive), the crack is at 45°; if the stress is positive (tensile), the crack is longitudinal. Depending on the notch sensitivity of the material, any surface defects will also affect the results by causing premature cracking. A typical surface defect is a **seam**: a longitudinal scratch, a string of inclusions, or a fold from prior working of the material.

Upsetting tests can be performed at various temperatures and strain rates. An optimal range for these parameters can then be specified for forging a particular material. Such tests can serve only as guidelines, since, in actual forging, the metal is subjected to a different state of stress as compared with a simple upsetting operation.

**Hot-twist test.**   In the hot-twist test, a round specimen is twisted continuously until it fails. The test is performed at various temperatures, and the number of turns that each specimen undergoes before failure is observed. The optimal forging temperature is then determined. The hot-twist test is particularly useful in determining the forgeability of steels, although upsetting tests can also be used for that purpose. Impurities in the metal or small changes in the composition of the metal can have a significant effect on forgeability.

**Effect of hydrostatic pressure on forgeability.**   As described in Section 2.2.8, hydrostatic pressure has a significant beneficial effect on the ductility of metals and nonmetallic materials. Experiments have indicated that the results of room-temperature forgeability tests, outlined previously, are improved (that is, cracking takes place at higher strain levels) if the tests are conducted in an environment of high *hydrostatic pressure*. Although such a test would be difficult to perform at elevated temperatures, techniques have been developed to forge metals in a high compressive environment in order to take advantage of this phenomenon. The pressure-transmitting medium is usually a low-strength ductile metal.

**Forgeability of various metals.**   Based on the results of various tests and observations, the forgeability of several metals and alloys has been determined. It is based on such considerations as the ductility and strength of the metals, forging temperature required, frictional behavior, and quality of the forgings obtained. In general, aluminum, magnesium, copper and their alloys, carbon and low-alloy steels have good forgeability; high-temperature materials such as superalloys, tantalum, molybdenum, and tungsten and their alloys have poor forgeability.

## 6.2.7  Die design

Proper design of forging dies and selection of die materials require considerable experience and knowledge concerning the strength and ductility of the workpiece material, its sensitivity to strain rate and temperature, and its frictional characteristics. Die *distortion* under high forging loads can also be an important consideration in die design, particularly if close dimensional tolerances are required.

Complex forgings are produced within a number of die cavities, as shown in Fig. 6.26. Starting with a round bar stock, the bar is (a) first *preformed* (**intermediate shape**) by some of the techniques previously described, (b) formed into the final shape in two additional forging operations, and (c) trimmed. Note how proper distribution of the material is essential to fill the die cavities to produce the part successfully. The reason for the intermediate steps can be best understood from Eq. (4.6), with the understanding that it is desirable to maximize die life and, therefore, minimize wear. The approach used is to design the forging operation so that a cavity encounters

(a) Blank (bar stock)

(b) Edging

(c) Blocking

(d) Finishing

(e) Trimming

**FIGURE 6.26**   Stages in forging a connecting rod for an internal combustion engine. Note the amount of flash that is necessary to fill the die cavities properly.

**FIGURE 6.27** Standard terminology for various features of a typical forging die.

either high sliding speeds or high pressures, but not both. Thus, edging produces large deformations on a relatively thick workpiece, while finishing produces fine details and involves small strains, but requires large pressures to obtain the desired part definition.

Care is essential in calculating the volume of the material required to properly fill the die cavities and to enable efficient material flow through the cavities. Computer techniques are now being developed to expedite these calculations. Fig. 6.27 provides the terminology used in die design. The significance of various parameters are discussed next.

The **parting line** is where the two dies meet. For simple symmetrical shapes, the parting line is a straight line at the center of the forging; for more complex shapes, the line may be offset and may not be in a single plane. Selection of the proper location for the parting line is based on the shape of the part, flow of the metal, balance of forces, and the flash. The significance of **flash** is described in Section 6.2.3. After lateral flow has been sufficiently constrained (by the length of the land), the flash is allowed to flow into a **gutter**; thus, the extra flash does not unnecessarily increase the forging load. A general guideline for flash clearance (between the dies) is 3% of the maximum thickness of the forging. The length of the land is usually five times that of the flash clearance.

**Draft angles** are necessary in almost all forgings to facilitate removal of the part from the die. Draft angles usually range between 3° and 10°. Because the forging shrinks in its radial direction (as well as in other directions) as it cools, internal draft angles are made larger than external ones. Typically, internal angles are about 7° to 10° and external angles about 3° to 5°. Proper selection of the *radii* for corners and fillets ensures smooth flow of the metal in the die cavity and improves die life. Small radii are generally not desirable, because of their adverse effect on metal flow and their tendency to wear rapidly from stress concentration and thermal cycling. Small radii in fillets can cause fatigue cracking in dies.

**Die materials.** Because most forgings, particularly large ones, are performed at elevated temperatures, die materials generally must have strength and toughness at elevated temperatures, hardenability, resistance to mechanical and thermal shock, and resistance to wear (particularly to abrasive wear, because of the presence of scale on heated forgings). Selection of die materials depends on the size of the die, the composition and properties of the workpiece, the complexity of the workpiece shape, the forging temperature, the type of operation, the cost of die material, the number of forgings required, and the heat-transfer and distortion characteristics of the die material. Common die materials are tool and die steels, which contain chromium, nickel, molybdenum, and vanadium (see Section 3.10.3).

**Forging temperatures and lubrication.** Table 6.3 provides the range of temperatures for hot forging of various metals. *Lubrication* plays an important role in

**TABLE 6.3**

| Hot-Forging Temperature Ranges for Various Metals | | | | | |
|---|---|---|---|---|---|
| Metal | °C | °F | Metal | °C | °F |
| Aluminium alloys | 400–450 | 750–850 | Alloy steels | 925–1260 | 1700–2300 |
| Copper alloys | 625–950 | 1150–1750 | Titanium alloys | 750–795 | 1400–1800 |
| Nickel alloys | 870–1230 | 1600–2250 | Refractory alloys | 975–1650 | 1800–3000 |

forging. It affects friction and wear and, consequently, the flow of metal into the die cavities. Lubricants can also serve as a *thermal barrier* between the hot forging and the relatively cool dies, thus slowing the rate of cooling of the workpiece. Another important role of a lubricant is to serve as a *parting agent*, that is, to prevent the forging from sticking to the dies. Lubricants can significantly affect the wear pattern in forging dies. A wide variety of metalworking fluids can be used in forging. For hot forging, graphite, molybdenum disulfide, and (sometimes) glass are commonly used as lubricants. For cold forging, mineral oils and soaps are common lubricants.

## 6.2.8  Equipment

Forging equipment of various designs, capacities, speeds, and speed-stroke characteristics is available (Fig. 6.28):

a) **Hydraulic presses.**  Hydraulic presses have a constant low speed and are **load limited**. Large amounts of energy can be transmitted to the workpiece by a constant load available throughout the stroke. Ram speed can be varied during the stroke. These presses are used for both open-die and closed-die forging operations. The largest hydraulic press in existence has a capacity of 670 MN (75,000 tons).

b) **Mechanical presses.**  Mechanical presses are **stroke limited**. They are of either the crank or the eccentric type, with speeds varying from a maximum at the center of the stroke to zero at the bottom. The force available depends on the stroke position and becomes extremely large at the bottom-dead-center position; thus, proper setup is essential to avoid breaking the dies or other equipment. The largest mechanical press has a capacity of 107 MN (12,000 tons).

c) **Screw presses.**  Screw presses derive their energy from a flywheel. The forging load is transmitted through a vertical screw. These presses are **energy limited** and can be used for many forging operations. They are particularly suitable for producing small quantities, for parts requiring precision (such as turbine blades),

Crank
(a)

Knucklejoint
(b)

Flywheel
Friction drive
Screw
Ram
Screw
(c)

Fluid
Ram
Hydraulic
(d)

**FIGURE 6.28**  Schematic illustration of various types of presses used in metalworking. The choice of the press is an important factor in the overall operation.

and for control of ram speed. The largest screw press has a capacity of 280 MN (31,500 tons).

d) **Hammers.** Hammers derive their energy from the *potential energy* of the ram, which is then converted to kinetic energy; thus, hammers are energy limited. The speeds of hammers are high; therefore, the low forming times minimize cooling of the hot forging, allowing the forging of complex shapes, particularly with thin and deep recesses. To complete the forging, several blows may have to be made on the part. In power hammers, the ram is accelerated in the downstroke by steam or air, in addition to gravity. The highest energy available in power hammers is 1150 kJ (850,000 ft-lb).

e) **Counterblow hammers.** Counterblow hammers have two rams that simultaneously approach each other to forge the part. They are generally of the mechanical–pneumatic or mechanical–hydraulic type. These machines transmit less vibration to the foundation than other hammers. The largest counterblow hammer has a capacity of 1200 kJ (900,000 ft-lb).

**Equipment selection.** Selection of forging equipment depends on the size and complexity of the forging, the strength of the material and its sensitivity to strain rate, and the degree of deformation required. Generally, presses are preferred for aluminum, magnesium, beryllium, bronze, and brass; hammers are preferred for copper, steels, titanium, and refractory alloys. Production rate is also a consideration in equipment selection. The number of strokes per minute ranges from a few for hydraulic presses to as many as 300 for power hammers.

## 6.3 | Rolling

*Rolling* is the process of reducing the thickness or changing the cross-section of a long workpiece by compressive forces applied through a set of rolls (Fig. 6.29), similar to rolling dough with a rolling pin to reduce its thickness. Rolling, which accounts for about 90% of all metals produced by metalworking processes, was first developed in the late 1500s. The basic operation is *flat rolling*, or simply *rolling*, where the rolled products are flat plate and sheet.

Plates, which are generally regarded as having a thickness greater than 6 mm ($\frac{1}{4}$ in.), are used for structural applications such as ship hulls, boilers, bridges, girders, machine structures, and nuclear vessels. Plates can be as much as 0.3 m (12 in.) thick for the supports for large boilers, 150 mm (6 in.) thick for reactor vessels, and 100–125 mm (4–5 in.) thick for battleships and tanks.

Sheets are generally less than 6 mm thick and are used for automobile bodies, appliances, containers for food and beverages, and kitchen and office equipment. Commercial-aircraft fuselages are usually made of about 1-mm-thick (0.040-in-thick) aluminum-alloy sheet, while beverage cans are made of 0.15-mm-thick (0.006-in-thick) aluminum sheet. Aluminum foil used to wrap candy and cigarettes has a thickness of 0.008 mm (0.0003 in.). Sheets are provided as flat pieces or as strip in coils to manufacturing facilities for further processing into products.

Traditionally, the initial form of material for rolling is an ingot. However, this practice is now being rapidly replaced by continuous casting and rolling, with their much higher efficiency and lower cost. Rolling is first carried out at elevated temperatures (hot rolling), wherein the coarse-grained, brittle, and porous structure of the ingot or continuously cast metal is broken down into a **wrought** structure, with finer grain size (Fig. 6.30).

**FIGURE 6.29** Schematic outline of various flat- and shape-rolling processes. *Source*: American Iron and Steel Institute.

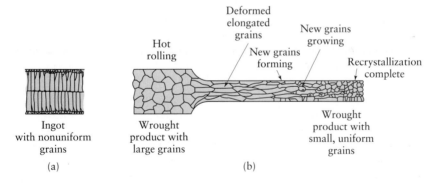

(a)

(b)

**FIGURE 6.30** Changes in the grain structure of cast or large-grain wrought metals during hot rolling. Hot rolling is an effective way to reduce grain size in metals for improved strength and ductility. Cast structures of ingots or continuous castings are converted to a wrought structure by hot working.

### 6.3.1 Mechanics of flat rolling

The flat-rolling process is shown schematically in Fig. 6.31. A strip of thickness $h_0$ enters the roll gap and is reduced to a thickness of $h_f$ by the rotating (powered) rolls. The surface speed of the roll is $V_r$. To keep the volume rate of metal flow constant, the velocity of the strip must increase as it moves through the roll gap (similar to fluid flow through a converging channel). At the exit of the roll gap, the velocity of the strip is $V_f$ (Fig. 6.32). Because $V_r$ is constant along the roll gap, sliding occurs between the roll and the strip.

At one point along the arc of contact, the two velocities are the same. This point is known as the **neutral point**, or **no-slip point**. To the left of this point, the roll moves

**FIGURE 6.31**    Schematic illustration of the flat-rolling process. A greater volume of metal is formed by rolling than by any other metalworking process.

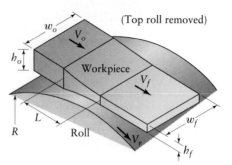

**FIGURE 6.32**    Relative velocity distribution between roll and strip surfaces. Note the difference in the direction of frictional forces. The arrows represent the frictional forces acting on the strip.

faster than the workpiece, and to the right the workpiece moves faster than the roll. Because of friction at the interfaces, the frictional forces (which oppose motion) act on the strip surfaces, as shown in Fig. 6.32.

Note the similarity between Figs. 6.32 and 6.1b. The frictional forces oppose each other at the neutral point. In upsetting, these forces are equal to each other, because of the symmetry of the operation. In rolling, the frictional force on the left of the neutral point must be greater than the frictional force on the right. This difference yields a net frictional force to the right, which makes the rolling operation possible by pulling the strip into the roll gap. Furthermore, the net frictional force and the surface velocity of the roll must be in the same direction in order to supply work to the system. Therefore, the neutral point should be located toward the exit in order to satisfy these requirements.

**Forward slip** in rolling is defined in terms of the exit velocity of the strip, $V_f$, and the surface speed of the roll, $V_r$, as

$$\text{Forward slip} = \frac{V_f - V_r}{V_r} \tag{6.23}$$

and is a measure of the relative velocities involved.

a) **Roll pressure distribution.** Although the deformation zone in the material is subjected to a state of stress similar to that in upsetting, the calculation of forces and stress distribution in flat rolling is more involved, because of the curved surface of contact. In addition, the material at the exit is strain hardened, so the flow stress at the exit is higher than that at the entry.

Figure 6.33 shows the stresses on an element in the entry and exit zones. Note that the only difference between the two elements is the direction of the friction force. Using the slab method of analysis for *plane strain* (described in Section 6.2.2), we may analyze the stresses in rolling as follows.

From the equilibrium of the horizontal forces on the element shown in Fig. 6.33, we have

$$(\sigma_x + d\sigma_x)(h + dh) - 2pR\,d\phi\,\sin\phi - \sigma_x h \pm 2\mu pR\,d\phi\,\cos\phi = 0.$$

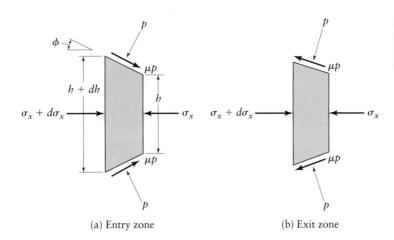

**FIGURE 6.33**  Stresses on an element in rolling: (a) entry zone and (b) exit zone.

(a) Entry zone          (b) Exit zone

When simplified, and if second-order terms are neglected, this expression reduces to

$$\frac{d(\sigma_x h)}{d\phi} = 2pR(\sin \phi \mp \mu \cos \phi).$$

In rolling, the angle $\alpha$ is only a few degrees; hence, we may assume that $\sin \phi = \phi$ and $\cos \phi = 1$. Thus,

$$\frac{d(\sigma_x h)}{d\phi} = 2pR(\phi \mp \mu). \tag{6.24}$$

Because the angles involved are small, we assume that $p$ is a *principal stress*, $\sigma_x$ being the other principal stress. The relationship between these two stresses and the flow stress, $Y_f$, of the material is given by Eq. (2.45) for plane strain, or

$$p - \sigma_x = \frac{2}{\sqrt{3}} Y_f = Y_f'. \tag{6.25}$$

For strain-hardening material, the flow stress, $Y_f$, in these expressions corresponds to the strain that the material has undergone at that particular location in the gap. Rewriting Eq. (6.24), we have

$$\frac{d[(p - Y_f')h]}{d\phi} = 2pR(\phi \mp \mu),$$

or

$$\frac{d}{d\phi}\left[ Y_f'\left(\frac{p}{Y_f'} - 1\right)h \right] = 2pR(\phi \mp \mu),$$

which, upon differentiation, becomes

$$Y_f' h \frac{d}{d\phi}\left(\frac{p}{Y_f'}\right) + \left(\frac{p}{Y_f'} - 1\right)\frac{d}{d\phi}(Y_f' h) = 2pR(\phi \mp \mu).$$

The second term is very small, because as $h$ decreases, $Y_f'$ increases, due to cold working, thus making the product nearly a constant, and its derivative thus becomes zero. We now have

$$\frac{\dfrac{d}{d\phi}\left(\dfrac{p}{Y_f'}\right)}{\dfrac{p}{Y_f'}} = \frac{2R}{h}(\phi \mp \mu). \tag{6.26a}$$

Letting $h_f$ be the final thickness, we have

$$h = h_f + 2R(1 - \cos \phi),$$

or, approximately,

$$h = h_f + R\phi^2. \tag{6.26b}$$

Substituting this expression for $h$ in Eq. (6.26a) and integrating, we obtain

$$\ln \frac{p}{Y_f'} = \ln \frac{h}{R} \mp 2\mu \sqrt{\frac{R}{h_f}} \tan^{-1} \sqrt{\frac{R}{h_f}} \phi + \ln C,$$

or

$$p = C Y_f' \frac{h}{R} e^{\mp \mu H},$$

where

$$H = 2\sqrt{\frac{R}{h_f}} \tan^{-1} \left( \sqrt{\frac{R}{h_f}} \phi \right). \tag{6.27}$$

At entry, $\phi = \alpha$; hence, $H = H_0$, with $\phi$ replaced by $\alpha$. At exit, $\phi = 0$; hence, $H = H_f = 0$. Also, at entry and exit, $p = Y_f'$. Thus, in the **entry zone**,

$$C = \frac{R}{h_f} e^{\mu H_i}$$

and

$$p = Y_f' \frac{h}{h_0} e^{\mu(H_0 - H)}. \tag{6.28}$$

In the **exit zone**, we have

$$C = \frac{R}{h_f}$$

and

$$p = Y_f' \frac{h}{h_f} e^{\mu H}. \tag{6.29}$$

Note that the pressure $p$ is a function of $h$ and the angular position $\phi$ along the arc of contact. These expressions also indicate that the pressure increases with increasing strength of the material, increasing coefficient of friction, and increasing $R/h_f$ ratio. The $R/h_f$ ratio in rolling is equivalent to the $a/b$ ratio in upsetting. (See Section 6.2.2.)

The dimensionless theoretical pressure distribution in the roll gap is shown in Fig. 6.34. Note the similarity of this curve to that in Fig. 6.5 (friction hill). Also note that the neutral point shifts toward the exit as friction decreases, which is to be expected. When friction approaches zero, the rolls begin to slip instead of pulling the strip in, so the neutral point must approach the exit. (This situation is similar to the spinning of automobile wheels when accelerating on wet pavement.)

Figure 6.35 shows the effect of reduction in thickness of the strip on the pressure distribution. As reduction increases, the length of contact in the roll gap increases, which in turn increases the peak pressure. The curves shown in Fig. 6.35 are theoretical; actual pressure distributions, as determined experimentally, have smoother curves, with their peaks rounded off.

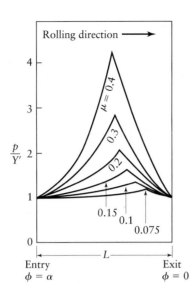

**FIGURE 6.34** Pressure distribution in the roll gap as a function of the coefficient of friction. Note that as friction increases, the neutral point shifts toward the entry. Without friction, the rolls slip, and the neutral point shifts completely to the exit.

**FIGURE 6.35** Pressure distribution in the roll gap as a function of reduction in thickness. Note the increase in the area under the curves with increasing reduction in thickness, thus increasing the roll force.

b) **Determination of the neutral point.** The neutral point can be determined simply by equating Eqs. (6.28) and (6.29); thus, at the neutral point,

$$\frac{h_0}{h_f} = \frac{e^{\mu H_0}}{e^{2\mu H_n}} = e^{\mu(H_0 - 2H_n)},$$

or

$$H_n = \frac{1}{2}\left(H_0 - \frac{1}{\mu}\ln\frac{h_0}{h_f}\right). \qquad (6.30)$$

Substituting Eq. (6.30) into Eq. (6.27), we obtain

$$\phi_n = \sqrt{\frac{h_f}{R}}\tan\left(\sqrt{\frac{h_f}{R}}\cdot\frac{H_n}{2}\right). \qquad (6.31)$$

**FIGURE 6.36** Pressure distribution as a function of front and back tension. Note the shifting of the neutral point and the reduction in the area under the curves with increasing tension.

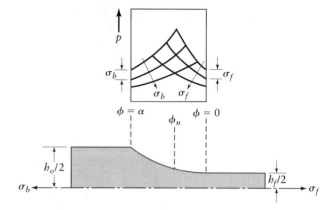

c) **Front and back tension.** The roll force, $F$, can be reduced by various means, such as lowering friction, using rolls of smaller radii, taking smaller reductions, and raising workpiece temperature. A particularly effective method is to reduce the apparent compressive yield stress of the material by applying longitudinal *tension*. Recall from our discussion of yield criteria (Section 2.11) that if tension is applied to a strip (Fig. 6.36), the yield stress normal to the strip surface drops, and hence the roll pressure will decrease.

Tensions in rolling can be applied either at the entry (**back tension, $\sigma_b$**) or at the exit (**front tension, $\sigma_f$**) of the strip or at both. Equations (6.28) and (6.29) can then be modified to include the effect of tension for the entry and exit zones as follows, respectively:

$$\text{Entry zone: } p = (Y'_f - \sigma_b)\frac{h}{h_0}e^{\mu(H_0 - H)} \tag{6.32}$$

and

$$\text{Exit zone: } p = (Y'_f - \sigma_f)\frac{h}{h_f}e^{\mu H}. \tag{6.33}$$

Depending on the relative magnitudes of the tensions applied, the neutral point may shift, as shown in Fig. 6.36. This shift affects the pressure distribution, torque, and power requirements in rolling. Front tension is controlled by the torque on the coiler (**delivery reel**) around which the rolled sheet is coiled. The back tension is controlled by a braking system in the uncoiler (**payoff reel**). Special instrumentation is available for these controls. Tensions are particularly important in rolling thin, high-strength materials, because such materials require high roll forces.

---

**EXAMPLE 6.3 Back tension required to cause roll slip**

It is to be expected that in rolling a strip, the rolls will begin to slip if the back tension, $\sigma_b$, is too high. Derive an expression for the magnitude of back tension required to make the rolls begin to slip.

**Solution.** Slipping of the rolls means that the neutral point has moved all the way to the exit in the roll gap; thus, the whole contact area becomes the entry zone, and Eq. (6.32) is applicable. Also, we know that when $\phi = 0$, $H = 0$; hence, the pressure at the exit is

$$p_{\phi=0} = (Y_f' - \sigma_b)\left(\frac{h_f}{h_0}\right)e^{\mu H_0}.$$

However, at the exit, the pressure is equal to $Y_f'$. Rearranging this equation, we obtain

$$\sigma_b = Y_f'\left[1 - \left(\frac{h_0}{h_f}\right)(e^{-\mu H_0})\right],$$

where $H_0$ is obtained from Eq. (6.27) with the condition $\phi = \alpha$. Since all the quantities are known, the magnitude of the back tension can be calculated.

d) **Roll forces.** The area under the pressure vs. contact–length curve multiplied by the strip width, $w$, is the *roll force*, $F$, (also called *roll–separating force*) on the strip. This area can be obtained graphically after the points are plotted. The force can also be calculated from the expression

$$F = \int_0^{\phi_n} wpR\, d\phi + \int_{\phi_n}^{\alpha} wpR\, d\phi. \tag{6.34}$$

A simpler method of calculating the roll force is to multiply the contact area by an average contact stress, as in

$$F = Lwp_{av}, \tag{6.35}$$

where $L$ is the *arc of contact*. The dimension $L$ can be approximated as the projected length by the equation

$$L = \sqrt{R\Delta h}, \tag{6.36}$$

where $R$ is the roll radius and $\Delta h$ is the difference between the original and final thicknesses of the strip (**draft**).

The estimation of $p_{av}$ depends on the $h/L$ ratio, where $h$ is now the *average thickness* of the strip in the roll gap. (See Figs. 6.12 and 6.13.) For large $h/L$ ratios (small reductions and/or large roll diameters), the rolls act as indenters in a hardness test. Friction is not significant, and $p_{av}$ is obtained from Fig. 6.12, where $a = L/2$. Small $h/L$ ratios (large reductions and/or large roll diameters) are equivalent to high $a/h$ ratios; thus, friction is predominant, and $p_{av}$ is obtained from Eq. (6.15). For strain-hardening materials, the appropriate flow stresses must be calculated.

As a rough approximation and for low frictional conditions, Eq. (6.35) can be simplified as

$$F = Lw\bar{Y}' \tag{6.37}$$

where $\bar{Y}'$ is the average flow stress in plane strain (see Fig. 2.39) of the material in the roll gap. For higher frictional conditions, we may write an expression similar to Eq. (6.15):

$$F = Lw\bar{Y}'\left(1 + \frac{\mu L}{2h_{av}}\right). \tag{6.38}$$

e) **Roll torque and power.** The *roll torque*, $T$, for each roll can be calculated analytically from the expression

$$T = \underbrace{\int_{\phi_n}^{\alpha} w\mu pR^2\, d\phi}_{\text{(entry zone)}} - \underbrace{\int_0^{\phi_n} w\mu pR^2\, d\phi}_{\text{(exit zone)}}. \tag{6.39}$$

Note that the minus sign indicates the change in direction of the friction force at the neutral point; thus, if the frictional forces are equal to each other, the torque is zero.

The torque in rolling can also be estimated by assuming that the roll force, $F$, acts in the *middle* of the arc of contact—that is, a moment arm of $0.5L$—and that $F$ is perpendicular to the plane of the strip. (Whereas $0.5L$ is a good estimate for hot rolling, $0.4L$ is a better estimate for cold rolling.)

The **torque per roll** is then

$$T = \frac{FL}{2}.$$

The **power required per roll** is

$$\text{Power} = T\omega, \tag{6.40}$$

where $\omega = 2\pi N$, and $N$ is the revolutions per minute of the roll. Consequently, the power per roll is

$$\text{Power} = \frac{\pi FLN}{60,000}\text{kW}, \tag{6.41}$$

where $F$ is in newtons, $L$ is in meters, and $N$ is the rpm of the roll. The power per roll can also be expressed as

$$\text{Power} = \frac{\pi FLN}{33,000}\text{hp}, \tag{6.42}$$

where $F$ is in lb and $L$ is in ft.

---

### EXAMPLE 6.4 Power required in rolling

A 9-in-wide 6061-O aluminum strip is rolled from a thickness of 1.00 in. to 0.80 in. If the roll radius is 12 in. and the roll rpm is 100, estimate the total horsepower required for this operation.

**Solution.** The power needed for a set of two rolls is given by Eq. (6.42) as

$$\text{Power} = \frac{2\pi FLN}{33,000}\text{hp},$$

where $F$ is given by Eq. (6.37) and $L$ by Eq. (6.36). Hence,

$$F = Lw\bar{Y}' \quad \text{and} \quad L = \sqrt{R\Delta h}.$$

Therefore,

$$L = \sqrt{(12)(1.0 - 0.8)} = 1.55 \text{ in.} = 0.13 \text{ ft} \quad \text{and} \quad w = 9 \text{ in.}$$

For 6061-O aluminum, $K = 30,000$ psi and $n = 0.2$ (from Table 2.3). The true strain in this operation is

$$\varepsilon_1 = \ln\left(\frac{1.0}{0.8}\right) = 0.223.$$

Thus, from Eq. (6.10), we have

$$\bar{Y} = \frac{(30,000)(0.223)^{0.2}}{1.2} = 18,500 \text{ psi}$$

and

$$\bar{Y}' = (1.15)(18,500) = 21,275 \text{ psi.}$$

Therefore,

$$F = (1.55)(9)(21,275) = 297,000 \text{ lb,}$$

and for two rolls

$$\text{Power} = \frac{(2\pi)(297,000)(0.13)(100)}{33,000} = 735 \text{ hp.}$$

f) **Forces in hot rolling.** Because ingots and slabs are usually hot rolled, the calculation of forces and torque in hot rolling is important. However, there are two major difficulties. One is the proper estimation of the coefficient of friction, $\mu$, at elevated temperatures. The other is the strain-rate sensitivity of metals at high temperatures.

The *average strain rate* in flat rolling can be obtained by dividing the strain by the time required for an element to undergo this strain in the roll gap. The time can be approximated as $L/V_r$, so

$$\dot{\varepsilon} = \frac{V_r}{L} \ln\left(\frac{h_0}{h_f}\right). \tag{6.43}$$

The flow stress, $Y_f$, of the material corresponding to this strain rate must first be obtained and then substituted into the proper equations. These calculations are approximate; variations in $\mu$ and the temperature within the strip in hot rolling further contribute to the difficulties in calculating forces accurately.

g) **Friction.** In rolling, although the rolls cannot pull the strip into the roll gap without some friction, forces and power requirements rise with increasing friction. In cold rolling, $\mu$ usually ranges between 0.02 and 0.3, depending on the materials and lubricants used. The low ranges for the coefficient of friction are obtained with effective lubricants and regimes approaching *hydrodynamic lubrication*, such as in cold rolling of aluminum at high speeds. In hot rolling, $\mu$ may range from about 0.2, with effective lubrication, to as high as 0.7, indicating sticking, which usually occurs with steels, stainless steels, and high-temperature alloys.

The maximum possible draft, that is, $h_0 - h_f$, in flat rolling is a function of friction and radius

$$\Delta h_{\max} = \mu^2 R. \tag{6.44}$$

Therefore, the higher the friction and the larger the roll radius, the greater is the maximum draft. As expected, the draft is zero when there is no friction. The maximum value of the angle $\alpha$ (**angle of acceptance**) in Fig. 6.32 is geometrically related to Eq. (6.44). From the simple model of a block sliding down an inclined plane, it can be seen that

$$\alpha_{\max} = \tan^{-1} \mu. \tag{6.45}$$

If $\alpha_{\max}$ is larger than this value, the rolls begin to slip, because the friction is not high enough to pull the material through the roll gap.

h) **Roll deflection and flattening.** Roll forces tend to bend the rolls, as shown in Fig. 6.37a, with the result that the strip is thicker at its center than at its edges (**crown**). The usual method of avoiding this problem is to grind the rolls

**FIGURE 6.37** (a) Bending of straight cylindrical rolls (exaggerated) because of the roll force. (b) Bending of rolls, ground with camber, that produce a sheet of uniform thickness during rolling.

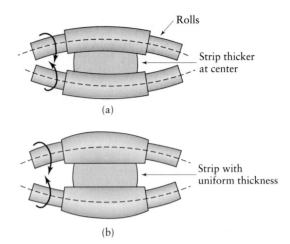

so that their diameter at the center is slightly larger than at the edges. This curvature is known as **camber**. In rolling sheet metals, the camber is generally less than 0.25 mm (0.01 in.) on the radius. Because of the heat generated during rolling, rolls can become slightly barrel-shaped, known as **thermal camber**. This effect can be controlled by varying the location of the coolant on the rolls.

When properly designed, such rolls produce flat strips, as shown in Fig. 6.37b. However, a particular camber is correct only for a certain load and width of strip. In hot rolling, uneven temperature distribution in the roll can also cause its diameter to vary along the length of the roll. In fact, camber can be controlled by varying the location of the coolant (lubricant) on the rolls in hot rolling.

Forces also tend to *flatten* the rolls elastically, much like the flattening of tires on automobiles. Flattening of a roll increases the roll's radius and hence yields a larger contact area for the same reduction in thickness; thus, the roll force, $F$, increases.

The new (distorted) roll radius $R'$ is

$$R' = R\left(1 + \frac{CF'}{h_0 - h_f}\right). \tag{6.46}$$

In Eq. (6.46), $C$ is $2.3 \times 10^{-2}$ mm²/kN ($1.6 \times 10^{-4}$ in²/klb) for steel rolls and $4.57 \times 10^{-2}$ ($3.15 \times 10^{-4}$) for cast-iron rolls. $F'$ is the **roll force per unit width** of strip, expressed in kN/mm (klb/in.). The higher the elastic modulus of the roll material, the less the roll distorts. Note from Eq. (6.46) that reducing the roll force, such as by using an effective lubricant or taking smaller reductions, also reduces roll flattening. We cannot determine $R'$ directly from Eq. 6.46, because it is a function of the roll force, $F$, which is itself a function of the roll radius. The solution is obtained by trial and error. (Note that $R$ in all previous equations should be replaced by $R'$ when significant roll flattening occurs).

Side view

Top view

**FIGURE 6.38** Increase in the width of a strip (spreading) in flat rolling. Spreading can be similarly observed when dough is rolled with a rolling pin.

i) **Spreading.** Although rolling plates and sheets with high width-to-thickness ratios is essentially a process of plane strain, with smaller ratios, such as a square cross-section, the width increases considerably during rolling. This increase in width is known as spreading (Fig. 6.38). Spreading decreases with increasing width-to-thickness ratios of the entering material, decreasing friction, and increasing ratios of roll radius to strip thickness. Spreading can be prevented by using vertical rolls that are in contact with the edges of the rolled product (**edger mill**).

## 6.3.2  Defects in rolling

Successful rolling practice requires balancing many factors, including material properties, process variables, and lubrication. There may be defects on the surfaces of the rolled plates and sheets, or there may be structural defects within the material. **Surface defects** may result from inclusions and impurities in the material, scale, rust, dirt, roll marks, and other causes related to the prior treatment and working of the material. In hot rolling blooms, billets, and slabs, the surface is usually preconditioned by various means, such as by torch (**scarfing**), to remove scale.

   **Structural defects** are defects that distort or affect the integrity of the rolled product. Figure 6.39 shows some typical defects. **Wavy edges** are caused by bending of the rolls; the edges of the strip are thinner than the center. Because the edges elongate more than the center and are restrained from expanding freely, they buckle. The cracks shown in Fig. 6.39b and c are usually caused by low ductility and barreling. **Alligatoring** is a complex phenomenon that results from inhomogeneous deformation of the material during rolling or from defects in the original cast ingot, such as piping.

**Residual stresses.**   Residual stresses can be generated in rolled sheets and plates because of inhomogeneous plastic deformation in the roll gap. Small-diameter rolls or small reductions tend to work the metal plastically at its surfaces (similarly to shot peening or roller burnishing). This working generates compressive residual stresses on the surfaces and tensile stresses in the bulk (Fig. 6.40a). Large-diameter rolls and high reductions, however, tend to deform the bulk to a greater extent than the surfaces, because of frictional constraint at the surfaces along the arc of contact. This situation generates residual stresses that are opposite to those of the previous case, as shown in Fig. 6.40b.

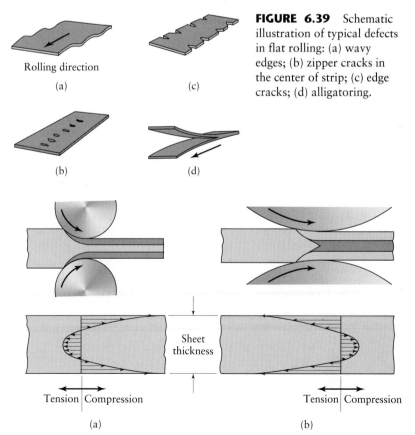

**FIGURE 6.39**  Schematic illustration of typical defects in flat rolling: (a) wavy edges; (b) zipper cracks in the center of strip; (c) edge cracks; (d) alligatoring.

Rolling direction

(a)          (c)

(b)          (d)

**FIGURE 6.40**  The effect of roll radius on the type of residual stresses developed in flat rolling: (a) small rolls, or small reduction in thickness; and (b) large rolls, or large reduction in thickness.

Sheet thickness

Tension | Compression          Tension | Compression

(a)          (b)

### 6.3.3 Vibration and chatter in rolling

*Vibration* and *chatter* in metalworking processes can have significant effects on product quality and productivity of the operation. Chatter, generally defined as self-excited vibration, can occur in rolling as well as in extrusion, drawing, machining, and grinding operations. (See Sections 8.11 and 9.6.8.) In rolling, it leads to random variations in the thickness of the rolled sheet and its surface finish and, consequently, can lead to excessive scrap. Chatter in rolling has been found to occur predominantly in *tandem mills* (see Fig. 6.42e); it is a very complex phenomenon and results from interactions between the structural dynamics of the mill stand and the dynamics of the rolling operation. Rolling speed and lubrication are found to be the two most important parameters. Chatter is very important to productivity; it has been estimated, for example, that modern rolling mills could operate at speeds up to 50% higher were it not for chatter. Considering the very high cost of rolling mills, this issue is a major economic concern.

The vibration modes commonly encountered in rolling are classified as torsional, third-octave, and fifth-octave chatter (*octave* meaning a frequency on the eighth degree from a given frequency).

1. **Torsional chatter** is characterized by a low resonant frequency (approximately 5–15 Hz) and is usually a result of forced vibration (see Section 8.11), although it can also occur simultaneously with third-octave chatter. Torsional chatter leads to minor variations in gage thickness and surface finish and is usually not significant unless improper conditions exist in the machinery, such as malfunctioning speed controls, broken gear teeth, and misaligned shafts.

2. **Third-octave chatter** occurs in the frequency range of 125–240 Hz (the third musical octave is 128–256 Hz) and is *self-excited*; that is, energy from the rolling mill is transformed into vibratory energy, regardless of the presence of any external forcing (forced vibration). This mode of vibration is the most serious in rolling, leading to significant gage variations and fluctuations in strip tension between the stands and often resulting in strip breakage. Third-octave chatter is generally controlled by reducing the rolling speed, $V$. Although not always practical to implement, it has been suggested that this type of chatter can also be reduced by increasing the distance between the stands of the rolling mill; increasing the strip width, $w$; incorporating *dampers* (see Section 8.11.1) in the roll supports; decreasing the reduction per pass (draft); increasing the roll radius $R$; and increasing the coefficient of strip-roll friction, $\mu$.

3. **Fifth-octave chatter** occurs in the frequency range of 550–650 Hz; it has been attributed to chatter marks that develop on work rolls and are caused by extended operation at certain critical speeds, surface defects on a backup roll or in the incoming strip, and/or improperly ground rolls. The chatter marks or striations are then imprinted onto the rolled sheet at any speed, adversely affecting its surface appearance. In addition, fifth-octave chatter is undesirable because of the objectionable noise generated. This type of chatter can be controlled by modulating the mill speed, using backup rolls of progressively larger diameters in the stands of a tandem mill, avoiding chatter in roll grinding, and eliminating the sources of other vibrations in the mill stand.

### 6.3.4 Flat-rolling practice

The initial breaking down of an ingot (or continuous casting) by hot rolling converts the coarse-grained, brittle, and porous cast structure to a wrought structure. (See Fig. 6.30.) This structure has finer grains and enhanced ductility, resulting from the breaking up of brittle grain boundaries and the closing up of internal defects, such as

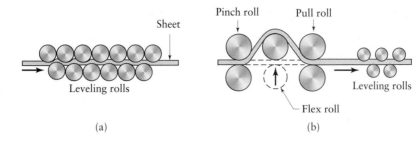

**FIGURE 6.41** Schematic illustration of methods of roller leveling. These processes are used to flatten rolled sheets.

porosity. Traditional methods of rolling ingots are now being rapidly replaced by continuous casting. Temperature ranges for hot rolling are similar to those for forging. (See Table 6.3.)

The product of the first hot-rolling operation is called a **bloom** or **slab**. (See Fig. 6.29.) A bloom usually has a square cross-section, at least 150 mm (6 in.) on the side; a slab is usually rectangular in cross-section. Blooms are processed further by shape rolling into structural shapes, such as I-beams and railroad rails. Slabs are rolled into plates and sheet. **Billets** are usually square, with a cross-sectional area smaller than that of blooms, and are rolled into various shapes, such as round rods and bars, using shaped rolls. (Hot-rolled round rods are used as the starting material for rod and wire drawing and are called **wire rods**.)

In hot rolling blooms, billets, and slabs, the surface of the material is usually **conditioned** (i.e., prepared for a subsequent operation) prior to rolling. Conditioning is done by various means, such as with a torch (scarfing) to remove heavy scale, or by rough grinding to smoothen surfaces. Prior to cold rolling, the scale developed during hot rolling or other defects may be removed by pickling with acids; by mechanical means, such as blasting with water; or by grinding.

**Pack rolling** is a flat-rolling operation in which two or more layers of metal are rolled together, thus improving productivity. Aluminum foil, for example, is pack rolled in two layers. One side of aluminum foil is matte, and the other side is shiny. The foil-to-foil side has a matte and satiny finish, and the foil-to-roll side is shiny and bright, because it has been in contact with the polished roll.

Mild steel, when stretched during sheet-forming operations, undergoes **yield-point elongation**. This phenomenon causes surface irregularities called **stretcher strains** or **Lueder's bands**. (See also Section 7.2.) To avoid this situation, the sheet metal is subjected to a light pass of 0.5–1.5% reduction, known as **temper rolling** (*skin pass*).

A rolled sheet may not be sufficiently flat as it leaves the roll gap, because of variations in the material or in the processing parameters during rolling. To improve flatness, the strip is then passed through a series of **leveling rolls**. Each roll is usually driven separately with individual electric motors. The strip is flexed in opposite directions as it passes through the sets of rollers. Several different roller arrangements are used (Fig. 6.41).

Sheet thickness is identified by a **gage number**; the smaller the number, the thicker the sheet is. There are different numbering systems for different sheet metals. Rolled sheets of copper and brass are also identified by thickness changes during rolling, such as $\frac{1}{4}$ hard, $\frac{1}{2}$ hard, and so on.

**Lubrication.** Ferrous alloys are usually hot rolled without a lubricant, although graphite may be used. Aqueous solutions are used to cool the rolls and break up the scale on the workpiece. Nonferrous alloys are hot rolled with a variety of compounded oils, emulsions, and fatty acids. Cold rolling is done with low-viscosity lubricants, including mineral oils, emulsions, paraffin, and fatty oils.

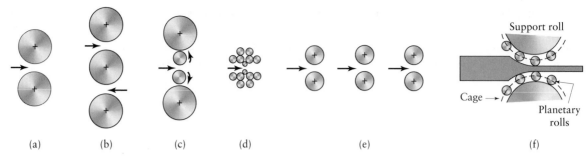

(a)  (b)  (c)  (d)  (e)  (f)

**FIGURE 6.42** Schematic illustration of various roll arrangements: (a) two high; (b) three high; (c) four high; (d) cluster; (e) tandem rolling with three stands; (f) planetary.

**Equipment.** A wide variety of rolling equipment is available with a number of roll arrangements, the major types of which are shown in Fig. 6.42. Small-diameter rolls are preferable, because the smaller the roll radius, the lower the roll force is. However, small rolls deflect under roll forces and have to be supported by other rolls (Fig. 6.42c and 6.42d).

Two-high or three-high (developed in the mid-1800s) rolling mills typically are used for initial breakdown passes on cast ingots (primary roughing) with roll diameters ranging up to 1400 mm (55 in.). The **cluster mill** (**Sendzimir**, or **Z, mill**) is particularly suitable for cold rolling thin strips of high-strength metals. Figure 6.43 shows one arrangement for the rolls. The **work roll** (the smallest roll) can be as small as 6 mm ($\frac{1}{4}$ in.) in diameter and is usually made of tungsten carbide for rigidity, strength, and wear resistance. The rolled product obtained in a cluster mill can be as wide as 5000 mm (200 in.) and as thin as 0.0025 mm (0.0001 in.)

In **tandem rolling** (Fig. 6.42e) the strip is rolled continuously through a number of **stands**; a group of stands is called a **train**. Control of the gage and of the speed at which the strip travels through each roll gap is critical. Flat rolling can also be carried out with front tension only, using idling rolls (**Steckel rolling**); for this case, the torque on the roll is zero, assuming frictionless bearings.

*Stiffness* in rolling mills is important for controlling dimensions. Mills can be highly automated, with rolling speeds as high as 25 m/s (5000 ft/min). Requirements for roll materials are mainly strength and resistance to wear. Three common roll materials are cast iron, cast steel, and forged steel. For hot rolling, roll surfaces are generally rough and may even have notches or grooves in order to pull the metal through the roll gap at high reductions. Rolls for cold rolling are ground to a fine finish and, for special applications, are also polished.

**FIGURE 6.43** Schematic illustration of a cluster (*Sendzimir*) mill. These mills are very rigid and are used in rolling thin sheets of high-strength materials, with good control of dimensions.

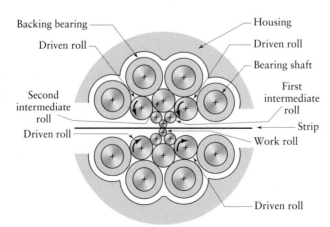

**Minimills.**  In minimills, scrap metal is melted in electric-arc furnaces, cast continuously, and rolled directly into specific lines of products. Each minimill produces essentially one kind of rolled product (rod, bar, or structural shapes) from one type of metal or alloy and is usually oriented to markets within the mill's particular geographic area. The scrap metal, which is obtained locally to reduce transportation costs, is usually old machinery, automobiles, and farm equipment.

**Integrated mills.**  Integrated mills are large facilities that involve complete activities, from the production of hot metal in a blast furnace to casting and rolling of finished products that are ready to be shipped to the customer.

### 6.3.5  Miscellaneous rolling operations

a) **Shape rolling.**  Straight structural shapes such as bars of various cross-sections, channel sections, I-beams, and railroad rails are rolled by passing the stock through a number of pairs of specially designed rollers (Fig. 6.44); these processes were first developed in the late 1700s. The original material is usually a bloom (Fig. 6.29). Designing a series of rolls (**roll-pass design**) requires experience in order to avoid defects and hold dimensional tolerances, although some defects may already exist in the material being rolled. The material elongates as it is reduced in cross-section. However, for a shape such as a channel, the reduction is different in various locations within the section; thus, elongation is not uniform, which can cause the product to warp or crack. Airfoil shapes can also be produced by shape-rolling techniques.

b) **Ring rolling.**  In ring rolling, a small-diameter, thick ring is expanded into a larger-diameter, thinner ring. The ring is placed between two rolls, one of which is driven (Fig. 6.45). The thickness is reduced by bringing the rolls closer as they rotate. The reduction in thickness is compensated for by an increase in the diameter of the ring. A great variety of cross-sections can be ring rolled with shaped rolls. This process can be carried out at room or elevated temperatures, depending on the size and strength of the product. The advantages of ring rolling, compared with other processes for making the same part, are short production runs, material savings, close dimensional tolerances, and favorable

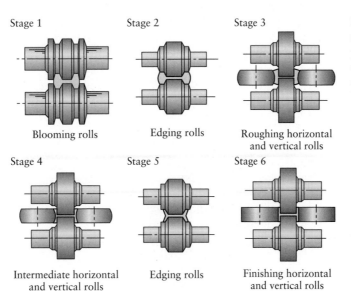

Stage 1   Stage 2   Stage 3

Blooming rolls   Edging rolls   Roughing horizontal and vertical rolls

Stage 4   Stage 5   Stage 6

Intermediate horizontal and vertical rolls   Edging rolls   Finishing horizontal and vertical rolls

**FIGURE 6.44**  Stages in shape rolling of an H-section part. Various other structural sections, such as channels and I-beams, are also rolled by this process.

(a)

(b)

**FIGURE 6.45**    (a) Schematic illustration of a ring-rolling operation. Reducing the thickness results in an increase in the part's diameter. (b) Examples of cross-sections that can be formed by ring rolling.

grain-flow direction. Typical applications of ring rolling include large rings for rockets and turbines, gearwheel rims, ball- and roller-bearing races, flanges and reinforcing rings for pipes, and pressure vessels.

c) **Thread and gear rolling.** In the cold-forming thread-and-gear-rolling process, threads are formed on round rods or workpieces by passing them between reciprocating or rotating dies (Fig. 6.46a). Typical products made by this process include screws, bolts, and similar threaded parts. With flat dies, the threads are rolled on the rod or wire with each stroke of the reciprocating die. Production rates are very high, but depend on the diameter of the product. With products of small diameter, the rates can be as high as eight pieces per second, and with products of larger diameters (as much as 25 mm [1 in.]), rates are about one piece per second. Two- or three-roller thread-rolling machines are also available. In another design (Fig. 6.46b), threads are formed using a rotary die, with production rates as high as 80 pieces per second.

**FIGURE 6.46**    Thread-rolling processes: (a) flat dies and (b) two-roller dies. These processes are used extensively in making threaded fasteners at high rates of production.

(a)

(b)

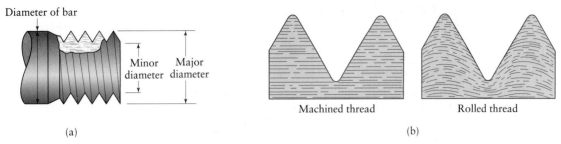

(a)                                                                              (b)

**FIGURE 6.47**  (a) Schematic illustration of threads. (b) Grain-flow lines in machined and rolled threads. Unlike machined threads, which are cut through the grains of the metal, rolled threads follow the grains and are stronger, because of the cold working involved.

The thread-rolling process generates threads without any loss of metal and with greater strength, because of the cold working involved. The surface finish is very smooth, and the process induces compressive residual stresses on the surfaces, which improve fatigue life. The product is superior to that made by thread cutting and is used in the production of almost all externally threaded fasteners.

Because of volume constancy in plastic deformation, a rolled thread requires a round stock of smaller diameter to produce the same major diameter as that of a machined thread (Fig. 6.47). Also, whereas machining removes material by cutting through the grain-flow lines of the material, rolled threads have a grain-flow pattern that improves the strength of the thread, because of the cold working involved. Thread rolling can also be carried out internally with a fluteless forming tap. The process is similar to external thread rolling and produces accurate threads with good strength. In all thread-rolling processes, it is essential that the material have sufficient ductility and that the rod or wire be of proper size. Lubrication is also important for good surface finish and to minimize defects.

Spur and helical gears are also produced by cold-rolling processes similar to thread rolling. The process may be carried out on solid cylindrical blanks or on precut gears. Helical gears can also be made by a direct extrusion process, using specially shaped dies. Cold rolling of gears has many applications in automatic transmissions and power tools.

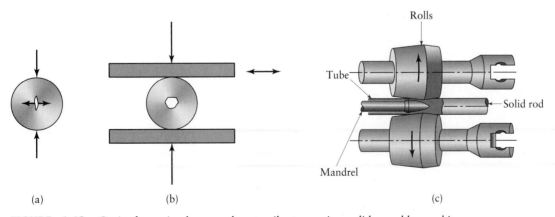

(a)                              (b)                                             (c)

**FIGURE 6.48**  Cavity formation by secondary tensile stresses in a solid round bar and its use in the rotary-tube-piercing process. This is the principle of the Mannesmann mill for seamless tube making. The mandrel is held in place by the long rod, although techniques have been developed in which the mandrel remains in place without the rod.

d) **Rotary tube piercing.** A hot-working process, rotary tube piercing is used to make long, thick-walled seamless tubing, as shown in Fig. 6.48. The process is based on the principle that when a round bar is subjected to radial compression in the manner shown in Fig. 6.48a, tensile stresses develop at the center of the roll. When the rod is subjected to cycling compressive stresses, as shown in Fig. 6.48b, a cavity begins to form at the center of the rod.

The rotary-tube-piercing process (the **Mannesmann process**, developed in the 1880s) is carried out by an arrangement of rotating rolls, as shown in Fig. 6.48c. The axes of the rolls are skewed in order to pull the round bar through the rolls by the longitudinal-force component of their rotary action. A mandrel assists the operation by expanding the hole and sizing the inside diameter of the tube. Because of the severe deformation that the metal undergoes in this process, high-quality, defect-free bars must be used.

e) **Tube rolling.** The diameter and thickness of tubes and pipe can be reduced by *tube rolling* using shaped rolls, either with or without mandrels. In the **pilger mill**, the tube and an internal mandrel undergo a reciprocating motion, and the tube is advanced and rotated periodically.

## 6.4 | Extrusion

In the basic extrusion process, developed in the late 1700s for lead pipe, a round billet is placed in a chamber and forced through a die opening by a ram. The die may be round or of various other shapes. There are four basic types of extrusion: direct, indirect, hydrostatic, and impact (Figs. 6.49 and 6.50).

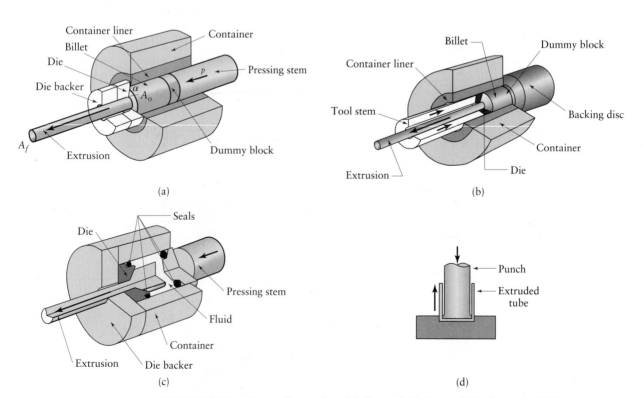

**FIGURE 6.49**    Types of extrusion: (a) direct; (b) indirect; (c) hydrostatic; (d) impact.

(a)                                      (b)

**FIGURE 6.50** (a) Extrusions and examples of products made by sectioning off extrusions. *Source*: Kaiser Aluminum. (b) Examples of extruded cross sections. *Source*: © 2002. Photo courtesy of Plymouth Extruded Shapes, 1.270.886.6631.

**Direct extrusion** (*forward extrusion*) is similar to forcing the toothpaste through the opening of a toothpaste tube. The billet slides relative to the container wall; the wall friction increases the ram force considerably.

In **indirect extrusion** (*reverse, inverted,* or *backwards extrusion*), the die moves toward the billet; thus, except at the die, there is no relative motion at the billet–container interface. This process is especially advantageous for materials with high billet–container friction.

In **hydrostatic extrusion**, the chamber is filled with a fluid that transmits the pressure to the billet, which is then extruded through the die. There is no friction along the container walls.

**Impact extrusion** is a form of indirect extrusion and is particularly suitable for hollow shapes.

Extrusion processes can be carried out hot or cold. Because a chamber is involved, each billet is extruded individually, and hence extrusion is basically a *batch* process. **Cladding** by extrusion can also be carried out with **coaxial** billets (such as copper clad with silver), provided that the flow stresses of the two metals are similar.

The **circumscribing-circle diameter** (CCD), which is the diameter of the smallest circle into which the extruded cross-section will fit, is a parameter that describes the shape of the extruded product. Thus, the CCD for a square cross-section is the cross-section's diagonal dimension. The complexity of an extrusion is a function of the ratio of the perimeter to the cross-sectional area of the part, known as the **shape factor**; thus, a solid, round extrusion is the simplest shape.

## 6.4.1 Metal flow in extrusion

Because the billet is forced through a die, with a substantial reduction in its cross-section, the *metal flow pattern* in extrusion is an important factor in the overall process. A common technique for investigating the flow pattern is to halve the round billet lengthwise and mark one face with a grid pattern. The two halves are placed together in the container (they may also be fastened together, or *brazed* (Section 12.12), to keep the two halves intact) and extruded. They are then taken apart and inspected.

**FIGURE 6.51**   Schematic illustration of three different types of metal flow in direct extrusion.

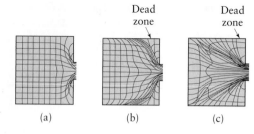

(a)        (b)        (c)

Figure 6.51 shows three typical results of direct extrusion with **square dies**. The conditions under which these different flow patterns are obtained are described as follows:

1. The most homogeneous flow pattern is obtained when there is no friction at the billet–container–die interfaces (Fig. 6.51a). This type of flow occurs when the lubricant is very effective or with indirect extrusion.

2. When friction along all interfaces is high, a **dead-metal zone** develops (Fig. 6.51b). Note the high-shear area as the material flows into the die exit, somewhat like liquid flowing into a funnel. This configuration may indicate that the billet surfaces (with their oxide layer and lubricant) could enter this high-shear zone and be extruded, causing defects in the extruded product.

3. In the third configuration, the high-shear zone extends farther back (Fig. 6.51c). This extension can result from high container-wall friction, which retards the flow of the billet, or from materials in which the flow stress drops rapidly with increasing temperature. In hot working, the material near the container walls cools rapidly and hence increases in strength; thus, the material in the central regions flows toward the die more easily than that at the outer regions. As a result, a large dead-metal zone forms, and the flow is inhomogeneous. This flow pattern leads to a defect known as a *pipe*, or *extrusion, defect*.

Thus, the two factors that greatly influence metal flow in extrusion are the frictional conditions at billet–container–die interfaces and thermal gradients in the billet.

### 6.4.2 Mechanics of extrusion

The ram force in direct extrusion can be calculated in the following ways for different situations:

a) **Ideal deformation.**   The extrusion ratio, $R$, is defined as

$$R = \frac{A_o}{A_f},$$   (6.47)

where $A_o$ is the cross-sectional area of the billet and $A_f$ is the area of the extruded product (Fig. 6.49). The absolute value of the true strain is then

$$\varepsilon_1 = \ln\left(\frac{A_o}{A_f}\right) = \ln\left(\frac{L_f}{L_o}\right) = \ln R,$$   (6.48)

where $L_o$ and $L_f$ are the lengths of the billet and the extruded product, respectively. For a perfectly plastic material with a yield stress, $Y$, the energy dissipated in plastic deformation per unit volume, $u$, is

$$u = Y\varepsilon_1.$$   (6.49)

Hence, the work done on the billet is

$$\text{Work} = (A_o)(L_o)(u). \tag{6.50}$$

This work is supplied by the ram force, $F$, which travels a distance $L_o$. Therefore,

$$\text{Work} = FL_o = pA_oL_o, \tag{6.51}$$

where $p$ is the extrusion pressure at the ram. Equating the work of plastic deformation to the external work done, we find that

$$p = u = Y \ln\left(\frac{A_o}{A_f}\right) = (Y)(\ln R). \tag{6.52}$$

For strain-hardening materials, $Y$ should be replaced by the *average flow stress*, $\bar{Y}$. Note that Eq. (6.52) is equal to the area under the true-stress–true-strain curve for the material.

b) **Ideal deformation and friction.** Equation (6.52) pertains to ideal deformation without any friction. Based on the slab method of analysis, when friction at the die–billet interface (but not the container-wall friction) is included and for small die angles, the pressure $p$ is given by

$$p = Y\left(1 + \frac{\tan\alpha}{\mu}\right)[R^{\mu\cot\alpha} - 1]. \tag{6.53}$$

Based on the assumption that the frictional stress is equal to the shear yield stress $k$, and that because of the dead zone formed, the material flows along a 45° "die angle," an estimate of $p$ can be given as

$$p = Y\left(1.7R + \frac{2L}{D_o}\right). \tag{6.54}$$

For strain-hardening materials, $Y$ in these expressions should be replaced by $\bar{Y}$. Note that as the billet is extruded farther, $L$ decreases, and thus the ram force decreases (Fig. 6.52), whereas in indirect extrusion, the ram force is not a function of billet length.

c) **Actual forces.** The derivation of analytical expressions, including those for friction, the die angle, and the redundant work due to inhomogeneous deformation of the material, can be difficult. Furthermore, there are difficulties in estimating the coefficient of friction, the flow stress of the material, and the redundant work in a particular operation. Consequently, a convenient *empirical* formula has been developed. The formula is

$$p = Y(a + b\ln R), \tag{6.55}$$

where $a$ and $b$ are experimentally determined constants. An approximate value for $a$ is 0.8, and $b$ ranges approximately from 1.2 to 1.5. Again, note that for strain-hardening materials, $Y$ is replaced by $\bar{Y}$.

d) **Optimum die angle.** The die angle has an important effect on forces in extrusion. Its relationship to work is as follows:

1. The **ideal work** of deformation is independent of the die angle (Fig. 6.53), because the work is a function only of the extrusion ratio.
2. The **frictional work** increases with decreasing die angle, because the length of contact at the billet–die interface increases, thus requiring more work.
3. The **redundant work** caused by inhomogeneous deformation increases with die angle.

**FIGURE 6.52** Schematic illustration of typical extrusion pressure as a function of ram travel: (a) direct extrusion and (b) indirect extrusion. The pressure in direct extrusion is higher because of frictional resistance in the chamber as the billet moves toward the die.

**FIGURE 6.53** Schematic illustration of extrusion force as a function of die angle: (a) total force; (b) ideal force; (c) force required for redundant deformation; (d) force required to overcome friction. Note that there is an optimum die angle where the total extrusion force is a minimum.

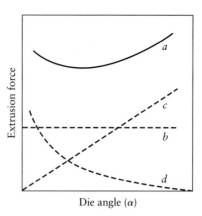

Extrusion force

Die angle ($\alpha$)

Because the total ram force is the sum of these three components, there is an angle at which this force is a *minimum* (Fig. 6.53). Unless the behavior of each component as a function of the die angle is known, determination of this *optimum angle* is difficult.

---

**EXAMPLE 6.5  Strain rate in extrusion**

Determine the true strain rate in extruding a round billet of radius $r_o$ as a function of distance $x$ from the entry of the die.

**Solution.**  From the geometry in the die gap, we have

$$\tan \alpha = \frac{(r_o - r)}{x}, \tag{6.56}$$

or

$$r = r_o - x \tan \alpha.$$

The incremental true strain can be defined as

$$d\varepsilon = \frac{dA}{A},$$

where $A = \pi r^2$. Therefore, $dA = 2\pi r \, dr$, and hence

$$d\varepsilon = \frac{2 \, dr}{r},$$

where $dr = -\tan \alpha \, dx$. We also know that

$$\dot{\varepsilon} = \frac{d\varepsilon}{dt} = -\left(\frac{2 \, \tan \alpha}{r}\right)\left(\frac{dx}{dt}\right).$$

However, $dx/dt = V$, which is the velocity of the material at any location $x$ in the die. Hence, we have

$$\dot{\varepsilon} = -\frac{2V \tan \alpha}{r}.$$

Because the flow rate is constant, we can write

$$V = \frac{V_o r_o^2}{r^2}$$

and hence we have

$$\dot{\varepsilon} = -\frac{2V_o r_o^2 \tan \alpha}{r^3}$$

$$= -\frac{2V_o r_o^2 \tan \alpha}{(r_o - x \tan \alpha)^3} \qquad (6.57)$$

The negative sign is due to the fact that true strain is defined in terms of the cross-sectional area, which decreases as $x$ increases.

e) **Forces in hot extrusion.** Because of the strain-rate sensitivity of metals at elevated temperatures, forces in hot extrusion are difficult to calculate. The average true-strain rate, $\dot{\bar{\varepsilon}}$ is

$$\dot{\bar{\varepsilon}} = \frac{6V_o D_o^2 \tan \alpha}{D_o^3 - D_f^3} \ln R, \qquad (6.58)$$

where $V_0$ is the ram velocity. Note from this equation that for high extrusion ratios ($D_o \gg D_f$) and for $\alpha = 45°$, as may be the case with a square die (thus developing a dead zone) and poor lubrication, the strain rate reduces to

$$\dot{\bar{\varepsilon}} = \frac{6V_o}{D_o} \ln R. \qquad (6.59)$$

Figure 6.54 shows the effect of ram speed and temperature on extrusion pressure. As expected, pressure increases rapidly with ram speed, especially at elevated temperatures. As extrusion speed increases, the rate of work done per unit time also increases. Because work is converted into heat, the heat generated at high speeds may not be dissipated fast enough. The subsequent rise in temperature can cause incipient melting of the workpiece material and cause defects. Circumferential surface cracks caused by **hot shortness** may also develop (see Section 3.4); in extrusion, this phenomenon is known as **speed cracking**. These problems can be eliminated by reducing the extrusion speed.

A convenient parameter that is used to estimate forces in extrusion is an experimentally determined extrusion constant $K_e$, which includes various factors and can be determined from

$$p = K_e \ln R. \qquad (6.60)$$

Figure 6.55 gives some typical values of $K_e$ for various materials.

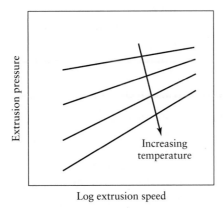

Log extrusion speed

**FIGURE 6.54**  Schematic illustration of the effect of temperature and ram speed on extrusion pressure. Compare with Fig. 2.11.

**FIGURE 6.55** Extrusion constant, $K_e$, for various materials as a function of temperature. *Source*: After P. Loewenstein, ASTME Paper SP63-89.

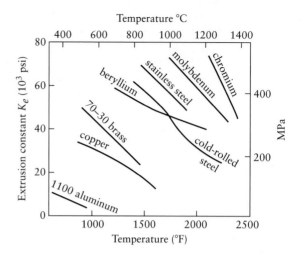

---

**EXAMPLE 6.6** Force in hot extrusion

A copper billet 5 in. in diameter and 10 in. long is extruded at 1500°F at a speed of 10 in./s. Using square dies and assuming poor lubrication, estimate the force required in this operation if the final diameter is 2 in.

**Solution.** The extrusion ratio is

$$R = \frac{5^2}{2^2} = 6.25.$$

The average true-strain rate, from Eq. (6.58), is

$$\dot{\varepsilon} = \frac{(6)(10)}{(5)} \ln 6.25 = 22 s^{-1}$$

From Table 2.5, let's assume that $C = 19,000$ psi (an average value) and $m = 0.06$. Then we have

$$\sigma = C\dot{\varepsilon}^m = (19,000)(22)^{0.06} = 22,870 \text{ psi}.$$

Assuming that $\bar{Y} = \sigma$, we have, from Eq. (6.54),

$$p = \bar{Y}\left(1.7 \ln R + \frac{2L}{D_o}\right) = (22,870)\left[(1.7)(1.83) + \frac{(2)(10)}{(5)}\right] = 162,630 \text{ psi}.$$

Hence,

$$F = (p)(A_o) = (162,630)\frac{(\pi)(5)^2}{4} = 3.2 \times 10^6 \text{ lb}.$$

---

### 6.4.3 Miscellaneous extrusion processes

a) **Cold extrusion.** *Cold extrusion* is a general term often used to denote a combination of processes, such as direct and indirect extrusion and forging (Fig. 6.56). Many materials can be extruded into various configurations, with the billet either at room temperature or at a temperature of a few hundred degrees. Cold extrusion has gained wide acceptance in industry, because of the following advantages over hot extrusion:

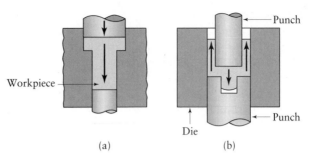

**FIGURE 6.56** Examples of cold extrusion. Arrows indicate the direction of material flow. These parts may also be considered as forgings.

1. Improved mechanical properties resulting from strain hardening, provided that the heat generated by plastic deformation and friction does not recrystallize the extruded metal;
2. Good control of dimensional tolerances, thus requiring a minimum of machining operations;
3. Improved surface finish, provided that lubrication is effective;
4. Lack of oxide layers;
5. High production rates and relatively low costs.

However, the stresses on tooling in cold extrusion are very high, especially with steel workpieces. The levels on tooling are on the order of the hardness of the material, that is, at least three times its flow stress. The design of tooling and selection of appropriate tool materials are therefore crucial to success in cold extrusion. The hardness of tooling usually ranges between 60 and 65 HRC for the punch and 58 and 62 HRC for the die. Punches are a critical component; they must have sufficient strength, toughness and resistance to wear and fatigue.

Lubrication also is crucial, especially with steels, because of the generation of new surfaces and the possibility of seizure between the metal and the tooling, caused by the breakdown of lubrication. The most effective lubrication is provided by phosphate conversion coatings on the workpiece and soap (or wax in some cases) as the lubricant. Temperature rise in cold extrusion is an important factor, especially at high extrusion ratios. The temperature may be sufficiently high to initiate and complete the recrystallization process of the cold-worked metal, thus reducing the advantages of cold working.

b) **Impact extrusion.** A process often included in the category of cold extrusion, *impact extrusion* (Fig. 6.49d) is similar to indirect extrusion. In impact extrusion, the punch descends at a high speed and strikes the *blank (slug)*, extruding it upward. The thickness of the extruded tubular section is a function of the clearance between the punch and the die cavity.

The impact-extrusion process usually produces tubular sections with wall thicknesses that are small in relation to their diameters. This ratio can be as small as 0.005. The concentricity between the punch and the blank is important for uniform wall thickness. A typical example of impact extrusion is the production of collapsible tubes, such as for toothpaste (Fig. 6.57a). The amount of distance traveled by the punch is determined by the setting of the press. A variety of nonferrous metals are impact extruded in this manner into various shapes (Fig. 6.57b), using vertical presses at production rates as high as two parts per second.

c) **Hydrostatic extrusion.** In *hydrostatic extrusion*, the pressure required for extrusion is supplied through a fluid medium surrounding the billet (Fig. 6.49c). Consequently, there is no container-wall friction. Pressures are usually about 1400 MPa (200 ksi). The high pressure in the chamber transmits some of the

**FIGURE 6.57** (a) Impact extrusion of a collapsible tube (Hooker process). (b) Two examples of products made by impact extrusion. These parts may also be made by casting, forging, and machining, depending on the dimensions and materials involved and the properties desired. Economic considerations are also important in final process selection.

(a)

(b)

fluid to the die surfaces, thus significantly reducing friction and forces (Fig. 6.58). Hydrostatic extrusion, which was developed in the early 1950s, has been improved by extruding the part into a second pressurized chamber, which is under lower pressure (**fluid-to-fluid extrusion**). This operation reduces the number of defects in the extruded product.

Because the hydrostatic pressure increases the ductility of the material, brittle materials can be extruded successfully by this method. However, the main reasons for this success appear to be low friction and use of low die angles and high extrusion ratios. Most commercial hydrostatic-extrusion operations use ductile materials. However, a variety of metals and polymers, solid shapes, tubes and other hollow shapes, and honeycomb and clad profiles have been extruded successfully.

Hydrostatic extrusion is usually carried out at room temperature, typically using vegetable oils as the fluid, particularly castor oil, because it is a good lubricant and its viscosity is not influenced significantly by pressure. For elevated-temperature extrusion, waxes, polymers, and glass are used as the fluid; these materials also serve as thermal insulators and help maintain the billet temperature during extrusion. In spite of the success obtained, hydrostatic extrusion has had limited industrial applications, largely because of the somewhat complex nature of tooling, the experience required in working with high pressures and design of specialized equipment, and the long cycle times required.

**FIGURE 6.58** Extrusion pressure as a function of the extrusion ratio for an aluminum alloy. (a) Direct extrusion, $\alpha = 90°$. (b) Hydrostatic extrusion, $\alpha = 45°$. (c) Hydrostatic extrusion, $\alpha = 22.5°$. (d) Ideal homogeneous deformation, calculated. *Source*: After H. Li., D. Pugh, and K. Ashcroft.

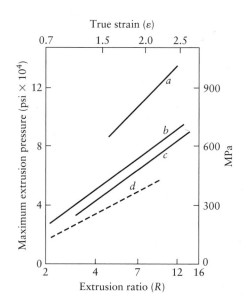

### 6.4.4  Defects in extrusion

There are three principal defects in extrusion: surface cracking, extrusion defect, and internal cracking. These defects are described as follows:

a) **Surface cracking.** If the extrusion temperature, friction, or extrusion speed is too high, surface temperatures rise significantly, which can lead to surface cracking and tearing (*fir-tree cracking* or *speed cracking*). These cracks are intergranular and are usually the result of hot shortness; they occur especially with aluminum, magnesium, and zinc alloys, but are also observed with other metals, such as molybdenum alloys. This situation can be avoided by using lower temperatures and speeds.

   However, surface cracking may also occur at low temperatures and has been attributed to periodic sticking of the extruded product along the die land (**stick-slip**) during extrusion. When the product being extruded sticks to the die land, the extrusion pressure increases rapidly; shortly thereafter, the product moves forward again, and the pressure is released. The cycle is then repeated. Because of its appearance, this defect is known as a **bamboo defect**. This type of defect is especially common with hydrostatic extrusion where the pressure is sufficient to significantly increase the viscosity of the fluid. This increase in viscosity allows a thick lubricant film to develop; the billet then surges forward, which, in turn, relieves the pressure of the fluid and increases friction. Thus, the change in the physical properties of the metalworking fluid/lubricant is responsible for the stick-slip phenomenon. In such circumstances, it has been found that proper selection of a fluid is critical, but also that increasing the extrusion speed can eliminate stick-slip.

b) **Extrusion defect.** The type of metal flow observed in Fig. 6.51c tends to draw surface oxides and impurities toward the center of the billet, much like a funnel. This defect is known as *extrusion defect*, **pipe**, **tailpipe**, and **fishtailing**. A considerable portion of the material can be rendered useless as an extruded product because of it—as much as one third the length of the extrusion. This defect can be reduced by modifying the flow pattern to a less inhomogeneous one, such as by controlling friction and minimizing temperature gradients. Another method is to machine the surface of the billet prior to extrusion to eliminate scale and impurities. The extrusion defect can also be avoided by using a dummy block (Fig. 6.49a) that is smaller in diameter than the container, thus leaving a thin shell along the container wall as extrusion progresses.

c) **Internal cracking.** The center of an extruded product can develop cracks (variously known as **centerburst**, **center cracking**, **arrowhead fracture**, and **chevron cracking**), as shown in Fig. 6.59. These cracks are attributed to a state of *hydrostatic tensile stress* (also called *secondary tensile stress*) at the centerline of the deformation zone in the die. This situation is similar to the necked region in a uniaxial tensile-test specimen.

   The major variables affecting hydrostatic tension are the die angle, the extrusion ratio (reduction in cross-sectional area), and friction; we can understand them best by observing the extent of inhomogeneous deformation in extrusion. Experimental results indicate that, for the same reduction, as the die angle becomes larger, the deformation across the part becomes more inhomogeneous. In addition to the die angle, another factor in internal cracking is the die contact length. The smaller the die angle, the longer is the contact length. This situation is similar to a hardness test with a flat indenter. The size and depth of the deformation zone increases with increasing contact length. This condition is illustrated in Fig. 6.59a.

(a)

(b)

**FIGURE 6.59** (a) Deformation zone in extrusion, showing rigid and plastic zones. Note that the plastic zones do not meet, leading to chevron cracking. The same observations are also made in drawing round bars through conical dies and drawing flat sheet or plate through wedge-shaped dies. *Source*: After B. Avitzur. (b) Chevron cracking in round steel bars during extrusion. Unless the part is inspected properly, such internal detects may remain undetected and possibly cause failure of the part in service.

As shown in Fig. 6.12, $h/L$ ratio is an important parameter: The higher this ratio, the more inhomogeneous is the deformation. High ratios mean small reductions and large die angles. Inhomogeneous deformation indicates that the center of the billet is not in a fully plastic state. (It is rigid.) The reason is that the plastic deformation zones under the die contact lengths do not reach each other (Fig. 6.59a). Likewise, small reductions and high die angles retard the flow of the material at the surfaces, while the central portions are freer to move through the die.

The high $h/L$ ratios described generate hydrostatic tensile stresses in the center of the billet, causing the type of defects shown in Fig. 6.59b. These defects form more readily in materials with impurities, inclusions, and voids, because they serve as nucleation sites for defect formation. As for the role of friction, high friction in extrusion apparently delays formation of these cracks. Such cracks have also been observed in tube extrusion and in spinning of tubes. The cracks appear on the inside surfaces for the reasons already given. In summary, the tendency for center cracking increases with increasing die angles and levels of impurities and decreases with increasing extrusion ratios. These observations are also valid for the drawing of rod and wire. (See Section 6.5.)

### 6.4.5 Extrusion practice

Numerous materials can be extruded to a wide variety of cross-sectional shapes and dimensions. Extrusion ratios can range from about 10 to over 100. Ram speeds may be up to 0.5 m/s (100 ft/min). Generally, slower speeds are preferable for aluminum, magnesium, and copper and higher speeds for steels, titanium, and refractory alloys. Presses for hot extrusion are generally hydraulic and horizontal; presses for cold extrusion are usually vertical.

**Hot extrusion.** In addition to the strain-rate sensitivity of the material at elevated temperatures, hot extrusion requires other special considerations. Cooling of the billet in the container can result in highly inhomogeneous deformation. Furthermore, because the billet is heated prior to extrusion, it is covered with an oxide layer (unless heated in an inert atmosphere). The resulting different frictional properties can affect the flow of the material and may produce an extrusion covered with an oxide layer.

In order to avoid this problem, the diameter of the dummy block ahead of the ram (Fig. 6.49a) is made a little smaller than that of the container. A thin cylindrical shell (skull), consisting mainly of the oxidized layer, is thus left in the container, and the extruded product is free of oxides. Temperature ranges for hot extrusion are similar to those for forging. (See Table 6.3.)

**Lubrication.** Lubrication is important in hot extrusion. For steels, stainless steels, and high-temperature materials, glass is an excellent lubricant (**Séjournet process**). Glass maintains its viscosity at elevated temperatures, has good wetting characteristics, and acts as a thermal barrier between the billet, the container, and the die, thus minimizing cooling. A circular glass pad is usually placed at the die entrance. This pad softens and melts away slowly as extrusion progresses and forms an optimal die geometry. The viscosity–temperature index of the glass is an important factor in this application. Solid lubricants such as graphite and molybdenum disulfide are also used in hot extrusion. Nonferrous metals are usually extruded without a lubricant, although graphite may be used.

For materials that have a tendency to stick to the container and the die, the billet can be enclosed in a jacket of a softer metal, such as copper or mild steel (**jacketing** or **canning**). Besides providing a low-friction interface, canning prevents contamination of the billet by the environment, or prevents the billet material from contaminating the environment if the material is toxic or radioactive. The canning technique is also used for processing metal powders.

**Die design.** Designing dies requires considerable experience, because of the variety of products extruded. Dies with angles of 90° (**square dies**, or **shear dies**) can also be used in extrusion of nonferrous metals, especially aluminum. Tubing is also extruded with a ram fitted with a mandrel (Fig. 6.60). For billets with a pierced hole, the mandrel may be attached to the ram. If the billet is solid, it must first be pierced in the container by the mandrel.

Hollow shapes (Fig. 6.61a) can also be extruded by **welding-chamber methods**, using various special dies known as *spider, porthole*, and *bridge dies* (Fig. 6.61b).

**FIGURE 6.60**  Extrusion of a seamless tube. The hole in the billet may be prepunched or pierced, or it may be generated during extrusion.

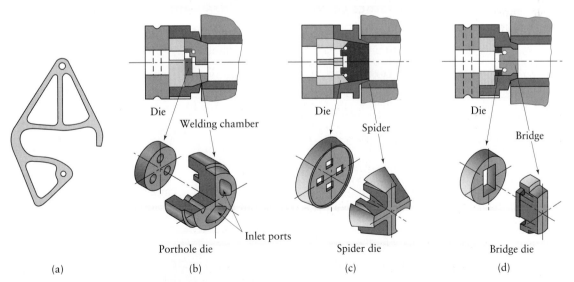

(a)                    (b)                    (c)                    (d)

**FIGURE 6.61**  (a) An extruded 6063-T6 aluminum ladder lock for aluminum extension ladders. This part is 8 mm ($\frac{5}{16}$ in.) thick and is sawed from the extrusion. (See Fig. 6.50a.) (b) Components of various dies for extruding intricate hollow shapes. *Source* for (b)-(d): K. Laue and H. Stenger, *Extrusion-Processes, Machinery, Tooling*, ASM International, 1981. Used with permission.

The metal flows around the arms of the die into strands, which are then rewelded under the high die pressures at the exit. This process is suitable only for aluminum and some of its alloys, because of their capacity for pressure welding. Lubricants cannot be used, as they prevent rewelding during extrusion.

Die materials for hot extrusion are usually hot-work die steels. To extend die life, various coatings may be applied to the dies. (see Section 4.5.)

**Equipment.** The basic equipment for extrusion is a hydraulic press, usually horizontal; these presses are designed for a variety of extrusion operations. Crank-type mechanical presses are also used for cold extrusion and piercing and for mass production of steel tubing. The largest hydraulic press for extrusion has a capacity of 160 MN (16,000 tons).

## 6.5 | Rod, Wire, and Tube Drawing

Drawing is an operation in which the cross-sectional area of a bar or tube is reduced by pulling it through a converging die (Fig. 6.62). The die opening may be any shape. Wire drawing involves materials of smaller diameter than those used for rod drawing, with sizes as small as 0.025 mm (0.001 in.). The drawing process, which was an established art by the 11th century AD, is somewhat similar to extrusion, except that in drawing, the bar is under tension, whereas in extrusion it is under compression.

*Rod and wire drawing* are usually finishing processes. The product is either used as produced or further processed into other shapes, usually by bending or machining. Rods are used for various applications, such as small pistons, tension-carrying structural members, shafts, and spindles, and as the raw material for fasteners such as bolts and screws. Wire and wire products have a wide range of applications, such as electrical and electronic equipment and wiring, cables, springs, musical instruments, fencing, bailing, wire baskets, and shopping carts.

### 6.5.1 Mechanics of rod and wire drawing

The major variables in the drawing process are reduction in cross-sectional area, die angle, and friction; these variables are illustrated in Fig. 6.62.

a) **Ideal deformation.** For a round rod or wire, the drawing stress $\sigma_d$ for the simplest case of ideal deformation (no friction or redundant work) can be obtained by the same approach as that used for extrusion. Thus, we have

$$\sigma_d = Y \ln \left( \frac{A_o}{A_f} \right). \tag{6.61}$$

Note that this expression is the same as Eq. (6.52) and that it also represents the energy per unit volume, $u$. For strain-hardening materials, $Y$ is replaced by an

**FIGURE 6.62**    Variables in drawing round rod or wire.

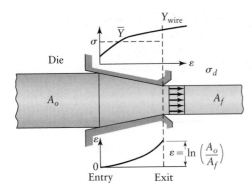

**FIGURE 6.63**   Variation in strain and flow stress in the deformation zone in drawing. Note that the strain increases rapidly toward the exit. The reason is that when the exit diameter is zero, the true strain reaches infinity. The point $Y_{wire}$ represents the yield stress of the drawn wire.

average flow stress, $\bar{Y}$ in the deformation zone (Fig. 6.63). Therefore, for a material that exhibits the true-stress–true-strain behavior of

$$\sigma = K\varepsilon^n,$$

the average flow stress, $\bar{Y}$ is given as

$$\bar{Y} = \frac{K\varepsilon_1^n}{n + 1}. \tag{6.62}$$

The drawing force, $F$, is then given as

$$F = \bar{Y}A_f \ln\left(\frac{A_o}{A_f}\right). \tag{6.63}$$

Note that the greater the reduction in cross-sectional area and the stronger the material, the higher is the drawing force.

b) **Ideal deformation and friction.**   In drawing, friction increases the drawing force, because work has to be supplied to overcome that friction. Using the slab method of analysis and on the basis of Fig. 6.64, we can obtain the following expression for the drawing stress, $\sigma_d$:

$$\sigma_d = Y\left(1 + \frac{\tan \alpha}{\mu}\right)\left[1 - \left(\frac{A_f}{A_o}\right)^{\mu \cot \alpha}\right]. \tag{6.64}$$

For a strain-hardening material, the yield stress, $Y$, is replaced by the average flow stress, $\bar{Y}$, as shown in Fig. 6.63. Investigations have shown that, even though Eq. (6.64) does not include the redundant work, it is in good agreement with experimental data for small die angles and for a wide range of reductions.

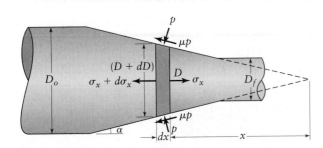

**FIGURE 6.64**   Stresses acting on an element in drawing of a solid cylindrical rod or wire through a conical converging die.

c) **Redundant work of deformation.** Depending on the die angle and reduction, the material in drawing undergoes *inhomogeneous deformation*, much
as it does in extrusion; therefore, the redundant work of deformation also has
to be included in the expression for the drawing stress. One such expression is

$$\sigma_d = \bar{Y}\left\{ \left(1 + \frac{\tan \alpha}{\mu}\right)\left[1 - \left(\frac{A_f}{A_o}\right)^{\mu \cot \alpha}\right] + \frac{4}{3\sqrt{3}}\alpha^2\left(\frac{1-r}{r}\right) \right\}, \quad (6.65)$$

where $r$ is the fractional reduction of area and $\alpha$ is the die angle in radians. The
first term in Eq. (6.65) represents the ideal and frictional components of work.
The second term represents the redundant-work component, which, as expected, is a function of the die angle. The larger the angle is, the greater is the inhomogeneous deformation and, hence, the greater is the redundant work.

Another expression for the drawing stress for small die angles (including
all three components of work) is

$$\sigma_d = \bar{Y}\left[\left(1 + \frac{\mu}{\alpha}\right)\ln\left(\frac{A_o}{A_f}\right) + \frac{2}{3}\alpha\right]. \quad (6.66)$$

The last term in this expression is the redundant-work component; this component of work increases linearly with the die angle, as shown in Fig. 6.53. Because redundant deformation is a function of the $h/L$ ratio (see Figs. 6.12 and
6.13), an **inhomogeneity** factor $\Phi$ can replace the last term in Eq. (6.66). For
drawing of round sections this factor is, approximately,

$$\Phi = 1 + 0.12(h/L). \quad (6.67)$$

A simple expression for the drawing stress is then

$$\sigma_d = \Phi\bar{Y}\left(1 + \frac{\mu}{\alpha}\right)\ln\left(\frac{A_o}{A_f}\right). \quad (6.68)$$

Equations (6.64), (6.65), (6.66), and (6.68), while not always agreeing
with experimental data, approximate reasonably well the stresses required for
wire drawing. More important, they identify the effects of the various parameters involved. With good lubrication, the value of $\mu$ in these equations ranges
from about 0.03 to 0.1.

d) **Die pressure.** The *die pressure, p*, at any diameter along the die contact length
can be conveniently obtained from

$$p = Y_f - \sigma \quad (6.69)$$

where $\sigma$ is the tensile stress in the deformation zone at any diameter. (Thus, $\sigma$
is equal to $\sigma_d$ at the exit and is zero at the entry). $Y_f$ is the flow stress of the material at any diameter. Equation (6.69) is based on yield criteria for an element
subjected to the stress system shown in Fig. 2.35b, where the compressive stresses in the two principal directions are both equal to $p$. Note that as the tensile
stress increases toward the exit, the die pressure drops toward the exit. Tensile-
stress and die-pressure distributions are shown qualitatively in Fig. 6.65.

**FIGURE 6.65** Variation in
the drawing stress and die
contact pressure along the
deformation zone. Note that
as the drawing stress increases,
the die pressure decreases. This
condition can be observed
from the yield criteria,
described in Section 2.11. Note
the effect of back tension.

(a)

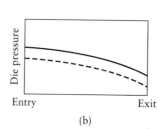

(b)

**EXAMPLE 6.7  Power and die pressure in rod drawing**

A round rod of annealed 302 stainless steel is being drawn from a diameter of 10 mm to a diameter of 8 mm at a speed of 0.5 m/s. Assume that the frictional and redundant work together constitute 40% of the ideal work of deformation. (1) Calculate the power required in this operation. (2) Calculate the die pressure at the exit of the die.

**Solution.**  The true-strain in this operation is

$$\varepsilon_1 = \ln\left(\frac{10^2}{8^2}\right) = 0.446.$$

From Table 2.3, we know that $K = 1300$ MPa and $n = 0.30$. Hence,

$$\bar{Y} = \frac{K\varepsilon_1^n}{n+1} = \frac{(1300)(0.446)^{0.30}}{1.30} = 785 \text{ MPa}.$$

From Eq. (6.63), we have

$$F = \bar{Y}A_f \ln\left(\frac{A_o}{A_f}\right),$$

where

$$A_f = \frac{(\pi)(0.008)^2}{4} = 5 \times 10^{-5} \text{ m}^2.$$

Hence,

$$F = (785)(5 \times 10^{-5})(0.446) = 0.0175 \text{ MN}.$$

So,

$$\begin{aligned}
\text{Power} &= (F)(V_f) = (0.0175)(0.5) \\
&= 0.00875 \text{ MN} \cdot \text{m/s} \\
&= 0.00875 \text{ MW} = 8.75 \text{ kW},
\end{aligned}$$

and

$$\text{Actual power} = (1.4)(8.75) = 12.25 \text{ kW}.$$

The die pressure can be obtained by noting from Eq. (6.69) that

$$p = Y_f - \sigma$$

where $Y_f$ represents the flow stress of the material at the exit of the die. Thus,

$$Y_f = K\varepsilon_1^n = (1300)(0.446)^{0.30} = 1020 \text{ MPa}.$$

In this equation, $\sigma$ is the drawing stress, $\sigma_d$. Hence, using the actual force, we have

$$\sigma_d = \frac{F}{A_f} = \frac{(1.4)(0.0175)}{0.00005} = 490 \text{ MPa}.$$

Therefore, the die pressure at the exit is

$$p = 1020 - 490 = 530 \text{ MPa}.$$

e) **Drawing at elevated temperatures.** At elevated temperatures, the flow stress of metals is a function of the strain rate. The *average true-strain rate $\dot{\bar{\varepsilon}}$* in the deformation zone in drawing is

$$\dot{\bar{\varepsilon}} = \frac{6V_o}{D_o} \ln\left(\frac{A_o}{A_f}\right). \tag{6.70}$$

[Note that this expression is the same as Eq. (6.59).] For a particular drawing operation at an elevated temperature, we first calculate the average strain rate; from that quantity, we can determine the flow stress and the average flow stress, $\bar{Y}$, of the material.

f) **Optimum die angle.** Because of the various effects of the die angle on the three components of work (i.e., ideal, friction, and redundant work), there is an *optimum die angle* at which the extrusion force is a minimum. Drawing involves a similar type of deformation, so there is an optimum die angle in rod and wire drawing also, as shown in Fig. 6.53. Figure 6.66 shows a typical example from an experimental study. The optimum die angle for the minimum force increases with reduction. Note that the optimum angles in Fig. 6.66 are rather small.

g) **Maximum reduction per pass.** With greater reduction, the drawing stress increases, but, obviously, the magnitude of the drawing stress has a limit. If it reaches the yield stress of the material, the material will simply continue to yield further as it leaves the die. This result is unacceptable, because the product will undergo further deformation; thus, the maximum possible drawing stress is the yield stress of the exiting material.

In the ideal case of a perfectly plastic material with a yield stress Y, the limiting condition is

$$\sigma_d = Y \ln\left(\frac{A_o}{A_f}\right) = Y, \tag{6.71}$$

or

$$\ln\left(\frac{A_o}{A_f}\right) = 1.$$

**FIGURE 6.66** The effect of reduction in cross-sectional area on the optimum die angle. *Source:* After J. G. Wistreich.

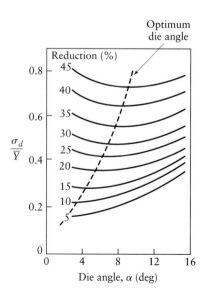

Hence,

$$\frac{A_o}{A_f} = e,$$

and therefore,

Maximum reduction per pass $= \dfrac{A_o - A_f}{A_o} = 1 - \dfrac{1}{e} = 0.63 = 63\%.$   (6.72)

The effects of friction and die angle on maximum reduction per pass are similar to those shown in Fig. 6.53. Because both friction and redundant work increase the drawing stress, the maximum reduction per pass will be lower than the ideal. In other words, the exiting material must have a larger cross-sectional area to sustain the higher drawing forces. With strain hardening, the exiting material is stronger than the rest of the material, and the maximum reduction per pass increases.

---

**EXAMPLE 6.8  Maximum reduction per pass for a strain-hardening material**

Obtain an expression for the maximum reduction per pass for a material with a true-stress–true-strain curve of $\sigma = K\varepsilon^n$. Ignore friction and redundant work.

**Solution.**  From Eq. (6.61), we have for this material

$$\sigma_d = \bar{Y} \ln\left(\frac{A_o}{A_f}\right) = \bar{Y}\varepsilon_1,$$

where

$$\bar{Y} = \frac{K\varepsilon_1^n}{n + 1}$$

and $\sigma_d$, for this problem, can have a maximum value equal to the flow stress at $\varepsilon_1$, or

$$\sigma_d = Y_f = K\varepsilon_1^n.$$

Hence, we can write Eq. (6.71) as

$$K\varepsilon_1^n = \frac{K\varepsilon_1^n}{n + 1}\varepsilon_1,$$

or

$$\varepsilon_1 = n + 1.$$

With $\varepsilon_1 = \ln(A_o/A_f)$ and a maximum reduction of $(A_o - A_f)/A_o$, these expressions reduce to

Maximum reduction per pass $= 1 - e^{-(n+1)}.$   (6.73)

Note that when $n = 0$ (perfectly plastic material), this expression reduces to Eq. (6.72). As $n$ increases, the maximum reduction per pass increases.

---

h) **Drawing of flat strip.**  Whereas drawing of round sections is an axisymmetric problem, drawing of flat strips with high width-to-thickness ratios can be regarded as a **plane-strain** problem. The dies are wedge shaped, and thus the process is somewhat similar to that of rolling wide strips. There is little or no

**FIGURE 6.67** Various methods of tube drawing.

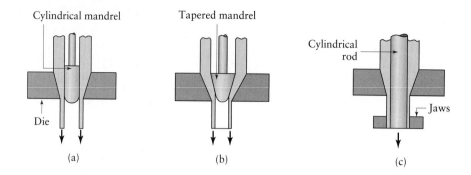

change in the width of the strip during drawing. This process, although not of any industrial significance in and of itself, is the fundamental deformation mechanism in **ironing**, as described in Section 7.12.

The treatment of this subject, as far as forces and maximum reductions are concerned, is similar to that for round sections. The drawing stress for the ideal condition is

$$\sigma_d = Y' \ln\left(\frac{t_o}{t_f}\right), \tag{6.74}$$

where $Y'$ is the yield stress of the material in plane strain and $t_o$ and $t_f$ are the original and final thicknesses of the strip, respectively. The effects of friction and redundant deformation in strip drawing are similar to those for round sections.

The maximum reduction per pass can be obtained by equating the drawing stress [Eq. (6.74)] to the *uniaxial* yield stress of the material, because the drawn strip is subjected only to simple tension. Thus,

$$\sigma_d = Y' \ln\left(\frac{t_o}{t_f}\right) = Y, \qquad \ln\left(\frac{t_o}{t_f}\right) = \frac{Y}{Y'} = \frac{\sqrt{3}}{2}, \qquad \text{and} \qquad \frac{t_o}{t_f} = e^{\sqrt{3}/2}.$$

It can then be shown that

$$\text{Maximum reduction per pass} = 1 - \frac{1}{e^{\sqrt{3}/2}} = 0.58 = 58\%. \tag{6.75}$$

b) **Drawing of tubes.** Tubes produced by extrusion or other processes can be reduced in thickness or diameter (**tube sinking**) by the tube-drawing processes illustrated in Fig. 6.67. A variety of mandrels are used to produce different shapes. Shape changes can also be imparted by using dies and mandrels with various profiles. Drawing forces, die pressures, and the maximum reduction per pass in tube drawing can be calculated by methods similar to those described for round rods.

### 6.5.2 Defects in drawing

Defects in drawing are similar to those observed in extrusion, especially center cracking. The factors influencing these internal cracks are the same, namely, that the tendency for cracking increases with increasing die angle, with decreasing reduction per pass, with friction, and with the presence of inclusions in the material.

Another type of defect is the formation of **seams**, which are longitudinal scratches or folds in the material. Such defects can open up during subsequent forming operations by upsetting, heading, thread rolling, or bending of the rod or wire. Various surface defects can also result from improper selection of process parameters and lubrication.

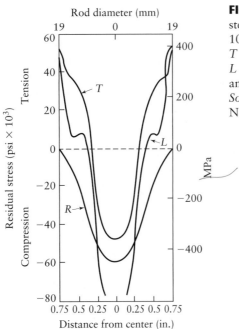

**FIGURE 6.68** Residual stesses in cold-drawn AISI 1045 carbon steel round rod: $T$ = transverse direction, $L$ = longitudinal direction, and $R$ = radial direction. *Source*: After E. S. Nachtman.

Because of inhomogeneous deformation, a cold-drawn rod, wire, or tube usually contains residual stresses. Typically, as shown in Fig. 6.68, a wide range of residual stresses is present within the rod in three principal directions. However, for very light reductions, the surface residual stresses are compressive. Light reductions are equivalent to shot peening or surface rolling, thus improving fatigue life. In addition to their effect on fatigue life, residual stresses can be significant in stress-corrosion cracking over a period of time and in warping of the component when a layer is subsequently removed, as by machining or grinding.

### 6.5.3  Drawing practice

Successful drawing operations require careful selection of process parameters and consideration of many factors. A typical die design for drawing, with its characteristic features, is shown in Fig. 6.69. Die angles usually range from 6° to 15°. The purpose of the land is to size and set the final diameter of the product. Also, when the die is reground after use, the land maintains the exit dimension of the die opening. Reductions in cross-sectional area per pass range from about 10% to 45%; usually, the smaller the cross-section, the smaller is the reduction per pass. Reductions per pass greater than 45% may result in breakdown of lubrication and deterioration of surface finish. Light reductions may also be made (**sizing pass**) on rods to improve surface finish and dimensional accuracy.

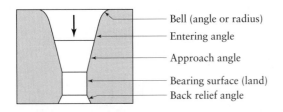

**FIGURE 6.69** Terminology for a typical die for drawing round rod or wire.

A rod or wire is fed into the die by first **pointing** it by swaging (i.e., forming the tip of the rod into a conical shape, as discussed in Section 6.6). After the rod or wire is placed in the die, the tip is clamped into the jaws of the wire-drawing machine, and the rod or wire is drawn continuously through the die. In most wire-drawing operations, the wire passes through a series of dies (**tandem drawing**). In order to avoid excessive tension in the exiting wire, it is wound one or two turns around a capstan between each pair of dies. The speed of the capstan is adjusted so that it supplies not only tension, but also a small back tension to the wire entering the next die. Back tension reduces the die pressure and extends die life.

Drawing speeds depend on the material and cross-sectional area. They may be as low as 0.15 m/s (30 ft/min) for heavy sections and as high as 50 m/s (10,000 ft/min) for very fine wire. The temperature can rise substantially at high drawing speeds.

Rods and tubes that are not sufficiently straight (or are supplied in a coiled form) can be straightened by passing them through pairs of rolls placed at different axes. The rolls subject the material to a series of bending and unbending operations, similar to the method shown in Fig. 6.41 for rolling.

Large cross-sections can be drawn at elevated temperatures. In cold drawing, because of strain hardening, intermediate annealing between passes may be necessary to maintain sufficient ductility. Steel wires for springs and musical instruments are made by a heat-treatment process that precedes or follows the drawing operation (**patenting**). These wires have ultimate tensile strengths as high as 4800 MPa (700,000 psi), with tensile reduction of area of about 20%.

**Bundle drawing.** In this process, many wires (as much as several thousand) are drawn simultaneously as a bundle. To prevent sticking, the wires are separated from each other by a suitable material. The cross-section of the wires is somewhat polygonal.

**Dies.** Die materials are usually alloy tool steels, carbides, or diamond. A diamond die (used for drawing fine wires) may be a single crystal or a polycrystalline diamond in a metal matrix. Carbide and diamond dies are made as inserts or nibs, which are then supported in a steel casing. Figure 6.70 shows a typical wear pattern on a drawing die. Note that die wear is highest at the entry. Although the die pressure is highest in this region and may be partially responsible for wear, other factors are involved, including variation in the diameter of the entering wire and vibration (thus subjecting the entry contact zone to fluctuating stresses), and the presence of abrasive scale on the surface of the entering wire.

In addition to rigid dies, a set of idling rolls is also used in drawing of rods or bars of various shapes. This arrangement is known as **Turk's head** and is more versatile than using ordinary dies, since the rolls can be adjusted to various positions for different products.

**Lubrication.** Proper lubrication is essential in rod, tube, and wire drawing, regardless of whether the process is dry or wet drawing. In **dry drawing**, the surface of the wire is coated with various lubricants, depending on its strength and frictional characteristics.

**FIGURE 6.70** Schematic illustration of a typical wear pattern in a wire-drawing die.

A common lubricant used is soap. The rod to be drawn is first surface treated by **pickling**. This treatment removes the surface scale that could lead to surface defects and, being quite abrasive, could considerably reduce die life. The soap is picked up by the wire as it goes through a box filled with soap powder (*stuffing box*).

With high-strength materials, such as steels, stainless steels, and high-temperature alloys, the surface of the rod or wire may be coated either with a softer metal or with a **conversion coating**. Copper or tin can be chemically deposited on the surface of the metal. This thin layer of softer metal acts as a solid lubricant during drawing. Conversion coatings may consist of sulfate or oxalate coatings on the rod, which typically are then coated with soap, as a lubricant. Polymers are also used as solid lubricants, such as in drawing of titanium. In **wet drawing**, the dies and rod are completely immersed in a lubricant. Typical lubricants include oils and emulsions containing fatty or chlorinated additives, and various chemical compounds.

The technique of **ultrasonic vibration** of the dies and mandrels is also used successfully in drawing. When carried out properly, this technique reduces drawing forces, improves surface finish and die life, and allows higher reductions per pass.

**Equipment.** Two types of equipment are used in drawing. A **draw bench** is similar to a long horizontal tensile-testing machine. It is used for single draws of straight rods with large cross-sections and tubes with lengths up to 30 m (100 ft) with a hydraulic or chain-drive mechanism. Rod and wire of smaller cross-section are drawn by a **bull block** (**capstan**), which is a rotating drum around which the wire is wrapped. The tension in this setup provides the force required to draw the wire.

## 6.6 | Swaging

In *swaging*, also known as **rotary swaging** or **radial forging**, a solid rod or a tube is reduced in diameter by the reciprocating radial movement of two or four dies (Fig. 6.71). The die movements are generally obtained by means of a set of rollers in a cage. The

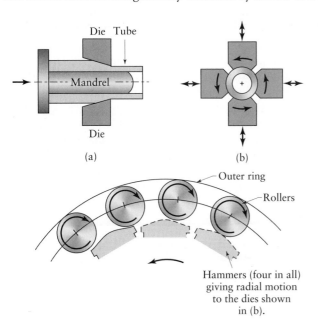

(a)

(b)

(c)

Outer ring

Rollers

Hammers (four in all)
giving radial motion
to the dies shown
in (b).

**FIGURE 6.71** Schematic illustration of the swaging process: (a) side view and (b) front view. (c) Schematic illustration of roller arrangement, curvature on the four radial hammers (that give motion to the dies), and the radial movement of a hammer as it rotates over the rolls.

**FIGURE 6.72** Reduction of outer and inner diameters of tubes by swaging. (a) Free sinking without a mandrel. The ends of solid bars and wire are tapered (pointing) by this process in order to feed the material into the conical die. (b) Sinking on a mandrel. Coaxial tubes of different materials can also be swaged in one operation.

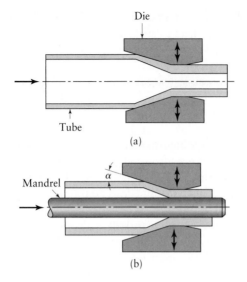

**FIGURE 6.73** (a) Typical cross-sections produced by swaging tube blanks with a constant wall thickness on shaped mandrels. Rifling of small gun barrels can also be made by swaging, using a specially shaped mandrel. The formed tube is then removed by slipping it out of the mandrel. (b) These parts can also be made by swaging.

internal diameter and the thickness of the tube can be controlled with or without mandrels, as shown in Fig. 6.72. Mandrels can also be made with longitudinal grooves (similar in appearance to a splined shaft); thus, internally shaped tubes can be swaged (Fig. 6.73a). The rifling in gun barrels is made by swaging a tube over a mandrel with spiral grooves. Externally shaped parts can also be made by swaging (Fig. 6.73b).

The swaging process is usually limited to workpiece diameters of about 50 mm (2 in.), although special machinery has been built to swage gun barrels of larger diameter. The length of the product in this process is limited only by the length of the mandrel (if needed). Die angles are usually only a few degrees and may be compound, that is, the die may have more than one angle, for more favorable material flow during swaging. Lubricants are used for improved surface finish and longer die life. The process is generally carried out at room temperature. Parts produced by swaging have improved mechanical properties and good dimensional accuracy.

## 6.7 | Die-Manufacturing Methods

Several manufacturing methods, either singly or in combination, are used in making dies for metal working operations, as well as for the molds used in the various processes described in Chapters 5, 7, 10, and 11. These processes include casting, forging, machining, grinding, and electrical and electrochemical methods of die sinking and finishing operations, such as honing, polishing, and coating. The selection of a die-manufacturing method depends primarily on the following parameters:

a) The particular process for which the die will be used;

b) The size and shape of the die;

c) The bulk and surface quality required;

d) Lead time to produce the die (e.g., some dies require months to prepare); and

e) Cost.

Economic considerations often dictate the process selected because tool and die costs can be significant in manufacturing operations. For example, the cost of a set of dies for pressworking automotive body panels (Figs. 1.5 and 7.68) may run up to $2 million; even small and simple dies can cost hundreds of dollars. On the other hand, because a large number of parts are typically produced by using the same die, the die cost per piece made is generally only a small fraction of a part's manufacturing cost (Chapter 16).

Dies may be classified as *male* and *female*; they may also be classified by their size. Small dies generally have a surface area of $10^3$–$10^4$ mm$^2$ (2–15 in.$^2$), whereas large dies have a surface area of 1 m$^2$ (9 ft$^2$) or larger. Dies of various sizes and shapes can be made from steels, cast irons, and nonferrous alloys, as well as carbides, ceramics, and diamond (Chapters 8 and 11).

The processes used for producing die materials typically include *casting*, followed by various primary processing (forging, rolling, and extrusion) and secondary processing (machining, grinding, polishing and coating). Small dies and tools can also be made by the *powder metallurgy* techniques described in Chapter 11. (See also *rapid tooling*, Section 10.12). Sand casting is used for large dies weighing many tons and shell molding for small dies. Cast steels are generally preferred as die materials, because of their strength and toughness and the ease with which their composition, grain size, and properties can be controlled and modified. Depending on how they solidify, cast dies, unlike dies made of wrought metals, may not have directional properties; thus, they can exhibit isotropic properties on all working surfaces. However, because of shrinkage, the control of dimensional accuracy in cast-metal dies can be difficult compared with that in machined and finished dies, thus requiring further processing.

Most commonly, dies are *machined* from forged *die blocks* by processes such as milling, turning, grinding, electrical and electrochemical machining, and polishing (Chapters 8 and 9). Traditionally, dies have been machined by milling on an automatic copy machine that traces a master model (*pattern*), typically made of wood, epoxy resins, aluminum, or gypsum. Increasingly, however, dies are now machined with *computer-controlled machine tools* and *machining centers* using various software packages (Fig. 1.8 and Sections 8.10 and 14.3). For improved production, the cutter tool path is optimized.

Conventional machining can, however, be difficult and time consuming for high-strength, hard, tough, and wear-resistant die materials. Consequently, advanced machining processes such as *chemical* and *electrical-discharge machining* are used extensively (Sections 9.10 through Section 9.14), particularly for small or medium-sized dies; more recent developments include *hard machining* of dies (*hard turning*, Section 8.8.2). These processes are generally faster and more economical than traditional machining, and the dies may not require additional finishing. However, one must always consider any possible adverse effects that these processes may have on the properties of the die (including its fatigue life) because of possible surface damage and cracks.

Small dies with shallow cavities may also be produced by *hubbing* (Section 6.2.4). Diamond dies for drawing fine wire are manufactured by producing holes with a thin rotating needle, coated with diamond powder suspended in oil.

To improve their hardness, wear resistance, and strength, die steels (Section 3.10.3) are usually *heat treated* (Section 5.11). Improper heat treatment, however, is one of the most common causes of die failure, as described in Section 6.8. Particularly important are the condition and composition of the die surfaces; thus, the proper selection of temperatures and atmospheres for heat treatment, quenching media, quenching practice, tempering procedures, and handling is important. Dies may also be subjected to various surface treatments, including *coatings* (Sections 4.5.1 and 8.6.5) for improved frictional and wear characteristics, as described in Chapter 4.

After heat treatment, tools and dies are subjected to operations, such as grinding and polishing, to obtain the desired surface finish and dimensional accuracy. Note that heat treatment may distort dies because of microstructural changes or uneven thermal cycling. Also, if not controlled properly, the grinding process can cause surface damage by emitting excessive heat and inducing detrimental tensile residual stresses on the surface of the die, thus reducing its fatigue life. Scratches on a die's structure can act as stress raisers as well, and commonly used die-making and finishing processes, such as electrical-discharge machining (Section 9.13), can cause surface damage and cracks, unless the process parameters are carefully controlled. Finish grinding, deburring, and polishing may be carried out, either by hand or with programmable *industrial robots* (Section 14.7).

## 6.8 | Die Failures

Failure of dies in metalworking operations generally results from one or more of the following causes:

a) Improper die design;

b) Defective die materials;

c) Improper heat treatment and finishing operations;

d) Improper installation, assembly, and alignment;

e) Overheating and heat checking (i.e., cracking caused by thermal cycling);

f) Excessive wear, and

g) Overloading, misuse, and improper handling.

The proper *design* of dies is as important as the appropriate selection of die materials. With rapid advances in *computer modeling and simulation* of processes and systems (Section 15.7), die design and optimization has now become an advanced technology. Basic design guidelines include the consideration that, in order to withstand the forces in manufacturing processes, a die must have appropriate cross sections and clearances. Sharp corners, radii, and fillets, as well as abrupt changes in cross section, act as stress raisers and can have detrimental effects on the life of the die. For improved strength, dies may be made in segments and may be prestressed during their assembly. Also, dies may be designed and constructed with inserts that can be replaced when worn or cracked. Overloading of dies can cause premature failure; for example, a common cause of failure of cold-extrusion dies (Section 6.4.3) is the operator's forgetting to remove an extruded part from the die before loading it with another blank. This type of failure can also occur in robots that automatically load and unload parts in dies.

In spite of their hardness and resistance to abrasion, die materials such as heat-treated steels, carbides, and diamond are susceptible to cracking and chipping from *impact forces* (as occur in mechanical presses and forging hammers) or from *thermal stresses* (caused by temperature gradients within the die in hot-working operations); thus, die surface preparation and finishing are important. Even metalworking fluids, often necessary, can adversely affect tool and die materials. For example, sulfur and chlorine additives in lubricants and coolants can attack the cobalt binder in tungsten carbide and lower its strength and toughness (*leaching*, Section 3.9.7).

During use, dies may also undergo *heat checking* from thermal cycling. To reduce heat checking (which has the appearance of parched land) and eventual breakage in hot-working operations, dies are usually preheated to temperatures of about 150 to 250°C (300 to 500°F). Cracked or worn dies may be repaired by welding and metal-deposition techniques. Even if they are manufactured properly, dies are subjected to stresses and temperature during their normal use, causing wear and, hence, changes in their shape and surface finish. When the shape of a die changes, the parts, in turn, have improper dimensions and surface finish; hence, the economics of the manufacturing operation are adversely affected.

Die failure and fracture in manufacturing plants can be hazardous to employees. It is not unusual for a set of dies resting on the plant floor or a shelf to disintegrate suddenly, because of their high internally stressed condition. The broken pieces are propelled at high speed and can cause serious injury or death. Highly stressed dies and tooling should always be surrounded by metal shielding, which should be properly designed and sufficiently strong to contain the fractured pieces in the event of sudden die failure.

---

**CASE STUDY** | Manufacturing Solid Rocket-Motor Case Segments for the Space Shuttle

The U.S. Space Shuttle is a very complex spacecraft that has required extensive engineering to design and manufacture. The components on this space-launch vehicle are highly specialized and are subjected to extreme conditions during its flight. To lift the orbiter into space, the Space Shuttle employs two large, strap-on cylindrical solid rocket boosters, as seen in Fig. 6.74, which shows the launching of the Space Shuttle *Atlantis*. Each booster provides a 3.3-million-lb thrust force for the initial liftoff and assent. The rocket boosters consist of steel case segments (Fig. 6.75), which contain the solid rocket-fuel propellant and act as very large pressure vessels during launch.

The manufacture of the steel case segments is very challenging and requires a series of special processes and equipment (Fig. 6.76). The case segments are each approximately 3 m (120 in.) in diameter, 3 m (120 in.) long, and less than 13 mm (0.5 in.) thick; the dimensional tolerance on thickness is 0.25 mm (0.010 in.). A very thin and tightly controlled wall section is required for this pressure-vessel application. However, the case segments are so large that the weight of the final component is very close to the weight limit for the melting and casting of specialty-grade steel into ingots to be used as starting stock (blanks). Thus, it is imperative that very little material be wasted or discarded during the manufacturing process.

The mechanical-property requirements for these components dictate that ultraclean steel be used to eliminate melt-related material defects. (See Sections 5.3.3 and 5.5 and Fig. 3.24.) Ladish D6AC steel, with an ultimate tensile strength of 1500–1700 MPa (220–250 ksi), is first cast into large premium-quality vacuum-melted ingots. Because the weight of the steel ingots is so close to the weight of the finished components, material-efficient manufacturing steps have to be incorporated. To transform the cast ingot into final cylinder components, several different manufacturing steps are used, namely, (a) upsetting, (b) extrusion, (c) ring

**FIGURE 6.74**  The Space Shuttle *Atlantis* is launched by two strap-on solid rocket boosters. *Source*: NASA.

**FIGURE 6.75**  Assembly of steel case segments to form a solid rocket booster. Note that most of the rocket casing is below the platform level. *Source*: NASA.

(a) As cast blank

Sheared section
(b) Reverse extruded

(c) Ring rolled

(d) Roll-formed casing

**FIGURE 6.76**  The forming processes involved in the manufacture of solid rocket casings for the U.S. Space Shuttle.

rolling, (d) roll forming, (e) machining, and (f) heat treating. Figure 6.76 shows a schematic of the forming steps used to make a case segment.

Indirect, or backward, extrusion (Section 6.4) is used first to produce a rough cylindrical shape for the case segments from the cast ingot. In this process, the ingot is upset (see Fig. 6.1), and then a large central punch is pressed into the upset steel blank. As the punch presses into the center of the heated metal, the metal is extruded backward between the moving punch and the container. The rough cylinder is removed from the container, and the thin slug of material that remains underneath the punch at the end of the operation is removed by hot shearing.

The hot cylinder is then placed on a large ring-rolling machine, where it is ring rolled (see Fig. 6.45a) to achieve the proper shape; this process reduces the thickness of the wall section of the cylinder while increasing its diameter. The length of the cylinder is maintained by containment or pinch rollers on the end faces. The rolled cylinder has a contour that contains the end flanges and the central pressure vessel. After this operation is completed, the cylinder is machined to remove all scale (from hot working) and to produce a smooth, uniform, bright surface finish. This step is designed to remove a minimum amount of material in order to produce a shape that can be ready for the final forming operation.

The machined cylinder is further processed by roll forming (also known as *flow forming* or *shear forming*, similar to spinning; see Section 7.8) to produce the final pressure-vessel walls. This process is performed at room temperature and by means of two opposing rollers, whereby the wall section is reduced, the diameter remains constant, and the length of the cylinder is increased. The cylinder wall is thus cold reduced to a net final thickness; it is then heat treated by normalizing, austenitizing, quenching, and tempering. (See Section 5.11.) Only the end flanges of the case segments require some final machining for proper assembly.

The aforementioned manufacturing steps have been optimized for the special requirements and constraints of this application, including (a) a clean, high-tensile-strength material; (b) a defect-free and inspectable component geometry; (c) a minimal loss of input material; and (d) a stable, robust, and reproducible process.

*Source*: D. Furrer, Ladish Co., Inc.

# SUMMARY

- Bulk deformation processes, which generally consist of forging, rolling, extrusion, and drawing, involve major changes in the dimensions of the workpiece. Bulk properties as well as surface characteristics of the material are important. (Section 6.1)

- Forging denotes a family of processes by which deformation of the workpiece is carried out by compressive forces applied through dies. Forging is capable of producing a wide variety of parts with favorable characteristics of strength, roughness, dimensional accuracy, and reliability in service. Among important factors in forging are friction, material flow in dies, heat transfer, and lubrication. Various defects can develop, depending on die geometry, quality of the billet materials, and preform. Several forging machines are available, with various characteristics and capabilities. (Section 6.2)

- Currently used methods of analysis of the stresses, strains, strain rates, temperature distribution, and loads in forging and other bulk-deformation processes include the slab method and the finite-element method. (Section 6.2)

- Rolling is the process of reducing the thickness or changing the cross-section of a long workpiece by compressive forces applied through a set of rolls. It is a continuous operation and involves several material and process variables, including the size of the roll relative to the thickness of the material, the amount of reduction per pass, speed, lubrication, and temperature. Rolled products include plate, sheet, foil, rod, pipe, and tube, as well as shape-rolled products such as I-beams, structural shapes, rails, and bars of various cross-sections. (Section 6.3)

- Extrusion is the process of forcing a material through the opening of a die, much like squeezing toothpaste from a tube. This process is capable of producing lengths of solid or hollow sections with constant cross-sectional area. Important factors include die design, extrusion ratio, lubrication, billet temperature, and extrusion speed. Cold extrusion, which is a combination of extrusion and forging operations, is capable of economically producing parts with good mechanical properties. (Section 6.4)

- Rod, wire, and tube drawing involve pulling the material through one or more dies. Proper die design and appropriate selection of materials and lubricants are essential to obtaining products of high quality and with good surface finish and product integrity. The major process variables in the mechanics of drawing are die angle, friction, and amount of reduction per pass. (Section 6.5)

- In swaging, a solid rod or a tube is reduced in diameter by the reciprocating radial movement of two or four dies. This process is suitable for producing short or long lengths of bar or tubing with various internal or external profiles. (Section 6.6)

- Because die failure has a major economic impact, die design, material selection, and the manufacturing methods employed are of major importance. A wide variety of die materials and manufacturing methods is available, including advanced material-removal processes and subsequent treatment, coating, and finishing operations. (Sections 6.7, 6.8)

# SUMMARY OF EQUATIONS

### Forging

Pressure in plane-strain compression, $p = Y'e^{2\mu(a-x)/h}$

Average pressure in plane-strain compression, $p_{av} \simeq Y'\left(1 + \dfrac{\mu a}{h}\right)$

Pressure in axisymmetric compression, $p = Ye^{2\mu(r-x)/h}$

Average pressure in axisymmetric compression, $p_{av} \simeq Y\left(1 + \dfrac{2\mu r}{3h}\right)$

Average pressure in plane-strain compression, sticking, $p_{av} = Y'\left(1 + \dfrac{a-x}{h}\right)$

Pressure in axisymmetric compression, sticking, $p = Y\left(1 + \dfrac{r-x}{h}\right)$

### Rolling

Roll pressure in entry zone, $p = Y'_f\dfrac{h}{h_0}e^{\mu(H_0-H)}$

with back tension, $p = (Y'_f - \sigma_b)\dfrac{h}{h_0}e^{\mu(H_0-H)}$

Roll pressure in exit zone, $p = Y'_f\dfrac{h}{h_f}e^{\mu H}$

with front tension, $p = (Y'_f - \sigma_f)\dfrac{h}{h_f}e^{\mu H}$

Parameter $H = 2\sqrt{\dfrac{R}{h_f}}\tan^{-1}\left(\sqrt{\dfrac{R}{h_f}}\phi\right)$

Roll force, $F = \displaystyle\int_0^{\phi_n} wpR\,d\phi + \int_{\phi_n}^{\alpha} wpR\,d\phi$

Roll force, approximate, $F = Lw\bar{Y}'\left(1 + \dfrac{\mu L}{2h_{av}}\right)$

Roll torque, $T = \displaystyle\int_{\phi_n}^{\alpha} w\mu pR^2\,d\phi - \int_0^{\phi_n} w\mu pR^2\,d\phi$

Roll contact length, approximate, $L = \sqrt{R\Delta h}$

Power per roll $= \dfrac{\pi FLN}{60{,}000 \text{ kW}} = \dfrac{\pi FLN}{33{,}000 \text{ hp}}$

Maximum draft, $\Delta h_{max} = \mu^2 R$

Maximum angle of acceptance, $\alpha_{max} = \tan^{-1}\mu$.

### Extrusion

Extrusion ratio, $R = A_o/A_f$

Extrusion pressure, ideal, $p = Y\ln R$

Extrusion pressure, with friction, $p = Y\left(1 + \dfrac{\tan\alpha}{\mu}\right)[R^{\mu\cot\alpha} - 1]$

### Rod and wire drawing

Drawing stress, ideal, $\sigma_d = Y\ln\left(\dfrac{A_o}{A_f}\right)$

Drawing stress, with friction, $\sigma_d = Y\left(1 + \dfrac{\tan\alpha}{\mu}\right)\left[1 - \left(\dfrac{A_f}{A_o}\right)^{\mu\cot\alpha}\right]$

Die pressure, $p = Y_f - \sigma$

# BIBLIOGRAPHY

Altan, T., S.-I. Oh, and H. Gegel, *Metal Forming—Fundamentals and Applications*, ASM International, 1983.

*ASM Handbook*, Vol. 14: *Forming and Forging*, ASM International, 1988.

Avitzur, B., *Handbook of Metalforming Processes*, Wiley-Interscience, 1983.

Blazynski, T. Z., *Plasticity and Modern Metal-Forming Technology*, Elsevier, 1989.

Boer, C. R., N. Rebelo, H. Rydstad, and G. Schroder, *Process Modelling of Metal Forming and Thermomechanical Treatment*, Springer, 1986.

Byrer, T. G. (ed.), *Forging Handbook*, ASM International, 1985.

Davis, J. R. (ed.), *Tool Materials*, ASM International, 1995.

Dieter, G. E., *Mechanical Metallurgy*, 3d. ed., McGraw-Hill, 1986.

—— (ed.), *Workability Testing Techniques*, ASM International, 1984.

*Forging Design Handbook*, ASM International, 1972.

Frost, H. J., and M. F. Ashby, *Deformation-Mechanism Maps*, Pergamon, 1982.

Ginzburg, V. B., *High-Quality Steel Rolling: Theory and Practice*, Dekker,1993.

——, *Steel-Rolling Technology: Theory and Practice*, Dekker, 1989.

Hoffman, H. (ed.), *Metal Forming Handbook*, Springer Verlag, 1998.

Hosford, W. F., and R. M. Caddell, *Metal Forming, Mechanics and Metallurgy*, 2d. ed., Upper Saddle River, NJ: Prentice Hall, 1993.

Inoue, N., and M. Nishihara (eds.), *Hydrostatic Extrusion: Theory and Applications*, Elsevier, 1985.

Kobayashi, S., S.-I. Oh, and T. Altan, *Metal Forming and the Finite-Element Method*, Oxford, 1989.

Lange, K. (ed.), *Handbook of Metal Forming*, McGraw-Hill, 1985.

Laue, K., and H. Stenger, *Extrusion—Processes, Machinery, Tooling*, ASM International, 1981.

Lenard, J. G., M. Pietrzyk, and L. Cser, *Mathematical and Physical Simulation of the Properties of Hot Rolled Products*, Elsevier, 1999.

Mielnik, E. M., *Metalworking Science and Engineering*, McGraw-Hill, 1991.

Nachtman, E. S., and S. Kalpakjian, *Lubricants and Lubrication in Metalworking Operations*, Dekker, 1985.

Pietrzyk, M., and J. G. Lenoard, *Thermal-Mechanical Modelling of the Flat Rolling Process*, Springer, 1991.

Prasad, Y. V. R. K., and S. Sasidhara (eds.), *Hot Working Guide: A Compendium of Processing Maps*, ASM International, 1997.

*Product Design Guide for Forging*, Forging Industry Association, 1997.

Roberts, W. L., *Cold Rolling of Steel*, Dekker, 1978.

——, *Hot Rolling of Steel*, Dekker, 1983.

Sabroff, A. M., F. W. Boulger, and H. J. Henning, *Forging Materials and Practices*, Reinhold, 1968.

Saha, P. K., *Aluminum Extrusion Technology*, ASM International, 2000.

Sheppard, T., *Extrusion of Aluminum Alloys*, Chapman & Hall, 1998.

*Tool and Manufacturing Engineers Handbook*, Vol. II: *Forming*, Society of Manufacturing Engineers, 1984.

Wagoner, R. H., and Chenot, J. L., *Fundamentals of Metal Forming*, Wiley, 1996.

——, *Metal Forming Analysis*, Cambridge, 2001.

# QUESTIONS

## Forging

**6.1**   How can you tell whether a certain part is forged or cast? Describe the features that you would investigate to arrive at a conclusion.

**6.2**   Why is the control of volume of the blank important in closed-die forging?

**6.3**   What are the advantages and limitations of a cogging operation? Of die inserts in forging?

**6.4**   Explain why there are so many different kinds of forging machines.

**6.5**   Devise an experimental method whereby you can measure the force required for forging only the flash in impression-die forging. (See Fig. 6.15a.)

**6.6**   A manufacturer is successfully hot forging a certain part, using material supplied by Company *A*. A new supply of material is obtained from Company *B*, with the same nominal composition of the major alloying elements as that of the material from Company *A*. However, it is found that the new forgings are cracking even though the same procedure is followed as before. What is the probable reason?

**6.7**   Explain why there might be a change in the density of a forged product as compared to that of the blank.

**6.8**   Describe the role of surface oxide layers on a blank as it is being deformed by impression-die forging.

**6.9**   Since glass is a good lubricant for hot extrusion, would you use glass for impression-die forging as well? Explain.

**6.10**   Describe and explain the factors that influence spread in cogging operations on square billets.

**6.11**   In deriving the formula for forging, we had to use the yield criteria described in Chapter 2. Explain why.

**6.12**   Why are end grains generally undesirable in forged products? Give examples of such products.

**6.13**   Explain why one cannot obtain a finished forging in one press stroke, starting with a blank.

**6.14**   List the advantages and disadvantages of using a lubricant in forging.

## Rolling

**6.15**   Explain how a cast structure is converted into a wrought structure by hot rolling.

**6.16**   As stated in Section 6.3.1, three factors that influence spreading in rolling are the width-to-thickness ratio of the strip, friction, and the ratio of the radius of the roll to the thickness of the strip. Explain how these factors affect spreading.

**6.17**   Explain how the residual stress patterns in Fig. 6.40 become reversed when the roll radius or reduction per pass is changed.

**6.18**   Explain how you would go about applying front and back tensions to sheet metals during rolling.

**6.19**   It was noted that rolls tend to flatten under roll forces. Which property(ies) of the roll material can be increased to reduce flattening?

**6.20**   Describe the methods by which roll flattening can be reduced.

**6.21**   Explain the technical and economic reasons for taking larger rather than smaller reductions per pass in flat rolling.

**6.22**   Surface roughness in hot-rolled products is higher than in cold-rolled products. Explain why.

**6.23**   List and explain the methods that can be used to reduce the roll force.

**6.24**   Explain the advantages and limitations of using small-diameter rolls in flat rolling.

**6.25**  Is it possible to adjust back and front tensions simultaneously so that the neutral point remains in the same position as in ordinary rolling without tensions? Explain.

**6.26**  A ring-rolling operation is successful for the production of bearing races. However, when the bearing race diameter is changed, the operation results in very poor surface finish. List the possible causes and the investigation you would perform to identify the parameters involved and correct the problem.

**6.27**  Describe the importance of controlling roll speed, roll gap, temperature, and other process variables in a tandem-rolling operation.

**6.28**  Is it possible to have a negative forward slip? Explain.

**6.29**  Rolling reduces the thickness of plates and sheets. However, it is also possible to reduce the thickness by simply stretching the material. Would this process be feasible? Explain.

**6.30**  In Fig. 6.34, explain why the neutral point moves towards the roll-gap entry as friction increases.

## Extrusion

**6.31**  The extrusion ratio, die geometry, extrusion speed, and billet temperature all affect the extrusion pressure. Explain why.

**6.32**  How would you go about avoiding centerburst defects in extrusion? Explain why your methods would be effective.

**6.33**  Assume that you are reducing the diameter of two round rods, one by simple tension and the other by indirect extrusion. Which method will require more force? Why?

**6.34**  How would you go about making a stepped extrusion that has increasingly larger cross-sections along its length? Is it possible? Would your process be economical and suitable for high production runs? Explain.

**6.35**  It has been stated in this chapter that the temperature ranges for extruding various metals are similar to those for forging. (See Table 6.3.) Describe the consequences of extruding at a temperature (a) below and (b) above these ranges.

**6.36**  Note from Eq. (6.52) that, for low values of the extrusion ratio, such as $R = 2$, the ideal extrusion pressure $p$ can be lower than the yield stress, $Y$, of the material. Explain whether or not this phenomenon is logical.

**6.37**  Explain why the three variables in Eq. (6.59) should influence the strain rate in extrusion.

**6.38**  In hydrostatic extrusion, complex seals are used between the ram and the container, but not between the extrusion and the die. Explain why.

## Drawing

**6.39**  What changes would you expect in the strength, hardness, ductility, and anisotropy of annealed metals after they have been cold drawn through dies? Why?

**6.40**  We have seen that in rod and wire drawing, the maximum die pressure is at the die entry. Why?

**6.41**  Describe the conditions under which wet drawing and dry drawing, respectively, are desirable.

**6.42**  Name the important process variables in drawing, and explain how they affect the drawing process.

**6.43**  Assume that a rod-drawing operation can be carried out either in one pass or in two passes in tandem. If all die angles are the same and the total reduction is the same, will the drawing forces be different? Explain.

**6.44**  In Fig. 6.62, assume that the reduction in the cross-section is taking place by pushing the rod through the die instead of pulling it. Assuming that the material is perfectly plastic, sketch the die-pressure distribution, for the following situations: (a) frictionless; (b) with friction; (c) frictionless, but with front tension. Explain your answers.

**6.45**  In deriving Eq. (6.72), no mention is made of the ductility of the original material being drawn. Explain why.

**6.46**  Explain why the die pressure in drawing decreases toward the exit of the die.

**6.47**  What is the value of the die pressure at the exit when a drawing operation is being carried out at the maximum reduction per pass?

**6.48**  Explain why the maximum reduction per pass in drawing should increase as the strain-hardening exponent, $n$, increases.

**6.49**  Explain why the maximum reduction per pass is higher in drawing round bars (63%) as compared with plane-strain drawing of flat sheet or plate (58%).

**6.50**  If, in deriving Eq. (6.72), we include friction, will the maximum reduction per pass be the same (63%), higher, or lower? Explain.

**6.51**  Explain what effects back tension has on the die pressure in wire or rod drawing, and discuss why these effects occur.

**6.52**  Explain why the inhomogeneity factor $\phi$ in rod and wire drawing depends on the ratio $h/L$, as plotted in Fig. 6.12.

**6.53**  Describe the reasons for the development of the swaging process.

**6.54**  Occasionally, steel-wire drawing will take place within a sheath of a soft metal such as copper or lead. Why would this procedure be useful?

**6.55**  Recognizing that it is very difficult to manufacture a die with a submillimeter diameter, how would you produce a 10-$\mu$m-diameter wire?

## General

**6.56**  With respect to the topics covered in this chapter, list and explain specifically two examples each where (a) friction is desirable and (b) friction is not desirable.

**6.57**  Take any three topics from Chapter 2 and, with a specific example for each, show their relevance to the topics covered in this chapter.

**6.58**  Take any three topics from Chapter 3 and, with a specific example for each, show their relevance to the topics covered in this chapter.

**6.59**  List and explain the reasons that there are so many different types of die materials used for the processes described in this chapter.

**6.60**  Explain the reasons that we should be interested in residual stresses developed in workpieces made by the forming processes described in this chapter.

**6.61**  Make a summary of the types of defects found in the processes described in this chapter. Indicate, for each type, methods of reducing or eliminating the defects.

# PROBLEMS

## Forging

**6.62**  In the free-body diagram in Fig. 6.4b, we have shown the incremental stress $d\sigma_x$ on the element, pointing to the left. Yet it would appear that, because of the direction of frictional stresses $\mu p$, the incremental stress should point to the right in order to balance the horizontal forces. Show that the same answer for the forging pressure is obtained regardless of the direction of this incremental stress.

**6.63**  Plot the force vs. reduction in height curve in open-die forging of a cylindrical, annealed copper specimen 2 in. high and 1 in. in diameter, up to a reduction of 70%, for the cases of (a) no friction between the flat dies and the specimen, (b) $\mu = 0.2$, and (c) $\mu = 0.4$. Ignore barreling. Use average-pressure formulas.

**6.64**  Use Fig. 6.9b to provide the answers to Problem 6.63.

**6.65**  Calculate the work done for each case in Problem 6.63.

**6.66**  Determine the temperature rise in the specimen for each case in Problem 6.63, assuming that the process is adiabatic and the temperature is uniform throughout the specimen.

**6.67**  To determine forgeability, a hot-twist test is performed on a round bar 30 mm in diameter and 300 mm long. It is found that bar underwent 200 turns before it fractured. Calculate the shear strain at the outer surface of the bar at fracture.

**6.68**  Derive an expression for the average pressure in plane-strain compression with sticking friction.

**6.69**  What is the value of $\mu$ when, for plane-strain compression, the forging load with sliding friction is equal to the load with sticking friction? Use average-pressure formulas.

**6.70**  Note that in cylindrical upsetting, the frictional stress cannot be greater than the shear yield stress $k$ of the material. Thus, there may be a distance $x$ in Fig. 6.8 where a transition occurs from sliding to sticking friction. Derive an expression for $x$ in terms of $r$, $h$, and $\mu$ only.

**6.71**  Assume that a workpiece is being pushed to the right by a lateral force $F$ while being compressed between flat dies. (See the accompanying figure.) (a) Make a sketch of the die-pressure distribution for the condition for which $F$ is not large enough to slide the workpiece to the right. (b) Make a similar sketch, except with $F$ now being large enough so that the workpiece slides to the right while being compressed.

**6.72**  For the sticking example in Fig. 6.10, derive an expression for the lateral force $F$ required to slide the workpiece to the right while the workpiece is being compressed between flat dies.

**6.73**  Two solid cylindrical specimens $A$ and $B$, made of a perfectly plastic material, are being forged with friction and isothermally at room temperature to a reduction in height of 50%. Specimen $A$ has a height of 2 in. and a cross-sectional area of 1 in$^2$, and specimen $B$ has a height of is 1 in. and a cross-sectional area of 2 in$^2$. Will the work done be the same for the two specimens? Explain.

**6.74** In Fig. 6.6, does the pressure distribution along the four edges of the workpiece depend on the particular yield criterion used? Explain.

**6.75** Under what conditions would you have a normal pressure distribution in forging a workpiece as shown in the accompanying figure? Explain.

**6.76** Derive the average die-pressure formula given by Eq. (6.15). (*Hint*: Obtain the volume under the friction hill over the surface by integration, and divide that quantity by the cross-sectional area of the workpiece.)

**6.77** Take two solid cylindrical specimens of equal diameter, but different heights, and compress them (frictionless) to the same percent reduction in height. Show that the final diameters will be the same.

**6.78** A rectangular workpiece has the following original dimensions: $2a = 120$ mm, $h = 30$ mm, and width $= 20$ mm. The metal has a strength coefficient of 400 MPa and a strain-hardening exponent of 0.3. It is being forged in plane strain with $\mu = 0.2$. Calculate the force required for the height to be reduced by 20%. Do not use average-pressure formulas.

**6.79** Assume that in upsetting a solid cylindrical specimen between two flat dies with friction, the dies are rotated at opposite directions to each other. How, if at all, will the forging force change from that for nonrotating dies? (*Hint*: Note that the dies will now require torque, because of the change in the direction of frictional forces at the die–workpiece interfaces.)

**6.80** A solid cylindrical specimen, made of a perfectly plastic material, is being upset between flat dies with no friction. The process is being carried out by a falling weight, as in a drop hammer. The downward velocity of the hammer is at a maximum when it first contacts the workpiece and becomes zero when the hammer stops at a certain height of the specimen. Establish quantitative relationships between workpiece height and velocity, and make a qualitative sketch of the velocity profile of the hammer. (*Hint*: The loss in the kinetic energy of the hammer is the plastic work of deformation; thus, there is a direct relationship between workpiece height and velocity.)

**6.81** How would you go about estimating the swaging force acting on each die?

**6.82** We note that in swaging a tube, the thickness of the tube increases. How would you go about calculating the thickness change without measuring it? You are allowed to make any other measurements.

**6.83** A mechanical press is powered by a 30-hp motor and operates at 40 strokes per minute. It uses a flywheel, so that the rotational speed of the crankshaft does not vary appreciably during the stroke. If the stroke length is 6 in., what is the maximum contact force that can be exerted over the entire stroke length? To what height can a 5052-O aluminum cylinder with a diameter of 1 in. and a height of 2 in. be forged before the press stalls?

**6.84** Estimate the force required to upset a 0.125-in.-diameter C74500 brass rivet in order to form a 0.25-in.-diameter head. Assume that the coefficient of friction between the brass and the tool-steel die is 0.25 and that the head is 0.125 in. in thickness.

**6.85** Using the slab method of analysis, derive Eq. (6.17).

## Rolling

**6.86** In Example 6.4, what is the velocity of the strip leaving the rolls?

**6.87** With appropriate sketches, explain the changes that occur in the roll-pressure distribution if one of the rolls is idling, i.e., power is shut off to that roll.

**6.88** It can be shown that it is possible to determine $\mu$ in flat rolling without measuring torque or forces. By inspecting equations for rolling, describe an experimental procedure to do so. Note that you are allowed to measure any quantity other than torque or forces.

**6.89** Derive a relationship between back tension $\sigma_b$ and front tension $\sigma_f$ in rolling such that when both tensions are increased, the neutral point remains in the same position.

**6.90** Take an element at the center of the deformation zone in flat rolling. Assuming that all the stresses acting on this element are principal stresses, indicate the stresses qualitatively, and state whether they are tension or compression stresses. Explain the reasoning behind your devices. Is it possible for all three principal stresses to be equal to each other in magnitude? Explain.

**6.91** It has been stated that in rolling a flat strip, the roll force is reduced about twice as effectively by back tension as it is by front tension. Explain this difference, using appropriate sketches. (*Hint*: Note the position of the neutral point.)

**6.92** It can be seen that in rolling a strip, the rolls will begin to slip if the back tension, $\sigma_b$, is too high. Derive an analytical expression for the magnitude of the back tension in order to make the powered rolls begin to slip. Use the same terminology as applied in the text.

**6.93**    Prove Eq. (6.44).

**6.94**    In Steckel rolling, the rolls are idling, and thus there is no net torque, assuming frictionless bearings. Where, then, is the energy coming from to supply the work of deformation in rolling? Explain with appropriate sketches, and state the conditions that have to be satisfied.

**6.95**    Derive an expression for the tension required in Steckel rolling of a flat sheet, without friction, for a workpiece whose true-stress–true-strain curve is given by $\sigma = a + b\varepsilon$.

**6.96**    (a) Make a neat sketch of the roll-pressure distribution in ordinary rolling with powered rolls. (b) Assume now that the power to both rolls is shut off and that rolling is taking place by front tension only, i.e., Steckel rolling. Superimpose on your diagram the new roll-pressure distribution, explaining your reasoning clearly. (c) After completing part (b), further assume that the roll bearings are getting rusty and deprived of lubrication while rolling is still taking place by front tension only. Superimpose a third roll-pressure distribution diagram for this condition, explaining your reasoning.

**6.97**    Derive Eq. (6.26b), based on the equation preceding it. Comment on how different the $h$ values are as the angle $\phi$ increases.

**6.98**    In Fig. 6.35, assume that $L = 2L_2$. Is the roll force, $F$, for $L$ twice or more than twice the roll force for $L_2$? Explain.

**6.99**    A flat-rolling operation is being carried out where $h_o = 0.15$ in., $h_f = 0.100$ in., $w_o = 10$ in., $R = 10$ in., $\mu = 0.2$, and the average flow stress of the material is 40,000 psi. Estimate the roll force and the torque. Include the effects of roll flattening.

**6.100**    A rolling operation takes place under the conditions shown in the accompanying figure. What is the position $x_n$ of the neutral point? Note that there are a front and back tension that have not been specified. Additional data are as follows: Material is 5052-O aluminum; $\mu = 0.1$; hardened steel rolls; surface roughness of the rolls $= 0.02$ $\mu$m; rolling temperature $= 210°C$.

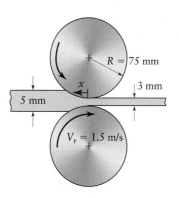

**6.101**    Estimate the roll force and power for annealed low carbon steel strip 400 mm wide and 10 mm thick, rolled to a thickness of 7 mm. The roll radius is 200 mm, and the roll rotates at 200 rpm. Use $\mu = 0.1$.

## Extrusion

**6.102**    Derive an expression for the true-strain rate, $\dot{\varepsilon}$, in the deformation zone in extrusion of a square billet, in terms of the entering velocity $V_o$, the die angle $\alpha$, an initial billet dimension $t_o$, and the distance $x$ from entry.

**6.103**    Calculate the force required in direct extrusion of 1100-O aluminum from a diameter of 8 in. to a diameter of 2 in. Assume that the redundant work is 30% of the ideal work of deformation, and the friction work is 25% of the total work of deformation.

**6.104**    Prove Eq. (6.56).

**6.105**    Calculate the theoretical temperature rise in the extruded material in Example 6.6, assuming that there is no heat loss. (See Section 3.9 for information on the physical properties of the material.)

**6.106**    Using the same approach as that shown in Section 6.5 for wire drawing, show that the extrusion pressure is given by the expression

$$p = Y\left(1 + \frac{\tan \alpha}{\mu}\right)\left[1 - \left(\frac{A_o}{A_f}\right)^{\mu \cot \alpha}\right],$$

where $A_o$ and $A_f$ are the original and final workpiece diameters, respectively.

**6.107**    Derive Eq. (6.54).

**6.108**    A planned extrusion operation involves steel at 800°C, with an initial diameter of 100 mm and a final diameter of 20 mm. Two presses, one with a capacity of 20 MN and the other of 10 MN, are available for this operation. Obviously, the larger press requires greater care and more expensive tooling. Is the smaller press sufficient for the operation? If not, what recommendations would you make to allow use of the smaller press?

## Drawing

**6.109**    Calculate the power required in Example 6.7 if the workpiece material is annealed 70–30 brass.

**6.110**    Using Eq. (6.61), make a plot similar to Fig. 6.66 for the following conditions: $K = 75$ MPa; $n = 0.3$; $\mu = 0.04$.

**6.111**    Using the same approach as that described in Section 6.5 for wire drawing, show that the drawing stress, $\sigma_d$, in plane-strain drawing of a flat sheet or plate is given by the expression

$$\sigma_d = Y'\left(1 + \frac{\tan \alpha}{\mu}\right)\left[1 - \left(\frac{h_f}{h_o}\right)^{\mu \cot \alpha}\right],$$

where $h_o$ and $h_f$ are the original and final thickness, respectively, of the workpiece.

**6.112**   Derive an analytical expression for the die pressure in wire drawing, without friction or redundant work, as a function of the instantaneous diameter in the deformation zone.

**6.113**   A linearly strain-hardening material with a true-stress–true-strain curve $\sigma = 5,000 + 25,000\varepsilon$ psi is being drawn into a wire. If the original diameter of the wire is 0.2 in., what is the minimum possible diameter at the exit of the die? Assume that there is no redundant work and that the frictional work is 25% of the ideal work of deformation. (*Hint*: The yield stress of the exiting wire is the point on the true-stress–true-strain curve that corresponds to the total strain that the material has undergone.)

**6.114**   In Fig. 6.68, assume that the longitudinal residual stress at the center is −60,000 psi. Using the distortion-energy criterion, calculate the minimum yield stress that this particular steel must have in order to sustain these levels of residual stresses.

**6.115**   What is the answer to Problem 6.114 for the maximum-shear-stress criterion?

**6.116**   Derive an expression for the die-separating force in frictionless wire drawing of a perfectly plastic material. Use the same terminology as in the text.

**6.117**   A material with a true-stress–true-strain curve $\sigma = 10,000\varepsilon^{0.3}$ is used in wire drawing. Assuming that the friction and redundant work compose a total of 60% of the ideal work of deformation, calculate the maximum reduction in cross-sectional area per pass that is possible.

**6.118**   Derive an expression for the maximum reduction per pass for a material of $\sigma = K\varepsilon^n$, assuming that the friction and redundant work contribute a total of 20% to the ideal work of deformation.

**6.119**   Prove that the true-strain rate $\dot{\varepsilon}$ in drawing or extrusion in plane strain with wedge-shaped dies is given by the expression

$$\dot{\varepsilon} = -\frac{2 \tan \alpha V_o t_o}{(t_o - 2x \tan \alpha)^2},$$

where $\alpha$ is the die angle, $t_o$ is the original thickness, and $x$ is the distance from die entry (*Hint*: Note that $d\varepsilon = dA/A$.)

**6.120**   In drawing a strain-hardening material with $n = 0.3$, what should be the percentage of friction plus redundant work, in terms of ideal work, so that the maximum reduction per pass is 63%?

**6.121**   Assume that you are asked to give a quiz to students on the contents of this chapter. Prepare three quantitative problems and three qualitative questions, and supply the answers.

# DESIGN

**6.122**   Forging is one method of producing turbine blades for jet engines. Study the design of such blades and the relevant technical literature, and then prepare a step-by-step procedure for making the blades. Comment on the potential difficulties that may be encountered.

**6.123**   In comparing forged parts with cast parts, we have noted that the same part may be made by either process. Comment on the pros and cons of each process, considering factors such as part size, shape complexity, and design flexibility if a particular design has to be modified.

**6.124**   Review the technical literature, and make a detailed list of the manufacturing steps involved in the manufacture of hypodermic needles.

**6.125**   Figure 6.50a shows examples of products that can be obtained by slicing long extruded sections into small discrete pieces. Name several other products that can be made in a similar manner.

**6.126**   Make an extensive list of products that either are made of or have one or more components of (a) wire, (b) very thin wire, and (c) rods of various cross-sections.

**6.127**   Although extruded products typically are straight, it is possible to design dies whereby the extrudate has a constant radius of curvature. (a) What applications could you think of for such products? (b) Describe your ideas as to the shape that such a die should have.

**6.128**   Survey the technical literature, and describe the design features of the various roll arrangements shown in Fig. 6.42.

**6.129**   We have indicated the beneficial effects of using ultrasonic vibration to reduce friction in some of the processes described in this chapter. Survey the technical literature, and offer design concepts to apply such vibrations.

# CHAPTER 7

# Sheet-Metal Forming Processes

## 7.1 | Introduction

It was noted in Chapter 6 that workpieces can be characterized by the ratio of their surface area to their volume or thickness. **Sheet forming**, unlike bulk-deformation processes, involves workpieces with a high ratio of surface area to thickness. A sheet thicker than 6 mm ($\frac{1}{4}$ in.) is generally called a **plate**. Although relatively thick plates, such as those used in boilers, bridges, ships, and nuclear power plants, may have smaller ratios of surface area to thickness, sheet forming usually involves relatively thin materials.

Typical examples of applications that use sheets include metal desks, appliance bodies, hubcaps, aircraft fuselage, beverage cans, car bodies, and kitchen utensils; thus, sheet forming, also called **pressworking** or **stamping**, is among the most important of metalworking processes. It dates back to as early as 5000 BC, when household utensils, jewelry, and other objects were made by hammering and stamping metals such as gold, silver, and copper. The products made by sheet-forming processes include a large variety of shapes and sizes, ranging from simple bends to double curvatures with shallow or deep recesses.

Sheet metal is produced by a rolling process, as described in Section 6.3. If the sheet is thin, it is generally coiled after rolling; if thick, it is available as flat sheets or plates, which may have been decoiled and flattened. Before a part is formed, a blank of suitable dimensions is first cut or removed from a large sheet. Removal is usually done by a shearing process; however, there are several other methods for cutting sheet and plates. Laser cutting has become an important process and is used with computer-controlled equipment to cut consistently a variety of shapes.

This chapter describes the principles and the technology of the wide variety of sheet-forming processes available to manufacturing industries. As shown in Table 7.1, each process has specific characteristics and uses different types of hard tools and dies, as well as soft tooling, such as rubber. The sources of energy typically involve mechanical means, but also include various other sources of energy, such as hydraulic, magnetic, and explosive.

7.1 Introduction
7.2 Sheet-Metal Characteristics
7.3 Shearing
7.4 Bending of Flat Sheet and Plate
7.5 Stretch Forming
7.6 Bulging
7.7 Rubber Forming and Hydroforming
7.8 Spinning
7.9 High-Energy-Rate Forming
7.10 Superplastic Forming
7.11 Various Forming Methods
7.12 Deep Drawing
7.13 Formability of Sheet Metals
7.14 Equipment for Sheet-Metal Forming
7.15 Design Considerations
7.16 Economics of Sheet-Metal Forming

## 7.2 | Sheet-Metal Characteristics

Sheet metals are generally characterized by a high ratio of surface area to thickness. Forming of sheet metals generally is carried out by tensile forces in the plane of the sheet; otherwise, the application of compressive forces could lead to buckling, folding, and wrinkling of the sheet. In bulk-deformation processes such as forging, rolling,

**TABLE 7.1**

## General Characteristics of Sheet-Metal Forming Processes

| Process | Characteristics |
|---|---|
| Roll forming | Long parts with constant complex cross-sections; good surface finish; high production rates; high tooling costs. |
| Stretch forming | Large parts with shallow contours; suitable for low-quantity production; high labor costs; tooling and equipment costs depend on part size. |
| Drawing | Shallow or deep parts with relatively simple shapes; high production rates; high tooling and equipment costs. |
| Stamping | Includes a variety of operations, such as punching, blanking, embossing, bending, flanging, and coining; simple or complex shapes formed at high production rates; tooling and equipment costs can be high, but labor costs are low. |
| Rubber forming | Drawing and embossing of simple or complex shapes; sheet surface protected by rubber membranes; flexibility of operation; low tooling costs. |
| Spinning | Small or large axisymmetric parts; good surface finish; low tooling costs, but labor costs can be high unless operations are automated. |
| Superplastic forming | Complex shapes, fine detail, and close tolerances; forming times are long, and hence production rates are low; parts not suitable for high-temperature use. |
| Peen forming | Shallow contours on large sheets; flexibility of operation; equipment costs can be high; process is also used for straightening parts. |
| Explosive forming | Very large sheets with relatively complex shapes, although usually axisymmetric; low tooling costs, but high labor costs; suitable for low-quantity production; long cycle times. |
| Magnetic-pulse forming | Shallow forming, bulging, and embossing operations on relatively low-strength sheets; most suitable for tubular shapes; high production rates; requires special tooling. |

extrusion, and wire drawing (described in Chapter 6), the thickness or the lateral dimensions of the workpiece are intentionally changed, whereas in most sheet-forming processes, any change in thickness is due to stretching of the sheet under tensile stresses (*Poisson's effect*). Thickness decreases in sheet forming generally should be avoided, as they can lead to necking and failure.

Because the mechanics of all sheet forming basically consists of the processes of stretching and bending, certain factors significantly influence the overall operation. As described next, the major factors are elongation, yield-point elongation, anisotropy, grain size, residual stresses, springback, and wrinkling.

## 7.2.1 Elongation

Although, as we shall see, sheet-forming operations rarely involve simple uniaxial stretching, the observations made in regard to simple tensile testing can be useful in understanding the behavior of sheet metals in forming operations. Recall from Fig. 2.2 that a specimen subjected to tension first undergoes **uniform elongation** (up to the UTS, after which necking begins); this elongation is then followed by additional nonuniform elongation (**postuniform elongation**) until fracture occurs. Because the material is being stretched during forming, high uniform elongation is, of course, desirable for good formability.

It was shown in Section 2.2.3 that, for a material that has a true-stress–true-strain curve that can be represented by the equation

$$\sigma = K\varepsilon^n, \tag{7.1}$$

the strain at which necking begins (**instability**) is given by

$$\varepsilon = n. \tag{7.2}$$

Thus, the true uniform strain in a simple stretching operation (uniaxial tension) is numerically equal to the strain-hardening exponent, $n$. A large value of $n$ indicates large uniform elongation, and, thus, it is desirable for sheet forming.

Necking of a sheet specimen generally takes place at an angle $\phi$ to the direction of tension, as shown in Fig. 7.1a (**localized necking**). For an *isotropic* sheet specimen under simple tension, the Mohr's circle is constructed as shown in Fig. 7.1b. The strain $\varepsilon_1$ is the longitudinal strain, and $\varepsilon_2$ and $\varepsilon_3$ are the two lateral strains. Poisson's ratio in the plastic range is 0.5; hence, the lateral strains have the value $-\varepsilon_1/2$. The narrow neck band in Fig. 7.1a is in **plane strain**, because it is constrained by the material above and below the neck band.

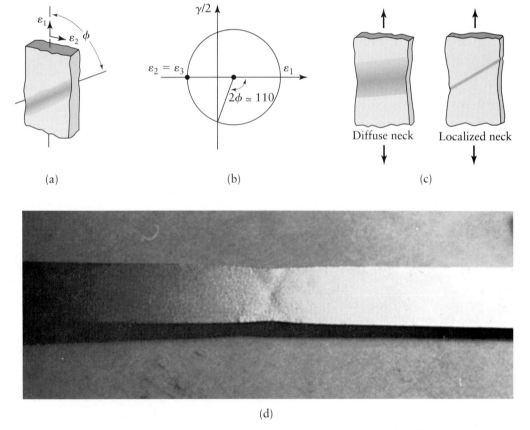

(a)                    (b)                    (c)

(d)

**FIGURE 7.1**   (a) Localized necking in a sheet specimen under tension. (b) Determination of the angle of neck from the Mohr's circle for strain. (c) Schematic illustrations for diffuse and localized necking. (d) Localized necking in an aluminum strip stretched in tension. Note the double neck.

The angle $\phi$ can be determined from the Mohr's circle by a rotation (either clockwise or counterclockwise) of $2\phi$ from the $\varepsilon_1$ position (Fig. 7.1b). This angle is about 110°; thus, the angle $\phi$ is about 55°. Note that although the length of the neck band essentially remains constant during the test, its thickness decreases (because of volume constancy), and the specimen eventually fractures. The angle $\phi$ will be different for materials that are anisotropic in the plane of the sheet (planar anisotropy).

Whether necking is *localized* or **diffuse** (Fig. 7.1c) depends on the strain-rate sensitivity, $m$, of the material, as given by the equation

$$\sigma = C\dot{\varepsilon}^m. \tag{7.3}$$

As described in Section 2.2.7, the higher the value of $m$, the more diffuse the neck becomes. Figure 7.1d shows an example of localized necking on an aluminum strip in tension; note also the double localized neck. In other words, $\phi$ can be in either the clockwise or counterclockwise position in Fig. 7.1a.

In addition to uniform elongation, the total elongation of a tension-test specimen [at a gage length of 50 mm (2 in.)] is also a significant factor in formability of sheet metals. Note that total elongation is the sum of uniform elongation and postuniform elongation. As stated previously, uniform elongation is governed by the strain-hardening exponent, $n$, whereas postuniform elongation is governed by the strain-rate sensitivity index, $m$. The higher the $m$ value, the more diffuse the neck is, and hence the greater the postuniform elongation becomes before fracture; thus, the total elongation of the material increases with increasing values of both $n$ and $m$. The following are important factors that affect the sheet-metal forming process:

a) **Yield-point elongation.** Low-carbon steels exhibit a behavior called *yield-point elongation*, exhibiting *upper* and *lower* yield points, as shown in Fig. 7.2a; yield-point elongation is usually on the order of a few percent. This behavior indicates that, after the material yields, it stretches further in certain regions along the specimen, with no increase in the lower yield point, while other regions have not yet yielded. When the overall elongation reaches the yield-point elongation, the entire specimen has been deformed uniformly. The magnitude of the yield-point elongation depends on the strain rate (with higher rates, the elongation generally increases) and the grain size of the sheet metal. As grain size decreases, yield-point elongation increases.

(a)                    (b)                    (c)

**FIGURE 7.2**   (a) Yield point elongation and Lueders bands in tension testing. (b) Lueder's bands in annealed low-carbon steel sheet. (c) Stretcher strains at the bottom of a steel can for household products. *Source:* (b) Reprinted Courtesy of Caterpillar Inc.

This behavior of low-carbon steels produces **Lueder's bands** (also called **stretcher strain marks** or *worms*) on the sheet, as shown in Fig. 7.2b. These bands are elongated depressions on the surface of the sheet and can be objectionable in the final product, because of its surface appearance. These bands can also cause difficulty in subsequent coating and painting operations. Stretcher strain marks can be seen on the curved bottom of steel cans for common household products, as shown in Fig. 7.2c; aluminum cans do not exhibit this behavior.

The usual method of avoiding this problem is to eliminate or to reduce yield-point elongation by reducing the thickness of the sheet 0.5% to 1.5% by cold rolling, known as **temper rolling** or **skin rolling**. However, because of strain aging (see Section 3.8.1), yield-point elongation reappears after even a few days at room temperature, or after a few hours at higher temperatures. Thus, the sheet metal should be formed within a certain period of time (from one to three weeks for rimmed steel; see Section 5.7.1) to avoid the reappearance of stretcher strains.

b) **Anisotropy.** Another important factor influencing sheet-metal forming is *anisotropy*, or **directionality**, of the sheet metal. Anisotropy is acquired during the thermomechanical-processing history of the sheet. Recall from Section 3.5 that there are two types of anisotropy: **crystallographic anisotropy** (from preferred grain orientation) and **mechanical fibering** (from alignment of impurities, inclusions, voids, and the like throughout the thickness of the sheet during processing). Anisotropy may be present not only in the plane of the sheet, but also in its thickness direction; the former is called **planar anisotropy**, and the latter is called **normal** or **plastic anisotropy**. These topics are particularly important in deep drawing of sheet metals, as described in Section 7.12.

c) **Grain size**    *Grain size* of the sheet metal is important for two reasons: first, because of its effect on the mechanical properties of the material, and second, because of its effect on the surface appearance of the formed part. The coarser the grain, the rougher the surface appears (**orange peel**). An ASTM grain size of No. 7 or finer is preferred for general sheet-metal forming. (See Section 3.4.1.)

d) **Residual stresses.**    *Residual stresses* can develop in sheet-metal parts because of the nonuniform deformation that the sheet undergoes during forming. When disturbed, such as by removing a portion of it, the part may distort. Tensile residual stresses on surfaces can also lead to **stress-corrosion cracking** of the part (Fig. 7.3) unless it is properly stress relieved. (See also Section 3.8.2.)

e) **Springback**    Because they generally are thin and are subjected to relatively small strains during forming, sheet-metal parts are likely to experience considerable *springback*. This effect is particularly significant in bending and other forming operations where the bend radius-to-thickness ratio is high, such as in automotive body parts.

f) **Wrinkling.**    Although, in sheet forming, the metal is generally subjected to tensile stresses, the method of forming may be such that compressive stresses are developed in the plane of the sheet; an example is the *wrinkling* of the flange in deep drawing (Section 7.12) because of circumferential *compression*. Other terms used to describe similar phenomena are **buckling, folding,** and **collapsing**. The tendency for wrinkling increases with (a) the unsupported or unconstrained length or surface area of the sheet, (b) decreasing thickness, and (c) nonuniformity of the thickness of the sheet. Lubricants that trapped or distributed unevenly between the surface of the sheet and the dies also can contribute to the initiation of wrinkling.

**FIGURE 7.3** Stress-corrosion cracking in a deep-drawn brass part for a light fixture. The cracks developed over a period of time. Brass and austenitic (300 series) stainless steels are among metals that are susceptible to stress-corrosion cracking.

g) **Coated sheet.** Sheet metals, especially steel, can be **precoated** with a variety of organic coatings, films, and laminates; such coatings are used primarily for appearance and eye appeal, but they also offer corrosion resistance. Coatings are applied to the coil stock on continuous lines, at thicknesses generally ranging from 0.0025 to 0.2 mm (0.0001 to 0.008 in.). Coatings are available with a wide range of properties, such as flexibility, durability, hardness, resistance to abrasion and chemicals, color, texture, and gloss. Coated sheet metals are subsequently formed into products such as TV cabinets, appliance housings, paneling, shelving, residential siding, and metal furniture. Zinc is used extensively as a coating on sheet steel (**galvanized** steel; see Sections 3.9.7 and 4.5) to protect it from corrosion, particularly in the automotive industry. Galvanizing of steel can be done by hot dipping, electrogalvanizing, or galvannealing processes. (See also Section 4.5.1.)

## 7.3 | Shearing

The *shearing* process involves cutting sheet metal by subjecting it to shear stresses, usually between a **punch** and a **die**, much like a paper punch (Fig. 7.4). The punch and die may be of any shape; for example, they may be circular or straight blades, similar to a pair of scissors. The major variables in the shearing process are the punch force, $F$, the speed of the punch, lubrication, the edge condition of the sheet, the punch and die materials, the corner radii of the punch and die, and the clearance between the punch and die.

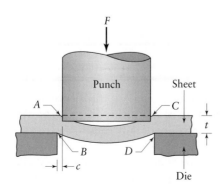

**FIGURE 7.4** Schematic illustration of the shearing process with a punch and die. This process is a common method of producing various openings in sheet metals.

**FIGURE 7.5**

Characteristic features of (a) a punched hole and (b) the punched slug. Note that the slug has been scaled down as compared with the hole.

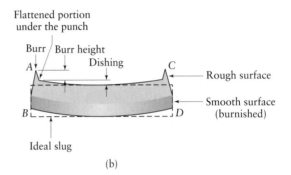

Figure 7.5 shows the overall features of a typical sheared edge for the two sheared surfaces (the sheet and the slug); note that the edges are neither smooth nor perpendicular to the plane of the sheet. The **clearance** $c$ (Fig. 7.4) is the major factor that determines the shape and quality of the sheared edge. As shown in Fig. 7.6a, as clearance increases, the edges become rougher and the deformation zone becomes larger. Furthermore, the material is pulled into the clearance area, and the edges of the sheared zone become more and more rounded. In fact, if the clearance is too large, the sheet metal is bent and subjected to tensile stresses, instead of undergoing a shearing deformation. As Fig. 7.6b shows, sheared edges can undergo severe cold working, because of the high strains involved; this result can adversely affect the formability of the sheet during subsequent operations.

Observation of the shearing mechanism reveals that shearing usually starts with the formation of cracks on both the top and bottom edges of the workpiece ($A$ and $B$ in Fig. 7.4). These cracks eventually meet, and complete separation takes place. The **rough fracture surface** in the slug in Fig. 7.5 is caused by these cracks. The **smooth, shiny, and burnished surface** is from the contact and rubbing of the sheared edge against the walls of the die. In the slug shown in Fig. 7.5b, the burnished surface is in the lower region, because this region is the section that rubs against the die wall. On the other hand, inspection of the sheared surface on the sheet itself reveals that the burnished surface is on the upper region in the sheared edge and results from rubbing against the punch.

The ratio of the burnished to rough areas on the sheared edge increases with increasing ductility of the sheet metal; it decreases with increasing material thickness

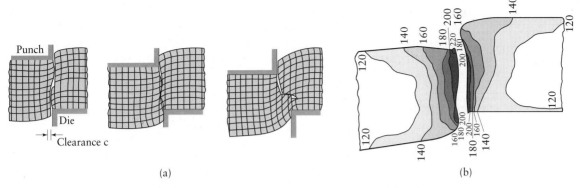

**FIGURE 7.6**   (a) Effect of clearance $c$ between the punch and die on the deformation zone in shearing. As clearance increases, the material tends to be pulled into the die, rather than being sheared. In practice, clearances usually range between 2% and 10% of the thickness of the sheet. (b) Microhardness (HV) contours for a 6.4-mm (0.25-in.) thick AISI 1020 hot-rolled steel in the sheared region. *Source*: After H. P. Weaver and K. J. Weinmann.

and clearance. The amount of punch travel required to complete the shearing process depends on the maximum shear strain that the material can undergo before fracture. Hence, a brittle metal or a material that is highly cold worked requires little travel of the punch to complete shearing.

Note from Fig. 7.6 that the deformation zone is subjected to high shear strains. The width of this zone depends on the rate of shearing, that is, the punch speed. With increasing punch speed, the heat generated by plastic deformation is confined to a smaller zone (approaching a narrow adiabatic zone), and the sheared surface is smoother.

Note also the formation of a **burr** in Fig. 7.5. *Burr height* increases with increasing clearance and increasing ductility of the metal. Tools with dull edges are also a major factor in burr formation. (See also the upcoming discussion of *slitting*.) The height, shape, and size of the burr can significantly affect many subsequent forming operations, such as in flanging, where a burr can lead to cracks. Furthermore, burrs on parts may become dislodged under external forces during their movements in mechanisms and interfere with the operation or contaminate lubricants.

**Punch force.**   The *punch force*, $F$, is basically the product of the shear strength of the sheet metal and the cross-sectional area being sheared. However, *friction* between the punch and the workpiece can increase this force substantially. Because the sheared zone is subjected to cracks, plastic deformation, and friction, the punch-force–stroke curves can have various shapes. Figure 7.7 shows one typical curve for a ductile material; the area under the curve is the *total work* done in shearing.

An approximate empirical formula for estimating the **maximum punch force** $F_{max}$ is

$$F_{max} = 0.7(UTS)(t)(L), \qquad (7.4)$$

where UTS is the ultimate tensile strength of the sheet metal, $t$ is its thickness, and $L$ is the total length of the sheared edge. Thus, for example, for a round hole of diameter $D$, $L = \pi D$. In addition to the basic punch force, a force is also required to strip the punch from the sheet during its return stroke. This force is difficult to calculate, however, because of the many factors involved, especially *friction* between the punch and the sheet.

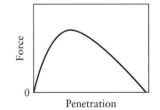

**FIGURE 7.7**   Typical punch–penetration curve in shearing. The area under the curve is the work done in shearing. The shape of the curve depends on process parameters and material properties.

**EXAMPLE 7.1** Calculation of maximum punch force

Estimate the force required in punching a 1-in. (25 mm) diameter hole through a $\frac{1}{8}$-in. (3.2 mm) thick annealed titanium-alloy Ti-6Al-4V sheet at room temperature.

**Solution.** The force is estimated from Eq. (7.4), and the UTS for this alloy is found from Table 3.13 to be 1000 MPa, or 140,000 psi. Therefore,

$$F = 0.7\left(\frac{1}{8}\right)(\pi)(1)(140,000) = 38,500\,\text{lb}$$

$$= 19.25\,\text{tons} = 0.17\,\text{MN}.$$

## 7.3.1 Shearing operations

In this section, we describe various operations that are based on the shearing process. First, we must define two terms: in **punching**, the sheared slug is discarded (Fig. 7.8a); in **blanking**, the slug is the part itself, and the rest is scrap. The following processes are common shearing operations:

a) **Die cutting.** *Die cutting* typically consists of the following operations (Fig. 7.8b): (1) **perforating**, or punching a number of holes in a sheet; (2) **parting**, or shearing the sheet into two or more pieces, usually when the adjacent blanks do not have a matching contour; (3) **notching**, or removing pieces or various shapes from the edges; (4) **slitting** (see also the upcoming discussion), and (5) **lancing**, or leaving a tab on the sheet without removing any material. Parts produced by these processes have various uses, particularly in assembly with other components.

b) **Fine blanking.** Very smooth and square edges can be produced by *fine blanking* (Fig. 7.9a). One basic die design is shown in Fig. 7.9b, in which a V-shaped *stinger*, or *impingement*, locks the sheet metal tightly in place and prevents the type of distortion of the material shown in Fig. 7.6. This process involves clearances on the order of 1% of the sheet thickness, as compared with as much as 8% in ordinary shearing operations; the thickness of the sheet may range from 0.13 to 13 mm (0.005 to 0.5 in.). This operation is usually carried out on triple-action hydraulic presses where the movements of the punch, pressure pad, and die are controlled individually. Fine blanking usually involves parts that have

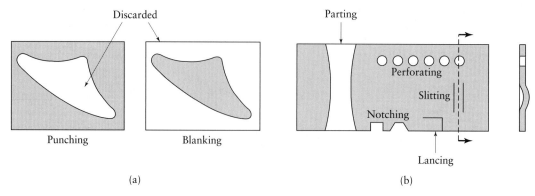

**FIGURE 7.8** (a) Punching (piercing) and blanking. (b) Examples of various shearing operations on sheet metal.

(a)                                (b)

**FIGURE 7.9** (a) Comparison of sheared edges by conventional (left) and fine-blanking (right) techniques. (b) Schematic illustration of the setup for fine blanking. *Source*: Feintool International Holding.

holes, which are punched simultaneously with blanking. Suitable sheet hardness is typically in the range of 50 to 90 HRB.

c) **Slitting.** Shearing operations can be carried out with a pair of circular blades (*slitting*), similar to the blade of a can opener (Fig. 7.10). The blades follow either a straight line or a circular or curved path. A slit edge normally has a burr, which may be removed by rolling it over the sheet's edge. There are two types of slitting equipment. In the *driven* type, the blades are powered; in the *pull-through* type, the strip is pulled through idling blades. Slitting operations, if not performed properly, may cause various planar distortions of the slit part or strip.

d) **Steel rules.** Soft metals, paper, leather, and rubber can be blanked with *steel-rule dies*. Such a die consists of a thin strip of hardened steel bent to the shape to be sheared (similar to a cookie cutter) and held on its edge on a flat wooden base. The die is pressed against the sheet, which rests on a flat surface, and cuts the sheet to the shape of the steel rule.

e) **Nibbling.** In *nibbling*, a machine called a *nibbler* moves a straight punch up and down rapidly into a die. The sheet metal is fed through the gap, and a number of overlapping holes are made. (Thus, the movements are similar to that of a sewing machine into which cloth is fed.) This operation is similar to making a large elongated hole by successively punching holes with a paper punch. Sheets can be cut along any desired path by manual control. The process is economical for small production runs, since no special dies are required; intricate slots and notches can be produced, using standard punches.

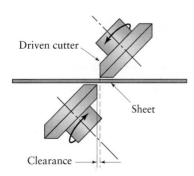

**FIGURE 7.10** Slitting with rotary knives. This process is similar to opening cans.

**Scrap in shearing.** The amount of *scrap* produced, or **trim loss**, in shearing operations can be significant. On large stampings, it can be as high as 30% of the original sheet. A significant factor in manufacturing cost, scrap can be reduced significantly by proper arrangement of the shapes on the sheet to be cut, called **layout and nesting** (see Section 7.15). Computer-aided design techniques have been developed to minimize scrap in shearing operations.

### 7.3.2 Shearing dies

Because the formability of a sheared part can be influenced by the quality of its sheared edges, clearance control is important. In practice, clearances usually range between 2% and 8% of the sheet's thickness; generally, the thicker the sheet, the larger is the clearance (as much as 10%). However, the smaller the clearance, the better is the quality of the edge. In a process called **shaving** (Fig. 7.11), the extra material from a rough sheared edge is trimmed by cutting. Some common shearing dies are described as follows:

a) **Punch and die shapes.** Note in Fig. 7.4 that the surfaces of the punch and die are flat; thus, the punch force builds up rapidly during shearing, because the entire thickness of the sheet is sheared at the same time. However, the area being sheared at any moment can be controlled by beveling the punch and die surfaces, a shown in Fig. 7.12. The shape of a beveled punch is similar to that of a paper punch, which you can observe by looking closely at its tip. This geometry is particularly suitable for shearing thick blanks, because it reduces the total shearing force (since it is an incremental process); it also reduces the noise level during punching. Note also that because there will be a net lateral force in the punch shape shown in Fig. 7.12b, the punch and the press must have sufficient lateral rigidity.

b) **Compound dies.** Several operations on the same strip may be performed in one stroke with a *compound die* in one station. Although they have higher productivity than the simple punch and die processes, these operations are usually limited to relatively simple shearing, because they are somewhat slow, and the dies are more expensive than those for individual shearing operations.

**FIGURE 7.11** Schematic illustrations of shaving on a sheared edge. (a) Shaving a sheared edge. (b) Shearing and shaving, combined in one stroke.

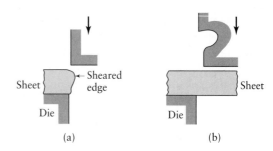

**FIGURE 7.12** Examples of the use of shear angles on punches and dies.

(a)                                                    (b)

**FIGURE  7.13**   (a) Schematic illustration of the making of a washer in a progressive die. (b) Forming of the top piece of an aerosol spray can in a progressive die. Note that the part is attached to the strip until the last operation is completed.

c) **Progressive dies.**  Parts requiring multiple operations, such as punching, bending, and blanking, are made at high production rates in *progressive dies*. The sheet metal is fed through a coil strip, and a different operation is performed at the same station with each stroke of a series of punches (Fig. 7.13a). Figure 7.13b shows an example of a part made in progressive dies.

d) **Transfer dies.**  In a *transfer-die* setup, the sheet metal undergoes different operations at different stations, which are arranged along a straight line or a circular path. After each operation, the part is transferred (hence the name) to the next station for additional operations.

e) **Tool and die materials.**  Tool and die materials for shearing are generally tool steels and, for high production rates, carbides. (See Table 3.5.) Lubrication is important for reducing tool and die wear and for improving edge quality.

## 7.3.3  Other methods of cutting sheet metal.

The following list describes several other methods of cutting sheets and, particularly, plates:

a) The sheet or plate may be cut with a **band saw**; see Section 8.9.5.

b) Particularly for thick plates, **oxyfuel-gas (flame) cutting** may be employed, as used widely in shipbuilding and heavy-construction industries. (See Section 9.14.2.)

c) **Laser-beam cutting** has become an important process; it is now used widely with computer-controlled equipment. (See Section 9.14.1.)

d) **Friction sawing** involves the use of a disk or blade that rubs against the sheet or plate at high surface speed. (See Section 8.9.5.)

e) **Water-jet** and **abrasive water-jet cutting** are effective on metals as well as on nonmetallic materials. (See Section 9.15.)

## 7.3.4  Tailor-welded blanks

In sheet-metal forming processes, the blank is typically made of one piece, is of a uniform thickness, and is usually cut from a large sheet. However, an important trend now involves laser butt welding (see Section 12.4) of pieces of sheet metal of different shapes and thicknesses (*tailor-welded blanks*, or *TWB*); the welded sheet is subsequently formed into a final shape. The most commonly used technique is laser-beam welding. (See Section 12.4.) Laser-welding techniques are now highly developed, and, as a consequence, weld joints are very strong. However, because of the small thicknesses involved, proper alignment of the sheets prior to welding is important. This method is becoming increasingly important, particularly in the automotive industry. Because each welded piece can have a different thickness (as guided by design considerations such as stiffness), grade of sheet metal, coating, or other characteristics, these blanks possess the needed characteristics in the desired locations of the formed part. As a result, productivity is increased, the need for subsequent spot welding of the product (e.g., a car body) is reduced or eliminated, and dimensional control is improved.

---

**EXAMPLE 7.2 Use of tailor-welded blanks in the automotive industry**

An example of the use of tailor-welded blanks is the production of an automobile outer side panel, shown in Fig. 7.14a. Note that five different pieces are first

**FIGURE 7.14**
(a) Production of an outer side panel of a car body by laser welding and stamping. (b) Examples of laser-welded and stamped automotive body components. *Source*: After M. Geiger and T. Nakagawa.

blanked; four of the pieces are 1 mm (0.04 in.) thick, and one is 0.8 mm (0.03 in.) thick. The pieces are then laser butt welded and stamped into the final shape. Thus, the blanks can be tailored for a particular application, including not only sheets of different shape and thickness, but also sheets of different quality, and with or without coatings on one or both surfaces. This trend of welding and forming of sheet-metal pieces allows significant flexibility in product design, structural stiffness and crash behavior (*crashworthiness*), formability, and use of different materials in one component, as well as weight savings and cost reduction in materials, scrap, equipment, assembly, and labor.

There is a growing number of applications for this type of production in automotive companies. The various other components shown in Fig. 7.14b use the advantages outlined above. For example, note that the required strength and stiffness for supporting the shock absorber is achieved by welding a round piece on the surface of the large sheet. The sheet thickness in these components varies, depending on its location and contribution to characteristics such as stiffness and strength, while allowing significant weight and cost savings. *Source*: After M. Geiger and T. Nakagawa.

## 7.4 | Bending of Flat Sheet and Plate

One of the most common metalworking operations is *bending*. This process is used not only to form parts such as flanges, curls, seams, and corrugations, but also to impart stiffness to a part by increasing its moment of inertia. Note, for instance, that a long strip of metal is much less rigid when it is flat than when it is formed into a *V* cross-section, as can be demonstrated with a piece of paper.

Figure 7.15a depicts the terminology used for bending. **Bend allowance** is the length of the *neutral axis* in the bend area and is used to determine the blank length for a bent part. However, the radial position of the neutral axis depends on the bend radius and bend angle, as described in texts on the mechanics of solids. An approximate formula for the bend allowance, $L_b$, is given by

$$L_b = \alpha(R + kt),$$

where $\alpha$ is the bend angle (in radians), $R$ is the bend radius, $k$ is a constant, and $t$ is the sheet thickness. For the ideal case, the neutral axis remains at the center, and hence $k = 0.5$. In practice, $k$ values usually range from 0.33 for $R < 2t$ to 0.5 for $R > 2t$.

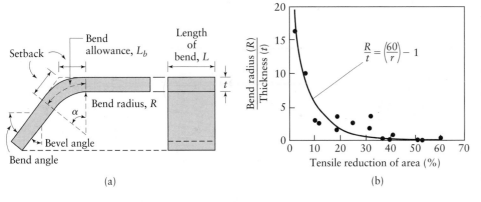

(a)

(b)

**FIGURE 7.15** (a) Bending terminology. The bend radius is measured to the inner surface of the bend. Note that the length of the bend is the width of the sheet. Also note that the bend angle and the bend radius (sharpness of the bend) are two different variables. (b) Relationship between the ratio of bend radius to sheet thickness and tensile reduction of area for various materials. Note that sheet metal with a reduction of area of about 50% can be bent and flattened over itself without cracking. *Source*: After J. Datsko and C. T. Yang.

### 7.4.1 Minimum bend radius

The outer fibers of the part being bent are in tension, and the inner fibers are in compression. Theoretically, the strains at the outer and inner fibers are equal in magnitude and are given by the equation

$$e_o = e_i = \frac{1}{(2R/t) + 1}. \tag{7.5}$$

However, due to the *shifting of the neutral axis* toward the inner surface, the *length of bend* (the sketch on the right in Fig. 7.15a) is smaller in the outer region than in the inner region. (This phenomenon can easily be observed by bending a rectangular eraser.) Consequently, the outer and inner strains are different, and the difference increases with decreasing $R/t$ ratio.

It can be seen from Eq. (7.5) that, as the $R/t$ ratio decreases, the tensile strain at the outer fiber increases, and the material may crack after a certain strain is reached. The radius, $R$, at which a crack appears on the outer surface of the bend is called the *minimum bend radius*. The minimum bend radius to which a part can be bent safely is normally expressed in terms of its thickness, such as $2t$, $3t$, $4t$, and so on. Thus, for example, a bend radius of $3t$ indicates that the smallest radius to which the sheet can be bent without cracking is three times the thickness of the sheet. The minimum bend radii for various materials have been determined experimentally and are available in various handbooks; some typical results are given in Table 7.2.

Studies have also been conducted to establish a relationship between the minimum $R/t$ ratio and a given mechanical property of the material. One such analysis is based on the following assumptions: (1) The true strain at cracking on the outer fiber in bending is equal to the true strain at fracture, $\varepsilon_f$, of the material in a simple tension test; (2) the material is homogeneous and isotropic; and (3) the sheet is bent in a state of plane stress—that is, its $L/t$ ratio is small.

The true strain at fracture in tension is

$$\varepsilon_f = \ln\left(\frac{A_0}{A_f}\right) = \ln\left(\frac{100}{100 - r}\right),$$

TABLE 7.2

**Minimum Bend Radii for Various Materials at Room Temperature**

| MATERIAL | MATERIAL CONDITION | |
|---|---|---|
| | SOFT | HARD |
| Aluminum alloys | 0 | 6t |
| Beryllium copper | 0 | 4t |
| Brass, low leaded | 0 | 2t |
| Magnesium | 5t | 13t |
| Steels | | |
|    austenitic stainless | 0.5t | 6t |
|    low carbon, low alloy, and HSLA | 0.5t | 4t |
| Titanium | 0.7t | 3t |
| Titanium alloys | 2.6t | 4t |

where $r$ is the percent reduction of area of the sheet in a tension test. From Section 2.2.2, we have for true strain

$$\varepsilon_o = \ln(1 + e_o) = \ln\left(1 + \frac{1}{(2R/t) + 1}\right) = \ln\left(\frac{R + t}{R + (t/2)}\right).$$

Equating the two expressions and simplifying, we obtain

$$\text{Minimum } \frac{R}{t} = \frac{50}{r} - 1. \tag{7.6}$$

The experimental data are shown in Fig. 7.15b; note that the curve that best fits the data is

$$\text{Minimum } \frac{R}{t} = \frac{60}{r} - 1. \tag{7.7}$$

The $R/t$ ratio approaches zero (complete **bendability**—that is, the material can be folded over itself) at a tensile reduction of area of 50%. Interestingly, this percent is the same value obtained for spinnability of metals (described in Section 7.8); that is, a material with 50% reduction of area is found to be completely spinnable. A curve similar to that in Fig. 7.15b also is obtained when the data are plotted against the total elongation (such as percent in 50-mm gage length) of the material.

**Factors affecting bendability.** The bendability of a metal may be increased by increasing its tensile reduction of area, either by *heating* or by the application of *hydrostatic pressure*. Other techniques may also be employed to increase the compressive environment in bending, such as applying compressive forces in the plane of the sheet during bending to minimize tensile stresses in the outer fibers of the bend area.

As the length of the bend increases, the state of stress at the outer fibers changes from a uniaxial stress to a *biaxial stress*. The reason for this change is that the bend length $L$ tends to become smaller due to stretching of the outer fibers, but it is constrained by the material around the bend area. Biaxial stretching tends to reduce ductility, that is, the amount of strain to fracture; thus, as $L$ increases, the minimum bend radius increases (Fig. 7.16). However, at a bend length of about $10t$, the minimum bend radius increases no further, and a *plane-strain condition* is fully developed. As

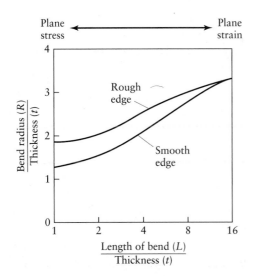

**FIGURE 7.16**   The effect of length of bend and edge condition on the ratio of bend radius to thickness for 7075-T aluminum. *Source:* After G. Sachs and G. Espey.

**FIGURE 7.17** (a) and (b) The effect of elongated inclusions (stringers) on cracking as a function of the direction of bending with respect to the original rolling direction of the sheet. This example shows the importance of the direction of cutting from large sheets in workpieces that are subsequently bent to make a product. (c) Cracks on the outer radius of an aluminum strip bent to an angle of 90°.

the $R/t$ ratio decreases, narrow sheets (smaller length of bend) crack at the edges, and wider sheets crack at the center, where the biaxial stress is the highest.

Bendability also depends on the **edge condition** of the sheet being bent; because rough edges are points of stress concentration, bendability decreases as edge roughness increases. Another important factor is the amount of **cold working** that the edges undergo during shearing, as can be observed from microhardness tests in the sheared region (Fig. 7.6b). Removal of the cold-worked regions, such as by shaving, machining, or heat treating, greatly improves the resistance to edge cracking during bending.

Another significant factor in edge cracking is the amount and shape of *inclusions* in the sheet metal. Inclusions in the form of **stringers** are more detrimental than globular-shaped inclusions; thus, anisotropy of the sheet is also important in bendability. As depicted in Fig. 7.17, cold rolling of sheets produces **anisotropy**, because of the alignment of impurities, inclusions, and voids (*mechanical fibering*), and thus transverse ductility is reduced, as shown in Fig. 7.17c; see also Fig. 3.15. In bending such a sheet, caution should be exercised in cutting or slitting the blank in the proper direction of the rolled sheet, although this may not always be possible in practice.

## 7.4.2 Springback

Because all materials have a finite modulus of elasticity, plastic deformation is followed by **elastic recovery** upon removal of the load; in bending, this recovery is known as *springback*. As shown in Fig. 7.18, the final bend angle after springback is smaller and the final bend radius is larger than before. This phenomenon can easily be observed by bending a piece of wire or a short strip metal. Springback occurs not only in flat sheets or plate, but also in bending bars, rod, and wire of any cross-section.

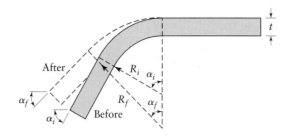

**FIGURE 7.18**
Terminology for springback in bending. Springback is caused by the elastic recovery of the material upon unloading. In this example, the material tends to recover toward its originally flat shape. However, there are situations where the material bends farther upon unloading (negative springback), as shown in Fig. 7.20.

A quantity characterizing springback is the **springback factor**, $K_s$, which is determined as follows. Because the bend allowance (see Fig. 7.15a) is the same before and after bending, the relationship obtained for pure bending is

$$\text{Bend allowance} = \left(R_i + \frac{t}{2}\right)\alpha_i = \left(R_f + \frac{t}{2}\right)\alpha_f. \tag{7.8}$$

From this relationship, $K_s$ is defined as

$$K_s = \frac{\alpha_f}{\alpha_i} = \frac{(2R_i/t) + 1}{(2R_f/t) + 1}, \tag{7.9}$$

where $R_i$ and $R_f$ are the initial and final bend radii, respectively. Note from this expression that $K_s$ depends only on the $R/t$ ratio. A factor of $K_s = 1$ indicates that there is no springback, and $K_s = 0$ indicates complete elastic recovery (Fig. 7.19), as in a leaf spring. Thus, a true spring always has a springback factor of zero.

Recall from Fig. 2.3 that the amount of elastic recovery depends on the stress level and the modulus of elasticity, $E$, of the material; hence, elastic recovery increases with the stress level and with decreasing elastic modulus. Based on this observation, an approximate formula has been developed to estimate springback:

$$\frac{R_i}{R_f} = 4\left(\frac{R_i Y}{Et}\right)^3 - 3\left(\frac{R_i Y}{Et}\right) + 1. \tag{7.10}$$

In this equation, $Y$ is the uniaxial yield stress of the material at 0.2% offset. (See Figure 2.2b.)

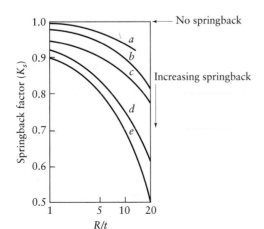

**FIGURE 7.19** Springback factor $K_s$ for various materials: (a) 2024-0 and 7075-0 aluminum; (b) austenitic stainless steels; (c) 2024-T aluminum; (d) $\frac{1}{4}$-hard austenitic stainless steels; (e) $\frac{1}{2}$-hard to full-hard austenitic stainless steels. *Source*: After G. Sachs.

**FIGURE 7.20** Schematic illustration of the stages in bending round wire in a V-die. This type of bending can lead to negative springback, which does not occur in air bending (shown in Fig. 7.24a). *Source*: After K. S. Turke and S. Kalpakjian.

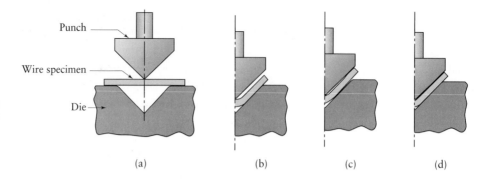

(a)    (b)    (c)    (d)

**Negative springback.** The springback observed in Fig. 7.18 can be called *positive springback*. However, under certain conditions, *negative springback* is also possible. In other words, the bend angle in such cases becomes larger after the bend has been completed and the load is removed. This phenomenon is generally associated with V-die bending (Fig. 7.20).

The development of negative springback can be explained by observing the sequence of deformation in Fig. 7.20. If we remove the bent piece at stage (b), it will undergo positive springback. At stage (c), the ends of the piece are touching the male punch. Note that between stages (c) and (d), the part is actually being bent in the direction opposite to that between stages (a) and (b). Note also the lack of conformity of the punch radius and the inner radius of the part in both stage (b) and stage (c); in stage (d), however, the two radii are the same. Upon unloading, the part in stage (d) will spring back inwardly, because it is being *unbent* from stage (c), both at the tip of the punch and in the arms of the part. The amount of this inward (negative) springback can be greater than the amount of positive springback, because of the large strains that the material has undergone in the small bend area in stage (b). The net result is negative springback.

**Compensation for springback.** In practice, springback is usually compensated for by using various techniques:

1. **Overbending** the part in the die (Figs. 7.21a and b) can compensate for springback; overbending can also be achieved by the **rotary bending** technique shown in Fig. 7.21e. The upper die has a cylindrical rocker (with an angle of $< 90°$) and is free to rotate; as it travels downward, the sheet is clamped and bent by

**FIGURE 7.21** Methods of reducing or eliminating springback in bending operations. *Source*: V. Cupka, T. Nakagawa, and H. Tyamoto.

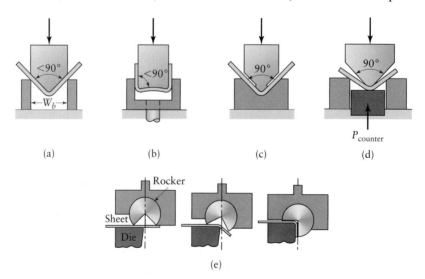

(a)    (b)    (c)    (d)

(e)

the rocker over the lower die (die anvil). A relief angle in the lower die allows overbending of the sheet at the end of the stroke, thus compensating for springback.

2. **Coining** the bend region by subjecting it to high localized compressive stresses between the tip of the punch and the die surface (Figs. 7.21c and 7.21d), known as **bottoming**.

3. **Stretch bending**, in which the part is subjected to tension while being bent, may be applied. The bending moment required to make the sheet deform plastically will be reduced as the combined tension (due to bending of the outer fibers and the applied tension) in the sheet increases. Therefore, springback, which is the result of nonuniform stresses due to bending, will also decrease. This technique is used to limit springback in stretch forming of shallow automotive bodies. (See Section 7.5.)

4. Because springback decreases as yield stress decreases [see Eq. (7.10)], all other parameters being the same, bending may also be carried out at *elevated temperatures* to reduce springback.

---

**EXAMPLE 7.3** **Estimating springback**

A 20-gage steel sheet is bent to a radius of 0.5 in. Assuming that its yield stress is 40,000 psi, calculate the radius of the part after it is bent.

**Solution.** The appropriate formula is Eq. (7.10), where

$$R_i = 0.5 \text{ in.}, \quad Y = 40,000 \text{ psi}, \quad E = 29 \times 10^6 \text{ psi},$$

and $t = 0.0359$ in. for 20-gage steel, as found in handbooks. Thus, we have

$$\frac{R_i Y}{Et} = \frac{(0.5)(40,000)}{(29 \times 10^6)(0.0359)} = 0.0192,$$

and

$$\frac{R_i}{R_f} = 4(0.0192)^3 - 3(0.0192) + 1 = 0.942.$$

Hence,

$$R_f = \frac{0.5}{0.942} = 0.531 \text{ in.}$$

---

## 7.4.3 Forces

Bending forces can be estimated by assuming that the process is that of the simple bending of a rectangular beam. Thus, the bending force is a function of the strength of the material, the length and thickness of the part, and the width $W$ of the die opening, as shown in Fig. 7.22. Excluding friction, the general expression for the **maximum bending force**, $F_{max}$, is

$$F_{max} = k \frac{(\text{UTS})Lt^2}{W}, \tag{7.11}$$

where the factor $k$ ranges from about 1.2 to 1.33 for a $V$ die. The $k$ values for wiping and for $U$ dies are 0.25 and 2 times, respectively, that of $k$ for $V$ dies. The effect

**FIGURE 7.22** Common die-bending operations, showing the die-opening dimension $W$ used in calculating bending forces. [See Eq. (7.11).]

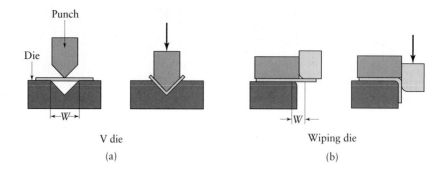

V die
(a)

Wiping die
(b)

of various factors on the bending force, such as friction, is included in the factor $k$. Equation (7.11) applies well to situations in which the punch radius and sheet thickness are small compared with the size of the die opening, $W$.

The bending force is also a function of punch travel; It increases from zero to a maximum and may decrease as the bend is completed. The force then increases sharply as the punch bottoms in the case of die bending. In *air bending*, or *free bending* (see Fig. 7.24a), the force does not increase again after it begins to decrease.

### 7.4.4 Common bending operations

In this section, we describe common bending operations. Some of these processes are performed on discrete sheet-metal parts; others are done continuously, as in roll forming of coiled sheet stock. The operations are described as follows:

a) **Press-brake forming.** Sheet metal or plate can be bent easily with simple fixtures, using a press. Parts that are long, i.e., 7 m (20 ft) or more, and relatively narrow are usually bent in a *press brake*. This machine uses long dies in a mechanical or hydraulic press and is suitable for small production runs. The tooling is simple and adaptable to a wide variety of shapes (Fig. 7.23); furthermore, the process can easily be automated. Die materials may range from hardwood (for low-strength materials and small production runs) to carbides; for most applications, carbon-steel or gray-iron dies are generally used.

**FIGURE 7.23** Schematic illustrations of various bending operations in a press brake.

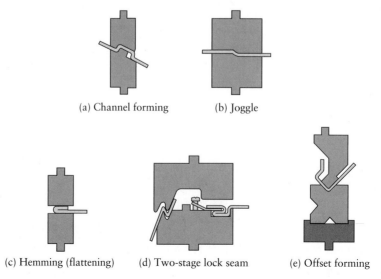

(a) Channel forming　　(b) Joggle

(c) Hemming (flattening)　　(d) Two-stage lock seam　　(e) Offset forming

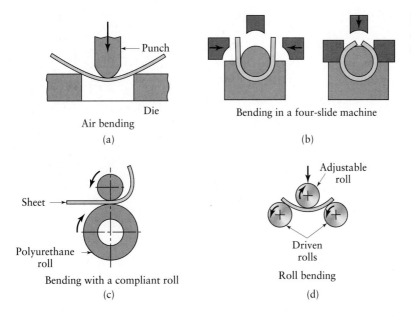

**FIGURE 7.24** Examples of various bending operations.

b) **Other bending operations.** Sheet metal may be bent by a variety of processes (Fig. 7.24). In **air bending** (*free bending*), there is only one die. Bending of sheet metals can be carried out with a *pair of rolls*, the larger one of which is flexible and made of polyurethane. The upper roll pushes the sheet into the flexible lower roll and imparts a curvature to the sheet, the shape of which depends on the degree of indentation into the flexible roll. Hence, by controlling this depth, it is possible to form a sheet with a variety of curvatures on it. Plates are bent by the **roll-bending** process shown in Fig. 7.24d. Adjusting the distance between the three rolls produces various curvatures. Bending of relatively short pieces, such as a bushing, can also be done on highly automated **four-slide machines**, a process similar to that shown in Fig. 7.24b.

c) **Beading.** In *beading*, the edge of the sheet metal is bent into the cavity of a die (Fig. 7.25). The bead gives stiffness to the part by increasing the moment of inertia of the edges. Also, it improves the appearance of the part and eliminates exposed sharp edges, which may be a safety hazard.

d) **Flanging.** Flanging is a process of bending the edges of sheet metals, usually to 90°. In **shrink flanging** (Fig. 7.26a), the flange is subjected to compressive hoop stresses, which, if excessive, can cause the edges of the flange to wrinkle. The wrinkling tendency increases with decreasing radius of curvature of the flange. In **stretch flanging**, the flange edges are subjected to tensile stresses, which, if excessive, can lead to cracking at the edges.

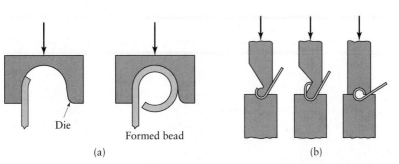

(a)

(b)

**FIGURE 7.25** (a) Bead forming with a single die. (b) Bead forming with two dies in a press brake.

**FIGURE 7.26** Various flanging operations. (a) Flanges on flat sheet. (b) Dimpling. (c) Piercing sheet metal to form a flange. In this operation, a hole does not have to be prepunched before the punch descends. Note, however, the rough edges along the circumference of the flange. (d) Flanging of a tube. Note the thinning of the edges of the flange.

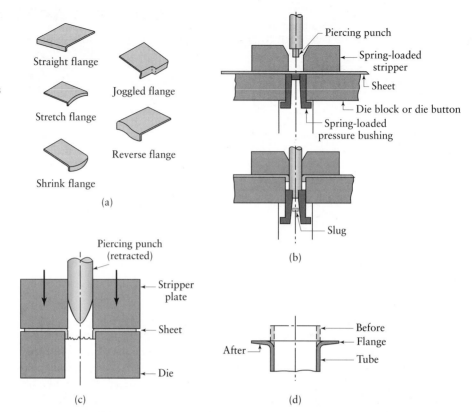

e) **Dimpling.** In this operation (Fig. 7.26b), a hole is first punched and then expanded into a flange, or a shaped punch pierces the sheet metal and expands the hole. Flanges may be produced by **piercing** with a bullet-shaped punch (Fig. 7.26c). The ends of tubes are flanged by a similar process (Fig. 7.26d). When the angle of bend is less than 90°, as in fittings with conical ends, the process is called **flaring**. The condition of the edges is important in these operations; stretching the material causes high tensile stresses at the edges, which may lead to cracking and tearing of the flange. As the ratio of flange to hole diameter increases, the strains increase proportionately. The rougher the edge, the greater will be the tendency for cracking. Sheared or punched edges may be shaved with a sharp tool (see Fig. 7.11) to improve the surface finish of the edge and reduce the tendency for cracking.

f) **Hemming.** In the *hemming* process (also called **flattening**), the edge of the sheet is folded over itself. (See Fig. 7.23c.) Hemming increases the stiffness of the part, improves its appearance, and eliminates sharp edges. *Seaming* (Fig. 7.23d) involves joining two edges of sheet metal by hemming. Double seams are made by a similar process, using specially shaped rollers, for watertight and airtight joints, such as in food and beverage containers.

g) **Roll forming.** *Roll forming* is used for bending continuous lengths of sheet metal and for large production runs (**contour roll forming**, or *cold roll forming*). The metal strip is bent in stages by passing it through a series of rolls (Fig. 7.27). Typical products formed by this method include channels, gutters, siding, panels, frames (Fig. 7.28), and pipes and tubing with lock seams. The length of the part is limited only by the amount of material supplied from the coiled stock. The parts are usually sheared and stacked continuously. The thickness of the sheet used typically ranges from about 0.125 to 20 mm (0.005 to 0.75 in.). Forming speeds are generally below 1.5 m/s (300 ft/min), although they can be much higher for special applications.

**FIGURE 7.27** The roll-forming process.

**FIGURE 7.28** Stages in roll forming of a sheet-metal door frame. In Stage 6, the rolls may be shaped as in *A* or *B*. *Source*: Oehler, G., *Bending*, Carl Hanser Verlag Munich, 1963.

Proper design and sequencing of the rolls, which usually are mechanically driven, requires considerable experience. Dimensional tolerances, springback, and tearing and buckling of the strip have to be considered. The rolls are generally made of carbon steel or gray iron and may be chromium plated, for better surface finish of the product and wear resistance of the rolls. Lubricants may be used to improve roll life and surface finish and to cool the rolls and the workpiece.

## 7.4.5 Tube bending

Bending and forming tubes and other hollow sections requires special tooling to avoid buckling and folding. The oldest and simplest method of bending a tube or pipe is to pack the inside with loose particles (such as sand) and bend the part in a suitable fixture. This technique does not allow the tube wall to buckle inward. After the tube is bent, the sand is shaken out. Tubes can also be *plugged* with various flexible internal mandrels, such as those shown in Fig. 7.29, which also illustrates various bending

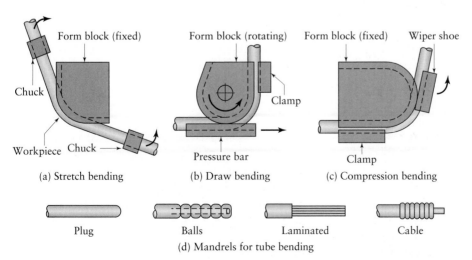

**FIGURE 7.29** Methods of bending tubes. Using internal mandrels, or filling tubes with particulate materials such as sand, is often necessary to prevent collapsing of the tubes during bending. Solid rods and structural shapes are also bent by these techniques.

**FIGURE 7.30** A method of forming a tube with sharp angles, using axial compressive forces. Compressive stresses are beneficial in forming operations because they delay fracture. Note that the tube is supported internally with rubber or fluid to avoid collapsing during forming. *Source*: After J. L. Remmerswaal and A. Verkaik.

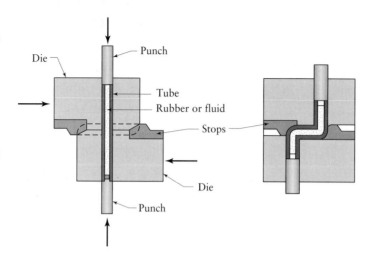

methods and fixtures for tubes and sections. A relatively thick tube that has a large bend radius can be bent without filling it with particulates or using plugs, since it will have less of a tendency to buckle inward. The beneficial effect of forming metals under highly compressive stresses is demonstrated in Fig. 7.30 with respect to the bending of a tube with relatively sharp corners. Note that in this operation, the tube is subjected to longitudinal compressive stresses, which reduce the stresses in the outer fibers in the bend area, thus improving the bendability of the material.

## 7.5 | Stretch Forming

In *stretch forming*, the sheet metal is clamped around its edges and stretched over a die or form block, which moves upward, downward, or sideways, depending on the particular machine (Fig. 7.31). Stretch forming is used primarily to make aircraft-wing skin panels, automobile door panels, and window frames. Aluminum skins for the Boeing 767 and 757 aircraft are made by stretch forming, with a tensile force of 9 MN (2 million lb). The rectangular sheets are 12 m × 2.5 m × 6.4 mm (40 ft × 8.3 ft × 0.25 in.). Although this process is generally used for low-volume production, it is versatile and economical.

In most operations, the blank is a rectangular sheet, clamped along its narrower edges and stretched lengthwise, thus allowing the material to shrink in its width; however, controlling the amount of stretching is important to avoid tearing. Stretch forming cannot produce parts with sharp contours or re-entrant corners (depressions of the surface of the die). Dies for stretch forming are generally made of zinc alloys, steel, hard plastics, or wood, and most applications require little or no lubrication. Var-

**FIGURE 7.31** Schematic illustration of a stretch-forming process. Aluminum skins for aircraft can be made by this process. *Source*: Cyril Bath Co.

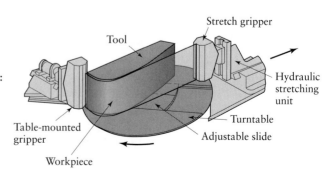

ious accessory equipment can be used in conjunction with stretch forming, including additional forming with both male and female dies while the part is in tension.

---

**EXAMPLE 7.4  Work done in stretch forming**

A 15-in.–long workpiece with a cross-sectional area of 0.5 in². (see the accompanying figure ) is stretched with a force, $F$, until $\alpha = 20°$. The material has a true-stress–true-strain curve $\sigma = 100{,}000\,\varepsilon^{0.3}$. (a) Find the total work done, ignoring end effects or bending. (b) What is $\alpha_{max}$ before necking begins?

**Solution.**

a) This situation is equivalent to stretching a workpiece from 15 in. to a length of $a + b$. For $\alpha = 20°$, the final length is $L_f = 16.8$ in. The true strain is then

$$\varepsilon = \ln\left(\frac{L_f}{L_o}\right) = \ln\left(\frac{16.8}{15}\right) = 0.114.$$

The work done per unit volume is

$$u = \int_0^{0.114} \sigma\,d\varepsilon = 10^5 \int_0^{0.114} \varepsilon^{0.3}\,d\varepsilon = 10^5 \left[\frac{\varepsilon^{1.3}}{1.3}\right]_0^{0.114} = 4570\ \text{in}\cdot\text{lb/in}^3.$$

The volume of the workpiece is

$$V = (15)(0.5) = 7.5\ \text{in}^3.$$

Hence,

$$\text{Work} = (u)(V) = 34{,}275\ \text{in.-lb.}$$

b) The necking limit for uniaxial tension is given by Eq. (7.2), the problem being similar to a stretch-forming operation. Thus,

$$L_{max} = L_o\varepsilon^n = 15\varepsilon^{0.3} = 20.2\ \text{in.}$$

Therefore, $a + b = 20.2$ in., and from similarities in triangles, we obtain

$$a^2 - 10^2 = b^2 - 5^2,$$

or

$$a^2 = b^2 + 75.$$

Therefore, $a = 12$ in. and $b = 8.2$ in. Hence,

$$\cos\alpha = \tfrac{10}{12} = 0.833, \quad \text{or} \quad \alpha_{max} = 33.6°.$$

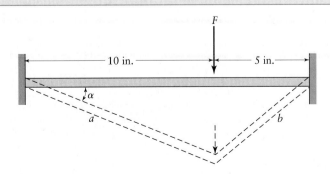

## 7.6 | Bulging

The basic forming process of *bulging* involves placing a tubular, conical, or curvilinear hollow part in a split female die and expanding it with, say, a rubber or polyurethane plug (Fig. 7.32a). The punch is then retracted, the plug returns to its original shape, and the part is removed by opening the dies. Typical products made by this process include coffee and water pitchers, barrels, and beads on drums. For parts with complex shapes, the plug, instead of being cylindrical, may be shaped in order to apply greater pressure at critical points. The major advantage of using polyurethane plugs is that they are very resistant to abrasion, wear, and lubricants, and they do not damage the surface finish of the part being formed.

Formability in bulging processes can be enhanced by the application of longitudinal compressive stresses. *Hydraulic pressure* may also be used in bulging, although this technique requires sealing and needs hydraulic controls (Fig. 7.32b). *Bellows* are manufactured by a bulging process, as shown in Fig. 7.32c. After the tube is bulged at several equidistant locations, it is compressed axially to collapse the bulged regions, thus forming bellows. The tube material must, of course, be able to undergo the large strains involved during the collapsing process. **Segmented dies** that are expanded and retracted mechanically may also be used for bulging operations. These dies are relatively inexpensive and can be used for large production runs.

**Embossing.** This process involves forming a number of shallow shapes, such as numbers, letters, or designs, on a metal sheet. Parts may be embossed with male and female dies or by other means described throughout this chapter. The process is used principally for decorative purposes.

**FIGURE 7.32** (a) Bulging of a tubular part with a flexible plug. Water pitchers can be made by this method. (b) Production of fittings for plumbing by expanding tubular blanks with internal pressure. The bottom of the piece is then punched out to produce a "T." *Source*: Schey J. A., *Introduction to Manufacturing Processes*, 2d ed., New York: McGraw-Hill, 1987. (c) Manufacturing of bellows.

# 7.7 | Rubber Forming and Hydroforming

In describing the processes in the preceding sections, we noted that the dies are made of solid materials. However, in *rubber forming*, one of the dies in a set is made of flexible material, such as a rubber or polyurethane membrane. Polyurethanes are used widely because of their resistance to abrasion, long fatigue life, and resistance to damage by burrs or sharp edges of the sheet blank.

In bending and embossing sheet metal by the rubber-forming method, as shown in Fig. 7.33, the female die is replaced with a rubber pad. Note that the outer surface of the sheet is now protected from damage or scratches, because it is not in contact with a hard metal surface during forming. Parts can also be formed with laminated sheets of various nonmetallic materials or coatings. Pressures are typically on the order of 10 MPa (1500 psi).

In *hydroforming*, or **fluid-forming process** (Fig. 7.34), the pressure over the rubber membrane is controlled throughout the forming cycle, with maximum pressures reaching 100 MPa (15,000 psi). This procedure allows close control of the part during forming to prevent wrinkling or tearing. Deeper draws are obtained than in conventional deep drawing (see Section 7.12), because the pressure around the rubber membrane forces the cup against the punch. The friction at the punch–cup interface reduces the longitudinal tensile stresses in the cup being formed and thus delays fracture. The control of frictional conditions in rubber forming as well as in other sheet-forming operations can be a critical factor in making parts successfully. Selection of proper lubricants and application methods is important. In **tube hydroforming** (Fig. 7.35), steel or other metal tubing is formed in a die and pressurized by a fluid. This procedure can form simple tubes (Fig. 7.35a), or it can form intricate hollow tubes (Fig. 7.35b). Applications of tube-hydroformed parts include automotive exhaust and structural components.

When selected properly, rubber-forming and hydroforming processes have the advantages of (a) low tooling cost, (b) flexibility and ease of operation, (c) low die wear, (d) no damage to the surface of the sheet, and (e) capability to form complex shapes.

(a)    (b)    (c)

**FIGURE 7.33** Examples of bending and embossing sheet metal with a metal punch and a flexible pad serving as the female die. *Source*: Polyurethane Products Corporation.

Pressure control valve
Forming cavity (oil filled)
Rubber diaphragm
Draw ring
Punch
Blank
Part→

(a)    (b)    (c)    (d)

**FIGURE 7.34** The hydroform, or fluid-forming, process. Note that, unlike in the ordinary deep-drawing process, the dome pressure forces the cup walls against the punch. The cup travels with the punch, and thus deep drawability is improved.

(a)

(b)

**FIGURE 7.35** (a) Schematic illustration of the tube-hydroforming process. (b) Example of tube-hydroformed parts. Automotive exhaust and structural components, bicycle frames, and hydraulic and pneumatic fittings are produced through tube hydroforming. *Source*: Schuler GmBH.

## 7.8 | Spinning

*Spinning* involves the forming of axisymmetric parts over a rotating mandrel, using rigid tools or rollers. There are three basic types of spinning processes: conventional (or manual), shear, and tube spinning. The equipment used for these processes is similar to a lathe (see Section 8.8.2) with various special features.

### 7.8.1 Conventional spinning

In the *conventional-spinning* process, a circular blank of flat or preformed sheet metal is held against a rotating mandrel while a rigid tool deforms and shapes the material over the mandrel (Fig. 7.36a). The tools may be actuated either manually or by a hydraulic mechanism. The operation involves a sequence of passes and requires considerable skill. Figure 7.37 shows some typical shapes made by conventional spinning; note that this process is particularly suitable for conical and curvilinear shapes, which would otherwise be difficult or uneconomical to form by other methods. Part diameters may range up to 6 m (20 ft). Although most spinning is performed at room temperature, thick parts or metals with low ductility or high strength require spinning at

**FIGURE 7.36** Schematic illustration of spinning processes: (a) conventional spinning and (b) shear spinning. Note that in shear spinning, the diameter of the spun part, unlike in conventional spinning, is the same as that of the blank. The quantity $f$ is the feed (in mm/rev or in./rev).

(a)

(b)

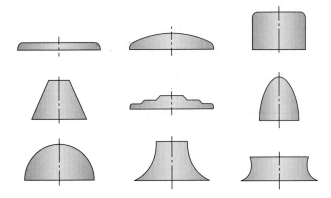

**FIGURE 7.37** Typical shapes produced by the conventional-spinning process. Circular marks on the external surfaces of components usually indicate that the parts have been made by spinning. Examples include aluminum kitchen utensils and light reflectors.

elevated temperatures. Tooling costs in spinning are relatively low; however, because the operation requires multiple passes to form the final part, it is economical for relatively small production runs only. (See Section 7.16.)

## 7.8.2 Shear spinning

In the *shear-spinning* process (also known as **power spinning, flow turning, hydrospinning**, and **spin forging**), an axisymmetric conical or curvilinear shape is generated in a manner whereby the diameter of the part remains constant. (See Fig. 7.36b.) Parts typically made by this process include rocket-motor casings and missile nose cones. Although a single roller can be used, two rollers are desirable in order to balance the radial forces acting on the mandrel. Parts up to about 3 m (10 ft) in diameter can be spun to close dimensional tolerances. Little material is wasted, and the operation is completed in a relatively short time. If the operation is carried out at room temperature, the spun part has a higher yield strength than the original material, but lower ductility and toughness. A great variety of shapes can be spun with relatively simple tooling, generally made of tool steel. Because of the large plastic deformation involved, the process generates considerable heat; it is usually carried away by a coolant-type fluid applied during spinning.

In shear spinning over a conical mandrel, the thickness, $t$, of the spun part is simply

$$t = t_o \sin \alpha, \tag{7.12}$$

where $t_o$ is the original sheet thickness. The force primarily responsible for supplying the energy required is the *tangential force*, $F_t$. For an *ideal* case in shear spinning of a cone, this force is

$$F_t = u\, t_o\, f \sin \alpha, \tag{7.13}$$

where $u$ is the specific energy of deformation, from Eq. (2.59), and $f$ is the feed. As described in Section 2.12, $u$ is the area under the true-stress–true-strain curve that corresponds to a true strain related to the shear strain by

$$\varepsilon = \frac{\gamma}{\sqrt{3}} = \frac{\cot \alpha}{\sqrt{3}}. \tag{7.14}$$

(See also Section 2.11.7.) The actual force can be as much as 50% higher than that given by Eq. (7.13), because of factors such as redundant work and friction. (See Section 6.4.2.)

An important factor in shear spinning is the **spinnability** of the metal, that is, the maximum percent reduction to which a part can be spun without fracture. To determine spinnability, a simple test method has been developed (Fig. 7.38), in which a

**FIGURE 7.38** Schematic illustration of a shear-spinnability test. As the roller advances, the part thickness is reduced. The reduction in thickness at fracture is called the *maximum spinning reduction per pass. Source*: After R. L. Kegg.

**FIGURE 7.39** Experimental data showing the relationship between maximum spinning reduction per pass and the tensile reduction of area of the original material. Note that once a material has about 50% reduction of area in a tension test, any further increase in the ductility of the original material does not improve the material's spinnability. *Source*: S. Kalpakjian.

circular blank is spun over an ellipsoid mandrel, according to Eq. (7.12). As the thickness is eventually reduced to zero, all materials will fail at some thickness $t_f$. The maximum spinning reduction in thickness is defined as

$$\text{Maximum reduction} = \frac{t_o - t_f}{t_o} \times 100\% \qquad (7.15)$$

and is plotted against the tensile reduction of area of the material tested (Fig. 7.39). Note that if a metal has a tensile reduction of area of about 50% or higher, it can be reduced in thickness by 80% by spinning in one pass. Note also that any further increase in the ductility of the original material does not improve its spinnability. Process variables such as feed and speed do not appear to have a significant effect on spinnability. In comparing Figs. 7.15b and 7.39, note that, interestingly, maximum bendability and spinnability both coincide with a tensile reduction of area of about 50%. For materials with low ductility, spinnability can usually be improved by forming at elevated temperatures, which improves the reduction of area.

### 7.8.3 Tube spinning

In the *tube-spinning* process, tubes are reduced in thickness by spinning them on a mandrel, using rollers. The operation may be carried out externally or internally (Fig. 7.40). Also, the part may be spun *forward* or *backward*, similar to a drawing or a backward extrusion process. (See Section 6.4.) In either case, the reduction in wall thickness results in a longer tube (because of volume constancy). Various internal

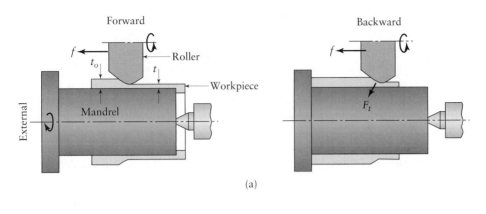

Forward                                          Backward

**FIGURE 7.40**   Examples of external and internal tube spinning and the process variables involved.

(a)

(b)

(a)

(b)

(c)

(d)

**FIGURE 7.41**   Stages in tube and shear spinning of a compressor shaft for the jet engine of the supersonic Concorde aircraft. Economic analysis indicated that the best method of manufacturing this part was to spin a preformed (forged and machined) tubular blank.

and external profiles can be produced by controlling the path of the roller during its travel along the mandrel. Tube spinning can be used for making pressure vessels, automotive components, and rocket and missile parts. It can also be combined with shear spinning, as is the case for the production of the compressor shaft for the Olympus engine (Rolls Royce) of the Concorde supersonic aircraft, as shown in Fig. 7.41.

Spinnability in this process is determined by a test method similar to that for shear spinning. Maximum reduction per pass in tube spinning is related to the tensile reduction of area of the material (see Fig. 7.39), with results very similar to those of shear spinning. As in shear spinning, ductile metals fail in tension after the reduction in thickness has taken place, whereas less ductile metals fail in the deformation zone *under* the roller.

The *ideal tangential force*, $F_t$, in *forward tube spinning* can be expressed as

$$F_t = \bar{Y}(t_o - t)f, \tag{7.16}$$

where $\bar{Y}$ is the *average flow stress* of the material. For *backward spinning*, the ideal tangential force is roughly twice that given by Eq. (7.16). Furthermore, friction and redundant work of deformation cause the actual forces to be about twice those calculated by this formula.

## 7.9 | High-Energy-Rate Forming

This section describes sheet-metal forming processes that use chemical, electrical, or magnetic sources of energy. They are called *high-energy-rate processes*, because the energy is released in a very short period of time.

### 7.9.1 Explosive forming

Figure 7.42 depicts the most common *explosive-forming* process. In this process, the workpiece is clamped over a die, the air in the die cavity is evacuated, and the whole assembly is lowered into a tank filled with water. An explosive charge is then placed at a certain distance from the sheet surface and detonated. The rapid conversion of the explosive into gas generates a shock wave; the pressure of this wave is sufficiently high to force the metal into the die cavity.

The peak pressure generated in water is given by the expression

$$p = K\left(\frac{\sqrt[3]{W}}{R}\right)^a, \tag{7.17}$$

where $p$ is the peak pressure in psi, $K$ is a constant that depends on the type of explosive (e.g., 21,600 for TNT), $W$ is the weight of the explosive in pounds, $R$ is the

**FIGURE 7.42** Schematic illustration of the explosive-forming process. Although explosives are generally used for destructive purposes, their energy can be controlled and employed in forming large parts that would otherwise be difficult or expensive to produce by other methods.

**FIGURE 7.43** Influence of the standoff distance and type of energy-transmitting medium on the peak pressure obtained using 1.8 kg (4 lb) of TNT. To be effective, the pressure-transmitting medium should have high density and low compressibility. In practice, water is a commonly used medium.

distance of the explosive from the workpiece (*standoff*) in feet; and *a* is a constant that is generally taken as 1.15.

An important factor in determining peak pressure is the **compressibility** of the energy-transmitting medium (such as water) and the medium's **acoustic impedance** (the product of mass density and sound velocity in the medium). The lower the compressibility of the medium and the higher its density, the higher is the peak pressure (Fig. 7.43). Detonation speeds are typically 6700 m/s (22,000 ft/s), and the speed at which the metal is formed is estimated to be on the order of 30 to 200 m/s (100 to 600 ft/s).

A variety of shapes can be formed explosively, provided that the material is sufficiently ductile at high strain rates. Depending on the number of parts to be formed, dies used in this process may be made of aluminum alloys, steel, ductile iron, zinc alloys, reinforced concrete, wood, plastics, or composite materials. The final properties of the parts made by this process are basically the same as the properties of parts made by conventional methods. Safety is an important aspect in explosive forming.

Requiring only one die and being versatile, explosive forming is particularly suitable for small-quantity production runs of large parts. Steel plates 25 mm (1 in.) thick and 3.6 m (12 ft) in diameter have been formed by this method. Also, tubes with a wall thickness of 25 mm have been bulged by explosive-forming techniques. Another explosive-forming method, which uses a cartridge as the source of energy, is shown in Fig. 7.44; no other energy-transmitting medium is used. The process can be used for bulging and expanding of thin-walled short tubes; however, die wear and die failure can be a significant problem in this operation.

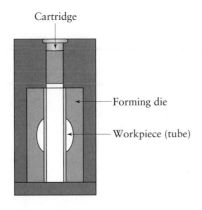

**FIGURE 7.44** Schematic illustration of the confined method of explosive bulging of tubes. Thin-walled tubes of nonferrous metals can be formed to close tolerances by this process.

**EXAMPLE 7.5**  Peak pressure in explosive forming

(a) Calculate the peak pressure in water for 0.1 lb of TNT at a standoff of 1 ft. (b) Is this pressure sufficiently high for forming sheet metals?

**Solution.**

a)  Using Eq. (7.17), we find that

$$p = (21,600)\left(\frac{\sqrt[3]{0.1}}{1}\right)^{1.15} = 9000 \text{ psi.}$$

b)  This pressure is sufficiently high to form sheet metals. Note in Example 2.6, for instance, that the pressure required to expand a thin-walled spherical shell of a material similar to soft aluminum alloys is only 400 psi. Also note that a process such as hydroforming (Section 7.7) has a maximum hydraulic pressure in the dome of about 100 MPa (15,000 psi). Other rubber-forming processes use pressures from about 1500 psi to 7500 psi. Therefore, the pressure obtained in this problem is sufficient for most sheet-forming processes.

## 7.9.2  Electrohydraulic forming

In the *electrohydraulic-forming* process, also called **underwater-spark** or **electric-discharge forming**, the source of energy is a spark from two electrodes connected with a thin wire (Fig. 7.45). The energy is stored in a bank of condensers charged with direct current. The rapid discharge of this energy through the electrodes generates a shock wave, which is strong enough to form the part. This process is essentially similar to explosive forming, except that it utilizes a lower level of energy and is used with smaller workpieces; it is also a safer operation than explosive forming.

## 7.9.3  Magnetic-pulse forming

In the *magnetic-pulse-forming* process, the energy stored in a capacitor bank is discharged rapidly through a magnetic coil. In a typical example, a ring-shaped coil is placed over a tubular workpiece to be formed over another solid piece, thus making an integral part (Fig. 7.46). The magnetic field produced by the coil crosses the metal tube, generating an *eddy currents* in the tube; this current, in turn, produces its own magnetic field. The forces produced by the two magnetic fields oppose each other; thus, there is a repelling force between the coil and the tube, and the high forces generated collapse the tube over the inner piece.

The higher the electrical conductivity of the workpiece, the higher is the magnetic force. Note, however, that the metal does not need to have any special magnetic properties. Magnetic-pulse forming is used for a variety of operations, such as swaging of thin-walled tubes over rods, cables, and plugs (such as placing end fittings onto torque tubes for the Boeing 777 aircraft); bulging; and flaring. Flat coils are also used for forming sheet metal, as in embossing and shallow drawing operations.

**FIGURE 7.45**  Schematic illustration of the electrohydraulic-forming process.

(a)  (b)

**FIGURE 7.46**
(a) Schematic illustration of the magnetic-pulse-forming process. The part is formed without physical contact with any object.
(b) Aluminum tube collapsed over a hexagonal plug by the magnetic-pulse-forming process.

# 7.10 | Superplastic Forming

Section 2.2.7 described the *superplastic behavior* of some very fine-grained alloys (normally less than 10 to 15 $\mu$m), where very large elongations (up to 2000%) are obtained at certain temperatures and low strain rates. These alloys, such as zinc–aluminum and titanium, can be formed into complex shapes by employing traditional metalworking or polymer-processing techniques. Commonly used die materials in superplastic forming include low-alloy steels, cast tool steels, ceramics, graphite, and plaster of paris; selection of the material depends on the forming temperature and strength of the superplastic alloy.

The high ductility and relatively low strength of superplastic alloys present the following advantages in superplastic forming:

1. Lower strength of tooling, because of the low strength of the material at forming temperatures, and, hence, lower tooling costs;
2. Ability to form complex shapes in one piece, with fine detail, close dimensional tolerances, and elimination of secondary operations;
3. Weight and material savings, because of the good formability of superplastic materials;
4. Little or no residual stresses in the formed parts.

The limitations are:

1. The material must not be superplastic at service temperatures;
2. Because of the extreme strain-rate sensitivity of the superplastic material, the material must be formed at sufficiently low rates (typically at strain rates of $10^{-4}$/s to $10^{-2}$/s). Forming times range anywhere from a few seconds to several hours, and thus cycle times are much longer than in conventional forming processes. Superplastic forming is therefore a batch-forming process.

Superplastic alloys (particularly Zn-22Al and Ti-6Al-4V) can be formed by bulk deformation processes, such as compression molding, closed-die forging, coining, hubbing, and extrusion. Sheet forming of these materials can also be done using operations such as thermoforming, vacuum forming, and blow molding.

**FIGURE 7.47** Two types of structures made by diffusion bonding and superplastic forming of sheet metal. Such structures have a high stiffness-to-weight ratio. *Source*: Rockwell Automation, Inc.

An important aspect is the ability to fabricate sheet-metal structures by combining **diffusion bonding** (see Section 12.11) *with* superplastic forming (SPF/DB). Typical structures in which flat sheets are diffusion bonded and formed are shown in Fig. 7.47. After diffusion bonding of selected locations of the sheets, the unbonded regions (*stop off*) are expanded into a mold by air pressure. These structures are thin and have high stiffness-to-weight ratios; hence, they are particularly important in aircraft and aerospace applications. This process improves productivity by eliminating mechanical fasteners and produces parts with good dimensional accuracy and low residual stresses. The technology is now well advanced for titanium structures (typically Ti-6Al-4V alloy) for aerospace applications. Structures made of 7475-T6 aluminum alloy continue to be developed using this technique; other metals suitable for superplastic forming include Inconel 100 and Incoloy 718 nickel alloys and iron-base, high-carbon alloys.

## 7.11 | Various Forming Methods

a) **Peen forming** is used to produce curvatures on thin sheet metals by **shot peening** one surface of the sheet. (See Section 4.5.1.) Peening is done with cast-iron or steel shot, discharged either from a rotating wheel or with an air blast from a nozzle. Peen forming is used by the aircraft industry to generate complex and smooth curvatures on aircraft-wing skins (Fig. 7.48). Cast-steel shot about 2.5 mm (0.1 in.) in diameter at speeds of 60 m/s (200 ft/s) has been used to form wing panels 25 m (80 ft) long. For heavy sections, shot diameters as large as 6 mm $\left(\frac{1}{4} \text{ in.}\right)$ may be used.

In peen forming, the surface of the sheet is subjected to compressive stresses, which tend to expand the surface layer. Since the material below the peened surface remains rigid, the surface expansion causes the sheet to develop a curvature. The process also induces compressive surface residual stresses, which improve the fatigue strength of the sheet. The peen-forming process is also used for straightening twisted or bent parts; out-of-round rings, for example, can be straightened by this method.

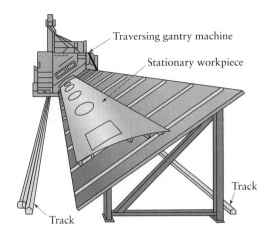

Traversing gantry machine

Stationary workpiece

Track

Track

**FIGURE 7.48** Peen-forming machine to form a large sheet-metal part, such as an aircraft-skin panel. The sheet is stationary, and the machine traverses it. *Source*: Metal Improvement Company.

b) **Gas mixtures** in a closed container have been used as an energy source in sheet-metal forming. When the gas is ignited, the pressures generated are sufficiently high to form parts. The principle is similar to the generation of pressure in an internal combustion engine.

c) **Liquefied gases,** such as liquid nitrogen, may be used to develop pressures high enough to form sheet metals. When allowed to reach room temperature in a closed container, liquified nitrogen becomes gaseous and expands, developing the necessary pressure to form the part.

d) **Laser forming** is a more recent technology that uses localized heating to induce thermal stress gradients through the thickness of the sheet. These stresses are sufficiently high to cause localized plastic deformation of the sheet without the use of external forces and result in, for example, a bent sheet.

e) **Laser-assisted forming** uses lasers as a local heat source in order to reduce the flow stress of the material at specific locations and to improve formability, thus increasing process flexibility. Examples of its uses include straightening, bending, embossing, and forming of complex flat or tubular components.

f) **Microforming** is a more recent term; it describes processes that are used to produce very small metallic parts and components. Examples of recently miniaturized products include a wristwatch with an integrated digital camera and a one-gigabyte hard-disk drive that fits into an eggshell. As discussed in greater detail in Chapter 13, several processes are involved in the fabrication of micromechanical devices. Through the use of specialized tooling, machines, and handling systems, microforming consists of miniaturization of traditional as well as more recent forming and machining processes, including the use of lasers. Other, similar terminology includes *microshaping, microcasting,* and *microsintering.* Typical components made by microforming include small shafts for micromotors; springs; screws; and a variety of cold-headed, extruded, bent, embossed, coined, punched, or deep-drawn parts, with dimensions typically in the submillimeter range and weights on the order of milligrams.

## 7.11.1 Manufacturing honeycomb structures

There are two principal methods of manufacturing honeycomb materials. In the **expansion** process (Fig. 7.49a), which is the most common method, sheets are cut from a coil, and an adhesive is applied at intervals (node lines). The sheets are stacked and cured in an oven, whereby strong bonds develop at the adhesive joints. The block is then cut into slices of the desired dimensions and stretched to produce a honeycomb

**FIGURE 7.49** Methods of making honeycomb materials: (a) expansion process and (b) corrugation process. *Source: Materials Engineering.* Reprinted with permission. (c) Making a honeycomb sandwich.

structure. This procedure is similar to expanding folded paper structures into the shape of decorative objects such as lanterns.

In the **corrugation** process (Fig. 7.49b), sheets pass through a pair of specially designed rolls, which make them into corrugated sheets; the corrugated sheets are then cut into desired lengths. Again, adhesive is applied to the node lines, and the block is cured. Note that no expansion process is involved. The honeycomb material is then made into a sandwich structure (Fig. 7.49c). Face sheets are subsequently joined with adhesives to the top and bottom surfaces.

Honeycomb structures are most commonly made of 3000-series aluminum, but are also made of titanium, stainless steels, and nickel alloys. Because of their light weight and high resistance to bending forces, these structures are used for aircraft and aerospace components, as well as buildings and transportation equipment.

# 7.12 | Deep Drawing

In *deep drawing*, a flat sheet-metal blank is formed into a cylindrical or box-shaped part by means of a punch that presses the blank into the die cavity (Fig. 7.50a). Although the process is generally called *deep drawing*, meaning *forming deep parts*, the basic operation also produces parts that are shallow or have moderate depth. Deep drawing, first developed in the 1700s, has been studied extensively and has become an important metalworking process. Typical parts produced by this method include beverage cans, pots and pans, containers of all shapes and sizes, kitchen sinks, and automobile panels.

Figure 7.50b shows the basic parameters in deep drawing a cylindrical cup. A circular sheet blank with a diameter $D_o$ and thickness $t_o$ is placed over a die opening with a corner radius $R_d$. The blank is held in place with a **blankholder,** or **hold-down ring,** with a certain force. A punch with a diameter $D_p$ and a corner radius $R_p$ moves downward and pushes the blank into the die cavity, thus forming a cup. The significant independent variables in deep drawing are

1. Properties of the sheet metal;
2. The ratio of the blank diameter to the punch diameter;
3. The thickness of the sheet;
4. The clearance between the punch and the die;
5. The corner radii of the punch and die;
6. The blankholder force;
7. Friction and lubrication at the punch, die, and workpiece interfaces;
8. The speed of the punch.

At an intermediate stage during the deep-drawing operation, the workpiece is subjected to the states of stress shown in Fig. 7.51. On element $A$ in the blank, the radial tensile stress is due to the blank being pulled into the cavity, and the compressive stress normal to the element is due to the blankholder pressure. With a free-body

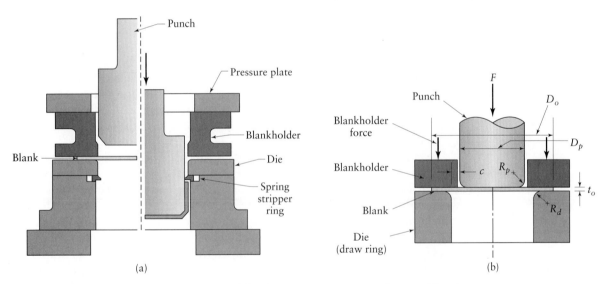

(a)                                                                  (b)

**FIGURE 7.50**   (a) Schematic illustration of the deep-drawing process. This procedure is the first step in the basic process by which aluminum beverage cans are produced today. The stripper ring facilitates the removal of the formed cup from the punch. (b) Variables in deep drawing of a cylindrical cup. Only the punch force in this illustration is a dependent variable; all others are independent variables, including the blankholder force.

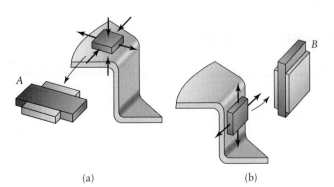

(a)                                        (b)

**FIGURE 7.51**
Deformation of elements in (a) the flange and (b) the cup wall in deep drawing of a cylindrical cup.

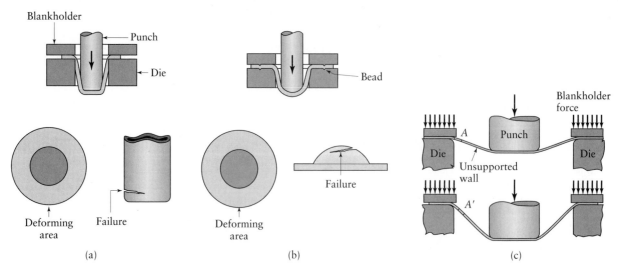

**FIGURE 7.52** Examples of drawing operations: (a) pure drawing and (b) pure stretching. The bead prevents the sheet metal from flowing freely into the die cavity. (c) Possibility of wrinkling in the unsupported region of a sheet in drawing. *Source*: After W. F. Hosford and R. M. Caddell.

diagram of the blank along its diameter, it can be shown that the radial tensile stresses lead to compressive hoop stresses on element $A$. (It is these hoop stresses that tend to cause the flange to wrinkle during drawing, thus requiring a blankholder under a certain force.) Under this state of stress, element $A$ contracts in the hoop direction and elongates in the radial direction.

The cup wall, which is already formed, is subjected principally to a longitudinal tensile stress, as shown in element $B$ in Fig. 7.51. The punch transmits the drawing force, $F$, (see Fig. 7.50b) through the walls of the cup and to the flange that is being drawn into the die cavity. The tensile hoop stress on element $B$ is caused by the cup being held tightly on the punch because of its contraction under tensile stresses in the cup wall. Note that a thin-walled tube, when subjected to longitudinal tension, becomes smaller in its diameter, as can be shown from the generalized *flow rule equations*; see Eq. (2.43). Thus, because it is constrained by the rigid punch, element $B$ tends to elongate in the longitudinal (axial) direction, with no change in its width.

An important aspect of drawing is determining how much **stretching** and how much **pure drawing** is taking place (Fig. 7.52). Note that either with a high blankholder force or with the use of **draw beads** (Figs. 7.52b and 7.53), the blank can be prevented from flowing freely into the die cavity. The deformation of the sheet metal takes place mainly under the punch, and the sheet begins to *stretch*, eventually resulting in necking and tearing. Whether necking is localized or diffused depends on (a) the strain-rate sensitivity exponent, $m$, of the sheet metal (the higher the $m$ value, the more diffuse the neck is), (b) geometry of the punch, and (c) lubrication.

Conversely, a low blankholder force will allow the blank to flow freely into the die cavity (*pure drawing*), whereby the blank diameter is reduced as drawing progresses. The deformation of the sheet is mainly in the flange, and the cup wall is subjected only to elastic stresses. However, these stresses increase with an increasing $D_o/D_p$ ratio and can eventually lead to failure when the cup wall cannot support the load required to draw in the flange (Fig. 7.52a). Also note that, in pure drawing, element $A$ in Fig. 7.51a will tend to increase in thickness as it moves toward the die cavity (because it is being reduced in diameter).

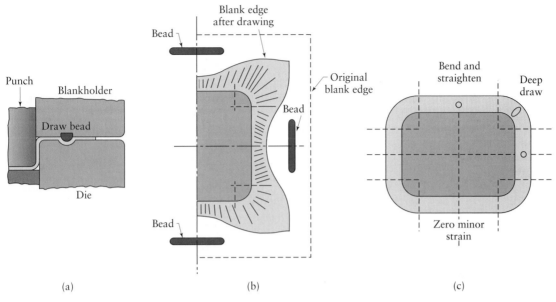

(a)                                   (b)                                   (c)

**FIGURE 7.53**  (a) Schematic illustration of a draw bead. (b) Metal flow during drawing of a box-shaped part, using beads to control the movement of the material. (c) Deformation of circular grids in drawing. (See Section 7.13.) *Source*: After S. Keeler.

In drawing operations, the length of the unsupported wall is significant in that it can lead to **wrinkling**. As shown in Fig. 7.52c, element *A* in the sheet is being pulled into the die cavity as the punch descends. Note, however, that the blank is becoming smaller in diameter and the circumference at the element is becoming smaller as the element moves to position *A'*. Thus, at this position, the element is being subjected to circumferential compressive strains and is unsupported by any tooling, unlike an element between the blankholder and the die surface. Because the sheet is thin and cannot support circumferential compressive strains to any significant extent, it tends to wrinkle in the unsupported region. This situation is particularly common in pure drawing, whereas it is less so as the process becomes more pure stretching and the blank diameter remains basically the same.

**Ironing.**  If the thickness of the sheet as it enters the die cavity is more than the clearance between the punch and the die, the thickness will have to be reduced; this effect is known as *ironing*. Ironing produces a cup with constant wall thickness (Fig. 7.54); thus, the smaller the clearance, the greater is the amount of ironing. Obviously, because of volume constancy, an ironed cup will be longer than a cup produced with a large clearance. (See also the case study at the end of this chapter.)

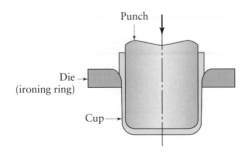

**FIGURE 7.54**  Schematic illustration of the ironing process. Note that the cup wall is thinner than its bottom. All beverage cans without seams (known as two-piece cans) are ironed, generally in three steps, after being deep drawn into a cup. (Cans with separate tops and bottoms are known as three-piece cans.)

### 7.12.1 Deep drawability (limiting drawing ratio)

The *limiting drawing ratio* (LDR) is defined as the maximum ratio of blank diameter to punch diameter that can be drawn without failure, or $D_o/D_p$. Many attempts have been made to correlate this ratio with various mechanical properties of the sheet metal. In an ordinary deep-drawing process, failure generally occurs by *thinning* in the cup wall under high longitudinal tensile stresses. By observing the movement of the material into the die cavity (see Fig. 7.51), we note that the material should be capable of undergoing a reduction in width (by being reduced in diameter), yet it should resist thinning under the longitudinal tensile stresses in the cup wall.

The ratio of width strain to thickness strain (Fig. 7.55) is defined as

$$R = \frac{\varepsilon_w}{\varepsilon_t} = \frac{\ln\left(\dfrac{w_o}{w_f}\right)}{\ln\left(\dfrac{t_o}{t_f}\right)}, \tag{7.18}$$

where $R$ is known as the **normal anisotropy** of the sheet metal (also called **plastic anisotropy** or the **strain ratio**). The subscripts $o$ and $f$ refer to the original and final dimensions, respectively. An $R$ value of unity indicates that the width and thickness strains are equal to each other; that is, the material is isotropic.

As sheet metals are generally thin compared with their surface area, errors in the measurement of small thicknesses is possible. Equation (7.18) can be modified, based on volume constancy, to

$$R = \frac{\ln\left(\dfrac{w_o}{w_f}\right)}{\ln\left(\dfrac{w_f \ell_f}{w_o \ell_o}\right)}, \tag{7.19}$$

where $\ell$ refers to the gage length of the sheet specimen. The final length and width in a test specimen are generally measured at an elongation of 15% to 20% or, for materials with lower ductility, below the elongation, where necking begins.

Rolled sheets generally have **planar anisotropy**; thus, the $R$ value of a specimen cut from a rolled sheet (Fig. 7.55) will depend on the specimen's orientation with respect to the rolling direction of the sheet. (See also Fig. 3.15.) In this case, an *average* $R$ value, $\bar{R}$, is calculated as

$$\bar{R} = \frac{R_0 + 2R_{45} + R_{90}}{4}, \tag{7.20}$$

where the subscripts 0, 45, and 90 refer to angular orientation (in degrees) of the test specimen with respect to the rolling direction of the sheet. Note that an isotropic

**FIGURE 7.55**   Definition of the normal anisotropy, $R$, in terms of width and thickness strains in a tensile-test specimen cut from a rolled sheet. Note that the specimen can be cut in different directions with respect to the length, or rolling direction, of the sheet.

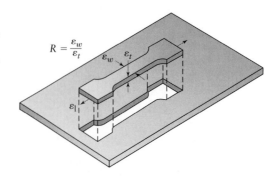

TABLE 7.3

| Typical Range of the Average Normal Anisotropy Ratio, $\bar{R}$, for Various Sheet Metals | |
| --- | --- |
| Zinc alloys | 0.4–0.6 |
| Hot-rolled steel | 0.8–1.0 |
| Cold-rolled rimmed steel | 1.0–1.4 |
| Cold-rolled aluminum-killed steel | 1.4–1.8 |
| Aluminum alloys | 0.6–0.8 |
| Copper and brass | 0.6–0.9 |
| Titanium alloys ($\alpha$) | 3.0–5.0 |
| Stainless steels | 0.9–1.2 |
| High-strength low-alloy steels | 0.9–1.2 |

material has an $\bar{R}$ value of unity. Table 7.3 gives some typical $\bar{R}$ values. Although hexagonal close-packed metals usually have high $\bar{R}$ values, the low value for zinc is the result of its high $c/a$ ratio in the crystal lattice. (See Fig. 3.2c.) The $\bar{R}$ value also depends on the grain size and texture of the sheet metal. For cold-rolled steels, for example, $\bar{R}$ decreases as grain size decreases (Fig. 7.56; note that grain size and grain number are inversely related). For hot-rolled sheet steels, $\bar{R}$ is approximately unity, because the texture developed has a random orientation.

Experiments with sheet steels have also shown that $\bar{R}$ is related to the average modulus of elasticity, $\bar{E}$, as shown in Fig. 7.57; thus, a high value of $\bar{R}$ is associated with a high value of $\bar{E}$. The average modulus is determined in the same manner as $\bar{R}$ in Eq. 7.20. Techniques and instruments are now available to obtain the modulus of elasticity of sheet specimens by measuring their natural frequencies, from which the elastic modulus can be calculated. (See Problem 3.47.)

Figure 7.58 shows the direct relationship between $\bar{R}$ and LDR, as determined experimentally. In spite of its scatter, no other mechanical property of sheet metal indicates as consistent a relationship with LDR as $\bar{R}$ does. For an isotropic material, and based on ideal deformation, the maximum LDR is equal to $e = 2.718$.

The **planar anisotropy** of a sheet, $\Delta R$, can also be defined in terms of directional $R$ values as

$$\Delta R = \frac{R_0 - 2R_{45} + R_{90}}{2}, \tag{7.21}$$

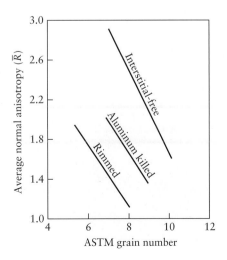

**FIGURE 7.56** Effect of grain size on the average normal anisotropy for various low-carbon steels. *Source*: After D. J. Blickwede.

**FIGURE 7.57**
Relationship between the average normal anisotropy, $\bar{R}$, and the average modulus of elasticity, $\bar{E}$, for steel sheet. *Source*: After P. R. Mould and T. R. Johnson, Jr.

**FIGURE 7.58**   Effect of average normal anisotropy, $\bar{R}$, on limiting drawing ratio (LDR) for a variety of sheet metals. Zinc has a high $c/a$ ratio (see Figure 3.2c), whereas titanium has a low ratio. *Source*: After M. Atkinson.

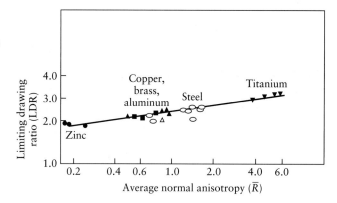

which is the difference between the average of the $R$ values in the 0° and 90° directions to rolling and the $R$ value at 45°. In addition, $\Delta R$ is related to $\Delta E$, where $\Delta E$ is determined in the same manner as $\Delta R$ is in Eq (7.21), with $R$ replaced by $E$.

**Earing.**   Planar anisotropy causes **ears** to form in drawn cups (Fig. 7.59), producing a wavy edge. The number of ears produced may be four, six, or eight. The height of the ears increases with increasing $\Delta R$. When $\Delta R = 0$, no ears form. Ears are objectionable, because they have to be trimmed off, thus also wasting material. The ear length in a typical drawn aluminum beverage can is about 1–2% of the length of the can.

**FIGURE 7.59**   Earing in a drawn steel cup, caused by the planar anisotropy of the sheet metal.

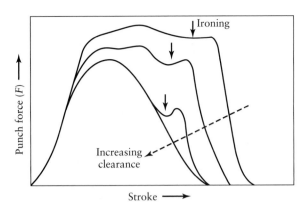

**FIGURE 7.60** Schematic illustration of the variation of punch force with stroke in deep drawing. Note that ironing does not begin until after the punch has traveled a certain distance and the cup is formed partially. Arrows indicate the beginning of ironing.

Deep drawability is thus enhanced with a high $\bar{R}$ and a low $\Delta R$. Generally, however, sheet metals with a high $\bar{R}$ also have a high $\Delta R$. Attempts continue to be made to develop textures in sheet metals to improve their drawability. The controlling parameters in processing metals have been found to be alloying elements, additives, processing temperatures, annealing cycles after processing, thickness reduction in rolling, and cross (biaxial) rolling of plates when rolling them into sheets.

**Maximum punch force.** The *maximum punch force*, $F_{max}$, supplies the work required in deep drawing. As in other deformation processes, the work consists of ideal work of deformation, redundant work, friction work, and, when present, the work required for ironing (Fig. 7.60). Because of the many variables involved in this operation, and because deep drawing is not a steady-state process, calculating the punch force is difficult. Various expressions have been developed; one simple and approximate formula for the punch force is

$$F_{max} = \pi D_p t_o \, (\text{UTS}) \left( \frac{D_o}{D_p} - 0.7 \right). \tag{7.22}$$

Note that Eq. (7.22) does not include friction, the corner radii of the punch and die, or the blankholder force. However, this empirical equation makes approximate provisions for these factors.

The punch force is supported basically by the cup wall; if this force is excessive, **tearing** occurs, as shown in the lower part of Fig. 7.52a. Note that the cup was drawn to a considerable depth before failure occurred. This result can be expected by observing Fig. 7.60, where it can be seen that the punch force does not reach a maximum until after the punch has traveled a certain distance. The punch corner radius and die radius (if they are greater than 10 times the sheet thickness) do not significantly affect the maximum punch force.

---

**EXAMPLE 7.6 Estimating the limiting drawing ratio**

Estimate the limiting drawing ratio (LDR) that you would expect from a sheet metal that, when stretched by 23% in length, decreases in thickness by 10%.

**Solution.** From volume constancy of the test specimen, we have

$$w_o t_o l_o = w_f t_f l_f \qquad \text{or} \qquad \frac{w_f t_f l_f}{w_o t_o l_o} = 1.$$

From the information given, we obtain

$$\frac{\ell_f - \ell_o}{\ell_o} = 0.23 \quad \text{or} \quad \frac{\ell_f}{\ell_o} = 1.23$$

and

$$\frac{t_f - t_o}{t_o} = -0.10 \quad \text{or} \quad \frac{t_f}{t_o} = 0.90.$$

Hence,

$$\frac{w_f}{w_o} = 0.903.$$

From Eq. (7.18), we have

$$R = \frac{\ln\left(\dfrac{w_o}{w_f}\right)}{\ln\left(\dfrac{t_o}{t_f}\right)} = \frac{\ln 1.107}{\ln 1.111} = 0.965.$$

If the sheet has planar isotropy. then $R = \bar{R}$, and from Fig. 7.58 we estimate that

$$\text{LDR} = 2.4.$$

---

### EXAMPLE 7.7  Theoretical limiting drawing ratio

Show that the theoretical limiting drawing ratio is 2.718.

**Solution.**  In reviewing Fig. 7.50b and assuming that, ideally, there is no thickness change of the sheet during drawing, we note that the diametral change from a blank to a cup involves a true strain of

$$\varepsilon_{max} = \ln(\pi D_o / \pi D_p) = \ln(D_o / D_p).$$

The work required for plastic deformation is supplied by the radial stress on element $A$ in Fig. 7.51a. Let's now refer to Fig. 6.62 for wire drawing and assume that the cross-section being drawn is thin and rectangular, with the thickness direction perpendicular to the page, and a thickness that ideally remains constant. From Eq. (6.61), we then note that the drawing stress (the radial stress in Fig. 7.51a), which we denote as $\sigma_d$, is given by

$$\sigma_d = Y \varepsilon_{max},$$

where $Y$ is the yield stress of a perfectly plastic (ideal) material. Since, in the limit, the drawing stress is equal to the yield stress [see also the derivation of Eq. (6.72)], we have

$$Y = Y \varepsilon_{max},$$

or $\varepsilon_{max} = 1$. Consequently, $\ln(D_o / D_p) = 1$, and hence

$$D_o / D_p = 2.718.$$

---

### EXAMPLE 7.8  Estimating cup diameter and earing

A steel sheet has $R$ values of 0.9, 1.3, and 1.9 for the 0°, 45°, and 90° directions to rolling, respectively. If a round blank is 100 mm in diameter, estimate the smallest cup diameter to which it can be drawn. Will ears form?

**Solution.** Substituting the given values into Eq (7.20), we obtain

$$\bar{R} = \frac{0.9 + (2)(1.3) + 1.9}{4} = 1.35.$$

The limiting drawing ratio (LDR) is defined as the maximum ratio of the diameter of the blank to the diameter of the punch that can be drawn without failure, i.e., $D_o/D_p$. From Fig. 7.58, we estimate the LDR for this steel to be approximately 2.5.

To determine whether or not earing will occur, we substitute the $R$ values into Eq (7.21). Thus, we obtain

$$\Delta R = \frac{0.9 - (2)(1.3) + 1.9}{2} = 0.1.$$

We know that ears will not form if $\Delta R = 0$, and since this is not the case here, ears will form in deep drawing this material.

## 7.12.2 Drawing practice

a) **Clearances and radii.** Generally, *clearances* are 7–14% greater than the original thickness of the sheet. The selection of clearance depends on the thickening of the cup wall (which is a function of the drawing ratio). As the clearance decreases, ironing increases. If the clearance is too small, the blank may simply be pierced and sheared by the punch. The *corner radii* of the punch and the die are important. If they are too small, they can cause fracture at the corners (Fig. 7.61). If they are too large, the unsupported area wrinkles. Wrinkling in this region (and from the flange area) causes a defect on the cup wall called **puckering**.

b) **Draw beads.** Draw beads (see Figs. 7.52b and 7.53) are useful in controlling the flow of the blank into the die cavity. They are essential in drawing box-shaped or nonsymmetric parts. Draw beads also help in reducing the blankholder forces required, because of the stiffness imparted to the flange by the bent regions along the beads. Proper design and location of draw beads requires considerable experience.

c) **Blankholder pressure.** Blankholder pressure is generally 0.7% to 1.0% of the sum of the yield and the ultimate tensile strength of the sheet metal. Too high a blankholder force increases the punch load (because of friction) and leads to tearing of the cup wall; on the other hand, if the force is too low, the flange wrinkles. Since the blankholder force controls the flow of the sheet metal within the die, new presses have been designed that are capable of applying a **variable blankholder force**; that is, the blankholder force can be programmed and varied with punch stroke. Additional features include the control of metal flow, which is achieved by using separate die cushions that allow the *local* blankholder force to be applied at the proper location on the sheet in order to improve drawability.

**FIGURE 7.61** Effect of die and punch corner radii on fracture in deep drawing of a cylindrical cup. (a) Die corner radius too small. The die corner radius should generally be 5 to 10 times the sheet thickness. (b) Punch corner radius too small. Because friction between the cup and the punch aids in the drawing operation, excessive lubrication of the punch is detrimental to drawability.

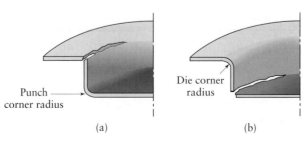

Punch corner radius

Die corner radius

(a)          (b)

**FIGURE 7.62** Reducing the diameter of drawn cups by redrawing operations: (a) conventional redrawing and (b) reverse redrawing. Small-diameter deep containers undergo many redrawing operations.

(a) Conventional redrawing       (b) Reverse redrawing

**FIGURE 7.63** Deep drawing without a blankholder, using a *tractrix* die profile. The tractrix is a special curve, the construction for which can be found in texts on analytical geometry or in handbooks.

(a)          (b)          (c)

d) **Redrawing.** Containers or shells that are too difficult to draw in one operation are generally *redrawn*, as shown in Fig. 7.62a. In **reverse redrawing**, shown in Fig. 7.62b, the metal is subjected to bending in the direction opposite to its original bending configuration. This reversal in bending results in **strain softening** and is another example of the Bauschinger effect (see Section 2.3.2); this operation requires lower forces than direct redrawing, and the material behaves in a more ductile manner.

e) **Drawing without a blankholder.** Deep drawing may be carried out without a blankholder, provided that the sheet metal is sufficiently thick to prevent wrinkling. The dies are specially contoured for this operation; one example is shown in Fig. 7.63. An approximate limit for drawing without a blankholder is given by

$$D_o - D_p < 5t_o. \tag{7.23}$$

It is apparent from Eq (7.23) that the process can be used with thin materials for shallow draws. Although the punch stroke is longer than ordinary drawing, a major advantage of this process is the reduced cost of tooling and equipment.

f) **Tooling and equipment.** The most commonly used tool materials for deep drawing are tool steels and alloyed cast iron; other materials, including carbides and plastics, may also be used, depending on the particular application. (See Table 3.5.) A double-action mechanical press is generally used for deep drawing, although hydraulic presses are also used. The double-action press controls the punch and blankholder independently and forms the part at a constant speed. Punch speeds generally range between 0.1 and 0.3 m/s (20 and 60 ft/min). Speed is generally not important in drawability, although lower speeds are used for high-strength metals.

g) **Lubrication.** Lubrication in deep drawing is important in lowering forces, increasing drawability, reducing tooling wear, and reducing part defects. In general, lubrication of the punch should be minimized, as friction between the punch and the cup improves drawability. For general applications, commonly used lubricants include mineral oils, soap solutions, and heavy-duty emulsions. For more difficult applications, coatings, wax, and solid lubricants are used.

## 7.13 | Formability of Sheet Metals

The formability of sheet metals has been of great and continued interest, because of its technological and economic significance. *Sheet-metal formability* is generally defined as the ability of a sheet to undergo the desired shape changes without failure such as necking, tearing, or splitting. Three factors have a major influence on formability: (a) properties of the sheet metal; (b) lubrication at various interfaces between the sheet, the dies, and the tooling; (c) characteristics of the equipment, tools, dies used in forming the sheet. Several techniques have been developed to test the formability of sheet metals. New developments include the ability to predict formability by modeling the particular forming process, using several inputs, including the crystallographic texture of the sheet metal.

### 7.13.1 Testing for formability

There are several tests used to determine the formability of sheet metal:

a) **Tension tests.** The *tension test* (see Section 2.2) is the most basic and common test used to evaluate formability. This test determines important properties of the sheet metal, such as total elongation of the sheet specimen at fracture, the strain hardening exponent, $n$, the planar anisotropy, $\Delta R$, and the normal anisotropy, $R$.

b) **Cupping tests.** Because, as we have seen, sheet forming is basically a biaxial stretching process, the earliest tests developed to determine or to predict formability were *cupping tests*, such as the **Erichsen** and **Olsen** tests (*stretching*) and the **Swift** and **Fukui** tests (*drawing*). In the Erichsen test, a sheet-metal specimen is clamped over a circular flat die with a load of 1000 kg. A 20-mm-diameter steel ball is then hydraulically pushed into the sheet metal until a crack appears on the stretched specimen or until the punch force reaches a maximum. The distance $D$, in mm, is the Erichsen number. The greater the value of $D$, the greater is the formability of the sheet. Cupping tests measure the capability of the material to be stretched before fracturing and are relatively easy to perform; however, because the stretching under the ball is axisymmetric, they do not at all simulate the exact conditions of actual forming operations.

**FIGURE 7.64** Schematic illustration of the punch–stretch test on sheet specimens with different widths, clamped at the edges. The narrower the specimen, the more uniaxial is the stretching. A large square specimen stretches biaxially under the hemispherical punch. (See also Fig. 7.65.)

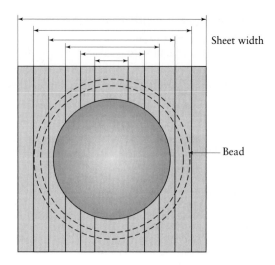

c) **Bulge test.** Equal (balanced) biaxial stretching of sheet metals is also performed in the *bulge test*, which has been used extensively to simulate sheet-forming operations. In this test, a circular blank is clamped at its periphery and is bulged by *hydraulic* pressure, thus replacing the punch. The process is one of pure stretch forming, and no friction is involved, as would be in using a punch. Bulge tests can be used to obtain effective-stress–effective-strain curves for biaxial loading under frictionless conditions. The biaxial bulge limit (depth) is also a sensitive measure of sheet cleanliness (quality).

d) **Forming-limit diagrams.** An important development in testing formability of sheet metals is the construction of *forming-limit diagrams* (FLDs). In the punch–stretch test, the sheet blank is marked with a grid pattern of circles, typically 2.5 to 5 mm (0.1 to 0.2 in.) in diameter, or similar patterns, using chemical-etching or photoprinting techniques. The blank is then stretched over a punch (Fig. 7.64), and the deformation of the circles is observed and measured in regions where failure (necking and tearing) occurs. Note that the sheet is fixed by a draw bead, which prevents the sheet from being drawn into the die. For improved accuracy of measurement, the circles on the sheet are made as small as possible and the grid lines (if used) as narrow as possible. Because lubricants can significantly influence the test results, lubrication conditions should be stated in reporting the test results. In this manner, different lubricants can be evaluated for the same sheet metal, or different metals can be evaluated using the same lubricant. This test can also be run with an unlubricated punch.

In order to vary the deformation *strain path*, the specimens are prepared with varying widths, as shown in Fig. 7.64b. Thus, the center of the square specimen produces **equal (balanced) biaxial stretching** under the punch, whereas a specimen with a small width approaches a state of **uniaxial stretching** (simple tension). After a series of such tests is performed on a particular type of sheet metal (Fig. 7.65), the boundaries between *safe* and *failed* regions are plotted on the forming-limit diagram (Fig. 7.66a). Figure 7.67 shows another example of a failed part with round and square grid patterns. Note in Fig. 7.66a the various strain paths that are straight lines: (a) The 45° line on the right represents **equal biaxial strain**; (b) the vertical line at the center of the diagram represents **plane strain**, because the minor planar strain is zero (although the sheet does become thinner); (c) the **simple tension** line on the left has a 2:1 slope, because the Poisson's ratio in the plastic range is 0.5 (note that the minor strain is one half of the major strain); (d) the **pure-shear** line has a negative 45° slope, because of the nature of deformation, as can also be seen in Figure 2.21.

**FIGURE 7.65**  Bulge-test results on steel sheets of various widths. The first specimen (farthest left) stretched farther before cracking than the last specimen. From left to right, the state of stress changes from almost uniaxial to biaxial stretching. *Source*: Courtesy of Ispat Inland Inc.

(a)

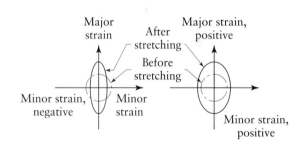

(b)

**FIGURE 7.66**  (a) Forming-limit diagram (FLD) for various sheet metals. The major strain is always positive. The region above the curves is the failure zone; hence, the state of strain in forming must be such that it falls below the curve for a particular material; $R$ is the normal anisotropy. (b) Note the definition of positive and negative minor strains. If the area of the deformed circle is larger than the area of the original circle, the sheet is thinner than the original, because the volume remains constant during plastic deformation. *Source*: After S. S. Hecker and A. K. Ghosh.

**FIGURE 7.67**   An example of the use of grid marks (circular and square) to determine the magnitude and direction of surface strains in sheet-metal forming. Note that the crack (tear) is generally perpendicular to the major (positive) strain.
*Source*: After S. P. Keeler.

In the forming-limit diagram, the **major strain** and the **minor strain** are obtained as follows: Note in Fig. 7.66b that after stretching, the original circle has deformed into an ellipse. The major axis of the ellipse represents the major direction and magnitude of stretching; the major strain plotted in Fig. 7.66a is the *engineering strain* (in percentage terms) in this direction. Likewise, the minor axis of the ellipse represents the magnitude of *stretching* (positive minor strain) or *shrinking* (negative minor strain) in the transverse direction. (The major strain is always positive, because forming sheet metal takes place by stretching in at least one direction.)

As an example, if we draw a circle in the center and on the surface of a sheet-metal tensile-test specimen and then stretch it plastically, the specimen becomes narrower; thus, the minor strain is negative. Because of volume constancy in the plastic range [see Eq. (2.48)], and assuming that the normal anisotropy is $R = 1$ (that is, the width and thickness strains are equal), we have $\varepsilon_w = -0.5\,\varepsilon_\ell$. As an experimental model, this phenomenon can easily be demonstrated by placing a circle on a wide rubber band (even though the deformation will only be in the elastic range) and stretching it. However, if we place a circle on a spherical rubber balloon and inflate the balloon, the minor strain is positive and equal in magnitude to the major strain; the circle simply becomes a larger circle. By observing the difference in surface area between the original circle and the ellipse, we can also determine whether the *thickness* of the sheet has changed. If the area of the ellipse is larger than that of the original circle, the sheet has become thinner (because volume remains constant in plastic deformation).

From Fig. 7.66a, it can be observed that, as expected, different materials have different forming-limit diagrams; note that the higher the curve, the better is the formability of the material. Also, it is important to note that for the same minor strain, say 20%, a compressive (negative) minor strain is associated with a higher major strain before failure than is a tensile (positive) minor strain. In other words, it is desirable for the minor strain to be *negative*, that is, to allow shrinking in the minor direction during sheet-forming operations. Special tooling has been designed for forming sheet metals and tubing that take advantage of the beneficial effect of negative minor strains on extending formability. (See, for example, Fig. 7.30.) The possible effect of the *rate of deformation* on forming-limit diagrams should also be assessed for each material. (See also Section 2.2.7.)

The effect of sheet-metal *thickness* on forming-limit diagrams is to *raise* the curves in Fig. 7.66a. Hence, the thicker the sheet, the higher its curve, and thus the more formable the sheet is. In actual forming operations, however, a thick blank may not bend easily around small radii and may develop cracks.

*Friction* at the punch–metal interface can also be significant in test results. With well-lubricated interfaces, the strains are more uniformly distributed over the punch. Also, depending on the notch sensitivity of the sheet metal, surface scratches, deep gouges, and blemishes can reduce formability and cause premature tearing and failure during testing and in actual forming operations. With effective lubrication, the coefficient of friction in sheet-metal forming generally ranges from about 0.05 to 0.1 for cold forming, and from 0.1 to 0.2 for elevated-temperature forming. (See also Table 4.1.)

e) **Limiting dome-height test.**  Recall that in the punch–stretch test (illustrated in Fig. 7.64), the major and minor strains were measured and plotted on the forming-limit diagram. In the *limiting dome-height* (LDH) test, performed with similarly prepared specimens, the *height* of the dome at failure of the sheet, or when the punch force reaches a maximum, is measured. Because the specimens are clamped, the LDH test indicates the capability of the material to stretch without failure. It has been shown that high LDH values are related to sheet-metal properties such as high $n$ and $m$ values, as well as high total elongation (in percentage terms) of the sheet metal.

---

**EXAMPLE 7.9**  **Estimating diameter of expansion**

A thin-walled spherical shell made of the aluminum alloy whose forming-limit diagram is shown in Fig. 7.66a is being expanded by internal pressure. If the original diameter is 200 mm, what is the maximum diameter to which the shell can safely be expanded?

**Solution.**  Because the material is being stretched in a state of equal (balanced) biaxial tension, we find from Fig. 7.66a that the maximum allowable engineering strain is about 40%. Thus,

$$e = \frac{\pi D_f - \pi D_o}{\pi D_o} = \frac{D_f - 200}{200} = 0.40$$

Hence,

$$D_f = 280 \text{ mm.}$$

**FIGURE 7.68**   Major and minor strains in various regions of an automobile body.

---

**EXAMPLE 7.10   Strains in the sheet-metal body of an automobile**

Computer programs have been developed that compute the major and minor strains and their orientations from the measured distortions of grid patterns placed on the surface of sheet metal. Figure 7.68 shows one such application for the various panels of an automobile body. Note that the trunk lid and the roof are subjected mainly to plane strain (see the vertical line in Fig. 7.66a), whereas the front door and front fender are subjected to biaxial strains. The numbers in the strain paths indicate the frequency of occurrence. *Source*: After T. J. Nihill and W. R. Thorpe.

---

### 7.13.2   Dent resistance of sheet-metal parts

In certain applications involving sheet-metal parts, such as automotive body panels, appliances, and office furniture, an important consideration is the dent resistance of the sheet-metal panel. A *dent* is a small, but permanent, biaxial deformation. The factors significant in dent resistance are the yield stress, $Y$, the thickness, $t$, and the shape of the panel. *Dent resistance* is expressed by a combination of material and geometrical parameters as

$$\text{Dent resistance} \propto \frac{Y^2 t^4}{S},\tag{7.24}$$

where $S$ is the panel stiffness, which, in turn, is defined as

$$S = (E)(t^a)(\text{shape}),\tag{7.25}$$

where the value of $a$ ranges from 1 to 2 for most panels. As for shape, the smaller the curvature (and hence, the flatter the panel), the greater the dent resistance is, because of the sheet's flexibility. Hence, dent resistance (a) increases with increasing strength and thickness of the sheet, (b) decreases with increasing elastic modulus and stiffness, and (c) decreases with decreasing curvature. Note that for $a = 2$, dent resistance is proportional to $t^2$. Dents are usually caused by *dynamic* forces, such as forces developed by falling objects or other objects that hit the surface of the panel from different directions. In typical automotive panels, for instance, impact velocities range up to 45 m/s (150 ft/s). Thus, the *dynamic yield stress* (that is, at high strain rates) rather than the static yield stress is the significant strength parameter. However, denting under quasistatic forces could also be important.

For materials in which yield stress increases with strain rate, denting requires higher energy levels than under static conditions. Furthermore, dynamic forces tend to cause more *localized* dents than do static forces, which tend to spread the dented area. Because a portion of the energy goes into elastic deformation, the modulus of resilience of the sheet metal [see Eq. (2.5)] is an additional factor to be considered.

### 7.13.3 Modeling of sheet-forming processes

In Section 6.2.2, we outlined the techniques for studying bulk-deformation processes and described *modeling* of impression-die forging, using the finite-element method (see Fig. 6.11). Such mathematical modeling is now also being applied to sheet-forming processes. The ultimate goal of such simulation techniques is rapid analysis of stresses, strains, flow patterns, wrinkling, and springback as functions of parameters such as material characteristics, friction, anisotropy, deformation speed, and temperature.

Through such interactive analyses, we can, for example, determine the optimum tool and die geometry to make a certain part, thus reducing or eliminating costly die tryouts. In addition, these simulation techniques can be used to determine the size and shape of blanks (including intermediate shapes), press characteristics, and process parameters to optimize the forming operation. The application of computer modeling, although requiring powerful computers and extensive software, has already proven to be a cost-effective technique in sheet-metal forming operations, particularly in the automotive industry.

## 7.14 | Equipment for Sheet-Metal Forming

The basic equipment for most pressworking operations consists of mechanical, hydraulic, pneumatic, or pneumatic–hydraulic presses. (See also Section 6.2.8 and Fig. 6.28.) The traditional *C-frame* press structure, with an open front, has been widely used for ease of tool and workpiece accessibility. However, the *box-type* (O-type) pillar and double-column frame structures are stiffer, and advances in automation, industrial robots, and computer controls have made accessibility less important.

*Press selection* includes factors such as (1) the type of operation, (2) the size and shape of the workpiece, (3) the length of stroke of the slide(s), (4) the number of strokes per minute, (5) the press speed, (6) the *shut height* (i.e., the distance from the top of the press bed to the bottom of the slide, with the stroke down), (7) the number of slides (e.g., double action and triple action), (8) the press capacity and tonnage, (9) the type of controls, and (10) the safety features.

In addition, because changing dies in presses can involve a significant amount of effort and time, rapid die-changing systems have been developed. Called **single-minute exchange of dies** (SMED), such systems use automated hydraulic or pneumatic equipment to reduce die-changing times from hours down to as little as 10 minutes.

**FIGURE 7.69** Efficient nesting of parts for optimum material utilization in blanking. *Source*: Society of Manufacturing Engineers.

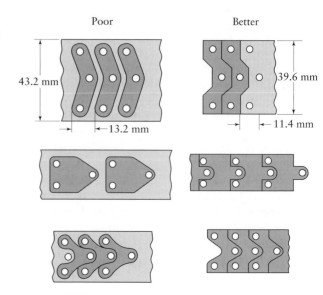

## 7.15 | Design Considerations

As with most other processes described throughout this text, certain design guidelines and practices have evolved with time. Careful design, using the best established design practices, computational tools, and manufacturing techniques, is the optimum approach for achieving high-quality designs and realizing cost savings. The following guidelines identify the most significant design issues of sheet-metal forming operations:

a) **Blank design.** Material scrap is the primary concern in blanking operations. Poorly designed parts will not nest, and there will be considerable amounts of scrap between successive blanking operations (Fig. 7.69). Some restrictions on the blank shape are made by the design application, but, when possible, blanks should be designed to reduce scrap to a minimum.

b) **Bending.** In bending operations, the main concerns are fracture of the material, wrinkling, and inability to form the bend. As shown in Fig. 7.70, bending a sheet-metal part with a flange that is to be bent will force the flange to undergo compression, which can cause buckling. This problem can be controlled by using a relief notch to limit the stresses from bending, as shown in Fig. 7.70. Right-angle bends have similar difficulties, and relief notches can be used to avoid tearing (Fig. 7.71).

**FIGURE 7.70** Control of tearing and buckling of a flange in a right-angle bend. *Source*: Society of Manufacturing Engineers.

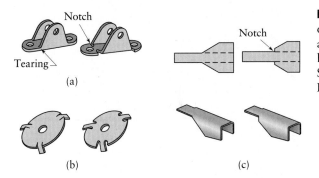

Notch

Tearing

(a)

(b)

Notch

(c)

**FIGURE 7.71** Application of notches to avoid tearing and wrinkling in right-angle bending operations. *Source*: Society of Manufacturing Engineers.

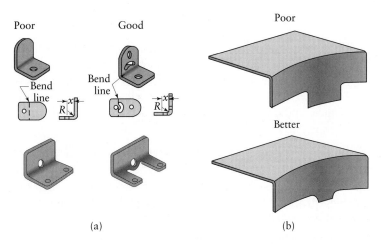

Poor

Bend line

Good

Bend line

$R$

(a)

Poor

Better

(b)

**FIGURE 7.72** Stress concentrations near bends. (a) Use of a crescent or ear for a hole near a bend. (b) Reduction of the severity of a tab in a flange. *Source*: Society of Manufacturing Engineers.

The bend radius should be recognized as a highly stressed area, and all stress concentrations should be removed from the bend-radius location. An example is the placement of holes near bends; it is normally advantageous to remove the hole from the stress concentration, but when this is not possible, a crescent slot or ear can be used (Fig. 7.72a). Similarly, when bending flanges, tabs and notches should be avoided, since these stress concentrations will greatly reduce the formability. When tabs are necessary, large radii should be used to reduce the stress concentration (Fig. 7.72b).

When both bending and the production of notches are to be implemented, it is important to orient the notches properly with respect to the grain direction. As shown in Fig. 7.17, bends should ideally be perpendicular to the rolling direction (or oblique, if perpendicularity is not possible) in order to avoid cracking. Bending of sharp radii can be accomplished through scoring or embossing (Fig. 7.73), but it should be recognized that this procedure can result in fracture. Burrs should not be present in a bend allowance, since they are brittle (because of strain hardening) and can fracture, leading to a stress concentration that propagates the crack into the rest of the sheet.

c) **Stamping and progressive-die operations.** In progressive dies, the cost of the tooling and the number of stations are determined by the number of features and spacing of the features on a part. It is therefore advantageous to hold the number of features to a minimum in order to minimize tooling costs. Note that closely spaced features may provide insufficient clearance for punches and would thereby require two punches. Narrow cuts and protrusions are also problematic for forming with a single punch and die.

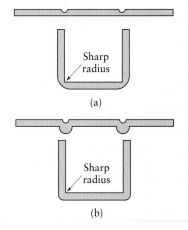

Sharp radius

(a)

Sharp radius

(b)

**FIGURE 7.73** Application of (a) scoring or (b) embossing to obtain a sharp inner radius in bending. However, unless properly designed, these features can lead to fracture. *Source*: Society of Manufacturing Engineers.

**FIGURE 7.74** Cost comparison for manufacturing a round sheet-metal container by conventional spinning and deep drawing. Note that for small quantities, spinning is more economical.

d) **Deep drawing.** After a deep-drawing operation, a cup will invariably spring back towards its original shape. For this reason, designs that use a vertical wall in a deep-drawn cup are difficult to form. Relief angles of at least 3° on each wall are easier to produce. Cups with sharp internal radii are difficult to produce, and deep cups often require additional ironing operations.

## 7.16 | Economics of Sheet-Metal Forming

Sheet-metal forming involves economic considerations similar to those for other metal working processes described in Chapter 6. Sheet-forming operations compete with each other, as well as with other processes, more than other processes compete with each other. As noted earlier, sheet-forming operations are versatile, and a number of different processes can be used to produce the same part. For example, a cup-shaped part can be formed by deep drawing, spinning, rubber forming, or explosive forming. Similarly, a cup can be formed by impact extrusion or casting, or by fabrication from different pieces.

As described in Section 16.5.5, almost all manufacturing processes produce some scrap, which can range up to 60% of the original material for machining operations and up to 25% for hot forging; in contrast, sheet-forming operations produce scrap on the order of 10% to 25% of the original material. The scrap should be recycled, although scrap with residual lubricant layers is less desirable than dry scrap.

The part shown in Fig. 7.74 can be made either by deep drawing or by conventional spinning; however, the die costs for the two processes are significantly different. Deep-drawing dies have many components and cost much more than the relatively simple mandrels and tools employed in spinning. Consequently, the die cost per part in drawing will be high if few parts are needed. On the other hand, this part can also be made by deep drawing in a much shorter time (seconds) than by spinning (minutes), even if the latter operation is highly automated. Furthermore, spinning requires more skilled labor. Considering all these factors, it is found that the *break-even point* for this particular part is at about 700 parts. Hence, deep drawing is more economical for quantities greater than that number.

---

**CASE STUDY** | Can Manufacturing

Can manufacturing is a major and competitive industry worldwide, with approximately 100 billion beverage cans and 30 billion food cans produced each year in the United States alone. These containers are strong and lightweight, typically weighing less than 0.5 oz.; they can reliably sustain 90 psi of internal pressure without leakage. In addition, there are stringent requirements on their surface finish, with brightly decorated and shiny cans preferred over dull, drab containers. Metal cans possessing these features are quite inexpensive: Can makers charge approximately $40 per 1000 cans, or around 4 cents per can.

The most common food and beverage can designs are two-piece and three-piece cans. These cans can be produced in a number of ways. Two-piece cans consist of a body and a lid; because the body has been *drawn and ironed*, the industry refers to this style as D&I (after drawn and ironed; Section 7.12) cans. Three-piece cans are produced by attaching a lid and a bottom to a cylindrical sheet-metal body. This case study describes the production of D&I cans, although many of the manufacturing considerations apply equally to three-piece cans.

Drawn and ironed can bodies are produced from a number of alloys, but the two most common are 3004-H19 aluminum and electrolytic tin-plated (Section 4.5.1) ASTM A623 steel. The can lids are produced from 5182 aluminum, sometimes in the H19 condition and sometimes precoated in the H48 condition. Aluminum lids are used for both aluminum and steel cans. The reason can be seen by inspecting Fig. 7.75. The lids have an extremely demanding set of design requirements: Not only must the can top be easily scored, but an integral rivet is stretch formed and headed in the lid to hold the tab in place. The 5182 aluminum alloy has the unique characteristic of being sufficiently formable to make the integral rivet, as well as being able to be scored. Basically, the lids are stamped from 5182 aluminum sheet, then the pop top is scored (to provide a controlled fracture path when tearing it), and, finally, a plastic seal is placed around the periphery of the lid. The polymer seals the can's contents after the lid is seamed to the can body, as described below. Note that the pop top opens by means of a Mode III fracture (i.e., tearing), as shown in Fig. 3.30.

The traditional method of manufacturing the can bodies consists of several processes, as shown in the sequence of operations depicted in Fig. 7.76: (a) First, 5.5-in.-diameter *blanks* are produced from rolled sheet stock. (b) Then the blanks are *deep drawn* to a diameter of around 3.5 in. (c) Next, the cans are *redrawn* to the final diameter of around 2.6 in. (See also Fig. 7.62.) (d) After that, the redrawn cans are *ironed* through two or three ironing rings in one pass. (See also Fig. 7.54.) (e) Finally, the can bottom is *domed*. The first two operations are carried out in a short-stroke mechanical press (Fig. 6.28); the last three operations are performed in a special type of press that typically produces cans at speeds of over 400 strokes

Integral rivet        Pop-top cantilever

Scored region

**FIGURE 7.75**   The lid of an aluminum beverage container.

**FIGURE 7.76** The metal-forming processes used to manufacture two-piece beverage cans.

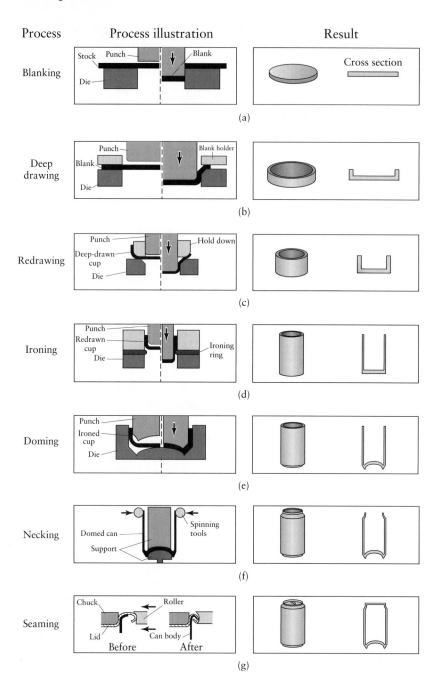

per minute. Effective cooling and lubrication is necessary, and emulsion lubricants (Section 4.4.4) are widely used for this purpose.

Following this series of operations, a number of additional processes take place. *Necking* of the can body is performed either through spinning (Fig. 7.76f) or by die necking (an operation similar to that shown in Fig. 6.72a, in which the top of the can is pushed into a die). Then the body is spin-flanged. The reason for necking the can top is that the 5182 aluminum used for the lid is relatively expensive. By tapering the top of the can, a smaller volume of metal is required, thereby reducing the material cost. The cost of a can is often calculated to millionths of a dollar; hence, any design feature that reduces its cost will be exploit-

**FIGURE 7.77** Aluminum two-piece beverage cans. Note the fine surface finish.

ed by this competitive industry. For example, reducing the blank diameter for the lid by 10% reduces the surface area, and hence the amount of metal, by $(1 - 0.9^2)/1 = 19\%$.

The interior surface of the can is not suitable for contact with food after forming, because the metal will rapidly adulterate its contents, either from metal-particulate contamination (resulting from the forming operations) or by metal–food interaction. To avoid contamination, a polymer solution is applied to the can's interior. However, lubricants used in the metal-forming operations must first be removed, because they can prevent adhesion of the coating to the metal. Cans are, therefore, subjected to a solvent wash operation (Section 4.5.2), after which the inner surfaces are coated with a thermoplastic or a thermosetting resin that is safe upon contact with food. Coating is performed by dissolving the polymer in a carrier fluid and applying it to the interior of the can.

After this application, an oven-curing operation allows the resin to bond to the interior surface, while the toxic carrier fluid boils off and is released as a volatile organic compound (VOC). In order to make this harmful by-product environmentally safe, an afterburner is used to reduce it to water vapor and carbon dioxide. Recent developments in coatings include the use of water-based carriers for the polymer coating and using prelaminated steels in the canmaking operation; these approaches eliminate the environmentally harmful VOCs without significantly affecting the cost of the can.

*Decoration* of the cans is a critical aspect of the overall manufacturing process. Aluminum cans are produced with an extremely smooth surface (Fig. 7.77) so that they appear shiny and vibrant. Lacquers and inks are printed directly onto the aluminum body, resulting in an aesthetically pleasing and attractive product. The final operation on a can is *seaming* the top over the body (Fig. 7.76g; see also Fig. 12.50). Note that, unlike aluminum cans, steel cans either have a base coating applied prior to the final decoration or have a paper or plastic label attached to the can.

*Source*: Courtesy of J. E. Wang, Texas A&M University.

# SUMMARY

- Sheet-metal forming processes involve workpieces that have a high ratio of surface area to thickness. In these processes, tensile stresses are usually applied in the plane of the sheet through various punches and dies. The material is generally prevented from being reduced in thickness in order to avoid necking and subsequent tearing and fracture. Material properties and behavior important to these processes include elongation, yield-point elongation, anisotropy, and grain size. (Sections 7.1, 7.2)

- Blanks for subsequent forming operations are generally sheared from large sheets, using various processes. The quality of the sheared edge is important in subsequent processing, and important process parameters include the clearance between the dies, sharpness, and lubrication. (Section 7.3)

- Bending of sheet, plate, and tubing is a commonly performed operation, and important parameters include the minimum bend radius (without causing cracking in the bend area) and springback. Several techniques are available for bending and for minimizing springback. (Section 7.4)

- Discrete sheet-metal bodies can be made by stretch forming, bulging, rubber forming, and hydroforming, whereby the workpiece is shaped by stretching in different directions. In rubber forming and hydroforming, in contrast to the other two methods, one of the dies can be made of a flexible material such as rubber or polyurethane. (Sections 7.5, 7.6, 7.7)

- Spinning involves the forming of axisymmetric parts over a rotating mandrel, using rigid tools or rollers, and comprises conventional, shear, and tube-spinning processes. Various external and internal profiles can be produced. (Section 7.8)

- High-energy-rate forming processes use chemical, electrical, or magnetic means of energy. They include explosive forming, usually of large sheets; electrohydraulic forming; and magnetic-pulse forming of relatively small parts. (Section 7.9)

- Superplastic-forming methods utilize the superplastic behavior of materials and are capable of producing complex shapes with high strength-to-weight ratios. They are particularly effective when combined with diffusion bonding of two or more sheets prior to forming. (Section 7.10)

- Various other sheet-forming processes include peen forming, based on the principle of shot peening; use of gas mixtures or liquified gases; laser forming and laser-assisted forming; microforming; and manufacturing of honeycomb structures. (Section 7.11)

- Deep drawing, which uses a punch and a die, is one of the most important sheet-forming processes, because it can produce container-shaped parts with high depth-to-diameter ratios. Deep drawability has been found to depend on the normal anisotropy of the sheet metal. Earing also occurs in deep drawing, because of the planar anisotropy of the metal. Buckling and wrinkling in drawing can be significant problems; techniques have been developed to eliminate or minimize these problems. (Section 7.12)

- Tests have been developed to determine the formability of sheet metals in forming processes. The most comprehensive test is the punch–stretch (bulge) test, which enables the construction of a forming-limit diagram, indicating safe and failure regions as a function of the planar strains involved. Each type of metal, its condition and its thickness, has its own curve. Limiting dome-height tests have also been developed for determining formability. (Section 7.13)

- Several types of equipment are available for sheet-metal forming processes. Selection among these pieces of equipment depends on various factors concerning workpiece size and shape and equipment characteristics and their control features. Rapid die-changing techniques are important in the productivity and overall economics of sheet-metal forming. (Section 7.14)

- Design considerations in sheet-metal forming are based on factors such as shape of the blank, the type of operation to be performed, and the nature of the deformation. Several design rules and practices have been established to eliminate or minimize production problems. (Section 7.15)

- As in all manufacturing processes, economic considerations in sheet-forming operations often determine the specific processes and practices to be employed. Many processes can be competitive; consequently, a thorough knowledge of process characteristics and capabilities is essential for economical production. (Section 7.16)

## SUMMARY OF EQUATIONS

Maximum punch force in shearing, $F_{max} = 0.7(\text{UTS})(t)(L)$

Strain in bending, $e_o = e_i = \dfrac{1}{(2R/t) + 1}$

Minimum $\dfrac{R}{t} = \dfrac{50}{r} - 1$

Springback, $\dfrac{R_i}{R_f} = 4\left(\dfrac{R_i Y}{Et}\right)^3 - 3\left(\dfrac{R_i Y}{Et}\right) + 1$

Maximum bending force, $F_{max} = k\dfrac{(\text{UTS})Lt^2}{W}$

Normal anisotropy, $R = \dfrac{\varepsilon_w}{\varepsilon_t}$

Average, $\bar{R} = \dfrac{R_0 + 2R_{45} + R_{90}}{4}$

Planar anisotropy, $\Delta R = \dfrac{R_0 - 2R_{45} + R_{90}}{2}$

Maximum punch force in drawing, $F_{max} = \pi D_p t_o(\text{UTS})\left(\dfrac{D_o}{D_p} - 0.7\right)$

Pressure in explosive forming, $p = K\left(\dfrac{\sqrt[3]{W}}{R}\right)^a$

## BIBLIOGRAPHY

*ASM Handbook*, Vol. 14: *Forming and Forging*, ASM International, 1988.

Benson, S. D., *Press Brake Technology*, Society of Manufacturing Engineers, 1997.

Davis, J. R. (ed.), *Tool Materials*, ASM International, 1995.

*Fundamentals of Tool Design*, 4th ed., Society of Manufacturing Engineers, 1998.

Eary, D. F., and E. A. Read, *Techniques for Pressworking Sheet Metal*, 2d. ed. Prentice-Hall, 1974.

Hosford, W. F., and R. M. Caddell, *Metal Forming, Mechanics and Metallurgy*, 2d. ed., Prentice Hall, 1993.

Koistinen, D. P., and N. M. Wang (eds.), *Mechanics of Sheet Metal Forming*, Plenum, 1978.

Marciniak, Z., and J. L. Duncan, *The Mechanics of Sheet Metal Forming*, Edward Arnold, 1992.

Morgan, E., *Tinplate and Modern Canmaking Technology*, Pergamon, 1985.

Nachtman, E. S., and S. Kalpakjian, *Lubricants and Lubrication in Metalworking Operations*, Dekker, 1985.

Pacquin, J. R., and R. E. Crowley, *Die Design Fundamentals*, Industrial Press, 1987.

Pearce, R., *Sheet Metal Forming*, Adam Hilger, 1991.

*Progressive Dies*, Society of Manufacturing Engineers, 1994.

Sachs, G., *Principles and Methods of Sheet Metal Fabricating*, 2d. ed., Reinhold, 1966.

Smith, D. A. (ed.), *Die Design Handbook*, 3d. ed., Society of Manufacturing Engineers, 1990.

Suchy, I., *Handbook of Die Design*, McGraw-Hill, 1997.

*Tool and Manufacturing Engineers Handbook*, 4th ed., Vol. 2: *Forming*, Society of Manufacturing Engineers, 1984.

Wagoner, R. H., K. S. Chan, and S. P. Keeler (eds.), *Forming Limit Diagrams*, The Minerals, Metals and Materials Society, 1989.

# QUESTIONS

**7.1** Take any three topics from Chapter 2, and, with specific examples for each, show their relevance to the topics covered in this chapter.

**7.2** Do the same as for Question 7.1, but for Chapter 3.

**7.3** Describe (a) the similarities and (b) the differences between the bulk-deformation processes described in Chapter 6 and the sheet-metal forming processes described in this chapter.

**7.4** Discuss the material and process variables that influence the shape of the curve of punch force vs. stroke for shearing, such as that shown in Fig. 7.7, including its height and width.

**7.5** In preparing a large number of blanks for sheet-forming operations, the blanks are spaced (*layout and nesting*) such that material waste is minimized. Make some sketches of blanks, and describe the considerations involved in this procedure, with particular emphasis on subsequent forming operations that may be performed.

**7.6** Describe your observations concerning Figs. 7.5 and 7.6.

**7.7** Inspect a common paper punch, and comment on the shape of the tip of the punch as compared with those shown in Fig. 7.12.

**7.8** Explain how you would go about estimating the temperature rise in the shear zone in a shearing operation.

**7.9** As a practicing engineer in manufacturing, why would you be interested in the shape of the curve in Fig. 7.7? Explain.

**7.10** Do you think the presence of burrs can be beneficial in certain applications? Give specific examples.

**7.11** Explain the reasons that there are so many different types of die materials used for the processes described in this chapter.

**7.12** Describe the differences between compound, progressive, and transfer dies.

**7.13** It has been stated that the quality of the sheared edges can influence the formability of sheet metals. Explain why.

**7.14** Explain why and how various factors influence springback in bending of sheet metals.

**7.15** Can the hardness of a sheet metal have an effect on the metal's springback in bending? Explain.

**7.16** We note in Fig. 7.16 that the state of stress shifts from plane stress to plane strain as the ratio of length of bend to sheet thickness increases. Explain why.

**7.17** Describe the material properties that have an effect on the relative position of the curves shown in Fig. 7.19.

**7.18** In Table 7.2, we note that hard materials have higher $R/t$ ratios than soft ones. Explain why.

**7.19** Why do tubes buckle when bent? Experiment with a soda straw, and describe your observations.

**7.20** If you wish to bend a piece of copper tubing without collapsing it, you could go to a hardware store and buy a helical springlike gadget. Visit a hardware store, find out what this gadget is, and explain why you can use it successfully to bend a tube.

**7.21** Based on Fig. 7.22, sketch the shape of a *U*-die used to produce channel-shaped bends.

**7.22** Explain why negative springback does not occur in air bending of sheet metals.

**7.23** Give examples of products in which the presence of beads is beneficial or even necessary.

**7.24** Explain why cupping tests do not always predict the behavior of sheet metals in actual forming operations.

**7.25** Assume that you are carrying out a sheet-forming operation and you find that the material is not sufficiently ductile. Make suggestions to improve its ductility.

**7.26** Referring to Fig. 7.39, explain why a further increase in the ductility of the material (beyond about 50% tensile reduction of area) does not improve the maximum spinning reduction per pass.

**7.27** Many missile components are made by spinning. What other methods would you use to make missile components if spinning processes were not available? Describe the relative advantages and limitations of each method.

**7.28** In deep drawing of a cylindrical cup, is it always necessary for there to be tensile circumferential stresses on the element in the cup wall? (See Fig. 7.51b.) Explain.

**7.29** When comparing the hydroforming process with the deep-drawing process, it has been stated that deeper draws are possible in the former method. With appropriate sketches, explain why.

**7.30** We note in Fig. 7.51a that element *A* in the flange is subjected to compressive circumferential (hoop) stresses. Using a simple free-body diagram, explain why.

**7.31** From the topics covered in this chapter, list and explain specifically several examples where (a) friction is desirable and (b) friction is not desirable.

**7.32** Explain why increasing the normal anisotropy, $R$, improves the deep drawability of sheet metals.

**7.33** What is the reason for the negative sign in the numerator of Eq. (7.21)?

**7.34** If you had a choice whereby you could control the state of strain in a sheet-forming operation, would you rather work on the left or the right side of the forming-limit diagram? Explain.

**7.35**   Comment on the effect of lubrication of the punch on the limiting drawing ratio in deep drawing.

**7.36**   Comment on the size of the circles placed on the surfaces of sheet metals in determining the metals' formability. Are square grid patterns, as shown in Fig. 7.67, useful? Explain.

**7.37**   Make a list of the independent variables that influence the punch force in deep drawing of a cylindrical cup, and explain why and how the variables influence this force.

**7.38**   Explain why the simple tension line in the forming-limit diagram in Fig. 7.66a states that it is for $R = 1$, where $R$ is the normal anisotropy of the sheet.

**7.39**   What are the reasons for developing forming-limit diagrams? Do you have any specific criticisms of such diagrams? Explain.

**7.40**   Explain the reasoning behind Eq. (7.20), for normal anisotropy, and Eq. (7.21), for planar anisotropy, respectively.

**7.41**   Describe why earing occurs. How would you avoid it?

**7.42**   What is the significance of the size of the grid patterns used in the study of sheet-metal formability?

**7.43**   We have stated in Section 7.13.1 that the thicker the sheet metal, the higher is the curve in the forming-limit diagram. Explain why.

**7.44**   Inspect the earing shown in Fig. 7.59, and estimate the direction in which the blank was cut.

**7.45**   Describe the factors that influence the size and length of beads in sheet-metal forming operations.

**7.46**   It is known that the strength of metals depends on the metals' grain size. Would you then expect strength to influence the $R$ value of sheet metals? Explain.

**7.47**   Equation (7.23) gives a general rule for dimensional relationships for successful drawing without a blankholder. Explain what happens if this limit is exceeded.

**7.48**   Explain why the three broken lines (simple tension, plane strain, and equal biaxial stretching) in Fig. 7.66a have those particular slopes.

**7.49**   Identify specific parts on a typical automobile, and explain which of the processes described in

Chapters 6 and 7 can be used to make that part. Then choose only one method for each part, and explain your reasoning.

**7.50**   We have seen that bendability and spinnability have a common aspect as far as properties of the workpiece material are concerned. Describe this common aspect.

**7.51**   Explain the reasons that such a wide variety of sheet-forming processes has been developed and used over the years.

**7.52**   Make a summary of the types of defects found in sheet-metal forming processes, and include brief comments on the reason(s) for each defects.

**7.53**   Which of the processes described in this chapter use only one die? What are the advantages of using only one die?

**7.54**   It has been suggested that deep drawability can be increased by (a) heating the flange and/or (b) chilling the punch by some suitable means. Comment on how these methods could improve drawability.

**7.55**   Offer designs whereby the suggestions given in Question 7.54 can be implemented. Would the required production rate affect your designs? Explain.

**7.56**   In the manufacture of automotive body panels from carbon-steel sheet, stretcher strains (Lueder's bands) are observed, which detrimentally affect surface finish. How can the stretcher strains be eliminated?

**7.57**   In order to improve its ductility, a coil of sheet metal is placed in a furnace and annealed. However, it is observed that the sheet has a lower limiting drawing ratio than it had before being annealed. Explain the reasons for this behavior.

**7.58**   What effects does friction have on a forming-limit diagram?

**7.59**   Why are lubricants generally used in sheet-metal forming? Explain, using examples.

**7.60**   Through changes in clamping, a sheet-metal forming operation can allow the material to undergo a negative minor strain. Explain how this effect can be advantageous.

**7.61**   How would you produce the part shown in Fig. 7.35 other than by tube hydroforming?

# PROBLEMS

**7.62**   Referring to Eq. (7.5), it is stated that actual values of $e_o$ are considerably higher than values of $e_i$, due to the shifting of the neutral axis during bending. With an appropriate sketch, explain this phenomenon.

**7.63**   Note in Eq. (7.11) that the bending force is a function of $t^2$. Why? (*Hint*: Consider bending-moment equations in mechanics of solids.)

**7.64**    Calculate the minimum tensile true fracture strain that a sheet metal should have in order to be bent to the following $R/t$ ratios: (a) 2, (b) 3, and (c) 5. (See Table 7.2.)

**7.65**    Estimate the maximum bending force required for a $\frac{1}{8}$-in. thick and 10-in. wide Ti-5Al-2.5Sn titanium alloy in a $V$-die with a width of 8 in.

**7.66**    In Example 7.4, calculate the work done by the force–distance method, i.e., work is the integral product of the vertical force, $F$, and the distance it moves.

**7.67**    What would be the answer to Example 7.4 if the tip of the force, $F$, were fixed to the strip by some means, thus maintaining the lateral position of the force? (*Hint*: Note that the left portion of the strip will now be strained more than the right portion.)

**7.68**    Calculate the magnitude of the force $F$ in Example 7.4 for $\alpha = 15°$.

**7.69**    How would the force in Example 7.4 vary if the workpiece were made of a perfectly plastic material?

**7.70**    Calculate the press force needed in punching 0.1-mm-thick 5052-0 aluminum foil in the shape of a square hole 30 mm on each side.

**7.71**    A straight bead is being formed on a 1-mm-thick aluminum sheet in a 20-mm-diameter die, as shown in the accompanying figure. (See also Fig. 7.25a.) Let $Y = 120$ MPa. Considering springback, calculate the outside diameter of the bead after it is formed and unloaded from the die.

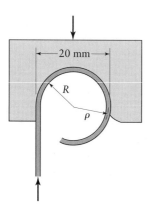

**7.72**    Inspect Eq. (7.10), and, substituting in some numerical values, show whether the first term can be neglected without significant error in calculating springback.

**7.73**    In Example 7.5, calculate the amount of TNT required to develop a pressure of 15,000 psi on the surface of the workpiece. Use a standoff of one foot.

**7.74**    Estimate the limiting drawing ratio (LDR) for the materials listed in Table 7.3.

**7.75**    For the same material and thickness as in Problem 7.65, estimate the force required for deep drawing with a blank of diameter 12 in. and a punch of diameter 9 in.

**7.76**    A cup is being drawn from a sheet metal that has a normal anisotropy of 3. Estimate the maximum ratio of cup height to cup diameter that can be drawn successfully in a single draw. Assume that the thickness of the sheet throughout the cup remains the same as the original blank thickness.

**7.77**    Obtain an expression for the curve shown in Fig. 7.58 in terms of the LDR and the average normal anisotropy, $\bar{R}$. (*Hint*: See Fig. 2.7b).

**7.78**    A steel sheet has $R$ values of 1.0, 1.5, and 2.0 for the 0°, 45°, and 90° directions to rolling, respectively. If a round blank is 200 mm in diameter, estimate the smallest cup diameter to which it can be drawn.

**7.79**    In Problem 7.78, explain whether ears will form and, if so, why.

**7.80**    A 1-mm-thick isotropic sheet metal is inscribed with a circle 5 mm in diameter. The sheet is then stretched uniaxially by 25%. Calculate (a) the final dimensions of the circle and (b) the thickness of the sheet at this location.

**7.81**    Make a search of the literature, and obtain the equation for a tractrix curve, as used in Fig. 7.63.

**7.82**    In Example 7.4, assume that the stretching is done by two equal forces $F$, each at 5 in. from the ends of the workpiece. (a) Calculate the magnitude of this force for $\alpha = 10°$. (b) If we want stretching to be done up to $\alpha_{max} = 50°$ without necking, what should be the minimum value of $n$ of the material?

**7.83**    Derive Eq. (7.5).

**7.84**    Estimate the maximum power in shear spinning a 0.5-in. thick annealed 304 stainless-steel plate that has a diameter of 15 in. on a conical mandrel of $\alpha = 45°$. The mandrel rotates at 100 rpm, and the feed is $f = 0.1$ in./rev.

**7.85**    Assume that you are asked to give a quiz to students on the contents of this chapter. Prepare five quantitative problems and five qualitative questions, and supply the answers.

**7.86**    Obtain an aluminum beverage can, and cut it in half lengthwise with a pair of tin snips. Using a micrometer, measure the thickness of the bottom of the can and of the wall of the can. Estimate the thickness reductions in ironing and the diameter of the original blank.

**7.87**    What is the force required to punch a square hole, 100 mm on each side, from a 1-mm-thick 5052-O aluminum sheet, using flat dies? What would be your answer if beveled dies were used instead?

# DESIGN

**7.88** Consider several shapes to be blanked from a large sheet (such as oval, triangle *L*-shape, and so forth) by laser-beam cutting, and sketch a nesting layout to minimize scrap generation.

**7.89** Estimate the percent scrap in producing round blanks if the clearance between blanks is one tenth of the radius of the blank. Consider single and multiple-row blanking, as sketched in the accompanying figure.

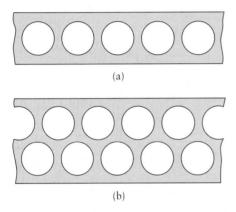

(a)

(b)

**7.90** Give several structural applications in which diffusion bonding and superplastic forming are used jointly.

**7.91** On the basis of experiments, it has been suggested that concrete, either plain or reinforced, can be a suitable material for dies in sheet-metal forming operations. Describe your thoughts regarding this suggestion, considering die geometry and any other factors that may be relevant.

**7.92** Metal cans are of either the two-piece variety (in which the bottom and sides are integral) or the three-piece variety (in which the sides, the bottom, and the top are each separate pieces). For a three-piece can, should the seam be (a) in the rolling direction, (b) normal to the rolling direction, or (c) oblique to the rolling direction? Explain your answer, using equations from solid mechanics.

**7.93** Find and inspect a brass drum cymbal, and describe the manufacturing processes used in its production.

**7.94** Investigate methods for determining optimum shapes of blanks for deep-drawing operations. Sketch the optimally shaped blanks for drawing rectangular cups, and optimize their layout on a large sheet of metal.

**7.95** The design shown in the accompanying figure is proposed for a metal tray, the main body of which is made from cold-rolled sheet steel. Noting its features and that the sheet is bent in two different directions, comment on various manufacturing considerations. Include factors such as anisotropy of the rolled sheet, the sheet's surface texture, the bend directions, the nature of the sheared edges, and the way the handle is snapped in for assembly.

**7.96** Design a box that will contain a 4-in. × 6-in. × 3-in. volume. The box should be produced from two pieces of sheet metal and require no tools or fasteners for assembly.

**7.97** In opening a can using an electric can opener, you will note that the lid develops a scalloped periphery. (a) Explain why scalloping occurs. (b) What design changes for the can opener would you recommend in order to minimize or eliminate, if possible, this scalloping effect? (c) Since lids typically are discarded or recycled, do you think it is necessary or worthwhile to make such design changes? Explain.

# CHAPTER 8

# Material-Removal Processes: Cutting

8.1 Introduction
8.2 Mechanics of Chip Formation
8.3 Tool Wear and Failure
8.4 Surface Finish and Integrity
8.5 Machinability
8.6 Cutting-Tool Materials
8.7 Cutting Fluids
8.8 Cutting Processes and Machine Tools for Producing Round Shapes
8.9 Cutting Processes and Machine Tools for Producing Various Shapes
8.10 Machining and Turning Centers
8.11 Vibration and Chatter
8.12 Machine-Tool Structures
8.13 Design Considerations
8.14 Economics of Machining

## 8.1 | Introduction

Parts manufactured by casting, forming, and various shaping processes often require further processing or finishing operations to impart specific characteristics, such as dimensional accuracy and surface finish, before the product is ready for use. This chapter and the following chapter describe in detail the various operations that are employed to obtain these characteristics; these processes are generally classified as **material-removal processes.** Because they are capable of producing shapes competitive with those produced by other methods, as described in the preceding chapters, critical choices have often to be made about the extent of shaping and forming versus the extent of machining to be done on a workpiece to produce an acceptable part.

Although **machining** is the general term used to describe *removal of material* from a workpiece, it covers several processes, which are usually divided into the following broad categories, the first of which is the subject of this chapter:

- **Cutting,** which generally involves single-point or multipoint cutting tools and processes, such as turning, boring, drilling, tapping, milling, sawing, and broaching;
- **Abrasive processes,** such as grinding, honing, lapping, and ultrasonic machining (Chapter 9);
- **Advanced machining processes,** which use electrical, chemical, thermal, hydrodynamic, and optical sources of energy to remove material from the workpiece surface (Chapter 9).

As described throughout this chapter, material-removal processes are desirable or even necessary in manufacturing operations for the following reasons:

1. Closer **dimensional accuracy** may be required than is available from casting, forming, or shaping processes alone. For example, in a forged crankshaft, the bearing surfaces and the holes cannot be produced with good dimensional accuracy and surface finish by forming and shaping processes alone.

2. Parts may possess external and internal **geometric features,** as well as sharp corners, flatness, etc., that cannot be produced by forming and shaping processes.

3. Some parts are heat treated for improved hardness and wear resistance. Since heat-treated parts may undergo distortion and surface discoloration, they generally require additional **finishing operations,** such as grinding, to obtain the desired characteristics.

4. Special **surface characteristics** or textures that cannot be produced by other means may be required on all or selected areas of the surfaces of the product. Copper mirrors with very high reflectivity, for example, are made by machining with a diamond cutting tool.

5. It may be more **economical** to machine the part than to manufacture it by other processes, particularly if the number of parts desired is relatively small.

Against these advantages, material-removal processes have certain limitations:

1. Removal processes inevitably **waste material** and generally require more energy, capital, and labor than forming and shaping operations; thus, they should be avoided or minimized whenever possible.

2. Removing a volume of material from a workpiece generally takes **longer time** than does shaping it by other processes.

3. Unless carried out properly, material-removal processes can have **adverse effects** on the surface integrity of the product.

In spite of these limitations, material-removal processes and machines, called **machine tools**, are indispensable to manufacturing technology. Ever since lathes were introduced in the 1700s, these processes have developed continuously; we now have available a wide variety of computer-controlled machines, as well as new techniques that use various sources of energy. It is important to view machining, as well as all manufacturing operations, as a *system* consisting of the workpiece, the tool, and the machine. As will be evident in reading this chapter, machining operations cannot be carried out efficiently and economically without a fundamental knowledge of the complex interactions among these elements. The importance of these considerations is even more evident from the estimate that, in the United States, labor and overhead costs alone for machining are about $300 billion per year.

# 8.2 | **Mechanics of Chip Formation**

Cutting processes (Fig. 8.1) remove material from the surface of the workpiece by producing chips. The basic mechanics of chip formation is essentially the same for all cutting operations and is represented by the two-dimensional model shown in Fig. 8.2. In this model, a tool moves along the workpiece at a certain velocity, $V$, and a depth of cut, $t_o$. A **chip** is produced just ahead of the tool by *shearing* of the material continuously along the shear plane.

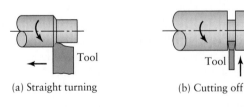

(a) Straight turning          (b) Cutting off

**FIGURE 8.1**   Examples of cutting processes.

(c) Slab milling          (d) End milling

**FIGURE 8.2** Schematic illustration of a two-dimensional cutting process, also called *orthogonal cutting*.

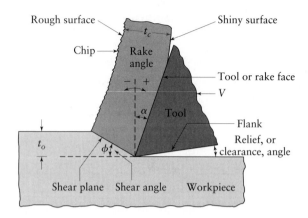

In the cutting process, the major *independent variables*, i.e., the variables that we can change and control directly, are

- Tool material and its condition;
- Tool shape, surface finish, and sharpness;
- Workpiece material and its condition and temperature;
- Cutting conditions, such as speed, feed, and depth of cut;
- Type of cutting fluids, if used;
- Characteristics of the machine tool, such as its structure, stiffness, and damping;
- Workholding and fixturing devices.

The *dependent variables*, which are the variables that are influenced by changes in the independent variables, are:

- Type of chip produced;
- Force and energy dissipated in the cutting process;
- Temperature rise in the workpiece, the chip, and the tool;
- Wear and failure of the tool;
- Surface finish and integrity of the workpiece after machining.

In order to appreciate the importance of studying the complex interrelationships among these variables, let's briefly ask the following questions: If, for example, the surface finish of the workpiece being cut is poor and unacceptable, which of the foregoing independent variables do we change first? The tool geometry? If so, which feature? If the workpiece becomes too hot, what do we do? If the tool wears and becomes dull rapidly, do we change the cutting speed, the depth of cut, or the tool material itself? If the cutting tool begins to vibrate and chatter, what should we do to eliminate this problem?

Although almost all cutting processes are three dimensional in nature, the two-dimensional model shown in Fig. 8.2 is very useful in studying the basic mechanics of cutting. In this model, known as **orthogonal cutting** (i.e., the tool edge is perpendicular to the movement of the tool), the tool has a **rake angle**, $\alpha$, (positive, as shown in the figure) and a **relief**, or **clearance, angle**. Note that the sum of the rake, relief, and included angles of the tool is 90°.

Microscopic examinations reveal that chips are produced by the **shearing** process shown in Fig. 8.3a. Shearing takes place along a **shear plane**, which makes an angle $\phi$, called the **shear angle**, with the surface of the workpiece. Note that this shearing

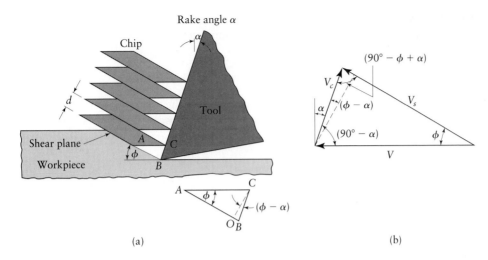

process is like cards in a deck sliding against each other. Below the shear plane, the workpiece is undeformed (though it does undergo some elastic distortion), and above the shear plane, we have a chip already formed and moving up the face of the tool as cutting progresses. Because of the movement of the chip, there is friction between the chip and the rake face of the tool.

Note that we can determine the thickness of the chip, $t_c$, if we know $t_o$, $\alpha$, and $\phi$. The ratio of $t_o$ to $t_c$ is known as the **cutting ratio**, $r$, which can be expressed as

$$r = \frac{t_o}{t_c} = \frac{\sin \phi}{\cos(\phi - \alpha)}. \tag{8.1}$$

A review of this process indicates that the chip thickness is always greater than the depth of cut; hence, the value of $r$ is always less than unity. The reciprocal of $r$ is known as the **chip compression ratio** and is a measure of how thick the chip has become compared with the depth of cut; thus, the chip compression ratio is always greater than unity.

On the basis of Fig. 8.3a, we can write the **shear strain**, $\gamma$, that the material undergoes as

$$\gamma = \frac{AB}{OC} = \frac{AO}{OC} + \frac{OB}{OC}, \tag{8.2}$$

or

$$\gamma = \cot \phi + \tan(\phi - \alpha). \tag{8.3}$$

Note from this equation that large shear strains are associated with low shear angles and low or negative rake angles. Shear strains of 5 or higher have been observed in actual cutting operations; thus, compared with forming and shaping processes (see Chapter 6), the material undergoes greater deformation during cutting.

From Fig. 8.2, note that since chip thickness, $t_c$, is greater than the depth of cut, $t_o$, the *velocity of the chip*, $V_c$, has to be less than the cutting speed, $V$. Since mass continuity has to be maintained, we have

$$V t_o = V_c t_c, \qquad \text{or} \qquad V_c = Vr. \tag{8.4}$$

Hence,

$$V_c = V \frac{\sin \phi}{\cos(\phi - \alpha)}. \tag{8.5}$$

We can construct a velocity diagram such as that shown in Fig. 8.3b, and from trigonometric relationships, we obtain the equations

$$\frac{V}{\cos(\phi - \alpha)} = \frac{V_s}{\cos \alpha} = \frac{V_c}{\sin \phi},$$

(8.6)

where $V_s$ is the velocity at which shearing takes place in the shear plane. The **shear-strain rate** is the ratio of $V_s$ to the thickness $d$ of the sheared element (shear zone), or

$$\dot{\gamma} = \frac{V_s}{d}.$$

(8.7)

Experimental evidence indicates that the magnitude of $d$ is on the order of $10^{-2}$ mm to $10^{-3}$ mm ($10^{-3}$ in. to $10^{-4}$ in.). This range means that, even at low cutting speeds, the shear-strain rate is very high, on the order of $10^3$/s to $10^6$/s. A knowledge of the shear-strain rate is essential, because of its effects on the strength and ductility of the material (see Section 2.2.7) and the type of chip.

### 8.2.1 Chip morphology

The type of chips produced (*chip morphology*) significantly influences surface finish and integrity, as well as the overall cutting operation. When actual chips are observed with a microscope and under different cutting conditions, there are some deviations from the ideal model shown in Figs. 8.2 and 8.3a. Figure 8.4 shows some types of metal chips commonly observed in practice; they are described in this subsection in the following order:

1. Continuous;
2. Built-up edge;
3. Serrated, or segmented;
4. Discontinuous.

Let's first note that a chip has two surfaces: One is in contact with the tool face (rake face), and the other is from the original surface of the workpiece. The tool side of the chip surface is shiny, or *burnished* (Fig. 8.5), caused by rubbing of the chip as it climbs up the tool face. The other surface of the chip does not come into contact with any solid body; this surface has a jagged, steplike appearance (see Fig. 8.4a), which is due to the shearing mechanism of chip formation. (See also Figs. 3.4 and 3.6 for an analogy.)

The basic types of chips produced in metal-cutting operations are described as follows:

1. **Continuous chips.** *Continuous chips* are usually formed at high cutting speeds and/or high rake angles (Fig. 8.4a); the deformation of the material takes place along a very narrow shear zone, called the **primary shear zone**. Continuous chips may develop a **secondary shear zone** at the tool–chip interface (Fig. 8.4b), caused by friction. As expected, the secondary zone becomes deeper as tool–chip friction increases. In continuous chips, deformation may also take place along a wide primary shear zone with *curved boundaries* (Fig. 8.4d). Note that the lower boundary is *below* the machined surface, which subjects the machined surface to distortion, as depicted by the distorted vertical lines to the right of the tool tip. This situation occurs particularly in machining soft metals at low speeds and low rake angles; it can produce poor surface finish and induce residual surface stresses, which may be detrimental to the properties of the machined part.

**FIGURE 8.4** Basic types of chips produced in metal cutting and their micrographs: (a) continuous chip with narrow, straight primary shear zone; (b) secondary shear zone at the tool–chip interface; (c) continuous chip with built-up edge; (d) continuous chip with large primary shear zone; (e) segmented or nonhomogeneous chip; and (f) discontinuous chip. *Source*: After M. C. Shaw, P. K. Wright, and S. Kalpakjian.

**FIGURE 8.5** Shiny (burnished) surface on the tool side of a continuous chip produced in turning.

Although they generally produce good surface finish, continuous chips are not always desirable, particularly in automated machine tools, because they tend to get tangled around the tool, and the operation has to be stopped to clear away the chips. This problem can be alleviated with **chip breakers** (discussed later).

As a result of strain hardening (caused by the shear strains to which the chip is subjected), the chip usually becomes harder, less ductile, and stronger than the original workpiece material. The increase in hardness and strength of the chip depends on the shear strain; as the rake angle decreases, the shear strain increases, and thus the chip becomes stronger and harder. With increasing strain, the chip tends to behave more like a rigid, perfectly plastic material. (See Fig. 2.8b.)

2. **Built-up-edge chips.** A *built-up edge* (BUE) may form at the tip of the tool during cutting (Fig. 8.4c); it consists of layers of material from the workpiece that are gradually deposited on the tool (hence the term *built-up*). As it becomes larger, the BUE becomes unstable and eventually breaks up; a portion of the BUE material is carried away on the tool side of the chip, and the rest is deposited randomly on the workpiece surface. The process of BUE formation and breakup is repeated continuously during the cutting operation. Although BUE is generally unwanted, a thin, stable BUE is usually regarded as desirable, because it protects the tool's surface.

The built-up edge is commonly observed in practice and is one of the important factors that adversely affects surface finish and integrity in cutting, as Figs. 8.4 and 8.6 show. A built-up edge, in effect, changes the geometry of cutting. Note, for example, the large tip radius of the BUE and the rough surface finish produced. Because of work hardening and deposition of successive layers of material, BUE hardness increases significantly (Fig. 8.6a).

The exact mechanism of formation of the BUE is not yet clearly understood; however, investigations have identified two distinct mechanisms that contribute to its formation. One is the adhesion of the workpiece material to the rake face of the tool; the strength of this bond depends on the affinity of the workpiece and tool materials. The second mechanism is the growth of the adhered metal layers. One of the important factors in forming the BUE is the tendency of the workpiece material for strain hardening; the higher the strain-hardening exponent, $n$, the greater is the probability for BUE formation. Experimental evidence indicates that as the cutting speed increases, the BUE decreases or is eliminated. In addition to making appropriate adjustments to the aforementioned factors, the tendency for BUE to form can be reduced by (a) decreasing the depth of cut, (b) increasing the rake angle, (c) using a tool with a small tip radius, and (d) using an effective cutting fluid.

3. **Serrated chips.** *Serrated chips*, also called **segmented** or *nonhomogeneous* chips, are semicontinuous chips, with zones of low and high shear strain (Fig. 8.4e). The chips have the appearance of saw teeth. Metals with low thermal conductivity and strength that decreases sharply with temperature, such as titanium, exhibit this behavior.

4. **Discontinuous chips.** *Discontinuous chips* consist of segments that may be either firmly or loosely attached to each other (Fig. 8.4f). These chips usually form under the following conditions: (a) when the workpiece materials are brittle, because they do not have the capacity to undergo the high shear strains in

**FIGURE 8.6**    (a) Hardness distribution in the cutting zone for 3115 steel. Note that some regions in the built-up edge are as much as three times harder than the bulk workpiece. (b) Surface finish in turning 5130 steel with a built-up edge. (c) Surface finish on 1018 steel in face milling. Magnifications: 15×. *Source*: Courtesy of TechSolve, formerly the Institute of Advanced Manufacturing Sciences, Inc.

cutting; (b) when the workpiece materials contain hard inclusions and impurities (see Figs. 3.24 and 3.25) or have structures such as the graphite flakes in gray cast iron (Fig. 5.20a); (c) when the cutting speeds are very low or very high; (d) when the depths of cut are large or the rake angles are low; (e) when there is low stiffness and poor damping of the machine tool's structure; and (f) when there is a lack of an effective cutting fluid.

Impurities and hard particles act as sites for cracks, thereby producing discontinuous chips; as expected, a large depth of cut increases the probability of such defects being present in the cutting zone. Higher cutting speeds mean higher temperatures; hence, the material is more ductile and has less of a tendency to form discontinuous chips. Another factor in the formation of discontinuous chips is the magnitude of the compressive stresses on the shear plane; recall from Fig. 2.22 that the maximum shear strain at fracture increases with increasing compressive stress. Thus, if the normal stress is not sufficiently high, the material will be unable to undergo the large shear strain required to form a continuous chip.

Because of the nature of discontinuous-chip formation, forces continually vary during cutting. Consequently, the stiffness of the cutting-tool holder and the machine tool is important in cutting with discontinuous-chip as well as serrated-chip formation. If not sufficiently stiff, the machine tool may begin to vibrate and chatter. (See Section 8.11.) This effect, in turn, adversely affects the surface finish and dimensional accuracy of the machined component and may cause damage or excessive wear of the cutting tool.

**Chip formation in cutting nonmetallic materials.** Much of the discussion thus far for metals is also generally applicable to nonmetallic materials. A variety of chips is obtained in cutting thermoplastics, depending on the type of polymer and the process parameters, such as depth of cut, tool geometry, and cutting speed. Because they are brittle, thermosetting plastics and ceramics generally produce discontinuous chips. (See also the discussion on *ductile regime cutting* in Section 8.8.2.)

**Chip curl.** *Chip curl* (Figs. 8.5 and 8.7a) is common to all cutting operations with metals and nonmetallic materials such as plastics and wood. The reasons for chip curl are not clearly understood. Among possible factors contributing to it are the distribution of stresses in the primary and secondary shear zones, thermal gradients, the work-hardening characteristics of the workpiece material, and the geometry of the rake face of the tool. Process variables also affect chip curl; generally, the radius of curvature decreases (i.e., the chip becomes curlier) with decreasing depth of cut, increasing rake angle, and decreasing friction at the tool–chip interface. The use of cutting fluids and various additives in the workpiece material also influences chip curl.

**FIGURE 8.7**
(a) Schematic illustration of the action of a chip breaker. Note that the chip breaker decreases the radius of curvature of the chip.
(b) Chip breaker clamped on the rake face of a cutting tool. (c) Grooves in cutting tools, acting as chip breakers.

(a)

(b)

(c)

(d)

**FIGURE 8.8**  Various chips produced in turning: (a) tightly curled chip; (b) chip hits workpiece and breaks; (c) continuous chip moving away from workpiece; and (d) chip hits tool shank and breaks off. *Source*: G. Boothroyd, *Fundamentals of Metal Machining and Machine Tools.*

**Chip breakers.**  Long, continuous chips are undesirable, because they tend to become entangled and interfere with cutting operations and can be a safety hazard. This situation is especially troublesome in high-speed automated machinery. The usual procedure to avoid the formation of continuous chips is to break the chip intermittently with a *chip breaker*. Although the chip breaker has traditionally been a piece of metal clamped to the rake face of the tool (Fig. 8.7b), it is now an integral part of the tool itself (Fig. 8.7c). Chips can also be broken by changing the tool geometry, thus controlling chip flow, as in the turning operations shown in Fig. 8.8.

A wide variety of cutting tools and inserts with chip-breaker features are available. However, with soft workpiece materials, such as commercially pure aluminum and copper, chip breakers are generally not as effective; the remedy is usually to change the process parameters. In interrupted cutting operations, such as milling, chip breakers are generally not necessary, since the chips already have finite lengths, due to the intermittent nature of the operation. (See Fig. 8.54a.)

## 8.2.2  Mechanics of oblique cutting

Unlike the two-dimensional situations described thus far, the majority of cutting operations involve tool shapes that are three-dimensional (**oblique**). Figure 8.9a illustrates the basic difference between two-dimensional and oblique cutting. We have seen that in orthogonal cutting, the tool edge is perpendicular to the movement of the tool, and the chip slides directly up the rake face of the tool. In oblique cutting, however, the cutting edge is at an angle $i$, called the **inclination angle** (Fig. 8.9b). Note that the chip in Fig. 8.9a flows up the rake face of the tool at an angle $\alpha_c$ (**chip flow angle**), measured in the plane of the tool face. (Note that this situation is similar to an angled snowplow blade, which throws the snow sideways.) Angle $\alpha_n$ is known as the **normal rake angle**, which is a basic geometric property of the tool. The normal rake angle is the angle between the normal $oz$ to the workpiece surface and the line $oa$ on the tool face.

The workpiece material approaches the tool at a velocity $V$ and leaves the workpiece surface (as a chip) with a velocity $V_c$. The effective rake angle, $\alpha_e$, in the plane of these two velocities can be calculated as follows: Assuming that the chip flow angle, $\alpha_c$, is equal to the inclination angle, $i$ (an assumption that has experimentally been found to be approximately correct), the effective rake angle, $\alpha_e$, is given by

$$\alpha_e = \sin^{-1}(\sin^2 i + \cos^2 i \sin \alpha_n). \tag{8.8}$$

Since $i$ and $\alpha_n$ can both be measured directly, we can use this formula to calculate the value of the effective rake angle. Note that as $i$ increases, the effective rake angle increases, and the chip becomes thinner and longer. Figure 8.9c illustrates the effect of the inclination angle on chip shape.

Figure 8.10 shows a typical single-point turning tool used on a lathe. Note the various angles, each of which has to be selected properly for efficient cutting. Various three-dimensional cutting tools are described in greater detail in Sections 8.8 and 8.9, including tools for drilling, tapping, milling, planing, shaping, broaching, sawing, and filing.

**Shaving and skiving.** Thin layers of material can be removed from straight or curved surfaces by a process similar to using a plane to shave wood. *Shaving* is particularly useful in improving the surface finish and dimensional accuracy of punched slugs or holes. (See Fig. 7.11.) Parts that are long or have a combination of shapes are shaved by *skiving* with a specially shaped cutting tool.

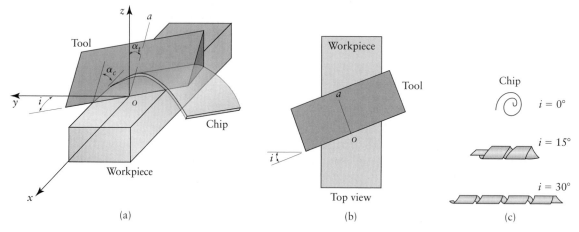

**FIGURE 8.9** (a) Schematic illustration of cutting with an oblique tool. (b) Top view, showing the inclination angle, $i$. (c) Types of chips produced with different inclination angles.

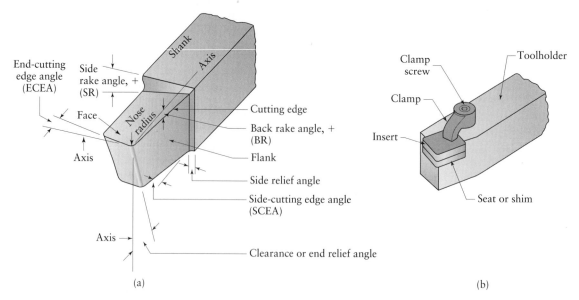

**FIGURE 8.10** (a) Schematic illustration of a right-hand cutting tool. Although these tools have traditionally been produced from solid tool-steel bars, they have been largely replaced by carbide or other inserts of various shapes and sizes, as shown in (b).

## 8.2.3 Forces in orthogonal cutting

Knowledge of the forces and power involved in cutting operations is essential for the following reasons:

1. **Power requirements** must be determined so that a machine of suitable capacity can be selected;
2. Data on forces are necessary for the proper **design of machine tools** so that they are sufficiently stiff, thus maintaining desired dimensional accuracy for the machined part and avoiding excessive distortion of the machine components and vibration and chatter;
3. The workpiece must be able to withstand the cutting forces without excessive **distortion** in order to maintain dimensional tolerances.

The following list discusses the various factors that influence the forces and power involved in orthogonal cutting:

a) **Cutting forces.** Figure 8.11 depicts the forces acting on the tool in orthogonal cutting. **The cutting force, $F_c$,** acts in the direction of the cutting speed, $V$, and supplies the energy required for cutting. The **thrust force, $F_t$,** acts in the direction normal to the cutting velocity, that is, perpendicular to the workpiece. These two forces produce the **resultant force, $F_r$.** Note that the resultant force can be resolved into two components on the tool face: a **friction force, $F$,** along the tool–chip interface and a **normal force, $N$,** perpendicular to the interface. From Fig. 8.11, it can be shown the friction force is

$$F = R \sin \beta \qquad (8.9)$$

and that the normal force is

$$N = R \cos \beta. \qquad (8.10)$$

Note also that the resultant force is balanced by an equal and opposite force along the shear plane and is resolved into a **shear force, $F_s$,** and a **normal force, $F_n$.**

From Fig. 8.11, the cutting force can be shown to be

$$F_c = R \cos(\beta - \alpha) = \frac{wt_o\tau \cos(\beta - \alpha)}{\sin \phi \cos(\phi + \beta - \alpha)}, \qquad (8.11)$$

where $\tau$ is the *average shear stress* along the shear plane.

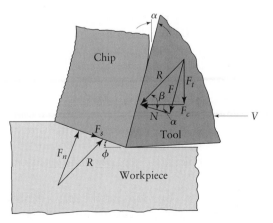

**FIGURE 8.11**   Forces acting on a cutting tool in two-dimensional cutting.

The ratio of $F$ to $N$ is the **coefficient of friction**, $\mu$, at the tool–chip interface (see also Section 4.4.1), and the angle $\beta$ is the **friction angle**. The coefficient of friction can be expressed as

$$\mu = \tan\beta = \frac{F_t + F_c \tan\alpha}{F_c - F_t \tan\alpha}. \tag{8.12}$$

In metal cutting, $\mu$ generally ranges from about 0.5 to 2, thus indicating that the chip encounters considerable frictional resistance while climbing up the rake face of the tool.

Although the magnitude of forces in actual cutting operations is generally on the order of a few hundred newtons, the *local* stresses in the cutting zone and the normal stresses on the rake face of the tool are very high, because the contact areas are very small. The tool–chip contact length (Fig. 8.2), for example, is typically on the order of 1 mm (0.04 in.); thus, the tool is subjected to very high stresses.

b) **Thrust force and its direction.** A knowledge of the magnitude of the thrust force in cutting is important, because the tool holder, the workholding and fixturing devices, and the machine tool must be sufficiently stiff to minimize deflections caused by this force. For example, if the thrust force (see Fig. 8.11) is too high, or if the machine tool is not sufficiently stiff, the tool will be pushed away from the workpiece surface. This movement will, in turn, reduce the actual depth of cut, leading to loss of dimensional accuracy of the machined part and possibly to vibration and chatter.

Note in Fig. 8.11, that the cutting force must always be in the direction of cutting in order to supply energy for cutting. Although the thrust force does not contribute to the total work done, it is important to know its magnitude. Also note from Fig. 8.11 that the thrust force is *downward*. However, it can be shown that this force can be upward (negative), by first observing that

$$F_t = R \sin(\beta - \alpha) \tag{8.13}$$

or

$$F_t = F_c \tan(\beta - \alpha). \tag{8.14}$$

Because $F_c$ is always positive (as shown in Fig. 8.11), the sign of $F_t$ can be either positive or negative. When $\beta > \alpha$ the sign of $F_t$ is positive (downward), and when $\beta < \alpha$, it is negative (upward). Thus, it is possible to have an upward thrust force at low friction at the tool-chip interface and/or high rake angles.

This situation can be visualized by inspecting Fig. 8.11 and noting that when $\mu = 0$, $\beta = 0$, and consequently the resultant force, $R$, coincides with the normal force, $N$. In this case, then, $F_r$ will have an upward thrust-force component; also note that for $\alpha = 0$ and $\beta = 0$, the thrust force is zero.

These observations have been verified experimentally, as shown in Fig. 8.12. The influence of the depth of cut is obvious, because as $t_o$ increases, $R$ must also increase so that $F_c$ will increase as well. This action supplies the additional energy required to remove the extra material produced by the increased depth of cut. The change in direction and magnitude of the thrust force can play a significant role. Within a certain range of operating conditions, it can lead to *instability* problems in machining, particularly if the machine tool is not sufficiently stiff.

c) **Observations on cutting forces.** In addition to being a function of the strength of the workpiece material, forces in cutting are influenced by other

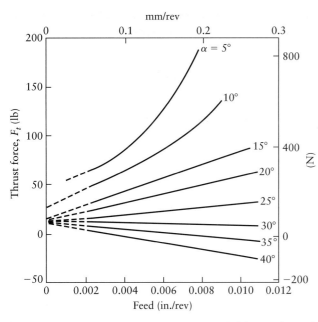

**FIGURE 8.12** Thrust force as a function of rake angle and feed in orthogonal cutting of AISI 1112 cold-rolled steel. Note that at high rake angles, the thrust force is negative. A negative thrust force has important implications in the design of machine tools and in controlling the stability of the cutting process. *Source:* After S. Kobayashi and E. G. Thomsen.

variables. Extensive data are availables such as those shown in Tables 8.1 and 8.2, which show that the cutting force, $F_c$, increases with increasing depth of cut, decreasing rake angle, and decreasing speed. From analysis of the data in Table 8.2, we can attribute the effect of cutting speed to the fact that as speed decreases, the shear angle decreases and the coefficient of friction increases. Both effects increase the cutting force.

## TABLE 8.1

### Data on Orthogonal Cutting of 4130 Steel*

| $\alpha$ | $\phi$ | $\gamma$ | $\mu$ | $\beta$ | $F_c$ (lb) | $F_t$ (lb) | $u_t$ (in.-lb/in$^3$) $\times 10^3$) | $u_s$ | $u_f$ | $u_f/u_t$ (%) |
|---|---|---|---|---|---|---|---|---|---|---|
| 25° | 20.9° | 2.55 | 1.46 | 56 | 380 | 224 | 320 | 209 | 111 | 35 |
| 35 | 31.6 | 1.56 | 1.53 | 57 | 254 | 102 | 214 | 112 | 102 | 48 |
| 40 | 35.7 | 1.32 | 1.54 | 57 | 232 | 71 | 195 | 94 | 101 | 52 |
| 45 | 41.9 | 1.06 | 1.83 | 62 | 232 | 68 | 195 | 75 | 120 | 62 |

*$t_o$ = 0.0025 in.; $w$ = 0.475 in.; $V$ = 90 ft/min; tool: high-speed steel.
*Source:* After E. G. Thomsen.

## TABLE 8.2

### Data on Orthogonal Cutting of 9445 Steel*

| $\alpha$ | $V$ | $\phi$ | $\gamma$ | $\mu$ | $\beta$ | $F_c$ | $F_t$ | $u_t$ | $u_s$ | $u_f$ | $u_f/u_t$ (%) |
|---|---|---|---|---|---|---|---|---|---|---|---|
| +10 | 197 | 17 | 3.4 | 1.05 | 46 | 370 | 273 | 400 | 292 | 108 | 27 |
| | 400 | 19 | 3.1 | 1.11 | 48 | 360 | 283 | 390 | 266 | 124 | 32 |
| | 642 | 21.5 | 2.7 | 0.95 | 44 | 329 | 217 | 356 | 249 | 107 | 30 |
| | 1186 | 25 | 2.4 | 0.81 | 39 | 303 | 168 | 328 | 225 | 103 | 31 |
| −10 | 400 | 16.5 | 3.9 | 0.64 | 33 | 416 | 385 | 450 | 342 | 108 | 24 |
| | 637 | 19 | 3.5 | 0.58 | 30 | 384 | 326 | 415 | 312 | 103 | 25 |
| | 1160 | 22 | 3.1 | 0.51 | 27 | 356 | 263 | 385 | 289 | 96 | 25 |

*$t_o$ = 0.037 in.; $w$ = 0.25 in.; tool: cemented carbide.
*Source:* After M. E. Merchant.

Another factor that can significantly influence the magnitude of the cutting force is the tip radius of the tool: The larger the radius (and hence the duller the tool), the higher is the force. Experimental evidence indicates that, for depths of cut on the order of five times the tip radius or higher, the effect of tool dullness on the cutting forces is negligible. (See also Section 8.4.)

d) **Shear and normal stresses in the cutting zone.** The stresses in the shear plane and at the tool–chip interface can be analyzed by assuming that they are uniformly distributed. The forces in the shear plane can be resolved into shear and normal forces and stresses. First, note that the area of the shear plane, $A_s$, is

$$A_s = \frac{wt_o}{\sin \phi}. \tag{8.15}$$

Therefore, the *average shear stress* in the shear plane is

$$\tau = \frac{F_s}{A_s} = \frac{F_s \sin \phi}{wt_o}, \tag{8.16}$$

and the *average normal stress* is

$$\sigma = \frac{F_n}{A_s} = \frac{F_n \sin \phi}{wt_o}. \tag{8.17}$$

Figure 8.13 provides some data pertaining to these average stresses. The rake angle is a parameter, and the shear plane area is increased by increasing the depth of cut. The following conclusions can be drawn from these curves:

1. The shear stress on the shear plane is independent of the rake angle.
2. The normal stress on the shear plane decreases with increasing rake angle.
3. Consequently, the normal stress in the shear plane has no effect on the magnitude of the shear stress; this phenomenon has also been verified by other mechanical tests. However, normal stress strongly influences the magnitude of the shear strain in the shear zone. The maximum shear strain to fracture increases with the normal compressive stress, as shown in Fig. 2.22.

Determining the stresses on the rake face of the tool presents considerable difficulties; one problem is accurately determining the length of contact at the tool–chip interface. This length increases with decreasing shear angle, indicating that the contact length is a function of rake angle, cutting speed, and

**FIGURE 8.13** Shear force and normal force as a function of the area of the shear plane and the rake angle for 85–15 brass. Note that the shear stress in the shear plane is constant, regardless of the magnitude of the normal stress. Thus, normal stress has no effect on the shear flow stress of the material. *Source*: After S. Kobayashi and E. G. Thomsen, *J. Eng. Ind.*, 81: 251–262, 1959.

(a)

(b)

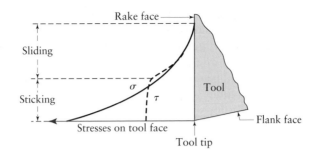

**FIGURE 8.14**    Schematic illustration of the distribution of normal and shear stresses at the tool–chip interface (rake face). Note that, whereas the normal stress increases continuously toward the tip of the tool, the shear stress reaches a maximum and remains at that value (a phenomenon known as *sticking*; see Section 4.4.1).

friction at the tool–chip interface. Another problem is that the stresses are not uniformly distributed on the rake face. *Photoelastic* studies have shown that the actual stress distribution is as shown in Fig. 8.14. Note that the stress normal to the tool face is a maximum at the tip and decreases rapidly toward the end of the contact length. The shear stress has a similar trend, except that it levels off after about the center of the contact length. This behavior indicates that *sticking* is taking place, whereby the shear stress has reached the shear yield strength of the workpiece material. (See Section 4.4.1.) Sticking regions have been experimentally observed on some chips. Sticking regions also can be observed in various other metal-forming processes; see, for example, Section 6.2.2.

e) **Measuring cutting forces.**    Cutting forces can be measured by using suitable force **dynamometers** (with resistance-wire strain gages) or force **transducers** (such as piezoelectric crystals) mounted on the machine tool. Forces can also be calculated from the amount of **power consumption** during cutting, often measured by a power monitor, provided that the mechanical efficiency of the machine tool can be determined.

## 8.2.4  Shear-angle relationships

Because the shear angle and the shear zone have great significance in the mechanics of cutting, a great deal of effort has been expended to determine the relationship of the shear angle to material properties and process variables. One of the earliest analyses (by M. E. Merchant) is based on the assumption that the shear angle adjusts itself so that the cutting force is a minimum, or so that the maximum shear stress occurs in the shear plane. From the force diagram in Fig. 8.11, we can express the shear stress in the shear plane as

$$\tau = \frac{F_s}{A_s} = \frac{F_c \sec(\beta - \alpha) \cos(\phi + \beta - \alpha) \sin \phi}{wt_o}. \tag{8.18}$$

If we assume that $\beta$ is independent of $\phi$, we can find the shear angle corresponding to the maximum shear stress by differentiating Eq. (8.18) with respect to $\phi$ and equating to zero. Thus, we have

$$\frac{d\tau}{d\phi} = \cos(\phi + \beta - \alpha) \cos \phi - \sin(\phi + \beta - \alpha) \sin \phi = 0, \tag{8.19}$$

and hence,

$$\tan(\phi + \beta - \alpha) = \cot \phi = \tan(90° - \phi),$$

or

$$\phi = 45° + \frac{\alpha}{2} - \frac{\beta}{2}. \tag{8.20}$$

Note that Eq. (8.20) indicates that, as the rake angle decreases and/or as the friction at the tool–chip interface increases, the shear angle decreases, and the chip is thus thicker. This result is to be expected, because decreasing $\alpha$ and increasing $\beta$ tend to lower the shear angle.

A second method of determining $\phi$ is based on slip-line analysis (by E. H. Lee and B. W. Shaffer) and gives an expression for the shear-plane angle as

$$\phi = 45° + \alpha - \beta. \tag{8.21}$$

Note that this expression is similar to Eq. (8.20) and indicates the same trends, although numerically it gives different values.

In another study (by T. Sata and M. Mizuno), the following simple relationships have been proposed and serve as a simple guide for estimating the shear angle for practical purposes:

$$\phi = \alpha \quad \text{for} \quad \alpha > 15°; \tag{8.22}$$

$$\phi = 15° \quad \text{for} \quad \alpha < 15°. \tag{8.23}$$

Several other expressions, based on various models and different assumptions, have been obtained for the shear angle; many of these expressions do not agree well with experimental data over a wide range of conditions (Fig. 8.15a), largely because shear rarely occurs in a thin plane. However, the shear angle always decreases with increasing $\beta - \alpha$, as shown in Fig. 8.15b. More recent and comprehensive studies appear to accurately predict the shear angle analytically, especially for continuous chips.

### 8.2.5 Specific energy

We note from Fig. 8.11 that the **total power** input in cutting is

$$\text{Power} = F_c V.$$

The **total energy per unit volume** of material removed (**specific energy**), $u_t$, is then

$$u_t = \frac{F_c V}{w t_o V} = \frac{F_c}{w t_o}, \tag{8.24}$$

where $w$ is the width of the cut. In other words, $u_t$ is simply the ratio of the cutting force to the projected area of the cut. Figures 8.3 and 8.11 also show that the power

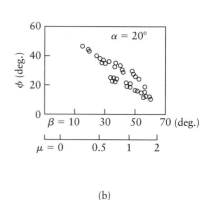

**FIGURE 8.15**    (a) Comparison of experimental and theoretical shear-angle relationships. More recent analytical studies have resulted in better agreement with experimental data. (b) Relation between the shear angle and the friction angle for various alloys and cutting speeds. *Source*: S. Kobayashi et al., *Trans. ASME, J. Eng. Ind.*, 82: 333–347, 1960.

required to overcome friction at the tool–chip interface is the product of $F$ and $V_c$, or, in terms of **specific energy for friction,** $u_f$,

$$u_f = \frac{FV_c}{wt_o V} = \frac{Fr}{wt_o} = \frac{(F_c \sin \alpha + F_t \cos \alpha)r}{wt_o} \qquad (8.25)$$

Likewise, the power required for shearing along the shear plane is the product of $F_s$ and $V_s$. Hence, the **specific energy for shearing,** $u_s$, is

$$u_s = \frac{F_s V_s}{wt_o V}. \qquad (8.26)$$

The **total specific energy,** $u_t$, is the sum of the two energies, or

$$u_t = u_f + u_s. \qquad (8.27)$$

There are two additional sources of energy in cutting. One is the **surface energy** resulting from the formation of *two new surfaces* (the rake-face side of the chip and the newly machined surface) when a layer of material is removed by cutting. However, this energy represents a very small amount as compared with the shear and frictional energies involved. The other source is the energy associated with the **momentum change** as the volume of metal being removed suddenly crosses the shear plane. (This situation is similar to the forces involved in a turbine blade from momentum changes of the liquid or gas.) Although in ordinary metal-cutting operations, the momentum energy is negligible, it can be significant at very high cutting speeds, such as above 125 m/s (25,000 ft/min).

Tables 8.1 and 8.2 provide experimental data on specific energies. Note that as the rake angle increases, the frictional specific energy remains more or less constant, whereas the shear specific energy is rapidly reduced; thus, the ratio $u_f/u_t$ increases significantly as $\alpha$ increases. This trend can also be predicted by obtaining an expression for the energy ratio as follows:

$$\frac{u_f}{u_t} = \frac{FV_c}{F_c V} = \frac{R \sin \beta}{R \cos(\beta - \alpha)} \cdot \frac{Vr}{V} = \frac{\sin \beta}{\cos(\beta - \alpha)} \cdot \frac{\sin \phi}{\cos(\phi - \alpha)}. \qquad (8.28)$$

Experimental observations have indicated that as $\alpha$ increases, both $\beta$ and $\phi$ increase, and inspection of Eq. (8.28) also indicates that the ratio $u_f/u_t$ should increase with $\alpha$. Obviously, $u_f$ and $u_s$ are related. Although $u_f$ is not necessary for the cutting

**TABLE 8.3**

## Approximate Specific-Energy Requirements in Cutting Operations

| | SPECIFIC ENERGY* | |
|---|---|---|
| MATERIAL | W-s/mm$^3$ | hp-min/in$^3$ |
| Aluminum alloys | 0.4–1.1 | 0.15–0.4 |
| Cast irons | 1.6–5.5 | 0.6–2.0 |
| Copper alloys | 1.4–3.3 | 0.5–1.2 |
| High-temperature alloys | 3.3–8.5 | 1.2–3.1 |
| Magnesium alloys | 0.4–0.6 | 0.15–0.2 |
| Nickel alloys | 4.9–6.8 | 1.8–2.5 |
| Refractory alloys | 3.8–9.6 | 1.1–3.5 |
| Stainless steels | 3.0–5.2 | 1.1–1.9 |
| Steels | 2.7–9.3 | 1.0–3.4 |
| Titanium alloys | 3.0–4.1 | 1.1–1.5 |

*At drive motor, corrected for 80% efficiency; multiply the energy by 1.25 for dull tools.

action to take place, it affects the magnitude of $u_s$. The reason is that as friction increases, the shear angle decreases; a decreasing shear angle, in turn, increases the magnitude of $u_s$.

The calculation of all the parameters involved in these specific energies in cutting presents considerable difficulties; good theoretical computations are available, but they are difficult to perform. Consequently, the reliable prediction of cutting forces and energies is still based largely on experimental data (as shown in Tables 8.1, 8.2, and 8.3). The wide range of values in Table 8.3 can be attributed to differences in strength within each material group and to the effects of various other variables, such as friction and operating conditions.

---

**EXAMPLE 8.1 Relative energies in cutting**

An orthogonal cutting operation is being carried out in which $t_o = 0.005$ in., $V = 400$ ft/min, $\alpha = 10°$, and the width of cut = 0.25 in. It is observed that $t_c = 0.009$ in., $F_c = 125$ lb, and $F_t = 50$ lb. Calculate the percentage of the total energy that is dissipated in friction at the tool–chip interface.

**Solution.** The percentage of energy can be expressed as

$$\frac{\text{Friction energy}}{\text{Total energy}} = \frac{FV_c}{F_c V} = \frac{Fr}{F_c},$$

where

$$r = \frac{t_o}{t_c} = \frac{5}{9} = 0.555,$$

$$F = R \sin \beta,$$

$$F_c = R \cos(\beta - \alpha)$$

and

$$R = \sqrt{F_t^2 + F_c^2} = \sqrt{50^2 + 125^2} = 135 \text{ lb.}$$

Thus,

$$125 = 135 \cos(\beta - 10°),$$

from which we find that

$$\beta = 32° \quad \text{and} \quad F = 135 \sin 32° = 71.5 \text{ lb.}$$

Therefore, the percentage of friction energy is calculated as

$$\text{Percentage} = \frac{(71.5)(0.555)}{125} = 0.32 = 32\%,$$

and similarly, the percentage of shear energy is calculated as 68%.

---

**EXAMPLE 8.2 Comparison of forming and machining energies**

You are given two pieces of annealed 304 stainless-steel rods, each with a diameter of 0.500 in. and a length of 6 in., and you are asked to reduce their diameters to 0.480 in., (a) for one piece by *pulling* it in tension and (b) for the other by *machining* it on a lathe (see Fig. 8.8) in one pass. Calculate the respective amounts of work involved, and explain the reasons for the difference in the energies dissipated.

**Solution.** (a) The work done in pulling the rod (see Section 2.12) is

$$W_{\text{tension}} = (u)(\text{volume}),$$

where

$$u = \int_0^{\varepsilon_1} \sigma \, d\varepsilon.$$

The true strain is found from the expression

$$\varepsilon_1 = \ln\left(\frac{0.500}{0.480}\right)^2 = 0.0816.$$

From Table 2.3, we obtain the following values for $K$ and $n$ for this material:

$$K = 1275 \text{ MPa} = 185,000 \text{ psi} \quad \text{and} \quad n = 045.$$

Thus,

$$u = \frac{K\varepsilon_1^{n+1}}{n+1} = \frac{(185,000)(0.0816)^{1.45}}{1.45} = 3370 \text{ in.} \cdot \text{lb/in}^3$$

and thus

$$W_{\text{tension}} = (3370)(\pi)(0.25)^2(6) = 3970 \text{ in.-lb.}$$

(b) From Table 8.3, let's estimate an average value for the specific energy in machining stainless steels as 1.5 hp-min/in$^3$. The volume of material machined is

$$\text{Volume} = \frac{\pi}{4}[(0.5)^2 - (0.480)^2](6) = 0.092 \text{ in}^3.$$

The specific energy, in appropriate units, is calculated as

$$\text{Specific energy} = (1.5)(33,000)(12) = 594,000 \text{ in.-lb/in}^3.$$

Hence, the work done in machining is

$$W_{\text{mach}} = (594,000)(0.092) = 54,650 \text{ in.-lb.}$$

Note that the work done in machining is about 14 times higher than that for tension. The reasons for the difference between the two energies are that tension involves very little strain and that there is no friction. Machining, however, involves friction, and the material removed (even though relatively small in volume) has undergone much higher strains than the bulk material undergoing tension. Assuming, from Tables 8.1 and 8.2, an average shear strain of 3, which is equivalent to an effective strain of 1.7 [see Eq. (2.58)], the material in machining is subjected to a strain of $1.7/0.0816 = 21$ times higher than that in tension.

These differences explain why machining consumes much more energy than reducing the diameter of this rod by stretching. It can be shown, however, that as the diameter of the rod decreases, the difference between the two energies becomes *smaller*, assuming that the same depth of material is to be removed. This result can be explained by noting the changes in the relative volumes involved in machining vs. tension as the diameter of the rod decreases.

### 8.2.6 Temperature

As in all metalworking operations, the energy dissipated in cutting operations is converted into heat, which, in turn, raises the temperature in the cutting zone. Knowledge of the temperature rise in cutting is important, because increases in temperature:

- Adversely affect the strength, hardness, and wear resistance of the cutting tool,
- Cause dimensional changes in the part being machined, making control of dimensional accuracy difficult, and
- Can induce thermal damage to the machined surface, adversely affecting its properties and service life.

In addition, the machine tool itself may be subjected to temperature gradients, causing distortion of the machine.

Because of the work done in shearing and in overcoming friction on the rake face of the tool, the main sources of heat generation are the primary shear zone and the tool–chip interface. In addition, if the tool is dull or worn, heat is also generated by the tool tip rubbing against the machined surface.

**Variables affecting temperature.** Various studies have been made of temperatures in cutting, based on heat transfer and dimensional analysis, using experimental data. Although Fig. 8.16 shows that there are severe temperature gradients in the cutting zone, a simple and approximate expression for the *mean temperature* for orthogonal cutting is

$$T = \frac{1.2Y_f}{\rho c} \sqrt[3]{\frac{Vt_o}{K}}, \qquad (8.29)$$

where $T$ is the mean temperature of the tool–chip interface (°F); $Y_f$ is the flow stress of the material (psi); $V$ is the cutting speed (in./s); $t_o$ is the depth of cut (in.); $\rho c$ is the volumetric specific heat of the workpiece (in.-lb/in³-°F); and $K$ is the *thermal diffusivity* (ratio of thermal conductivity to volumetric specific heat) of the workpiece material (in²/s). Note that because some of the parameters in Eq. (8.29) depend on temperature, it is important to use appropriate values that are compatible with the predicted temperature range.

**FIGURE 8.16**  Typical temperature distribution in the cutting zone. Note that the maximum temperature is about halfway up the rake face of the tool and that there is a steep temperature gradient across the thickness of the chip. Some chips may become red hot, causing safety hazards to the operator and thus necessitating the use of safety guards. *Source*: After G. Vieregge.

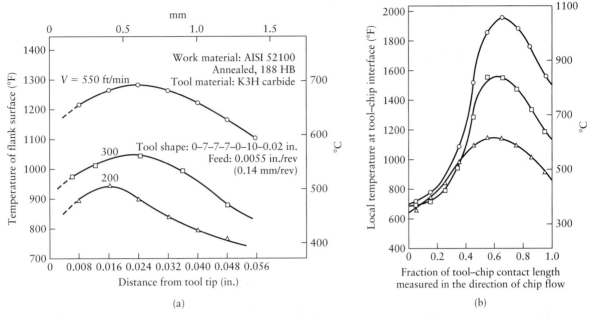

**FIGURE 8.17**   Temperature distribution in turning: (a) flank temperature for tool shape (see Fig. 8.41); (b) temperature of the tool–chip interface. Note that the rake face temperature is higher than that at the flank surface. *Source:* After B. T. Chao and K. J. Trigger.

The temperature generated in the shear plane is a function of the specific energy for shear, $u_s$, and the specific heat of the material. Hence, temperature rise is highest in cutting materials with high strength and low specific heat, as Eq. (2.65) indicates. The temperature rise at the tool–chip interface is, of course, also a function of the coefficient of friction. Flank wear (see Section 8.3) is another source of heat, caused by rubbing of the tool on the machined surface.

Figure 8.17 shows results from one experimental measurement of temperature (using thermocouples). Note that the maximum temperature is at a point away from the tip of the tool and that it increases with cutting speed. As the speed increases, there is little time for the heat to be dissipated, and hence temperature rises. As cutting speed increases, a larger proportion of the heat is carried away by the chip, as shown in Fig. 8.18. The chip is a good heat sink, in that it carries away most of the heat generated. (This action of the chip is somewhat similar to *ablation*, where layers of metal melt away from a surface, thus carrying away the heat.) It has been shown that thermal properties of the tool material (see Section 8.6) are relatively unimportant compared with those of the workpiece.

Note that Eq. (8.29) also indicates that the mean temperature increases with the strength of the workpiece material (because more energy will be required) and the depth of cut (because of the ratio of surface area to thickness of the chip; a thin chip cools faster than a thick chip). However, the depth of cut has negligible influence on the mean temperature for depths exceeding twice the tip radius of the tool. (See Fig. 8.29.)

Based on Eq. (8.29), another expression for the mean temperature in turning (on a lathe; see Fig. 8.42) is given as

$$T \propto V^a f^b, \tag{8.30}$$

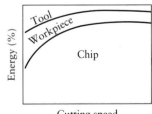

**FIGURE 8.18**   Typical plot of energy distribution as a function of cutting speed. Note that most of the cutting energy is carried away by the chip (in the form of heat), particularly as speed increases. For dimensional accuracy during cutting, it is important not to allow the workpiece temperature to rise significantly.

**FIGURE 8.19**

Terminology used in a turning operation on a lathe, where *f* is the feed (in in./rev or mm/rev) and *d* is the depth of cut. Note that feed in turning is equivalent to the depth of cut in orthogonal cutting (Fig. 8.2), and the depth of cut in turning is equivalent to the width of cut in orthogonal cutting. See also Fig. 8.42.

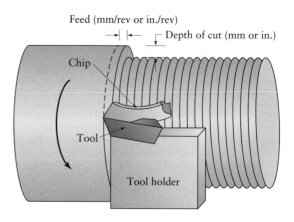

where *a* and *b* are constants, *V* is the cutting speed, and *f* is the feed of the tool (Fig. 8.19). Approximate values for *a* and *b* are given in the following table:

| TOOL | $a$ | $b$ |
|------|-----|-----|
| Carbide | 0.2 | 0.125 |
| High-speed steel | 0.5 | 0.375 |

**Techniques for measuring temperature.** Temperatures and their distribution in the cutting zone may be determined by using **thermocouples** embedded in the tool or the workpiece. This technique has been used successfully, although it involves considerable effort. A simpler technique for determining the average temperature is by the measuring **thermal emf** (electromotive force) at the tool–chip interface, which acts as a hot junction between two different (tool and chip) materials. **Infrared radiation** from the cutting zone may also be monitored with a *radiation pyrometer*; however, this technique indicates only surface temperatures, and the accuracy of the results depends on the emissivity of the surfaces, which is difficult to determine precisely.

## 8.3 | Tool Wear and Failure

We have seen that cutting tools are subjected to forces, temperature, and sliding; all these conditions induce wear. (See Section 4.4.2.) Because of its effects on the quality of the machined surface and the economics of machining, *tool wear* is one of the most important and complex aspects of machining operations. Although cutting speed is an independent variable, the forces and temperatures generated in cutting are dependent variables and are functions of numerous parameters. Similarly, wear depends on tool and workpiece materials (i.e., their physical, mechanical, and chemical properties), tool geometry, properties of cutting fluids (if used), and various other operating parameters. The types of wear on a tool depend on the relative roles of the foregoing variables. Analytical studies of tool wear present considerable difficulties, and therefore our knowledge of wear is based largely on experimental observations.

The basic wear behavior of a cutting tool is shown in Fig. 8.20c for a two-dimensional cut and in Figs. 8.20a and 8.21b for a three-dimensional cut in which the position of the tool is shown in Fig. 8.19. The various regions of wear are identified as *flank wear*, *crater wear*, *nose wear*, and *chipping* of the cutting edge. The tool

Thermal cracks in interrupted cutting

Chamfer

Carbide

High-speed steel

Ceramic

1. Flank wear (wear land)
2. Crater wear
3. Primary groove or depth-of-cut line

4. Secondary groove (oxidation wear)
5. Outer-metal chip notch
6. Inner chip notch

(a)

Chamfer

High-speed steel tool, thermal softening and plastic flow

Ceramic tool, chipping and fracture

1. Flank wear
2. Crater wear
3. Failure face

4. Primary groove or depth-of-cut line
5. Outer-metal chip notch
6. Plastic flow around failure face

(b)

Rake face

Crater wear depth ($KT$)

Tool

Flank wear

Flank face

Rake face

Crater wear

$R$ Nose radius

Depth-of-cut line

Flank wear

Depth-of-cut line

$VB_{max}$ $VB$

Flank face

(c)

**FIGURE 8.20** (a) Types of wear observed in cutting tools. The thermal cracks shown are usually observed in interrupted cutting operations, such as in milling. (b) Catastrophic failure of tools. (c) Features of tool wear in a turning operation. The *VB* indicates average flank wear. *Source*: (a) and (b) After V. C. Venkatesh. (c) Terms and definitions reproduced with the permission of the International Organization for Standardization, ISO. Copyright remains with ISO.

(a)

(b)

**FIGURE 8.21** (a) Crater wear and (b) flank wear on a carbide tool. *Source*: J. C. Keefe, Lehigh University.

profile is of course, altered by these various wear and fracture processes, and this change influences the cutting operation. In addition to wear, *plastic deformation* of the tool can also take place to some extent, such as in softer tools at elevated temperatures. Gross chipping of the tool is called *catastrophic failure*, whereas wear is generally a gradual process. Because of the complex interactions among the numerous material and process variables involved, the proper interpretation of wear patterns and a careful study of the wear mechanisms are essential.

### 8.3.1 Flank wear

*Flank wear* has been studied extensively and is generally attributed to

1. Sliding of the tool along the machined surface, causing adhesive and/or abrasive wear, depending on the materials involved, and

2. Temperature rise, because of its influence on the properties of the tool material.

Following an extensive study by F. W. Taylor published in 1907, a relationship was established for cutting various steels:

$$VT^n = C. \qquad (8.31)$$

In this equation, $V$ is the cutting speed, $T$ is the time (in minutes) that it takes to develop a flank wear land ($VB$ in Fig. 8.20c), $n$ is an exponent that depends on cutting conditions, and $C$ is a constant. Equation (8.31) is a simple version of the many relationships developed by Taylor among the variables involved; each combination of workpiece and tool materials and each cutting condition has its own $n$ value and a different constant $C$.

**Tool-life curves.** *Tool-life curves* are plots of experimental data obtained in cutting tests (Fig. 8.22). Note the rapid decrease in tool life as cutting speed increases and the

| | Hardness (HB) | Ferrite | Pearlite |
|---|---|---|---|
| a. As cast | 265 | 20% | 80% |
| b. As cast | 215 | 40 | 60 |
| c. As cast | 207 | 60 | 40 |
| d. Annealed | 183 | 97 | 3 |
| e. Annealed | 170 | 100 | — |

(a)

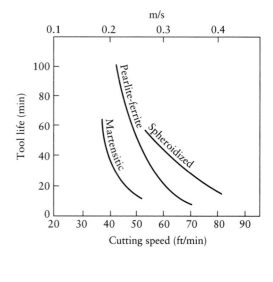

(b)

**FIGURE 8.22** Effect of workpiece microstructure on tool life in turning. Tool life is given in terms of the time (in minutes) required to reach a flank wear land of a specified dimension. (a) Ductile cast iron. (b) Steels, with identical hardness. Note in both figures the rapid decrease in tool life as the cutting speed increases.

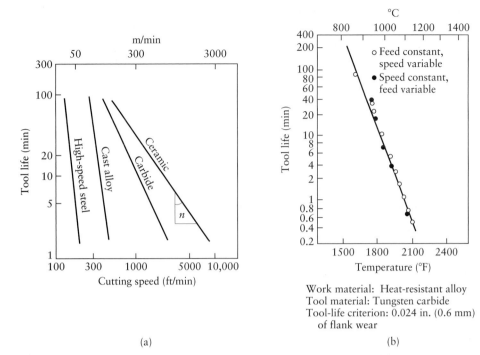

(a)                                                    (b)

Work material: Heat-resistant alloy
Tool material: Tungsten carbide
Tool-life criterion: 0.024 in. (0.6 mm)
of flank wear

**FIGURE 8.23** (a) Tool-life curves for a variety of cutting-tool materials. The negative inverse of the slope of these curves is the exponent $n$ in tool-life equations. (b) Relationship between measured temperature during cutting and tool life (flank wear). Note that high cutting temperatures severely reduce tool life. See also Eq. (8.30). *Source*: After H. Takeyama and Y. Murata.

strong influence of the condition of the workpiece material on tool life; also note the large difference in tool life for different workpiece microstructures. Heat treatment is important largely because of increasing workpiece hardness; for example, ferrite has a hardness of about 100 HB, pearlite 200 HB, and martensite 300 HB to 500 HB. (See Section 5.11.) Impurities and hard constituents in the workpiece material are also important considerations, because they reduce tool life by their abrasive action.

Tool-life curves are usually plotted on log–log paper, from which we can easily determine the exponent $n$ (Fig. 8.23a). Table 8.4 provides the range of $n$ values that has been determined experimentally. Tool-life curves are usually linear over a certain range of cutting speeds, but are rarely so over a wide range. Moreover, the exponent $n$ can indeed become negative at low cutting speeds; thus, tool-life curves may actually reach a maximum and then curve downward. Consequently, caution should be exercised when using tool-life equations beyond the range of cutting speeds for which they were developed.

**TABLE 8.4**

| Range of $n$ Values for Various Cutting Tools | |
|---|---|
| High-speed steels | 0.08–0.2 |
| Cast alloys | 0.1–0.15 |
| Carbides | 0.2–0.5 |
| Ceramics | 0.5–0.7 |

Because of the major influence of temperature on the physical and mechanical properties of materials, we would expect that wear is strongly influenced by temperature. Experimental investigations have shown that there is indeed a direct relationship between flank wear and the temperature generated during cutting (Fig. 8.23b). Although cutting speed has been found to be the most significant process variable in tool life, depth of cut and feed rate are also important; thus, Eq. (8.31) can be modified as

$$VT^n d^x f^y = C, \tag{8.32}$$

where $d$ is the depth of cut and $f$ is the feed rate (in mm/rev or in./rev) in turning. The exponents $x$ and $y$ must be determined experimentally for each cutting condition. For example, taking $n = 0.15$, $x = 0.15$, and $y = 0.6$ as typical values encountered in practice, we can see that cutting speed, feed rate, and depth of cut are of decreasing order of importance.

Equation (8.32) can be rewritten as

$$T = C^{1/n} V^{-1/n} d^{-x/n} f^{-y/n}, \tag{8.33}$$

or

$$T \simeq C^7 V^{-7} d^{-1} f^{-4}. \tag{8.34}$$

For a constant tool life, the following observations can be made from Eq. (8.34):

1. If the feed rate or the depth of cut is increased, the cutting speed must be decreased, and vice versa;

2. Depending on the exponents, a reduction in speed can then result in an increase in the volume of the material removed, because of the increased feed rate and/or depth of cut.

**Allowable wear land.** The *allowable wear land* (VB in Fig. 8.20c) for various conditions is given in Table 8.5; for improved dimensional accuracy and surface finish, the allowable wear land may be made smaller than the values given in the table. The recommended cutting speed for a high-speed-steel tool is generally the one that gives a tool life of 60–120 min (30–60 min for carbide tools).

**Optimum cutting speed.** We have stated previously that as cutting speed increases, tool life is rapidly reduced. On the other hand, if cutting speeds are low, tool life is long, but the rate at which material is removed is also low; thus, there is an *optimum* cutting speed.

TABLE 8.5

**Allowable Average Wear Land for Cutting Tools for Various Operations**

| | Allowable Wear Land (mm) | |
|---|---|---|
| Operation | High-Speed Steels | Carbides |
| Turning | 1.5 | 0.4 |
| Face milling | 1.5 | 0.4 |
| End milling | 0.3 | 0.3 |
| Drilling | 0.4 | 0.4 |
| Reaming | 0.15 | 0.15 |

The effect of cutting speed on the volume of metal removed between tool resharpenings or replacements can be appreciated by analyzing Fig. 8.22a. Let's assume that we are machining the material whose condition is represented by curve "a"—that is, as cast with a hardness of 265 HB. If the cutting speed is 1 m/s (200 ft/min), the tool life is about 40 min. Thus, the tool travels a distance of (1 m/s)(60 s/min)(40 min) = 2400 m before it is resharpened or replaced. If we now change the cutting speed to 2 m/s, tool life will be about 5 min, and the tool travels (2)(60)(5) = 600 m. Since the volume of material removed is directly proportional to the distance the tool has traveled, we see that by decreasing the cutting speed, we can remove more material between tool changes. Note, however, that the lower the cutting speed, the longer is the time required to machine a part; these variables have important economic impact. (See Section 8.14.)

---

**EXAMPLE 8.3   Increasing tool life by reducing the cutting speed**

Using the Taylor equation [Eq. (8.31)] for tool life and letting $n = 0.5$ and $C = 400$, calculate the percentage increase in tool life when the cutting speed is reduced by 50%.

**Solution.** Since $n = 0.5$, the Taylor equation can be rewritten as $V\sqrt{T} = 400$. Letting $V_1$ be the initial speed and $V_2$ the reduced speed, we note that, for this problem, $V_2 = 0.5V_1$. Because $C$ is a constant, we have the relationship

$$0.5V_1\sqrt{T_2} = V_1\sqrt{T_1}.$$

Simplifying this expression, we find that

$$\frac{T_2}{T_1} = \frac{1}{0.25} = 4.0.$$

This relation indicates that the tool-life change is

$$\frac{T_2 - T_1}{T_1} = \left(\frac{T_2}{T_1}\right) - 1 = 4 - 1 = 3,$$

or that it is increased by 300%. Note that the reduction in cutting speed has resulted in a major increase in tool life and that, in this problem, the magnitude of $C$ is not relevant.

---

### 8.3.2   Crater wear

The factors affecting flank wear also influence *crater wear*, but the most significant factors in crater wear are temperature and the degree of chemical affinity between the tool and the workpiece. As shown earlier, the rake face of the tool is subjected to high levels of stress and temperature, in addition to sliding at relatively high speeds. As shown in Fig. 8.17b, peak temperatures, for instance, can be on the order of 1100°C (2000°F). Interestingly, the location of *maximum depth* (KT) of crater wear generally coincides with the location of maximum temperature.

Experimental evidence indicates a direct relation between crater-wear rate and tool–chip interface temperature (Fig. 8.24); note the sharp increase in crater wear

**FIGURE 8.24**   Relation between crater-wear rate and average tool–chip interface temperature in turning: (a) high-speed-steel tool; (b) C-1 carbide; (c) C-5 carbide. Note that crater wear increases rapidly within a narrow range of temperature. *Source*: After K. J. Trigger and B. T. Chao.

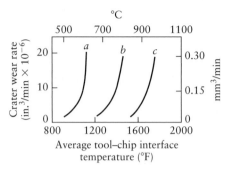

after a certain temperature range has been reached. Figure 8.25 shows the cross-section of the tool–chip interface in cutting steel at high speeds. Note the location of the crater-wear pattern and the discoloration of the tool (loss of temper) as a result of high temperatures; note also how well the discoloration profile agrees with the temperature profile shown in Fig. 8.16.

The effect of temperature on crater wear has been described in terms of a **diffusion** mechanism (the movement of atoms across the tool–chip interface). Diffusion depends on the tool–workpiece material combination and on temperature, pressure, and time; as these quantities increase, the diffusion rate increases. Unless these factors are favorable, crater wear will not take place by diffusion; wear will then be caused by other factors, such as those outlined for flank wear. High temperatures may also cause softening and *plastic deformation* of the tool, because of the decrease in the tool's yield strength and hardness with temperature. This type of deformation is generally observed in machining high-strength metals.

### 8.3.3 Chipping

The term *chipping* is used to describe the breaking away of a piece from the cutting edge of the tool. The chipped pieces may be very small (*microchipping* or *macrochipping*), or they may involve relatively large fragments (*gross chipping* or

**FIGURE 8.25**   Interface of chip (left) and rake face of tool (right) and crater wear in cutting AISI 1004 steel at 3 m/s (585 ft/min). Discoloration of the tool indicates high temperature (loss of temper). Note how the crater-wear pattern coincides with the discoloration pattern. Compare this pattern with the temperature distribution shown in Fig. 8.16. *Source*: Courtesy of P. K. Wright.

*fracture*). Unlike wear, which is a more gradual process, chipping is a phenomenon that results in sudden loss of tool material. Two main causes of chipping are **mechanical shock** and **thermal fatigue** (both by interrupted cutting, as in milling operations).

Chipping by mechanical shock may occur in a region in the tool where a small crack or defect already exists. High positive rake angles also can contribute to chipping, because of the small included angle of the tool tip. Thermal cracks, which are generally perpendicular to the cutting edge (Fig. 8.20a), are caused by the thermal cycling of the tool in interrupted cutting. Thus, chipping can be reduced by selecting tool materials with high impact and thermal-shock resistance. Furthermore, crater wear may also progress toward the tool tip and weaken it, causing chipping.

### 8.3.4  General observations on tool wear

Because of the many factors involved, including the characteristics of the machine tool, the wear behavior of cutting tools varies significantly. In addition to the wear processes already described, other phenomena also occur in tool wear (Fig. 8.20a). The wear **groove** or **notch** on cutting tools (Fig. 8.20) has been attributed to the fact that this region is the boundary where the chip is no longer in contact with the tool. This boundary, or **depth-of-cut line** (DOC), oscillates, because of inherent variations in the cutting operation, and accelerates the wear process. Furthermore, this region is in contact with the machined surface from the previous cut. Since a machined surface may develop a thin work-hardened layer, this contact could contribute to the formation of the wear groove.

Because they are hard and abrasive, scale and oxide layers on the surface of a workpiece increase wear. In such cases, the depth of cut, $d$, (see Fig. 8.19) should be greater than the thickness of the oxide film or the work-hardened layer. In other words, light cuts should not be taken on rusted workpieces.

### 8.3.5  Tool-condition monitoring

With rapidly increasing use of computer-controlled machine tools and implementation of highly automated manufacturing, the reliable and repeatable performance of cutting tools has become an important consideration. Once programmed properly, most modern machine tools operate with little direct supervision by an operator; consequently, the failure of a cutting tool can have serious detrimental effects on the quality of the machined part as well as on the efficiency and economics of the overall machining operation. It is therefore essential to continuously and indirectly monitor the condition of the cutting tool, i.e., for wear, chipping, or gross failure. Most state-of-the-art machine controls are now equipped with tool-condition monitoring systems.

Techniques for condition monitoring typically fall into two general categories: direct and indirect. The **direct** method for observing the condition of a cutting tool involves *optical* measurement of wear, such as by periodically observing changes in the profile of the tool. This method is the most commonly employed and reliable technique and is usually done using a microscope (*toolmakers' microscope*). Continued research on computer vision systems (see Section 14.8.1) for this task is being conducted, mainly because conventional optical measurement techniques require that the cutting operation be stopped.

**Indirect** methods of measuring wear involve correlating the tool condition with process variables such as forces, power, temperature rise, surface finish, and vibrations. One such technique is **acoustic emission**, which uses a piezoelectric transducer attached to a tool holder. The transducer picks up acoustic-emission signals (typically above 100 kHz) resulting from the stress waves generated during cutting. By analyzing the signals, tool wear and chipping can be monitored (Fig. 8.26). This technique is

**FIGURE 8.26**

Relationship between mean flank wear, maximum crater wear, and acoustic emission (noise generated during cutting) as a function of machining time. This technique is being developed as a means for monitoring wear rate in various cutting processes without interrupting the operation. *Source*: After M. S. Lan and D. A. Dornfeld.

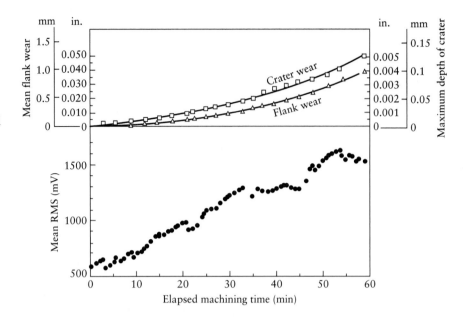

particularly effective in precision machining operations, where, because of the very small amounts of material removed, cutting forces are low.

A similar indirect tool-condition monitoring system consists of **transducers** that are installed in original machine tools or retrofitted on existing machines. They continually monitor spindle *torque* and tool *forces* in various directions in a variety of machining processes. The signals are preamplified, and a microprocessor analyzes the content of the signals and interprets it. This system is capable of differentiating the signals that come from tool breakage, tool wear, a missing tool, overloading of the machine, or collision of machine components such as spindles; it can also automatically compensate for tool wear.

The design of the transducers must be such that they are (a) nonintrusive to the machining operation, (b) accurate and repeatable in signal detection, (c) resistant to abuse and shop-floor environment, and (d) cost effective. Continued progress is being made in the development of such sensors (see Section 14.8), including the use of *infrared* and *fiber optic* techniques for temperature measurement in machining operations.

## 8.4 | Surface Finish and Integrity

Surface finish influences not only the dimensional accuracy of machined parts, but also the properties of the parts, especially fatigue strength. Whereas **surface finish** describes the *geometric* features of surfaces, **surface integrity** pertains to *properties* such as fatigue life and corrosion resistance, which are strongly influenced by the type of surface produced. The factors that influence surface integrity are (a) temperatures generated during processing, (b) residual stresses, (c) metallurgical (phase) transformations, and (d) plastic deformation, tearing, and cracking of the surface. Figure 8.27 provides ranges of surface roughness for machining and other processes.

The built-up edge, with its significant effect on tool profile, has the greatest influence on surface roughness of all the factors. Figure 8.28 shows surfaces obtained

**FIGURE 8.27** Range of surface roughnesses obtained in various machining processes. Note the wide range within each group. (See also Fig. 9.27).

(a)                    (b)

**FIGURE 8.28** Surfaces produced on steel by cutting, as observed with a scanning electron microscope: (a) turned surface and (b) surface produced by shaping. *Source*: J. T. Black and S. Ramalingam.

**FIGURE 8.29** Schematic illustration of a dull tool in orthogonal cutting (exaggerated). Note that at small depths of cut, the rake angle can effectively become negative. In such cases, the tool may simply ride over the workpiece surface, burnishing it.

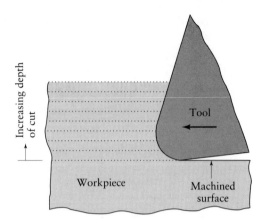

in two different cutting operations; note the considerable damage to the surfaces from BUE. Ceramic and diamond tools generally produce better surface finish than other tools, largely because of their much lower tendency to form BUE.

A tool that is not sharp has a large radius along its edges (see Fig. 8.20c), just as a dull pencil or knife does. Figure 8.29 illustrates the relationship between the radius of the cutting edge and depth of cut in orthogonal cutting. Note that at small depths of cut, the rake angle can effectively become negative, and thus the tool may simply ride over the workpiece surface and not remove any chips. You can simulate this behavior by trying to scrape the surface of a stick of butter along its length with a dull knife; note that you will not be able to remove a thin layer of butter, whereas if the knife is sharp, you will be able to do so. If this radius, not to be confused with radius $R$ in Fig. 8.20c, is large in relation to the depth of cut, the tool will rub over the machined surface. Rubbing in cutting generates heat and induces surface residual stresses, which, in turn, may cause surface damage such as tearing and cracking. Thus, the depth of cut should generally be greater than the radius on the cutting edge.

**Feed marks.** In turning, as in other cutting operations, the tool leaves a spiral profile (*feed marks*) on the machined surface as it moves across the workpiece. (See Fig. 8.19.) The higher the feed, $f$, and the smaller the radius, $R$, the more prominent are these marks. Although not significant in rough machining operations, the marks are important in finish machining. In turning, the peak-to-valley roughness, $R_t$, can be expressed as

$$R_t = \frac{f^2}{8R}, \tag{8.35}$$

where $f$ is the feed and $R$ is the nose radius. If the nose radius is much smaller than the feed, the roughness is given by

$$R_t = \frac{f}{\tan \alpha_s + \cot \alpha_e}, \tag{8.36}$$

where $\alpha_s$ and $\alpha_e$ are the side and edge cutting angles, respectively (see Fig. 8.41). For face milling (see Fig. 8.57), the roughness is given by

$$R_t = \frac{f^2}{16(D \pm (2fn/\pi))}, \tag{8.37}$$

where $D$ is the cutter diameter, $f$ is the feed per tooth, and $n$ is the number of inserts on the cutter. Equations (8.35) to (8.37) are derived considering geometry only and do not account for workpiece cracking (see Fig. 8.6c), thermal distortion, or chatter. Vibration and chatter are described in greater detail in Section 8.11. For now, we should recognize that if the tool vibrates or chatters during cutting, surface finish will be adversely affected, the reason being that a vibrating tool changes the dimensions of the cut periodically. Excessive chatter can also cause chipping and premature failure of the more brittle cutting-tool materials, such as ceramics and diamond.

# 8.5 | Machinability

The *machinability* of a material is usually defined in terms of the following factors: (1) surface finish and integrity of the machined part, (2) tool life obtained, (3) force and power requirements, and (4) chip control. Thus, good machinability indicates *good surface finish and integrity, long tool life, low force and power* requirements, and easy collection of *chips* that does not interfere with the cutting operation. (As stated earlier, long, thin, and curled chips, if not broken up, can become entangled in the cutting zone.)

Because of the complex nature of machining operations, it is difficult to establish relationships to define quantitatively the machinability of a material. However, in manufacturing plants, tool life and surface roughness are generally considered to be the most important factors in machinability. Although not used much anymore, approximate **machinability ratings** are available. Machinability ratings are usually based on a tool life of $T = 60$ min. The standard material is AISI 1112 steel, which is given a rating of 100. Therefore, for a tool life of 60 min, this steel should be machined at a cutting speed of 100 ft/min (0.5 m/s); higher speeds will reduce tool life, and lower speeds will increase it. Note also that as the hardness of a material increases, its machinability rating generally decreases proportionately.

For example, (a) 3140 steel has a machinability rating of 55, which means that its tool life will be 60 min if it is machined at a cutting speed of 55 ft/min (0.275 m/s), and (b) nickel has a rating of 200, indicating that it should be machined at 200 ft/min (1 m/s) to obtain a tool life of 60 min. Machinability ratings for various other materials are as follows: free-cutting brass, 300; 2011 wrought aluminum, 200; pearlitic gray iron, 70; Inconel, 30; and precipitation-hardening 17–7 steel, 20.

## 8.5.1 Machinability of steels

Because steels are among the most important engineering materials, their machinability has been studied extensively. Their machinability has been improved mainly by adding lead and sulfur to obtain so-called **free-machining steels**.

**Leaded steels.** Lead is added to the molten steel and takes the form of dispersed fine lead particles. (See Fig. 5.2a.) Lead is insoluble in iron, copper, and aluminum and their alloys; thus, during cutting, the lead particles are sheared and smeared over the tool–chip interface. Because of their low shear strength, the lead particles act as a solid lubricant. This behavior has been verified by the presence of high concentrations of lead on the tool-side face of chips in machining leaded steels. In addition to this effect, it is believed that lead probably lowers the shear stress in the primary shear zone, thus reducing cutting forces and power consumption. Lead may be used in either nonsulfurized or resulfurized steels; however, because of environmental concerns, the trend now is towards eliminating the use of leaded steels in favor of bismuth and tin (*lead-free steels*). Leaded steels are identified by the letter L between the second and third numerals—for example, 10L45. However, similar use of the letter L in identifying stainless steels refers to a low carbon content, which improves the steels' resistance to corrosion.

(a) (b) (c)

**FIGURE 8.30** Photomicrographs showing various types of inclusions in low-carbon, resulfurized free-machining steels. (a) Manganese-sulfide inclusions in AISI 1215 steel. (b) Manganese-sulfide inclusions and glassy manganese-silicate-type oxide (dark) in AISI 1215 steel. (c) Manganese sulfide with lead particle as tails in AISI 12L14 steel. *Source:* Courtesy of Ispat Inland Inc.

**Resulfurized and rephosphorized steels.** Sulfur in steels forms manganese-sulfide inclusions (second-phase particles, see Fig. 8.30), which act as stress raisers in the primary shear zone. As a result, the chips produced are small and break up easily, thus improving machinability. The shape, orientation, distribution, and concentration of these inclusions significantly influence machinability. Elements such as tellurium and selenium (both chemically similar to sulfur) in resulfurized steels act as inclusion modifiers. Phosphorus in steels improves machinability by virtue of strengthening the ferrite, thereby increasing the hardness of the steels.

**Calcium-deoxidized steels.** In these steels, oxide flakes of calcium aluminosilicate (CaO, $SiO_2$, and $Al_2O_3$) are formed that, in turn, reduce the strength of the secondary shear zone, thus decreasing tool–chip interface friction and wear and, hence, temperature. Consequently, these steels cause less crater wear of the tool, especially at high cutting speeds.

**Effects of other elements in steels on machinability.** The presence of aluminum and silicon in steels is always harmful, because these elements combine with oxygen and form aluminum oxide and silicates. These compounds are hard and abrasive, thus increasing tool wear and reducing machinability.

Carbon and manganese have various effects on the machinability of steels, depending on their composition. As the carbon content increases, machinability decreases; however, plain low-carbon steels (less than 0.15% C) can produce poor surface finish by forming a built-up edge. Cast steels are more abrasive, although their machinability is similar to that of wrought steels. Tool and die steels are very difficult to machine and usually require annealing prior to machining. Machinability of most steels is generally improved by cold working, which reduces the tendency for built-up edge formation.

Other alloying elements, such as nickel, chromium, molybdenum, and vanadium, which improve the properties of steels, generally reduce machinability. The effect of boron is negligible. The role of gaseous elements such as oxygen, hydrogen, and nitrogen has not been clearly established; any effect that they may have would depend on the presence and amount of other alloying elements.

In selecting various elements to improve machinability, the possible detrimental effects of the elements on the properties and strength of the machined part in service also should be considered. At elevated temperatures, for example, lead causes embrittlement of steels (*hot shortness*; see Section 3.4.2), although at room temperature, it has no effect on mechanical properties. Sulfur can severely reduce hot workability of steels, because of the presence of iron sulfide, unless sufficient manganese is present to prevent the formation of iron sulfide. At room temperature, the mechanical properties of resulfurized steels depend on the orientation of the deformed manganese-sulfide inclusions (anisotropy). Rephosphorized steels are significantly less ductile than resulfurized steels and are produced solely to improve machinability.

**Stainless steels.** Austenitic (300 series) stainless steels are generally difficult to machine. Chatter could be a problem, which necessitates the use of machine tools with high stiffness and damping capacity; however, ferritic stainless steels (also 300 series) have good machinability. Martensitic (400 series) stainless steels are abrasive, tend to form built-up edge, and require tool materials with high hot hardness and resistance to crater wear. Precipitation-hardening stainless steels are strong and abrasive and thus require hard and abrasion-resistant tool materials.

## 8.5.2 Machinability of various other metals

*Aluminum* is generally easy to machine; however, the softer grades tend to form built-up edge. High cutting speeds, high rake angles, and high relief angles are recommended. Wrought alloys with high silicon content and cast-aluminum alloys may be abrasive and, hence, require harder tool materials. Dimensional control may be a problem in machining aluminum, because of its low elastic modulus and a relatively high thermal coefficient of expansion.

*Beryllium* is similar to cast irons, but is more abrasive and toxic; hence, it requires machining in a controlled environment.

*Gray cast irons* are generally machinable, but are abrasive. Free carbides in castings reduce the castings' machinability and cause tool chipping or fracture, thus requiring tools with high toughness. Nodular and malleable irons are machinable with hard tool materials.

*Cobalt-base alloys* are abrasive and highly work hardening; they require sharp and abrasion-resistant tool materials and low feeds and speeds.

*Wrought copper* can be difficult to machine, because of built-up edge formation, although cast-copper alloys are easy to machine. *Brasses* are easy to machine as well, especially those to which lead has been added (*leaded free-machining brass*). *Bronzes* are more difficult to machine than brass.

*Magnesium* is very easy to machine, has good surface finish, and prolongs tool life; however, care should be exercised when using it, because of its high rate of oxidation (*pyrophoric*) and the danger of fire.

*Molybdenum* is ductile and work hardening; hence, it can produce poor surface finish, thereby requiring the use of sharp tools.

*Nickel-base alloys* are work hardening, abrasive, and strong at high temperatures; their machinability is similar to that of stainless steels.

*Tantalum* is very work hardening, ductile, and soft; hence, it produces a poor surface finish, and tool wear is high.

*Titanium* and its alloys have poor thermal conductivity (the lowest of all metals), thus causing significant temperature rise and built-up edge; hence, it can be difficult to machine.

*Tungsten* is brittle, strong, and very abrasive; hence, its machinability is low. However, its machinability improves greatly at elevated temperatures.

*Zirconium* has good machinability; however, a coolant-type cutting fluid must be used, because of the danger of explosion and fire.

### 8.5.3  Machinability of various materials

*Graphite* is abrasive; thus, it requires hard, abrasion-resistant sharp tools.

*Thermoplastics* generally have low thermal conductivity, a low elastic modulus, and a low softening temperature. Consequently, machining them requires tools with a positive rake angle to reduce cutting forces; large relief angles; small depths of cut and feed; relatively high speeds; and proper support of the workpiece, because of lack of stiffness. Tools should be sharp, and external cooling of the cutting zone may be necessary to keep the chips from becoming "gummy" and sticking to the tools. Cooling can usually be done with a jet of air, vapor mist, or water-soluble oils. Residual stresses may develop during machining. To relieve residual stresses, machined parts can be annealed at temperatures ranging from 80°C to 160°C (175°F to 315°F) for a period of time and then cooled slowly and uniformly to room temperature.

*Thermosetting plastics* are brittle and sensitive to thermal gradients during cutting; their machinability is generally similar to that of thermoplastics.

Because of the fibers present, *reinforced plastics* are generally very abrasive and are difficult to machine. Fiber tearing and pulling is a significant problem. Furthermore, machining of these materials requires careful removal of machining debris to avoid human contact with and inhalation of the fibers.

*Metal-matrix* and *ceramic-matrix composites* can be difficult to machine, depending on the properties of the individual components in the material.

The machinability of *ceramics* has improved steadily, especially with the development of *nanoceramics* (see Section 11.8.1) and with the selection of appropriate machining parameters, such as in *ductile-regime cutting* (Section 8.8.2).

### 8.5.4  Thermally assisted machining

Materials that are difficult to machine at room temperature can be machined more easily at elevated temperatures, thus lowering cutting forces and increasing tool life. In *thermally assisted machining* (**hot machining**), the source of heat is a torch, high-energy beam (such as a laser or electron beam) or plasma arc, focused to an area just ahead of the cutting tool. Most applications for hot machining are in turning and milling. The process of heating to and maintaining a uniform temperature distribution within the workpiece may be difficult to control; also, the original microstructure of the workpiece may be adversely affected. Except in isolated cases, thermally assisted machining offers no significant advantage over machining at room temperature with the use of appropriate cutting tools and fluids. Some success has been obtained in laser-assisted machining of silicon-nitride ceramics and *Stellite*.

## 8.6  |  Cutting-Tool Materials

The proper selection of cutting-tool materials is among the most important factors in machining operations, as is the selection of mold and die materials for forming and shaping processes. We noted previously that in cutting, the tool is subjected to high temperatures, contact stresses, rubbing on the workpiece surface, and the effects of the chip climbing up the rake face of the tool. Consequently, a cutting tool must have the following characteristics in order to produce good-quality and economical parts:

- **Hardness**, particularly at elevated temperatures (**hot hardness**), so that the hardness and strength of the tool are maintained at the temperatures encountered in cutting operations (Fig. 8.31);

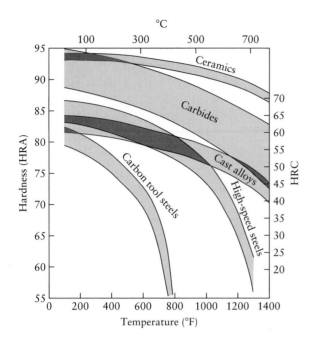

**FIGURE 8.31** Hardness of various cutting-tool materials as a function of temperature (hot hardness). The wide range in each group of materials results from the variety of tool compositions and treatments available for that group.

- **Toughness,** so that impact forces on the tool in interrupted cutting operations, such as milling or turning a splined shaft, do not chip or fracture the tool;

- **Wear resistance,** so that an acceptable tool life is obtained before the tool is resharpened or replaced;

- **Chemical stability** or **inertness** with respect to the workpiece material, so that any adverse reactions contributing to tool wear are avoided or minimized.

Several cutting-tool materials having a wide range of these characteristics are now available. (See Table 8.6.) Tool materials are usually divided into the following general categories, which are listed in the approximate chronological order in which they were developed and implemented:

1. Carbon and medium-alloy steels;

2. High-speed steels;

3. Cast-cobalt alloys;

4. Carbides;

5. Coated tools;

6. Alumina-base ceramics;

7. Cubic boron nitride;

8. Silicon-nitride-base ceramics;

9. Diamond;

10. Whisker-reinforced and nanocrystalline tool materials.

Many of these materials are also used for dies and molds (Chapters 5, 6, 7, 10 and 11).

This section describes the characteristics, applications, and limitations of these tool materials; also discussed are characteristics such as hot hardness, toughness, impact strength, wear resistance, resistance to thermal shock, and costs, as well as the range of cutting speeds and depth of cut for optimum performance.

TABLE 8.6

**Typical Range of Properties of Various Tool Materials**

| PROPERTY | HIGH-SPEED STEEL | CAST ALLOYS | CARBIDES WC | CARBIDES TiC | CERAMICS | CUBIC BORON NITRIDE | SINGLE-CRYSTAL DIAMOND* |
|---|---|---|---|---|---|---|---|
| Hardness | 83–86 HRA | 82–84 HRA<br>46–62 HRC | 90–95 HRA<br>1800–2400 HK | 91–93 HRA<br>1800–3200 HK | 91–95 HRA<br>2000–3000 HK | 4000–5000 HK | 7000–8000 HK |
| Compressive strength | | | | | | | |
| MPa | 4100–4500 | 1500–2300 | 4100–5850 | 3100–3850 | 2750–4500 | 6900 | 6900 |
| psi $\times 10^3$ | 600–650 | 220–335 | 600–850 | 450–560 | 400–650 | 1000 | 1000 |
| Transverse rupture strength | | | | | | | |
| MPa | 2400–4800 | 1380–2050 | 1050–2600 | 1380–1900 | 345–950 | 700 | 1350 |
| psi $\times 10^3$ | 350–700 | 200–300 | 150–375 | 200–275 | 50–135 | 105 | 200 |
| Impact strength | | | | | | | |
| J | 1.35–8 | 0.34–1.25 | 0.34–1.35 | 0.79–1.24 | <0.1 | <0.5 | <0.2 |
| in.-lb | 12–70 | 3–11 | 3–12 | 7–11 | <1 | <5 | <2 |
| Modulus of elasticity | | | | | | | |
| GPa | 200 | — | 520–690 | 310–450 | 310–410 | 850 | 820–1050 |
| psi $\times 10^6$ | 30 | — | 75–100 | 45–65 | 45–60 | 125 | 120–150 |
| Density | | | | | | | |
| kg/m$^3$ | 8600 | 8000–8700 | 10,000–15,000 | 5500–5800 | 4000–4500 | 3500 | 3500 |
| lb/in$^3$ | 0.31 | 0.29–0.31 | 0.36–0.54 | 0.2–0.22 | 0.14–0.16 | 0.13 | 0.13 |
| Volume of hard phase (%) | 7–15 | 10–20 | 70–90 | - | 100 | 95 | 95 |
| Melting or decomposition temperature | | | | | | | |
| °C | 1300 | — | 1400 | 1400 | 2000 | 1300 | 700 |
| °F | 2370 | — | 2550 | 2550 | 3600 | 2400 | 1300 |
| Thermal conductivity, W/mK | 30–50 | — | 42–125 | 17 | 29 | 13 | 500–2000 |
| Coefficient of thermal expansion, $\times 10^{-6}$/°C | 12 | — | 4–6.5 | 7.5–9 | 6–8.5 | 4.8 | 1.5–4.8 |

* The values for polycrystalline diamond are generally lower, except impact strength, which is higher.

442

## 8.6.1  Carbon and medium-alloy steels

*Carbon steels* are the oldest of tool materials and have been used widely for drills, taps, broaches, and reamers since the 1880s. Low-alloy and medium-alloy steels were developed later for similar applications, but with longer tool life. Although inexpensive and easily shaped and sharpened, these steels do not have sufficient hot hardness and wear resistance for cutting at high speeds, where, as we have seen, the temperature rises significantly. Note in Fig. 8.31, for example, how rapidly the hardness of carbon steels decreases as the temperature increases. Consequently, the use of these steels is limited to very low-speed cutting operations.

## 8.6.2  High-speed steels

*High-speed-steel* (HSS) tools are so named because they were developed to cut at speeds higher than previously possible. First produced in the early 1900s, high-speed steels are the most highly alloyed of the tool steels. They can be hardened to various depths, have good wear resistance, and are relatively inexpensive. Because of their high toughness and resistance to fracture, high-speed steels are especially suitable for high positive-rake-angle tools (small included angle), for interrupted cuts, and for use on machine tools with low stiffness that are subject to vibration and chatter. High-speed steels account for the largest tonnage of tool materials used today, followed by various die steels and carbides. They are used in a wide variety of cutting operations that require complex tool shapes such as drills, reamers, taps, and gear cutters. Their basic limitation is the relatively low cutting speeds when compared with those of carbide tools.

There are two basic types of high-speed steels: **molybdenum** (*M* series) and **tungsten** (*T* series). The *M* series contains up to about 10% molybdenum, with chromium, vanadium, tungsten, and cobalt as alloying elements. The *T* series contains 12% to 18% tungsten, with chromium, vanadium, and cobalt as alloying elements. The *M* series generally has higher abrasion resistance than the *T* series, undergoes less distortion during heat treating, and is less expensive. Consequently, 95% of all high-speed steel tools produced in the United States are made of *M*-series steels.

High-speed-steel tools are available in wrought, cast, and sintered (powder-metallurgy) forms. They can be **coated** for improved performance (see Section 8.6.5), and they may be subjected to surface treatments, such as case hardening for improved hardness and wear.

## 8.6.3  Cast-cobalt alloys

Introduced in 1915, *cast-cobalt alloys* have the following ranges of composition: 38% to 53% cobalt, 30% to 33% chromium, and 10% to 20% tungsten. Because of their high hardness, typically 58 HRC to 64 HRC, they have good wear resistance and maintain their hardness at elevated temperatures. Commonly known as *Stellite* tools, these alloys are cast and ground into relatively simple tool shapes. They are not as tough as high-speed steels and are sensitive to impact forces; consequently, they are less suitable than high-speed steels for interrupted cutting operations. Tools made of cast-cobalt alloys are now used only for special applications that involve deep, continuous roughing operations at relatively high feeds and speeds, as much as twice the rates possible with high-speed steels.

### 8.6.4 Carbides

The three groups of tool materials described previously have sufficient toughness, impact strength, and thermal shock resistance, but have important limitations on characteristics such as strength and hardness, particularly hot hardness. Consequently, they cannot be used as effectively where high cutting speeds, and hence high temperatures, are involved; thus, their tool life can be short. To meet the challenge of higher speeds for higher production rates, *carbides* (also known as **cemented** or **sintered carbides**) were introduced in the 1930s. Because of their high hardness over a wide range of temperatures (see Fig. 8.31), high elastic modulus and thermal conductivity, and low thermal expansion, carbides are among the most important, versatile, and cost-effective tool and die materials for a wide range of applications. However, stiffness of the machine tool is important, and light feeds, low cutting speeds, and chatter can be detrimental. The two basic groups of carbides used for machining operations are tungsten carbide and titanium carbide. In order to differentiate them from coated tools (see Section 8.6.5), plain carbide tools are usually referred to as **uncoated** carbides. (See also the discussion of **micrograin carbides** in Section 8.6.10.)

a) **Tungsten carbide.** *Tungsten carbide* (WC) is a composite material consisting of tungsten-carbide particles bonded together in a cobalt matrix; hence, it is also called *cemented carbide*. Tungsten-carbide tools are manufactured by powder-metallurgy techniques (Chapter 11) in which WC particles are combined together with cobalt in a mixer, resulting in a cobalt matrix surrounding the WC particles. These particles, which are $1-5$ $\mu$m ($40-200$ $\mu$in.) in size, are then pressed and sintered (hence tungsten carbide is also called *sintered carbide*) into the desired insert shapes. (See Example 11.6.) WC is frequently compounded with carbides of titanium and niobium to impart special properties to the carbide.

The amount of cobalt significantly affects the properties of carbide tools. As the cobalt content increases, the strength, hardness, and wear resistance decrease, while the toughness increases, because of the higher toughness of cobalt. Tungsten-carbide tools are generally used for cutting steels, cast irons, and abrasive nonferrous materials and have largely replaced HSS tools in many applications, due to the better performance of the tungsten-carbide tools.

b) **Titanium carbide.** *Titanium carbide* (TiC) has higher wear resistance than that of tungsten carbide, but is not as tough. With a nickel–molybdenum alloy as the matrix, TiC is suitable for machining hard materials, mainly steels and cast irons, and for cutting at speeds higher than those for which tungsten carbide may be used.

c) **Inserts.** Carbon-steel and high-speed-steel cutting tools can be shaped in one piece and ground to various shapes (see Fig. 8.33); other such tools include drills and milling cutters. However, after the cutting edge wears and becomes dull, the tool has to be removed from its holder and reground. The need for a more effective method has led to the development of *inserts*, which are individual cutting tools with a number of cutting edges (Fig. 8.32). Inserts are available in a variety of shapes, such as square, triangle, diamond, and round. A square insert, for example, has eight cutting edges, and a triangular insert has six.

Inserts are usually clamped on the tool *shank* with various locking mechanisms (Fig. 8.32b and c). Less frequently, inserts are *brazed* to the tool shank (Fig. 8.32d); however, because of the difference in thermal expansion between the insert and the tool-shank materials, brazing must be done carefully in order to avoid cracking or warping. Clamping is the preferred method because each insert has a number of cutting edges, and after one edge is worn, it is *indexed* (i.e., rotated in its holder) so that another edge can be used. In addition to these examples, a wide variety of other toolholders is available for specific applications, including toolholders with quick insertion and removal features for efficient operation.

The strength of the cutting edge of an insert depends on the shape of the insert; the smaller the included angle (Fig. 8.33), the lower is the strength of the edge. In order to further improve edge strength and prevent chipping, insert

(a)                (b)

(c)                (d)

**FIGURE 8.32** (a) Typical carbide inserts with various shapes and chip-breaker features. Round inserts are also available. The holes in the inserts are standardized for interchangeability. *Source*: Courtesy of Kyocera Engineered Ceramics, Inc., and *Manufacturing Engineering*, Society of Manufacturing Engineers. (b) Methods of attaching inserts to a tool shank by clamping, (c) with wing lockpins, and (d) with a brazed insert on a shank.

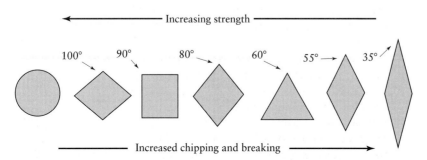

**FIGURE 8.33** Relative edge strength and tendency for chipping and breaking of inserts with various shapes. Strength refers to that of the cutting edge shown by the included angles. *Source*: Kennametal, Inc.

**FIGURE 8.34** Edge preparation of inserts to improve edge strength. *Source*: Kennametal, Inc.

Increasing edge strength

edges are usually honed, chamfered, or are produced with a negative land (Fig. 8.34). Most inserts are honed to a radius of about 0.025 mm (0.001 in.). Inserts are available with a wide variety of **chip-breaker** features for controlling chip flow and reducing vibration and the heat generated. Optimum chip-breaker geometries are now being developed using computer-aided design and finite-element techniques.

### 8.6.5 Coated tools

As described in Chapter 3, new alloys and engineered materials are being developed continuously, particularly since the 1960s. These materials have high strength and toughness, but are generally abrasive and highly chemically reactive with tool materials. The difficulty of machining these materials efficiently and the need to improve the performance in machining the more common engineering materials have led to important developments in coated tools. Because of their unique properties, coated tools can be used at high cutting speeds, thus reducing the time required for machining operations and, hence, costs. Note in Fig. 8.35 how the cutting time has been reduced by a factor of more than 100 since 1900. Coated tools can improve tool life by as much as 10 times that of uncoated tools.

Commonly used *coating materials* include titanium nitride, titanium carbide, titanium carbonitride, and aluminum oxide ($Al_2O_3$). Generally in the thickness range of $2-10 \ \mu$m ($80-400 \ \mu$in.), these coatings are applied on tools and inserts by **chemical vapor deposition** (CVD) and **physical-vapor deposition** (PVD) techniques, described in Section 4.5. The CVD process is the most commonly used coating application method for carbide tools with multiphase and ceramic coatings. The PVD-coated carbides with TiN coatings, on the other hand, have higher cutting-edge strength, lower friction, lower tendency to form a built-up edge, and are smoother and more

**FIGURE 8.35** Relative time required to machine with various cutting-tool materials, with indication of the year the tool materials were introduced. *Source*: Sandvik Coromant.

uniform in thickness, which is generally in the range of $2-4\,\mu$m ($80-160\,\mu$in.). A more recent technology, particularly for multiphase coatings, is **medium-temperature chemical-vapor deposition** (MTCVD); it provides higher resistance to crack propagation than do CVD coatings.

Coatings should have the following general characteristics:

- High hardness at elevated temperatures,
- Chemical stability and inertness to the workpiece material,
- Low thermal conductivity,
- Good bonding to the substrate to prevent flaking or spalling, and
- Little or no porosity.

The effectiveness of coatings, in turn, is enhanced by hardness, toughness, and high thermal conductivity of the substrate, which may be carbide or high-speed steel. Honing (see Section 9.8) of the cutting edges is an important procedure to maintain the strength of the coating; otherwise, the coating may chip off at sharp edges and corners. Some different types of coatings are discussed as follows:

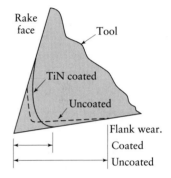

**FIGURE 8.36** Wear patterns on high-speed-steel uncoated and titanium-nitride-coated tools. Note that flank wear is lower for the coated tool.

a) **Titanium nitride.** *Titanium nitride* (TiN) coatings have a low coefficient of friction, high hardness, high resistance to high temperature, and good adhesion to the substrate. Consequently, they greatly improve the life of high-speed-steel tools, as well as carbide tools, drills, and cutters. Titanium-nitride-coated tools (gold in color) perform well at higher cutting speeds and feeds. Flank wear is significantly lower than for uncoated tools (Fig. 8.36), and flank surfaces can be reground after use; regrinding does not remove the coating on the rake face of the tool. However, coated tools do not perform as well at low cutting speeds as uncoated tools, because the coating can be worn off by chip adhesion. Hence, the use of appropriate cutting fluids to discourage adhesion is important.

b) **Titanium carbide.** *Titanium carbide* (TiC) coatings on tungsten-carbide inserts have high resistance to flank wear in machining abrasive materials.

c) **Titanium carbonitride.** *Titanium carbonitride* (TiCN), which is deposited by physical-vapor deposition techniques, is harder and tougher than TiN. It can be used on carbides and high-speed-steel tools and is particularly effective in cutting stainless steels.

d) **Ceramic coatings.** Because of their resistance to high temperature, chemical inertness, low thermal conductivity, and resistance to flank and crater wear, ceramics are suitable coatings for tools. The most commonly used ceramic coating is aluminum oxide ($Al_2O_3$). However, because ceramic coatings are very stable (i.e., not chemically reactive), oxide coatings generally bond weakly with the substrate and thus may have a tendency to peel off.

e) **Multiphase coatings.** The desirable properties of the types of coating just described can be combined and optimized with the use of *multiphase coatings* (Fig. 8.37). Carbide tools are available with two or three layers of such coatings and are particularly effective in machining cast irons and steels.

The first layer over the substrate is TiC, followed by $Al_2O_3$, and then TiN. The first layer should bond well with the substrate; the outer layer should resist wear and have low thermal conductivity; the intermediate layer should bond well and be compatible with both layers. Typical applications of multiple-coated tools are listed as follows:

1. High-speed, continuous cutting: TiC/$Al_2O_3$.
2. Heavy-duty, continuous cutting: TiC/$Al_2O_3$/TiN.
3. Light, interrupted cutting: TiC/TiC + TiN/TiN.

**FIGURE 8.37** Multiphase coatings on a tungsten-carbide substrate. Three alternating layers of aluminum oxide are separated by very thin layers of titanium nitride. Inserts with as many as 13 layers of coatings have been made. Coating thicknesses are typically in the range of $2-10 \mu m$ ($80-400 \mu in.$). *Source*: Courtesy of Kennametal, Inc., and *Manufacturing Engineering*, Society of Manufacturing Engineers.

A more recent development in coatings is **alternating multiphase layers**, with layers that are thinner than in regular multiphase coatings (Fig. 8.37). The thickness of these layers is in the range of $2-10 \mu m$ ($80-400 \mu in.$). The reason for using thinner coatings is that coating hardness increases with decreasing grain size, a phenomenon that is similar to the increase in strength of metals with decreasing grain size (see Section 3.4.1); hence, thinner layers are harder than thicker layers.

f) **Diamond coatings.** A more recent development concerns the use of poly-crystalline diamond as coatings for cutting tools, particularly tungsten-carbide and silicon-nitride inserts. However, difficulties exist regarding adherence of the diamond film to the substrate and the difference in thermal expansion between diamond and substrate materials (see Table 8.6).

*Thin-film diamond-coated* inserts are available, as are thick-film diamond-brazed-tip cutting tools. Thin films are deposited on substrates by PVD and CVD techniques, whereas thick films are obtained by growing a large sheet of pure diamond, which is then laser cut to shape and brazed to a carbide shank. Diamond-coated tools are particularly effective in machining abrasive materials, such as aluminum alloys that contain silicon, fiber-reinforced and metal-matrix composite materials, and graphite. Improvements in tool life of as much as tenfold have been obtained over that of other coated tools.

g) **Other coating materials.** While titanium-nitride coatings made by chemical-vapor deposition processes are still common, advances are being made in developing and testing new coating materials. **Titanium aluminum nitride** (TiAlN) is effective in machining aerospace alloys. Chromium-based coatings such as **chromium carbide** (CrC) have been found to be effective in machining softer metals that tend to adhere to the cutting tool, such as aluminum, copper, and titanium. Other new coating materials include **zirconium nitride** (ZrN) and **hafnium nitride** (HfN). Much experimental data are still needed before these coatings and their behavior can be fully assessed. More recent developments include (a) **nanocoatings** with carbide, boride, nitride, oxide, or some combination thereof, and (b) **composite coatings**, using a variety of materials.

h) **Ion implantation.** In this process, ions are introduced into the surface of the cutting tool. The process does not change the dimensions of tools, but improves their surface properties. (See Section 4.5.1.) **Nitrogen-ion**-implanted carbide tools have been used successfully on alloy steels and stainless steels. **Xenon-ion** implantation of tools is also under development.

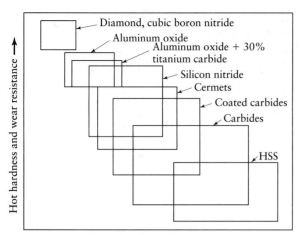

**FIGURE 8.38** Ranges of properties for various groups of tool materials. (See also various tables in this chapter.)

### 8.6.6 Alumina-base ceramics

Ceramic tool materials, introduced in the early 1950s, consist primarily of fine-grained, high-purity **aluminum oxide**. They are pressed into insert shapes at room temperature and under high pressure, sintered at high temperature, and called **white**, or **cold-pressed, ceramics**. (See also Section 11.9.3.) Additions of titanium carbide and zirconium oxide improve properties such as toughness and resistance to thermal shock.

   *Alumina-base ceramic* tools have very high abrasion resistance and hot hardness (Fig. 8.38). Chemically, they are more stable than high-speed steels and carbides; thus, they have less of a tendency to adhere to metals during cutting and, hence, less of a tendency to form a built-up edge. Consequently, good surface finish is obtained with ceramic tools in cutting cast irons and steels. However, ceramics lack toughness, which results in premature tool failure by chipping or catastrophic failure. (See Fig. 8.20.) The shape and setup of ceramic tools are important. Negative rake angles, and hence large included angles, are generally preferred in order to avoid chipping due to poor tensile strength. The occurrence of tool failure can be reduced by increasing the stiffness and damping capacity of machine tools, mountings, and workholding devices, thus reducing vibration and chatter (see Section 8.11).

**Cermets.**   *Cermets* (from *cer*amic and *met*al), which are also called **black**, or **hot-pressed, ceramics** (carboxides), were introduced in the 1960s. They typically contain 70% aluminum oxide and 30% titanium carbide; other cermets may contain molybdenum carbide, niobium carbide, or tantalum carbide. The brittleness and high cost of cermets have been a problem when using them as cutting tools; however, further refinements have resulted in improved strength, toughness, and reliability. Their performance is between that of ceramics and carbides. Chip-breaker features are important when using cermets. Although cermets can be coated, the benefits of coatings are somewhat controversial, as the improvement in wear resistance appears to be marginal.

### 8.6.7  Cubic boron nitride

Next to diamond, *cubic boron nitride* (cBN) is the hardest material presently available. The cBN cutting tool was introduced in 1962; it is made by bonding a 0.5- to 1-mm (0.02- to 0.04-in.) layer of *polycrystalline cubic boron nitride* to a carbide substrate by sintering under pressure (Fig. 8.39). While the carbide provides shock

**FIGURE 8.39**

Construction of polycrystalline cubic-boron-nitride or diamond layer on a tungsten-carbide insert.

resistance, the cBN layer provides very high wear resistance and cutting-edge strength. Cubic-boron-nitride tools are also made in small sizes without a substrate. At elevated temperatures, cBN is chemically inert to iron and nickel, and its resistance to oxidation is high. It is therefore particularly suitable for cutting hardened ferrous and high-temperature alloys. (See the description of *hard turning* in Section 8.8.2.) Cubic boron nitride is also used as an abrasive. (See Chapter 9.) Because cBN tools are brittle, stiffness and damping capacity of the machine tool and fixturing is important to avoid vibration and chatter.

### 8.6.8  Silicon-nitride-base ceramics

Developed in the 1970s, *silicon-nitride* (SiN) *-base ceramic* tool materials consist of silicon nitride with various additions of aluminum oxide, yttrium oxide, and titanium carbide. These tools have high toughness, hot hardness, and good thermal-shock resistance. An example of an SiN-base material is **sialon**, so called after the elements *si*licon, *al*uminum, *o*xygen, and *n*itrogen in its composition. It has higher resistance to thermal shock than silicon nitride and is recommended for machining cast irons and nickel-base superalloys at intermediate cutting speeds. Because of their chemical affinity to steels, SiN-base tools are not suitable for machining steels.

### 8.6.9  Diamond

The hardest substance of all known materials is *diamond*, a crystalline form of carbon. (See also Section 11.13.) As a cutting tool, it has low tool–chip friction, high wear resistance, and the ability to maintain a sharp cutting edge. It is used when good surface finish and dimensional accuracy are required, particularly with soft nonferrous alloys and abrasive nonmetallic materials. **Single-crystal diamond** of various carats is used for special applications, such as machining copper-front high-precision optical mirrors. Because diamond is brittle, tool shape and sharpness are important; low rake angles (large included angles) are normally used to provide a strong cutting edge. Special attention should be given to proper mounting and crystal orientation in order to obtain optimum useage. Diamond wear may occur by microchipping (caused by thermal stresses and oxidation) and transformation to carbon (caused by the heat generated during cutting).

Single-crystal (single-point) diamond tools have been largely replaced by **polycrystalline diamond** tools (*compacts*), which are also used as wire-drawing dies for fine wire. These materials consist of very small synthetic crystals, fused by a high-pressure, high-temperature process to a thickness of about 0.5–1 mm (0.02–0.04 in.), and bonded to a carbide substrate, similar to cBN tools. (See Fig. 8.39.) The random orientation of the diamond crystals prevents the propagation of cracks through the structure, thus significantly improving its toughness.

Diamond tools can be used satisfactorily at almost any speed, but are suitable mostly for light, uninterrupted finishing cuts. In order to minimize tool fracture, the single-crystal diamond must be resharpened as soon as it becomes dull. Because of its strong chemical affinity, diamond is not recommended for machining plain-carbon steels and titanium, nickel, and cobalt-base alloys. Diamond is also used as an abrasive in grinding and polishing operations (see Chapter 9) and as coatings (see Section 4.5).

### 8.6.10  Whisker-reinforced and nanocrystalline tool materials

In order to further improve the performance and wear resistance of cutting tools, particularly in machining new workpiece materials and composites that are under continued development, new tool materials are being developed with enhanced properties such as (1) high fracture toughness, (2) resistance to thermal shock, (3) cutting-edge strength, and (4) hot hardness.

More recent developments consist of using *whiskers* (see Section 3.8.3) as reinforcing fibers in composite tool materials. Examples of **whisker-reinforced** materials include silicon-nitride-base tools reinforced with silicon-carbide (SiC) whiskers, and aluminum-oxide-base tools reinforced with silicon-carbide whiskers, sometimes with the addition of *zirconium oxide* ($ZrO_2$). However, the low inertness of silicon carbide to ferrous metals makes SiC-reinforced tools unsuitable for machining irons and steels. Progress in nanomaterials (Section 3.11.9) has led to the development of cutting tools made of very fine-grained (**micrograin**) carbides of tungsten, titanium, and tantalum. These materials are stronger, harder, and more wear resistant than are traditional carbides.

### 8.6.11  Cryogenic treatment of cutting tools

Studies are continuing on the beneficial effect of cryogenic treatment of tools and other metals on their performance in cutting operations. (See Section 5.11.6 for details.) In this procedure, the tool is cooled very slowly to temperatures of around $-180°C$ ($-300°F$) and slowly returned to room temperature; it is then tempered. Claims have been made that, depending on the combinations of tool and workpiece material involved, tool-life increases of up to 300% can be achieved with this procedure.

## 8.7  |  Cutting Fluids

Also called *lubricants* and *coolants, cutting fluids* are used extensively in machining operations to

- Reduce friction and wear, hence improving tool life and surface finish;
- Reduce forces and energy consumption;
- Cool the cutting zone, thus reducing workpiece temperature and distortion and improving tool life;
- Wash away chips;
- Protect the newly machined surfaces from environmental corrosion.

A cutting fluid can interchangeably be a **coolant** and a **lubricant** (see Section 4.4.3); its effectiveness in cutting operations depends on a number of factors, such as the method of application, the temperature, the cutting speed, and the type of machining operation. As described in Section 8.2.6, temperature increases as cutting speed increases; therefore, *cooling* of the cutting zone is of major importance at high cutting speeds. Water is an excellent coolant; however, it causes rusting of workpieces and machine components and is a poor lubricant. On the other hand, if the cutting speed is low, such as in broaching or tapping, *lubrication*, not cooling, is the important factor. Lubrication reduces the tendency for the formation of built-up edge and thus improves surface finish.

There are situations, however, in which the use of cutting fluids can be detrimental. For example, in interrupted cutting operations, such as milling, the cooling action of the cutting fluid increases the extent of alternate heating and cooling (*thermal cycling*) to which the cutter teeth are subjected. This condition can lead to thermal cracks (*thermal fatigue* or *thermal shock*). Cutting fluids may also cause the chip to become more curled, thus concentrating the stresses near the tool tip. These stresses, in turn, concentrate the heat closer to the tool tip and reduce tool life.

Cutting fluids can be expensive and represent a **biological** and **environmental hazard** (see also Section 4.4.4) that requires proper recycling and disposal, thus adding to the cost of the machining operation. For these reasons, **dry cutting**, or **dry machining**, has become an increasingly important approach; in dry machining, no coolant or lubricant is used. Even though this approach would suggest that higher temperatures and more rapid tool wear would occur, some tool materials and coatings display a reasonable tool life at the higher temperatures. Dry cutting has sometimes been associated with high-speed machining, because higher cutting speeds transfer a greater amount of heat conveyed from cutting to the chip (see Fig. 8.18), which is a natural strategy for reducing the need for a coolant. (See also Section 3.9.7 on possible detrimental effects of cutting fluids on some cutting tools, called *selective leaching*, such as carbide tools with cobalt binders.)

### 8.7.1 Types of cutting fluids and methods of application

There are generally four types of cutting fluids commonly used in machining operations: **oils, emulsions, semisynthetics,** and **synthetics.** The general characteristics of these fluids are described in Section 4.4.4. Cutting-fluid recommendations for specific machining operations are given throughout the rest of this chapter. In selecting an appropriate cutting fluid, considerations should be given to its possibly detrimental effects on the workpiece material (e.g., corrosion, stress-corrosion cracking, staining), the components of the machine tool, biological and environmental effects, and recycling and disposal.

The most common method of applying cutting fluid is **flood cooling**. Flow rates range from 10 L/min (3 gal/min) for single-point tools to 225 L/min (60 gal/min) per cutter for multiple-tooth cutters, such as in milling. In operations such as gun drilling and end milling, fluid pressures of 700–14,000 kPa (100–2000 psi) are used to wash away the chips.

**Mist cooling** is another method of applying cutting fluids and is used particularly with water-base fluids. Although it requires venting (to prevent inhalation of fluid particles by the machine operator and others nearby) and has limited cooling capacity, mist cooling supplies fluid to otherwise inaccessible areas and provides better visibility of the workpiece being machined. It is particularly effective in grinding operations (see Chapter 9) and at air pressures of 70–600 kPa (10–80 psi).

With increasing speed and power of machine tools, heat generation in machining operations has become a significant factor. More recent developments include the use of **high-pressure refrigerated coolant** systems to improve the rate of heat removal from the cutting zone, as well as to reduce costs and avoid adverse environmental effects. High pressures as high as 35 MPa (5000 psi) are also used in delivering the cutting fluid to the cutting zone via specially designed nozzles that aim a powerful jet of fluid to the zone. This action breaks up the chips (thus acting as a chip breaker) in situations where the chips produced would otherwise be long and continuous and hence interfere with the cutting operation.

## 8.7.2 Cryogenic machining

Largely in the interest of reducing or eliminating the adverse environmental impact of using metalworking fluids, a recent technology is the use of *liquid nitrogen* as a coolant in machining, as well as in grinding. (See Section 9.6.9.) With appropriate small-diameter nozzles, liquid nitrogen at around −200°C (−320°F) is injected into the tool–workpiece interface, thus reducing its temperature. As a result, tool hardness and, hence, tool life is enhanced, thus allowing for higher cutting speeds; furthermore, the chips become more brittle and, thus, easier to remove from the cutting zone. Because no fluids are involved and the liquid nitrogen simply evaporates, the chips can be recycled more easily, thus also improving the economics of machining operations, without any adverse effects on the environment.

---

**EXAMPLE 8.4  Effect of cutting fluids on machining**

A machining operation is being carried out with a cutting fluid that is effective as a lubricant. Explain the changes in the mechanics of the cutting operation and total energy consumption if the fluid is shut off.

**Solution.**  Since the cutting fluid is a good lubricant, the friction at the tool–chip interface will increase when the fluid is shut off. The following chain of events then takes place:

1. Fluid is shut off.
2. Friction at the tool–chip interface increases.
3. The shear angle decreases.
4. The shear strain increases.
5. The chip is thicker.
6. A built-up edge is likely to form.

As a consequence,

1. The shear energy in the primary zone increases.
2. The friction energy in the secondary zone increases.
3. Hence the total energy increases.
4. Surface finish is likely to deteriorate.
5. The temperature in the cutting zone increases, and hence tool wear increases and workpiece surface may lose its integrity.
6. Dimensional tolerances may be difficult to maintain, because of the increased temperature and expansion of the workpiece during machining.

# 8.8 | Cutting Processes and Machine Tools for Producing Round Shapes

This section describes the processes that produce parts which are basically *round* in shape (Table 8.7); typical products made include parts as small as miniature screws for eyeglass-frame hinges and as large as shafts, pistons, cylinders, gun barrels, and turbines for hydroelectric power plants. The processes are usually performed by turning the workpiece on a lathe. *Turning* means that the part is rotating while it is being machined. The starting material is usually a workpiece that has been made by other processes, such as casting, shaping, forging, extrusion, and drawing. Turning processes are versatile and capable of producing a wide variety of shapes, as outlined in Fig. 8.40. The various types of turning processes are described as follows:

- **Turning** straight, conical, curved, or grooved workpieces, such as shafts, spindles, pins, handles, and various machine components;

- **Facing,** to produce a flat surface at the end of the part, such as for parts that are attached to other components, or to produce grooves for O-ring seats;

- Producing various shapes by **form tools,** such as for functional purposes or for appearance;

TABLE 8.7

## General Characteristics of Machining Processes

| Process | Characteristics | Commercial tolerances (±mm) |
|---|---|---|
| Turning | Turning and facing operations are performed on all types of materials; requires skilled labor; low production rate, but medium to high rates can be achieved with turret lathes and automatic machines, requiring less skilled labor. | Fine: 0.05–0.13 Rough: 0.13 Skiving: 0.025–0.05 |
| Boring | Internal surfaces or profiles, with characteristics similar to those produced by turning; stiffness of boring bar is important to avoid chatter. | 0.025 |
| Drilling | Round holes of various sizes and depths; requires boring and reaming for improved accuracy; high production rate, labor skill required depends on hole location and accuracy specified. | 0.075 |
| Milling | Variety of shapes involving contours, flat surfaces, and slots; wide variety of tooling; versatile; low to medium production rate; requires skilled labor. | 0.13–0.25 |
| Planing | Flat surfaces and straight contour profiles on large surfaces; suitable for low-quantity production; labor skill required depends on part shape. | 0.08–0.13 |
| Shaping | Flat surfaces and straight contour profiles on relatively small workpieces; suitable for low-quantity production; labor skill required depends on part shape. | 0.05–0.13 |
| Broaching | External and internal flat surfaces, slots, and contours with good surface finish; costly tooling; high production rate; labor skill required depends on part shape. | 0.025–0.15 |
| Sawing | Straight and contour cuts on flats or structural shapes; not suitable for hard materials unless the saw has carbide teeth or is coated with diamond; low production rate; requires only low labor skill. | 0.8 |

**FIGURE 8.40** Various cutting operations that can be performed on a lathe.

(a) Straight turning     (b) Taper turning     (c) Profiling

(d) Turning and external grooving     (e) Facing     (f) Face grooving

(g) Cutting with a form tool     (h) Boring and internal grooving     (i) Drilling

(j) Cutting off     (k) Threading     (l) Knurling

- **Boring**, to enlarge a hole made by a previous process or in a tubular workpiece, or to produce internal grooves;
- **Drilling**, to produce a hole, which may be followed by tapping or boring to improve the accuracy of the hole and its surface finish;
- **Parting**, also called **cutting off**, to cut a piece from the end of a part, as in making slugs or blanks for additional processing into discrete parts;
- **Threading**, to produce external and internal threads in workpieces;
- **Knurling**, to produce a regularly shaped geometric roughness on cylindrical surfaces, as in making knobs.

These operations may be performed at various rotational speeds of the workpiece, depths of cut, $d$, and feed, $f$, (see Fig. 8.19), depending on the workpiece and tool materials, the surface finish and dimensional accuracy required, and the capacity of the machine tool. Cutting speeds are usually in the range of 0.15–4 m/s (30–800 ft/min). **Roughing cuts**, which are performed for large-scale material removal, usually involve depths of cut greater than 0.5 mm (0.02 in.) and feeds on the order of 0.2–2 mm/rev (0.008–0.08 *in./rev*). **Finishing cuts** usually involve lower depths of cut and feed.

## 8.8.1 Turning parameters

The majority of turning operations involve single-point cutting tools. Figure 8.41 shows the geometry of a typical right-hand simple cutting tool for turning; such tools

**FIGURE 8.41**
(a) Designations and symbols for a right-hand cutting tool; solid high-speed-steel tools have a similar designation. The designation "right hand" means that the tool travels from right to left, as shown in Fig. 8.19. (b) Square insert in a right-hand toolholder for a turning operation. A wide variety of toolholders is available for holding inserts at various angles. Thus, the angles shown in (a) can easily be achieved by selecting an appropriate insert and toolholder. *Source:* Kennametal, Inc.

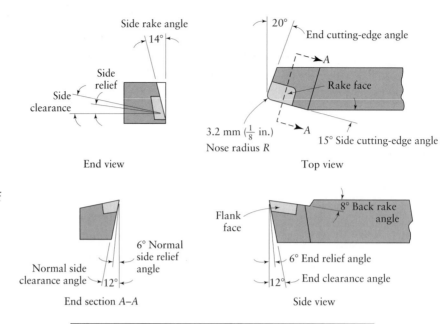

| Tool Signature | Dimensions | Abbreviation |
|---|---|---|
| 8 .............. | Back rake angle | BR |
| 14 .............. | Side rake angle | SR |
| 6 .............. | End relief angle | ER |
| 12 .............. | End clearance angle | ... |
| 6 .............. | Side relief angle | SRF |
| 12 .............. | Side clearance angle | ... |
| 20 .............. | End cutting-edge angle | ECEA |
| 15 .............. | Side cutting-edge angle | SCEA |
| $\frac{1}{8}$ .............. | Nose radius | NR |

(a)

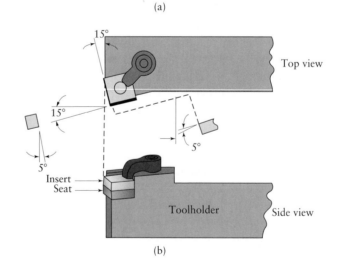

(b)

are identified by a standardized nomenclature. Each group of tool and workpiece materials has an optimum set of tool angles, which have been developed over many years, largely through experience. Data on tool geometry can be found in the references given in the bibliography at the end of this chapter.

a) **Tool geometry.** The various angles on a tool have important functions in cutting operations. **Rake angles** are important in controlling the direction of chip flow and the strength of the tool tip. Positive angles improve the cutting operation by reducing forces and temperatures; however, positive angles also produce a small included angle of the tool tip. (See Fig. 8.2.) Depending on the toughness of the tool material, this included angle may cause premature tool chipping and failure. **Side rake angle** is more important than **back rake angle**, although the latter usually controls the direction of chip flow. **Relief angles** control interference and rubbing at the tool–workpiece interface; if the relief angle is too large, the tool may chip off, and if it is too small, flank wear may be excessive. **Cutting-edge angles** affect chip formation, tool strength, and cutting forces to various degrees. **Nose radius** affects surface finish and tool-tip strength; also, the smaller the nose radius, the rougher will be the surface finish of the workpiece, and the lower will be the strength of the tool. On the other hand, large nose radii can lead to tool chatter.

b) **Material removal rate.** The *material removal rate* (MRR) is the volume of material removed per unit time, such as $mm^3/min$ or $in^3/min$. Referring to Fig. 8.42a, note that for each revolution of the workpiece, we remove a ring-shaped layer of material whose cross-sectional area is the product of the distance the tool travels in one revolution (i.e., the feed $f$) and the depth of cut, $d$. The volume of this ring is the product of the cross-sectional area–that is, $fd$–and the average circumference of the ring–that is, $\pi D_{avg}$, where $D_{avg} = (D_o + D_f)/2$. For light cuts on large-diameter workpieces, the average diameter can be replaced by $D_o$. The rotational speed of the workpiece is $N$.

Thus, the material removal rate per revolution will be $\pi D_{avg}df$. Since we have $N$ revolutions per minute, the removal rate is given by

$$MRR = \pi D_{avg}dfN. \tag{8.38}$$

The dimensional accuracy of this equation can be checked by substituting dimensions into the right-hand side; hence, $(mm)(mm)(mm/rev)(rev/min) = mm^3/min$, which indicates the volume rate of removal. Similarly, the cutting time $T$ for a workpiece of length $\ell$ can be calculated by noting that the tool travels at a feed rate of $fN = (mm/rev)(rev/min) = mm/min$. Since the distance traveled is $\ell$ mm, the cutting time is

$$t = \frac{\ell}{fN}. \tag{8.39}$$

This time does not include the time required for *tool approach* and *retraction*. Because the time spent in *noncutting* cycles of a machining operation is

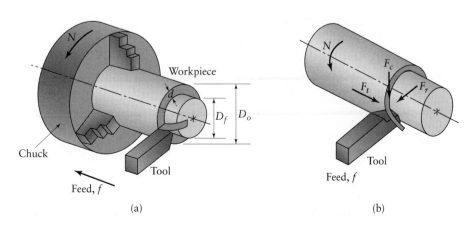

(a)

(b)

**FIGURE 8.42**
(a) Schematic illustration of a turning operation showing depth of cut, $d$, and feed, $f$. Cutting speed is the surface speed of the workpiece at the tool tip. (b) Forces acting on a cutting tool in turning. $F_c$ is the cutting force; $F_t$ is the thrust or feed force (in the direction of feed); and $F_r$ is the radial force that tends to push the tool away from the workpiece being machined. Compare this figure with Fig. 8.11 for a two-dimensional cutting operation.

nonproductive and affects the overall economics of the process, the time involved in approaching and retracting tools to and from the workpiece is an important consideration. Machine tools are now designed and built to minimize this time, using computer controls; a typical method employed is first to rapidly traverse the tools and then to slow the movement as the tool engages the workpiece.

c) **Forces in turning.** Figure 8.42b illustrates the three forces acting on a cutting tool; these forces are important in the design of machine tools as well as in the deflection of tools for precision machining operations. The *cutting force*, $F_c$, acts downward on the tool tip and thus tends to deflect the tool downward. The cutting force is the force that supplies the energy required for the cutting operation. As can be seen in Example 8.5, the cutting force can be calculated from the energy per unit volume, described in Section 8.2.5, and by using the data in Table 8.3. The *thrust force*, $F_t$, acts in the longitudinal direction; this force is also called the *feed force*, because it is in the feed direction. The *radial force*, $F_r$, is in the radial direction and thus tends to push the tool away from the workpiece. The thrust and radial forces are difficult to calculate, because of the many factors involved in the cutting process, so they are determined experimentally.

d) **Tool materials, feeds, and cutting speeds.** Section 8.6 describes the general characteristics of cutting-tool materials. Figure 8.43 gives a broad range of cutting speeds and feeds applicable for these tool materials. Specific recommendations for cutting speeds for turning various workpiece materials and cutting tools are given in Table 8.8. In this table, **roughing** means cutting material without much consideration to dimensional tolerances and surface finish, and **finishing** means cutting to obtain final dimensions with acceptable dimensional tolerances and surface finish.

e) **Cutting fluids.** Although many metals and nonmetallic materials can be machined without a cutting fluid (dry machining), the application of a cutting fluid can improve the operation significantly in many cases. Specific recommendations for cutting fluids for various workpiece materials and operations can be found in the references in the bibliography at the end of this chapter.

**FIGURE 8.43** The range of applicable cutting speeds and feeds for a variety of tool materials. *Source*: Valenite, Inc.

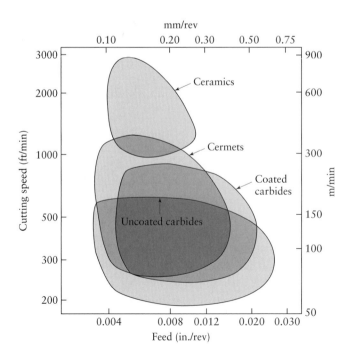

TABLE 8.8

**Approximate Ranges of Recommended Cutting Speeds for Turning Operations**

| WORKPIECE MATERIAL | CUTTING SPEED | |
| --- | --- | --- |
| | m/min | ft/min |
| Aluminum alloys | 200–1000 | 650–3300 |
| Cast iron, gray | 60–900 | 200–3000 |
| Copper alloys | 50–700 | 160–2300 |
| High-temperature alloys | 20–400 | 65–1300 |
| Steels | 50–500 | 160–1600 |
| Stainless steels | 50–300 | 160–1000 |
| Thermoplastics and thermosets | 90–240 | 300–800 |
| Titanium alloys | 10–100 | 30–330 |
| Tungsten alloys | 60–150 | 200–500 |

*Note:* (a) The speeds given in this table are for carbides and ceramic cutting tools. Speeds for high-speed-steel tools are lower than indicated. The higher ranges are for coated carbides and cermets. Speeds for diamond tools are significantly higher than any of the values indicated in the table.
(b) Depths of cut, $d$, are generally in the range of 0.5–12 mm (0.02–0.5 in.).
(c) Feeds, $f$, are generally in the range of 0.15–1 mm/rev (0.006–0.040 in./rev).

**EXAMPLE 8.5** Material removal rate and cutting force in turning

A 6-in.-long, 0.5-in.-diameter 304 stainless-steel rod is being reduced in diameter to 0.480 in. by turning on a lathe. The spindle rotates at $N = 400$ rpm, and the tool is traveling at an axial speed of 8 in./min. Calculate the cutting speed, material removal rate, cutting time, power dissipated, and cutting force.

**Solution.** The cutting speed is the tangential speed of the workpiece. The maximum cutting speed is at the outer diameter, $D_o$, and is obtained from the expression

$$V = \pi D_o N.$$

Thus,

$$V = (\pi)(0.500)(400) = 628 \text{ in./min} = 52 \text{ ft/min}.$$

The cutting speed at the machined diameter is

$$V = (\pi)(0.480)(400) = 603 \text{ in./min} = 50 \text{ ft/min}.$$

From the information given, note that the depth of cut is

$$d = (0.500 - 0.480)/2 = 0.010 \text{ in.,}$$

and the feed is

$$f = 8/400 = 0.02 \text{ in./rev.}$$

Thus, according to Eq. (8.38), the material removal rate is

$$\text{MRR} = (\pi)(0.490)(0.010)(0.02)(400) = 0.123 \text{ in}^3/\text{min.}$$

The actual time to cut, according to Eq. (8.39), is

$$t = 6/(0.02)(400) = 0.75 \text{ min.}$$

We can calculate the power required by referring to Table 8.3 and taking an average value for stainless steel as 4 W-s/mm$^3$ = 4/2.73 = 1.47 hp-min/in$^3$. Therefore, the power dissipated is

$$\text{Power} = (1.47)(0.123) = 0.181 \text{ hp},$$

and since 1 hp = 396,000 in.-lb/min, the power dissipated is 71,700 in.-lb/min. The cutting force, $F_c$, is the tangential force exerted by the tool. Since power is the product of torque, $T$, and rotational speed in radians per unit time, we have

$$T = \frac{(71,700)}{(400)(2\pi)} = 29 \text{ lb-in.}$$

Since $T = (F_c)(D_{avg}/2)$, we obtain

$$F_c = \frac{(29)(2)}{(0.490)} = 118 \text{ lb.}$$

### 8.8.2 Lathes and lathe operations

Lathes are generally considered to be the oldest machine tools. Although wood-working lathes were first developed during the period 1000–1 BC, metalworking lathes with lead screws were not built until the late 1700s. The most common type of lathe, shown in Fig. 8.44, was originally called an **engine lathe**, because it was powered with overhead pulleys and belts from nearby engines. Lathes are now equipped with various features, including computer controls, described as follows:

a) **Lathe components.** Lathes are equipped with a variety of components and accessories (Fig. 8.44). The *bed* supports all the other major components of the lathe. The *carriage*, or *carriage assembly*, which slides along the ways, consists

**FIGURE 8.44** The components of a lathe. *Source*: Courtesy of MAKINO GmbH.

of an assembly of the *cross slide*, *tool post*, and *apron*. The cutting tool is mounted on the *tool post*, usually with a *compound rest* that swivels for tool positioning and adjustments. The *headstock*, which is fixed to the bed, is equipped with motors, pulleys, and V-belts that supply power to the *spindle* at various rotational speeds. Headstocks have a hollow spindle to which workholding devices, such as *chucks* and *collets*, are attached. The *tailstock*, which can slide along the ways and be clamped at any position, supports the other end of the workpiece. The *feed rod*, which is powered by a set of gears from the headstock, rotates during operation of the lathe and provides movement to the carriage and the cross slide. The *lead screw*, which is used for cutting threads accurately, is engaged with the carriage by closing a split nut around the lead screw.

A lathe is usually specified by its *swing*, that is, the maximum diameter of the workpiece that can be machined; by the maximum distance between the headstock and tailstock centers; and by the length of the bed. A variety of lathes is available, including *bench lathes*, *toolroom lathes*, *engine lathes*, *gap lathes*, and *special-purpose lathes*.

*Tracer lathes*, also called *duplicating lathes* or *contouring lathes*, are capable of turning parts with various contours; the cutting tool follows a path that duplicates the contour of a template through a hydraulic or electrical system. *Automatic lathes*, also called *chucking machines* or *chuckers*, are used for machining individual pieces of regular or irregular shapes. *Turret lathes* are capable of performing multiple cutting operations on the same workpiece, such as turning, boring, drilling, thread cutting, and facing. Several cutting tools are mounted on the hexagonal *main turret*. The lathe usually has a *square turret* on the cross slide, with as many as four cutting tools mounted on it.

b) **Computer-controlled lathes.** In the most advanced lathes, movement and control of the machine and its components are actuated by **computer numerical controls** (CNCs); the features of such a lathe are shown in Fig. 8.45. These lathes are usually equipped with one or more turrets; each turret is equipped with a variety of tools and performs several operations on different surfaces of the workpiece. These machines are highly automated, the operations are repetitive and maintain the desired dimensional accuracy, and less skilled labor is required (after the machine is set up) than with common lathes. They are suitable for low to medium volumes of production. The details of computer controls are presented in Chapters 14 and 15. (See also Section 8.10.)

Chuck

CNC unit

Round turret for
OD operations

End turret for
ID operations

Tailstock

**FIGURE 8.45** A computer-numerical-control lathe. Note the two turrets on this machine. *Source*: Jones & Lamson.

67.4 mm
(2.654")

87.9 mm
(3.462")

98.4 mm
(3.876")

85.7 mm (3.375")
32 threads per in.

Material: Titanium Alloy
Number of tools: 7
Total machining time
(two operations):
5.25 minutes

235.6 mm
(9.275")

78.5 mm
(3.092")

Material: 52100 Alloy Steel
Number of tools: 4
Total machining time
(two operations):
6.32 minutes

50.8 mm
(2")

23.8 mm
(0.938")

53.2 mm
(2.094")

Material: 1020 Carbon Steel
Number of tools: 8
Total machining time
(two operations):
5.41 minutes

(a) Housing base

(b) Inner bearing race

(c) Tube reducer

**FIGURE  8.46**    Typical parts made on computer-numerical-control machine tools.

**EXAMPLE 8.6  Typical parts made on CNC turning machine tools**

Figure 8.46 illustrates the capabilities of CNC turning machine tools. Workpiece materials, the number of cutting tools used, and machining times are indicated for each part. Although not as effectively or as consistently so, these parts can also be made on manual or turret lathes.
*Source*: Monarch Machine Tool Company.

c) **Turning process capabilities.** Table 8.9 provides relative *production rates* for turning, as well as for other cutting operations described in the rest of this chapter; these rates have an important bearing on productivity in machining operations. Note that there are major differences in the production rates of these processes. These differences are due not only to the inherent characteristics of the processes and machine tools, but also to various other factors, such as setup times and the types and sizes of the workpieces involved. The proper selection of a process and the machine tool for a particular product is important for minimizing production costs, as is also discussed in Section 8.14 and Chapter 16.

As stated previously, the ratings in Table 8.9 are relative, and there can be significant variations in specialized applications. For example, heat-treated, high-carbon cast-steel rolls (for rolling mills; see Section 6.3) can be machined on special lathes at material removal rates as high as 6000 cm$^3$/min (370 in$^3$/min), using multiple cermet tools. The important factor in this operation (also called **high-removal-rate machining**) is the very high stiffness of the machine tool (to avoid tool breakage due to chatter; see Section 8.11) and the machine tool's high power, which can be up to 450 kW (600 hp).

The surface finish and dimensional accuracy obtained in turning and related operations depend on such factors as the characteristics and condition of the machine tool, stiffness, vibration and chatter, process parameters, tool geometry and wear, cutting fluids, machinability of the workpiece material, and operator skill. As a result, a wide range of surface finishes can be obtained, as shown in Fig. 8.27. (See also Fig. 9.27.)

TABLE 8.9

| Typical Production Rates for Various Cutting Operations | |
| --- | --- |
| OPERATION | RATE |
| Turning | |
| Engine lathe | Very low to low |
| Tracer lathe | Low to medium |
| Turret lathe | Low to medium |
| Computer-control lathe | Low to medium |
| Single-spindle chuckers | Medium to high |
| Multiple-spindle chuckers | High to very high |
| Boring | Very low |
| Drilling | Low to medium |
| Milling | Low to medium |
| Planing | Very low |
| Gear cutting | Low to medium |
| Broaching | Medium to high |
| Sawing | Very low to low |

*Note*: Production rates indicated are relative: *Very low* is about one or more parts per hour; *medium* is approximately 100 parts per hour; *very high* is 1000 or more parts per hour.

d) **High-speed machining.** With increasing demands for higher productivity and lower production costs, investigations have been carried out since the late 1950s to increase the material removal rate, particularly for applications in the aerospace and automotive industries. One obvious possibility is to increase the cutting speed. The term *high speed* is relative; as shown in Table 8.8, general recommendations for typical applications state that aluminum alloys, for example, should be machined at much higher speeds than titanium alloys or stainless steels. As a general guide, however, an approximate range of speeds may be defined as follows: **High speed**, 600–1800 m/min (2000–6000 ft/min); **very high speed**, 1800–18,000 m/min (6000–60,000 ft/min); **ultrahigh speed**: greater than 18,000 m/min (60,000 ft/min).

Much research is being carried out on *high-speed machining* (turning, milling, boring, and drilling), particularly on aluminum alloys, titanium alloys, steels, and superalloys. Much data have been collected regarding the effects of high speeds on the type of chips produced, the cutting forces, the temperatures generated, the tool wear, the surface finish, and the economics of the process. These studies have indicated that high-speed machining can be economical for certain applications, and consequently, it is now implemented for machining aircraft-turbine components and automotive engines, with 5 to 10 times the productivity of traditional machining. Important factors in these operations include (a) the selection of an appropriate cutting tool, (b) the power of the machine tools and their stiffness and damping capacity, (c) the stiffness of toolholders and the workholding devices, (d) special spindle designs for high power and high rotational speeds, (e) the inertia of the machine-tool components, (f) fast feed drives, and (g) the level of automation.

It is important to note, however, that high-speed machining should be considered mainly for situations in which *cutting time* is a significant portion of the floor-to-floor time of the operation. Other factors, such as *noncutting time* and labor costs, are important in the overall assessment of the benefits of high-speed machining for a particular application.

e) **Ultraprecision machining.** Beginning in the 1960s, increasing demands have been made for precision manufacturing of components for computer, electronics, nuclear energy, and defense applications. Examples include optical mirrors, computer memory disks, and drums for photocopying machines, with surface finish in the range of tens of nanometers [$10^{-9}$ m, or $0.001\mu$m ($0.04$ $\mu$in.)] and form accuracies in the $\mu$m and sub-$\mu$m range. The cutting tool for these *ultraprecision machining* applications is exclusively a single-crystal diamond (hence, the process is also called **diamond turning**), with a polished cutting-edge that has a radius as small as in the range of tens of nanometers. Wear of the diamond can be a significant problem; recent advances include **cryogenic diamond turning**, in which the tooling system is cooled by liquid nitrogen to a temperature of about $-120°$C ($-184°$F). (See also Section 8.7.2.)

The workpiece materials for ultraprecision machining to date include copper alloys, aluminum alloys, silver, gold, electroless nickel, infrared materials, and plastics (acrylics). The depths of cut involved are in the nanometer range. In this range, hard and brittle materials produce continuous chips (known as **ductile-regime cutting**; see also the discussion of *ductile-regime grinding* in Section 9.5.3); deeper cuts produce discontinuous chips.

The machine tools for ultraprecision machining are built with very high precision and high machine, spindle, and workholding-device stiffness. These machines, some parts of which are made of structural materials with low thermal expansion and good dimensional stability, are located in a dust-free environment (*clean rooms*) where the temperature is controlled within a fraction of one degree. Vibrations from external and internal sources are avoided as much as possible. Feed and position controls are made by laser metrology, and the machines are equipped with highly advanced computer control systems and with thermal and geometric error-compensating features.

f) **Hard turning.** As described in Chapter 9, there are several other mechanical (particularly grinding) processes and nonmechanical methods of removing material economically from hard or hardened metals. However, it is still possible to apply traditional cutting processes to hard metals and alloys by selecting an appropriate tool material and a machine tool with high stiffness. One common example is finish machining of heat-treated-steel (45 HRC to 65 HRC) machine and automotive components, using polycrystalline cubic-boron-nitride (PcBN) cutting tools. Called *hard turning*, this process produces machined parts with good dimensional accuracy, surface finish, and surface integrity. It can compete successfully with grinding of the same components, from both technical and economic aspects. A comparative example of hard turning vs. grinding is presented in Example 9.4.

g) **Cutting screw threads.** Threads are produced primarily by forming (*thread rolling*; see Fig. 6.46), which constitutes the process by which the largest quantity of threaded parts is produced, and cutting. *Casting* of threaded parts is also possible, although dimensional accuracy, surface finish, and production rate are not as good as obtained by other processes. Figure 8.40k shows that turning operations are capable of producing threads on round bar stock. When threads are produced externally or internally by cutting with a lathe-type tool, the process is called *thread cutting*, or *threading*. When the threads are cut internally with a special threaded tool (*tap*), the process is called *tapping*. External threads may also be cut with a die or by milling. Although it adds considerably to the cost of the operation, threads may be ground for improved accuracy and surface finish.

**Automatic screw machines** are designed for high-production-rate machining of screws and similar threaded parts. Because such machines are also capable of producing other components, they are now called *automatic bar machines*. All operations on these machines are performed automatically, with tools attached to a special turret. The bar stock is fed forward automatically through the headstock after each screw or part is machined to finished dimensions and cut off. The machines may be equipped with single or multiple spindles, and capacities range from 3- to 150-mm ($\frac{1}{8}$- to 6-in.)-diameter bar stock.

### 8.8.3  Boring and boring machines

The basic boring operation, as carried out on a lathe, is shown in Fig. 8.40h. *Boring* consists of producing circular internal profiles in hollow workpieces or on a hole made by drilling or another process and is carried out using cutting tools that are similar to those used in turning. However, because the boring bar has to reach the full length of the bore, tool deflection, and hence maintenance of dimensional accuracy, can be a significant problem. The boring bar must be sufficiently stiff—that is, made of a material with high elastic modulus, such as carbides—to minimize deflection and avoid vibration and chatter. Boring bars have been designed with capabilities for damping vibrations.

Although boring operations on relatively small workpieces can be carried out on a lathe, **boring mills** are used for large workpieces. These machines are either vertical or horizontal and are capable of performing operations such as turning, facing, grooving, and chamfering. A *vertical boring machine* (Fig. 8.47) is similar to a lathe, but with a vertical axis of workpiece rotation. In *horizontal boring machines*, the workpiece is mounted on a table that can move horizontally, both axially and radially. The cutting tool is mounted on a spindle that rotates in the headstock, which is capable of both vertical and longitudinal movements. Drills, reamers, taps, and milling cutters can also be mounted on the spindle.

### 8.8.4  Drilling, reaming, and tapping

One of the most common machining processes is *drilling*. **Drills** usually have a high length-to-diameter ratio (Fig. 8.48) and thus are capable of producing deep holes. However, they are somewhat flexible, depending on their diameter, and should be used with care in order to drill holes accurately and to prevent the drill from breaking. Moreover, chips are produced within the workpiece, and the chips have to move in a direction opposite to the axial movement of the drill. Consequently, chip disposal and the effectiveness of cutting fluids are important.

BCross rail

Tool head

Workpiece

Worktable

Bed

Column

**FIGURE 8.47**   Schematic illustration of the components of a vertical boring mill.

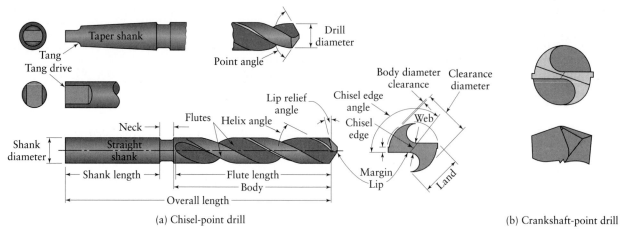

(a) Chisel-point drill

(b) Crankshaft-point drill

**FIGURE 8.48** (a) Standard chisel-point drill, with various features indicated. (b) Crankshaft-point drill.

The most common drill is the standard-point **twist drill** (Fig. 8.48). The main features of the drill point are a *point angle*, a *lip-relief angle*, a *chisel-edge angle*, and a *helix angle*. The geometry of the drill tip is such that the normal rake angle and velocity of the cutting edge vary with the distance from the center of the drill. Other types of drills include the *step drill*, *core drill*, *counterboring* and *countersinking drills*, *center drill*, and *spade drill*, as illustrated in Fig. 8.49. *Crankshaft drills* have good centering ability, and because the chips they produce tend to break up easily, they are suitable for drilling deep holes. In *gun drilling*, a special drill is used for drilling deep holes; depth-to-diameter ratios of the holes produced can be 300 or higher. In the *trepanning* technique, a cutting tool produces a hole by removing a disk-shaped piece of material (*core*), usually from flat plates; thus, a hole is produced without reducing all the material to chips. The process can be used to make disks up to 150 mm (6 in.) in diameter from flat sheet or plate.

The *material removal rate* in drilling is the ratio of volume removed to time; thus, it can be expressed as

$$\text{MRR} = \frac{\pi D^2}{4} fN, \tag{8.40}$$

where $D$ is the drill diameter, $f$ is the feed, and $N$ is the rpm of the drill. (See Table 8.10 for recommendations for speed and feed.)

**FIGURE 8.49** Various types of drills and drilling operations.

TABLE 8.10

## General Recommendations for Speeds and Feeds in Drilling

| WORKPIECE MATERIAL | SURFACE SPEED | | FEED, mm/rev (in./rev) DRILL DIAMETER | | RPM | |
|---|---|---|---|---|---|---|
| | m/min | ft/min | 1.5 mm (0.060 in.) | 12.5 mm (0.5 in.) | 1.5 mm (0.060 in.) | 12.5 mm (0.5 in.) |
| Aluminum alloys | 30–120 | 100–400 | 0.025 (0.001) | 0.30 (0.012) | 6400–25,000 | 800–3000 |
| Magnesium alloys | 45–120 | 150–400 | 0.025 (0.001) | 0.30 (0.012) | 9600–25,000 | 1100–3000 |
| Copper alloys | 15–60 | 50–200 | 0.025 (0.001) | 0.25 (0.010) | 3200–12,000 | 400–1500 |
| Steels | 20–30 | 60–100 | 0.025 (0.001) | 0.30 (0.012) | 4300–6400 | 500–800 |
| Stainless steels | 10–20 | 40–60 | 0.025 (0.001) | 0.18 (0.007) | 2100–4300 | 250–500 |
| Titanium alloys | 6–20 | 20–60 | 0.010 (0.0004) | 0.15 (0.006) | 1300–4300 | 150–500 |
| Cast irons | 20–60 | 60–200 | 0.025 (0.001) | 0.30 (0.012) | 4300–12,000 | 500–1500 |
| Thermoplastics | 30–60 | 100–200 | 0.025 (0.001) | 0.13 (0.005) | 6400–12,000 | 800–1500 |
| Thermosets | 20–60 | 60–200 | 0.025 (0.001) | 0.10 (0.004) | 4300–12,000 | 500–1500 |

*Note*: As hole depth increases, speeds and feeds should be reduced. Selection of speeds and feeds also depends on the specific surface finish required.

The thrust force in drilling is the force that acts in the direction of the hole axis. If this force is excessive, it can cause the drill to break or bend. The thrust force depends on factors such as the strength of the workpiece material, feed, rotational speed, cutting fluids, drill diameter, and drill geometry. Although some attempts have been made, accurate calculation of the thrust force on the drill has proven to be difficult. Experimental data are available as an aid in the design and use of drills and drilling equipment. Thrust forces in drilling range from a few newtons for small drills to as high as 100 kN (22.5 klb) in drilling high-strength materials with large drills. Similarly, drill torque can range up to as high as 4000 N-m (3000 ft-lb).

The torque during drilling also is difficult to calculate; it can be obtained by using the data in Table 8.3 and by noting that the power dissipated during drilling is the product of torque and rotational speed. Therefore, by first calculating the MRR, we can calculate the torque on the drill. **Drill life**, as well as tap life, is usually measured by the number of holes drilled before the drill becomes dull and forces increase.

*Drilling machines* are used for drilling holes, tapping, reaming, and other general-purpose, small-diameter boring operations; they are generally vertical, the most common type being a **drill press**. The workpiece is placed on an adjustable table, either by clamping it directly into the slots and holes on the table or by using a vise that can be clamped to the table. The drill is then lowered manually by hand wheel or by power feed at preset rates. In order to maintain proper cutting speeds, the spindle speed on drilling machines has to be adjustable to accommodate different sizes of drills. Drill presses are usually designated by the largest workpiece diameter that can be accommodated on the table; sizes typically range from 150 to 1250 mm (6 to 50 in.).

**Reaming and reamers.** *Reaming* is an operation that makes an existing hole dimensionally more accurate than can be achieved by drilling alone; it also improves the hole's surface finish. The most accurate holes are produced by the following sequence of operations: centering, drilling, boring, and reaming. For even better accuracy and surface finish, holes may be internally ground and honed (Chapter 9). A *reamer* (Fig. 8.50) is a multiple-cutting-edge tool with straight or helically fluted edges that removes very little material. The shanks may be straight or tapered, as in drills. The basic types of reamers are *hand* and *machine* (*chucking*) reamers; other types include *rose* reamers with cutting edges that have wide margins and no relief and *fluted, shell, expansion*, and *adjustable* reamers.

**FIGURE 8.50**
Terminology for a helical
reamer.

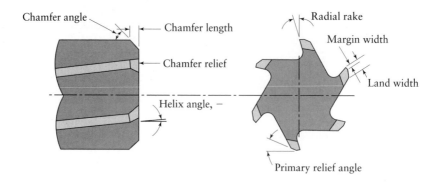

**FIGURE 8.51**
Terminology for a tap.

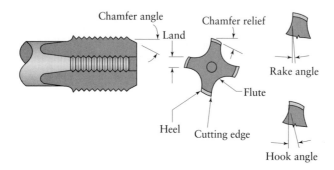

**Tapping and taps.** Internal threads in workpieces can be produced by tapping. A **tap** is basically a threading tool with multiple cutting teeth (Fig. 8.51). Taps are generally available with three or four flutes; three-fluted taps are stronger, because of the larger amount of material available in the flute. *Tapered* taps are designed to reduce the torque required for tapping through holes, and *bottoming* taps are designed for tapping blind holes to their full depth. *Collapsible* taps are used for large-diameter holes; after tapping is completed, the tap is mechanically collapsed and removed from the hole, without having to be rotated. Tap sizes range up to 100 mm (4 in.).

## 8.9 | Cutting Processes and Machine Tools for Producing Various Shapes

Several cutting processes and machine tools are capable of producing complex shapes with the use of multitooth, as well as single-point, cutting tools (Fig. 8.52 and Table 8.7). We begin with one of the most versatile processes, called *milling*, in which a multitooth cutter rotates along various axes with respect to the workpiece. The planing, shaping, and broaching are then described, wherein flat and shaped surfaces are produced, and the tool or workpiece travels along a straight path. This section ends with discussion of sawing and filing processes.

### 8.9.1 Milling operations

*Milling* includes a number of versatile machining operations that use a **milling cutter,** a multitooth tool that produces a number of chips per revolution, to produce a variety of configurations. Parts such as the ones shown in Fig. 8.52 can be machined efficiently and repeatedly using various milling cutters.

The main types of milling operations are described as follows:

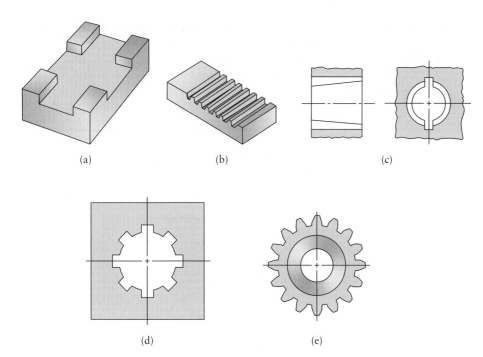

**FIGURE 8.52**  Typical parts and shapes produced by the cutting processes described in Section 8.9.

a)  **Slab milling.**  In *slab milling*, also called **peripheral milling**, the axis of cutter rotation is parallel to the surface of the workpiece to be machined (Fig. 8.53a). The cutter, generally made of high-speed steel, has a number of teeth along its circumference, each tooth acting like a single-point cutting tool called a *plain mill*. Cutters used in slab milling may have *straight* or *helical* teeth, producing orthogonal or oblique cutting action. Figure 8.1c shows the helical teeth on a cutter. In **conventional milling**, also called *up milling*, the maximum chip thickness is at the end of the cut (Fig. 8.53b). The advantages of conventional milling are that tooth engagement is not a function of workpiece surface characteristics, and contamination or scale on the surface does not affect tool life; this process is the dominant method of milling. The cutting process is smooth, provided that the cutter teeth are sharp. However, there is a tendency for the tool to chatter, and the workpiece has a tendency to be pulled upward; thus, proper clamping is important.

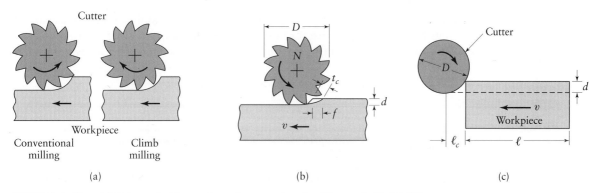

**FIGURE  8.53**  (a) Schematic illustration of conventional milling and climb milling. (b) Slab-milling operation, showing depth of cut, $d$; feed per tooth, $f$; chip depth of cut, $t_c$; and workpiece speed, $v$. (c) Schematic illustration of cutter travel distance $\ell_c$ to reach full depth of cut.

In **climb milling**, also called *down milling*, cutting starts with the thickest location of the chip. The advantage of climb milling is that the downward component of cutting forces holds the workpiece in place, particularly for slender parts. However, because of the resulting high-impact forces when the teeth engage the workpiece, this operation must have a rigid setup, and backlash must be eliminated in the table feed mechanism. Climb milling is not suitable for machining workpieces that have surface scale, such as hot-worked metals, forgings, and castings; the scale is hard and abrasive and causes excessive wear and damage to the cutter teeth, thus reducing tool life. Climb milling is recommended, in general, for maximum cutter life when CNC machine tools are used; a typical application is in finishing cuts on aluminum.

The cutting speed, $V$, in milling is the peripheral speed of the cutter, or

$$V = \pi D N, \tag{8.41}$$

where $D$ is the cutter diameter and $N$ is the rotational speed of the cutter (Fig. 8.53b). Note that the thickness of the chip in slab milling varies along its length, because of the relative longitudinal motion between cutter and workpiece. For a straight-tooth cutter, the approximate *undeformed chip thickness*, $t_c$ (*chip depth of cut*) can be determined from the equation

$$t_c = 2f\sqrt{\frac{d}{D}}, \tag{8.42}$$

where $f$ is the feed per tooth of the cutter, measured along the workpiece surface (that is, the distance the workpiece travels per tooth of the cutter, in mm/tooth or in./tooth), and $d$ is the depth of cut. As $t_c$ becomes greater, the force on the cutter tooth increases.

Feed per tooth is determined from the equation

$$f = \frac{v}{Nn}, \tag{8.43}$$

where $v$ is the linear speed (feed rate) of the workpiece and $n$ is the number of teeth on the cutter periphery. We can check the dimensional accuracy of this equation by substituting appropriate units for the individual terms; thus, (mm/tooth) = (m/min) ($10^3$ mm/m)/(rev/min)(number of teeth/rev), which checks out correctly. The cutting time, $t$, is given by the expression

$$t = \frac{(\ell + \ell_c)}{v}, \tag{8.44}$$

where $\ell$ is the length of the workpiece (Fig. 8.53c) and $\ell_c$ is the extent of the cutter's first contact with the workpiece. Based on the assumption that $\ell_c \ll \ell$ (although this relationship is not generally so), the *material removal rate* is

$$\text{MRR} = \frac{\ell w d}{t} = wdv, \tag{8.45}$$

where $w$ is the width of the cut, which, for a workpiece narrower than the length of the cutter, is the same as the width of the workpiece. The distance that the cutter travels in the noncutting-cycle operation is an important economic consideration and should be minimized.

Although we can calculate the *power requirement* in slab milling, the *forces* acting on the cutter (tangential, radial, and axial; see also Fig. 8.42b) are difficult to calculate, because of the many variables involved, particularly tool geometry. These forces can be measured experimentally for a variety of conditions. The *torque* on the cutter spindle can be calculated from the power.

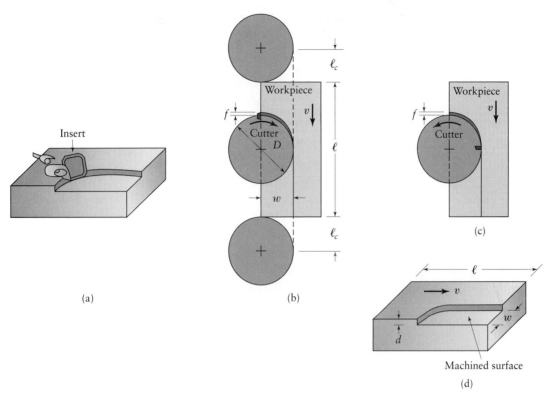

**FIGURE 8.54**    Face-milling operation showing (a) action of an insert in face milling; (b) climb milling; (c) conventional milling; (d) dimensions in face milling. The width of cut, *w*, is not necessarily the same as the cutter radius. *Source*: Courtesy of The Ingersoll Cutting Tool Company.

Although the torque is the product of the tangential force on the cutter and the cutter's radius, the tangential force per tooth depends on how many teeth are engaged during the cut.

b) **Face milling.** In *face milling*, the cutter is mounted on a spindle that has an axis of rotation perpendicular to the workpiece surface, and removes material in the manner shown in Fig. 8.54a. The cutter rotates at a rotational speed $N$, and the workpiece moves along a straight path at a linear speed, $v$. When the cutter rotates in the direction shown in Fig. 8.54b, the operation is *climb* milling; when it rotates in the opposite direction (Fig. 8.54c), it is *conventional* milling.

Because of the relative motion between the cutting teeth and the workpiece, a face-milling cutter leaves *feed marks* on the machined surface, much as in turning operations. Surface roughness depends on insert corner geometry and feed per tooth. [see also Eqs. (8.35) to (8.37).]

Figure 8.55 provides the terminology for a face-milling cutter and the various angles. The side view is shown in Fig. 8.56, where we note that, as in turning operations, the *lead angle* of the insert in face milling has a direct influence on the *undeformed chip thickness*. As the lead angle (positive as shown) increases, the undeformed chip thickness (thus, also the thickness of the actual chip) decreases, and the length of contact increases. The range of lead angles for most face-milling cutters is from 0° to 45°, thus influencing the axial and tangential forces. Note that the cross-sectional area of the undeformed chip remains constant. The lead angle also influences forces in milling. It can be seen

**FIGURE 8.55**
Terminology for a face-
milling cutter.

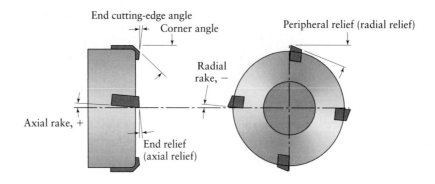

**FIGURE 8.56**   The effect
of lead angle on the
undeformed chip thickness in
face milling. Note that as the
lead angle increases, the
undeformed chip thickness
(and hence the thickness of
the actual chip) decreases,
but the length of contact
(and hence the width of the
chip) increases. The insert
must be sufficiently large to
accommodate the increase in
contact length.

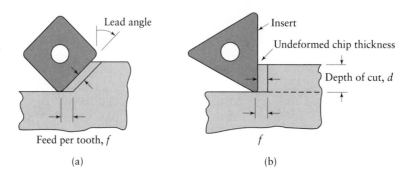

that as the lead angle decreases, there is less and less of a vertical force (axial force on the cutter spindle) component.

A wide variety of milling cutters is available. The cutter diameter should be chosen in such a manner that it will not interfere with fixtures and other components in the setup. In a typical face-milling operation, the ratio of the cutter diameter, $D$, to the width of cut, $w$, should be no less than $3:2$. The cutting tools are usually carbide or high-speed-steel inserts and are mounted on the cutter body. (See Fig. 8.55.)

The relationship of cutter diameter and insert angles and their position relative to the surface to be milled is important, because they determine the angle at which an insert enters and exits the workpiece. Note in Fig. 8.54b for climb milling that if the insert has zero axial and radial rake angles (see Fig. 8.55), then the rake face of the insert engages the workpiece directly. However, as can be seen in Fig. 8.57a and b, the same insert engages the workpiece at different angles, depending on the relative positions of the cutter and the workpiece. In Fig. 8.57a, the tip of the insert makes the first contact, and hence there is potential for the cutting edge to chip off. In Fig. 8.57b, on the other hand, the contacts (i.e., at entry, reentry, and the two exits) are at an angle and away from the tip of the insert. There is, therefore, less of a tendency for the insert to fail, because the force on the insert increases and decreases gradually. Note from Fig. 8.55 that the radial and axial rake angles will also have an effect on this situation.

Figure 8.57c shows the exit angles for various cutter positions. Note that in the first two examples, the insert exits the workpiece at an angle, whereby the force on the insert diminishes to zero at a slower rate (desirable) than in the third example, where the insert exits suddenly (undesirable).

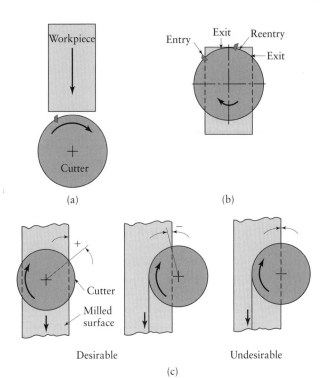

**FIGURE 8.57** (a) Relative position of the cutter and insert as it first engages the workpiece in face milling, (b) insert positions toward the end of cut, and (c) examples of exit angles of insert, showing desirable (positive or negative angle) and undesirable (zero angle) positions. In all figures, the cutter spindle is perpendicular to the page.

---

**EXAMPLE 8.7 Calculation of material removal rate, power required, and cutting time in face milling**

Refer to Fig. 8.54, and assume that $D = 150$ mm, $w = 60$ mm, $\ell = 500$ mm, $d = 3$ mm, $v = 0.6$ m/min, and $N = 100$ rpm. The cutter has 10 inserts, and the workpiece material is a high-strength aluminum alloy. Calculate the material removal rate, cutting time, and feed per tooth, and estimate the power required.

**Solution.** The cross-section of the cut is $wd = (60)(3) = 180$ mm$^2$. Since the workpiece speed $v$ is 0.6 m/min = 600 mm/min, we calculate the material removal rate as

$$\text{MRR} = (180)(600) = 108,000 \text{ mm}^3/\text{min}.$$

The cutting time is given by

$$t = \frac{(\ell + 2\ell_c)}{v}.$$

Note from Fig. 8.54 that for this problem, $\ell_c = D/2$, or 75 mm. Thus, the cutting time is

$$t = \frac{(500 + 150)(60)}{600} = 65 \text{ s} = 1.08 \text{ min}.$$

We obtain the feed per tooth from Eq. (8.43). Noting that $N = 100$ rpm = 1.67 rev/s, we find

$$f = \frac{10}{(1.67)(10)} = 0.6 \text{ mm/tooth}.$$

For this material, the unit power can be taken from Table 8.3 to be 1.1 W-s/mm$^3$; hence, the power can be estimated as

$$\text{Power} = (1.1)(1800) = 1980 \text{ W} = 1.98 \text{ kW}.$$

c) **End milling.** The cutter in *end milling* is shown in Fig. 8.1d; it has either straight or tapered shanks for smaller and larger cutter sizes, respectively. The cutter usually rotates on an axis perpendicular to the workpiece, although it can be tilted to machine tapered surfaces as well. Flat surfaces as well as various profiles can be produced by end milling. The end faces of some end mills have cutting teeth, and these teeth can be used as a drill to start a cavity; end mills are also available with hemispherical ends for producing curved surfaces, as in making dies. *Hollow end mills* have *internal* cutting teeth and are used for machining the cylindrical surface of solid round workpieces, as in preparing stock with accurate diameters for automatic machines. End mills are made of high-speed steels or have carbide inserts.

Section 8.8.2 has described high-speed machining and its applications. One of the more common applications is **high-speed milling,** using an end mill and with the same general provisions regarding the stiffness of machines, work-holding devices, etc., that have been described earlier. A typical application is milling aluminum-alloy aerospace components and honeycomb structures, with spindle speeds on the order of 20,000 rpm. Another application is in *die sinking*, i.e., producing cavities in die blocks. Chip collection and disposal can be a significant problem in these operations, because of the high rate of material removal.

d) **Other milling operations and cutters.** Several other types of milling operations and cutters are used to machine various surfaces. In *straddle milling*, two or more cutters are mounted on an arbor and are used to machine two *parallel* surfaces on the workpiece (Fig. 8.58a). *Form milling* is used to produce curved profiles, using cutters that have specially shaped teeth (Fig. 8.58b); such cutters are also used for cutting gear teeth (see Section 8.9.7).

*Circular cutters* are used for slotting and slitting. The teeth may be staggered slightly, as in a saw blade (see Section 8.9.5), to provide clearance for the cutter in making deep slots. *Slitting saws* are relatively thin, usually less than 5 mm $\left(\frac{3}{16} \text{ in.}\right)$. *T-slot cutters* are used to mill T-slots, such as those in machine-tool worktables for clamping of workpieces. The slot is first milled with an end mill; a T-slot cutter then cuts the complete profile of the slot in one pass. *Key-seat cutters* are used to make the semicylindrical (Woodruff) key sets for shafts. *Angle-milling cutters*, either with a single angle or double angles, are used to produce tapered surfaces with various angles.

*Shell mills* are hollow inside and are mounted on a shank, thus allowing the same shank to be used for different-sized cutters. The use of shell mills is

**FIGURE 8.58** Cutters for (a) straddle milling and (b) form milling.

(a) Straddle milling          (b) Form milling

similar to that of end mills. Milling with a single cutting tooth mounted on a high-speed spindle is known as *fly cutting* and is generally used in simple face-milling and boring operations. The tool can be shaped as a single-point tool and can be placed in various radial positions on the spindle.

e) **Toolholders.** Milling cutters are classified into two basic types. *Arbor cutters* are mounted on an arbor, such as for slab, face, straddle, and form milling; in *shank-type cutters*, the cutter and the shank are one piece. The most common examples of shank cutters are end mills. Whereas small end mills have straight shanks, larger end mills have tapered shanks for better clamping to resist the higher forces and torque involved. Cutters with straight shanks are mounted in collet chucks or special end-mill holders; cutters with tapered shanks are mounted in tapered toolholders. In addition to mechanical means, hydraulic tool holders and arbors are also available. The stiffness of cutters and toolholders is important for surface quality and reducing vibration and chatter in milling operations. Conventional tapered toolholders have a tendency to wear and bellmouth under the radial forces in milling.

f) **Milling machines and process capabilities.** In addition to the various characteristics of milling processes described thus far, milling capabilities include parameters such as production rate, surface finish, dimensional tolerances, and cost considerations. Table 8.11 gives the typical conventional ranges of feeds and cutting speeds for milling. Cutting speeds vary over a wide range, from 30 to 3000 m/min (90 to 10,000 ft/min), depending on workpiece material, cutting-tool material, and process parameters. Because of their capability to perform a variety of operations, milling machines, which were first built in 1876, are among the most versatile and useful machine tools, and a wide selection of machines with numerous features is available. Used for general-purpose milling operations, **column-and-knee-type** machines are the most common milling machines. The spindle, to which the milling cutter is attached, may be *horizontal* (Fig. 8.59a) for slab milling or *vertical* for face and end milling, boring, and drilling operations (Fig. 8.59b). The various components are moved manually or by power.

In **bed-type** machines, the worktable is mounted directly on the bed, which replaces the knee, and can move only longitudinally. These milling

## TABLE 8.11

**Approximate Range of Recommended Cutting Speeds for Milling Operations**

| WORKPIECE MATERIAL | CUTTING SPEED | |
| --- | --- | --- |
| | m/min | ft/min |
| Aluminum alloys | 300–3000 | 1000–10,000 |
| Cast iron, gray | 90–1300 | 300–4200 |
| Copper alloys | 90–1000 | 300–3300 |
| High-temperature alloys | 30–550 | 100–1800 |
| Steels | 60–450 | 200–1500 |
| Stainless steels | 90–500 | 300–1600 |
| Thermoplastics and thermosets | 90–1400 | 300–4500 |
| Titanium alloys | 40–150 | 130–500 |

*Note*: (a) These speeds are for carbides, ceramic, cermets, and diamond cutting tools. Speeds for high-speed-steel tools are lower than those indicated in this table.
(b) Depths of cut, *d*, are generally in the range of 1–8 mm (0.04–0.3 in.).
(c) Feeds per tooth, *f*, are generally in the range of 0.08–0.46 mm/rev (0.003–0.018 in./rev).

**FIGURE 8.59**  (a) Schematic illustration of a horizontal-spindle column-and-knee-type milling machine. (b) Schematic illustration of a vertical-spindle column-and-knee-type milling machine. *Source*: G. Boothroyd, *Fundamentals of Machining and Machine Tools.*

machines are not as versatile as others, but have high stiffness and are used for high-production-rate work. The spindles may be horizontal or vertical and of duplex or triplex types, that is, with two or three spindles for simultaneously milling two or three workpiece surfaces, respectively. **Planer-type** machines, which are similar to bed-type machines, are equipped with several heads and cutters to mill various surfaces. They are used for heavy workpieces and are more efficient than planers (see Section 8.9.2) when used for similar operations. *Rotary-table machines* are similar to vertical milling machines and are equipped with one or more heads for face milling. *Profile-milling machines* have five-axis movements. Using tracer fingers, *duplicating machines* (copy-milling machines) reproduce parts from a master model; they are used in the automotive and aerospace industries for machining complex parts and dies (die sinking), although they are being replaced rapidly with CNC machines. These machine tools are versatile and capable of milling, drilling, boring, and tapping with repetitive accuracy.

### 8.9.2 Planing and planers

*Planing* is a relatively simple cutting process by which flat surfaces, as well as various cross-sections with grooves and notches, are produced along the length of the workpiece. Planing is usually done on large workpieces, as large as 25 m × 15 m (75 ft × 45 ft). In a *planer*, the workpiece is mounted on a table that travels along a straight path. A horizontal cross rail, which can be moved vertically along the ways in the column, is equipped with one or more tool heads. The cutting tools are attached to the heads, and machining is done along a straight path. Because of the reciprocating motion of the workpiece, elapsed noncutting time during the return stroke is significant, both in planing and in shaping (see Section 8.9.3). Consequently, these operations are not efficient or economical, except for in low-quantity production. Efficiency of the operation can be improved by equipping planers with toolholders and tools that cut in both directions of table travel.

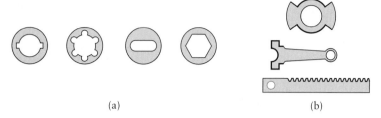

**FIGURE 8.60**  (a) Typical parts made by internal broaching. (b) Parts made by surface broaching. The heavy lines indicate broached surfaces. *Source*: General Broach and Engineering Company.

(a)                          (b)

### 8.9.3  Shaping and shapers

Cutting by *shaping* is basically the same as in planing. In a *horizontal shaper*, the tool travels along a straight path, and the workpiece is stationary; the cutting tool is attached to the tool head, which is mounted on the ram. The ram has a reciprocating motion, and in most machines, cutting is done during the forward movement of the ram (*push cut*); in others, however, it is done during the return stroke of the ram (*draw cut*). *Vertical shapers* (*slotters*) are used for machining notches and keyways. Shapers are also capable of producing complex shapes, such as cutting a helical impeller, in which the workpiece is rotated during the cut through a master cam. Because of their low production rate, shapers are now generally used only in toolrooms, job shops, and for repair work.

### 8.9.4  Broaching and broaching machines

The *broaching* operation is similar to shaping and is used to machine internal and external surfaces (Fig. 8.60), such as holes of circular, square, or irregular section; keyways; teeth of internal gears; multiple spline holes; and flat surfaces. A *broach* (Fig. 8.61) is, in effect, a long multitooth cutting tool that makes successively deeper cuts; hence, the total depth of material removed in one stroke is the sum of the depths of cut of each tooth. A broach can remove material as deep as 6 mm (0.25 in.) in one stroke. Broaching can produce parts with good surface finish and dimensional accuracy; thus, it competes favorably with other processes to produce similar shapes. Although broaches can be expensive, the cost is justified, because of their use for high-quantity production runs.

Figure 8.61b provides the terminology for a broach. The rake (hook) angle depends on the material being cut, as in turning and other cutting operations, and usually ranges between 0° and 20°. The clearance angle is usually 1° to 4°; finishing teeth have smaller angles. Too small a clearance angle causes rubbing of the teeth

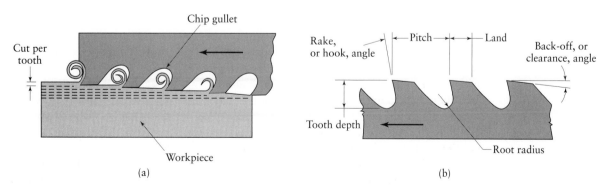

**FIGURE 8.61**  (a) Cutting action of a broach, showing various features. (b) Terminology for a broach.

**FIGURE 8.62**
Terminology for a pull-type internal broach used for enlarging long holes.

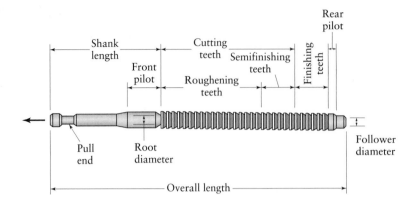

against the broached surface. The pitch of the teeth depends on factors such as length of the workpiece (length of cut), tooth strength, and size and shape of chips. The tooth depth and pitch must be sufficiently large to accommodate the chips produced during broaching, particularly for long workpieces, but at least two teeth should be in contact with the workpiece at all times (similar to sawing).

Broaches are available with various tooth profiles, including some that have chip breakers. Broaches are also made round with circular cutting teeth and are used to enlarge holes (Fig. 8.62). Note that the cutting teeth on broaches have three regions: roughing, semifinishing, and finishing. Irregular internal shapes are usually broached by starting with a round hole in the workpiece, produced by drilling or boring.

In **turn broaching** of crankshafts, the workpiece rotates between centers, and the broach is equipped with multiple inserts and passes tangentially across the part; the process is thus a combination of broaching and skiving. There are also machines that broach a number of crankshafts simultaneously; main bearings for engines are broached in this way.

*Broaching machines* either pull or push the broaches and are made to be horizontal or vertical. *Push broaches* are usually shorter, generally in the range of 150–350 mm (6–14 in.) long. *Pull broaches* tend to straighten the hole, whereas pushing permits the broach to follow any irregularity of the leader hole. Horizontal machines are capable of longer strokes than are vertical machines. Many types of broaching machines are available, some with multiple heads, allowing a variety of shapes and parts to be produced, including helical splines and rifled gun barrels. Sizes range from machines for making needlelike parts to those used for broaching gun barrels. The capacities of broaching machines are as high as 0.9 MN (100 tons) of pulling force.

### 8.9.5 Sawing and saws

*Sawing* is a cutting operation in which the cutting tool is a blade that has a series of small teeth, with each tooth removing a small amount of material. Saws are used for all metallic and nonmetallic materials that are machinable by other cutting processes, and they are capable of producing various shapes. The width of cut (**kerf**) in sawing is usually narrow, and hence sawing wastes little material. Figure 8.63 illustrates typical saw-tooth and saw-blade configurations. Tooth spacing is usually in the range of 0.08 to 1.25 teeth per mm (2 to 32 teeth per in.).

Saw blades are made from carbon and high-speed steels; steel blades with tips of carbide steel or high-speed steel are used for sawing harder materials (Fig. 8.64). In order to prevent the saw from binding and rubbing during cutting, the teeth are alternately set in opposite directions so that the kerf is wider than the blade

**FIGURE 8.63**
(a) Terminology for saw teeth. (b) Types of saw teeth, staggered to provide clearance for the saw blade to prevent binding during sawing.

**FIGURE 8.64**   (a) High-speed-steel teeth welded on a steel blade. (b) Carbide inserts brazed to blade teeth.

(Fig. 8.63b). At least two or three teeth should always be engaged with the workpiece in order to prevent snagging (catching of the saw tooth on the workpiece); this requirement is the reason that sawing thin materials satisfactorily is difficult. Cutting speed in sawing usually ranges up to 1.5 m/s (300 ft/min), with lower speeds for high-strength metals; cutting fluids are generally used to improve the quality of cut and the life of the saw.

*Hacksaws* have straight blades and reciprocating motions; they may be manually or power operated. *Circular saws* (*cold saws*) are generally used for high-production-rate sawing of large cross-sections. *Band saws* have long, flexible, continuous blades and provide continuous cutting action. Blades and high-strength wire can be coated with diamond powder (*diamond-edged blades* and *diamond saws*). They are suitable for sawing hard metallic, nonmetallic, and composite materials.

**Friction sawing.** *Friction sawing* is a process in which a mild-steel blade or disk rubs against the work-piece at speeds of up to 125 m/s (25,000 ft/min); the frictional energy is converted into heat, which rapidly softens a narrow zone in the workpiece. The action of the blade or disk, which is sometimes provided with teeth or notches, pulls and ejects the softened metal from the cutting zone. The heat generated in the workpiece produces a *heat-affected zone* (see also Section 12.5) on the cut surfaces; thus, their properties can be adversely affected. Because only a small portion of the blade is engaged with the workpiece at any time, the blade cools rapidly as it passes through the air. Friction sawing is suitable for hard ferrous metals and fiber-reinforced plastics, but not for nonferrous metals, because they have a tendency to stick to the blade. Disks for friction sawing as large as 1.8 m (6 ft) in diameter are used to cut off large steel sections in rolling mills. Friction sawing can also be used to remove flash from castings.

### 8.9.6  Filing

*Filing* is small-scale removal of material from a surface, corner, edge, or hole. First developed in about 1000 BC, files are usually made of hardened steels and are available in a variety of cross-sections, including flat, round, half round, square, and triangular. Files have many tooth forms and grades of coarseness, such as smooth cut, second cut, and bastard cut. Although filing is usually done by hand, various machines with automatic features are available for high production rates, with files reciprocating at up to 500 *strokes/min*. *Band files* consist of file segments, each about 75 mm (3 in.) long, that are riveted to flexible steel bands and used in a manner similar to band saws. *Disk-type files* are also available. *Rotary files* and **burs** are used for special applications; they are usually conical, cylindrical, or spherical in shape and have various tooth profiles. The rotational speeds range from 1500 rpm, for cutting steel with large burs, to as high as 45,000 rpm, for cutting magnesium using small burs.

### 8.9.7  Gear manufacturing by cutting

Gears can be manufactured by casting, forging, extrusion, drawing, thread rolling, powder metallurgy, and blanking (for making thin gears such as those used in watches and small clocks); however, most gears are machined for better dimensional control and surface finish. Nonmetallic gears are usually made by injection molding and casting.

In **form cutting**, the cutting tool is similar to a form-milling cutter (see Fig. 8.58b) in the shape of the space between the gear teeth. The cutter travels axially along the length of the gear tooth at the appropriate depth to produce the gear tooth. After each tooth is cut, the cutter is withdrawn, the gear blank is rotated (indexed), and the cutter proceeds to cut another tooth. The process continues until all teeth are cut. *Broaching* can also be used to produce gear teeth and is particularly applicable to internal teeth. The process is rapid and produces fine surface finish with high dimensional accuracy. However, because broaches are expensive and a separate broach is required for each size of gear, this method is suitable mainly for high-quantity production.

In **gear generating**, the tool may be (a) a *pinion-shaped cutter*, (b) a *rack-shaped straight cutter*, or (c) a *hob*. The pinion-shaped cutter can be considered as one of the gears in a conjugate pair, where the other is the gear blank (Fig. 8.65a). This type of cutter is used in gear generating on machines called **gear shapers** (Fig. 8.65b). The cutter has an axis parallel to that of the gear blank and rotates slowly with the blank at the same pitch-circle velocity with an axial reciprocating motion. A train of gears provides the required relative motion between the cutter shaft and the gear-blank shaft. Cutting may take place at either the downstroke or the upstroke of the machine. Because the clearance required for cutter travel is small, gear shaping is suitable for gears that are located close to obstructing surfaces such as flanges (as in the gear blank in Fig. 8.65b). The process can be used for low-quantity as well as high-quantity production.

On a **rack shaper**, the generating tool is a segment of a rack (Fig. 8.65c), which reciprocates parallel to the axis of the gear blank. Because it is not practical to have more than 6 to 12 teeth on a rack cutter, the cutter must be disengaged at suitable intervals and returned to the starting point; the gear blank, meanwhile, remains fixed. A gear-cutting **hob** is basically a worm, or screw, that has been made into a gear-generating tool by machining a series of longitudinal slots, or gashes, into it to produce the cutting teeth (Fig. 8.65d). When hobbing a spur gear, the angle between the hob and gear-blank axes is 90° minus the lead angle at the hob threads. All motions in hobbing are rotary, and the hob and gear blank rotate continuously as in two gears meshing until all teeth are cut.

**FIGURE 8.65** (a) Schematic illustration of gear generating with a pinion-shaped gear cutter. (b) Schematic illustration of gear generating in a gear shaper, using a pinion-shaped cutter. Note that the cutter reciprocates vertically. (c) Gear generating with a rack-shaped cutter. (d) Schematic illustration of three views of gear cutting with a hob. *Source*: After E. P. DeGarmo and Society of Manufacturing Engineers.

Through gear production by any of the processes previously described, the surface finish and dimensional accuracy of gear teeth may not be sufficiently accurate for certain specific applications; thus, several *finishing* operations are also performed, including shaving, burnishing, grinding, honing, and lapping. (See Chapter 9.) Modern gear-manufacturing machines are computer controlled, and multiaxis computer-controlled machines have capabilities of machining many types of gears, using indexable cutters.

# 8.10 | Machining and Turning Centers

In the descriptions of individual machining processes and machine tools in the preceding chapters, we noted that each machine, regardless of how highly it is automated, is designed to perform basically one type of operation. Previous chapters have also shown that in manufacturing operations, most parts generally require a number of different machining operations on their various surfaces. Many parts have a variety of features and surfaces that require different types of machining operations, such as milling, facing, boring, drilling, reaming, and threading, to certain specified dimensional tolerances and surface finishes.

None of the processes and machine tools described thus far could individually produce these parts. Traditionally, these operations have been performed by moving the part from one machine tool to another until all machining is completed. This process is a viable manufacturing method that can be highly automated and is the principle behind **transfer lines**, consisting of numerous machine tools arranged in a sequence (described in Section 14.2.4). The workpiece, such as an engine block, moves from station to station, with a particular machining operation performed at each station; the part is then transferred automatically to the next machine for another operation, and so on. Transfer lines are commonly used in high-volume or mass production.

There are situations and products, however, for which such lines are not feasible or economical, particularly when the types of products to be machined change rapidly. An important concept, developed in the late 1950s, is machining centers. A **machining center** is a computer-controlled machine tool capable of performing a variety of cutting operations on different surfaces and in different orientations on a workpiece (Fig. 8.66). In general, the workpiece is stationary, and the cutting tools rotate, such as in milling and drilling. The development of machining centers is intimately related to advances in **computer control** of machine tools, details of which are described in Chapter 14. Recall that, as an example of advances in modern lathes, Fig. 8.45 illustrated a numerically controlled lathe (**turning center**), with two turrets that carry several cutting tools for operations such as turning, facing, boring, and threading. Because of the high productivity of machining centers, large amounts of chips are produced and must be collected and disposed of properly.

**FIGURE 8.66** A horizontal-spindle machining center, equipped with an automatic tool changer. Tool magazines can store 200 cutting tools. *Source*: Courtesy of Cincinnati Machine, a UNOVA Company.

The workpiece in a machining center is placed on a **pallet** that can be oriented in three principal directions, as well as rotated around one or more axes on the pallet. Thus, after a particular cutting operation is completed, the workpiece does not have to be moved to another machine for subsequent operations, say, drilling, reaming, and tapping. In other words, the tools and the machine are brought to the workpiece. After all the cutting operations are completed, the pallet automatically moves away, carrying the finished workpiece, and another pallet containing another workpiece to be machined is brought into position by **automatic pallet changers**. All movements are guided by computer control, and pallet-changing cycle times are on the order of 10–30 s. Pallet stations are available with multiple pallets that serve the machining center. These machines can also be equipped with various automatic parts, such as loading and unloading devices.

Machining centers are equipped with a **programmable automatic tool changer**. Depending on the design, as many as 200 cutting tools can be stored in a magazine, drum, or chain (*tool storage*); auxiliary tool storage is available on some special machining centers for complex operations. The cutting tools are automatically selected, with random access for the shortest route to the machine spindle. A common design is a **tool-exchange arm** that swings around to pick up a particular tool (each tool has its own toolholder) and places it in the spindle. Tools are identified by coded tags, bar codes, or memory chips applied directly to the toolholders. Tool changing times are typically on the order of a few seconds.

Machining centers may be equipped with a **tool-checking** and/or **part-checking station** that feeds information to the computer numerical control to compensate for any variations in tool settings or tool wear. **Touch probes** are available that can be automatically loaded into a toolholder for determining reference surfaces of the workpiece, for selection of tool setting, and for on-line inspection of parts being machined. Several surfaces can be contacted, and their relative positions are determined and stored in the database of the computer software. The data are then used for programming tool paths and compensating for tool length and diameter, as well as compensating for tool wear in more advanced systems.

## 8.10.1  Types of machining and turning centers

Although there are several designs for machining centers, the two basic types are vertical spindle and horizontal spindle; many machines have the capability of using both axes. The maximum dimensions that the cutting tools can reach around a workpiece in a machining center are known as the *work envelope*; this term was first used in connection with industrial robots. (See Section 14.7.2.)

1. **Vertical-spindle machining centers**, or *vertical machining centers*, are suitable for performing various machining operations on flat surfaces with deep cavities, such as in mold and die making. Because the thrust forces in vertical machining are directed downward, these machines have high stiffness and produce parts with good dimensional accuracy. Also, they are generally less expensive than horizontal-spindle machines.

2. **Horizontal-spindle machining centers**, or *horizontal machining centers*, (see Fig. 8.66) are suitable for large or tall workpieces that require machining on a number of their surfaces. The pallet can be rotated on different axes and to various angular positions. Another category of horizontal-spindle machines is **turning centers**, which are computer-controlled *lathes* with several features. Figure 8.67 shows a three-turret computer-numerical-control turning center. This machine is designed with two horizontal spindles and three turrets, which are equipped with a variety of cutting tools to perform several operations on a rotating workpiece.

**FIGURE 8.67** Schematic illustration of a three-turret, two-spindle computer-numerical-control turning center. *Source*: Hitachi Seiki USA, Inc.

1st Turret head

2nd Turret head

#1 Spindle head

#3 Turret head

#2 Spindle head

3. **Universal machining centers** are equipped with both vertical and horizontal spindles. They have a variety of features and are capable of machining all surfaces of a workpiece (vertical, horizontal, and diagonal), hence the term *universal*.

Frequent product changes are a major consideration in computer-integrated manufacturing. Consequently, a more recent approach is the design of **reconfigurable machines**, which have flexibility. This flexibility is achieved by using **modular components** that can be arranged and rearranged in different ways. Those machines use advanced computer hardware and software and communications systems. (See also Chapters 14 and 15.)

## 8.10.2 Characteristics and capabilities of machining centers

The major characteristics of machining centers are described as follows:

- They are capable of handling a variety of part sizes and shapes efficiently, economically, and with repetitively high dimensional tolerances, on the order of ±0.0025 mm (0.0001 in.).

- The machines are versatile, having as many as six axes of linear and angular movements and the capability of quick changeover from one type of product to another. Thus, the need for a variety of machine tools and a large amount of floor space is significantly reduced.

- The time required for loading and unloading workpieces, changing tools, gaging, and troubleshooting is reduced, thus improving productivity, reducing labor requirements (particularly for skilled labor), and minimizing overall costs.

- Machining centers are highly automated and relatively compact, so that one operator can attend to two or more machines at the same time.

- The machines are equipped with tool-condition monitoring devices for detecting tool breakage and wear, as well as probes for compensation for tool wear and for tool positioning.

- In-process and postprocess gaging and inspection of machined workpieces are now standard features of machining centers.

Machining centers are available in a wide variety of sizes and with a large selection of features, and their cost ranges from about $50,000 to $1 million and higher. Typical capacities range up to 75 kW (100 hp), and maximum spindle speeds are usually in the range of 4000–8000 rpm, but some are as high as 75,000 rpm for special applications and with small-diameter cutters. Some pallets are capable of supporting workpieces that weigh as much as 7,000 kg (15,000 lb), although higher capacities are available for special applications.

**Machine-tool selection.** Machining centers can require significant capital expenditures; therefore, to be cost effective, they have to be in operation for at least two shifts per day. Consequently, there must be sufficient and continued demand for products made on machining centers to justify this cost. However, because of their inherent versatility, machining centers can be used to produce a wide range of products, particularly in just-in-time manufacturing (see Section 15.11).

The selection of a particular type and size of machining center depends on several factors, among which are

- Type of products, their size, and their shape complexity;
- Type of machining operations to be performed and the type and number of tools required;
- Dimensional accuracy of the products;
- Production rate.

Although versatility is the key factor in selecting machining centers, these considerations must be weighed against the high capital investment required and compared with to those of manufacturing the same products by using a number of the more traditional machine tools.

**EXAMPLE 8.8   Machining outer bearing races on a turning center**

Outer bearing races (see Fig. 8.68) are machined on a turning center. The starting material is hot-rolled 52100 steel tube with an outside diameter of

1. Finish turning of outside diameter
2. Boring and grooving on outside diameter
3. Internal grooving with a radius-form tool
4. Finish boring of internal groove and rough boring of internal diameter
5. Internal grooving with form tool and chamfering
6. Cutting off finished part; inclined bar picks up bearing race

**FIGURE 8.68** Machining outer bearing races on a turning center.

91 mm (3.592 in.) and an inside diameter of 75.5 mm (2.976 in.). The cutting speed is 95 m/min (313 ft/min) for all operations. All tools are carbide, including the cutoff tool (the last operation), which is 3.18 mm $\left(\frac{1}{8}\text{ in.}\right)$ wide instead of 4.76 mm $\left(\frac{3}{16}\text{ in.}\right)$, for the high-speed-steel cutoff tool that was formerly used. The material saved by this change is significant, because the width of the race is small. The turning center was able to machine these races at high speeds and with repeatable dimensional tolerances of $\pm0.025$ mm (0.001 in.).

# 8.11 | Vibration and Chatter

In describing cutting processes and machine tools, we pointed out that machine stiffness is important in controlling dimensional accuracy and surface finish of parts. In this section, we describe the adverse effects of low stiffness on product quality and machining operations. Low stiffness affects the level of vibration and chatter in tools and machines. If uncontrolled, vibration and chatter can result in the following effects:

- Poor surface finish (right central region in Fig. 8.69);
- Loss of dimensional accuracy of the workpiece;
- Premature wear, chipping, and failure of the cutting tool, which is crucial with brittle tool materials, such as ceramics, some carbides, and diamond;
- Damage to machine-tool components from excessive vibrations;
- Objectionable noise generated, particularly if it is of high frequency, such as the squeal heard in turning brass on a lathe.

Extensive studies have shown that vibration and chatter in cutting are complex phenomena. Cutting operations cause two basic types of vibration: forced vibration and self-excited vibration. **Forced vibration** is generally caused by some periodic force present in the machine tool, such as from gear drives, imbalance of the machine-tool components, misalignment, or motors and pumps. In processes such as milling or turning a splined shaft or a shaft with a keyway, forced vibrations are caused by the periodic engagement of the cutting tool with the workpiece surface, including its exit from the surface.

**FIGURE 8.69** Chatter marks (right of center of photograph) on the surface of a turned part. *Source*: General Electric Company.

The basic solution to forced vibrations is to isolate or remove the forcing element. If the forcing frequency is at or near the natural frequency of a component of the machine-tool system, one of the frequencies may be raised or lowered. The amplitude of vibration can be reduced by increasing the stiffness or damping the system. Although changing the process parameters generally does not appear to greatly influence forced vibrations, changing the cutting speed and the tool geometry can be helpful.

Generally called **chatter, self-excited vibration** is caused by the interaction of the chip-removal process and the structure of the machine tool; these vibrations usually have very high amplitude. Chatter typically begins with a disturbance in the cutting zone. Such disturbances include lack of homogeneity in the workpiece material or its surface condition, changes in chip morphology, or a change in frictional conditions at the tool–chip interface, as also influenced by cutting fluids and their effectiveness. Self-excited vibrations can generally be controlled by increasing the dynamic stiffness of the system and by damping. **Dynamic stiffness** is the ratio of the amplitude of the force applied to the amplitude of vibration. Because a machine tool has different stiffnesses at different frequencies, changes in cutting parameters, such as cutting speed, can influence chatter.

The most important type of self-excited vibration in machining is **regenerative chatter,** which is caused when a tool cuts a surface that has a roughness or disturbances left from the previous cut. Because the depth of cut thus varies, the resulting variations in the cutting force subject the tool to vibrations; the process continues repeatedly, hence the term *regenerative*. You may observe this type of vibration while driving over a rough road (the so-called *washboard effect*).

## 8.11.1  Damping

The term *damping* is defined as the rate at which vibrations decay. Damping is an important factor in controlling machine-tool vibration and chatter.

**Internal damping of structural materials.**  Damping results from the energy loss within materials during vibration. Thus, for example, steel has less damping than gray cast iron, and composite materials have more damping than gray iron (Fig. 8.70). You can observe this difference in the damping capacity of materials by striking them with a gavel and listening to the sound; try, for example, striking pieces of steel, concrete, and wood.

**Joints in the machine-tool structure.**  Although less significant than internal damping, bolted joints in the structure of a machine tool are also a source of damping. Because friction dissipates energy, small relative movements along dry (unlubricated) joints dissipate energy and thus improve damping. In joints where oil is present, the internal friction of the oil layers also dissipates energy, thereby contributing to

(a)

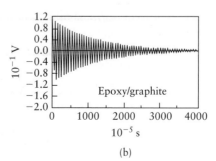

(b)

**FIGURE 8.70**  Relative damping capacity of gray cast iron and epoxy–granite composite material. The vertical scale is the amplitude of vibration, and the horizontal scale is time. *Source*: Courtesy of Cincinnati Machine, a UNOVA Company.

**FIGURE 8.71** Damping of vibrations as a function of the number of components on a lathe. Joints dissipate energy; thus, the greater the number of joints, the higher the damping. *Source*: J. Peters, CIRP Annals, 1965.

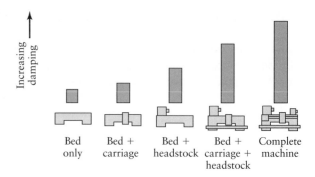

damping. In describing the machine tools for various cutting operations, we noted that all machines consist of a number of large and small components, assembled into a structure. Consequently, this type of damping is cumulative, owing to the presence of a number of joints in a machine tool. Note in Fig. 8.71 how damping increases as the number of components on a lathe and their contact area increase; thus, the more joints, the greater is the amount of energy dissipated, and the higher is the damping.

**External damping.** External damping is accomplished with external dampers, which are similar to shock absorbers on automobiles. Special vibration absorbers have been developed and can be installed on machine tools for this purpose.

### 8.11.2 Factors influencing chatter

Studies have indicated that the tendency for a particular workpiece to chatter during cutting is proportional to the cutting forces and the depth and width of cut. Consequently, because cutting forces increase with strength (hardness), the tendency to chatter generally increases as the hardness of the workpiece material increases. Thus, for example, aluminum and magnesium alloys have less tendency to chatter than do martensitic and precipitation-hardening stainless steels, nickel alloys, and high-temperature and refractory alloys. An important factor in chatter is the type of chip produced during cutting operations. As we have described, continuous chips involve steady cutting forces. Consequently, they generally do not cause chatter. Discontinuous chips and serrated chips, on the other hand, may do so. These chips are produced periodically, and the resulting variations in force during cutting can cause chatter.

## 8.12 | Machine-Tool Structures

The design and construction of machine tools are important aspects of manufacturing engineering. This section describes the *material* and *design* aspects of machine tools as structures with certain desired characteristics. The proper design of machine-tool structures requires a knowledge of the materials available for construction, their forms and properties, the dynamics of the particular machining process, and the forces involved. *Stiffness* and *damping* are important factors in machine-tool structures. Stiffness involves both the dimensions of the structural components and the elastic modulus of the materials used. Damping involves the type of materials used, as well as the number and nature of the joints in the structure.

**Materials and design.** Traditionally, the base and some of the major components of machine tools have been made of gray or nodular cast iron, which have the advantages of low cost and good damping capacity, but are heavy. Lightweight designs are desirable, because of ease of transportation, higher natural frequencies, and lower inertial forces of moving members. Lightweight designs and design flexibility require fabrication processes such as (a) mechanical fastening (bolts and nuts) of individual components and (b) welding. However, this approach to fabrication increases labor and material costs, because of the preparations involved.

Wrought steels are the likely choice for such lightweight structures, because of their low cost; availability in various section sizes and shapes (such as channels, angles, and tubes); desirable mechanical properties; and favorable characteristics, such as formability, machinability, and weldability. Tubes, for example, have high stiffness-to-weight ratios. On the other hand, the benefit of the higher damping capacity of castings and composites is not available with steels.

In addition to stiffness, another factor that contributes to lack of precision of the machine tool is the *thermal expansion* of its components, causing distortion. The source of heat may be internal, such as bearings, ways, motors, and heat generated from the cutting zone, or external, such as nearby furnaces, heaters, sunlight, and fluctuations in cutting fluid and ambient temperatures. Equally important in machine-tool precision are *foundations*, particularly their mass and how they are installed in a plant. For example, in installing a large grinding machine for grinding marine-propulsion gears 2.75 m (9 ft) in diameter with high precision, the concrete foundation is 6.7 m (22 ft deep). This large mass of concrete and the machine base reduce the amplitude of vibrations and their adverse effects.

Significant developments concerning the materials used for machine-tool bases and components have taken place. For example, **acrylic concrete** is a mixture of concrete and polymer (polymethylmethacrylate); it can easily be cast into desired shapes for machine bases and various components; several compositions are being developed. It can also be used for sandwich construction with cast iron, thus combining the advantages of each type of material. **Granite-epoxy composite** is a castable composite, with a typical composition of about 93% crushed granite and 7% epoxy binder. First used in precision grinders in the early 1980s, this composite material has several favorable properties: (1) good castability, which allows design versatility in machine tools; (2) high stiffness-to-weight ratio; (3) thermal stability; (4) resistance to environmental degradation; and (5) good damping capacity (see Fig. 8.70).

With the goal of improving the stiffness of machine tools and, therefore, attaining extremely fine dimensional tolerances on machined components, innovative experimental designs for machine-tool structures are continually being developed. Although still largely in a development stage, these designs are self-contained octahedral (eight-sided) machine frames, variously called **hexapods**, **parallel kinematic linked machines**, and **prism-shaped machines**. They are based on a mechanism called the **Stewart platform** (an invention used to position aircraft cockpit simulators), and the machines have high stiffness and high flexibility of operation. The workpiece is fixtured on a fixed table, and pairs of telescoping tubes (struts or legs), equipped with ballscrews and each with its own motor, are used to maneuver a rotating cutting tool. During machining, the machine controller shortens some tube while lengthening others, so that the cutting tool follows a specified contour around the workpiece. Design considerations include the coordinate system for machine movements and inertia effects resulting from rapid acceleration and deceleration of machine components.

# 8.13 | Design Considerations

Major design considerations for the processes described in this chapter are outlined as follows:

a) **Turning.** Blanks to be machined should be as close to the final dimensions (such as by net or near-net-shape forming) as possible so as to reduce machining cycle time. The machinability of raw materials should be properly considered. Parts should be designed so that they can be fixtured and clamped in work-holding devices with relative ease. Thin and slender workpieces may be difficult to support properly. Dimensional accuracy and surface finish specified should be as broad as permissible for the part to function as desired. Sharp corners, tapers, and major dimensional variations along the part should be avoided. Design features should be such that commercially available standard cutting tools, inserts, and toolholders can be used.

b) **Screw thread cutting.** Designs should allow the termination of threads before the threads reach a shoulder. Internal threads in blind holes should have an unthreaded length at the bottom. Shallow blind-threaded holes should be avoided. Chamfers should be specified to avoid burr formation. Threaded sections should not be interrupted with holes, slots, or other discontinuities. Thread design should be such that standard threading tooling and inserts can be used. Parts should be designed so that all cutting operations can be completed in one setup.

c) **Drilling, reaming, and tapping.** Holes should be specified on flat surfaces and perpendicular to the direction of drill movement for ease of drilling. Hole bottoms should match standard drill-point angles, and holes with interrupted surfaces should be avoided. Through holes should be preferred to blind holes, and blind holes should be drilled deeper than subsequent reaming or tapping operations require. Part design and fixturing should be such that all operations can be performed in one setup.

d) **Boring.** Whenever possible, through holes rather than blind holes should be specified. The greater the length-to-diameter ratio of the hole, the more difficult is the boring operation. Interrupted internal surfaces should be avoided.

e) **Milling.** Designs should be such that standard milling cutters can be used. Chamfers should be preferred over radii, and internal cavities and pockets with sharp inner corners should be avoided. Parts should be sufficiently stiff, i.e., have a high section modulus, to minimize deflections from clamping as well as from cutting forces and to minimize vibration and chatter.

f) **Broaching.** Parts should be designed such that they can be fixtured securely and should have sufficient stiffness to withstand the forces involved. Chamfers should be preferred over radii. Holes, sharp corners, dovetail splines, and flat surfaces should be avoided.

g) **Gear machining.** Blank design is important for proper fixturing and to ease cutting operations. Wide gears are more difficult to machine than narrow gears. Designs should allow for sufficient clearance between gear teeth and flanges. Design of the part should allow the use of standard gear cutters. Dimensional accuracy and surface finish specified should be as broad as possible for economical production.

# 8.14 | Economics of Machining

Section 8.1 outlined the advantages and limitations of machining, and the technical considerations were described in detail in various sections in this chapter. These considerations should be viewed in light of the competitive nature of manufacturing processes. Thus, for example, as compared with forming and shaping processes, machining involves more time and wastes more material; however, it can produce parts with dimensional control and surface finish better than that provided by other processes.

This section deals with the *economics* of machining. The two most important parameters are the *minimum cost per part* and the *maximum production rate*. It can be seen that the **total cost per piece** consists of four items, or

$$C_p = C_m + C_s + C_\ell + C_t, \tag{8.46}$$

where $C_p$ is the cost (say, in dollars) per piece; $C_m$ is the machining cost; $C_s$ is the cost of setting up for machining, such as mounting the cutter and fixtures and preparing the machine for the particular operation; $C_\ell$ is the cost of loading, unloading, and machine handling; and $C_t$ is the tooling cost, which includes tool changing, regrinding, and depreciation of the cutter. The **machining cost** is given by

$$C_m = T_m(L_m + B_m), \tag{8.47}$$

where $T_m$ is the machining time per piece; $L_m$ is the labor cost of production operator per hour; and $B_m$ is the burden rate, or overhead charge, of the machine, including depreciation, maintenance, indirect labor, and the like. The **setup cost** is a fixed figure in dollars per piece. The **loading, unloading,** and **machine-handling cost** is given by

$$C_\ell = T_\ell(L_m + B_m), \tag{8.48}$$

where $T_\ell$ is the time involved in loading and unloading the part, changing speeds, changing feed rates, and so on. The **tooling cost** is expressed as

$$C_t = \frac{1}{N_p}[T_c(L_m + B_m) + T_g(L_g + B_g) + D_c], \tag{8.49}$$

where $N_p$ is the number of parts machined per tool grind; $T_c$ is the time required to change the tool; $T_g$ is the time required to grind the tool; $L_g$ is the labor cost of tool-grinder operator per hour; $B_g$ is the burden rate of tool grinder per hour; and $D_c$ is the depreciation of the tool in dollars per grind.

The **time** needed to produce one part is

$$T_p = T_\ell + T_m + \frac{T_c}{N_p}, \tag{8.50}$$

where $T_m$ has to be calculated for each particular operation. For example, let's consider a turning operation; the machining time (see Section 8.8.1) is

$$T_m = \frac{L}{fN} = \frac{\pi LD}{fV}, \tag{8.51}$$

where $L$ is the length of cut, $f$ is the feed, $N$ is the rpm of the workpiece, $D$ is the workpiece diameter, and $V$ is the cutting speed. Note that appropriate units must be used in all these equations. From the Taylor tool-life equation, we have

$$VT^n = C.$$

Hence,

$$T = \left(\frac{C}{V}\right)^{1/n}, \tag{8.52}$$

where $T$ is the time, in minutes, required to reach a flank wear of certain dimension, after which the tool has to be reground or changed. The number of pieces per tool grind is thus simply

$$N_p = \frac{T}{T_m}. \tag{8.53}$$

The combination of Eqs. (8.51)–(8.53) gives

$$N_p = \frac{fC^{1/n}}{\pi LDV^{(1/n)-1}}. \tag{8.54}$$

The **cost per piece,** $C_p$, in Eq. (8.46) can now be defined in terms of several variables. To find the optimum cutting speed and the optimum tool life for **minimum cost,** we differentiate $C_p$ with respect to $V$ and set it to zero. Thus, we have

$$\frac{\partial C_p}{\partial V} = 0. \tag{8.55}$$

We then find that the optimum cutting speed, $V_o$, is

$$V_o = \frac{C(L_m + B_m)^n}{\{[(1/n) - 1][T_c(L_m + B_m) + T_g(L_g + B_g) + D_c]\}^n}, \tag{8.56}$$

and the optimum tool life, $T_o$, is

$$T_o = [(1/n) - 1]\frac{T_c(L_m + B_m) + T_g(L_g + B_g) + D_c}{L_m + B_m}. \tag{8.57}$$

To find the optimum cutting speed and the optimum tool life for **maximum production,** we differentiate $T_p$ with respect to $V$ and set the result to zero. Thus, we have

$$\frac{\partial T_p}{\partial V} = 0. \tag{8.58}$$

The **optimum cutting speed** now becomes

$$V_o = \frac{C}{\{[(1/n) - 1]T_c\}^n}, \tag{8.59}$$

and the **optimum tool life** is

$$T_o = [(1/n) - 1]T_c. \tag{8.60}$$

Qualitative plots of the minimum cost per piece and the minimum time per piece, that is, the maximum production rate, are given in Fig. 8.72. The cost of a machined surface also depends on the degree of finish required (see Section 9.17); the machining cost increases rapidly with finer surface finish.

**FIGURE 8.72** Graphs showing (a) cost per piece and (b) time per piece in machining. Note the optimum speeds for both cost and time. The range between the two optimum speeds is known as the *high-efficiency machining range.*

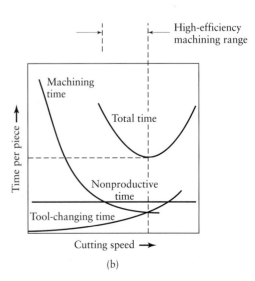

This analysis indicates the importance of identifying all relevant parameters in a machining operation, determining various cost factors, obtaining relevant tool-life curves for the particular operation, and properly measuring the various time intervals involved in the overall operation. The importance of obtaining accurate data is clearly shown in Fig. 8.72, as small changes in cutting speed can have a significant effect on the minimum cost or time per piece.

---

**CASE STUDY** | High-Speed Dry Machining of Cast-Iron Engine Blocks

High-speed machining has clear economic advantages as long as spindle damage and chatter are avoided. Similarly, dry machining is recognized as a promising approach to solving environmental problems associated with cutting fluids. However, it has often been perceived that the two approaches cannot be applied simultaneously, because the higher speeds without the use of cutting fluids would result in low tool life. However, high-speed dry machining is possible with aluminum and cast iron, and we examine it here for roughing, finishing, and hole-making operations on a cast-iron cylinder head for a 4.3-liter engine.

High-speed dry machining is a demanding operation. Heat dissipation without a liquid coolant requires high-performance coatings and heat-resistant tool materials. The cutting-tool materials must be either ceramic or cBN, because of the intense heat generated in the process. In addition, machine tools with high power are required, and spindles with high stiffness and power ratings are needed. Chatter can be a problem with high-speed machining; hence, CNC programs must be developed that maintain process parameters within a window of operation that is known to be chatter free.

The key to dry machining the cast-iron head is the use of two advanced techniques. The first strategy is *red-crescent machining*, where high temperatures are generated in the workpiece in front of the tool, creating a visible red arc (hence the name of the process). The second strategy is flush fine machining (see later in this case study), using in-spindle high-pressure air to remove chips that normally would be carried away by a cutting fluid. Red-crescent machining refers to milling operations that take place at high feed rates and high spindle speeds. This combination reduces the thrust force against the workpiece as compared with a standard milling operation. Red-crescent machining concentrates heat in the material in front of the tool and the workpiece can reach temperature ranges of 600–700°C (1100–1300°F) in some cases. This situation occurs because, at the high speeds, the heat generated is localized in the shear zone and does not have time to be conducted into the workpiece or the tool. This high-temperature zone generates a red glow. Red-crescent machining is beneficial with cast irons because the heat increases the ductility and reduces the strength of the cast iron, allowing for machining at much higher efficiency. Part distortion does not occur, because most of the heat is retained in the chip and removed before it can be conducted into the workpiece. Table 8.12 provides a comparison of red-crescent machining with conventional machining for this particular application.

Finish milling is accomplished by using low-thrust machining strategies, usually by using cutters with fewer inserts per cutter, at a lower feed rate per revolution and higher surface speeds than for conventional machining. Low-thrust-force machining has an additional advantage in that work-holder design is simplified, often allowing better access to complicated parts. cBN inserts are used for finish machining of the cast-iron engine blocks. Although cBN is more expensive than the alternatives of carbide or ceramic inserts, tool life is longer, and fewer

**TABLE 8.12**

**Comparison of Conventional vs. Red-Crescent Machining Conditions**

| Parameter | Conventional | Red crescent |
|---|---|---|
| Surface speed (ft/min) | 250 | 5000 |
| Spindle speed (rpm) | 160 | 3200 |
| Feed (in./rev) | 0.020 | 0.005 |
| Number of inserts | 10 | 5 |
| Radial thrust force (lb) | 4300 | 262 |
| Metal removal rate (in$^3$/min) | 32 | 80 |

**FIGURE 8.73** A high-speed tool for single-point milling, chamfering, counterboring, and threading of holes. *Source*: Courtesy of Makino, Inc.

setups are needed; thus the insert cost per machined part can be comparable with or lower than that for the conventional tool materials.

*Flush fine machining* is a process that uses high-pressure through-spindle air for chip disposal, removing heat from the cutting tool as well. (See also Section 8.7.1.) The air pressure required is in the range of 1100–1380 kPa (160–200 psi) in order to reduce the temperature of the workpiece by approximately 50%. Flush fine machining is suitable for most shallow holes (depth-to-diameter ratios of five or less), which accounts for most automotive applications. Deeper holes are produced by using more conventional drilling operations. For shallow holes, single-tool milling, chamfering, counterboring, and threading of holes can be accomplished by using tools such as shown in Fig. 8.73. The process uses high-speed, high-accuracy CNC control to move a small-diameter tool in a helical pattern to shape and thread the holes. In addition, this technique allows a single tool to machine holes of different diameters.

*Source*: Adapted from G. Hyatt, "High Speed, Dry Machining Can Cut Cycle Times and Cost," *Manufacturing Engineering*, 119(3), 1997.

# SUMMARY

- Material removal processes are often necessary in order to impart the desired dimensional accuracy, geometric features, and surface finish characteristics to components, particularly those with complex shapes that cannot be produced economically or properly by other shaping techniques. On the other hand, these processes waste some material in the form of chips, generally cause the manufacturing process to take longer, and may have adverse effects on the surfaces produced. (Section 8.1)

- Important process variables include tool shape and material; cutting conditions such as speed, feed, and depth of cut; use of cutting fluids; and the characteristics of the machine tool and the workpiece material. Parameters influenced by these variables include forces and power consumption, tool wear, surface finish and integrity, temperature, and dimensional accuracy of the workpiece. The commonly observed chip types are continuous, built-up edge, discontinuous, and serrated. (Section 8.2)

- Temperature rise is an important consideration, as it can have adverse effects on tool life as well as on dimensional accuracy and surface integrity of the machined part. (Section 8.2)

- Tool wear depends primarily on workpiece and tool-material characteristics, cutting speed, and cutting fluids. Feeds, depth of cut, and machine-tool characteristics also have an effect. Two major types of wear are flank wear and crater wear. (Section 8.3)

- Surface finish of machined components is an important consideration, as it can adversely affect product integrity. Important variables that affect surface finish include the geometry and condition of the cutting tool, chip morphology, and process variables. (Section 8.4)

- Machinability is usually defined in terms of surface finish, tool life, force and power requirements, and chip type. Machinability of materials depends not only on their intrinsic properties and microstructure, but also on proper selection and control of process variables. (Section 8.5)

- A variety of cutting-tool materials has been developed, the most common ones being high-speed steels, carbides, ceramics, and cubic boron nitride. The materials have a broad range of mechanical and physical properties, such as hot hardness, toughness, chemical stability and inertness, and resistance to chipping and wear. Various tool coatings have also been developed, resulting in major improvements in tool life. (Section 8.6)

- Cutting fluids are important in machining operations. They reduce friction, forces, and power requirements and improve tool life. Generally, slower operations with high tool pressures require a fluid with good lubricating characteristics, whereas in high-speed operations, where temperature rise can be significant, fluids with cooling capacity are preferred. (Section 8.7)

- The cutting processes that produce external and internal circular profiles are turning, boring, drilling, tapping, and thread cutting. Because of the three-dimensional nature of these operations, chip movement and control are important considerations, since chips can interfere with the operation. Optimization of each process requires an understanding of the interrelationships among design and processing parameters. (Section 8.8)

- High-speed machining, ultraprecision machining, and hard turning are among more recent developments in cutting. They can help reduce machining costs, and they produce parts with exceptional surface finish and dimensional accuracy. (Section 8.8)

- Various complex shapes can be machined by slab, face, and end milling; broaching; and sawing. These processes use multitooth tools and cutters at various axes with respect to the workpiece. The machine tools are now mostly computer controlled, have various features and attachments, and possess considerable flexibility in operation. (Section 8.9)

- Because of their versatility and capability of performing a variety of cutting operations, machining and turning centers are among the most important developments in machine tools. Their selection depends on factors such as part complexity required, the number and type of cutting operations to be performed, the number of cutting tools needed, the dimensional accuracy required, and the production rate required. (Section 8.10)

- Vibration and chatter in machining are important considerations for workpiece dimensional accuracy and surface finish, as well as tool life. Stiffness and damping capacity of machine tools are important factors in controlling vibration and chatter. New materials are being developed and used for constructing machine-tool structures. (Sections 8.11 and 8.12)

- Several design guidlines have been developed for parts to be produced by machining. (Section 8.13)

- The economics of machining processes depend on various costs. Optimum cutting speeds can be determined for minimum machining time per piece and minimum cost per piece, respectively. (Section 8.14)

# SUMMARY OF EQUATIONS

Cutting ratio, $r = \dfrac{t_o}{t_c} = \dfrac{\sin\phi}{\cos(\phi - \alpha)}$

Shear strain, $\gamma = \cot\phi + \tan(\phi - \alpha)$

Velocity relationships: $\dfrac{V}{\cos(\phi - \alpha)} = \dfrac{V_s}{\cos\alpha} = \dfrac{V_c}{\sin\phi}$

Friction force, $F = R\sin\beta$

Normal force, $N = R\cos\beta$

Coefficient of friction, $\mu = \tan\beta = \dfrac{F_t + F_c\tan\alpha}{F_c - F_t\tan\alpha}$

Thrust force, $F_t = R\sin(\beta - \alpha) = F_c\tan(\beta - \alpha)$

Shear-angle relationships, $\phi = 45° + \dfrac{\alpha}{2} - \dfrac{\beta}{2}$

$\phi = 45° + \alpha - \beta$

Total cutting power $= F_c V$

Specific energy, $u_t = \dfrac{F_c}{wt_o}$

Frictional specific energy, $u_f = \dfrac{Fr}{wt_o}$

Shear specific energy, $u_s = \dfrac{F_s V_s}{wt_o V}$

Mean temperature, $T = \dfrac{1.2Y_f}{\rho c} \sqrt[3]{\dfrac{Vt_o}{K}}$

$$T \propto V^a f^b$$

Tool life, $VT^n = C$

Material removal rate (MRR) in turning $= \pi D_{avg} df N$

in drilling $= \pi \left(\dfrac{D^2}{4}\right) fN$

in milling $= wdv$

# BIBLIOGRAPHY

Armarego, E. J. A., and R. H. Brown, *The Machining of Metals*, Prentice Hall, 1969.

Arnone, M., *High Performance Machining*, Hanser, 1998.

*ASM Handbook*, Vol. 1: *Properties and Selection: Irons, Steels, and High-Performance Alloys*, ASM International, 1990.

*ASM Handbook*, Vol. 16: *Machining*, ASM International, 1989.

*ASM Specialty Handbook: Tool Materials*, ASM International, 1995.

Astakhov, V. P., *Metal Cutting Mechanics*, CRC Press, 1998.

Boothroyd, G., and W. A. Knight, *Fundamentals of Machining and Machine Tools*, 2d. ed., Dekker, 1989.

Brown, J., *Advanced Machining Technology Handbook*, McGraw-Hill, 1998.

Byers, J. P. (ed.), *Metalworking Fluids*, Dekker, 1994.

DeVries, W. R., *Analysis of Material Removal Processes*, Springer, 1992.

Ewert, R. H., *Gears and Gear Manufacture: The Fundamentals*, Chapman & Hall, 1997.

*Handbook of High-Speed Machining Technology*, Chapman & Hall, 1985.

Hoffman, E. G., *Jig and Fixture Design*, 4th ed., Industrial Press, 1996.

Juneja, B. L., *Fundamentals of Metal Cutting and Machine Tools*, Wiley, 1987.

Komanduri, R., "Tool Materials," in *Kirk-Othmer Encyclopedia of Chemical Technology*, 4th ed., Vol. 24, Wiley, 1997.

Krar, S. F., and A. F. Check, *Technology of Machine Tools*, 5th ed., Glencoe Macmillan/McGraw-Hill, 1996.

*Machinery's Handbook*, Industrial Press, revised periodically.

*Machining Data Handbook*, 2 vols., 3d. ed., Machinability Data Center, 1980.

*Metal Cutting Tool Handbook*, 7th ed., Industrial Press, 1989.

Nachtman, E. S., and S. Kalpakjian, *Lubricants and Lubrication in Metalworking Operations*, Dekker, 1985.

Nakazawa, H., *Principles of Precision Engineering*, Oxford, 1994.

Oxley, P. L. B., *Mechanics of Machining: An Analytical Approach to Assessing Machinability*, Wiley, 1989.

Rechetov, D. N., and V. T. Portman, *Accuracy of Machine Tools*, ASME, 1989.

Rivin, E. I., *Stiffness and Damping in Mechanical Design*, Dekker, 1999.

Roberts, G. A., G. Krauss, and R. Kennedy, *Tool Steels*, 5th ed., ASM International, 1997.

Schey, J. A., *Tribology in Metalworking—Friction, Wear and Lubrication*, ASM International, 1983.

Schneider, A. F., *Mechanical Deburring and Surface Finishing Technology*, Dekker, 1990.

Shaw, M. C., *Metal Cutting Principles*, Oxford, 1984.

Sluhan, C. (ed.), *Cutting and Grinding Fluids: Selection and Application*, Society of Manufacturing Engineers, 1992.

Stephenson, D., and J. S. Agapiou, *Metal Cutting: Theory and Practice*, Dekker, 1996.

Stout, K. J, E. J. Davis, and P. J. Sullivan, *Atlas of Machined Surfaces*, Chapman & Hall, 1990.

*Tool and Manufacturing Engineers Handbook*, 4th ed., Vol. 1: *Machining*, Society of Manufacturing Engineers, 1983.

Townsend, D. P., *Dudley's Gear Handbook: The Design, Manufacturing, and Application of Gears*, 2d ed., McGraw-Hill, 1991.

Trent, E. M., and P. K. Wright, *Metal Cutting*, 4th ed., Butterworth Heinemann, 1999.

Venkatesh, V. C., and H. Chandrasekaran, *Experimental Techniques in Metal Cutting*, rev. ed. Prentice-Hall, 1987.

Walsh, R. A., *McGraw-Hill Machining and Metalworking Handbook*, McGraw-Hill, 1994.

Weck, M., *Handbook of Machine Tools*, 4 vols., Wiley, 1984.

# QUESTIONS

**8.1** In metal cutting, it has been observed that the shear strains obtained are higher than those calculated from the properties of the material. To what factors would you attribute this difference? Explain.

**8.2** Describe the effects of material properties and process variables on serrated chip formation.

**8.3** Explain why the cutting force, $F_c$, increases with increasing depth of cut and decreasing rake angle.

**8.4** What are the effects of performing a cutting operation with a dull tool tip?

**8.5** Describe the trends that you have observed in Tables 8.1 and 8.2.

**8.6** To what factors would you attribute the large difference in the specific energies within each group of materials shown in Table 8.3?

**8.7** Describe the effects of cutting fluids on chip formation. Explain why and how they influence the cutting operation.

**8.8** Under what conditions would you discourage the use of cutting fluids?

**8.9** Give reasons that pure aluminum and copper are generally classified as easy to machine.

**8.10** Can you offer an explanation as to why the maximum temperature in cutting is located at about the middle of the tool–chip interface? (*Hint*: Note that there are two principal sources of heat: the shear plane and the tool–chip interface.)

**8.11** State whether or not the following statements are true, explaining your reasons: (a) For the same shear angle, there are two rake angles that give the same cutting ratio. (b) For the same depth of cut and rake angle, the type of cutting fluid used has no influence on chip thickness. (c) If the cutting speed, shear angle, and rake angle are known, the chip velocity can be calculated. (d) The chip becomes thinner as the rake angle increases. (e) The function of a chip breaker is to decrease the curvature of the chip.

**8.12** It is generally undesirable to allow temperatures to rise too high in cutting operations. Explain why.

**8.13** Explain the reasons that the same tool life may be obtained at two different cutting speeds.

**8.14** Inspect Table 8.6, and identify tool materials that would not be particularly suitable for interrupted cutting operations. Explain your choices.

**8.15** Explain the possible disadvantages of a cutting operation in which the type of chip produced is discontinuous.

**8.16** We have noted that tool life can be almost infinite at low cutting speeds. Would you then recommend that all machining be done at low speeds? Explain.

**8.17** How would you explain the effect of cobalt content on the properties of carbides?

**8.18** Explain why studying the types of chips produced is important in understanding machining operations.

**8.19** How would you expect the cutting force to vary for the case of serrated-chip formation? Explain.

**8.20** Wood is a highly anisotropic material; that is, it is orthotropic. Explain the effects of orthogonal cutting of wood at different angles to the grain direction on chip formation.

**8.21** Describe the advantages of oblique cutting.

**8.22** Explain why it is possible to remove more material between tool resharpenings by lowering the cutting speed.

**8.23** Explain the significance of Eq. (8.8).

**8.24** Why are there so many different types of cutting-tool materials? Explain, with appropriate examples.

**8.25** How would you go about measuring the hot hardness of cutting tools?

**8.26** Describe the reasons for making cutting tools with multiphase coatings of different materials.

**8.27** Explain the advantages and limitations of inserts. Why were they developed?

**8.28** Make a list of alloying elements used in high-speed-steel cutting tools. Explain why they are used.

**8.29** What are the purposes of chamfers on cutting tools?

**8.30** Why does temperature have such an important effect on cutting-tool performance?

**8.31** Ceramic and cermet cutting tools have certain advantages over carbide tools. Why, then, are they not replacing carbide tools to a greater extent?

**8.32** Why are chemical stability and inertness important in cutting tools?

**8.33** What precautions would you take in machining with brittle tool materials?

**8.34** Why do cutting fluids have different effects at different cutting speeds? Is the control of cutting-fluid temperature important? Explain.

**8.35** Which of the two materials, diamond or cubic boron nitride, is more suitable for cutting steels? Why?

**8.36** Describe the properties that the substrate for multiphase cutting tools should have.

**8.37** List and explain the considerations involved in determining whether a cutting tool should be reconditioned, recycled, or discarded after use.

**8.38** List the parameters that influence the temperature in metal cutting, and explain why and how they do so.

**8.39** List and explain factors that contribute to poor surface finish in machining operations.

**8.40** Explain the functions of different angles on a single-point lathe cutting tool. How does the chip thickness vary as the side cutting-edge angle is increased?

**8.41** The helix angle for drills is different for different groups of workpiece materials. Why?

**8.42** A turning operation is being carried out on a long, round bar at a constant depth of cut. Explain what changes, if any, may occur in the machined diameter from one end of the bar to the other. Give reasons for any changes that may occur.

**8.43** Describe the relative characteristics of climb milling and up milling.

**8.44** In Fig. 8.64a, high-speed-steel cutting teeth are welded to a steel blade. Would you recommend that the whole blade be made of high-speed steel? Explain your reasons.

**8.45** Explain the technical requirements that led to the development of machining centers. What are the distinctive features of these centers? Why can their spindle speeds be varied over a wide range? Explain.

**8.46** Describe the adverse effects of vibrations and chatter in machining.

**8.47** Make a list of components of machine tools that could be made of ceramics, and explain why ceramics would be a suitable material for these components.

**8.48** In Fig. 8.12, why do the thrust forces start at a finite value when the feed is zero?

**8.49** Do you think temperature rise in cutting is related to the hardness of the workpiece material? Explain.

**8.50** Describe the effects of tool wear on the workpiece and on the machining operation in general.

**8.51** Explain whether or not it is desirable to have a high or low $n$ value in the Taylor tool-life equation.

**8.52** Are there cutting operations that cannot be performed on machining centers? Explain.

**8.53** What is the importance of the cutting ratio?

**8.54** Emulsion cutting fluids typically consist of 95% water and 5% soluble oil and chemical additives. Why is the ratio so unbalanced? Is the oil needed at all?

**8.55** It is possible for the $n$ value in the Taylor tool-life equation to be negative. Explain how.

**8.56** Assume that you are asked to estimate the cutting force in slab milling with a straight-tooth cutter, but without running a test. Describe the procedure that you would follow.

**8.57** Explain the possible reasons that a knife cuts better when it is moved back and forth. Consider factors such as the material being cut, interfacial friction, and the dimensions of the knife.

**8.58** What are the effects of lowering the friction at the tool–chip interface, say with an effective cutting fluid, on the mechanics of cutting operations? Explain, giving several examples.

**8.59** Why is it not always advisable to increase cutting speed in order to increase production rate? Explain.

**8.60** It has been observed that the shear-strain rate in metal cutting is high even though the cutting speed may be relatively low. Why?

**8.61** We note from the exponents in Eq. (8.30) that the cutting speed has a greater influence on temperature than does the feed. Why?

**8.62** Describe the consequences of exceeding the allowable wear land (see Table 8.5) for cutting tools.

**8.63** Explain whether it is desirable to have a high or low C value in the Taylor tool-life equation.

**8.64** Comment on and explain your observations regarding Figs. 8.34, 8.38, and 8.43.

**8.65** The tool-life curve for ceramic tools in Fig. 8.23a is to the right of those for other tools. Why?

**8.66** In Fig. 8.18, we note that the percentage of the energy carried away by the chip increases with cutting speed. Why?

**8.67** How would you go about measuring the effectiveness of cutting fluids? Explain.

**8.68** Describe the conditions that are critical in using the capabilities of diamond and cubic-boron-nitride cutting tools.

**8.69** In Table 8.6, the last two properties listed can be important to the life of the cutting tool. Why? Which of the properties listed are, in your opinion, the least important in machining? Explain.

**8.70** In Fig. 8.31, we note that tool materials, especially carbides, have a wide range of hardnesses at a particular temperature. Why?

**8.71** How would you go about recycling used cutting tools? Explain.

**8.72** As you can see, there is a wide range of tool materials available and used successfully today, yet much research and development continues to be carried out on these materials. Why?

**8.73** Drilling, boring, and reaming of large holes is generally more accurate than just drilling and reaming. Why?

**8.74** A badly oxidized and uneven round bar is being turned on a lathe. Would you recommend a relatively small or large depth of cut? Explain your reasons.

**8.75** Does the force or torque in drilling change as the hole depth increases? Explain.

**8.76** Explain the advantages and limitations of producing threads by forming and cutting, respectively.

**8.77** Describe your observations regarding the contents of Tables 8.8, 8.10, and 8.11.

**8.78** The footnote to Table 8.10 states that as the depth of the hole increases, speeds and feeds should be reduced. Why?

**8.79** List and explain the factors that contribute to poor surface finish in machining operations.

**8.80** Make a list of the machining operations described in this chapter, according to the difficulty of the operation and the desired effectiveness of cutting fluids. (*Example*: Tapping of holes is a more difficult operation than turning straight shafts.)

**8.81** Are the feed marks left on the workpiece by a face-milling cutter true segments of a true circle? Explain with appropriate sketches.

**8.82** What determines the selection of the number of teeth on a milling cutter? (See, for example, Figs. 8.53 and 8.55.)

**8.83** Explain the technical requirements that led to the development of machining and turning centers. Why do their spindle speeds vary over a wide range?

**8.84** In addition to the number of components, as shown in Fig. 8.71, what other factors influence the rate at which damping increases in a machine tool? Explain.

**8.85** Why is thermal expansion of machine-tool components important? Explain, with examples.

**8.86** Would using the machining processes described in this chapter be difficult on nonmetallic or rubberlike materials? Explain your thoughts, commenting on the influence of various physical and mechanical properties of workpiece materials, the cutting forces involved, the parts geometries, and the fixturing required.

**8.87** The accompanying illustration shows a part that is to be machined from a rectangular blank. Suggest the type of operations required and their sequence, and specify the machine tools that are needed.

**8.88** Select a specific cutting-tool material, and estimate the machining time for the parts shown in the accompanying figure: (a) pump shaft, stainless steel; (b) ductile (nodular) iron crankshaft; (c) 304 stainless-steel tube with internal rope thread.

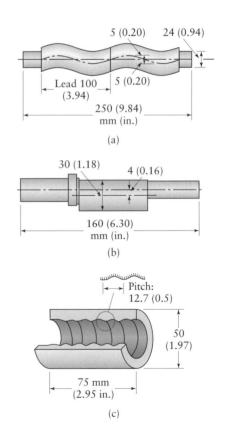

**8.89** Why is the machinability of alloys difficult to assess?

**8.90** What are the advantages and disadvantages of dry machining?

**8.91** Can high-speed machining be performed without the use of cutting fluids? Explain.

**8.92** If the rake angle is 0°, then the frictional force is perpendicular to the cutting direction and, therefore, does not consume machine power. Why, then, is there an increase in the power dissipated when machining with a rake angle of, say, 20°?

**8.93** Would you recommend broaching a keyway on a gear blank before or after the teeth are machined? Explain.

## PROBLEMS

**8.94** Assume that you are an instructor covering the topics in this chapter, and you are giving a quiz on the quantitative aspects to test the understanding of the students. Prepare several numerical problems, and supply the answers to them.

**8.95** Prove Eq. (8.1).

**8.96** With a simple analytical expression to prove the validity of the last paragraph in Example 8.2.

**8.97** Using Eq. (8.3), make a plot of the shear strain, $\gamma$, vs. the shear angle, $\phi$, with the rake angle, $\alpha$, as a parameter. Describe your observations.

**8.98** Assume that in orthogonal cutting, the rake angle is 15° and the coefficient of friction is 0.25. Determine the percentage change in chip thickness when the friction is doubled.

**8.99** Derive Eq. (8.11).

**8.100** Determine the shear angle in Example 8.1. Is this calculation exact or an estimate? Explain.

**8.101** Taking carbide as an example and using Eq. (8.30), determine how much the feed should be changed in order to keep the mean temperature constant when the cutting speed is tripled.

**8.102** With appropriate diagrams, show how the use of a cutting fluid can affect the magnitude of the thrust force, $F_t$, in orthogonal cutting.

**8.103** An 8-in-diameter stainless-steel bar is being turned on a lathe at 600 rpm and at a depth of cut, $d = 0.1$ in. If the power of the motor is 4 hp, what is the maximum feed that you can have before the motor stalls?

**8.104** Using the Taylor equation for tool wear and letting $n = 0.4$, calculate the percentage increase in tool life if the cutting speed is reduced by (a) 20% and (b) 50%.

**8.105** Determine the $n$ and $C$ values for the four tool materials shown in Fig. 8.23a.

**8.106** Using Eq. (8.30) and referring to Fig. 8.18a, estimate the magnitude of the coefficient $a$.

**8.107** (a) Estimate the machining time required in rough turning a 1.25-m-long, annealed aluminum-alloy round bar that is 75 mm in diameter, using a high-speed-steel tool. (b) Estimate the time for a carbide tool. Use a feed of 2 mm/rev.

**8.108** A 200-mm-long, 75-mm-diameter titanium-alloy rod is being reduced in diameter to 65 mm by turning on a lathe. The spindle rotates at 400 rpm, and the tool is traveling at an axial velocity of 250 mm/min. Calculate the cutting speed, material removal rate, time of cut, power required, and cutting force.

**8.109** Calculate the same quantities as in Example 8.5 for high-strength cast iron at $N = 300$ rpm.

**8.110** A 0.75-in-diameter drill is being used on a drill press operating at 200 rpm. If the feed is 0.005 in./rev, what is the material removal rate? What is the MRR if the drill diameter is tripled?

**8.111** A hole is being drilled in a block of magnesium alloy with a 10-mm drill at a feed of 0.1 mm/rev. The spindle is running at 800 rpm. Calculate the material removal rate, and estimate the torque on the drill.

**8.112** Show that the distance $\ell_c$ in slab milling is approximately equal to $\sqrt{Dd}$ for situations where $D \gg d$.

**8.113** Calculate the chip depth of cut in Example 8.7.

**8.114** In Example 8.7, which of the quantities will be affected when the spindle speed is increased to 200 rpm?

**8.115** A slab-milling operation is being carried out on a 30-in.-long, 6-in.-wide high-strength-steel block at a feed of 0.01 in./tooth and a depth of cut of 0.15 in. The cutter has a diameter of 3 in, has eight straight cutting teeth, and rotates at 150 rpm. Calculate the material removal rate and the cutting time, and estimate the power required.

**8.116** Referring to Fig. 8.54, assume that $D = 250$ mm, $w = 50$ mm, $\ell = 600$ mm, $d = 2$ mm, $v = 1$ mm/s, and $N = 200$ rpm. The cutter has 10 inserts, and the workpiece material is 304 stainless steel. Calculate the material removal rate, cutting time, and feed per tooth, and estimate the power required.

**8.117** Estimate the time required for face milling an 8-in.-long, 3-in.-wide brass block with an 8-in-diameter cutter that has 10 high-speed-steel teeth.

**8.118** A 10-in-long, 2-in-thick plate is being cut on a band saw at 100 ft/min. The saw has 12 teeth per in. If the feed per tooth is 0.003 in., how long will it take to saw the plate along its length?

**8.119** A single-thread hob is used to cut 40 teeth on a spur gear. The cutting speed is 150 ft/min, and the hob has a diameter of 4 in. Calculate the rotational speed of the spur gear.

**8.120** In deriving Eq. (8.20), we assumed that the friction angle, $\beta$, was independent of the shear angle, $\phi$. Is this assumption valid? Explain.

**8.121** An orthogonal cutting operation is being carried out under the following conditions: depth of cut = 0.15 mm, width of cut = 5 mm, chip thickness = 0.2 mm, cutting speed = 2 m/s, rake angle = 15°, cutting force = 500 N, and thrust force = 200 N. Calculate the percentage of the total energy that is dissipated in the shear plane during cutting.

**8.122** An orthogonal cutting operation is being carried out under the following conditions: depth of cut = 0.020 in., width of cut = 0.1 in., cutting ratio = 0.3, cutting speed = 300 ft/min, rake angle = 0°, cutting force = 200 lb, thrust force = 150 lb, workpiece density = 0.26 lb/in³, and workpiece specific heat = 0.12 BTU/lb°F. Assume that (a) the sources of heat are the shear plane and the tool–chip interface; (b) the thermal conductivity of the tool is zero, and there is no heat loss to the environment; (c) the temperature of the chip is uniform throughout. If the temperature rise in the chip is 155°F, calculate the percentage of the energy dissipated in the shear plane that goes into the workpiece.

**8.123** It can be shown that the angle $\psi$ between the shear plane and the direction of maximum grain elongation (see Fig. 8.4a) is given by the expression

$$\psi = 0.5 \cot^{-1}\left(\frac{\gamma}{2}\right),$$

where $\gamma$ is the shear strain, as given by Eq. (8.3). Assume that you are given a piece of the chip obtained from orthogonal cutting of an annealed metal. The rake angle and cutting speed are also given, but you have not seen the setup on which the chip was produced. Outline the procedure that you would follow to estimate the power required in producing this chip. Assume that you have access to a fully equipped laboratory and a technical library.

**8.124** In Figs. 8.16 and 8.17b, we note that the maximum temperature is about halfway up the face of the tool. We have also described the adverse effects of temperature on various tool materials. Considering the mechanics of cutting operations, describe your thoughts on the technical and economic merits of embedding a small insert, made of materials such as ceramic or carbide, halfway up the rake face of a tool made of a material with lower resistance to temperature than ceramic or carbide.

**8.125** A lathe is set up to machine a taper on a bar stock 100 mm in diameter; the taper is 1 mm per 10 mm. A cut is made with an initial depth of cut of 2 mm at a feed rate of 0.250 mm/rev and at a spindle speed of 150 rpm. Calculate the average metal removal rate.

**8.126** Obtain an expression for optimum feed rate that minimizes the cost per piece if the tool life is as described by Eq. (8.34).

**8.127** Assuming that the coefficient of friction in cutting is 0.3, calculate the maximum depth of cut for turning a hard aluminum alloy on a 20-hp lathe (mechanical efficiency of 80%) at a width of cut of 0.25 in, a rake angle of 10°, and a cutting speed of 300 ft/min. What is your estimate of the material's shear strength?

**8.128** Assume that, using a carbide cutting tool, you measure the temperature in a cutting operation at a speed of 250 ft/min and feed of 0.0025 in./rev as 1200°F. What is the approximate temperature if the speed is doubled? What speed is required to lower the maximum temperature to 800°F?

**8.129** A 4-in-diameter gray cast-iron cylindrical part is to be turned on a lathe at 600 rpm, with a depth of cut of 0.25 in. and a feed of 0.02 in./rev. What should be the minimum horsepower of the lathe?

**8.130** (a) A 6-in-diameter aluminum cylinder with a length of 10 in. is to have its diameter reduced to 5 in. Estimate the machining time if an uncoated carbide tool is used. (b) What if a TiN-coated tool is used?

# DESIGN

**8.131** Tool life could be greatly increased if an effective means of cooling and lubrication were developed. Design methods of delivering a cutting fluid to the cutting zone, and discuss the advantages and shortcomings of your design.

**8.132** Devise an experimental setup whereby you can perform an orthogonal cutting operation using a round tubular workpiece on a lathe.

**8.133** Cutting tools are sometimes designed so that the chip–tool contact length is controlled by recessing it at the rake face some distance away from the tool tip. Explain the possible advantages of such a tool.

**8.134** The accompanying illustration shows drawings for a cast-steel valve body before (left) and after (right) machining. Identify the surfaces that are to be machined. What type of machine tool would be suitable to machine this part? What type of machining operations are involved, and what should be the sequence of these operations? (Note that not all surfaces are to be machined.)

100 mm

|←—100 mm —→|

(a) Casting          (b) After machining

**8.135** Make a comprehensive table of the process capabilities of the machining processes described in this chapter. Use several columns to describe the machines involved, the type of tools and tool materials used, the shapes of blanks and parts produced, the typical maximum and minimum sizes produced, the surface finish produced, the dimensional tolerances produced, and the production rates achieved.

**8.136** A large bolt is to be produced from hexagonal bar stock by placing the hex stock into a chuck and machining the cylindrical shank of the bolt by turning on a lathe. List and explain the difficulties that may be presented by this operation.

**8.137** The part shown in the accompanying figure is a power-transmitting shaft; it is to be produced on a lathe. List the operations that are needed to make this part, and estimate the machining time.

Dimensions in inches

**8.138** Design appropriate fixtures and describe the machining operations required to produce the aluminum piston shown in Fig. 5.43.

# CHAPTER

# 9

# Material-Removal Processes: Abrasive, Chemical, Electrical, and High-Energy Beams

9.1  Introduction
9.2  Abrasives
9.3  Bonded Abrasives
9.4  Mechanics of Grinding
9.5  Grinding-Wheel Wear
9.6  Grinding Operations and Machines
9.7  Ultrasonic Machining
9.8  Finishing Operations
9.9  Deburring
9.10 Chemical Machining
9.11 Electrochemical Machining
9.12 Electrochemical Grinding
9.13 Electrical-Discharge Machining
9.14 High-Energy-Beam Machining
9.15 Water-Jet, Abrasive Water-Jet, and Abrasive-Jet Machining
9.16 Design Considerations
9.17 Process Economics

## 9.1 | Introduction

In all the cutting processes described in Chapter 8, the tool is made of a certain material and has a clearly defined geometry. Furthermore, the cutting process is carried out by chip removal, the mechanics of which are reasonably well understood. There are many situations in manufacturing, however, where the workpiece material is either too *hard* or too *brittle*, or its *shape* is difficult to produce with sufficient accuracy by any of the cutting methods described previously. One of the best methods for producing such parts is to use **abrasives**. An abrasive is a small, hard particle that has sharp edges and an irregular shape, unlike the cutting tools described earlier. Abrasives are capable of removing small amounts of material from a surface by a cutting process that produces tiny chips.

**Abrasive machining** processes are generally among the last operations performed on manufactured products. These processes, however, are not necessarily confined to fine or small-scale material removal. They are also used for large-scale removal operations and can indeed compete economically with some machining processes, such as milling and turning. Because they are hard, abrasives are also used in finishing very hard or heat-treated parts; shaping hard nonmetallic materials, such as ceramics and glasses; removing unwanted weld beads; cutting off lengths of bars, structural shapes, masonry, and concrete; and cleaning surfaces, for which jets of air or water that contain abrasive particles are used.

There are situations, however, where none of these processes is satisfactory or economical, for one or more of the following reasons:

1. The hardness and strength of the material is very high, typically above 400 HB.
2. The workpiece is too flexible or slender to support the cutting or grinding forces, or parts are difficult to clamp in workholding devices.
3. The shape of the part is complex—for example, it has internal and external profiles or small-diameter holes.

4. Surface finish and dimensional accuracy better than those obtainable by other processes are required.

5. Temperature rise or residual stresses in the workpiece is undesirable or unacceptable.

These requirements led to the development of chemical, electrical, and other means of material removal in the 1940s; these methods are now called **advanced machining**, but have also been called **nontraditional** or **unconventional machining**. When selected and applied properly, these processes offer significant economic and technical advantages over the traditional machining methods described in the preceding chapter.

## 9.2 | Abrasives

The abrasives commonly used in manufacturing are as follows:

1. **Conventional abrasives:**

   - *Aluminum oxide*;
   - *Silicon carbide*.

2. **Superabrasives:**

   - *Cubic boron nitride*;
   - *Diamond*.

Abrasives are considerably harder than conventional cutting-tool materials (see Table 8.6 and Table 9.1); the last two of the four aforementioned abrasives are the two hardest materials known, hence the term *superabrasives*. In addition to hardness, an important characteristic of abrasives is **friability**, that is, the ability of abrasive grains to fracture (break down) into smaller pieces; this property gives abrasives *self-sharpening* characteristics, which are important in maintaining the sharpness of the abrasives in their use. High friability indicates low strength or low fracture resistance of the abrasive; thus, a highly friable abrasive grain fragments more rapidly under grinding forces than an abrasive grain with low friability. Aluminum oxide has lower friability than silicon carbide, hence it has less tendency to fragment.

The **shape** and **size** of the abrasive grain also affect its friability; blocky grains, for example, which may be analogous to negative-rake-angle cutters, are less friable than platelike grains. Furthermore, because the probability of defects in smaller grains is lower (due to the *size effect*; Section 3.8.3), they are stronger and less friable than larger grains. The importance of friability in abrasive processes is described in Section 9.5.

**Types of abrasives.** Abrasives found in nature include *emery*, *corundum* (*alumina*), *quartz*, *garnet*, and *diamond*. However, natural abrasives contain unknown amounts

**TABLE 9.1**

| Knoop Hardness Range for Various Materials and Abrasives | | | |
|---|---|---|---|
| Common glass | 350–500 | Titanium nitride | 2000 |
| Flint, quartz | 800–1100 | Titanium carbide | 1800–3200 |
| Zirconium oxide | 1000 | Silicon carbide | 2100–3000 |
| Hardened steels | 700–1300 | Boron carbide | 2800 |
| Tungsten carbide | 1800–2400 | Cubic boron nitride | 4000–5000 |
| Aluminum oxide | 2000–3000 | Diamond | 7000–8000 |

of impurities and possess nonuniform properties; consequently, their performance is inconsistent and unreliable. As a result, aluminum oxides and silicon carbides used as abrasives are now made synthetically. These two abrasives and the two superabrasives are described as follows:

a) Synthetic **aluminum oxide** ($Al_2O_3$), first made in 1893, is obtained by fusing bauxite, iron filings, and coke. Aluminum oxides are divided into two groups: fused and unfused. **Fused** aluminum oxides are categorized as *white* (very friable), *dark* (less friable), and *monocrystalline*. **Unfused** alumina, also known as *ceramic aluminum oxides*, can be harder than fused alumina. The purest form of fused alumina is seeded gel. First introduced in 1987, **seeded gel** has a particle size on the order of $0.2 \mu$m ($8 \mu$in.), which is much smaller than that of commonly used abrasive grains. These particles are sintered to form larger sizes. Because of their hardness and relatively high friability, seeded gels maintain their sharpness and are therefore used for difficult-to-grind materials.

b) **Silicon carbide** (SiC), first discovered in 1891, is made with silica sand, petroleum coke, and small amounts of sodium chloride. Silicon carbides are divided into *green* (more friable) and *black* (less friable) types and generally have higher friability than aluminum oxides; hence, they have a higher tendency to fracture and remain sharp.

c) **Cubic boron nitride** (cBN) was first discovered in the 1970s. Its properties and characteristics are described in Chapters 8 and 11.

d) **Diamond** was first used as an abrasive in 1955. It is also produced synthetically, whereby it is known as *synthetic* or *industrial diamond*. Its properties and characteristics are described in Chapters 8 and 11.

**Grain size.** As used in manufacturing processes, abrasives are generally very small compared with the size of the cutting tools and inserts described in Chapter 8. Also, abrasives have sharp edges, thus allowing the removal of very small quantities of material from the workpiece surface. Consequently, very fine surface finish and dimensional accuracy can be obtained. (See Figs. 8.27 and 9.27.)

The size of an abrasive grain is identified by a **grit number**, which is a function of sieve size. The smaller the sieve size, the larger is the grit number. For example, grit number 10 is rated as very coarse, 100 as fine, and 500 as very fine. Sandpaper and emery cloth are also identified in this manner, with the grit number printed on the back of the abrasive paper or cloth.

## 9.3 | Bonded Abrasives

Because each abrasive grain usually removes only a very small amount of material at a time, high rates of material removal can be obtained only if a large number of these grains act together; this is accomplished by using *bonded abrasives* (as opposed to loose abrasives), typically in the form of a **grinding wheel**. A simple grinding wheel is shown schematically in Fig. 9.1. The abrasive grains are held together by a **bonding material** (various types of which are described in Section 9.3.1), which acts as supporting posts or braces between the grains. Some porosity is essential in bonded wheels to provide clearance for the minute chips being produced and to provide cooling; otherwise, the chips would interfere with the grinding process. (It is thus impossible to use a grinding wheel that is fully dense and solid with no porosity.) Porosity can easily be observed simply by looking at the surface of any grinding

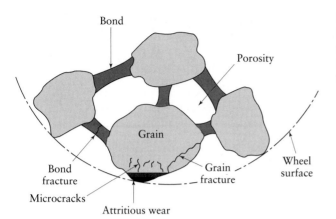

**FIGURE 9.1** Schematic illustration of a physical model of a grinding wheel, showing its structure and wear and fracture patterns.

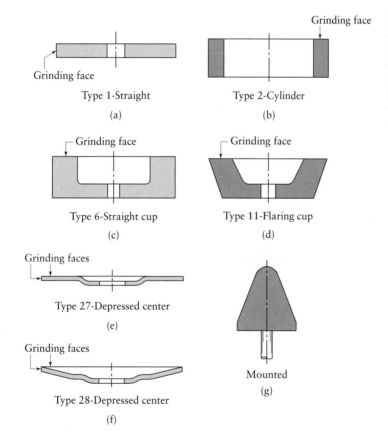

**FIGURE 9.2** Some common types of grinding wheels made with conventional abrasives. Note that each wheel has a specific grinding face. Grinding on other surfaces is improper and unsafe.

wheel. Other features of the grinding wheel shown in Fig. 9.1 are described in Sections 9.4 and 9.5.

Some of the more commonly used types of grinding wheels are shown in Fig. 9.2 for conventional abrasives and in Fig. 9.3 for superabrasives. Note that due to the high cost of the latter, only a small percentage of the wheels used is of the superabrasives variety. Bonded abrasives are marked with a standardized system of letters and numbers, indicating the type of abrasive, grain size, grade, structure, and bond type.

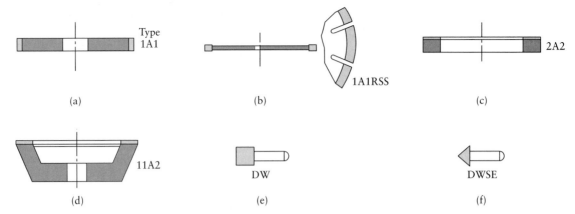

**FIGURE 9.3** Examples of superabrasive-wheel configurations. The annular regions (rim) are superabrasive grinding surfaces, and the wheel itself (core) is generally made of metal or composites. Note that the basic numbering of wheel type is the same as shown in Fig. 9.2. The bonding materials for the superabrasives are (a), (d), and (e) resinoid, metal, or vitrified; (b) metal; (c) vitrified; and (f) resinoid.

Figure 9.4a shows the marking system for aluminum-oxide and silicon-carbide bonded abrasives, and Fig. 9.4b shows the marking system for diamond and cubic-boron-nitride bonded abrasives.

### 9.3.1 Bond types

The common bond types for bonded abrasives are vitrified, resinoid, rubber, and metal. These bonds are used for conventional abrasives as well as for superabrasives (except rubber) and are described as follows:

1. **Vitrified.** Essentially a glass, *vitrified bond* is also called a *ceramic bond*, particularly outside the United States; it is the most common and widely used bond. The raw materials consist of feldspar (a crystalline mineral) and clays. These materials are mixed with the abrasives, moistened, and molded under pressure into the shape of grinding wheels. These "green" products, which are similar to powder-metallurgy parts, are then fired slowly, up to a temperature of about 1250°C (2300°F), to fuse the glass and develop structural strength. The wheels are then cooled slowly to avoid thermal cracking, finished to size, inspected for quality and dimensional accuracy, and tested for defects. Vitrified bonds produce wheels that are strong, stiff, porous, and resistant to oils, acids, and water. Because the wheels are brittle, they lack resistance to mechanical and thermal shock. However, vitrified wheels are also available with steel backing plates or cups for better structural support during their use.

2. **Resinoid.** Resinoid bonding materials are *thermosetting resins* (see Section 10.4) and are available in a wide range of compositions and properties. Because the bond is an organic compound, wheels with *resinoid bonds* are also called **organic wheels**. The basic manufacturing technique consists of mixing the abrasive with liquid or powdered phenolic resins and additives, pressing the mixture into the shape of a grinding wheel, and curing it at temperatures of about 175°C (350°F). Because the elastic modulus of thermosetting resins is lower than that of glasses, resinoid wheels are more flexible than vitrified wheels. **Reinforced resinoid wheels** are widely used, in which one or more layers of fiberglass mats of various mesh sizes provide the reinforcement. The main purpose of the reinforcements is to retard the disintegration of the wheel should it break for

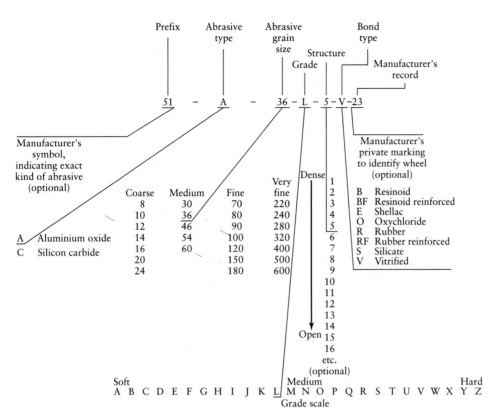

**FIGURE 9.4a** Standard marking system for aluminum-oxide and silicon-carbide bonded abrasives.

**FIGURE 9.4b** Standard marking system for diamond and cubic-boron-nitride bonded abrasives.

| Prefix | Abrasive type | Grit size | Grade | Diamond concentration | Bond modification | | Diamond depth (in.) |
|---|---|---|---|---|---|---|---|
| B Cubic boron nitride | | 20 | A (soft) | 25 (low) | B | Resinoid | 1/16 |
| D Diamond | | 24 | | | M | Metal | 1/8 |
| | | 30 | | 50 | V | Vitrified | 1/4 |
| | | 36 | to | 75 | | | Absence of depth |
| | | 46 | | 100 (high) | | | symbol indicates |
| | | 54 | | | | | solid diamond |
| | | 60 | Z (hard) | | | | |
| | | 80 | | | | | |
| | | 90 | | | | | |
| | | 100 | | | | | |
| | | 120 | | | | | |
| | | 150 | | | | | |
| Manufacturer's | | 180 | | | | | |
| symbol to | | 220 | | A letter, numeral, | | | |
| indicate type of | | 240 | | or combination used | | | |
| diamond | | 280 | | here indicates | | | |
| | | 320 | | a variation from | | | |
| | | 400 | | standard bond | | | |
| | | 500 | | | | | |
| | | 600 | | | | | |
| | | 800 | | | | | |
| | | 1000 | | | | | |

some reason, rather than to improve its strength. Large-diameter wheels can be supported additionally with one or more internal rings made of round steel bar, which are inserted during molding of the wheel.

3. **Rubber.** The most flexible bond used in abrasive wheels is *rubber*. The manufacturing process consists of mixing crude rubber, sulfur, and the abrasive grains together, rolling the mixture into sheets, cutting out circles, and heating the circles under pressure to vulcanize the rubber. Thin wheels can be made in this manner and are used like saws for cutting-off operations (cut-off blades).

4. **Metal bonds.** The abrasive grains, which are usually diamond or cubic boron nitride, are bonded, using powder-metallurgy techniques, to the periphery of a metal wheel, to depths of 6 mm (0.25 in.) or less. (See Fig. 9.3.) *Metal bonding* is carried out under high pressure and temperature. The wheel itself (core) may be made of aluminum, bronze, steel, ceramics, or composite materials, depending on requirements for the wheel, such as strength, stiffness, and dimensional stability.

5. **Other bonds.** In addition to those described previously, other types of bonds include *silicate*, *shellac*, and *oxychloride* bonds; however, these bonds have limited uses, and are not discussed further here. A new development is the use of *polyimide* (see Section 10.6) as a substitute for the phenolic in resinoid wheels. Polyimide is tough and has resistance to high temperatures. Superabrasive wheels may also be layered so that a single abrasive layer is plated or brazed to a metal wheel with a particular desired shape. Such wheels are lower in cost and are used for small production quantities.

### 9.3.2 Wheel grade and structure

The *grade* of a bonded abrasive is a measure of the bond's strength; it includes both the *type* and the *amount* of bond in the wheel. Because strength and hardness are directly related, the grade is also referred to as the *hardness* of a bonded abrasive. Thus, a hard wheel has a stronger bond and/or a larger amount of bonding material between the grains than a soft wheel. The *structure* is a measure of the *porosity* (the spacing between the grains in Fig. 9.1) of the bonded abrasive. Some porosity is essential to provide clearance for the grinding chips; otherwise, they would interfere with the grinding process. The structure of bonded abrasives ranges from dense to open.

## 9.4 | Mechanics of Grinding

Grinding is basically a *chip removal* process in which the cutting tool is an individual abrasive grain. The following are major factors that differentiate the action of a single grain from that of a single-point tool:

a) The individual grain has an irregular geometry and is spaced randomly along the periphery of the wheel (Fig. 9.5).

b) The average rake angle of the grains is highly negative: $-60°$ or even lower. Consequently, the shear angles are very low.

c) The radial positions of the grains in a grinding wheel vary.

d) The cutting speeds of grinding wheels are very high (Table 9.2), typically 30 m/s (6000 ft/min).

Figure 9.6 shows an example of chip formation by an abrasive grain; note the negative rake angle, the low shear angle, and the small size of the chip. A variety of

**FIGURE 9.5**    The grinding surface of an abrasive wheel (A46–J8V), showing grains, porosity, wear flats on grains (see also Fig. 9.9), and metal chips from the workpiece adhering to the grains. Note the random distribution and shape of the abrasive grains. Magnification: 50×.

TABLE 9.2

**Typical Ranges of Speeds and Feeds for Abrasive Processes**

| Process Variable | Conventional Grinding | Creep-Feed Grinding | Buffing | Polishing |
|---|---|---|---|---|
| Wheel speed (m/min) | 1500–3000 | 1500–3000 | 1800–3600 | 1500–2400 |
| Work speed (m/min) | 10–60 | 0.1–1 | – | – |
| Feed (mm/pass) | 0.01–0.05 | 1–6 | – | – |

**FIGURE 9.6**    Grinding chip being produced by a single abrasive grain: (A) chip, (B) workpiece, (C) abrasive grain. Note the large negative rake angle of the grain. The inscribed circle is 0.065 mm (0.0025 in.) in diameter. *Source*: M. E. Merchant.

metal chips can be observed in grinding. (Chips are easily collected on a piece of adhesive tape held against the sparks of a grinding wheel.)

The mechanics of grinding and the variables involved can best be studied by analyzing the *surface-grinding* operation shown in Fig. 9.7. In this figure, a grinding wheel of diameter $D$ is removing a layer of metal at a depth $d$, known as the **wheel depth of cut**. An individual grain on the periphery of the wheel is moving at a tangential velocity $V$ (*up*, or *conventional*, *grinding*, as shown in Fig. 9.7), and the workpiece is moving at a velocity $v$. The grain is removing a chip whose *undeformed thickness*

**FIGURE 9.7** Variables in surface grinding. In actual grinding, the wheel depth of cut, $d$, and contact length, $\ell$, are much smaller than the wheel diameter, $D$. The dimension $t$ is called the *grain depth of cut*.

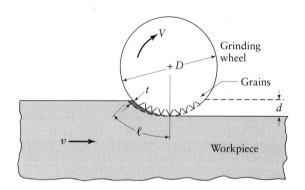

(**grain depth of cut**) is $t$ and the undeformed length is $\ell$. For $v \ll V$, the *undeformed-chip length*, $\ell$ is approximately

$$\ell \simeq \sqrt{Dd}. \tag{9.1}$$

For *external (cylindrical) grinding* (see Section 9.6),

$$\ell = \sqrt{\frac{Dd}{1 + (D/D_w)}}, \tag{9.2}$$

and for *internal grinding*,

$$\ell = \sqrt{\frac{Dd}{1 - (D/D_w)}}, \tag{9.3}$$

where $D_w$ is the diameter of the workpiece.

The relationship between $t$ and other process variables can be derived as follows: Let $C$ be the number of cutting points per unit area of wheel surface; $v$ and $V$ are the surface speeds of the workpiece and the wheel, respectively (Fig. 9.7). If we let the width of the workpiece be unity, the number of chips produced per unit time is $VC$, and the volume of material removed per unit time is $vd$. Letting $r$ be the ratio of the chip width $w$ to the average chip thickness, we find that the volume of a chip with a rectangular cross-sectional area and constant width is

$$\text{Vol}_{\text{chip}} = \frac{wt\ell}{2} = \frac{rt^2\ell}{4}. \tag{9.4}$$

The volume of material removed per unit time, then, is the product of the number of chips produced per unit time and the volume of each chip, or

$$VC\frac{rt^2\ell}{4} = vd,$$

and because $\ell = \sqrt{Dd}$, the undeformed chip thickness in surface grinding is

$$t = \sqrt{\frac{4v}{VCr}\sqrt{\frac{d}{D}}}. \tag{9.5}$$

Experimental observations indicate the value of $C$ to be roughly on the order of 0.1 per mm$^2$ to 10 per mm$^2$ ($10^2$ per in$^2$ to $10^3$ per in$^2$); the finer the grain size of the wheel, the larger is this number. The magnitude of $r$ is between 10 and 20 for most grinding operations. If we substitute typical values for a grinding operation into Eqs. (9.1) through (9.5), we obtain very small quantities for $\ell$ and $t$. For example, typical values for $t$ are in the range of 0.3–0.4 $\mu$m (12–160 $\mu$in.).

**EXAMPLE 9.1   Chip dimensions in grinding**

Estimate the undeformed-chip length and the undeformed chip thickness for a typical surface-grinding operation.

**Solution.** The formulas for undeformed length and thickness, respectively, are

$$\ell = \sqrt{Dd} \quad \text{and} \quad t = \sqrt{\frac{4v}{VCr}}\sqrt{\frac{d}{D}}.$$

From Table 9.2, we select the following values:

$$v = 0.5 \text{ m/s} \quad \text{and} \quad V = 30 \text{m/s}.$$

We also assume that

$$d = 0.05 \text{mm} \quad \text{and} \quad D = 200 \text{ mm}$$

and let

$$C = 2 \text{ per mm}^2 \quad \text{and} \quad r = 15.$$

Therefore,

$$\ell = \sqrt{(200)(0.05)} = 3.2 \text{ mm} = 0.126 \text{ in.}$$

and

$$t = \sqrt{\frac{(4)(0.5)}{(30)(2)(15)}}\sqrt{\frac{0.05}{200}} = 0.006 \text{ mm} = 2.3 \times 10^{-4} \text{ in.}$$

Because of plastic deformation, the actual length of the chip is shorter and the thickness greater than these values. (See Fig. 9.6.)

## 9.4.1 Forces

A knowledge of forces is essential not only in the design of grinding machines, but also in determining the deflections that the workpiece and the machine will undergo, as well as in the design and use of workholding devices. Deflections, in turn, adversely influence dimensional accuracy and are thus critical in precision grinding. If we assume that the *force* on the grain (see the discussion of cutting force, $F_c$, in Section 8.2.3) is proportional to the cross-sectional area of the undeformed chip, we can show that the **relative grain force** is given by

$$\text{Relative grain force} \propto \frac{v}{VC}\sqrt{\frac{d}{D}}. \tag{9.6}$$

The actual force is the product of the relative grain force and the strength of the metal being ground.

The **specific energy** consumed in producing a grinding chip consists of three components:

$$u = u_{\text{chip}} + u_{\text{plowing}} + u_{\text{sliding}}. \tag{9.7}$$

In this equation, $u_{\text{chip}}$ is the specific energy required for chip formation by plastic deformation, and $u_{\text{plowing}}$ is the specific energy required for plowing, which is plastic

**FIGURE 9.8** Chip formation and plowing of the workpiece surface by an abrasive grain.

**FIGURE 9.9** Schematic illustration of chip formation by an abrasive grain. Note the negative rake angle, the small shear angle, and the wear flat on the grain.

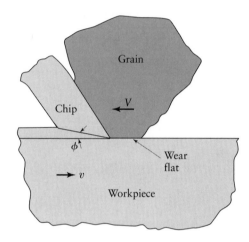

deformation without chip removal (Fig. 9.8). The last term, $u_{sliding}$, can best be understood by observing the grain in Fig. 9.9. The grain develops a **wear flat** as a result of the grinding operation (similar to flank wear in cutting tools). The wear flat slides along the surface being ground and, because of friction, requires energy for sliding; the larger the wear flat, the higher the grinding force is.

Table 9.3 provides typical specific-energy requirements in grinding. Note that these energy levels are much higher than those in cutting operations with single-point tools. (See Table 8.3.) This difference has been attributed to the following factors:

1. **Size effect**: As previously stated, the size of grinding chips is quite small compared with chips produced by other cutting operations, by about two orders of magnitude. As described in Section 3.8.3, the smaller the size of a piece of metal, the greater is its strength; consequently, grinding involves higher specific energy than cutting. Studies have indicated that extremely high dislocation densities (see Section 3.3.3) occur in the shear zone during chip formation, thus influencing the grinding energies involved.

**TABLE 9.3**

## Approximate Specific-Energy Requirements for Surface Grinding

| Workpiece Material | Hardness | Specific Energy | |
| --- | --- | --- | --- |
| | | W-s/mm$^3$ | hp-min/in$^3$ |
| Aluminum | 150 HB | 7–27 | 2.5–10 |
| Cast iron (class 40) | 215 HB | 12–60 | 4.5–22 |
| Low-carbon steel (1020) | 110 HB | 14–68 | 5–25 |
| Titanium alloy | 300 HB | 16–55 | 6–20 |
| Tool steel (T15) | 67 HRC | 18–82 | 6.5–30 |

2. **Wear flat**: Because a wear flat (see Fig. 9.9) requires frictional energy for sliding, this energy can contribute substantially to the total energy consumed. The size of the wear flat in grinding is much larger than the grinding chip, unlike in metal cutting by a single-point tool, where flank wear land is small compared with the size of the chip.

3. **Chip morphology**: Because the average rake angle of a grain is highly negative (see Fig. 9.6), the shear strains are very large. This indicates that the energy required for plastic deformation to produce a grinding chip is higher than in other cutting processes. Furthermore, plowing consumes energy without contributing to chip formation.

---

**EXAMPLE 9.2** Forces in surface grinding

Assume that you are performing a surface-grinding operation on low-carbon steel with a wheel of diameter $D = 10$ in. that rotates at $N = 4000$ rpm, with a width of cut $w = 1$ in. The depth of cut is $d = 0.002$ in. and the feed rate $v$ of the workpiece is 60 in./min Calculate the cutting force, $F_c$ (the force tangential to the wheel), and the thrust force, $F_n$ (the force normal to the workpiece).

**Solution.** We first determine the material removal rate as follows:

$$\text{MRR} = dwv = (0.002)(1)(60) = 0.12 \text{ in}^3/\text{min.}$$

The power consumed is given by

$$\text{Power} = (u)(\text{MRR}),$$

where $u$ is the specific energy, as obtained from Table 9.3. For low-carbon steel, let's estimate $u$ to be 15 hp-min/in$^3$. Hence, we have

$$\text{Power} = (15)(0.12) = 1.8 \text{ hp.}$$

By noting that 1 hp = 33,000 ft-lb/min = 396,000 in.-lb/min, we obtain

$$\text{Power} = (1.8)(396,000) = 712,800 \text{ in.-lb/min.}$$

Since power is defined as

$$\text{Power} = T\omega,$$

where $T$ is the torque and is equal to $(F_c)(D/2)$ and $\omega$ is the rotational speed of the wheel in radians per minute, we also have $\omega = 2\pi N$. Thus,

$$712,800 = (F_c)\left(\frac{10}{2}\right)(2\pi)(4000),$$

and therefore, $F_c = 57$ lb. The thrust force, $F_n$, can be calculated by noting from experimental data in the technical literature that it is about 30% higher than the cutting force, $F_c$. Consequently,

$$F_n = (1.3)(57) = 74 \text{ lb.}$$

---

## 9.4.2 Temperature

*Temperature rise* in grinding is an important consideration, because it can adversely affect the surface properties and cause residual stresses on the workpiece. Furthermore, temperature gradients in the workpiece cause distortions by differential thermal expansion and contraction. When a portion of the heat generated is conducted into the workpiece, the heat expands the part being ground, thus making it difficult to control dimensional accuracy. The work expended in grinding is mainly converted into heat. The *surface temperature rise*, $\Delta T$, has been found to be a function of the ratio of the total energy input to the surface area ground. Thus, in *surface grinding*, if $w$ is the width and $L$ is the length of the surface area ground, then

$$\Delta T \propto \frac{uwLd}{wL} \propto ud.$$ (9.8)

If we introduce size effect and assume that $u$ varies inversely with the undeformed-chip thickness $t$, then the temperature rise is

$$\Delta T \propto \frac{d}{t} \propto d^{3/4} \sqrt{\frac{VC}{v}\sqrt{D}}.$$ (9.9)

The *peak temperatures* in chip generation during grinding may be as high as $1650°C\,(3000°F)$. However, the time involved in producing a chip is extremely short (on the order of microseconds); hence, melting may or may not occur. Because, as in metal cutting, the chips carry away much of the heat generated, only a fraction of the heat generated is conducted to the workpiece. Experiments indicate that in grinding, as much as one half the energy is conducted to the workpiece; this percentage is higher than that in metal cutting (see Section 8.2). The heat generated by sliding and plowing is conducted mostly into the workpiece.

**Sparks.** The sparks observed in metal grinding are actually glowing chips; the glowing occurs because of the exothermic reaction of the hot chips with oxygen in the atmosphere. Sparks have not been observed with any metal ground in an oxygen-free environment. The color, intensity, and shape of the sparks depend on the composition of the metal being ground. If the heat generated by the exothermic reaction is sufficiently high, the chip may melt and, because of surface tension, acquire a round shape and solidify as a shiny spherical particle. Observation of these particles under scanning electron microscopy has revealed that they are hollow and have a fine dendritic structure (see Fig. 5.8), indicating that they were once molten (by exothermic oxidation of hot chips in air) and resolidified rapidly. It has been suggested that some of the spherical particles may also be produced by plastic deformation and rolling of chips at the grit–workpiece interface during grinding.

### 9.4.3 Effects of temperature

The following list describes briefly the major effects of temperature in grinding:

a) **Tempering.** Excessive temperature rise caused by grinding can temper and soften the surfaces of steel components, which are often ground in the hardened state. Grinding process parameters must therefore be chosen carefully to avoid excessive temperature rise. The use of grinding fluids can effectively control temperatures.

b) **Burning.** If the temperature is excessive, the surface may burn; burning produces a bluish color on steels, which indicates oxidation at high temperatures. A burn may not be objectionable in itself; however, the surface layers may undergo metallurgical transformations, with martensite formation in high-carbon steels from reaustenization followed by rapid cooling (see Section 5.11). This effect is known as *metallurgical burn*, which also is a serious problem with nickel-base alloys. High temperatures in grinding may also lead to thermal cracking of the surface of the workpiece, known as **heat checking**. (See also

(a)                                        (b)

**FIGURE 9.10** Residual stresses developed on the workpiece surface in grinding tungsten: (a) effect of wheel speed and (b) effect of grinding fluid. Tensile residual stresses on a surface are detrimental to the fatigue life of ground components. The variables in grinding can be controlled to minimize residual stresses, a process known as *low-stress grinding*. *Source*: After N. Zlatin et al., 1963.

Section 5.10.3.) Cracks are usually perpendicular to the grinding direction; however, under severe grinding conditions, parallel cracks may also develop.

c) **Residual stresses.** Temperature change and gradients within the workpiece are mainly responsible for residual stresses in grinding. Other contributing factors are the physical interactions of the abrasive grain in chip formation and the sliding of the wear flat along the workpiece surface, causing plastic deformation of the surface. Figure 9.10 shows two examples of residual stresses in grinding, demonstrating the effects of wheel speed and the type of grinding fluid used. The method and direction of the application of grinding fluid can also have a significant effect on residual stresses. Because of the deleterious effect of tensile residual stresses on fatigue strength, process parameters should be chosen carefully. Residual stresses can usually be lowered (**low-stress**, or **gentle, grinding**) by using softer grade wheels (*free-cutting wheels*), lower wheel speeds, and higher work speeds.

## 9.5 | Grinding-Wheel Wear

Grinding-wheel wear is an important consideration, because it adversely affects the shape and accuracy of ground surfaces, as is the case with cutting tools. Grinding wheels wear by three different mechanisms: attritious grain wear, grain fracture, and bond fracture. These mechanisms are described as follows:

1. **Attritious wear.** In *attritious wear*, the cutting edges of a sharp grain become dull by attrition, developing a *wear flat* (see Fig. 9.9) that is similar to flank wear in cutting tools. Wear is caused by the interaction of the grain with the workpiece material, involving both physical and chemical reactions. These reactions are

complex and involve diffusion, chemical degradation or decomposition of the grain, fracture at a microscopic scale, plastic deformation, and melting.

Attritious wear is low when the two materials are chemically inert with respect to each other, much like with the use of cutting tools. The more inert the materials, the lower will be the tendency for reaction and adhesion to occur between the grain and the workpiece. For example, because aluminum oxide is relatively inert with respect to iron, its rate of attritious wear when it is used to grind steels is much lower than that for silicon carbide and diamond. On the other hand, silicon carbide can dissolve in iron, and hence it is not suitable for grinding steels. Cubic boron nitride has a higher inertness with respect to steels, and hence it is suitable for use as an abrasive. The selection of the type of abrasive for low attritious wear is, therefore, based on the reactivity of the grain and the workpiece and their relative mechanical properties, such as hardness and toughness. The environment and the type of grinding fluid used also have an influence on grit–workpiece-material interactions.

2. **Grain fracture.** Because abrasive grains are brittle, their fracture characteristics in grinding are important. If the wear flat caused by attritious wear is excessive, the grain becomes dull, and grinding becomes inefficient and produces undesirably high temperatures. Optimally, the grain should fracture or fragment at a moderate rate, so that new sharp cutting edges are produced continuously during grinding. This process is equivalent to breaking a piece of stone or dull chalk into two or more pieces in order to expose new sharp edges. Section 9.2 has described *friability* of abrasives, which gives them their self-sharpening characteristics, an important consideration in effective grinding.

   The selection of grain type and size for a particular application also depends on the attritious-wear rate. A grain–workpiece-material combination with high attritious wear and low friability causes dulling of grains and development of a large wear flat. Grinding then becomes inefficient, and surface damage such as burning is likely to occur.

   The following combinations are generally recommended:

   1. *Aluminum oxide*: for steels, ferrous alloys, and alloy steels.
   2. *Silicon carbide*: for cast iron, nonferrous metals, and hard and brittle materials, such as carbides, ceramics, marble, and glass.
   3. *Diamond*: for cemented carbide ceramics and some hardened steels.
   4. *Cubic boron nitride*: for steels and cast irons at 50 HRC or above and for high-temperature superalloys.

3. **Bond fracture.** The strength of the bond (*grade*) is a significant parameter in grinding. If the bond is too strong, dull grains cannot be dislodged so that other, sharp grains along the circumference of the grinding wheel can begin to contact the workpiece and remove chips; thus, the grinding process becomes inefficient. On the other hand, if the bond is too weak, the grains are dislodged easily, and the wear rate of the wheel increases; consequently, maintaining dimensional accuracy becomes difficult. In general, softer bonds are recommended for harder materials and for reducing residual stresses and thermal damage to the workpiece. Hard-grade wheels are used for softer materials and for removing large amounts of material at high rates.

## 9.5.1 Dressing, truing, and shaping of grinding wheels

*Dressing* is the process of conditioning worn grains on the surface of a grinding wheel in order to produce sharp new grains and for truing an out-of-round wheel. Dressing is necessary when excessive attritious wear dulls the wheel (a phenomenon called **glazing**, because of the shiny appearance of the wheel surface) or when the wheel becomes loaded. **Loading** occurs when the porosities on the grinding surfaces of the wheel (see Fig. 9.5) become filled or clogged with chips. Loading can occur in grinding soft materials or by improper selection of the grinding wheel, such as a wheel with low porosity, and improper selection of process parameters. A loaded wheel cuts very inefficiently, generating much frictional heat, causing surface damage and loss of dimensional accuracy.

Dressing is done by any of the following techniques:

1. A specially shaped diamond-point tool or diamond cluster is moved across the width of the grinding face of a rotating wheel, removing a small layer from the wheel surface with each pass. This method can be done either dry or wet, depending on whether the wheel is to be used dry or wet, respectively.

2. A set of star-shaped steel disks is manually pressed against the wheel. Material is removed from the wheel surface by crushing the grains. This method produces a coarse grinding surface on the wheel and is used only for rough grinding operations on bench or pedestal grinders.

3. Abrasive sticks may be held against the grinding surface of the wheel.

4. More recent developments in dressing techniques, for metal-bonded diamond wheels, involve the use of electrical-discharge and electrochemical material-removal techniques, which erode away very small layers of the metal bond, exposing new diamond cutting edges.

Dressing for *form grinding* involves **crush dressing**, or *crush forming*, and consists of pressing a metal roll on the surface of the grinding wheel, which is usually a vitrified wheel. The roll, which may be made of high-speed steel, tungsten carbide, or boron carbide, has a machined or ground profile and thus reproduces a replica of this profile on the surface of the grinding wheel being dressed. Dressing techniques and the frequency with which the wheel surface is dressed are significant, affecting grinding forces and surface finish. Modern computer-controlled grinders are equipped with automatic dressing features, which dress the wheel as grinding continues. Dressing can also be done to generate a certain shape or form on a grinding wheel for the purpose of grinding profiles on workpieces. (See Section 9.6.2.)

*Truing* is an operation by which a wheel is restored to its original shape; a round wheel is dressed to make its circumference a true circle (hence the word *truing*). Grinding wheels can also be *shaped* to the form to be ground on the workpiece. The grinding face on the Type 1 straight wheel shown in Fig. 9.2a is cylindrical and thus produces a flat surface; however, this surface can be shaped by dressing it to various forms. Although templates have been used for this purpose, new grinders are equipped with computer-controlled shaping features, whereby the diamond dressing tool traverses the wheel face automatically along a certain prescribed path (Fig. 9.11). Note that the axis of the diamond dressing tool remains normal to the wheel face at the point of contact.

## 9.5.2 Grinding ratio

Grinding-wheel wear is generally correlated with the amount of material ground by a parameter called the *grinding ratio*, $G$, which is defined as

$$G = \frac{\text{Volume of material removed}}{\text{Volume of wheel wear}}. \qquad (9.10)$$

In practice, grinding ratios vary widely, ranging from 2 to 200 and higher, depending on the type of wheel, the workpiece material, the grinding fluid, and process

**FIGURE 9.11** Shaping the grinding face of a wheel by dressing it with computer controlled shaping features. Note that the diamond dressing tool is normal to the surface at point of contact. *Source*: OKUMA America Corporation.

parameters such as depth of cut and speeds of the wheel and workpiece. Attempting to obtain a high grinding ratio in practice isn't always desirable, because high ratios may indicate grain dulling and possible surface damage. A lower ratio may be quite acceptable when an overall economic analysis justifies it.

**Soft- or hard-acting wheels.** During a grinding operation, a particular wheel may *act soft* (wear rate is high) or *hard* (wear rate is low), regardless of its grade. Note, for example, that an ordinary pencil acts soft when you write on rough paper and acts hard when you write on soft paper. This behavior is a function of the force on the grain. The greater the force, the greater is the tendency for the grains to fracture or be dislodged from the wheel surface, and hence the higher is the wheel wear and the lower is the grinding ratio. Equation (9.6) shows that the grain force increases with the strength of the workpiece material, the work speed, and the depth of cut and decreases with increasing wheel speed and wheel diameter. Thus, a wheel acts soft when $v$ and $d$ increase, or when $V$ and $D$ decrease.

---

**EXAMPLE 9.3** Action of a grinding wheel

A surface-grinding operation is being carried out with the wheel running at a constant spindle speed. Will the wheel act soft or hard as the wheel wears down over a period of time?

**Solution.** Referring to Eq. (9.6), we note that the parameters that change with time in this operation are $v$ and the wheel diameter $D$ (assuming that the diameter remains constant and the wheel is dressed periodically). Hence, as $D$ becomes smaller, the relative grain force increases, and therefore the wheel acts softer. Some grinding machines are equipped with variable-speed spindle motors to accommodate these changes and to make provisions for wheels of different diameter.

---

### 9.5.3 Wheel selection and grindability of materials

Selection of a grinding wheel for a certain application greatly influences the quality of surfaces produced, as well as the economics of the operation. The selection involves not only the shape of the wheel with respect to the shape of the part to be produced, but the characteristics of the workpiece material as well. The *grindability* of materials, like machinability (see Section 8.5), is difficult to define precisely. It is a general indication of how easy it is to grind a material and includes considerations such as the quality of the surface produced, surface finish, surface integrity, wheel wear, cycle time, and overall economics. It should also be pointed out that just as with machinability, grindability of a material can be enhanced greatly by proper selection of process parameters, the type of wheel, and grinding fluids, as well as machine characteristics and fixturing methods.

Grinding practices have been well established for a wide variety of metallic and nonmetallic materials, including newly developed materials for aerospace applications. Specific recommendations for selecting wheels and process parameters can be found in various handbooks. Some examples of such recommendations are C60-L6V for cast irons, A60-M6V for steels, C60-I9V or D150-R75B for carbides, and A60-K8V for titanium. Ceramics can be ground with relative ease by using diamond wheels, as well as by carefully selecting process parameters; a typical wheel selection for ceramics is D150-N50M.

It has been shown that with light passes and rigid machines with good damping capacity, it is possible to obtain continuous chips in grinding ceramics. (See Figs. 9.8 and 9.9.) This procedure is known as **ductile regime grinding**; it produces good workpiece surface integrity. However, ceramic chips are typically $1-10 \ \mu$m ($40-400 \ \mu$in.) in size and hence are more difficult to remove from grinding fluids than metal chips, thus requiring fine filters and other effective methods.

## 9.6 | Grinding Operations and Machines

Grinding operations are carried out with a variety of wheel–workpiece configurations. The selection of a grinding process for a particular application depends on part shape, part size, ease of fixturing, and the production rate required. The basic types of grinding operations are surface, cylindrical, internal, and centerless grinding. The relative movement of the wheel may be along the surface of the workpiece (*traverse grinding*, *through feed grinding*, or *cross-feeding*), or it may be radially into the workpiece (*plunge grinding*). Surface grinders constitute the largest percentage of grinders in use in industry, followed by bench grinders (usually with two wheels), cylindrical grinders, and tool and cutter grinders. The least used type of grinder is the internal grinder.

Grinding machines are available for various workpiece geometries and sizes. Modern grinding machines may be computer controlled, with features such as automatic workpiece loading and unloading, clamping, cycling, gaging, dressing, and wheel shaping. Grinders can also be equipped with probes and gages for determining the relative position of the wheel and workpiece surfaces, as well as with tactile sensing features whereby diamond-dressing-tool breakage, if any, can be detected during the dressing cycle. Because grinding wheels are brittle and are operated at high speeds, certain procedures must be carefully followed in their handling, storage, and use. Failure to follow these rules and the instructions and warnings printed on individual wheel labels may result in serious injury or death.

### 9.6.1 Surface grinding

*Surface grinding* involves grinding flat surfaces and is one of the most common grinding operations (Fig. 9.12). Typically, the workpiece is secured on a magnetic chuck attached to the worktable of a *surface grinder* (Fig. 9.13). Nonmagnetic materials generally are held by vises, special fixtures, vacuum chucks, or double-sided adhesive tapes. A straight wheel is mounted on the *horizontal spindle* of the grinder. Traverse grinding is done as the table reciprocates longitudinally and feeds laterally after each stroke. In *plunge grinding*, the wheel is moved radially into the workpiece, as in

Horizontal-spindle surface grinder: traverse grinding

(a)

Horizontal-spindle surface grinder: plunge grinding

(b)

Rotary table

Worktable

(c)

**FIGURE 9.12** Schematic illustrations of surface-grinding operations. (a) Traverse grinding with a horizontal-spindle surface grinder. (b) Plunge grinding with a horizontal-spindle surface grinder, producing a groove in the workpiece. (c) Vertical-spindle rotary-table grinder (also known as the *Blanchard-type* grinder).

**FIGURE 9.13** Schematic illustration of a horizontal-spindle surface grinder. The majority of grinding operations are done on such machines.

grinding a groove (Fig. 9.12b). The size of a surface grinder is determined by the surface dimensions that can be ground on the machine. Other types of setups for surface grinding have *vertical spindles* and *rotary tables* (Fig. 9.12c), also referred to as *Blanchard-type* grinders. These configurations allow a number of pieces to be ground in one setup.

### 9.6.2 Cylindrical grinding

In *cylindrical grinding*, also called *center-type grinding*, the workpiece's external cylindrical surfaces and shoulders are ground. Typical applications include crankshaft bearings, spindles, pins, bearing rings, and rolls for rolling mills. The rotating cylindrical workpiece reciprocates laterally along its axis, although in grinders used for large and long workpieces, the grinding wheel reciprocates. The latter design is called a *roll grinder* and is capable of grinding rolls as large as 1.8 m (72 in.) in diameter.

The workpiece in cylindrical grinding is held between centers or in a chuck, or it is mounted on a faceplate in the headstock of the grinder. For straight cylindrical surfaces, the axes of rotation of the wheel and workpiece are parallel. Separate motors drive the wheel and workpiece at different speeds. Long workpieces with two or more diameters are ground on cylindrical grinders. Cylindrical grinding can produce shapes in which the wheel is dressed to the form to be ground on the workpiece (*form grinding* and *plunge grinding*).

Cylindrical grinders are identified by the maximum diameter and length of the workpiece that can be ground, similar to engine lathes. In *universal grinders*, both the workpiece and the wheel axes can be moved and swiveled around a horizontal plane,

**FIGURE 9.14**   Thread grinding by (a) traverse and (b) plunge grinding.

thus permitting the grinding of tapers and other shapes. These machines are equipped with computer controls, thereby reducing labor and producing parts accurately and repetitively. Cylindrical grinders can also be equipped with computer-controlled features so that *noncylindrical* parts (such as cams) can be ground on rotating workpieces. The workpiece spindle speed is synchronized such that the distance between the workpiece and wheel axes is varied continuously to produce a particular shape.

**Thread grinding** is done on cylindrical grinders, as well as on centerless grinders (see Section 9.6.4), with specially dressed wheels that match the shape of the threads (Fig. 9.14). Although this process is costly, it produces threads more accurately than any other manufacturing process, and the threads have a very fine surface finish. The workpiece and wheel movements are synchronized to produce the pitch of the thread, usually in about six passes.

### 9.6.3  Internal grinding

In *internal grinding* (Fig. 9.15), a small wheel is used to grind the inside diameter of a part, such as bushings and bearing races. The workpiece is held in a rotating chuck, and the wheel rotates at 30,000 rpm or higher. Internal profiles can also be ground with profile-dressed wheels that move radially into the workpiece. The headstock of internal grinders can be swiveled on a horizontal plane to grind tapered holes.

### 9.6.4  Centerless grinding

*Centerless grinding* is a high-production process for continuously grinding cylindrical surfaces in which the workpiece is supported not by centers (hence the term *centerless*) or chucks, but by a blade (Figs. 9.16 and 9.17). Typical parts made by centerless grinding include roller bearings, piston pins, engine valves, camshafts, and similar components. This continuous-production process requires little operator skill. Parts with diameters as small as 0.1 mm (0.004 in.) can be ground using this process. Centerless grinders are now capable of wheel surface speeds on the order of 10,000 m/min (35,000 ft/min), using cubic-boron-nitride abrasive wheels.

In *through-feed grinding*, the workpiece is supported on a work-rest blade and is ground between two wheels. Grinding is done by the larger wheel, while the smaller wheel regulates the axial movement of the workpiece. The *regulating wheel*, which is rubber bonded, is tilted and runs at speeds of only about $\frac{1}{20}$ of those of the grinding wheel.

(a) Traverse grinding     (b) Plunge grinding     (c) Profile grinding

**FIGURE 9.15**   Schematic illustrations of internal-grinding operations.

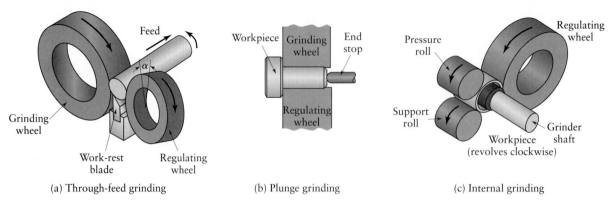

(a) Through-feed grinding          (b) Plunge grinding          (c) Internal grinding

**FIGURE 9.16**    Schematic illustrations of centerless-grinding operations.

**FIGURE 9.17**  A computer-numerical-control centerless grinding machine. The movement of the workpiece is perpendicular to the page. *Source:* Courtesy of Cincinnati Milacron, Inc.

Parts with variable diameters, such as bolts, valve tappets, and distributor shafts, can be ground by centerless grinding; called *infeed*, or *plunge*, *grinding* ((Fig. 9.16b), the process is similar to plunge or form grinding with cylindrical grinders. Tapered pieces are centerless ground by *end-feed grinding*. High-production-rate thread grinding can be done with centerless grinders, using specially dressed wheels. In *internal centerless grinding*, the workpiece is supported between three rolls and is internally ground. Typical applications include sleeve-shaped parts and rings.

### 9.6.5  Other grinders

A variety of special-purpose grinders is available. *Bench grinders* are used for routine offhand grinding of tools and small parts. They are usually equipped with two wheels mounted on the two ends of the shaft of an electric motor. One wheel is usually coarse for rough grinding, and the other is fine for finish grinding. *Pedestal*, or *stand*, *grinders* are placed on the floor and used similarly to bench grinders.

*Universal tool and cutter grinders* are used for grinding single-point or multi-point tools and cutters. They are equipped with special workholding devices for accurate positioning of the tools to be ground. *Tool-post grinders* are self-contained units and are usually attached to the tool post of a lathe. The workpiece is mounted on the headstock and is ground by moving the tool post. These grinders are versatile, but the lathe should be protected from abrasive debris.

*Swing-frame grinders* are used in foundries for grinding large castings. Rough grinding of castings is called *snagging* and is usually done on floorstand grinders, using wheels as large as 0.9 m (36 in.) in diameter. *Portable grinders*, either air or electrically driven, or with a flexible shaft connected to an electric motor or gasoline engine, are available for operations such as grinding off weld beads and *cutting-off* operations using thin abrasive disks, usually on large workpieces.

### 9.6.6  Creep-feed grinding

Grinding has traditionally been associated with small rates of material removal and fine finishing operations; however, grinding can also be used for large-scale metal removal operations, similar to milling, broaching, and planing. In *creep-feed grinding*, developed in the late 1950s, the wheel depth of cut, *d*, is as much as 6 mm (0.25 in.), and the workpiece speed is low (Fig. 9.18). The wheels are mostly softer grade resin bonded with open structure to keep temperatures low and improve surface finish. Grinders with capabilities for continuously dressing the grinding wheel with a diamond roll are available. The machines used for creep-feed grinding have special features, such as high power of up to 225 kW (300 hp), high stiffness (because of the high forces due to the depth of material removed), high damping capacity, variable and well-controlled spindle and worktable speeds, and ample capacity for grinding fluids.

Its overall economics and competitive position with respect to other material removal processes indicate that creep-feed grinding can be economical for specific applications, such as in grinding shaped punches, key seats, twist-drill flutes, the roots of turbine blades (Fig. 9.18c), and various complex superalloy parts. The wheel is dressed to the shape of the workpiece to be produced. Consequently, the workpiece does not have to be previously milled, shaped, or broached; thus, near-net-shape castings and forgings are suitable parts for creep-feed grinding. Although a single pass generally is sufficient, a second pass may be necessary for improved surface finish.

### 9.6.7  Heavy stock removal by grinding

Grinding processes can be used for large-scale material removal by varying process parameters. However, this is a rough grinding operation, with possibly detrimental effects on the workpiece surface and its integrity. It can be economical in specific

$d = 1$–6 mm

Low workpiece speed, *v*

(a)                                 (b)                                 (c)

**FIGURE  9.18**    (a) Schematic illustration of the creep-feed grinding process. Note the large wheel depth of cut. (b) A shaped groove produced on a flat surface in one pass by creep-feed grinding. Groove depth can be on the order of a few mm. (c) An example of creep-feed grinding with a shaped wheel. *Source*:  Courtesy of Blohm, Inc., and *Manufacturing Engineering*, Society of Manufacturing Engineers.

applications and compete favorably with machining processes, particularly milling, but also turning and broaching as well. In this operation, surface finish is of secondary importance, and the grinding wheel (or belt) is used to its maximum extent for minimum cost per piece. The geometric tolerances in this process are on the same order as those obtained by other machining processes. (See also Fig. 9.27.)

### 9.6.8 Grinding chatter

*Chatter* is particularly significant in grinding, because it adversely affects surface finish and wheel performance. Vibrations during grinding may be caused by bearings, spindles, and unbalanced wheels, as well as external sources, such as from nearby machinery; also, the grinding process can itself cause regenerative chatter. The analysis of chatter in grinding is similar to that for machining operations and involves *self–excited vibration* and *regenerative chatter* (see Section 8.11). Thus, the important variables are (a) stiffness of the machine tool and workholding devices and (b) damping. Additional factors that are unique to grinding chatter include nonuniformities in the grinding wheel itself, the dressing techniques used, and uneven wheel wear.

Because the foregoing variables produce characteristic **chatter marks** on ground surfaces, a study of these marks can often lead to the source of the problem. General guidelines have been established to reduce the tendency for chatter in grinding, such as (a) using soft-grade wheels, (b) dressing the wheel frequently, (c) changing dressing techniques, (d) reducing the material removal rate, and (e) supporting the workpiece rigidly. Advanced techniques include the use of sensors to monitor grinding chatter and control of it by chatter suppression systems.

### 9.6.9 Grinding fluids

The functions of grinding fluids are similar to those for cutting fluids (see also Section 8.7). Although grinding and other abrasive-removal processes can be performed dry, the use of a fluid is important and usually preferred, as the fluid prevents temperature rise in the workpiece and improves the part's surface finish and dimensional accuracy. Fluids also improve the efficiency of the operation by reducing wheel loading and wear and lowering power consumption.

Grinding fluids are typically *water-base emulsions* and *chemicals* and *synthetics; oils* may be used for thread grinding. The fluids may be applied as a stream (flood) or as mist, which is a mixture of fluid and air. Because of the high surface speeds involved, an air stream or air blanket around the periphery of the wheel usually prevents the fluid from reaching the cutting zone. Special *nozzles* that mate with the shape of the cutting surface of the grinding wheel have been designed for effective application of the grinding fluid under high pressure.

The temperature of water-base grinding fluids can rise significantly during their use as they remove heat from the grinding zone, in which case the workpiece can expand, making it difficult to control dimensional tolerances. The common method employed to maintain even temperature is to use refrigerating systems (chillers) through which the fluid is circulated. Also, as discussed with respect to cutting fluids in Section 8.7, the biological and ecological aspects, treatment, recycling, and disposal of grinding fluids are among the important considerations in their selection and use. The practices employed must comply with federal, state, and local laws and regulations.

**Cryogenic grinding.** As described in Section 8.7.2, a more recent technology is the use of liquid nitrogen as a coolant in grinding operations, largely in the interest of reducing or eliminating the adverse environmental impact of the use of metalworking fluids. With appropriate small-diameter nozzles, liquid nitrogen at around $-200°C$ ($-320°F$) is injected into the wheel–workpiece grinding zone, thus reducing its temperature; as stated previously, high temperatures can have significant adverse effects on the part, such as poor surface finish and integrity. Experimental studies have indicated that, as compared with the use of traditional grinding fluids, use of cryogenic grinding is associated with a reduced level of surface (metallurgical) burn and oxidation, improved surface finish, lower tensile residual stress levels, and less loading of the grinding wheel (hence, less need for dressing). Cryogenic grinding appears to be particularly suitable for materials such as titanium, which has low thermal conductivity (see Table 3.2), low specific heat, and high reactivity. Other effects, such as on fatigue life of the ground pieces, as well as the economic advantages of cryogenic grinding, continue to be investigated.

**EXAMPLE 9.4  Grinding vs. hard turning**

Section 8.8.2 describes hard turning, an example of which is the machining of heat-treated steels (usually above 45 HRC), using a single-point polycrystalline cubic-boron-nitride cutting tool. In view of the discussions presented thus far in this chapter, it is evident that grinding and hard turning can be competitive in specific applications; consequently, there has been considerable debate regarding their respective merits.

Hard turning continues to be increasingly competitive with grinding, and dimensional tolerances and surface finish produced by hard turning are beginning to approach those for grinding. Turning requires much less energy than grinding (compare, for example, Tables 8.3 and 9.3), and with turning, thermal and other damage to the workpiece surface is less likely to occur, cutting fluids may not be necessary, and the machine tools are less expensive. In addition, finishing the part while it is still chucked in the lathe eliminates the need for material handling and setting the part in the grinder. On the other hand, workholding devices for large and slender workpieces for hard turning can present problems, since the cutting forces are higher than in grinding. Furthermore, tool wear and its control in hard turning can be a significant problem as compared with the automatic dressing of grinding wheels. It is thus evident that the competitive position of hard turning versus grinding must be evaluated individually for each application and in terms of product surface integrity, quality, and overall economics.

# 9.7 | Ultrasonic Machining

In *ultrasonic machining* (UM), material is removed from a surface by microchipping or erosion with abrasive particles. The tip of the tool (Fig. 9.19a), called a *sonotrode*, vibrates at low amplitudes [0.05 to 0.125 mm (0.002 to 0.005 in.)] and at high frequency (20 kHz). This vibration, in turn, transmits a high velocity to fine abrasive grains between the tool and the surface of the workpiece. The grains are usually boron carbide, but aluminum oxide and silicon carbide are also used; grain size ranges from 100 (for roughing) to 1000 (for finishing). The grains are in a water slurry with concentrations ranging from 20% to 60% by volume; the slurry also carries away the debris from the cutting area.

Ultrasonic machining is best suited for hard and brittle materials, such as ceramics, carbides, glass, precious stones, and hardened steels. The tip of the tool is usually made of low-carbon steel and undergoes wear; it is attached to a transducer

**FIGURE 9.19**
(a) Schematic illustration of the ultrasonic-machining process by which material is removed through microchipping and erosion. (b) and (c) Typical examples of holes produced by ultrasonic machining. Note the dimensions of cut and the types of workpiece materials.

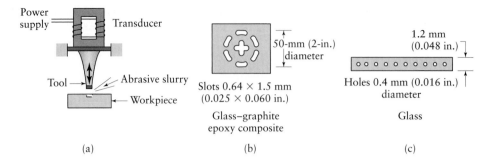

(a)

Slots 0.64 × 1.5 mm
(0.025 × 0.060 in.)

Glass–graphite
epoxy composite

(b)

Holes 0.4 mm (0.016 in.)
diameter

Glass

(c)

through the toolholder. With fine abrasives, dimensional tolerances of 0.0125 mm (0.0005 in.) or better can be held in this process. Figures 9.19b and c show two applications of ultrasonic machining.

*Microchipping* in ultrasonic machining is possible because of the high stresses produced by particles striking a surface. The time of contact between the particle and the surface is very short (10 to 100 $\mu$s), and the area of contact is very small. The time of contact, $t_o$, can be expressed as

$$t_o \simeq \frac{5r}{c_o}\left(\frac{c_o}{v}\right)^{1/5},$$ (9.11)

where $r$ is the radius of a spherical particle, $c_o$ is the elastic wave velocity in the workpiece $\left(c_o = \sqrt{E/\rho}\right)$, and $v$ is the velocity with which the particle strikes the surface. The force $F$ of the particle on the surface is obtained from the rate of change of momentum. That is,

$$F = \frac{d(mv)}{dt},$$ (9.12)

where $m$ is the mass of the particle. The *average force*, $F_{ave}$, of a particle striking the surface and rebounding is

$$F_{ave} = \frac{2mv}{t_o}.$$ (9.13)

Substitution of numerical values into Eq. (9.13) indicates that even small particles can exert significant forces and, because of the very small contact area, produce very high stresses. In brittle materials, these stresses are sufficiently high to cause microchipping and surface erosion. (See also the discussion of *abrasive-jet machining* in Section 9.15.)

**Rotary ultrasonic machining.** In this process, the abrasive slurry is replaced by a tool with metal-bonded diamond abrasives that have been either impregnated or electroplated on the tool surface. The tool is rotated and ultrasonically vibrated, and the workpiece is pressed against it at a constant pressure; the process is similar to a face milling operation. (See Section 8.9.1.) This process is particularly effective in producing deep holes in ceramics and operating at high material removal rates.

## 9.8 | Finishing Operations

In addition to the abrasive processes described thus far, several processes generally are used on workpieces as the final finishing operation. These processes use mainly abrasive grains. This section describes commonly used finishing operations, in the

**FIGURE 9.20** Schematic illustration of the structure of a coated abrasive. Sandpaper, developed in the 16th century, and emery cloth are common examples of coated abrasives.

order of improved surface finish produced. Finishing operations can contribute significantly to production time and product cost; they should therefore be specified with due consideration to their costs and benefits.

The commonly used finishing operations are described as follows:

1. **Coated abrasives.** Typical examples of *coated abrasives* include sandpaper and emery cloth. The grains used in coated abrasives are more pointed than those used for grinding wheels. The grains are electrostatically deposited on flexible backing materials, such as paper or cloth (Fig. 9.20), with their long axes perpendicular to the plane of the backing. The matrix (coating) is made of resins. Coated abrasives are available as sheets, belts, and disks and usually have a much more open structure than the abrasives used on grinding wheels. Coated abrasives are used extensively in finishing flat or curved surfaces of metallic and nonmetallic parts, in finishing metallographic specimens and in wood-working. The surface finish obtained depends primarily on grain size.

   Coated abrasives are also used as *belts* for high-rate material removal. **Belt grinding** has become an important production process and in some cases has replaced conventional grinding operations such as grinding of camshafts, with 8 to 16 lobes (of approximately elliptic shapes) per shaft. Belt speeds are usually in the range of 700–1800 m/min (2500–6000 ft/min) Machines for abrasive-belt operations require proper belt support and rigid construction to minimize vibrations.

   A more recent development is **microreplication,** in which aluminum-oxide abrasives in the shape of tiny pyramids are placed in a predetermined orderly arrangement on the belt surface. When they are used on stainless steels and superalloys, their performance is more consistent, and the temperature rise is lower then when using other coated abrasives. Typical applications include surgical implants, turbine blades, and medical and dental instruments.

2. **Wire brushing.** In the *wire-brushing* process, the workpiece is held against a circular wire brush that rotates at high speed. The tips of the wire produce longitudinal scratches on the workpiece surface. This process is used to produce a fine surface texture. Some efforts are taking place in developing wire brushing as a light material removal process.

3. **Honing.** *Honing* is an operation used primarily to give holes a fine surface finish. The honing tool (Fig. 9.21) consists of a set of aluminum-oxide or silicon-carbide bonded abrasives, called *stones*. The stones are mounted on a mandrel

**FIGURE 9.21** Schematic illustration of a honing tool to improve the surface finish of bored or ground holes.

**FIGURE 9.22** Schematic illustration of the superfinishing process for a cylindrical part: (a) cylindrical microhoning; (b) centerless microhoning.

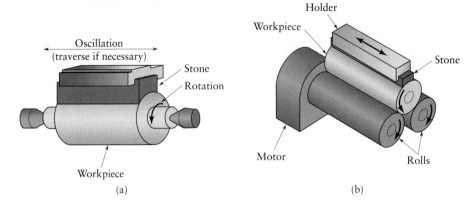

(a)

(b)

that rotates in the hole, applying a radial force with a reciprocating axial motion, thus producing a crosshatched pattern. The stones can be adjusted radially for different hole sizes. The fineness of surface finish can be controlled by the type and size of abrasive used, the speed of rotation and the pressure applied. A fluid is used to remove chips and to keep temperatures low. If not implemented properly, honing can produce holes that are not straight and cylindrical, but rather that are bellmouthed, wavy, barrel shaped, or tapered. Honing is also done on external cylindrical or flat surfaces and to remove sharp edges on cutting tools and inserts. (See Fig. 8.34.)

In **superfinishing**, the pressure applied is very light, and the motion of the honing stone has a short stroke. The process is controlled so that the grains do not travel along the same path along the surface of the workpiece. Figure 9.22 shows examples of external superfinishing of a round part.

4. **Electrochemical honing.** This process combines the fine abrasive action of honing with electrochemical action. Although the equipment is costly, the process is as much as 5 times faster than conventional honing, and the tool lasts longer by as much as 10 times. Electrochemical honing is used primarily for finishing internal cylindrical surfaces.

5. **Lapping.** *Lapping* is a finishing operation used on flat or cylindrical surfaces. The lap (Fig. 9.23a) is usually made of cast iron, copper, leather, or cloth. The abrasive particles are embedded in the lap, or they may be carried through a slurry. Dimensional tolerances on the order of $\pm 0.0004$ mm ($\pm 0.000015$ in.) can be obtained with the use of fine abrasives, up to size 900. Surface finish can be as smooth as $0.025-0.1$ $\mu$m($1-4$ $\mu$in.). Production lapping on flat or cylindrical pieces is done on machines such as those shown in Figs. 9.23b and c. Lapping

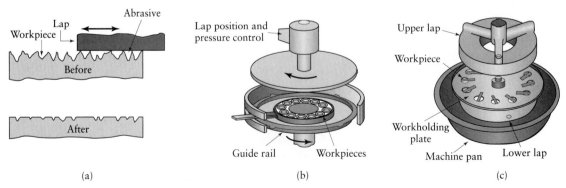

(a)

(b)

(c)

**FIGURE 9.23** (a) Schematic illustration of the lapping process. (b) Production lapping on flat surfaces. (c) Production lapping on cylindrical surfaces.

is also done on curved surfaces, such as spherical objects and glass lenses, using specially shaped laps. Running-in of mating gears can be done by lapping. Depending on the hardness of the workpiece, lapping pressures range from 7–140 kPa (1–20 psi).

6. **Polishing.** *Polishing* is a process that produces a smooth, lustrous surface finish. Two basic mechanisms are involved in the polishing process: (1) fine-scale abrasive removal and (2) softening and smearing of surface layers by frictional heating. The shiny appearance of polished surfaces results from the smearing action. Polishing is done with disks or belts made of fabric, leather, or felt and coated with fine powders of aluminum oxide or diamond. Parts with irregular shapes, sharp corners, deep recesses, and sharp projections are difficult to polish.

7. **Buffing.** *Buffing* is similar to polishing, with the exception that very fine abrasives are used on soft disks made of cloth or hide. The abrasive is supplied externally from a stick of abrasive compound. Polished parts may be buffed to obtain an even finer surface finish.

8. **Electropolishing.** Mirrorlike finishes can be obtained on metal surfaces by *electropolishing*, a process that is the reverse of electroplating (see Section 4.5.1). Because there is no mechanical contact with the workpiece, this process is particularly suitable for polishing irregular shapes. The electrolyte attacks projections and peaks on the workpiece surface at a higher rate than for the rest of the surface, thus producing a smooth surface. Electropolishing is also used for deburring operations.

9. **Chemical mechanical polishing.** *Chemical mechanical polishing* is a process in which a chemically reactive surface is polished with a ceramic slurry in a sodium-hydroxide solution. Most of the material removal is due to chemical action, with the slurry serving mainly to ensure a good surface finish. A major application of this process is the polishing of silicon wafers. (See Section 13.9.)

10. **Polishing processes using magnetic fields.** A more recent development in polishing involves the use of magnetic fields to support abrasive slurries in finishing (polishing) ceramic balls and bearing rollers. **Magnetic float polishing** of ceramic balls is illustrated schematically in Fig. 9.24a. In this process, a

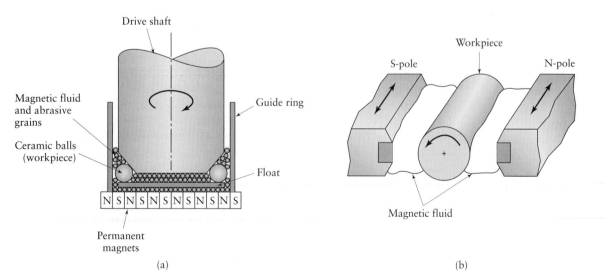

(a)　　　　　　　　　　　　(b)

**FIGURE 9.24** Schematic illustration of the use of magnetic fields to polish balls and rollers: (a) magnetic float polishing of ceramic balls; (b) magnetic-field-assisted polishing of rollers. *Source*: R. Komanduri, M. Doc, and M. Fox.

magnetic fluid, containing abrasive grains and extremely fine ferromagnetic particles in a carrier fluid such as water or kerosene, fills the chamber within a guide ring. The ceramic balls are located between a drive shaft and a float. The abrasive grains, ceramic balls, and the float (which is made of a nonmagnetic material) are all suspended by magnetic forces. The balls are pressed against the rotating drive shaft and are polished by the abrasive action. The forces applied by the abrasive particles on the balls are extremely small and controllable, and hence the polishing action is very fine. Polishing times are much lower than those for other polishing methods; thus, the process is very economical, and the surfaces produced have little or no defects.

   **Magnetic-field-assisted polishing** of ceramic rollers is illustrated in Fig. 9.24b. In this process, a ceramic or steel roller (the workpiece) is clamped and rotated on a spindle. The magnetic poles are oscillated, introducing a vibratory motion to the magnetic–abrasive conglomerate. This action polishes the cylindrical roller surface. Bearing steels of 63 HRC have been mirror finished in 30 seconds by this process.

# 9.9 | Deburring

**Burrs** are thin ridges, usually triangular in shape, that develop along the edges of a workpiece from shearing sheet materials (see Fig. 7.5), trimming forgings and castings, and machining. Burrs may interfere with the assembly of parts and can cause jamming of parts, misalignment, and short circuits in electrical components; furthermore, burrs may reduce the fatigue life of components. Because they are usually sharp, burrs can be a safety hazard to personnel. The need for deburring may be reduced by adding chamfers to sharp edges on parts. On the other hand, burrs on thin drilled or tapped components, such as tiny parts in watches, can provide extra thickness and hence improve the holding torque of screws.

   Several *deburring processes* are available: (1) manually with files; (2) mechanically by cutting; (3) wire brushing; (4) flexible abrasive finishing, using rotary nylon brushes with the nylon filaments embedded with abrasive grits; (5) abrasive belts; (6) ultrasonics; (7) electropolishing; (8) electrochemical machining; (9) vibratory finishing; (10) shot blasting; (11) abrasive-flow machining; and (12) thermal energy. The last four processes are described as follows, and the other processes are covered in this and in other chapters:

1. **Vibratory** and **barrel-finishing** processes are used to improve the surface finish and remove burrs from large numbers of relatively small workpieces. In this batch-type operation, specially shaped *abrasive pellets* or media are placed in a container along with the parts to be deburred. The container is either vibrated or tumbled. The impact of individual abrasives and metal particles removes sharp edges and burrs from the parts. Depending on the application, this process is performed dry or wet, and liquid compounds may be added for requirements such as degreasing and providing resistance to corrosion.

2. **In shot blasting** (also called *grit blasting*), abrasive particles (usually sand) are propelled by a high-velocity jet of air or by a rotating wheel onto the surface of the workpiece. Shot blasting is particularly useful in deburring both metallic and nonmetallic materials and stripping, cleaning, and removing surface oxides. The surface produced has a matte finish. Small-scale polishing and etching can also be done by this process on bench-type units (*microabrasive blasting*).

3. **In abrasive-flow machining**, abrasive grains, such as silicon carbide or diamond, are mixed in a puttylike matrix, which is then forced back and forth through

the openings and passageways in the workpiece. The movement of the abrasive matrix under pressure erodes away burrs and sharp corners and polishes the part. The process is particularly suitable for workpieces with internal cavities that are inaccessible by other means. Pressures applied range from 0.7 MPa to 22 MPa (100 psi to 3200 psi). External surfaces can also be deburred using this process by containing the workpiece within a fixture that directs the abrasive media to the edges and areas to be deburred.

4. **The thermal-energy** method of deburring consists of placing the part in a chamber that is then injected with a mixture of natural gas and oxygen. This mixture is ignited, whereby a heat wave is produced with a temperature of 3300°C (6000°F). The burrs heat up instantly and are melted away, while the temperature of the part reaches only about 150°C (300°F). The process is effective in many applications on noncombustible parts. However, larger burrs or flashes tend to form beads after melting; the process can distort thin and slender parts; and the process does not polish or buff the workpiece surfaces as do many of the other deburring processes.

**Robotic deburring.**   Deburring and flash removal from castings and forgings are now being performed increasingly by *programmable robots* (see Section 14.7), using a force-feedback system for control. This method eliminates tedious and costly manual labor and results in more consistent deburring.

## 9.10 | Chemical Machining

We know that certain chemicals attack metals and etch them, thereby removing small amounts of material from the surface. Thus, *chemical machining* (CM) was developed (Table 9.4), whereby material is removed from a surface by chemical dissolution, using chemical **reagents**, or **etchants**, such as acids and alkaline solutions. Chemical machining is the oldest of the nontraditional machining processes and has been used for many years for engraving metals and hard stones and, more recently, in the production of printed-circuit boards and microprocessor chips. Parts can also be deburred by chemical means.

### 9.10.1 Chemical milling

In *chemical milling*, shallow cavities are produced on sheets, plates, forgings, and extrusions for overall reduction of weight (Fig. 9.25). Chemical milling has been used on a wide variety of metals, with depths of removal to as much as 12 mm (0.5 in.). Selective attack by the chemical reagent on different areas of the workpiece surfaces is controlled by removable layers of *maskant* (Fig. 9.26) or by partial immersion in the reagent.

Chemical milling is used in the aerospace industry for removing shallow layers of material from large aircraft, missile skin panels, and extruded parts for airframes. Tank capacities for reagents are as large as 3.7 m × 15 m (12 ft × 50 ft). The process is also used to fabricate microelectronic devices (see Sections 13.8 and 13.13.) Figure 9.27 shows the range of surface finish and dimensional tolerances obtained by chemical machining and other machining processes. Some surface damage may result from chemical milling, because of preferential etching and intergranular attack (see Section 3.4), which adversely affect surface properties. Chemical milling of welded and brazed structures may produce uneven material removal. Chemical milling of castings may result in uneven surfaces caused by porosity and nonuniformity of the structure.

**TABLE 9.4**

## General Characteristics of Advanced Machining Processes

| Process | Characteristics | Process Parameters and Typical Material Removal Rate or Cutting Speed |
|---------|-----------------|--------------------------------------------|
| Chemical machining (CM) | Shallow removal (up to 12 mm) on large flat or curved surfaces; blanking of thin sheets; low tooling and equipment cost; suitable for low production runs. | 0.025–0.1 mm/min. |
| Electrochemical machining (ECM) | Complex shapes with deep cavities; highest rate of material removal; expensive tooling and equipment; high power consumption; medium to high production quantity. | V: 5–25 dc; A: 1.5–8 A/mm$^2$; 2.5–12 mm/min, depending on current density. |
| Electrochemical grinding (ECG) | Cutting off and sharpening hard materials, such as tungsten-carbide tools; also used as a honing process; higher material removal rate than grinding. | A: 1–3 A/mm$^2$; typically 1500 mm$^3$/min per 1000 A. |
| Electrical-discharge machining (EDM) | Shaping and cutting complex parts made of hard materials; some surface damage may result; also used for grinding and cutting; versatile; expensive tooling and equipment. | V: 50–380; A: 0.1–500; typically 300 mm$^3$/min. |
| Wire EDM | Contour cutting of flat or curved surfaces; expensive equipment. | Varies with workpiece material and its thickness. |
| Laser-beam machining (LBM) | Cutting and hole making on thin materials; heat-affected zone; does not require a vacuum; expensive equipment; consumes much energy; extreme caution required in use. | 0.50–7.5 m/min. |
| Electron-beam machining (EBM) | Cutting and hole making on thin materials; very small holes and slots; heat-affected zone; requires a vacuum; expensive equipment. | 1–2 mm$^3$/min. |
| Water-jet machining (WJM) | Cutting all types of nonmetallic materials to 25 mm (1 in.) and greater in thickness; suitable for contour cutting of flexible materials; no thermal damage; environmentally safe process. | Varies considerably with workpiece material. |
| Abrasive water-jet machining (AWJM) | Single or multilayer cutting of metallic and nonmetallic materials. | Up to 7.5 m/min. |
| Abrasive-jet machining (AJM) | Cutting, slotting, deburring, flash removal, etching, and cleaning of metallic and nonmetallic materials; tends to round off sharp edges; some hazard because of airborne particulates. | Varies considerably with workpiece material. |

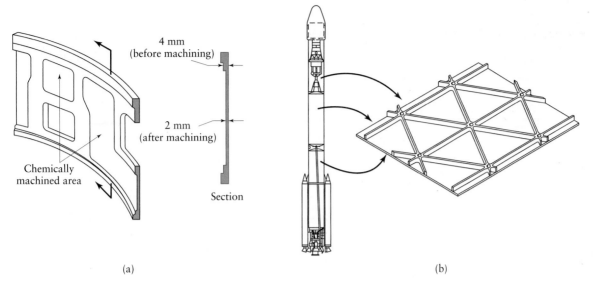

(a)                    (b)

**FIGURE 9.25** (a) Missile skin-panel section contoured by chemical milling to improve the stiffness-to-weight ratio of the part. (b) Weight reduction of space launch vehicles by chemical milling of aluminum-alloy plates. These panels are milled after the plates have first been formed into shape, such as by roll forming or stretch forming. The design of the chemically machined rib patterns can readily be modified at minimal cost. *Source*: *Advanced Materials and Processes*, ASM International, December 1990, p. 43.

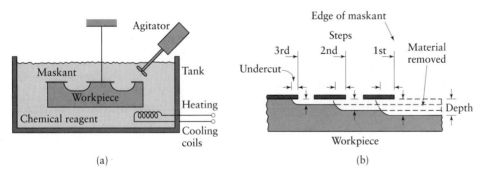

(a)                    (b)

**FIGURE 9.26** (a) Schematic illustration of the chemical-machining process. Note that no forces or machine tools are involved in this process. (b) Stages in producing a profiled cavity by chemical machining.

## 9.10.2 Chemical blanking

*Chemical blanking* is similar to blanking of sheet metal in that it is used to produce features that penetrate through the thickness of the material (see Fig. 7.8), with the exception that material is removed by chemical dissolution rather than by shearing. Typical applications for chemical blanking include burr-free etching of printed-circuit boards and the production of decorative panels, thin sheet-metal stampings, and complex or small shapes.

## 9.10.3 Photochemical blanking

Also called *photoetching*, *photochemical blanking* is a modification of chemical milling; material is removed, usually from flat thin sheet, by photographic techniques. Complex burr-free shapes can be blanked (Fig. 9.28) on metals as thin as 0.0025 mm (0.0001 in.); the process is also used for etching. Typical applications for photochemical blanking include fine screens, printed-circuit cards, electric-

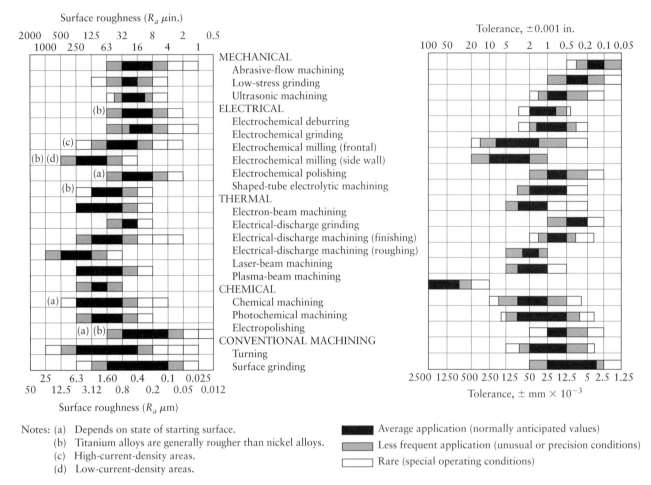

Surface roughness ($R_a$ μin.)

2000 500 125 32 8 2 0.5
1000 250 63 16 4 1

MECHANICAL
Abrasive-flow machining
Low-stress grinding
Ultrasonic machining
ELECTRICAL
Electrochemical deburring
Electrochemical grinding
Electrochemical milling (frontal)
Electrochemical milling (side wall)
Electrochemical polishing
Shaped-tube electrolytic machining
THERMAL
Electron-beam machining
Electrical-discharge grinding
Electrical-discharge machining (finishing)
Electrical-discharge machining (roughing)
Laser-beam machining
Plasma-beam machining
CHEMICAL
Chemical machining
Photochemical machining
Electropolishing
CONVENTIONAL MACHINING
Turning
Surface grinding

25 6.3 1.60 0.4 0.1 0.025
50 12.5 3.12 0.8 0.2 0.05 0.012

Surface roughness ($R_a$ μm)

Tolerance, ±0.001 in.

100 50 20 10 5 2 1 0.5 0.2 0.1 0.05

2500 1250 500 250 125 50 25 12.5 5 2.5 1.25

Tolerance, $\pm$ mm $\times$ $10^{-3}$

Notes: (a) Depends on state of starting surface.
(b) Titanium alloys are generally rougher than nickel alloys.
(c) High-current-density areas.
(d) Low-current-density areas.

■ Average application (normally anticipated values)
▨ Less frequent application (unusual or precision conditions)
▢ Rare (special operating conditions)

**FIGURE 9.27** Surface roughness and tolerances obtained in various machining processes. Note the wide range within each process. (See also Fig. 8.27.) *Source*: Courtesy of TechSolve, formerly the Institute of Advanced Manufacturing Sciences'.

**FIGURE 9.28** Various parts made by chemical blanking. Note the fine detail. *Source*: Courtesy of Buckabee-Mears St. Paul.

motor laminations, flat springs, and masks for color television. Although skilled labor is required for photochemical blanking, tooling costs are low, the process can be automated, and it is economical for medium to high production volume. Photochemical blanking is capable of forming very small parts for which traditional blanking dies (Section 7.3) are difficult to make. Also, the process is effective for blanking fragile workpieces and materials. Handling of chemical reagents requires precautions and special safety considerations to protect personnel against exposure to both liquid chemicals and volatile chemicals. Furthermore, disposal of chemical by-products from this process is a major consideration, although some by-products can be recycled.

## 9.11 | Electrochemical Machining

*Electrochemical machining* (ECM) is basically the reverse of electroplating. An electrolyte (Fig. 9.29) acts as current carrier, and the high rate of electrolyte movement in the tool–workpiece gap washes metal ions away from the workpiece (*anode*) before they have a chance to plate onto the tool (*cathode*). Note that the cavity produced is the female mating image of the tool. Modifications of this process are used for turning, facing, slotting, trepanning, and profiling operations in which the electrode becomes the cutting tool.

The shaped tool is generally made of brass, copper, bronze, or stainless steel. The **electrolyte** is a highly conductive inorganic salt solution, such as sodium chloride mixed in water or sodium nitrate; it is pumped at a high rate through the passages in the tool. A dc power supply in the range of 5–25 V maintains current densities, which for most applications are 1.5–8 A/mm² (1000–5000 A/in²) of active machined surface. Machines that have current capacities as high as 40,000 A and as small as 5 A are available. Because the metal removal rate is only a function of ion exchange rate, it is not affected by the strength, hardness, or toughness of the workpiece, which must be electrically conductive.

The material removal rate by electrochemical machining can be calculated from the equation

$$\text{MRR} = CI\eta, \qquad (9.14)$$

where MRR $= \text{mm}^3/\text{min}$; $I =$ current in amperes; and $\eta =$ current efficiency, which typically ranges from 90% to 100%. $C$ is a material constant in $\text{mm}^3$/A-min and, for pure metals, depends on valence; the higher the valence, the lower is its value. If a

Pump for circulating electrolyte

dc power supply

(−)

Insulating coating

(+)

Tool

Tank

Workpiece     Electrolyte

**FIGURE 9.29**   Schematic illustration of the electrochemical-machining process. This process is the reverse of electroplating, described in Section 4.5.1.

**FIGURE 9.30** Typical parts made by electrochemical machining. (a) Turbine blade made of a nickel alloy, 360 HB. *Source: Metal Handbook*, 9th ed., Vol. 3, Materials Park, OH: ASM International, 1980, p. 849. (b) Thin slots on a 4340-steel roller-bearing cage. (c) Integral airfoils on a compressor disk.

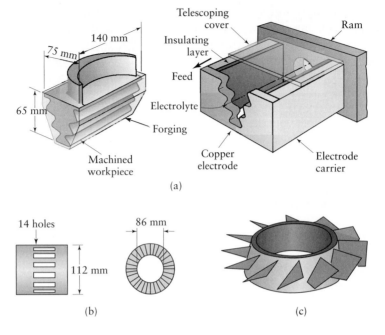

cavity of uniform cross-sectional area $A_o$ mm$^2$ is being electrochemically machined, the feed rate, $f$, in mm/min would be

$$f = \frac{\text{MRR}}{A_o}. \tag{9.15}$$

Note that the feed rate is the speed at which the electrode is penetrating the workpiece. For most metals, the value of $C$ typically ranges between about 1 and 2.

Electrochemical machining is generally used for machining complex cavities in high-strength materials, particularly in the aerospace industry for mass production of turbine blades, jet-engine parts, and nozzles (Fig. 9.30). It is also used for machining forging-die cavities and producing small holes. The ECM process leaves a burr-free surface; in fact, it can also be used as a deburring process. It does not cause any thermal damage to the part, and the lack of tool forces prevents distortion of the part. Furthermore, there is no tool wear, and the process is capable of producing complex shapes as well as machining hard materials. However, the mechanical properties of components made by ECM should be compared carefully with those produced by other material removal methods. Electrochemical-machining systems are available as numerically controlled machining centers, with the capability for high production rates, high flexibility, and maintenance of close dimensional tolerances.

**Pulsed electrochemical machining.** *Pulsed electrochemical machining* (PECM) is a refinement of ECM; it uses very high current densities (on the order of 100 A/cm$^2$), but the current is pulsed rather than direct. The purpose of pulsing is to eliminate the need for high electrolyte flow rates, which limits the usefulness of ECM in die and mold making (die sinking). Investigations have shown that PECM improves fatigue life, and the process has been proposed as a possible method for eliminating the recast layer left on die and mold surfaces by electrical discharge machining. (See Section 9.13.)

# 9.12 | Electrochemical Grinding

*Electrochemical grinding* (ECG) combines electrochemical machining with conventional grinding (Table 9.4). The equipment used in electrochemical grinding is similar to a conventional grinder, except that the wheel is a rotating cathode with abrasive particles (Fig. 9.31a). The wheel is metal bonded with diamond or aluminum-oxide abrasives and rotates at a surface speed of 1200–2000 m/min (4000–7000 ft/min.) The abrasives serve as insulators between the wheel and the workpiece and mechanically remove electrolytic products from the working area. A flow of electrolyte, usually sodium nitrate, is provided for the electrochemical-machining phase of the operation. The majority of metal removal in ECG is by electrolytic action, and typically less than 5% of metal is removed by the abrasive action of the wheel. Therefore, wheel wear is very low, and current densities range from 1–3 $A/mm^2$ (500–2000 $A/in^2$). Finishing cuts are usually made by the grinding action, but only to produce a surface with good finish and dimensional accuracy.

The material removal rate in electrochemical grinding can be calculated from the equation

$$\text{MRR} = \frac{GI}{\rho F} \qquad (9.16)$$

where MRR is in $mm^3/min$, $G$ = equivalent weight in grams, $I$ = current in amperes, $\rho$ = density in $g/mm^3$, and $F$ = *Faraday's constant* in Coulombs. The speed of penetration, $V_s$, of the grinding wheel into the workpiece is given by the equation

$$V_s = \left(\frac{G}{\rho F}\right)\left(\frac{E}{gK_p}\right)K \qquad (9.17)$$

where $V_s$ is in $mm^3/min$; $E$ = cell voltage in volts; $g$ = wheel–workpiece gap in mm; $K_p$ = coefficient of loss, which is in the range of 1.5 to 3; and $K$ = electrolyte conductivity in $\Omega^{-1}mm^{-1}$.

Electrochemical grinding is suitable for applications similar to those for milling, grinding, and sawing (Fig. 9.31b). It is not adaptable to cavity-sinking operations, such as die making; it has been successfully applied to carbides and high-strength alloys. The process offers a distinct advantage over traditional diamond-wheel grinding when processing very hard materials, for which wheel wear can be high. ECG machines are available with numerical controls, thus improving dimensional accuracy and providing repeatability and increased productivity.

(a)                                    (b)

**FIGURE 9.31**    (a) Schematic illustration of the electrochemical-grinding process. (b) Thin slot produced on a round nickel-alloy tube by this process.

**EXAMPLE 9.5 Machining time in electrochemical machining vs. drilling**

A round hole 12.5 mm in diameter is being produced in a titanium-alloy block by electrochemical machining. Using a current density of 6 A/mm², estimate the time required for machining a 20-mm-deep hole. Assume that the efficiency is 90%. Compare this time with that required for ordinary drilling.

**Solution.** We note from Eqs. (9.14) and (9.15) that the feed rate can be expressed by the equation

$$f = \frac{CI\eta}{A_o}.$$

Letting $C = 1.6 \text{ mm}^3/\text{A} \cdot \text{min}$ and $I/A_o = 6 \text{ A/mm}^2$, we find that the feed rate is $f = (1.6)(6)(0.9) = 8.64 \text{ mm/min}$. Since the hole is 20 mm deep,

$$\text{Machining time} = \frac{20}{8.64} = 2.3 \text{ min.}$$

To determine the drilling time, we refer to Table 8.10 and note the data for titanium alloys. Let's select the following values for a 12.5-mm drill: drill rpm = 300 and feed = 0.15 mm/rev. From these values, we calculate that the feed rate will be (300 rev/min)(0.15 mm/rev) = 45 mm/min. Since the hole is 20 mm deep,

$$\text{Drilling time} = \frac{20}{45} = 0.45 \text{ min,}$$

which is about 20% of the time for ECM.

## 9.13 | Electrical-Discharge Machining

The principle of *electrical-discharge machining* (EDM), also called *electrodischarge* or *sparkerosion machining*, is based on erosion of metals by spark discharges (Table 9.4). We know that when two current-conducting wires are allowed to touch each other, an arc is produced. If we look closely at the point of contact between the two wires, we note that a small portion of the metal has been eroded away, leaving a small crater. Although this phenomenon has been known since the discovery of electricity, it was not until the 1940s that a machining process based on this principle was developed. The EDM process has become one of the most important and widely accepted production technologies in manufacturing industries.

The EDM system (Fig. 9.32) consists of a shaped tool (**electrode**) and the workpiece, which are connected to a dc power supply and placed in a **dielectric** (electrically

**FIGURE 9.32** Schematic illustration of the electrical-discharge-machining process.

nonconducting) **fluid**. When a voltage is applied to the tool, a magnetic field causes suspended particles in the dielectric fluid to concentrate, eventually forming a bridge for current to flow to the workpiece. An intense electrical arc is then generated, causing sufficient heating to melt a portion of the workpiece and usually some of the tooling material as well. In addition, the dielectric fluid is heated rapidly, causing evaporation of the fluid in the arc gap; this evaporation, in turn, increases the resistance of the interface, until the arc can no longer be maintained. Once the arc is interrupted, heat is removed from the gas bubble by the surrounding dielectric fluid, and the bubble collapses (cavitates). The associated shock wave and flow of dielectric fluid flush debris from the surface and entrain any molten workpiece material into the dielectric fluid. The capacitor discharge is repeated at rates of between 50 kHz and 500 kHz, with voltages usually ranging between 50 V and 380 V and currents from 0.1 A to 500 A.

The dielectric fluid (1) acts as an insulator until the potential is sufficiently high, (2) acts as a flushing medium and carries away the debris in the gap, and (3) provides a cooling medium. The gap between the tool and the workpiece (called *overcut*) is critical; thus, the downward feed of the tool is controlled by a servomechanism, which automatically maintains a constant gap. The most common dielectric fluids are mineral oils, although kerosene and distilled and deionized water may be used in specialized applications. The workpiece is fixtured within the tank containing the dielectric fluid, and its movements are controlled by numerically controlled systems. The machines are equipped with a pump and filtering system for the dielectric fluid.

The EDM process can be used on any material that is an electrical conductor. The melting point and latent heat of melting are important physical properties that determine the volume of metal removed per discharge; as these values increase, the rate of material removal slows. The volume of material removed per discharge is typically in the range of $10^{-6}$ to $10^{-4}$ mm$^3$ ($10^{-10}$ to $10^{-8}$ in$^3$). Since the process doesn't involve mechanical energy, the hardness, strength, and toughness of the workpiece material do not necessarily influence the removal rate. The frequency of discharge or the energy per discharge is usually varied to control the removal rate, as are the voltage and current. The rate and surface roughness increase with increasing current density and decreasing frequency of sparks.

*Electrodes* for EDM are usually made of graphite, although brass, copper, or copper–tungsten alloy may be used. The tools are shaped by forming, casting, powder metallurgy, or machining. Electrodes as small as 0.1 mm (0.005 in.) in diameter have been used. *Tool wear* is an important factor, since it affects dimensional accuracy and the shape produced. Tool wear can be minimized by reversing the polarity and using copper tools, a process called **no-wear EDM**.

Electrical-discharge machining has numerous applications, such as producing die cavities for large automotive-body components (die-sinking machining centers); narrow slots; turbine blades; various intricate shapes (Figs. 9.33a and b); and small-diameter deep holes (Fig. 9.33c), using tungsten wire as the electrode. Stepped cavities can be produced by controlling the relative movements of the workpiece in relation to the electrode (Fig. 9.34).

The material removal rate in electrical-discharge machining is basically a function of the current and the melting point of the workpiece material, although other process variables also have an effect. The following approximate empirical relationship can be used as a guide to estimate the metal removal rate in EDM:

$$\text{MRR} = 4 \times 10^4 I T_w^{-1.23}. \tag{9.18}$$

In this equation, MRR is in mm$^3$/min, $I$ is the current in amperes, and $T_w$ is the melting point of the workpiece in °C.

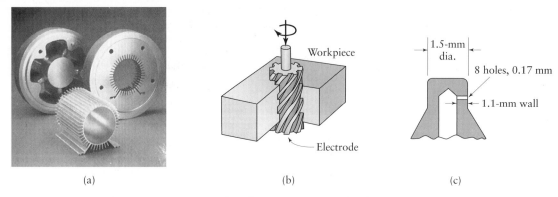

(a) (b) (c)

**FIGURE 9.33** (a) Examples of cavities produced by the electrical-discharge-machining process, using shaped electrodes. The two round parts (rear) are the set of dies for extruding the aluminum piece shown in front. *Source*: Courtesy of AGIE USA Ltd. (b) A spiral cavity produced by a rotating electrode. *Source*: *American Machinist*. (c) Holes in a fuel-injection nozzle made by electrical-discharge machining. Material: Heat-treated steel.

**FIGURE 9.34** Stepped cavities produced with a square electrode by EDM. The workpiece moves in the two principal horizontal directions, and its motion is synchronized with the downward movement of the electrode to produce various cavities. Also shown is a round electrode capable of producing round or elliptical cavities. *Source*: Courtesy of AGIE USA Ltd.

The wear rate of the electrode, $W_t$, can be estimated from the empirical equation

$$W_t = 1.1 \times 10^{11} I T_t^{-2.38} \tag{9.19}$$

where $W_t$ is mm$^3$/min and $T_t$ is the melting point of the electrode material in °C. The wear ratio of the workpiece to electrode, $R$, can be estimated from the expression

$$R = 2.25\, T_r^{-2.38} \tag{9.20}$$

where $T_r$ is the ratio of workpiece to electrode melting points. In practice, the value of $R$ varies over a wide range, between 0.2 and 100. (Note that in abrasive processing, grinding ratios also are in the same range; see Section 9.5.2.)

Metal removal rates usually range from 2 to 400 mm$^3$/min. High rates produce a very rough finish, with a molten and resolidified (recast) structure with poor surface integrity and low fatigue properties; thus, finishing cuts are made at low metal removal rates, or the recast layer is removed later by finishing operations. A more recent technique is to use an oscillating electrode, which provides very fine surface finish, requiring significantly less benchwork to improve lustrous cavities.

**EXAMPLE 9.6** Machining time in electrical-discharge machining vs. drilling

Calculate the machining time for producing the hole in Example 9.5 by EDM, and compare the time with that for drilling and for ECM. Assume that the titanium alloy has a melting point of 1600°C (see Table 3.2) and that the current is 100 A. Also calculate the wear rate of the electrode, assuming that the melting point of the electrode is 1100°C.

**Solution.**  Using Eq. (9.18), we find that

$$MRR = (4 \times 10^4)(100)(1600^{-1.23}) = 458 \text{ mm}^3/\text{min}.$$

The volume of the hole is

$$V = \pi \left[ \frac{(12.5)^2}{4} \right](20) = 2454 \text{ mm}^3.$$

Hence, the machining time for EDM is $2454/458 = 5.4$ min; this time is 2.35 times that for ECM and 11.3 times that for drilling. Note, however, that if the current is increased to 300 A, then the machining time for EDM will only be 1.8 minutes, which is less than the time for ECM. The wear rate of the electrode is calculated using Eq. (9.19). Thus,

$$W_t = (11 \times 10^3)(100)(1100^{-2.38}) = 0.064 \text{ mm}^3/\text{min}.$$

## 9.13.1  Electrical-discharge grinding

The grinding wheel in *electrical-discharge grinding* (EDG) is made of graphite or brass and contains no abrasives. Material is removed from the surface of the workpiece by repetitive spark discharges between the rotating wheel and the workpiece. The material removal rate can be estimated from the equation

$$MRR = KI, \tag{9.21}$$

where MRR is in $\text{mm}^3/\text{min}$, $I$ = current in amperes, and $K$ = workpiece material factor in $\text{mm}^3/\text{A-min}$; for example, $K = 16$ for steel and 4 for tungsten carbide.

The EDG process can be combined with electrochemical grinding; the process is then called **electrochemical-discharge grinding** (ECDG). Material is removed by chemical action, with the electrical discharges from the graphite wheel breaking up the oxide film, which is washed away by the electrolyte flow. The process is used primarily for grinding carbide tools and dies, but can also be used for grinding fragile parts, such as surgical needles, thin-walled tubes, and honeycomb structures. The ECDG process is faster than EDG, but power consumption for ECDG is higher.

In **sawing** with EDM, a setup similar to a band or circular saw (but without any teeth) is used with the same electrical circuit as in EDM. Narrow cuts can be made at high rates of metal removal. Because the cutting forces are negligible, the process can be used on slender components.

## 9.13.2  Wire EDM

A variation of EDM is *wire EDM* (Fig. 9.35), or *electrical-discharge wire cutting*. In this process, which is similar to contour cutting with a band saw (Section 8.9.5), a slowly moving wire travels along a prescribed path, cutting the workpiece, with the discharge sparks acting like cutting teeth. This process is used to cut plates as thick

**FIGURE 9.35** Schematic illustration of the wire EDM process. As much as 50 hours of machining can be performed with one reel of wire, which is then discarded or recycled.

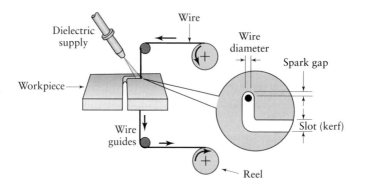

as 300 mm (12 in.) and for making punches, tools, and dies from hard metals; it can also cut intricate components for the electronics industry (Table 9.4).

The wire is usually made of brass, copper, or tungsten and is typically about 0.25 mm (0.01 in.) in diameter, making narrow cuts possible; zinc- or brass-coated and multicoated wires are also used. The wire should have sufficient tensile strength and fracture toughness, as well as high electrical conductivity and capacity to flush away the debris produced during cutting. The wire is generally used only once, as it is relatively inexpensive. It travels at a constant velocity in the range of 0.15 to 9 m/min (6 to 360 in./min), and a constant gap (*kerf*) is maintained during the cut. The cutting speed is generally given in terms of the cross-sectional area cut per unit time; some typical examples are 18,000 mm$^2$/hr (28 in$^2$/hr) for 50-mm-thick (2-in-thick) D2 tool steel, and 45,000 mm$^2$/hr (70 in$^2$/hr) for 150-mm-thick (6-in-thick) aluminum. These removal rates indicate a linear cutting speed of 18,000/50 = 360 mm/hr = 6 mm/min, and 45,000/150 = 300 mm/hr = 5 mm/min, respectively.

The material removal rate for wire EDM can be obtained from the expression

$$\text{MRR} = V_f h b, \tag{9.22}$$

where MRR is in mm$^3$/min; $V_f$ = feed rate of the wire into the workpiece in mm/min; $h$ = workpiece thickness or height in mm; and the kerf (see Fig. 9.35) is denoted as $b = d_w + 2s$, where $d_w$ = wire diameter in mm and $s$ = gap, in mm, between wire and workpiece during machining.

Modern wire-EDM machines (**multiaxis EDM wire-cutting machining centers**) are equipped with (1) computer controls to control the cutting path of the wire, (2) automatic self-threading features in case of wire breakage, (3) multiheads for cutting two parts at the same time, (4) controls that prevent wire breakage, and (5) programmed machining strategies. Two-axis computer-controlled machines can produce cylindrical shapes in a manner similar to that of a turning operation or cylindrical grinding.

## 9.14 | High-Energy-Beam Machining

### 9.14.1 Laser-beam machining

In *laser-beam machining* (LBM), the source of energy is a laser (an acronym for *Light Amplification by Stimulated Emission of Radiation*), which focuses optical energy on the surface of the workpiece (Fig. 9.36a). The highly focused, high-density energy melts and evaporates portions of the workpiece in a controlled manner. This process, which does not require a vacuum, is used to machine a variety of metallic and non-metallic materials. There are several types of lasers used in manufacturing operations:

**FIGURE 9.36**
(a) Schematic illustration of the laser-beam-machining process. (b) and (c) Examples of holes produced in nonmetallic parts by LBM.

$CO_2$ (pulsed or continuous wave), Nd:YAG (neodymium: yttrium–aluminum–garnet), Nd:glass, ruby, and excimer (from the words *excited* and di*mer*, meaning two mers, or two molecules of the same chemical composition). The typical applications of laser-beam machining are outlined in Tables 9.4 and 9.5.

Important physical parameters in LBM are the *reflectivity* and *thermal conductivity* of the workpiece surface and its specific heat and latent heats of melting and evaporation; the lower these quantities, the more efficient is the process. The cutting depth $t$ may be expressed as

$$t \propto \frac{P}{vd}, \tag{9.23}$$

where $P$ is the power input, $v$ is the cutting speed, and $D$ is the laser-beam spot diameter. The surface produced by LBM is usually rough and has a heat-affected zone that, in critical applications, may have to be removed or heat treated. Kerf width is an important consideration, as it is in other cutting processes such as sawing, wire EDM, and electron-beam machining.

**TABLE 9.5**

| General Applications of Lasers in Manufacturing | |
|---|---|
| Application | Laser Type |
| Cutting | |
|    Metals | $PCO_2$; $CWCO_2$; Nd:YAG; ruby |
|    Plastics | $CWCO_2$ |
|    Ceramics | $PCO_2$ |
| Drilling | |
|    Metals | $PCO_2$; Nd:YAG; Nd:glass; ruby |
|    Plastics | Excimer |
| Marking | |
|    Metals | $PCO_2$; Nd:YAG |
|    Plastics | Excimer |
|    Ceramics | Excimer |
| Surface treatment (metals) | $CWCO_2$ |
| Welding (metals) | $PCO_2$; $CWCO_2$; Nd:YAG; Nd:glass; ruby |

*Note*: P = pulsed; CW = continuous wave.

Laser beams may be used in combination with a gas stream, such as oxygen, nitrogen, or argon (*laser-beam torch*), for cutting thin sheet materials. High-pressure inert-gas-assisted (e.g., nitrogen) laser cutting is used for stainless steel and aluminum; it leaves an oxide-free edge that can improve weldability. Gas streams also have the important function of blowing away molten and vaporized material from the workpiece surface.

Laser-beam machining is used widely in cutting and drilling metals, nonmetals, and composite materials (Figs. 9.36b and c). The abrasive nature of composite materials and the cleanliness of the operation have made laser-beam machining an attractive alternative to traditional machining methods. Holes as small as 0.005 mm (0.0002 in.), with hole depth-to-diameter ratios of 50 to 1, have been produced in various materials, although a more practical minimum is 0.025 mm (0.001 in.). Extreme caution should be exercised with lasers, as even low-power lasers can cause damage to the retina of the eye if proper precautions are not observed.

Laser-beam machining is being used increasingly in the electronics and automotive industries. Bleeder holes for fuel-pump covers and lubrication holes in transmission hubs, for example, are being drilled; the cooling holes in the first-stage vanes of the Boeing 747 jet engines are also produced by lasers. Significant cost savings have been achieved by laser-beam machining, such that it now competes with electrical-discharge machining in this respect.

Laser beams are also used for (1) *welding* (see Section 12.4), (2) small-scale *heat treating* of metals and ceramics to modify their surface mechanical and tribological properties (see Section 4.4), and (3) *marking* of parts with letters, numbers, codes, etc. Marking can be done by processes such as with ink; with mechanical devices such as punches, pins, styli, scroll rolls, or stamps; by etching; and with lasers. Although the equipment is more expensive than for other methods, marking and engraving with lasers has become increasingly common, due to the process' accuracy, reproducibility, flexibility, ease of automation, and on-line application in manufacturing.

The inherent flexibility of the laser-beam cutting process with fiber-optic-beam delivery, simple fixturing, low setup times, the availability of multi-kW machines and 2D and 3D computer-controlled systems are attractive features; thus, laser cutting can compete successfully with cutting of sheet metal using the traditional punching processes described in Section 7.3. There are now efforts to combine the two processes for improved overall efficiency.

### 9.14.2 Electron-beam machining and plasma-arc cutting

The source of energy in *electron-beam machining* (EBM) is high-velocity electrons, which strike the surface of the workpiece and generate heat (Fig. 9.37). The applications of this process are similar to those of laser-beam machining, except that EBM

**FIGURE 9.37** Schematic illustration of the electron-beam-machining process. Unlike LBM, this process requires a vacuum, and hence workpiece size is limited.

requires a vacuum. These machines use voltages in the range of 50–200 kV to accelerate the electrons to speeds of 50% to 80% of the speed of light. The interaction of the electron beam with the workpiece surface produces hazardous X rays; consequently, the equipment should be used only by highly trained personnel. Electron-beam machining is used for very accurate cutting of a wide variety of metals. Surface finish is better and kerf width is narrower than that for other thermal-cutting processes. (See also the discussion in Section 12.4 on electron-beam welding.)

In **plasma-arc cutting** (PAC), *plasma beams* (ionized gas) are used for rapid cutting of nonferrous and stainless-steel plates. The temperatures generated are very high (9400°C, or 17,000°F, in the torch for oxygen as a plasma gas); consequently, the process is fast, kerf width is small, and the surface finish is good. Material removal rates are much higher than in the EDM and LBM processes, and parts can be machined with good reproducibility. Plasma-arc cutting is highly automated today, through the use of programmable controls.

A more traditional method is **oxyfuel-gas cutting** (OFC), using a torch as in welding. It is particularly useful for cutting steels, cast irons, and cast steels. Cutting occurs mainly by oxidation and burning of the steel, with some melting taking place as well. Kerf widths are usually in the range of 1.5–10 mm (0.06–0.4 in.).

# 9.15  Water-Jet, Abrasive Water-Jet, and Abrasive-Jet Machining

We know that when we put our hand across a jet of water or air, we feel a considerable concentrated force acting on our hand. This force results from the momentum change of the stream and, in fact, is the principle on which the operation of water or gas turbines is based. This principle is also used in water-jet, abrasive water-jet, and abrasive-jet machining, described as follows:

1. **Water-jet machining** (WJM). In water-jet machining, also called *hydrodynamic machining* (Fig. 9.38a), the force from the jet is used in cutting and deburring operations. The water jet acts like a saw and cuts a narrow groove

(a)   (b)

**FIGURE 9.38**   (a) Schematic illustration of water-jet machining. (b) Examples of various nonmetallic parts cut by a water-jet machine. *Source*: Jet Edge.

in the material. Although pressures as high as 1400 MPa (200 ksi) can be generated, a pressure level of about 400 MPa (60 ksi) is generally used for efficient operation. Jet-nozzle diameters usually range between 0.05 mm and 1 mm (0.002 in. and 0.040 in.).

A variety of materials can be cut with this technique, including plastics, fabrics, rubber, wood products, paper, leather, insulating materials, brick, and composite materials (Fig. 9.38b). Thicknesses range up to 25 mm (1 in.) and higher. Vinyl and foam coverings for some automobile dashboards, for example, are cut using multiple-axis, robot-guided water-jet machining equipment. Because it is an efficient and clean operation compared with other cutting processes, WJM is also used in the food-processing industry for cutting and slicing food products (Table 9.4).

The advantages of this process are that cuts can be started at any location without the need for predrilled holes; no heat is produced; no deflection of the rest of the workpiece takes place (hence the process is suitable for flexible materials); little wetting of the workpiece takes place; and the burr produced is minimal. It is also an environmentally safe manufacturing process.

2. **Abrasive water-jet machining.**  In *abrasive water-jet machining* (AWJM), the water jet contains abrasive particles such as silicon carbide or aluminum oxide, thus increasing the material removal rate over that of water-jet machining. Metallic, nonmetallic, and advanced composite materials of various thicknesses can be cut in single or multiplelayers, particularly heat-sensitive materials that cannot be machined by processes in which heat is produced. Cutting speeds are as high as 7.5 m/min (2.5 ft/min) for reinforced plastics, but much lower for metals. With multiple-axis and robotic-control machines, complex three-dimensional parts can be machined to finish dimensions by this process. Nozzle life is improved by making the nozzle from rubies, sapphires, and carbide-base composite materials.

3. **Abrasive-Jet Machining.**  In *abrasive-jet machining* (AJM), a high-velocity jet of dry air, nitrogen, or carbon dioxide containing abrasive particles is aimed at the workpiece surface under controlled conditions (Fig. 9.39). The impact of the particles develops sufficient concentrated force (see also Section 9.7) to perform operations such as (1) cutting small holes, slots, or intricate patterns in very hard or brittle metallic and nonmetallic materials; (2) deburring or removing small flash from parts; (3) trimming and beveling; (4) removing oxides and other surface films; and (5) general cleaning of components with irregular surfaces.

The gas supply pressure is on the order of 850 kPa (125 psi), and the abrasive-jet velocity can be as high as 300 m/s (100 ft/s) and is controlled by a valve. The handheld nozzles are usually made of tungsten carbide or sapphire.

**FIGURE 9.39** Schematic illustration of the abrasive-jet-machining process.

The abrasive size is in the range of 10–50 $\mu$m (400–2000 $\mu$in.). Because the flow of the free abrasives tends to round off corners, designs for abrasive-jet machining should avoid sharp corners; also, holes made in metal parts tend to be tapered. Because of airborne particulates, there is some hazard involved in using this process.

# 9.16 | Design Considerations

The important design considerations for the processes and operations described in this chapter are outlined as follows:

1. **Abrasive processes.** Parts to be ground should be designed so that they can be held securely in workholding devices without causing part distortion. For high dimensional accuracy, interrupted surface features should be avoided, as they could cause vibration and chatter. In cylindrical grinding, parts should be balanced, and long, slender designs should be avoided to minimize deflections. Fillets and corner radii should be as large as possible, or relief should be provided for by prior machining. Designs requiring accurate form grinding should be kept as simple as possible to avoid frequent wheel dressing. Deep and small holes, and blind holes requiring internal grinding, should be avoided, or they should include a relief. In ultrasonic machining, sharp profiles should be avoided, and the bottom of brittle parts should be supported to avoid chipping; some taper should be expected for holes made by this process.

2. **Chemical, electrochemical, and electrical processes.** In chemical machining, sharp corners and deep and narrow cavities should be avoided; undercuts may occur under the edges of the maskant; the bulk of the workpiece should preferably be preshaped by other processes to reduce machining time; and size changes in artwork due to humidity and temperature can affect dimensional accuracy. In electrochemical machining, sharp corners or flat bottoms should be avoided; designs should make provision for a small taper for holes and cavities; and irregular cavities may not be produced to the desired dimensions. In electrical-discharge machining, the bulk of the workpiece should preferably be preshaped by other processes to reduce machining time; deep slots and narrow openings should be avoided; and, for economic production, the surface finish specified should not be too fine.

3. **High-energy-beam processes.** Any adverse effects of the processes on surface finish and integrity should be considered. In laser-beam machining, dull and unpolished workpiece surfaces are preferable, and designs should avoid deep cuts or very sharp profiles. In electron-beam machining, part size and shape are limited by the size of the vacuum chamber, and if a part requires machining on only a small portion, consideration should be given to making the part in a number of small components.

# 9.17 | Process Economics

This chapter has shown that grinding may be used both as a finishing operation and for large-scale removal operations (as in creep-feed grinding). The use of grinding as a finishing operation is often necessary, because forming and machining processes

**FIGURE 9.40** Increase in the cost of machining and finishing a part as a function of the surface finish required.

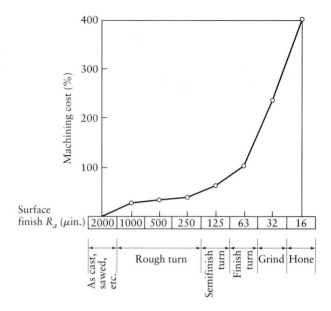

alone usually cannot produce parts with the desired dimensional accuracy and surface finish. However, because it is an additional operation, grinding contributes significantly to product cost. Creep-feed grinding, on the other hand, has proved to be an economical alternative to machining operations such as milling, even though wheel wear is high.

All finishing operations contribute to product cost. On the basis of the discussion thus far, it can be seen that as the surface finish improves, more operations are required, and hence the cost increases. Note in Fig. 9.40 how rapidly cost increases as surface finish is improved by processes such as grinding and honing.

Much progress has been made in automating the equipment involved in finishing operations, including computer controls. Consequently, labor costs and production times have been reduced, even though such machinery may require significant capital investment. If finishing is likely to be an important factor in manufacturing a particular product, the conceptual and design stages should involve an analysis of the degree of surface finish and dimensional accuracy required. Furthermore, all processes that precede finishing operations should be analyzed for their capability to produce a more acceptable surface finish and dimensional accuracy. This can be accomplished through proper selection of tools and process parameters and the characteristics of the machine tools involved.

The economic production run for a particular process depends on the cost of tooling and equipment, the material removal rate, operating costs, and the level of operator skill required, as well as the cost of secondary and finishing operations that may be necessary. In chemical machining, the costs of reagents, maskants, and disposal, together with the cost of cleaning the parts, are important factors; in electrical-discharge machining, the cost of electrodes and the need to replace them periodically are significant. The rate of material removal, and hence the production rate, can vary significantly in these processes. The cost of tooling and equipment also varies significantly, as does the operator skill required. If possible, the high capital investment for machines such as for electrical and high-energy-beam machining should be justified in terms of the production runs they are able to generate and the lack of feasibility of manufacturing the same part by other means.

## CASE STUDY | Manufacture of Stents

Heart attacks, strokes, and other cardiovascular diseases claim one life every 33 seconds in the United States alone.[1] These illnesses are attributed mostly to coronary artery disease, which is the gradual buildup of fat (cholesterol) within the artery wall causing the coronary arteries to become narrowed or blocked. This condition reduces the blood flow to the heart muscle and eventually leads to a heart attack, stroke, or other cardiovascular diseases. One of the most popular methods today for keeping blocked arteries open is to implant a stent into the artery. The MULTI-LINK TETRA™, shown in Fig. 9.41, consists of a tiny mesh tube that is expanded using a balloon dilatation catheter and implanted into a blocked or partially blocked coronary artery. A stent serves as a scaffolding or mechanical brace to keep the artery open. A stent offers the patient a minimally invasive method for treating coronary heart disease for which the alternative is usually open-heart bypass surgery, a procedure with greater risk, pain, rehabilitation time, and cost to the patient.

Stent manufacturing is extremely demanding for which the accuracy and precision of its design are paramount for proper performance. The manufacture of the stent must allow satisfaction of all of the design constraints and provide an extremely reliable device; thus, strict quality control procedures must be in place throughout the manufacturing process. When designing a stent there are many material-selection factors to consider, such as radial strength, corrosion resistance, endurance limit, flexibility, and biocompatibility. Radial strength is important because the stent must withstand the pressure the artery exerts on the stent upon expansion. Because the stent is implantable in the body it must be corrosion resistant; it must also be able to withstand the stress undulations caused by heartbeats, and consequently, the material must also have high resistance to fatigue. In addition, the stent must have the appropriate wall thickness in order to be flexible enough to negotiate through the tortuous anatomy of the heart. Above all, the material must be biocompatible to the body, because it remains there for the rest

Proximal and distal markers
indicate position of stent on radiograph

8–38 mm (0.315–1.50 in.)

Guide wire
0.356 mm
(0.014 in.) max

2.5–4.0 mm
(0.010–0.16 in.)

Catheter and balloon
used for stent expansion

Variable Thickness Strut (VTS™)
3-3-3 Pattern

**FIGURE  9.41**   The Guidant MULTI-LINK TETRA™ coronary stent system.

[1] 2002 Heart and Stroke Statistical Update, American Heart Association, p. 4

**FIGURE 9.42** Detail of the 3-3-3 MULTI-LINK TETRA™ pattern.

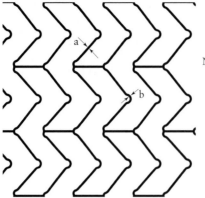

Notes:
a. 0.12 mm (0.0049 in.) section thickness to provide radiopacity
b. 0.091 mm (0.0036 in.) thickness for flexibility

of the patient's life. A standard material for stents, one that satisfies all of these requirements, is 316L stainless steel (Section 3.10.2).

The pattern of a stent also affects its performance. A finite-element analysis of the stent's design is first performed to determine how the stent will perform under the stresses induced from the beating heart. The strut pattern, tube thickness, and strut-leg width all affect the performance of the stent. Many different patterns are possible, such as the MULTI-LINK TETRA™ stent, which is shown in Figure 9.42 including some of the critical dimensions.

A stent starts out as a drawn stainless-steel tube with outer diameter that matches the final stent dimension, and a wall thickness selected to provide proper strength when expanded. The drawn tube is then laser machined (Section 9.14.1) to achieve the desired pattern (Fig. 9.43a). This method has proven to be very effective because of the small, intricate pattern of the stent and the tight dimensional tolerances that it must have. As the laser cuts the pattern of the stent, it leaves small metal slugs that need to be removed, therefore, it is very important that the laser cut all the way through the tube wall. A thick oxide layer or slag inevitably develops on the stainless steel due to the thermal and chemical attack from the air; in addition, weld splatter, burrs, and other surface defects result from laser machining. Finishing operations are therefore required to remove the splatter and oxide layer.

The first finishing operation after laser machining involves chemical etching to remove as much of the slag on the stent as possible; this operation is typically done in an acid solution, and the resulting surface is shown in Fig. 9.43b. Once the slag has been sufficiently removed, any residual burrs from the laser machining process need to be removed and the stent surface must be finished. It is very important that the surface finish of the stent be smooth to prevent possible forma-

(a)

(b)

(c)

**FIGURE 9.43** Evolution of the stent surface. (a) MULTI-LINK TETRA™ after lasing. Note that a metal slug is still attached. (b) After removal of slag. (c) After electropolishing.

tion of clots (thrombi) on the stent. To remove the sharp points, the stent is electropolished (Section 9.8) by passing an electrical current through an electrochemical solution, to ensure proper surface finish (Fig. 9.43c) that is both shiny and smooth. The stent is then placed onto a balloon catheter assembly, and sterilized and packaged for delivery to the surgeon.

*Source*: Courtesy of K. L. Graham, Guidant Corporation.

# SUMMARY

- Abrasive machining and advanced machining processes are often necessary and economical when workpiece hardness is high, the materials are brittle or flexible, part shapes are complex, and surface finish and dimensional tolerance requirements are high. (Section 9.1)

- Conventional abrasives consist basically of aluminum oxide and silicon carbide, and superabrasives consist of cubic boron nitride and diamond. Friability of abrasive grains is an important factor, as are shape and size of the grains. (Section 9.2)

- Grinding wheels consist of a combination of abrasive grains and bonding agents. Important characteristics of wheels include abrasive-grain and bond types, grade, and hardness. Wheels may be reinforced to maintain wheel integrity. (Section 9.3)

- The mechanics of grinding processes are studied in order to establish quantitative relationships regarding chip dimensions, grinding forces and energy requirements, temperature rise, residual stresses, and any adverse effects on surface integrity of ground components. (Section 9.4)

- Wheel wear is an important consideration in ground surface quality and product integrity. Wear is usually monitored in terms of the grinding ratio, defined as the ratio of material removed to the volume of wheel wear. Dressing and truing of wheels are done by various techniques, with the aid of computer controls. (Section 9.5)

- A variety of abrasive machining processes and equipment is available for surface, external, and internal grinding, as well as for large-scale material removal processes. The proper selection of abrasives, process variables, and grinding fluids is important for obtaining the desired surface finish and dimensional accuracy and for avoiding damage such as burning, heat checking, harmful residual stresses, and chatter. (Section 9.6)

- Ultrasonic machining removes material by microchipping and is best suited for hard and brittle materials. (Section 9.7)

- Several finishing operations are available to improve surface finish. Because finishing processes can contribute significantly to product cost, proper selection and implementation of finishing operations are important. (Section 9.8)

- Deburring may be necessary for certain finished components. Commonly used deburring methods include vibratory and barrel finishing and shot blasting, although thermal-energy processes and other methods can also be used. (Section 9.9)

- Advanced machining processes involve chemical and electrical means and high-energy beams; they are particularly suitable for hard materials and complex shapes. The effects of these processes on surface integrity must be investigated, as they can damage surfaces considerably, thus reducing fatigue life. (Sections 9.10–9.14)

- Water-jet, abrasive water-jet, and abrasive-jet machining processes can be used effectively for cutting as well as deburring operations. Because they do not use hard tooling, they have inherent flexibility of operation. (Sections 9.15)

- As in all manufacturing processes, certain design guidelines should be followed for effective use of advanced machining processes. (Section 9.16)
- The processes described in this chapter have unique capabilities; however, because they involve different types of machinery, process variables, and cycle times, the competitive aspects of each process with respect to a particular product must be studied individually. (Section 9.17)

## SUMMARY OF EQUATIONS

Undeformed chip length, $\ell$:

Surface grinding, $\ell = \sqrt{Dd}$

External grinding, $\ell = \sqrt{\dfrac{Dd}{1 + D/D_w}}$

Internal grinding, $\ell = \sqrt{\dfrac{Dd}{1 - D/D_w}}$

Undeformed-chip thickness, surface grinding, $t = \sqrt{\dfrac{4v}{VCr}\sqrt{\dfrac{d}{D}}}$

Relative grain force $\propto \dfrac{v}{VC}\sqrt{\dfrac{d}{D}}$

Temperature rise in surface grinding, $\Delta T \propto \dfrac{d}{t} \propto d^{3/4}\sqrt{\dfrac{VC}{v}}\sqrt{D}$

Grinding ratio, $G = \dfrac{\text{volume of material removed}}{\text{volume of wheel wear}}$

Average force of particle in ultrasonic machining, where $F_{ave} = \dfrac{2mv}{t_o}$ where $t_o = \dfrac{5r}{c_o}\left(\dfrac{c_o}{v}\right)^{1/5}$

Material removal rate in

electrochemical machining, MRR $= CI\eta$

electrochemical grinding, MRR $= \dfrac{GI}{\rho F}$

electrical-discharge machining, MRR $= 4 \times 10^4 IT_w^{-1.23}$

electrical-discharge grinding, MRR $= KI$

wire EDM, MRR $= V_f h b$, where $b = d_w + 2s$

Penetration rate in electrochemical grinding, $V_s = \left[\dfrac{G}{dF}\right]\left[\dfrac{E}{gK_p}\right]K$

Wear rate of electrode in electrical-discharge machining, $W_t = 1.1 \times 10^{11} IT_t^{-2.38}$

Laser-beam cutting time, $t = \dfrac{P}{vd}$

## BIBLIOGRAPHY

Andrew, C., T. D. Howes, and T. R. A. Pearce, *Creep-Feed Grinding*, Industrial Press, 1985.

*ASM Handbook*, Vol. 16: *Machining*, ASM International, 1989.

Borkowski, J., and A. Szymanski, *Uses of Abrasives and Abrasive Tools*, Ellis Horwood, 1992.

Brown, J., *Advanced Machining Technology Handbook*, McGraw-Hill, 1998.

Chryssolouris, G., and P. Sheng, *Laser Machining, Theory & Practice*, Springer, 1991.

Crafer, R. C., and P. J. Oakley, *Laser Processing in Manufacturing*, Chapman & Hall, 1993.

Farago, F. T., *Abrasive Methods Engineering*, Industrial Press, Vol. 1, 1976; Vol. 2, 1980.

Gillespie, L. K., *Deburring and Edge Finishing Handbook*, Society of Manufacturing Engineers/American Society of Mechanical Engineers, 2000.

Guitran, E. B., *The EDM Handbook*, Hanser-Gardner, 1997.

Hwa, L. S., *Chemical Mechanical Polishing in Silicon Processing*, Academic Press, 1999.

Jain, V. K., and P. C. Pandey, *Theory and Practice of Electrochemical Machining*, Wiley, 1993.

King, R. I., and R. S. Hahn, *Handbook of Modern Grinding Technology*, Chapman & Hall/Methuen, 1987.

Krar, S., and E. Ratterman, *Superabrasives: Grinding and Machining with CBN and Diamond*, McGraw-Hill, 1990.

Krar, S., *Grinding Technology*, 2d. ed., Delmar Publishers, 1995.

*Machining Data Handbook*, 3d. ed., 2 vols, Machinability Data Center, 1980.

Malkin, S., *Grinding Technology: Theory and Applications of Machining with Abrasives*, Wiley, 1989.

Marinescu, I. D. (ed.), *Handbook of Ceramics Grinding & Polishing*, Noyes, 1998.

Maroney, M. L., *A Guide to Metal and Plastic Finishing*, Industrial Press, 1991.

McGeough, J. A., *Advanced Methods of Machining*, Chapman & Hall, 1988.

McKee, R. L., *Machining with Abrasives*, Van Nostrand Reinhold, 1982.

Metzger, J. L., *Superabrasive Grinding*, Butterworths, 1986.

Momber, A. W., and R. Kovacevic, *Principles of Abrasive Water Jet Machining*, Springer, 1998.

Nachtman, E. S., and S. Kalpakjian, *Lubricants and Lubrication in Metalworking Operations*, Dekker, 1985.

Powell, J., $CO_2$ *Laser Cutting*, Springer, 1991.

Salmon, S. C., *Modern Grinding Process Technology*, McGraw-Hill, 1992.

Schneider, A. F., *Mechanical Deburring and Surface Finishing Technology*, Dekker, 1990.

Shaw, M. C., *Metal Cutting Principles*, Oxford, 1984.

————, *Principles of Abrasive Processing*, Oxford, 1996.

Sluhan, C. (ed.), *Cutting and Grinding Fluids: Selection and Application*, Society of Manufacturing Engineers, 1992.

Sommer, C., and S. Sommer, *Wire EDM Handbook*, Technical Advanced Publishing Co., 1997.

Steen, W. M., *Laser Material Processing*, Springer, 1991.

Szymanski, A., and J. Borkowski, *Technology of Abrasives and Abrasive Tools*, Ellis Horwood, 1992.

Taniguchi, N. (ed.), *Nanotechnology*, Oxford, 1996.

————, *Energy-Beam Processing of Materials*, Oxford, 1989.

*Tool and Manufacturing Engineers Handbook*, 4th ed., Vol. 1: *Machining*, Society of Manufacturing Engineers, 1983.

Webster, J. A., *Abrasive Processes: Theory, Technology, and Practice*, Dekker, 1996.

# QUESTIONS

**9.1**  Why are grinding operations necessary for parts that have been machined by other processes?

**9.2**  Explain why there are so many different types and sizes of grinding wheels.

**9.3**  Explain the reasons for the large differences between the specific energies involved in grinding (Table 9.3) and in machining (Table 8.3).

**9.4**  Describe the advantages of superabrasives over conventional abrasives.

**9.5**  Give examples of applications for the grinding wheels shown in Fig. 9.2.

**9.6**  Explain why the same grinding wheel may act soft or hard.

**9.7**  Describe your understanding of the role of friability of abrasive grains on grinding-wheel performance.

**9.8**  Explain the factors involved in selecting the appropriate type of abrasive for a particular grinding operation.

**9.9**  What are the effects of wear flat on the grinding operation?

**9.10**  The grinding ratio, $G$, depends on the following factors: (1) type of grinding wheel, (2) workpiece hardness, (3) wheel depth of cut, (4) wheel and workpiece speeds, and (5) type of grinding fluid. Explain how.

**9.11**  List and explain the precautions you would take when grinding with high precision. Comment on the role of the machine, process parameters, the grinding wheel, and grinding fluids.

**9.12**  Describe the methods you would use to determine the number of active cutting points per unit surface area on the periphery of a straight (i.e., Type 1; see Fig. 9.2a) grinding wheel. What is the significance of this number?

**9.13**  Describe and explain the difficulties involved in grinding parts made of thermoplastics.

**9.14** Explain why ultrasonic machining is not suitable for soft and ductile metals.

**9.15** It is generally recommended that a soft-grade grinding wheel be used for hardened steels. Explain why.

**9.16** Explain the reasons that the processes described in this chapter may adversely affect the fatigue strength of materials.

**9.17** Describe the factors that may cause chatter in grinding operations, and give reasons that these factors cause chatter.

**9.18** Outline the methods that are generally available for deburring parts. Discuss the advantages and limitations of each.

**9.19** In which of the processes described in this chapter are physical properties of the workpiece material important? Why?

**9.20** Give all possible technical and economic reasons that the material removal processes described in this chapter may be preferred, or even required, over those described in Chapter 8.

**9.21** What processes would you recommend for die sinking in a die block? Explain.

**9.22** The proper grinding surfaces for each type of wheel are shown with arrows in Fig. 9.2. Explain why grinding on other surfaces of the wheel is improper and unsafe.

**9.23** Note that wheel (b) in Fig. 9.3 has serrations along its periphery, somewhat similar to circular metal saws. What does this design offer in grinding?

**9.24** In Fig. 9.10, we note that wheel speed and grinding fluids can have a major effect on the type and magnitude of residual stresses developed in grinding. Explain the possible reasons for these phenomena.

**9.25** Explain the consequences of allowing the workpiece temperature to rise too high in grinding operations.

**9.26** Comment on any observations you have regarding the contents of Table 9.4.

**9.27** Why has creep-feed grinding become an important process? Explain.

**9.28** There has been a trend in manufacturing industries to increase the spindle speed of grinding wheels. Explain the possible advantages and limitations of such an increase.

**9.29** Why is preshaping or premachining of parts generally desirable in the advanced machining processes described in this chapter?

**9.30** Why are finishing operations sometimes necessary? How could they be minimized? Explain, with examples.

**9.31** Why has the wire-EDM process become so widely accepted in industry?

**9.32** Make a list of material removal processes described in this chapter that may be suitable for the following workpiece materials: (1) ceramics, (2) cast iron, (3) thermoplastics, (4) thermosets, (5) diamond, and (6) annealed copper.

**9.33** Explain why producing sharp corners and profiles using some of the processes described in this chapter can be difficult.

**9.34** How do you think specific energy, $u$, varies with respect to wheel depth of cut and hardness of the workpiece material? Explain.

**9.35** In Example 9.2, it is stated that the thrust force in grinding is about 30% higher than the cutting force. Why is it higher?

**9.36** Why should we be interested in the magnitude of the thrust force in grinding? Explain.

**9.37** Why is the material removal rate in electrical-discharge machining a function of the melting point of the workpiece material?

**9.38** Inspect Table 9.4, and, for each process, list and describe the role of various mechanical, physical, and chemical properties of the workpiece material that may affect performance.

**9.39** Which of the processes listed in Table 9.4 would not be applicable to nonmetallic materials? Explain.

**9.40** Why does the machining cost increase so rapidly as surface finish requirements become finer?

**9.41** Which of the processes described in this chapter are particularly suitable for workpieces made of (a) -ceramics, (b) thermoplastics, and (c) thermosets? Why?

**9.42** Other than cost, is there a reason that a grinding wheel intended for a hard workpiece cannot be used for a softer workpiece?

**9.43** How would you grind the facets on a diamond, as for an engagement ring, since diamond is the hardest material known?

**9.44** Define dressing and truing, and describe the difference between them.

# PROBLEMS

**9.45** In a surface-grinding operation, calculate the chip dimensions for the following process variables: $D = 10$ in., $d = 0.001$ in., $v = 50$ ft/min, $V = 5000$ ft/min, $C = 500$ per in$^2$, and $r = 20$.

**9.46** If the workpiece strength in grinding is doubled, what should be the percentage decreases in the wheel depth of cut, $d$, in order to maintain the same grain force, all other variables being the same?

**9.47** Taking a thin Type 1 grinding wheel as an example and referring to texts on stresses in rotating bodies, plot the tangential stress, $\sigma_t$, and radial stress, $\sigma_r$, as a function of radial distance, i.e., from the hole to the periphery of the wheel. Note that because the wheel is thin, this problem can be regarded as a plane-stress problem. How would you determine the maximum combined stress and its location in the wheel?

**9.48** Derive a formula for the material removal rate (MRR) in surface grinding in terms of process parameters. Use the same terminology as in the text.

**9.49** Assume that a surface-grinding operation is being carried out under the following conditions: $D = 250$ mm, $d = 0.1$ mm, $v = 0.5$ m/s, and $V = 50$ m/s. These conditions are then changed to the following: $D = 150$ mm, $d = 0.1$ mm, $v = 0.3$ m/s, and $V = 25$ m/s. What is the difference in the temperature rise from the initial condition?

**9.50** For a surface-grinding operation, derive an expression for the power dissipated in imparting kinetic energy to the chips. Comment on the magnitude of this energy. Use the same terminology as in the text.

**9.51** The shaft of a Type 1 grinding wheel is attached to a flywheel only, which is rotating at a certain initial rpm. With this setup, a surface-grinding operation is being carried out on a long workpiece and at a constant workpiece speed, $v$. Obtain an expression for estimating the linear distance ground on the workpiece before the wheel comes to a stop. Ignore wheel wear.

**9.52** Calculate the average impact force on a steel plate by a spherical aluminum-oxide abrasive particle with a 0.1-mm diameter, dropped from heights of (1) 1 m, (2) 2 m, and (3) 10 m.

**9.53** A 50-mm-deep hole, 30 mm in diameter, is being produced by electrochemical machining. A high production rate is more important than the quality of the machined surface. Estimate the maximum current and the time required to perform this operation.

**9.54** If the operation in Problem 9.53 were performed on an electrical-discharge machine, what would be the estimated machining time?

**9.55** A cutting-off operation is being performed with a laser beam. The workpiece being cut is $\frac{3}{8}$ in. thick and 3 in. long. If the kerf width is $\frac{1}{6}$ in., estimate the time required to perform this operation.

**9.56** Refering to Table 3.2, identify two metals or metal alloys that, when used as workpiece and electrode, respectively, in EDM would give the (1) lowest and (2) highest wear ratios, $R$. Calculate these quantities.

**9.57** In Section 9.5.2, it was stated that, in practice, grinding ratios typically range from 2 to 200. Based on the information given in Section 9.13, estimate the range of wear ratios in electrical-discharge machining, and then compare them with grinding ratios.

**9.58** Assume that you are an instructor covering the topics in this chapter, and you are giving a quiz on the quantitative aspects to test the understanding of the students. Prepare several numerical problems, and supply the answers.

**9.59** It is known that heat checking occurs when grinding with a spindle speed of 4000 rpm, a wheel diameter of 10 in., and a depth of cut of 0.0015 in. for a feed rate of 50 ft/min. For this reason, the spindle speed should be kept at 3500 rpm. If a new, 8-in-diameter wheel is used, what spindle speed can be employed before heat checking occurs? What spindle speed should be used to keep the same grinding temperatures as those encountered with the existing operating conditions?

**9.60** It is desired to grind a hard aerospace aluminum alloy. A depth of 0.003 in. is to be removed from a cylindrical section 10 in. long and with a 4-in. diameter. If each part is to be ground in not more than one minute, what is the approximate power requirement for the grinder? What if the material is changed to a hard titanium alloy?

**9.61** A grinding operation is taking place with a 10-in. grinding wheel at a spindle rotational speed of 3000 rpm. The workpiece feed rate is 60 ft/min, and the depth of cut is 0.002 in. Contact thermometers record an approximate maximum temperature of 1800°F. If the workpiece is steel, what is the temperature if the spindle speed is increased to 4000 rpm? What if it is increased to 10,000 rpm?

**9.62** The regulating wheel of a centerless grinder is rotating at a surface speed of 50 ft/min and is inclined at an angle of 5°. What is the feed rate of material past the grinding wheel?

**9.63** Using some typical values, explain what changes, if any, take place in the magnitude of impact force of a particle in ultrasonic machining as the temperature of the workpiece is increased. Assume that the workpiece is made of hardened steel.

# DESIGN

**9.64** Would you consider designing a machine tool that combines, in one machine, two or more of the processes described in this chapter? Explain. For what types of parts would such a machine be useful? Make preliminary sketches for such machines.

**9.65** With appropriate sketches, describe the principles of various fixturing methods and devices that can be used for each of the processes described in this chapter.

**9.66** Surface finish can be an important consideration in the design of products. Describe as many parameters as you can that could affect the final surface finish in grinding, including the role of process parameters as well as the setup and the equipment used.

**9.67** A somewhat controversial subject in grinding is size effect. (See Section 9.4.1.) Design a setup and a series of experiments whereby size effect can be studied.

**9.68** Describe how the design and geometry of the workpiece affects the selection of an appropriate shape and type of a grinding wheel.

**9.69** Prepare a comprehensive table of the capabilities of abrasive machining processes, including the shapes of parts ground, types of machines involved, typical maximum and minimum workpiece dimensions, and production rates.

**9.70** How would you produce a thin, large-diameter round disk with a thickness that decreases linearly from the center outward?

**9.71** Marking surfaces of manufactured parts with letters and numbers can be done not only with labels and stickers, but also by various mechanical and non-mechanical means. Make a list of some of these methods, explaining their advantages and limitations.

# Properties and Processing of Polymers and Reinforced Plastics; Rapid Prototyping and Rapid Tooling

10.1 Introduction
10.2 The Structure of Polymers
10.3 Thermoplastics: Behavior and Properties
10.4 Thermosets: Behavior and Properties
10.5 Thermoplastics: General Characteristics and Applications
10.6 Thermosets: General Characteristics and Applications
10.7 High-Temperature Polymers, Electrically Conducting Polymers, and Biodegradable Plastics
10.8 Elastomers (Rubbers): General Characteristics and Applications
10.9 Reinforced Plastics
10.10 Processing of Plastics
10.11 Processing of Polymer-Matrix-Reinforced Plastics
10.12 Rapid Prototyping and Rapid Tooling
10.13 Design Considerations
10.14 Economics of Processing Plastics

## 10.1 | Introduction

Although the word **plastics**, first used around 1909, is a commonly used synonym for **polymers**, plastics are one of numerous polymeric materials and have extremely large molecules (*macromolecules*). Consumer and industrial products made of polymers include food and beverage containers, packaging, signs, housewares, textiles, medical devices, foams, paints, safety shields, and toys. Compared with metals, polymers are generally characterized by low density, low strength and stiffness (Table 10.1), low electrical and thermal conductivity, good resistance to chemicals, and a high coefficient of thermal expansion. However, the useful temperature range for most polymers is generally low, up to about 350°C(660°F), and they are not as dimensionally stable in service, over a period of time, as metals.

The word *plastics* is from the Greek word *plastikos*, meaning "it can be molded and shaped." Plastics can be machined, cast, formed, and joined into many shapes with relative ease. Minimal or no additional surface-finishing operations are required, which is an important advantage over metals. Plastics are commercially available as sheet, plate, film, rods, and tubing of various cross-sections. The word *polymer* was first used in 1866. The earliest polymers were made of **natural organic materials** from animal and vegetable products, cellulose being the most common example. With various chemical reactions, cellulose is modified into cellulose acetate, used in making photographic films (celluloid), sheets for packaging, and textile fibers; cellulose is also modified into cellulose nitrate for plastics, explosives, rayon (a cellulose textile fiber), and varnishes. The earliest **synthetic polymer** was a phenol formaldehyde, a thermoset developed in 1906 and called *Bakelite* (a trade name, after L. H. Baekeland, 1863–1944).

The development of modern plastics technology began in the 1920s, when raw materials necessary for making polymers were extracted from coal and petroleum products. Ethylene was the first example of such a raw material, as it became the building block for polyethylene; it is the product of the reaction between acetylene and hydrogen, and acetylene is the product of the reaction between coke and methane.

**TABLE 10.1**

**Approximate Range of Mechanical Properties for Various Engineering Plastics at Room Temperature**

| Material | UTS (MPa) | E (GPa) | Elongation in 50 mm (%) | Poisson's Ratio ($\nu$) |
|---|---|---|---|---|
| ABS | 28–55 | 1.4–2.8 | 75–5 | – |
| ABS (reinforced) | 100 | 7.5 | – | 0.35 |
| Acetals | 55–70 | 1.4–3.5 | 75–25 | – |
| Acetals (reinforced) | 135 | 10 | – | 0.35–0.40 |
| Acrylics | 40–75 | 1.4–3.5 | 50–5 | – |
| Cellulosics | 10–48 | 0.4–1.4 | 100–5 | – |
| Epoxies | 35–140 | 3.5–17 | 10–1 | – |
| Epoxies (reinforced) | 70–1400 | 21–52 | 4–2 | – |
| Fluorocarbons | 7–48 | 0.7–2 | 300–100 | 0.46–0.48 |
| Nylon | 55–83 | 1.4–2.8 | 200–60 | 0.32–0.40 |
| Nylon (reinforced) | 70–210 | 2–10 | 10–1 | – |
| Phenolics | 28–70 | 2.8–21 | 2–0 | – |
| Polycarbonates | 55–70 | 2.5–3 | 125–10 | 0.38 |
| Polycarbonates (reinforced) | 110 | 6 | 6–4 | – |
| Polyesters | 55 | 2 | 300–5 | 0.38 |
| Polyesters (reinforced) | 110–160 | 8.3–12 | 3–1 | – |
| Polyethylenes | 7–40 | 0.1–0.14 | 1000–15 | 0.46 |
| Polypropylenes | 20–35 | 0.7–1.2 | 500–10 | – |
| Polypropylenes (reinforced) | 40–100 | 3.6–6 | 4–2 | – |
| Polystyrenes | 14–83 | 1.4–4 | 60–1 | 0.35 |
| Polyvinyl chloride | 7–55 | 0.014–4 | 450–40 | – |

Likewise, commercial polymers including polypropylene, polyvinyl chloride, poly-methylmethacrylate, polycarbonate, and others are all made in a similar manner; these materials are known as *synthetic organic polymers*. Figure 10.1 illustrates an outline of the basic process of making various synthetic polymers.

This chapter also covers the characteristics and processing of **polymer-matrix reinforced plastics** (also called *composite materials*), which exhibit a wide range of properties such as stiffness, strength, resistance to creep, and high strength-to-weight and stiffness-to-weight ratios. Their applications include numerous consumer and industrial products, as well as products in the automotive and aerospace industries. The properties and processing of metal-matrix and ceramic-matrix composites are covered in Chapter 11. Another important advance in manufacturing is **rapid prototyping and tooling,** also called *desktop manufacturing* or *free-form fabrication*. This method is a family of processes by which a solid physical model of a part is made directly from a three-dimensional CAD drawing, involving a time span that is much shorter than that for traditional prototype manufacturing. As described in detail in Section 10.12, rapid prototyping entails several different material-consolidation techniques, including resin curing, deposition, solidification, and sintering.

# 10.2 | The Structure of Polymers

Many of a polymer's properties depend largely on (a) the structure of individual poly-mer molecules, (b) the shape and size of the molecules, and (c) how the molecules are arranged to form a polymer structure. Polymer molecules are characterized by their extraordinary size, a feature that distinguishes them from other organic chemical

**FIGURE 10.1** Basic structure of polymer molecules: (a) ethylene molecule; (b) polyethylene, a linear chain of many ethylene molecules; (c) molecular structure of various polymers. These molecules are examples of the basic building blocks for plastics.

compositions. Polymers are *long-chain molecules*, also called **macromolecules** or **giant molecules**, that are formed by polymerization, that is, by linking and cross-linking different monomers. A **monomer** is the basic building block of polymers. The word **mer**, from the Greek *meros*, meaning "part," indicates the smallest repetitive unit, similar to the term *unit cell* used in connection with crystal structures of metals (see Section 3.2).

The term **polymer** means "many mers or units," generally repeated hundreds or thousands of times in a chainlike structure. Most monomers are organic materials in which carbon atoms are joined in *covalent* (electron-sharing) bonds with other atoms, such as hydrogen, oxygen, nitrogen, fluorine, chlorine, silicon, and sulfur. An ethylene molecule is a simple monomer consisting of carbon and hydrogen atoms (Fig. 10.1a).

## 10.2.1 Polymerization

Monomers can be linked in repeating units to make longer and larger molecules by a chemical reaction known as the *polymerization reaction*. Polymerization processes are complex and are described only briefly here. Although there are numerous variations, two basic polymerization processes are condensation polymerization and addition polymerization:

1. In **condensation polymerization**, polymers are produced by the formation of bonds between two types of reacting mers. One characteristic of this reaction is that reaction by-products such as water are condensed out, hence the term

*condensation*. This process is also known as *step-growth* or *step-reaction polymerization*, because the polymer molecule grows step by step until all of one reactant is consumed.

2. In **addition polymerization**, also known as *chain-growth* or *chain-reaction polymerization*, bonding takes place without reaction by-products; it is called *chain-reaction polymerization* because of the high rate at which long molecules form simultaneously, usually within a few seconds. This rate is much higher than that for condensation polymerization. In this reaction, an initiator is added to open the double bond between the carbon atoms and begins the linking process by adding many more monomers to a growing chain. For example, ethylene monomers (Fig. 10.1a) link to produce polyethylene (Fig. 10.1b); other examples of addition-formed polymers are shown in Fig. 10.1c.

Some basic characteristics and types of polymers are described below.

a) **Molecular weight.** The sum of the molecular weight of the mers in the polymer chain is the *molecular weight* of the polymer; the higher the molecular weight in a given polymer, the greater is the length of the chain. Because polymerization is a random event, the polymer chains produced are not all of equal length; however, the chain lengths produced fall into a traditional distribution curve (see Fig. 11.4). The average molecular weight of a polymer is determined and expressed on a statistical basis by averaging; the spread observed in the weight distribution is referred to as the **molecular weight distribution** (MWD). Molecular weight and MWD have a strong influence on the properties of the polymer. For example, tensile and impact strength, resistance to cracking, and viscosity in the molten state all increase with increasing molecular weight (Fig. 10.2). Most commercial polymers have a molecular weight between 10,000 and 10,000,000.

b) **Degree of polymerization.** In some cases, it is more convenient to express the size of a polymer chain in terms of the *degree of polymerization* (DP), defined as the ratio of the molecular weight of the polymer to the molecular weight of the repeat unit. In terms of polymer processing (see Section 10.10), the higher the DP, the higher is the polymer's viscosity, or resistance to flow (Fig. 10.2), which affects ease of shaping and overall cost of processing.

c) **Bonding.** During polymerization, the monomers are linked together in a *covalent bond*, forming a polymer chain. Because of their strength, covalent bonds are also called **primary bonds**. The polymer chains are, in turn, held

**FIGURE 10.2**    Effect of molecular weight and degree of polymerization on the strength and viscosity of polymers.

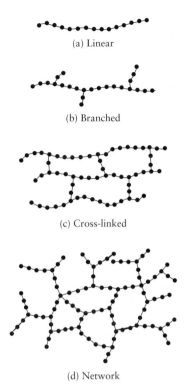

(a) Linear

(b) Branched

(c) Cross-linked

(d) Network

**FIGURE 10.3**    Schematic illustration of polymer chains. (a) Linear structure; thermoplastics such as acrylics, nylons, polyethylene, and polyvinyl chloride have linear structures. (b) Branched structure, such as in polyethylene. (c) Cross-linked structure; many rubbers and elastomers have this structure. Vulcanization of rubber produces this structure. (d) Network structure, which is basically highly cross-linked; examples include thermosetting plastics, such as epoxies and phenolics.

together by the secondary bonds, such as van der Waals bonds, hydrogen bonds, and ionic bonds. **Secondary bonds** are much weaker than primary bonds, by one to two orders of magnitude. In a given polymer, the increase in strength and viscosity with molecular weight comes in part from the fact that the longer the polymer chain, the greater is the energy needed to overcome the strength of the secondary bonds. For example, ethylene mers that have a DP of 1, 6, 35, 140, and 1350 are, respectively, in the form of gas, liquid, grease, wax, and hard plastic at room temperature.

d) **Linear polymers.**    The chainlike polymers shown in Fig. 10.1 are called *linear polymers*, because of their linear structure (Fig. 10.3a). A linear molecule is not straight in shape. In addition to those shown in Fig. 10.3, other linear polymers include polyamides (nylon 6,6) and polyvinyl fluoride. Generally, a polymer consists of more than one type of structure; thus, a linear polymer may contain some branched and cross-linked chains. (See parts e and f of this list.) As a result of branching and cross-linking, the polymer's properties change.

e) **Branched polymers.**    The properties of a polymer depend not only on the type of monomers in the polymer, but also on their arrangement in the molecular structure. In *branched polymers* (Fig. 10.3b), side-branch chains are attached to the main chain during the synthesis of the polymer. Branching interferes with the relative movement of the molecular chains; as a result, resistance to defor-mation increases and stress cracking resistance is affected. Also, the density of branched polymers is lower than that of linear-chain polymers, as branches interfere with the packing efficiency of polymer chains. The behavior of branched polymers can be compared with that of linear-chain polymers by making an analogy with a pile of tree branches (branched polymers) and a bundle of straight logs (linear-chain polymers). Note that it is more difficult to move a branch within the pile of branches than to move a log in its bundle. The three-dimensional entanglements of branches make movements more difficult, a phenomenon akin to increased strength.

**FIGURE 10.4** Behavior of polymers as a function of temperature and (a) degree of crystallinity and (b) cross-linking. The combined elastic and viscous behavior of polymers is known as *viscoelasticity*.

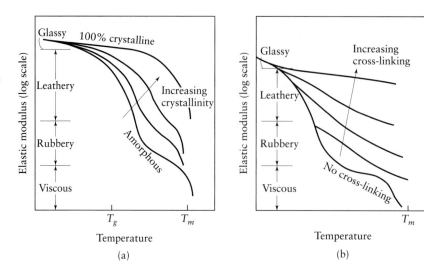

f) **Cross-linked polymers.** Generally three-dimensional in structure, *cross-linked polymers* have adjacent chains linked by covalent bonds (Fig. 10.3c). Polymers with a cross-linked chain structure are called **thermosets** or **thermosetting plastics**, such as epoxies, phenolics, and silicones. Cross-linking has a major influence on the properties of polymers (generally imparting hardness, strength, stiffness, brittleness, and better dimensional stability; see Fig. 10.4), as well as in the **vulcanization** of rubber. **Network polymers** consist of spatial (three-dimensional) networks of three active covalent bonds (Fig. 10.3d). A highly cross-linked polymer is also considered a network polymer. Thermoplastic polymers that have already been formed or shaped can be cross-linked to obtain greater strength by being subject to high-energy radiation, such as ultraviolet light, X rays, or electron beams. However, excessive radiation can cause *degradation* of the polymer.

g) **Copolymers and terpolymers.** If the repeating units in a polymer chain are all of the same type, the molecule is called a **homopolymer**. However, as with solid-solution metal alloys (see Section 5.2.1), two or three different types of monomers can be combined to impart certain special properties and characteristics to the polymer, such as improvement of both the strength and toughness as well as the formability of the polymer. *Copolymers* contain two types of polymers, such as styrene–butadiene, used widely for automobile tires. *Terpolymers* contain three types, such as ABS (acrylonitrile–butadiene–styrene), used for helmets, telephones, and refrigerator liners.

---

**EXAMPLE 10.1  Degree of polymerization in polyvinyl chloride (PVC)**

Determine the molecular weight of a polyvinyl-chloride mer. If a PVC polymer has an average molecular weight of 50,000, what is the degree of polymerization?

**Solution.** From Fig. 10.1c, note that each PVC mer has three hydrogen atoms, two carbon atoms, and one chlorine atom. Since the atomic number of these elements are 1, 12, and 35.5, respectively, the weight of a PVC mer is $(3)(1) + (2)(12) + (1)(35.5) = 62.5$; thus, the degree of polymerization is $50,000/62.5 = 800$.

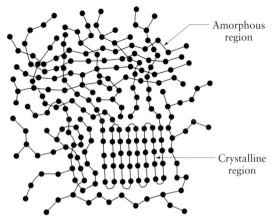

**FIGURE 10.5** Amorphous and crystalline regions in a polymer. The crystalline region (crystallite) has an orderly arrangement of molecules. The higher the crystallinity, the harder, stiffer, and less ductile is the polymer.

Amorphous region

Crystalline region

## 10.2.2 Crystallinity

Polymers such as polymethylmethacrylate, polycarbonate, and polystyrene are generally *amorphous*; that is, the polymer chains exist without long-range order. (See also the discussion of *amorphous alloys* in Section 3.11.9.) The amorphous arrangement of polymer chains is often described as similar to a bowl of spaghetti, or worms in a bucket, all intertwined with each other. However, in some polymers, it is possible to impart some crystallinity and thereby modify their characteristics. This may be done either during the synthesis of the polymer or by deformation during its subsequent processing.

The crystalline regions in polymers are called **crystallites** (Fig. 10.5). The crystals are formed when the long molecules arrange themselves in an orderly manner, similar to folding a fire hose in a cabinet or facial tissue in a box. Thus, a partially crystalline (semicrystalline) polymer can be regarded as a two-phase material, one phase being crystalline and the other amorphous. By controlling the rate of solidification during cooling and chain structure, it is possible to impart different **degrees of crystallinity** to polymers, although a polymer can never be 100% crystalline. Crystallinity ranges from an almost complete crystal (up to about 95% by volume in the case of polyethylene) to slightly crystallized, but mostly amorphous polymers. The degree of crystallinity is also affected by branching. A linear polymer can become highly crystalline, but a highly branched polymer cannot. Although the latter may develop some low level of crystallinity, it will never achieve a high crystalline content, because the branches interfere with the alignment of the chains into a regular crystal array.

**Effects of crystallinity.** The mechanical and physical properties of polymers are greatly influenced by the degree of crystallinity; as crystallinity increases, polymers become stiffer, harder, less ductile, more dense, less rubbery, and more resistant to solvents and heat (Fig. 10.4). The increase in density with increasing crystallinity is caused by crystallization shrinkage and a more efficient packing of the molecules in the crystal lattice. For example, the highly crystalline form of polyethylene, known as high-density polyethylene (HDPE), has a specific gravity in the range of 0.941 to 0.970 (80% to 95% crystalline) and is stronger, stiffer, tougher, and less ductile than low-density polyethylene (LDPE), which is about 60% to 70% crystalline and has a specific gravity of about 0.910 to 0.925.

Optical properties also are affected by the degree of crystallinity. The reflection of light from the boundaries between the crystalline and amorphous regions in the polymer causes opaqueness. Furthermore, because the index of refraction is proportional to density, the greater the density difference between the amorphous and crystalline phases, the greater is the opaqueness of the polymer. Polymers that are completely amorphous can be transparent, such as polycarbonate and acrylics.

**FIGURE 10.6**   Specific volume of polymers as a function of temperature. Amorphous polymers, such as acrylic and polycarbonate, have a glass-transition temperature, $T_g$, but do not have a specific melting point, $T_m$. Partly crystalline polymers, such as polyethylene and nylons, contract sharply at their melting points during cooling.

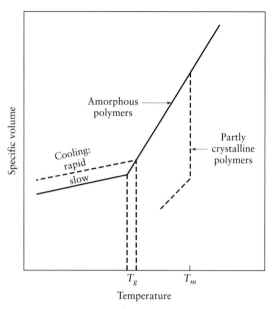

### 10.2.3  Glass-transition temperature

Amorphous polymers do not have a specific melting point, but they undergo a distinct change in their mechanical behavior across a narrow range of temperature. At low temperatures, they are hard, rigid, brittle, and glassy, and at high temperatures, they are rubbery or leathery. The temperature at which this transition occurs is called the *glass-transition temperature*, $T_g$, or the *glass point* or *glass temperature*. The term *glass* is included in this definition because glasses, which are amorphous solids (see Section 3.11.9), behave in the same manner (as you can see by holding a glass rod over a flame and observing its behavior). Although most amorphous polymers exhibit this behavior, there are some exceptions, such as polycarbonate, which is not rigid or brittle below its glass-transition temperature. Polycarbonate is tough at ambient temperature and is therefore used for safety helmets and shields.

To determine $T_g$, the specific volume of the polymer is measured and plotted against temperature to find the sharp change in the slope of the curve (Fig. 10.6). However, in the case of highly cross-linked polymers, the slope of the curve changes gradually near $T_g$, making it difficult to determine $T_g$ for these polymers. The glass-transition temperature varies for different polymers (Table 10.2); for example, room tem-

**TABLE 10.2**

| Glass-Transition and Melting Temperatures of Some Polymers | | |
|---|---|---|
| Material | $T_g(°C)$ | $T_m(°C)$ |
| Nylon 6,6 | 57 | 265 |
| Polycarbonate | 150 | 265 |
| Polyester | 73 | 265 |
| Polyethylene | | |
|    High density | −90 | 137 |
|    Low density | −110 | 115 |
| Polymethylmethacrylate | 105 | – |
| Polypropylene | −14 | 176 |
| Polystyrene | 100 | 239 |
| Polytetrafluoroethylene (Teflon) | −90 | 327 |
| Polyvinyl chloride | 87 | 212 |
| Rubber | −73 | – |

perature is above $T_g$ for some polymers and below it for others. Unlike amorphous polymers, partly crystalline polymers have a distinct melting point, $T_m$ (Fig. 10.6; see also Table 10.2). Because of the structural changes (first-order changes) that occur the specific volume of the polymers drops rapidly as their temperature is reduced.

## 10.2.4 Polymer blends

To improve the brittle behavior of amorphous polymers below their glass-transition temperature, they can be blended, usually with small quantities of an *elastomer* (see Section 10.8). These tiny particles are dispersed throughout the amorphous polymer, enhancing its toughness and impact strength by improving its resistance to crack propagation; these blended polymers are known as **rubber-modified polymers**. Blending may involve several components (**polyblends**) that use the favorable properties of different polymers. Advances have been made in **miscible blends** (mixing without separation of two phases), a process similar to alloying of metals, enabling polymer blends to become more ductile. Polymer blends account for about 20% of all polymers produced.

## 10.2.5 Additives in polymers

In order to impart certain specific properties, polymers are usually compounded with *additives*. The additives modify and improve certain characteristics of the polymers, such as their stiffness, strength, color, weatherability, flammability, arc resistance for electrical applications, and ease of subsequent processing.

1. **Fillers** used are generally wood flour (fine sawdust), silica flour (fine silica powder), clay, powdered mica, and short fibers of cellulose, glass, and asbestos. Because of their low cost, fillers are important in reducing the overall cost of polymers. Depending on their type, fillers improve the strength, hardness, toughness, abrasion resistance, dimensional stability, and/or stiffness of plastics. These properties are greatest at various percentages of different types of polymer–filler combinations. As in reinforced plastics (Section 10.9), a filler's effectiveness depends on the nature and strength of the bond between the filler material and the polymer chains.

2. **Plasticizers** are added to some polymers to impart flexibility and softness by lowering their glass-transition temperature. Plasticizers are low-molecular-weight solvents with high boiling points (nonvolatile). They reduce the strength of the secondary bonds between the long-chain molecules, thus making the polymer soft and flexible. The most common use of plasticizers is in polyvinyl chloride (PVC), which remains flexible during its many uses. Other applications of plasticizers include thin sheet, film, tubing, shower curtains, and clothing materials.

3. Most polymers are adversely affected by **ultraviolet radiation** (sunlight) and oxygen, which weaken and break the primary bonds, resulting in the scission (splitting) of the long-chain molecules; the polymer then degrades and becomes brittle and stiff. On the other hand, degradation may be beneficial, as in the disposal of plastic objects by subjecting them to environmental attack. (See also Section 10.7.3.) A typical example of imparting protection against ultraviolet radiation is the compounding of certain plastics and rubber with *carbon black* (soot); the carbon black absorbs a high percentage of the ultraviolet radiation. Protection against degradation by oxidation, particularly at elevated temperatures, is achieved by adding antioxidants to the polymer. Applying various coatings is another means of protecting polymers against environmental attack.

4. The wide variety of colors available in plastics is obtained by adding **colorants**; these materials are either organic (dyes) or inorganic (pigments). The selection of a colorant depends on the polymer's service temperature and exposure to light. Pigments are dispersed particles and generally have greater resistance than dyes to temperature and light.

5. If the temperature is sufficiently high, most polymers will ignite. The **flammability** (ability to support combustion) of polymers varies considerably, depending on their composition (such as the chlorine and fluorine content). The flammability of polymers can be reduced either by making them from less flammable raw materials or by adding of **flame retardants**, such as compounds of chlorine, bromine, and phosphorus. Examples of polymers with different burning characteristics include the following: (a) Fluorocarbons (e.g., *Teflon*) do not burn; (b) carbonate, nylon, and vinyl chloride burn, but are self-extinguishing; and (c) acetal, acrylic, ABS, polyester, polypropylene, and styrene burn and are not self-extinguishing.

6. **Lubricants** may be added to polymers to reduce friction during their subsequent processing into useful products and to prevent parts from sticking to the molds. Lubrication is also important in preventing thin polymer films from sticking to each other.

## 10.3 | Thermoplastics: Behavior and Properties

We noted earlier that the bonds between adjacent long-chain molecules (secondary bonds) are much weaker than the covalent bonds (primary bonds) within each molecule. Furthermore, it is the strength of the secondary bonds that determines the overall strength of the polymer; linear and branched polymers have weak secondary bonds. As the temperature is raised above the glass-transition temperature or the melting point certain polymers become easier to form or mold into desired shapes. The increased temperature weakens the secondary bonds (due to thermal vibration of the long molecules), and the adjacent chains can thus move more easily under the shaping forces. If the polymer is then cooled, it returns to its original hardness and strength; in other words the effects of the process are reversible. Polymers that exhibit this behavior are known as *thermoplastics*; typical examples include acrylics, cellulosics, nylons, polyethylenes, and polyvinyl chloride. Although the effects of the process are reversible, the repeated heating and cooling of thermoplastics can cause **degradation (thermal aging)**.

The behavior of thermoplastics, like their structure and composition, depends on numerous variables. Among the most important are temperature and rate of deformation. Below the glass-transition temperature, most polymers are glassy (described as rigid, brittle, or hard), and they behave like an elastic solid. If the load exceeds a certain critical value, the polymer fractures, just as a piece of glass does at room temperature. In the glassy region, the relationship between stress and strain is linear, or

$$\sigma = E\varepsilon. \tag{10.1}$$

If the polymer is tested in torsion, then

$$\tau = G\gamma. \tag{10.2}$$

The glassy behavior can be represented by a spring that has a stiffness equivalent to the modulus of elasticity of the polymer (Fig. 10.7a). Note that the strain is completely recovered when the load is removed at time $t_1$. When the applied stress is increased, the polymer eventually fractures, just as a piece of glass does at ambient

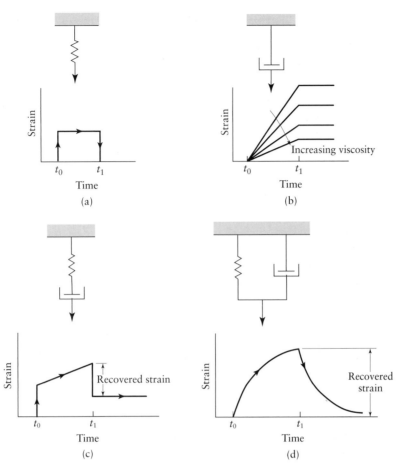

**FIGURE 10.7** Various deformation modes for polymers: (a) elastic; (b) viscous; (c) viscoelastic (Maxwell model); and (d) viscoelastic (Voigt or Kelvin model).

temperature. The mechanical properties of several polymers listed in Table 10.1 indicate that thermoplastics are about two orders of magnitude less stiff than metals. Their ultimate tensile strength is about one order of magnitude lower than that of metals. (See Table 2.1.) Figure 10.8 shows typical stress–strain curves for some thermoplastics and thermosets at room temperature. Note that these plastics exhibit different behaviors, which can be described as rigid, soft, brittle, flexible, and so on. Plastics, like metals, undergo fatigue and creep phenomena as well.

**FIGURE 10.8** General terminology describing the behavior of three types of plastics. PTFE (polytetrafluoroethylene) is Teflon, a trade name. *Source*: R. L. E. Brown.

Some of the major characteristics of thermoplastics are described as follows.

a) **Effects of temperature and deformation rate.** The typical effects of temperature on the strength and elastic modulus of thermoplastics are similar to those for metals (see Sections 2.2.6 and 2.2.7); thus, with increasing temperature, the strength and modulus of elasticity decrease, and the toughness increases (Fig. 10.9). The effect of temperature on impact strength is shown in Fig. 10.10; note the large difference in the impact behavior of various polymers.

If we raise the temperature of a thermoplastic polymer above its $T_g$, it first becomes leathery and then rubbery with increasing temperature (Fig. 10.4). Finally, at higher temperatures, e.g., above $T_m$ for crystalline thermoplastics, it becomes a viscous fluid, with viscosity decreasing with increasing temperature and strain rate. Since viscosity is not constant, polymers display **viscoelastic** behavior, as demonstrated by the spring and dashpot models in Figs. 10.7c and d, known as the *Maxwell* and *Kelvin* (or *Voigt*) models, respectively. When a constant load is applied, the polymer first stretches at a high strain rate and then continues to elongate over a period of time, because of its viscous behavior. Note in the models shown in Fig. 10.7 that the elastic portion of the elongation is reversible (elastic recovery), but the viscous portion is not.

**FIGURE 10.9**  Effect of temperature on the stress–strain curve for cellulose acetate, a thermoplastic. Note the large drop in strength and increase in ductility with a relatively small increase in temperature. *Source*: After T. S. Carswell and H. K. Nason.

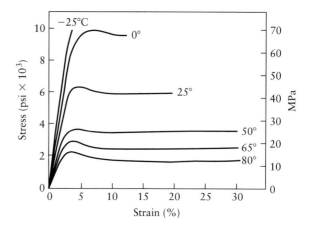

**FIGURE 10.10**  Effect of temperature on the impact strength of various plastics. Small changes in temperature can have a significant effect on impact strength. *Source*: P. C. Powell.

**FIGURE 10.11**
Parameters used to describe
viscosity; see Eq. (10.3).

**FIGURE 10.12**    Viscosity of common thermoplastics as a function of (a) temperature and
(b) shear rate. *Source*:  From D. H. Morton-Jones, *Polymer Processing*, Chapman and Hall,
1989.

The viscous behavior is expressed by

$$\tau = \eta\left(\frac{dv}{dy}\right) = \eta\dot{\gamma},\qquad(10.3)$$

where $\eta$ is the *viscosity* and $dv/dy$ is the *shear-strain rate, $\dot{\gamma}$*, as shown in
Fig. 10.11. Figure 10.12 provides the viscosities of some polymers. When the
shear stress $\tau$ is directly proportional to the shear-strain rate, the behavior of the
thermoplastic polymer is known as *Newtonian*. For many polymers, Newtonian
behavior is not a good approximation, and Eq. (10.3) will give poor predictions
of process performance. For example, polyvinyl chloride, polyethylene (low
density and high density), and polypropylene have viscosities that decrease
markedly with increasing strain rate (*pseudoplastic behavior*). Their viscosity as
a function of strain rate can be expressed as

$$\eta = A\dot{\gamma}^{1-n},\qquad(10.4a)$$

where $A$ is the *consistency index* and $n$ is the *power-law index* for the polymer.
    Note from the foregoing equations that the viscous behavior of thermo-
plastics is similar to the strain-rate sensitivity of metals (Section 2.2.7) and is
given by the expression

$$\sigma = C\dot{\varepsilon}^{m},\qquad(10.4b)$$

**FIGURE 10.13**

(a) Load–elongation curve for polycarbonate, a thermoplastic. *Source*: R. P. Kambour and R. E. Robertson. (b) High-density polyethylene tension-test specimen, showing uniform elongation (the long, narrow region in the specimen).

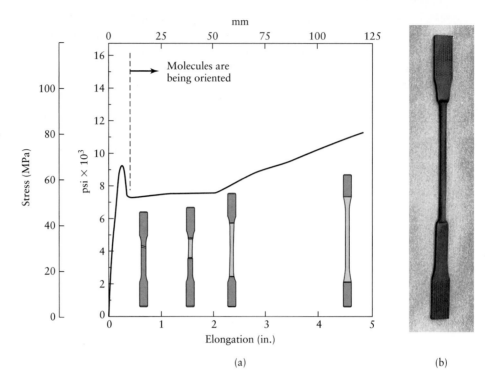

(a)    (b)

where, for Newtonian behavior, $m = 1$. Thermoplastics have high $m$ values, indicating that they can undergo large uniform deformations in tension before fracture. Note how, unlike with ordinary metals, the necked region elongates considerably (Fig. 10.13). This behavior can easily be demonstrated by stretching a piece of the plastic (polypropylene) holder for six-pack beverage cans. This characteristic, which is the same as in superplastic metals (see Section 2.2.7), enables thermoplastics to be formed into complex shapes, such as bottles for soft drinks, cookie and meat trays, and lighted signs, as described in Section 10.10. Note also that the effect of increasing the rate of load application is that the strength of the polymer is increased.

Between $T_g$ and $T_m$, thermoplastics exhibit leathery and rubbery behavior, depending on their structure and degree of crystallinity, as shown in Fig. 10.4. A term combining the strains caused by elastic behavior ($e_e$) and viscous flow ($e_v$) is called the *viscoelastic modulus*, $E_r$, and is expressed as

$$E_r = \frac{\sigma}{(e_e + e_v)}. \tag{10.5}$$

This modulus essentially represents a time-dependent elastic modulus.

The viscosity $\eta$ of polymers (a measure of the resistance of their molecules in sliding along each other) depends on temperature and on the polymer's structure, molecular weight, and pressure. The effect of temperature can be represented by

$$\eta = \eta_0 e^{E/kT}, \tag{10.6}$$

where $\eta_0$ is a material constant, $E$ is the activation energy (the energy required to initiate a reaction), $k$ is Boltzmann's constant (the thermal-energy constant, or $13.8 \times 10^{-24}$ J/K), and $T$ is the temperature (in K). Thus, as the temperature of the polymer is increased, $\eta$ decreases, because of the higher mobility of the molecules. Note that viscosity may be regarded as the inverse of fluidity of molten metals in casting. Increasing the molecular weight (and hence increasing the length of the chain) increases $\eta$, because of the greater number of secondary bonds present. As the molecular-weight distribution widens, there are more shorter chains, and $\eta$ decreases. The effect of increasing pressure is that viscosity is increased, because of reduced *free volume*, or *free space*, defined as the volume in excess of the true volume of the crystal in the crystalline regions of the polymer.

Based on experimental observations that, at the glass-transition temperature, $T_g$, polymers have a viscosity $\eta$ of about $10^{12}$ Pa-s, an empirical relationship between viscosity and temperature has been developed for linear thermoplastics:

$$\log \eta = 12 - \frac{17.5\Delta T}{52 + \Delta T}. \tag{10.7}$$

In this equation, $\Delta T = T - T_g$, in K or °C; thus, we can estimate the viscosity of the polymer at any temperature.

b) **Creep and stress relaxation.** The terms *creep* and *stress relaxation* have been defined in Section 2.8. Because of their viscoelastic behavior, thermoplastics are particularly susceptible to these phenomena. Referring back to the dashpot models in Figs. 10.7b, c, and d, note that under a constant load, the polymer undergoes further strain, and thus it creeps. The recovered strain depends on the stiffness of the spring and hence the modulus of elasticity. Stress relaxation in polymers occurs over a period of time, according to the relation

$$\tau = \tau_0 \, e^{-t/\lambda}, \tag{10.8}$$

where $\tau_0$ is the shear stress at time zero, $\tau$ is the stress at time $t$, and $\lambda$ is the *relaxation time*. This expression may also be given in terms of the normal stresses $\sigma$ and $\sigma_0$. The relaxation time is defined as

$$\lambda = \frac{\eta}{G}, \tag{10.9}$$

where $G$ is the shear modulus of the polymer and $\gamma = \tau/G$; thus, the relaxation time is a function of the material and its temperature. Because both $\eta$ and $G$ depend on temperature to varying degrees, the relaxation time also depends on temperature.

c) **Orientation.** When thermoplastics are permanently deformed, say by stretching, the long-chain molecules align in the general direction of elongation. This process is called *orientation*, and, just as with metals, the polymer becomes *anisotropic* (see Section 3.5). The specimen becomes stronger and stiffer in the elongated (stretched) direction than in the transverse direction. This process is an important technique to enhance the strength and toughness of polymers; however, orientation weakens the polymer in the transverse direction.

d) **Crazing.** Some thermoplastics, such as polystyrene and polymethylmethacrylate, develop localized, wedge-shaped narrow regions of highly deformed material when subjected to tensile stresses or to bending. This phenomenon is called *crazing*. Although they may appear to be like cracks, crazes are areas of spongy material, typically containing about 50% voids. With increasing tensile load on the specimen, the voids coalesce to form a crack and eventually lead to fracture of the polymer. Crazing has been observed in transparent glassy polymers as well as in other types of polymers. The environment and the presence of solvents, lubricants, and water vapor enhance the formation of crazes (**environmental stress cracking** and **solvent crazing**). Residual stresses in the material contribute to crazing and cracking of the polymer. Radiation and ultraviolet radiation also can adversely affect the crazing behavior in certain polymers.

A related phenomenon is **stress whitening.** When subjected to tensile stresses, such as by folding or bending, the plastic becomes lighter in color. This phenomenon is usually attributed to the formation of microvoids in the material. As a result, the material becomes less translucent (i.e., transmits less light), or more opaque; you can easily demonstrate this result by bending plastic components commonly found in household products and toys.

e) **Water absorption.** An important limitation of some polymers, such as nylons, is their ability to absorb water (*hygroscopy*). Water acts as a plasticizing agent—that is, it makes the polymer more plastic; thus, in a sense, it lubricates the chains in the amorphous region. Typically, with increasing moisture absorption, the glass-transition temperature, the yield stress, and the elastic modulus of the polymer are lowered severely. Dimensional changes also occur because of water absorption, such as in a humid environment.

f) **Thermal and electrical properties.** Compared with metals, plastics are generally characterized by low thermal and electrical conductivity, low specific gravity (ranging from 0.90 to 2.2), and a relatively high coefficient of thermal expansion—about one order of magnitude higher (see Table 3.2). Because polymers generally have low electrical conductivity, they can be used for electrical and electronic components. However, polymers can be made to be electrically conductive, as described in Section 10.7.2.

---

**EXAMPLE 10.2** Lowering the viscosity of a polymer

In processing a batch of polycarbonate at 170°C to make a certain part, it is found that the polycarbonate's viscosity is twice that desired. Determine the temperature at which this polymer should be processed.

**Solution.** From Table 10.2, we know that the $T_g$ for polycarbonate is 150°C. Its viscosity at 170°C is determined by Eq. (10.7):

$$\log \eta = 12 - \frac{17.5(20)}{52 + 20} = 7.14.$$

Hence, $\eta = 13.8$ MPa-s, and because this magnitude is twice what we want, the new viscosity should be 6.9 MPa-s. We substitute this new value to find the new temperature:

$$\log(6.9 \times 10^6) = 12 - \frac{17.5(\Delta T)}{52 + \Delta T}.$$

Thus, $\Delta T = 21.7$, or the new temperature is $150 + 21.7 = 171.7$°C. Note that the numbers for temperature are rounded and that viscosity is very sensitive to temperature.

**EXAMPLE 10.3**  Stress relaxation in a thermoplastic member under tension

A long piece of thermoplastic is stretched between two rigid supports at a stress level of 5 MPa. After 30 days, the stress level is found to have decayed to one half the original level. How long will it take for the stress level to reach one tenth of the original value?

**Solution.**  Substituting these data into Eq. (10.8) as modified for normal stress, we have

$$2.5 = 5e^{-30/\lambda}.$$

Hence,

$$\ln\left(\frac{2.5}{5}\right) = \frac{-30}{\lambda}, \quad \text{or} \quad \lambda = 43.3 \text{ days}.$$

Therefore,

$$\ln\left(\frac{0.5}{5}\right) = \frac{-t}{43.3}, \quad \text{or} \quad t = 99.7 \text{ days}.$$

# 10.4 | Thermosets: Behavior and Properties

When the long-chain molecules in a polymer are cross-linked in a three-dimensional arrangement, the structure in effect becomes one giant molecule with strong covalent bonds. As previously stated, such polymers are called **thermosetting polymers,** or *thermosets*, because during polymerization, the network is completed, and the shape of the part is permanently set. This **curing** (*cross-linking*) reaction, unlike that of thermoplastics, is irreversible. The response of a thermosetting plastic to temperature can be likened to baking a cake or boiling an egg; once the cake is baked and cooled, or the egg is boiled and cooled, reheating it will not change its shape.

Some thermosets, such as epoxy, polyester, and urethane, cure at room temperature. Although curing takes place at ambient temperature, the heat of the exothermic reaction cures the plastic. Thermosetting polymers do not have a sharply defined glass-transition temperature. The polymerization process for thermosets generally takes place in two stages: (1) The first stage occurs at the chemical plant, where the molecules are partially polymerized into linear chains, and (2) the second stage occurs at the parts-producing plant, where cross-linking is completed under heat and pressure during the molding and shaping of the part (Section 10.10).

Because of the nature of the bonds, strength and hardness of thermosets, unlike those of thermoplastics, are not affected by temperature or rate of deformation. A typical thermoset is phenolic, which is a product of the reaction between phenol and formaldehyde; common products made from this polymer are the handles and knobs on cooking pots and pans, and components of light switches and outlets. Thermosetting plastics (Section 10.6) generally possess better mechanical, thermal, and chemical properties, electrical resistance, and dimensional stability than do thermoplastics. However, if the temperature is increased sufficiently, the thermosetting polymer begins to burn up, degrade, and char.

# 10.5 | Thermoplastics: General Characteristics and Applications

This section outlines the general characteristics and typical applications of major thermoplastics, particularly as they relate to manufacturing plastic products and their service life. Table 10.3 gives general recommendations for various plastics applications. The major thermoplastics are described as follows.

**TABLE 10.3**

**General Recommendations for Plastic Products**

| Design Requirement | Typical Applications | Plastics |
|---|---|---|
| Mechanical strength | Gears, cams, rollers, valves, fan blades, impellers, pistons. | Acetals, nylon, phenolics, polycarbonates, polyesters, polypropylenes, epoxies, polyimides. |
| Wear resistance | Gears, wear strips and liners, bearings, bushings, roller-skate wheels. | Acetals, nylon, phenolics, polyimides, polyurethane, ultrahigh-molecular-weight polyethylene. |
| Frictional properties | | |
|   High | Tires, nonskid surfaces, footware, flooring. | Elastomers, rubbers. |
|   Low | Sliding surfaces, artificial joints. | Fluorocarbons, polyesters, polyethylene, polyimides. |
| Electrical resistance | All types of electrical components and equipment, appliances, electrical fixtures. | Polymethylmethacrylate, ABS, fluorocarbons, nylon, polycarbonate, polyester, polypropylenes, ureas, phenolics, silicones, rubbers. |
| Chemical resistance | Containers for chemicals, laboratory equipment, components for chemical industry, food and beverage containers. | Acetals, ABS, epoxies, polymethylmethacrylate, fluorocarbons, nylon, polycarbonate, polyester, polypropylene, ureas, silicones. |
| Heat resistance | Appliances, cookware, electrical components. | Fluorocarbons, polyimides, silicones, acetals, polysulfones, phenolics, epoxies. |
| Functional and decorative features | Handles, knobs, camera and battery cases, trim moldings, pipe fittings. | ABS, acrylics, cellulosics, phenolics, polyethylenes, polpropylenes, polystyrenes, polyvinyl chloride. |
| Functional and transparent features | Lenses, goggles, safety glazing, signs, food-processing equipment, laboratory hardware. | Acrylics, polycarbonates, polystyrenes, polysulfones. |
| Housings and hollow shapes | Power tools, housings, sport helmets, telephone cases. | ABS, cellulosics, phenolics, polycarbonates, polyethylenes, polypropylene, polystyrenes. |

1. **Acetals** (from acetic and alcohol) have good strength; stiffness; and resistance to creep, abrasion, moisture, heat, and chemicals. Typical applications include mechanical parts and components where high performance is required over a long period: bearings, cams, gears, bushings, rollers, impellers, wear surfaces, pipes, valves, showerheads, and housings. Common trade name: *Delrin*.

2. **Acrylics** (such as **polymethylmethacrylate,** or PMMA) possess moderate strength, good optical properties, and weather resistance. They are transparent, but can be made opaque and are generally resistant to chemicals and have good electrical resistance. Typical applications include lenses, lighted signs, displays, window glazing, skylights, bubble tops, automotive lenses, windshields, lighting fixtures, and furniture. Common trade names: *Orlon, Plexiglas,* and *Lucite*.

3. **Acrylonitrile–butadiene–styrene** (ABS) is dimensionally stable and rigid and has good resistance to impact, abrasion, chemicals, and electricity, strength and toughness, and low-temperature properties. Typical applications include pipes, fittings, chrome-plated plumbing supplies, helmets, tool handles, automotive components, boat hulls, telephones, luggage, housing, appliances, refrigerator liners, and decorative panels.

4. **Cellulosics** have a wide range of mechanical properties, depending on their composition. They can be made rigid, strong, and tough. However, they weather poorly and are affected by heat and chemicals. Typical applications include tool handles, pens, knobs, frames for eyeglasses, safety goggles, machine guards, helmets, tubing and pipes, lighting fixtures, rigid containers, steering wheels, packaging film, signs, billiard balls, toys, and decorative parts.

5. **Fluorocarbons** possess good resistance to temperature, chemicals, weather, and electricity; they also have unique nonadhesive properties and low friction. Typical applications include linings for chemical-processing equipment, nonstick coatings for cookware, electrical insulation for high-temperature wire and cable, gaskets, low-friction surfaces, bearings, and seals. Common trade name: *Teflon*.

6. **Polyamides** (from the words *poly, amine,* and *carboxyl acid*) are available in two main types—nylons and aramids:

   a) **Nylons** (a coined word) have good mechanical properties and abrasion resistance. They are self-lubricating and resistant to most chemicals. All nylons are hygroscopic (i.e., they absorb water). Moisture absorption reduces mechanical properties and increases part dimensions. Typical applications include gears, bearings, bushings, rollers, fasteners, zippers, electrical parts, combs, tubing, wear-resistant surfaces, guides, and surgical equipment.

   b) **Aramids** (aromatic polyamides) have very high tensile strength and stiffness. Typical applications include fibers for reinforced plastics (composite materials), bulletproof vests, cables, and radial tires. Common trade name: *Kevlar*.

7. **Polycarbonates** are versatile and have good mechanical and electrical properties; they also have high impact resistance and can be made resistant to chemicals. Typical applications include safety helmets, optical lenses, bullet-resistant window glazing, signs, bottles, food-processing equipment, windshields, load-bearing electrical components, electrical insulators, medical apparati, business-machine components, guards for machinery, and parts requiring dimensional stability. Common trade name: *Lexan*.

8. **Polyesters** (thermoplastics; see also Section 10.6) have good mechanical, electrical, and chemical properties; good abrasion resistance; and low friction. Typical applications include gears, cams, rollers, load-bearing members, pumps, and electromechanical components. Common trade names: *Dacron, Mylar,* and *Kodel*.

9. **Polyethylenes** possess good electrical and chemical properties. Their mechanical properties depend on their composition and structure. The three major classes of polyethylenes are low density (LDPE), high density (HDPE), and ultrahigh molecular weight (UHMWPE). Typical applications for LDPE and HDPE include housewares, bottles, garbage cans, ducts, bumpers, luggage, toys, tubing, bottles, and packaging material. UHMWPE is used in parts requiring high-impact toughness and abrasive wear resistance; examples include artificial knee and hip joints.

10. **Polyimides** have the structure of a thermoplastic, but the nonmelting characteristic of a thermoset; see Section 10.6. Common trade name: *Torlon*.

11. **Polypropylenes** have good mechanical, electrical, and chemical properties and good resistance to tearing. Typical applications include automotive trim and components, medical devices, appliance parts, wire insulation, TV cabinets, pipes, fittings, drinking cups, dairy-product and juice containers, luggage, ropes, and weather stripping.

12. **Polystyrenes** are inexpensive, have generally average properties, and are somewhat brittle. Typical applications include disposable containers; packaging; trays for meats, cookies, and candy; foam insulation; appliances; automotive, radio, and TV components; housewares; and toys and furniture parts (as a wood substitute).

13. **Polysulfones** have excellent resistance to heat, water, and steam and are highly resistant to some chemicals, but are attacked by organic solvents. Typical applications include steam irons, coffeemakers, hot-water containers, medical equipment that requires sterilization, power-tool and appliance housings, aircraft cabin interiors, and electrical insulators.

14. **Polyvinyl chloride** (PVC) has a wide range of properties, is inexpensive and water resistant, and can be made rigid or flexible. It is not suitable for applications that require strength and heat resistance. **Rigid PVC** is tough and hard and is used for signs and in the construction industry, such as for pipes and conduits. **Flexible PVC** is used in wire and cable coatings, low-pressure flexible tubing and hose, footwear, imitation leather, upholstery, records, gaskets, seals, trim, film, sheet, and coatings. Common trade names: *Saran* and *Tygon*.

## 10.6 | Thermosets: General Characteristics and Applications

This section outlines the general characteristics and typical applications of major thermosetting plastics:

1. **Alkyds** (from *alkyl*, meaning alcohol, and *acid*) possess good electrical insulating properties, impact resistance, and dimensional stability and have low water absorption. Typical applications include electrical and electronic components.

2. **Aminos** (**urea** and **melamine**) have properties that depend on composition. Generally, aminos are hard and rigid and are resistant to abrasion, creep, and electrical arcing. Typical applications include small appliance housings, countertops, toilet seats, handles, and distributor caps. Urea is used for electrical and electronic components, and melamine is used for dinnerware.

3. **Epoxies** have excellent mechanical and electrical properties, dimensional stability, strong adhesive properties, and good resistance to heat and chemicals. Typical applications include electrical components that require mechanical strength and

high insulation, tools and dies, and adhesives. *Fiber-reinforced epoxies* have excellent mechanical properties and are used in pressure vessels, rocket motor casings, tanks, and similar structural components.

4. **Phenolics**, although brittle, are rigid; dimensionally stable; and have high resistance to heat, water, electricity, and chemicals. Typical applications include knobs, handles, laminated panels, telephones, bond material to hold abrasive grains together in grinding wheels, and electrical components, such as wiring devices, connectors, and insulators.

5. **Polyesters** (thermosetting plastics; see also Section 10.5) have good mechanical, chemical, and electrical properties. Polyesters are generally reinforced with glass or other fibers. Typical applications include boats, luggage, chairs, automotive bodies, swimming pools, and material for impregnating cloth and paper. Polyesters are also available as casting resins.

6. **Polyimides** possess good mechanical, physical, and electrical properties at elevated temperatures; they also have creep resistance and low friction and wear characteristics. Polyimides have the nonmelting characteristics of a thermoset, but the structure of a thermoplastic. Typical applications include pump components (bearings, seals, valve seats, retainer rings, and piston rings); electrical connectors for high-temperature use; aerospace parts; high-strength, impact-resistant structures; sports equipment; and safety vests.

7. **Silicones** have properties that depend on composition. Generally, they weather well, possess excellent electrical properties over a wide range of humidities and temperatures, and resist chemicals and heat. (See also Section 10.8.) Typical applications include electrical components that require strength at elevated temperatures, oven gaskets, heat seals, and waterproof materials.

# 10.7 | High-Temperature Polymers, Electrically Conducting Polymers, and Biodegradable Plastics

This section describes three important trends in the development of polymers.

## 10.7.1 High-temperature polymers

During the past several years, there have been important advances in polymers and polymer blends for high-temperature applications, particularly in the aerospace industry. Polymers with properties such as high strength-to-weight ratio and fatigue strength are being further enhanced with high-temperature resistance. This resistance, as in metals, may be short term at relatively high temperatures or long term at lower temperatures. (See, for example, Section 3.11.6 for a similar consideration regarding titanium alloys.)

Short-term exposure to high temperatures generally requires *ablative* materials to dissipate the heat, such as phenolic–silicone copolymers, which have been used for rocket and missile components at temperatures of thousands of degrees. Long-term exposure for polymers is presently confined to temperatures on the order of 260°C (500°F ). High-temperature thermoplastic polymers include fluorine-containing thermoplastics, polyketones, and polyimides. High-temperature thermosetting polymers include phenolic resins, epoxy resins, silicone-based thermosetting polymers, and phenolic–fiberglass systems. Several other polymers with complex structures are being developed for high-temperature applications.

### 10.7.2 Electrically conducting polymers

Most polymers are electrical insulators and hence are used as insulators for electrical equipment (e.g., light switches, outlets, and circuit breakers) and as packaging material for electronic components. In addition, the electrical conductivity of some polymers can be increased by **doping**, i.e., introducing certain impurities in the polymer, such as metal powder, salts, and iodides. The conductivity of polymers increases with moisture absorption, and the electronic properties of polymers can be altered by irradiation as well. Applications of electrically conducting polymers include microelectronic devices, rechargeable batteries, capacitors, fuel cells, catalysts, fuel-level sensors, de-icer panels, antistatic coatings, and as conducting adhesives for surface-mount technologies (see Section 12.13.4).

### 10.7.3 Biodegradable plastics

One third of plastics produced are in the disposable-products sector, such as bottles, packaging, and garbage bags. With the growing use of plastics and increasing concern over environmental issues regarding disposal of plastic products and limited landfills, major efforts are underway to develop biodegradable plastics. Three different **biodegradable plastics** are currently available: starch-based bioplastics, lactic-based bioplastics, and bioplastics made from the fermentation of sugar. These plastics degrade over various periods of time, from a few months to a few years, and have different degradability characteristics. They are designed to degrade completely when exposed to microorganisms in soil or water, without producing toxic by-products. It should also be noted that the costs of biodegradable plastics are substantially higher than those of synthetic polymers.

Three types of biodegradable plastics are described as follows:

a) The **starch-based system** is the farthest along of the three types of bioplastics in terms of production capacity. Starch may be extracted from potatoes, wheat, rice, and corn. In this system, starch granules are processed into a powder, which is heated and becomes a sticky liquid; various additives and binders are also blended in. The liquid is then cooled, formed into pellets, and processed in conventional plastic-processing equipment (Section 10.10). Various additives are blended with the starch to impart special characteristics to the bioplastic materials.

b) In the **lactic-based system**, fermenting corn or other feedstocks produce lactic acid, which is then polymerized to form a polyester resin.

c) In the third system, **organic acids** are added to a **sugar** feedstock. Through a specially developed process, the resulting reaction produces a highly crystalline and very stiff polymer that, after further processing, behaves in a manner similar to polymers developed from petroleum.

**Recycling.** Because the development of biodegradable plastics is relatively new, their long-range performance, both during their useful life cycle as products and in landfills, has not been fully assessed. There is concern that emphasis on biodegradability will divert attention from the issue of *recyclability* of plastics and the efforts for *conservation* of materials and energy. Recycled plastics are being used increasingly for a variety of products, including automotive-body components. Plastic products now carry the following numerals within a triangular mark, indicating the corresponding type of plastic: 1—PETE (polyethylene), 2—HDPE (high-density polyethylene), 3—V (vinyl), 4—LDPE (low-density polyethylene), 5—PP (polypropylene), 6—PS (polystyrene), 7—others.

# 10.8 | Elastomers (Rubbers): General Characteristics and Applications

The terms *rubber* and *elastomer* are often used interchangeably. Generally, however, an **elastomer** is defined as being capable of recovering substantially in shape and size after a load has been removed; **rubber** is defined as being capable of recovering from large deformations quickly.

Elastomers comprise a large family of amorphous polymers that have a low glass-transition temperature. They have the characteristic ability to undergo large elastic deformations without rupture, and they are soft and have a low elastic modulus. The term *elastomer* is derived from the words *elastic* and *mer*. The structure of these polymers is highly kinked (tightly twisted or curled). They stretch, but then return to their original shape after the load is removed (Fig. 10.14). They can be cross-linked; the best example is the elevated-temperature **vulcanization** of rubber with sulfur, discovered by Charles Goodyear in 1839 and named for Vulcan, the Roman god of fire. Once the elastomer is cross-linked, it cannot be reshaped. For example, an automobile tire, which is one giant molecule, cannot be softened and reshaped.

The hardness of elastomers, which is measured with a durometer (see Section 2.6.7), increases with increasing cross-linking of the molecular chains. A variety of additives can be blended with elastomers to impart specific properties, as with plastics. Elastomers have a wide range of applications, such as high-friction and nonskid surfaces, protection against corrosion and abrasion, electrical insulation, and shock and vibration insulation. Specific examples include tires, hoses, weather stripping, footwear, linings, gaskets, seals, printing rolls, and flooring. A characteristic of elastomers is their hysteresis loss in stretching or compression (Fig. 10.14). The clockwise loop in Fig. 10.14 indicates energy loss, whereby mechanical energy is converted into heat; this property is desirable for absorbing vibrational energy (damping) and sound deadening.

Some of the major types of elastomers are described below:

a) **Natural rubber.** The base for natural rubber is **latex**, a milklike sap obtained from the inner bark of a tropical tree. It has good resistance to abrasion and fatigue and high frictional properties, but low resistance to oil, heat, ozone, and sunlight. Typical applications include tires, seals, shoe heels, couplings, and engine mounts.

b) **Synthetic rubbers.** Further developed than natural rubbers, synthetic rubbers include synthetic natural rubber, butyl, styrene butadiene, polybutadiene, and ethylene propylene. Compared with natural rubbers, synthetic rubbers have improved resistance to heat, gasoline, and chemicals and a higher useful-temperature range. Examples of synthetic rubbers that are resistant to oil include neoprene, nitrile, urethane, and silicone. Typical applications of synthetic rubbers include tires, shock absorbers, seals, and belts.

c) **Silicones.** Silicones (see also Section 10.6) have the highest useful temperature range of all elastomers, up to 315°C (600°F), but their other properties (such as strength and resistance to wear and oils) are generally inferior to those of other elastomers. Typical applications include seals, gaskets, thermal insulation, high-temperature electrical switches, and electronic apparati.

d) **Polyurethane.** This elastomer has very good overall properties of high strength, stiffness, and hardness and exceptional resistance to abrasion, cutting, and tearing. Typical applications include seals, gaskets, cushioning, diaphragms for rubber forming of sheet metals (see Section 7.7), and auto-body parts.

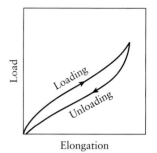

**FIGURE 10.14** Typical load–elongation curve for rubbers. The clockwise loop, indicating loading and unloading paths, is the hysteresis loss. Hysteresis gives rubbers the capacity to dissipate energy, damp vibration, and absorb shock loading, as in automobile tires and vibration dampeners placed under machinery.

## 10.9 | Reinforced Plastics

Among the major developments in materials are *reinforced plastics* (**composite materials**); in fact, composites are now one of the most important classes of **engineered materials**, as they offer several outstanding properties as compared with conventional materials. These materials can be defined as a combination of two or more chemically distinct and insoluble phases whose properties and structural performance are superior to those of the constituents acting independently. It has been shown, for example, that plastics possess mechanical properties (particularly strength, stiffness, and creep resistance) that are generally inferior to those of metals and alloys.

These properties can be improved by embedding reinforcements of various types, such as glass or graphite fibers, to produce reinforced plastics. As shown in Table 10.1, reinforcements improve the strength, stiffness, and creep resistance of plastics, as well as their strength-to-weight and stiffness-to-weight ratios. Reinforced plastics have found increasingly wider applications in aircraft, space vehicles, offshore structures, piping, electronics, automobiles, boats, ladders, and sporting goods.

The oldest example of composites is the addition of straw to clay for making mud huts and bricks for structural use, dating back to 4000 BC. In that application, the straws are the reinforcing fibers, and the clay is the matrix. Another example of a composite material is the reinforcing of masonry and concrete with iron rods, begun in the 1800s. In fact, concrete itself is a composite material, consisting of cement, sand, and gravel. In reinforced concrete, steel rods impart the necessary tensile strength to the composite, since concrete is brittle and generally has little or no useful tensile strength.

### 10.9.1 Structure of polymer-matrix-reinforced plastics

Reinforced plastics consist of **fibers** (the *discontinuous* or *dispersed* phase) in a polymer **matrix** (the *continuous* phase). Commonly used fibers include glass, graphite, aramids, and boron (Table 10.4). These fibers are strong and stiff and have a high

TABLE 10.4

| **Typical Properties of Reinforcing Fibers** | | | | |
|---|---|---|---|---|
| Type | Tensile Strength (MPa) | Elastic Modulus (GPa) | Density ($kg/m^3$) | Relative Cost |
| Boron | 3500 | 380 | 2600 | Highest |
| Carbon | | | | |
|   High strength | 3000 | 275 | 1900 | Low |
|   High modulus | 2000 | 415 | 1900 | Low |
| Glass | | | | |
|   E type | 3500 | 73 | 2480 | Lowest |
|   S type | 4600 | 85 | 2540 | Lowest |
| Kevlar | | | | |
|   29 | 2800 | 62 | 1440 | High |
|   49 | 2800 | 117 | 1440 | High |
|   129 | 3200 | 85 | 1440 | High |
| Nextel | | | | |
|   312 | 1630 | 135 | 2700 | High |
|   610 | 2770 | 328 | 3960 | High |
| Spectra | | | | |
|   900 | 2270 | 64 | 970 | High |
|   1000 | 2670 | 90 | 970 | High |

*Note*: These properties vary significantly, depending on the material and method of preparation. Strain to failure for these fibers is typically in the range of 1.5% to 5.5%.

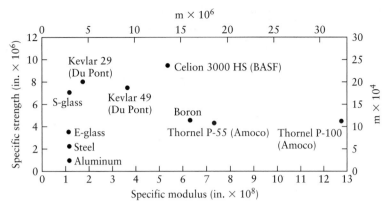

**FIGURE 10.15** Specific tensile strength (ratio of tensile strength to density) and specific tensile modulus (ratio of modulus of elasticity to density) for various fibers used in reinforced plastics. Note the wide range of specific strengths and stiffnesses available.

specific strength (strength-to-weight ratio) and specific modulus (stiffness-to-weight ratio), as shown in Fig. 10.15. However, they are generally brittle and abrasive and lack toughness; thus, the fibers, by themselves, have little structural value. The plastic matrix is less strong and less stiff, but tougher than the fibers; reinforced plastics combine the advantages of each of the two constituents (Table 10.5). When more than one type of fiber is used in a reinforced plastic, the composite is called a **hybrid**, which generally has even better properties than a reinforced plastic with only one type of fiber.

In addition to high specific strength and specific modulus, reinforced plastic structures have improved fatigue resistance, greater toughness, and higher creep resistance than those of unreinforced plastics. These structures are relatively easy to design, fabricate, and repair. The percentage of fibers (by volume) in reinforced plastics usually ranges between 10% and 60%. Practically, the percentage of fiber in a matrix

## TABLE 10.5

**Types and General Characteristics of Reinforced Plastics and Metal-Matrix and Ceramic-Matrix Composites**

| Material | Characteristics |
|---|---|
| FIBER | |
| Glass | High strength, low stiffness, high density; E (calcium aluminoborosilicate) and S (magnesia–aluminosilicate) types are commonly used; lowest cost. |
| Graphite | Available typically as high modulus or high strength; less dense than glass; low cost. |
| Boron | High strength and stiffness; has tungsten filament at its center (coaxial); highest density; highest cost. |
| Aramids (Kevlar) | Highest strength-to-weight ratio of all fibers; high cost. |
| Other | Nylon, silicon carbide, silicon nitride, aluminum oxide, boron carbide, boron nitride, tantalum carbide, steel, tungsten, and molybdenum; see Chapters 3, 8, 9, and 10. |
| MATRIX | |
| Thermosets | Epoxy and polyester, with the former most commonly used; others are phenolics, fluorocarbons, polyethersulfone, silicon, and polyimides. |
| Thermoplastics | Polyetheretherketone; tougher than thermosets, but lower resistance to temperature. |
| Metals | Aluminum, aluminum–lithium alloy, magnesium, and titanium; fibers used are graphite, aluminum oxide, silicon carbide, and boron. |
| Ceramics | Silicon carbide, silicon nitride, aluminum oxide, and mullite; fibers used are various ceramics. |

is limited by the average distance between adjacent fibers or particles. The highest practical fiber content is 65%; higher percentages generally result in diminished structural properties.

## 10.9.2 Reinforcing fibers: characteristics and manufacture

The major types of reinforcing fibers are described as follows:

1. **Polymer fibers.** Most polymer fibers are used in the textile industry and are mass produced. The most common reinforcing fibers are **aramids** (such as *Kevlar*); they are among the toughest fibers and have very high specific strength (Fig. 10.15 and Table 10.4). They can undergo some plastic deformation before fracture and thus have higher toughness than that of brittle fibers; however, aramids absorb moisture, which reduces their properties and complicates their application, as hygrothermal stresses must be considered. Other polymer reinforcements include rayon, nylon, and acrylics.

   A more recent high-performance polyethylene fiber is *Spectra* (a trade name); it has ultrahigh molecular weight and high molecular-chain orientation. The fiber has better abrasion resistance and flexural fatigue than the aramid fiber and at a similar cost. In addition, because of its lower density, it has higher specific strength and specific stiffness than the aramid fiber. However, a low melting point and poor adhesion characteristics as compared with those of other polymers are its major limitations. New polymer fibers are continually being introduced, including *Nextel*.

   Polymers are either melt spun or dry spun to make fibers. **Melt spinning** involves extruding a liquid polymer through small die holes, or *spinnerettes*. The fibers are then cooled before being gathered and wound onto bobbins. The fibers may be stretched to further orient the polymer in the fibers. In **dry spinning**, the polymer is dissolved in a liquid solution to form a partially oriented liquid-crystal form. As the polymer passes through the spinnerette, it is further oriented, and at this point, the fibers are washed, dried, and wound. Aramids are oriented in solution and are fully oriented when they pass through the spinnerette and therefore do not need to be further drawn.

2. **Glass fibers.** Glass fibers are the most widely used and least expensive of all fibers. (See also Section 11.10.2.) The composite material is called *glass-fiber reinforced plastic* (GFRP) and may contain between 30% and 60% glass fibers by volume. Glass fibers are made by drawing molten glass through small openings in a platinum die. The molten glass is then mechanically elongated, cooled, and wound on a roll. A protective coating or sizing may be applied to facilitate the passage of the glass fibers through machinery. The glass fibers are treated with **silane** (a silicon hydride) for improved wetting and bonding between the fiber and the matrix.

   The principal types of glass fibers are (1) the **E type**, a calcium aluminoborosilicate glass, which is used most often; (2) the **S type**, a magnesia–aluminosilicate glass, which has higher strength and stiffness and is more expensive than the other two types and (3) the **E-CR type**, which offers higher resistance to elevated temperature and acid corrosion than the other two types.

3. **Graphite fibers.** Graphite fibers (Fig. 10.16a), although more expensive than glass fibers, have a combination of low density, high strength, and high stiffness; the product is called *carbon-fiber reinforced plastic* (CFRP). Graphite fibers are made by pyrolysis of organic **precursors**, commonly polyacrylonitrile (PAN), because of its lower cost; rayon and pitch (the residue from catalytic crackers in petroleum refining) also can be used as precursors. **Pyrolysis** is the term used

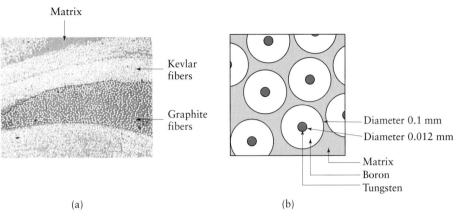

**FIGURE 10.16**   (a) Cross-section of a tennis racket, showing graphite and aramid (Kevlar) reinforcing fibers. *Source*: J. Dvorak, Mercury Marine Corporation, and F. Garrett, Wilson Sporting Goods Co. (b) Cross-section of boron-fiber-reinforced composite material.

to describe the process of inducing chemical changes by heat, such as burning a length of yarn, which becomes carbon and black in color. With PAN, the fibers are partially cross-linked at a moderate temperature in order to prevent melting during subsequent processing steps, and the fibers are simultaneously elongated. At this point, the fibers are carburized, that is, they are exposed to elevated temperature to expel the hydrogen (dehydrogenation) and nitrogen (denitrogenation) from the PAN. The temperatures range up to about 1500°C (2730°F) for carbonizing and to 3000°C (5400°F ) for graphitizing.

The difference between carbon and graphite, although the terms are often used interchangeably, depends on the temperature of pyrolysis and the purity of the material. Carbon fibers are generally 80% to 95% carbon, and graphite fibers are usually more than 99% carbon; the rest is graphite and carbon, respectively. The fibers are classified by the magnitude of their elastic modulus, which typically ranges from 35 GPa to 800 GPa; tensile strengths typically range from 250 MPa to 2600 MPa.

**Conductive graphite fibers** are being produced to make it possible to enhance the electrical and thermal conductivity of reinforced plastics, such as in electromagnetic and radio-frequency shielding and lightning protection. These fibers are coated with a metal, usually nickel, typically to a thickness of $0.5\ \mu m$ ($20\ \mu in.$) on a graphite-fiber core $7\ \mu m$ ($280\ \mu in.$) in diameter, using a continuous electroplating process.

4. **Boron fibers.**   Boron fibers consist of boron deposited (by chemical vapor-deposition techniques) on tungsten fibers (Fig. 10.16b), although boron can also be deposited on carbon fibers. These fibers have favorable properties, such as high strength and stiffness in tension and compression and resistance to high temperatures. However, because of the use of tungsten, they have high density and are expensive, thus increasing the cost and weight of the reinforced plastic component.

5. **Other fibers.**   Other fibers that are used for composite materials include silicon carbide, silicon nitride, aluminum oxide, sapphire, steel, tungsten, molybdenum, boron carbide, boron nitride, and tantalum carbide. (See also Chapter 11.) Metallic fibers are drawn as described in Section 6.5, although at the smaller diameters, the wires are drawn in bundles. **Whiskers** (see Section 3.8.3) also are used as reinforcing fibers. Whiskers are tiny needlelike single crystals that grow to $1–10\ \mu m$ ($40–400\ \mu in.$) in diameter and have aspect ratios (defined as the ratio of fiber length to diameter) ranging from 100 to 15,000. Because of their small size, either the fibers are free of imperfections or the imperfections they contain do not significantly affect their strength, which approaches the theoretical strength of the material (Section 3.3.2).

### 10.9.3 Fiber size and length

The mean diameter of fibers used in reinforced plastics is usually less than 0.01 mm (0.0004 in.). The fibers are very strong and rigid in tension. The reason is that the molecules in the fibers are oriented in the longitudinal direction, and their cross-sections are so small that the probability is low that any defects exist in the fiber. Glass fibers, for example, can have tensile strengths as high as 4600 MPa (650 ksi), whereas the strength of glass in bulk form is much lower; thus, glass fibers are stronger than steel.

Fibers are classified as **short** or **long** fibers, also called **discontinuous fibers** or **continuous fibers**, respectively. Short fibers generally have an *aspect ratio* between 20 and 60, and long fibers have an aspect ratio between 200 and 500. The short- and long-fiber designations are, in general, based on the following observations: In a given fiber, if the mechanical properties improve as a result of increasing the fiber length, then the fiber is denoted as a short fiber. When no additional improvement in properties occurs, the fiber is denoted as a long fiber. In addition to the discrete fibers just described, reinforcements in composites may be in the form of continuous *roving* (slightly twisted strand of fibers), *woven* fabric (similar to cloth), *yarn* (twisted strand), and *mats* of various combinations. Reinforcing elements may also be in the form of particles and flakes.

### 10.9.4 Matrix materials

The matrix in reinforced plastics has three functions:

1. Support and transfer the stresses to the fibers, which carry most of the load;
2. Protect the fibers against physical damage and the environment;
3. Reduce propagation of cracks in the composite by virtue of the ductility and toughness of the plastic matrix.

Matrix materials are usually epoxy, polyester, phenolic, fluorocarbon, polyethersulfone, or silicon. The most commonly used matrix materials are epoxies (80% of all reinforced plastics) and polyesters, which are less expensive than epoxies. Polyimides, which resist exposure to temperatures in excess of 300°C (575°F), are being developed for use with graphite fibers. Some thermoplastics, such as polyetheretherketone (PEEK), also are used as matrix materials; they generally have higher toughness than thermosets, but their resistance to temperature is lower, being limited to the range of 100–200°C (200–400°F).

### 10.9.5 Properties of reinforced plastics

The properties of reinforced plastics depend on the type, shape, and orientation of the reinforcing material; the length of the fibers; and the volume fraction (percentage) of the reinforcing material. Short fibers are less effective than long fibers (Fig. 10.17), and their properties are strongly influenced by time and temperature. Long fibers transmit the load through the matrix better and thus are commonly used in critical applications, particularly at elevated temperatures. Fiber reinforcement also affects many other properties of composites.

A critical factor in reinforced plastics is the strength of the bond between the fiber and the polymer matrix, since the load is transmitted through the fiber–matrix interface. Weak bonding causes **fiber pullout** and **delamination** of the structure, particularly under adverse environmental conditions. Adhesion at the interface can be improved by special surface treatments, such as coatings and coupling agents. Glass fibers, for example, are treated with *silane* (see Section 10.9.2) for improved wetting

**FIGURE 10.17** Effect of the percentage of reinforcing fibers and fiber length on the mechanical properties of reinforced nylon. Note the significant improvement with increasing percentage of fiber reinforcement. *Source*: Courtesy of Wilson Fiberfill International.

(a)  (b)

**FIGURE 10.18**
(a) Fracture surface of glass-fiber-reinforced epoxy composite. The fibers are $10\mu$m ($400\mu$in.) in diameter and have random orientation. (b) Fracture surface of a graphite-fiber-reinforced epoxy composite. The fibers, $9-11\mu$m in diameter, are in bundles and are all aligned in the same direction. *Source*: L. J. Broutman.

and bonding between the fiber and the matrix. The importance of proper bonding can be appreciated by inspecting Fig. 10.18, which shows the fracture surfaces of a reinforced plastic.

Generally, the highest stiffness and strength in reinforced plastics is obtained when the fibers are aligned in the direction of the tension force; this composite is, of course, highly anisotropic (Fig. 10.19). As a result, other properties of the composite, such as stiffness, creep resistance, thermal and electrical conductivity, and thermal expansion, are also anisotropic. The transverse properties of such a unidirectionally reinforced structure are much lower than the longitudinal properties. Note, for example, how easily you can split fiber-reinforced packaging tape, yet how strong it is when you pull on it (tension).

For a specific service condition, we can give a reinforced-plastic part an optimal configuration. For example, if the reinforced-plastic part is to be subjected to forces in different directions (such as thin-walled, pressurized vessels), the fibers are

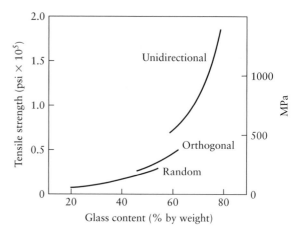

**FIGURE 10.19** The tensile strength of glass-reinforced polyester as a function of fiber content and fiber direction in the matrix. *Source*: R. M. Ogorkiewicz, *The Engineering Properties of Plastics*, Oxford, U.K.: Oxford University Press, 1977.

crisscrossed in the matrix. (See the discussion of *filament winding* in Section 10.11.2.) Reinforced plastics may also be made with various other materials and shapes of the polymer matrix in order to impart specific properties (such as permeability and dimensional stability), to make processing easier, and to reduce costs.

**Strength and elastic modulus of reinforced plastics.** The strength of a reinforced plastic with longitudinal fibers can be determined in terms of the strength of the fibers and matrix, respectively, and the volume fraction of fibers in the composite. In the equations given next, $c$ refers to the composite, $f$ to the fiber, and $m$ to the matrix. The total load $F_c$ on the composite is shared by the fiber load $F_f$ and the matrix load $F_m$; thus,

$$F_c = F_f + F_m, \tag{10.10}$$

which can be written as

$$\sigma_c A_c = \sigma_f A_f + \sigma_m A_m, \tag{10.11}$$

where $A_c$, $A_f$, and $A_m$ are the cross-sectional areas of the composite, fiber, and matrix, respectively, and $A_c = A_f + A_m$. Let's now denote $x$ as the area fraction of the fibers in the composite. (Note that $x$ also represents the volume fraction, because the fibers are uniformly longitudinal in the matrix.) Equation (10.11) can now be rewritten as

$$\sigma_c = x\sigma_f + (1 - x)\sigma_m. \tag{10.12}$$

We can now calculate the fraction of the total load carried by the fibers. First, note that in the composite under a tension load, the strains sustained by the fibers and the matrix are the same (that is, $e_c = e_f = e_m$), and then recall from Section 2.2 that

$$e = \frac{\sigma}{E} = \frac{F}{AE}.$$

Consequently,

$$\frac{F_f}{F_m} = \frac{A_f E_f}{A_m E_m}. \tag{10.13}$$

Since we know the relevant quantities for a specific case, we can use Eq. (10.10) to determine the fraction $F_f/F_c$. Then, using the foregoing relationships, we can also calculate the elastic modulus $E_c$ of the composite by replacing $\sigma$ in Eq. (10.12) with $E$. Hence,

$$E_c = xE_f + (1 - x)E_m. \tag{10.14}$$

**EXAMPLE 10.4**  **Properties of a graphite–epoxy-reinforced plastic**

Let's assume that a graphite–epoxy-reinforced plastic with longitudinal fibers contains 20% graphite fibers with a strength of 2500 MPa and an elastic modulus of 300 GPa. The strength of the epoxy matrix is 120 MPa, and it has an elastic modulus of 100 GPa. Calculate the elastic modulus of the composite and the fraction of the load supported by the fibers.

**Solution.**  The data given are $x = 0.2, E_f = 300$ GPa, $E_m = 100$ GPa, $\sigma_f = 2500$ MPa, and $\sigma_m = 120$ MPa. Using Eq. (10.14), we have

$$E_c = 0.2(300) + (1 - 0.2)100 = 60 + 80 = 140 \text{ GPa}.$$

The load fraction $F_f/F_m$ is obtained from Eq. (10.13):

$$\frac{F_f}{F_m} = \frac{(0.2)(300)}{(0.8)(100)} = 0.75.$$

Since

$$F_c = F_f + F_m$$

and

$$F_m = \frac{F_f}{0.75},$$

we obtain

$$F_c = F_f + \frac{F_f}{0.75} = 2.33 F_f,$$

or

$$F_f = 0.43 F_c.$$

Hence, the fibers support 43% of the load, even though they occupy only 20% of the cross-sectional area (and hence volume) of the composite.

## 10.9.6  Applications

The first application of reinforced plastics (in 1907) was for an acid-resistant tank, made of a phenolic resin with asbestos fibers. Formica, commonly used for countertops, was developed in the 1920s. Epoxies were first used as a matrix in the 1930s. Beginning in the 1940s, boats were made with fiberglass, and reinforced plastics were used for aircraft, electrical equipment, and sporting goods. Major developments in composites began in the 1970s, and these materials are now called **advanced composites**. Glass- or carbon-fiber-reinforced hybrid plastics are being developed for high-temperature applications, with continuous use ranging up to about 300°C (550°F).

Reinforced plastics are typically used in military and commercial aircraft and rocket components, helicopter blades, automotive bodies, leaf springs, drive shafts, pipes, ladders, pressure vessels, sporting goods, sports and military helmets, boat hulls, and various other structures. Applications of reinforced plastics include components in the McDonnell Douglas DC-10 aircraft, the Lockheed Martin L-1011 aircraft, and the Boeing aircraft. The Boeing 777 is made of about 9% composites by total weight, which is triple the composite content of prior Boeing transport aircraft. The floor beams and panels and most of the vertical and horizontal tail are made of composite materials.

By virtue of the resulting weight savings, reinforced plastics have reduced aircraft fuel consumption by about 2%. The newly designed Airbus jumbo jet A380, with a capacity of 550 to 700 passengers and to be in service in 2006, will have horizontal stabilizers, ailerons, wing boxes and leading edges, secondary mounting brackets of the fuselage, and the deck structure made of composites with carbon fibers, thermosetting resins, and thermoplastics. The upper fuselage will be made of alternating layers of aluminum and glass-fiber-reinforced epoxy prepregs.

Substitution of aluminum in large commercial aircraft with graphite–epoxy-reinforced plastics could reduce both weight and production costs by 30%, with improved fatigue and corrosion resistance. The structure of the Lear Fan 2100 passenger aircraft is almost totally made of graphite–epoxy-reinforced plastic. Nearly 90% of the structure of the lightweight Voyager aircraft, which circled the Earth without refueling, was made of carbon-reinforced plastic. The contoured frame of the Stealth bomber is made of composites consisting of carbon and glass fibers, epoxy-resin matrices, high-temperature polyimides, and other materials. Boron-fiber-reinforced composites are used in military aircraft, golf-club shafts, tennis rackets, fishing rods, and sailboards.

The processing of polymer-matrix-reinforced plastics (Section 10.11) presents significant challenges; several innovative techniques have been developed to manufacture both large and small parts by molding, forming, cutting, and assembly. Careful inspection and testing of reinforced plastics is essential in critical applications in order to ensure that good bonding between the reinforcing fiber and the matrix has been obtained throughout. In some instances, the cost of inspection can be as high as one quarter of the total cost of the composite product.

# 10.10  Processing of Plastics

The processing of plastics involves operations similar to those used to form and shape metals, as described in the preceding chapters. Plastics can be molded, cast, and formed, as well as machined and joined, into many shapes with relative ease and with few or no additional operations required (Table 10.6). Plastics melt or cure at relatively low temperatures (Table 10.2) and thus, unlike metals, are easy to handle and require less energy to process. However, because the properties of plastic parts and components are influenced greatly by the method of manufacture and the processing conditions, the proper control of these conditions is important for part quality. Plastics are usually shipped to manufacturing plants as pellets or powders and are melted just before the shaping process. Plastics are also available as sheet, plate, rod, and tubing, which may be formed into a variety of products. Liquid plastics are especially used in making reinforced plastic parts.

## 10.10.1 Extrusion

In *extrusion*, raw materials in the form of thermoplastic pellets, granules, or powder are placed into a hopper and fed into the extruder barrel (Fig. 10.20). The barrel is equipped with a *screw* that blends and conveys the pellets down the barrel. The internal friction from the mechanical action of the screw, along with heaters around the extruder's barrel, heats the pellets and liquefies them. The screw action also builds up pressure in the barrel.

Screws have three distinct sections: (1) a *feed section*, which conveys the material from the hopper area into the central region of the barrel; (2) a *melt*, or *transition*, *section*, where the heat generated from shearing of the plastic causes melting to begin;

**TABLE 10.6**

| **Characteristics of Processing Plastics and Reinforced Plastics** | |
| --- | --- |
| Process | Characterisics |
| Extrusion | Long, uniform, solid or hollow, simple or complex cross-sections; wide range of dimensional tolerances; high production rates; low tooling cost. |
| Injection molding | Complex shapes of various sizes and with fine detail; good dimensional accuracy; high production rates; high tooling cost. |
| Structural foam molding | Large parts with high stiffness-to-weight ratio; low production rates; less expensive tooling than in injection molding. |
| Blow molding | Hollow thin-walled parts of various sizes; high production rates and low cost for making beverage and food containers. |
| Rotational molding | Large hollow shapes of relatively simple design; low production rates; low tooling cost. |
| Thermoforming | Shallow or deep cavities; medium production rates; low tooling costs. |
| Compression molding | Parts similar to impression-die forging; medium production rates; relatively inexpensive tooling. |
| Transfer molding | More complex parts than in compression molding, and higher production rates; some scrap loss; medium tooling cost. |
| Casting | Simple or intricate shapes, made with flexible molds; low production rates. |
| Processing of reinforced plastics | Long cycle times; dimensional tolerances and tooling costs depend on the specific process. |

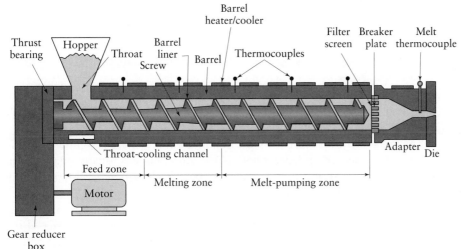

**FIGURE 10.20** Schematic illustration of a typical extruder. *Source: Encyclopedia of Polymer Science and Engineering*, 2d ed. Copyright © 1985. This material is used by permission of John Wiley & Sons, Inc.

and (3) a *pumping section*, where additional shearing and melting occurs, with pressure buildup at the die. The lengths of these sections can be changed to accommodate the melting characteristics of different plastics.

a) **Mechanics of polymer extrusion.** The pumping section of the screw determines the rate of polymer flow through the extruder. Consider a uniform screw geometry with narrow *flights* and small clearance with the barrel (Fig. 10.21). At any point in time, the molten plastic is in the shape of a helical ribbon, and this ribbon is being conveyed towards the extruder outlet by the screw flights. If the pressure is constant along the pumping zone, then the volume flow rate of plastic out of the extruder, or *drag flow*, is given by

$$Q_d = \frac{vHW}{2},$$   (10.15)

**FIGURE 10.21** Geometry
of the pumping section of an
extruder screw.

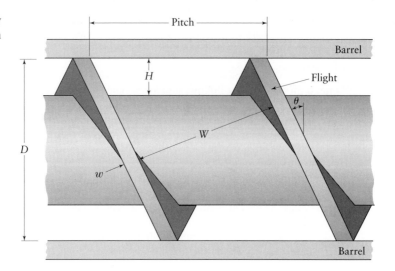

where $v$ is the velocity of the flight in the extrusion direction, $H$ is the channel depth, and $W$ is the width of the polymer ribbon. From the geometry defined in Fig. 10.21, we can write

$$V = \omega \cos \theta = \pi DN \cos \theta \tag{10.16}$$

and

$$W = \pi D \sin \theta - w, \tag{10.17}$$

where $\omega$ is the angular velocity of the screw, $D$ is the screw diameter, $N$ is the shaft speed (usually in rev/min), $\theta$ is the flight angle, and $w$ is the flight width. If the flight width can be considered as negligibly small, then the drag flow can be simplified as

$$Q_d = \frac{\pi^2 HD^2\, N\sin \theta \cos \theta}{2}. \tag{10.18}$$

The actual flow rate can be larger than this value if there is significant pressure buildup in the feed or melt sections of the screw, but it is usually smaller, because of the large pressure at the die end of the barrel. Therefore, the flow rate through the extruder can be taken as

$$Q = Q_d - Q_p, \tag{10.19}$$

where $Q_p$ is a flow correction due to pressure. For Newtonian fluids (see Section 10.3), $Q_p$ can be taken as

$$Q_p = \frac{WH^3 p}{12\eta(l/\sin \theta)} = \frac{p\pi DH^3 \sin^2 \theta}{12\eta l}. \tag{10.20}$$

where $l$ is the length of the pumping section. Hence, Eq. (10.19) becomes

$$Q = \frac{\pi^2 HD^2 N \sin \theta \cos \theta}{2} - \frac{p\pi DH^3 \sin^2 \theta}{12\eta l}. \tag{10.21}$$

Equation (10.21) is known as the *extruder characteristic*. If the power-law index $n$ from Eq. 10.4a is known, then the extruder characteristic is given by the following approximate relationship (after Rauwendaal, 1984):

$$Q = \left(\frac{4 + n}{10}\right)(\pi^2 HD^2 N\sin \theta \cos \theta) - \frac{p\pi DH^3 \sin^2 \theta}{(1 + 2n)4\eta}. \tag{10.22}$$

The die plays a major role in determining the output of the extruder. The *die characteristic* is the expression relating flow to the pressure drop across the die and in general form is written as

$$Q_{\text{die}} = Kp \tag{10.23}$$

where $Q_{\text{die}}$ is the flow through the die, $p$ is the pressure at the die inlet, and $K$ is a function of the die's geometry. The determination of $K$ is usually complicated and difficult to obtain analytically, although computer-based tools are becoming increasingly available for predictions. More commonly, $K$ is determined experimentally, but one closed-form solution for extruding solid round cross-sections is given by

$$K = \frac{\pi D_d^4}{128 \eta \ell_d}, \tag{10.24}$$

where $D_d$ is the die opening diameter and $\ell_d$ is the die land. If both the die characteristic and the extruder characteristic are known, we have two simultaneous algebraic equations that can be solved for the pressure and flow rate during the operation.

b) **Process characteristics.** Once the extruded product exits the die, it is cooled, either by air or by passing through a water-filled channel. Controlling the rate and uniformity of cooling is important for minimizing product shrinkage and distortion. The extruded product can also be drawn (sized) by a puller after it has cooled; the extruded product is then coiled or cut into desired lengths. Complex shapes with constant cross-section can be extruded with relatively inexpensive tooling. This process is also used to extrude elastomers.

Because this operation is continuous, long products with cross-sections of solid rods, channels, pipe, window frames, and architectural components, as well as sheet, are extruded through dies of various cross-sectional shapes. Plastic-coated wire, cable, and strips for electrical applications are also extruded and coated by this process. The wire is fed into the die opening at a controlled rate with the extruded plastic.

**Pellets,** which are used for other plastic-processing methods described in this chapter, are also made by extrusion. In this case, the extruded product is a small-diameter rod and is chopped into short lengths, or pellets, as it is extruded. With some modifications, extruders can also be used as simple melters for other shaping processes, such as injection molding and blow molding.

Because the material being extruded is still soft as it leaves the die and the pressure is relieved, the cross-section of the extruded product is different than the shape of the die opening; this effect is known as **die swell.** (See Fig. 10.47b.) Thus, the diameter of a round extruded part, for example, is larger than the die diameter, the difference between the diameters depends on the type of polymer. Considerable experience is required to design proper dies for extruding complex cross-sections of desired shape and dimensions.

Extruders are generally rated by the diameter of the barrel and by the length-to-diameter ($L/D$) ratio of the barrel. Typical commercial units are 25 to 200 mm (1 to 8 in.) in diameter, with $L/D$ ratios of 5 to 30. Production-size extrusion equipment costs \$30,000–\$80,000, with an additional \$30,000 cost for equipment for downstream cooling and winding of the extruded product; therefore, large production runs are generally required to justify such an expenditure.

c) **Sheet and film extrusion.** Polymer sheet and film can be produced using a flat extrusion die. The polymer is extruded by forcing it through a specially designed die, after which the extruded sheet is taken up first on water-cooled rolls and then by a pair of rubber-covered pull-off rolls.

**FIGURE 10.22** Schematic illustration of production of thin film and plastic bags from a tube produced by an extruder, and then blown by air. *Source*: D. C. Miles and J. H. Briston, *Polymer Technology*, Chemical Publishing Co., 1979. Reproduced by permission of Chemical Publishing Co., Inc.

Thin polymer films and common plastic bags are made from a tube produced by an extruder (Fig. 10.22). In this process, a thin-walled tube is extruded vertically and expanded into a balloon shape by blowing air through the center of the extrusion die until the desired film thickness is reached. The balloon is usually cooled by air from a cooling ring around it, which can also act as a barrier to further expansion of the balloon. **Blown film** is sold as wrapping film (after the cooled bubble has been slit) or as bags (where the bubble is pinched and cut off). Film is also produced by shaving solid round billets of plastics, especially polytetrafluoroethylene (PTFE), by *skiving* (shaving with specially designed knives).

---

### EXAMPLE 10.5  Blown film

Assume that a typical plastic shopping bag, made by blown film, has a lateral (width) dimension of 400 mm. (a) What should be the extrusion die diameter? (b) These bags are relatively strong. How is this strength achieved?

**Solution.**

a) The perimeter of the bag is $(2)(400) = 800$ mm. Since the original cross-section of the film was round, the blown diameter should be $\pi D = 800$, or $D = 255$ mm. Recall that in this process a tube is expanded 1.5 to 2.5 times the extrusion die diameter. Taking the maximum value of 2.5, we calculate the die diameter as $255/2.5 = 100$ mm.

b) Note in Fig. 10.22 that, after being extruded, the bubble is being pulled upward by the pinch rolls. Thus, in addition to diametral stretching and the attendant molecular orientation, the film is stretched and oriented in the longitudinal direction as well. The biaxial orientation of the polymer molecules significantly improves the strength and toughness of the blown film.

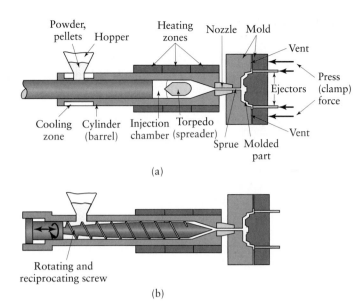

**FIGURE 10.23** Injection molding with (a) a plunger and (b) a reciprocating rotating screw. Telephone receivers, plumbing fittings, tool handles, and housings are made by injection molding.

## 10.10.2 Injection molding

*Injection molding* is essentially the same process as hot-chamber die casting (see Fig. 5.28). The pellets or granules are fed into a heated cylinder, where they are melted. The melt is then forced into a split-die chamber (Fig. 10.23a), either by a hydraulic plunger or by the rotating screw system of an extruder. Newer equipment is of the *reciprocating-screw* type (Fig. 10.23b). As the pressure builds up at the mold entrance, the rotating screw begins to move backward under pressure to a predetermined distance, thus controlling the volume of material to be injected. The screw then stops rotating and is pushed forward hydraulically, forcing the molten plastic into the mold cavity. Injection-molding pressures usually range from 70 MPa to 200 MPa (10,000 psi to 30,000 psi).

Typical injection-molded products include cups, containers, housings, tool handles, knobs, electrical and communication components (such as cell phones), toys, and plumbing fittings. Although the molds are relatively cool for thermoplastics, thermosets are molded in heated molds, where *polymerization* and *cross-linking* take place. In either case, after the part is sufficiently cooled (for thermoplastics) or set or cured (for thermosets), the molds are opened, and the part is ejected. The molds are then closed, and the process is repeated automatically. Elastomers also are injection molded by these processes. Molds with moving and unscrewing mandrels are also used; they allow the molding of parts with multiple cavities and internal and external threads.

Because the material is molten when injected into the mold, complex shapes and good dimensional accuracy can be achieved. However, as in metal casting, the molded part shrinks during cooling. (Note also that plastics have higher thermal expansion than metals.) Shrinkage in plastics molding is compensated for by making the molds a little larger. Linear shrinkage for plastics typically ranges between 0.005 and 0.025 mm/mm or in./in., and the volumetric shrinkage is typically in the range of 1.5% to 7%. (See also Table 5.1 and Section 11.4.)

Various aspects of injection molding are discussed below.

a) **Molds.** Injection molds have several components, depending on part design, such as runners (as in metal-casting dies), cores, cavities, cooling channels, inserts, knockout pins, and ejectors. There are three basic types of molds: (1) *cold-*

*runner two-plate* mold, which is the basic and simplest mold design; (2) *cold-runner three-plate* mold, in which the runner system is separated from the part when the mold opens; and (3) *hot-runner* mold, also called *runnerless* mold, in which the molten plastic is kept hot in a heated runner plate. In cold-runner molds, the solidified plastic in the channels (that connect the mold cavity to the end of the barrel) must be removed, usually by trimming. This scrap can be chopped and recycled. In hot-runner molds, which are more expensive, there are no gates, runners, or sprues attached to the part. Cycle times are shorter, because only the injection-molded part must be cooled and ejected.

Metallic components, such as screws, pins, and strips, can also be placed in the mold cavity to become an integral part of the injection-molded product (**insert molding**; see Fig. 10.24); the most common examples are electrical components. *Multicomponent* injection molding (also called *coinjection* or *sandwich* molding) allows the forming of parts with a combination of colors and shapes. Examples include multicolor molding of rear-light covers for automobiles and ball joints made of different materials. Also, printed film can be placed in the mold cavity; thus, parts need not be decorated or labeled after molding.

Injection molding is a high-rate production process, with good dimensional control. Typical cycle times range from 5 to 60 seconds, but can be several minutes for thermosetting materials. The molds, generally made of tool steels or beryllium–copper, may have multiple cavities so that more than one part can be made in one cycle. Proper mold design and control of the flow of material in the die cavities are important factors in the quality of the product; other factors that affect quality include injection pressure, temperature, and condition of the resin. *Computer models* have been developed to study the flow of material in dies and thereby improve die design and establish appropriate process parameters. Modern machines used in injection molding are equipped with microprocessors and microcomputers in a control panel and monitor all aspects of the operation.

b) **Machines.** Injection-molding machines are usually horizontal, and the clamping force on the dies is supplied generally by hydraulic means, although electrical types are also available. Electrically driven models weigh less and are quieter than hydraulic machines. Vertical machines are used for making small, close-tolerance parts and for insert molding. Injection-molding machines are rated according to the capacity of the mold and the clamping force. Although in most machines, this force generally ranges from 0.9 MN to 2.2 MN (100 tons to 250 tons), the largest machine in operation has a capacity of 45 MN (5000 tons) and can

**FIGURE 10.24** Products made by insert injection molding. Metallic components are embedded in these parts during molding. *Source*: Rayco Mold and Mfg. LLC.

produce parts weighing up to 25 kg (55 lb); however, parts typically weigh 100 g to 600 g (3 oz to 20 oz). Because of the high cost of dies, typically ranging from $20,000 to $200,000, high-volume production is required to justify such an expenditure.

c) **Overmolding and ice-cold molding.** *Overmolding* is a technique for producing hinge joints and ball-and-socket joints in one operation. Two different plastics are used to ensure that no bond will form between the two parts. *Ice-cold molding* uses the same kind of plastic to form both components, such as in a door hinge. It is done in one cycle and involves a two-cavity mold, using cooling inserts positioned such that no bond forms between the two pieces.

d) **Reaction-injection molding.** In the *reaction-injection-molding* (RIM) process, a mixture of two or more reactive fluids is forced into the mold cavity (Fig. 10.25). Chemical reactions take place rapidly in the mold, and the polymer solidifies, producing a thermoset part. Major applications include automotive bumpers and fenders, thermal insulation for refrigerators and freezers, and stiffeners for structural components. Various reinforcing fibers, such as glass or graphite, may also be used to improve the product's strength and stiffness. Mold costs are low for this process.

e) **Structural foam molding.** The *structural-foam-molding* process is used to make plastic products that have a solid skin and a cellular inner structure. Typical products include furniture components, TV cabinets, business-machine housings, and storage-battery cases. Although there are several foam-molding processes, they are basically similar to injection molding or extrusion. Both thermoplastics and thermosets can be used for foam molding, but thermosets are in the liquid-processing form, similar to polymers for reaction-injection molding.

f) **Injection foam molding.** In *injection foam molding*, also called *gas-assist molding*, thermoplastics are mixed with a **blowing agent** (usually an inert gas, such as nitrogen, or a chemical agent that produces gas during molding), which expands the material. The core of the part is cellular, and the skin is rigid. The thickness of the skin can be as much as 2 mm (0.08 in.), and part densities are

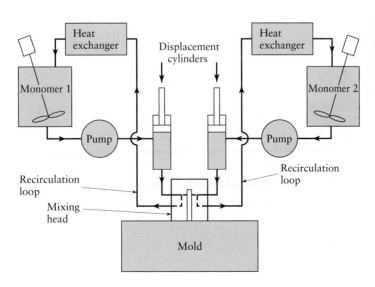

**FIGURE 10.25** Schematic illustration of the reaction-injection-molding process. *Source*: *Modern Plastics Encyclopedia.*

as low as 40% of the density of the solid plastic; thus, parts have a high stiffness-to-weight ratio and can weigh as much as 55 kg (120 lb).

---

**EXAMPLE 10.6**  **Injection molding of gears**

A 250-ton injection-molding machine is to be used to make 4.5-in. diameter spur gears that are 0.5 in.-thick. The gears have a fine-tooth profile. How many gears can be injection molded in one set of molds? Does the thickness of the gears influence the answer?

**Solution.**  Because of the fine detail involved, the pressures required in the mold cavity will probably be on the order of 100 MPa (15 ksi). The cross-sectional (projected) area of the gear is $\pi(4.5)^2/4 = 15.9$ in$^2$. If we assume that the parting plane of the two halves of the mold is in the midplane of the gear, the force required is $(15.9)(15,000) = 238,500$ lb. The capacity of the machine is 250 tons, so we have $(250)(2000) = 500,000$ lb of clamping force available. Therefore, the mold can accommodate two cavities, thus producing two gears per cycle. Because it does not influence the cross-sectional area of the gear, the thickness of the gear does not directly influence the pressures involved, and hence the answer is the same for different gear thicknesses.

---

### 10.10.3  Blow molding

*Blow molding* is a modified extrusion and injection-molding process. In **extrusion blow molding**, a tube is extruded (usually vertically), clamped into a mold with a cavity much larger than the tube's diameter, and then blown outward to fill the mold cavity (Fig. 10.26a). Blowing is usually done with an air blast, at a pressure of 350 kPa to 700 kPa (50 psi to 100 psi). In some operations, the extrusion is continuous, and the molds move with the tubing. The molds close around the tubing, close off both ends (thereby breaking the tube into sections), and then move away as air is injected into the tubular piece. The part is then cooled and ejected. Corrugated pipe and tubing are made by continuous blow molding, whereby the pipe or tube is extruded horizontally and blown into moving molds. (See also Section 10.10.1).

In **injection blow molding**, a short tubular piece (**parison**) is first injection molded (Fig. 10.26b). The dies then open, and the parison is transferred to a blow-molding die. Hot air is injected into the parison, which then expands and fills the mold cavity. Typical products made by this process include plastic beverage bottles and hollow containers.

**Multilayer blow molding** involves the use of coextruded tubes or parisons, thus allowing the use of multilayer structures. Typical examples of multilayer structures include plastic packaging for food and beverages with characteristics such as odor and permeation barrier, taste and aroma protection, resistance to scuffing, printing capability, and ability to be filled with hot fluids. Other applications are in the cosmetics and pharmaceutical industries.

### 10.10.4  Rotational molding

Most thermoplastics and some thermosets can be formed into large hollow parts by *rotational molding*. The thin-walled metal mold is made of two pieces (*split female mold*) and is designed to be rotated about two perpendicular axes (Fig. 10.27). A premeasured quantity of powdered plastic material is placed inside a warm mold. The powder is obtained from a polymerization process that precipitates the powder from a liquid. The mold is then heated, usually in a large oven, while it is rotated about

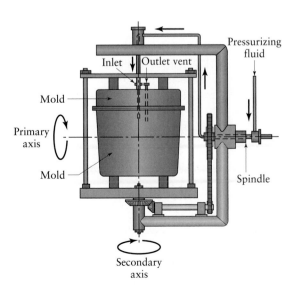

**FIGURE 10.26**   Schematic illustrations of (a) the blow-molding process for making plastic beverage bottles and (b) a three-station injection-blow-molding machine. *Source*: *Encyclopedia of Polymer Science and Engineering*, 2d ed. Wiley, Copyright © 1985. This material is used by permission of John Wiley & Sons, Inc.

**FIGURE 10.27**   The rotational molding (rotomolding or rotocasting) process. Trash cans, buckets, and plastic footballs can be made by this process.

the two axes. This action tumbles the powder against the mold, where the heat fuses the powder without melting it. In some parts, a chemical cross-linking agent is added to the powder, and cross-linking occurs after the part is formed in the mold by continued heating. Typical parts made by rotational molding include tanks of various sizes, trash cans, boat hulls, buckets, housings, toys, carrying cases, and footballs. Various metallic or plastic *inserts* may also be molded into the parts made by this process.

Liquid polymers, called **plastisols** (vinyl plastisols being the most common), can also be used in a process called **slush molding**. The mold is simultaneously heated and rotated, and the particles of plastic material are forced against the inside walls of the heated mold by the tumbling action. Upon contact, the material melts and coats the walls of the mold. The part is cooled while still rotating and is then removed by opening the mold.

Rotational molding can produce parts with complex hollow shapes, with wall thicknesses as small as 0.4 mm (0.016 in.). Parts as large as 1.8 m × 1.8 m × 3.6 m (6 ft × 6 ft × 12 ft) have been formed using this method. The outer surface finish of the part is a replica of the surface finish of the mold walls. Cycle times are longer than in other processes, but equipment costs are low. Quality-control considerations usually involve proper weight of powder placed in the mold, proper rotation of the mold, and the temperature–time relationship during the oven cycle.

### 10.10.5 Thermoforming

*Thermoforming* is a family of processes for forming thermoplastic sheet or film over a mold with the application of heat and pressure differentials (Fig. 10.28). In this process, a sheet is heated in an oven to the *sag* (softening) *point*, but not to the melting point. The sheet is then removed from the oven, placed over a mold, and pulled against the mold through the application of a vacuum. Since the mold is usually at room temperature, the shape of the plastic is set upon contacting the mold. Because of the low strength of the materials formed, the pressure differential caused by the vacuum is usually sufficient for forming, although air pressure or mechanical means are also applied for some parts. The sheets used in thermoforming are made by sheet extrusion.

Typical parts made using this method include advertising signs, refrigerator liners, packaging, appliance housings, and panels for shower stalls; however, parts with openings or holes cannot be formed by this method, because the necessary pressure

| 1. Straight vacuum forming | 2. Drape vacuum forming | 3. Force above sheet | 4. Plug and ring forming |

*a.* Heater     *d.* Mold
*b.* Clamp      *e.* Vacuum line
*c.* Plastic sheet

**FIGURE 10.28**   Various thermoforming processes for thermoplastic sheet. These processes are commonly used in making advertising signs, cookie and candy trays, panels for shower stalls, and packaging.

differential cannot be developed during forming. Because thermoforming is a combination of drawing and stretching operations, much like sheet-metal forming (see Chapter 7), the material should exhibit high uniform elongation; otherwise, it will neck and fail. Thermoplastics have a high capacity for uniform elongation by virtue of their high strain-rate sensitivity exponent, *m*. (See Section 2.2.7.)

Hollow parts can be produced by using twin sheets. In this operation, the two mold halves come together with the heated sheets between them. The sheets are drawn to the mold halves through a combination of vacuum from the mold and compressed air that is introduced between the sheets. The molds also fuse the sheet around the mold cavity. (See the discussion on *hot–plate welding* in Section 12.15.1.)

Molds for thermoforming are usually made of aluminum, since high strength is not a requirement. The holes in the molds are usually less than 0.5 mm (0.02 in.) in diameter in order not to leave any marks on the formed sheets. Tooling is inexpensive. Quality considerations include tears, nonuniform wall thickness, improperly filled molds, and poor part definition (surface details).

## 10.10.6  Compression molding

In *compression molding*, a preshaped charge of material, a premeasured volume of powder, or a viscous mixture of liquid resin and filler material is placed directly in a heated mold cavity. Forming is done under pressure with a plug or the upper half of the die (Fig. 10.29). Compression molding results in flash formation; the flash is removed by trimming or some other means. Typical parts made by this method include dishes, handles, container caps, fittings, electrical and electronic components, washing-machine agitators, and housings. Fiber-reinforced parts with long

**FIGURE 10.29** Types of compression molding, a process similar to forging: (a) positive, (b) semipositive, and (c) flash. The flash in part (c) has to be trimmed off. (d) Die design for making a compression-molded part with undercuts. Such designs are also used in other molding and shaping operations.

chopped fibers are formed exclusively by this process. Elastomers also are shaped by compression molding.

Compression molding is used mainly with thermosetting plastics, with the original material in a partially polymerized state. Cross-linking is completed in the heated die, with curing times typically ranging from 0.5 min to 5 min, depending on the material, part geometry, and thickness of the part. The thicker the material, the longer it will take to cure. Because of their relative simplicity, dies used in compression molding generally cost less than dies used in injection molding. Three types of compression molds are available: (1) **flash type** for shallow or flat parts, (2) **positive** for high-density parts, and (3) **semipositive** for high-quality production. Undercuts in parts are not recommended; however, dies can be designed to open sideways (Fig. 10.29d) to allow removal of the part. In general, the part complexity is less than with injection molding, and the dimensional control produced by compression molding is better.

### 10.10.7  Transfer molding

*Transfer molding* represents a further development of the compression-molding process. The uncured thermosetting material is placed in a heated transfer pot or chamber (Fig. 10.30). After the material is heated, it is injected into heated, closed molds. Depending on the type of machine used, a ram, plunger, or rotating screw feeder forces the material to flow through the narrow channels into the mold cavity. This flow generates considerable internal heat, which raises the temperature of the material and homogenizes it. Curing takes place by cross-linking. Because the resin is molten as it enters the molds, the complexity of the part and dimensional control approach those for injection molding.

Typical parts made by transfer molding include electrical and electronic components and rubber and silicone parts. The process is particularly suitable for intricate shapes that have varying wall thicknesses. The molds for this process tend to be more expensive than those for compression molding, and material is wasted in the channels of the mold during filling. (See also the discussion of *resin transfer molding* in Section 10.11.1.)

**FIGURE 10.30**  Sequence of operations in transfer molding for thermosetting plastics. This process is particularly suitable for intricate parts with varying wall thicknesses.

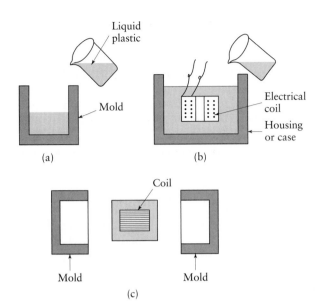

**FIGURE 10.31**
Schematic illustration of (a) casting, (b) potting, and (c) encapsulation of plastics.

## 10.10.8 Casting

Some thermoplastics, such as nylons and acrylics, and thermosetting plastics, such as epoxies, phenolics, polyurethanes, and polyester, can be *cast* in rigid or flexible molds into a variety of shapes (Fig. 10.31a). Typical parts cast include large gears, bearings, wheels, thick sheets, and components that require resistance to abrasive wear. In casting thermoplastics, a mixture of monomer, catalyst, and various additives is heated and poured into the mold. The part is formed after polymerization takes place at ambient pressure. Intricate shapes can be formed with flexible molds, which are then peeled off. *Centrifugal casting* (see Section 5.10.4) is also used with plastics, including reinforced plastics with short fibers. Thermosets are cast in a similar manner. Typical parts produced are similar to those made by thermoplastic castings. Degassing may be necessary for maintaining product integrity.

**Potting and encapsulation.** A variation of casting that is important to the electrical and electronics industry is potting and encapsulation. This process involves casting the plastic around an electrical component, thus embedding it in the plastic. *Potting* (Fig. 10.31b) is done in a housing or case, which is an integral part of the product. In *encapsulation* (Fig. 10.31c), the component is covered with a layer of the solidified plastic. In both applications, the plastic serves as a dielectric (nonconductor). Structural members, such as hooks and studs, may also be partly encapsulated.

**Foam molding.** Products such as *styrofoam* cups and food containers, insulating blocks, and shaped packaging materials (such as for cameras, appliances, and electronics) are made by *foam molding*. The material is made of expandable polystyrene, produced by placing **polystyrene beads** (obtained by polymerization of styrene monomer), containing a blowing agent, in a mold and exposing them to heat, usually by steam. As a result of their exposure to heat, the beads expand to as much as 50 times their original size and take the shape of the mold. The amount of expansion can be controlled through temperature and time. A common method of molding is using *preexpanded beads*, in which the beads are expanded by steam (or hot air, hot water, or an oven) in an open-top chamber. They are then placed in a storage bin and allowed to stabilize for a period of 3 to 12 hours. The beads can then be molded into shapes as described previously.

Polystyrene beads are available in three sizes: small for cups, medium for molded shapes such as containers, and large for molding of insulating blocks (which can then be cut to size); thus, the bead size chosen depends on the minimum wall thickness of the product. Beads can also be colored prior to expansion, or integrally colored beads can be used.

**Polyurethane foam processing**, for products such as cushions and insulating blocks, involves several processes that basically consist of mixing two or more chemical components. The reaction forms a cellular structure, which solidifies in a mold. Various low-pressure and high-pressure machines, with computer controls for proper mixing, are available.

### 10.10.9  Cold forming and solid-phase forming

Processes that are used in cold working of metals, such as rolling, deep drawing, extrusion, closed-die forging, coining, and rubber forming (Chapters 6 and 7), can also be used to form many thermoplastics at room temperature (*cold forming*). Typical materials formed include polypropylene, polycarbonate, ABS, and rigid PVC. The important considerations are that (a) the material be sufficiently ductile at room temperature (hence, polystyrenes, acrylics, and thermosets cannot be formed) and (b) the material's deformation must be nonrecoverable (to minimize springback and creep).

The advantages of cold forming of plastics over other methods of shaping are listed as follows:

1. Strength, toughness, and uniform elongation are increased.
2. Plastics with high molecular weight can be used to make parts with superior properties.
3. Forming speeds are not affected by the thickness of the part, since there is no heating or cooling involved.
4. Typical cycle times are shorter than those for molding processes.

*Solid-phase forming* is carried out at a temperature about 10°C to 20°C (20°F to 40°F) below the melting temperature of the plastic, if the plastic is a crystalline polymer, and it is formed while still in a solid state. The advantages of solid-phase forming over cold forming are that forming forces and springback effects are lower for the former. These processes are not as widely used as hot-processing methods and are generally restricted to special applications.

### 10.10.10  Processing elastomers

Elastomers can be formed by a variety of processes used for shaping thermoplastics. Thermoplastic elastomers are commonly shaped by extrusion and injection molding, extrusion being the most economical and the fastest process. In terms of its processing characteristics, a thermoplastic elastomer is a polymer; in terms of its function and performance, it is a rubber (Section 10.8). These polymers can also be formed by blow molding and thermoforming. Thermoplastic polyurethane can be shaped by all conventional methods; it can also be blended with thermoplastic rubbers, polyvinyl chloride compounds, ABS, and nylon. Dryness of the materials is important. For extrusion, the temperatures are in the range of 170°C to 230°C (340°F to 440°F), and for molding, they range up to 60°C (140°F). Typical extruded products include tubing, hoses, moldings, and inner tubes. Injection-molded products cover a broad range of applications, such as components for automobiles and appliances.

Rubber and some thermoplastic sheets are formed by the **calendering** process (Fig. 10.32), wherein a warm mass of the compound is fed through a series of rolls

**FIGURE 10.32**  Schematic illustration of calendering. Sheets produced by this process are subsequently used in thermoforming.

(*masticated*) and is then stripped off in the form of a sheet. The rubber may also be formed over both surfaces of a fabric liner. Discrete rubber products, such as gloves, are made by dipping a form repeatedly into a liquid compound that adheres to the form; the material is then *vulcanized*, usually in steam, and stripped from the form.

# 10.11 | Processing of Polymer-Matrix-Reinforced Plastics

As described in Section 10.9, reinforced plastics are among the most important materials and can be engineered to meet specific design requirements such as high strength-to-weight and stiffness-to-weight ratios and creep resistance. Because of their unique structure, reinforced plastics require special methods to shape them into useful products (Fig. 10.33).

The care required and the several steps involved in manufacturing reinforced plastics make processing costs substantial, so this process generally is not competitive with processing traditional materials and shapes at high production rates. This situation has necessitated the careful assessment and integration of the design and manufacturing processes (*concurrent engineering*) in order to minimize costs while maintaining product integrity and production rate. An important environmental concern with respect to reinforced plastics is the dust generated during processing, such as airborne carbon fibers, which are known to remain in the work area long after fabrication of parts has been completed.

Reinforced plastics can usually be fabricated by the methods described in this chapter, with some provision for the presence of more than one type of material in the composite. The reinforcement may be chopped fibers, woven fabric or mat, roving or yarn (slightly twisted fiber), or continuous lengths of fiber. In order to obtain good bonding between the reinforcing fibers and the polymer matrix, as well as to protect the fibers during subsequent processing, the fibers are surface treated by *impregnation* (*sizing*). Short fibers are commonly added to thermoplastics for injection

**FIGURE 10.33**
Reinforced-plastic components for a Honda motorcycle. The parts shown are front and rear forks, a rear swing arm, a wheel, and brake disks.

(a)                                                              (b)

**FIGURE 10.34**    (a) Manufacturing process for polymer-matrix composite.
*Source*: T.-W. Chou, R. L. McCullough, and R. B. Pipes. (b) Boron–epoxy prepreg tape.
*Source*: Textron Systems.

molding; milled fibers can be used for reaction-injection molding; and longer chopped fibers are used primarily in compression molding of reinforced plastics.

When the impregnation is done as a separate step, the resulting partially cured sheets are referred to by various terms:

1.  **Prepregs.**   The continuous fibers are aligned (Fig. 10.34a) and subjected to surface treatment to enhance their adhesion to the polymer matrix. They are then coated by being dipped in a resin bath and made into a *sheet* or *tape* (Fig. 10.34b). Finally, individual pieces of the sheet are assembled into laminated structures, such as the horizontal stabilizer for the F-14 fighter aircraft. Special computer-controlled tape-laying machines have been developed for this purpose. Typical products include flat or corrugated architectural paneling, panels for construction and electric insulation, and structural components of aircraft that require good retention of properties and fatigue strength under hot or wet conditions.

2.  **Sheet-molding compound** (SMC).   Continuous strands of reinforcing fiber are chopped into short fibers (Fig. 10.35) and deposited over a layer of resin paste, usually a polyester mixture, carried on a polymer film such as polyethylene. A second layer of resin paste is deposited on top, and the sheet is pressed between rollers. The product is gathered into rolls, or placed into containers in layers, and stored until it undergoes a *maturation period*, reaching the desired

**FIGURE 10.35**
Manufacturing process for producing reinforced-plastic sheets. The sheet is still viscous at this stage and can later be shaped into various products. *Source*: T.-W. Chou, R. L. McCullough, and R. B. Pipes.

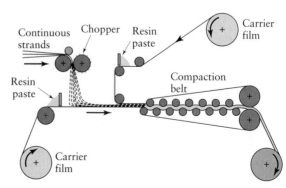

molding viscosity. The maturing process is under controlled temperature and humidity and usually takes one day. The matured SMC, which has a leatherlike feel and is tack free, has a shelf life of about 30 days and must be processed within this period. Alternatively, the resin and the fibers can be mixed together only at the time they are placed in the mold.

3. **Bulk-molding compound** (BMC). These compounds are in the shape of billets, generally up to 50 mm (2 in.) in diameter; they are made in the same manner as SMCs, that is, by the extrusion process. When processed into products, BMCs have flow characteristics similar to that of dough; hence, they are called *dough-molding compounds* (DMCs).

4. **Thick-molding compound** (TMC). This compound combines the characteristics of BMCs (lower cost) and SMCs (higher strength); it is usually injection molded, using chopped fibers of various lengths. One of its applications is for electrical components, because of the high dielectric strength of TMCs.

---

**EXAMPLE 10.7**  Tennis rackets made of composite materials

In order to impart certain desirable characteristics, such as light weight and stiffness, composite-material tennis rackets are being manufactured with graphite, fiberglass, boron, ceramic (silicon carbide), and Kevlar as the reinforcing fibers. Rackets have a foam core; some have unidirectional reinforcement, and others have braided reinforcement. Rackets with boron fibers have the highest stiffness, followed by rackets with graphite (carbon), glass, and Kevlar fibers. The racket with the lowest stiffness has 80%-fiberglass, whereas the stiffest racket has 95%-graphite and 5%-boron fibers; thus, the latter has the highest percentage of inexpensive reinforcing fiber and the smallest percentage of the most expensive fiber.

---

## 10.11.1 Molding

There are four basic methods for molding reinforced plastics, described as follows.

1. In **compression molding,** the material is placed between two molds, and pressure is applied. Depending on the material, the molds either are at room temperature or are heated to accelerate hardening. The material may be in bulk form (bulk-molding compound), which is a viscous, sticky mixture of polymers, fibers, and additives. It is generally shaped into a log, which is cut into the desired mass. Fiber lengths generally range from 3 mm to 50 mm (0.125 in. to 2 in.), although longer fibers [75 mm (3 in.)] may be used as well. Sheet-molding compounds can also be used in molding; these compounds are similar to BMCs, except that the resin–fiber mixture is laid between plastic sheets to make a sandwich that can be easily handled. The sheets are removed when the SMC is placed in the mold.

2. In **vacuum-bag molding** (Fig. 10.36), prepregs are laid in a mold to form the desired shape. The pressure required to form the shape and to develop good bonding is obtained by covering the layup with a plastic bag and creating a vacuum. If additional heat and pressure are desired, the entire assembly is put in an *autoclave*. Care should be exercised to maintain fiber orientation, if specific fiber orientations are desired. In materials with chopped fibers, no specific orientation is intended. In order to prevent the resin from sticking to the vacuum bag, and to facilitate removal of any excess, several sheets of various materials (*release cloth* or *bleeder cloth*) are placed on top of the prepreg sheets. The

**FIGURE 10.36**   (a) Vacuum-bag forming. (b) Pressure-bag forming. *Source*: T. H. Meister.

**FIGURE 10.37**   Manual methods of processing reinforced plastics: (a) hand lay-up and (b) spray-up. These methods are also called *open-mold processing*.

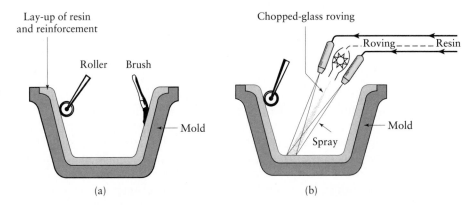

molds can be made of metal, usually aluminum, but more often are made from the same resin (with reinforcement) as the material to be cured. This condition eliminates any problem with differential thermal expansion between the mold and the part.

3. **Contact-molding** processes use a single male or female mold (Fig. 10.37) made of materials such as reinforced plastics, wood, or plaster. Contact molding is used in making products with high surface-area-to-thickness ratios, such as swimming pools, boats, tub and shower units, and housings. This process is a "wet" method, in which the reinforcement is impregnated with the resin at the time of molding. The simplest method is called **hand lay-up**. In this method, the materials are placed and formed in the mold by hand (Fig. 10.37a), and the squeezing action expels any trapped air and compacts the part.

   Molding may also be done by spraying (**spray-up**; see Fig. 10.37b). Although spraying can be automated, these processes are relatively slow, and labor costs are high. However, they are simple, and the tooling is inexpensive. Only the mold-side surface of the part is smooth, and the choice of materials is limited. Many types of boats are made by this process.

4. **Resin transfer molding** is based on transfer molding (Section 10.10.7) whereby a resin, mixed with a catalyst, is forced by a piston-type positive-displacement pump into a mold cavity filled with fiber reinforcement. The process is a viable alternative to hand lay-up, spray-up, and compression molding for low- or intermediate-volume production.

5. **Transfer/injection molding** is an automated process that combines compression molding, injection molding, and transfer molding. It combines the advantages of each process and produces parts with enhanced properties.

(a)

(b)

**FIGURE 10.38**
(a) Schematic illustration of the filament-winding process. (b) Fiberglass being wound over aluminum liners for slide-raft inflation vessels for the Boeing 767 aircraft. *Source*: Advanced Technical Products Group, Inc., Lincoln Composites.

## 10.11.2 Filament winding, pultrusion, and pulforming

**Filament winding.** *Filament winding* is a process whereby the resin and fibers are combined at the time of curing. Axisymmetric parts, such as pipes and storage tanks, as well as asymmetric parts are produced on a rotating mandrel. The reinforcing filament, tape, or roving is wrapped continuously around the form. The reinforcements are impregnated by passing through a polymer bath (Fig. 10.38a). The process can be modified by wrapping the mandrel with prepreg material.

The products made by filament winding are very strong, because of their highly reinforced structure. Filament winding has also been used for strengthening cylindrical or spherical pressure vessels (Fig. 10.38b) made of materials such as aluminum or titanium. The presence of a metal inner lining makes the part impermeable. Filament winding can be used directly over solid-rocket propellant forms. Seven-axis computer-controlled machines (see Chapter 14) have been developed for making asymmetric parts that automatically dispense several unidirectional prepregs. Typical asymmetric parts made include aircraft engine ducts, fuselages, propellers, blades, and struts.

**Pultrusion.** Long shapes with various constant profiles, such as rods, structural profiles, and tubing (similar to drawn metal products), are made by the *pultrusion* process. Typical products made by this process include golf clubs; drive shafts; and structural members such as ladders, walkways, and handrails. In this process, developed in the early 1950s, the continuous reinforcement (roving or fabric) is pulled through a thermosetting-polymer bath, and then through a long heated steel die (Fig. 10.39). The product is cured during its travel through the die and then cut into desired lengths. The most common material used in pultrusion is polyester with glass reinforcements.

**FIGURE 10.39** Schematic illustration of the pultrusion process. This figure shows a microwave heating arrangement, although curing can be performed in a heated die.

**Pulforming.** Continuously reinforced products other than profiles with a constant cross-section are made by *pulforming*. After being pulled through the polymer bath, the composite is clamped between the two halves of a die and cured into a finished product. The dies recirculate and shape the products successively. Common examples of products made by this method include glass-fiber-reinforced hammer handles and curved automotive leaf springs.

### 10.11.3  Product quality

The major quality considerations for the processes described thus far involve internal voids and gaps between successive layers of material. Volatile gases that develop during processing must be allowed to escape from the lay-up through the vacuum bag to avoid porosity due to trapped gases within the lay-up. Microcracks may develop due to improper curing or during transportation and handling. These defects can be detected using ultrasonic scanning and other techniques. (See Section 4.8.1.)

## 10.12 | Rapid Prototyping and Rapid Tooling

Making a prototype, that is, a first full-scale model of a product (see Fig. 1.2), has traditionally involved actual manufacturing processes, using a variety of tooling and machines and usually taking weeks or months. An important advance in manufacturing is *rapid prototyping*, also called **desktop manufacturing** or **free-form fabrication**, a process by which a solid physical model of a part is made directly from a three-dimensional CAD drawing. Developed in the mid-1980s, rapid prototyping entails several different consolidation techniques: resin curing, deposition, solidification, and sintering.

The importance and economic impact of rapid prototyping can be appreciated by noting the following:

a) The conceptual design is viewed in its entirety and at different angles on the monitor through a three-dimensional CAD system, as described in Section 15.4;

b) A prototype from various nonmetallic and metallic materials is manufactured and studied thoroughly from functional, technical, and esthetic aspects; and

c) Prototyping is accomplished in a much shorter time and at lower cost than by traditional methods.

It should be noted, however, that modern CNC machine tools (Chapters 8 and 15) also provide the capability of producing complex shapes quickly and are a viable option for rapid prototyping.

This section emphasizes use of **additive manufacturing**, that is, processes that build parts in layers. This technique inherently involves integrated computer-driven hardware and software. As an example, Fig. 10.40 shows the computational steps involved in producing a part. Part production in rapid prototyping takes hours for most parts. Therefore, these processes are not viable for large production runs, especially since the workpiece material is often an expensive polymer or laminate. Table 10.7 summarizes characteristics of the rapid-prototyping process, and Table 10.8 lists typical material properties that are attainable via this method.

### 10.12.1  Stereolithography

The *stereolithography* (STL) process is based on the principle of curing (hardening) a *liquid photopolymer* into a specific shape. Consider what happens when a laser beam is focused on and translated across the surface of a liquid photopolymer. The laser serves to cure the polymer, providing the energy necessary for polymerization.

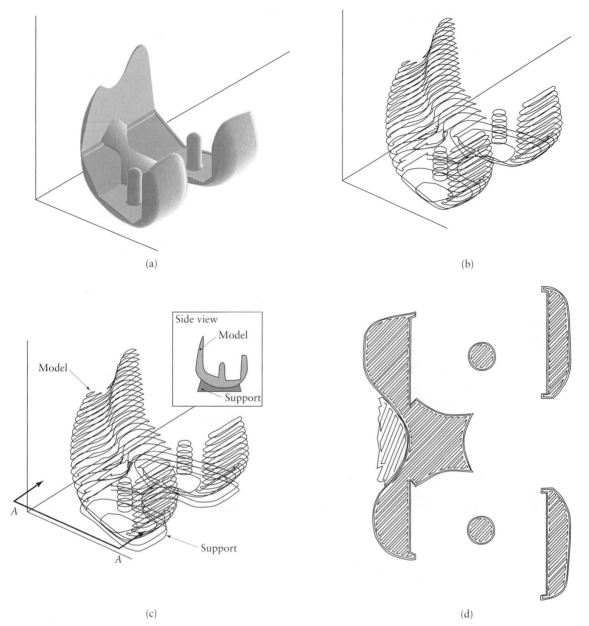

(a)

(b)

(c)

(d)

**FIGURE 10.40** The computational steps in producing a stereolithography file. (a) Three-dimensional decription of the part. (b) The part is divided into slices. (Only 1 in 10 is shown.) (c) Support material is planned. (d) A set of tool directions is determined for manufacturing each slice. Shown is the extruder path at section *A–A* from (c), for a fused-deposition modeling operation.

Laser energy is absorbed by the polymer, but according to the Beer–Lambert law, the exposure decreases exponentially with depth according to the rule

$$E(z) = E_0^{-z/D_p}, \tag{10.25}$$

where $E$ is the exposure in energy per area, $E_0$ is the exposure at the resin surface $(z = 0)$, and $D_p$ is the "penetration depth" at the laser wavelength and is a property of the resin. At the cure depth, the polymer is sufficiently exposed for it to gel, or

$$E_c = E_0^{-C_d/D_p}, \tag{10.26}$$

TABLE 10.7

**Characteristics of Rapid-Prototyping Processes**

| Supply Phase | Process | Layer Creation Technique | Phase-Change Type | Materials |
|---|---|---|---|---|
| Liquid | Stereolithography | Liquid-layer curing | Photopolymerization | Photopolymers (acrylates, epoxies, colorable resins, and filled resins) |
| | Solid-base curing | Liquid-layer curing and milling | Photopolymerization | Photopolymers |
| | Fused-deposition modeling | Extrusion of melted plastic | Solidification by cooling | Thermoplastics (ABS, polycarbonate, and elastomer) and wax |
| | Ballistic-particle manufacturing | Droplet deposition | Solidification by cooling | Polymers and wax |
| Powder | Three-dimensional printing | Binder-droplet deposition onto powder layer | No phase change | Ceramic; polymer and metal powder with binder |
| | Selective laser sintering | Layer of powder | Laser-driven sintering or melting | Polymers, metals with binder, metals, ceramics, and sand with binder |
| Solid | Laminated-object manufacturing | Deposition of sheet material | No phase change | Paper and polymers |

TABLE 10.8

**Mechanical Properties of Selected Materials for Rapid Prototyping**

| Process | Material | Tensile Strength (MPa) | Elastic Modulus (GPa) | Elongation in 50 mm (%) |
|---|---|---|---|---|
| Stereo-lithography | SL 5180[a] | 55–65 | 2.4–2.6 | 6–11 |
| | SL 5195[a] | 46.5 | 2.1 | 11 |
| | SL 5510[b] | 73 | 2.8 | 7.9 |
| | SL 7940[b] | 37–39 | 1.3 | 18–21 |
| Fused-deposition modeling | Polycarbonate | 62 | – | 100 |
| | ABS | 35 | 2.5 | 50 |
| Selective laser sintering | Nylon | 36 | 1.4 | 32 |
| | Polycarbonate | 23.4 | 1.2 | 5 |
| | Polyamide | 44 | 1.6 | 9 |
| | SOMOS 201 | 17.3 | 14 | 130 |
| | ST-100[c] | 305 | 137 | 10 |

[a.] After a 90-min UV cure.
[b.] After a 90-min UV cure at 80°C.
[c.] Sintered and bronze-infiltrated steel powder.

where $E_c$ is the critical threshold exposure and $C_d$ is the cure depth. Thus, the cure depth is given by

$$C_d = D_p \ln\left(\frac{E_0}{E_c}\right). \tag{10.27}$$

The cure depth represents the thickness in which the resin has polymerized into a gel, but this state does not have high strength. Recognizing this condition, the controlling

**FIGURE 10.41**   Schematic illustration of the stereolithography process. *Source*:  Courtesy of A. S. Alpert, 3D Systems.

software slightly overlaps the cured volumes, but curing under fluorescent lamps is often necessary as a finishing operation.

The polymer at the periphery of the laser spot does not receive sufficient exposure to polymerize. It can be shown that the cured line width $L_w$ at the surface is given by

$$L_w = B\sqrt{\frac{C_d}{2D_p}}, \tag{10.28}$$

where $B$ is the diameter of the laser-beam spot.

Stereolithography (Fig. 10.41) involves a vat that contains a mechanism whereby a platform can be lowered and raised vertically and that is filled with a photocurable liquid acrylate polymer. The liquid is a mixture of acrylic monomers, oligomers (polymer intermediates), and a photoinitiator. When the platform is at its highest position, the layer of liquid above it is shallow. A *laser*, generating an ultraviolet beam, is then focused along a selected surface area of the photopolymer at surface *a* and moved in the *x–y*-direction.

The beam cures that portion of the photopolymer (say, a ring-shaped portion), producing a thin solid body. The platform is then lowered enough to cover the cured polymer with another layer of liquid polymer, and the sequence is repeated. In Fig. 10.41, the process is repeated until level *b* is reached. Hence, thus far we have a cylindrical part with a constant wall thickness. Note that the platform is now lowered by a vertical distance *ab*.

At level *b*, the *x–y* movements of the beam are wider, so that we now have a flange-shaped piece that is being produced over the previously formed part. After the proper thickness of the liquid has been cured, the process is repeated, producing another cylindrical section between levels *b* and *c*. Note that the surrounding liquid polymer is still fluid, because it has not been exposed to the ultraviolet beam, and that the part has been produced from the bottom up in individual "slices." The unused portion of the liquid polymer can be used again to make another part or another prototype. Note that the word *stereolithography*, as used to describe this process, comes from the fact that the movements are three dimensional and that the process is similar to lithography, in which the image to be printed on a flat surface is ink receptive and the blank areas are ink repellent. After completion, the part is removed from the platform, blotted, cleaned ultrasonically and with an alcohol bath, and subjected to a final curing cycle.

By controlling the movements of the beam and the platform through a servo-control system, a variety of parts can be formed by this process. Total cycle times range from a few hours to a day. Progress is being made toward improvements in (1) accuracy and dimensional stability of the prototypes produced, (2) less expensive liquid modeling materials, (3) CAD interfaces to transfer geometrical data to model-making systems, and (4) strength so that the prototypes made by this process can be truly

**FIGURE 10.42** Schematic illustration of the fused-deposition modeling process. *Source*: Courtesy of Stratasys, Inc.

considered prototypes and models in the traditional sense. Sloping surfaces tend to be relatively rough, because of the manner in which layer-by-layer curing takes place.

Depending on capacity, the cost of machines for stereolithography ranges from $100,000 to $500,000, while the cost of the liquid polymer is on the order of $300 per gallon. Maximum part size is 0.5 m × 0.5 m × 0.6 m (20 in. × 20 in. × 24 in.). The automotive, aerospace, electronics, and medical industries are among the industries that use stereolithography as a rapid and inexpensive method of producing prototypes and thus reducing product-development-cycle times. One major application of stereolithography is in the area of making molds and dies for casting and injection molding.

### 10.12.2 Fused-deposition modeling

In the *fused-deposition modeling* (FDM) process (Fig. 10.42), a gantry-robot-controlled extruder head moves in two principal directions over a table. The table can be raised and lowered as needed. A thermoplastic or wax filament is extruded through the small orifice of a heated die. The initial layer is placed on a foam foundation by extruding the filament at a constant rate while the extruder head follows a predetermined path (Fig. 10.40d). When the first layer is completed, the table is lowered so that subsequent layers can be superposed.

Occasionally, complicated parts are required, such as the part shown in Fig. 10.43a. This part is difficult to manufacture directly, because once the part has been constructed up to height *a*, the next slice would require the filament to be placed

**FIGURE 10.43** (a) A part with a protruding section that requires support material. (b) Common support structures used in rapid-prototyping machines. *Source*: Figure reprinted with permission of the Society of Manufacturing Engineers, from Rapid Prototyping & Manufacturing: Fundamentals of StereoLithography, copyright 1992.

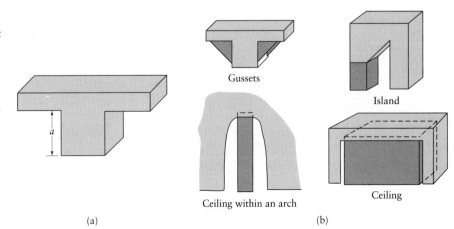

Gussets

Island

Ceiling within an arch

Ceiling

(a)

(b)

in a location where no material exists to support it. The solution is to extrude a support material separately from the modeling material, so that a filament can be placed safely in the center of the part. The support material is extruded with a less dense spacing of filament on a layer, so that it is weaker than the model material and can be broken off or dissolved after the part is completed.

In the FDM process, the extruded layer's thickness is determined by the extruder-die diameter, the diameters typically range from 0.33 mm to 0.12 mm (0.013 in. to 0.005 in.). This thickness represents the best achievable dimensional tolerance in the vertical direction. In the $x$–$y$-plane, dimensional accuracy can be as fine as 0.025 mm (0.001 in.), as long as a filament can be extruded into the feature. Close examination of an FDM-produced part will indicate that a stepped surface exists on oblique exterior planes. If this surface roughness is unacceptable, chemical-vapor polishing or a heated tool can be used to smoothen the surface, or a coating can be applied, often in the form of a polishing wax. However, the overall dimensional tolerances are then compromised, unless care is taken in these finishing operations.

## 10.12.3  Selective laser sintering

*Selective laser sintering* (SLS) is a process based on the sintering of polymer (or, less commonly, metallic) powders selectively into an individual object. (See also Chapter 11.) The basic elements in this process are shown in Fig. 10.44. The bottom of the processing chamber is equipped with two cylinders: a part-build cylinder, which is lowered incrementally to where the sintered part is formed, and a powder-feed cylinder, which is raised incrementally to supply powder to the part-build cylinder through a roller mechanism. A thin layer of powder is first deposited in the part-build cylinder. A laser beam, guided by a process-control computer (using instructions generated by the 3D CAD program of the desired part), is then focused on that layer, tracing and melting (or, for metals, *sintering*; see Section 11.4) a particular cross-section, which then quickly resolidifies into a solid mass (after the laser beam moves to another section). The powder in other areas remains loose, yet supports the solid portion. Another layer of powder is then deposited; this cycle is repeated continuously until the entire three-dimensional part has been produced. The loose particles are then shaken off, and the part is recovered.

A variety of materials can be used in this process, including polymers (ABS, PVC, nylon, polyester, polystyrene, and epoxy), wax, and metals, as well as ceramics

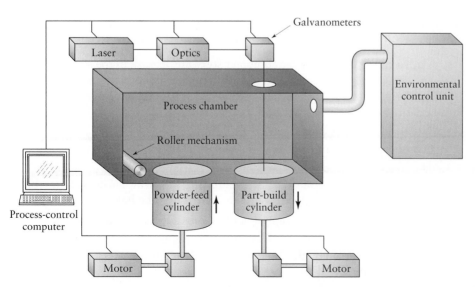

**FIGURE 10.44** Schematic illustration of the selective-laser-sintering process. *Source*: C. Deckard and P. F. McClure.

with appropriate binders. It is most common to use polymers, because of the smaller, less expensive, and less complicated lasers required for sintering. With ceramics and metals, it is common to sinter only a polymer binder that has been blended with the ceramic or metal powders and then to complete the sintering in a furnace.

### 10.12.4  Laminated-object manufacturing

Lamination implies a laying down of layers that are adhesively bonded to one another. *Laminated-object manufacturing* (LOM) uses layers of paper or plastic sheets with a heat-activated glue on one side to produce parts. The desired shapes are burned into the sheet with a laser beam, and the parts are built layer by layer. Once the part is completed, the excess material must be removed manually. This process is simplified by programming the laser beam to burn perforations in crisscrossed patterns; the resulting grid lines make the part appear as if it had been constructed from gridded paper. LOM uses sheets as thin as 0.05 mm (0.002 in.), although a thickness of 0.5 mm (0.02 in.) is more commonly used, so that it can achieve dimensional tolerances similar to those available from stereolithography and fused-deposition modeling. The compressed paper has the appearance and strength of soft wood, and it is often mistaken for elaborate wood carvings. The paper parts are subsequently easy to finish or coat.

### 10.12.5  Ballistic-particle manufacturing

In the *ballistic-particle-manufacturing* process, a stream of a material such as a plastic, a ceramic, a metal, or wax, is ejected through a small orifice at a surface (target), by use of an ink-jet-style mechanism. The mechanism uses a piezoelectric pump, which operates when an electric charge is applied, generating a shock wave that propels $50$-$\mu$m (0.002-in.) droplets at a rate of 10,000/s. The operation is repeated, in a manner similar to other processes, to form the part, with layers of wax deposited on top of each other. The ink-jet head is guided by a three-axis robot.

### 10.12.6  Solid-base curing

The *solid-base-curing* process, also called *solid-ground curing*, is unique in that entire slices of a part are manufactured at one time; as a result, a large throughput is achieved compared with that from other rapid-prototyping processes. This process is, however, among the most expensive, so its adoption has been less common than for other types of rapid prototyping.

The basic approach consists of the following steps:

1. Once a slice is created by the computer software, a mask of the slice is printed on a glass sheet, by use of an electrostatic printing process similar to that used in laser printers. A mask is required because the area of the slice where solid material is desired remains transparent.

2. While the mask is being prepared, a thin layer of photoreactive polymer is deposited on the work surface and spread evenly.

3. The mask is placed over the work surface, and an ultraviolet floodlight is projected through the mask. Wherever the mask is clear, the light shines through to cure the polymer and causes the desired slice to be hardened.

4. The unaffected resin, still liquid, is vacuumed off the surface.

5. Water-soluble liquid wax is spread across the work area, filling the cavities previously occupied by the unexposed liquid polymer. Since the workpiece is on a chilling plate and the workspace remains cool, the wax hardens quickly.

6. The layer is then milled (see Section 8.9.1) to achieve the correct thickness and flatness.

7. The process is repeated, layer by layer, until the part is completed.

Solid-base curing has the advantage of high production rate, both because entire slices are produced at once and because two glass screens are used concurrently. That is, while one mask is being used to expose the polymer, the next mask is already being prepared, and it is ready as soon as the milling operation is completed. The wax support is water soluble; it may be removed immediately, or it may remain in place as protection during shipping of the part. Use of the wax support means that any shape can be manufactured, including "ship-in-the-bottle" type of parts, for which removal of a solid support would be difficult. Usually, no finishing operation is required, but occasionally the polymer is cleaned in a slightly acidic solution to bleach the surface and to obtain a smoother finish.

### 10.12.7  Direct manufacturing and rapid tooling

The polymer parts that can be obtained from various rapid-prototyping operations are useful not only for design evaluation and troubleshooting, but also occasionally for manufacturing marketable products directly. Unfortunately, it is often desirable, for functional reasons, to use metallic parts, while the best developed and most available rapid-prototyping operations involve polymeric workpieces.

The solution is to use the components manufactured through rapid prototyping as aids in further processing. As an example, Fig. 10.45 shows an approach for investment casting. Here, the individual patterns are made in a rapid-prototyping operation (in this case, stereolithography) and then used as the blanks in assembling a tree for investment casting. It should be noted that this approach requires a polymer that will melt and burn from the ceramic mold completely; such polymers are available for all forms of polymer rapid-prototyping operations. Furthermore, the parts as drawn in CAD programs are usually software modified to account for shrinkage; it is the modified part that is produced in the rapid-prototyping machinery.

Rapid-prototyping processes are now being used to produce *engineering components* (*direct manufacturing*), but not economically for large quantities. The advantage of rapid prototyping is that it makes traditionally costly processes economical for very short production runs, often for merely one part. Several approaches have been devised for the rapid production of *tooling* by means of rapid-prototyping processes. For example, in the sand-casting operation, the pattern plates can be manufactured via rapid-prototyping approaches, which can then be applied to conventional sand casting (see Section 5.8.1). The advantage to this approach is the short time needed to produce the pattern plate compared with that in conventional manufacturing. The main shortcoming is the reduced pattern life, as compared with that obtained from machined high-strength metals.

Another common application is injection molding (see Section 10.10.2), where the mold (or, more typically, a *mold insert*) is manufactured via rapid prototyping and is used in injection-molding machines. Two approaches are used for the production of the tooling: Either a high-melting-point thermoplastic or a stable thermoset is used for low-temperature injection molding, or an investment-cast insert is produced.

Molds for slip casting of ceramics (see Section 11.9.1) can also be produced in this manner. To produce individual molds, rapid-prototyping processes are used directly, but the molds will be shaped with the desired permeability. For example, in fused-deposition modeling, this requirement mandates that the filaments be placed onto the individual slices with a small gap between adjacent filaments. These filaments are later positioned at right angles in adjacent layers. The advantage of rapid

Pattern creation

(a)

Tree assembly

(b)

Insert into
flask

(c)

Fill with
investment

(d)

Heat

Wax melt-out/
burnout

(e)

Crucible

Molten metal

Fill mold
with metal

(f)

Cool

(g)

Finish

(h)

**FIGURE 10.45**   Manufacturing steps for investment casting that uses rapid-prototyped wax parts as blanks. This approach uses a flask for the investment, but a shell method can also be used. *Source*: 3D Systems, Inc.

tooling is the capability to produce a mold (or mold insert) that can be used for the manufacture of plastic components without the time lag (several months) traditionally required for the procurement of tooling. Furthermore, the design is simplified, because the designer need only analyze a CAD file of the desired part. Software then produces the tool geometry and automatically compensates for shrinkage.

---

**EXAMPLE 10.8  Rapid prototyping of an injection manifold**

The Rover Group, a subsidiary of British Aerospace, is well known for four-wheel-drive vehicles and performance automobiles. An innovative new injection-manifold design (Fig. 10.46), based on a plenum chamber for efficient movement of air into the cylinders, was designed for a new Rover power unit. However, the chamber is very intricate and has large trapped volumes, and consequently a prototype had to be constructed from one piece to provide accurate flow characteristics.

**FIGURE 10.46** Rapid prototyped model of an injection-manifold design, produced through stereolithography. *Source:* 3D Systems.

Normally, Rover would have relied on traditional patternmaking followed by complex core-making and casting operations. However, this process is expensive and has a typical turnaround time of around 16 weeks. Using stereolithography, Rover produced a prototype multitrack injection manifold from a CAD file in only 39 hours, at less than 10% of the cost of the conventional prototype. More importantly, due to the model's strength and accuracy, it could be attached to an engine test bed, and its volumetric efficiency, i.e., its ability to allow the maximum air volume into the cylinder during one stroke under different conditions, could be directly measured. This procedure allowed optimization of the complex inlet tracts and plenum chamber and allowed Rover to incorporate design iterations into the product before committing to the purchase of expensive tooling.

*Source*: Courtesy of 3D Systems, Inc., and The Rover Group.

## 10.13 | Design Considerations

The design considerations for forming and shaping plastics are to a large extent similar to those for processing metals. However, the mechanical and physical properties of plastics should carefully be considered during design, material, and process selection. Selection of an appropriate material from an extensive list requires consideration of mechanical and physical properties; service requirements; possible long-range effects of processing on properties and behavior, such as dimensional stability and degradation; and ultimate disposal at the end of the product's life cycle. Compared with metals, plastics have lower strength and stiffness, although the strength-to-weight and stiffness-to-weight ratios for reinforced plastics are higher than for many metals. Thus, section sizes should be selected accordingly, with a view to maintaining a sufficiently high section modulus (the ratio of moment of inertia to the distance from the neutral axis to the surface of the part) for desired stiffness. Improper part design or assembly can lead to warping and shrinking (Fig. 10.47a). Plastics can easily be molded around metallic parts and inserts; however, their compatibility with metals when so assembled should be assessed.

The overall part shape often determines the choice of the particular forming or molding process. Even after a particular process is selected, the design of the part

**FIGURE 10.47** Examples of design modifications to eliminate or minimize distortion of plastic parts. (a) Suggested design changes to minimize distortion. *Source*: F. Strasser. (b) Die design (exaggerated) for extrusion of square sections. Without this design, product cross-sections would not have the desired shape, because of the recovery of the material, an effect known as *die swell*. (c) Design change in a rib to minimize pull-in caused by shrinkage during cooling. (d) Stiffening of the bottom of thin plastic containers by doming, which is similar to the process used to make the bottoms of

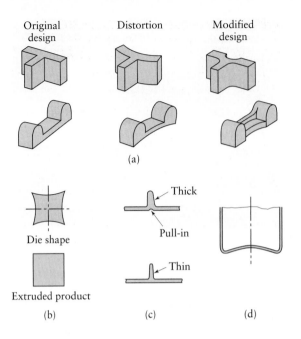

and die should be such that it will not cause problems concerning shape generation (Fig. 10.47b), dimensional control, and surface finish. As in casting metals and alloys (Chapter 5), material flow in the mold cavities should be properly controlled. The effects of molecular orientation during processing should also be considered, especially in extrusion, thermoforming, and blow molding.

Large variations in cross-section (Fig. 10.47c) and abrupt changes in geometry should be avoided in order to achieve better product quality and increased mold life. Furthermore, contraction in large cross-sections can cause porosity in plastic parts. Conversely, because of a lack of stiffness, removing thin sections from molds after shaping may be difficult. The low elastic modulus of plastics further requires that shapes be properly selected for improved stiffness of the component (Fig. 10.47d), particularly when it is important to save material. These considerations are similar to those in designing metal castings and forgings.

The properties of the final product depend on the original material and its processing history. Cold working of polymers improves their strength and toughness; on the other hand, because of the nonuniformity of deformation (even in simple rolling), residual stresses develop, as they do in metals. These stresses can also be generated by thermal cycling of the part. Also, the magnitude and direction of residual stresses, however the stresses are produced, can be important factors, as these stresses can relax over a period of time and cause distortion of the part during its service life, or during subsequent processing such as drilling; in addition, they can affect the part's resistance to adverse environmental conditions. (See Section 10.3.)

**Reinforced plastics.** The processing of reinforced plastics depends on the nature of the fibers, whether long, short, or chopped, as well as on the type of fibers and matrix materials. The presence of fibers makes processing of these composites more complex than that of thermoplastics and thermosets, including control of dimensional tolerances, and involves long cycle times. A major design advantage of reinforced plastics is the directional nature of the strength of the material. (See, for example, Fig. 10.19.) Forces applied to the material are transferred by the resin matrix to the fibers, which are much stronger and stiffer than the matrix. When fibers are all oriented in one direction, the resulting material is exceptionally strong in the fiber direction. This property is often used in designing reinforced plastic structures. For strength in two principal directions, the unidirectional materials are often laid at different angles to each other. If strength in the third (thickness) direction is desired, a different type of material is used to form a sandwich structure.

# 10.14 | Economics of Processing Plastics

As in all processes, design and manufacturing decisions are ultimately based on performance and cost, including the costs of equipment, tooling, and production; the final selection of a process depends greatly on production volume. High equipment and tooling costs in plastics processing can be acceptable only if the production run is large, as is the case in casting and forging. Table 10.9 provides a general guide to economical processing of plastics and composite materials. Several types of equipment are used in plastics forming and shaping processes; the most expensive is injection-molding machines, with costs being directly proportional to the clamping force. A machine with a 2000-kN (225-ton) clamping force costs about $100,000, and one with a 20,000-kN (2250-ton) clamping force costs about $450,000. For composite materials, equipment and tooling costs for most molding operations are generally high, and production rates and economic production quantities vary widely.

The optimum number of cavities in the die for making the product in one cycle is an important consideration, as in die casting (Section 5.10.3). For small parts, a number of cavities can be made in a die, with runners to each cavity. If the part is large, then only one cavity may be accommodated. As the number of cavities increases, so does the cost of the die; larger dies may be considered for larger numbers of cavities, thus increasing die cost even further. On the other hand, more parts will be produced per machine cycle on a larger die, thereby increasing the production rate. Hence, a detailed analysis has to be made to determine the optimum number of cavities, die size, and machine capacity.

**TABLE 10.9**

**Comparative Costs and Production Volumes for Processing of Plastics**

| | Equipment Capital Cost | Production Rate | Tooling Cost | Typical Production Volume, Number of Parts | | | | | | |
|---|---|---|---|---|---|---|---|---|---|---|
| | | | | 10 | $10^2$ | $10^3$ | $10^4$ | $10^5$ | $10^6$ | $10^7$ |
| Machining | Med | Med | Low | ⊢————————⊣ | | | | | | |
| Compression molding | High | Med | High | | | ⊢—————————————⊣ | | | | |
| Transfer molding | High | Med | High | | | ⊢—————————⊣ | | | | |
| Injection molding | High | High | High | | | | | ⊢—————————⊣ | | |
| Extrusion | Med | High | Low | * | | | | | | |
| Rotational molding | Low | Low | Low | | ⊢————⊣ | | | | | |
| Blow molding | Med | Med | Med | | | ⊢——————————⊣ | | | | |
| Thermoforming | Low | Low | Low | | ⊢————⊣ | | | | | |
| Casting | Low | Very low | Low | ⊢————⊣ | | | | | | |
| Forging | High | Low | Med | ⊢————⊣ | | | | | | |
| Foam molding | High | Med | Med | | | ⊢————————⊣ | | | | |

* Continuous process.

*Source*: After R. L. E. Brown, *Design and Manufacture of Plastic Parts*. Copyright © 1980 by John Wiley & Sons, Inc. Reprinted by permission of John Wiley & Sons, Inc.

**CASE STUDY** | The EPOCH Hip

Each year, over one million osteoarthritic joints are replaced with artificial joints worldwide; these devices generally eliminate pain and greatly improve the quality of life of their recipients. A total hip prosthesis replaces both the ball and the socket of the natural joint, in most cases, with a femoral component inserted in the upper end of the femur and an acetabular component placed in the pelvis. These implants have been produced since the 1950s, and can last for years before a revision is needed. Most modern implants are constructed from (a) Ti-6Al-4V alloy, (b) ASTM F-75 cobalt-chrome alloy, or (c) 316 stainless steel alloy, with a separate head made of an appropriate alloy and attached via a tapered seat on the femoral component, and a Ti-6Al-4V shell with a polymer insert for the acetabular component.

Total hips may be classified in two basic types: cemented and press fit. Cemented implants use a polymethyl methacrylate-based grout for fixation, whereas press-fit implants have a rough, porous surface intended for bone to grow into. They are fixed directly to bone. This rough surface can be manufactured by plasma spray operations, by sintering metal beads to the surface, or by diffusion bonding wires to the implant. These kinds of implants are often used in patients that are younger and more active, since their bone is better able to grow than those of older, more sedentary patients.

One of the problems that has been encountered with total hip replacements is that of bone resorption (loss of bone). Healthy bone must be stressed to maintain its strength and vigor. However, metal implants that are very stiff tend to transfer loads to the femur differently than the patient's natural hip does, and much of the bone toward the top of the femur experiences very little stress. The body reacts to this low-stress state by resorbing the bone. This loss of bone mass can leave the implant in an unsupported loading environment. This unsupported condition may limit the service life of the implant, leading to revision sooner than would otherwise be necessary. The revision may also require the use of bone graft to replace some of the resorbed bone and support the revision of the implant.

The main cause of stress shielding is that the stiffness of the metals used for the implant is much greater than that of bone. For example, the elastic modulus of Ti-6Al-4V is around 105 GPa, whereas bone has a modulus nearer to 20 GPa. Many polymers have a stiffness close to that of bone, but their fatigue strengths are not high enough to be used directly as implants.

A new and innovative implant design is the EPOCH hip system shown in Fig. 10.48. It consists of three layers: (a) an outermost layer comprised of a

**FIGURE 10.48** The EPOCH hip stem. This design uses a poly(aryletherketone) (PAEK) layer and bone ingrowth pad around a cobalt–chrome core in order to maximize bone ingrowth. The EPOCH Hip Stem is not available for sale in the United States. *Source:* Zimmer, Inc.

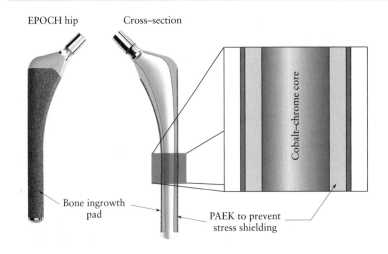

diffusion-bonded titanium-alloy mesh for bone ingrowth, (b) a thermoplastic polymer layer intended to provide a stiffness similar to bone, and (c) a cobalt-alloy core to provide static and fatigue strength. The polymer used in the Epoch hip is from the poly(aryletherketone) (PAEK) family, and the layers are proportioned to mimic bone stiffness while maximizing fatigue strength.

The Epoch hip is manufactured through an insert injection-molding operation (Fig. 10.49; see also Fig. 10.24). A vertical injection-molding machine is used (as opposed to a horizontal machine) in order to facilitate proper placement of the inserts. Two flexible pads are placed into seats on the mold halves. A machined cobalt-alloy insert is preheated and placed in the bottom half of the mold. The mold is then closed and the polymer is injected into the cavity defined by the mold halves, the flexible pads, and the insert. The polymer adheres to the cobalt-alloy insert and partially permeates the porous pad, integrally locking all components together. After the part is ejected from the mold, the flashing is trimmed and the hip stem is finished to obtain desired surface finish. A cross-section of the resulting structure is shown in Fig. 10.48.

In laboratory testing, the EPOCH hip has demonstrated flexibility near 20 GPa while meeting established strength requirements. This contribution of properties should assist in maintaining bone density without compromising performance expectations.

*Source*: M. Hawkins, Zimmer, Inc.

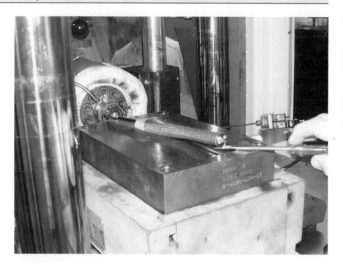

**FIGURE 10.49** An EPOCH hip is removed from the mold after an insert-injection-molding operation. The EPOCH Hip Stem is not available for sale in the United States. *Source*: Zimmer, Inc.

# SUMMARY

- Polymers are an important class of materials, because they possess a very wide range of mechanical, physical, chemical, and optical properties. Compared with metals, plastics are generally characterized by lower density, strength, elastic modulus, and thermal and electrical conductivity, and a higher coefficient of thermal expansion. (Section 10.1)

- Plastics are composed of polymer molecules and various additives. The smallest repetitive unit in a polymer chain is called a mer. Monomers are linked by polymerization processes to form larger molecules. The properties of the polymer depend on the molecular structure (linear, branched, cross-linked, or network), the degree of crystallinity, and additives (which perform various functions, such as improving strength, flame retardation, and lubrication, as well as imparting flexibility, color, and stability against ultraviolet radiation and oxygen). The glass-transition temperature separates the brittle and ductile regions of polymers. (Section 10.2)

- Thermoplastics become soft and are easy to form at elevated temperatures. Their mechanical behavior can be characterized by various spring and dashpot models and includes phenomena such as creep and stress relaxation, crazing, and water absorption. (Section 10.3)
- Thermosets, which are obtained by cross-linking polymer chains, do not become soft to any significant extent with increasing temperature. (Section 10.4)
- Thermoplastics and thermosets have a very wide range of consumer and industrial applications with a variety of characteristics. (Sections 10.5, 10.6)
- Among important developments in polymers is biodegradable plastics; several formulations have been developed. (Section 10.7)
- Elastomers have the characteristic ability to undergo large elastic deformations and return to their original shape when unloaded. Consequently, they have important applications as tires, seals, footwear, hoses, belts, and shock absorbers. (Section 10.8)
- Reinforced plastics are an important class of materials that have superior mechanical properties and are lightweight. The reinforcing fibers are usually glass, graphite, aramids, or boron; epoxies commonly serve as a matrix material. The properties of reinforced plastics depend on the composition of the plastic and the orientation of the fibers. (Section 10.9)
- Plastics can be formed and shaped by a variety of processes, such as extrusion, molding, casting, and thermoforming. Thermosets are generally molded or cast. (Section 10.10)
- Reinforced plastics are processed into structural components, using liquid plastics, prepregs, and bulk- and sheet-molding compounds. Fabricating techniques include various molding methods, filament winding, and pultrusion. The type and orientation of the fibers, and the strength of the bond between fibers and matrix and between layers of materials, are important factors. (Section 10.11)
- Rapid prototyping has become an increasingly important technology, because of its inherent flexibility and the much shorter times required for making prototypes. Typical techniques include stereolithography, fused-deposition modeling, selective laser sintering, laminated-object manufacturing, ballistic-particle manufacturing, and solid-base curing. Rapid production of tooling is a further development of these processes. (Section 10.12)
- The design of plastic parts includes considerations of their relatively low strength and stiffness, as well as high thermal expansion and generally low resistance to temperature. (Section 10.13)
- Because of the variety of low-cost materials and processing techniques available, the economics of processing plastics and reinforced plastics is an important consideration, particularly when compared with metal components, and includes die costs, cycle times, and production volume as important parameters. (Section 10.14)

## SUMMARY OF EQUATIONS

Linearly elastic behavior
  in tension, $\sigma = E\varepsilon$
  in shear, $\tau = G\gamma$

Viscous behavior, $\tau = \eta\left(\dfrac{dv}{dy}\right) = \eta\dot{\gamma}$

Strain-rate sensitivity, $\sigma = C\dot{\varepsilon}^m$

Viscoelastic modulus, $E_r = \sigma/(e_e + e_v)$

Viscosity,
$$\eta = \eta_0 e^{E/kT}$$

Extruder Characteristic,
$$Q = \frac{\pi^2 HD^2 N \sin\theta \cos\theta}{2} - \frac{p\pi DH^3 \sin^2\theta}{12\eta l}.$$

Die Characteristic, $Q_{\text{die}} = Kp$

For round die, $K = \dfrac{\pi D_d^4}{128\eta \ell_d}$,

Cure depth in stereolithography, $C_d = D_p \ln\left(\dfrac{E_0}{E_c}\right)$.

**10.19** What products have you personally seen that are made of reinforced plastics? How can you tell that they are reinforced?

**10.20** Identify metals and alloys that have strengths comparable with those of reinforced plastics.

**10.21** Compare the advantages and disadvantages of metal-matrix composites, reinforced plastics, and ceramic-matrix composites.

**10.22** You have studied the many advantages of composite materials in this chapter. What limitations or disadvantages do these materials have? What suggestions would you make to overcome these limitations?

**10.23** A hybrid composite is defined as a material containing two or more different types of reinforcing fibers. What advantages would such a composite have over other composites.

**10.24** Why are fibers capable of supporting a major portion of the load in composite materials?

**10.25** Assume that you are manufacturing a product in which all the gears are made of metal. A salesperson visits you and asks you to consider replacing some of the metal gears with plastic ones. Make a list of the questions that you would raise before making such a decision.

**10.26** Review the three curves in Fig. 10.8, and name applications for each type of behavior. Explain your choices.

**10.27** Repeat Question 10.26 for the curves in Fig. 10.10.

**10.28** Do you think that honeycomb structures could be used in passenger cars? If so, where? Explain.

**10.29** Other than those described in this chapter, what materials can you think of that can be regarded as composite materials?

**10.30** What applications for composite materials can you think of in which high thermal conductivity would be desirable?

**10.31** Make a survey of a variety of sports equipment, and identify the components that are made of composite materials. Explain the reasons for and advantages of using composites for these specific applications.

**10.32** We have described several material combinations and structures in this chapter. In relative terms, identify those that would be suitable for applications involving one of the following: (1) very low temperatures; (2) very high temperatures; (3) vibrations; (4) high humidity.

**10.33** Explain how you would go about determining the hardness of the reinforced plastics and composite materials described in this chapter. Are hardness measurements for these types of materials meaningful? Does the size of the indentation make a difference in your answer? Explain.

**10.34** Describe the advantages of applying traditional metalworking techniques to the formation of plastics.

**10.35** Describe the advantages of cold forming of plastics over other processing methods.

**10.36** Explain the reasons that some forming processes are more suitable for certain plastics than for others.

**10.37** Would you use thermosetting plastics for injection molding? Explain.

**10.38** By inspecting plastic containers, such as for baby powder, you can see that the lettering on them is raised rather than sunk. Can you offer an explanation as to why they are molded in that way?

**10.39** Give examples of several parts that are suitable for insert molding. How would you manufacture these parts if insert molding were not available?

**10.40** What manufacturing considerations are involved in making a metal beverage container versus a plastic one?

**10.41** Inspect several electrical components, such as light switches, outlets, and circuit breakers, and describe the process or processes used in making them.

**10.42** Inspect several similar products that are made of metals and plastics, such as a metal bucket and a plastic bucket of similar shape and size. Comment on their respective thicknesses, and explain the reasons for their differences, if any.

**10.43** Make a list of processing methods used for reinforced plastics. Identify which of the following fiber orientation and arrangement capabilities each has: (1) uniaxial, (2) cross-ply, (3) in-plane random, and (4) three-dimensional random.

**10.44** Some plastic products have lids with integral hinges; that is, no other material or part is used at the junction of the two parts. Identify such products, and describe a method for making them.

**10.45** Explain why operations such as blow molding and film-bag making are done vertically and why buildings that house equipment for these operations have ceilings 10 m to 15 m (35 ft to 50 ft) high.

**10.46** Consider the case of a coffee mug being produced by rapid prototyping. How can the top of the handle be manufactured, since there is no material directly beneath the arch?

**10.47** Make a list of the advantages and disadvantages of each of the rapid-prototyping operations.

**10.48** Explain why finishing operations are needed for rapid-prototyping operations. If you are making a prototype of a toy automobile, list the post-rapid-prototyping finishing operations you would perform.

**10.49** A current topic of research involves producing parts from rapid-prototyping operations and then using them in experimental stress analysis, in order to infer the strength of final parts produced by means of conventional manufacturing operations. List your concerns with this approach, and outline means of addressing these concerns.

**10.50** Because of relief of residual stresses during curing, long unsupported overhangs in parts from stereolithography will tend to curl. Suggest methods of controlling or eliminating this curl.

**10.51** One of the major advantages of stereolithography is that it can use semitransparent polymers, so that internal details of parts can be readily discerned. List parts for which this feature is valuable.

**10.52** Based on the processes used to make fibers, explain how you would produce carbon foam. How would you make a metal foam?

**10.53** Die swell in extrusion is radially uniform for circular cross-sections, but is not uniform for other cross-sections. Recognizing this fact, make a qualitative sketch of a die profile that will produce (a) square and (b) triangular cross-sections of extruded polymer.

**10.54** What are the advantages of using whiskers as a reinforcing material?

**10.55** By incorporating small amounts of blowing agent, it is possible to manufacture polymer fibers with gas cores. List some applications for such fibers.

**10.56** With injection-molding operations, it is common practice to remove the part from its runner and then to place the runner into a shredder and recycle the resultant pellets. List the concerns you would have in using such recycled pellets as opposed to so-called virgin pellets.

# PROBLEMS

**10.57** Calculate the areas under the stress–strain curve (toughness) for the material in Fig. 10.9, plot them as a function of temperature, and describe your observations.

**10.58** Note in Fig. 10.9 that, as expected, the elastic modulus of the polymer decreases as temperature increases. Using the stress–strain curves in the figure, make a plot of the modulus of elasticity versus temperature.

**10.59** Calculate the percentage increase in mechanical properties of reinforced nylon from the data shown in Fig. 10.17.

**10.60** A rectangular cantilever beam 100 mm high, 25 mm wide, and 1 m long is subjected to a concentrated force of 100 N at its end. Select two different unreinforced and reinforced materials from Table 10.1, and calculate the maximum deflection of the beam. Then select aluminum and steel, and for the same beam dimensions, calculate the maximum deflection. Compare the results.

**10.61** In Sections 10.5 and 10.6, we listed several plastics and their applications. Rearrange this information, respectively, by making a table of products and the type of plastics that can be used to make the products.

**10.62** Determine the dimensions of a tubular steel drive shaft for a typical automobile. If you now replace this shaft with shafts made of unreinforced and reinforced plastic, respectively, what should be the shaft's new dimensions to transmit the same torque for each

case? Choose the materials from Table 10.1, and assume a Poisson's ratio of 0.4.

**10.63** Calculate the average increase in the properties of the plastics listed in Table 10.1 as a result of their reinforcement, and describe your observations.

**10.64** In Example 10.4, what would be the percentage of the load supported by the fibers if their strength is 1250 MPa and the matrix strength is 240 MPa? What if the strength is unaffected, but the elastic modulus of the fiber is 600 GPa while the matrix is 50 GPa?

**10.65** Estimate the die clamping force required for injection molding ten identical 2-in.-diameter disks in one die. Include the runners of appropriate length and diameter.

**10.66** A two-liter plastic beverage bottle is made from a parison with the same diameter as the threaded neck of the bottle and has a length of 5 in. Assuming uniform deformation during blow molding, estimate the wall thickness of the tubular section.

**10.67** Estimate the consistency index and power-law index for the polymers in Fig. 10.12.

**10.68** An extruder has a barrel diameter of 75 mm. The screw rotates at 100 rpm, has a channel depth of 6 mm, and a flight angle of 17.5°. What is the largest flow rate of polypropylene that can be achieved?

**10.69**    The extruder in Problem 10.68 has a pumping section that is 3 m long and is used to extrude round polyethylene rod. The die has a land of 1 mm and a diameter of 5 mm. If the polyethylene is at a mean temperature of 250°C, what is the flow rate through the die? What if the die diameter is 10 mm?

**10.70**    An extruder has a barrel diameter of 4 in., a channel depth of 0.2 in., a flight angle of 18°, and a pumping zone that is 6 ft long. It is used to pump a plastic with a viscosity of $100 \times 10^{-4}$ lb-s/in$^2$. If the die characteristic is experimentally determined as $Q_x = (0.00210 \text{ in}^5/\text{lb-s})\, p$, what screw speed is needed to achieve a flow rate of 6 in$^3$/s from the extruder?

**10.71**    What flight angle should be used on a screw so that a flight translates a distance equal to the barrel diameter with every revolution?

**10.72**    For a laser providing 12.5 kJ of energy to a spot with diameter of 0.25 mm, determine the cure depth and the cured line width in stereolithography. Use $E_c = 6.36 \times 10^{10}$ J/m$^2$ and $D_p = 100\mu$m.

**10.73**    For the stereolithography system described in Problem 10.72, estimate the time required to cure a layer defined by a 50-mm (2 in.) circle if adjacent lines overlap each other by 10% and the power available is 10 MW.

**10.74**    The extruder head in a fused-deposition-modeling setup has a diameter of 1.25 mm (0.05 in.) and produces layers that are 0.25 mm (0.01 in.) thick. If the velocities of the extruder head and polymer extrudate are both 50 mm/s, estimate the production time for the generation of a 50-mm (2-in.) solid cube. Assume that there is a 15-s delay between layers as the extruder head is moved over a wire brush for cleaning.

# DESIGN

**10.75**    Discuss the design considerations involved in replacing a metal beverage container with a container made of plastic.

**10.76**    Using specific examples, discuss the design issues involved in products made of plastics versus reinforced plastics.

**10.77**    Make a list of products, parts, or components that are not currently made of plastics, and offer some reasons that they are not.

**10.78**    In order to use a steel or aluminum container to hold an acidic material, such as tomato juice or sauce, the inside of the container is coated with a polymeric barrier. Describe the methods of producing such a container. (See also Chapter 7.)

**10.79**    Using the information given in this chapter, develop special designs and shapes for possible new applications of composite materials.

**10.80**    Would a composite material with a strong and stiff matrix and soft and flexible reinforcement have any practical uses? Explain.

**10.81**    Make a list of products for which the use of composite materials could be advantageous because of their anisotropic properties.

**10.82**    Name several product designs in which both specific strength and specific stiffness are important.

**10.83**    Describe designs and applications in which strength in the thickness direction of a composite is important.

**10.84**    Design and describe a test method to determine the mechanical properties of reinforced plastics in their thickness direction.

**10.85**    As described in this chapter, reinforced plastics can be adversely affected by environmental factors, such as moisture, chemicals, and temperature variations. Design and describe test methods to determine the mechanical properties of composite materials under these conditions.

**10.86**    As with other materials, the mechanical properties of composites are obtained by preparing appropriate specimens and testing them. Explain what problems you might encounter in preparing specimens for testing and in the actual testing process itself.

# Properties and Processing of Metal Powders, Ceramics, Glasses, Composites, and Superconductors

11.1 Introduction
11.2 Production of Metal Powders
11.3 Compaction of Metal Powders
11.4 Sintering
11.5 Secondary and Finishing Operations
11.6 Design Considerations for Powder Metallurgy
11.7 Economics of Powder Metallurgy
11.8 Ceramics: Structure, Properties, and Applications
11.9 Shaping Ceramics
11.10 Glasses: Structure, Properties, and Applications
11.11 Forming and Shaping Glass
11.12 Design Considerations for Ceramics and Glasses
11.13 Graphite and Diamond
11.14 Processing Metal-Matrix and Ceramic-Matrix Composites
11.15 Processing Superconductors

## 11.1 | Introduction

In the manufacturing processes described in the preceding chapters, the raw materials used are either in a molten state or in solid form. This chapter describes how metal parts can be made by compacting metal powders in tool-steel or carbide dies and sintering them (heating without melting). This process is called **powder metallurgy (P/M)**. One of its first uses was in the early 1900s to make the tungsten filaments for incandescent light bulbs. The availability of a wide range of powder compositions, the capability to produce parts to net dimensions (**net-shape forming**), and the economics of the overall operation make this process attractive for many applications.

Typical products made by powder-metallurgy techniques include gears, cams, bushings, cutting tools, porous products such as filters and oil-impregnated bearings, and automotive components such as piston rings, valve guides, connecting rods, and hydraulic pistons (Fig. 11.1). Advances in technology now permit structural parts of aircraft, such as landing gears, engine-mount supports, engine disks, impellers, and engine nacelle frames, to be made by P/M.

Powder metallurgy has become competitive with processes such as casting, forging, and machining, particularly for relatively complex parts made of high-strength and hard alloys. Nearly 70% of P/M part production is for automotive applications. Parts made by this process have good dimensional accuracy, and their sizes range from tiny balls for ball-point pens to parts weighing about 50 kg (100 lb), although most parts made by P/M weigh less than 2.5 kg (5 lb). A typical family car now contains, on average, 13 kg (33 lb) of precision metal parts made by powder metallurgy, an amount that has been increasing by about 10% annually.

This chapter also describes the *structure, properties, and processing of ceramics, glasses, graphite, diamond, and metal-matrix and ceramic-matrix composites.* Because of the large number of possible combinations of elements, a great variety of these materials is now available for a wide range of consumer, industrial, and aerospace applications. There are several methods of producing these materials, as well as a number of techniques for subjecting them to additional processing for improved properties, dimensional tolerances, and surface finish.

(a)                                    (b)

**FIGURE 11.1**
(a) Examples of typical parts made by powder-metallurgy processes. (b) Upper trip lever for a commercial irrigation sprnkler, made by P/M. This part is made of unleaded brass alloy; it replaces a die-cast part, with a 60% cost savings. *Source*: Reproduced with permission from *Success Stories on P/M Parts*, Princeton, NJ: Metal Powder Industries Federation, 1998.

# 11.2 | Production of Metal Powders

The powder-metallurgy process basically consists of the following operations:

- Powder production,
- Blending,
- Compaction,
- Sintering, and
- Finishing operations.

For improved quality and dimensional accuracy, or for special applications, additional processing such as coining, sizing, machining, infiltration, and resintering may be carried out.

## 11.2.1 Methods of powder production

There are several methods of producing metal powder. Metal sources are generally bulk metals and alloys, ores, salts, and other compounds. Most metal powders can be produced by more than one method, the choice of which depends on the requirements of the end product. Particle sizes produced range from 0.1 $\mu$m to 1000 $\mu$m (0.04 $\mu$in. to 4 $\mu$in.). The shape, size distribution, porosity, chemical purity, and bulk and surface characteristics of the particles depend on the particular process used (Fig. 11.2). These characteristics are important, because they significantly affect flow during compaction and reactivity in subsequent sintering operations.

The methods of powder production are:

1. **Atomization.** *Atomization* produces a liquid-metal stream by injecting molten metal through a small orifice (Fig. 11.3a). The stream is broken up by jets of inert gas, air, or water. The size of the particles formed depends on the temperature of the metal, the rate of flow, nozzle size, and jet characteristics. Melt-atomization methods are widely used for production of powders for P/M. In one variation of this method, a consumable electrode is rotated rapidly in a helium-filled chamber (Fig. 11.3b). The centrifugal force breaks up the molten tip of the electrode, producing metal particles.

2. **Reduction.** *Reduction* of metal oxides (i.e., removal of oxygen) uses gases, such as hydrogen and carbon monoxide, as reducing agents; thus, very fine metallic oxides are reduced to the metallic state. The powders produced by this method are spongy and porous and have uniformly sized spherical or angular shapes.

**FIGURE 11.2** Particle shapes in metal powders and the processes by which they are produced. Iron powders are produced by many of these processes.

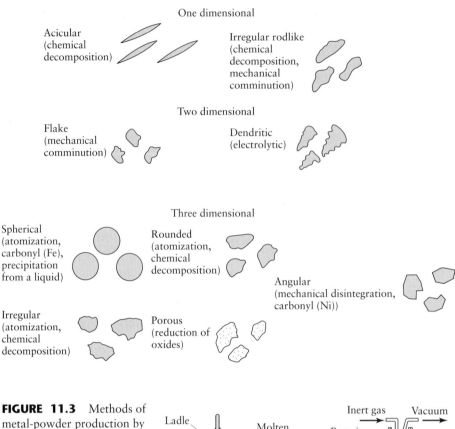

**FIGURE 11.3** Methods of metal-powder production by atomization: (a) melt atomization and (b) atomization with a rotating consumable electrode.

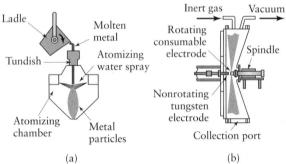

3. **Electrolytic deposition.** *Electrolytic deposition* uses either aqueous solutions or fused salts. The powders produced are among the purest of all metal powders.

4. **Carbonyls.** Metal *carbonyls*, such as iron carbonyl [$Fe(CO)_5$] and nickel carbonyl [$Ni(CO)_4$], are formed by letting iron or nickel react with carbon monoxide. The reaction products are then decomposed to iron and nickel, producing small, dense, and uniform spherical particles of high purity.

5. **Comminution.** Mechanical *comminution* (*pulverization*) involves crushing, milling in a *ball mill*, or grinding brittle or less ductile metals into small particles. A ball mill is a machine with a rotating hollow cylinder that is partly filled with steel or white cast-iron balls. When made from brittle materials, the powder particles have angular shapes, whereas when they are made from ductile metals, they are flaky and are not particularly suitable for powder-metallurgy applications.

6. **Mechanical alloying.** In *mechanical alloying*, powders of two or more pure metals are mixed in a ball mill. Under the impact of the hard balls, the powders repeatedly fracture and weld together by diffusion, forming alloy powders.

7. **Other methods.** Other, less commonly used methods include (a) **precipitation** from a chemical solution, (b) production of fine metal chips by **machining**, and (c) **vapor condensation**. New developments include techniques based on high-temperature *extractive metallurgical processes.* Metal powders are being produced using high-temperature processing techniques based on (a) the reaction of volatile halides (a compound of halogen and an electropositive element) with liquid metals and (b) the controlled reduction and reduction/carburization of solid oxides.

**Nanopowders.**  More recent developments include the production of *nanopowders* of various metals such as copper, aluminum, iron, and titanium. (See also the discussion of *nanomaterials* in Section 3.11.9.) When the metals are subjected to large plastic deformation at stress levels of 5500 MPa (800 ksi), their particle size is reduced, and the material becomes pore free and thus possesses enhanced properties.

**Microencapsulated powders.**  *Microencapsulated powders* are completely coated with a binder and are compacted by warm pressing. (See also the discussion of *metal injection molding* in Section 11.3.4.) In electrical applications, the binder acts like an insulator, preventing electricity from flowing and thus reducing eddy-current losses.

## 11.2.2  Particle size, distribution, and shape

Particle *size* is measured usually by screening, that is, by passing the metal powder through screens (sieves) of various mesh sizes. Screening is achieved by using a vertical stack of screens with increasing mesh size as the powder flows downward through the screens. The larger the mesh size, the smaller is the opening in the screen. For example, a mesh size of 30 has openings of 600 $\mu$m, size 100 has openings of 150 $\mu$m, and size 400 has openings of 38 $\mu$m. (This method is similar to the technique used for numbering abrasive grains—the larger the number, the smaller is the size of the abrasive particle; see Section 9.2.)

In addition to screen analysis, several other methods are also used for particle-size analysis, particularly for powders finer than 45 $\mu$m (0.0018 in.):

1. **Sedimentation,** which involves measuring the rate at which particles settle in a fluid;

2. **Microscopic analysis,** including the use of transmission and scanning electron microscopy;

3. **Light scattering** from a laser that illuminates a sample consisting of particles suspended in a liquid medium—the particles cause the light to be scattered, which is then focused on a detector that digitizes the signals and computes the particle-size distribution;

4. **Optical means,** such as having particles blocking a beam of light, which is then sensed by a photocell;

5. **Suspension of particles** in a liquid and subsequent detection of particle size and distribution by *electrical sensors.*

The *size distribution* of particles is an important consideration, since it affects the processing characteristics of the powder. The distribution of particle size is given

**FIGURE 11.4** (a) A plot of the distribution of particle size, given as weight percentage. The most populous size is termed the *mode*. In this case, it is between 75 $\mu$m and 90 $\mu$m. (b) Cumulative particle-size distribution as a function of weight. *Source*: Reprinted with permission from Randall M. German, *Powder Metallurgy Science*, Princeton, NJ: Metal Powder Industries Federation, 1984.

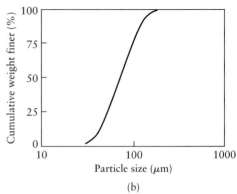

(a)       (b)

in terms of a *frequency-distribution* plot, as shown in Fig. 11.4a. (See Section 4.9.1 and Fig. 4.21a for a detailed description of this type of plot.) In this plot, note that the highest percentage of the particles (by weight) has a size in the range of 75 $\mu$m to 90 $\mu$m. This maximum is called the *mode size*. The same particle-size data are also plotted in the form of a *cumulative distribution* (Fig. 11.4b); thus, for example, about 75% of the particles, by weight, have a size finer than 100 $\mu$m. As expected from the distribution plot in Fig. 11.4b, the cumulative weight reaches 100% at a particle size of about 200 $\mu$m.

The *shape* of particles has a major influence on the particles' processing characteristics. The shape is usually described in terms of aspect ratio or shape index. *Aspect ratio* is the ratio of the largest dimension to the smallest dimension of the particle. This ratio ranges from 1 for a spherical particle to about 10 for flakelike or needlelike particles. *Shape index*, or **shape factor (SF)**, is a measure of the surface area to the volume of the particle with reference to a spherical particle of equivalent diameter; thus, the shape factor for a flake is higher than that for a sphere.

---

**EXAMPLE 11.1 Particle shape-factor determination**

Determine the shape factor for (1) a spherical particle, (2) a cubic particle, and (3) a cylindrical particle with a length-to-diameter ratio of 2.

**Solution.**

(1) In general, the ratio of surface area $A$ to volume $V$ can be expressed as

$$\frac{A}{V} = \frac{k}{D_{eq}},$$

where $k$ is the shape factor and $D_{eq}$ is the diameter of a sphere that has the same volume as the particle or object being considered. In the case of a spherical particle, its diameter $D = D_{eq}$, its area $A = \pi D^2$, and its volume $V = \pi D^3/6$. Thus,

$$k = \left(\frac{A}{V}\right)D_{eq} = \left[\frac{(\pi D^2)}{(\pi D^3/6)}\right]D = 6.$$

(2) The surface area of a cube of length $L$ is $6L^2$, and the cube's volume is $L^3$. Hence,

$$\frac{A}{V} = \frac{6L^2}{L^3} = \frac{6}{L}.$$

The particle's equivalent diameter is

$$D_{eq} = \left(\frac{6V}{\pi}\right)^{1/3} = \left(\frac{6L^3}{\pi}\right)^{1/3} = 1.24\,L.$$

Therefore, the shape factor is

$$k = \left(\frac{A}{V}\right)D_{eq} = \left(\frac{6}{L}\right)(1.24L) = 7.44.$$

(3) The surface area of a cylindrical particle with length-to-diameter ratio of 2 is

$$A = \left(\frac{2\pi D^2}{4}\right) + \pi DL = \left(\frac{\pi D^2}{2}\right) + 2\pi D^2 = 2.5\pi D^2.$$

The particle's volume is

$$V = \left(\frac{\pi D^2}{4}\right)(L) = \left(\frac{\pi D^2}{4}\right)(2D) = \frac{\pi D^3}{2},$$

and its equivalent diameter is

$$D_{eq} = \left(\frac{6V}{\pi}\right)^{1/3} = \left(\frac{6\pi D^3}{2\pi}\right)^{1/3} = 1.442D.$$

Hence, its shape factor is

$$k = \left(\frac{A}{V}\right)D_{eq} = \left[\frac{(2.5\pi D^2)}{(\pi D^3/2)}\right](1.442D) = 7.21.$$

## 11.2.3 Blending metal powders

*Blending* (mixing) powders is the second step in powder-metallurgy processing and is carried out for the following purposes:

a) Because the powders made by various processes may have different sizes and shapes, they must be mixed to obtain uniformity. The ideal mix is one in which all the particles of each material are distributed uniformly.

b) Powders of different metals and other materials may be mixed in order to impart special physical and mechanical properties and characteristics to the P/M product.

c) Lubricants may be mixed with the powders to improve the powders' flow characteristics. Such blends result in reduced friction between the metal particles, improved flow of the powder metals into the dies, and longer die life. The lubricants typically used are stearic acid and zinc stearate, in proportions of 0.25% to 5% by weight.

Powder mixing must be carried out under controlled conditions in order to avoid contamination and deterioration. Deterioration is caused by excessive mixing, which may alter the shape of the particles and work harden them, thus making the subsequent compacting operation more difficult. Powders can be mixed in air, in inert atmospheres (to avoid oxidation), or in liquids, which act as lubricants and make the mix more uniform; several types of blending equipment are available. These operations are being increasingly controlled by microprocessors to improve and maintain quality.

Because of their high surface-area-to-volume ratio, metal powders are explosive, particularly aluminum, magnesium, titanium, zirconium, and thorium powders. Great care must be exercised both during blending and during storage and handling. Precautions include grounding equipment, preventing the creation of sparks (by using nonsparking tools and avoiding the use of friction as a source of heat), and avoiding dust clouds, open flames, and chemical reactions.

## 11.3 | Compaction of Metal Powders

*Compaction* is the step in which the blended powders are pressed into shapes in dies (Figs. 11.5a and b), using presses that are either hydraulically or mechanically actuated. The purposes of compaction are to obtain the required shape, density, and particle-to-particle contact and to make the part strong enough to be processed further. The as-pressed powder is known as a **green compact**. The powder must flow easily to feed properly into the die cavity. Pressing is generally carried out at room temperature, although it can be done at elevated temperatures as well.

In powder metallurgy, one refers to *density* at three different stages: (1) as loose powder, (2) as a green compact, and (3) after sintering. The particle shape, average size, and size distribution dictate the packing density of loose powder. Spherical powder with a wide size distribution gives a high packing density. But a compact made from such powder has poor green strength, and, hence, such powder is unsuitable for P/M parts to be made by the die-pressing method (which is the most commonly used method of fabricating P/M parts). For P/M parts production, powders with some irregularity of shape (for example, water-atomized ferrous powders) are preferred, even though the fill density of powder in the die is somewhat lower than that of spherical powder. For reproducibility of part dimensions, the fill density of the powder should

**FIGURE 11.5**
(a) Compaction of metal powder to form a bushing. The pressed-powder part is called *green compact*.
(b) Typical tool and die set for compacting a spur gear. *Source*: Reprinted with the permission of the Metal Powder Industries Federation.

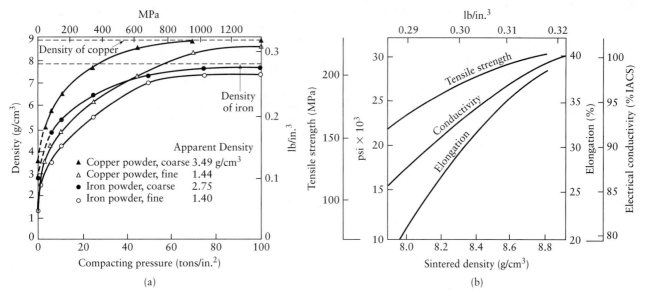

**FIGURE 11.6** (a) Density of copper- and iron-powder compacts as a function of compacting pressure. Density greatly influences the mechanical and physical properties of P/M parts. *Source*: F. V. Lenel, *Powder Metallurgy: Principles and Applications*, Princeton, NJ: Metal Powder Industries Federation, 1980. Reprinted by permission of Metal Powder Industries Federation, Princeton, NJ. (b) Effect of density on tensile strength, elongation, and electrical conductivity of copper powder. IACS means International Annealed Copper Standard for electrical conductivity.

be consistent from one powder lot to the next. Spherical powder is preferred for hot isostatic pressing, a special technique that is described in Section 11.3.3.

The density after compaction (**green density**) depends primarily on the compaction pressure, the metal powder composition, and the hardness of the powder (Fig. 11.6a). Higher pressure and softer powder give a higher green density. Furthermore, pure-iron powder will compact to a higher density than powders composed of alloyed steels. The green density and its uniformity within a compact improve with the addition of a small quantity of admixed (blended-in) lubricant.

To understand the effect of particle shape on green density, consider two powder grades with the same chemical composition and hardness, one with a spherical particle shape and the other with an irregular particle shape. The spherical grade will have a higher apparent (or fill) density, but after compaction at a higher pressure, compacts from both grades will have similar green densities. When comparing two similar powders pressed under some standard conditions, the powder that gives a higher green density is said to have a higher *compressibility*. (See also the discussion of sintered density in Section 11.4.)

The higher the density, the higher will be the strength and elastic modulus of the part (Fig. 11.6b). The reason is that the higher the density, the higher will be the amount of solid metal in the same volume, hence the greater will be the part's resistance to external forces. Because of friction between the metal particles in the powder and between the punches and the die walls, the density can vary considerably within the part. This variation can be minimized by proper punch and die design and by friction control; for example, it may be necessary to use multiple punches, with separate movements in order to ensure that the density is more uniform throughout the part (Fig. 11.7).

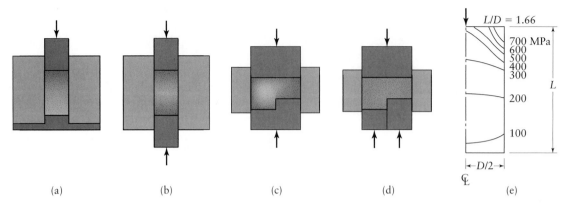

**FIGURE 11.7** Density variation in compacting metal powders in different dies: (a) and (c) single-action press; (b) and (d) double-action press. Note in (d) the greater uniformity of density in pressing with two punches with separate movements as compared with (c). Generally, uniformity of density is preferred, although there are situations in which density variation, and hence variation of properties, within a part may be desirable. (e) Pressure contours in compacted copper powder in a single-action press. *Source*: P. Duwez and L. Zwell.

---

**EXAMPLE 11.2  Density of metal-powder–lubricant mix**

Zinc stearate is a lubricant that commonly is mixed with metal powders prior to compaction, in proportions up to 2% by weight. Calculate the theoretical and apparent densities of an iron-powder–zinc-stearate mix, assuming that (1) 1000 g of iron powder is mixed with 20 g of lubricant, (2) the density of the lubricant is $1.10 \text{ g/cm}^3$, (3) the theoretical density of the iron powder is $7.86 \text{ g/cm}^3$ (from Table 3.3), and (4) the apparent density of the iron powder is $2.75 \text{ g/cm}^3$ (from Fig. 11.6a).

**Solution**  The volume of the mixture is

$$V = \left(\frac{1000}{7.86}\right) + \left(\frac{20}{1.10}\right) \simeq 145.41 \text{ cm}^3.$$

The combined weight of the mixture is 1020 g, so the mixture's theoretical density is $1020/145.41 = 7.01 \text{ g/cm}^3$. The apparent density of the iron powder is $2.75 \text{ g/cm}^3$, so it is $(2.75/7.86)100 = 35\%$ of the theoretical density. Assuming a similar percentage for the mixture, we can estimate the apparent density of the mix as $(0.35)(7.01) = 2.45 \text{ g/cm}^3$. (*Note*: Although g and $\text{g/cm}^3$ are not SI units, they are commonly used in the powder-metallurgy industry.)

---

### 11.3.1  Pressure distribution in powder compaction

Figure 11.7e shows that, in single-action pressing, the pressure decays rapidly toward the bottom of the compact. The pressure distribution along the length of the compact can be determined by using the slab method of analysis of deformation processes, described in Section 6.2.2. As done in Fig. 6.4, we first describe the operation in terms of its coordinate system, as shown in Fig. 11.8: $D$ is the diameter of the compact, $L$ is the compact's length, and $p_0$ is the pressure applied by the punch. We take an element $dx$ thick and place on it all the relevant stresses, namely, the compacting pressure, $p_x$, the die-wall pressure, $\sigma_r$, and the frictional stress, $\mu\sigma_r$. Note that the frictional stresses act upward on the element, because the punch movement is downward.

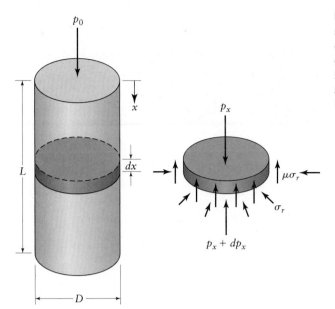

**FIGURE 11.8** Coordinate system and stresses acting on an element in compaction of powders. The pressure is assumed to be uniform across the cross-section. (See also Fig. 6.4.)

Balancing the vertical forces acting on this element, we have

$$\left(\frac{\pi D^2}{4}\right) p_x - \left(\frac{\pi D^2}{4}\right)(p_x + dp_x) - (\pi D)(\mu \sigma_r)\, dx = 0, \qquad (11.1)$$

which we can simplify to

$$D\, dp_x + 4\mu \sigma_r\, dx = 0.$$

We have one equation, but two unknowns ($p_x$ and $\sigma_r$). Let's now introduce the factor $k$, which is a measure of the interparticle friction during compaction:

$$\sigma_r = k p_x.$$

If there is no friction between the particles, $k = 1$, the powder behaves like a fluid, and we have $\sigma_r = p_x$, signifying a state of hydrostatic pressure. We now have the expression

$$dp_x + \frac{4\mu k p_x\, dx}{D} = 0,$$

or

$$\frac{dp_x}{p_x} = -\frac{4\mu k\, dx}{D}.$$

This expression is similar to that given in Section 6.2.2 for upsetting. In the same manner as in that section, we integrate this expression, noting that the boundary condition in this case is $p_x = p_0$ when $x = 0$:

$$p_x = p_0 e^{-4\mu k x/D}. \qquad (11.2)$$

Thus, the pressure within the compact decays as the coefficient of friction, the parameter $k$, and the length-to-diameter ratio increase.

**EXAMPLE 11.3** Pressure decay in compaction

Assume that a powder mix has $k = 0.5$ and $\mu = 0.3$. At what depth will the pressure in a straight cylindrical compact 10 mm in diameter become (a) zero and (b) one half the pressure at the punch?

**Solution.** For case (a) from Eq. 11.2, $p_x = 0$. Consequently, we have the expression

$$0 = p_0 e^{-(4)(0.3)(0.5)x/10}, \quad \text{or} \quad e^{-0.06x} = 0.$$

The value of $x$ must approach $\infty$ for the pressure to decay to 0. For case (b), we have $p_x/p_0 = 0.5$. Therefore,

$$e^{-0.06x} = 0.5, \quad \text{or} \quad x = 11.55 \text{ mm}.$$

In a practical case, a pressure drop of 50% is severe, as the compact density will then be unacceptably low. This example shows that, under the conditions assumed, uniaxial compaction of even a cylinder of length-to-diameter ratio of about 1.2 will be unsatisfactory.

## 11.3.2 Equipment

The pressure required for pressing metal powders ranges from 70 MPa (10 ksi) for aluminum to 800 MPa (120 ksi) for high-density iron parts (Table 11.1). The compacting pressure required depends on the characteristics and shape of the particles, the method of blending, and the lubrication.

Press capacities are on the order of 1.8–2.7 MN (200–300 tons), although presses with much higher capacities are used for special applications. Most applications require less than 100 tons. For small tonnage, crank or eccentric-type mechanical presses are used; for higher capacities, toggle or knucklejoint presses are employed. (See Fig. 6.28.) Hydraulic presses with capacities as high as 45 MN (5000 tons) can be used for large parts.

**TABLE 11.1**

**Compacting Pressures for Various Metal Powders**

| Metal | Pressure | |
|---|---|---|
| | MPa | psi $\times 10^3$ |
| Aluminum | 70–275 | 10–40 |
| Brass | 400–700 | 60–100 |
| Bronze | 200–275 | 30–40 |
| Iron | 350–800 | 50–120 |
| Tantalum | 70–140 | 10–20 |
| Tungsten | 70–140 | 10–20 |
| Other Materials | | |
| Auminum oxide | 110–140 | 16–20 |
| Carbon | 140–165 | 20–24 |
| Cemented carbides | 140–400 | 20–60 |
| Ferrites | 110–165 | 16–24 |

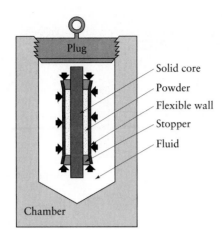

**FIGURE 11.9**  Schematic illustration of cold isostatic pressing as applied to formation of a tube. The powder is enclosed in a flexible container around a solid core rod. Pressure is applied isostatically to the assembly inside a high-pressure chamber. *Source*: Reprinted with permission from Randall M. German, *Powder Metallurgy Science*, Princeton, NJ: Metal Powder Industries Federation, 1984.

The selection of the press depends on part size and configuration, density requirements, and production rate. The higher the pressing speed, the greater is the tendency to trap air in the die cavity. Therefore, good die design, including provision of vents, is important so that trapped air does not hamper proper compaction.

### 11.3.3  Isostatic pressing

Compaction can also be carried out or improved by using a number of additional processes, such as isostatic pressing, rolling, and forging. Because the density of die-compacted powders can vary significantly, powders are subjected to *hydrostatic pressure* in order to achieve more uniform compaction. This process is similar to pressing your cupped hands together when making snowballs.

In **cold isostatic pressing** (CIP), the metal powder is placed in a flexible rubber mold made of neoprene rubber, urethane, polyvinyl chloride, or other elastomers (Fig. 11.9). The assembly is then pressurized hydrostatically in a chamber, usually with water. The most common pressure is 400 MPa (60 ksi), although pressures of up to 1000 MPa (150 ksi) have been used. The applications of CIP and other compacting methods, in terms of size and complexity of part, are shown in Fig. 11.10.

**FIGURE 11.10**
Capabilities of part size and shape complexity according to various P/M operations. *P/F* means powder forging. *Source*: Metal Powder Industries Federation.

**FIGURE 11.11** Schematic illustration of hot isostatic pressing. The pressure and temperature variation versus time are shown in the diagram. *Source*: Reprinted with permission from Randall M. German, *Powder Metallurgy Science*, Princeton, NJ: Metal Powder Industries Federation, 1984.

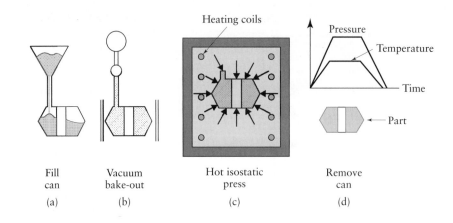

Fill can

(a)

Vacuum bake-out

(b)

Hot isostatic press

(c)

Remove can

(d)

In **hot isostatic pressing** (HIP), the container is usually made of a high-melting-point sheet metal, and the pressurizing medium is inert gas or vitreous (glasslike) fluid (Fig. 11.11). A common condition for HIP is 100 MPa (15 ksi) at 1100°C (2000°F), although the trend is toward higher pressures and temperatures. The main advantage of HIP is its ability to produce compacts with essentially 100% density, good metallurgical bonding among the particles, and good mechanical properties. The HIP process is relatively expensive and is used mainly in making superalloy components for the aerospace industry. It is routinely used as a final densification step for tungsten-carbide cutting tools and P/M tool steels. The process is also used to close internal porosity and improve properties in superalloy and titanium-alloy castings for the aerospace industry.

The main advantage of isostatic pressing is that, because of the uniformity of pressure from all directions and the absence of die-wall friction, it produces compacts of practically uniform grain structure and density, irrespective of shape. Parts with high length-to-diameter ratios have been produced by this method, with very uniform density, high strength, high toughness, and good surface details. The limitations of this method include wider dimensional tolerances produced, higher cost and time required, and relatively small production quantities output.

### 11.3.4 Other compacting and shaping processes

Other compacting and shaping processes used in powder metallurgy include:

a) **Metal injection molding.** In *metal injection molding* (MIM), very fine metal powders (generally <45 $\mu$m, and often <10 $\mu$m) are blended with a polymer or a wax-based binder. The mixture then undergoes a process similar to injection molding of plastics. (See Section 10.10.2.) The molded green parts are placed in a low-temperature oven to burn off the plastic, or else the binder is partially removed by solvent extraction, and then the green parts are sintered in a furnace. The term *powder injection molding* (PIM) is used with a broader reference to molding of both metal and ceramic powders.

   The major advantage of injection molding over conventional compaction is that complex shapes, with wall thicknesses as small as 0.5 mm (0.02 in.), can be molded and removed easily from the dies. The process is particularly well suited for small, complex parts required in large quantities; most commercially produced parts fabricated by this method weigh from a fraction of a gram to about 250 g (0.55 lb) each. Typical parts made by this method include components for guns, surgical instruments, automobiles, and watches. In spite of the high cost of fine metal powders, parts made by MIM are competitive with investment-cast parts (see Section 5.9.2), small forgings (Section 6.2), and complex machined parts (Chapter 8).

b) **Rolling.** In *powder rolling*, also called *roll compaction*, the powder is fed to the roll gap in a two-high rolling mill (see Fig. 6.42a) and is compacted into a continuous strip at speeds of up to 0.5 m/s (100 ft/min). The process can be carried out at room or elevated temperatures. Sheet metal for electrical and electronic components and for coins can be made by powder rolling.

c) **Extrusion.** Powders may be compacted by *hot extrusion*; in this process, the powder is encased in a metal container and extruded. (See Section 6.4.) Super-alloy powders, for example, are hot extruded for improved properties. Preforms made from the extrusions may be reheated and forged in a closed die to their final shape.

d) **Pressureless compaction.** In *pressureless compaction*, the die is filled with metal powder by gravity, and the powder is sintered directly in the die. Because of the resulting low density, pressureless compaction is used principally for porous parts, such as filters.

e) **Ceramic molds.** Molds for shaping metal powders are made by the technique used in investment casting. After the ceramic mold is made, it is filled with metal powder and placed in a steel container; the space between the mold and the container is filled with particulate material. The container is then evacuated, sealed, and subjected to hot isostatic pressing. Titanium-alloy compressor rotors for missile engines have been made by this process.

f) **Spray deposition.** This method is a shape-generation process in which the basic components are an atomizer, a spray chamber with inert atmosphere, and a mold for producing preforms; although there are several variations, the best known is the *Osprey* process (Fig. 11.12). After the metal is atomized, it is deposited on a cooled preform mold, usually made of copper or ceramic, where it solidifies. The metal particles weld together, developing a density that is normally above 99% of the solid metal density. The mold may be composed of various shapes, such as billets, tubes, disks, and cylinders. Spray-deposited preforms may be subjected to additional shaping and consolidation processes, such as forging, rolling, and extrusion. The grain size of the part is fine, and the part's mechanical properties are comparable with those of wrought products of the same alloy.

Induction-heated ladle
Atomizer (nitrogen gas)
Recipient substrate
Deposition chamber
Particle injector (optional)
Tube

**FIGURE 11.12** Spray casting (Osprey process) in which molten metal is sprayed over a rotating mandrel to produce seamless tubing and pipe. *Source*: After J. Szekeley, "Can Advanced Technology Save the U.S. Steel Industry," *Scientific American*, July 1987, © George V. Kelvin/Scientific American.

### 11.3.5 Punch and die materials

The selection of punch and die materials for P/M depends on the abrasiveness of the powder metal and the number of parts to be made. The most commonly used die materials are air- or oil-hardening tool steels, such as D2 or D3, with a hardness range of 60–64 HRC. Because of their greater hardness and wear resistance, tungsten-carbide dies are used for more severe applications. Punches are generally made of similar materials as those for dies. Close control of die and punch dimensions is essential for proper compaction and die life. Too large a clearance between the punch and the die will allow the metal powder to enter the gap, interfere with the operation, and result in eccentric parts. Diametral clearances are generally less than $25\ \mu m$ (0.001 in.). Die and punch surfaces must be lapped or polished in the direction of tool movements for improved die life and overall performance.

## 11.4 | Sintering

*Sintering* is the process whereby compressed metal powder is heated in a controlled-atmosphere furnace to a temperature below its melting point, but sufficiently high to allow bonding (fusion) of the individual particles. Prior to sintering, the compact is brittle, and its strength, known as **green strength**, is low. The nature and strength of the bond between the particles, and hence of the sintered compact, depend on the mechanisms of diffusion, plastic flow, evaporation of volatile materials in the compact, recrystallization, grain growth, and pore shrinkage.

The sintered density of a part depends mainly on the part's green density and on the sintering conditions, in terms of temperature, time, and furnace atmosphere. The sintered density increases with increasing values of temperature and time and usually with a more deoxidizing type of furnace atmosphere. For structural P/M parts, a higher sintered density is very desirable, as it leads to better mechanical properties. At the same time, in conventional P/M parts production, it is preferable to minimize the increase in density during sintering, for better dimensional accuracy. Better properties and accuracy can be achieved by using a powder with a high compressibility, i.e., a powder that will give a high green density and keep the sintering temperature moderate (not high). Such a powder also has another important benefit in that larger parts can be produced with a specific press tonnage. Therefore, a considerable amount of development work by powder producers is aimed at producing powder grades with higher compressibility.

Sintering temperatures (Table 11.2) are generally within 70% to 90% of the melting point of the metal or alloy. Sintering times range from a minimum of about 10 minutes for iron and copper alloys to as much as eight hours for tungsten and

TABLE 11.2

| **Sintering Temperature and Time for Various Metal Powders** | | |
|---|---|---|
| Material | Temperature (°C) | Time (min) |
| Copper, brass, and bronze | 760–900 | 10–45 |
| Iron and iron graphite | 1000–1150 | 8–45 |
| Nickel | 1000–1150 | 30–45 |
| Stainless steels | 1100–1290 | 30–60 |
| Alnico alloys (for permanent magnets) | 1200–1300 | 120–150 |
| Ferrites | 1200–1500 | 10–600 |
| Tungsten carbide | 1430–1500 | 20–30 |
| Molybdenum | 2050 | 120 |
| Tungsten | 2350 | 480 |
| Tantalum | 2400 | 480 |

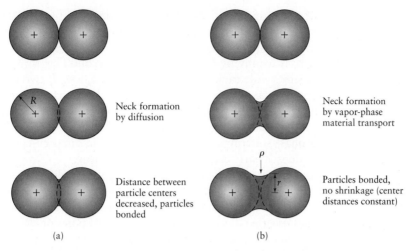

**FIGURE 11.13** Schematic illustration of two mechanisms for sintering metal powders: (a) solid-state material transport; and (b) liquid-phase material transport. $R$ = particle radius, $r$ = neck radius, and $\rho$ = neck profile radius.

tantalum. Continuous-sintering furnaces are used for most production today. These furnaces have three chambers: (1) a burn-off chamber to volatilize the lubricants in the green compact in order to improve bond strength and prevent cracking; (2) a high-temperature chamber for sintering; and (3) a cooling chamber.

Proper control of the furnace atmosphere is essential for successful sintering and to obtain optimum properties. An oxygen-free atmosphere is necessary in order to control the carburization and decarburization of iron and iron-base compacts and to prevent oxidation of powders. Oxide inclusions have a detrimental effect on mechanical properties. For the same volume of inclusions, the smaller inclusions have a larger effect because there are more of them per unit volume of the part. A vacuum is generally used for sintering refractory metal alloys and stainless steels. The gases most commonly used for sintering a variety of other metals are hydrogen, dissociated or burned ammonia, partially combusted hydrocarbon gases, and nitrogen.

Sintering mechanisms are complex and depend on the composition of the metal particles as well as the processing parameters (Fig. 11.13). As the temperature increases, two adjacent particles begin to form a bond by **diffusion** (*solid-state bonding*). As a result, the strength, density, ductility, and thermal and electrical conductivities of the compact increase (Fig. 11.14). At the same time, however, the compact shrinks; hence, allowances should be made for shrinkage, as in casting. (See Section 5.12.1.)

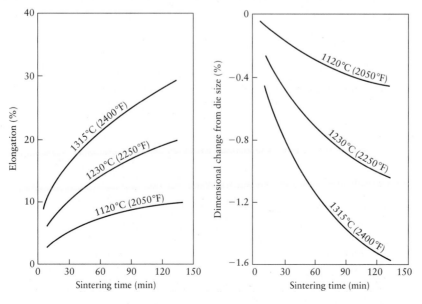

**FIGURE 11.14** Effect of sintering temperature and time on elongation and dimensional change during sintering of type 316L stainless steel. *Source*: ASM International.

A second sintering mechanism is **vapor-phase transport**. Because the material is heated very close to its melting temperature, metal atoms will release to the vapor phase from the particles. At convergent geometries (the interface of two particles), the melting temperature is locally higher, and the vapor resolidifies; thus, the interface grows and strengthens, while each particle shrinks as a whole. If two adjacent particles are of different metals, *alloying* can take place at the interface of the two particles. One of the particles may have a lower melting point than the other. In that case, one particle may melt and, because of surface tension, surround the particle that has not melted (**liquid-phase sintering**). An example is cobalt in tungsten-carbide tools and dies. Stronger and denser parts can be obtained in this way.

Depending on temperature, time, and processing history, different structures and porosities can be obtained in a sintered compact. However, porosity cannot be completely eliminated, because voids remain after compaction, and gases evolve during sintering. Porosities can consist of either a network of interconnected pores or closed holes. Their presence is an important consideration in making P/M filters and bearings.

Another method, which is at an experimental stage, is **spark sintering**. In this process, loose metal powders are placed in a graphite mold, heated by an electric current, subjected to a high-energy discharge, and compacted, all in one step. The rapid discharge strips contaminants or any oxide coating, such as that found on aluminum, from the surfaces of the particles and thus encourages good bonding during compaction at elevated temperatures.

Table 11.3 gives typical mechanical properties for several sintered P/M alloys. Note the effect of heat treating on the properties of metals. To evaluate the differences between the properties of P/M, wrought, and cast metals and alloys, compare this table with those in other chapters. Table 11.4 shows the effects of various processes on the mechanical properties of a particular titanium alloy. Note that hot isostatically pressed (HIP) titanium has properties that are similar to those for cast and forged titanium. It should be remembered, however, that forged components are likely to require additional machining processes (unless the components are precision forged to net shape), while P/M components may not. Therefore, powder metallurgy can be a competitive alternative to casting and forging.

---

**EXAMPLE 11.4  Shrinkage in sintering**

In solid-state bonding during sintering of a powder-metal green compact, the linear shrinkage is 4%. If the desired sintered density is 95% of the theoretical density of the metal, what should be the density of the green compact? Ignore the small changes in mass that occur during sintering.

**Solution.** Linear shrinkage is defined as $\Delta L/L_o$, where $L_o$ is the original part length. The volume shrinkage during sintering can then be expressed as

$$V_{\text{sint}} = V_{\text{green}}\left(1 - \frac{\Delta L}{L_o}\right)^3. \qquad (11.3)$$

The volume of the green compact must be larger than that of the sintered part. However, the mass does not change during sintering, so we can rewrite this expression in terms of the density $\rho$ as

$$\rho_{\text{green}} = \rho_{\text{sint}}\left(1 - \frac{\Delta L}{L_o}\right)^3. \qquad (11.4)$$

Therefore,

$$\rho_{\text{green}} = 0.95(1 - 0.04)^3 = 0.84, \qquad \text{or} \qquad 84\%.$$

## TABLE 11.3

### Typical Mechanical Properties of Selected P/M Materials

| Designation | MPIF type | Condition | Ultimate Tensile Strength (MPa) | Yield Stress (MPa) | Hardness | Elongation in 25 mm (%) | Elastic Modulus (GPa) |
|---|---|---|---|---|---|---|---|
| FERROUS | | | | | | | |
| FC-0208 | N | AS | 225 | 205 | 45 HRB | <0.5 | 70 |
| | | HT | 295 | — | 95 HRB | <0.5 | 70 |
| | R | AS | 415 | 330 | 70 HRB | 1 | 110 |
| | | HT | 550 | — | 35 HRC | <0.5 | 110 |
| | S | AS | 550 | 395 | 80 HRB | 1.5 | 130 |
| | | HT | 690 | 655 | 40 HRC | <0.5 | 130 |
| FN-0405 | S | AS | 425 | 240 | 72 HRB | 4.5 | 145 |
| | | HT | 1060 | 880 | 39 HRC | 1 | 145 |
| | T | AS | 510 | 295 | 80 HRB | 6 | 160 |
| | | HT | 1240 | 1060 | 44 HRC | 1.5 | 160 |
| ALUMINUM | | | | | | | |
| 601 AB, | | AS | 110 | 48 | 60 HRH | 6 | — |
| pressed bar | | HT | 252 | 241 | 75 HRH | 2 | — |
| BRASS | | | | | | | |
| CZP-0220 | T | — | 165 | 76 | 55 HRH | 13 | — |
| | U | — | 193 | 89 | 68 HRH | 19 | — |
| | W | — | 221 | 103 | 75 HRH | 23 | — |
| TITANIUM | | | | | | | |
| Ti-6AI-4V | | HIP | 917 | 827 | — | 13 | — |
| SUPERALLOYS | | | | | | | |
| Stellite 19 | | — | 1035 | — | 49 HRC | <1 | — |

*Note*: MPIF = Metal Powder Industries Federation; AS = as sintered; HT = heat treated; HIP = hot isostatically pressed.

## TABLE 11.4

### Mechanical Property Comparisons for Ti-6AI-4V Titanium Alloy

| Process | Density (%) | Yield Stress (MPa) | Ultimate Tensile Strength (MPa) | Elongation (%) | Reduction of Area (%) |
|---|---|---|---|---|---|
| Cast | 100 | 840 | 930 | 7 | 15 |
| Cast and forged | 100 | 875 | 965 | 14 | 40 |
| Powder metallurgy | | | | | |
| Blended elemental (P + S)[*] | 98 | 786 | 875 | 8 | 14 |
| Blended elemental (HIP)[*] | >99 | 805 | 875 | 9 | 17 |
| Realloyed (HIP)[*] | 100 | 880 | 975 | 14 | 26 |

(*) P + S = pressed and sintered; HIP = hot isostatically pressed.
*Source*: R. M. German.

**EXAMPLE 11.5** **Production of automotive-engine main bearing caps by powder metallurgy**

Main bearing caps (MBCs), as shown in Figure 11.15, are essential structural elements in internal combustion engines. Most MBCs are made from gray cast iron (see Section 5.6), which is cast and then followed by extensive machining operations. Grey cast iron is a very versatile material that exhibits excellent castability with good machinability (due to its low hardness and flake graphite in the microstructure), compressive strength, and sound-dampening properties. However, it has low fatigue strength and is very brittle, which limits its use for highly stressed applications.

Powder-metallurgy MBCs were used for the first time in 1993 in General Motors 3100 and 3800 V6 engines. P/M parts were chosen to replace the previous cast-iron parts because of the former's design flexibility and functional advantages, and, because of its net-shape capability, P/M eliminated the need for major investments in machining lines required for finishing the cast parts to final shape and dimensions. A high-strength, low-cost P/M alloy steel was developed for this application, involving liquid-phase sintering and using an optimized low-alloy combination in a carbon-steel matrix, designated as Zenith Material 833 (ZM833).

The P/M process molds individual MBCs into net shape, except for the new side bolt holes, which are drilled and tapped. Critical to the success of the P/M approach was the molding of the long bolt holes, which would be costly and difficult to machine. This molding step requires "upright" compaction, i.e., vertical bolt holes, which presents significant powder compaction, challenges due to the irregular shape conferred by the bearing arch. The advantages of upright compaction, however, include the ability to press in the bearing notch and the features of the top face, which provide identification and orientation for assembly. Upright compaction also permits use of a much smaller compacting press than for "flat" pressing, i.e., horizontal bolt holes, with associated cost savings. In addition, the 3800 engine's rear MBC has several features that would be impossible to compact in the flat orientation; thus, the use of upright compaction was inevitable. Principal features of the MBCs produced by this method include recessed bolt-hole bosses, a rectangular oil-drain hole, vertical side

**FIGURE 11.15** Powder-metal main bearing caps for 3.8- and 3.1-liter GM engines. *Source*: Courtesy of Zenith Sintered Products, Inc., Milwaukee, WI.

grooves, a double arch to accept an oil seal, and a raised-arch feature for added strength.

Once the desired fatigue strength was achieved, the ZM833 was used to produce actual MBCs that were subjected to severe engine-endurance testing in both engines. It was found that the cast-iron engine block generally broke before the P/M cap cracked, and often several engine blocks had to be sacrificed to obtain a fatigue-test point, but eventually an *S–N* curve (see Fig. 2.28) was generated for the P/M material so that a safety margin could be determined. The P/M was proven to be a robust design by repeated successful fatigue testing, simulating 1.8 times the most severe running conditions predicted for the engine.

Material ductility is a functional benefit, since it is possible to induce cracks in a very brittle material during press-fit installation. The ZM833 P/M material had the following properties: (1) density of $6.6 \, g/cm^3$; (2) tensile strength of 450 MPa (65,000 psi); (3) hardness of 70 HRB; (4) modulus of elasticity of $10^7$ GPa ($15.5 \times 10^6$ psi)—the P/M material developed for this application has a lower modulus of elasticity than that of cast iron, resulting in significantly lower initial tensile stresses in highly stressed areas of the MBC; (5) tensile elongation of 3% to 4%, which compares with that for gray iron, at 0.5%, and that for ductile iron, typically at 3%; and (6) it was found that the P/M alloy's fatigue endurance limit, determined by flexural bending fatigue tests, was intermediate between those of the gray and ductile irons, being double the gray iron's level.

Adapted from "Powder Metallurgy Main Bearing Caps for Automotive Engines," by T. M. Cadle, M. A. Jarrett, and W. L. Miller, *Int. Jr. of Powder Metallurgy*, 30(3): 275–282, 1994.

# 11.5 | Secondary and Finishing Operations

In order to further improve the properties of sintered P/M products or to give them special characteristics, several additional operations may be carried out after sintering:

1. **Re-pressing**, also called **coining** and **sizing**, is an additional compaction operation, performed under high pressure in presses. The purposes of this operation are to impart dimensional accuracy to the sintered part and to improve the part's strength and surface finish by additional densification.

2. **Forging** involves the use of unsintered or sintered alloy-powder preforms that are subsequently hot forged in heated, confined dies to the desired final shapes; the preforms may also be shaped by impact forging. The process is usually referred to as *powder-metallurgy forging* (P/F), and when the preform is sintered, the process is usually referred to as *sinter forging*. Ferrous and nonferrous powders can be processed in this manner, resulting in full-density products (on the order of 99.9% of the theoretical density of the material). The deformation of the preform may consist of the following two modes: *upsetting* and *re-pressing*. In upsetting, the material flows laterally outward, as shown in Fig. 6.1. The preform is subjected to compressive and shear stresses during deformation; any interparticle residual oxide films are broken up, as a result of which the part has better toughness, ductility, and fatigue strength. In *re-pressing*, the material flow is mainly in the direction of the movement of the punch, with little lateral flow; as a result, the mechanical properties of the part are not as high as those produced by upsetting.

P/M forged products have good surface finish, good dimensional tolerances, little or no flash formation, and uniform fine grain size. The superior properties obtained make this technology particularly suitable for applications such as highly stressed components for automotive (e.g., connecting rods), jet-engine, military, and off-road equipment.

3. The inherent porosity of powder-metallurgy components can be taken advantage of by **impregnating** the components with a fluid. A typical application is to impregnate the sintered part with oil, which is usually done by immersing the part in heated oil. Bearings and bushings that are internally lubricated, with up to 30% oil by volume, are made by this method. Internally lubricated components have a continuous supply of lubricant during their service lives. Universal joints are now being made with grease-impregnated P/M techniques; as a result, these parts no longer requiring grease fittings.

4. **Infiltration** is a process whereby a metal slug with a lower melting point than that of the part is placed against the sintered part, and the assembly is heated to a temperature sufficient to melt the slug. The molten metal infiltrates the pores by *capillary action*, resulting in a relatively pore-free part with good density and strength. The most common application is the infiltration of iron-base compacts with copper. The advantages of this technique are that hardness and tensile strength are improved and the pores of the part are filled, thus preventing moisture penetration, which could cause corrosion. Infiltration may also be done with lead, whereby, because of the low shear strength of lead, the infiltrated part has lower frictional characteristics than the uninfiltrated one; some bearing materials are formed in this manner.

5. Powder-metal parts may be subjected to other operations, including the following:

   a. **Heat treating** (quench and temper; steam treat), for improved strength, hardness, and resistance to wear:

   b. **Machining** (turning, milling, drilling, tapping, and grinding), for production of undercuts and slots, improved surface finish and dimensional accuracy, and production of threaded holes and other surface features;

   c. **Finishing** (deburring, burnishing, mechanical finishing, plating, and coating), for improved surface characteristics, corrosion and fatigue resistance, and appearance.

---

**EXAMPLE 11.6 Production of tungsten carbide for tools and dies**

Tungsten carbide is an important tool and die material, mainly because of its hardness, strength, and wear resistance to wear over a wide range of temperatures. Powder-metallurgy techniques are used in making these carbides. First, powders of tungsten and carbon are blended together in a ball mill or rotating mixer. The mixture (typically 94% tungsten and 6% carbon, by weight) is heated to approximately 1500°C (2800°F) in a vacuum-induction furnace. As a result of this process, the tungsten is carburized, forming tungsten carbide in fine-powder form. A binding agent (usually cobalt) is then added to the tungsten carbide (with an organic fluid such as hexane), and the mixture is ball milled to produce a uniform and homogeneous mix, a process that can take several hours, or even days.

The mixture is then dried and consolidated, usually by cold compaction, at pressures in the range of 200 MPa (30,000 psi). Finally, it is sintered in a hydrogen-atmosphere or vacuum furnace at a temperature of 1350°C to 1600°C (2500°F to 2900°F) depending on its composition. At this temperature, the cobalt is in a liquid phase and acts as a binder for the carbide particles. (Powders may also be hot pressed at the sintering temperature, using graphite dies.) During sintering, the tungsten carbide undergoes a linear shrinkage of about 16%, corresponding to a volume shrinkage of about 40%; thus, control of size and shape is important for producing tools with accurate dimensions. A combination of other carbides, such as titanium carbide and tantalum carbide, can also be produced, using mixtures made by the method described in this example.

# 11.6 | Design Considerations for Powder Metallurgy

Because of the unique properties of metal powders, their flow characteristics in the die, and the brittleness of green compacts, there are certain design principles that should be followed when using them (Figs. 11.16–11.18):

- The shape of the compact must be kept as simple and uniform as possible. Sharp changes in contour, thin sections, variations in thickness, and high length-to-diameter ratios should be avoided.

- Provision must be made for ejection of the green compact from the die without damaging the compact; thus, holes or recesses should be parallel to the axis of punch travel. Chamfers should also be provided.

- As is the case for parts made by most other processes, P/M parts should be made with the widest dimensional tolerances, consistent with their intended applications, in order to increase tool and die life and reduce production costs.

- Walls should not be less than 1.5 mm (0.060 in.) thick, although walls as thin as 0.34 mm (0.0135 in.) have been pressed successfully on components 1 mm (0.04 in.) in length. Walls with length-to-thickness ratios greater than 8:1 are difficult to press, and density variations are virtually unavoidable.

- Simple steps can be produced if their size doesn't exceed 15% of the overall part length. Larger steps can be pressed, but require more complex, multiple-motion tooling.

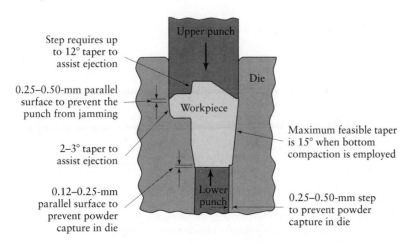

**FIGURE 11.16** Die geometry and design features for powder-metal compaction. *Source*: Metal Powder Industries Federation.

**FIGURE 11.17** Examples of P/M parts, showing poor and good designs. Note that sharp radii and reentry corners should be avoided, and that threads and transverse holes have to be produced separately by additional machining operations. *Source*: Metal Powder Industries Federation.

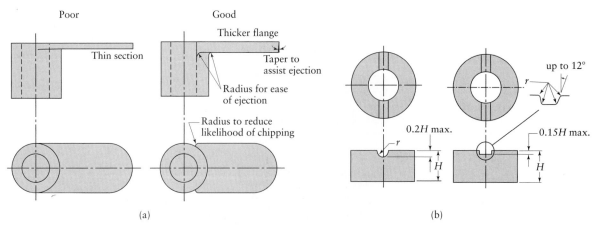

**FIGURE 11.18** (a) Design features for use with unsupported flanges. (b) Design features for use with grooves. *Source*: Metal Powder Industries Federation.

- Letters can be pressed if they are oriented perpendicular to the direction of pressing; letters can also be raised or recessed. However, raised letters are more susceptible to damage in the green stage and prevent stacking during sintering.

- Flanges or overhangs can be produced by a step in the die. However, long flanges may be broken upon ejection and may require more elaborate tooling. A long flange should incorporate a draft around the flange, a radius at the bottom edge, and a radius at the juncture of the flange and/or component body to reduce stress concentrations and the likelihood of fracture.

- A true radius cannot be pressed into the edge of a part, because it would require the punch to be feathered to a zero thickness (Fig. 11.17c). Chamfers or flats are preferred for pressing; a common design approach is to use a 45° angle and 0.25-mm (0,010-in.) flat (Fig. 11.17e).

- Keys, keyways, and holes, used for transmitting torques on gears and pulleys, can be formed during compaction. Bosses can be produced if proper drafts are used and their length is small compared with the overall size of the component.

- Notches and grooves can be made if they are oriented perpendicular to the pressing direction (Fig. 11.17b). It is recommended that circular grooves not exceed a depth of 20% of the overall depth of the component. (Rectangular grooves shouldn't exceed a depth of 15% of the overall depth of the component - see Fig. 11.18b.)

- Parts produced through MIM have similar design constraints as injection molding. With MIM, wall thicknesses should be uniform in order to minimize distortion during sintering.

Dimensional tolerances of sintered P/M parts are usually on the order of $\pm 0.05 - 0.1$ mm ($\pm 0.002 - 0.004$ in.). Tolerances improve significantly with additional operations such as sizing, machining, and grinding. (See Table 16.1.)

# 11.7 | Economics of Powder Metallurgy

As we have seen, powder metallurgy can produce parts at or near net shape, thus eliminating many secondary manufacturing and assembly operations; consequently, P/M has become increasingly competitive with cast, forged, extruded, or machined parts (Table 11.5). Although some larger parts also are produced, the vast majority of conventional P/M parts weigh from a few grams to about a kilogram each. Many conventional die-pressed and sintered parts are used without any machining, although some may require minor finishing operations. Even in highly complex parts, finish machining of P/M parts involves relatively simple operations, such as drilling and tapping holes or grinding some surfaces; furthermore, a single P/M part may replace an assembly of several parts made by other manufacturing processes.

The net- or near-net-shape capability of P/M is an important factor because of its cost-effectiveness in numerous applications. Major cost elements in P/M parts production are the costs for metal powders, compacting dies, and equipment for compacting and sintering. The unit cost of metal powders is much higher than that for molten metal for casting or wrought bar stock for machining; the cost also depends on the method of powder production. The least expensive powder metal is iron, followed by aluminum, zinc, copper, chromium, stainless steel, molybdenum, tungsten, cobalt, niobium, zirconium, and tantalum. Due to the high initial cost of punches, dies,

**TABLE 11.5**

| Competitive Features of P/M and Some Other Manufacturing Processes | | |
|---|---|---|
| Process | Advantages Over P/M | Limitations as Compared With P/M |
| Casting | Wide range of part shapes and sizes produced; generally low mold and setup cost. | Some waste of material in processing; some finishing required; may not be feasible for some high-temperature alloys. |
| Forging (hot) | High production rate of a wide range of part sizes and shapes; high mechanical properties through control of grain flow. | Some finishing required; some waste of material in processing; die wear; relatively poor surface finish and dimensional control. |
| Extrusion (hot) | High production rate of long parts; complex cross-sections may be produced. | Only a constant cross-sectional shape can be produced; die wear; poor dimensional control. |
| Machining | Wide range of part shapes and sizes; short lead time; flexibility; good dimensional control and surface finish; simple tooling. | Waste of material in the form of chips; relatively low productivity. |

and various equipment for P/M processing, production volume must be sufficiently high to warrant this major expenditure, typically upwards of 50,000 parts per year. Today's equipment is highly automated, and P/M production is ideal for many automotive parts made in millions per year, for such large quantities, the labor cost per part is significantly low.

The cost-effectiveness of specialized P/M methods results from special considerations that are different from those for conventional die-press and sinter methods. For example, P/M forging is used mainly for critical applications in which full density and the accompanying superior fatigue resistance are essential. The technique is especially well suited for high-quantity production. For specific applications, P/M forging competes with processes such as conventional forging and casting, in terms of both product properties and production cost. Automotive connecting rods can be produced by both P/M forging and casting. A more recently developed process is metal injection molding (MIM), which uses finer, and therefore more expensive, powder; it also requires more production steps, such as feedstock preparation, molding, debinding, and sintering. Because of these higher cost elements, MIM is cost effective mainly for small (generally weighing less than 100 g), but highly complex, parts required in large quantities.

Relatively large aerospace parts can be produced in small quantities when there are critical property requirements or special metallurgical considerations. For example, some nickel-based superalloys are so highly alloyed that they are prone to segregation during casting and can be processed only by P/M methods; the powders may first be consolidated by hot extrusion and then hot forged in dies maintained at a high temperature. All beryllium processing is also based on powder metallurgy and involves either cold isostatic pressing and sintering or hot isostatic pressing. Many of these materials are difficult to process and may require much finish machining even when the parts are made by P/M methods; this is in sharp contrast to the numerous ferrous press-and-sinter P/M parts that require practically no finish machining. To a large extent, these special and rather expensive P/M methods are used for aerospace applications because no other production alternatives are technically feasible.

# 11.8 | Ceramics: Structure, Properties, and Applications

*Ceramics* are compounds of metallic and nonmetallic elements. The term ceramics refers both to the material and to the ceramic product itself. In Greek, the word *keramos* means potter's clay, and *keramikos* means clay products. Because of the large number of possible combinations of elements, a great variety of ceramics is now available for widely different consumer and industrial applications.

The earliest use of ceramics was in pottery and bricks, dating back to before 4000 BC. Ceramics have been used for many years in automotive spark plugs as an electrical insulator and for high-temperature strength; they are becoming increasingly important in heat engines and various other applications (Table 11.6), as well as in tool

**TABLE 11.6**

| Types and General Characteristics of Ceramics and Glasses | |
|---|---|
| Type | General Characteristics |
| Oxide Ceramics | |
|    Alumina | High hot hardness and abrasion resistance, moderate strength and toughness; most widely used ceramic; used for cutting tools, abrasives, and electrical and thermal insulation. |
|    Zirconia | High strength and toughness; resistance to thermal shock, wear, and corrosion; partially-stabilized zirconia and transformation-toughened zirconia have better properties; suitable for heat-engine components. |
| Carbides | |
|    Tungsten carbide | High hardness, strength, toughness, and wear resistance, depending on cobalt binder content; commonly used for dies and cutting tools. |
|    Titanium carbide | Not as tough as tungsten carbide, but has a higher wear resistance; has nickel and molybdenum as the binder; used as cutting tools. |
|    Silicon carbide | High-temperature strength and wear resistance, used for heat engines and as abrasives. |
| Nitrides | |
|    Cubic boron nitride | Second hardest substance known, after diamond; high resistance to oxidation; used as abrasives and cutting tools. |
|    Titanium nitride | Used as coatings on tools, because of its low frictional characteristics. |
|    Silicon nitride | High resistance to creep and thermal shock; high toughness and hot hardness; used in heat engines. |
| Sialon | Consists of silicon nitrides and other oxides and carbides; used as cutting tools. |
| Cermets | Consist of oxides, carbides, and nitrides; high chemical resistance but is somewhat brittle and costly; used in high-temperature applications. |
| Nanophase ceramics | Stronger and easier to fabricate and machine than conventional ceramics; used in automotive and jet-engine applications. |
| Silica | High temperature resistance; quartz exhibits piezoelectric effects; silicates containing various oxides are used in high-temperature, nonstructural applications. |
| Glasses | Contain at least 50% silica; amorphous structure; several types available, with a wide range of mechanical, physical, and optical properties. |
| Glass ceramics | High crystalline component to their structure; stronger than glass; good thermal-shock resistance; used for cookware, heat exchangers, and electronics. |
| Graphite | Crystalline form of carbon; high electrical and thermal conductivity; good thermal-shock resistance; also available as fibers, foam, and buckyballs for solid lubrication; used for molds and high-temperature components. |
| Diamond | Hardest substance known; available as single-crystal or polycrystalline form; used as cutting tools and abrasives and as die insert for fine wire drawing; also used as coatings. |

and die materials. More recent applications of ceramics are in automotive components, such as exhaust-port liners, coated pistons, and cylinder liners, with the desirable properties of strength and corrosion resistance at high operating temperatures.

Some properties of ceramics are significantly better than those of metals, particularly their hardness and thermal and electrical resistance. Ceramics are available as a single crystal or in polycrystalline form, consisting of many grains. Grain size has a major influence on the strength and properties of ceramics; the finer the grain size, the higher are the strength and toughness—hence, the term **fine ceramics**. Ceramics are general divided into the categories of **traditional ceramics** (whiteware, tiles, brick, pottery, and abrasive wheels) and **industrial ceramics**, also called **engineering** or **high-tech ceramics** (heat exchangers, cutting tool, semiconductors, and prosthetics).

### 11.8.1  Structure and types of ceramics

The structure of ceramic crystals is among the most complex of all materials, containing various elements of different sizes. The bonding between these atoms is generally covalent (electron sharing, hence strong bonds) and ionic (primary bonding between oppositely charged ions, hence strong bonds).

Among the oldest *raw materials* for ceramics is *clay*, a fine-grained sheetlike structure, the most common example being *kaolinite* (from Kaoling, a hill in China). It is a white clay, consisting of silicate of aluminum with alternating weakly bonded layers of silicon and aluminum ions (Fig. 11.19). When added to kaolinite, water attaches itself to the layers (adsorption), makes them slippery, and gives wet clay its well-known softness and plastic properties (*hydroplasticity*) that make it formable. Other major raw materials for ceramics that are found in nature are *flint* (rock of very fine-grained silica, $SiO_2$) and *feldspar* (a group of crystalline minerals consisting of aluminum silicates, potassium, calcium, or sodium). In their natural state, these raw materials generally contain impurities of various kinds, which have to be removed prior to further processing of the materials into useful products with reliable performance. Highly refined raw materials produce ceramics with improved properties.

The various types of ceramics are:

1. **Oxide ceramics**

   **Alumina.**  Also called *corundum* or *emery*, *alumina* (aluminum oxide, $Al_2O_3$) is the most widely used *oxide ceramic*, either in pure form or as a raw material to be mixed with other oxides. It has high hardness and moderate strength. Although alumina exists in nature, it contains varying amounts of impurities and possesses nonuniform properties in natural form. As a result, its behavior is unreliable. Aluminum oxide, as well as silicon carbide and many other ceramics, is now almost totally manufactured synthetically, so that its quality can be controlled.

   First made in 1893, synthetic aluminum oxide is obtained by the fusion of molten bauxite (an aluminum-oxide ore that is the principal source of alu-

**FIGURE 11.19** The crystal structure of kaolinite, commonly known as clay.

Silicon ions

Oxygen ions

Aluminum ions

OH ions

minum), iron filings, and coke in electric furnaces. The material is then crushed and graded by size by passing the particles through standard screens. Parts made of aluminum oxide are cold pressed and sintered (*white ceramics*). Their properties are improved by minor additions of other ceramics, such as titanium oxide and titanium carbide. Structures containing various alumina and other oxides are known as *mullite* and *spinel* and are used as refractory materials for high-temperature applications. The mechanical and physical properties of alumina are particularly suitable for applications such as electrical and thermal insulation, cutting tools, and abrasives.

**Zirconia.** *Zirconia* (zirconium oxide, $ZrO_2$), white in color, has good toughness; good resistance to thermal shock, wear, and corrosion; low thermal conductivity, and a low friction coefficient. A more recent development is **partially stabilized zirconia** (PSZ), which has high strength and toughness and performs more reliably than zirconia. It is obtained by doping the zirconia with oxides of calcium, yttrium, or magnesium. This process forms a material with fine particles of tetragonal zirconia in a cubic lattice. Typical applications include dies for hot extrusion of metals and zirconia beads for grinding and dispersion media for aerospace coatings, automotive primers and topcoats, and fine glossy print on flexible food packaging.

Another important characteristic of PSZ is the fact that its coefficient of thermal expansion is only about 20% lower than that of cast iron, and its thermal conductivity is about one third that of other ceramics. Consequently, it is very suitable for heat-engine components, such as cylinder liners and valve bushings, to keep the cast-iron engine assembly intact. New developments to further improve upon the properties of PSZ include **transformation-toughened zirconia** (TTZ), which has higher toughness than that of PSZ, because of dispersed tough phases in the ceramic matrix.

## 2. Other ceramics

**Carbides.** Typical examples of *carbides* are tungsten and titanium, used as cutting tools and die materials, and silicon carbide, used as abrasives (e.g., in grinding wheels):

a) **Tungsten carbide** (WC) consists of tungsten-carbide particles with cobalt as a binder. The amount of binder has a major influence on the material's properties: Toughness increases with the cobalt content, whereas hardness, strength, and resistance decrease.

b) **Titanium carbide** (TiC) has nickel and molybdenum as the binder and is not as tough as tungsten carbide.

c) **Silicon carbide** (SiC) has good wear, thermal shock, and corrosion resistance. It has a low friction coefficient and retains its strength at elevated temperatures. It is suitable for high-temperature components in heat engines and is also used as an abrasive. Synthetic silicon carbide is made from silica sand, coke, and small amounts of sodium chloride and sawdust. The process is similar to making synthetic aluminum oxide.

**Nitrides.** Another important class of ceramics is the *nitrides*:

a) **Cubic boron nitride** (cBN), the second hardest known substance, after diamond, has special applications, such as abrasives in grinding wheels and as cutting tools. It does not exist in nature and was first made synthetically in the 1970s, with techniques similar to those used for making synthetic diamond.

b) **Titanium nitride** (TiN) is used extensively as coatings on cutting tools; it improves tool life, because of its low frictional characteristics.

c) **Silicon nitride** ($Si_3N_4$) has high resistance to creep at elevated temperatures, low thermal expansion, and high thermal conductivity, and hence it resists thermal shock. It is suitable for high-temperature structural applications, such as in automotive engine and gas-turbine components, cam-follower rollers, bearings, sand-blast nozzles, and components for the paper industry.

**Sialon.** *Sialon* consists of silicon nitride, with various additions of aluminum oxide, yttrium oxide, and titanium carbide. (See Section 8.6.8.) The word *sialon* is derived from the words *silicon, aluminum, oxygen,* and *nitrogen*. Sialon has higher strength and thermal-shock resistance than silicon nitride does and thus far is used primarily as a cutting-tool material.

**Cermets.** *Cermets* are combinations of ceramics bonded with a metallic phase. Introduced in the 1960s and also called **black ceramics** or *hot-pressed ceramics*, they combine the high-temperature oxidation resistance of ceramics and the toughness, thermal-shock resistance, and ductility of metals. An application of cermets is cutting tools, a typical composition being 70% aluminum oxide and 30% titanium carbide; other cermets contain various oxides, carbides, and nitrides. They have been developed for high-temperature applications such as nozzles for jet engines and aircraft brakes. Cermets can be regarded as composite materials and can be used in various combinations of ceramics and metals bonded by powder-metallurgy techniques. (See also Section 11.14.)

### 3. Silica

Abundant in nature, *silica* is a polymorphic material; that is, it can have different crystal structures. The cubic structure is found in refractory bricks used for high-temperature furnace applications. Most glasses contain more than 50% silica. The most common form of silica is **quartz**, which is a hard, abrasive hexagonal crystal. It is used extensively as oscillating crystals of fixed frequency in communications applications, since it exhibits the piezoelectric effect (See Section 3.9.6.)

**Silicates** are products of the reaction of silica with oxides of aluminum, magnesium, calcium, potassium, sodium, and iron. Examples include clay, asbestos, mica, and silicate glasses. **Lithium aluminum silicate** has very low thermal expansion and thermal conductivity and good thermal shock resistance. However, it has very low strength and fatigue life, and thus it is suitable only for nonstructural applications, such as catalytic converters, regenerators, and heat-exchanger components.

### 4. Nanophase ceramics

A more recent development is *nanophase ceramics*, which consist of atomic clusters containing a few thousand atoms. Nanophase ceramics exhibit ductility at significantly lower temperatures than for conventional ceramics and are stronger and easier to fabricate, with fewer flaws; however, control of particle size and distribution and of contamination is important. Applications of nanophase ceramics in the automotive industry, for example, include valves, rocker arms, turbocharger rotors, and cylinder liners; they are also used in jet-engine components. Nanocrystalline second-phase particles, on the order of 100 nm or less, and fibers are also being used as reinforcement in composites (see Section 11.14); they have enhanced properties such as tensile strength and creep resistance (See also the discussion of *nanomaterials*, in Section 3.11.9.)

## 11.8.2  General properties and applications of ceramics

Compared with metals, ceramics have the following relative characteristics: brittleness, high strength and hardness at elevated temperatures, high elastic modulus, low toughness, low density, low thermal expansion, and low thermal and electrical conductivity. However, because of the wide variety of ceramic material composition and grain size, the mechanical and physical properties of ceramics vary significantly. For example, the electrical conductivity of ceramics can be modified from poor to good, which is the principle behind semiconductors. Because of their sensitivity to flaws, defects, and cracks (surface or internal); the presence of different types and levels of impurities; and different methods of manufacturing, ceramics can have a wide range of properties.

a) **Mechanical properties.** Table 11.7 presents the mechanical properties of several engineering ceramics. Note that the ceramics' strength in tension (transverse rupture strength) is approximately one order of magnitude lower than their compressive strength, because of their sensitivity to cracks, impurities, and porosity. Such defects lead to the initiation and propagation of cracks under tensile stresses, severely reducing tensile strength. Therefore, reproducibility and reliability (acceptable performance over a specified period of time) are important aspects in the service life of ceramic components. The tensile strength of polycrystalline ceramic parts increases with decreasing grain size.

Tensile strength is empirically related to porosity as

$$UTS \simeq UTS_0 e^{-nP}, \qquad (11.5)$$

where $P$ is the volume fraction of pores in the solid, $UTS_0$ is the tensile strength at zero porosity, and the exponent $n$ ranges between 4 and 7.

The modulus of elasticity is likewise affected by porosity, as given by

$$E \simeq E_0(1 - 1.9P + 0.9P^2), \qquad (11.6)$$

**TABLE 11.7**

### Approximate Range of Properties of Various Ceramics at Room Temperature

| Material | Symbol | Transverse Rupture Strength (MPa) | Compressive Strength (MPa) | Elastic Modulus (GPa) | Hardness (HK) | Poisson's Ratio ($v$) | Density ($kg/m^3$) |
|---|---|---|---|---|---|---|---|
| Aluminum oxide | $Al_2O_3$ | 140–240 | 1000–2900 | 310–410 | 2000–3000 | 0.26 | 4000–4500 |
| Cubic boron nitride | cBN | 725 | 7000 | 850 | 4000–5000 | – | 3480 |
| Diamond | – | 1400 | 7000 | 830–1000 | 7000–8000 | – | 3500 |
| Silica, fused | $SiO_2$ | – | 1300 | 70 | 550 | 0.25 | – |
| Silicon carbide | SiC | 100–750 | 700–3500 | 240–480 | 2100–3000 | 0.14 | 3100 |
| Silicon nitride | $Si_3N_4$ | 480–600 | – | 300–310 | 2000–2500 | 0.24 | 3300 |
| Titanium carbide | TiC | 1400–1900 | 3100–3850 | 310–410 | 1800–3200 | – | 5500–5800 |
| Tungsten carbide | WC | 1030–2600 | 4100–5900 | 520–700 | 1800–2400 | – | 10,000–15,000 |
| Partially stabilized zirconia | PSZ | 620 | – | 200 | 1100 | 0.3 | 5800 |

*Note*: These properties vary widely, depending on the condition of the material.

where $E_0$ is the modulus of elasticity at zero porosity. Equation (11.5) is valid up to 50% porosity. Common earthenware has a porosity ranging between 10% and 15%, whereas the porosity of hard porcelain is about 3%.

Although there are exceptions, ceramics, unlike most metals and thermoplastics, generally lack impact toughness and thermal-shock resistance because of their inherent lack of ductility. Once initiated, a crack propagates rapidly. In addition to undergoing fatigue failure under cyclic loading, ceramics (particularly glasses) exhibit a phenomenon called **static fatigue**: When subjected to a static tensile load over a period of time, these materials may suddenly fail. This phenomenon occurs in environments where water vapor is present. Static fatigue, which does not occur in a vacuum or dry air, has been attributed to a mechanism similar to stress-corrosion cracking of metals.

Ceramic components that are to be subjected to tensile stresses may be **prestressed**, much like prestressed concrete. Prestressing shaped ceramic components subjects them to compressive stresses. Methods used include (1) heat treatment and chemical tempering; (2) laser treatment of surfaces; (3) coating with ceramics with different degrees of thermal expansion; and (4) surface-finishing operations, such as grinding, in which compressive residual stresses are induced on the surfaces. Significant advances are being made in improving the toughness and other properties of ceramics, including **machinable ceramics**. Among these advances are proper selection and processing of raw materials, control of purity and structure, and use of reinforcements, with particular emphasis during design on advanced methods of stress analysis in ceramic components.

b) **Physical properties.**  Most ceramics have relatively low specific gravity, ranging from about 3 to 5.8 for oxide ceramics, compared with 7.86 for iron. They have very high melting or decomposition temperatures. The thermal conductivity of ceramics varies by as much as three orders of magnitude, depending on their composition, whereas the thermal conductivity of metals varies by one order. The thermal conductivity of ceramics, as well as of other materials, decreases with increasing temperature and porosity, because air is a poor thermal conductor.

The thermal conductivity $k$ is related to porosity by

$$k = k_0(1 - P), \qquad (11.7)$$

where $k_0$ is the thermal conductivity at zero porosity.

The thermal-expansion characteristics of ceramics are shown in Fig. 11.20. Thermal expansion and thermal conductivity induce thermal stresses that can lead to thermal shock or thermal fatigue. The tendency for *thermal cracking* (called *spalling* when a piece or a layer from the surface breaks off) is lower with low thermal expansion and high thermal conductivity. For example, fused silica has high thermal shock resistance because of its virtually zero thermal expansion.

A familiar example that illustrates the importance of low thermal expansion is heat-resistant ceramics for cookware and stove tops. These ceramics can sustain high thermal gradients, from hot to cold, and vice versa. Moreover, the relative thermal expansion of ceramics and metals is an important reason for the use of ceramic components in heat engines. The fact that the thermal conductivity of partially stabilized zirconia components is close to that of the cast iron in engine blocks is a further advantage in the use of PSZ in heat engines. An additional characteristic is the **anisotropy of thermal expansion** exhibited by oxide ceramics, whereby thermal expansion varies in different directions of the ceramic. This behavior causes thermal stresses that can lead to cracking of the ceramic component.

The *optical properties* of ceramics can be controlled by various formulations and control of structure, imparting different degrees of transparency and

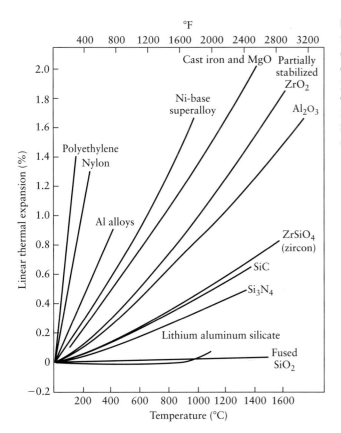

°F

colors. For example, single-crystal sapphire is completely transparent, zirconia is white, and fine-grained polycrystalline aluminum oxide is a translucent gray. Porosity influences the optical properties of ceramics, much like trapped air in ice cubes, which makes the ice less transparent and gives it a white appearance.

Although ceramics basically are resistors, it has been shown that they can be made to be electrically conductive by alloying them with certain elements, thus making the ceramic act like a semiconductor or even a superconductor.

c) **Applications.**  As shown in Table 11.6, ceramics have numerous consumer and industrial applications. Several types of ceramics are used in the electrical and electronics industry, because of their high electrical resistivity, dielectric strength (the voltage required for electrical breakdown per unit thickness), and magnetic properties suitable for applications such as magnets for speakers. An example of such a ceramic is **porcelain**, which is a white ceramic composed of kaolin, quartz, and feldspar. Certain ceramics also have good piezoelectric properties.

The capability of ceramics to maintain their strength and stiffness at elevated temperatures (Figs. 11.21 and 11.22) makes them very attractive for high-temperature applications. Their high resistance to wear makes them suitable for applications such as cylinder liners, bushings, seals, and bearings. The higher operating temperatures made possible by the use of ceramic components enable more efficient fuel burning and reduced emissions. Currently, internal combustion engines are only about 30% efficient, but with the use of ceramic components, their operating performance can be improved by at least 30%. Ceramics that have been used successfully, especially in gasoline and diesel-engine components and as rotors, are silicon nitride, silicon carbide, and partially stabilized zirconia. Ceramics are also used to coat metal, which may be done to reduce wear, prevent corrosion, and provide a thermal barrier.

**FIGURE 11.21** Effect of temperature on the strength of various engineering ceramics. Note that much of the strength is maintained at high temperatures.

**FIGURE 11.22** Effect of temperature on the modulus of elasticity for several ceramics. *Source*: D. W. Richerson, *Modern Ceramic Engineering*, New York: Marcel Dekker, Inc., 1982.

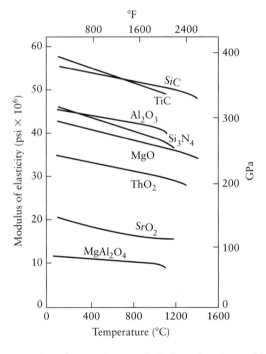

Other attractive properties of ceramics are their low density and high elastic modulus, which make it possible to reduce engine weight; also, in other applications that use ceramics, the inertial forces generated by moving parts are lower than with metals. High-speed components for machine tools, for example, are candidates for ceramics. Furthermore, the higher elastic modulus of ceramics makes them attractive for improving the stiffness, while reducing the weight, of machines. Silicon-nitride ceramics are also used as ball bearings and rollers. Because of their strength and inertness, ceramics are used as *biomaterials* to replace joints in the human body, as prosthetic devices, and for dental work. Commonly used bioceramics include aluminum oxide, silicon nitride, and various compounds of silica. Furthermore, ceramics can be made to be porous, thus allowing bone to grow into the porous surface and develop a strong mechanical bond.

**EXAMPLE 11.7** Effect of porosity on properties

If a fully dense ceramic has the properties of $UTS_0 = 100$ MPa, $E_0 = 400$ GPa, and $k_0 = 0.5$ W/m·K, what are these properties at 10% porosity? Let $n = 5$ and $P = 0.1$.

**Solution.** Using Eqs. 11.5–11.7, we have

$$UTS = 100e^{-(5)(0.1)} = 61 \text{ MPa},$$
$$E = 400[1 - (1.9)(0.1) + (0.9)(0.1)^2] = 328 \text{ GPa},$$

and

$$k = 0.5(1 - 0.1) = 0.45 \text{ W/m·K}.$$

# 11.9 | Shaping Ceramics

Several techniques have been developed for shaping ceramics into useful products (Table 11.8). Generally, the procedure involves the following steps: crushing or grinding the raw materials into very fine particles; mixing the particles with additives to impart certain desirable characteristics; and shaping, drying, and firing the material. *Crushing* (also called *comminution* or *milling*) of the raw materials is generally done in a ball mill (Fig. 11.23b), either dry or wet. Wet crushing is more effective, because

**TABLE 11.8**

**General Characteristics of Ceramics Processing**

| Process | Advantages | Limitations |
|---|---|---|
| Slip casting | Large parts; complex shapes; low equipment cost. | Low production rate; limited dimensional accuracy. |
| Extrusion | Hollow shapes and small diameters; high production rate. | Parts have constant cross-section; limited thickness. |
| Dry pressing | Close tolerances; high production rate with automation. | Density variation in parts with high length-to-diameter ratios; dies require high abrasive-wear resistance; equipment can be costly. |
| Wet pressing | Complex shapes; high production rate. | Limited part size and dimensional accuracy; tooling costs can be high. |
| Hot pressing | Strong, high-density parts. | Protective atmospheres required; die life can be short. |
| Isostatic pressing | Uniform density distribution. | Equipment can be costly. |
| Jiggering | High production rate with automation; low tooling cost. | Limited to axisymmetric parts; limited dimensional accuracy. |
| Injection molding | Complex shapes; high production rate. | Tooling costs can be high. |

(a)       (b)       (c)

**FIGURE 11.23** Methods of crushing ceramics to obtain very fine particles: (a) roll crushing, (b) ball milling, and (c) hammer milling.

it keeps the particles together and prevents the suspension of fine particles in air. The ground particles are then mixed with *additives*, the function of which is one or more of the following:

1. *Binder* for the ceramic particles;

2. *Lubricant* for mold release and to reduce internal friction between particles during molding;

3. *Wetting agent* to improve mixing;

4. *Plasticizer* to make the mix more plastic and formable;

5. *Deflocculent* to make the ceramic–water suspension uniform. The additive changes the electrical charges on the particles of clay so that they repel instead of attract each other. Water is added to make the mixture more pourable and less viscous. Typical deflocculants include $Na_2CO_3$ and $Na_2SiO_3$ in amounts of less than 1%.;

6. Various agents to *control foaming* and *sintering*.

The three basic shaping processes for ceramics are casting, plastic forming, and pressing, and are described in the following three subsections.

### 11.9.1 Casting

The most common casting process is **slip casting**, also called *drain casting* (Fig. 11.24). A **slip** is a suspension of ceramic particles in a liquid, generally water. In this process, the slip is poured into a porous mold made of plaster of paris. The slip must have sufficient fluidity and low viscosity to flow easily into the mold, much like the fluidity of molten metals. After the mold has absorbed some of the water from the outer layers of the suspension, it is inverted, and the remaining suspension is poured out (for making hollow objects, as in slush casting of metals - see Fig. 5.16). The top of the part is then trimmed, the mold is opened, and the part is removed.

Large and complex parts, such as plumbing ware, art objects, and dinnerware, can be made by slip casting. Although dimensional control is limited and the production rate is low, mold and equipment costs are also low. In some applications, components of the product (such as handles for cups and pitchers) are made separately and then joined, using the slip as an adhesive.

Thin sheets of ceramics, less than 1.5 mm (0.06 in.) thick, can be made by a casting technique called the **doctor-blade process**. In this operation, the slip is cast over a moving plastic belt, and its thickness is controlled by a blade. Other processes include *rolling* the slip between pairs of rolls and *casting* the slip over a paper tape, which is then burned off during firing. For solid ceramic parts, the slip is supplied continuously into the mold to replenish the absorbed water; the suspension is not drained from the mold. At this stage, the part is a soft solid or semirigid. The higher the concentration of solids in the slip, the less water has to be removed. The part, called *green*, as in powder metallurgy, is then fired.

**FIGURE 11.24** Sequence of operations in slip casting a ceramic part. After the slip has been poured, the part is dried and fired in an oven to give it strength and hardness. *Source*: F. H. Norton, *Elements of Ceramics*.

(a)          (b)          (c)          (d)          (e)

## 11.9.2 Plastic forming

*Plastic forming* (also called *soft*, *wet*, or *hydroplastic forming*) can be done by various methods, such as extrusion, injection molding, or molding and jiggering (as done on a potter's wheel). Plastic forming tends to orient the layered structure of clays along the direction of material flow. This orientation leads to anisotropic behavior of the material, both in subsequent processing and in the final properties of the ceramic product. In **extrusion**, the clay mixture, containing 20% to 30% water, is forced through a die opening by screw-type equipment. The cross-section of the extruded product is constant, but there are limitations on the wall thickness for hollow extrusions. Tooling costs are low, and production rates are high. The extruded products may be subjected to additional shaping operations.

## 11.9.3 Pressing

The various methods of pressing are:

a) **Dry pressing.** Similar to powder-metal compaction, *dry pressing* is used for relatively simple shapes. Typical parts made by this method include whiteware, refractories, and abrasive products. The process has the same high production rates and close control of dimensional tolerances as in P/M. The moisture content of the mixture is generally below 4%, but may be as high as 12%. Organic and inorganic binders, such as stearic acid, wax, starch, and polyvinyl alcohol, are usually added to the mixture; they also act as lubricants. The pressure for pressing is between 35 MPa and 200 MPa (5 ksi and 30 ksi). Modern presses used for dry pressing are highly automated. Dies, usually made of carbides or hardened steel, must have high wear resistance to withstand the abrasive ceramic particles and can be expensive.

   Density can vary greatly in dry-pressed ceramics (Fig. 11.25), because of friction between particles and at the mold walls, as in P/M compaction. Density variations cause warping during firing. Warping is particularly severe for parts that have high length-to-diameter ratios; the recommended maximum ratio is 2 : 1. Several methods may be used to minimize density variations. Design of tooling is important in this respect. Vibratory pressing and impact forming are used, particularly for nuclear-reactor fuel elements. Isostatic pressing also reduces density variations.

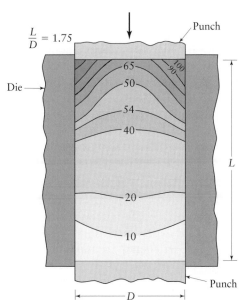

**FIGURE  11.25**   Density variation in pressed compacts in a single-action press. The variation increases with increasing *L/D* ratio. See also Fig. 11.7e. *Source*: Adapted from W. D. Kingery et al., *Ceramic Fabrication Processes*, MIT Press.

**FIGURE 11.26** Extruding and jiggering operations. *Source*: R. F. Stoops.

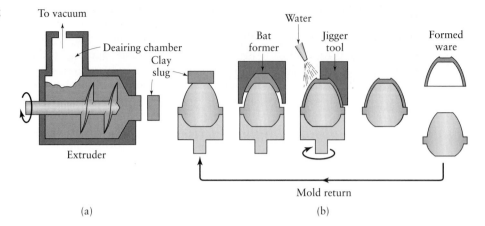

(a)  (b)

b) **Wet pressing.** In *wet pressing*, the part is formed in a mold while under high pressure in a hydraulic or mechanical press. This process is generally used to make intricate shapes. The moisture content of the part usually ranges from 10% to 15%. Production rates are high, but part size is limited; dimensional control is difficult, because of shrinkage during drying; and tooling costs can be high.

c) **Isostatic pressing.** Used extensively in powder metallurgy, as you have seen, *isostatic pressing* is also used for ceramics in order to obtain uniform density distribution throughout the part. Automotive spark-plug insulators are made by this method. Silicon-nitride vanes for high-temperature use are also made by hot isostatic pressing.

d) **Jiggering.** A combination of processes is used to make ceramic plates. Clay slugs are first extruded, then formed into a *bat* over a plaster mold, and finally jiggered on a rotating mold (Fig. 11.26). *Jiggering* is an operation in which the clay bat is formed with templates or rollers. The part is then dried and fired. The process is limited to axisymmetric parts and has limited dimensional accuracy, but the operation can be automated.

e) **Injection molding.** The advantages of *injection molding* of plastics and powder metals were indicated earlier; this process is now being used extensively for precision forming of ceramics for high-technology applications, as in rocket-engine components. The raw material is mixed with a binder, such as a thermoplastic polymer (polypropylene, low-density polyethylene, ethylene vinyl acetate, or wax). The binder is usually removed by pyrolysis, and the part is sintered by firing. This process can produce thin sections, typically less than 10–15 mm (0.4–0.6 in.), using most engineering ceramics, such as alumina, zirconia, silicon nitride, and silicon carbide. Thicker sections require careful control of the materials used and of processing parameters in order to avoid internal voids and cracks, such as those due to shrinkage.

f) **Hot pressing.** In *hot pressing*, also called *pressure sintering*, pressure and temperature are applied simultaneously. This method reduces porosity, making the part denser and stronger. *Hot isostatic pressing* (see Section 11.3.3) may also be used in this operation, particularly to improve the quality of high-technology ceramics. Because of the presence of both pressure and temperature, die life in hot pressing can be short. Protective atmospheres are usually employed, and graphite is a commonly used punch and die material.

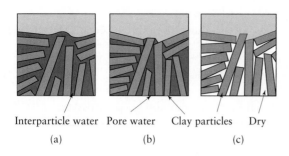

Interparticle water   Pore water   Clay particles   Dry

(a)            (b)            (c)

**FIGURE 11.27**   Shrinkage of wet clay caused by removal of water during drying. Shrinkage may be as much as 20% by volume. *Source*: After F. H. Norton.

## 11.9.4  Drying and firing

After the ceramic has been shaped by any of the methods described in this section, the next step is to dry and fire the part to give it the proper strength. *Drying* is a critical stage, because of the tendency for the part to warp or crack from variations in moisture content and thickness within the part and the complexity of its shape. Control of atmospheric humidity and temperature is important in order to reduce warping and cracking. Loss of moisture results in shrinkage of the part by as much as 15% to 20% of the original moist size (Fig. 11.27). In a humid environment, the evaporation rate is low, and consequently the moisture gradient across the thickness of the part is lower than that in a dry environment. The low moisture gradient, in turn, prevents a large, uneven gradient in shrinkage from the surface to the interior during drying. The dried part (called *green*, as in powder metallurgy) can be machined relatively easily to bring it closer to its final shape, although it must be handled carefully.

*Firing*, also called *sintering*, involves heating the part to an elevated temperature in a controlled environment, similar to sintering in powder metallurgy. Some shrinkage occurs during firing. Firing gives the ceramic part its strength and hardness. The improvement in properties results from (1) development of a strong bond between the complex oxide particles in the ceramic and (2) reduced porosity. *Nanophase ceramics* (see Section 11.8.1) can be sintered at lower temperatures than those for conventional ceramics. Nanophase ceramics are also easier to fabricate and can be compacted at room temperature to high densities, hot pressed to theoretical density, and formed into net-shape parts without binders or sintering aids.

## 11.9.5  Finishing operations

After firing, additional operations may be performed to give the part its final shape, remove surface flaws, and improve surface finish and dimensional tolerances. The processes that can be used are (a) grinding, (b) lapping, (c) ultrasonic machining, (d) electrical-discharge machining, (e) laser-beam machining, (f) abrasive water-jet machining, and (g) tumbling to remove sharp edges and grinding marks. The choice of the process is important in view of the brittle nature of most ceramics and the additional costs involved in these processes. The effect of the finishing operation on the properties of the product must also be considered; because of notch sensitivity, the finer the finish, the higher is the part's strength. To improve appearance and strength, and to make them impermeable, ceramic products are often coated with a **glaze** material, which forms a glassy coating after firing.

**EXAMPLE 11.8**  **Dimensional changes during shaping of ceramic components**

A solid cylindrical ceramic part is to be made whose final length must be $L = 20$ mm. It has been established that for this material, linear shrinkages during drying and firing are 7% and 6%, respectively, based on the dried dimension $L_d$. Calculate (1) the initial length $L_o$ of the part and (2) the dried porosity $P_d$ if the porosity of the fired part, $P_f$, is 3%.

**Solution.**

(1) On the basis of the information given, and remembering that firing is preceded by drying, we can write

$$\frac{(L_d - L)}{L_d} = 0.06,$$

or

$$L = (1 - 0.06)L_d;$$

hence,

$$L_d = \frac{20}{0.94} = 21.28 \text{ mm}$$

and

$$L_o = (1 + 0.07)L_d = (1.07)(21.28) = 22.77 \text{ mm}.$$

(2) Since the final porosity is 3%, the actual volume $V_a$ of the ceramic material is

$$V_a = (1 - 0.03)V_f = 0.97V_f,$$

where $V_f$ is the fired volume of the part. Since the linear shrinkage during firing is 6%, we can determine the dried volume $V_d$ of the part as

$$V_d = \frac{V_f}{(1 - 0.06)^3} = 1.2V_f.$$

Hence,

$$\frac{V_a}{V_d} = \frac{0.97}{1.2} = 0.81 = 81\%.$$

Therefore, the porosity $P_d$ of the dried part is 19%.

# 11.10  Glasses: Structure, Properties, and Applications

*Glass* is an amorphous solid with the structure of a liquid. In other words, it has been *supercooled*, that is, cooled at a rate too high for crystals to form. Generally, we define glass as an inorganic product of fusion that has cooled to a rigid condition without crystallizing. Glass has no distinct melting or freezing point; thus, its behavior is similar to that of amorphous polymers. At present, there are some 750 different types of commercially available glasses. The uses of glass range from windows, bottles, and cookware to glasses with special mechanical, electrical, high-temperature, chemical-resistant, corrosion-resistant, and optical characteristics. Special glasses are used in fiber optics for communication, with little loss in signal power, and in glass fibers with very high strength for reinforced plastics.

All glasses contain at least 50% silica, which is known as a *glass former*. The composition and properties of glasses, except strength, can be modified greatly by the addition of oxides of aluminum, sodium, calcium, barium, boron, magnesium, titanium, lithium, lead, and potassium. Depending on their function, these oxides are known as *intermediates* or *modifiers*. Glasses are generally resistant to chemical attack and are ranked by their resistance to acid, alkali, or water corrosion.

## 11.10.1  Types of glasses

Almost all commercial glasses are categorized by type (Table 11.9):

1. **Soda-lime glass** (the most common type);
2. **Lead-alkali glass**;
3. **Borosilicate glass**;
4. **Aluminosilicate glass**;
5. **96% silica glass**;
6. **Fused silica.**

Glasses are also classified as colored, opaque (white and translucent), multiform (variety of shapes), optical, photochromatic (darkens when exposed to light, as in some types of sunglasses), photosensitive (changes from clear to opal), fibrous (drawn into long fibers, as in fiberglass), and foam or cellular glass (containing bubbles, hence a good thermal insulator). Glasses are also referred to as **hard** or **soft**, usually in the sense of a thermal property rather than a mechanical property, as in hardness; thus, a soft glass softens at a lower temperature than does a hard glass. Soda-lime and lead-alkali glasses are considered soft, and the rest of the types of glass are considered hard.

**TABLE 11.9**

### General Characteristics of Various Types of Glasses

|  | Soda-lime Glass | Lead Glass | Borosilicate Glass | 96% Silica Glass | Fused Silica |
|---|---|---|---|---|---|
| Density | High | Highest | Medium | Low | Lowest |
| Strength | Low | Low | Moderate | High | Highest |
| Resistance to thermal shock | Low | Low | Good | Better | Best |
| Electrical resistivity | Moderate | Best | Good | Good | Good |
| Hot workability | Good | Best | Fair | Poor | Poorest |
| Heat treatability | Good | Good | Poor | None | None |
| Chemicals resistance | Poor | Fair | Good | Better | Best |
| Impact abrasion resistance | Fair | Poor | Good | Good | Best |
| Ultraviolet-light transmission | Poor | Poor | Fair | Good | Good |
| Relative cost | Lowest | Low | Medium | High | Highest |

### 11.10.2 Mechanical properties

For all practical purposes, we regard the behavior of glass, as for most ceramics, as perfectly elastic and brittle. The range of modulus of elasticity for most commercial glasses is 55–90 GPa (8–13 million psi), and the Poisson's ratio of glasses ranges from 0.16–0.28. The hardness of glasses, as a measure of resistance to scratching, ranges from 5–7 on the Mohs scale, equivalent to a range of approximately 350–500 HK.

Glass in *bulk* form has a strength of less than 140 MPa (20 ksi). The relatively low strength of bulk glass is attributed to the presence of small flaws and microcracks on the surface of the glass, some or all of which may be introduced during normal handling of the glass by inadvertent abrading. These defects reduce the strength of glass by two to three orders of magnitude, compared with its ideal (defect-free) strength. Glasses can be strengthened by thermal or chemical treatments to obtain high strength and toughness.

The strength of glass can theoretically reach as high as 35 GPa (5 million psi). When molten glass is freshly drawn into fibers (*fiberglass*), its tensile strength ranges from 0.2 GPa to 7 GPa (30 ksi to 1000 ksi), with an average value of about 2 GPa (300 ksi). Thus, glass fibers are stronger than steel and are used to reinforce plastics in applications such as boats, automobile bodies, furniture, and sports equipment. The strength of glass is usually measured by bending it. The surface of the glass is first thoroughly abraded (roughened) to ensure that the test gives a reliable strength level in actual service under adverse conditions. The phenomenon of *static fatigue* observed in ceramics (see Section 11.8.2) is also exhibited by glasses. If a glass item must withstand a load for 1000 hours or longer, the maximum stress that can be applied to it is approximately one third the maximum stress that the same item can withstand during the first second of loading.

### 11.10.3 Physical properties

Glasses have low thermal conductivity and high electrical resistivity and dielectric strength. Their thermal expansion coefficient is lower than those for metals and plastics and may even approach zero. Titanium-silicate glass (a clear, synthetic high-silica glass), for example, has a near-zero coefficient of thermal expansion. Fused silica, a clear, synthetic amorphous silicon dioxide of very high purity, also has a near-zero coefficient of expansion. (See Fig. 11.20.) Optical properties of glasses, such as reflection, absorption, transmission, and refraction, can be modified by varying their composition and treatment.

### 11.10.4 Glass ceramics

Although glasses are amorphous, *glass ceramics* (such as Pyroceram, a trade name) have a high crystalline component to their microstructure. Glass ceramics contain large proportions of several oxides, and hence their properties are a combination of those for glass and ceramics. Most glass ceramics are stronger than glass. These products are first shaped and then heat treated, with **devitrification** (recrystallization) of the glass occurring. Unlike most glasses, which are clear, glass ceramics are generally white or gray in color.

The hardness of glass ceramics ranges approximately from 520 HK to 650 HK. They have a near-zero coefficient of thermal expansion; hence, they have good thermal shock resistance and high strength, because of the absence of the porosity usually found in conventional ceramics. The properties of glass ceramics can be improved by modifying their composition and by heat-treatment techniques. First developed in 1957, glass ceramics are suitable for cookware, heat exchangers for gas-turbine engines, radomes (housings for radar antenna), and electrical and electronics applications.

# 11.11 | Forming and Shaping Glass

Glass products can generally be categorized as

- Flat sheet or plate, ranging in thickness from about 0.8 mm to 10 mm (0.03 in. to 0.4 in.), such as window glass, glass doors, and glass tabletops;

- Rods and tubing used for chemicals, neon lights, and decorative artifacts;

- Discrete products, such as bottles, vases, headlights, and television tubes;

- Glass fibers to reinforce composite materials and for fiber optics.

All glass forming and shaping processes begin with molten glass, which has the appearance of red-hot viscous syrup, supplied from a melting furnace or tank.

Traditionally, **flat-sheet glass** has been made by drawing or rolling from the molten state and, more recently, by the **float** method, all of which are continuous processes. In the **drawing** process, the molten glass passes through a pair of rolls, similar to an old-fashioned clothes wringer. The solidifying glass is squeezed between the rolls, forming a sheet, which is then moved forward over a set of smaller rolls. In the **rolling** process, the molten glass is squeezed between rollers, forming a sheet. The surfaces of the glass can be embossed with a pattern by shaping the surfaces of the rollers accordingly. Glass sheet produced by drawing and rolling has a rough surface appearance. In making plate glass, both surfaces have to be ground parallel and polished.

In the **float** method (Fig. 11.28), developed in the 1950s, molten glass from the furnace is fed into a bath of molten tin, under controlled atmosphere; the glass floats on the tin bath. The glass then moves over rollers into another chamber (called a *lehr*) and solidifies. *Float glass* has a smooth (*fire-polished*) surface and needs no further grinding or polishing.

Glass **tubing** is manufactured by the process shown in Fig. 11.29. In this process, molten glass is wrapped around a rotating hollow cylindrical or cone-shaped mandrel and is drawn out by a set of rolls. Air is blown through the mandrel to keep the

**FIGURE 11.28** The float method of forming sheet glass. *Source*: Corning Glass Works.

**FIGURE 11.29** Manufacturing process for glass tubing. Air is blown through the mandrel to keep the tube from collapsing. *Source*: Corning Glass Works.

glass tube from collapsing. The machines used in this process may be horizontal, vertical, or slanted downward. Glass **rods** are made in a similar manner, but air is not blown through the mandrel; thus, the drawn product becomes a solid rod.

Continuous **fibers** are drawn through multiple (200 to 400) orifices in heated platinum plates, at speeds as high as 500 m/s (1700 ft/s). Fibers as small as 2 $\mu$m (80 $\mu$in.) in diameter can be produced by this method. In order to protect their surfaces, the fibers are subsequently coated with chemicals. Short glass fibers, used as thermal insulating material (**glass wool**) or for acoustic insulation, are made by a *centrifugal spraying process* in which molten glass is fed into a rotating head.

### 11.11.1 Making discrete glass products

Several processes are used for making discrete glass objects. These processes include blowing, pressing, centrifugal casting, and sagging.

**Blowing.** The *blowing* process is used to make hollow, thin-walled glass items, such as bottles and flasks, and is similar to blow molding of thermoplastics. Figure 11.30 shows the steps involved in the production of an ordinary glass bottle by the blowing

**FIGURE 11.30** Stages in manufacturing a common glass bottle. *Source*: F. H. Norton.

**FIGURE 11.31**
Manufacturing a glass item by pressing in a mold. *Source*: Corning Glass Works.

Stage 1 — Empty mold    Stage 2 — Loaded mold    Stage 3 — Glass pressed    Stage 4 — Finished piece

process. In this process, blown air expands a hollow gob of heated glass against the walls of a mold. The mold is usually coated with a parting agent, such as oil or emulsion, to prevent the part from sticking to it. The surface finish of products made by the blowing process is acceptable for most applications. Although it is difficult to control the wall thickness of the product, this process is used for high rates of production. Lightbulbs are made in automatic blowing machines, at a rate of over 1000 bulbs per minute.

**Pressing.**  In *pressing*, a gob of molten glass is placed in a mold and is pressed into shape with the use of a plunger. The mold may be made in one piece (Fig. 11.31), or it may be a split mold (Fig. 11.32). After pressing, the solidifying glass acquires the shape of the cavity between the mold and the plunger. Because of the confined environment, the product has greater dimensional accuracy than can be obtained with blowing. However, pressing cannot be used on thin-walled items or for parts such as bottles from which the plunger cannot be retracted.

**Centrifugal casting.**  Also known as *spinning* in the glass industry, the *centrifugal-casting* process is similar to that for metals. In this process, the centrifugal force pushes the molten glass against the mold wall, where it solidifies. Typical products include picture tubes for televisions and missile nose cones.

**Sagging.**  Shallow dish-shaped or lightly embossed glass parts can be made by the *sagging* process. In this process, a sheet of glass is placed over the mold and is heated. The glass sags by its own weight and takes the shape of the mold. The process is similar to thermoforming with thermoplastics, but without pressure or a vacuum. Typical applications include dishes, lenses for sunglasses, mirrors for telescopes, and lighting panels.

## 11.11.2  Techniques for treating glass

As produced by the methods described in this chapter, glass can be strengthened by thermal tempering, chemical tempering, and laminating. Glass products may also be subjected to annealing and other finishing operations. These treatment techniques are described as follows.

Step 1 — Empty mold    Step 2 — Loaded mold    Step 3 — Glass pressed    Step 4 — Finished product

**FIGURE 11.32**  Pressing glass in a split mold. *Source*: Shand, E. B., Glass Engineering Handbook. New York, McGraw-Hill, 1958.

**FIGURE 11.33** Residual stresses in tempered glass plate, and stages involved in inducing compressive surface residual stresses for improved strength.

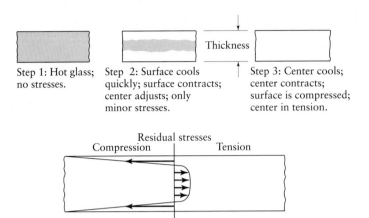

Step 1: Hot glass; no stresses.

Step 2: Surface cools quickly; surface contracts; center adjusts; only minor stresses.

Thickness

Step 3: Center cools; center contracts; surface is compressed; center in tension.

Residual stresses

Compression          Tension

a)  In **thermal tempering** (also called *physical tempering* or *chill tempering*), the surfaces of the hot glass are cooled rapidly (Fig. 11.33). As a result, the surfaces shrink, and because the bulk is still hot, tensile stresses develop on the surfaces. As the bulk of the glass begins to cool, it contracts. The solidified surfaces are now forced to contract, thus developing residual compressive surface stresses and interior tensile stresses. Compressive surface stresses improve the strength of the glass, as they do in other materials. Note that the higher the coefficient of thermal expansion of the glass and the lower its thermal conductivity, the higher is the level of residual stresses developed and, hence, the stronger the glass becomes. Thermal tempering takes a relatively short time (measured in minutes) and can be applied to most glasses. Because of the large amount of energy stored from residual stresses, **tempered glass** shatters into a large number of pieces when broken.

b)  In **chemical tempering**, the glass is heated in a bath of molten $KNO_3$, $K_2SO_4$, or $NaNO_3$, depending on the type of glass. Ion exchange takes place, with larger atoms replacing the smaller atoms on the surface of the glass. As a result, residual compressive stresses develop on the surface. This condition is similar to that created by forcing a wedge between two bricks in a wall. The time required for chemical tempering is longer (about one hour) than for thermal tempering. Chemical tempering may be performed at various temperatures. At low temperatures, distortion of the part is minimal, and complex shapes can be treated. At elevated temperatures, there may be some distortion of the part, but the product can be used at higher temperatures without loss of strength.

c)  Another strengthening method is to make **laminated glass**, which is two pieces of flat glass with a thin sheet of tough plastic between them; the process is also called **laminate strengthening**. When laminated glass is broken, its pieces are held together by the plastic sheet. You probably have seen this phenomenon in a shattered automobile windshield.

**Finishing operations.**  As in metal products, residual stresses can develop in glass products if the products are not cooled slowly enough. In order to ensure that the product is free from these stresses, it is *annealed* by a process similar to stress-relief annealing of metals. In this process, the glass is heated to a certain temperature and cooled gradually. Depending on the size, thickness, and type of glass, annealing times may range from a few minutes to as long as 10 months, as in the case of a 600-mm (24-in.) mirror for a telescope. In addition to annealing, glass products may be subjected to further operations, such as cutting, drilling, grinding, and polishing. Sharp edges and corners can be smoothened by grinding, as in glass tops for desks and shelves, or by holding a torch against the edges (**fire polishing**), which rounds them by localized softening and surface tension.

## 11.12 | Design Considerations for Ceramics and Glasses

Ceramic and glass products require careful selection of composition, processing methods, finishing operations, and methods of assembly into other components. Limitations such as general lack of tensile strength, sensitivity to defects, and low impact toughness are important. These limitations have to be balanced against desirable characteristics, such as their hardness, resistance to scratching, compressive strength at room and elevated temperatures, and diverse physical properties. Control of processing parameters and the quality of and level of impurities in the raw materials used are important. As in all design decisions, there are priorities, limitations, and various factors that should be considered, including the number of parts needed and the costs of tooling, equipment, and labor.

The possibilities of dimensional changes, warping and cracking during processing are significant factors in selecting methods for shaping these materials. When a ceramic or glass component is part of a larger assembly, compatibility with other components is another important consideration. Particularly important in such cases are thermal expansion (as in seals) and the type of loading. The potential consequences of part failure are always a significant factor in designing ceramic products.

## 11.13 | Graphite and Diamond

### 11.13.1 Graphite

*Graphite* is a crystalline form of carbon that has a *layered structure* of basal planes or sheets of close-packed carbon atoms. Consequently, graphite is weak when sheared along the layers. This characteristic, in turn, gives graphite its low frictional properties as a solid lubricant. However, its frictional properties are low only in an environment of air or moisture; graphite is abrasive and a poor lubricant in a vacuum.

Although brittle, graphite has high electrical and thermal conductivity and resistance to thermal shock and high temperature, although it begins to oxidize at 500°C, (930°F). It is therefore an important material for applications such as electrodes, heating elements, brushes for motors, high-temperature fixtures and furnace parts, mold materials such as crucibles for melting and casting of metals, and seals (because of its low friction and high wear resistance). Unlike other materials, the strength and stiffness of graphite increases with temperature.

An important use of graphite is as **fibers** in composite materials and reinforced plastics (Chapter 10). A characteristic of graphite is its resistance to chemicals; hence, it is used as filters for corrosive fluids. Furthermore, its low thermal-neutron-absorption cross-section and high scattering cross-section make graphite suitable for nuclear applications. Ordinary pencil "lead" is a mixture of graphite and clay.

Graphite is generally graded in terms of decreasing order of grain size: *industrial, fine grain,* and *micrograin*. As with ceramics, the mechanical properties of graphite improve with decreasing grain size. Micrograin graphite can be impregnated with copper and is used as electrodes for electrical-discharge machining and for furnace fixtures. Amorphous graphite is known as *lampblack* (black soot) and is used as a pigment. Graphite is usually processed by molding or forming, oven baking, and then machining to the final shape. It is available commercially in square, rectangular, or round shapes of various sizes.

A more recent development is the production of soccer-ball-shaped carbon molecules, called *buckyballs* (from *Buckminster*, after Buckminster Fuller, 1895–1983, inventor of the geodesic dome). Also called **fullerenes** (after *Fuller*), these chemically inert spherical molecules are produced from soot and act much like solid-lubricant particles. Fullerenes become superconductors when mixed with metals. Another development is **microcellular carbon foam**, which has uniform porosity and isotropic-strength characteristics. It is used, for example, as reinforcing components in aerospace structures that can be shaped directly.

### 11.13.2 Diamond

The second principal form of carbon is *diamond*, which has a covalently bonded structure. It is the hardest substance known (7000 HK to 8000 HK). This characteristic makes diamond an important cutting-tool material (Section 8.6.9), either as a single crystal or in polycrystalline form, as an abrasive in grinding wheels (Section 9.2); and for dressing of grinding wheels (sharpening of abrasive grains). Diamond is also used as a die material for drawing thin wire, with a diameter of less than 0.06 mm (0.0025 in.). Diamond is brittle and it begins to decompose in air at about 700°C (1300°F); in nonoxidizing environments it resists higher temperatures.

**Synthetic diamond**, or **industrial diamond**, was first made in 1955 and is used extensively for industrial applications. One method of manufacturing it is to subject graphite to a hydrostatic pressure of 14 GPa (2 million psi) and a temperature of 3000°C (5400°F). Synthetic diamond is identical to natural diamond for industrial applications and has superior properties because of its lack of impurities. It is available in various sizes and shapes, the most common abrasive grain size being 0.01 mm (0.004 in.) in diameter. *Gem-quality synthetic diamond* is now being made with electrical conductivity 50 times higher than that for natural diamond and 10 times more resistance to laser damage; potential applications include heat sinks for computers, products in the telecommunications and integrated-circuit industries, and windows for high-power lasers. Diamond particles can also be *coated* (with nickel, copper, or titanium) for improved performance in grinding operations.

A more recent development is **diamondlike carbon** (DLC). It is used as a coating, as described in Section 4.5.1.

## 11.14 | Processing Metal-Matrix and Ceramic-Matrix Composites

New developments in composite materials are continually taking place, with a wide range and form of polymeric, metallic, and ceramic materials being used both as fibers and as matrix materials. Research and development activities in this area are concerned with improving strength, toughness, stiffness, resistance to high temperatures, and reliability in service.

### 11.14.1 Metal-matrix composites

The advantage of a metal matrix over a polymer matrix is the former's higher resistance to elevated temperatures and higher ductility and toughness. The limitations of the metal matrix are higher density and greater difficulty in processing components. The matrix materials in these composites are usually aluminum, aluminum–lithium, magnesium, and titanium, although other metals are also being investigated for this purpose. The fiber materials are graphite, aluminum oxide, silicon carbide, and boron, with beryllium and tungsten as other possibilities.

Because of their high specific stiffness, light weight, and high thermal conductivity, boron fibers in aluminum matrix have been used for structural tubular supports

TABLE 11.10

**Metal-Matrix Composite Materials and Typical Applications**

| Fiber | Matrix | Typical Applications |
|---|---|---|
| Graphite | Aluminum | Satellite, missile, and helicopter structures |
| | Magnesium | Space and satellite structures |
| | Lead | Storage-battery plates |
| | Copper | Electrical contacts and bearings |
| Boron | Aluminum | Compressor blades and structural supports |
| | Magnesium | Antenna structures |
| | Titanium | Jet-engine fan blades |
| Alumina | Aluminum | Superconductor restraints in fusion power reactors |
| | Lead | Storage-battery plates |
| | Magnesium | Helicopter transmission structures |
| Silicon carbide | Aluminum, titanium | High-temperature structures |
| | Superalloy (cobalt base) | High-temperature engine components |
| Molybdenum, tungsten | Superalloy | High-temperature engine components |

in the Space Shuttle Orbiter. Other applications include bicycle frames and sporting goods. Studies of techniques for optimum bonding of fibers to the metal matrix are in progress. Current applications of metal-matrix composites are in gas turbines, electrical components, and various structural components (Table 11.10).

There are three methods of manufacturing these composites into near-net-shape parts: liquid-phase processes, solid-phase processes, and two-phase (liquid–solid) processes. These methods are described as follows:

a) **Liquid-phase processing** consists basically of casting the liquid matrix and the solid reinforcement, using either conventional casting processes or pressure-infiltration casting. In the latter process, pressurized gas is used to force the liquid matrix metal into a perform (usually as sheet or wire) made of the reinforcing fibers.

b) **Solid-phase processing** consists basically of powder-metallurgy techniques, including cold and hot isostatic pressing. Proper mixing for homogeneous distribution of the fibers is important. A sample use of this technique, employed in tungsten-carbide tool and die manufacturing with cobalt as the matrix material, is discussed in Example 11.6. In making complex MMC parts with whisker or fiber reinforcement, die geometry and control of process variables are very important in ensuring proper distribution and orientation of the fibers within the part. MMC parts made by powder-metallurgy processes are generally heat treated for optimum properties.

c) The techniques used for **two-phase processing** consist of *rheocasting*, described in Section 5.10.6, and *spray atomization and deposition*. In the latter processes, the reinforcing fibers are mixed with a matrix that contains both liquid and solid phases.

**EXAMPLE 11.9**  Metal-matrix-composite brake rotors and cylinder liners

Brake rotors are currently being made of composites consisting of an aluminum matrix with 20% silicon-carbide (SiC) particles. The particles are stirred into molten aluminum alloys and cast into ingots. The ingots are then remelted and cast into shapes, such as brake rotors and drums, using processes such as green sand, bonded sand, investment, permanent mold, and squeeze casting. (See Chapter 5.) The

MMC brake rotors weigh about half as much as do gray-cast-iron rotors, conduct heat three times faster, exhibit the stiffness and wear-resistance of ceramics, and reduce noise and vibration.

To improve the wear- and heat-resistance of cast-iron cylinder liners in aluminum engine blocks, aluminum-matrix liners are being developed. The MMC layer consists of 12% aluminum-oxide ($Al_2O_3$) fiber and 9% graphite fiber, in a thickness that ranges from 1.5 mm to 2.5 mm (0.06 in. to 0.1 in.).

### 11.14.2 Ceramic-matrix composites

Ceramic-matrix composites are another important recent development in engineered materials. As described earlier, ceramics are strong and stiff and resist high temperatures, but generally lack toughness. Silicon carbide, silicon nitride, aluminum oxide, and mullite (a compound of aluminum, silicon, and oxygen) are new matrix materials that retain their strength to 1700°C (3100°F). Under development are carbon–carbon-matrix composites that retain much of their strength up to 2500°C (4500°F), although they lack oxidation resistance at high temperatures. Present applications for ceramic-matrix composites are in jet and automotive engines, equipment for deep-sea mining, pressure vessels, and various structural components.

There are several processes used to make these composites; three commons ones are briefly described next. In addition to these processes, new techniques such as melt infiltration, controlled oxidation, and hot-press sintering, which are still largely in the experimental stage, are being developed for improving the properties and performance of these composites. The three common processes for making ceramic-matrix composites are described as follows:

a) **Slurry infiltration,** the most common process, involves the preparation of a fiber preform that is hot pressed and then impregnated with a slurry that contains the matrix powder, a carrier liquid, and an organic binder. A further improvement is *reaction bonding* or *reaction sintering* of the slurry. High strength, toughness, and uniform structure are obtained by the slurry-infiltration process, but the product has limited high-temperature properties, due to the low melting temperature of the matrix materials used.

b) **Chemical-synthesis** processes involve the solgel and polymer-precursor techniques. In the *solgel* process, a sol (a colloidal fluid with the liquid as the continuous phase) containing fibers is converted to a gel, which is then subjected to heat treatment to produce a ceramic-matrix composite. The *polymer-precursor* method is analogous to the process used in making ceramic fibers.

c) In **chemical-vapor infiltration,** a porous fiber preform is infiltrated with the matrix phase, using the chemical vapor deposition technique (see Section 4.5.1). The product has very good high-temperature properties, but the process is costly and time consuming.

### 11.14.3 Other composites

Composites may also consist of coatings of various kinds on base metals or substrates. Examples include plating of aluminum and other metals over plastics for decorative purposes and enamels, dating to before 1000 BC, or applying similar vitreous (glasslike) coatings on metal surfaces for various functional or ornamental purposes.

Composites such as cemented carbides, usually tungsten carbide and titanium carbide, with cobalt and nickel, respectively, as a binder, are made into tools and dies. Other composites include grinding wheels made of aluminum-oxide, silicon-carbide, diamond, or cubic-boron-nitride abrasive particles, held together with various organic, inorganic, or metallic binders. Another category of composites is cermets.

A composite developed relatively recently is granite particles in an epoxy matrix. (See Section 8.12.) It has high strength, good vibration-damping capacity (better than that of gray cast iron), and good frictional characteristics. This composite is used as machine-tool beds for some precision grinders.

# 11.15 | Processing Superconductors

Although superconductors (Section 3.9.6) have major energy-saving potential in the generation, storage, and distribution of electrical power, their processing into useful shapes and sizes for practical applications presents significant difficulties. Two basic types of superconductors are metals [**low-temperature superconductors** (LTSCs), including combinations of niobium, tin, and titanium] and ceramics [**high-temperature superconductors** (HTSCs), including various copper oxides]. Here, *high temperature* means closer to *ambient* temperature, hence the HTSCs are of more practical use.

Ceramic superconducting materials are available in powder form. The fundamental difficulty in manufacturing them is their inherent brittleness and anisotropy, which make it difficult to align the grains in the proper direction for high efficiency; the smaller the grain size, the more difficult it is to align the grains.

The basic manufacturing process for superconductors consists of the following steps:

a)  Preparing the powder, mixing it, and grinding it in a ball mill to a grain size of 0.5 mm to 10 mm,

b)  Forming it into shape, and

c)  Heat treating to improve grain alignment.

The most common forming process is **oxide powder in tube** (OPIT). In this process, the powder is packed into silver tubes (because silver has the highest electrical conductivity of all metals) and sealed at both ends. The tubes are then mechanically worked, by such deformation processes as swaging, drawing, extrusion, isostatic pressing, and rolling, into final shapes, which may be wire, tape, coil, or bulk.

Other principal superconductor-shaping processes include (1) coating of silver wire with superconducting material, (2) deposition of superconductor films by laser ablation, (3) the doctor-blade process (see Section 11.9.1), (4) explosive cladding (see Section 12.10), and (5) chemical spraying.

---

**CASE STUDY** | Production of High-Temperature Superconducting Tapes

In recent years, significant progress has been made in understanding high-temperature superconducting materials and their potential use as conductors. Two bismuth-based oxides ($Bi_2Sr_2Ca_1Cu_2O_x$, also known as Bi-2212, and $Bi_2Sr_2Ca_2Cu_3O_x$, or Bi-2223) are superconducting, ceramic materials of choice for various military and commercial applications, such as electrical propulsion for ships and submarines, shallow-water and ground minesweeping systems, transmission cable generators, and superconducting magnetic-energy storage (SMES). However, wide variations in the electrical performance of test magnet coils using *monofilament* superconducting strips indicate that *multifilament* conductors are essential in reliably achieving the required critical current densities, so that property variations are averaged over a length. Tape stacking is a new approach to manufacturing multifilament tape. In this method, 7 to 10 monofilament tapes or strips are stacked, heated, and pressed together; the resulting laminate is then rolled to final thickness.

**FIGURE 11.34**    Schematic illustration of the powder-in-tube process. *Source*: Copyright © Concurrent Technologies Corporation (CTC) 1998 used with permission.

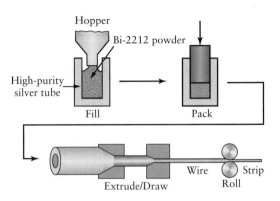

The quality and performance of multifilament tapes are directly related to that of the constituents. A variety of processes have been explored to produce wires and tapes, using Bi-2212 and Bi-2223. The powder-in-tube process (Fig. 11.34) has been used successfully to fabricate long lengths of Bi-based wires and tapes with desired properties. This case study discusses a prior implementation of this method for the production of high-temperature superconducting multifilament tapes. Although this case study concerns only Bi-2212, similar processing can be used for Bi-2223.

Production of the monofilament tapes involves the following steps:

a)  Production of a composite billet, using a silver casing and Bi-2212 powder;

b)  Extrusion and drawing of the billet to reduce the diameter and increase the powder density;

c)  Rolling to final dimensions.

**Composite-billet preparation.**   The composite billets were approximately 254 mm (10 in.) long, with an outside diameter of 6.35 mm (0.25 in.), and the wall thickness of the silver tube was 1 mm. The as-received high-purity silver tubes were cleaned and annealed at 400°C (750°F) for one hour to improve ductility. They were subsequently filled with fine-grained Bi-2212 powder (with a median particle size of 2.78 $\mu$m, or 110 $\mu$in.) that had been baked at 105°C (220°F) for an hour to remove moisture. The powder was poured into the silver container in an inert atmosphere and compacted to a 30% relative density, using a steel ram. In order to minimize density gradients (see Fig. 11.7), about one gram of powder was added to the billet for each stroke of the ram. Each billet was weighed and measured to verify the initial packing density. The ends of each billet were then sealed with a silver alloy to avoid contamination during subsequent deformation processing.

**Wire drawing.**   A swaging machine was used for preparing the end of the composite wire for drawing. To measure elongation, two circumferential score markers, 25.4 mm apart, were placed on each billet and measured after each pass. Billets were drawn down to a final diameter of 1.63 mm on a drawing bench, using 12 passes of 20.7% reduction per pass. The drawing speed was approximately 1.4 m/min, and the dies had a semicone angle of 8°. A semisoluble oil and zinc-stearate spray were used as a lubricant.

Figure 11.35a shows the evolution of the silver-to-oxide (area) ratio with the drawing pass number while Fig. 11.35b shows the same for the powder density. Note that both quantities increase rapidly during the initial passes and tend to saturate towards the end of the process.

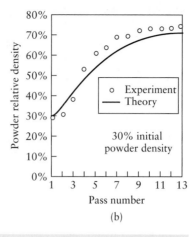

In addition to measuring these quantities directly from tests, a process model was developed using the finite-element method. This model enables prediction of stresses, strains and other quantities of interest resulting from the complex deformation the material undergoes. In the current case, it was used to model both the wire-drawing and tape-rolling (described later) processes. The evolution of the powder relative density and silver-to-oxide ratio obtained from the computer-based models is shown in Fig. 11.35. The models accurately predict these quantities, demonstrating their potential to conduct parametric studies and prescribe an optimum reduction schedule to achieve the desired makeup. The model can also be used to study the influence of die geometry, friction conditions, initial powder density, powder type, and initial silver tube thickness on dependent variables such as the drawing force, wire geometry, final core density, potential for sheath fracture, and mechanical properties of the superconducting wire. This information can reduce the number of experimental trial-and-error attempts needed for optimizing the process, thereby reducing the cost of the process.

**Rolling.** Following the drawing process, the wire was progressively transformed into tape, using a single-stand rolling mill in two-high and four-high configurations. (See Figs. 6.42a and c.) For the four-high case, the diameter of the backup rolls (which are the work rolls for the two-high configuration) was 213 mm, and the diameter of the work rolls was 63.5 mm. The mill was also equipped with payoff and take-up reel systems to maintain front and back tension, respectively, on the tape. Short samples were cut after each rolling pass for metallographic examination, including optical micrographs of polished, longitudinal, and transverse cross-sections, as shown in Fig. 11.36. The final tape was 100–200 $\mu$m thick and 2–3 mm wide, with a Bi-2212 core ranging from 40–80 $\mu$m in thickness and 1.0–1.5 mm in width.

The results depicted in Figs. 11.36 reflect a manufacturing sequence that was experimentally optimized with guidance from mathematical models. The performance of the high-temperature superconductor is strongly affected by the density distribution of the oxide core, the texture of the core, the integrity of the silver–oxide interface, the absence of microcracking, and variations in filament thickness (sausaging), all of which are strongly dependent on the work-roll diameters, rolling-reduction schedule, and frictional conditions.

Typically, the oxide's microstructure is dense, elongated, and well aligned only in the high-shear regions near the silver–oxide interface. The key to producing high-$J_c$ tapes is to enhance the texture/grain structure and reduce the porosity in the oxide core during wire drawing and tape rolling. This case study has characterized four rolling schedules: two-high and four-high rolling configurations and

**FIGURE 11.36** Selected cross-sections of rolled filament. *Source*: Copyright © Concurrent Technologies Corporation (CTC) 1998 used with permission.

Pass No. 5

Pass No. 11

Pass No. 17

Pass No. 22

10% and 25% reductions per pass. The longitudinal sections of tapes from the four rolling schedules show that the four-high + 10%-reduction schedule has the smoothest silver–oxide interface.

*Source*: Courtesy of S. Vaze and M. Pandheeradi, Concurrent Technologies Corporation.

## SUMMARY

- The powder-metallurgy process is capable of economically producing relatively complex parts in net-shape form to close dimensional tolerances, from a wide variety of metal and alloy powders. (Section 11.1)

- The steps in powder-metallurgy processing are powder production, blending, compaction, sintering, and additional processing to improve dimensional accuracy, surface finish, mechanical or physical properties, or appearance. (Sections 11.2– 11.5)

- Design considerations in powder metallurgy include the shape of the compact, the ejection of the green compact from the die, and the acceptable dimensional tolerances of the application. (Section 11.6)

- From an economic standpoint, the powder-metallurgy process is suitable for medium- to high-volume production runs and for relatively small parts; it has certain competitive advantages over other processes. (Sections 11.7)

- Ceramics have typical characteristics such as brittleness, high strength and hardness at elevated temperatures, high modulus of elasticity, low toughness, low density, low thermal expansion, and low thermal and electrical conductivity. (Section 11.8)

- Three basic shaping processes for ceramics are casting, plastic forming, and pressing; the resulting product is dried and fired to give it the proper strength. Finishing operations, such as machining and grinding, may be performed to give the part its final shape, or the part may be subjected to surface treatments to improve specific properties. (Section 11.9)

- Almost all commercial glasses are categorized as one of six types that indicate their composition. Glass in bulk form has relatively low strength, but it can be strengthened by thermal or chemical treatments to obtain high strength and toughness. (Section 11.10)

- Continuous methods of glass processing are drawing, rolling, and floating. Discrete glass products can be manufactured by blowing, pressing, centrifugal casting, and sagging. After initial processing, glasses can be strengthened by thermal or chemical tempering or by laminating. (Section 11.11)

- Design considerations for ceramics and glasses are guided by factors such as general lack of tensile strength and toughness, and sensitivity to external and internal defects. Warping and cracking are important considerations, as are the methods employed for production and assembly. (Section 11.12)

- Graphite, fullerenes, and diamond are forms of carbon that display unusual combinations of properties. These materials have unique and emerging applications. (Section 11.13)

- Metal-matrix and ceramic-matrix composites possess useful combinations of properties and have increasingly broad applications. Metal-matrix composites are processed via liquid-phase, solid-phase, and two-phase processes; ceramic-matrix composites can be processed via slurry infiltration, chemical synthesis, and chemical-vapor infiltration. (Sections 11.14)

- Manufacture of superconductors into useful products is a challenging area, because of the anisotropy and inherent brittleness of the materials involved. Although other processes are being developed, the basic and common practice involves packing the powder into a silver tube and deforming it plastically into desired shapes. (Section 11.15)

## SUMMARY OF EQUATIONS

Shape factor of particles, $k = \left(\dfrac{A}{V}\right)D_{eq}$

Pressure distribution in compaction, $p_x = p_0 e^{-4\mu k x/D}$

Volume, $V_{sint} = V_{green}\left(1 - \dfrac{\Delta L}{L_0}\right)^3$

Density, $\rho_{green} = \rho_{sint}\left(1 - \dfrac{\Delta L}{L_0}\right)^3$

Tensile strength, $UTS \simeq UTS_0 e^{-nP}$
Elastic modulus, $E \simeq E_0(1 - 1.9P + 0.9P^2)$
Thermal conductivity, $k = k_0(1 - P)$

## BIBLIOGRAPHY

*ASM Engineered Materials Handbook, Desk Edition*, ASM International, 1995.

*ASM Handbook*, Vol. 7: *Powder Metal Technologies and Applications*, ASM International, 1998.

*ASM Handbook*, Vol. 21: *Composites*, ASM International, 2001.

Barsoum, M. W., *Fundamentals of Ceramics*, McGraw-Hill, 1996.

Belitskus, D. L., *Fiber and Whisker Reinforced Ceramics for Structural Applications*, Dekker, 1993.

Brook, R. J., (ed.), *Concise Encyclopedia of Advanced Ceramic Materials*, Pergamon, 1991.

Buchanan, R. C., *Ceramic Materials for Electronics: Processing, Properties, and Applications*, 2d. ed., Dekker, 1991.

Burns, M., *Automated Fabrication: Improving Productivity in Manufacturing*, PTR Prentice Hall, 1993.

Chawla, K. K., *Composite Materials*, 2d. ed., Springer, 1998.

Cheremisinoff, N. P., *Handbook of Ceramics and Composites*, 3 vols., Dekker, 1991.

Chiang, Y.-M., *Physical Ceramics: Principles for Ceramics Science and Engineering*, Wiley, 1995.

Clyne, T. W., and P. J. Withers, *An Introduction to Metal Matrix Composites*, Cambridge University Press, 1993.

*Concise Encyclopedia of Advanced Ceramics*, The MIT Press, 1991.

Cranmer, D. C., and D. W. Richerson, *Mechanical Testing Methodology for Ceramic Design and Reliability*, Dekker, 1998.

Doresmus, R. H., *Glass Science*, 2d. ed., Wiley, 1994.

Dowson, G., *Powder Metallurgy: The Process and Its Products*, Adam Hilger, 1991.

*Engineered Materials Handbook*, Vol. 4: *Ceramics and Glasses*, ASM International, 1991.

German, R. M., *Powder Metallurgy Science*, 2d. ed., Metal Powder Industries Federation, 1994.

_____, *Sintering Technology and Practice*, Wiley, 1996.

German, R. M., and A. Bose, *Injection Molding of Metals and Ceramics*, Metal Powder Industries Federation, 1997.

Gutowski, T. G. (ed.), *Advanced Composites Manufacturing*, Wiley, 1997.

Harper, C. A. (ed), *Handbook of Ceramics, Glasses, and Diamonds*, McGraw-Hill, 2001.

Hausner, H. H., and M. K. Mal, *Handbook of Powder Metallurgy*, 2d. ed., Chemical Publishing Co., 1982.

Hlavac, J., *The Technology of Glass and Ceramics*, Elsevier, 1983.

Hoa, S. V., *Computer-Aided Design for Composite Structures*, Dekker, 1996.

Holand, W., and G. H. Beall, *Design and Properties of Glass-Ceramics*, American Chemical Society, 2001.

Jahanmir, S., *Friction and Wear of Ceramics*, Dekker, 1994.

Jones, J. T., and M. F. Berard, *Ceramics—Industrial Processing and Testing*, 2d. ed., Iowa State University Press, 1993.

Karlsson, L. (ed.), *Modeling in Welding, Hot Powder Forming and Casting*, ASM International, 1997.

Kelly, A. (ed.), *Concise Encyclopedia of Composite Materials*, Pergamon, 1989.

King, A. G., *Ceramics Processing and Technology*, Noyes Pub., 2001.

Kingery, W. D., H. K. Bowen, and D. R. Uhlmann, *Introduction to Ceramics*, 2d. ed., Wiley, 1976.

Kubicki, B., *Sintered Machine Elements*, Prentice Hall, 1995.

Kuhn, H. A., and B. L. Ferguson, *Powder Forging*, Metal Powder Industries Federation, 1990.

Lawley, A., *Atomization: The Production of Metal Powders*, Metal Powder Industries Federation, 1992.

Lenel, F. V., *Powder Metallurgy: Principles and Applications*, Metal Powder Industries Federation, 1980.

Lewis, M. H. (ed.), *Glasses and Glass-Ceramics*, Chapman & Hall, 1989.

Loewenstein, K. L., *The Manufacturing Technology of Continuous Glass Fibers*, 2d. ed., Elsevier, 1983.

Lu, H. Y., *Introduction to Ceramic Science*, Dekker, 1996.

Mallick, P. K. (ed.), *Composites Engineering Handbook*, Dekker, 1997.

*Nanoceramics*, Institute of Materials, 1993.

Norton, F. H., *Elements of Ceramics*, 2d. ed., Addison-Wesley, 1974.

Ochiai, S., *Mechanical Properties of Metallic Composites*, Dekker, 1994.

Phillips, G. C., *A Concise Introduction to Ceramics*, Van Nostrand Reinhold, 1991.

*Powder Metallurgy Design Guidebook*, American Powder Metallurgy Institute, revised periodically.

*Powder Metallurgy Design Manual*, 3d. ed., Metal Powder Industries Federation, 1998.

Prelas, M. A., G. Popovici, and L. K. Bigelow (eds.), *Handbook of Industrial Diamonds and Diamond Films*, Dekker, 1997.

Rahaman, M. N., *Ceramic Processing Technology and Sintering*, Dekker, 1996.

Reed, J. S., *Principles of Ceramics Processing*, 2d. ed., Wiley, 1995.

Rhodes, M. J. (ed.), *Principles of Powder Technology*, Wiley, 1990.

Richerson, D. W., *Modern Ceramic Engineering: Properties, Processing, and Use in Design*, 2d. ed., Dekker, 1992.

Schwartz, M. M., *Handbook of Composite Ceramics*, McGraw-Hill, 1992.

_____, *Handbook of Structural Ceramics*, McGraw-Hill, 1992.

Thummler, F., and K. Obercker, *An Introduction to Powder Metallurgy*, Institute of Materials, 1994.

Tooley, F. V. (ed.), *The Handbook of Glass Manufacture*, Ashlee Pub. Co., 1984.

Upadhyaya, G. S., *Sintering Metallic and Ceramic Materials: Preparation, Properties and Applications*, Wiley, 2000.

Vincenzini, P., *Fundamentals of Ceramic Engineering*, Elsevier, 1991.

Wilks, J., and E. Wilks, *Properties and Applications of Diamond*, Butterworth-Heinemann, 1991.

Yosomiya, R., *Adhesion and Bonding in Composites*, Dekker, 1990.

# QUESTIONS

## Powder metallurgy

**11.1**  Explain the advantages of blending metal powders.

**11.2**  Is green strength important in powder-metal processing? Explain.

**11.3**  Explain why density differences may be desirable within some P/M parts. Give specific examples.

**11.4**  Give the reasons that injection molding of metal powders has become an important process.

**11.5**  Describe the events that occur during sintering.

**11.6**  What is mechanical alloying, and what are its advantages over conventional alloying of metals?

**11.7**  It is possible to infiltrate P/M parts with various resins, as well as with metals. What possible benefits would result from infiltration?

**11.8**  What concerns would you have when electroplating P/M parts?

**11.9**  Describe the effects of different shapes and sizes of metal powders in P/M processing, commenting on the magnitude of the shape factor (SF) of the particles.

**11.10**  Comment on the shapes of the curves and their relative positions shown in Fig. 11.6.

**11.11**  Should green compacts be brought up to the sintering temperature slowly or rapidly? Explain the advantages and limitations of both methods.

**11.12**  Explain the effects of using fine powders and coarse powders, respectively, in making P/M parts.

**11.13**  Are the requirements for punch and die materials in powder metallurgy different than those for forging and extrusion, described in Chapter 6? Explain.

**11.14**  Describe the relative advantages and limitations of cold and hot isostatic pressing, respectively.

**11.15**  Why do mechanical and physical properties depend on the density of P/M parts? Explain with appropriate sketches.

**11.16**  What type of press is required to compact parts by the set of punches shown in Fig. 11.7d? (See Section 6.2.8.)

**11.17**  Explain the difference between impregnation and infiltration. Give some applications of each.

**11.18**  Tool steels are now being made by P/M techniques. Explain the advantages of this method over traditional methods such as casting and subsequent metal-working techniques.

**11.19**  Why do the compacting pressure and sintering temperature depend on the type of powder metal used?

## Ceramics and other materials

**11.20**  Compare the major differences between ceramics, metals, thermoplastics, and thermosets.

**11.21**  Explain why ceramics are weaker in tension than in compression.

**11.22**  Explain why the mechanical and physical properties of ceramics decrease with increasing porosity.

**11.23**  What engineering applications could benefit from the fact that, unlike metals, ceramics generally maintain their modulus of elasticity at elevated temperatures?

**11.24**  Explain why the mechanical-property data in Table 11.7 have such a broad range. What is the significance of this range in engineering applications?

**11.25**  List the factors that you would consider when replacing a metal component with a ceramic component. Give examples of such substitutions.

**11.26**  How are ceramics made tougher? Explain.

**11.27**  Describe applications in which static fatigue can be important.

**11.28**  Explain the difficulties involved in making large ceramic components. What recommendations would you make to improve the process?

**11.29**  Explain why ceramics are effective cutting-tool materials. Would ceramics also be suitable as die materials for metal forming? Explain.

**11.30**  Describe applications in which the use of a ceramic material with a zero coefficient of thermal expansion would be desirable.

**11.31**  Give reasons for the development of ceramic-matrix components. Name some present and possible future applications of these components.

**11.32**  List the factors that are important in drying ceramic components, and explain why they are important.

**11.33**  It has been stated that the higher the coefficient of thermal expansion of the glass and the lower the glass' thermal conductivity, the higher is the level of residual stresses developed during processing. Explain why.

**11.34**  What types of finishing operations are performed on ceramics? Why are they done?

**11.35**  What should be the property requirements for the metal balls used in a ball mill? Explain why these conditions are required.

**11.36**  Which properties of glasses allow them to be expanded and shaped into bottles by blowing?

**11.37**  What properties should plastic sheet have when used in laminated glass? Why?

**11.38** Consider some ceramic products that you are familiar with, and outline a sequence of processes performed to manufacture each of them.

**11.39** Explain the difference between physical and chemical tempering of glass.

**11.40** What do you think is the purpose of the operation shown in Fig. 11.24d?

**11.41** As you have seen, injection molding is a process that is used for plastics and powder metals as well as for ceramics. Why is it used for all these materials?

**11.42** Are there any similarities between the strengthening mechanisms for glass and those for other metallic and nonmetallic materials described throughout this text? Explain.

**11.43** Describe and explain the differences in the manner in which each of the following flat surfaces would fracture when struck with a large piece of rock: (a) ordinary window glass, (b) tempered glass, and (c) laminated glass.

**11.44** Describe the similarities and the differences between the processes described in this chapter and in Chapters 5 through 10.

# PROBLEMS

**11.45** Estimate the number of particles in a 500-g sample of iron powder, if the particle size is 100 $\mu$m.

**11.46** Assume that the surface of a copper particle is covered with a 0.1-$\mu$m-thick oxide layer. What is the volume occupied by this layer if the copper particle itself is 50 $\mu$m in diameter? What would be the role of this oxide layer in subsequent processing of the powders?

**11.47** Determine the shape factor for a flakelike particle with a ratio of surface cross-sectional area to thickness of 15 × 15 × 1 and for an ellipsoid with an axial ratio of 5 × 2 × 1.

**11.48** We stated in Section 3.3 that the energy in brittle fracture is dissipated as surface energy. We also noted that the comminution process for powder preparation generally involves brittle fracture. What are the relative energies involved in making spherical powders of diameters 1, 10, and 100 $\mu$m, respectively?

**11.49** Referring to Fig. 11.6a, what should be the volume of loose, fine iron powder in order to make a solid cylindrical compact 20 mm in diameter and 10 mm high?

**11.50** In Fig. 11.7e, we note that the pressure is not uniform across the diameter of the compact. Explain the reasons for this lack of uniformity.

**11.51** Plot the family of pressure-ratio $p_x/p_0$ curves as a function of $x$ for the following ranges of process parameters: $\mu$ = 0 to 1, $k$ = 0 to 1, and $D$ = 5 mm to 50 mm.

**11.52** Derive an expression similar to Eq. (11.2) for compaction in a square die with dimensions $a$ by $a$.

**11.53** For the ceramic in Example 11.7, calculate (a) the porosity of the dried part if the porosity of the fired part is to be 9%, and (b) the initial length $L_0$ of the part if the linear shrinkages during drying and firing are 8% and 7%, respectively.

**11.54** Plot the UTS, $E$, and $k$ values for ceramics as a function of porosity $P$, and describe and explain the trends that you observe in their behavior.

**11.55** Plot the total surface area of a 1-g sample of aluminum as a function of the natural log of particle size.

**11.56** How large is the grain size of metal powders that can be produced in atomization chambers? Conduct a literature search to determine the answer.

**11.57** A coarse copper powder is compacted in a mechanical press at a pressure of 30 tons/in². During sintering, the green part shrinks an additional 9%. What is the final density of the part?

**11.58** A gear is to be manufactured from iron powder. It is desired that it have a final density that is 90% of that of cast iron, and it is known that the shrinkage in sintering will be approximately 7%. For a gear 2.5 in. in diameter and a 0.75-in. hub, what is the required press force?

**11.59** What volume of powder is needed to make the gear in Problem 11.58?

**11.60** The axisymmetric part shown in the accompanying figure is to be produced from fine copper powder and is to have a tensile strength of 175 GPa. Determine the compacting pressure and the initial volume of powder needed.

Dimensions in mm

**11.61** What techniques, other than the powder-in-tube process, could be used to produce superconducting monofilaments?

**11.62** Describe other methods of manufacturing the parts shown in Fig. 11.1a. Comment on the advantages and limitations of these methods over P/M.

# DESIGN

**11.63** Describe the design features of several P/M products in which density variations would be desirable.

**11.64** Compare the design considerations for P/M products with those for (a) products made by casting and (b) products made by forging. Describe your observations.

**11.65** It is known that in the design of P/M gears, the outside diameter of the hub should be as far as possible from the root of the gear. Explain the reasons for this design consideration.

**11.66** How are the design considerations for ceramics different, if at all, than those for the other materials described in this chapter?

**11.67** Are there any shapes or design features that are not suitable for production by powder metallurgy? By ceramics processing? Explain.

**11.68** What design changes would you recommend for the part shown in Problem 11.60.

**11.69** The axisymmetric parts shown in the accompanying figure are to be produced through P/M. Describe the design changes that you would recommend.

**11.70** Assume that in a particular design, a metal beam is to be replaced with a beam made of ceramics. Discuss the differences in the behavior of the two beams, such as with respect to strength, stiffness, deflection, and resistance to temperature and to the environment.

**11.71** Describe any special design considerations in products that use ceramics with a near-zero coefficient of thermal expansion.

**11.72** Describe your thoughts regarding designs of internal combustion engines using ceramic pistons.

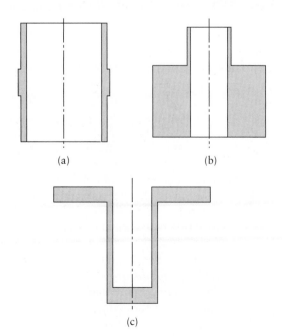

(a)          (b)

(c)

# 12

# Joining and Fastening Processes

## 12.1 | Introduction

**Joining** is a general term covering numerous processes that are important aspects of manufacturing operations. The necessity of joining and assembly of components can be appreciated simply by inspecting various products around us such as automobile bodies, bicycle frames, machines, pipes and tubing, appliances, pots and pans, and printed circuit boards. Joining may be preferred or necessary for one or more of the following reasons:

12.1 Introduction
12.2 Arc-Welding Processes: Consumable Electrode
12.3 Arc-Welding Processes: Nonconsumable Electrode
12.4 High-Energy-Beam Welding
12.5 The Fusion-Welded Joint
12.6 Cold Welding
12.7 Ultrasonic Welding
12.8 Friction Welding
12.9 Resistance Welding
12.10 Explosion Welding
12.11 Diffusion Bonding
12.12 Brazing and Soldering
12.13 Adhesive Bonding
12.14 Mechanical Fastening
12.15 Joining Plastics, Ceramics, and Glasses
12.16 Design Considerations in Joining
12.17 Economic Considerations

- The product is impossible or uneconomical to manufacture as a single piece;
- The product is easier to manufacture in individual components, which are then assembled, than as a single piece;
- The product may have to be taken apart for repair or maintenance during its service life;
- Different properties may be desirable for functional purposes of the product—for example, surfaces subjected to friction and wear, or corrosion and environmental attack, typically require characteristics different than those of the component's bulk;
- Transportation of the product in individual components and subsequent assembly may be easier and more economical than transportation of the product as a single piece.

Although there are different ways of categorizing the wide variety of joining and fastening processes, this chapter will follow the sequence shown in the section headings listed to the left. Joining methods can be categorized in terms of their common principle of operation. **Fusion welding** involves melting together and coalescing materials by means of heat, usually supplied by electrical or high-energy means; filler metals may or may not be used. Fusion welding is a major category of welding and consists of consumable- and nonconsumable-electrode *arc welding*, and *high-energy-beam welding* processes. The high-energy processes are also used for cutting and machining applications, as described in Section 9.14. Because they generally are not used in a modern industrial setting, but are used mainly for maintenance and repair work, oxyfuel-gas welding and thermit welding processes are only briefly described in this text.

In **solid-state welding**, joining takes place without fusion; thus, there is no liquid (molten) phase in the joint. The basic categories of solid-state welding are *cold*, *ultrasonic*, *friction*, *resistance*, *explosion welding*, and *diffusion bonding*; the latter, combined with superplastic forming, has become an important manufacturing process for complex shapes. The joints produced by most of the aforementioned welding processes undergo metallurgical and physical changes; these changes, in turn, have major effects on the properties and performance of the welded components.

| (a) Butt joint | (b) Corner joint | (c) T joint | (d) Lap joint | (e) Edge joint |

**FIGURE 12.1**   Examples of joints that can be made through the various joining processes described in Chapter 12.

**Brazing** and **soldering** use filler metals and involve lower temperatures than used in welding; the heat required is supplied externally. **Adhesive bonding** has been developed into an important technology, because of its various unique advantages for applications requiring strength, sealing, insulation, vibration damping, and resistance to corrosion between dissimilar or similar metals. Included in this category are electrically conducting adhesives for surface-mount technologies. **Mechanical-fastening** processes typically involve traditional methods, using a wide variety of fasteners, bolts, nuts, screws, and rivets. **Joining nonmetallic materials** can be accomplished by such means as adhesive bonding, fusion by various external or internal heat sources, diffusion, preplating with metal, and mechanical fastening.

An inspection of welded components in numerous household and industrial products will clearly indicate that a common feature of all the components is the type of joint. Although there are some variations, the basic and common joint designs are outlined in Fig. 12.1. The individual groups of joining processes briefly outlined previously have, as in all manufacturing processes, several important characteristics, such as joint design, size and shape of the parts to be joined, strength and reliability of the joint, cost and maintenance of equipment, and skill level of labor. These characteristics are outlined, in relative terms, in Tables 12.1 and 12.2 as a guide to process selection, as described in greater detail throughout this chapter.

**TABLE 12.1**

## Comparison of Various Joining Methods

| | | | | | Characteristics | | | | |
|---|---|---|---|---|---|---|---|---|---|
| Method | Strength | Design Variability | Small Parts | Large Parts | Tolerances | Reliability | Ease of Maintenance | Visual Inspection | Cost |
| Arc welding | 1 | 2 | 3 | 1 | 3 | 1 | 2 | 2 | 2 |
| Resistance welding | 1 | 2 | 1 | 1 | 3 | 3 | 3 | 3 | 1 |
| Brazing | 1 | 1 | 1 | 1 | 3 | 1 | 3 | 2 | 3 |
| Bolts and nuts | 1 | 2 | 3 | 1 | 2 | 1 | 1 | 1 | 3 |
| Riveting | 1 | 2 | 3 | 1 | 1 | 1 | 3 | 1 | 2 |
| Fasteners | 2 | 3 | 3 | 1 | 2 | 2 | 2 | 1 | 3 |
| Seaming, crimping | 2 | 2 | 1 | 3 | 3 | 1 | 3 | 1 | 1 |
| Adhesive bonding | 3 | 1 | 1 | 2 | 3 | 2 | 3 | 3 | 2 |

*Note*: 1, very good; 2, good; 3, poor.

**TABLE 12.2**

## General Characteristics of Joining Processes

| Joining Process | Operation | Advantage | Skill Level Required | Welding Position | Current Type | Distortion* | Cost of Equipment |
|---|---|---|---|---|---|---|---|
| Shielded metal arc | Manual | Portable and flexible | High | All | ac, dc | 1 to 2 | Low |
| Submerged arc | Automatic | High deposition | Low to medium | Flat and horizontal | ac, dc | 1 to 2 | Medium |
| Gas metal arc | Semiautomatic or automatic | Works with most metals | Low to high | All | dc | 2 to 3 | Medium to high |
| Gas tungsten arc | Manual or automatic | Works with most metals | Low to high | All | ac, dc | 2 to 3 | Medium |
| Flux-cored arc | Semiautomatic or automatic | High deposition | Low to high | All | dc | 1 to 3 | Medium |
| Oxyfuel | Manual | Portable and flexible | High | All | – | 2 to 4 | Low |
| Electron beam, laser beam | Semiautomatic or automatic | Works with most metals | Medium to high | All | – | 3 to 5 | High |

* 1, highest; 5, lowest

# 12.2 | Arc-Welding Processes: Consumable Electrode

In *arc welding*, developed in the mid-1800s, the heat required is obtained through electrical energy. Through use of either a *consumable* or *nonconsumable electrode* (rod or wire), an arc is produced between the tip of the electrode and the parts to be welded, using ac or dc power supplies. Arc welding includes several processes (Table 12.2), as described in the upcoming subsections.

## 12.2.1 Shielded metal arc-welding

*Shielded metal-arc welding* (SMAW) is one of the oldest, simplest, and most versatile joining processes; currently, about 50% of all industrial and maintenance welding is performed by this process. The electric arc (hence the term *electric-arc welding*) is generated by touching the tip of a coated electrode to the workpiece and then withdrawing it quickly to a distance sufficient to maintain the arc (Fig. 12.2a). The electrodes are in the shape of thin, long sticks (see Section 12.2.7); hence, this process is also known as **stick welding**. The heat generated melts a portion of the tip of the electrode, its coating, and the base metal in the immediate area of the arc. A weld forms after the molten metal (a mixture of the base metal, electrode metal, and substances from the coating on the electrode) solidifies in the weld area. The electrode coating deoxidizes and provides a shielding gas in the weld area to protect it from oxygen in the environment.

A bare section at the end of the electrode is clamped to one terminal of the power source, while the other terminal is connected to the part being welded (Fig. 12.2b). The current usually ranges between 50 A and 300 A, with power requirements generally less than 10 kW; too low a current causes incomplete fusion, and

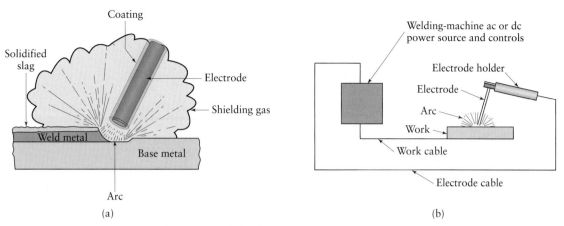

**FIGURE 12.2**    (a) Schematic illustration of the shielded metal-arc-welding process. About 50% of all large-scale industrial welding operations use this process. (b) Schematic illustration of the shielded metal-arc-welding operation, also known as *stick welding*, because the electrode is in the shape of a stick.

too high a current can damage the electrode coating and reduce its effectiveness. The current may be ac or dc; for sheet-metal welding, dc is preferred, because of the steady arc produced. The **polarity** (direction of current flow) of the dc current can be important, depending on factors such as the type of electrode, the metals to be welded, and the depth of the heated zone. In **straight polarity**, the workpiece is positive and the electrode negative; it is preferred for sheet metals, because of its shallow penetration, and for joints with very wide gaps. In **reverse polarity**, where the electrode is positive and the workpiece negative, deeper weld penetration is possible. When the current is ac, the arc pulsates rapidly; this condition is suitable for welding thick sections and using large-diameter electrodes at maximum currents.

The SMAW process requires a relatively small variety of electrodes, and the equipment consists of a power supply, power cables, and an electrode holder. This process is commonly used in general construction, shipbuilding, and for pipelines, as well as for maintenance work, since the equipment is portable and can be easily maintained. It is especially useful for work in remote areas, where portable fuel-powered generators can be used as the power supply. The SMAW process is best suited for workpiece thicknesses of 3–19 mm (0.12–0.75 in.), although this range can easily be extended by using multiple-pass techniques (Fig. 12.3). This process requires that *slag* (compounds of oxides, fluxes, and electrode coating materials) be cleaned after each weld bead; unless removed completely, the solidified slag can cause severe corrosion of the weld area and can lead to failure of the weld. Slag should also be completely removed, such as by wire brushing, before another weld is applied for multiple-pass welding; thus labor costs are high, as are the material costs.

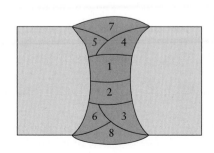

**FIGURE 12.3**    A weld zone showing the build-up sequence of individual weld beads in deep welds.

**FIGURE 12.4** Schematic illustration of the submerged arc-welding process and equipment. Unfused flux is recovered and reused.

### 12.2.2 Submerged arc welding

In *submerged arc welding* (SAW), the weld arc is shielded by granular flux, consisting of lime, silica, manganese oxide, calcium fluoride, and other elements; the flux is fed into the weld zone by gravity flow through a nozzle (Fig. 12.4). The thick layer of flux completely covers the molten metal and prevents spatter and sparks and suppresses the intense ultraviolet radiation and fumes. The flux also acts as a thermal insulator, allowing deep penetration of heat into the workpiece; the unfused flux can be recovered (using a *recovery tube*), treated, and reused.

The consumable electrode is a coil of bare wire 1.5–10 mm $\left(\frac{1}{16}-\frac{3}{8}\text{ in.}\right)$ in diameter and is fed automatically through a tube (**welding gun**). Electric currents usually range between 300 A and 2000 A; the power supplies are usually connected to standard single- or three-phase power lines with a primary rating up to 440 V. Because the flux is fed by gravity, the SAW process is somewhat limited to welds in a flat or horizontal position, with a backup piece. Circular welds can be made on pipes, provided that the pipes are rotated during welding.

The SAW process can be automated for greater economy and is used to weld a variety of carbon- and alloy-steel and stainless-steel sheet or plate, at speeds as high as 5 m/min (16 ft/min). Typical applications include thick-plate welding for shipbuilding and fabrication of pressure vessels. The quality of the weld is very high, with good toughness, ductility, and uniformity of properties. The process provides very high welding productivity, depositing 4 to 10 times the amount of weld metal per hour as deposited by the SMAW process.

### 12.2.3 Gas metal-arc welding

In *gas metal-arc welding* (GMAW), the weld area is shielded by an external source, such as argon, helium, carbon dioxide, or various other gas mixtures (Fig. 12.5a). In addition, deoxidizers are usually present in the electrode metal itself, in order to prevent oxidation of the molten weld puddle. The consumable bare wire is fed automatically through a nozzle into the weld arc (Fig. 12.5b), and multiple weld layers can be deposited at the joint. In this process, formerly called *MIG welding* (for *metal inert gas*), the metal can be transferred in one of three ways:

a) In **spray transfer**, small droplets of molten metal from the electrode are transferred to the weld area, at rates of several hundred droplets per second; the transfer is spatter free and very stable. High dc current, high voltages, and large-diameter electrodes are used, with argon or argon-rich gas mixtures used as the shielding gas. The average current required in this process can be reduced by

**FIGURE 12.5**    (a) Gas metal-arc-welding process, formerly known as MIG welding (for *metal inert gas*). (b) Basic equipment used in gas metal-arc-welding operations.

using *pulsed arcs*, which are high-amplitude pulses superimposed over a low, steady current. The process can be used for all welding positions.

b)  In **globular transfer**, carbon-dioxide rich gases are used, and globules are propelled by the forces of the electric arc that transfer the metal, thus resulting in considerable spatter. High welding currents are used, with greater weld penetration and welding speed than in spray transfer. Heavier sections are commonly joined by this method.

c)  In **short circuiting**, the metal is transferred in individual droplets, at rates of more than 50/s, as the electrode tip touches the molten weld pool and short circuits. Low currents and voltages are used, with carbon-dioxide rich gases and electrodes made of small-diameter wire. The power required is about 2 kW, and the temperatures involved are relatively low. Hence, this method is suitable only for thin sheets and sections [less than 6 mm (0.25 in.) thick]; with thicker materials, incomplete fusion may occur. This process is very easy to use and may be the most popular method for welding ferrous metals in thin sections; however, pulsed-arc systems are gaining wider usage for thin ferrous and nonferrous metals.

Developed in the 1950s, the GMAW process is suitable for a variety of ferrous and nonferrous metals and is used extensively in the metal-fabrication industry. The process is rapid, versatile, and economical; its welding productivity is double that of the SMAW process, and it can easily be automated and lends itself readily to robotics and flexible manufacturing systems (see Chapters 14 and 15).

## 12.2.4  Flux-cored arc welding

The *flux-cored arc welding* (FCAW) process (Fig. 12.6) is similar to gas metal-arc welding, with the exception that the electrode is tubular in shape and is filled with flux (hence the term *flux cored*). Cored electrodes produce a more stable arc, improve weld contour, and improve the mechanical properties of the weld metal. The flux in these electrodes is much more flexible than the brittle coating used on SMAW

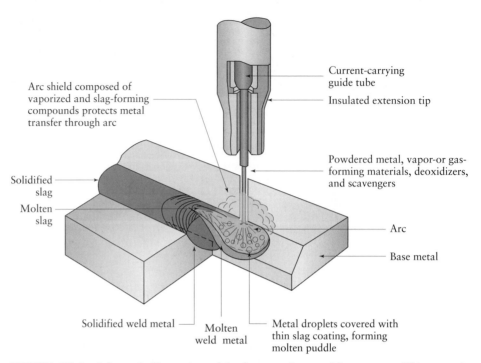

Arc shield composed of vaporized and slag-forming compounds protects metal transfer through arc

Current-carrying guide tube

Insulated extension tip

Powdered metal, vapor-or gas-forming materials, deoxidizers, and scavengers

Solidified slag

Molten slag

Arc

Base metal

Solidified weld metal

Molten weld metal

Metal droplets covered with thin slag coating, forming molten puddle

**FIGURE 12.6**   Schematic illustration of the flux-cored arc-welding process. This operation is similar to gas metal-arc welding, shown in Fig. 12.5.

electrodes; thus, the tubular electrode can be provided in long coiled lengths. The power required is about 20 kW.

The electrodes are usually 0.5–4 mm (0.020–0.15 in.) in diameter; small-diameter electrodes have made welding of thinner materials not only possible, but often desirable. Also, small-diameter electrodes make it relatively easy to weld parts out of position, and the flux chemistry enables welding of many base metals. Self-shielded cored electrodes are also available; these electrodes do not require external gas shielding, because they contain emissive fluxes that shield the weld area against the surrounding atmosphere. Advances in manufacturing of electrodes for FCAW, as well as in the chemistry of the flux, have made this process the fastest growing method in welding.

The flux-cored arc-welding process combines the versatility of SMAW with the continuous and automatic electrode-feeding feature of GMAW. It is economical and is used for welding a variety of joints, mainly with steels, stainless steels, and nickel alloys. The higher weld-metal deposition rate of FCAW over that of GMAW has led to the use of the former in joining sections of all thicknesses. The availability of tubular electrodes with very small diameters has extended the use of this process to smaller section sizes. A major advantage of FCAW is the ease with which specific weld-metal chemistries can be developed; by adding alloys to the flux core, virtually any alloy composition can be developed. This process is easy to automate and is readily adaptable to flexible manufacturing systems and robotics.

### 12.2.5 Electrogas welding

*Electrogas welding* (EGW) is used primarily for welding the edges of sections vertically in one pass with the pieces placed edge to edge (butt welding); it is classified as a machine-welding process because it requires special equipment (Fig. 12.7). The weld metal is deposited into a weld cavity between the two pieces to be joined; the space is enclosed by two water-cooled copper dams (*shoes*) to prevent the molten

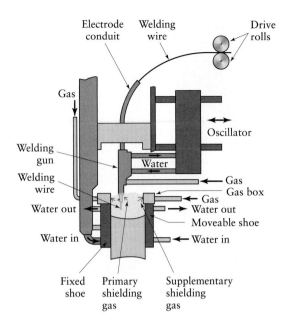

**FIGURE 12.7** Schematic illustration of the electrogas-welding process. *Source:* Courtesy of the American Welding Society.

slag from running off. Mechanical drives move the shoes upward. Circumferential welds such as on pipes are also possible, with rotation of the workpiece.

Single or multiple electrodes are fed through a conduit, and a continuous arc is maintained, using flux-cored electrodes at currents up to 750 A, or solid electrodes at 400 A; power requirements are about 20 kW. Shielding is provided by an inert gas, such as carbon dioxide, argon, or helium, depending on the type of material being welded. The gas may be provided from an external source, or it may be produced from a flux-cored electrode, or both. Weld thickness ranges from 12 mm to 75 mm (0.5 in. to 3 in.) on steels, titanium, and aluminum alloys. Typical applications are in the construction of bridges, pressure vessels, thick-walled and large-diameter pipes, storage tanks, and ships.

## 12.2.6  Electroslag welding

*Electroslag welding* (ESW) was developed in the 1950s. Its applications are similar to those for electrogas welding; the main difference between the two processes is that in electroslag welding, the arc is started between the electrode tip and the bottom of the part to be welded (Fig. 12.8). Flux is added and melted by the heat of the arc. After the molten slag reaches the tip of the electrode, the arc is extinguished; energy is supplied continuously through the electrical resistance of the molten slag. However, because the arc is extinguished, ESW is not strictly an arc-welding process. Single or multiple solid as well as flux-cored electrodes may be used, and the guide may be nonconsumable (conventional method) or consumable.

Electroslag welding is capable of welding plates with thicknesses ranging from 50 mm to more than 900 mm (2 in. to more than 36 in.). Welding is done in one pass. The current required is about 600 A at 40–50 V, although higher currents are used for thick plates. The travel speed of the weld is 12–36 mm/min (0.5–1.5 in./min). The weld quality is good, and the process is used for heavy structural steel sections, such as heavy machinery and nuclear-reactor vessels.

## 12.2.7  Electrodes for arc welding

The *electrodes* for the consumable-electrode arc-welding processes described thus far are classified according to the strength of the deposited weld metal, the current (ac or dc), and the type of coating. Electrodes are identified by numbers and letters or by

**FIGURE 12.8** Equipment used for electroslag-welding operations. *Source*: Courtesy of the American Welding Society.

color code, particularly if they are too small to imprint with identification. Typical coated-electrode dimensions are a length of 150–460 mm (6–18 in.) and a diameter of 1.5–8 mm $\left(\frac{1}{16}-\frac{5}{16}\text{ in.}\right)$. The thinner the sections to be welded and the lower the current required, the smaller the diameter of the electrode should be.

The specifications for electrodes and filler metals, including tolerances, quality-control procedures, and processes, have been provided by the American Welding Society (AWS) and the American National Standards Institute (ANSI) and are stated in the Aerospace Materials Specifications (AMS) by the Society of Automotive Engineers (SAE). The specifications require, for examples that the diameter of the wire not vary more than 0.05 mm (0.002 in.) from nominal size and that the coatings be concentric with the wire. Electrodes are sold by weight and are available in a wide variety of sizes and specifications. Information on selection and recommendations for electrodes for a particular metal and its application can be found in supplier literature and various handbooks and references listed in the bibliography at the end of this chapter.

**Electrode coatings.** Electrodes are *coated* with claylike materials that include silicate binders and powdered materials such as oxides, carbonates, fluorides, metal alloys, and cellulose (cotton cellulose and wood flour). The coating, which is brittle and has complex interactions during welding, has the following basic functions:

a) Stabilize the arc;

b) Generate gases to act as a shield against the surrounding atmosphere (the gases produced are carbon dioxide and water vapor, and carbon monoxide and hydrogen in small amounts);

c) Control the rate at which the electrode melts;

d) Act as a flux to protect the weld against the formation of oxides, nitrides, and other inclusions and, with the slag produced, thus protect the molten weld pool;

e) Add alloying elements to the weld zone to enhance the properties of the weld, including deoxidizers to prevent the weld from becoming brittle.

The deposited electrode coating or slag must be removed after each pass in order to ensure a good weld. A manual or powered wire brush can be used for this purpose. Bare electrodes and wires made of stainless steels and aluminum alloys are also available; they are used as filler metals in various welding operations.

# 12.3 | Arc-Welding Processes: Nonconsumable Electrode

Unlike the arc-welding processes that use consumable electrodes, described in the previous section, nonconsumable-electrode arc-welding processes typically use a **tungsten electrode**. As one pole of the arc, the electrode generates the heat required for welding; a shielding gas is supplied from an external source. The nonconsumable-electrode arc-welding process was first developed using carbon electrodes; hence, it was called *carbon-arc welding* (CAW). The three basic processes are described in the upcoming subsections.

## 12.3.1 Gas tungsten-arc welding

In *gas tungsten-arc welding* (GTAW), formerly known as *TIG welding* (for *tungsten inert gas*), a filler metal is typically supplied from a **filler wire** (Fig. 12.9a), although welding may be done without filler metals, as in welding close-fit joints. The composition of filler metals is similar to that of the metals to be welded, and flux is not used; the shielding gas is usually argon or helium, or a mixture of the two. Because the tungsten electrode is not consumed in this operation, a constant and stable arc gap is maintained at a constant level of current.

The power supply (Fig. 12.9b) ranges from 8 kW to 20 kW and is either dc at 200 A or ac at 500 A, depending on the metals to be welded. In general, ac is preferred for aluminum and magnesium, because the cleaning action of ac removes oxides and improves weld quality. Thorium or zirconium may be used in the tungsten electrodes to improve their electron-emission characteristics. Contamination of the tungsten electrode by the molten metal can be a significant problem, particularly in critical applications, as it can cause discontinuities in the weld; therefore, contact of the electrode with the molten metal pool should be avoided.

(a)                                    (b)

**FIGURE 12.9**    (a) Gas tungsten-arc-welding process, formerly known as TIG welding (for *tungsten inert gas*). (b) Equipment for gas tungsten-arc-welding operations.

The GTAW process is used for a wide variety of applications and metals, particularly aluminum, magnesium, titanium, and refractory metals, and it is especially suitable for thin metals. The cost of the inert gas makes this process more expensive than SMAW, but it provides welds with very high quality and surface finish.

### 12.3.2  Atomic-hydrogen welding

In *atomic-hydrogen welding* (AHW), an arc is generated between two tungsten electrodes in a shielding atmosphere of flowing hydrogen gas. The arc is maintained independently of the workpiece or parts being welded. The hydrogen gas is normally diatomic ($H_2$), but near the arc, where the temperatures are over 6000°C (11000° F), the hydrogen breaks down into its atomic form, simultaneously absorbing a large amount of heat from the arc. When the hydrogen strikes a relatively cold surface, i.e., the weld zone, it recombines into its diatomic form and releases the stored heat very rapidly. The energy in AHW can easily be varied by changing the distance between the arc stream and the workpiece surface. This process was previously much more commonly used than it is now; shielded arc-welding processes have become more popular, mainly because of the availability of inexpensive inert gases.

### 12.3.3  Plasma-arc welding

In *plasma-arc welding* (PAW), developed in the 1960s, a concentrated plasma arc is produced and aimed at the weld area; the arc is stable and reaches temperatures as high as 33,000°C (60,000° F). A *plasma* is ionized hot gas, composed of nearly equal numbers of electrons and ions. The plasma is initiated between the tungsten electrode and the orifice, using a low-current pilot arc. Unlike the arc in other processes, the plasma arc is concentrated, because it is forced through a relatively small orifice. Operating currents for this process are usually below 100 A, but they can be higher for special applications. When a filler metal is used, it is fed into the arc, as in GTAW. Arc and weld-zone shielding is supplied through an outer shielding ring by gases such as argon, helium, or mixtures of these gases.

There are two methods of plasma-arc welding. In the **transferred-arc** method (Fig. 12.10a), the workpiece being welded is part of the electrical circuit. The arc thus transfers from the electrode to the workpiece, hence the term *transferred*. In the **nontransferred** method (Fig. 12.10b), the arc is between the electrode and the nozzle, and the heat is carried to the workpiece by the plasma gas; this technique is also used for *thermal spraying*. (See Section 4.5.1.)

Compared with other arc-welding processes, plasma-arc welding has a greater energy concentration (hence, deeper and narrower welds can be made); better arc stability; less thermal distortion; and higher welding speeds, such as 120–1000 mm/min (5–40 in./min ). A variety of metals can be welded, generally with part thicknesses less than 6 mm (0.25 in.). The high heat concentration can penetrate completely through

**FIGURE 12.10**  Two types of plasma-arc-welding processes: (a) transferred and (b) nontransferred. Deep and narrow welds are made by this process at high welding speeds.

(a)                     (b)

the joint (**keyhole technique**), to thicknesses as much as 20 mm (0.75 in.) for some titanium and aluminum alloys. In the keyhole technique, the force of the plasma arc displaces the molten metal and produces a hole at the leading edge of the weld pool. Plasma-arc welding, rather than the GTAW process, is often used for butt and lap joints, because of the former's higher energy concentration, better arc stability, and higher welding speeds.

# 12.4 | High-Energy-Beam Welding

Joining with high-energy beams, principally laser-beam and electron-beam welding, has increasingly important applications in modern manufacturing, because of its high quality and technical and economic advantages. The general characteristics of high-energy beams and their unique applications in machining operations are described in Section 9.14.

## 12.4.1 Electron-beam welding

In *electron-beam welding* (EBW), heat is generated by high-velocity, narrow-beam electrons; the kinetic energy of the electrons is converted into heat as the electrons strike the workpiece. This process requires special equipment to focus the beam on the workpiece and in a vacuum; the higher the vacuum, the more the beam penetrates and the greater is the depth-to-width ratio. The level of vacuum is specified as HV (high vacuum) or MV (medium vacuum), both of which are used commonly, or NV (no vacuum), which can be effective on some materials. Almost any metals, similar or dissimilar, can be butt or lap welded using this process, with thicknesses ranging from foil to plates as thick as 150 mm (6 in.). The intense energy is also capable of producing holes in the workpiece. Generally, no shielding gas, flux, or filler metal is required. Capacities of electron-beam guns range to 100 kW.

Developed in the 1960s, EBW has the capability to make high-quality welds that are almost parallel sided, are deep and narrow, and have small heat-affected zones. (See Section 12.5.) Depth-to-width ratios range between 10 and 30 (Fig. 12.11). By using servo controls, parameters can be precisely controlled, with welding speeds as high as 12 m/min (40 ft/min). Distortion and shrinkage in the weld area are minimal, and the weld quality is good, with very high purity. Typical applications of EBW include welding of aircraft, missile, nuclear, and electronic components, as well as gears and shafts in the automotive industry. Electron-beam welding equipment generates X rays, and hence proper monitoring and periodic maintenance are important.

(a)

(b)

**FIGURE 12.11**
Comparison of the size of weld beads in (a) electron-beam or laser-beam welding with that in (b) conventional (tungsten-arc) welding. *Source*: American Welding Society, *Welding Handbook*, 8th ed., 1991.

## 12.4.2 Laser-beam welding

*Laser-beam welding* (LBW) uses a high-power laser beam as the source of heat (see Fig. 9.36 and Table 9.5) to produce a fusion weld; because the beam can be focused to a very small area, it has high-energy density and, therefore, has deep penetrating capability. The beam can be directed, shaped, and focused precisely on the workpiece; consequently, this process is particularly suitable for welding deep and narrow joints (Fig. 12.11a), with depth-to-width ratios typically ranging from 4 to 10. The laser beam may be **pulsed** (milliseconds) for applications such as spot welding of thin materials, with power levels up to 100 kW; **continuous** multi-kW laser systems are used for deep welds on thick sections. The efficiency of this process decreases with increasing reflectivity of the workpiece materials; to improve the process' performance, oxygen may be used for steels, and inert gases for nonferrous metals.

The LBW process can be automated and used successfully on a variety of materials with thicknesses of up to 25 mm (1 in.); it is particularly effective on thin workpieces. Typical metals and alloys welded by this process include aluminum, titanium, ferrous metals, copper, superalloys, and refractory metals. Welding speeds range from 2.5 m/min (8 ft/min) to as high as 80 m/min (250 ft/min) for thin metals. Because of the nature of the process, welding can be done in otherwise inaccessible locations. Safety is a particularly important consideration in laser-beam welding, because of the extreme hazards to the eye as well as the skin; solid-state (YAG) lasers are particularly harmful.

Laser-beam welding produces welds of good quality, with minimum shrinkage and distortion. Laser welds have good strength and are generally ductile and free of porosity. In the automotive industry, welding of transmission components is a widespread application of this process; among numerous other applications is the welding of thin parts for electronic components. As shown in Fig 7.14, another important application of laser welding is the forming of butt-welded sheet metal, particularly for automotive-body panels. (See the discussion of **tailor-welded blanks** in Section 7.3.4.)

The major advantages of LBW over electron-beam welding are:

- The laser beam can be transmitted through air, so a vacuum is not required, unlike in electron-beam welding.

- Because laser beams can be shaped, manipulated, and focused optically (using fiber optics), the process can easily be automated.

- The laser beams do not generate X rays, while the electron beams do.

- The quality of the weld is better with laser-beam welding, with less tendency for incomplete fusion, spatter, porosity, and distortion.

---

**EXAMPLE 12.1**  Laser-beam welding of razor blades

Figure 12.12 shows a close-up of the Gillette Sensor razor cartridge. Each of the two narrow, high-strength blades has 13 pinpoint welds, 11 of which can be seen, as darker spots about 0.5 mm in diameter, on each blade in the photograph. The welds are made with an Nd : YAG laser equipped with fiber-optic delivery, thus providing flexible beam manipulation to the exact locations along the length of the blade. With a set of these machines, production runs at a rate of 3 million welds per hour, with accurate and consistent weld quality.

*Source*: Courtesy of Lumonics Corporation, Industrial Products Division.

**FIGURE 12.12** Gillette Sensor razor cartridge, with laser-beam welds.

# 12.5 | The Fusion-Welded Joint

This section describes the following aspects of *fusion-welding* processes:

- The nature, properties, and quality of the welded joint;
- Weldability of metals;
- Testing welds.

The mechanical properties of a welded joint depend on many factors; for example, the rate of heat application and the thermal properties of metals are important, because they control the magnitude and distribution of temperature in a joint during welding. The microstructure and grain size of the welded joint depend on the magnitude of heat applied and of the temperature rise, the degree of prior cold work of the metals, and the rate of cooling after the weld is made. Weld quality depends on factors such as the geometry of the weld bead and the presence of cracks, residual stresses, inclusions, and oxide films; control of these factors is essential for producing reliable welds that have acceptable mechanical properties.

Figure 12.13 shows a typical fusion-welded joint, where three distinct zones can be identified:

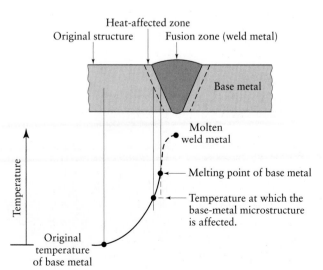

**FIGURE 12.13**
Characteristics of a typical fusion-weld zone in oxyfuel-gas and arc welding.

**FIGURE 12.14** Grain structure in (a) a deep weld and (b) a shallow weld. Note that the grains in the solidified weld metal are perpendicular to their interface with the base metal. In a good weld, the solidification line at the center in the deep weld shown in (a) has grain migration, thus developing uniform strength in the weld bead.

(a)        (b)

1. The **base metal**, that is, the metal to be joined;
2. The **heat-affected zone** (HAZ);
3. The **fusion zone**, that is, the region that has melted during welding.

The metallurgy and properties of the second and third zones depend greatly on the metals joined (single phase, two phase, dissimilar), the welding process, the filler metals used (if any), and the process variables. A joint produced without a filler metal is called **autogenous**, and the weld zone is composed of the *resolidified* base metal. A joint made with a filler metal has a central zone called the **weld metal**, which is composed of a mixture of the base and filler metals.

**Solidification of the weld metal.** After heat has been applied and filler metal, if any, has been introduced into the weld area, the molten weld joint is allowed to cool to ambient temperature. The *solidification* process is similar to that in casting and begins with the formation of columnar (dendritic) grains; these grains are relatively long and form parallel to the heat flow. (See Fig. 5.9.) Because metals are much better heat conductors than the surrounding air, the grains lie parallel to the plane of the two plates or sheets being welded (Fig. 12.14a); the grains developed in a shallow weld are shown in Figs. 12.14b. Grain structure and size depend on the specific alloy, the welding process, and the filler metal used.

The weld metal is basically a *cast* structure, and, because it has cooled somewhat slowly, it typically has coarse grains; consequently, this structure has relatively low strength, hardness, toughness, and ductility (Fig. 12.15). However, with proper selection of filler-metal composition or with subsequent heat treatments, the joint's mechanical properties can be improved. The results depend on the particular alloy, its composition, and the thermal cycling to which the joint is subjected. Cooling rates may, for example, be controlled and reduced by *preheating* the weld area prior to welding; preheating is particularly important for metals with high thermal conductivity, such as aluminum and copper, as otherwise the heat during welding rapidly dissipates.

**FIGURE 12.15** (a) Weld bead on a cold-rolled nickel strip produced by a laser beam. (b) Microhardness profile across the weld bead. Note the softer condition of the weld bead compared with the base metal. *Source*: IIT Research Institute.

(a)

(b)

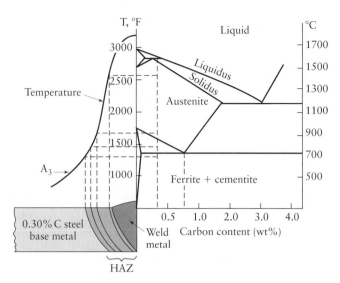

**FIGURE 12.16** Schematic illustration of various regions in a fusion-weld zone and the corresponding phase diagram for 0.30% C steel. *Source*: Courtesy of the American Welding Society.

**Heat-affected zone.** The *heat-affected zone* is within the base metal itself. It has a microstructure different than that of the base metal before welding, because it has been subjected to elevated temperatures for a period of time during welding. The portions of the base metal that are far away from the heat source do not undergo any structural changes during welding. The properties and microstructure of the HAZ depend on (a) the rate of heat input and cooling and (b) the temperature to which this zone has been raised during welding. Figure 12.16 shows the heat-affected zone and the corresponding phase diagram for 0.30% C steel. In addition to metallurgical factors such as original grain size, grain orientation, and degree of prior cold work, the specific heat and thermal conductivity of the metals also influence the HAZ's size and characteristics.

The strength and hardness of the heat-affected zone depend partly on how the original strength and hardness of the particular alloy were developed prior to welding. As described in Chapters 3 and 5, they may have been developed by cold working, solid-solution strengthening, precipitation hardening, or various heat treatments. Of metals strengthened by these various methods, the simplest to analyze is base metal that has been cold worked, say, by cold rolling or forging.

The heat applied during welding *recrystallizes* the elongated grains (preferred orientation) of the cold-worked base metal. Grains that are away from the weld metal will recrystallize into fine equiaxed grains. However, grains close to the weld metal, having been subjected to elevated temperatures for a longer period of time, will grow. (See also Section 3.6.) This grain growth will result in a region that is softer and has less strength, and such a joint will be weakest in its heat-affected zone. The grain structure of such a weld exposed to corrosion by chemical reaction is shown in Fig. 12.17; the central vertical line is where the two workpieces originally met (butt welded). It is apparent that the effects of heat during welding on the HAZ for joints made with dissimilar metals, and for alloys strengthened by other mechanisms, are complex, and they are beyond the scope of this text.

**FIGURE 12.17** Intergranular corrosion of a weld in ferritic stainless-steel welded tube after exposure to a caustic solution. The weld line is at the center of the photograph. Scanning electron micrograph at 20×. *Source*: Courtesy of Allegheny Ludlum Corp.

## 12.5.1 Weld quality

Because of a history of thermal cycling and attendant microstructural changes, a welded joint may develop *discontinuities*. Welding discontinuities can also be caused by inadequate or careless application of established welding technologies or substandard operator training. The major discontinuities that affect weld quality are described as follows:

a) **Porosity.** *Porosity* in welds is caused by trapped *gases* that are released during melting of the weld zone, but trapped during solidification; by chemical reactions during welding; or by *contaminants*. Most welded joints contain some porosity, which is generally spherical in shape or in the form of elongated pockets. (See also Section 5.4.6.) The distribution of porosity in the weld zone may be random, or it may be concentrated in a certain region. Particularly important is the presence of hydrogen, which may be due to the use of damp fluxes or to environmental humidity; the result is porosity in aluminum-alloy welds and hydrogen embrittlement in steels. (See also Section 3.8.2.)

   Porosity in welds can be reduced by the following methods:

   1. Proper selection of electrodes and filler metals;
   2. Improvement of welding techniques, such as by preheating the weld area or increasing the rate of heat input;
   3. Proper cleaning of the weld zone and prevention of contaminants from entering the weld zone;
   4. Slowing the welding speed to allow time for gas to escape.

b) **Slag inclusions.** *Slag inclusions* are compounds, such as oxides, fluxes, and electrode coating materials, that are trapped in the weld zone. If the shielding gases used are not effective during welding, contamination from the environment may also contribute to such inclusions. Maintenance of proper welding conditions is also important, and when proper techniques are used, the molten slag will float to the surface of the molten weld metal and not be entrapped. Slag inclusions, or *slag entrapment*, may be prevented by the following methods:

   1. Cleaning the weld bead surface before the next layer is deposited (see Fig. 12.3), by using a hand or power wire brush;
   2. Providing adequate shielding gas;
   3. Redesigning the joint (see Section 12.16) to permit sufficient space for proper manipulation of the puddle of molten weld metal.

c) **Incomplete fusion and penetration.** *Incomplete fusion* (*lack of fusion*) produces poor weld beads, such as those shown in Fig. 12.18. A better weld can be obtained by the following methods:

   1. Raising the temperature of the base metal;
   2. Cleaning the weld area prior to welding;
   3. Changing the joint design and type of electrode;
   4. Providing adequate shielding gas.

   *Incomplete penetration* occurs when the depth of the welded joint is insufficient. Penetration can be improved by the following methods:

   1. Increasing the heat input;
   2. Lowering travel speed during welding;
   3. Modifying the design of the joint;
   4. Ensuring that the surfaces to be joined fit properly.

d) **Weld profile.** *Weld profile* is important not only because of its effects on the strength and appearance of the weld, but also because it can indicate incomplete

Incomplete fusion from oxide or dross at
the center of a joint, especially in aluminum

(b)

Incomplete fusion in fillet welds. *B* is often
termed *bridging.*

(a)

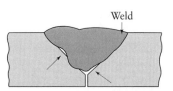

Incomplete fusion in a groove weld

(c)

**FIGURE  12.18**  Examples
of various discontinuities in
fusion welds.

fusion or the presence of slag inclusions in multiple-layer welds. **Underfilling** results when the joint is not filled with the proper amount of weld metal (Fig. 12.19a); **undercutting** results from melting away of the base metal and subsequent generation of a groove in the shape of a sharp recess or notch. If deep or sharp, an undercut can act as a stress raiser and reduce the fatigue strength of the joint, which may lead to premature failure. **Overlap** (Fig. 12.19b) is a surface discontinuity generally caused by poor welding practice and selection of the wrong materials. A proper weld bead is shown in Fig. 12.19c.

e) **Cracks.**  *Cracks* may occur in various locations and directions in the weld area. The typical types of cracks are longitudinal, transverse, crater, underbead, and toe cracks (Fig. 12.20). These cracks generally result from a combination of the following factors:

1.  Temperature gradients that cause thermal stresses in the weld zone;
2.  Variations in the composition of the weld zone that cause different contractions;
3.  Embrittlement of grain boundaries by segregation of elements, such as sulfur, to the grain boundaries (see Section 3.4.2) as the solid–liquid boundary moves when the weld metal begins to solidify;
4.  Hydrogen embrittlement (see Section 3.8.2);

(a)

(b)

(c)

**FIGURE  12.19**  Examples of various defects in fusion welds.

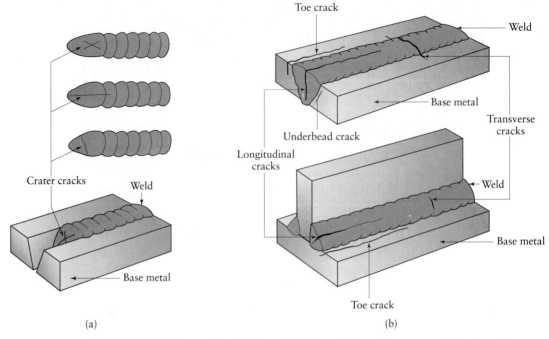

(a)                                        (b)

**FIGURE 12.20**   Types of cracks in welded joints. The cracks are caused by thermal stresses that develop during solidification and contraction of the weld bead and the welded structure: (a) crater cracks; (b) various types of cracks in butt and T joints.

5.  Inability of the weld metal to contract during cooling (Fig. 12.21), a situation similar to the development of *hot tears* in castings, due to excessive restraint of the workpiece (see also Section 5.4.6).

Cracks are classified as follows: **Hot cracks** occur while the joint is still at elevated temperatures; **cold cracks** develop after the weld metal has solidified. Some crack-prevention measures are listed as follows:

1.  Modify the design of the joint to minimize thermal stresses from shrinkage during cooling;
2.  Change welding-process parameters, procedures, and sequence;
3.  Preheat components being welded;
4.  Avoid rapid cooling of the components after welding.

**FIGURE 12.21**   Crack in a weld bead. The two components were not allowed to contract after the weld was completed. *Source*: Courtesy of Packer Engineering.

f) **Lamellar tears.** In describing the anisotropy of plastically deformed metals (see Section 3.5), we stated that because of the alignment of nonmetallic impurities and inclusions (stringers), the workpiece is weaker when tested in its thickness direction than in other directions (*mechanical fibering*). This condition is particularly evident in rolled plates and structural shapes. In welding such components, *lamellar tears* may develop, because of shrinkage of the restrained members in the structure during cooling. Such tears can be avoided by providing for shrinkage of the members or by modifying the joint design to make the weld bead penetrate the weaker member more deeply.

g) **Surface damage.** During welding, some of the metal may spatter and be deposited as small droplets on adjacent surfaces; also, in arc welding, the electrode may inadvertently contact the parts being welded at places outside of the weld zone (**arc strikes**). Such surface discontinuities may be objectionable for reasons of appearance or subsequent use of the welded part. If severe, they may adversely affect the properties of the welded structure, particularly for notch-sensitive metals. Thus, following proper welding procedures is important in avoiding surface damage.

h) **Residual stresses.** Because of localized heating and cooling during welding, expansion and contraction of the weld area cause *residual stresses* in the workpiece. (See also Section 2.10.) Residual stresses can cause the following detrimental effects:

1. Distortion, warping, and buckling of the welded parts (Fig. 12.22);
2. Stress-corrosion cracking (see Section 3.8.2);
3. Further distortion if a portion of the welded structure is subsequently removed, say, by machining or sawing;
4. Reduced fatigue life.

The type and distribution of residual stresses in welds are best described by reference to Fig. 12.23a. When two plates are being welded, a long, narrow region is subjected to elevated temperatures, whereas the plates as a whole are essentially at ambient temperature. As the weld is completed and time elapses, the heat from the weld area dissipates laterally to the plates. The plates thus begin to expand longitudinally while the welded length begins to contract; these two opposing effects cause residual stresses that are typically distributed as shown in Fig. 12.23b. Note that the magnitude of compressive residual stresses in the plates diminishes to zero at a point away from the weld area. Because no external forces are acting on the welded plates, the tensile and compressive forces represented by these residual stresses must balance each other.

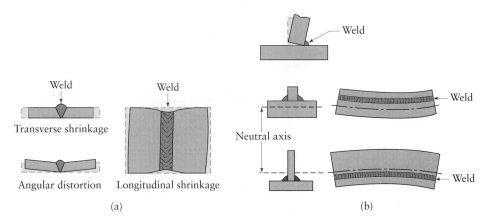

(a)                                      (b)

**FIGURE 12.22** Distortion of parts after welding: (a) butt joints and (b) fillet welds. Distortion is caused by differential thermal expansion and contraction of different regions of the welded assembly. Warping can be reduced or eliminated by proper weld design and fixturing prior to welding.

**FIGURE 12.23** Residual stresses developed in a straight butt joint. *Source*: Courtesy of the American Welding Society.

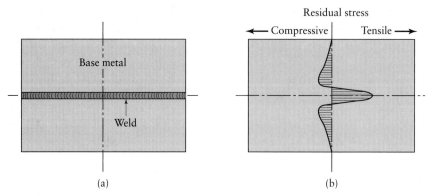

(a)                                                    (b)

In complex welded structures, residual-stress distributions are three dimensional and hence difficult to analyze. The preceding example involves only two plates that are not restrained from movement; in other words, the plates are not an integral part of a larger structure. If the plates are restrained, reaction stresses will be generated, because the plates are not free to expand or contract. This situation arises particularly in structures with high stiffness.

**Stress relieving of welds.** The problems caused by residual stresses, such as distortion, buckling, or cracking, can be reduced by *preheating* the base metal or the parts to be welded. Preheating reduces distortion by lowering the cooling rate and the magnitude of thermal stresses (by reducing the elastic modulus of the metals being welded); this technique also reduces shrinkage and possible cracking of the joint. Preheating is also effective in welding hardenable steels for which rapid cooling of the weld area would otherwise be harmful. The workpieces may be heated in a furnace or electrically or inductively; thin sections may be heated by radiant lamps or a hot-air blast. For optimum results, preheating temperatures and cooling rates must be controlled in order to attain acceptable strength and toughness in the welded structure.

Residual stresses can be reduced by *stress relieving* or *stress-relief annealing* the welded structure; the temperature and time required for stress relieving depend on the type of material and magnitude of the residual stresses developed. Other methods of stress relieving include peening, hammering, and surface rolling the weld bead area. (See Section 4.5.1.) These processes induce compressive residual stresses, thus reducing or eliminating tensile residual stresses in the weld. For multilayer welds, the first and last layers should not be peened, because peening may cause damage to those layers.

Residual stresses can also be relieved, or reduced, by plastically deforming the structure by a small amount. (See Fig. 2.34.) This technique can be used in some welded structures, such as pressure vessels, by pressurizing the vessels internally (*proof stressing*). In order to reduce the possibility of sudden fracture under high internal pressure, the weld must be free from notches or discontinuities, which could act as points of stress concentration.

**Heat treatment of welds.** In addition to stress relieving, welds may also be *heat treated* by various techniques in order to modify and improve properties such as strength and toughness. These techniques include (a) annealing, normalizing, or quenching and tempering of steels; and (b) solution treatment and aging of precipitation-hardenable alloys (see Section 5.11).

## 12.5.2 Weldability

*Weldability* of a metal is generally defined as the metal's capacity to be welded into a specific structure that has certain properties and characteristics and that will satisfactorily meet service requirements. As can be expected, weldability involves a large number of variables, making generalizations difficult. Material characteristics, such as alloying elements, impurities, inclusions, grain structure, and processing history of the base metal and filler metal, are important. Because of the melting, solidification, and microstructural changes involved, a thorough knowledge of the phase diagram and the response of the metal or alloy to elevated temperatures over a period of time also is essential. Other factors that influence weldability are the mechanical and physical properties of strength, toughness, ductility, notch sensitivity, elastic modulus, specific heat, melting point, thermal expansion, surface-tension characteristics of the molten metal, and corrosion.

Preparation of surfaces for welding is important, as are the nature and properties of surface oxide films and adsorbed gases. The welding process employed significantly affects the temperatures developed and their distribution in the weld zone. Other factors that influence weldability are shielding gases, fluxes, moisture content of the coatings on electrodes, welding speed, welding position, cooling rate, preheating, and postwelding techniques (such as stress relieving and heat treating).

The following list briefly summarizes, in alphabetic order, the general weldability of specific metals. The weldability of these materials can vary significantly, with some requiring special welding techniques and good control of processing parameters.

1. *Aluminum alloys*: weldable at a high rate of heat input; aluminum alloys containing zinc or copper generally are considered unweldable.
2. *Cast irons*: generally weldable.
3. *Copper alloys*: similar to that of aluminum alloys.
4. *Lead*: easy to weld.
5. *Magnesium alloys*: weldable with the use of protective shielding gas and fluxes.
6. *Molybdenum*: weldable under well-controlled conditions.
7. *Nickel alloys*: weldable by various processes.
8. *Niobium (columbium)*: weldable under well-controlled conditions.
9. *Stainless steels*: weldable by various processes.
10. *Steels, galvanized and prelubricated*: The presence of zinc coating and lubricant layer adversely affects weldability.
11. *Steels, high-alloy*: generally good weldability under well-controlled conditions.
12. *Steels, low-alloy*: fair to good weldability.
13. *Steels, plain-carbon*: excellent weldability for low-carbon steels; fair to good weldability for medium-carbon steels; poor weldability for high-carbon steels.
14. *Tantalum*: weldable under well-controlled conditions.
15. *Tin*: easy to weld.
16. *Titanium alloys*: weldable with the use of proper shielding gases.
17. *Tungsten*: weldable under well-controlled conditions.
18. *Zinc*: difficult to weld; soldering preferred.
19. *Zirconium*: weldable with the use of proper shielding gases.

## 12.5.3 Testing welded joints

As in all manufacturing processes, the *quality* of a welded joint is established by testing. Several standardized tests and test procedures have been established and are available from organizations such as the American Society for Testing and Materials (ASTM), the American Welding Society (AWS), the American Society of Mechanical

Engineers (ASME), the American Society of Civil Engineers (ASCE), and various federal agencies. Welded joints may be tested either destructively or nondestructively (see Section 4.8); each technique has its capabilities, sensitivity, limitations, reliability, and need for special equipment and operator skill.

a) **Destructive testing techniques.** Five methods of destructively testing welded joints are commonly used. A brief review of each method is given below.

**Tension test.**   Longitudinal and transverse tension tests are performed on specimens removed from actual welded joints and from the weld-metal area; stress–strain curves are then obtained by the procedures described in Section 2.2. These curves indicate the yield strength ($Y$), ultimate tensile strength (UTS), and ductility of the welded joint in different locations and directions. Ductility is measured in terms of percentage elongation and percentage reduction of area. Tests of weld hardness may also be used to indicate weld strength and microstructural changes in the weld zone.

**Tension-shear test.**   The specimens in the tension-shear test (Fig. 12.24a) are specially prepared to simulate actual welded joints and procedures. The specimens are subjected to tension, and the shear strength of the weld metal and the location of fracture are determined.

**Bend test.**   Several bend tests have been developed to determine the ductility and strength of welded joints. In one test, the welded specimen is bent around a fixture (*wraparound bend test*; Fig. 12.24b). In another test, the specimens are tested in *three-point transverse bending* (Fig. 12.24c; see also Fig. 2.23). These tests help establish the relative ductility and strength of welded joints.

**Fracture toughness test.**   Fracture toughness tests commonly use impact-testing techniques, described in Section 2.9. Charpy V-notch specimens are prepared and tested for toughness; other toughness tests include the drop-weight test, in which the energy is supplied by a falling weight.

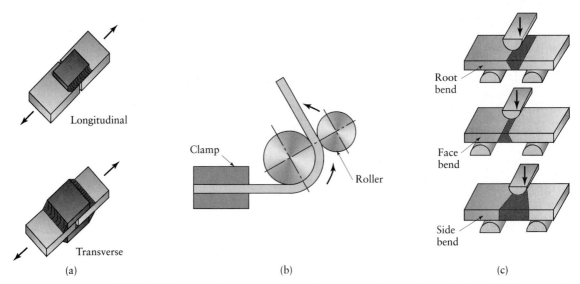

(a)                                    (b)                                    (c)

**FIGURE  12.24**   (a) Types of specimens for tension-shear testing of welds. (b) Wraparound bend test method. (c) Three-point bending of welded specimens. (See also Fig. 2.23.)

**FIGURE  12.25**
(a) Tension-shear test for
spot welds. (b) Cross-tension
test. (c) Twist test.
(d) Peel test.

**Corrosion and creep tests.**   In addition to mechanical tests, welded joints
may be tested for corrosion and creep resistance. Because of the difference in the
composition and microstructure of the materials in the weld zone, preferential
corrosion may take place in this zone. (See Fig. 12.17.) Creep tests are essential
in determining the behavior of welded joints at elevated temperatures.

**Testing of spot welds.**   Spot-welded joints may be tested for weld-nugget
strength (Fig. 12.25), using the (1) *tension-shear*, (2) *cross-tension*, (3) *twist*,
and (4) *peel tests*. Because they are easy to perform and inexpensive, tension-
shear tests are commonly used in fabricating facilities. The cross-tension and
twist tests are capable of revealing flaws, cracks, and porosity in the weld area.
The peel test is commonly used for thin sheets. After the joint has been bent and
peeled, the shape and size of the torn-out weld nugget are evaluated.

b) **Nondestructive testing techniques.** Welded structures often have to be
tested nondestructively, particularly for critical applications where weld failure
can be catastrophic, such as in pressure vessels, pipelines, load-bearing structural
members, and power plants. Nondestructive testing techniques for welded joints
usually consist of visual, radiographic, magnetic-particle, liquid-penetrant, and
ultrasonic testing methods (see Section 4.8.1).

## 12.5.4  Welding-process selection

In addition to the material characteristics described thus far, selection of an appropriate welding process involves the following considerations:

a) Configuration of the parts or structure to be welded and their thickness and
size;

b) The methods used to manufacture component parts;

c) Service requirements, such as the type of loading and stresses generated;

d) Location, accessibility, and ease of welding;

e) Effects of distortion and discoloration;

f) Appearance;

g) Costs involved in edge preparation, welding, and postprocessing of the weld, including machining and finishing operations.

# 12.6 | Cold Welding

In *cold welding* (CW), pressure is applied to the surfaces of the parts, through either dies or rolls. Because of the resulting plastic deformation, it is necessary that at least one, but preferably both, of the mating parts be sufficiently ductile. The interface is usually cleaned by wire brushing prior to welding. However, in joining two dissimilar metals that are mutually soluble, brittle *intermetallic compounds* may form (see Section 5.2.2), resulting in a weak and brittle joint. An example is the bonding of aluminum and steel, where a brittle intermetallic compound is formed at the interface. The best bond strength and ductility are obtained with two similar materials. Cold welding can be used to join small workpieces made of soft, ductile metals; applications include electrical connections and wire stock.

**Roll bonding.** In *roll bonding*, or *roll welding* (ROW), the pressure required for cold welding of long pieces or continuous strips can be applied through a pair of rolls (Fig. 12.26). This process, developed in the 1960s, is used for manufacturing some U.S. coins. (See Example 12.2.) Surface preparation is important for improved interfacial strength. Roll bonding can also be carried out at elevated temperatures (*hot roll bonding*) for improved interfacial strength; typical examples include *cladding* (Section 4.5.1) pure aluminum over precipitation-hardened aluminum-alloy sheet (Section 5.11.2) and cladding stainless steel over mild steel for corrosion resistance. Another common application of roll bonding is the production of bimetallic strips for thermostats and similar controls, using two layers of materials with different coefficients of thermal expansion. (See also Section 12.7.) Bonding in only selected regions in the interface can be achieved by depositing a parting agent such as graphite or ceramic (called *stop-off*). The use of this technique in *superplastic forming* of sheet-metal structures is shown in Fig. 7.47.

**FIGURE 12.26** Schematic illustration of the roll-bonding, or cladding, process.

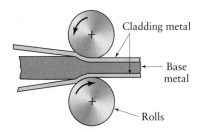

---

**EXAMPLE 12.2** Roll bonding of the U.S. quarter

The technique used for manufacturing composite U.S. quarters is roll bonding of two outer layers of 75% copper-25% nickel (cupronickel), each 1.2 mm (0.048 in.) thick, with an inner layer of pure copper 5.1 mm (0.20 in.) thick. To obtain good bond strength, the mating surfaces (*faying surfaces*) are first chemically cleaned and wire brushed. The strips are then rolled to a thickness of 2.29 mm (0.090 in.); a second rolling operation reduces the final thickness to 1.36 mm (0.0535 in.). The strips thus undergo a total reduction in thickness of 82%. Because volume constancy is maintained in plastic deformation, there is a major increase in the surface area between the layers, thus generating clean interfacial surfaces and hence better bonding. This extension in surface area under the high pressure of the rolls, combined with the solid solubility of nickel in copper (see Section 5.2.1), produces a strong bond.

---

## 12.7 | Ultrasonic Welding

In *ultrasonic welding* (USW), the faying surfaces of the two members are subjected to a static normal force and oscillating shearing (tangential) stresses. The shearing stresses are applied by the tip of a **transducer** (Fig. 12.27a) similar to that used for ultrasonic machining (see Fig. 9.19a); the frequency of oscillation generally ranges from 10 kHz to 75 kHz. The energy required increases with the thickness and hardness of the materials being joined. Proper coupling between the transducer and the tip (called *sonotrode*, from the word *sonic*) is important for effective operation. The welding tip can be replaced with rotating disks (Fig. 12.27b) for seam-welding structures (similar to those shown in Fig. 12.33 for resistance seam welding), one component of which can be a sheet or foil.

The shearing stresses cause small-scale plastic deformation at the workpiece interfaces, breaking up oxide films and contaminants, thus allowing good contact and producing a strong solid-state bond. Temperatures generated in the weld zone are usually in the range of one third to one half the melting point (absolute scale) of the

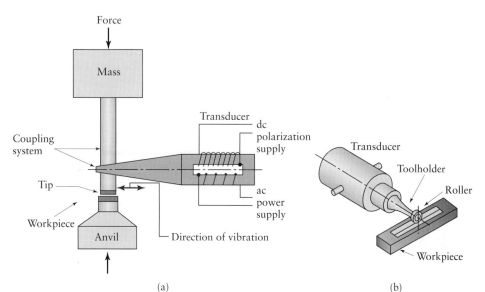

(a)                           (b)

**FIGURE 12.27**
(a) Components of an ultrasonic-welding machine for lap welds. The lateral vibrations of the tool tip cause plastic deformation and bonding at the interface of the workpieces.
(b) Ultrasonic seam welding using a roller.

metals joined; therefore, no melting and fusion take place. In certain situations, however, temperatures can be sufficiently high to cause metallurgical changes in the weld zone.

The ultrasonic-welding process is reliable and versatile; it can be used with a wide variety of metallic and nonmetallic materials, including dissimilar metals (*bimetallic strips*). It is used extensively in joining plastics and in the automotive and consumer-electronics industries for lap welding of sheet, foil, and thin wire and in packaging with foils. The mechanism of joining thermoplastics (see Section 12.15.1) by ultrasonic welding is different than that for metals, and interfacial melting takes place, due to the much lower melting temperature of plastics (see Table 10.2).

# 12.8 | Friction Welding

In the joining processes described thus far, the energy required for welding is supplied externally. In *friction welding* (FRW), the heat required for welding is, as the name implies, generated through friction at the interface of the two members being joined; thus, the source of energy is mechanical. You can demonstrate the significant rise in temperature from friction simply by rubbing your hands together fast or sliding down a rope rapidly.

In the friction-welding process, one of the members remains stationary while the other is placed in a chuck or collet and rotated at a high constant speed; the two members to be joined are then brought into contact under an axial force (Fig. 12.28). The surface speed of rotation may be as high as 900 m/min (3000 ft/min). The rotating member must be clamped securely to resist both torque and axial forces without slip. After sufficient interfacial contact is established, the rotating member is brought to a sudden stop, so that the weld is not destroyed by shearing, while the axial force is increased. The pressure at the interface and the resulting friction produce sufficient heat for a strong joint to take place.

The weld zone is usually confined to a narrow region, depending on (a) the amount of heat generated, (b) the thermal conductivity of the materials, and (c) the mechanical properties of the materials at elevated temperature. The shape of the welded joint depends on the rotational speed and the axial force applied (Fig. 12.29); these factors must be controlled to obtain a uniformly strong joint. The radially outward movement of the hot metal at the interface pushes oxides and other contaminants out of the interface.

Developed in the 1940s, friction welding can be used to join a wide variety of materials, provided that one of the components has some rotational symmetry. Solid as well as tubular parts can be joined by this method, with good joint strength. Solid steel bars up to 100 mm (4 in.) in diameter and pipes up to 250 mm (10 in.) outside diameter have been welded successfully using this process. Because of the combined

**FIGURE 12.28** Sequence of operations in the friction-welding process. (a) The part on the left is rotated at high speed. (b) The part on the right is brought into contact under an axial force. (c) The axial force is increased; flash begins to form. (d) The part on the left stops rotating. The weld is completed. Flash can be removed by machining or grinding.

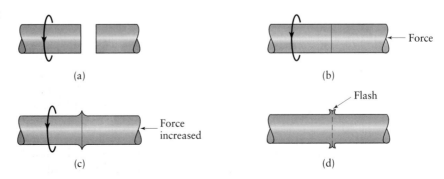

(a)    (b)    Force

(c)    Force increased    Flash    (d)

**FIGURE 12.29**    Shape of the fusion zone in friction welding as a function of the force applied and the rotational speed.

(1) High pressure        (2) Low pressure        (3) Optimum
    or low speed             or high speed

heat and pressure, the interface in FRW develops a flash by plastic deformation (upsetting) of the heated zone; if objectionable, this flash can easily be removed by subsequent machining or grinding.

Variations on the friction-welding process are described as follows:

a) **Inertia friction welding.** *Inertia friction welding* is a modification of FRW, although the term has been used interchangeably with *friction welding.* The energy required in this process is supplied through the kinetic energy of a flywheel. The flywheel is accelerated to the proper speed, the two members are brought into contact, and an axial force is applied; as friction at the interface begins to slow down the flywheel, the axial force is increased. The weld is completed when the flywheel comes to a stop. The timing of this sequence is important; if it is not properly controlled, the weld quality will be poor. The rotating mass of inertia friction-welding machines can be adjusted for applications requiring different levels of energy, which depend on workpiece size and its properties. In one application of inertia friction welding, a shaft 10 mm (0.4 in.) in diameter is welded to automotive turbocharger impellers at a rate of one every 15 s.

b) **Linear friction welding.** *Linear friction welding* is another development that involves subjecting the interface of the two parts to be joined to a linear reciprocating motion of at least one of the components to be joined; thus, the parts do not have to be circular or tubular in cross-section. In this process, one part is moved across the face of the other part, using a balanced reciprocating mechanism. Linear friction welding is capable of joining square or rectangular components, as well as round parts, made of metals or plastics. In one application, a rectangular titanium-alloy part was friction welded at a linear frequency of 25 Hz, an amplitude of ±2 mm (0.08 in.), and a pressure of 100 MPa (15,000 psi) acting on a 240-mm$^2$ (0.38-in$^2$) interface. Various other metal parts have been welded successfully, with rectangular cross-sections as large as 50 mm × 20 mm (2 in. × 0.8 in.).

c) **Friction stir welding.** This technique is a more recent process, originally intended for welding of aerospace alloys, especially aluminum extrusions, although current research is being directed at extending this process towards polymers and composite materials. Whereas in conventional friction welding, heating of interfaces is achieved through friction by rubbing two contacting surfaces, in the *friction stir-welding* (FSW) process, a third body is rubbed against the two surfaces to be joined. The rotating tool is a small (5 mm to 6 mm in diameter, 5 mm in height) rotating member that is plunged into the joint (Fig. 12.30). The contact pressures cause frictional heating, raising the temperature to the range of 230°C to 260°C (450°F to 500°F). The probe at the tip of the rotating tool forces heating and mixing or stirring of the material in the joint.

The welding equipment can be a conventional, vertical-spindle milling machine (see Fig. 8.59b), and the process is relatively easy to implement. The thickness of the welded material can be as little as 1 mm (0.04 in.) and as much as 30 mm (1.2 in.) or so. The weld quality is high, with minimal pores and with

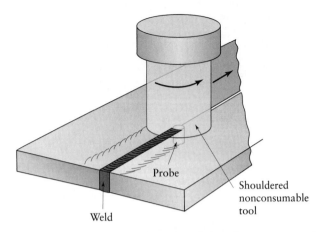

**FIGURE 12.30** The principle of the friction stir-welding process. Aluminum-alloy plates up to 75 mm (3 in.) thick have been welded by this process. *Source*: TWI, Cambridge, United Kingdom.

Probe

Shouldered nonconsumable tool

Weld

uniform material structure. The welds are produced with low heat input and, therefore, low distortion and little microstructural changes; furthermore, there are no fumes or spatter produced, and the process is suitable for automation.

## 12.9 | Resistance Welding

*Resistance welding* (RW) covers a number of processes in which the heat required for welding is produced by means of the *electrical resistance* between the two members to be joined; hence, these processes' major advantages include the fact that they don't require consumable electrodes, shielding gases, or flux. The heat generated in resistance welding is given by the general expression

$$H = I^2Rt, \tag{12.1}$$

where $H$ = heat generated, in joules (watt-seconds); $I$ = current, in amperes; $R$ = resistance, in ohms; and $t$ = flow time of the current, in seconds. This equation is often modified to represent the actual heat energy available in the weld by including a factor $K$ that represents the energy losses through radiation and conduction. Thus, Eq. (12.1) becomes $H = I^2RtK$, where the value of $K$ is less than unity.

The total resistance in these processes, such as in the resistance spot welding shown in Fig. 12.31, is the sum of the following:

1. Resistance of the electrodes;
2. Electrode–workpiece contact resistances;
3. Resistances of the individual parts to be welded;
4. Workpiece–workpiece contact resistances (faying surfaces).

The magnitude of the current in resistance-welding operations may be as high as 100,000 A, although the voltage is typically only 0.5–10 V. The actual temperature rise at the joint depends on the specific heat and thermal conductivity of the metals to be joined; consequently, because they have high thermal conductivity, metals such as aluminum and copper require high heat concentrations. Similar as well as dissimilar metals can be joined by resistance welding. Electrode materials should have high thermal conductivity and strength at elevated temperatures; they are typically made of copper alloys.

Developed in the early 1900s, resistance-welding processes require specialized machinery, and many are now operated with programmable computer control. The machinery is generally not portable, and the process is most suitable for use in

Pressure applied     Current on     Current off, pressure on     Pressure released

(a)

Electrodes

Weld nugget

Lap joint

Electrode

Electrode tip

Indentation

Sheet separation

Weld nugget

Heat-affected zone

Electrode

(b)

**FIGURE  12.31**
(a) Sequence in the resistance spot-welding process. (b) Cross-section of a spot weld, showing weld nugget and light indentation by the electrode on sheet surfaces. This method is one of the most common processes used in sheet-metal fabrication and automotive-body assembly.

manufacturing plants and machine shops. There are five basic methods of resistance welding: spot, seam, projection, flash, and upset welding. Lap joints are used in the first three processes and butt joints in the last two.

## 12.9.1  Resistance spot welding

In *resistance spot welding* (RSW), the tips of two opposing solid cylindrical electrodes contact the lap joint of two sheet metals, and resistance heating produces a spot weld (Fig. 12.31a). In order to obtain a good bond in the **weld nugget** (Fig. 12.31b), pressure is continually applied until the current is turned off; accurate control and timing of the electric current and pressure are essential. Currents usually range from 3000 A to 40,000 A, depending on the materials being welded and their thickness. Modern equipment for spot welding is computer controlled for optimum timing of current and pressure, and the spot-welding guns are manipulated by programmable robots (Section 14.7). Spot welding is widely used for fabricating sheet-metal products. Examples of its applications range from attaching handles to stainless-steel cookware to rapid spot welding of automobile bodies, using multiple electrodes; a typical automobile may have as many as 10,000 spot welds.

The strength of the weld depends on the surface roughness and cleanliness of the mating surfaces; thus oil, paint, and thick oxide layers should be removed before welding. However, the presence of uniform, thin oxide layers and other contaminants is not critical. The weld nugget is generally up to 10 mm (0.375 in.) in diameter. The surface of the weld spot has a slightly discolored indentation.

Spot welding is the simplest and most commonly used resistance-welding process. Welding may be performed by means of single or multiple electrodes, and the pressure required is supplied through mechanical or pneumatic means. The *rocker-arm type* of spot-welding machines is generally used for smaller parts, whereas the *press-type* machines is used for larger workpieces. The shape and surface condition of the electrode tip and accessibility of the area to be welded are important factors in spot welding; a variety of electrode shapes is used to spot weld areas that are difficult to reach (Fig. 12.32).

**FIGURE 12.32** Types of special electrodes designed for easy access in spot-welding operations for complex shapes.

(a)                                    (b)

---

**EXAMPLE 12.3** Heat generated in resistance spot welding

Assume that two 1-mm (0.04-in.)-thick steel sheets are being spot welded at a current of 5000 A and current flow time $t = 0.1$ s. Using electrodes 5 mm (0.2 in.) in diameter, estimate the amount of heat generated and its distribution in the weld zone.

**Solution.** Let's assume that the effective resistance in this operation is 200 $\mu\Omega$. Then, according to Eq. (12.1),

$$\text{Heat} = (5000)^2(0.0002)(0.1) = 500 \text{ J}.$$

From the information given, we estimate the weld-nugget volume to be 30 mm$^3$ (0.0018 in$^3$). If we assume a density for steel of 8000 kg/m$^3$ (0.008 g/mm$^3$), the weld nugget has a mass of 0.24 g. Since the heat required to melt 1 g of steel is about 1400 J, the heat required to melt the weld nugget is $(1400)(0.24) = 336$ J. Consequently, the remaining heat (164 J), or 33%, is dissipated into the metal surrounding the nugget.

---

## 12.9.2 Resistance seam welding

*Resistance seam welding* (RSEW) is a modification of resistance spot welding wherein the electrodes are replaced by rotating wheels or rollers (Fig. 12.33). With a continuous ac power supply, the electrically conducting rollers produce continuous spot welds whenever the current reaches a sufficiently high level in the ac cycle; these welds are actually overlapping spot welds and can produce a joint that is liquid tight and gas tight (Fig. 12.33b). The typical welding speed is 1.5 m/min (60 in./min) for thin sheet. With intermittent application of current to the rollers, a series of spot welds at

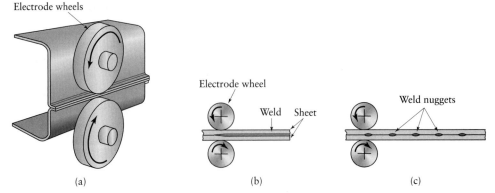

Electrode wheels

Electrode wheel                          Weld nuggets

Weld    Sheet

(a)                        (b)                        (c)

**FIGURE 12.33** (a) Seam-welding process, with rolls acting as electrodes. (b) Overlapping spots in a seam weld. (c) Roll spot welds.

various intervals can be made along the length of the seam (Fig. 12.33c), a procedure called **roll spot welding**. The RSEW process is used to make the longitudinal (side) seam of cans for household products, mufflers, gasoline tanks, and other containers. In *mash-seam welding*, the overlapping welds are about one to two times the sheet thickness; tailor-welded blanks may be made by this process.

High-frequency resistance welding (HFRW) is similar to resistance seam welding, except that high-frequency (up to 450 kHz) current is employed. A typical application is making butt-welded tubing; structural sections such as I-beams can also be fabricated by HFRW, and spiral pipe and tubing, finned tubes for heat exchangers, and wheel rims may be made by these techniques as well. In **high-frequency induction welding** (HFIW), the tube or pipe to be welded is subjected to high-frequency induction heating, using an induction coil ahead of the rollers.

### 12.9.3 Resistance projection welding

In *resistance projection welding* (RPW), high electrical resistance at the joint is developed by embossing one or more projections (dimples) on one of the surfaces to be welded (Fig. 12.34). The projections may be round or oval for design or strength purposes. High localized temperatures are generated at the projections, which are in contact with the flat mating part. The electrodes, typically made of copper-base alloys and water cooled to keep their temperature low, are large and flat. Weld nuggets, similar to those produced in spot welding, are formed as the electrodes exert pressure to compress the projections.

Spot-welding equipment can be used for RPW by modifying the electrodes. Although embossing workpieces is an added expense, this process produces a number of welds in one stroke, extends electrode life, and is capable of welding metals with different thicknesses. Nuts and bolts are also welded to sheet and plate by this process, with projections that may be produced by machining or forging. The process used for joining a network of wires, such as metal baskets, grills, oven racks, and shopping carts, is considered as resistance projection welding as well, because of the small contact area between crossing wires (grids).

### 12.9.4 Flash welding

In *flash welding* (FW), also called **flash butt welding**, heat is generated from the arc as the ends of the two members begin to make contact, developing an electrical resistance at the joint (Fig. 12.35). Because of the arc's presence, this process is also classified as arc welding. After the proper temperature is reached and the interface begins to soften, an axial force is applied at a controlled rate, and a weld is formed by plastic deformation of the joint—a process of *hot upsetting*, hence the term **upset welding** (UW). Some metal is expelled from the joint as a shower of sparks during the flashing process. Because impurities and contaminants are squeezed out during this

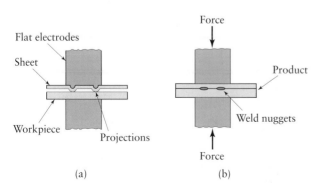

(a)                          (b)

**FIGURE 12.34** Schematic illustration of resistance projection welding: (a) before and (b) after. The projections are produced by embossing operations, as described in Section 7.6.

**FIGURE 12.35** Flash-welding process for end-to-end welding of solid rods or tubular parts.

(a)   (b)

operation, the quality of the weld is good, although a significant amount of material may be burned off during the welding process. The joint may later be machined to improve its appearance. The machines for FW are usually automated and large, with a variety of power supplies ranging from 10 kVA to 1500 kVA.

The flash-welding process is suitable for end-to-end or edge-to-edge joining of similar or dissimilar metals 1–75 mm (0.05–3 in.) in diameter and sheet and bars 0.2–25 mm (0.01–1 in.) thick. Thinner sections have a tendency to buckle under the axial force applied during flash welding. Rings made by forming processes, such as those shown in Figs. 7.24b and c, can be flash butt welded as well. This process is also used to repair broken band-saw blades (Section 8.9.5), with fixtures that are attached to the band-saw frame. Flash welding can be automated for reproducible welding operations; typical applications include joining pipe and tubular shapes for metal furniture and windows. It is also used for welding high-speed steels to steel shanks and for welding the ends of coils of sheet or wire for continuous operation of rolling mills and wire-drawing equipment.

### 12.9.5 Stud arc welding

*Stud arc welding* (SW) is similar to flash welding. The stud, such as a small piece, bolt or threaded rod, hook, or hanger, serves as one of the electrodes while being joined to another member, which is typically a flat plate (Fig. 12.36). In order to properly concentrate the heat generated, prevent oxidation, and retain the molten metal in the weld zone, a disposable ceramic ring (*ferrule*) is placed around the joint. The equipment for stud welding can be automated, with various controls for arcing and applying pressure; portable equipment is also available. The process has numerous extensive applications in the automotive, construction, appliance, electrical, and shipbuilding industries. **Stud welding** is a more general term for joining a metal stud to a workpiece and can be done by various joining processes such as resistance welding, friction welding, or any other suitable means described in this chapter; shielding gases may or may not be used.

In **capacitor-discharge stud welding**, a dc arc is produced from a capacitor bank. No ferrule or flux is required, because the welding time is very short, on the order of one to six milliseconds. The process is capable of stud welding on thin metal sheets with coated or painted surfaces. The choice between this process and stud arc welding depends on factors such as the metals to be joined, part thickness, stud diameter, and shape of the joint.

**FIGURE 12.36** Sequence of operations in stud arc welding, which is used for welding bars, threaded rods, and various fasteners on metal plates.

(a)   (b)   (c)   (d)

### 12.9.6 Percussion welding

The resistance-welding processes described thus far require a transformer to supply power requirements; however, the electrical energy for welding may also be stored in a capacitor. *Percussion welding* (PEW) uses this technique, in which the power is discharged in a very short period of time (1–10 milliseconds), thus developing high localized heat at the joint. This process is useful where heating of the components adjacent to the joint is to be avoided, such as in electronic components.

## 12.10 | Explosion Welding

In *explosion welding* (EXW), pressure is applied by detonating a layer of explosive placed over one of the members being joined, called the *flyer plate* (Fig. 12.37). The contact pressures developed are extremely high, and the plate's kinetic energy striking the mating member produces a turbulent wavy interface (vortexes); this impact mechanically interlocks the two surfaces (Fig. 12.38). In addition, cold pressure welding by plastic deformation also takes place. The flyer plate is placed at an angle, and any oxide films present at the interface are broken up and propelled from the interface; as a result, bond strength in explosion welding is very high.

The explosive may be in the form of flexible plastic sheet, cord, granular solid, or liquid that is cast or pressed onto the flyer plate. Detonation speeds are usually 2400–3600 m/s (8000–12,000 ft/s), depending on the type of explosive, thickness of the explosive layer and its packing density. There is a minimum detonation speed for

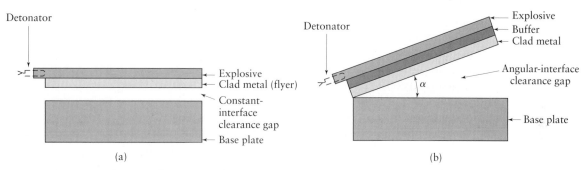

**FIGURE 12.37** Schematic illustration of the explosion-welding process: (a) constant-interface clearance gap and (b) angular-interface clearance gap.

(a)

(b)

**FIGURE 12.38** Cross-sections of explosion-welded joints: (a) titanium (top) on low-carbon steel (bottom) and (b) incoloy 800 (iron–nickel-base alloy) on low-carbon steel. The wavy interfaces shown improve the shear strength of the joint. Some combinations of metals, such as tantalum and vanadium, produce a much less wavy interface. If the two metals have little metallurgical compatibility, an interlayer may be added that has compatibility with both metals. *Source:* Courtesy of DuPont Company.

welding to occur in this process. Detonation is carried out using a standard commercial blasting cap. This process is particularly suitable for cladding plates and slab with dissimilar metals, such as for the chemical industry. Plates as large as 6 m × 2 m (20 ft × 7 ft) have been explosively clad; the resulting material may then be rolled into thinner sections. Tube and pipe are often joined to the holes in header plates of boilers and tubular heat exchangers by placing the explosive inside the tube; the explosion expands the tube and seals it against the plate.

# 12.11 | Diffusion Bonding

*Diffusion bonding*, or **diffusion welding** (DFW), is a solid-state joining process in which the strength of the joint results primarily from diffusion (movement of atoms across the interface) and, to a lesser extent, some plastic deformation of the faying surfaces. This process requires temperatures of about $0.5T_m$ (where $T_m$ is the melting point of the metal on the absolute scale) in order to have a sufficiently high diffusion rate between the parts to be joined. The bonded interface in DFW has essentially the same physical and mechanical properties as those of the base metal. Its strength depends on pressure, temperature, time of contact, and the cleanliness of the faying surfaces; these requirements can be lowered by using filler metal at the interfaces. The principle of diffusion bonding actually dates back centuries, when goldsmiths bonded gold over copper. To produce this material, called **filled gold**, a thin layer of gold foil is first made by hammering; the foil is placed over copper, and a weight is placed on top of it. The assembly is then placed in a furnace and left until a good bond is obtained; this process is also called *hot pressure welding* (HPW). The pressure required can also be achieved by using mechanisms with different coefficients of thermal expansion.

In diffusion bonding, the parts are usually heated in a furnace or by electrical resistance; pressure may be applied by dead weights, by a press, by using differential gas pressure, or from the relative thermal expansion of the parts to be joined. High-pressure autoclaves are also used for bonding complex parts. The process is generally most suitable for dissimilar metal pairs; however, it is also used for reactive metals, such as titanium, beryllium, zirconium, and refractory metal alloys. Diffusion bonding is also important in sintering in powder metallurgy and for processing composite materials (Chapter 11). Because diffusion involves migration of the atoms across the joint, the process is slower than other welding methods. Although DFW is used for fabricating complex parts in small quantities for the aerospace, nuclear, and electronics industries, it has been automated to make it suitable and economical for moderate-volume production.

**Diffusion bonding/superplastic forming.** An important development is the ability to fabricate complex sheet-metal structures by combining *diffusion bonding* with *superplastic forming* (DB/SPF; see also Section 7.10). After diffusion bonding of selected locations of the sheets, the unbonded regions are expanded into a mold by air pressure. Typical structures made by this process are shown in Fig. 12.39. The structures are thin and have high stiffness-to-weight ratios; hence, they are particularly important in aircraft and aerospace applications. This process improves productivity by eliminating mechanical fasteners, reducing the number of parts required, producing parts with good dimensional accuracy and low residual stresses, and reducing labor costs and lead times. First developed in the 1970s, this technology is well advanced for titanium structures (typically Ti-6Al-4V alloy) for aerospace applications; structures made of 7475-T6 aluminum alloy and various other alloys are also being developed.

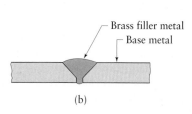

**FIGURE 12.39** (a) Sequence of operations in diffusion bonding and superplastic forming of a structure with three flat sheets. *Source:* After D. Stephen and S. J. Swadling. (b) Typical structures fabricated. *Source:* Rockwell Automation, Inc.

## 12.12 | Brazing and Soldering

There are two joining processes that involve lower temperatures than those required for welding: *brazing* and *soldering*. Brazing temperatures are higher than those for soldering, and the brazed joint has higher strength than thus the soldered joint; brazing and soldering are arbitrarily distinguished by temperature.

### 12.12.1 Brazing

In *brazing*, a process first used as far back as 3000 BC–2000 BC, a filler metal is placed at or between the faying surfaces to be joined, and the temperature is raised to melt the filler metal, but not the workpieces (Fig. 12.40a). The molten metal fills the closely fitting space by *capillary* action; upon cooling and solidification of the filler metal, a strong joint is developed. There are two types of brazing processes: (a) **brazing**, as we have already described, and (b) **braze welding**, in which the filler metal is deposited at the joint (Fig. 12.40b).

**FIGURE 12.40**
(a) Brazing and (b) braze-welding operations.

Filler metals for brazing generally melt above 450°C (840°F), but below the melting point (*solidus temperature*; see Fig. 5.3) of the metals to be joined; thus, this process is unlike liquid-state welding processes, in which the workpieces must melt in the weld area for fusion to occur. Any difficulties associated with heat-affected zones, warping, and residual stresses are, therefore, reduced in brazing. The strength of the brazed joint depends on joint design of the joint and adhesion at the interfaces of the workpiece and filler metal. Consequently, the surfaces to be brazed should be chemically or mechanically cleaned to ensure full capillary action; thus, the use of a flux is important.

The *clearance* between mating surfaces is an important parameter, as it directly affects the strength of the brazed joint. The smaller the gap, the higher is the *shear strength* of the joint; also, there is an optimum gap to achieve maximum *tensile strength*. The typical joint clearance ranges from 0.025 mm to 0.2 mm (0.001 in. to 0.008 in.), and because clearances are very small, surface roughness of the mating surfaces is also important.

**Filler metals.** Several *filler metals* (**braze metals**) are available, with a range of brazing temperatures (Table 12.3); they come in a variety of shapes such as wire, strip, rings, shims, preforms, and filings or powder. The choice of filler metal and its composition is important in order to avoid *embrittlement* of the joint (by grain-boundary penetration of liquid metal; see Section 3.4.2), formation of brittle inter-metallic compounds at the joint, and galvanic corrosion in the joint. Note that filler metals for brazing, unlike those for other welding operations, generally have significantly different compositions than those of the metals to be joined.

Because of diffusion between the filler metal and the base metal, mechanical and metallurgical properties of joints can change during the service life of brazed components or in subsequent processing of the brazed parts. For example, when titanium is brazed with pure-tin filler metal, it is possible for the tin to completely diffuse into the titanium base metal by subsequent aging or heat treatment; when that happens, the joint no longer exists. The filler metal selected must also prevent galvanic corrosion of the brazed joint.

**Fluxes.** The use of a *flux* is essential in brazing in order to prevent oxidation and to remove oxide films from workpiece surfaces. Brazing fluxes are generally made of borax, boric acid, borates, fluorides, and chlorides and are available as paste, slurry, or powder. *Wetting agents* may also be added to improve both the wetting characteristics of the molten filler metal and capillary action. Surfaces to be brazed must be clean and free from rust, oil, lubricants, and other contaminants; clean surfaces are

**TABLE 12.3**

**Typical Filler Metals for Brazing Various Metals and Alloys**

| Base Metal | Filler Metal | Brazing Temperature (°C) |
|---|---|---|
| Aluminum and its alloys | Aluminum–silicon | 570–620 |
| Magnesium alloys | Magnesium–aluminum | 580–625 |
| Copper and its alloys | Copper–phosphorus | 700–925 |
| Ferrous and nonferrous alloys (except aluminum and magnesium) | Silver and copper alloys, copper–phosphorus | 620–1150 |
| Iron-, nickel-, and cobalt-base alloys | Gold | 900–1100 |
| Stainless steels, nickel- and cobalt-base alloys | Nickel–silver | 925–1200 |

essential to obtain the proper wetting and spreading characteristics of the molten filler metal in the joint, as well as maximum bond strength. Sand blasting may also be used to improve surface finish of the faying surfaces. Because they are corrosive, fluxes should be removed after brazing, especially in hidden crevices, usually by washing vigorously with hot water.

### 12.12.2  Brazing methods

The *heating methods* used in brazing also identify the various processes, as described in the upcoming list. A variety of special fixtures may be used to hold the parts together during brazing, some with provision for allowing thermal expansion and contraction.

a) **Torch brazing.**  The heat source in *torch brazing* (TB) is oxyfuel gas with a carburizing flame. Brazing is performed by first heating the joint with the torch, and then depositing the brazing rod or wire in the joint. Suitable part thicknesses are usually in the range of 0.25–6 mm (0.01–0.25 in.). More than one torch may be used in this process. Although it can be automated as a production process, torch brazing is difficult to control and requires skilled labor. This process can also be used for repair work.

b) **Furnace brazing.**  As the name suggests, *furnace brazing* (FB) is carried out in a furnace. The parts are precleaned and preloaded with brazing metal in appropriate configurations before being placed in the furnace (Fig. 12.41), and the whole assembly is heated uniformly in the furnace. Furnaces may be batch type for complex shapes or continuous type for high production runs, especially for small parts with simple joint designs. **Vacuum furnaces** or neutral atmospheres are used for metals that react with the environment.

c) **Induction brazing.**  The source of heat in *induction brazing* (IB) is induction heating by high-frequency ac current. (See Section 5.5.) Parts are preloaded with filler metal and are placed near the induction coils for rapid heating. Unless a protective atmosphere is used, fluxes are generally needed. Part thicknesses are usually less than 3 mm (0.125 in.). Induction brazing is particularly suitable for brazing parts continuously.

d) **Resistance brazing.**  In *resistance brazing* (RB), the source of heat is through electrical resistance of the components to be brazed. Electrodes are used for this purpose, as in resistance welding. Either parts are preloaded with filler metal or the filler metal is supplied externally during brazing. Parts that are commonly brazed by this process have a thickness of 0.1–12 mm (0.004–0.5 in.). As in induction brazing, the process is rapid, heating zones can be confined to very small areas, and the process can be automated to produce uniform quality.

e) **Dip brazing.**  *Dip brazing* (DB) is carried out by dipping the assemblies to be brazed into either a molten filler-metal bath or a molten salt bath (at a temperature just above the melting point of the filler metal), which serves as the heat

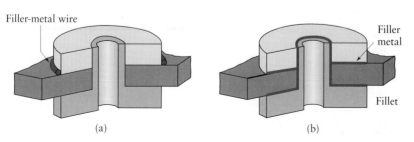

Filler-metal wire

Filler metal

Fillet

(a)          (b)

**FIGURE 12.41**  An example of furnace brazing: (a) before and (b) after. Note that the filler metal is a shaped wire.

726 Chapter 12 Joining and Fastening Processes

source. All workpiece surfaces are thus coated with the filler metal. Dip brazing in metal baths is used only for small parts, such as sheet, wire, and fittings, usually of less than 5 mm (0.2 in.) in thickness or diameter. Molten salt baths, which also act as fluxes, are used for complex assemblies of various thicknesses. Depending on the size of the parts and the bath, as many as 1000 joints can be made at one time by dip brazing.

f) **Infrared brazing.** The heat source in *infrared brazing* (IRB) is a high-intensity quartz lamp. This process is particularly suitable for brazing very thin components, usually less than 1 mm (0.04 in.) thick, including honeycomb structures (see Fig. 7.49). The radiant energy is focused on the joint, and the process can be carried out in a vacuum. *Microwave* heating may also be used.

g) **Diffusion brazing.** *Diffusion brazing* (DFB) is carried out in a furnace, where, with proper control of temperature and time, the filler metal diffuses into the faying surfaces of the components to be joined. The brazing time required may range from 30 min to 24 hrs. Diffusion brazing is used for strong lap or butt joints and for difficult joining operations. Because the rate of diffusion at the interface does not depend on the thickness of the components, part thicknesses may range from foil to as much as 50 mm (2 in.).

h) **High-energy beams.** For specialized high-precision applications and with high-temperature metals and alloys, *electron-beam* and *laser-beam* heating may also be used.

i) **Braze welding.** The joint in *braze welding* is prepared as in fusion welding. Through use of an oxyacetylene torch with an oxidizing flame, filler metal is deposited at the joint, rather than by capillary action, as in brazing; thus, considerably more filler metal is used as compared with brazing. However, temperatures in braze welding are generally lower than in fusion welding, and part distortion is minimal. The use of a flux is essential in this process. The principal use of braze welding is to maintain and repair parts, such as ferrous castings and steel components; however, the process can be automated and used for mass production as well.

Figure 12.42 shows examples of typical brazed joints. In general, dissimilar metals can be assembled with good joint strength, including carbide drill bits (*masonry drill*) and carbide inserts on steel shanks (see Fig. 8.32d). The shear strength of brazed joints can reach 800 MPa (120 ksi), using brazing alloys containing silver (*silver solder*). Intricate, lightweight shapes can be joined rapidly and with little distortion by braze welding.

**FIGURE 12.42** Joint designs commonly used in brazing operations.

**FIGURE 12.43**  Joint designs commonly used for soldering.

(a) Flanged T          (b) Flush lap          (c) Flanged corner

(d) Line contact       (e) Flat lock seam     (f) Flanged bottom

### 12.12.3  Soldering

In soldering, the filler metal (**solder**) melts below 450°C (840°F); as in brazing, the solder fills the joint by capillary action between closely fitting or closely placed components (Fig. 12.43). Soldering with copper–gold and tin–lead alloys was first practiced as long ago as 4000 BC–3000 BC Heat sources for soldering are usually soldering irons, torches, or ovens. Soldering can be used to join various metals and part thicknesses. Although manual soldering operations require skill and are time consuming, soldering speeds can be high with automated equipment.

**Types of solders and fluxes.**  Solders are usually tin–lead alloys in various proportions. For better joint strength and special applications, other solder compositions that can be used are tin–zinc, lead–silver, cadmium–silver, and zinc–aluminum alloys (Table 12.4). Because of the toxicity of lead and its adverse effects on the environment, **lead-free solders** are continually being developed and used; these solders are essentially *tin-based* solders, typical compositions being 96.5% Sn-3.5% Ag, and 42% Sn-58% Bi.

Fluxes are used in soldering as they are in welding and brazing and for the same purposes. Fluxes are generally of two types:

1.  Inorganic acids or salts, such as zinc ammonium chloride solutions, which clean the surface rapidly. After soldering, the flux residues should be removed by washing thoroughly with water to avoid corrosion.

2.  Noncorrosive resin-based fluxes, used in electrical applications.

Soldering is used extensively in the electronics industry. Unlike in brazing, temperatures in soldering are relatively low; consequently, a soldered joint has very limited use at elevated temperatures. Moreover, because solders do not generally have much strength, they are not used for load-bearing structural members. Because of the small faying surfaces, butt joints are rarely made with solders; in other situations, joint strength is improved by mechanical interlocking of the joint. (See also Section 12.16.)

**TABLE 12.4**

| Types of Solders and their Applications | |
| --- | --- |
| Tin–lead | General purpose |
| Tin–zinc | Aluminum |
| Lead–silver | Strength at higher than room temperature |
| Cadmium–silver | Strength at high temperatures |
| Zinc–aluminum | Aluminum; corrosion resistance |
| Tin–silver | Electronics |
| Tin–bismuth | Electronics |

**Solderability.** *Solderability* may be defined in a manner similar to weldability, as described in Section 12.5.2. Copper and precious metals, such as silver and gold, are easy to solder; iron and nickel are more difficult to solder. Aluminum and stainless steels are difficult to solder, because of their strong, thin oxide film (see Section 4.2); however, these and other metals can be soldered by using special fluxes that modify surfaces. Other materials, such as cast irons, magnesium, and titanium, as well as nonmetallic materials such as graphite and ceramics, may be soldered by first plating the parts with metallic elements. (See also Section 12.15.3 for a discussion of the use of a similar technique for joining ceramics.) A common sheet metal is *tinplate*, which is steel coated with tin and used for food containers; the coating makes it very easy to solder.

**Soldering methods.** There are several soldering methods, which are similar to brazing methods:

a) **Torch soldering** (TS);

b) **Furnace soldering** (FS);

c) **Iron soldering** (INS), using a soldering iron;

d) **Induction soldering** (IS);

e) **Resistance soldering** (RS);

f) **Dip soldering** (DS);

g) **Infrared soldering** (IRS);

h) **Ultrasonic soldering,** in which a transducer subjects the molten solder to ultrasonic cavitation, which removes the oxide films from the surfaces to be joined—the need for a flux is thus eliminated;

i) **Wave soldering** (WS), used for automated soldering of printed circuit boards;

j) **Reflow (paste) soldering** (RS).

The last two techniques, which are significantly different from the other soldering methods, are described in more detail as follows:

a) **Reflow (paste) soldering.** Solder pastes are solder-metal particles bound together by flux and by binding and wetting agents; the pastes are semisolid in consistency. They have high viscosity, but are able to maintain a solid shape for relatively long periods of time—in this property, they are similar to greases and cake frostings. The paste is placed directly onto the joint, or on flat objects for finer detail, and it can be applied via a *screening* or *stenciling* process (Fig. 12.44). Stenciling is very commonly used during the mounting of electrical components to printed circuit boards. An additional benefit of this method is that the surface tension of the paste helps keep surface-mount packages aligned on their pads; this feature improves the reliability of these solder joints.

**FIGURE 12.44** Screening solder paste onto a printed circuit board in reflow soldering. *Source:* V. Solberg, Design Guidelines for Surface Mount and Fine-Pitch Technology, 2d ed. New York, McGraw-Hill, 1996.

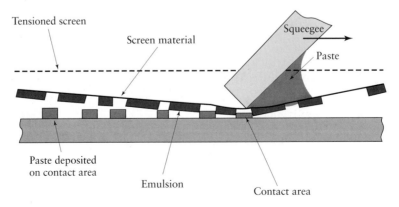

Once the paste has been placed and the joint assembled, the paste is heated in a furnace, and reflow soldering takes place. In *reflow soldering*, the product is heated in a controlled manner, so that the following sequence of events occurs:

1. Solvents present in the paste are evaporated.
2. The flux in the paste is activated, and fluxing action occurs.
3. The components are carefully preheated.
4. The solder particles are melted and wet the joint.
5. The assembly is cooled at a low rate to prevent thermal shock to and fracture of the solder joint.

While this process appears to be straightforward, there are several process variables for each stage, and good control over temperatures and exposures must be maintained at each stage to ensure proper joint strength.

b) **Wave soldering.** Wave soldering is a very common method for attaching circuit components to their boards (Section 13.12). To understand wave soldering, it is imperative to appreciate the fact that molten solder does not wet all surfaces; indeed, solder will not stick to most polymer surfaces and can easy be removed while molten. Also, as can be shown with a simple hand-soldering iron, the solder only wets metal surfaces and forms a good bond when the metal is preheated to a certain temperature. Therefore, wave soldering requires separate fluxing and preheating operations before it can be successfully completed.

Wave soldering is schematically illustrated in Fig. 12.45a. A standing laminar wave of molten solder is generated by a pump. Preheated and prefluxed circuit boards are conveyed over the wave; the solder wets the exposed metal surfaces, but it does not stay attached to the polymer package for the integrated circuits, and it does not stick to the polymer-coated circuit boards. An air knife (basically a high-velocity jet of hot air) blows excess solder from the joint, to prevent bridging between adjacent leads. When surface-mount packages are to be wave soldered, they must first be adhesively bonded to the circuit board before the soldering can commence. This bonding is usually accomplished by screening or stenciling epoxy onto the boards, placing the components in their proper locations, curing the epoxy, inverting the board, and performing wave soldering. Fig. 12.45b shows an SEM photograph of a typical surface-mount joint.

(a)                                                              (b)

**FIGURE 12.45**   (a) Schematic illustration of the wave-soldering process. (b) SEM image of a wave-soldered joint on a surface-mount device.

# 12.13 | Adhesive Bonding

Numerous components and products can be joined and assembled with an *adhesive*, rather than using any of the joining methods described thus far. **Adhesive bonding** has been a common method of joining and assembly for applications such as labeling, packaging, bookbinding, home furnishings, and footwear. Plywood, developed in 1905, is a typical example of adhesive bonding of several layers of wood with glue. Adhesive bonding has been gaining increased acceptance in manufacturing ever since its first use on a large scale in assembling load-bearing components in military aircraft manufactured during World War II (1939–1945). Figure 12.46 shows examples of adhesively bonded joints. Adhesives are available in various forms such as liquids, pastes, solutions, emulsions, powder, tape, and film; when applied, adhesives typically are on the order of 0.1 mm (0.004 in.) thick.

## 12.13.1 Adhesives

Numerous types of adhesives are available, and continue to be developed, that provide adequate joint strength, including fatigue strength. The three basic types of adhesives are listed as follows:

1. **Natural adhesives**, such as starch, dextrin (a gummy substance obtained from starch), soya flour, and animal products;
2. **Inorganic adhesives**, such as sodium silicate and magnesium oxychloride;
3. **Synthetic organic adhesives**, which may be thermoplastics (for nonstructural and some structural bonding) or thermosetting polymers (primarily for structural bonding).

Because of their cohesive strength, synthetic organic adhesives are the most important adhesives in manufacturing processes, particularly for load-bearing applications (*structural adhesives*). They are generally classified as follows:

1. **Chemically reactive**, such as polyurethanes, silicones, epoxies, cyanoacrylates, modified acrylics, phenolics, and polyimides—also included are anaerobics,

**FIGURE 12.46** Various configurations for adhesively bonded joints: (a) single lap, (b) double lap, (c) scarf, and (d) strap.

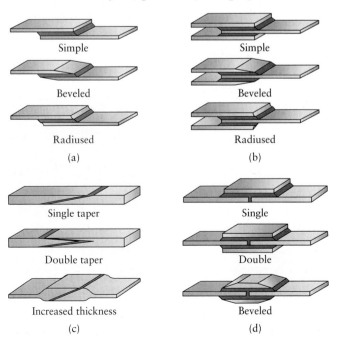

which cure in the absence of oxygen (such as Loctite® for threaded fasteners; see the case study at end of this chapter);

2. **Pressure-sensitive**, including are natural rubber, styrene–butadiene rubber, butyl rubber, nitrile rubber, and polyacrylates;

3. **Hot melt**, which are thermoplastics such as ethylene–vinyl-acetate copolymers, polyolefins, polyamides, polyester, and thermoplastic elastomers;

4. **Reactive hot-melt**, which have a thermoset portion based on the urethane's chemistry, with improved properties;

5. **Evaporative or diffusion**, including vinyls, acrylics, phenolics, polyurethanes, synthetic rubbers, and natural rubbers;

6. **Film** and **tape**, such as nylon–epoxies, elastomer–epoxies, nitrile–phenolics, vinyl–phenolics, and polyimides;

7. **Delayed tack**, including styrene–butadiene copolymers, polyvinyl acetates, polystyrenes, and polyamides;

8. **Electrically** and **thermally conductive**, including epoxies, polyurethanes, silicones, and polyimides (see Section 12.13.4).

*Adhesive systems* may be classified based on their specific chemistries:

1. **Epoxy-based systems.** These systems have high strength and high-temperature properties, to as high as 200°C (400°F); typical applications include automotive brake linings and as bonding agents for sand molds for casting. (See Section 5.8.1.)

2. **Acrylics.** Suitable for applications with substrates that are not clean.

3. **Anaerobic systems.** Curing of these adhesives is by oxygen deprivation, and the bond is usually hard and brittle; curing times can be reduced by external heat or ultraviolet (UV) radiation.

4. **Cyanoacrylate.** The bond lines are thin, and the bond sets within 5 s to 40 s.

5. **Urethanes.** High toughness and flexibility at room temperature; widely used as sealants.

6. **Silicones.** Highly resistant to moisture and solvents, silicones have high impact and peel strength; however, curing times are typically in the range of one to five days.

Many of the adhesives listed previously can be combined to optimize their properties; examples include epoxy–silicon, nitrile–phenolic, and epoxy–phenolic. The least expensive adhesives are epoxies and phenolics, followed by polyurethanes, acrylics, silicones, and cyanoacrylates. Adhesives for high-temperature applications at up to about 260°C (500°F), such as polyimides and polybenzimidazoles, are generally the most expensive adhesives.

Depending on the particular application, an adhesive generally must have one or more of the following properties (Table 12.5):

- Strength (shear and peel);
- Toughness;
- Resistance to various fluids and chemicals;
- Resistance to environmental degradation, including heat and moisture;
- Capability to wet the surfaces to be bonded.

Adhesive joints are designed to withstand shear, compressive, and tensile forces, but they should not be subjected to peeling forces (Fig. 12.47). Note, for example, how

**TABLE 12.5**

**Typical Properties and Characteristics of Chemically Reactive Structural Adhesives**

|  | Epoxy | Polyurethane | Modified Acrylic | Cyanocrylate | Anaerobic |
|---|---|---|---|---|---|
| Impact resistance | Poor | Excellent | Good | Poor | Fair |
| Tension-shear strength, MPa ($10^3$ psi) | 15.4 (2.2) | 15.4 (2.2) | 25.9 (3.7) | 18.9 (2.7) | 17.5 (2.5) |
| Peel strength, N/m (lb/in.) | < 525 (3) | 14,000 (80) | 5250 (30) | < 525 (3) | 1750 (10) |
| Substrates bonded | Most | Most smooth, nonporous | Most smooth, nonporous | Most nonporous metals or plastics | Metals, glass, thermosets |
| Service temperature range, °C (°F) | −55 to 120 (−70 to 250) | −160 to 80 (−250 to 175) | −70 to 120 (−100 to 250) | −55 to 80 (−70 to 175) | −55 to 150 (−70 to 300) |
| Heat cure or mixing required | Yes | Yes | No | No | No |
| Solvent resistance | Excellent | Good | Good | Good | Excellent |
| Moisture resistance | Excellent | Fair | Good | Poor | Good |
| Gap limitation, mm (in.) | None | None | 0.75 (0.03) | 0.25 (0.01) | 0.60 (0.025) |
| Odor | Mild | Mild | Strong | Moderate | Mild |
| Toxicity | Moderate | Moderate | Moderate | Low | Low |
| Flammability | Low | Low | High | Low | Low |

*Source*: *Advanced Materials & Processes*, July 1990, ASM International.

(a)          (b)

**FIGURE 12.47** Characteristic behavior of (a) brittle and (b) tough adhesives in a peeling test. This test is similar to peeling adhesive tape from a solid surface. Adhesive joints should not be subjected to this type of loading in service.

easily an adhesive tape can be peeled from a surface. During peeling, the behavior of an adhesive may be brittle, or it may be ductile and tough, requiring large forces to peel it.

## 12.13.2 Surface preparation and application

Surface preparation is very important in adhesive bonding, as joint strength depends greatly on the absence of dirt, dust, oil, and various other contaminants; note, for example, that is virtually impossible to apply an adhesive tape over a dusty or oily surface. Contaminants also affect the wetting ability of the adhesive, and they can prevent spreading of the adhesive evenly over the interface; thick, weak, or loose oxide films on workpieces also are detrimental to adhesive bonding. On the other hand, a porous or thin and strong oxide film may be desirable, particularly one with some surface roughness to improve adhesion, by mechanical locking. However, the roughness must not be too high, or else air may be trapped and joint strength is reduced. Various compounds and primers are available to modify surfaces to increase the strength of adhesive bonds. Liquid adhesives may be applied by brushes, sprayers, and rollers.

### 12.13.3  Process capabilities

A wide variety of similar and dissimilar metallic and nonmetallic materials and components with different shapes, sizes, and thicknesses can be bonded to each other by adhesives. Adhesive bonding can be combined with mechanical fastening methods (see Section 12.14) to further improve the strength of the bond. An important consideration in the use of adhesives in production is curing time, which can range from a few seconds at high temperatures to many hours at room temperature, particularly for thermosetting adhesives; thus, production rates may be low. Furthermore, adhesive bonds for structural applications are rarely suitable for service above 250°C (500°F). Joint design and bonding methods require care and skill, and special equipment, such as fixtures, presses, tooling, and autoclaves and ovens for curing, is also usually needed.

Nondestructive inspection of the quality and strength of adhesively bonded components can be difficult; however, some of the techniques described in Section 4.8.1, such as acoustic impact (tapping), holography, infrared detection, and ultrasonic testing, are effective nondestructive testing methods. Major industries that use adhesive bonding extensively are aerospace, automotive, appliances, and building products. Applications include attaching rearview mirrors to windshields, automotive brake lining assemblies, laminated windshield glass (see Section 11.11.2), various appliances, helicopter blades, honeycomb structures, and aircraft bodies and control surfaces.

The advantages of adhesive bonding are listed as follows:

1. It provides a bond at the interface, either for structural strength or for non-structural applications such as sealing, insulating, preventing electrochemical corrosion between dissimilar metals, and reducing vibration and noise through internal damping at the joints.

2. It distributes the load at an interface, thus eliminating localized stresses that typically result from joining the components with welds or mechanical fasteners, such as bolts and screws. Moreover, since no holes are required, structural integrity of the sections is maintained.

3. The external appearance of the joined components is unaffected.

4. Very thin and fragile components can be bonded without contributing significantly to weight.

5. Porous materials and materials with very different properties and sizes can be joined.

6. Because adhesive bonding is usually carried out between room temperature and about 200°C (400°F), there is no significant distortion of the components or change in their original properties; this factor is particularly important for materials that are heat sensitive.

The limitations of adhesive bonding are:

1. The service temperatures are relatively low.

2. The bonding time can be long.

3. Surface preparation is necessary.

4. It is difficult to test bonded joints nondestructively, particularly with large structures.

5. The reliability of adhesively bonded structures during their service life and under hostile environmental conditions (*degradation* by temperature, oxidation, radiation, stress corrosion, and dissolution) may be a significant concern.

### 12.13.4 Electrically conducting adhesives

Although the majority of adhesive bonding is for mechanical strength and structural integrity, a more recent advance is the development and application of *electrically conducting adhesives* to replace lead-base solder alloys, particularly in the electronics industry, as well as for economic considerations. These adhesives require curing or setting temperatures that are lower than that for soldering. Electrical conductivity in adhesives is obtained by the addition of *fillers*, such as silver (used most commonly, with up to 85% silver), copper, aluminum, nickel, gold, and graphite. (See also Section 10.7.2.) Recent developments in fillers include the use of polymeric particles, such as polystyrene, coated with thin films of silver or gold. Fillers are generally in the form of flakes or particles; fillers that improve electrical conductivity also improve thermal conductivity. The polymer is the matrix, and there is a minimum volume concentration of fillers in order to make the adhesive electrically conducting, typically in the range of 40%–70%. Matrix materials are generally epoxies, although various thermoplastics are also used.

The size, shape, and distribution of the conducting particles, as well as how heat and pressure are applied and the nature of contact among the individual particles, can be controlled so as to impart isotropic or anisotropic electrical conductivity to the adhesive. Applications of electrically conducting adhesives include calculators, remote controls, and control panels, as well as high-density use in electronic assemblies, liquid-crystal displays, pocket TVs, and electronic games.

## 12.14 | Mechanical Fastening

Two or more components may have to be joined or fastened in such a way that they can be taken apart sometime during the product's service life or life cycle. Countless products, including mechanical pencils, watches, computers, bicycles, engines, and aircraft have components that are fastened mechanically. *Mechanical fastening* may be preferred over other methods because of its

1. Ease of manufacturing;

2. Ease of assembly and transportation;

3. Ease of parts replacement, maintenance, and repair;

4. Ease in creating designs that require movable joints, such as hinges, sliding mechanisms for drawers and doors, and adjustable components and fixtures;

5. Lower overall cost of manufacturing the product.

The most common method of mechanical fastening is by using bolts, nuts, screws, rivets, pins, and a wide variety of other **fasteners**; these techniques are also known as **mechanical assembly**. Most mechanical fastening typically requires that the components have holes through which fasteners are inserted. These assemblies may be structural (load-bearing) components and thus may be subjected to both shear and tensile stresses; consequently, they should be designed to resist these forces.

## 12.14.1 Hole preparation

*Hole preparation* is an important aspect of mechanical fastening. A hole in a solid body can be produced by punching, drilling, chemical and electrical means, and high-energy beams (Chapters 7, 8, and 9), depending on the type of material, its properties, and its thickness. Recall also from Chapters 5, 6, and 11 that holes may be produced as an *integral part* of the product during casting, forging, extrusion, and powder metallurgy. For improved accuracy and surface finish, many of these holemaking operations may be followed by finishing operations, such as shaving, deburring, reaming, and honing.

Because of their fundamental differences, each type of holemaking operation produces a hole with different surface finish and properties and dimensional characteristics. The most significant influence of a hole in a solid body is to act as a stress concentration, and thus the hole has a tendency to reduce the component's fatigue life. For holes, fatigue life can best be improved by inducing compressive residual stresses on the cylindrical surface of the hole. These stresses are usually developed by pushing a round rod (*drift pin*) through the hole and expanding it by a very small amount, a process that plastically deforms the surface layers of the hole in a manner similar to shot peening or roller burnishing (see Section 4.5.1).

## 12.14.2 Threaded fasteners

Bolts, screws, and nuts are among the most commonly used *threaded fasteners*. References on machine design describe in detail numerous standards and specifications, including thread dimensions, tolerances, pitch, strength, and the quality of materials used to make these fasteners. Bolts and screws may be secured with nuts or they may be *self-tapping*, whereby the screw either cuts or forms the thread into the part to be fastened; the latter method is particularly effective and economical in plastic products. If the joint is to be subjected to vibration, such as in aircraft and various types of engines and high-speed machinery, several specially designed nuts and lock washers are available. They increase the frictional resistance in the torsional direction, thus preventing vibrational loosening of the fasteners.

## 12.14.3 Rivets

The most common method of permanent or semipermanent mechanical joining is by *riveting* (Fig. 12.48). Hundreds of thousands of rivets may be used in the construction and assembly of large commercial aircraft. There are several types of rivets, and some may be solid or hollow; blind rivets are inserted from one side only. Installing a rivet basically consists of placing the rivet in the hole and deforming the end of its shank by upsetting (*heading*, similar to the process shown in Fig. 6.17). The operation may be performed by hand or by mechanized means, including the use of robots. Riveting may be done either at room temperature or hot, using special tools, or by using explosives in the cavity of the rivet.

(a)          (b)          (c)          (d)

**FIGURE 12.48**   Examples of rivets: (a) solid, (b) tubular, (c) split, or bifurcated and (d) compression.

Standard loop

Flat clinch

Metal channel

Nonmetal

Roll–pin loop — Pin

(a)            (b)            (c)            (d)

**FIGURE 12.49**    Examples of metal stitching.

## 12.14.4 Other methods of fastening

Several other fastening methods are also used in joining and assembly applications; the most common types are described as follows:

a) **Metal stitching or stapling.**  The process of *metal stitching* or *stapling* (Fig. 12.49) is much like that of ordinary stapling of papers. This operation is fast and particularly suitable for joining thin metallic and nonmetallic materials, and it does not require holes to be made in the components; a common example is the stapling of cardboard containers for various products. In *clinching* (Fig. 12.49b), the fastening material should be sufficiently thin and ductile to withstand the large localized deformations during sharp bending.

b) **Seaming.**  *Seaming* is based on the simple principle of folding two thin pieces of material together, much like folding two or more pieces of paper at their corners in the absence of a paper clip or a staple. The most common examples of seaming (*lock seams*) are the lids of beverage cans and containers for food and household products (Fig. 12.50). In seaming, the materials should be capable of undergoing bending and folding at very small radii (see Section 7.4.1); otherwise, they may crack, and the seams will not be airtight or watertight. The performance and airtightness of lock seams may be improved with adhesives, coatings, polymeric seals, or by soldering.

c) **Crimping.**  The *crimping* process is a method of joining without using fasteners. It can be done with beads or dimples (Fig. 12.51), which can be produced by *shrinking* or *swaging* operations. Crimping can be used on both tubular and flat parts. Caps on glass bottles are attached by crimping, as are some connectors for electrical wiring.

**FIGURE 12.50**    Stages in forming a double-lock seam.

(a)            (b)            (c)            (d)

**FIGURE 12.51**    Two examples of mechanical joining by crimping.

(a)            (b)

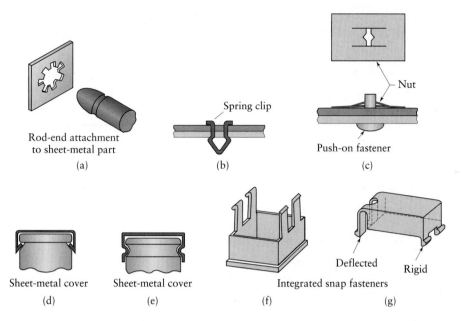

**FIGURE 12.52** Examples of spring and snap-in fasteners to facilitate assembly.

d) **Snap-in fasteners.** Figure 12.52 shows various spring and *snap-in fasteners*. Such *snap-fit* joints are widely used in automotive bodies and household appliances; they are economical and permit easy and rapid component assembly, particularly for plastic products.

e) **Shrink and press fits.** Components may also be assembled by shrink fitting and press fitting. *Shrink fitting* is based on the principle of the differential thermal expansion and contraction of two components; typical applications include the assembly of die components and mounting gears and cams on shafts. In *press fitting*, one component is forced over another, resulting in high joint strength.

# 12.15 | Joining Plastics, Ceramics, and Glasses

## 12.15.1 Joining thermoplastics

As described in Section 10.3, thermoplastics soften and melt as temperature increases; consequently, they can be joined by means whereby heat is generated at the interface, either through external means or internally. The heat softens the thermoplastic at the interface to a viscous or molten state and, with the application of pressure to ensure a good bond, allows *fusion* to take place. Filler materials of the same type of polymer may also be used.

Oxidation can be a problem in joining some polymers, such as polyethylene, causing degradation; typically, an inert shielding gas such as nitrogen is used to prevent oxidation. Because of the low thermal conductivity of thermoplastics, the heat source, if applied at too high a rate, may burn or char the surfaces of the components to be joined, possibly causing difficulties in obtaining sufficiently deep fusion for joint strength. *External heat sources* may consist of the following, depending on the compatibility of the polymers to be joined:

1. Hot air or gases, or infrared radiation using high-intensity focused quartz heat lamps;

2. Heated tools and dies, used in a process known as *hot-tool* or *hot-plate welding*, that contact and heat the surfaces to be joined by interdiffusion of molecular chains—this process is commonly used to butt-weld pipe and tubing;

3. Radio-frequency or dielectric heating, particularly useful for thin films;

4. Electrical-resistance wire or braids, or carbon-based tapes, sheet, and ropes, which are placed at the interface and pass electrical current, a process known as *resistive-implant welding*—these elements at the interface may be subjected to a radio-frequency field (*induction welding*) and must be compatible with the intended use of the joined product, since they remain in the weld zone;

5. Lasers, using defocused beams at low power to prevent degradation of the polymer.

*Internal heat sources* are developed through the following means:

1. Ultrasonic welding (see Section 12.7), which is the most commonly used process for thermoplastics, particularly amorphous polymers such as ABS and high-impact polystyrene;

2. Friction welding (see Section 12.8), also called *spin welding* for polymers—it includes linear friction welding, also called *vibration welding*, which is particularly useful for joining polymers with a high degree of crystallinity (see Section 10.2.2), such as acetal, polyethylene, nylon, and polypropylene;

3. *Orbital welding*, which is similar to friction welding, except that the rotary motion of one part is in an orbital path (see also Fig. 6.16a).

**Other joining methods.** *Adhesive bonding* of polymers is a versatile process, One of the best examples is the process for joining sections of polyvinyl chloride (PVC) pipe, used extensively in home plumbing systems, or ABS pipe, used in drain, waste, and vent systems. The liquid adhesive is applied to the connecting sleeve and pipe surfaces, sometimes using a primer to improve adhesion (much like using primers in painting). Adhesive bonding of polyethylene, polypropylene, and polytetrafluoroethylene (Teflon) can be difficult, because these polymers have low surface energy, and hence adhesives do not stick readily to their surfaces; the surfaces usually have to be treated chemically to improve bond strength. The use of adhesive primers or double-sided adhesive tapes can also be effective. Bonding may also be achieved by using solvents (*solvent bonding*).

Coextruded multilayer food wrappings consist of various types of films bonded by the *heat* generated during *extrusion*. Each film has a different function, such as to keep out moisture and oxygen or to facilitate heat sealing during packaging. Some wrappings have as many as seven layers, all bonded together during production of the film (*cocuring*).

Thermoplastics may also be joined by *mechanical* means, including the use of fasteners and self-tapping screws. The strength of the joint depends on the particular technique used and the inherent toughness and resilience of the plastic to resist tearing at the holes present for mechanical fastening. *Integrated snap fasteners* (Fig. 12.52f and g) are gaining wide acceptance for their simplified and cost-effective assembly operations.

Bonding may also be achieved by *magnetic* means, by embedding tiny [on the order of $1\,\mu$m ($40\,\mu$in.)] particles in the polymer. A high-frequency field then causes induction heating of the polymer and melts it at the interfaces to be joined (*electromagnetic bonding*).

## 12.15.2  Joining thermosets

Because they do not soften or melt with increasing temperature, thermosetting plastics, such as epoxy and phenolics, are usually joined (1) by using threaded or other molded-in inserts (see, for example, Fig. 10.24), (2) by using mechanical fasteners, and (3) by solvent bonding. Bonding with solvents basically involves the following sequence: (a) roughening the surfaces of the thermoset parts with an abrasive cloth or paper, (b) wiping the surfaces with a solvent, and (c) pressing the surfaces together and maintaining the pressure until sufficient bond strength is developed.

## 12.15.3  Joining ceramics and glasses

A wide variety of types of ceramics and glasses is now available with unique and important properties; these ceramics and glasses are used either as products or components of products, or as tools, molds, and dies. As any other material used in products, these materials also have to be assembled into components, either with the same type of material or with different materials. These joints have to meet specific requirements, depending on the particular application. Nonmetallic parts can, of course, be joined using adhesive bonding; a typical example is assembling broken ceramic pieces, using two-component epoxy, mixed just prior to application. Other methods include mechanical means, such as fasteners, and shrink or press fitting. The relative thermal expansion of the two materials should be taken into consideration to avoid damage if the assembly is subjected to elevated temperature. (See Section 3.9.5.)

**Ceramics.**  As described in Section 8.6.4, *tungsten carbide* has a matrix (binder) of cobalt, and *titanium carbide* has matrix of nickel–molybdenum alloy. Consequently, with both binders being metal, carbides can easily be brazed to other metals. Common applications include brazing carbide cutting tools to steel toolholders (see Fig. 8.32d), brazing carbide tips to masonry drills, and brazing cubic-boron-nitride or diamond tips to carbide inserts (see Fig. 8.39).

A commonly used technique that is effective in joining difficult-to-bond material combinations consists of first applying a coating of a material that bonds itself easily to one or both parts, thus acting as a bonding agent, just as an adhesive does between a piece of wood and a metal that otherwise would not bond to each other. Thus, the surface of *alumina ceramics*, for example, can be metallized (see Section 4.5.1); in this technique, known as the Mo–Mn process, the ceramic part is first coated with a slurry that, after being fired, forms a glassy layer. This layer is then plated with nickel, and, now that the part has a metallic surface, it can brazed to a metal surface. Depending on their particular structure, ceramics and metals can be joined by diffusion bonding (Section 12.11); it may be necessary to place a metallic layer at the joint to make the joint stronger.

It is also apparent that ceramic parts can be joined or assembled together during their shaping process (see Section 11.9); a common example is attaching the handle of a coffee mug. Thus, in a sense, the shaping of the product is done integrally, rather than as an additional operation after the part is already made.

**Glasses.**  As evidenced by the availability of numerous glass objects, glasses can easily be bonded to each other, which is accomplished by first softening the surfaces to be joined and then pressing the two pieces together and cooling them. Bonding of glass to metals is possible because of the diffusion of metal ions into the amorphous surface structure of glass.

# 12.16 | Design Considerations in Joining

Basic design guidelines for the variety of joining processes described throughout this chapter are outlined as follows:

## 12.16.1 Design for welding

As in all manufacturing processes, the optimum choice in welding is the one that satisfies all design and service requirements at minimum cost. Some examples of weld characteristics are shown in Fig. 12.53. General design guidelines for welding may be summarized as follows:

1. Product design should minimize the number of welds, as welding can be costly when it is not automated.
2. The weld location should be selected to avoid excessive stresses or stress concentrations in the welded structure, as well as for appearance.
3. The weld location should be selected so as not to interfere with subsequent processing of the part or with its intended use and appearance.
4. Parts should fit properly before welding; the method employed to produce the edges (sawing, machining, shearing, and flame cutting) can affect weld quality.
5. Modification of the design may avoid the need for edge preparation.
6. Weld-bead size should be kept to a minimum to conserve weld metal.
7. Mating surfaces for some processes may require uniform cross-sections at the joint, such as is shown for flash welding in Fig. 12.54.

**FIGURE 12.53** Design guidelines for welding. *Source*: Bralla, J. G. (ed.) *Handbook of Product Design for Manufacturing*, 2d ed. New York, McGraw-Hill, 1999.

**FIGURE 12.54** Design guidelines for flash welding.

**EXAMPLE 12.4** Weld design selection

Figure 12.55 shows three different types of weld design. In Fig. 12.55a, the two vertical joints can be welded either externally or internally. Obviously, full-length external welding takes a considerable amount of time and requires more weld material than the alternative design, which consists of intermittent internal welds. Moreover, in the alternative method, the appearance of the structure is improved, and distortion is reduced. Although both designs require the same amount of weld material and welding time (Fig. 12.55b), further analysis shows that the design on the right can carry three times the moment *M* than can the one on the left. In Fig. 12.55c, the weld on the left requires about twice the amount of weld material than that required by the design on the right. Note also that because more material must be machined, the design on the left will require more time for edge preparation, and more base metal will be wasted.

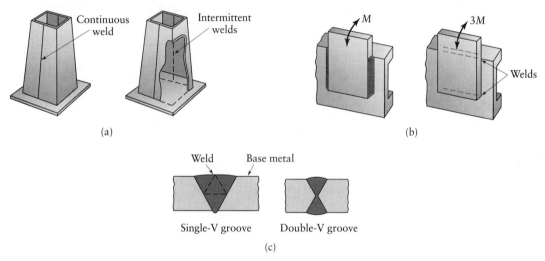

**FIGURE 12.55** Weld designs for Example 12.4.

## 12.16.2 Design for brazing and joining

Some general design guidelines for *brazing* are given in Fig. 12.56; note that strong joints require a greater contact area for brazing than for welding. Design guidelines for *soldering* are similar to those for brazing. Figures 12.42 and 12.43 shows some examples of frequently used joint designs; note again the importance of large contact surfaces to the development of sufficient joint strength.

## 12.16.3 Design for adhesive bonding

Designs for *adhesive bonding* should ensure that joints are subjected to compressive, tensile, or shear forces, but not peeling or cleavage (see Fig. 12.47). Figure 12.57 shows several joint designs for adhesive bonding. They vary considerably in their strength; hence, selection of the appropriate design is important and should include considerations such as the type of loading and the environment to which the bonded structure will be subjected, particularly over a period of time.

Butt joints require large bonding surfaces, and lap joints tend to distort under tension, because of the force couple developed at the joint (see Fig. 12.25a). The coefficients of thermal expansion of the components to be bonded should preferably be

**FIGURE 12.56** Examples of good and poor designs for brazing.

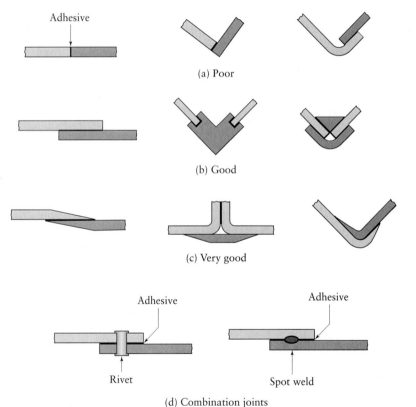

| Good | Poor | Comments |
|---|---|---|
| | | Too little joint area in shear |
| | | Improved design when fatigue loading is a factor to be considered |
| | | Insufficient bonding |

**FIGURE 12.57** Various joint designs in adhesive bonding. Note that good design requires large contact areas between the members to be joined.

Adhesive

(a) Poor

(b) Good

(c) Very good

Adhesive                    Adhesive

Rivet                    Spot weld

(d) Combination joints

close, in order to avoid internal stress during adhesive bonding. Also, situations where thermal cycling can cause differential movement across the joint should be avoided.

## 12.16.4 Design for mechanical fastening

The design of *mechanical joints* requires consideration of (a) the type of loading, such as shear, tension, and moments, to which the structure will be subjected; and (b) the size and spacing of the holes. Also, the compatibility of the fastener material with the

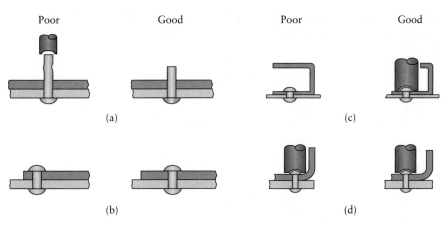

Poor          Good          Poor          Good

(a)                                        (c)

(b)                                        (d)

**FIGURE 12.58**  Design guidelines for riveting. *Source*: Bralla, J. G. (ed.) *Handbook of Product Design for Manufacturing*, 2d ed. New York, McGraw-Hill, 1999.

components to be joined is important; incompatibility may lead to galvanic corrosion, also known as *crevice corrosion*. In a system where, for example, a steel bolt or rivet is used to fasten copper sheets, the bolt is anodic and the copper plate cathodic, thus leading to rapid corrosion and subsequent loss of joint strength. Similar reactions take place when aluminum or zinc fasteners are used on copper products. Figure 12.58 illustrates some design guidelines for *riveting*.

Other general design guidelines for mechanical joining include the following (see also Section 14.10):

1. It is generally less costly to use fewer, but larger, fasteners than to use a large number of smaller fasteners.

2. Part assembly should be accomplished with a minimum number of fasteners.

3. The fit between parts to be joined should be as loose as permissible to reduce costs and facilitate the assembly process.

4. Fasteners of standard size should be used whenever possible.

5. Holes should not be too close to edges or corners in order to avoid the possibility that the material will tear when subjected to external forces.

## 12.17 | Economic Considerations

We have included details on the general economic considerations for each particular joining process in our description of the process' capabilities; it can be seen that because of the wide variety of joining methods available, it is difficult to make generalizations regarding costs. The overall relative costs for these processes have been included in the last columns in Tables 12.1 and 12.2. Note in Table 12.1, for example, that because of the preparations involved, brazing and mechanical fastening can be the most costly methods. On the other hand, because of their highly automated nature, resistance welding and seaming and crimping processes are the least costly of the eight methods listed.

The cost of adhesive bonding depends on the particular operation. However, in many applications, the overall economics of the process makes adhesive bonding an attractive joining process; sometimes, it is the only process that is feasible or practical for a particular application.

Table 12.2 lists the cost of equipment involved for major categories of joining processes. Note that because of the specialized equipment required, laser-beam welding is the most expensive process in this respect, whereas the traditional oxyfuel-gas welding and shielded metal-arc welding are the least expensive processes. Typical ranges of equipment costs for some processes are listed as follows in alphabetical order, although the cost of some equipment exceeds these values significantly, depending on the size of the equipment and the level of automation and controls implemented:

> Diffusion brazing: $50,000 to $300,000;
> Electron-beam welding: $75,000 to over $1 million;
> Electroslag welding: $15,000 to $25,000;
> Flash welding: $5000 to $1 million;
> Friction welding: $75,000 to $300,000;
> Furnace brazing: $2000 to $300,000;
> Gas metal-arc and flux-cored arc welding: $1000 to $3000;
> Gas tungsten-arc welding: $1000 to $5000;
> Laser-beam welding: $40,000 to $1 million;
> Plasma-arc welding: $3000 to $6000;
> Resistance welding: $20,000 to $50,000;
> Shielded metal-arc welding: $1500;

An important aspect of the economics of joining is the major progress being made in automation of processes. (See Chapters 14 and 15.) These developments include computer control and optimization of processing parameters, as well as improvement in the manner in which various components in the operation are manipulated and controlled. Particularly important has been the extensive and effective use of industrial robots, programmed in such a manner that they can accurately track complex weld paths (such as in tailor-welded blanks), using machine vision systems and closed-loop controls. Consequently, the repeatability and accuracy of welding processes have improved greatly. The operations are optimized so that, unlike in manual operations, little time is wasted in manipulating various components in the system, just as such optimization has reduced the noncutting time in machining operations with advanced machine tools and machining centers. These developments have significantly reduced costs and have had a major impact on the overall economics of joining operations.

## CASE STUDY | Light-Curing Acryclic Adhesives for Medical Products

Cobe Cardiovascular, Inc., is a leading manufacturer of blood-collection and blood-processing systems, as well as extracorporeal systems for cardiovascular surgery. In 1993, the company, like many other device manufacturers, used solvents for bonding several device components and subassemblies. However, several federal agencies began to encourage avoiding the use of solvents, and Cobe particularly wanted to eliminate its use of methylene chloride, for environmental and occupational safety reasons. To accomplish this goal, the company began to redesign most of its assemblies to accept light-curing (ultraviolet or visible) adhesives. Most of the company's devices were made of transparent plastics; consequently, its engineers determined that clear adhesive bonds were required for aesthetic purposes and because they have no tendency for stress cracking or crazing (see Section 10.3).

As an example of a typical product, Cobe's blood-salvage, or blood-collection, reservoir is an oval polycarbonate device that is approximately 300 mm (12 in.) tall, 200 mm (8 in.) in major diameter and 100 mm (4 in.) deep (Fig. 12.59). The reservoir is a one-time-use, or disposable, device; its purpose is to collect and hold blood during open-heart and chest surgery and arthroscopic and emergency-room procedures. Up to 3000 cc of blood may be stored in the reservoir while the blood awaits passage into a 250-cc centrifuge, which cleans the blood and returns it to the patient after the surgical procedure is completed. The collection reservoir consists of a clear, polycarbonate lid joined to a polycarbonate bucket. The joint is a tongue-and-groove

**FIGURE 12.59** The Cobe Laboratories blood reservoir. The lid is bonded to the bowl with an airtight adhesive joint and tongue-in-groove joint. *Source*: Cobe Laboratories.

configuration; the goal was to create a strong, elastic joint that could withstand repeated stresses with no chance of leakage.

Light-cure acrylic adhesives offer a range of performance properties that make them well suited for this application (see the discussion of *anaerobic adhesives* in Section 12.13.1). First and foremost, they achieve high bond strength to the thermoplastics typically used to form medical-device housings. Loctite® 3211, for example, achieves shear strengths of 11 MPa (1600 psi) on polycarbonate. As important as the initial shear strength may be, it is even more important that the adhesive be able to maintain the high bond strength after sterilization; fortunately, disposable medical devices typically are subjected to very few sterilization cycles during manufacture. Also, these adhesives can endure a limited number of cycles of gamma irradiation, electron-beam irradiation, autoclaving, and ethylene-oxide or chemical immersion.

Another consideration that makes light-cure adhesives well suited for this application is their availability in formulations that allow them to withstand large strains prior to yielding; Loctite® 3211 yields at elongations in excess of 200%. This flexibility is critical, because the bonded joints are typically subjected to large amounts of bending and flexing when the devices are pressurized during qualification testing and use. If an adhesive is too rigid, it will fail this type of testing, even if it offers higher shear strength than that of a comparable and more flexible adhesive. Finally, light-cure acrylics are widely available in formulations that meet international quality standard certification (ISO; see Section 4.9.4), which means that when processed properly, they will not cause biocompatibility problems in the final assembly.

While these performance features are attractive, the adhesive must also meet certain processing characteristics during manufacturing. Light-cure acrylic adhesives have found wide use in medical-device assembly/joining operations because their processing characteristics are compatible with the high-speed automated manufacturing processes employed. These adhesives are available in a wide variety of viscosities and are easily dispensed through either pressure–time or positive-displacement dispensing systems. Once dispensed on the part, they can remain in contact with even highly-stressed plastic parts for several minutes or longer without causing stress cracking or degradation of the plastic. Liquid Loctite® 3211, for example, can remain in contact with polycarbonate that has been bent so as to induce stresses up to 17 MPa (2500 psi) for more than 15 minutes without stress cracking. Finally, the adhesive can be completely converted from liquid to solid state in seconds when exposed to light of the proper intensity and wavelength.

Since Loctite® 3211 absorbs light in the visible as well as the ultraviolet range, it can be used successfully on plastics that contain UV blockers, such as many grades of polycarbonate. The ability to have a long open time during which parts can be positioned, and yet cure the adhesive on demand, is a unique benefit of light-curing adhesives and dramatically reduces scrap costs. The equipment used to irradiate the part with high-intensity light typically requires a space of 10 ft² to 20 ft² on a production line, which is generally much less than that required for the ovens used by heat-cure adhesives and the racking shelves required for slower curing adhesives. Since floor space carries a cost premium in clean-room environments, this factor is a significant benefit.

It was also important that the joint be designed for maximum performance. If the enclosure were bonded with a joint consisting of two flat faces in intimate contact, peel stresses (see Fig. 12.47) would act on the bond when the vessel is pressurized. Peel stresses are the most difficult type of stress for an adhesive joint to withstand, due to the fact that the entire load is concentrated on the leading edge of the joint. The tongue-and-groove design used addressed this concern; in this design, the groove acts as a reservoir for holding the adhesive during the dispensing operation. When the parts are mated and the adhesive is cured, this design allows much of the load on the joint (when the device is pressurized) to be translated into shear and tensile forces, which the adhesive is much better suited to withstand. The gap between the tongue and the groove can vary widely, because most light-cure adhesives can quickly be cured to depths in excess of 5 mm (0.20 in.). This feature allows the manufacturer to have a robust joining process (meaning that wide dimensional tolerances can be accommodated).

With the new design and by using Loctite® 3211, Cobe thus eliminated the environmental concerns and the issues associated with solvent bonding while gaining the benefit of a safer, faster, and more consistent bond. The light-curing adhesive provided the aesthetic bond line the company wanted—one that was clear and barely perceptible; it also provided the structural strength needed and thus enabled Cobe to maintain a competitive edge in the marketplace.

*Source*: Courtesy of P. J. Courtney, Loctite Corporation.

# SUMMARY

- Almost all products are an assemblage of many components; therefore, joining and fastening processes are an important and often necessary aspect of manufacturing, as well as of servicing and transportation. (Section 12.1)
- A major category of joining processes is fusion welding, in which the two pieces to be joined melt together and coalesce by means of heat; filler metals may or may not be used. Two common methods are consumable-electrode arc welding and nonconsumable-electrode are welding. Their selection of the particular method depends on factors such as workpiece material; shape complexity, size, and thickness; and type of joint. (Sections 12.2 and 12.3)
- Electron-beam and laser-beam welding are major categories of joining using high-energy beams. They produce small and high-quality weld zones and thus have important and unique applications in manufacturing. (Section 12.4)
- Because the joint undergoes important metallurgical and physical changes, the nature, properties, and quality of the weld joint are important factors to consider.

Weldability of metals, weld design, and process selection are factors that also should be considered. (Section 12.5)

- In solid-state welding, joining takes place without fusion, and pressure is applied either mechanically, using tools, electrodes, or rollers, or by explosives. Surface preparation and cleanliness can be important. Ultrasonic welding and resistance welding are two major examples of solid-state welding; these methods have a wide variety of applications, particularly for foil and sheet metals. (Sections 12.6–12.10)

- Diffusion bonding is another example of solid-state joining. Combined with superplastic forming, this method can be used to fabricate various complex sheet-metal structures with high strength-to-weight and stiffness-to-weight ratios. (Section 12.11)

- Brazing and soldering use filler metals at the interfaces to be joined. These processes require lower temperatures than those used in welding and are capable of joining dissimilar metals with intricate shapes and thicknesses. (Section 12.12)

- Adhesive bonding has gained wide acceptance in major industries such as aerospace, automotive, and building products. Adhesive-bonded joints have favorable characteristics such as strength, sealing, insulating, vibration damping, and resistance to corrosion between dissimilar metals. An important development is electrically conducting adhesives for surface-mount technologies. (Section 12.13)

- Mechanical fastening is among the oldest and most commonly used techniques. A wide variety of shapes and sizes of fasteners and fastening techniques have been developed for numerous permanent and semipermanent applications. (Section 12.14)

- With increasing use of thermoplastics and thermosetting plastics and various types of ceramics and glasses in a wide range of consumer and industrial applications, joining of these materials continue to undergo many advances. Several joining techniques are now available for these materials. (Section 12.15)

- General design guidelines have been established for the groups of processes described throughout this chapter. Some of these guidelines are applicable to a variety of processes, whereas other processes require special considerations, depending on their application. (Section 12.16)

- The economics of joining operations involves factors such as equipment cost, labor cost, and skill level required, as well as processing parameters such as time required, joint quality, and the need to meet specific requirements. The implementation of automation and computer controls can have a major impact on costs in these operations. (Section 12.17)

# BIBLIOGRAPHY

Adams, R. D., J. Comyn, and W. C. Wake, *Structural Adhesive Joints in Engineering*, Chapman & Hall, 1997.

*ASM Engineered Materials Handbook*, Vol. 3: *Adhesives and Sealants*, ASM International, 1993.

*ASM Handbook*, Vol. 6: *Welding, Brazing and Soldering*, ASM International, 1993.

Baghdachi, J., *Adhesive Bonding Technology*, Dekker, 1996.

Bickford, J. H., and S. Nassar (eds.), *Handbook of Bolts and Bolted Joints*, Dekker, 1998.

Blake, A., *What Every Engineer Should Know about Threaded Fasteners*, Dekker, 1983.

*Brazing Handbook*, 4th ed., American Welding Society, 1991.

Cary, H. B., *Modern Welding Technology*, 4th ed., Prentice Hall, 1997.

Davies, A. C., *The Science and Practice of Welding*, 10th ed., 2 vols., Cambridge, 1993.

Dawes, C. T., *Laser Welding*, Woodhead, 1992.

Easterling, K. E., *Introduction to the Physical Metallurgy of Welding*, Butterworth, 1983.

*Electronic Materials Handbook*, Vol. 1: *Packaging*, ASM International, 1989.

Evans, G. M., and N. Bailey, *Metallurgy of Basic Weld Metal*, Wooodhead, 1997.

Frear, D. R., W. B. Jones, and K. R. Kinsman (eds.), *Solder Mechanics: A State of the Art Assessment*, The Minerals, Metals & Materials Society, 1991.

Galyen, J., G. Sear., and C. Tuttle, *Welding: Fundamentals and Procedures*, Wiley, 1985.

Granjon, H., *Fundamentals of Welding Metallurgy*, Woodhead Publishing, 1991.

Grong, O., *Metallurgical Modeling of Welding*, The Institute of Metals, 1994.

*Handbook of Plastics Joining: A Practical Guide*, William Andrew, Inc., 1996.

Haviland, G. S., *Machinery Adhesives for Locking, Retaining, and Sealing*, Dekker, 1986.

Hicks, J. G., *Welded Joint Design*, 2d. ed., Abington, 1997.

Houldcroft, P. T., *Welding and Cutting: A Guide to Fusion Welding and Associated Cutting Processes*, Industrial Press, 1988.

Humpston, G., and D. M. Jacobson, *Principles of Soldering and Brazing*, ASM International, 1993.

Hwang, J. S., *Modern Solder Technology for Competitive Electronics Manufacturing*, McGraw-Hill, 1996.

*Introduction to the Nondestructive Testing of Welded Joints*, 2d. ed., American Society of Mechanical Engineers, 1996.

Jeffus, L. F., *Welding: Principles and Applications*, 4th ed., Delmar, 1997.

Jellison, R., *Welding Fundamentals*, Prentice Hall, 1995.

Karlsson, I. (ed.), *Modeling in Welding, Hot Powder Forming and Casting*, ASM International, 1997.

Koellhoffer, L., *Welding Processes and Practices*, Prentice Hall, 1988.

Kou, S., *Welding Metallurgy*, Wiley, 1987.

Lambert, L., *Soldering for Electronic Assemblies*, Dekker, 1988.

Lancaster, J. F., *The Metallurgy of Welding*, Chapman & Hall, 1993.

Landrock, A. H., *Adhesives Technology Handbook*, Noyes, 1985.

Lee, L.-H., *Adhesive Bonding*, Plenum, 1991.

Lincoln, B., K. J. Gomes, and J. F. Braden, *Mechanical Fastening of Plastics*, Dekker, 1984.

Linnert, J. E., *Welding Metallurgy*, 4th ed., Vol. 1, American Welding Society, 1994.

Manko, H. H., *Soldering Handbook for Printed Circuits and Surface Mounting*, Van Nostrand Reinhold, 1995.

_____, *Solders and Soldering—Materials, Design, Production, and Analysis for Reliable Bonding*, McGraw-Hill, 1992.

Messler, R. W., Jr., *Joining of Advanced Materials*, Butterworth-Heinemann, 1993.

Mouser, J. D., *Welding Codes, Standards, and Specifications*, McGraw-Hill, 1997.

Nicholas, M. G., *Joining Processes: Introduction to Brazing and Diffusion Bonding*, Chapman & Hall, 1998.

North, T. H., *Advanced Joining Technologies*, Chapman & Hall, 1990.

Parmley, R. O. (ed.), *Standard Handbook of Fastening and Joining*, 3d. ed., McGraw-Hill, 1997.

Pecht, M. G., *Soldering Processes and Equipment*, Wiley, 1993.

Petrie, E. M., *Handbook of Adhesives and Sealants*, McGraw-Hill, 1999.

Powell, J., *$CO_2$ Laser Cutting*, Springer, 1992.

Rahn, A., *The Basics of Soldering*, Wiley Interscience, 1993.

*Resistance Welding Manual*, 4th ed., Resistance Welder Manufacturers' Association, 1989.

Satas, D., *Handbook of Pressure-Sensitive Adhesive Technology*, 3d. ed., Satas & Associates, 1999.

Schultz, H., *Electron Beam Welding*, Woodhead, 1994.

Schwartz, M. M., *Brazing for the Engineering Technologist*, Chapman & Hall, 1995.

_____, *Ceramic Joining*, ASM International, 1990.

_____, *Joining of Composite-Matrix Materials*, ASM International, 1994.

Shields, J., *Adhesives Handbook*, 3d. ed., Butterworth, 1984.

Skeist, I., *Handbook of Adhesives*, 3d. ed., Van Nostrand Reinhold, 1990.

Speck, J. A., *Mechanical Fastening, Joining, and Assembly*, Dekker, 1997.

Steen, W. M., *Laser Material Processing*, 2d. ed., Springer, 1998.

Stout, R. D., *Weldability of Steels*, Welding Research Council, 1987.

Swenson, L.-E, *Control of Microstructures and Properties in Steel Arc Welds*, CRC Press, 1994.

*Tool and Manufacturing Engineers Handbook*, Vol. 4: "Quality Control and Assembly," Society of Manufacturing Engineers, 1986.

Tres, P. A., *Designing Plastic Parts for Assembly*, 3d. ed., Hanser-Gardner, 1998.

*Welding Handbook*, 8th ed., 3 vols., American Welding Society, 1987.

Woodgate, R. W., *Handbook of Machine Soldering*, Wiley, 1996.

# QUESTIONS

**12.1**  Explain the reasons that so many different welding processes have been developed.

**12.2**  List the advantages and disadvantages of mechanical fastening as compared with adhesive bonding.

**12.3**  What are the similarities and differences between consumable and nonconsumable electrodes?

**12.4**  What determines whether a certain welding process can be used for workpieces in horizontal, vertical, or upside-down positions, or for all types of positions? Explain, giving appropriate examples.

**12.5**  Comment on your observations regarding Fig. 12.3.

**12.6**  Discuss the need for and role of fixtures in holding workpieces in the welding operations described in this chapter.

**12.7**  Describe the factors that influence the size of the two weld beads in Fig. 12.11.

**12.8**  Why is the quality of welds produced by submerged arc welding very good?

**12.9**  Explain the factors involved in electrode selection in arc-welding processes.

**12.10**  Explain why the electroslag-welding process is suitable for thick plates and heavy structural sections.

**12.11**  What are the similarities and differences between consumable- and nonconsumable-electrode arc-welding processes?

**12.12**  In Table 12.2, there is a column on the distortion of welded components, ranging from lowest to highest. Explain why the degree of distortion varies among different welding processes.

**12.13**  Explain why the grains in Fig. 12.14 grow in the particular directions shown.

**12.14**  Prepare a table listing the processes described in this chapter and providing, for each process, the range of welding speeds as a function of workpiece material and thickness.

**12.15**  Explain what is meant by *solid-state welding*.

**12.16**  Describe your observations concerning Figs. 12.18, 12.19, and 12.20.

**12.17**  What advantages does friction welding have over the other joining methods described in this chapter?

**12.18**  Why is diffusion bonding, when combined with superplastic forming of sheet metals, an attractive fabrication process? Does it have any limitations?

**12.19**  Can roll bonding be applied to various part configurations? Explain.

**12.20**  Comment on your observations concerning Fig. 12.39.

**12.21**  If electrical components are to be attached to both sides of a circuit board, what soldering process(es) would you use? Explain.

**12.22**  Discuss the factors that influence the strength of (a) a diffusion-bonded component and (b) a cold-welded component.

**12.23**  Describe the difficulties you might encounter in applying explosion welding in a factory environment.

**12.24**  Inspect the edges of a U.S. quarter, and comment on your observations. Is the cross-section, i.e., the thickness of individual layers, symmetrical? Explain.

**12.25**  What advantages do resistance-welding processes have over others described in this chapter?

**12.26**  What does the strength of a weld nugget in resistance spot welding depend on?

**12.27**  Explain the significance of the magnitude of the pressure applied through the electrodes during resistance-welding operations.

**12.28**  Which materials can be friction stir welded, and which cannot? Explain your answer.

**12.29**  List the joining methods that would be suitable for a joint that will encounter high stresses and will need to be disassembled several times during the product life, and rank the methods.

**12.30**  Inspect Fig. 12.29, and explain why the particular fusion-zone shapes are developed as a function of pressure and speed. Comment on the influence of the properties of the material.

**12.31**  Which applications could be suitable for the roll spot-welding process shown in Fig. 12.33c? Give specific examples.

**12.32**  Give several examples concerning the bulleted items listed at the beginning of Section 12.1.

**12.33**  Could the projection-welded parts shown in Fig. 12.34 be made by any of the processes described in other parts of this text? Explain.

**12.34**  Describe the factors that influence flattening of the interface after resistance projection welding takes place.

**12.35**  What factors influence the shape of the upset joint in flash welding, as shown in Fig. 12.35b?

**12.36**    Explain how you would fabricate the structures shown in Fig. 12.39b with methods other than diffusion bonding and superplastic forming.

**12.37**    Make a survey of metal containers used for household products and foods and beverages. Identify those that have utilized any of the processes described in this chapter. Describe your observations.

**12.38**    Which process uses a solder paste? What are the advantages to this process?

**12.39**    Explain why some joints may have to be preheated prior to welding.

**12.40**    What are the similarities and differences between casting of metals (Chapter 5) and fusion welding?

**12.41**    Explain the role of the excessive restraint (stiffness) of various components to be welded on weld defects.

**12.42**    Discuss the weldability of several metals, and explain why some metals are easier to weld than others.

**12.43**    Must the filler metal be of the same composition as that of the base metal to be welded? Explain.

**12.44**    Describe the factors that contribute to the difference in properties across a welded joint.

**12.45**    How does the weldability of steel change as the steel's carbon content increases? Why?

**12.46**    Are there common factors among the weldability, solderability, castability, formability, and machinability of metals? Explain, with appropriate examples.

**12.47**    Assume that you are asked to inspect a weld for a critical application. Describe the procedure you would follow. If you find a flaw during your inspection, how would you go about determining whether or not this flaw is important for the particular application?

**12.48**    Do you think it is acceptable to differentiate brazing and soldering arbitrarily by temperature of application? Comment.

**12.49**    Loctite® is an adhesive (see the case study at the end of this chapter) used to keep metal bolts from vibrating loose; it basically glues the bolt to the nut once the nut is inserted in the bolt. Explain how this process is effective.

**12.50**    List the joining methods that would be suitable for a joint that will encounter high stresses and cyclic (fatigue) loading, and rank the methods in order of preference.

**12.51**    Why is surface preparation important in adhesive bonding?

**12.52**    Why have mechanical joining and fastening methods been developed? Give several specific examples of their applications.

**12.53**    Explain why hole preparation may be important in mechanical joining.

**12.54**    What precautions should be taken in mechanical joining of dissimilar metals?

**12.55**    What difficulties are involved in joining plastics? What about in joining ceramics? Why?

**12.56**    Comment on your observations concerning the numerous joints shown in the figures in Section 12.16.

**12.57**    How different is adhesive bonding from other joining methods? What limitations does it have?

**12.58**    Soldering is generally applied to thinner components. Why?

**12.59**    Explain why adhesively bonded joints tend to be weak in peeling.

**12.60**    Inspect various household products, and describe how they are joined and assembled. Explain why those particular processes were used.

**12.61**    Name several products that have been assembled by (a) seaming, (b) stitching, and (c) soldering.

**12.62**    Suggest methods of attaching a round bar made of thermosetting plastic perpendicularly to a flat metal plate.

**12.63**    Describe the tooling and equipment that are necessary to perform the double-lock seaming operation shown in Fig. 12.50, starting with flat sheet. (See also Fig. 7.23.)

**12.64**    What joining methods would be suitable to assemble a thermoplastic cover over a metal frame? Assume that the cover has to be removed periodically.

**12.65**    Repeat Question 12.64, but for a cover made of (a) a thermosetting plastic, (b) metal, and (c) ceramic. Describe the factors involved in your selection of methods.

**12.66**    Do you think the strength of an adhesively bonded structure is as high as that obtained by diffusion bonding? Explain.

**12.67**    Comment on workpiece size limitations, if any, for each of the processes described in this chapter.

**12.68**    Describe part shapes that cannot be joined by the processes described in this chapter. Gives specific examples.

**12.69** Give several applications of electrically conducting adhesives.

**12.70** Give several applications for fasteners in various household products, and explain why other joining methods have not been used instead.

**12.71** Comment on workpiece shape limitations, if any, for each of the processes described in this chapter.

**12.72** List and explain the rules that must be followed to avoid cracks in welded joints, such as hot tearing, hydrogen-induced cracking, lamellar tearing, etc.

**12.73** If a built-up weld is to be constructed (see Fig. 12.3), should all of it be done at once, or should it be done a little at a time, with sufficient time allowed for cooling between beads?

**12.74** Describe the reasons that fatigue failure generally occurs in the heat-affected zone of welds instead of through the weld bead itself.

**12.75** If the parts to be welded are preheated, is the likelihood that porosity will form increased or decreased? Explain.

# PROBLEMS

**12.76** Two 1-mm-thick, flat copper sheets are being spot welded using a current of 5000 A and a current flow time of $t = 0.18$ s. The electrodes are 5 mm in diameter. Estimate the heat generated in the weld zone.

**12.77** Calculate the temperature rise in Problem 12.76, assuming that the heat generated is confined to the volume of material directly between the two electrodes and that the temperature distribution is uniform.

**12.78** In Fig. 12.23, assume that most of the top portion of the top piece is cut horizontally with a sharp saw. Thus, the residual stresses will be disturbed, and, as described in Section 2.10, the part will undergo shape change. For this case, how will the part distort? Explain.

**12.79** The accompanying figure below shows a metal sheave that consists of two matching pieces of hot-rolled, low-carbon-steel sheets. These two pieces can be joined either by spot welding or by V-groove welding. Discuss the advantages and limitations of each process for this application.

**12.80** A welding operation takes place on an aluminum-alloy plate. A pipe 2.5 in. in diameter, with a 0.20-in. wall thickness and a 2-in. length, is butt welded onto an angle iron 6 in. by 6 in. by 0.25 in thick. The angle iron is of an $L$ cross-section and has a length of 1 ft. If the weld zone in a gas tungsten-arc welding process is approximately 0.5 in. wide, what would be the temperature increase of the entire structure due to the heat input from welding only? What if the process were an electron-beam welding operation, with a bead width of 0.08 in.? Assume that the electrode requires 1500 J and the aluminum alloy requires 1200 J to melt one gram.

**12.81** An arc-welding operation is taking place on carbon steel. The desired welding speed is around 1 in./sec. If the power supply is 10 V, what current is needed if the weld width is to be 0.25 in.?.

**12.82** The energy applied in friction welding is given by the formula $E = IS^2/C$, where $I$ is the moment of inertia of the flywheel, $S$ is the spindle speed in rpm, and $C$ is a constant of proportionality. ($C = 5873$ when the moment of inertia is given in lb-ft$^2$.) For a spindle speed of 600 rpm and an operation in which a steel tube with a 3.5-in. outside diameter and a 0.25 in.-wall thickness is welded to a flat frame, what is the required moment of inertia of the flywheel if all of the energy is used to heat the weld zone, approximated as the material $\frac{1}{4}$ in. deep and directly below the tube? Assume that 1.4 ft-lbm is needed to melt the electrode.

(a)

0.135 in.

$2\frac{7}{16}$ in.

Spot weld          V-groove weld

(b)                      (c)

# DESIGN

**12.83** Design a machine that can perform friction welding of two cylindrical pieces, as well as remove the flash from the welded joint. (See Fig. 12.28.)

**12.84** How would you modify your design in Problem 12.83 if one of the pieces to be welded is noncircular?

**12.85** Describe product designs that cannot be joined by friction-welding processes.

**12.86** Make a comprehensive outline of joint designs relating to the processes described in this chapter. Give specific examples of engineering applications for each type of joint.

**12.87** Review the two weld designs in Fig. 12.55a, and, based on the topics covered in courses on the strength of materials, show that the design on the right is capable of supporting a larger moment, as shown.

**12.88** In the building of large ships, there is a need to weld large sections of steel together to form a hull. For this application, consider each of the welding operations described in this chapter, and list the benefits and drawbacks of that operation for this product. Which welding process would you select? Why?

**12.89** Examine various household products, and describe how they are joined and assembled. Explain why those particular processes are used for these applications.

**12.90** Repeat Problem 12.89 for a cover made of (a) a thermoset, (b) a metal, and (c) a ceramic. Describe the factors involved in your selection of methods.

**12.91** A major cause of erratic behavior (hardware bugs) and failures of computer equipment is fatigue failure of the soldered joints, especially in surface-mount devices and devices with bond wires. (See Fig. 12.45.) Design a test fixture for cyclic loading of a surface-mount joint for fatigue testing.

**12.92** Using two strips of steel 1 in. wide and 8 in. long, design and fabricate a joint that gives the highest strength in a tension test in the longitudinal direction.

# Fabrication of Microelectronic and Micromechanical Devices

## 13.1 | Introduction

Although semiconducting materials have been used in electronics for many decades, it was the invention of the transistor in 1947 that set the stage for what would become one of the greatest technological advancements in all of history. Microelectronics has played an ever-increasing role since **integrated circuit** (IC) technology (Fig. 13.1) became the foundation for personal computers, cellular telephones, information systems, automotive control, and telecommunications.

(a)

(b)

(c)

**FIGURE 13.1** (a) The Pentium® 4 2.2GHz processor, as fabricated on a 300mm (11.8 in) wafer. (b) A Pentium® processor in a flip chip package with cover removed. (c) A Pentium® processor motherboard. *Source*: Courtesy Intel Corporation.

13.1 Introduction
13.2 Semiconductors and Silicon
13.3 Crystal Growing and Wafer Preparation
13.4 Films and Film Deposition
13.5 Oxidation
13.6 Lithography
13.7 Diffusion and Ion Implantation
13.8 Etching
13.9 Metallization and Testing
13.10 Wire Bonding and Packaging
13.11 Yield and Reliability of Chips
13.12 Printed Circuit Boards
13.13 Micromachining of MEMS Devices
13.14 The LIGA Microfabrication Process
13.15 Solid Free-Form Fabrication of Devices

**FIGURE 13.2** Cross-sectional views of the fabrication of a metal oxide semiconductor (MOS) transistor. *Source*: R. C. Jaeger.

The basic building block of a complex IC is the transistor. A transistor (Fig. 13.2) is a three-terminal device that acts as a simple on–off switch: If a positive voltage is applied to the "gate" terminal, a conducting channel is formed between the "source" and "drain" terminals, allowing current to flow between those two terminals (switched closed). If no voltage is applied to the gate, no channel is formed, and the source and drain are isolated from each other (switch opened). Figure 13.3 shows how the basic processing steps, described in this chapter, are combined to form a metal-oxide-semiconductor field effect transistor (MOSFET).

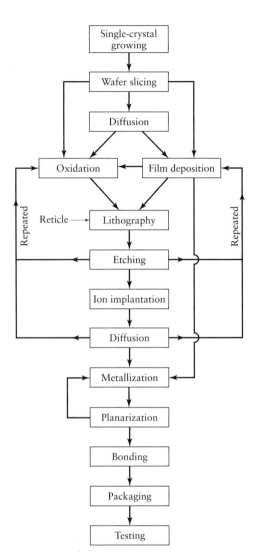

**FIGURE 13.3** General fabrication sequence for integrated circuits.

In addition to the metal-oxide semiconductor structure, the **bipolar junction transistor** (BJT) is also used, but to a lesser extent. While the actual fabrication steps for these transistors are very similar to those for both the MOSFET and MOS technologies, their circuit applications are different. Memory circuits, such as RAMs (random access memory), and microprocessors consist primarily of MOS devices, whereas linear circuits, such as amplifiers and filters, are more likely to contain bipolar transistors. Other differences between these two types of devices include the faster operating speeds, higher breakdown voltage, and greater current drive of the BJT, and the lower current of the MOSFET.

The major advantage of today's ICs is their high degree of complexity, reduced size, and low cost. As fabrication technology becomes more advanced, the size of devices decreases, and, consequently, more components can be put onto a **chip** (a small slice of semiconducting material on which the circuit is fabricated). In addition, mass processing and process automation have greatly helped reduce the cost of each completed circuit. The components fabricated include transistors, diodes, resistors, and capacitors. Typical chip sizes produced today range from 0.5 mm × 0.5 mm (0.02 in. × 0.02 in.) to more than 50 mm × 50 mm (2 in. × 2 in.). In the past, no more than 100 devices could be fabricated on a single chip, but technology now allows densities in the tens of millions of devices per chip, commonly referred to as

**very large-scale integration** (VLSI). The Pentium 4 processor, for example, contains 42 million transistors. Some of the most advanced ICs may contain more than 100 million devices, termed **ultralarge-scale integration** (ULSI).

Because of the minute scale of microelectronic devices, all fabrication must take place in an extremely clean environment. *Clean rooms* are used extensively for this purpose and are rated by the maximum number of 0.5-$\mu$m (20-$\mu$in.) particles per cubic foot; most modern clean rooms are class-1 to class-10 facilities.

This chapter begins with a description of the properties of common semiconductors, such as silicon, gallium arsenide, and polysilicon. It then describes in detail the current processes employed (Fig. 13.3) in fabricating microelectronic devices and integrated circuits, including IC testing, packaging, and reliability. The chapter also describes a more recent and potentially more important development concerning the manufacture of **microelectromechanical systems** (MEMS) which are combinations of electrical and mechanical systems with characteristic lengths of less than 1 mm (0.040 in.). These devices utilize many of the batch-processing technologies used for the manufacture of electronic devices, although a number of other, unique processes have been developed as well.

MEMS devices have the potential of dramatically changing society; applications as varied as precise and rapid sensors, microrobots for nanofabrication, medical delivery systems, and artificial organs have been proposed. Given that common MEMS devices such as ink-jet printers, accelerometers, and computer hard-disk heads are already widespread, it is clear that MEMS are having a fundamental effect on society. Although "MEMS" is a term that came into use around 1987, it has been applied to a wide variety of applications. Currently, there are only a few microelectromechanical systems in use, such as accelerometers and some pressure sensors with on-chip electronics. Often, "MEMS" is a label applied to devices, such as pressure sensors, valves, and mirrors. There are many more device applications than integrated systems; even so, MEMS sales were in the range of $3 to $10 billion in 2001.

# 13.2 | Semiconductors and Silicon

As the name suggests, **semiconductor materials** have electrical properties that lie between those of conductors and insulators; they exhibit resistivities between $10^{-3}\,\Omega$-cm and $10^{8}\,\Omega$-cm. Semiconductors have become the foundation for electronic devices, because their electrical properties can be altered by adding controlled amounts of selected impurity atoms into their crystal structures. These impurity atoms, also known as **dopants**, either have one more valence electron (*n*-type, or *negative dopant*) or one fewer valence electron (*p*-type, or *positive dopant*) than the atoms in the semiconductor lattice. For silicon, which is a Group IV element in the periodic table, typical *n*-type and *p*-type dopants include phosphorus and arsenic (Group V) and boron (Group III), respectively. The electrical operation of semiconductor devices is controlled by creating regions of different doping types and concentrations.

Although the earliest electronic devices were fabricated on *germanium*, **silicon** has become the industry standard. The abundance of silicon in its alternative forms in the earth's crust is second only to that of oxygen, thus making it economically attractive. Silicon's main advantage over germanium is its larger energy gap (1.1 eV) compared with that of germanium (0.66 eV); this larger energy gap allows silicon-based devices to operate at temperatures about 150°C (270°F) higher than the operating temperatures of devices fabricated on germanium [about 100°C (180°F).] Furthermore, the oxidized form of silicon, $SiO_2$, enables the production of metal-oxide-semiconductor (MOS) devices, which are the basis for MOS-Transistors. These materials are used in memory devices and processors and the like and account for by far the largest volume of semiconductor material produced worldwide.

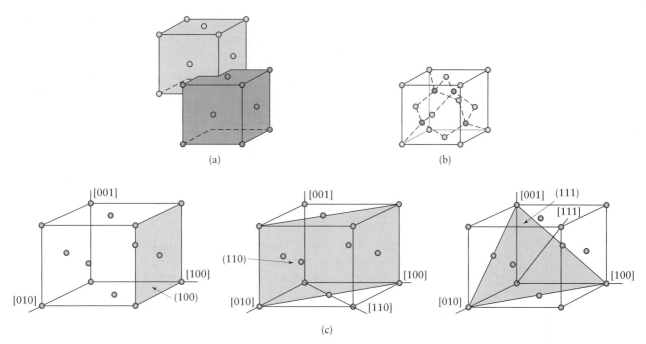

**FIGURE 13.4**  Crystallographic structure and Miller indices for silicon. (a) Construction of a diamond-type lattice from interpenetrating face-centered cubic cells (one of eight penetrating cells shown). (b) The diamond-type lattice of silicon. The interior atoms have been shaded darker than the surface atoms. (c) The Miller indices for a cubic lattice.

The crystallographic structure of silicon is a diamond-type fcc structure, as shown in Fig. 13.4, along with the Miller indices of an fcc material. Miller indices are a useful notation for identifying planes and directions within a unit cell. A crystallographic plane is defined by the reciprocal of its intercepts on the three axes. Since anisotropic etchants (see Section 13.8.1) preferentially remove material in certain crystallographic planes, the orientation of the silicon crystal in a wafer is important.

Silicon's important processing advantage is that its oxide (**silicon dioxide**) is an excellent insulator and is used for isolation and passivation purposes; conversely, germanium oxide is water soluble and unsuitable for electronic devices. However, silicon has some limitations, which has encouraged the development of compound semiconductors, specifically **gallium arsenide** (GaAs). Its major advantage over silicon is its capability for light emission [allowing fabrication of devices such as lasers and light-emitting diodes (LEDs)], in addition to its larger energy gap (1.43 eV) and, therefore, higher maximum operating temperature [to about 200°C (400°F).] Devices fabricated on gallium arsenide also have much higher operating speeds than those fabricated on silicon. Some disadvantages of gallium arsenide include its considerably higher cost, greater processing complications, and the difficulty of growing high-quality oxide layers, the need for which is emphasized throughout this chapter.

# 13.3 | Crystal Growing and Wafer Preparation

Silicon occurs naturally in the forms of silicon dioxide and various silicates. It must undergo a series of purification steps in order to become a high-quality, defect-free, and single-crystal material needed for semiconductor-device fabrication. The purification process begins by heating silica and carbon together in an electric furnace, resulting in 95%- to 98%-pure polycrystalline silicon. This material is converted to an alternative form, commonly trichlorosilane, which in turn is purified and decomposed in a high-temperature hydrogen atmosphere. The result is an extremely high-quality **electronic-grade silicon** (EGS).

Single-crystal silicon is usually obtained by using the **Czochralski, or CZ, process** (see Fig. 5.34); this method uses a seed crystal that is dipped into a silicon melt and then slowly pulled out while being rotated. At this point, controlled amounts of impurities can be added to the system to obtain a uniformly doped crystal. Typical pull rates are on the order of 10 $\mu$m/s (400 $\mu$in./s). The result of this growing technique is a cylindrical single-crystal ingot, typically 100–300 mm (4–12 in.) in diameter and over 1 m (40 in.) long. However, this technique does not allow precise control of the ingot's diameter; therefore, ingots are commonly grown to a diameter a few millimeters larger than required and then ground to a precise diameter.

Next, the crystal is sliced into individual **wafers** by using an inner-diameter blade. This method uses a rotating blade with its cutting edge on the inner ring. While the substrate depth needed for most electronic devices is no more than several microns, wafers are typically cut to a thickness of about 0.5 mm (0.02 in.). This thickness provides the necessary mass to absorb temperature variations and provide mechanical support during subsequent fabrication steps. Finally, the wafers must be polished and cleaned to remove surface damage caused by the sawing process.

**FIGURE 13.5**
Identification of single-crystal wafers of silicon.

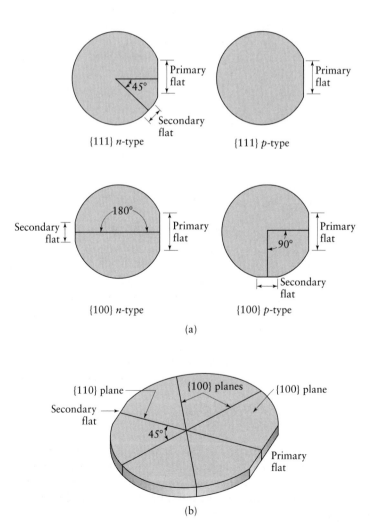

In order to properly control the manufacturing process, it is important to determine the orientation of the crystal in a wafer. The wafers, therefore, have notches and/or flats machined into them for identification, as shown in Fig. 13.5. Most commonly, the (100) or (111) plane of the crystal defines the wafer surface, although (110) surfaces can be used as well for micromachining applications. Wafers are also identified by a laser scribe marking produced by the manufacturer. Laser scribing of information can take place on the front or on the back side of the wafer. The front side of the wafer has an exclusion edge area from 3–10 mm in size reserved for the scribe information, including lot numbers, orientation, and a unique wafer identification code.

Device fabrication takes place over the entire wafer surface, and many identical circuits are generated at the same time. Wafers are typically kept in lots of 25 or 50 wafers with 6- to 8-in. diameters each or lots of 12–13 wafers with 12-in. diameters each, so that they can be easily handled and transferred during processing. Because of the small device size and large wafer diameter, thousands of individual circuits can be put on one wafer. Once processing is finished, the wafer is sliced into individual **chips**, each containing one complete integrated circuit.

# 13.4 | Films and Film Deposition

*Films* of many different types, particularly insulating and conducting, are used extensively in microelectronic-device processing. Common deposition films include polysilicon, silicon nitride, silicon dioxide, tungsten, titanium, and aluminum. In some instances, single-crystal silicon wafers serve merely as a mechanical support on which custom *epitaxial layers* are grown (see Section 13.5). These silicon epitaxial films are of the same lattice structure as the substrate, so they are also single crystal materials. The advantages of processing on these deposited films, instead of on the actual wafer surface, include the inclusion of fewer impurities (notably carbon and oxygen), improved device performance, and the attainment of tailored material properties not obtainable on the wafers themselves.

Some of the major functions of deposited films include **masking** for diffusion or implants and protection of the semiconductor's surface. In masking applications, the film must effectively inhibit the passage of dopants while also being able to be etched into patterns of high resolution. Upon completion of device fabrication, films are applied to protect the underlying circuitry. Films used for masking and protection include silicon dioxide, phosphosilicate glass (PSG), and silicon nitride; each of these materials has distinct advantages, and they are often used in combination.

Other films contain dopant impurities and are used as doping sources for the underlying substrate. Conductive films are used primarily for device interconnection; these films must have a low resistivity, be capable of carrying large currents, and be suitable for connecting to terminal–packaging leads with wire bonds. Aluminum and copper are generally used for this purpose. To date, aluminum is the more popular material, because it can be dry etched more easily. However, aluminum is difficult to use for small structures and high current densities, and for this reason, copper, with its low resistivity, has been of significant research interest. Increasing circuit complexity has required up to six levels of conductive layers, which must all be separated by insulating films.

Films may be deposited by a number of techniques, which involve a variety of pressures, temperatures, and vacuum systems:

a) **Evaporation.** One of the simplest and oldest methods of film deposition is *evaporation*, used primarily for depositing metal films. In this process, the metal is heated in a vacuum until vaporization; upon evaporation, the metal forms a thin layer on the surface. The heat of evaporation is usually provided by a heating filament or electron beam.

**FIGURE 13.6** Schematic diagrams of (a) a continuous, atmospheric-pressure CVD reactor and (b) a low-pressure CVD reactor. *Source*: Sze, S. M., ed., VLSI Technology. New York, McGraw-Hill, 1983. (c) Apparatus for ion implantation. *Source*: J. A. Schey.

b) **Sputtering.** Another method of metal deposition is *sputtering*, which involves bombarding a target with high-energy ions, usually argon ($Ar^+$), in a vacuum. (See Section 4.5.1.) Sputtering systems usually include a dc power source to obtain the energized ions. As the ions impinge on the target, atoms are knocked off and subsequently deposited on wafers mounted within the system. Although some argon may be trapped within the film, this technique can provide very uniform coverage. Advanced sputtering techniques include use of a radio-frequency power source (**RF sputtering**) and introduction of magnetic fields (**magnetron sputtering**).

c) **Chemical vapor deposition.** In one of the most common techniques, *chemical vapor deposition* (CVD), film deposition is achieved by the reaction and/or decomposition of gaseous compounds. (See Section 4.5.1.) In this technique, silicon dioxide is routinely deposited by the oxidation of silane or a chlorosilane. Figure 13.6a shows a continuous CVD reactor that operates at atmospheric pressure.

d) **Low-pressure chemical vapor deposition.** Figure 13.6b shows the apparatus for a method similar to CVD that operates at lower pressures, referred to as *low-pressure chemical vapor deposition* (LPCVD). Capable of coating hundreds of wafers at a time, this method has a much higher production rate that than of atmospheric-pressure CVD and provides superior film uniformity with less consumption of carrier gases. This technique is commonly used for depositing polysilicon, silicon nitride, and silicon dioxide.

e) **Plasma-enhanced chemical vapor deposition.** The *plasma-enhanced chemical vapor deposition* (PECVD) process involves placing wafers in radio-frequency (RF) plasma containing the source gases and offers the advantage of maintaining low wafer temperature during deposition. However, the films deposited by this method generally incorporate hydrogen and are of lower quality than those deposited by other methods.

Silicon **epitaxy** layers, in which the crystalline layer is formed using the substrate as a seed crystal, can be grown via a variety of methods. If the silicon is deposited from the gaseous phase, the process is known as **vapor-phase epitaxy** (VPE). In another variation, called **liquid-phase epitaxy** (LPE), the heated substrate is brought into contact with a liquid solution containing the material to be deposited.

Another high-vacuum process uses evaporation to produce a thermal beam of molecules that deposit on the heated substrate. This process, called **molecular-beam epitaxy** (MBE), offers a very high degree of purity. In addition, since the films are grown one atomic layer at a time, excellent control of doping profiles is achieved, which is especially important in gallium-arsenide technology. However, MBE suffers from relatively low growth rates compared with those of other conventional film-deposition techniques.

# 13.5 | Oxidation

Recall that the term *oxidation* refers to the growth of an oxide layer by the reaction of oxygen with the substrate material. Oxide films can also be formed by the deposition techniques described previously. Thermally grown oxides, described in this section, display a higher level of purity than deposited oxides, because the former are grown directly from the high-quality substrate. However, deposition methods must be used if the composition of the desired film is different than that of the substrate material.

Silicon dioxide is the most widely used oxide in integrated-circuit technology today, and its excellent characteristics are one of the major reasons for the widespread use of silicon. Aside from its functions of dopant masking and device isolation, silicon dioxide's most critical role is that of the *gate-oxide* material in MOSFETs. Silicon surfaces have an extremely high affinity for oxygen, and a freshly sawed silicon slice will quickly grow a native oxide of 30–40 Å (0.003–0.004 $\mu$m) in thickness. Modern IC technology requires oxide thicknesses from tens to thousands of angstroms.

a) **Dry oxidation.** Dry oxidation is a relatively simple process and is accomplished by elevating the substrate temperature, typically to 750°C–1100°C (1380°F–2020°F), in an oxygen-rich environment with variable pressure. Silicon dioxide is produced according to the chemical reaction

$$Si + O_2 \longrightarrow SiO_2. \tag{13.1}$$

Most oxidation is carried out in a batch process where up to 150 wafers are placed in a furnace. **Rapid thermal processing** (RTP) is a related process combining **rapid thermal oxidation** (RTO) with an anneal step [**rapid thermal annealing** (RTA)] and is used to produce thin oxides on a single wafer. As a layer of oxide forms, the oxidizing agents must be able to pass through the oxide layer and reach the silicon surface, where the actual reaction takes place. Thus, an oxide layer does not continue to grow on top of itself, but rather it grows from the silicon–silicon-dioxide interface outward. Some of the silicon substrate is consumed in the oxidation process (Fig. 13.7). The ratio of oxide thickness to the amount of silicon consumed is found to be 1:0.44. Thus, for example, to obtain an oxide layer 1000 Å (0.1 $\mu$m) thick, approximately 440 Å (0.044 $\mu$m) of silicon will be consumed; this condition does not present a problem, as substrates are always grown sufficiently thick.

**FIGURE 13.7** Growth of silicon dioxide, showing consumption of silicon. *Source*: Sze, S. M., Physics of Semiconductor Devices. New York: Wiley, 1981. This material is used by permission of John Wiley & Sons, Inc.

One important effect of this consumption of silicon is the rearrangement of dopants in the substrate near the interface. Some dopants deplete away from the oxide interface, while others pile up, and hence processing parameters have to be adjusted to compensate for this effect.

b) **Wet oxidation.** Another oxidizing technique uses a water-vapor atmosphere as the agent and is appropriately called *wet oxidation*. The chemical reaction involved in wet oxidation is

$$Si + 2H_2O \longrightarrow SiO_2 + 2H_2. \tag{13.2}$$

This method offers a considerably higher growth rate than that of dry oxidation, but it suffers from a lower oxide density and, therefore, a lower dielectric strength. The common practice in industry is to combine both dry and wet oxidation methods, growing an oxide in a three-part layer: dry, wet, and dry. This approach combines the advantages of the much higher growth rate in wet oxidation and the high quality obtained in dry oxidation.

c) **Selective oxidation.** The foregoing two oxidation methods are useful primarily for coating the entire silicon surface with oxide, but it is also necessary to oxidize only certain portions of the surface. The procedure used for this task, called *selective oxidation*, uses silicon nitride, which inhibits the passage of oxygen and water vapor. Thus, by masking certain areas with silicon nitride, the silicon under these areas remains unaffected while the uncovered areas are oxidized.

# 13.6 | Lithography

Table 13.1 provides a summary of lithographic techniques. *Lithography*, or *photolithography*, is the process by which the geometric patterns that define the devices are transferred from a **reticle** (also called a **photomask**, or **mask**) to the surface

**TABLE 13.1**

**General Characteristics of Lithography Techniques**

| Method | Wavelength (nm) | Finest feature size (nm) |
|---|---|---|
| Ultraviolet (Photolithoraphy) | 365 | 350 |
| Deep UV | 248 | 250 |
| Extreme UV | 10–20 | 30–100 |
| X ray | 0.01–1 | 20–100 |
| Electron beam | – | 80 |

*Source*: P. K. Wright, *21st Century Manufacturing*, Prentice Hall, Upper Saddle River, NJ, 2001.

of the chip. A reticle is a glass or quartz plate with a pattern of the chip deposited onto it usually with a chromium film, although iron oxide and emulsion are also used. The reticle image can be the same size as the desired structure on the chip, but it is often an enlarged image (usually 5× to 20× larger, although 10× magnification is most common). Enlarged images are then focused onto a wafer through a lens system in a process known as *reduction lithography*.

In current practice, the lithographic process is applied to microelectronic circuits several times, each time using a different reticle to define the different areas of the working devices. Typically designed at several thousand times their final size, reticle patterns undergo through a series of reductions before being applied permanently to a defect-free quartz plate. Computer-aided design (Section 15.4) has had a major impact on reticle design and generation. Cleanliness is especially important in lithography, and many manufacturers are turning to robotics and specialized wafer-handling apparati in order to minimize dust and dirt contamination.

Once the film deposition process is completed and the desired reticle patterns have been generated, the wafer is cleaned and coated with an organic **photoresist** (PR), which is sensitive to ultraviolet (UV) light. The wafers is then placed inside a resist spinner, and the photoresist is applied as a viscous liquid onto the wafer. Photoresist layers of $0.5–2.5 \ \mu m \ (20–100 \ \mu in.)$ thick are obtained by spinning at several thousand rpm for 30 to 60 s to give uniform coverage. The thickness of the resist is given by

$$t = \frac{k C^\beta \eta^\gamma}{\omega^\alpha}, \tag{13.3}$$

where $t$ is the resist thickness, $C$ is the polymer concentration in mass per volume, $\eta$ is the viscosity of the polymer, $\omega$ is the angular velocity during spinning, and $k, \alpha, \beta$, and $\gamma$ are constants for the particular spinning system. Where masking levels are considered critical, a **barrier antireflective layer** (BARL) or **barrier antireflective coating** (BARC) is applied either beneath or on top of the photoresist to provide line width control, especially over aluminum.

The next step in lithography is **prebaking** the wafer to remove the solvent from the photoresist and harden it. This step is carried out in a convection oven at around 100°C (273°F) for a period of 10 to 30 min. The wafer is then aligned under the desired reticle in a *stepper*. In this crucial step, called **registration**, the reticle must be aligned correctly with the previous layer on the wafer. Once the reticle is aligned, it is stepped across the wafer and exposed to UV radiation. Upon development and removal of the exposed photoresist, a duplicate of the reticle pattern will appear in the photoresist layer. As can be seen in Fig. 13.8, the reticle can be either a negative or a positive image of the desired pattern. A positive reticle uses the UV radiation to break down the chains in the organic film, so that these chains are preferentially removed by the developer. Positive masking is more commonly used than negative masking, because with negative masking, the photoresist can swell and distort, making it unsuitable for small geometries. Newer negative photoresist materials do not have this problem, however.

Following the exposure and development sequence, **postbaking** the wafer is implemented to drive off solvent and toughen and improve the adhesion of the remaining resist. In addition, a deep UV treatment, which consists of baking the wafer to 150°C–200°C (300°F–400°F) in ultraviolet light, can also be used to further strengthen the resist against high-energy implants and dry etches. The underlying film not covered by the PR is then implanted or etched away (Sections 13.7 and 13.8). Finally, the photoresist is stripped by exposing it to wet stripper or an oxygen plasma, a technique also referred to as **ashing**. The lithography process may be repeated as many as 25 times in the fabrication of the most advanced ICs.

**FIGURE 13.8**    Pattern transfer by lithography. Note that the mask in Step 3 can be either a positive or a negative image of the pattern. *Source*: After W. C. Till and J. T. Luxon.

One of the major issues in lithography is **linewidth**, which refers to the width of the smallest feature obtainable on the silicon surface. As circuit densities have escalated over the years, device sizes and features have become smaller and smaller. Today, minimum commercially feasible linewidths are between 0.13 $\mu$m and 0.21 $\mu$m, with considerable research being conducted at smaller linewidths.

As pattern resolution, and therefore device miniaturization, is limited by the wavelength of the radiation source used, the need has arisen to move to wavelengths shorter than those in the ultraviolet range, such as "deep" UV wavelengths, "extreme" UV wavelengths, electron beams, and X rays (see Table 13.1). In these technologies, the photoresist is replaced by a similar resist that is sensitive to a specific range of shorter wavelengths.

**Extreme-ultraviolet lithography.**    The pattern resolution in photolithography is limited by light diffraction. One method of reducing the effects of diffraction is to use ever shorter wavelengths. *Extreme-ultraviolet* (EUV) *lithography* uses light at a wavelength of 13 nm in order to obtain features around 30 to 100 nm in size. The waves are focused by highly reflective molybdenum/silicon mirrors (instead of glass lenses that absorb EUV light) through the mask to the wafer surface.

**X ray lithography.**    Although photolithography is the most widely used lithography technique, it has fundamental resolution limitations associated with light diffraction. *X ray lithography* is superior to photolithography, because of the shorter wavelength of the radiation and the very large depth of focus involved in the former. This characteristic allows much finer patterns to be resolved, and X ray lithography is far less susceptible to dust. Furthermore, the aspect ratio (defined as the ratio of depth to lateral dimension) can be more than 100 with X ray lithography, but is limited to around 10 with photolithography. However, in order to achieve this benefit, synchrotron radiation is needed, which is expensive and available at only a few research laboratories. Given the large capital investment required for a manufacturing facility, industry has preferred to refine and improve optical lithography instead of investing new capital into X ray-based production. X ray lithography is not currently widespread; however, the LIGA process (see Section 13.14) fully exploits the benefits of X ray lithography.

**Electron-beam and ion-beam lithography.** Like X ray lithography, *electron-beam* (**e-beam**) and *ion-beam* (*i-beam*) lithography are superior to photolithography with respect to the resolutions attainable. These methods involve high current density in narrow electron or ion beams (*pencil sources*) that scan a pattern, one pixel at a time, onto a wafer. The masking is done by controlling the point-by-point transfer of the stored pattern and therefore is performed by software. These techniques have the advantages of accurate control of exposure over small areas of the wafer, large depth of focus, and low defect densities. Resolutions are limited to around 10 nm, because of electron scatter, although 2-nm resolutions have been reported for some materials. It should be noted that the scan time increases significantly as the resolution increases, because more highly focused beams are required. The main drawback of these techniques is that electron and ion beams need to be maintained in a vacuum, which significantly increases equipment complexity and production cost; furthermore, the scan time for a wafer for this technique is much slower than that for other lithographic methods.

# 13.7 | Diffusion and Ion Implantation

As stated in Section 13.2, the operation of microelectronic devices often depends on regions of different doping types and concentrations. The electrical character of these regions can be altered by introducing dopants into the substrate, accomplished by *diffusion* and *ion implantation* processes. Since many different regions of microelectronic devices must be defined, this step in the fabrication sequence is repeated several times.

In the diffusion process, the movement of atoms results from thermal excitation. Dopants can be introduced to the substrate in the form of a deposited film, or the substrate can be exposed to a vapor containing the dopant source. This process takes place at elevated temperatures, usually 800°C–1200°C (1500°F–2200°F). Dopant movement within the substrate is a function of temperature, time, and the diffusion coefficient (or diffusivity) of the dopant species, as well as the type and quality of the substrate material. Because of the nature of diffusion, the dopant concentration is very high at the surface and drops off sharply away from the surface. To obtain a more uniform concentration within the substrate, the wafer is heated further to drive in the dopants, a process called **drive-in diffusion**. The fact that diffusion, desired or undesired, will always occur at high temperatures is invariably taken into account during subsequent processing steps. Although the diffusion process is relatively inexpensive, it is highly isotropic.

**Ion implantation** is a much more extensive process and requires specialized equipment (Fig. 13.6c). Implantation is accomplished by accelerating ions through a high-voltage beam of as much as one million volts and then choosing the desired dopant by means of a magnetic mass separator. In a manner similar to that used in cathode-ray tubes, the beam is swept across the wafer by sets of deflection plates, thus ensuring uniform coverage of the substrate. The complete implantation system must be operated in a vacuum.

The high-velocity impact of ions on the silicon surface damages the lattice structure, resulting in lower electron mobilities. This condition is undesirable, but the damage can be repaired by an annealing step, which involves heating the substrate to relatively low temperatures, usually 400°C–800°C (750°F–1500°F), for 15–30 min. This process provides the energy that the silicon lattice needs to rearrange and mend itself. Another important function of annealing is to allow the dopant to move from the interstitial to substitutional sites (see Fig. 3.8), where they are electrically active.

**FIGURE 13.9** Effect of residual stress on patterns etched in a 2 $\mu$m thick carbon layer: (a) Patterns destroyed by high residual stress; (b) Patterns etched after thermal curing to relieve residual stresses. *Source*: P. Hudek, et al., *Journal of Vac. Sci. and Technology*, B Nov. 17 (6), 1999, pp. 3127–3131.

(a)

(b)

## 13.8 | Etching

*Etching* is the process by which entire films, or particular sections of films or the substrate, are removed; it plays an important role in the fabrication sequence. One of the most important criteria in this process is **selectivity**, which refers to the ability to etch one material without etching another. Tables 13.2 and 13.3 provide a summary of etching processes. In silicon technology, an etching process must effectively etch the silicon-dioxide layer with minimal removal of the underlying silicon or the resist material. In addition, polysilicon and metals must be etched into high-resolution lines with vertical wall profiles and with minimal removal of the underlying insulating film. Typical etch rates range from tens to several thousands of nm/min, and selectivities (defined as the ratio of the etch rates of the two films) can range from 1 : 1 to 100 : 1.

It should be noted that etching is often performed in order to machine desired patterns into substrates or coatings. The resolution of the pattern is a critical concern, both for device performance and to satisfy the continuous goals of further device miniaturization. Residual stresses in the surface can be a major obstacle to pattern transfer. The effect of residual stresses can clearly be seen in Fig. 13.9a; as material is removed, residual stresses can cause slender features to warp or fracture (as also shown, for example, in Fig. 2.33). To reduce the likelihood of such defects, workpieces and coatings can be annealed or stress-relieved before etching, allowing good pattern transfer. When residual stresses are effectively controlled or eliminated, good pattern transfer can take place (Fig. 13.9b).

### 13.8.1 Wet etching

*Wet etching* involves immersing the wafers in a liquid, usually acidic, solution. The main drawback to most wet-etching operations is that they are *isotropic*, that is, they etch in all directions of the workpiece at the same rate. This condition results in undercuts beneath the mask material (see, for example, Fig. 13.10a) and limits the resolution of geometric features in the substrate.

Effective etching requires the following conditions:

1. Etchant transport to the surface;
2. A chemical reaction;
3. Transport of reaction products away from the surface;
4. Ability to stop the etching process rapidly in order to obtain superior pattern transfer (**etch stop**), usually using an underlying layer with high selectivity.

**TABLE 13.2**

## Comparison of Etch Rates

| Etchant | Target material | Etch rate (nm/min)[a] | | | | | | | |
|---|---|---|---|---|---|---|---|---|---|
| | | Polysilicon $n^+$ | Polysilicon, undoped | Silicon dioxide | Silicon nitride | Phosphosilicate glass, annealed | Aluminum | Titanium | Photoresist (OCG-820PR) |
| Wet etchants | | | | | | | | | |
| Concentrated HF (49%) | Silicon oxides | 0 | – | 2300 | 14 | 3600 | 4.2 | >1000 | 0 |
| 25:1 HF:$H_2O$ | Silicon oxides | 0 | 0 | 9.7 | 0.6 | 150 | – | – | 0 |
| 5:1 BHF[b] | Silicon oxides | 9 | 2 | 100 | 0.9 | 440 | 140 | >1000 | 0 |
| Silicon etchant (126 $HNO_3$:60 $H_2O$:5 $NH_4F$) | Silicon | 310 | 100 | 9 | 0.2 | 170 | 400 | 300 | 0 |
| Aluminum etchant (16 $H_3PO_4$:1 $HNO_3$:1 HAc:2 $H_2O$) | Aluminum | <1 | <1 | 0 | 0 | <1 | 660 | 0 | 0 |
| Titanium etchant (20 $H_2O$:1 $H_2O_2$:1 HF) | Titanium | 1.2 | – | 12 | 0.8 | 210 | >10 | 880 | 0 |
| Piranha (50 $H_2SO_4$:1 $H_2O_2$) | Cleaning off metals and organics | 0 | 0 | 0 | 0 | 0 | 180 | 240 | >10 |
| Acetone ($CH_3COOH$) | Photoresist | 0 | 0 | 0 | 0 | 0 | 0 | 0 | >4000 |
| Dry etchants | | | | | | | | | |
| $CF_4$ + $CHF_3$ + He, 450W | Silicon oxides | 190 | 210 | 470 | 180 | 620 | – | >1000 | 220 |
| $SF_6$ + He, 100W | Silicon nitrides | 73 | 67 | 31 | 82 | 61 | – | >1000 | 69 |
| $SF_6$, 125W | Thin silicon nitrides | 170 | 280 | 110 | 280 | 140 | – | >1000 | 310 |
| $O_2$, 400W | Ashing photoresist | 0 | 0 | 0 | 0 | 0 | 0 | 0 | 340 |

Notes:

a. Results are for fresh solutions at room temperature unless otherwise noted. Actual etch rates will vary with temperature and prior use of solution, area of exposure of film, other materials present, and film impurities and microstructure.

b. Buffered hydrofluoric acid, 33% NH4F and 8.3% HF by weight.

Source: After K. Williams and R. Muller, *J. Microelectromechanical Systems*, Vol. 5, 1996, pp. 256–269.

**TABLE 13.3**

## General Characteristics of Silicon Etching Operations

| | Temperature (°C) | Etch rate ($\mu$m/min) | {111}/{100} selectivity | Nitride etch rate (nm/min) | $SiO_2$ etch rate (nm/min) | $p++$ etch stop |
|---|---|---|---|---|---|---|
| **Wet etching** | | | | | | |
| $HF:HNO_3:CH_3COOH$ | 25 | 1–20 | – | Low | 10–30 | No |
| KOH | 70–90 | 0.5–2 | 100:1 | <1 | 10 | Yes |
| Ethylene-diamine pyrochatechol (EDP) | 115 | 0.75 | 35:1 | 0.1 | 0.2 | Yes |
| $N(CH_3)_4OH$ (TMAH) | 90 | 0.5–1.5 | 50:1 | <0.1 | <0.1 | Yes |
| **Dry (plasma) etching** | | | | | | |
| $SF_6$ | 0–100 | 0.1–0.5 | – | 200 | 10 | No |
| $SF_6/C_4F_8$ (DRIE) | 20–80 | 1–3 | – | 200 | 10 | No |

*Source*: Adapted from N. Maluf, *An Introduction to Microelectromechanical Systems Engineering*, Artech House, 2000.

If the first or third steps limit the speed of the process, agitation or stirring of the solution can increase etching rates. If the second step limits the speed of the process, the etching rate will depend strongly on temperature, etching material, and solution composition. Therefore, reliable etching requires both good temperature control and repeatable stirring capability.

**Isotropic etchants** are widely used for the following procedures:

1. Removal of damaged surfaces;
2. Rounding of sharp etched corners to avoid stress concentrations;
3. Reduction of roughness after anisotropic etching;
4. Creation of structures in single-crystal slices; and
5. Evaluation of defects.

Microelectronic devices and MEMS (see Sections 13.13 through 13.15) require accurate machining of structures, and this task is done through masking. However, masking is a challenge with isotropic etchants. The strong acids etch aggressively (at a rate of 0.1–1 $\mu$m/min) and produce rounded cavities. Furthermore, the etch rate is very sensitive to agitation, and, therefore, lateral and vertical features are difficult to control.

**FIGURE 13.10** Etching directionality. (a) Isotropic etching: Etch proceeds vertically and horizontally at approximately the same rate, with significant mask undercut. (b) Orientation-dependant etching (ODE): Etch proceeds vertically, terminating on {111} crystal planes with little mask undercut. (c) Vertical etching: Etch proceeds vertically with little mask undercut. *Source*: Courtesy of K. R. Williams.

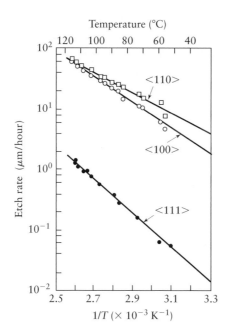

**FIGURE 13.11** Etch rates of silicon in different crystallographic orientations, using ethylene-diamine/pyrocatechol-in-water as the solution. *Source*: After H. Seidel et al., *J. Electrochemical Society*, 1990, pp. 3612–3626.

The size of the features in an integrated circuit determines its performance, and for this reason, there is a strong desire to produce well-defined, extremely small structures. Such small features cannot be attained through isotropic etching, because of the poor definition that results from undercutting of masks.

**Anisotropic etching** takes place when etching is strongly dependent on compositional or structural variations in the material. There are two basic kinds of anisotropic etching—*orientation-dependent etching* (ODE) and *vertical etching*—although most vertical etching is done with dry plasmas and is discussed in Section 13.8.2. Orientation-dependent etching commonly occurs in a single crystal when etching takes place at different rates in different directions, as shown in Fig. 13.10b. When ODE is performed properly, the etchants produce geometric shapes with walls defined by the crystallographic planes that resist the etchants. For example, Fig. 13.11 shows the vertical etch rate for silicon as a function of temperature. As can be seen, etching is more than one order of magnitude slower in the $[111]$ crystal direction than in other directions; therefore, well-defined walls can be obtained along the $[111]$ crystal direction.

The **anisotropy ratio**, $AR$, for etching is defined by

$$AR = \frac{E_1}{E_2},\qquad(13.4)$$

where $E$ is the etch rate and the subscripts refer to two crystallographic directions of interest. *Selectivity* is defined in a similar manner, but refers to the etch rates between materials of interest. The anisotropy ratio is unity for isotropic etchants and can be as high as 400/200/1 for (110)/(100)/(111) silicon. The {111} planes always etch the slowest, but the {100} and {110} planes can be controlled through etchant chemistry.

Masking is also a concern for anisotropic etching, but silicon oxide is less valuable as a mask material for different reasons than in isotropic etching. Anisotropic etching is slower than isotropic etching (typically 3 $\mu$m/min; 120 $\mu$in./min), and thus anisotropic etching through a wafer may take several hours. Silicon oxide may etch too rapidly to use as a mask, and hence a high-density silicon-nitride mask may be needed.

Often, it is important to rapidly halt the etching process, a technique referred to as *etch stop*. This situation is typically the case when thin membranes are to be manufactured or when features with very precise thicknesses are needed. Conceptually, this task can be accomplished by removing the wafer from the etching solution. However, etching depends to a great extent on the ability to circulate fresh etchants to the desired locations. Since the circulation varies across a wafer surface, this strategy for halting the etching process would lead to large variations in etch depth.

The most common approach for uniform feature sizes across a wafer is to use a boron etch stop, where a boron layer is diffused or implanted into silicon. Examples of common etch stops include the placement of a boron-doped layer beneath silicon and the placement of silicon oxide ($SiO_2$) beneath silicon nitride ($Si_3N_4$). Since anisotropic etchants do not attack boron-doped silicon as aggressively as they do undoped silicon, surface features or membranes can be created by **back etching**. Figure 13.12 shows an example of the boron etch-stop approach.

A large number of etchant formulations have been developed. Some of the more common wet etchants are summarized as follows:

1. Silicon dioxide is commonly etched with hydrofluoric (HF) acid solutions. The driving chemical reaction for pure HF etching is

$$SiO_2 + 6HF \longrightarrow H_2SiF_6 + 2H_2O. \tag{13.5}$$

It is rare that silicon dioxide is etched purely through the reaction in Eq. (13.5). Hydrofluoric acid is a weak acid, and it does not completely dissociate into hydrogen and fluorine ions in water. $HF_2^-$ is an additional ion that exists in hydrofluoric acid, and $HF_2^-$ attacks oxide roughly 4.5 times faster than does HF alone. The reaction involving the $HF_2^-$ ion is

$$SiO_2 + 3HF_2^- + H^+ \longrightarrow SiF_6^{2-} + 2H_2O. \tag{13.6}$$

**FIGURE 13.12**

Application of a boron etch stop and back etching to form a membrane and orifice. *Source*: After Brodie, I., and Murray, J. J., *The Physics of Microfabrication*, Plenum Press, 1982.

(a) Oxidation

(b) Lithography and development

(c) Boron diffusion

(d) Anisotropic etching

(e) Stripping and reoxidation

The pH value of the etching solution is critical, because acidic solutions have sufficient hydrogen ions to dissociate the $HF_2^-$ ions into HF ions. As HF and $HF_2^-$ are consumed, the etch rate decreases. For that reason, a buffer of ammonium fluoride ($NH_4F$) is used to maintain the pH and thus keep the concentrations of HF and $HF_2^-$ constant, stabilizing the etch rate. Such an etching solution is referred to as a *buffered hydrofluoric acid* (BHF) or *buffered oxide etch* (BOE) and has the reaction

$$SiO_2 + 4HF + 2NH_4F \longrightarrow (NH_4)_2SiF_6 + 2H_2O. \qquad (13.7)$$

2. Silicon nitride is etched with phosphoric acid ($H_3PO_4$), usually at an elevated temperature, typically 160°C (320°F.) The etch rate of phosphoric acid decreases with water content, so a *reflux system* is used to return condensed water vapor to the solution to maintain a constant etch rate.

3. Etching silicon often involves mixtures of nitric acid ($HNO_3$) and hydrofluoric acid (HF). Water can be used to dilute these acids; the preferred buffer is acetic acid because it preserves the oxidizing power of $HNO_3$; this system is referred to as an *HNA etching system*. A simplified description of this etching process is that the nitric acid oxidizes the silicon and then the hydrofluoric acid removes the silicon oxide. This two-step process is a common approach for chemical machining of metals. (See Section 9.10.) The overall reaction is

$$18HF + 4HNO_3 + 3Si \longrightarrow 3H_2SiF_6 + 4NO + 8H_2O. \quad (13.8)$$

The etch rate is limited by the silicon-oxide removal; therefore, a buffer of ammonium fluoride is usually used to maintain etch rates.

4. Anisotropic etching, or orientation-dependent etching, of single-crystal silicon can be done with solutions of potassium hydroxide, although other etchants have also be used. The reaction is

$$Si + 2OH^- + 2H_2O \longrightarrow SiO_2(OH)_2^{2-} + 2H_2. \qquad (13.9)$$

Note that this reaction does not require potassium to be the source of the OH ions; KOH attacks {111}-type planes much more slowly than other planes. Isopropyl alcohol is sometimes added to KOH solutions to reduce etch rates and increase uniformity of etching. KOH also is extremely valuable in that it stops etching when it contacts a very heavily doped *p*-type material (the boron etch stop described previously).

5. Aluminum is etched through a solution typically of 80% phosphoric acid ($H_3PO_4$), 5% nitric acid ($HNO_3$), 5% acetic acid ($CH_3COOH$), and 10% water. The nitric acid first oxidizes the aluminum, and the oxide is then removed by the phosphoric acid and water. This solution can be masked with a photoresist.

6. Wafer cleaning is accomplished through *Piranha solutions*, which have been in use for decades; these solutions are hot mixtures of sulfuric acid ($H_2SO_4$) and peroxide ($H_2O_2$) solutions. Piranha solutions strip photoresist and other organic coatings and remove metals on the surface, but do not affect silicon dioxide or silicon nitride, making it an ideal cleaning solution. Bare silicon forms a thin layer of hydrous silicon oxide, which is removed through a short dip in hydrofluoric acid.

7. Although photoresist can be removed through Piranha solutions, acetone is commonly used for this purpose instead. Acetone dissolves the photoresist, but it should be noted that if the photoresist is excessively heated during a process step, it will be significantly more difficult to remove with acetone; in such a case, the photoresist can be removed through an ashing plasma.

**EXAMPLE 13.1 Processing of a *p*-type region in *n*-type silicon**

Assume that we want to create a *p*-type region within a sample of *n*-type silicon. Draw cross-sections of the sample at each processing step in order to accomplish this task.

**Solution** See Fig. 13.13. This simple device is known as a *pn junction diode*, and the physics of its operation is the foundation for most semiconductor devices.

SOLUTION

| | Cross-section | Description |
|---|---|---|
| (a) | Silicon — *n* | Sample of *n*-type silicon. |
| (b) | Oxide — *n* | Grow silicon dioxide by oxidation. |
| (c) | Photo-resist — *n* | Apply photoresist. |
| (d) | UV light Mask — *n* | Expose photoresist, using appropriate lithographic mask. |
| (e) | *n* | Develop photoresist. |
| (f) | *n* | Etch silicon dioxide. |
| (g) | *n* | Remove photoresist. |
| (h) | *p* *n* | Implant boron. |
| (i) | *p* *n* | Remove silicon dioxide. |

**FIGURE 13.13** Processing of a *p*-type region in *n*-type silicon.

## 13.8.2 Dry etching

Modern integrated circuits are etched exclusively through *dry etching*, which involves the use of chemical reactants in a low-pressure system. In contrast to the wet process, dry etching can have a high degree of directionality, resulting in highly anisotropic etch profiles (Fig. 13.10c). Also, the dry process requires only small amounts of the reactant gases, whereas the solutions used in the wet process have to be refreshed periodically. Dry etching usually involves a plasma or discharge in areas of high electric and magnetic fields; any gases that are present are dissociated to form ions, electrons, or highly reactive molecules.

There are several specialized dry-etching techniques, described as follows:

a) **Sputter etching.** *Sputter etching* removes material by bombarding it with noble-gas ions, usually $Ar^+$. The gas is ionized in the presence of a cathode and anode (Fig. 13.14). If a silicon wafer is the target, the momentum transfer associated with bombardment of atoms causes bond breakage and material to be ejected, or sputtered. If the silicon chip is the substrate, then the material in the target is deposited onto the silicon after it has been sputtered by the ionized gas. Some of the concerns with sputter etching are listed as follows:

1. The ejected material can be redeposited onto the target, especially when the aspect ratios are large;
2. Sputter etching is not material selective (most materials sputter at about the same rate), and therefore masking is difficult;
3. Sputter etching is slow, with etch rates limited to tens of nm/min;
4. Sputtering can cause damage to or excessive erosion of the material;
5. The photoresist is difficult to remove.

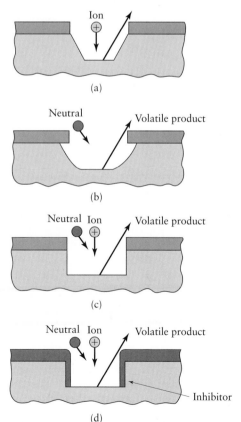

**FIGURE 13.14**
Machining profiles associated with different dry-etching techniques: (a) sputtering; (b) chemical; (c) ion-enhanced energetic; (d) ion-enhanced inhibitor. *Source*: After M. Madou.

b) **Reactive plasma etching.** Also referred to as *dry chemical etching, reactive plasma etching* involves chlorine or fluorine ions (generated by RF excitation) and other molecular species that diffuse to and chemically react with the substrate, forming a volatile compound that is removed by the vacuum system. The mechanism of reactive plasma etching is shown in Fig. 13.15. Here, a reactive species is produced, such as $CF_4$ dissociating upon impact with energetic electrons to produce fluorine atoms (step 1). The reactive species then diffuse to the surface (step 2), become adsorbed (step 3), and chemically react to form a volatile compound (step 4). The reactant then desorbs from the surface (step 5) and diffuses into the bulk gas, where it is removed by the vacuum system. Some reactants polymerize on the surface and require additional removal, either with oxygen in the plasma reactor or by an external ashing operation. The electrical charge of the reactive species is not great enough to cause damage through impact on the surface, so that no sputtering occurs. Thus, the etching is isotropic, and undercutting of the mask takes place (Fig. 13.10a). Table 13.2 lists some of the more common dry etchants, their target materials, and their typical etch rates.

c) **Physical–chemical etching.** Processes such as *reactive ion-beam etching* (RIBE) and *chemically assisted ion-beam etching* (CAIBE) combine the advantages of phys-

**FIGURE 13.15** (a) Schematic illustration of reactive plasma etching. *Source*: (a) After M. Madou. (b) Example of deep reactive ion etched trench. Note the periodic undercuts, or scalloping. (c) Near vertical sidewalls produced through DRIE with an anisotropic etching process. (d) An example of cryogenic dry etching, showing a145 $\mu$m deep structure etched into Si using a 2.0 $\mu$m thick oxide masking layer. The substrate temperature was −140°C during etching. *Source*: R. Kassing and I. W. Rangelow, University of Kassel, Germany.

ical and chemical etching. These processes use chemically reactive species to drive material removal, but this procedure is physically assisted by the impact of ions onto the surface. In RIBE, also known as *deep reactive ion etching* (DRIE), vertical trenches hundreds of micrometers deep can be produced by periodically interrupting the etch process and depositing a polymer layer. When done with an isotropic dry etching process, this results in scalloped sidewalls, as shown in Fig. 13.15b. Anisotropic DRIE can produce near vertical sidewalls, as shown in Fig. 13.15c.

In CAIBE, ion bombardment can assist dry chemical etching by

1. Making the surface more reactive;
2. Clearing the surface of reaction products and allowing the chemically reactive species access to the cleared areas;
3. Providing the energy to drive surface chemical reactions—however, the neutral species do most of the etching.

Physical–chemical etching is extremely useful, because the ion bombardment is directional, so that etching is anisotropic. Also, the ion bombardment energy is low and does not contribute much to mask removal. This factor allows generation of near vertical walls with very large aspect ratios. Since the ion bombardment does not directly remove material, masks can be used.

d) **Cryogenic dry etching.**  This procedure is an approach used to obtain very deep features with vertical walls. The workpiece is lowered to cryogenic temperatures, and then chemically assisted ion-beam etching takes place. The low temperatures involved ensure that insufficient energy is available for a surface chemical reaction to take place, unless ion bombardment is normal to the surface. Oblique impacts, such as occur on side walls in deep crevices, cannot drive the chemical reactions; therefore, very smooth vertical walls can be produced, as shown in Fig. 13.15d.

Since dry etching is not selective, etch stops cannot be directly applied. Dry-etch reactions must be terminated when the target film is removed. Optical emission spectroscopy is often used to determine the "end point" of a reaction. Filters can be used to capture the wavelength of light emitted during a particular reaction. A noticeable change in light intensity at the point of etching will be detected.

---

**EXAMPLE 13.2  Comparison of wet and dry etching**

Consider a case where a ⟨100⟩ wafer has an oxide mask placed on it in order to produce square or rectangular holes. The sides of the square mask are precisely oriented with the ⟨110⟩ direction (see Fig. 13.5) of the wafer surface, as shown in Fig. 13.16.

Isotropic etching results in the cavity shown in Fig. 13.16a. Since etching occurs at constant rates in all directions, a rounded cavity is produced that undercuts the mask. An orientation-dependent etchant produces the cavity shown in Fig. 13.16b. Since etching is much faster in the ⟨100⟩ and ⟨110⟩ directions than in the ⟨111⟩ direction, sidewalls defined by the {111} plane are generated. For silicon, these sidewalls are at an angle of 54.74° to the surface.

The effect of a larger mask or shorter etch time is shown in Fig. 13.16c. The resultant pit is defined by ⟨111⟩ sidewalls and a bottom in the ⟨100⟩ direction parallel to the surface. A rectangular mask and resultant pit are shown in Fig. 13.16d. Deep reactive ion etching is depicted in Fig. 13.16e. Note that a polymer layer is periodically deposited onto the hole sidewalls to allow for deep pockets, but scalloping (greatly exaggerated in the figure) is unavoidable. A hole from chemically reactive ion etching is shown in Fig. 13.16f.

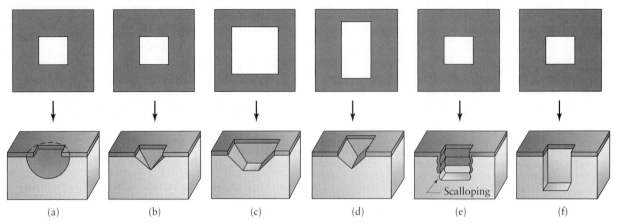

**FIGURE 13.16**    Different holes generated from a square mask in (a) isotropic (wet) etching; (b) ODE; (c) ODE with a larger hole; (d) ODE of a rectangular hole; (e) deep reactive ion etching; (f) vertical etching. *Source*:  After M. Madou.

## 13.9 | Metallization and Testing

The preceding sections have focused only on device fabrication; however, generation of a complete and functional integrated circuit requires that such devices be interconnected. **Interconnections** are made by metals that exhibit low electrical resistance and good adhesion to dielectric insulator surfaces. Aluminum and aluminum–copper alloys remain the most commonly used materials for this purpose in VLSI (very large scale integration) technology today. However, as device dimensions continue to shrink, electromigration has become more of a concern with aluminum interconnects.

**Electromigration** is the process by which metal atoms are physically moved by the impact of drifting electrons under high-current conditions. Low-melting-point metals such as aluminum are especially prone to electromigration. In extreme cases, this condition can lead to severed and/or shorted metal lines. Solutions to this problem include the addition of sandwiched metal layers such as tungsten and titanium, and more recently, the use of pure copper, which displays lower resistivity and significantly better antielectromigration performance than does aluminum.

Metals are deposited by standard deposition techniques, and interconnection patterns are generated by lithographic and etching processes. Modern ICs typically have one to four layers of metallization, in which case each layer of metal is insulated by a dielectric, either silicon oxide or borophosphosilicate glass. **Planarization** (producing a planar surface) of these interlayer dielectrics is critical to reducing metal shorts and linewidth variation of the interconnect. A common method for achieving a planar surface has been a uniform oxide etch process that smoothens out the peaks and valleys of the dielectric layer.

However, today's standard for planarizing high-density interconnects is **chemical mechanical polishing** (CMP; see Section 9.8), also called *chemical–mechanical planarization*. This process involves physically polishing the wafer surface similar to a disc or belt sander polishing the ridges on a piece of wood. A typical CMP process combines an abrasive medium with a polishing compound or slurry and can polish a wafer to within $0.03\ \mu\mathrm{m}\,(1.2\ \mu\mathrm{in.})$ of being perfectly flat, with a $R_q$ roughness on the order of 0.1 nm for a new, bare silicon wafer.

Different layers of metal are connected together by **vias**, and access to the devices on the substrate is achieved through **contacts** (Fig. 13.17). In recent years, as

(a)                                                                                            (b)

**FIGURE 13.17**    (a) Scanning electron microscope photograph of a two-level metal interconnect. Note the varying surface topography. (b) Schematic drawing of a two-level metal interconnect structure. *Source*:  R. C. Jaeger.

devices have become smaller and faster, the size and speed of some chips have become dominated by the resistance of the metallization process itself and the capacitance of the dielectric and the transistor gate. Wafer processing is completed upon application of a *passivation layer*, usually silicon nitride ($Si_3N_4$), which acts as an ion barrier for sodium ions and also provides excellent scratch resistance.

The next step is to test each of the individual circuits on the wafer (Fig. 13.18). Each chip, also referred to as a **die**, is tested with a computer-controlled platform that contains needlelike probes to access the bonding pads on the die. The probes are of two forms:

**FIGURE 13.18**  A probe checks for defects, and an ink mark is placed on each defective die. *Source*:  Intel Corp.

1. **Test patterns or structures.** The probe measures test structures, often outside of the active die, placed in the so-called *scribe line* (the empty space between dies). These structures consist of transistors and interconnect structures that measure various quantities such as resistivity, contact resistance, and electromigration.

2. **Direct probe.** This approach uses 100% testing on the bond pads of each die.

The platform steps across the wafer, testing whether each circuit functions properly, using computer-generated timing wave forms. If a defective chip is encountered, it is marked with a drop of ink. Up to one third of the cost of a micromachined part can be incurred during this testing.

After the wafer-level testing is complete, back grinding may be done to remove a large amount of the original substrate. Final die thickness is dependent on the packaging requirement, but anywhere from 25% to 75% of the wafer thickness can be removed. After back grinding, each die is separated from the wafer. Diamond sawing is a commonly used separation technique and results in very straight edges, with minimal chipping and cracking damage. The chips are then sorted, with the passing dice being sent on for packaging and the inked dice being discarded.

## 13.10 | Wire Bonding and Packaging

The working dice must be attached to a more rugged foundation to ensure reliability. One simple method is to *bond* a die to its packaging material with an *epoxy cement* (see Section 12.13); another method uses a *eutectic bond*, made by heating metal-alloy systems. One widely used mixture is 96.4% gold and 3.6% silicon, which has a eutectic point at 370°C (700°F). Once the chip has been bonded to its substrate, it must be electrically connected to the package leads. This task is accomplished by *wire bonding* very thin [15 $\mu$m (600 $\mu$in.) in diameter] gold wires from the package leads to bonding pads located around the perimeter or down the center of the die (Fig. 13.19a). The bonding pads on the die are typically drawn at 50 $\mu$m (2000 $\mu$in.) or more per side, and the bond wires are attached using thermocompression, ultrasonic, or thermosonic techniques (Fig. 13.20).

The connected circuit is now ready for final *packaging*. The packaging process largely determines the overall cost of each completed IC, since the circuits are mass produced on the wafer, but are then packaged individually. Packages are available in

(a)    (b)    (c)

**FIGURE 13.19**    (a) SEM photograph of wire bonds connecting package leads (left-hand side) to die bonding pads. (b) and (c) Detailed views of (a). *Source:* Courtesy of Micron Technology, Inc.

a variety of styles (Table 13.4), and selection of the appropriate one must take into account operating requirements. Consideration of a circuit's package includes consideration of chip size, the number of external leads, operating environment, heat dissipation, and power requirements. ICs that are used for military and industrial applications, for example, need packages of particularly high strength, toughness, hermicity and high-temperature resistance.

Packages are produced from polymers, metals, or ceramics. Metal containers are produced from alloys such as Kovar (an iron–cobalt–nickel alloy with a low coefficient of thermal expansion) and provide a hermetic seal and good thermal conductivity, but are limited in the number of leads that can be used. Ceramic packages are usually produced from $Al_2O_3$ and are hermetic and have good thermal conductivity, with higher lead counts than metal packages; however, they are also more expensive

**TABLE 13.4**

**Summary of Molded-Plastic IC Packages**

| Package | Abbreviation | Pins min. | Pins max. | Description |
|---|---|---|---|---|
| Through-hole mount | | | | |
|   Dual in-line | DIP | 8 | 64 | Two in-line rows of leads. |
|   Single in-line | SIP | 11 | 40 | One in-line row of leads. |
|   Zigzag in-line | ZIP | 16 | 40 | Two rows with staggered leads. |
|   Quad in-line package | QUIP | 16 | 64 | Four in-line rows of staggered leads. |
| Surface mount | | | | |
|   Small-outline IC | SOIC | 8 | 28 | Small package with leads on two sides. |
|   Thin small-outline package | TSOP | 26 | 70 | Thin version of SOIC. |
|   Small-outline J-lead | SOJ | 24 | 32 | Same as SOIC, with leads in a J-shape. |
|   Plastic leaded chip carrier | PLCC | 18 | 84 | J-shaped leads on four sides. |
|   Thin quad flat pack | TQFP | 32 | 256 | Wide, but thin, package with leads on four sides. |

than metal packages. Plastic packages are inexpensive, with high lead counts, but they have high thermal resistance and are not hermetic.

An older style of packaging is the **dual in-line package** (DIP), shown schematically in Fig. 13.21a. Characterized by low cost (if plastic) and ease of handling, DIP packages are made of thermoplastic, epoxy, or ceramic, and they can have from 2 to 500 external leads. Ceramic packages are designed for use over a broader temperature range and in military applications and cost considerably more than plastic packages. A flat ceramic package is shown in Fig. 13.21b, in which the package and all the leads are in the same plane. This package style does not offer the ease of handling or the modular design of the DIP package; therefore, it is usually affixed permanently to a multiple-level circuit board in which the low profile of the flat pack is essential.

**Surface-mount packages** have become the standard for today's integrated circuits. Some common examples are shown in Fig. 13.21c, where it can be seen that the main difference among them is in the shape of the connectors. The DIP connection to the surface board is by way of prongs that are inserted into corresponding holes, whereas a surface mount is soldered onto specially fabricated pad or land designs. Package size and land layouts are selected from standard patterns and usually require adhesive bonding of the package to the board, followed by wave soldering of the connections (see Section 12.12.3).

**FIGURE 13.21**   Schematic illustration of different IC packages: (a) dual in-line (DIP); (b) ceramic flat pack; (c) common surface-mount configurations; (d) ball-grid array (BGA). *Source*: After R. C. Jaeger, A. B. Glaser, and G. E. Subak-Sharpe. Coombs, C.F., Printed Circuits Handbook. New York, McGraw-Hill, 1996.

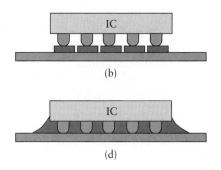

**FIGURE 13.22**
Illustration of flip-chip technology. (a) Flip-chip package with solder plated metal balls and pads on the printed circuit board; (b) flux application and placement; (c) reflow soldering; (d) encapsulation. *Source*: P. K. Wright, *21st Century Manufacturing*, Prentice Hall, Upper Saddle River, NJ, 2001.

Faster and more versatile chips require increasingly tightly spaced connections. **Pin-grid arrays** (PGAs) use tightly packed pins that connect by way of through-holes onto printed circuit boards. However, PGAs, and indeed other in-line and surface-mount packages, are extremely susceptible to plastic deformation of the wires and legs, especially with small-diameter, closely spaced wires. One way to achieve tight packing of connections and avoid the difficulties of slender connections is through **ball-grid arrays** (BGAs), shown in Fig. 13.21d. Such arrays have a plated solder coating on a number of closely spaced metal balls on the underside of the package. The spacing between balls can be as small as 50 $\mu$m (2000 $\mu$in.), but more commonly, the spacing is standardized as 1.0 mm (0.040 in.), 1.27 mm (0.050 in.), or 1.5 mm (0.060 in.).

BGAs can be designed with over 1000 connections, but such high numbers of connections are extremely rare; usually, 200–300 connections are sufficient for demanding applications. By using reflow soldering (Section 12.12.3), the solder serves to center the BGAs by surface tension, resulting in well-defined electrical connections for each ball. **Flip-chip technology** (FCT) refers to the attachment procedure of ball-grid arrays and is depicted in Fig. 13.22. The final encapsulation with an epoxy is necessary, not only to more securely attach the IC package to the printed circuit board, but also to evenly distribute thermal stresses during its operation.

After the chip has been sealed in the package, it undergoes final testing. Because one of the main purposes of packaging is isolation from the environment, testing at this stage usually encompasses heat, humidity, mechanical shock, corrosion, and vibration. Destructive tests are also performed to investigate the effectiveness of sealing.

# 13.11 | Yield and Reliability of Chips

*Yield* is defined as the ratio of functional chips to the total number of chips produced. The overall yield of the total IC manufacturing process is the product of the wafer yield, bonding yield, packaging yield, and test yield. This value can range from only a few percent for new processes to above 90% for mature manufacturing lines. Most loss of yield occurs during wafer processing, due to its complex nature; in this stage, wafers are commonly separated into regions of good and bad chips. Failures at this stage can arise from point defects (such as oxide pinholes), film contamination, or metal particles, as well as from area defects, such as uneven film deposition or etch nonuniformity.

A major concern about completed ICs is their **reliability** and **failure rate**. Since no device has an infinite lifetime, statistical methods are used to characterize the expected lifetimes and failure rates of microelectronic devices. The unit for failure rate is the FIT (*failure in time*), defined as the number of failures per one billion device-hours. However, complete systems may have millions of devices, so the overall failure rate in entire systems is correspondingly higher. Failure rates greater than 100 FIT are generally unacceptable.

Equally important in failure analysis is determination of the failure mechanism, that is, the actual process that causes the device to fail. Common failures due to processing involve (1) diffusion regions (leading to nonuniform current flow and junction breakdown), (2) oxide layers (dielectric breakdown and accumulation of surface charge), (3) lithography (uneven definition of features and mask misalignment), and (4) metal layers (poor contact and electromigration resulting from high current densities). Other failures can originate from improper chip mounting, degradation of wire bonds, and loss of package hermeticity. Wire-bonding and metallization failures account for over one half of all integrated-circuit failures.

Because device lifetimes are very long (10 years or more; see also Section 16.3), it is impractical to study device failure under normal operating conditions. One method of studying failures efficiently is by **accelerated life testing**, which involves accelerating the conditions whose effects are known to cause device breakdown. Cyclic variations in temperature, humidity, voltage, and current are used to stress the components. Statistical data taken from these tests are then used to predict device failure modes and device life under normal operating conditions. Chip mounting and packaging are strained by cyclical temperature variations.

## 13.12 | Printed Circuit Boards

Packaged integrated circuits are seldom used alone; rather, they are usually combined with other ICs to serve as building blocks of a yet larger system. A *printed circuit board* (PCB) is the substrate for the final interconnections among all completed chips and serves as the communication link between the outside world and the microelectronic circuitry within each packaged IC. In addition to the ICs, circuit boards also usually contain discrete circuit components, such as resistors and capacitors, that would take up too much "real estate" on the limited silicon surface, have special power dissipation requirements, or cannot be implemented on a chip. Other common discrete components include inductors, which cannot be integrated onto the silicon surface, and high-performance transistors, large capacitors, precision resistors, and crystals for frequency control.

A printed circuit board is basically a plastic resin material containing several layers of copper foil (Fig. 13.23). *Single-sided* PCBs have copper tracks on only one side of an insulating substrate, while *double-sided* boards have copper tracks on both sides. Multilayer boards also can be constructed from alternating copper and insulator layers. Single-sided boards are the simplest form of circuit board. Double-sided boards usually must have locations where electrical connectivity is established between the features on both sides of the board. This structure is accomplished with vias, as shown in Fig. 13.24. Multilayer boards can have partial, buried, or through-hole vias to allow for extremely flexible PCBs. Double and multilayer boards are beneficial in that IC packages can be bonded to both sides of the board, allowing more compact designs.

The insulating material is usually an epoxy resin 0.25 mm to 3 mm (0.01 in. to 0.12 in.) thick, reinforced with an epoxy/glass fiber, referred to as E-glass (see Section 10.9.2). They are produced by impregnating sheets of glass fiber with epoxy and then pressing the layers together between hot plates or rolls. The heat and pressure cure the board, resulting in a stiff and strong basis for printed circuit boards. Boards are

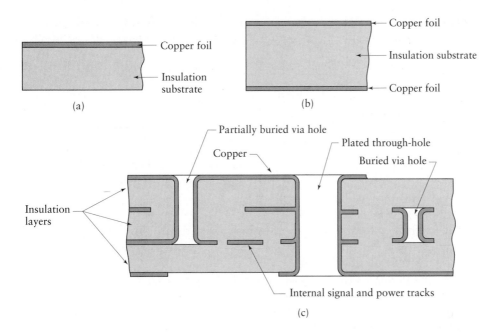

**FIGURE 13.23**  Types of circuit board structures: (a) single sided; (b) double sided; (c) multilayer, showing vias and pathways between layers. *Source*:  M. Groover, *Fundamentals of Modern Manufacturing*, Prentice Hall, Upper Saddle River, NJ, 1996.

sheared to a desired size, and roughly 3-mm-diameter locating holes are then drilled or punched into the board corners to permit alignment and proper location of the board within chip-insertion machines. Holes for vias and connections are punched or produced through CNC drilling; stacks of boards can be drilled simultaneously to increase production rates.

The conductive patterns on circuit boards are defined by lithography, although originally they were produced through screen-printing technologies, hence the terms *printed circuit board* and *printed wiring board* (PWB). In the *subtractive method*, a copper foil is bonded to the circuit board. The desired pattern on the board is then defined by a positive mask developed through photolithography, and the remaining copper is removed through wet etching. In the *additive method*, a negative mask is placed directly onto an insulator substrate to define the desired shape. Electroless plating and electroplating of copper serve to define the connections, tracks, and lands on the circuit board.

The ICs and other discrete components are then fastened to the board by soldering. This procedure is the final step in making the integrated circuits and

**FIGURE 13.24**  Design features in printed circuit boards. *Source*:  M. Groover, *Fundamentals of Modern Manufacturing*, Prentice Hall, Upper Saddle River, NJ, 1996.

the microelectronic devices they contain to be accessible. *Wave soldering* and *reflow paste soldering* are the preferred methods of soldering ICs onto circuit boards.

Some of the design considerations in laying out PCBs are:

1. Wave soldering should be used only on one side of the board; thus, all through-hole mounted components should be inserted from the same side of the board. Surface-mount devices placed on the insertion side of the board must be reflow soldered in place; surface-mount devices on the leg side can be wave soldered.

2. To allow good solder flow in wave soldering, IC packages should carefully be laid out on the printed circuit board (Fig. 13.25). Inserting the packages in the same direction is advantageous for automated placing, whereas random orientations can cause problems in flow of solder across all of the connections.

3. The spacing of ICs is determined mainly by the need to remove heat during operation. Sufficient clearance between packages and adjacent boards is required to allow forced air flow and heat convection.

4. There should be sufficient space around each IC package to allow for rework and repair without disturbing adjacent devices.

**FIGURE 13.25** Layout of IC packages and other components on circuit boards to facilitate wave soldering. *Source*: After G. Boothroyd, P. Dewhurst, and W. Knight, *Product Design for Manufacture and Assembly*, Marcel-Dekker, 1994.

Feed direction

(a) Preferred

(b) Acceptable

(c) Unacceptable

# 13.13 | Micromachining of MEMS Devices

Although lithography and etching can be used in the manufacture of MEMS devices (see Fig. 1.7) to obtain 2D or $2\frac{1}{2}$-D features on wafer surfaces, three-dimensional features are often required. The production of features from micrometers to millimeters in size is called *micromachining*. MEMS devices have been constructed from **polycrystalline silicon (polysilicon)** and single-crystal silicon because the technologies for integrated-circuit manufacture, described earlier in this chapter, are well developed and exploited for these devices, and other, new processes have been developed that are compatible with the existing processing steps. The use of anisotropic etching techniques allows the fabrication of devices with well-defined walls and high aspect ratios, and for this reason, some single-crystal silicon MEMS devices have been fabricated.

One of the recognized difficulties associated with using silicon for MEMS devices is the high adhesion encountered at small length scales and the associated rapid wear. Most commercial devices are designed to avoid friction by, for example, using flexing springs instead of bushings; however, this approach complicates designs and makes some MEMS devices unfeasible. Therefore, significant research is being conducted to identify materials and lubricants that provide reasonable life and performance. Silicon carbide, diamond, and metals such as aluminum, tungsten, and nickel have been investigated as potential MEMS materials. Lubricants have also been investigated; it is known that surrounding the MEMS device in a silicone oil, for example, practically eliminates adhesive wear (see Section 4.4.2), but it also limits the performance of the device. Self-assembling layers of polymers also are being investigated, as well as novel and new materials with self-lubricating characteristics; however, the tribology of MEMS devices remains a main technological barrier to expansion of their already widespread use.

An example of a commercial MEMS-based product is the digital pixel technology (DPT™) device illustrated in Fig. 13.26; this device uses an array of *digital micromirror devices* (DMD™) to project a digital image, such as in computer-driven projection systems. The aluminum mirrors can be tilted so that light is directed into or away from the optics that focus light onto a screen, enabling each mirror to represent a pixel of an image's resolution. The mirror allows light or dark pixels to be projected, but levels of gray can also be accommodated. Since the switching time is about 15 microseconds, which is much faster than the human eye response time, the mirror will switch between the on and off states in order to reflect the proper light dose to the optics.

An array of such mirrors represents a gray-scale screen; using three mirrors (one each for red, green, and blue light) for each pixel results in a color image with millions of discrete colors. Digital pixel technology is widely applied for digital projection systems, high-definition television, and other optical applications. However, production of the device shown in Fig. 13.26 requires much more than $2\frac{1}{2}$-D features, and full three-dimensional, multipart assemblies have to be manufactured.

## 13.13.1 Bulk micromachining

Until the early 1980s, bulk micromachining was the most common form of machining at micrometer scales. This process uses orientation-dependent etches on single-crystal silicon. (See Fig. 13.10b.) The approach is based on etching down into a surface, stopping on certain crystal faces, doped regions, and etchable films to form the desired structure. As an example of the process, consider the fabrication of the silicon cantilever shown in Fig. 13.27. By using the masking techniques described previously, a rectangular patch of the *n*-type silicon substrate is changed to *p*-type silicon

**FIGURE 13.26** The Texas Instruments digital pixel technology (DPT™) device.
(a) Exploded view of a single digital micromirror device (DMD™). (b) View of two
adjacent DMD pixels (c) Images of DMD arrays, with some mirrors removed for clarity.
Each mirror measures approximately 17 $\mu$m (680 $\mu$in.) on a side. (d) A typical DPT device,
used for digital projection systems, high-definition televisions, and other image-display
systems. The device shown contains 1,310,720 micromirrors and measures less than two
inches per side. *Source*: Texas Instruments Corp.

through boron doping. Recall that ODE etchants, such as potassium hydroxide, will
not be able to etch heavily boron-doped silicon; hence, this patch will not be etched.

A mask is then produced, such as with silicon nitride on silicon. When etched
with potassium hydroxide, the undoped silicon will be removed rapidly, while the
mask and the doped patch will essentially be unaffected. Etching progresses until the
(111) planes are exposed in the *n*-type silicon substrate, and they undercut the patch,
leaving a suspended cantilever, as shown in the figure.

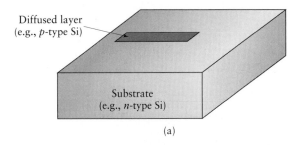

Diffused layer
(e.g., *p*-type Si)

Substrate
(e.g., *n*-type Si)

(a)

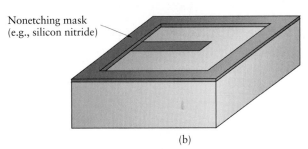

Nonetching mask
(e.g., silicon nitride)

(b)

Freestanding
cantilever

{111} planes

(c)

**FIGURE 13.27** Schematic illustration of bulk micromachining. (a) Diffuse dopant in desired pattern. (b) Deposit and pattern masking film. (c) Orientation-dependent etch, leaving behind a freestanding structure. *Source*: K. R. Williams.

## 13.13.2 Surface micromachining

The basic steps in surface micromachining are illustrated in Fig. 13.28 for silicon devices. A spacer, or sacrificial, layer is deposited onto a silicon substrate coated with a thin dielectric layer (isolation, or buffer, layer). Phosphosilicate glass deposited by chemical-vapor deposition is the most common material for a spacer layer, because it etches very rapidly in hydrofluoric acid. Figure 13.28b shows the spacer layer after application of masking and etching. At this point, a structural thin film is deposited onto the spacer layer; the film can be polysilicon, metal, metal alloy, or dielectric (Fig. 13.28c). The structural film is then patterned, usually through dry etching in order to maintain vertical walls and tight dimensional tolerances. Finally, wet etching of the sacrificial layer leaves a free-standing, three-dimensional structure (Fig. 13.28e). It should be noted that the wafer is usually annealed to remove the residual stresses in the deposited metal before it is patterned. If this procedure is not done, the structural film may severely warp once the spacer layer is removed.

Figure 13.29 shows a microlamp that emits a white light when current is passed through it. It has been produced through a combination of surface and bulk micromachining. The top patterned layer is a 2.2-$\mu$m-thick layer of plasma-etched tungsten forming a meandering filament and bond pad. The rectangular overhang is dry-etched silicon nitride. The steeply sloped layer is wet-HF-etched phosphosilicate glass. The substrate is silicon, which is ODE etched.

The etchant used to remove the spacer layer must be carefully chosen; it must preferentially etch the spacer layer while leaving the dielectric, silicon, and structural film as intact as possible. With large features and narrow spacer layers, this task

**FIGURE 13.28** Schematic illustration of the steps in surface micromachining: (a) deposition of a phosphosilicate glass (PSG) spacer layer; (b) etching of the spacer layer; (c) deposition of polysilicon; (d) etching of polysilicon; (e) selective wet etching of PSG, leaving the silicon substrate and deposited polysilicon unaffected.

**FIGURE 13.29** A microlamp produced from a combination of bulk and surface micromachining. *Source*: Courtesy of K. R. Williams.

**FIGURE 13.30** Stiction after wet etching: (a) unreleased beam; (b) released beam before drying; (c) released beam pulled to the surface by capillary forces during drying. Once contact is made, adhesive forces prevent the beam from returning to its original shape. *Source*: After B. Bhushan.

becomes very difficult to do, and etching can take hours. To reduce the etch time, additional etch holes can be designed into the microstructures to increase access to the spacer layer.

Another difficulty that must be overcome is **stiction** after wet etching. Consider the situation illustrated in Fig. 13.30: After the spacer layer has been removed, the liquid etchant is rinsed from the wafer surface. The rinse water meniscus formed between layers then results in capillary forces that can deform the film and cause it to contact the substrate as the liquid evaporates. Since adhesion forces are more significant at small length scales, it is possible that the film may permanently stick to the surface, and the desired three-dimensional features will not be produced.

Surface micromachining is a very widespread approach for the production of MEMS. Its applications include accelerometers, pressure sensors, micropumps, micromotors, actuators, and microscopic locking mechanisms. Often, these devices require very large vertical walls that cannot be directly manufactured, because the high vertical structure is difficult to deposit. This problem is overcome by machining large flat structures horizontally and then rotating or folding them into an upright position, as shown in Fig. 13.31.

(a)

(b)

**FIGURE 13.31** (a) SEM image of a deployed micromirror. (b) Detail of the micromirror hinge. *Source*: Sandia National Laboratories.

**EXAMPLE 13.3  Surface micromachining of a hinge for a mirror actuation system**

Figure 13.31a shows a micromirror that has been inclined with respect to the surface on which it was manufactured. Such systems can be used for reflecting light (that is oblique to a surface) onto detectors or towards other sensors. It is apparent that a device that has such depth and has the aspect ratio of the deployed mirror is very difficult to machine directly. Instead, it is easier to surface micromachine the mirror along with a linear actuator and then fold the mirror into a deployed position. In order to do so, special hinges, as shown in Fig. 13.31b, are integrated into the design.

Figure 13.32 shows the cross-section of a hinge during manufacture. The following steps are required to produce the hinge:

1. A 2-$\mu$m-thick layer of phosphosilicate glass (PSG) is deposited onto the substrate material.
2. A 2-$\mu$m-thick layer of polysilicon (Poly1 in Fig. 13.32a) is deposited onto the PSG, patterned by photolithography, and dry etched to form the desired structural elements, including hinge pins.
3. A second layer of sacrificial PSG, with a thickness of 0.5 $\mu$m, is deposited (Fig. 13.32b).
4. The connection locations are etched through both layers of PSG (Fig. 13.32c).
5. A second layer of polysilicon (Poly2 in Fig. 13.32d) is deposited, patterned, and etched.
6. The sacrificial layers of PSG are removed through wet etching.

**FIGURE 13.32**  Schematic illustration of the steps required to manufacture a hinge. (a) Deposition of a phosphosilicate glass (PSG) spacer layer and polysilicon layer. (See Fig. 13.28.) (b) Deposition of a second spacer layer. (c) Selective etching of the PSG. (d) Depostion of polysilicon to form a staple for the hinge. (e) After selective wet etching of the PSG, the hinge can rotate.

Hinges such as those discussed in this example have very high friction. If mirrors such as that shown in Fig. 13.31 are manually and carefully manipulated with probe needles, they will remain in position. Often, such mirrors will be combined with linear actuators to enable precise control of their deployment.

**SCREAM.** Another approach for making very deep MEMS structures is the *single-crystal silicon reactive etching and metallization* (SCREAM) process, depicted in Fig. 13.33. In this technique, standard lithography and etching processes produce trenches 10–50 $\mu$m (400–2000 $\mu$in.) deep, which are then protected by a layer of chemically vapor-deposited silicon oxide. An anisotropic etch step removes the oxide only at the bottom of the trench, and the trench is then extended through dry etching. An isotropic etch, using sulfur hexafluoride ($SF_6$), laterally etches the exposed sidewalls at the bottom of the trench; this undercut, when it overlaps adjacent undercuts, releases the machined structures.

**SIMPLE.** An alternative to SCREAM is the *silicon micromachining by single-step plasma etching* (SIMPLE) technique, as depicted in Fig. 13.34. This technique uses a chlorine-gas-based plasma etch process that machines *p*-doped, or lightly doped, silicon anisotropically, but heavily *n*-doped silicon isotropically. A suspended MEMS device can thus be produced in one plasma etching step, as shown in the figure.

Some of the concerns with the SIMPLE process are as follows:

1. The oxide mask is machined, although at a slower rate, by the chlorine-gas plasma; therefore, relatively thick oxide masks are needed.

2. The isotropic etch rate is low, typically 50 nm/min; consequently, this process is very slow.

**FIGURE 13.33** Steps involved in the SCREAM process. *Source*: After N. Maluf.

1. Deposit oxide and photoresist.

2. Lithography and oxide etch.

3. Silicon etch.

4. Coat sidewalls with PECVD oxide.

5. Remove oxide at bottom and etch silicon.

6. Plasma etch in $SF_6$ to release structures.

**FIGURE 13.34** Schematic illustration of silicon micromachining by the single-step plasma etching (SIMPLE) process. Reprinted with permission of Cambridge University Press.

1. Deposit oxide and photoresist on layered substrate.

2. Lithography and oxide etch.

3. Plasma etch $p^-$ doped silicon.

4. Isotropic etch of $n^+$ doped silicon.

3. The layer beneath the structures will contain deep trenches, which may affect the motion of free-hanging structures.

**Etching combined with fusion bonding.** Very tall structures can be produced in crystalline silicon through a combination of *silicon fusion bonding* and *deep reactive ion etching* (SFB–DRIE), as illustrated in Fig. 13.35. First, a silicon wafer is prepared with an insulating oxide layer, with the deep trench areas defined by a standard lithography procedure. This step is followed by conventional wet or dry etching to form a large cavity. A second layer of silicon is fusion bonded to this layer, which can then be ground and polished to the desired thickness, if necessary. At this stage, integrated circuitry is manufactured through the steps outlined in Fig. 13.3. A protective resist is applied and exposed, and the desired trenches are etched by deep reactive ion etching through to the cavity in the first layer of silicon.

**EXAMPLE 13.4 Operation and fabrication sequence for a thermal ink-jet printer**

Thermal ink-jet printers are perhaps the most successful application of MEMS to date. These printers operate by ejecting nano- or picoliters ($10^{-12}$ liters) of ink from a nozzle towards paper. Ink-jet printers use a variety of designs, but silicon machining technology is most applicable to high-resolution printers. It should be realized that a resolution of 1200 dots per inch (dpi) requires a nozzle pitch of approximately 20 $\mu$m (800 $\mu$in.).

Figure 13.36 depicts the mode of operation of an ink-jet printer. When an ink droplet is to be generated and expelled, a tantalum resistor below a nozzle is heated. This resistor heats a thin film of ink so that a bubble forms within five microseconds. The bubble expands rapidly, with internal pressures reaching 1.4 MPa (200 psi), and as a result, fluid is forced rapidly out of the nozzle. Within 24 microseconds, the tail of the ink droplet separates, because of surface tension. The heat source is then removed (turned off), and the bubble collapses inside the nozzle. Within 50 microseconds, sufficient ink has been drawn into the nozzle from a reservoir to form the desired meniscus for the next droplet.

Traditional ink-jet printer heads were produced with electroformed nickel nozzles, fabricated separately from the integrated circuitry, and required a bonding

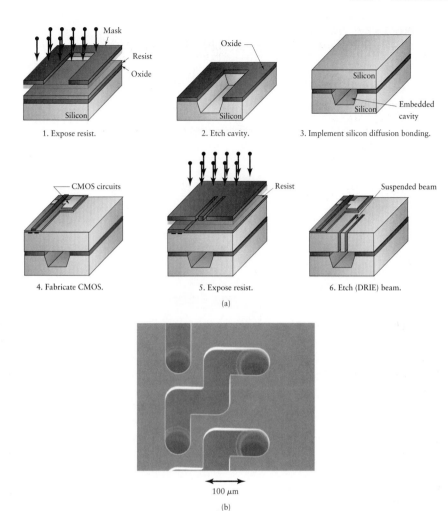

1. Expose resist.   2. Etch cavity.   3. Implement silicon diffusion bonding.

4. Fabricate CMOS.   5. Expose resist.   6. Etch (DRIE) beam.

(a)

(b)

**FIGURE 13.35** (a) Silicon fusion bonding combined with deep reactive ion etching to produce large suspended cantilevers. *Source*: After N. Maluf. (b) A micro–fluid-flow device manufactured by applying the DRIE process to two separate wafers and then aligning and silicon-fusion bonding them together. Afterward, a Pyrex layer (not shown) is anodically bonded over the top to provide a window for observing fluid flow. *Source*: K. R. Williams.

100 μm

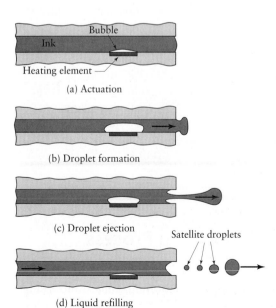

(a) Actuation

(b) Droplet formation

(c) Droplet ejection

(d) Liquid refilling

**FIGURE 13.36** Sequence of operation of a thermal ink-jet printer. (a) Resistive heating element is turned on, rapidly vaporizing ink and forming a bubble. (b) Within five microseconds, the bubble has expanded and displaced liquid ink from the nozzle. (c) Surface tension breaks the ink stream into a bubble, which is discharged at high velocity. The heating element is turned off at this time, so that the bubble collapses as heat is transferred to the surrounding ink. (d) Within 24 microseconds, an ink droplet (and undesirable satellite droplets) are ejected, and surface tension of the ink draws more liquid from the reservoir. *Source*: From F. G. Tseng, "Microdroplet Generators," in M. Gad-el-hak (ed.), *The MEMS Handbook*, CRC Press, 2002.

**FIGURE 13.37** The manufacturing sequence for producing thermal ink-jet printer heads. *Source*: From F. G. Tseng, "Microdroplet Generators," in M. Gad-el-hak (ed.), *The MEMS Handbook*, CRC Press, 2002.

(a) Silicon-nitride deposition

(b) Wet etch manifold; remove PSG

(c) Wet etch, enlarge chamber

(d) Heater and interconnection formulation

(e) Laser ablade nozzle

operation to attach these two components. With increasing printer resolution, however, it is more difficult to bond the components with a tolerance under a few micrometers. For this reason, single-component, or monolithic, fabrication is of interest.

Figure 13.37 shows the fabrication sequence for a monolithic ink-jet printer head. A silicon wafer is prepared and coated with a phosphosilicate-glass (PSG) pattern and low-stress silicon-nitride coating. The ink reservoir is obtained by isotropically etching the back side of the wafer, followed by PSG removal and then enlargement of the reservoir. The required CMOS controlling circuitry is then produced (this step is not shown in Fig. 13.37), and a tantalum heater pad is deposited. The aluminum interconnection between the tantalum pad and the CMOS circuit is formed, and the nozzle is produced through laser ablation. An array of such nozzles can be placed inside an ink-jet printing head, and resolutions of 2400 dpi or higher can be achieved.

## 13.14 | The LIGA Microfabrication Process

*LIGA* is a German acronym for the combined processes of X ray lithography, electrodeposition, and molding (**X ray Lithographie, Galvanoformung und Abformung**); a schematic illustration of this process is given in Fig. 13.38.

The LIGA process involves the following steps:

1. A very thick (up to hundreds of microns) resist layer of polymethylmethacrylate (PMMA) is deposited onto a primary substrate.
2. The PMMA is exposed to collimated X rays and developed.
3. Metal is electrodeposited onto the primary substrate.
4. The PMMA is removed or stripped, resulting in a freestanding metal structure.
5. Plastic injection molding is done in the metal structure, which acts as a mold.

Depending on the application, the final product from a LIGA process may be:

a) A freestanding metal structure, resulting from the electrodeposition process;
b) A plastic injection-molded structure;
c) An investment-cast metal part, where the injection molded structure was used as a blank;
d) A slip-cast ceramic part, produced with the injection-molded parts as the molds.

The substrate used in LIGA is a conductor or a conductor-coated insulator. Examples of primary substrate materials include austenitic steel plate; silicon wafers with a titanium layer; and copper plated with gold, titanium, or nickel. Metal-plated ceramic and glass have also been used. The surface may be roughened by grit blasting to encourage good adhesion of the resist material.

Resist materials must have high X ray sensitivity, dry- and wet-etching resistance when unexposed, and thermal stability. The most common resist material is polymethylmethacrylate, which has a very high molecular weight (more than $10^6$

**FIGURE 13.38** The LIGA (lithography, electrodeposition, and molding) technique. (a) Primary production of a metal final product or mold insert. (b) Use of the primary part for secondary operations, or *replication*. *Source:* Courtesy of IMM Institute für Mikrotechnik.

grams per mole; see Section 10.2.1). The X rays break the chemical bonds, leading to the production of free radicals and significantly reduced molecular weight in the exposed region. Organic solvents then preferentially dissolve the exposed PMMA in a wet-etching process. After development, the remaining three-dimensional structure is rinsed and dried, or it is spun and blasted with dry nitrogen.

Electrodeposition of metal usually involves electroplating of nickel (see Section 4.5.1 and Fig. 4.17). The nickel is deposited onto exposed areas of the substrate; it fills the PMMA structure and can even coat the resist (Fig. 13.38a). Nickel is the material of choice because of the relative ease in electroplating with well-controlled deposition rates and residual stress control. Electroless plating of nickel is also possible, and the nickel can be deposited directly onto electrically insulating substrates. However, because nickel displays high wear rates in MEMS, significant research has been directed towards the use of other materials or coatings.

After the metal structure has been deposited, precision grinding either removes the substrate material or a layer of the deposited nickel, a process referred to as *planarization* (see also Section 13.9). The need for planarization is clear when it is recognized that three-dimensional MEMS devices require micrometer tolerances on layers many hundreds of micrometers thick. Planarization is difficult to achieve; conventional lapping (Section 9.8) leads to preferential removal of the soft PMMA and smearing of the metal. Planarization is usually accomplished with a diamond lapping procedure referred to as *nanogrinding*. Here, a diamond slurry-loaded soft metal plate is used to remove material in order to maintain flatness within $1 \mu$m $(40 \mu$in.) over a 75-mm (3-in.) diameter substrate.

If cross-linked (Section 10.2.1), the PMMA resist is then exposed to synchrotron X ray radiation and removed by exposure to an oxygen plasma or through solvent extraction; the result is a metal structure, which may be used for further processing. Examples of freestanding metal structures produced through electrodeposition of nickel are shown in Fig. 13.39.

The processing steps used to make freestanding metal structures are extremely time consuming and expensive. The main advantage of LIGA is that these structures serve as molds for the rapid replication of submicron features through molding operations. Table 13.5 lists and compares the processes that can be used for producing micromolds; it can be seen that LIGA provides some clear advantages. Reaction injection molding, injection molding, and compression molding (Section 10.10) have also been used to make these micromolds.

(a)                                                    (b)

**FIGURE 13.39**    (a) Electroformed 200-$\mu$m-tall nickel structures. (b) Detail of 5-$\mu$m-wide nickel lines and spaces. *Source*: After T. Christenson, *The MEMS Handbook*, CRC Press, 2002.

**TABLE 13.5**

**Comparison of Micromold Manufacturing Techniques**

|  | Production technique | | |
|---|---|---|---|
|  | LIGA | Laser machining | EDM |
| Aspect ratio | 10–50 | 10 | up to 100 |
| Surface roughness | $<50$ nm | 100 nm | $0.3-1\ \mu m$ |
| Accuracy | $<1\ \mu m$ | $1-3\ \mu m$ | $1-5\ \mu m$ |
| Mask Required? | Yes | No | No |
| Maximum height | $1-500\ \mu m$ | $200-500\ \mu m$ | $\mu m$ to mm |

*Source*: L. Weber, W. Ehrfeld, H. Freimuth, M. Lacher, M. Lehr, and P. Pech, *SPIE Micromachining and Microfabrication Process Technology II*, Austin, TX, 1996.

---

**EXAMPLE 13.5**   Production of rare-earth magnets

A number of scaling issues in electromagnetic devices indicate that there is an advantage in using rare-earth magnets from the samarium cobalt (SmCo) and neodymium iron boron (NdFeB) families. These materials are available in powder form and are of interest because they can produce magnets that are an order of magnitude more powerful than conventional magnets (Table 13.6). Therefore, these materials can be used when effective miniature electromagnetic transducers are to be produced.

Figure 13.40 shows the processing steps used to manufacture these magnets. The polymethylmethacrylate mold is produced by exposure to X ray radiation and solvent extraction. The rare-earth powders are mixed with a binder of epoxy and then applied to the PMMA mold through a combination of calendering (see Fig. 10.32) and pressing. After curing in a press at a pressure around 70 MPa (10 ksi), the substrate is planarized. The substrate is then subjected to a magnetizing field of at least 35 kilooersteds in the desired orientation. Once the material has been magnetized, the PMMA substrate is dissolved, leaving behind the rare-earth magnets, as shown in Fig. 13.41.

*Source*: T. Christenson, Sandia National Laboratories.

---

**TABLE 13.6**

**Comparison of Properties of Permanent-Magnet Materials**

| Material | Energy product (Gauss–Oersted $\times\ 10^{-6}$) |
|---|---|
| Carbon steel | 0.20 |
| 36% cobalt steel | 0.65 |
| Alnico I | 1.4 |
| Vicalloy I | 1.0 |
| Platinum–cobalt | 6.5 |
| $Nd_2Fe_{14}B$, fully dense | 40 |
| $Nd_2Fe_{14}B$, bonded | 9 |

**FIGURE 13.40**
Fabrication process used to
produce rare-earth magnets
for microsensors. *Source*:
T. Christenson, Sandia
National Laboratories.

(a) Press

(b) Lap

(c) Magnetize

(e) Release

**FIGURE 13.41** SEM
images of $Nd_2Fe_{14}B$
permanent magnets. Powder
particle size ranges from 1 to
5 $\mu$m (40 to 200 $\mu$in.) and
the binder is a methylene-
chloride resistant epoxy.
Mild distortion is present in
the image, due to magnetic
perturbation of the imaging
electrons. Maximum-energy
products of 9 MGOe have
been obtained with this
process. *Source*:
T. Christenson, in Gad-el-
Hak, ed., The MEMS
Handbook, CRC Press,
2002.

(a)

(b)

**Multilayer X ray lithography.** The LIGA technique is very suitable for producing MEMS devices with large aspect ratios and reproducible shapes. Often, a multilayer stepped structure that cannot be made directly through LIGA is required. For
nonoverhanging geometries, direct plating can be applied. In this technique, a layer
of electrodeposited metal with surrounding PMMA is produced as described previously. A second layer of PMMA resist is then bonded to this structure and X ray
exposed with an aligned X ray mask.

It is often useful to have overhanging geometries within complex MEMS devices.
A batch diffusion bonding and release procedure has been developed for this purpose and is schematically illustrated in Fig. 13.42a. This process involves the preparation of two PMMA patterned and electroformed layers, with the PMMA being
subsequently removed. The wafers are then aligned face to face with guide pins that
press-fit into complimentary structures on the opposite surface. Next, the substrates

Alumina substrate

Alignment
gauge pin

Sacrificial layer

Nickel substrate (structural)

Diffusion bond and release

All-nickel structure

(a)

(b)

**FIGURE 13.42** (a) Multilevel MEMS fabrication through wafer-scale diffusion bonding. (b) A suspended ring structure for measurement of tensile strain, formed by two-layer wafer-scale diffusion bonding. *Source*: After T. Christenson, Sandia National Laboratories.

are joined in a hot press, and a sacrificial layer on one substrate is etched away, leaving one layer bonded to the other. Fig. 13.42b shows an example of such a structure.

# 13.15 | Solid Freeform Fabrication of Devices

*Solid freeform fabrication* is another term for rapid prototyping, described in Section 10.12. This method is unique in that complex three-dimensional structures are produced through additive manufacturing, as opposed to material removal. Many of the advances in rapid prototyping also are applicable to MEMS manufacture for processes with sufficiently high resolution. Recall that *stereolithography* involves curing of a liquid thermosetting polymer, using a photoinitiator and a highly focused light source. Conventional stereolithography uses layers between 75 $\mu$m and 500 $\mu$m (0.003 in. and 0.02 in.) in thickness, with a laser dot focused to a 0.25-mm (0.01 in.) diameter.

In *microstereolithography*, the process uses the same approach, but the laser is more highly focused, to a diameter as small as 1 $\mu$m (40 $\mu$in.), and layer thicknesses are around 10 $\mu$m (400 $\mu$in.). This technique has a number of cost advantages, but the MEMS devices are difficult to integrate with the controlling circuitry, because stereolithography produces nonconducting polymer structures.

*Instant masking* is a technique for producing MEMS devices (Fig. 13.43). The solid freeform fabrication of MEMS devices by instant masking is known as **electrochemical fabrication** (EFAB). A mask of elastomeric material is first produced through the conventional photolithography techniques described in Section 13.6. The mask is pressed against the substrate in an electrodeposition bath, so that the elastomer conforms to the substrate and excludes plating solution in contact areas. Electrodeposition takes place in areas that are not masked, eventually producing a mirror image of the mask. By using a sacrificial filler, made of a second material, complex three-dimensional shapes can be produced, complete with overhangs, arches, and other features, using instant masking technology.

**FIGURE 13.43** The instant masking process: (a) bare substrate; (b) during deposition, with the substrate and instant mask in contact; (c) the resulting pattern deposit. *Source*: After A. Cohen, MEMGen Corporation.

(a)                    (b)                    (c)

## CASE STUDY | Accelerometer for Automotive Air Bags

Accelerometers based on lateral resonators represent the largest commercial application of surface micromachining today and are widely used as sensors for automotive air-bag deployment systems. The sensor portion of such an accelerometer is shown in Fig. 13.44. A central mass is suspended over the substrate, but anchored through four slender beams that act as springs to center the mass under static-equilibrium conditions. An acceleration causes the mass to deflect, reducing or increasing the clearance between the fins on the mass and the stationary fingers on the substrate. By measuring the electrical capacitance between the mass and fins, the deflection of the mass, and therefore the acceleration or deceleration of the system, can be directly measured. Figure 13.44 shows an arrangement for the measurement of acceleration in one direction, but commercial sensors employ several masses so that accelerations can be measured in multiple directions simultaneously.

Figure 13.45 shows the 50-g surface-micromachined accelerometer (ADXL-50) with onboard signal conditioning and self-diagnostic electronics. The polysilicon sensing element (visible in the center of the die) occupies only 5% of the total die area, and the whole chip measures $500\,\mu m$ by $625\,\mu m$ ($20\,\mu in.$ by $25\,\mu in.$) The mass is approximately $0.3\,\mu g$, and the sensor has a measurement accuracy of 5% over the $\pm 50$-g range.

Fabrication of the accelerometer proved to be a challenge, since it required a complimentary metal-oxide semiconductor (CMOS) fabrication sequence to be closely integrated with a surface-micromachining approach. Analog Devices, Inc., was able to modify a CMOS production technique to directly incorporate surface micromachining. In the sensor design, $n+$ underpasses connect the sensor area to the electronic circuitry, replacing the usual heat-sensitive aluminum connect lines.

**FIGURE 13.44** Schematic illustration of a micro–acceleration sensor. *Source*: After N. Maluf.

**FIGURE 13.45**
Photograph of Analog Devices's ADXL-50 accelerometer with a surface-micromachined capacitive sensor (center), on-chip excitation, self-test, and signal-conditioning circuitry. The entire chip measures 0.500 by 0.625 mm. *Source*: From R. A. Core et al., *Solid State Technol.*, October 1993, pp. 39–47.

(a)

(b)

(c)

**FIGURE 13.46**
Preparation of IC chip for polysilicon. (a) Sensor area post-BPSG planarization and moat mask. (b) Blanket deposition of thin oxide and thin nitride layer. (c) Bumps and anchors made in LTO spacer layer. *Source*: From R. A. Core et al., *Solid State Technol.*, October 1993, pp. 39–47.

Most of the sensor processing is inserted into the fabrication process right after borosilicate-glass planarization. After planarization, a designated sensor region, or *moat*, is cleared in the center of the die (Fig. 13.46a). A thin oxide is then deposited to passivate the $n+$ underpass connects, followed by a thin, low-pressure chemical-vapor-deposited (LPCVD) nitride to act as an etch stop for the final polysilicon release etch (Fig. 13.46b). The spacer, or sacrificial oxide, used is a 1.6-$\mu$m (64-$\mu$in.)

densified low-temperature oxide (LTO) deposited over the whole die (Fig. 13.46c).

In a first etch, small depressions that will form bumps, or dimples, on the underside of the polysilicon sensor are created in the LTO layer. These bumps will limit adhesive forces and sticking in case the sensor comes in contact with the substrate. A subsequent etch cuts anchors into the spacer layer to provide regions of electrical and mechanical contact (Fig. 13.46c). The 2-$\mu$m-thick (80-$\mu$in.-thick) sensor polysilicon layer is then deposited, implanted, annealed, and patterned (Fig. 13.47a).

Metallization follows, which starts with the removal of the sacrificial spacer oxide from the circuit area, along with the LPCVD nitride and LTO layer. A low-temperature oxide is deposited on the polysilicon sensor part, and contact openings appear in the integrated circuit part of the die where platinum is deposited to form platinum silicide (Fig. 13.47b). The trimmable thin film material, TiW barrier metal, and Al/Cu interconnect metal are sputtered on and patterned in the integrated-circuit area.

The circuit area is then passivated in two separate deposition steps. First, plasma oxide is deposited and patterned (Fig. 13.47c), followed by a plasma nitride (Fig. 13.48a) to form a seal with the previously deposited LPCVD nitride. The nitride acts as a hydrofluoric-acid barrier in the subsequent etch release in surface micromachining. The plasma oxide left on the sensor acts as an etch stop for the removal of the plasma nitride (Fig. 13.48a). The sensor area is then prepared for the final release etch. The dielectrics are removed from the sensor, and the final protective resist mask is applied. The photoresist protects the circuit area from the long-term buffered oxide etch (Fig. 13.48b). The final device cross-section is shown in Fig. 13.48c.

*Source*: Adapted from M. Madou, *Fundamentals of Microfabrication*, 2d. ed., CRC Press, 2002.

**FIGURE 13.47**

Polysilicon deposition and IC metallization. (a) Cross-sectional view after polysilicon deposition, implant, anneal, and patterning. (b) Sensor area after removal of dielectrics from circuit area, contact mask, and platinum silicide. (c) Metallization scheme and plasma-oxide passivation and patterning. *Source*: From R. A. Core et al., *Solid State Technol.*, October 1993, pp. 39–47.

(a)

(b)

(c)

**FIGURE 13.48** Prerelease preparation and release. (a) Post-plasma-nitride passivation and patterning. (b) Photoresist protection of the IC. (c) Freestanding, released, polysilicon beam. *Source*: From R. A. Core et al., *Solid State Technol.*, October 1993, pp. 39–47.

# SUMMARY

- The microelectronics industry is well developed, but still rapidly changing. The possibilities for new device concepts and circuit designs appear to be endless. MEMS devices have become important technologies. (Section 13.1)

- Semiconductor materials, such as silicon, gallium arsenide, and polysilicon, have unique properties, and their electrical properties lie between those of conductors and insulators. (Section 13.2)

- Techniques have been developed for single crystal growing and wafer preparation; after fabrication, the wafer is sliced into individual chips, each containing one complete integrated circuit. (Section 13.3)

- The fabrication of microelectronic devices and integrated circuits involves many different types of processes. After bare wafers have been prepared, they undergo repeated oxidation, film deposition, lithographic, and etching steps to open windows in the oxide layer in order to provide access to the silicon substrate. (Sections 13.4–13.8)

- Dopants are introduced into various regions of the silicon structure to alter their electrical characteristics; this task is done by diffusion and ion implantation. (Section 13.9)

- Microelectronic devices are interconnected by multiple metal layers, and the completed circuit is packaged and made accessible through electrical connections. A variety of packages is available. (Section 13.10)

- Yield in devices is important for economic considerations, and reliability has become increasingly important, because of the long expected life of these devices. (Section 13.11)

- Requirements for more flexible and faster integrated circuits demand packaging that places many closely spaced contacts on a printed circuit board. (Section 13.12)

- MEMS devices are manufactured through techniques and with materials that, for the most part, have been pioneered in the microelectronics industry. Bulk and surface micromachining, LIGA, SCREAM, and fusion bonding of multiple layers are the most prevalent methods. (Sections 13.13–13.15.)

# BIBLIOGRAPHY

Bakoglu, H. B., *Circuits, Interconnections, and Packaging for VLSI*, Addison-Wesley, 1990.

Berger, L. I., *Semiconductor Materials*, CRC Press, 1997.

Brar, A. S., and P. B. Narayan, *Materials and Processing Failures in the Electronics and Computer Industries: Analysis and Prevention*, ASM International, 1993.

Campbell, S. A., *The Science and Engineering of Microelectronic Fabrication*, Oxford, 1996.

Chandrakasan, A., and R. Brodersen (eds.), *Low Power CMOS Design*, IEEE, 1998.

Chandrakasan, A., and R. Brodersen, *Low Power Digital CMOS Design*, Kluwer, 1995.

Chang, C.-Y., and S. M. Sze (eds.), *ULSI Devices*, Wiley-Interscience, 2000.

Colclaser, R. A., *Microelectronics: Processing and Device Design*, Wiley, 1980.

*Electronic Materials Handbook*, Vol. 1: *Packaging*, ASM International, 1989.

Elwenspoek, M., and H. Jansen, *Silicon Micromachining*, Cambridge University Press, 1998.

Elwenspoek, M., and R. Wiegerink, *Mechanical Microsensors*, Springer, 2001.

Gad-el-Hak, M. (ed.), *The MEMS Handbook*, CRC Press, 2002.

Ghandhi, S. K., *VLSI Fabrication Principles*, 2d. ed., Wiley-Interscience, 1993.

Griffin, P. B., J. D. Plummer, and M. D. Deal, *Silicon VLSI Technology: Fundamentals, Practice and Modeling*, Prentice Hall, 2000.

Harper, C. A. (ed.), *Electronic Packaging and Interconnection Handbook*, 3d. ed., McGraw-Hill, 2000.

_____, *High-Performance Printed Circuit Boards*, McGraw-Hill, 2000.

Hwang, J. S., *Modern Solder Technology for Competitive Electronics Manufacturing*, McGraw-Hill, 1996.

Javits, M. W. (ed.), *Printed Circuit Board Materials Handbook*, McGraw-Hill, 1997.

Judd, M., and K. Brindley, *Soldering in Electronics Assembly*, 2d. ed., Newnes, 1999.

Kovacs, G. T. A., *Micromachined Transducers Sourcebook*, McGraw-Hill, 1998.

Madou, M. J., *Fundamentals of Microfabrication*, CRC Press, 1997.

Mahajan, S., and K. S. S. Harsha, *Principles of Growth and Processing of Semiconductors*, McGraw-Hill, 1998.

Maluf, N., *An Introduction to Microelectromechanical Systems Engineering*, Artech House, 2000.

Manko, H. H., *Soldering Handbook for Printed Circuits and Surface Mounting*, Van Nostrand Reinhold, 1995.

Matisoff, B. S., *Handbook of Electronics Manufacturing*, 3d. ed., Chapman & Hall, 1996.

Nishi, Y., and R. Doering (eds.), *Handbook of Semiconductor Manufacturing Technologies*, Dekker, 2000.

Pierret, R. F., and G. W. Neudeck, *Semiconductor Device Fundamentals*, Addison-Wesley, 1996.

Quirk, M., and J. Serda, *Semiconductor Manufacturing Technology*, Prentice Hall, 2000.

Sze, S. M. (ed.), *Modern Semiconductor Devices Physics*, Wiley, 1997.

_____, *Semiconductor Devices: Physics and Technology*, Wiley, 2001.

Taur, Y., and T. H. Ning, *Fundamentals of Modern VLSI Devices*, Cambridge, 1998.

Wise, K. D. (ed.), *Proceedings of the IEEE: Special Issue on Integrated Sensors, Microactuators and Microsystems (MEMS)*, 1998.

Wolf, S., and R. N. Tauber, *Silicon Processing for the VSLI Era: Process Technology*, 2d. ed., Lattice Press, 1999.

van Zant, P., *Microchip Fabrication: A Practical Guide to Semiconductor Processing*, 4th ed., McGraw-Hill, 2000.

Varadan, V. K., X. Jiang, and V. V. Varadan, *Microstereolithography and Other Fabrication Techniques for 3D MEMS*, Wiley, 2001.

Yeap, G., *Practical Low Power Digital VLSI Design*, Kluwer, 1997.

# QUESTIONS

**13.1** Define the terms *wafer, chip, device*, and *integrated circuit*.

**13.2** Why is silicon the most commonly used semiconductor in IC technology?

**13.3** What do VLSI, IC, CVD, CMP, and DIP stand for?

**13.4** How do *n*-type and *p*-type dopants differ?

**13.5** How is epitaxy different than other forms of film deposition?

**13.6** Compare wet and dry etching.

**13.7** How is silicon nitride used in oxidation?

**13.8** What are the purposes of prebaking and postbaking in lithography?

**13.9** Define selectivity and isotropy and their importance in relation to etching.

**13.10** What do the terms *linewidth* and *registration* refer to?

**13.11** Compare diffusion and ion implantation.

**13.12** What is the difference between evaporation and sputtering?

**13.13** What is the definition of yield? How important is yield?

**13.14** What is accelerated life testing? Why is it practiced?

**13.15** What do BJT and MOSFET stand for?

**13.16** Explain the basic process of surface micromachining.

**13.17** What is LIGA? What are its advantages?

**13.18** What is the difference between isotropic and anisotropic etching?

**13.19** What is a mask? Of what materials is it composed?

**13.20** What is the difference between chemically reactive ion etching and dry plasma etching?

**13.21** Which process(es) in this chapter allow(s) fabrication of products from polymers?

**13.22** What is a PCB?

**13.23** Explain the process of thermosonic stitching.

**13.24** What is the difference between a die, a chip, and a wafer?

**13.25** Why are flats machined onto silicon wafers?

**13.26** What is a via? What is its function?

**13.27** What is a flip chip? What are its advantages over a surface-mount device?

**13.28** Explain how IC packages are attached to a printed circuit board if both sides will contain ICs.

**13.29** In a horizontal epitaxial reactor (see the accompanying figure), the wafers are placed on a stage (susceptor) that is tilted by a small amount, usually 1°–3°. Why is this procedure done?

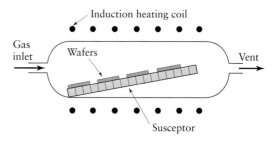

**13.30** The accompanying table describes three changes in the manufacture of a wafer: increase of the wafer diameter, reduction of the chip size, and increase of the process complexity. Complete the table by filling in the words "increase," "decrease," or "no change" to indicate the effect that each change would have on the wafer yield and on the overall number of functional chips.

**Effects of manufacturing changes**

| Change | Wafer yield | Number of functional chips |
|---|---|---|
| Increase wafer diameter | | |
| Reduce chip size | | |
| Increase process complexity | | |

**13.31** The speed of a transistor is directly proportional to the width of its polysilicon gate, with a narrower gate resulting in a faster transistor, and a wider gate resulting in a slower transistor. With the knowledge that the manufacturing process has a certain variation for the gate width, say $\pm 0.1\ \mu m$, how might a designer alter the gate size of a critical circuit in order to minimize its speed variation? Are there any penalties for making this change? Explain.

**13.32** A common problem in ion implantation is channeling, in which the high-velocity ions travel deep into the material through channels along the crystallographic planes before finally being stopped. What is one simple way to stop this effect?

**13.33** The MEMS devices discussed in this chapter apply macroscale machine elements, such as spur gears, hinges, and beams. Which of the following machine elements can and cannot be applied to MEMS, and why?

a. ball bearing;

b. bevel gears;

c. worm gears;

d. cams.

e. helical springs;

f. rivets;

g. bolts;

**13.34** Figure 13.5b shows the Miller indices on a wafer of (100) silicon. Referring to Fig. 13.4, identify the important planes for the other wafer types illustrated in Fig. 13.5a.

**13.35** Referring to Fig. 13.16, sketch the holes generated from a circular mask.

**13.36** Explain how you would produce a spur gear if its thickness were one tenth its diameter and its diameter were (a) 10 $\mu$m, (b) 100 $\mu$m, (c) 1 mm, (d) 10 mm, and (e) 100 mm.

# PROBLEMS

**13.37** A certain wafer manufacturer produces two equal-sized wafers, one containing 500 chips and the other containing 300 chips. After testing, it is observed that 50 chips on each wafer are defective. What are the yields of the two wafers? Can any relationship be drawn between chip size and yield?

**13.38** A chlorine-based polysilicon etch process displays a polysilicon:resist selectivity of 4:1 and a polysilicon:oxide selectivity of 50:1 How much resist and exposed oxide will be consumed in etching 3500 Å of polysilicon? What would the polysilicon:oxide selectivity have to be in order to lose only 40 Å of exposed oxide?

**13.39** During a processing sequence, four silicon-dioxide layers are grown by oxidation: 4000 Å, 1500 Å, 400 Å, and 150 Å. How much of the silicon substrate is consumed?

**13.40** A certain design rule calls for metal lines to be no less than 2 $\mu$m wide. If a 1-$\mu$m-thick metal layer is to be wet etched, what is the minimum photoresist width allowed (assuming that the wet etch is perfectly isotropic)? What would be the minimum photoresist width if a perfectly anisotropic dry-etch process were used?

**13.41** Obtain mathematical expressions for the etch rate as a function of temperature, using Fig. 13.11.

**13.42** If a square mask of side length 100 $\mu$m is placed on a {100} plane and oriented with a side in the ⟨110⟩ direction, how long will it take to etch a hole 4 $\mu$m deep at 80°C using thylene-diamine/pyrocatechol? Sketch the resulting profile.

**13.43** Obtain an expression for the width of the trench bottom as a function of time for the mask shown in Fig. 13.10b.

**13.44** Estimate the time of contact and average force when a fluorine atom strikes a silicon surface with a velocity of 1 mm/s. *Hint*: See Eqs. 9.11 and 9.13.

**13.45** Calculate the undercut in etching a 10-$\mu$m-deep trench if the anisotropy ratio is (a) 200, (b) 2, and (c) 0.5. What is the sidewall slope for these cases?

**13.46** Calculate the undercut in etching a 10-$\mu$m-deep trench for the wet etchants listed in Table 13.3. What would the undercut be if the mask were silicon oxide?

**13.47** Estimate the time required to etch a spur-gear blank from a 75-mm-thick slug of silicon.

**13.48** A resist is applied in a resist spinner spun operating at 2000 rpm, using a polymer resist with viscosity of 0.05 N·s/m². The measured resist thickness is 1.5 $\mu$m. What is the expected resist thickness at 6000 rpm? Assume that $\alpha = 1.0$ in Eq. (13.3).

# DESIGN

**13.49** The accompanying figure shows the cross-section of a simple *npn* bipolar transistor. Develop a process flow chart to fabricate this device.

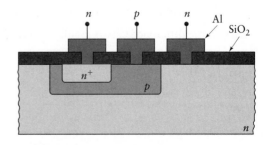

**13.50** Referring to the MOS transistor cross-section in the accompanying figure and the given table of design rules, what is the smallest obtainable transistor size $W$? Which design rules, if any, have no impact on $W$?

Table of Design Rules

| Rule # | Rule name | Value ($\mu$m) |
|---|---|---|
| R1 | Minimum polysilicon width | 0.50 |
| R2 | Minimum poly-to-contact spacing | 0.15 |
| R3 | Minimum enclosure of contact by diffusion | 0.10 |
| R4 | Minimum contact width | 0.60 |
| R5 | Minimum enclosure of contact by metal | 0.10 |
| R6 | Minimum metal-to-metal spacing | 0.80 |

**13.51** The accompanying figure shows a mirror that is suspended on a torsional beam and that can be inclined through electrostatic attraction by applying a voltage on either side of the mirror at the bottom of the trench. Make a flow chart of the manufacturing operations required to produce this device.

**13.52** Referring to Fig. 13.30, design an experiment to find the critical dimensions of an overhanging cantilever that will not stick to the substrate.

**13.53** Explain how you would manufacture the device shown in Fig. 13.27.

550-$\mu$m-diameter mirror
Torsion hinge
Moving comb teeth
Fixed comb teeth
Anchor

*Source*: R. Muller, University of California at Berkeley.

# 14

# Automation of Manufacturing Processes and Systems

## 14.1 | Introduction

Until the early 1950s, most manufacturing operations were carried out on traditional machinery, such as lathes, milling machines, and presses, which required skilled labor and lacked flexibility. Each time a different product was manufactured, the machinery had to be retooled, and the movement of materials had to be rearranged. Furthermore, the development of new products and parts with complex shapes required numerous trial-and-error attempts by the operator in order to set the proper processing parameters on the machine. Also, because of the human involvement, making parts that were exactly alike was difficult and time consuming. These circumstances meant that processing methods were generally inefficient and that labor costs were a significant portion of the overall production costs. The necessity for reducing the labor share of product cost gradually became apparent, as did the need to improve the efficiency and flexibility of manufacturing operations.

**Productivity** also became a major concern. Defined as the optimum use of all resources, such as materials, energy, capital, labor, and technology, or as output per employee per hour, productivity basically measures operating efficiency. **Mechanization** refers to the use of various mechanical, hydraulic, pneumatic, or electrical devices to run a process or operation. In mechanized systems, the operator still directly controls the particular process and must check each step of the machine's performance. For example, if a cutting tool breaks during machining, parts are overheated during heat treatment, surface finish begins to deteriorate during grinding, or dimensional tolerances become too large in sheet-metal forming, the operator must intervene and change one or more of the relevant process parameters.

Mechanization of machinery and operations had, by and large, reached its peak by the 1940s, although with rapid advances in the science and technology of manufacturing, the efficiency of manufacturing operations began to improve, and the percentage of total cost represented by labor costs began to decline. The next step in improving productivity was **automation**, from the Greek word *automatos*, meaning self-acting; this word was coined in the mid-1940s by the U.S. automobile industry to indicate automatic handling and processing of parts in production machines. During the past six decades, major advances and breakthroughs in the types and in the extent of automation have occurred, made possible largely through rapid advances in the capacity and sophistication of **computers and control systems.**

14.1 Introduction
14.2 Automation
14.3 Numerical Control
14.4 Programming for Numerical Control
14.5 Adaptive Control
14.6 Material Handling and Movement
14.7 Industrial Robots
14.8 Sensor Technology
14.9 Flexible Fixturing
14.10 Assembly, Disassembly, and Service
14.11 Design Considerations
14.12 Economic Considerations

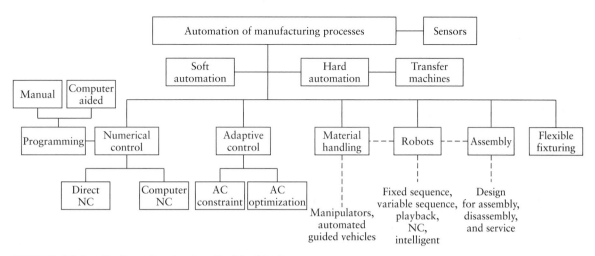

**FIGURE 14.1** Outline of topics described in this chapter.

This chapter follows the outline shown in Fig. 14.1. First, it reviews the history and principles of automation and how they have helped us integrate various operations and activities in manufacturing plants to improve productivity. Then, it introduces the important concept of control of machines and systems through **numerical control** and **adaptive-control** techniques. Next, an essential aspect of manufacturing is discussed: material handling—that is, the movement of raw materials and parts in various stages of completion throughout the plant. Material handling has been developed into various systems, particularly those that include the use of **industrial robots**. The subject of **sensor technology** is then described; this technology is an essential element in the control and optimization of machinery, processes, and systems. Other developments discussed include **flexible fixturing** and **assembly operations**. These methods enable us to take full advantage of advanced manufacturing technologies, particularly **flexible manufacturing systems** (described in Section 15.10), and of major developments in **computer-integrated manufacturing systems** and their impact on all aspects of manufacturing operations.

## 14.2 | Automation

*Automation* is generally defined as the process of having machines follow a predetermined sequence of operations with little or no human labor, using specialized equipment and devices that perform and control manufacturing processes. As described in Section 14.8 and in Chapter 15, automation is achieved through the use of a variety of devices, sensors, actuators, and specialized equipment that are capable of (a) observing all aspects of the manufacturing operation, (b) making decisions concerning the changes that should be made, and (c) controlling all aspects of the operation.

Automation is an *evolutionary* rather than a revolutionary concept. In manufacturing plants, it has been implemented successfully in the following basic areas of activity:

- **Manufacturing processes.** Machining, grinding, forging, cold extrusion, casting, and plastics molding are typical examples of processes that have been extensively automated.

- **Material handling.** Materials and parts in various stages of completion are moved throughout a plant by computer-controlled equipment, with little or no human guidance.

- **Inspection.** Parts are automatically inspected for quality, dimensional accuracy, and surface finish, especially as they are being made.
- **Assembly.** Individually manufactured parts are automatically assembled into subassemblies and, finally, into products.
- **Packaging.** Products are packaged automatically.

### 14.2.1 Evolution of automation

Some metalworking processes were developed as early as 4000 BC (see Table 1.1); it was not, however, until the beginning of the Industrial Revolution, in the 1750s, that automation began to be introduced in the production of goods. Machine tools, such as turret lathes, automatic screw machines, and automatic bottle-making equipment, were in development starting in the late 1890s. *Mass-production* techniques and transfer machines were developed in the 1920s; these systems had *fixed* automatic mechanisms and were designed to produce *specific products*. These developments were best represented in the automobile industry, which produced passenger cars at high production rates (mass production) and at low cost.

The major breakthrough in automation began with numerical control (NC) of machine tools, in the early 1950s. Since this historic development, rapid progress has been made in automating most aspects of manufacturing (Table 14.1), including the

**TABLE 14.1**

**Developments in the History of Automation and Control of Manufacturing Processes. (See also Table 1.1)**

| Date | Development |
|------|-------------|
| 1500–1600 | Water power for metalworking; rolling mills for coinage strips. |
| 1600–1700 | Hand lathe for wood; mechanical calculator. |
| 1700–1800 | Boring, turning, and screw cutting lathe, drill press. |
| 1800–1900 | Copying lathe, turret lathe, universal milling machine; advanced mechanical calculators. |
| 1808 | Sheet-metal cards with punched holes for automatic control of weaving patterns in looms. |
| 1863 | Automatic piano player (Pianola). |
| 1900–1920 | Geared lathe; automatic screw machine; automatic bottle-making machine. |
| 1920 | First use of the word *robot*. |
| 1920–1940 | Transfer machines; mass production. |
| 1940 | First electronic computing machine. |
| 1943 | First digital electronic computer. |
| 1945 | First use of the word *automation*. |
| 1947 | Invention of the transistor. |
| 1952 | First prototype numerical control machine tool. |
| 1954 | Development of the symbolic language APT (Automatically Programmed Tool); adaptive control. |
| 1957 | Commercially available NC machine tools. |
| 1959 | Integrated circuits; first use of the term *group technology*. |
| 1960 | Industrial robots. |
| 1965 | Large-scale integrated circuits. |
| 1968 | Programmable logic controllers. |
| 1970s | First integrated manufacturing system; spot welding of automobile bodies with robots; microprocessors; minicomputer-controlled robot; flexible manufacturing system; group technology. |
| 1980s | Artificial intelligence; intelligent robots; smart sensors; untended manufacturing cells. |
| 1990–2000s | Integrated manufacturing systems; intelligent and sensor-based machines; telecommunications and global manufacturing networks; fuzzy-logic devices; artificial neural networks; Internet tools; virtual environments; high-speed information systems. |

introduction of computers and their use in automation; computerized numerical control (CNC); adaptive control (AC); industrial robots; and computer-integrated manufacturing (CIM) systems, including computer-aided design, engineering, and manufacturing (CAD/CAE/CAM).

### 14.2.2 Goals of automation

Automation has the following primary goals:

a) **Integrate** various aspects of manufacturing operations so as to improve product quality and uniformity, minimize cycle times and effort involved, and, thus, reduce labor costs.

b) **Improve productivity** by reducing manufacturing costs through better control of production: Raw materials and parts are loaded, fed, and unloaded on machines faster and more efficiently; machines are used more effectively; and production is organized more effectively.

c) **Improve quality** by improving the repeatability of manufacturing processes.

d) **Reduce human involvement**, boredom, and thus the possibility of human error.

e) **Reduce workpiece damage** caused by manual handling of parts.

f) **Economize on floor space** in the manufacturing plant by arranging machines, material movement, and auxiliary equipment more efficiently.

g) **Raise the level of safety** for personnel, especially under hazardous working conditions.

**Automation and production quantity.** Production quantity is crucial in determining the type of machinery and the level of automation required to produce parts economically. Let's first define some basic production terms. **Total production quantity** is defined as the total number of parts to be made; this quantity can be produced in individual batches of various **lot sizes**. Lot size greatly influences the economics of production, as described in Chapter 16. **Production rate** is defined as the number of parts produced per unit time—for example, per day, per month, or per year. The approximate and generally accepted ranges of production volumes are shown in Table 14.2 for some typical applications. As expected, **experimental** and **prototype** products represent the lowest volume. (See also Section 10.12.)

Small quantities per year can be manufactured in **job shops** (Fig. 14.2), using various standard general-purpose machine tools (**stand-alone machines**) or **machining centers** (see Section 8.10). These operations have high part variety, meaning that different parts can be produced in a short time without extensive changes in tooling and in operations. On the other hand, machinery in job shops generally requires skilled labor, and the production quantity and rate are low; as a result, the cost per part can

**TABLE 14.2**

| Approximate Annual Volume of Production | | |
|---|---|---|
| Type of production | Number produced | Typical products |
| Experimental or prototype | 1–10 | All types |
| Piece or small batch | <5000 | Aircraft, machine tools, dies |
| Batch or high volume | 5000–100,000 | Trucks, agricultural machinery, jet engines, diesel engines, orthopedic devices |
| Mass production | 100,000+ | Automobiles, appliances, fasteners, bottles, food and beverage containers |

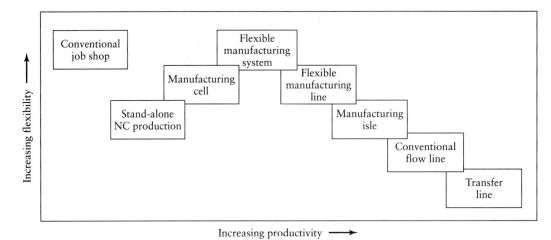

**FIGURE 14.2**  Flexibility and productivity of various manufacturing systems. Note the overlap between the systems, which is due to the various levels of automation and computer control that are possible in each group. See also Chapter 15 for more details. *Source*: U. Rembold et al., *Computer Integrated Manufacturing and Engineering*, Addison-Wesley, 1993.

be high (Fig. 14.3). When products involve a large labor component, their production is referred to as **labor intensive**. (See also Section 14.12.)

**Piece-part production** usually involves very small quantities and is suitable for job shops. The majority of piece-part production is in lot sizes of 50 or less. Quantities for **small-batch production** typically range from 10 to 100, using general-purpose machines and machining centers with various computer controls. **Batch production** usually involves lot sizes between 100 and 5000; it utilizes machinery similar to that used for small-batch production, but with specially designed fixtures for higher production rates.

**Mass production** often involves quantities over 100,000; it requires special-purpose machinery (**dedicated machines**) and automated equipment for transferring materials and parts. Although the machinery, equipment, and specialized tooling are expensive, both the labor skills required and the labor costs are relatively low, because of the high level of automation. However, these production systems are organized for a specific type of product and, hence, lack flexibility. Most manufacturing facilities operate with a variety of machines in combination and with various levels of automation and computer controls.

**FIGURE 14.3**  General characteristics of three types of production methods: job shop, batch production, and mass production.

## 14.2.3 Applications of automation

Automation can be applied to the manufacture of all types of goods, from raw materials to finished products, and in all types of production, from job shops to large manufacturing facilities. The decision to automate a new or existing facility requires that the following additional considerations be taken into account:

- Type of product manufactured;
- Quantity and the rate of production required;
- Particular phase of the manufacturing operation to be automated;
- Level of skill in the available workforce;
- Reliability and maintenance problems that may be associated with automated systems;
- Economics.

Because automation generally involves a high initial cost of equipment and requires a knowledge of the principles of operation and maintenance, a decision about the implementation of even low levels of automation must involve a study of the true needs of an organization. It is not unusual for a company to begin the implementation of automation with great enthusiasm and with high across-the-board goals, only to discover that the economic benefits of automation largely were illusory rather than real and that, in the final assessment, automation was not cost effective. In many situations, **selective automation**, rather than *total* automation, of a facility is desirable. Generally, the higher the level of skill available in the workforce, the lower is the need for automation, provided that labor costs are justified and that there is a sufficient number of workers available. Conversely, if a manufacturing facility is already automated, the skill level required is lower.

## 14.2.4 Hard automation

In *hard automation*, also called **fixed-position automation**, the production machines are designed to produce a standardized product, such as an engine block, a valve, a gear, or a spindle. Although product size and processing parameters (such as speed, feed, and depth of cut in machining) can be changed, these machines are specialized and thus lack flexibility. They cannot be modified to any significant extent to accommodate a variety of products with different shapes and dimensions. (See also the discussion of *group technology* in Section 15.8.) Because these machines are expensive to design and construct, their economical use requires the production of parts in very large quantities.

The machines that are used in hard-automation applications are usually fabricated on the **building-block (modular) principle**; they are generally called **transfer machines** and consist of two major components: power-head production units and transfer mechanisms. **Power-head production units** consist of a frame or bed, electric drive motors, gearboxes, and tool spindles and are self-contained; their individual components are commercially available in various standard sizes and capacities. Because of their intrinsic modularity, these units can easily be regrouped to produce a different part and thus have some adaptability and flexibility. Transfer machines consisting of two or more power-head units can be arranged on the shop floor in *linear, circular,* or U *patterns*. The weight and shape of the workpieces influence the arrangement selected. The arrangement is also important for continuity of operation in the event of tool failure or machine breakdown in one or more of the units. **Buffer storage** features are often incorporated in these systems to permit continued operation in the case of such an event.

**FIGURE 14.4** Two types of transfer mechanisms: (a) straight, and (b) circular patterns.

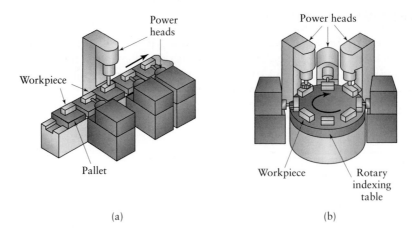

(a)  (b)

**Transfer mechanisms and transfer lines** are used to move the workpiece from one station to another in the machine, or from one machine to another, in order to enable various operations to be performed on the part. Workpieces are transferred by several methods: (1) *rails* along which the parts, usually placed on pallets, are pushed or pulled by various mechanisms (Fig. 14.4a); (2) *rotary indexing tables* (Fig. 14.4b); and (3) *overhead conveyors*. Transfer of parts from station to station is usually controlled by sensors and other devices. Tools on transfer machines can be easily changed in toolholders that have quick-change features. These machines can also be equipped with various automatic gaging and inspection systems; these systems are used between consecutive operations to ensure that a part produced in one station is within acceptable dimensional tolerances before it is transferred to the next station. (As described in Section 14.10, transfer machines are also used extensively in automatic assembly.)

Figure 14.5 shows the **transfer lines**, or **flow lines**, in a very large system for producing cylinder heads for engine blocks, consisting of a number of transfer machines. This system is capable of producing 100 cylinder heads per hour. Note the various machining operations performed: milling, drilling, reaming, boring, tapping, and honing, as well as washing, and gaging.

**FIGURE 14.5** A traditional transfer line for producing engine blocks and cylinder heads. *Source*: Ford Motor Company.

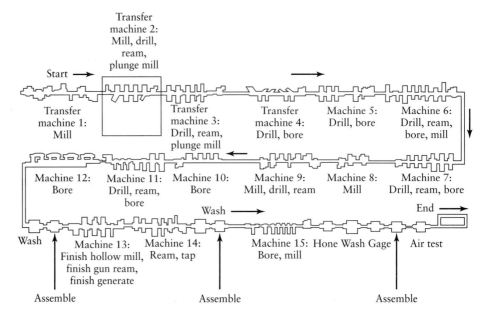

### 14.2.5  Soft automation

In contrast to hard automation, *soft automation*, also called *flexible automation* and *programmable automation*, has greater flexibility through the use of computer control of the machine and its functions by various programs; details are given in Sections 14.3 and 14.4. The machine can be easily and readily reprogrammed to produce a part that has a shape or dimensions different than those of the one produced just before it; because of this characteristic, soft automation can produce parts that have complex shapes. Advances in flexible automation have led to the development of **flexible manufacturing systems** (Section 15.10), which have high levels of efficiency and productivity.

### 14.2.6  Programmable controllers

The control of a manufacturing process in the proper sequence, especially a process that involves groups of machines and material-handling equipment, has traditionally been performed by timers, switches, relays, counters, and similar hard-wired devices based on mechanical, electromechanical, and pneumatic principles. Beginning in 1968, **programmable logic controllers** (PLCs) were introduced to replace these devices. Because PLCs eliminate the need for relay control panels and because they can be reprogrammed and take up less space than relay control panels, they have been widely adopted in manufacturing systems and operations. Their basic functions are (a) on–off motion, (b) sequential operations, and (c) feedback control. PLCs are also used in system control, with high-speed digital-processing and communication capabilities; they perform reliably in industrial environments and improve the overall efficiency of the operation.

Because of advances in numerical control machines, PLCs have become less popular in new installations, but they still represent a very large existing installation base. There is a growing trend toward using microcomputers instead of PLCs, because the former are less expensive, easier to program, and easy to network. This advance has been made possible by the new breeds of "real-time" operating systems such as Windows NT and SCADA (Supervisory Control and Data Acquisition Software).

### 14.2.7  Total productive maintenance

The management and maintenance of a wide variety of machines, equipment, and systems are among the significant aspects that affect the productivity of a manufacturing organization. Consequently, the concepts of *total productive maintenance* (TPM) and *total productive equipment management* (TPEM) are now being advanced. These concepts include continued analysis of such factors as (a) equipment breakdown and equipment problems; (b) monitoring and improving of equipment productivity; (c) implementation of preventive and predictive maintenance; (d) reduction of setup time, idle time, and cycle time; (d) full utilization of machinery and equipment and the improvement of their effectiveness; and (e) reduction of product defects. Teamwork— for example, as implemented by continuous improvement action teams—is an important component of this activity and involves the full cooperation of machine operators, maintenance personnel, engineers, and management of the organization.

# 14.3 | Numerical Control

*Numerical control* (NC) is a method of controlling the movements of machine components by directly inserting coded instructions, in the form of numbers and letters, into the system. The system automatically interprets these data and converts them to output signals. These signals, in turn, control various machine components, such as

(a) turning spindles on and off, (b) changing tools, (c) moving the workpiece or the tools along specific paths, and (d) turning cutting fluids on and off.

In order to appreciate the importance of numerical control of machines, let's briefly review how a process such as machining has traditionally been carried out. After studying the working drawings of a part, the operator sets up the appropriate process parameters, determines the sequence of the machining operations to be performed, clamps the workpiece in a workholding device (such as a chuck or collet), and proceeds with the fabrication of the part. Depending on the part's shape and on the dimensional accuracy specified, this approach usually requires skilled operators. The machining procedure followed may depend on the particular operator; because of the possibilities of human error, even parts produced by the same operator may not all be identical. Part quality may, therefore, depend on the particular operator—or, even with the same operator, on the hour of the day or on the day of the week. Because of increased concern with improving product quality and reducing manufacturing costs, such variability and its adverse effects on product quality are no longer acceptable; this situation can be eliminated by numerical control of the machining operation.

The importance of numerical control can be further illustrated by the following example: Assume that several holes are to be drilled on a part in the positions shown in Fig. 14.6. In the traditional manual method of machining this part, the operator positions the drill bit with respect to the workpiece, using reference points given by any of the three methods shown in the figure. The operator then proceeds to drill the holes. Let's first assume that 100 parts, all having exactly the same shape and dimensional accuracy, are to be drilled. Obviously, this operation is going to be tedious, because the operator has to go through the same motions repeatedly. Moreover, the probability is high that, for a variety of reasons, some of the parts machined will be different than others. Let's now assume that during this production run, the order for these parts is changed, and 10 of the parts now require holes in different positions. The machinist now has to reposition the worktable; this operation will thus be time consuming and is also subject to error.

Such operations can be performed easily by numerical control machines that are capable of producing parts repeatedly and accurately and of handling different parts by simply loading different part programs, as described in Section 14.4. Data concerning all aspects of the machining operation can now be stored on hard disks; specific information is then relayed to the machine tool's control panel. On the basis of input information, relays and other devices can be actuated to obtain a desired machine setup. Complex operations, such as turning a part that has various contours or die sinking in a milling machine, can be carried out easily.

### 14.3.1 Computer numerical control

In the next step in the development of numerical control, the control hardware (mounted on the NC machine) was converted to *local* computer control by software. Two types of computerized systems have been developed: direct numerical control and computer numerical control.

**FIGURE 14.6** Positions of drilled holes in a workpiece. Three methods of measurements are shown: (a) absolute dimensioning, referenced from one point at the lower left of the part; (b) incremental dimensioning, made sequentially from one hole to another; and (c) mixed dimensioning, a combination of both methods.

(a)                    (b)                    (c)

In **direct numerical control** (DNC), several machines are directly controlled, step by step, by a central mainframe computer. In this system, the operator has access to the central computer through a remote terminal, and the status of all machines in a manufacturing facility can be monitored and assessed from the central computer. However, DNC has a crucial disadvantage: If the computer shuts down, all the machines become inoperative. **Distributed numerical control** covers the use of a central computer serving as the control system over a number of individual computer numerical control machines that have onboard microcomputers. This system provides large memory and computational capabilities and offers flexibility while overcoming the disadvantage of direct numerical control.

*Computer numerical control* is a system in which a control microcomputer is an integral part of a machine or piece of equipment (onboard computer). The part program may be prepared at a remote site by the programmer, and it may incorporate information obtained from drafting software packages and from machining simulations, in order to ensure that the part program is bug free. The machine operator can, however, easily and manually program onboard computers; the operator can modify the programs directly, prepare programs for different parts, and store the programs. Because of the availability of small computers that have a large memory, microprocessor(s), and program-editing capabilities, as well as the increased flexibility, accuracy, and versatility of such computers, CNC systems are widely used today.

## 14.3.2 Principles of NC machines

Fig. 14.7 depicts the basic elements and operation of a typical NC machine. The functional elements involved are:

a) **Data input:** The numerical information is read and stored in computer memory.

b) **Data processing:** The programs are read into the machine control unit for processing.

c) **Data output:** This information is translated into commands (typically pulsed commands) to the servomotor (Fig. 14.8). The servomotor then moves the table (onto which the workpiece is mounted) to specific positions (through linear or rotary movements) by means of stepping motors, lead screws, and other devices.

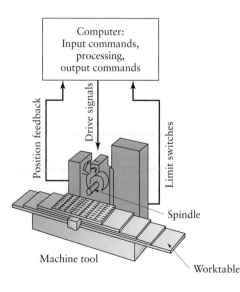

**FIGURE 14.7**  Schematic illustration of the major components of a numerical control machine tool.

Computer:
Input commands,
processing,
output commands

Position feedback

Drive signals

Limit switches

Spindle

Machine tool

Worktable

**FIGURE 14.8** Schematic illustration of the components of (a) an open-loop, and (b) a closed-loop control system for a numerical control machine. DAC means digital-to-analog converter.

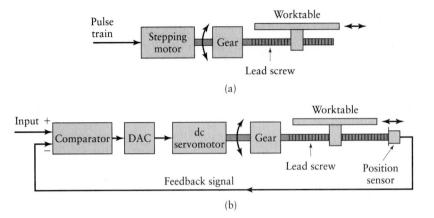

(a)

(b)

**Types of control circuits.** In the **open-loop** system (Fig. 14.8a), the signals are sent to the servomotor by the controller, but the movements and final positions of the worktable are not checked for accuracy. In contrast, the **closed-loop** system (Fig. 14.8b) is equipped with various transducers, sensors, and counters that accurately measure the position of the worktable. Through **feedback control**, the position of the worktable is compared against the signal; table movements terminate when the proper coordinates are reached. The closed-loop system is more complicated and more expensive than the open-loop system.

Position measurement in NC machines can be accomplished through direct or indirect methods. In *direct measuring systems*, a sensing device reads a graduated scale on the machine table or slide for linear movement (Fig. 14.9a). This system is more accurate than those used for indirect methods, because the scale is built into the machine, and backlash (the play between two adjacent mating gear teeth) in the mechanisms is not significant. In *indirect measuring systems*, **rotary encoders** or **resolvers** (Figs. 14.9b and c) convert rotary movement to translation movement. In this system, backlash can significantly affect measurement accuracy. Position feedback mechanisms use various sensors that are based mainly on magnetic and photoelectric principles.

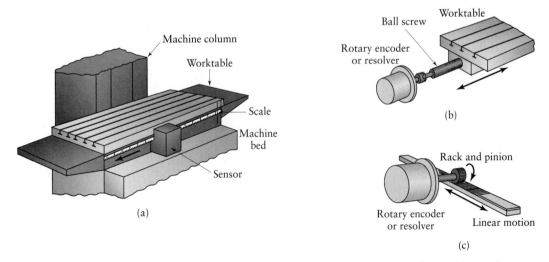

**FIGURE 14.9** (a) Direct measurement of the linear displacement of a machine-tool worktable. (b) and (c) Indirect measurement methods.

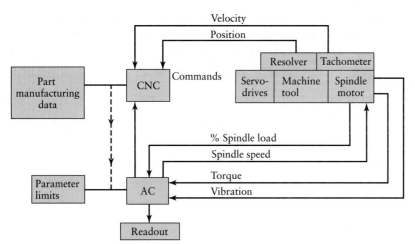

**FIGURE 14.13** Schematic illustration of the application of adaptive control (AC) for a turning operation. The system monitors such parameters as cutting force, torque, and vibrations; if they are excessive, it modifies process variables such as feed and depth of cut to bring them back to acceptable levels.

particular manufacturing process. In CNC machines, these parameters are held constant during a particular process cycle. In AC, on the other hand, the system is capable of automatic adjustments during processing, through closed-loop feedback control (Fig. 14.13).

**Principles and applications of adaptive control.**  The following are the basic functions common to adaptive-control systems:

1. Determine the operating conditions of the process, including measures of performance. This information is typically obtained by using sensors that measure process parameters, such as force, torque, vibration, and temperature.

2. Configure the process control in response to the operating conditions. Large changes in the operating conditions may provoke a decision to make a major switch in control strategy. More modest alterations may include the modification of process parameters, such as the speed of operation or of the feed in machining.

3. Continue to monitor the process, making further changes in the controller as needed.

In an operation such as turning on a lathe (Section 8.8.2), the adaptive-control system senses real-time cutting forces, torque, temperature, tool-wear rate, tool chipping or tool fracture, and surface finish of the workpiece. The system then converts this information into commands that modify the process parameters on the machine tool to hold the parameters constant (or within certain limits) or to optimize the cutting operation.

Those systems that place a constraint on a process variable (such as forces, torque, or temperature) are called **adaptive-control constraint** (ACC) systems. Thus, if the thrust force and the cutting force (and hence the torque) increase excessively (because, for example, of the presence of a hard region in a casting), the adaptive-control system changes the speed or the feed to lower the cutting force to an acceptable level (Fig. 14.14). Without adaptive control or without the direct intervention of the operator (as in traditional machining operations), high cutting forces may cause the tools to chip or break or cause the workpiece to deflect or distort excessively. As a result, the dimensional accuracy and surface finish deteriorate. Those systems that optimize an operation are called **adaptive-control optimization** (ACO) systems. Optimization may involve maximizing material-removal rate between tool changes (or resharpening) or improving surface finish. Currently, most systems are based on ACC, because the development and the proper implementation of ACO is complex.

*Response time* must be short for AC to be effective, particularly in high-speed operations (see Section 8.8.2). Assume, for example, that a turning operation is being

**FIGURE 14.14** An example of adaptive control in milling. As the depth of cut or the width of cut increases, the cutting forces and the torque increase. The system senses this increase and automatically reduces the feed to avoid excessive forces or tool breakage, in order to maintain cutting efficiency. *Source*: Y. Koren.

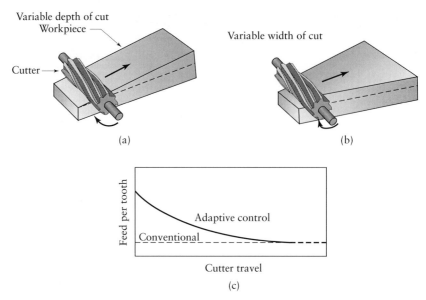

performed on a lathe at a spindle speed of 1000 rpm, and the tool suddenly breaks, adversely affecting the surface finish and dimensional accuracy of the part. In order for the AC system to be effective, the sensing system must respond within a very short time, as otherwise the damage to the workpiece will be extensive.

For adaptive control to be effective in manufacturing operations, *quantitative relationships* must be established and stored in the computer software as mathematical models. For example, if the tool-wear rate in a machining operation is excessive, the computer must be able to decide how much of a change in speed or feed is necessary (and whether to increase it or decrease it) in order to reduce the wear rate to an acceptable level. The system should also be able to compensate for dimensional changes in the workpiece due to such causes as tool wear and temperature rise (Fig. 14.15). If the operation is, for example, grinding (Chapter 9), the computer software must reflect the desired quantitative relationships among process variables (wheel and work speeds, feed, type of wheel) and such parameters as wheel wear, dulling of abrasive grains, grinding forces, temperature, surface finish, and part deflections. Similarly, for the bending of a sheet in a V-die (Section 7.4), data on the dependence of springback on punch travel and on other material and process variables must be stored in the computer memory.

It is apparent that, because of the many factors involved, mathematical equations for such quantitative relationships in manufacturing processes are difficult to establish. Compared to the various other parameters involved, forces and torque in machining have been found to be the easiest to monitor by AC. Various solid-state power controls are commercially available, in which power is displayed or interfaced with data-acquisition systems.

**FIGURE 14.15** In-process inspection of workpiece diameter in a turning operation. The system automatically adjusts the radial position of the cutting tool in order to produce the correct diameter.

# 14.6 | Material Handling and Movement

During a typical manufacturing operation, raw materials and parts are moved from storage to machines, from machine to machine, from inspection to assembly and to inventory, and finally to shipment. For example, a forging is mounted on a milling-machine bed for further processing; the machined forging may subsequently be ground for better surface finish and dimensional accuracy, and it is inspected prior to being assembled into a finished product. Similarly, cutting tools are mounted on lathes, either manually or automatically; dies are placed in presses or hammers; grinding wheels are mounted on spindles; and parts are mounted on special fixtures for dimensional measurement and inspection. *Material handling* is defined as the functions and systems associated with the transportation, storage, and control of materials and parts in the total manufacturing cycle of a product.

Material handling should be repeatable and reliable. Consider, for example, what happens if a part or workpiece is loaded improperly into a die, mold, or the chuck or collet on a lathe. The consequences of such action may well be broken dies and tools or parts that are improperly made or are out of dimensional tolerance. Moreover, this action can also present safety hazards and possibly cause injury to the operator and nearby personnel.

The total amount of time required for manufacturing a part depends on the part's size, complexity of shape, surface finish and dimensional tolerances specified, and the resulting set and sequence of operations required. Depending on the particular operations performed, idle time and the time required for transporting materials can constitute the majority of the production time consumed. **Plant layout** is an important aspect of the orderly flow of materials and components throughout the manufacturing cycle. Obviously, the time and distances required for moving raw materials and parts should be minimized, and storage areas and service centers should be organized accordingly. For parts requiring multiple operations, equipment should be grouped around the operator or the industrial robot. (See the discussion of *cellular manufacturing* in Section 15.9.) Material handling should, therefore, be an integral part of the planning, implementing, and controlling of manufacturing operations.

Important aspects of material handling are discussed as follows:

a) **Methods of material handling.** Several factors have to be considered in selecting a suitable material-handling method for a particular manufacturing operation:

1. Shape, weight, and characteristics of the parts;
2. Types and distances of movements, and the position and orientation of the parts during movement and at their final destination;
3. Conditions of the path along which the parts are to be transported;
4. Level of automation and control desired, and integration with other systems and equipment;
5. Operator skill required;
6. Economic considerations.

For small batch operations, raw materials and parts can be handled and transported by hand, but this method can be time consuming and hence costly. Furthermore, because it involves human beings, this practice can be unpredictable and unreliable; it can even be unsafe to the operator, because of the weight and shape of the parts to be moved and because of environmental factors, such as heat and smoke in older foundries and forging plants.

b) **Equipment.** Several types of equipment can be used to move materials, such as conveyors; rollers; self-powered monorails; carts; forklift trucks; and various mechanical, electrical, magnetic, pneumatic, and hydraulic devices and manipulators. **Manipulators** can be designed to be controlled directly by the operator, or they can be automated for repeated operations (such as the loading and unloading of parts from machine tools, presses, dies, and furnaces). Manipulators are capable of gripping and moving heavy parts and of orienting them as may be required in manufacturing and assembly operations. Machinery combinations that have the capability of conveying parts without the use of additional material-handling equipment are called **integral transfer devices.**

In order to use warehouse space efficiently and to reduce labor costs, **automated storage/retrieval systems** (AS/RS) have been developed. There are several such systems available, with varying degrees of automation and complexity. These systems are currently considered undesirable, because of the current focus on minimal or zero inventory and on just-in-time production methods (see Section 15.11).

c) **Automated guided vehicles.** Flexible material handling and movement, with real-time control, has become an integral part of modern manufacturing. *Automated guided vehicles* (AGVs) are used extensively in flexible manufacturing (Fig. 14.16); this transport system has high flexibility and is capable of random delivery to different workstations. The movements of AGVs can be planned so that they interface with automated storage/retrieval systems.

First developed in the 1950s, AGVs were put into extensive use in the 1970s to reduce direct-labor costs and improve productivity by automation. Although similar in their basic design and operation, there are several types of AGVs commonly in use, with a variety of designs, load-carrying capacities, and features for specific applications. Among these types of AGVs are (a) unit-load vehicles with pallets that can be handled from both sides (see Fig. 14.16); (b) light-load vehicles, equipped with bins or trays for light manufacturing; (c) tow vehicles; (d) forklift-style vehicles; and (e) assembly-line vehicles to carry

**FIGURE 14.16** A self-guided vehicle (Caterpillar Model SGC-M) carrying a machining pallet. The vehicle is aligned next to a stand on the floor. Instead of following a wire or stripe path on the factory floor, this vehicle calculates its own path and automatically corrects for any deviations. *Source*: Courtesy of Caterpillar Industrial, Inc.

subassemblies to final assembly. Pallet sizes vary, and loading and unloading are accomplished either manually or automatically, by using various transfer mechanisms. The load capacity of AGVs can range up to thousands of pounds.

AGVs are guided automatically along pathways with in-floor wiring (for magnetic guidance) or tapes or fluorescent painted strips (for optical guidance, called *chemical guide path*). Some systems may require additional operator guidance. More recent developments include *autonomous guidance* (without any wiring or tapes) and use of various optical, ultrasonic, inertial, and dead-reckoning techniques with onboard controllers. *Routing* of the AGV can be controlled such that the system optimizes the movement of materials and parts in case of congestion around workstations, machine breakdown, or the failure of one section of the manufacturing system. Proper traffic management on the factory floor is thus important; sensors and various controls on the vehicles are designed to ensure that collisions with other AGVs or machines do not occur.

d) **Coding systems.** Various *coding systems* have been developed to locate and identify parts and subassemblies throughout the manufacturing system and to correctly transfer them to their appropriate stations:

1. **Bar coding** is the most widely used and least costly system. The codes are printed on labels, which are attached to the parts themselves and read by fixed or portable bar code readers, using light pens.
2. **Magnetic strips,** such as those on the back of a credit card, constitute the second most common coding system.
3. The third system uses **RF** (radio frequency) **tags.** Although expensive, the tags do not require the clear line of sight needed by the two systems discussed previously; in addition, they have a long range and are rewritable.

Other identification systems are based on acoustic waves, optical character recognition, and machine vision. (See Section 14.8.1.)

# 14.7 | Industrial Robots

The word *robot* was coined in 1920 by the Czech author K. Čapek in his play *R.U.R.* (Rossum's Universal Robots); it is derived from the Czech word *robota*, meaning worker. An *industrial robot* has been defined as a reprogrammable multifunctional manipulator designed to move materials, parts, tools, or other devices by means of variable programmed motions and to perform a variety of other tasks. In a broader context, the term *robot* also includes manipulators that are activated directly by an operator. Introduced in the early 1960s, the first industrial robots were used in hazardous operations, such as the handling of toxic and radioactive materials and the loading and unloading of hot workpieces from furnaces and in foundries. Some simple rule-of-thumb applications for robots are the three D's (*dull, dirty, and dangerous, including demeaning, but necessary, tasks*) and the three H's (*hot, heavy, and hazardous.*) Industrial robots have become important components in manufacturing operations and systems and have greatly helped improve productivity and product quality and reduce labor costs.

## 14.7.1 Robot components

To appreciate the functions of robot components and their capabilities, we might simultaneously observe the flexibility and capability of the diverse movements of our arm, wrist, hand, and fingers in reaching for and grabbing an object from a shelf, in using a hand tool, or in operating a car or a machine. The basic components of an industrial robot are described as follows (Fig. 14.17a):

**FIGURE 14.17** (a) Schematic of a six-axis S-10 GMF robot. The payload at the wrist is 10 kg (22 lb.) and repeatability is ±0.2 mm (±0.008 in.) The robot has mechanical brakes on all its axes, which are coupled directly. (b) The work envelope of a robot, as viewed from the side. *Source*: GMFanuc Robotics Corporation.

(a)  (b)

1. **Manipulator.** Also called **arm and wrist,** the *manipulator* is a mechanical unit that provides motions (trajectories) similar to those of a human arm and hand; manipulation is carried out using devices such as linkages, gears, and various joints. The end of the wrist can reach a point in space with a specific set of coordinates and in a specific orientation. Most robots have six rotational joints. (See Fig. 14.17a.) There are also four-degrees-of-freedom (d.o.f.) and five-d.o.f. robots, but these kinds are not fully dexterous, because full dexterity, by definition, requires six d.o.f. Seven-d.o.f. (or *redundant*) robots for special applications are also available.

2. **End Effector.** The end of the wrist in a robot is equipped with an *end effector,* also called **end-of-arm tooling.** End effectors are generally custom made to meet special handling requirements. Mechanical grippers are the most commonly used end effectors and are equipped with two or more fingers. The selection of an appropriate end effector for a specific application depends on such factors as the *payload* (weight of the object to be lifted and moved; load-carrying capacity), environment, reliability, and cost. Depending on the type of operation, conventional end effectors may be equipped with any of the following:

   a) Grippers, hooks, scoops, electromagnets, vacuum cups, and adhesive fingers, for material handling (Fig. 14.18a);

   b) Spray guns, for painting;

   c) Various attachments, such as for spot and arc welding and for arc cutting;

   d) Power tools, such as drills, nut drivers, and burrs;

   e) Measuring instruments, such as dial indicators.

   **Compliant end effectors** are used to handle fragile materials or to facilitate assembly. They can use elastic mechanisms to limit the force that can be applied to a workpiece or part, and they can be designed with a specific required stiffness. For example, the end effector shown in Fig. 14.18b is stiff in the axial direction, but it is very compliant in lateral directions; this arrangement prevents damage to parts in those assembly operations in which slight misalignments can occur.

(a)

**FIGURE 14.18**
(a) Various devices and tools attached to end effectors to perform a variety of operations. (b) A system that compensates for misalignment during automated assembly. *Source*: ATI Industrial Automation.

(b)

3. **Power supply.** Each motion of the manipulator (in linear and rotational axes) is controlled and regulated by independent actuators that use an electrical, pneumatic, or hydraulic power supply. Each source of energy and each type of motor has its own characteristics, advantages, and limitations.

4. **Control system.** Whereas manipulators and end effectors are the robot's arms and hands, the control system is the *brain* of a robot. Also known as the **controller**, the control system is the communications and information-processing system that gives commands for the movements of the robot; it stores data to initiate and terminate movements of the manipulator. It also acts as the *nerves* of the robot; it interfaces with computers and other equipment such as manufacturing cells or assembly systems. **Feedback devices,** such as transducers, are an important part of the control system. Robots with a *fixed* set of motions have **open-loop control.** In this system, commands are given, and the robot arm goes through its motions; unlike in **closed-loop systems,** where feedback is given, accuracy of the movements is not monitored; consequently, this system does not have a self-correcting capability.

As in numerical control machines, the types of control in industrial robots are *point to point* and *continuous path*. (See also Section 14.3.3.) Depending on the particular task, the *positioning repeatability* required may be as small as 0.050 mm (0.002 in.), as in assembly operations for electronic printed circuitry. Specialized robots can reach such accuracy, although most robots are unable to do so. Accuracy and repeatability vary greatly with payload and with position within the *work envelope* (see Section 14.7.2) and, as such, are very difficult to quantify for most robots.

### 14.7.2 Classification of robots

Robots may be classified by basic type, as follows (Fig. 14.19):

a)  **Cartesian**, or **rectilinear**;

b)  **Cylindrical**;

c)  **Spherical**, or **polar**;

d)  **Articulated**, or **revolute, jointed**, or **anthropomorphic**.

Also, robots may be attached permanently to the floor of a manufacturing plant, move along overhead rails (**gantry robots**), or be equipped with wheels to move along the factory floor (**mobile robots**).

1.  **Fixed- and variable-sequence robots.** The *fixed-sequence robot* (also called a **pick-and-place robot**) is programmed for a specific sequence of operations. Its movements are from point to point, and the cycle is repeated continuously; these robots are simple and relatively inexpensive. The *variable-sequence robot* can be programmed for a specific sequence of operations, but can be reprogrammed to perform another sequence of operations.

2.  **Playback robot.** An operator leads or walks the *playback robot* and its end effector through the desired path; in other words, the operator *teaches* the robot by showing it what to do. The robot memorizes and records the path and sequence of motions and can then repeat them continually without any further action or guidance by the operator. The **teach pendant** uses handheld button boxes that are connected to the control panel; these button boxes are used to control and guide the robot and its tooling through the work to be performed. These movements are then registered in the memory of the controller and are automatically reenacted by the robot whenever needed.

3.  **Numerically controlled robot.** The *numerically controlled robot* is programmed and operated much like a numerically controlled machine; the robot is servocontrolled by digital data, and its sequence of movements can be changed with relative ease. As in NC machines, there are two basic types of controls: point to point and continuous path. Point-to-point robots are easy to program and have a higher payload and a larger **work envelope**, also called the **working**

**FIGURE 14.19**  Four types of industrial robots: (a) Cartesian (rectilinear); (b) cylindrical; (c) spherical (polar); and (d) articulated (revolute, jointed, or anthropomorphic).

(a)                (b)                (c)                (d)

(a) Rectangular

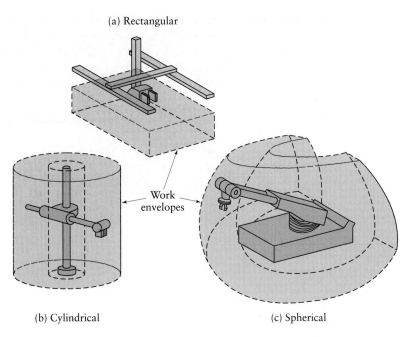

Work envelopes

(b) Cylindrical

(c) Spherical

**FIGURE 14.20** Work envelopes for three types of robots. The choice depends on the particular application (See also Fig. 14.17b.)

envelope (the maximum extent or reach of the robot hand or working tool in all directions; see Figs. 14.17b and 14.20). Continuous-path robots have a higher accuracy than point-to-point robots, but they have a lower payload. More advanced robots have a complex system of path control, enabling high-speed movements with high accuracy.

4. **Intelligent (sensory) robot.** The *intelligent robot* is capable of performing some of the functions and tasks carried out by human beings; it is equipped with a variety of sensors with *visual* (*computer vision*) and *tactile* (touching) capabilities. (See Section 14.8.) Much like humans, the robot observes and evaluates the immediate environment and its proximity to other objects, especially machinery in its path, by *perception* and *pattern recognition*; it then makes appropriate decisions for the next movement and proceeds accordingly. Because its operation is very complex, powerful computers are required to control this type of robot. Developments in intelligent robots include capabilities so that they will:

a) Behave more and more like humans, performing tasks such as moving among a variety of machines and equipment on the shop floor and avoiding collisions;

b) Recognize, select, and properly grip the correct raw material or workpiece;

c) Transport a part to a machine for further processing or inspection;

d) Assemble components into subassemblies or the final product.

## 14.7.3 Applications and selection of robots

Major applications of industrial robots include the following:

1. *Material handling* consists of the loading, unloading, and transferring of workpieces in manufacturing facilities. These operations can be performed reliably and repeatedly with robots, thereby improving product quality and reducing scrap losses. Some examples of material-handling applications are (a) casting and molding operations, in which molten metal, raw materials, lubricants, and parts in various stages of completion are handled without operator interference;

**FIGURE 14.21** Spot welding automobile bodies with industrial robots. *Source*: Courtesy of Cincinnati Milacron, Inc.

(b) heat treating, in which parts are loaded and unloaded from furnaces and quench baths; and (c) forming operations, in which parts are loaded and unloaded from presses and various other types of machinery.

2. *Spot welding* by robots unitizes automobile and truck bodies, producing welds of good and reliable quality (Fig. 14.21); robots also perform other, similar operations, such as arc welding, arc cutting, and riveting (see Chapter 12).

3. *Finishing operations* such as deburring, grinding, and polishing (Chapter 9) can be done by using appropriate tools attached to end effectors.

4. *Application of adhesives and sealants* by industrial robots has widespread usefulness—for example, in the automobile frame shown in Fig. 14.22.

5. *Spray painting*, particularly of parts with complex shapes, and cleaning operations are frequent applications of industrial robots, because the motions for one piece repeat very accurately for the next.

6. *Automated assembly*, a very repetitive operation, is performed by industrial robots (Fig. 14.23); see also Section 14.10.

7. *Inspection and gaging* in various stages of manufacture as performed by industrial robots allows speeds that are much higher than those achieved by humans.

Factors that influence the *selection of robots* in manufacturing are as follows: (a) payload, (b) speed of movement, (c) reliability, (d) repeatability, (e) arm configuration, (f) number of degrees of freedom, (g) control system, (h) program memory, and (i) work envelope.

**FIGURE 14.22** Sealing joints of an automobile body with an industrial robot. *Source*: Courtesy of Cincinnati Milacron, Inc.

**FIGURE 14.23**
Automated assembly operations using industrial robots and circular and linear transfer lines.

Robots

Remote center compliance

Linear transfer line

Torque sensor

Visual sensing

Circular transfer line

Programmable part feeder

# 14.8 | Sensor Technology

A *sensor* is a device that produces a signal in response to its detection or measurement of a certain quantity or property, such as position, force, torque, pressure, temperature, humidity, speed, acceleration, or vibration. Traditionally, sensors, actuators, and switches have been used to set limits on the performance of machines, such as stops on machine-tool slideways to restrict worktable movements, pressure and temperature gages with automatic shutoff features, and governors on engines to prevent excessive speed of operation. Sensor technology is an important aspect of manufacturing processes and systems, and sensors are essential for data acquisition, monitoring, communication, and computer control of machines and systems (Fig. 14.24). Because they convert one quantity to another, sensors are also often referred to as **transducers**. **Analog sensors** produce a signal, such as voltage, that is proportional to the measured quantity; **digital sensors** have numeric or digital outputs that can directly be transferred to computers. **Analog-to-digital converters** (ADCs) are used for interfacing analog sensors with computers.

## 14.8.1 Sensor classification

Sensors that are of interest in manufacturing may be classified generally as follows:

a) Mechanical sensors, which measure such quantities as position, shape, velocity, force, torque, pressure, vibration, strain, and mass;

Toolholder

Onboard electronics to process signals

Chuck

Drill

Inductive transmitter

Strain gages

**FIGURE 14.24**   A tool holder equipped with thrust-force and torque sensors (*smart tool holder*), capable of continuously monitoring the cutting operation. Such tool holders are necessary for adaptive control of manufacturing operations. (See Section 14.5). *Source*: Cincinnati Milacron, Inc.

b) **Electrical sensors**, which measure voltage, current, charge, and conductivity;

c) **Magnetic sensors**, which measure magnetic field, flux, and permeability;

d) **Thermal sensors**, which measure temperature, flux, conductivity, and specific heat;

e) **Other types of sensors**, such as acoustic, ultrasonic, chemical, optical, radiation, laser, and fiber-optic sensors.

Depending on its application, a sensor may be made of metallic, nonmetallic, organic, or inorganic materials and fluids, gases, plasmas, or semiconductors. Using the special characteristics of these materials, sensors convert the quantity or property measured to analog or digital output. The operation of a common mercury thermometer, for example, is based on the difference between the thermal expansion of mercury and that of glass. Similarly, a machine part or a physical obstruction or barrier in a space can be detected by a break a beam of light, sensed by a photoelectric cell. A proximity sensor, which senses and measures the distance between it and an object or a moving member of a machine, can be based on acoustics, magnetism, capacitance, or optics. Other actuators physically touch the object and take appropriate action, usually by electromechanical means. Sensors are essential to the control of intelligent robots; they continue to be developed with capabilities that resemble those of the senses of human beings (*smart sensors*; see later).

Sensors are also classified as follows:

1. **Tactile sensing** is the continuous sensing of varying contact forces, commonly by an *array* of sensors; such a system is capable of performing within an arbitrary three-dimensional space. Fragile parts, such as glass bottles and electronic devices, can be handled by robots with *compliant (smart) end effectors*. These effectors can sense the force applied to the object being handled, using piezoelectric devices, strain gages, magnetic induction, ultrasonics, and optical systems of fiber optics and light-emitting diodes. Tactile sensors that are capable of measuring and controlling gripping forces and moments in three axes have been designed (Fig. 14.25).

**FIGURE 14.25** A robot gripper with tactile sensors. In spite of their capabilities, tactile sensors are now being used less frequently, because of their high cost and their low durability in industrial applications. *Source*: Courtesy of Lord Corporation.

The gripping force of an end effector is sensed, monitored, and controlled through closed-loop feedback devices. Compliant grippers that have force-feedback capabilities and sensory perception can, however, be complicated and require powerful computers; thus, they can be costly. *Anthropomorphic end effectors* continue to be designed to simulate the human hand and fingers and to have the capability of sensing touch, force, movement, and pattern. The ideal tactile sensor must also sense slip (such as a heavy glass slipping between your fingers), a motion that human fingers sense so readily.

2. In **visual sensing (machine vision; computer vision)**, cameras optically sense the presence (and hence also the absence) and shape of the object (Fig. 14.26). A microprocessor then processes the image, usually in less than one second; the image is then measured, and the measurements are digitized (**image recognition**). There are two basic systems of machine vision: **Linear arrays** sense only one dimension, such as the presence of an object or some feature on its surface. **Matrix arrays** sense two, or even three, dimensions and are capable of detecting, for example, a properly inserted component in a printed circuit or a properly made solder joint, known as **assembly verification**. When used in automated inspection systems (Section 4.8.3), these sensors can also detect cracks and flaws.

Machine vision is particularly suitable for parts that are otherwise inaccessible, in hostile manufacturing environments, for measuring a large number of small features, and in situations where physical contact with the part may cause damage. Examples of applications of machine vision include (a) on-line,

**FIGURE 14.26** Examples of machine vision applications. (a) In-line inspection of parts. (b) Identification of parts with various shapes, and inspection and rejection of defective parts. (c) Use of cameras to provide positional input to a robot relative to the workpiece. (d) Painting of parts that have different shapes by means of input from a camera. The system's memory allows the robot to identify the particular shape to be painted and to proceed with the correct movements of a paint spray attached to the end effector.

real-time inspection in sheet-metal stamping lines; and (b) sensors for machine tools that can sense tool offset and tool breakage, verify part placement and fixturing, and monitor surface finish; see also Fig. 14.26. Machine vision is capable of in-line identification and inspection of parts and of rejection of defective parts. With visual sensing capabilities, end effectors can pick up parts and grip them in the proper orientation and location.

3. **Smart sensors** have the capability to perform a logic function, to conduct two-way communication, and to make decisions and take appropriate actions. The necessary input, and the knowledge required to make decisions, can be built into a smart sensor. For example, a computer chip with sensors can be programmed to turn a machine tool off in the event that a cutting tool fails. Likewise, a smart sensor can stop a mobile robot or a robot arm from accidentally coming in contact with an object or people by sensing quantities such as distance, heat, and noise.

The *selection* of a sensor for a particular application depends on such factors as (a) the particular quantity to be measured or sensed, (b) the sensor's interaction with other components in the system, (c) its expected service life, (d) the required level of its performance, (e) the difficulties associated with use of the sensor, and (f) its cost. Another important consideration is the type of environment in which the sensors are to be used. Rugged sensors have been developed to withstand extremes of temperature, shock and vibration, humidity, corrosion, dust and various contaminants, fluids, electromagnetic radiation, and other interferences. Industrial environments, in particular, require robust sensors. (See also the discussion of *robust design* in Section 16.2.3.)

## 14.8.2 Sensor fusion

*Sensor fusion* basically involves the integration of multiple sensors in such a manner that the individual data from each of the sensors (such as force, vibration, temperature, and dimensions) are combined to provide a higher level of information and reliability. It has been suggested that an example of sensor fusion occurs when someone drinks from a cup of hot tea or coffee; although we take such a common practice for granted, it can readily be seen that this process involves data input from the person's eyes, hands, lips, and tongue. Through our various senses (sight, hearing, smell, taste, and touch), there is real-time monitoring of temperature, movements, and positions. Thus, for example, if the fluid is too hot, the hand movement of the cup toward the lip is controlled accordingly.

The earliest applications of sensor fusion were in robot-movement control and in missile-flight tracking and similar military applications, primarily because these activities involve movements that mimic human behavior. In manufacturing, an example of sensor fusion is a machining operation in which a set of different, but integrated, sensors continuously monitors (a) the dimensions and surface finish of the workpiece; (b) cutting-tool forces, vibrations, wear, and fracture; (c) the temperatures in various regions of the tool–workpiece system; and (d) the spindle power.

An essential aspect in sensor fusion is **sensor validation**, in which the failure of one particular sensor is detected so that the control system retains high reliability. For this application, the receipt of redundant data from different sensors is essential. It can be seen that the receipt, integration, and processing of all data from various sensors can be a complex problem. With advances in sensor size, quality, and technology, and with new developments in computer-control systems, artificial intelligence, expert systems, and artificial neural networks (see Chapter 15), sensor fusion is becoming practical and available at relatively low cost.

# 14.9 | Flexible Fixturing

In describing workholding devices for the manufacturing operations covered in Chapters 6 through 9, the words *fixture*, *clamp*, and *jig* are often used interchangeably and sometimes in pairs, such as in *jigs and fixtures*. Common workholding devices include chucks, collets, and mandrels, many of which are usually operated manually. Other workholding devices are designed and operated at various levels of mechanization and automation, such as *power chucks*, which are driven by mechanical, hydraulic, or electrical means. **Fixtures** are generally designed for specific purposes; **clamps** are simple multifunctional devices; **jigs** have various reference surfaces and points for accurate alignment of parts and tools and are widely used in mass production (see also the discussion of *pallets* in Section 8.10). These devices may be used for actual manufacturing operations (in which case the forces exerted on the part must maintain the part's position in the machine without slipping or distortion), or they may be used to hold workpieces for purposes of measurement and inspection, where the part is not subjected to any forces.

Workholding devices have certain ranges of capacity; for example, (a) a particular collet can accommodate bars only within a certain range of diameters; (b) four-jaw chucks can accommodate square or prismatic workpieces that have certain dimensions; and (c) various other devices and fixtures are designed and made for specific workpiece shapes and dimensions and for specific tasks—such devices are called **dedicated fixtures**. It is very simple to reliably fixture a workpiece in the shape of, say, a brick, such as by clamping it between the parallel jaws of a vise. If the part has curved surfaces, it is possible to shape the contacting surfaces of the jaws themselves by machining them (*machinable jaws*) to conform to the workpiece surfaces.

The emergence of flexible manufacturing systems has necessitated the design and use of workholding devices and fixtures that have built-in **flexibility**. There are several methods of *flexible fixturing*, based on different principles (also called **intelligent fixturing systems**), and the term itself has been defined somewhat loosely. Basically, however, these devices are capable of quickly accommodating a range of part shapes and dimensions without the necessity of making extensive changes and adjustments or requiring operator intervention, both of which would adversely affect productivity.

A challenging example of the need for proper fixturing of a workpiece for further processing, such as grinding or deburring, is a series of turbine blades of various shapes and sizes. (See Fig. 3.1.) In addition to the use of some of the methods outlined previously, including the use of robots and computer controls for manipulating part orientation for proper clamping, other techniques for fixturing are described as follows:

1. **Modular fixturing.** *Modular fixturing* is often used for small or moderate lot sizes (Fig. 14.27), especially when the cost of dedicated fixtures and the time required to make them are difficult to justify. Complex workpieces can be located within machines through fixtures produced quickly from standard components and can be disassembled when a production run is completed. These modular fixtures are usually based on tooling plates or blocks configured with grid holes or T-slots upon which a fixture is constructed. A number of other standard components, such as locating pins, adjustable stops, workpiece supports, V-blocks, clamps, and springs, can be mounted onto the base plate or block to quickly produce a fixture. By computer-aided fixture planning for specific situations, such fixtures can be assembled and modified using robots. As compared with dedicated fixturing, modular fixturing has been shown to be

**FIGURE 14.27** Typical components of a modular workholding system. *Source*: Carr Lane Manufacturing Co.

low in cost, have a shorter lead time, provide greater ease of repair of damaged components, and offer a more intrinsic flexibility of application.

2. **Bed-of-nails device.** This fixture consists of a series of air-actuated pins that conform to the shape of the external surfaces of the part. Each pin moves as necessary to conform to the shape at its point of contact with the part; the pins are then mechanically locked against the part. The fixture is compact and has high stiffness, and it is reconfigurable.

3. **Adjustable-force clamping.** Figure 14.28 shows a schematic illustration of another flexible fixturing system. In this system, referred to as an *adjustable-force clamping* system, the strain gage attached to the clamp senses the magnitude of the clamping force; the system then adjusts this force to keep the workpiece securely clamped to the workpiece.

4. **Phase-change materials.** There are two methods that hold irregularly shaped or curved workpieces in a medium, other than hard tooling:

   a)  In the first, and older, method, a *low-melting-point metal* is used as the clamping medium. For example, an irregularly shaped workpiece is dipped

**FIGURE 14.28** Schematic illustration of an adjustable-force clamping system. The clamping force is sensed by the strain gage, and the system automatically adjusts this force. *Source*: P. K. Wright and D. A. Bourne, *Manufacturing Intelligence*, Reading, MA. Addison-Wesley, 1988.

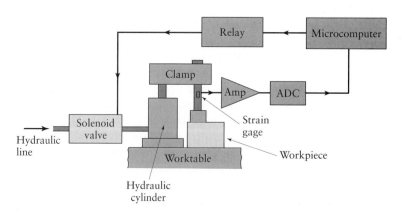

into molten lead and allowed to set (like the wooden stick in a Popsicle, a process similar to *insert molding*; see Section 10.10.2 and Fig. 10.24). After setting, the body of the solid metal itself is clamped in a simple fixture. This method is particularly common in the aerospace industry. Note, however, the possible adverse effects of such materials as lead (see the discussion of *liquid-metal embrittlement* in Section 3.4.2) on human health and the environment.

There is a similar application of this method in machining the honeycomb structures shown in Fig. 7.49. Note that because the walls of the hollow structure are very thin, the forces exerted by the cutting tool would easily distort it or damage it. One method of stiffening this structure is to fill the cavities with water and freeze it. Thus, the hexagonal cavities are now filled with ice, whose strength is sufficient to resist the cutting forces; after machining, the ice is allowed to melt away.

b)   In the second method, which is still in experimental stages, the supporting medium is a *magnetorheological* (MR) or *electrorheological* (ER) *fluid*. In the MR application, the particles, which have sizes measurable in micrometers or are nanoparticles (see Sections 3.11.9, 11.2.1, and 11.8.1), are ferromagnetic or paramagnetic in a nonmagnetic fluid; surfactants are added to maintain dispersal of powders. After the workpiece is immersed in the fluid, an external magnetic field is applied, whereby the particles are polarized and the behavior of the fluid changes from a liquid to a solid. Afterwards, the workpiece is retrieved by removing the external magnetic field. The process is particularly suitable for nonferrous metal parts. In the ER application, the fluid is a suspension of fine dielectric particles in a liquid with a low dielectric constant. After application of an electrical field, the liquid becomes a solid.

Studies continue on (a) improving the shear strength (currently around 80 kPa for MR) and yield stress (currently up to 60 kPa for ER) of the solidified fluid, (b) ensuring that the fluids do not react with the workpieces, (c) improving the speed of the operation, (d) maintaining the level of positioning accuracy that can be achieved in industrial applications, and (e) integration into CIM systems.

# 14.10 | Assembly, Disassembly, and Service

Some products are simple and have only two or three components to *assemble*; such an operation can be done with relative ease. Examples include the manufacture of an ordinary pencil with an eraser, a frying pan with a wooden handle, and an aluminum beverage can. Most products, however, consist of many parts, and their assembly requires considerable care and planning.

Traditionally, assembly has generally involved much *manual* work and, thus, has contributed significantly to product cost. The total assembly operation is usually broken into individual assembly operations (**subassemblies**), with an operator assigned to carry out each step. Assembly costs are now typically 25% to 50% of the total cost of manufacturing, with the percentage of workers involved in assembly operations ranging from 20% to 60%. In the electronics industries, for example, some 40% to 60% of total wages are paid to assembly workers. As costs increase, the necessity for **automated assembly** become obvious. Beginning with the hand assembly of muskets in the late 1700s (and the introduction of *interchangeable parts* in the early 1800s), assembly methods have been vastly improved over the years. The first application of large-scale modern assembly was for the flywheel magnetos for the Model T Ford; this activity eventually led to mass production of the automobile itself.

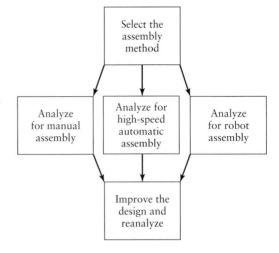

**FIGURE 14.29**   Stages in the design-for-assembly analysis. *Source*: *Product Design for Assembly*, 1989 edition, by G. Boothroyd and P. Dewhurst. Reproduced with permission.

The choice of an assembly method and system depends on the required production rate, the total quantity to be produced, the product's market life, labor availability, and cost.

As we have seen throughout this text, parts are manufactured within certain dimensional tolerance ranges. (See also Section 4.7.) Taking ball bearings as an example, we know that although they all have the same nominal dimensions, some balls in a particular lot will be smaller than others by a very small amount; likewise, some inner and outer races produced will be smaller than others in the lot. There are two methods of assembly for such high-volume products:

1. In *random assembly*, parts are put together by selecting them randomly from the lots produced.

2. In *selective assembly*, the balls and the races are segregated by groups of sizes, from smallest to largest. The parts are then selected to mate properly. Thus, the smallest diameter balls are mated with inner races that have the largest outside diameters and with outer races that have the smallest inside diameters.

### 14.10.1 Assembly systems

There are three basic methods of assembly: *manual*, *high-speed automatic*, and *robotic*. These methods can be used individually or, as is the case for most applications in practice, in combination. Before assembly operations commence, an analysis of the product design (Fig. 14.29) should be made to determine the most appropriate and economical method(s) of assembly.

a) **Manual assembly** uses simple tools and is generally economical for relatively small lots. Because of the dexterity of the human hand and fingers and their capability for feedback through various senses, workers can assemble even complex parts without much difficulty. Note, for example, that the alignment and placement of a simple square peg into a square hole, involving small clearances, can be difficult in automated assembly—yet the human hand is capable of doing this simple operation with relative ease. There can, however, be potential problems of *cumulative trauma disorders* (*carpal tunnel syndrome*) associated with this method.

b) **High-speed automated assembly** uses **transfer mechanisms** designed specially for assembly. Two examples of such assembly are shown in Fig. 14.30, in which individual assembly is carried out on products that are *indexed* for proper positioning during assembly. In **robotic assembly**, one or two general-purpose

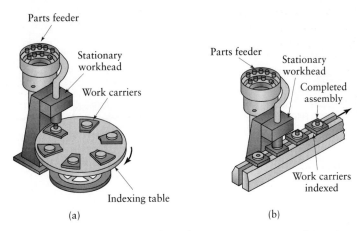

robots operate at a single workstation, or the robots operate at a multistation assembly system.

There are three basic types of assembly systems: synchronous, nonsynchronous, and continuous.

1. **In synchronous systems,** also called **indexing systems,** individual parts and components are supplied and assembled at a constant rate at fixed individual stations. The rate of movement is based on the station that takes the longest time to complete its portion of the assembly. This system is used primarily for high-volume, high-speed assembly of small products.

   Transfer systems move the partial assemblies from workstation to workstation by various mechanical means; two typical transfer systems are **rotary indexing** and **in-line indexing systems** (Fig. 14.30). These systems can operate in either a fully automatic mode or a semiautomatic mode. However, a breakdown of one station will shut down the whole assembly operation. The part feeders supply the individual parts to be assembled and place them on other components, which are secured on work carriers or fixtures. The feeders move the individual parts (by vibratory or other means) through delivery chutes and ensure their proper orientation by various ingenious means (Fig. 14.31). Proper orientation of parts and avoidance of jamming are essential in all automated assembly operations.

2. **In nonsynchronous systems,** each station operates independently, and any imbalance is accommodated in storage (**buffer**) between stations. The station continues operating until the next buffer is full or the previous buffer is empty. Furthermore, if one station becomes inoperative for some reason, the assembly line continues to operate until all the parts in the buffer have been used up. Nonsynchronous systems are suitable for large assemblies with many parts to be assembled. For types of assembly in which the times required for individual assembly operations vary widely, the rate of output will be constrained by the slowest station.

3. **In continuous systems,** the product is assembled while moving at a constant speed on pallets or similar workpiece carriers. The parts to be assembled are brought to the product by various workheads, and their movements are synchronized with the continuous movement of the product. Typical applications of this system are in bottling and packaging plants, although this method has also been used on mass-production lines for automobiles and appliances.

Assembly systems are generally set up for a certain product line; however, they can be modified for increased flexibility in order to be used in product lines that have a variety of models. Such *flexible assembly systems* (FAS) use computer controls,

**FIGURE 14.31** Various guides that ensure that parts are properly oriented for automated assembly. *Source*: G. Boothroyd.

interchangeable and programmable workheads and feeding devices, coded pallets, and automated guiding devices. The General Motors plant for the Saturn subcompact automobile, for example, is designed with a flexible assembly system. The system is capable of assembling up to a dozen different transmission and engine combinations and power-steering and air-conditioning units.

# 14.11 | Design Considerations

As in many aspects of manufacturing processes and systems, design is an integral part of the topics described in this chapter. Two of these topics require special design considerations, as described in this section.

## 14.11.1 Design for fixturing

The proper design, construction, and operation of flexible workholding devices and fixtures are essential to the operation and efficiency of advanced manufacturing systems. The following is a list of the major design issues involved:

1. Workholding devices must position the workpiece automatically and accurately; they must maintain its location precisely and with sufficient clamping force to withstand the particular manufacturing operation. Fixtures should also be able to accommodate parts placed repeatedly in the same position.

2. The fixtures must have sufficient stiffness to resist, without excessive distortion, the normal and shear stresses developed at the workpiece–fixture interfaces.

3. The presence of loose chips and other debris between the locating surfaces of the workpiece and the fixture can be a serious problem. Chips are most likely to be present where cutting fluids are used, because small chips tend to stick to the wet surfaces, due to surface-tension forces.

4. A flexible fixture should accommodate parts to be made by various processes and cases for which dimensions and surface features vary from part to part. These considerations are even more important when the workpiece (a) is fragile or made of a brittle material; (b) is made of a relatively soft and flexible material, such as plastic or rubber part; or (c) has a relatively soft coating on its contacting surfaces.

5. Fixtures and clamps should have low profiles, so as to avoid collision with cutting tools. Collision avoidance is also an important factor in programming tool paths in machining operations; see Sections 14.3 and 14.4.

6. Flexible fixturing must also meet special requirements in order to function properly in manufacturing cells and flexible manufacturing systems. The time required to load and unload parts on modern machinery, for example, should be minimal in order to reduce cycle times.

7. Workpieces should be designed to allow locating and clamping within a fixture. Flanges, flats, or other locating surfaces should be incorporated into the design in order to simplify fixture design and to aid in part transfer into machinery.

## 14.11.2  Design for assembly, disassembly, and service

**Design for assembly.**   Although the functions of a product and its design for manufacture have been matters of major interest for some time, *design for assembly* (DFA) has attracted special attention (particularly design for automated assembly), because of the need to reduce assembly costs. In **manual assembly**, a major advantage is that humans can easily pick the correct similar or different parts from bulk, such as from a nearby bin, and the human senses guide the hands for proper assembly. In **high-speed automated assembly**, however, automatic handling generally requires that parts be separated from the bulk, conveyed by hoppers or vibratory feeders (see Fig. 14.30), and assembled in the proper locations and orientations.

Based on analyses of assembly operations as well as on experience, general guidelines for DFA have been developed through the years and are summarized as follows:

1. Reduce the number and variety of parts in a product; incorporate multiple functions into a single part. Also, consider subassemblies that could serve as modules.

2. Parts should have a high degree of symmetry (such as round or square) or a high degree of asymmetry (such as oval or rectangular) so that they cannot be installed incorrectly and so that they do not require location, alignment, or adjustment. Design parts for easy insertion.

3. Designs should allow parts to be assembled without obstructions or the lack of a direct line of sight. Assemblies should not be turned over for insertion of parts.

4. Consider methods such as snap fits, and avoid the need for fasteners such as bolts, nuts, and screws. If such fasteners must be used, minimize the variety used, and space and locate them such that tools can be used without obstruction.

5. Part designs should consider such factors as size, shape, weight, flexibility, abrasiveness, and tangling with other parts.

6. Parts should be inserted from a single direction, preferably vertically (that is, from above) to take advantage of gravity; assembly from two or more directions can be difficult.

7. Products should be designed, or existing products redesigned, so that there are no physical obstructions to the free movement of the parts during assembly (see Fig. 1.3); thus, for example, sharp external and internal corners should be replaced with chamfers, tapers, or radii.

8. Color-code parts that may appear to be similar, but are different.

**Robotic-assembly** design guidelines have rules similar to those for manual and high-speed automated assembly. The development of *compliant end effectors* and *dexterous manipulators* has somewhat relaxed the inherent inflexibility of robots. Some distinguishing additional guidelines are as follows:

1. Parts should be designed so that they can be gripped and manipulated by the same gripper (end effector) of the robot (see Fig. 14.18); such a design avoids the need for different grippers. Also, parts should be made available to the gripper in the proper orientation.

2. Assembly that involves threaded fasteners, such as bolts, nuts, and screws, may be difficult for robots to perform; one exception is the use of self-threading screws for sheet metal, plastics, and wooden parts. Also, robots easily can handle snap fits, rivets, welds, and adhesives.

**Design for disassembly.** The manner and ease with which a product may be taken apart for maintenance or replacement of its parts is another important consideration in product design. Consider, for example, the difficulties one has in removing certain components from under the hood of some automobiles; similar difficulties exist in the disassembly of several other products. Although there is no established set of guidelines, the general approach to design for disassembly requires the consideration of factors that are similar to those outlined previously for design for assembly. Analysis of computer or physical models of products and their components with regard to disassembly generally indicates any potential problems, such as obstructions, narrow and long passageways, lack of line of sight, and the difficulty of firmly grasping and guiding objects.

An important aspect of design for disassembly is how, after its life cycle, a product is to be taken apart for *recycling*, especially with respect to the more valuable components. Note, for example, that depending on (a) their design and location, (b) the type of tools used for disassembly and (c) whether the tools are manual or power tools, rivets may take longer to remove than screws or snap fits, and that a bonded layer of valuable material on a component would be very difficult, if not impossible, to remove for recycling or reuse. Obviously, the longer it takes to disassemble components, the higher is the cost of doing so; it is then possible that this cost becomes prohibitive. Consequently, the time required for disassembly has been studied and measured; although it depends on the manner in which disassembly takes place, some examples are as follows: cutting wire, 0.25s; disconnecting wire, 1.5 s; removing snap fits and clips, 1–3 s; and removing screws and bolts, 0.15–0.6 s per revolution.

**Design for service.** Design for assembly and disassembly includes taking into account the ease with which a product can be serviced and, if necessary, repaired. Design for service is essentially based on the concept that the elements that are most likely to need servicing should be at the outer layers of the product. In this way, individual parts are easier to reach and to service, without the need to remove various other parts in order to do so.

# 14.12 | Economic Considerations

As described in greater detail in Chapter 16, and as we have seen throughout the preceding chapters, there are numerous considerations involved in determining the overall economics of production operations. Because all production systems are essentially a combination of machines and people, important factors influencing the final decisions include the types and cost of machinery, cost of its operation, the skill level and amount of labor required, and production quantity. We have seen that lot size and production rate greatly influence the economics of production. Small quantities per year can be manufactured in job shops; however, the type of machinery in job shops generally requires skilled labor to operate. Furthermore, production quantity and rate in job shops are low, and as a result, cost per part can be high. (See Fig. 14.3.)

At the other extreme, there is the production of very large quantities of parts, using conventional flow lines and transfer lines and involving special-purpose machinery and equipment, specialized tooling, and computer control systems. Although all these components constitute major investments, the level of labor skill required and the labor costs are both relatively low, because of the high level of automation implemented. However, these production systems are organized for a specific type of product and hence lack flexibility.

Because most manufacturing operations are between these two extremes, an appropriate decision has to be made regarding the desirable level of automation to be implemented. In many situations, *selective automation* rather than *total* automation of a facility has been found to be cost effective. Generally, the higher the level of skill available in the workforce, the lower is the need for automation, provided that higher labor costs are justified and that there is a sufficient number of qualified workers available. Conversely, if a manufacturing facility has already been automated, the skill level required is lower.

In addition, the manufacture of some products must have a large labor component, and thus their production is *labor intensive*; this is especially the case with products that require assembly. Examples of labor-intensive products include aircraft, software, bicycles, pianos and other musical instruments, furniture, toys, shoes, textiles, and garments. This high labor requirement is a major reason that so many household as well as high-tech products are now made or assembled in countries where labor costs are low, such as Mexico, China, the Philippines, and other Pacific-rim countries. (See Section 16.9).

In manual assembly, relatively simple tools are used, and the process is economical for relatively small lots. Because of the dexterity of the human hand and fingers and their capability for feedback through various senses, workers can manually assemble even complex parts without much difficulty. As described in Section 14.7, labor cost and benefit considerations are also significant aspects of robot selection and use. The increasing availability and reliability and the reduced cost of intelligent robots are having a major economic impact on manufacturing operations, causing human labor to be gradually replaced.

---

**CASE STUDY** | Development of a Modular Fixture

Figure 14.32 shows a round cast-iron housing that requires a fixture for milling the housing flat, counterboring the center hole, drilling the four corner holes, and machining the circular mounting area. These operations are to be performed in moderate lot sizes on a CNC milling machine; thus, a fixture that accurately and

**FIGURE 14.32** Cast-iron housing and the machining operations required.

consistently locates the workpiece is required. The lot size is not large enough to justify the design and fabrication of dedicated fixtures, and hence a modular fixture is constructed from the components illustrated in Fig. 14.33.

The first step in designing such a fixture is to select a tooling plate. Tooling plates with T-slots or gridded holes are the most common alternatives for modular fixtures. For this case, a rectangular plate with gridded holes is selected that has a sufficiently large surface area to accommodate both the workpiece and the fixturing elements.

The workpiece must not be clamped on the surface that is to be machined; however, there is a lower flange that is suitable for clamping. Clamping is generally accomplished by exerting a force with a clamp against a locating button. For this round workpiece, it is desirable to locate the lower flange at three points, spaced ap-

**FIGURE 14.33** Modular components used to construct the fixture for CNC machining of the cast-iron housing depicted in Fig. 14.32.

**FIGURE 14.34**
Completed modular fixture with cast-iron housing in place, as would be assembled for use in a machining center or CNC milling machine.

proximately 120° apart, for reasons of stability. The first locating elements selected are shoulder-type support cylinders. The flange rests on the shoulders, and the cylinders support the workpiece on its diameter and also raise the workpiece above the tooling plate. This method has the advantage of eliminating the adverse effects of chips (from machining) on the tooling plate, which could interfere with the orientation of the workpiece. The support cylinders are mounted onto the tooling plate, using locating screws. An adjustable extension support is then mounted to the tooling plate in a position where the correct diameter is set between the three locators. The extension support is positioned to contact only the workpiece's bottom edge, and sufficient space is allowed so that the locator will not bind the housing on the locating diameter.

The workpiece must be consistently oriented in order to maintain the required tolerances on the corner holes. An adjustable stop is utilized to locate the workpiece; this stop uses a threaded locating button that is adjusted to orient the workpiece. An extension support is used so that the stop is vertically located directly below the surface to be machined.

Strap clamps are then used to hold the workpiece in place. The springs and washers on the clamping stud permit the strap clamp to rise automatically when loosened, so that the clamp will not fall when the workpiece is removed. Figure 14.34 shows the modular fixture with the workpiece in place. Note that planar, concentric, and radial locations have all been accurately defined by the fixture.

*Source*: Courtesy of Carr Lane Manufacturing Company.

# SUMMARY

- There are several levels of automation, from simple automation to untended manufacturing cells. Automation has been implemented successfully in manufacturing processes, material handling, inspection, assembly, and packaging. Production quantity and production rate are major factors in selecting the most economic level of automation for a process. (Sections 14.1, 14.2)

- True automation began with the numerical control of machines, achieving flexibility of operation, lower cost, and ease of making different parts with lower operator skill. (Sections 14.3, 14.4)

- The quality and cost of manufacturing operations are further optimized by adaptive-control techniques, which continuously monitor the operation and automatically make appropriate adjustments in the processing parameters. (Section 14.5)

- Significant advances have been made in material handling, particularly with the implementation of industrial robots and automated guided vehicles. (Sections 14.6 and 14.7)

- Sensors are essential in the implementation of modern manufacturing technologies and computer-integrated manufacturing, and a wide variety of sensors based on various principles has been developed and successfully installed. (Section 14.8)

- Other advances include flexible fixturing and automated assembly techniques that reduce the need for worker intervention and lower manufacturing costs. The effective and economic implementation of these techniques requires that design for assembly, disassembly, and servicing be recognized as an important factor in the total design and manufacturing process. (Sections 14.9, 14.10)

- As in all manufacturing processes, there are certain design considerations regarding the implementation of the various topics described in this chapter. (Section 14.11)

- Economic considerations in automation include decisions regarding the level of automation to be implemented; such decisions, in turn, involve parameters such as the quantity and rate of production. (Section 14.12)

# BIBLIOGRAPHY

Amic, P. J., *Computer Numerical Control Programming*, Prentice Hall, 1996.

Astrom, K. J., and B. Wittenmark, *Adaptive Control*, 2d. ed., Addison-Wesley, 1994.

Bolhouse, V., *Fundamentals of Machine Vision*, Robotic Industries Association, 1997.

Boothroyd, G., *Assembly Automation and Product Design*, Dekker, 1991.

Boothroyd, G., P. Dewhurst, and W. Knight, *Product Design for Manufacture and Assembly*, 2d. ed., Dekker, 2001.

Brooks, R. R., and S. Iyengar, *Multi-Sensor Fusion: Fundamentals and Applications with Software*, Prentice Hall, 1997.

Burke, M., *Handbook of Machine Vision Engineering*, Chapman & Hall, 1999.

Busch-Vishniac, I., *Electromechanical Sensors and Actuators*, Springer, 1999.

Chow, W., *Assembly Line Design: Methodology and Applications*, Dekker, 1990.

Fraden, J., *Handbook of Modern Sensors: Physics, Designs, and Applications*, 2d. ed., Springer, 1996.

Galbiati, L. J., *Machine Vision and Digital Image Processing Fundamentals*, Prentice Hall, 1997.

Gibbs, D., and T. Crandell, *CNC: An Introduction to Machining and Part Programming*, Industrial Press, 1991.

Ioannu, P. A., *Robust Adaptive Control*, Prentice Hall, 1995.

Jain, R. (ed.), *Machine Vision*, Prentice Hall, 1995.

Lim, S. C. J., *Computer Numerical Control*, Delmar, 1994.

Lynch, M., *Computer Numerical Control for Machining*, McGraw-Hill, 1992.

Molloy, O., E. A. Warman, and S. Tilley, *Design for Manufacturing and Assembly: Concepts, Architectures and Implementation*, Kluwer, 1998.

Myler, H. R., *Fundamentals of Machine Vision*, Society of Photo-optical Instrumentation Engineers, 1998.

Nof, S. Y., W. E. Wilhelm, and H.-J. Warnecke, *Industrial Assembly*, Chapman & Hall, 1998.

Rampersad, H. K., *Integral and Simultaneous Design for Robotic Assembly*, Wiley, 1995.

Rehg, J. A., *Introduction to Robotics in CIM Systems*, 5th ed., Prentice Hall, 2002.

Sandler, B.-Z., *Robotics: Designing the Mechanisms for Automated Machinery*, 2d. ed., Prentice Hall, 1999.

Sava, M., and J. Pusztai, *Computer Numerical Control Programming*, Prentice Hall, 1997.

Seames, W., *Computer Numerical Control: Concepts and Programming*, 3d. ed., Delmar, 1995.

Smid, P., *CNC Programming Handbook*, Industrial Press, 2000.

Soloman, S., *Sensors Handbook*, McGraw-Hill, 1997.

Stenerson, J., and K. S. Curran, *Computer Numerical Control: Operation and Programming*, 2d. ed., Prentice Hall, 2000.

*Tool and Manufacturing Engineers Handbook*, 4th ed., Vol. 4: *Assembly, Testing, and Quality Control*, Society of Manufacturing Engineers, 1986.

_____, Vol. 9: *Material and Part Handling in Manufacturing*, Society of Manufacturing Engineers, 1998.

Valentino, J. V., and J. Goldenberg, *Introduction to Computer Numerical Control*, 2d. ed., Prentice Hall, 1999.

Williams, D. J., *Manufacturing Systems*, 2d. ed., Chapman & Hall, 1994.

Zuech, N., *Understanding and Applying Machine Vision*, 2d. ed., Dekker, 1999.

# QUESTIONS

**14.1** Describe the differences between mechanization and automation. Give several specific examples for each.

**14.2** Why is automation generally regarded as evolutionary rather than revolutionary?

**14.3** Are there activities in manufacturing operations that cannot be automated? Explain, and give specific examples.

**14.4** Explain the difference between hard and soft automation. Why are they named as such?

**14.5** Describe the principle of numerical control of machines. What factors led to the need for and development of numerical control? Name some typical applications of NC.

**14.6** Explain the differences between direct numerical control and computer numerical control. What are their relative advantages?

**14.7** Describe open-loop and closed-loop control circuits.

**14.8** What are the advantages of computer-aided NC programming?

**14.9** Describe the principle and purposes of adaptive control. Give some examples of present applications in manufacturing and other areas that you think can be implemented.

**14.10** What factors have led to the development of automated guided vehicles? Do automated guided vehicles have any limitations? Explain your answers.

**14.11** List and discuss the factors that should be considered in choosing a suitable material-handling system for a particular manufacturing facility.

**14.12** Make a list of the features of an industrial robot. Why are these features necessary?

**14.13** Discuss the principles of various types of sensors, and give two applications for each type.

**14.14** Describe the concept of design for assembly. Why has it become an important factor in manufacturing?

**14.15** Is it possible to have partial automation in assembly operations? Explain.

**14.16** Describe your thoughts on adaptive control in manufacturing operations.

**14.17** What are the two kinds of robot joints? Give applications for each.

**14.18** What are the advantages of flexible fixturing over other methods of fixturing? Are there any limitations to flexible fixturing? Explain.

**14.19** How are robots programmed to follow a certain path?

**14.20** Giving specific examples, discuss your observations concerning Fig. 14.2.

**14.21** What are the relative advantages and limitations of the two arrangements for power heads shown in Fig. 14.4?

**14.22** Discuss methods of on-line gaging of workpiece diameters in turning operations other than that shown in Fig. 14.15. Explain the relative advantages and limitations of the methods.

**14.23** Is drilling and punching the only application for the point-to-point system shown in Fig. 14.10a? Are there others? Explain.

**14.24** Describe possible applications for industrial robots not discussed in this chapter.

**14.25** What determines the number of robots in an automated assembly line such as that shown in Fig. 14.23?

**14.26** Describe situations in which the shape and size of the work envelope of a robot (see Fig. 14.20) can be critical.

**14.27** Explain the functions of each of the components of the robot shown in Fig. 14.17a. Comment on their degrees of freedom.

**14.28** Explain the difference between an automated guided vehicle and a self-guided vehicle.

**14.29** It has been commonly acknowledged that, at the early stages of development and implementation of industrial robots, the usefulness and cost effectiveness of the robots were overestimated. What reasons can you think of to explain this situation?

**14.30**   Describe the type of manufacturing operations (see Fig. 14.2) that are likely to make the best use of a machining center (see Section 8.10). Comment on the influence of product quantity and part variety.

**14.31**   Give a specific example of a situation in which an open-loop control system would be desirable, and give a specific example of a situation in which a closed-loop system would be desirable. Explain.

**14.32**   Why should the level of automation in a manufacturing facility depend on production quantity and production rate?

**14.33**   Explain why sensors have become so essential in the development of automated manufacturing systems.

**14.34**   Why is there a need for flexible fixturing for holding workpieces? Are there any disadvantages to such flexible fixturing? Explain.

**14.35**   Describe situations in manufacturing for which you would not want to apply numerical control. Explain your reasons.

**14.36**   Table 14.2 shows a few examples of typical products for each category of production by volume. Add several other examples to this list.

**14.37**   Describe situations for which each of the three positioning methods shown in Fig. 14.6 would be desirable.

**14.38**   Describe applications of machine vision for specific parts, similar to the examples shown in Fig. 14.26.

**14.39**   Add examples of guides other than those shown in Fig. 14.31.

**14.40**   Sketch the work envelope of each of the robots in Fig. 14.19. Describe its implications in manufacturing operations.

**14.41**   Give several applications for the types of robots shown in Fig. 14.19.

**14.42**   Name some applications for which you would not use a vibratory feeder. Explain why vibratory feeding is not appropriate for these applications.

**14.43**   Give an example of a metal-forming operation (from Chapters 6 and 7) that is suitable for adaptive control similar to that shown in Fig. 14.15.

**14.44**   Give some applications for the systems shown in Fig. 14.26a and c.

**14.45**   Comment on your observations regarding the system shown in Fig. 14.18b.

**14.46**   Give examples for which tactile sensors would not be suitable. Explain why tactile sensors are unsuitable for these applications.

**14.47**   Give examples for which machine vision cannot be applied properly and reliably. Explain why machine vision is inappropriate for these applications.

**14.48**   Comment of the effect of cutter wear on the profiles produced, such as those shown in Figs. 14.10b and 14.12.

**14.49**   Although future trends are always difficult to predict with certainty, describe your thoughts as to what new developments in the topics covered in this chapter could possibly take place as we move through the early 2000s.

**14.50**   Describe the circumstances under which a dedicated fixture, a modular fixture, and a flexible fixture would be preferable.

# DESIGN

**14.51**   Design two different systems of mechanical grippers for widely different applications.

**14.52**   For a system similar to that shown in Fig. 14.27, design a flexible fixturing setup for a lathe chuck. (See Figs. 8.42 and 8.44.)

**14.53**   Add other examples to those shown in Fig. 1.3.

**14.54**   Give examples of products that are suitable for the type of production shown in Fig. 14.3.

**14.55**   Choose one machine each from Chapters 6 through 12, and design a system for the machine in

which sensor fusion can be used effectively. How would you convince a prospective customer of the merits of such a system? Would the system be cost effective?

**14.56**   Does the type of material (metallic or nonmetallic) used for the parts shown in Fig. 14.31 have any influence on the effectiveness of the guides? Explain.

**14.57**   Think of a product, and design a transfer line for it similar to that shown in Fig. 14.5. Specify the types of and the number of machines required.

**14.58**   Section 14.9 has described the basic principles of flexible fixturing. Considering the wide variety of parts made, prepare design guidelines for flexible fixturing. Make simple sketches illustrating each guideline for each type of fixturing, describing its ranges of applications and limitations.

**14.59**   Describe your thoughts on the usefulness and applications of modular fixturing, consisting of various individual clamps, pins, supports, and attachments mounted on a base plate.

**14.60**   Inspect several household products, and describe the manner in which they have been assembled. Comment on any design changes you would make so that assembly, disassembly, and servicing are simpler and faster.

**14.61**   Review the last design shown in Fig. 14.18a, and design grippers that would be suitable for gripping the following products: (a) an egg; (b) an object made of soft rubber; (c) a metal ball with a very smooth and shiny surface; (d) a newspaper; and (e) tableware, such as forks, knives, and spoons.

**14.62**   Obtain an old toaster, and disassemble it. Make recommendations on its redesign, using the guidelines given in Section 14.11.2.

**14.63**   Comment on the design and materials used for the gripper shown in Fig. 14.25. Why do such grippers have low durability on the shop floor?

**14.64**   Comment on your observations regarding Fig. 14.27, and offer designs for similar applications in manufacturing. Also comment on the usefulness of the designs in actual production on the shop floor.

**14.65**   Review the various toolholders used in the machining operations described in Chapter 8, and design sensor systems similar for them to that shown in Fig. 14.24. Comment on the features of the sensor systems and discuss any problems that may be associated with their use on the factory floor.

**14.66**   Design a guide that operates in the same manner as those shown in Fig. 14.31, but to align U-shaped parts so that they are inserted with the open end down.

**14.67**   Design a flexible fixture that uses powered workholding devices for a family of parts, as in the case study, but that allows for a range of diameters and thicknesses.

**14.68**   Design a modular fixturing system for the part shown in the figure accompanying Problem 8.134.

# CHAPTER

# 15

# Computer-Integrated Manufacturing Systems

15.1 Introduction

15.2 Manufacturing Systems

15.3 Computer-Integrated Manufacturing

15.4 Computer-Aided Design and Engineering

15.5 Computer-Aided Manufacturing

15.6 Computer-Aided Process Planning

15.7 Computer Simulation of Manufacturing Processes and Systems

15.8 Group Technology

15.9 Cellular Manufacturing

15.10 Flexible Manufacturing Systems

15.11 Just-in-Time Production

15.12 Communications Networks in Manufacturing

15.13 Artificial Intelligence

15.14 The Factory of the Future

15.15 Economic Considerations

## 15.1 | Introduction

The implementation, benefits, and limitations of mechanization, automation, and computer control of various stages of manufacturing operations have been described in the preceding chapter; this chapter focuses its attention on the *integration of manufacturing activities*. This concept means that processes, equipment, operations, and their management are treated as a **manufacturing system**, one that makes possible total control of the manufacturing facility, thereby increasing productivity, product quality, and product reliability and reducing manufacturing costs. As described in this chapter, in **computer-integrated manufacturing** (CIM), the traditionally separate functions of product design, research and development, production, assembly, inspection, and quality control are all linked. Consequently, integration requires that quantitative relationships among product design, materials, manufacturing processes, process and equipment capabilities, and related activities be well understood. In this way, changes in, for example, material requirements, product types, or market demand can be properly accommodated.

The importance of product quality was emphasized in Section 4.9, along with the necessity of commitment of a company's management to total quality control. Recall the statements that (a) quality must be built into the product, (b) higher quality does not necessarily mean higher cost, and (c) marketing poor-quality products can indeed be very costly to the manufacturer. It has been shown that high quality is far more attainable and less expensive via the proper integration of design and manufacturing than if the two were separate activities. Integration is now successfully and effectively accomplished through **computer-aided design, engineering, manufacturing, process planning**, and simulation of processes and systems.

This chapter also emphasizes the importance of **flexibility** in machines, equipment, tooling, and production operations to the ability to respond to market and product changes and to ensure **on-time delivery** of high-quality products to the customer's satisfaction. As described in detail, important developments during the past 30 years or so have had a major impact on modern manufacturing, especially in an intensely competitive global marketplace. Among these advances are **group technology, cellular manufacturing, flexible manufacturing systems**, and **just-in-time production** (also called **zero inventory, stockless production**, or **demand scheduling**) with no excess production such as was previously stored in inventory for future sales. Furthermore, because of the extensive use of computer controls and hardware and software in integrated manufacturing, the planning and effective implementation of **communication networks** soon became an essential component of these activities.

The chapter concludes with a review of **artificial intelligence**, consisting of expert systems, natural-language processing, machine vision, artificial neural networks, and fuzzy logic, and how these developments impact manufacturing activities, as well as a discussion of the concept of **the factory of the future**, including its possibilities and existing concerns about it.

# 15.2 | Manufacturing Systems

As we have seen throughout this text, manufacturing consists of a large number of interdependent activities with distinct entities, such as materials, tools, machines, and controls; it should therefore be regarded as a *system*. Manufacturing is indeed a complex system, because it consists of numerous diverse physical and human elements. Some of these factors can be difficult to predict and to control, such as the supply and cost of raw materials, market changes, and global trends, as well as human behavior and performance.

Ideally, we should be able to represent a system by **mathematical and physical models**, which can identify the nature and extent of the interdependence of the variables involved. In a manufacturing system, a change or disturbance anywhere in the system requires that it adjust itself systemwide in order to continue functioning efficiently. For example, if the supply of a particular raw material is reduced (for example, by geopolitical maneuvers, wars, or strikes) and, consequently, its cost increases, alternative materials must be selected. This selection should be made only after careful consideration of the effect this change may have on processing conditions, production rate, product quality, and, especially, manufacturing costs.

Similarly, the demand for a product may fluctuate randomly and rapidly, or even unexpectedly, on account of its style, size, or capacity. Note, for example, the downsizing of automobiles during the 1980s in response to fuel shortages, and note as well the current popularity of sport-utility vehicles with comparatively low gas mileage. The manufacturing system must be capable of producing the modified product on a short **lead time** and, preferably, with relatively small major capital investment in machinery and tooling. Lead time is defined as the length of time between the creation of the product as a concept (or receipt of an order for the product) and the time that the product first becomes available in the marketplace (or is delivered to the customer).

Computer simulation and modeling of such a complex system can be difficult, because of a lack of comprehensive and reliable data on some of the numerous variables involved; furthermore, it is not always easy to predict correctly and to control some of these variables. The following are some examples of the problems that are encountered:

a) Machine tools' characteristics, performance, and response to random external disturbances cannot always be modeled precisely.

b) Raw-material costs over a period of time may be difficult to predict accurately.

c) Market demands and human behavior and performance are difficult to model reliably.

In spite of these challenges, however, much progress continue to be made in modeling and simulation of manufacturing systems. (See also Section 15.7.)

# 15.3 | Computer-Integrated Manufacturing

The various levels of automation in manufacturing operations, as described in Chapter 14, have been extended further by including *information-processing* functions, using an extensive network of interactive computers. The result is *computer-integrated manufacturing* (CIM), which is a broad term used to describe the computerized integration of product design, planning, production, distribution, and management. Computer-integrated manufacturing is a *methodology* and a *goal*, rather than merely an assemblage of equipment and computers.

The effectiveness of CIM greatly depends on the use of a large-scale **integrated communications system**, involving computers, machines, and their controls. The problems that may arise in such systems are described in Section 15.12. Furthermore, because CIM ideally should involve the total operation of a company, it requires an extensive database concerning the technical as well as business aspects of the operation. Consequently, if planned all at once, CIM can be prohibitively expensive, particularly for small and medium-size companies.

Implementation of CIM in existing plants may begin with the use of modules in selected phases of the operation; for new manufacturing plants, on the other hand, comprehensive and long-range strategic planning covering all phases of the operation is essential in order to benefit fully from CIM. Such plans must take into account considerations such as (a) the availability of resources; (b) the mission, goals, and culture of the organization; (c) existing as well as emerging technologies; and (d) the level of integration desired.

Computer-integrated manufacturing systems consist of *subsystems* that are integrated into a whole (Fig. 15.1); these subsystems consist of the following:

a) Business planning and support;

b) Product design;

c) Manufacturing process planning;

d) Process automation and control;

e) Factory-floor monitoring systems.

The subsystems are designed, developed, and implemented in such a manner that the output of one subsystem serves as the input of another subsystem, as shown by the various arrows in Fig. 15.1. Organizationally, the subsystems are generally divided into two functions: (a) **business-planning functions**, which include activities such as forecasting, scheduling, material-requirements planning, invoicing, and accounting; and (b) **business-execution functions**, which include production and process control, material handling, testing, and inspection.

The benefits of CIM include the following:

a) Responsiveness to shorter product life cycles (see Section 16.4), changing market demands, and global competition;

b) Emphasis on product quality and uniformity, implemented through better process control;

c) Better use of materials, machinery, and personnel and the reduction of *work-in-progress* (WIP) inventory, all of which improve productivity and lower product cost;

d) Better control of the production, scheduling, and management of the total manufacturing operation, resulting in lower product cost.

**FIGURE 15.1** A schematic illustration of a computer-integrated manufacturing system. *Source*: U. Rembold et al., *Computer-Integrated Manufacturing and Engineering*, Addison-Wesley, 1993.

## 15.3.1 Databases

An effective computer-integrated manufacturing system requires a single, large database that is shared by members of the entire organization. *Databases* consist of up-to-date, detailed, and accurate data on products, designs, machines, processes, materials, production, finances, purchasing, sales, marketing, and inventory. This vast array of data is stored in computer memory and randomly recalled or modified as necessary, either by individuals in the organization or by the CIM system itself while it is controlling various aspects of design and production.

A database generally consists of the following information:

a) **Product data,** such as part shape, dimensions, tolerances, and specifications;

b) **Data management attributes,** such as owner, revision level, and part number;

c) **Production data,** such as the manufacturing processes used in making parts and products;

d) **Operational data,** such as scheduling, lot sizes, and assembly requirements;

e) **Resources data,** such as capital, machines, equipment, tooling, and personnel, and the capabilities of these resources.

Databases are built by individuals and through the use of various sensors in the machinery and equipment employed in production. Data from the latter are collected automatically by a **data acquisition system** (DAS) that can report, for example, the number of parts being produced per unit time, dimensional accuracy, surface finish, and weight, at specified rates of sampling. The components of DASs include microprocessors, transducers, and analog-to-digital converters (ADCs). Data acquisition systems are also capable of analyzing the data and transferring them to other computers for such purposes as statistical analysis, data presentation, and forecasting of product demand.

Several factors are important in the use and implementation of databases:

a) Databases should be up to date, accurate, user friendly, and easily accessible and shared.

b) In the event that there is a malfunction in data retrieval, the original, correct data must be recovered and restored in the database.

c) Because it is used for many purposes and by many people, the database must be flexible and responsive to the needs of users with different backgrounds and interests.

d) Databases should be accessed by designers, manufacturing engineers, process planners, financial officers, and the management of the company by using appropriate access codes. Obviously, companies must protect data against tampering or unauthorized use.

# 15.4 | Computer-Aided Design and Engineering

*Computer-aided design* (CAD) involves the use of computers to create design drawings and geometric models of products and components (see also Fig. 1.2) and is usually associated with **interactive computer graphics**, known as a **CAD system**. *Computer-aided engineering* (CAE) simplifies the creation of the database by allowing several applications to share the information in the database. It basically consists of a comprehensive computer-based analysis and evaluation of engineering design. Its diverse applications include, for example, (a) finite-element analysis of stresses, strains, deflections, and temperature distribution in machine elements, load-bearing members, and structures; (b) generation, storage, and retrieval of numerical-control data for machining operations; and (c) design of integrated circuits and other microelectronic and micromechanical devices (see Chapter 13).

In CAD, the traditional drawing board is replaced by **workstations**, where the user can generate drawings or sections of a drawing on the computer monitor. The design is continuously displayed on the monitor and in different colors for its various components; the final drawing is printed or plotted on appropriate hardware connected to the computer. When using a CAD system, the designer can conceptualize the object to be designed on the graphics screen and can consider alternative designs or quickly modify a particular design to meet specific design requirements.

Through the use of powerful software, such as CATIA (after Computer-Aided Three-Dimensional Interactive Applications), the design can be subjected to *engineering analysis*, which can identify potential problems, such as an excessive load or deflection, and interference at mating surfaces during assembly. In addition to the design's geometric and dimensional features, other information, such as a list of materials, specifications, and manufacturing instructions, is stored in the CAD database. Using such information, the designer can also analyze the economics of alternative designs.

### 15.4.1 Exchange specifications

Because of the availability of a wide variety of CAD systems with different characteristics, supplied by different vendors, proper communication and exchange of data between these systems is essential. (See also Section 15.12.) **Drawing exchange format**

(DFX) was developed for use with Autodesk and has become a de facto standard, because of the long-term success of this software package. DFX is limited to transferring geometry information only. Similarly, **STL** (STereo Lithography) formats are used to export 3D geometries, initially to rapid prototyping systems; recently, however, STL has become a format for data exchange between CAD systems.

The need for a single, neutral format for better compatibility and for the transfer of more information than geometry alone is currently filled mainly by the **Initial Graphics Exchange Specification** (IGES). Vendors need only to provide translators for their own systems, to preprocess outgoing data into the neutral format, and to postprocess incoming data from the neutral format into their system. IGES is used for translation in two directions (into and out of a system) and is also used widely for translation of 3D line and surface data. Because IGES is evolving, there are many variations of IGES in existence.

Another specification is a solid-model-based standard called **Product Data Exchange Specification** (PDES), which is based on the Standard for the Exchange of Product Model Data (STEP) developed by the International Standards Organization. PDES allows information on shape, design, manufacturing, quality assurance, testing, maintenance, etc., to be transferred between CAD systems.

## 15.4.2 Elements of CAD systems

The design process in a CAD system consists of four stages, described as follows:

a) **Geometric modeling.** In *geometric modeling*, a physical object or any of its parts is described mathematically or analytically. The designer first constructs a geometric model by giving commands that create or modify lines, surfaces, solids, dimensions, and text that, together, compose an accurate and complete two- or three-dimensional representation of the object. The results of these commands are then displayed; the images can be moved around on the screen, and any section can be magnified in order to view its details. The models are stored in the database contained in computer memory.

The models can be presented in different ways, described as follows:

1. In **line representation** (also called **wire-frame representation**; see Fig. 15.2), all edges of the model are visible as solid lines. However, this type of image can be ambiguous, particularly for complex shapes; hence, various colors are generally used for different parts of the object. The three types of wire-frame representations are 2D, $2\frac{1}{2}$-D, and 3D. A 2D image shows the profile of the object, and a $2\frac{1}{2}$-D image can be obtained by *translational sweep*, that is, by moving the 2D object along the $z$-axis. For round objects, a $2\frac{1}{2}$-D model can be generated by simply *rotating* a 2D model around its axis.

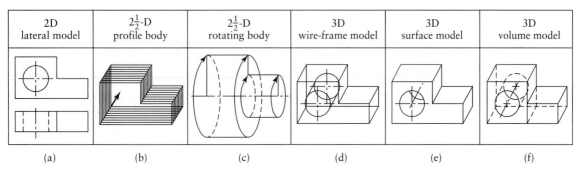

| 2D<br>lateral model | $2\frac{1}{2}$-D<br>profile body | $2\frac{1}{2}$-D<br>rotating body | 3D<br>wire-frame model | 3D<br>surface model | 3D<br>volume model |
|---|---|---|---|---|---|
| (a) | (b) | (c) | (d) | (e) | (f) |

**FIGURE 15.2** Various types of modeling for CAD.

(a)                              (b)                              (c)

**FIGURE 15.3**    Types of splines. (a) A Bezier curve passes through the first and last control point and generates a curve from the other points. Changing a control point modifies the entire curve. (b) A B-spline is constructed piecewise, so that changing a vertex affects the curve only in the vicinity of the changed control point. (c) A third-order piecewise Bezier curve constructed through two adjacent control points, with two other control points defining the curve slope at the end points. A third-order piecewise Bezier curve is continuous, but its slope may be discontinuous.

2. In the **surface model,** all visible surfaces are shown in the model. Surface models define surface features and edges of objects. Modern CAD programs use Bezier curves, B-splines, or nonuniform rational B-splines (NURBS) for surface modeling. Each of these methods uses control points to define a polynomial curve or surface. A Bezier curve passes through the first and last vertex and uses the other control points to generate a blended curve. The drawback to Bezier curves is that modification of one control point will affect the entire curve. B-splines are a blended piecewise polynomial curve, where modification of a control point affects the curve only in the area of the modification. Figure 15.3 shows examples of two-dimensional Bezier curves and B-splines. A NURBS is a special kind of B-spline where each control point has a weight associated with it.

3. In the **solid model,** all surfaces are shown, but the data also describe the interior volume. Solid models can be constructed from *swept volumes* (Figs. 15.2b and c) or by the techniques shown in Fig. 15.4. In **boundary representation** (BREP), surfaces are combined to develop a solid model (Fig. 15.4a); in **constructive solid geometry** (CSG), simple shapes (called **primitives**

**FIGURE 15.4**    (a) Boundary representation of solids, showing the enclosing surfaces and the generated solid model. (b) A solid model represented as compositions of solid primitives. (c) Three representations of the same part by CAD. *Source*: P. Ranky.

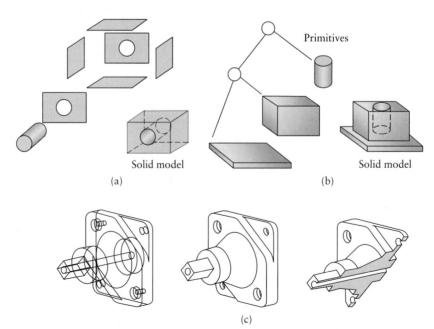

Solid model                              Solid model

(a)                              (b)

(c)

(a)                                                    (b)

**FIGURE 15.5**   An example of parametric design. Dimensions of part features can be easily modified to quickly obtain an updated solid model.

**of solids**) such as spheres, cubes, blocks, cylinders, and cones are combined to develop a solid model (Fig. 15.4b). The user selects any combination of primitives and their sizes and combines them into the desired solid model. Although solid models have certain advantages, such as ease of design analysis and ease of preparation for manufacture of the part, they require more computer memory and processing time than the wire-frame and surface models.

   A special kind of solid model is a **parametric model**, where a part not only is stored in terms of a BREP or CSG definition, but also is derived from the dimensions and constraints that define the features (Fig. 15.5). Whenever a change is made, the part is re-created from these definitions. This feature allows simple and straightforward updates and changes to be made to models.

4. Figure 15.6 shows the **octree representation** of a solid object; this type of model is basically a three-dimensional analog to pixels on a television screen or monitor. Just as any area can be broken down into quadrants, any volume

(a) One iteration

(b) Two iterations

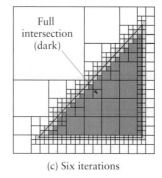

(c) Six iterations

**FIGURE 15.6**   The octree representation of a solid object. Any volume can be broken down into octants, which are then identified as solid, void, or partially filled. Shown is two-dimensional, or quadtree, version, for representation of shapes in a plane.

(a)

(b)

**FIGURE 15.7** (a) Illustration of the skeleton data structure for solid objects. The skeleton is the dashed line in the object's interior. (b) A skeleton model used for kinematic analysis of a clamp. *Source*: S. D. Lockhart and C. M. Johnson, *Engineering Design Communication*, Prentice Hall, 2000.

can be broken down into *octants*, which are then identified as solid, void, or partially filled. Partially filled *voxels* (from *vo*lume pi*xels*) are broken into smaller octants and reclassified. With increasing resolution, exceptional part detail can be achieved. This process may appear to be somewhat cumbersome, but it allows for accurate description of complex surfaces; it is used particularly in biomedical applications, such as modeling bone geometries.

5. A **skeleton** (Fig. 15.7) is commonly used for kinematic analysis of parts or assemblies. A skeleton is the family of lines, planes, and curves that describe a part without the detail of surface models. Conceptually, a skeleton can be constructed by fitting the largest circles (or spheres for three-dimensional objects) within the geometry; the skeleton is the set of points that connect the centers of the circles (or spheres). A current area of research involves using skeleton models instead of conventional surface or solid models.

b) **Design analysis and optimization.** After its geometric features have been determined, the design is subjected to an *engineering analysis*. This phase may consist of analyzing, for example, stresses, strains, deflections, vibrations, heat transfer, temperature distribution, or dimensional tolerances. Several software packages are available that have the capabilities to compute these quantities accurately and rapidly. Because of the relative ease with which such analyses can be done, designers can study a design more thoroughly before it moves on to production. Experiments and measurements in the field may nonetheless be necessary to determine and verify the actual effects of loads, temperature, and other variables on the designed components.

c) **Design review and evaluation.** An important design stage is *review and evaluation*, to check for any interference among various components. This step is necessary in order to avoid difficulties during assembly or use of the part and to determine whether moving members (such as linkages, gears, and cams) will operate and function as intended. Software with animation capabilities is available to identify potential problems with moving members and other dynamic situations. During the review and evaluation stage, the part is also precisely dimensioned and set within the full range of tolerance required for its production (Section 4.7).

d) **Documentation.** After the preceding stages have been completed, the design is reproduced by automated drafting machines, for documentation and reference. At this stage, detail and working drawings are also developed and printed. The CAD system is also capable of developing and drafting sectional views of the part, scaling the drawings, and performing transformations in order to present various views of the part.

e) **Database.** Many components are either standard components that are mass produced according to a given design specification (such as bolts or gears) or are identical to parts used in previous designs. Modern CAD systems therefore have a built-in database management system that allows designers to identify, view, and adopt parts from a library of stock parts. These parts can be parametrically modeled to allow cost-effective updating of the geometry. Some databases with extensive parts libraries are commercially available; many vendors make their part libraries available on the World Wide Web.

## 15.5 | Computer-Aided Manufacturing

*Computer-aided manufacturing* (CAM) involves the use of computers and computer technology to assist in all phases of manufacturing a product, including process and production planning, scheduling, manufacture, quality control, and management. Because of the increased benefits, computer-aided design and computer-aided manufacturing are often combined into **CAD/CAM systems.** This combination allows the transfer of information from the design stage to the planning stage for the manufacture of a product, without the need to manually reenter the data on part geometry. The database developed during CAD is stored; then it is processed further by CAM into the necessary data and instructions for operating and controlling production machinery and material-handling equipment and for performing automated testing and inspection for product quality (Section 4.8.3).

The emergence of CAD/CAM has had a major impact on manufacturing by standardizing product development and by reducing design effort, evaluation, and prototype work; it has made possible significant cost reductions and improved productivity. The two-engine Boeing 777 passenger airplane, for example, was designed completely by computer (**paperless design**), with 2000 workstations linked to eight computers. The plane is constructed directly from the CAD/CAM software developed (an enhanced CATIA system), and no prototypes or mock-ups were built, such as were required for previous models. The cost for this development was on the order of $6 billion.

An example of an important feature of CAD/CAM in machining is its capability to describe the *cutting-tool path* for various operations such as NC turning, milling, and drilling (see Section 14.4). The instructions (programs) are computer generated, and they can be modified by the programmer to optimize the tool path. The engineer or technician can then display and visually check the tool path for possible tool collisions with clamps or fixtures or for other interferences. The tool path can be modified at any time to accommodate other part shapes to be machined. CAD/CAM systems are also capable of *coding and classifying parts* into groups that have similar shapes, using alphanumeric coding. (See the discussion of *group technology* in Section 15.8.)

## 15.6 | Computer-Aided Process Planning

For a manufacturing operation to be efficient, all of its diverse activities must be planned and coordinated; this task has traditionally been done by process planners. *Process planning* involves selecting methods of production, tooling, fixtures, machinery,

**FIGURE 15.8**  An example of a simple routing sheet. These operation sheets may include additional information on materials, tooling, estimated time for each operation, processing parameters (such as cutting speeds and feeds), and other details. The routing sheet travels with the part from operation to operation. The current trend is to store all relevant data in computers and to affix to the part a bar code that serves as a key into the database of parts information.

ROUTING SHEET

CUSTOMER'S NAME: Midwest Valve Co.    PART NAME: Valve body

QUANTITY: 15    PART NO.: 302

| Operation no. | Description of operation | Machine |
|---|---|---|
| 10 | Inspect forging, check hardness | Rockwell tester |
| 20 | Rough machine flanges | Lathe No. 5 |
| 30 | Finish machine flanges | Lathe No. 5 |
| 40 | Bore and counterbore hole | Boring mill No. 1 |
| 50 | Turn internal grooves | Boring mill No. 1 |
| 60 | Drill and tap holes | Drill press No. 2 |
| 70 | Grind flange end faces | Grinder No. 2 |
| 80 | Grind bore | Internal grinder No. 1 |
| 90 | Clean | Vapor degreaser |
| 100 | Inspect | Ultrasonic tester |

the sequence of operations, the standard processing time for each operation, and methods of assembly; these choices are all documented on a **routing sheet** (Fig. 15.8). When done manually, this task is highly labor intensive and time consuming; also, it relies heavily on the experience of the process planner.

*Computer-aided process planning* (CAPP) accomplishes this complex task by viewing the total operation as an integrated system, so that the individual operations and steps involved in making each part are coordinated with each other and are performed efficiently and reliably. CAPP is thus an essential adjunct to CAD and CAM. Although it requires extensive software and good coordination with CAD/CAM as well as with other aspects of integrated manufacturing systems (described throughout the rest of this chapter), CAPP is a powerful tool for efficiently planning and scheduling manufacturing operations. It is particularly effective in small-volume, high-variety parts production requiring machining, forming, and assembly operations.

### 15.6.1 Elements of CAPP systems

There are two types of computer-aided process-planning systems: variant and generative process-planning systems. These types of systems are described as follows:

1. In the **variant system**, also called the **derivative system**, the computer files contain *a standard process plan* for a particular part to be manufactured. The search for a standard plan is made in the database, using a code number for the part; the plan is based on the part's shape and its manufacturing characteristics. (See Section 15.8.) The standard plan is retrieved, displayed for review, and printed as a routing sheet. The process plan includes information such as the types of tools and machines to be used, the sequence of operations to be performed, the speeds, the feeds, the time required for each sequence, and so on. Minor modifications to an existing process plan, which generally are necessary, can also be made. If the standard plan for a particular part is not in the computer files, a plan that is similar to it and that has a similar code number and an existing routing sheet is retrieved. If a routing sheet does not exist, one is prepared for the new part and stored in computer memory.

2. In the **generative system,** a process plan is automatically generated on the basis of the same logical procedures that would be followed by a traditional process planner in making that particular part. However, the generative system is complex, because it must contain comprehensive and detailed knowledge of the part's shape and dimensions, process capabilities, the selection of manufacturing methods, machinery, tools, and the sequence of operations to be performed. (Computers with such capabilities, known as **expert systems,** are described in Section 15.13.) The generative system is capable of creating a new plan instead of having to use or modify an existing plan, as the variant system must do. Although it currently is used less commonly than is the variant system, the generative system has such advantages as (a) flexibility and consistency for process planning for new parts and (b) higher overall planning quality, because of the capability of the decision logic in the system to optimize the planning and to use up-to-date manufacturing technology.

The process-planning capabilities of computers can be integrated into the planning and control of production systems. Process-planning activities are a subsystem of computer-integrated manufacturing, as described in Section 15.3. Several functions can be performed using these activities, such as **capacity planning** for plants to meet production schedules, control of *inventory*, *purchasing*, and *production scheduling*.

**Advantages of CAPP systems.**   The advantages of CAPP systems over traditional process-planning methods include the following:

a) The standardization of process plans improves the productivity of process planners, reduces lead times and costs of planning, and improves the consistency of product quality and reliability.

b) Process plans can be prepared for parts that have similar shapes and features, and the plans can be easily retrieved to produce new parts.

c) Process plans can be modified to suit specific needs.

d) Routing sheets can be prepared more quickly, and compared with the traditional handwritten routing sheets, computer printouts are neater and much more legible.

e) Other functions, such as cost estimation and work standards, can be incorporated into CAPP.

## 15.6.2  Material-requirements planning systems and manufacturing resource planning systems

Computer-based systems for managing inventories and delivery schedules of raw materials and tools are called *material-requirements planning* (MRP) systems. This activity, sometimes regarded as a method of inventory control, involves keeping complete records of inventories of materials, supplies, parts in various stages of production (work in progress), orders, purchasing, and scheduling. Several files of data are usually involved in a master production schedule. These files pertain to the raw materials required (**bill of materials**), product structure levels (i.e., individual items that compose a product, such as components, subassemblies, and assemblies), and scheduling.

A further development is *manufacturing resource planning* (MRP-II) systems, which, through feedback, control all aspects of manufacturing planning. Although the system is complex, MRP-II is capable of final production scheduling, monitoring actual results in terms of performance and output, and comparing those results against the master production schedule.

### 15.6.3 Enterprise resource planning

Beginning the 1990s, *enterprise resource planning* (ERP) became an important trend; it is basically an extension of MRP-II. Although there are variations, it generally has been defined as a method for effective planning and control of all the resources needed in a business enterprise (i.e., companies) to take orders for products, produce them, ship them to the customer, and service them. ERP thus attempts to coordinate, optimize, and dynamically integrate all information sources and the widely diverse technical and financial activities in a manufacturing organization, as outlined in Chapters 14, 15, and 16. Among its major goals are to improve productivity, reduce manufacturing cycle times, and optimize processes, benefiting not only the organization, but also the customer, in an increasingly competitive global marketplace.

However, truly effective implementation of ERP is a difficult and challenging task. The main reasons are the following:

a) Difficulties are encountered in attempting timely, effective, and reliable communication among all parties involved, especially in a global business enterprise—thus teamwork is essential;

b) There is a need for changing and evolving business practices in an age where information systems and e-commerce have become highly relevant to the successful future of business organizations;

c) Extensive and specific hardware and software requirements must be met for ERP.

## 15.7 | Computer Simulation of Manufacturing Processes and Systems

With increasing power and sophistication of computer hardware and software, one area that has grown rapidly is *computer simulation* of manufacturing processes and systems. Process simulation takes two basic forms:

a) It is a model of a specific operation, intended to determine the viability of a process or to optimize or improve its performance, and

b) It models multiple processes and their interactions, and it helps process planners and plant designers in the layout of machinery and facilities.

Individual processes have been modeled using various mathematical schemes. Finite-element analysis has been increasingly applied in software packages (**process simulation**) that are commercially available and inexpensive. Typical problems addressed by such models include **process viability** (such as assessing the formability and behavior of a certain sheet metal in a certain die) and **process optimization** (such as analyzing the metal-flow pattern in forging in a given die to identify potential defects, or improving mold design in casting to reduce or eliminate hot spots and minimize defects by promoting uniform cooling).

Simulation of an entire manufacturing system, involving multiple processes and equipment, helps plant engineers to organize machinery and to identify critical machinery elements. In addition, such models can assist manufacturing engineers with scheduling and routing by **discrete-event simulation**. Commercially available software packages are often used for these simulations, but the use of dedicated software programs written for a particular company and line of products is not unusual.

**EXAMPLE 15.1**  Simulation of plant-scale manufacturing

Several examples and case studies presented in this text have focused upon simulation of individual manufacturing processes. However, the proliferation of low-cost, high-performance computer systems and the development of advanced software have allowed the simulation of entire manufacturing systems and have led to the optimization of manufacturing and assembly operations.

As an example, Envision software, produced by the Delmia Corporation, allows the simulation of manufacturing processes in three dimensions, including the use of human manikins, to identify safety hazards, manufacturing problems, or bottlenecks; improve machining accuracy; or optimize tooling organization. Since simulation can be performed prior to building an assembly line, it can significantly reduce development times and cost. For example, Fig. 15.9 shows a simulation of a robotic welding line in an automotive plant, where a collision between a robot and a vehicle has been detected. The program for the robot can thereby be modified to prevent such collisions before the line is put into operation. While this example is a powerful demonstration of the utility of system simulation, the more common application is to optimize the sequence of operations and organization of machinery to reduce manufacturing costs.

Software also has the capability of conducting ergonomic analysis of various operations and machinery setups and therefore of identifying bottlenecks in the movement of parts, equipment, or personnel. The bottlenecks can then be relieved by the process planner by adjusting the automated or manual procedures at these locations. Using such techniques, a Daimler-Chrysler facility in Rastatt, Germany, was able to balance its production lines so that each worker is busy an average of 85% to 95% of the time.

Another application of systems simulation is planning of manufacturing operations to optimize production and to prepare for just-in-time production (Section 15.11). For example, if a car manufacturer needs to produce 1000 vehicles in a given period of time, production can be optimized by using certain strategies, such as by distributing the number of vehicles with sunroofs throughout the day

**FIGURE 15.9**  Simulation of a robotic welding station. A collision has been detected that production engineers can rectify before building the assembly line, thereby reducing development time and cost. *Source*: Mechanical Engineering, March 2001.

or by grouping vehicles by color, so that the number of paint changes in paint booths are minimized. With respect to just-in-time production, software such as that produced by ILOG Corporation can plan and schedule plant operations far enough in advance to order material as it is needed, thus eliminating stockpiled inventory.

The Nissan Corporation used the ILOG software to expand production at its Sunderland, England, facility. The plant was already producing 7,000 vehicles per week when Nissan decided to increase production by 3,000 vehicles per week. The conventional solution would have been to construct an additional assembly line, at great capital investment. However, it was noted that the existing scheduling software, based on the input of plant managers, did not take into account variables such as part availability and tool changeover, so that the plant adhered to its schedule only 3% of the time. The ILOG software, on the other hand, allows optimization of the manufacturing sequence and schedule, enabling cars to cross over from assembly lines to paint lines and tooling lines as needed to increase throughput. The result was that the Sunderland plant met the vehicle scheduling targets 85% of the time, eliminating much of the time-consuming task of rescheduling vehicles while they wait in storage buffers between the main sections of the plant. This approach led to increased production, and the cost of the software was recovered in three days of production.

*Source*: Adapted from *Mechanical Engineering*, March 2001.

## 15.8 | Group Technology

It can be seen throughout this text that many parts produced have certain similarities in their shape and in their method of manufacture. Traditionally, each part has been viewed as a separate entity, and each has been produced in individual batches. *Group technology* (GT) is a concept that seeks to take advantage of the **design and processing similarities** among the parts to be produced. This concept, first developed in Europe in the early 1900s, starts by categorizing the parts and recording them; the designs are then retrieved as needed. The term *group technology* was first used in 1959, but not until the use of interactive computers became widespread in the 1970s did this technology develop significantly.

The similarity in the characteristics of a group of parts (Fig. 15.10) suggests that benefits can be obtained by *classifying* and *coding* these parts into *families*. Surveys in manufacturing plants have repeatedly indicated the commonness of similarity in parts. Such surveys consist of breaking each product into its components and then identifying the similar parts; one survey, for instance, found that 90% of the 3000 parts made by a particular company fell into only five major families of parts. As an example, a pump can be broken into such basic components as the motor, the housing, the shaft, the seals, and the flanges. In spite of the variety of pumps manufactured, each of these components is basically the same in terms of design and manufacturing methods. Consequently, all shafts, for example, can be categorized into one family of shafts. Also, questions have been raised, for example, about why a particular product has so many different sizes of fasteners, such as bolts and nuts.

The group-technology approach becomes even more attractive in view of consumer demand for an ever-greater variety of products, each in smaller quantity, thus involving *batch production* (see Section 14.2.2). Maintaining high efficiency in batch operations can be difficult, and overall manufacturing efficiency can be adversely affected because nearly 75% of manufacturing today is batch production. The traditional

Part 1
10 parts per month
AISI 1020 steel
$R_q<0.01 \ \mu$m

Part 2
10,000 parts per month
Polyester
$R_q<0.1 \ \mu$m

(a)

(b)

**FIGURE 15.10** (a) Two parts with identical geometries but different manufacturing attributes. (b) Parts with similar manufacturing attributes but different geometries.

*product flow* in a batch manufacturing operation is shown in Fig. 15.11a. Note that machines of the same type are arranged in groups, that is, groups of lathes, of milling machines, of drill presses, and of grinders. In such a layout (**functional layout**), there is usually considerable random movement on the production floor, as shown by the arrows, which indicate movement of materials and parts. Such an arrangement is not efficient, because it wastes time and effort. The machines in *cellular manufacturing* (Section 15.9) are arranged in a more efficient product flow line (**group layout**; Fig. 15.11b).

## 15.8.1 Advantages of group technology

The major advantages of group technology are the following:

a) Group technology makes possible standardization of part designs and minimization of design duplication. New part designs can be developed using similar, but previously specified, designs, and in this way, a significant amount of time and effort can be saved. The product designer can quickly determine whether data on a similar part already exist in the computer files.

b) Data that reflect the experience of the designer and the process planner are stored in the database. A new and less experienced engineer can, therefore, quickly benefit from that experience by retrieving any of the previous designs and process plans.

c) Manufacturing costs can be estimated more easily, and the relevant statistics on materials, processes, number of parts produced, and other factors can be more easily obtained.

**FIGURE 15.11**
(a) Functional layout of machine tools in a traditional plant. Arrows indicate the flow of materials and parts in various stages of completion. (b) Group-technology (cellular) layout. Legend: L = lathe; M = milling machine; D = drilling machine; G = grinding machine; A = assembly. *Source*: M. P. Groover.

(a)

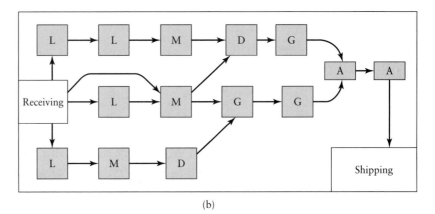

(b)

d) Process plans are standardized and scheduled more efficiently, orders are grouped for more efficient production, and machine utilization is improved. Further-more, setup times are reduced, and parts are produced more efficiently and with better and more consistent product quality. Also, similar tools, fixtures, and machinery are shared in the production of a family of parts, and programming for NC is more automated.

e) With the implementation of CAD/CAM, cellular manufacturing, and CIM, group technology is capable of greatly improving productivity and reducing costs in small-batch production (almost approaching those of mass produc-tion). Depending on the level of implementation, potential savings in each of the various design and manufacturing phases have been estimated to range from 5% to as much as 75%.

## 15.8.2 Classification and coding of parts

In group technology, parts are identified and grouped into families by **classification and coding (C/C) systems**. This process is a critical and complex first step in GT, and it is done according to the part's design and manufacturing attributes (see Fig. 15.10):

1. **Design attributes** pertain to similarities in geometric features and consist of the following:

   a)  External and internal shapes and dimensions;

   b)  Aspect ratio (length-to-width ratio or length-to-diameter ratio);

   c)  Dimensional tolerance;

   d)  Surface finish;

   e)  Part function.

2. **Manufacturing attributes** pertain to similarities in the methods and the sequence of the operations performed on the part. As described throughout this text, selection of a manufacturing process or processes depends on numerous factors, including the workpiece; consequently, manufacturing and design attributes are interrelated. The manufacturing attributes of a particular part or component consist of the following:

   a)  Primary processes;

   b)  Secondary and finishing processes;

   c)  Dimensional tolerances and surface finish;

   d)  Sequence of operations performed;

   e)  Tools, dies, fixtures, and machinery;

   f)  Production quantity and production rate.

In reviewing these lists, it can be appreciated that the coding can be time consuming and that it requires considerable experience in the design and manufacture of products. In its simplest form, coding can be done by viewing the shapes of the parts in a generic way and then classifying the parts accordingly, such as in the categories of parts having rotational symmetry, parts having rectilinear shape, and parts having large surface-to-thickness ratios. Parts may also be classified by studying their production flow throughout the total manufacturing cycle, called **production-flow analysis** (PFA). Recall from Section 15.6 that routing sheets clearly show process plans and the operations to be performed. One drawback of PFA is that a particular routing sheet does not necessarily indicate that the total operation is optimized. In fact, depending on the experience of the particular process planner, routing sheets for manufacturing the same part can be quite different. The benefits of computer-aided process planning as regards avoiding such problems are obvious.

## 15.8.3  Coding

Coding of parts may be based on a particular company's own coding system, or it may be based on one of several commercially available classification and coding systems. Because of widely varying product lines and organizational needs, none of the C/C systems has been universally adopted. Whether it was developed in-house or it was purchased, the system must be compatible with the company's other systems, such as NC machinery and CAPP systems. The *code structure* for part families typically consists of numbers, of letters, or of a combination of the two. Each specific component of a product is assigned a code; this code may pertain to design attributes only (generally less than 12 digits) or to manufacturing attributes only, although most advanced systems include both, using as many as 30 digits.

The three basic **levels of coding** outlined as follows vary in their degree of complexity:

1. **Hierarchical coding.** In this type of coding, also called **monocode**, the interpretation of each succeeding digit depends on the value of the preceding

**FIGURE 15.12** Decision-tree classification for a sheet-metal bracket. *Source:* G. W. Millar.

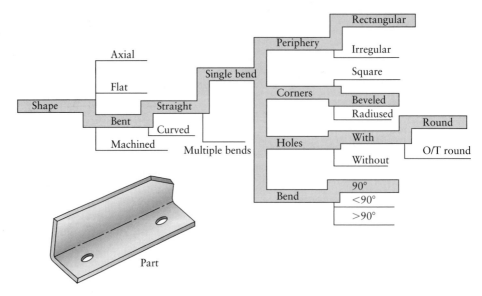

Part

digit. Each symbol amplifies the information contained in the preceding digit, so a single digit in the code cannot be interpreted alone. The advantage of this system is that a short code can contain a large amount of information. This method is, however, difficult to apply in a computerized system.

2. **Polycodes.** Each digit in this code, also known as **chain-type code**, has its own interpretation, which does not depend on the preceding digit. This structure tends to be relatively long, but it allows the identification of specific part attributes and is well suited to computer implementation.

3. **Decision-tree coding.** This system, also called **hybrid codes**, is the most advanced, and it combines both design and manufacturing attributes (Fig. 15.12).

### 15.8.4 Coding systems

Three major industrial coding systems are described as follows:

a)  The **Opitz system** was developed in the 1960s in Germany by H. Opitz (1905–1977), and it was the first comprehensive coding system ever presented. The basic code consists of nine digits (in the format 12345 6789) that represent design and manufacturing data (Fig. 15.13); four additional codes (in the format ABCD) may be used to identify the type and sequence of production operations. This system has two drawbacks: (a) It is possible to have different codes for parts that have similar manufacturing attributes, and (b) several parts with different shapes can have the same code.

b)  The **MultiClass system** was originally developed under the name MICLASS, for Metal Institute Classification System (Fig. 15.14), for the purpose of helping automate and standardize several design, production, and management functions in an organization. This system involves up to 30 digits; it is used interactively with a computer that asks the user a number of questions. On the basis of the answers given, the computer automatically assigns a code number to the part.

c)  The **KK-3 system** is a general-purpose classification and coding system for parts that are to be machined or ground. It uses a 21-digit decimal system. This code is much greater in length than the two coding systems described previously, but it classifies dimensions and dimensional ratios such as the length-to-diameter ratio of the part. The structure of a KK-3 system for rotational components is shown in Fig. 15.15.

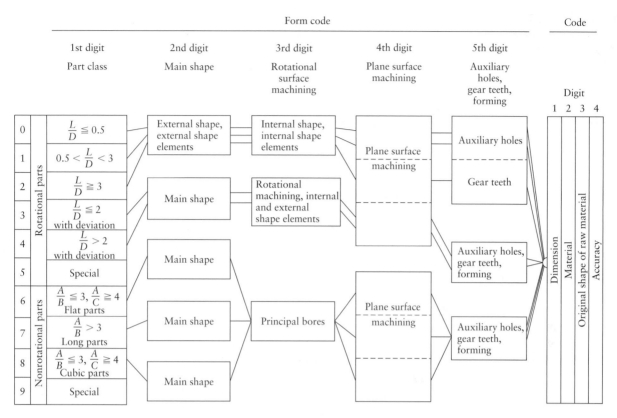

**FIGURE 15.13** A classification and coding scheme using the Opitz system, consisting of five digits and a supplementary code of four digits.

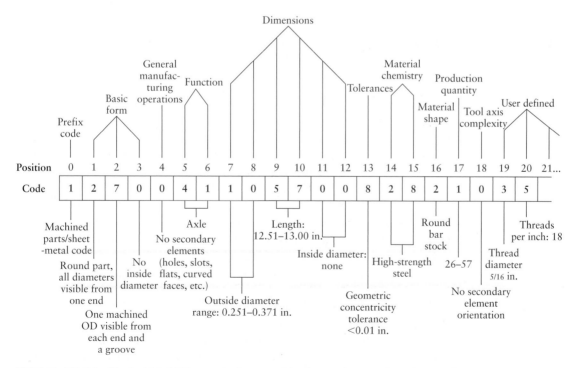

**FIGURE 15.14** Typical MultiClass code for a machined part. *Source:* Organization for Industrial Research.

**FIGURE 15.15** The structure of a KK-3 system for rotational components. *Source*: Japan Society for the Promotion of Machine Industry.

| Digit | Items | (Rotational component) |
|-------|-------|------------------------|
| 1 | Part name | General classification |
| 2 | | Detail classification |
| 3 | Materials | General classification |
| 4 | | Detail classification |
| 5 | Major dimensions | Length |
| 6 | | Diameter |
| 7 | Primary shapes and ratio of major dimensions | |
| 8 | Shape details and kinds of processes / External surface | External surface and outer primary shape |
| 9 | | Concentric screw-threaded parts |
| 10 | | Functional cut-off parts |
| 11 | | Extraordinary shaped parts |
| 12 | | Forming |
| 13 | | Cylindrical surface |
| 14 | Internal surface | Internal primary shape |
| 15 | | Internal curved surface |
| 16 | | Internal flat surface and cylindrical surface |
| 17 | End surface | |
| 18 | Nonconcentric holes | Regularly located holes |
| 19 | | Special holes |
| 20 | Noncutting process | |
| 21 | Accuracy | |

# 15.9 | Cellular Manufacturing

The concept of group technology can be effectively implemented in *cellular manufacturing*, which consists of one or more manufacturing cells. A **manufacturing cell** is a small unit, composed of one to several *workstations*, within a manufacturing system. A workstation usually contains either one machine (**single-machine cell**) or several machines (**group-machine cell**) that each perform a different operation on the part. The machines can be modified, retooled, and regrouped for different product lines within the same family of parts. Manufacturing cells are particularly effective in producing families of parts for which there is relatively constant demand.

Cellular manufacturing has thus far been used primarily in material removal (Chapters 8 and 9) and in sheet-metal forming operations (Chapter 7). The machine tools commonly used in manufacturing cells are lathes, milling machines, drills, grinders, electrical-discharge machines, and machining centers; for sheet-metal forming, the equipment consists of shearing, punching, bending, and other forming machines. Cellular manufacturing has some degree of automatic control for the following typical operations:

1. Loading and unloading of raw materials and workpieces at workstations;
2. Changing of tools and dies at workstations, thus reducing labor and time;
3. Transfer of workpieces and tools and dies between workstations;
4. Scheduling and controlling of the total operation in the manufacturing cell.

Central to these activities is a *material-handling system* for transferring materials and parts among workstations. In attended (manned) machining cells, materials can be moved and transferred manually by the operator or by an industrial robot (Section 14.7) located centrally in the cell. Automated inspection and testing equipment may also be components of such a cell.

The significant benefits of cellular manufacturing are the economics of reduced work in progress, improved productivity, and the fact that product quality problems are detected right away. Furthermore, because of the variety of the machines and the processes involved, the operator becomes multifunctional and is not subjected to the tedium experienced with always working on the same machine.

**Manufacturing cell design.** Because of the unique features of manufacturing cells, their design and implementation in traditional plants inevitably requires the reorganization of the plant and the rearrangement of existing product flow lines. The machines may be arranged along a line, a $U$ shape, an $L$ shape, or in a loop. For a group-machine cell, where the materials are handled by the operator, the $U$-shaped arrangement is convenient and efficient, because the operator can easily reach various machines. With mechanized or automated material handling, the linear arrangement and the loop layout are more efficient. Selecting the best arrangement of machines and material-handling equipment also involves taking into account such factors as the production rate, the type of product, and its shape, size, and weight.

**Flexible manufacturing cells.** In the introduction to this chapter and in Chapter 14, it was stressed that in view of rapid changes in market demand and the need for more product variety in smaller quantities, flexibility of manufacturing operations is highly desirable. Manufacturing cells can be made flexible by using CNC machines and machining centers (see Section 8.10) and by means of industrial robots or other mechanized systems for handling materials. Figure 15.16 shows an example of a *flexible manufacturing cell* (FMC) for machining operations.

Flexible manufacturing cells are usually untended, so their design and operation are more exacting than those for other cells. The selection of machines and robots, including the types and capacities of end effectors and their control systems, is important to the proper functioning of the FMC. The likelihood of significant change in demand for part families should be considered during design of the cell, to ensure that the equipment involved has the proper flexibility and capacity. As is the case for other flexible manufacturing systems (described in Section 15.10), the cost of flexible manufacturing cells is high. Cellular manufacturing generally requires more machine tools and thus increases manufacturing cost; however, this disadvantage is

**FIGURE 15.16** Schematic view of a flexible manufacturing cell, showing two machine tools, an automated part inspection system, and a central robot serving these machines. *Source*: P. K. Wright.

Machining center

Machining center

Anthropomorphic robot

Control unit

Vision-based inspection

outweighed by increased speed of manufacture, flexibility, and controllability. Proper maintenance of the tools and the machinery is essential, as is the implementation of two- or three-shift operation of the cells.

---

**EXAMPLE 15.2 Manufacturing cells in small shops**

Beginning in the mid-1990s, large companies have increasingly outsourced machining operations to their first-tier suppliers. This trend has created pressure for small job shops to develop the capability to perform sophisticated projects, and it also has created a demand for larger firms to take on lower volume, or turnkey, operations. These two basic market factors have driven a number of trends in the design and implementation of flexible manufacturing cells:

- Manufacturing cells are being developed with a smaller number of machines, but greater versatility. Instead of four or five machines, many new manufacturing cells consist of only one or two machines.
- Advanced CAD systems are being used by smaller firms. While high-end CAD systems running on computer workstations are available at larger corporations, increasing computational power and software accessibility has allowed smaller firms to use less expensive, but still very powerful, PC-based packages.
- PC controllers are becoming more popular and are driving a push to open architecture. Historically, machine manufacturers maintained full control over their control systems, and PC-based control was limited. Open architecture enables machine control to take place from a PC and therefore allows much greater flexibility in defining environments for machine operators.
- PC-based software is now available to translate solid-model part descriptions into surfaces for machining.

---

## 15.10 | Flexible Manufacturing Systems

A *flexible manufacturing system* (FMS) integrates all major elements of manufacturing into a highly automated system. First used in the late 1960s, FMS consists of a number of manufacturing cells, each containing an industrial robot (serving several CNC machines) and an automated material-handling system, all interfaced with a central computer. Different computer instructions for the manufacturing process can be downloaded for each successive part that passes through the workstation.

This system is highly automated and is capable of optimizing each step of the total manufacturing operation. These steps may involve one or more processes and operations (such as machining, grinding, cutting, forming, powder metallurgy, heat treating, and finishing), as well as handling of raw materials, inspection, and assembly. The most common applications of FMS to date have been in machining and assembly operations. A variety of FMS technology is available from machine-tool manufacturers. Flexible manufacturing systems represent the highest level of efficiency, sophistication, and productivity that has been achieved in manufacturing plants (Fig. 15.17). The flexibility of FMS is such that it can handle a variety of part configurations and produce them in any order.

FMS can be regarded as a system that combines the benefits of two other systems: (a) the highly productive, but inflexible, transfer lines (Section 14.2.4) and (b)

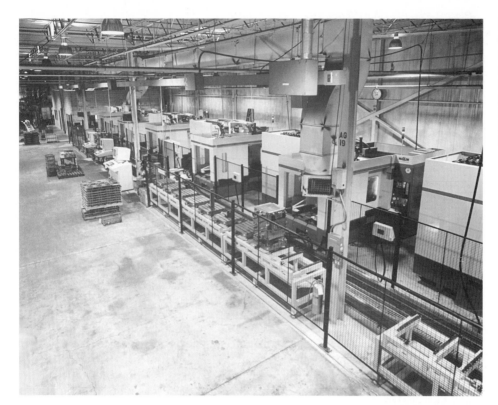

**FIGURE 15.17**   A general view of a flexible manufacturing system, showing several machine tools and an automated guided vehicle. *Source*: Courtesy of Cincinnati Machine, a UNOVA Company.

job-shop production, which can fabricate a large variety of products on stand-alone machines, but is inefficient. Table 15.1 compares some of the characteristics of transfer lines and FMS. Note that in FMS, the time required for changeover to a different part is very short. The quick response to product and market-demand variations is a major benefit of FMS.

**Elements of FMS.**   The basic elements of a flexible manufacturing system are (a) workstations, (b) automated handling and transport of materials and parts, and (c) control systems. The workstations are arranged to yield the greatest efficiency in production, with an orderly flow of materials, parts, and products through the system.

The types of machines in workstations depend on the type of production. For machining operations (Chapter 8), they usually consist of a variety of three- to five-

TABLE 15.1

**Comparison of the Characteristics of Transfer Lines and Flexible Manufacturing Systems**

| Characteristic | Transfer line | FMS |
|---|---|---|
| Types of parts made | Generally few | Infinite |
| Lot size | > 100 | 1–50 |
| Part changing time | $\frac{1}{2}$ to 8 hr | 1 min |
| Tool change | Manual | Automatic |
| Adaptive control | Difficult | Available |
| Inventory | High | Low |
| Production during breakdown | None | Partial |
| Efficiency | 60–70% | 85% |
| Justification for capital expenditure | Simple | Difficult |

axis machining centers, CNC lathes, milling machines, drill presses, and grinders. Also included is various other equipment, such as that for automated inspection (including coordinate measuring machines), assembly, and cleaning. Other types of operations suitable for FMS include sheet-metal forming, punching and shearing, and forging systems for these operations incorporate furnaces, forging machines, trimming presses, heat-treating facilities, and cleaning equipment.

Because of the flexibility of FMS, the proper operation of material-handling, storage, and retrieval systems is very important. Material handling is controlled by a central computer and performed by automated guided vehicles, conveyors, and various transfer mechanisms. The system is capable of transporting raw materials, blanks, and parts in various stages of completion to any machine, in any order, at any time. Prismatic parts are usually moved on specially designed **pallets**. Parts having rotational symmetry (such as parts for turning operations) are usually moved by mechanical devices and robots.

The computer control system of FMS is its brains and includes various software and hardware. This subsystem controls the machinery and equipment in workstations and the transportation of raw materials, blanks, and parts in various stages of completion from machine to machine. It also stores data and provides communication terminals that display the data visually.

**Scheduling.**  Because FMS involves a major capital investment, efficient machine utilization is essential: Machines must not stand idle. Consequently, proper scheduling and process planning are crucial. Scheduling for FMS is *dynamic*, unlike that in job shops, where a relatively rigid schedule is followed to perform a set of operations. The scheduling system for FMS specifies the types of operations to be performed on each part, and it identifies the machines or manufacturing cells to be used. Dynamic scheduling is capable of responding to quick changes in product type and thus is responsive to real-time decisions.

Because of the flexibility provided by FMS, no setup time is wasted in switching between manufacturing operations; the system is capable of performing different operations in different orders and on different machines. However, the characteristics, performance, and reliability of each unit in the system must be checked to ensure that parts moving from workstation to workstation are of acceptable quality and dimensional accuracy.

# 15.11 | Just-in-Time Production

The *just-in-time production* (JIT) concept was first implemented in Japan to eliminate waste of materials, machines, capital, manpower, and inventory throughout the manufacturing system. The JIT concept has the following goals:

1. Receive supplies just in time to be used;
2. Produce parts just in time to be made into subassemblies;
3. Produce subassemblies just in time to be assembled into finished products;
4. Produce and deliver finished products just in time to be sold.

In traditional manufacturing, the parts are made in batches, placed in inventory, and used whenever necessary. This approach is known as a **push system**, meaning that parts are made according to a schedule and are in inventory to be used if and when they are needed. In contrast, just-in-time manufacturing is a **pull system**, meaning that parts are produced to order, and the production is matched with demand for the final assembly of products. There are no stockpiles, and thus the extra motions and expenses involved in stockpiling parts and then retrieving them from storage are eliminated. Hence, the ideal production quantity is one product; therefore, this system is also called **zero inventory, stockless production**, and **demand scheduling**. Moreover, parts are inspected by the worker as they are being manufactured and are used within a short period of time. In this way, the worker maintains continuous production control, immediately identifying defective parts and reducing process variation in order to produce quality products. The worker takes pride in good product quality.

Implementation of the JIT concept requires that all aspects of manufacturing operations be continuously reviewed and monitored, so that all operations and resources that do not add value are eliminated. This approach emphasizes pride and dedication in producing high-quality products; elimination of idle resources; and the use of teamwork among workers, engineers, and management to quickly solve any problems that arise during production or assembly. The ability to detect production problems has been likened to the level of water (representing the inventory levels) in a lake covering a bed of boulders (representing production problems). When the water level is high (the high inventories associated with push production), the boulders are not exposed. By contrast, when the level is low (the low inventories associated with pull production), the boulders are exposed; consequently, the problems can be identified and removed. This analogy indicates that high inventory levels can mask quality and production problems involving parts that are already stockpiled.

An important aspect of the JIT concept is the delivery of supplies and parts from outside sources and from other divisions of the company, so as to significantly reduce in-plant inventory. As a result, major reductions in storage facilities can take place, and storage space can be reclaimed for productive purposes; in fact, the concept of building large warehouses for parts has now become obsolete. Suppliers are expected to deliver, often on a daily basis, preinspected goods as they are needed for production. Consequently, this approach requires reliable suppliers, close cooperation and trust between the company and its vendors, and a reliable system of transportation. Also important for smoother operation is reduction of the number of suppliers. In one example, an Apple Computer plant reduced the number of suppliers from 300 to 70.

**Kanban.** Although the basic concept of JIT originated in the United States decades ago, it was first demonstrated on a large scale in 1953 at the Toyota Motor Company, under the name *kanban*, meaning "visible record." These records usually consist of two types of cards (kanbans): (a) the **production card**, which authorizes the *production* of one container or cart of identical, specified parts at a workstation, and (b) the **conveyance card**, or **move card**, which authorizes the *transfer* of one container or cart of parts from that particular workstation to the workstation where the parts will be used next. The cards, which now consist of bar-coded plastic tags or other devices, contain information about the type of the part, the place of issue, the part number, and the number of items in the container. The number of containers in circulation at any time is completely controlled and can be scheduled as desired for maximum production efficiency.

**Advantages of JIT.** The advantages of just-in-time production may be summarized as follows:

- Low inventory-carrying costs;
- Fast detection of defects in the production or the delivery of supplies and, hence, low scrap loss;
- Reduced need for inspection and reworking of parts;
- Production of high-quality parts at low cost.

Although there are significant variations, implementation of just-in-time production has resulted in reductions estimated at 20% to 40% in product cost; 60% to 80% in inventory; up to 90% in rejection rates; 90% in lead times; and 50% in scrap, rework, and warranty costs. Furthermore, increases of 30% to 50% in direct labor productivity and of 60% in indirect labor productivity have also been achieved.

# 15.12 | Communications Networks in Manufacturing

In order to maintain a high level of coordination and efficiency of operation in integrated manufacturing, an extensive, high-speed, and interactive **communications network** is required. A major advance in communications technology is the **local area network** (LAN). In this hardware and software system, logically related groups of machines and equipment, such as those in a manufacturing cell, communicate with each other; the network links these groups to each other, bringing different phases of manufacturing into a unified operation. A local area network can be very large and complex, linking hundreds, or even thousands, of machines and devices in several buildings; various network layouts (Fig. 15.18) of fiber-optic or copper cables are typically used, over distances ranging from a few meters to as much as 32 km (20 mi). For longer distances, **wide area networks** (WANs) are used. Different types of networks can be linked or integrated through "gateways" and "bridges."

Access control to the network is important; otherwise, collisions can occur when several workstations are transmitting simultaneously. Thus, continuous scanning of the transmitting medium is essential. In the 1970s, a *carrier sense multiple access with collision detection* (CSMA/CD) system was developed and implemented in **Ethernet**. Other access-control methods are **token ring** and **token bus**, in which a "token" (special message) is passed from device to device; the device that has the token is allowed to transmit, while all other devices only receive.

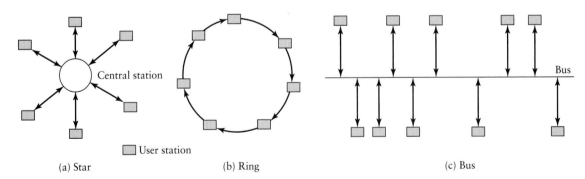

(a) Star                    (b) Ring                    (c) Bus

**FIGURE 15.18**    Three basic types of topology for a local area network (LAN). (a) The star topology is suitable for situations that are not subject to frequent configuration changes. All messages pass through a central station. Telephone systems in office buildings usually have this type of topology. (b) In the ring topology, all individual user stations are connected in a continuous ring. The message is forwarded from one station to the next until it reaches its assigned destination. Although the wiring is relatively simple, the failure of one station shuts down the entire network. (c) In the bus topology, all stations have independent access to the bus. This system is reliable and is easier to service than the other two. Because its arrangement is similar to the layout of the machines in a factory, its installation is relatively easy, and it can be rearranged when the machines are rearranged.

A new trend is the use of **wireless local area networks** (WLAN). Conventional LANs require routing of wires, often through masonry walls or other permanent structures, and require computers or machinery to remain stationary. WLANs allow equipment such as mobile test stands or data collection devices, such as bar code readers, to easily maintain a network connection. A communication standard (IEEE 802.11) currently defines frequencies and specifications of signals, and defines two radio frequency and one infrared methods. It should be noted that wireless networks are slower than those that are hard wired, but flexibility makes them desirable, especially for situations where slow tasks such as machine monitoring are the main application.

**Personal area networks** (PAN) are under development. PANs are based on communications standards such as Bluetooth, IrDA, and HomeRF, and are designed to allow data and voice communication over short distances. For example, a short-range Bluetooth device will allow communication over a 10 m (32 ft) distance. PANs are undergoing major changes, and communications standards are continually being refined.

## 15.12.1  Communications standards

Often, one manufacturing cell is built with machines and equipment purchased from one vendor, another cell with machines purchased from another vendor, a third cell with machines purchased from yet a third vendor, and so on. As a result, a variety of programmable devices is involved, driven by several computers and microprocessors purchased at various times from different vendors and having various capacities and levels of sophistication. Each cell's computers have their own specifications and proprietary standards, and they cannot communicate beyond the cell with others unless equipped with custom-built interfaces. This situation creates *islands of automation*, and in some cases, up to 50% of the cost of automation has been related to overcoming difficulties in the communications between individual manufacturing cells and other parts of the organization.

The existence of automated cells that could function only independently of each other, without a common base for information transfer, led to the need for a communications standard to improve communications and efficiency in computer-integrated manufacturing. After considerable effort and on the basis of existing national and international standards, a set of communications standards known as **manufacturing automation protocol** (MAP) was developed. The capabilities and effectiveness of MAP were first demonstrated in 1984 with the successful interconnection of devices from a number of vendors.

The International Organization for Standardization/Open System Interconnect (ISO/OSI) reference model is accepted worldwide. This model has a hierarchical structure in which communication between two users is divided into seven layers (Fig. 15.19). Each layer has a special task: (a) mechanical and electronic means of data transmission, (b) error detection and correction, (c) correct transmission of the message, (d) control of the dialog between users, (e) translation of the message into a common syntax, and (f) verification that the data transferred have been understood.

The operation of this system is complex. Basically, each standard-sized chunk of message or data to be transmitted from user *A* to user *B* moves sequentially through the successive layers at user *A*'s end, from layer 7 to layer 1. More information is added to the original message as it travels through each layer. The complete packet is transmitted through the physical communications medium to user *B* and then moves through the layers from layer 1 to layer 7 at user *B*'s end. The transmission takes place through fiber-optic cable, coaxial cable, microwaves, and similar devices.

Communication protocols have been extended to *office automation* as well with the development of **technical and office protocol** (TOP), which is based on the ISO/OSI reference model. In this way, total communication (MAP/TOP) can be established

**FIGURE 15.19** The ISO/OSI reference model for open communication. *Source*: U. Rembold et al., *Computer Integrated Manufacturing and Engineering*, Addison-Wesley, 1993.

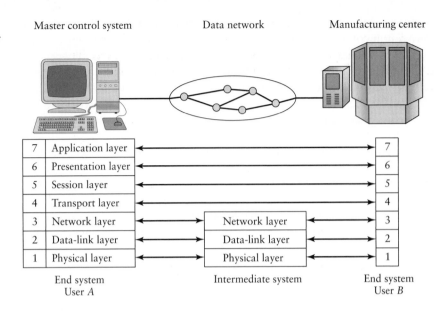

across the factory floor and among offices at all levels of an organization. **Internet tools** (hardware, software, and protocols) within a company now link all departments and functions into a self-contained and fully compatible **Intranet**. Several tools for implementing this linkage are available commercially; they are inexpensive and easy to install, integrate, and use.

## 15.13 | Artificial Intelligence

*Artificial intelligence* (AI) is an area of computer science concerned with systems that exhibit some characteristics which are usually associated with intelligence in human behavior, such as learning, reasoning, problem solving, and the understanding of language. The goal of AI is to simulate such human endeavors on the computer. The art of bringing relevant principles and tools of AI to bear on difficult application problems is known as **knowledge engineering**. Artificial intelligence is having a major impact on the design, the automation, and the overall economics of manufacturing operations, in large part because of advances in computer-memory expansion (VLSI chip design) and decreasing costs. Artificial-intelligence packages costing on the order of a few thousand dollars have been developed, many of which can now be run on personal computers; thus, AI has become accessible to office desks and shop floors.

Artificial-intelligence applications in manufacturing generally encompass the following:

a) Expert systems;

b) Natural language;

c) Machine (computer) vision;

d) Artificial neural networks;

e) Fuzzy logic.

### 15.13.1 Expert systems

Also called a **knowledge-based system**, an *expert system* (ES) is generally defined as an intelligent computer program that has the capability to solve difficult real-life problems, using **knowledge-base** and **inference** procedures (Fig. 15.20). The goal of an expert system is to develop the capability to conduct an intellectually demanding task in the way that a human expert would. The field of knowledge required to perform this task is called the **domain** of the expert system. Expert systems use a knowledge base containing facts, data, definitions, and assumptions. They also have the capacity to follow a **heuristic** approach, that is, to make good judgments on the basis of *discovery* and *revelation* and to make high-probability *guesses*, just as a human expert would. The knowledge base is expressed in computer codes, usually in the form of **if–then rules**, and can generate a series of questions; the mechanism for using these rules to solve problems is called an **inference engine**. Expert systems can also communicate with other computer software packages.

To construct expert systems for solving the complex design and manufacturing problems encountered, one needs (a) a great deal of knowledge and (b) a mechanism for manipulating this knowledge to create solutions. Because of the difficulty involved in accurately modeling the many years of experience of an expert or a team of experts and the complex inductive reasoning and decision-making capabilities of humans, including the capacity to learn from mistakes, the development of knowledge-based systems requires much time and effort.

Expert systems operate on a *real-time* basis, and their short reaction times provide rapid responses to problems. The programming languages most commonly used for this application are C++, LISP, and PROLOG; other languages can also be used. A significant development is expert-system software **shells**, or **environments**, also called **framework systems**. These software packages are essentially expert-system outlines that allow a person to write specific applications to suit special needs. Writing these programs requires considerable experience and time.

Several expert systems have been developed and used for such specialized applications as

1. Problem diagnosis in machines and equipment and determination of corrective actions;
2. Modeling and simulation of production facilities;
3. Computer-aided design, process planning, and production scheduling;
4. Management of a company's manufacturing strategy.

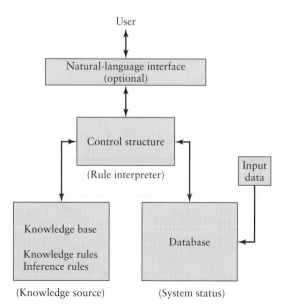

**FIGURE 15.20** The basic structure of an expert system. The knowledge base consists of knowledge rules (general information about the problem) and inference rules (the way conclusions are reached). The results may be communicated to the user through the natural-language interface. *Source*: K. W. Goff, *Mechanical Engineering*, October 1985.

### 15.13.2 Natural-language processing

Traditionally, obtaining information from a database in computer memory has required the use of computer programmers to translate questions expressed in natural language into "queries" in some machine language. Natural-language interfaces with database systems are in various stages of development; these systems allow a user to obtain information by entering English-language commands in the form of simple, typed questions. Natural-language software shells are used in such applications as scheduling of material flow in manufacturing and analyzing information in databases. Major progress continue to be made in the development of computer software that will have speech-synthesis and recognition (**voice recognition**) capabilities, thus eliminating the need to type commands on keyboards.

### 15.13.3 Machine vision

The basic features of *machine vision* are described in Section 14.8. In systems that incorporate machine vision, computers and software implementing artificial intelligence are combined with cameras and other optical sensors. These machines then perform such operations as inspecting, identifying, and sorting parts and guiding robots (*intelligent robots*; Fig. 15.21); in other words, operations that would otherwise require human involvement and intervention.

### 15.13.4 Artificial neural networks

Although computers are much faster than the human brain at *sequential* tasks, humans are much better at pattern-based tasks that can be attacked with *parallel processing*, such as recognizing features (e.g., in faces and voices, even under noisy conditions), assessing situations quickly, and adjusting to new and dynamic conditions. These advantages are also due partly to the ability of humans to use, in real time, their **senses** (i.e., sight, hearing, smell, taste, and touch) simultaneously, a process called *data fusion*. The branch of artificial intelligence called *artificial neural networks* (ANN) attempts to gain some of these human capabilities through computer imitation of the way that data are processed by the human brain.

The human brain has about 100 billion linked **neurons** (cells that are the fundamental functional units of nervous tissue) and more than a thousand times that many connections. Each neuron performs only one simple task: It receives input signals from a fixed set of neurons, and when those input signals are related in a certain way (specific to that particular neuron), it generates an electrochemical output signal that goes to a fixed set of neurons. It is now believed that human learning is accomplished by changes in the strengths of these signal connections between neurons.

**FIGURE 15.21** Expert system as applied to an industrial robot guided by machine vision.

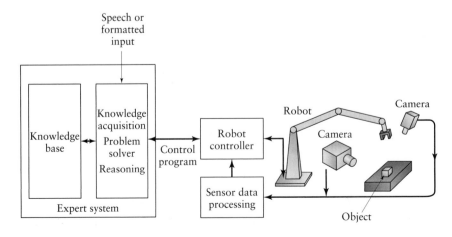

A fully developed, feed-forward network is the most common type of ANN, and it is built according to this principle from several layers of processing elements (simulating neurons). The elements in the first (input) layer are fed with input data—for example, forces, velocities, or voltages. Each element sums up all its inputs—one per element in the input layer, and many per element in succeeding layers. Each element in a layer then transfers the data (according to a transfer function) to all the elements in the next layer. Each element in that next layer, however, receives a different signal, because of the different connection weights between the elements.

The last layer is the output layer, within which each element is compared with the desired output—that of the process being simulated. The difference between the desired output and the calculated output (*error*) is fed back to the network by changing the weights of the connections in a way that reduces this error. After this procedure has been repeated several times, the network has been "trained," and it can thus be used on input data that have not previously been presented to the system.

Other kinds of artificial neural networks are (a) *associative memories*, (b) *self-organizing ANN*, and (c) *adaptive-resonance ANN*. The feature common to these neural networks is that they must be trained with concrete exemplars. It is, therefore, very difficult to formulate input–output relations mathematically and to predict an ANN's behavior with untrained inputs.

Artificial neural networks are used in such applications as noise reduction (in telephones), speech recognition, and process control in manufacturing. For example, they can be used to predict the surface finish of a workpiece obtained by end milling (see Fig. 8.1d) on the basis of input parameters such as cutting force, torque, acoustic emission, and spindle acceleration. Although this field still controversial, the opinion of many is that true artificial intelligence will evolve only through advances being made in ANN.

### 15.13.5 Fuzzy logic

An element of artificial intelligence that has important applications in control systems and pattern recognition is *fuzzy logic* (**fuzzy models**). Introduced in 1965, and based on the observation that people can make good decisions on the basis of imprecise and nonnumerical information, fuzzy models are mathematical means of representing vagueness and imprecise information (hence the term *fuzzy*). These models have the capability to recognize, represent, manipulate, interpret, and use data and information that are vague or lack precision. Fuzzy-logic methods deal with reasoning and decision making at a level higher than do neural networks. Typical linguistic examples of concepts used in fuzzy logic are the following words and terms: *few, very, more or less, small, medium, extremely*, and *almost all*.

Fuzzy-logic technologies and devices have been developed and successfully applied in areas such as robotics and motion control, image processing and machine vision, machine learning, and the design of intelligent systems. Some applications of fuzzy logic include the following: the automatic transmission of Lexus automobiles; a washing machine that automatically adjusts the washing cycle for load size, fabric type, and amount of dirt; and a helicopter that obeys vocal commands to go forward, up, left, and right; to hover; and to land.

## 15.14 | The Factory of the Future

On the basis of the advances made in all aspects of manufacturing technology and computer integration, we may envision the *factory of the future* as a fully automated facility in which human beings will not be directly involved with production on the shop floor (hence the term **untended factories**). All manufacturing, material handling, assembly, and inspection will be done by computer-controlled machinery and equipment. Similarly, activities such as the processing of incoming orders, production planning and scheduling, cost accounting, and various decision-making processes (usually performed by management) will also be done automatically by computers. The role of human beings will be confined to activities such as overall supervision, maintenance (especially preventive maintenance), and the upgrading of machines and equipment; the shipping and receiving of supplies and finished products; the provision of security for the plant facilities; the programming, upgrading, and monitoring of computer programs; and the monitoring, maintenance, and upgrading of computer hardware.

Some businesses in the food, petroleum, and chemical industries already operate automatically, with little human intervention. These industries involve continuous processes that are easier to automate fully than is piece-part manufacturing. Even so, the direct involvement of fewer people in the manufacturing of products is already apparent: Surveys of modern manufacturing facilities show that only 10% to 15% of the workforce is directly involved in production; the majority is involved in the gathering and processing of information. Virtually untended manufacturing cells already make products such as engine blocks, axles, and housings for clutches and air compressors. For large-scale flexible manufacturing systems, however, highly trained and skilled personnel will always be needed in order to plan, maintain, and oversee operations.

The reliability of machines, equipment, control systems, power supplies, and communications networks is crucial to full factory automation. Without rapid human intervention, a local or general breakdown in even one of these components can cripple production. The computer-integrated factory of the future must be capable of automatically rerouting materials and production flows to other machines and to the control of other computers in the case of such emergencies.

In the implementation of advanced concepts of manufacturing, with their attendant impact on the operation of the factory of the future, certain developments and improvements are bound to continue at a rapid pace:

a) Common communications networks, software, and standards for every aspect of manufacturing technology, from product design to manufacturing;

b) Reduction in the size and increase in the efficiency of knowledge-rich controllers, and their integration into machines instead of functioning as separate units;

c) Increasing availability of off-the-shelf control hardware;

d) Increase in the reliability of voice-recognition capabilities, so that machines can understand and implement spoken messages;

e) Tighter sensor fusion, employing more rugged sensors, and improvement of feedback methods for monitoring and diagnosing all aspects of manufacturing operations (both developments are augmented by advances in artificial-intelligence technology);

f) Cost-effective system integration at all levels of an organization.

**Concerns and opportunities.** An important consideration in fully automating factory is the nature and extent of its impact on employment, including all the social and political ramifications. Although economic forecasts indicate that there will be a decline in the number of machine-tool operators and tool-and-die workers, there will be major increases in the number of people working in service occupations (such as computer service technicians and maintenance electricians). Thus, the generally low-skilled, manual-effort labor force traditionally required in manufacturing will evolve into a knowledge-effort labor force that possesses the specialized training or retraining required in such activities as computer programming, information processing, implementation of CAD/CAM, application of information science and technology, and other high-technology tasks. The development of more user-friendly computer software is making retraining of the workforce much easier.

There are many widely divergent opinions about the impact of untended factories. Consequently, predicting the nature of future manufacturing strategies with any certainty is difficult. Although economic considerations and trade-offs are crucial, companies that do not install computer-integrated operations where they will be cost-effective are in obvious jeopardy. It is now widely recognized that, in a highly competitive global marketplace, rapid adaptability is crucial to the survival of a manufacturing organization.

The continuous advances in all aspects of the science, the engineering, and the technology of manufacturing are constantly being analyzed by corporate management, with a view to the near-term and long-term economic impact on their operations. Over the past few years, much has been written, discussed, and debated concerning the relative manufacturing technologies and industrial strengths among many industrialized nations. There are various complex issues stemming from the following considerations:

a) The characteristics of each industrialized nation—in particular, its history, its social and demographic structure, and its standard of living;

b) The nature, educational level, and loyalty of its workforce;

c) The relationships among management, labor, and government;

d) Interactions among companies, research universities, and social organizations;

e) Productivity and the effective utilization of capital equipment;

f) The missions and the operational philosophies of the management of industrial and business organizations;

g) Stockholder attitudes and preferences.

The term **world-class** is frequently used to describe a certain quality and level of manufacturing activities, signifying the fact that products must meet international standards and be acceptable worldwide. It should also be recognized that the designation of world-class, like quality, is not a fixed target for a manufacturing company or a country to reach, but rather a moving target, rising to higher and higher levels as time passes. Manufacturing organizations must be aware of this moving target and plan and execute their programs accordingly.

# 15.15 | **Economic Considerations**

The economic considerations in implementing the various computer-integrated activities described in this chapter are crucial in view of the complexities and the high costs involved. Installations of flexible manufacturing systems are very capital intensive; consequently, a thorough cost–benefit analysis must be conducted before a final decision is made. This analysis should include such factors as (a) the cost of capital, energy, materials, and labor; (b) the expected markets for which the products are to be produced; (c) anticipated fluctuations in market demand and product type; and (d) the time and effort required for installing and debugging the system.

Typically, an FMS system can take two to five years to install and at least six months to debug. Although FMS requires few, if any, machine operators, the personnel in charge of the total operation must be trained and highly skilled. These personnel include manufacturing engineers, computer programmers, and maintenance engineers. The most effective FMS applications have been in medium-volume batch production. (See Fig. 14.2.) When a variety of parts is to be produced, FMS is suitable for production volumes of 15,000–35,000 aggregate parts per year. For individual parts of the same configuration, production may reach 100,000 units per year. In contrast, high-volume, low-variety parts production is best obtained from transfer machines (dedicated equipment).

As indicated in Chapter 14, low-volume, high-variety parts production can best be done on conventional standard machinery (with or without NC) or by machining centers. Compared with conventional manufacturing systems, FMS has the following benefits:

1. Parts can be produced in any order, in batch sizes as small as one item, and at lower unit cost.

2. Direct labor and inventories are reduced, yielding major savings over conventional systems.

3. Lead times required for product changes are shorter.

4. Production is more reliable, because the system is self-correcting and thus product quality is uniform.

5. Work-in-progress inventories are reduced.

---

**CASE STUDY** | CAD Model Development for Automotive Components

CAD models are used extensively for a great variety of tasks, as described in Section 15.4. In the automotive industry, for example, it is especially important to have a detailed CAD model of a particular component in the product database in order to ensure that all those who will be working on it have all the data they need to perform their tasks. Special care is taken to build very precise CAD models of the automobile components that passengers will see and interact with on a regular basis; examples of such components are outer-body panels, handles, seats, and the instrument panel (Fig. 15.22). The quality of the *visible* (Class I) surfaces has a major impact on overall vehicle quality and customer perception of the look and feel of the automobile.

The route from product concept to full working CAD models of visible components involves several steps, as well as the efforts of a large number of highly skilled designers, engineers, and technicians. The personnel constructing the CAD models are in constant interaction with the workers who will be using them, from concept to final design, to support the activities that rely on CAD data throughout the life of the vehicle.

**FIGURE 15.22**   CAD description of an automobile. Every vehicle component, from body panels to knobs on the instrument panel, has a solid model associated with it. *Source*: Ford Motor Company.

**2D concept sketches.**  Stylists with a background and experience in industrial design and/or art first develop 2D concepts through a series of sketches. These sketches are most frequently drawn by hand, although software may be used, especially if the stylist starts with a photograph or a scanned drawing that needs to be modified. Concept sketches provide an overall feel for the aesthetics of the object and are frequently very detailed and show texture, color, and the areas at which individual surfaces on a vehicle should meet. Most often, stylists are given a set of packaging constraints such as (a) how the component should be assembled with other components, (b) what the size of the component should be, and (c) what the size and shape of any structure lying behind the visible surfaces should be. The time involved in producing a series of such concept sketches for an individual component or a set of components typically ranges from a few days to several weeks.

**3D concept model.**  It is difficult to have an accurate sense of how the vehicle or a component will look and feel from a 2D sketch alone; thus, the next step is to make a 3D model. Both physical and computer models are built. First, further sketches of the component (such as cross-sections, perspective views, and orthographic views) are drawn. In many cases, a rough digital surface model is built to better represent the 3D object. A skilled clay sculptor then constructs a full-scale mock-up of the component from clay (Fig. 15.23) that is spread over a foam core, either by hand, using the sketches as a guide, or by a combination of CNC milling driven by a computer model (see, for example, Fig. 1.8); this activity is then followed by hand finishing. Several iterations may be necessary to achieve the shape that the stylist intended; hence, constant interaction between the sculptor and stylist is essential.

Alternatively, the stylist or clay sculptor can use 3D sculpting software, such as FreeForm,[1] to construct a digital clay model of the component. This digital model is then projected onto a large screen so that designers can study the component in full scale, or the digital model is used to generate tool paths for CNC machining of a full-scale clay mock-up.

At this point, all information from the sketches, rough computer drawings, and the concept clay model begins to be fed to downstream processes, such as engineering analysis and manufacturing. Although the final surfaces have not yet been decided upon, the basic shape and size of the component can now be used to begin tooling design and to analyze component feasibility.

**FIGURE  15.23**  A clay sculptor working on an instrument-panel concept. *Source*: Ford Motor Company.

[1] FreeForm is a product of SensAble Technologies, Inc.

**3D surface model.** As a concept is being reviewed and refined, several very accurate surface models of the component are constructed. To start the surface model, a computer-controlled optical scanner scans the concept clay model, producing a *cloud of points* organized along scan lines. Depending on the size of the component, a point cloud may consist of hundreds of thousands to millions of points. The point cloud is read into point-processing software such as Paraform[2] or ICEM Surf[3] to further organize the points and filter out noise. Scanning can take anywhere from several hours to a day to be completed. If, however, a digital 3D clay model of the component is already available, it is converted to a point cloud organized into scan lines, without the need for physical scanning and point-cloud postprocessing.

Next, the scan lines from the point cloud are used to construct mathematical surfaces, using software such as ICEM Surf and Alias/Wavefront StudioTools.[4] First, to construct the surfaces, freeform NURBS (*nonuniform rational B-spline*, see Section 15.4.2) curves that interpolate or approximate the scan lines are constructed; then a NURBS surface patch is fit through the curves. An individual surface patch models a small region of a single component's face; several patches are constructed and joined smoothly at common edges to form the entire face. Faces join each other at common edges to model an entire component. A great deal of experience is required to determine how to divide a face into a collection of patches that can be fit with the simplest low-order surfaces possible and still meet smoothly at common boundaries. A surfacing specialist performs this task in conjunction with the stylist to ensure that the surfaces are of high quality and that they capture the stylist's intent. A single component may take as long as a week to model.

Surface models are passed along to the various downstream departments to be used for tooling design, feasibility checks, analysis, and for the design of *nonvisible* (Class II) surfaces. As the design evolves, the dimensional tolerances on the surfaces are gradually tightened. When the component surfaces are within 3 to 4 mm (0.12 to 0.16 in.) of their final shape, a great deal of engineering that relies on them has already been accomplished.

When designing outer-body panels, a major milestone is finalizing what is called *first flange and fillet*, in which the edges of the body panels are turned under, or hemmed (see Fig. 7.23c), to provide a connecting flange for the inner panel (Fig. 15.24). The shape of the flange and fillet affects the overall aesthetics and shape of the body panel, and hence their careful design is important. After the shape of the fillet and flange has been decided, inner-panel designs can be completed.

**Surface verification.** Once the final surface model is completed, it must be verified and evaluated for surface quality and aesthetics. NC tool paths are generated automatically from the surfaces, either by the surfacing software or through specialized machining software; the tool paths are then used to CNC-machine the sur-

**FIGURE 15.24** Schematic illustration of first flange and fillet, where the edges of the body panel are hemmed to provide a flange for connection. The shape of this flange and fillet is important for overall vehicle aesthetics.

[2] Paraform is a product of Paraform, Inc.
[3] ICEM Surf is a product of ICEM Technologies, Inc.
[4] StudioTools is a product of Alias/Wavefront.

faces in clay. If the component is small, then instead of machining a clay model, an STL file can be generated and the mock-up can be built on a rapid prototyping machine (see Section 10.12). It takes anywhere from several hours to several days to machine a component and perform any hand-finishing that may be required. Clay models can be coated with a thin layer of latex material and painted to make them look more realistic and to help evaluate surface quality. The coating may be modified to improve surface smoothness between patches or change the way that light reflects off the surface of the model. All changes made to the clay must be translated back to the digital surface model, either through rescanning and refitting the surface patches or by tweaking the shape of the existing patches. At this stage of surface verification, only very minor changes, to within $\pm 0.5$ mm (0.04 in.), can be made.

**Solid-model construction.** After the surface model is finalized, it is used to develop a solid model. For sheet-metal components, such as body panels, body specialists offset the surfaces to form a solid. For other components, such as instrument panels, door handles, and wheels, manufacturing features are added to the surface model, such as flanges, bosses, and ribs. Also, it is confirmed that all surfaces form a closed, watertight solid object. Suppliers are contacted and consulted to determine what features must be added and where they should be placed to ensure that the component can be fabricated from the desired materials and at the target cost.

In the process of making a solid model, it may be discovered that the surface model must be modified because (a) of changes in packaging constraints, (b) the component fails to meet minimal manufacturing requirements, or (c) surfaces do not match up properly at common edges with the appropriate smoothness. These changes are communicated back to the surface modeler and the stylist so that the surfaces can be modified and reverified. Finally, the solid model is entered into a product database, where it is now available to suppliers and engineers for further analysis and manufacturing.

*Source*: Courtesy of A. Marsan and P. Stewart, Ford Motor Company.

# SUMMARY

- Computer-integrated manufacturing systems have become the most important means of improving productivity, responding to changing market demands, and better controlling manufacturing operations and management functions. (Sections 15.1–15.3)

- With rapid developments in advanced and sophisticated software, together with the capability for computer simulation and analysis, computer-aided product design and engineering have become much more efficient, detailed, and comprehensive. (Section 15.4)

- Computer-aided manufacturing is often combined with computer-aided design to transfer information from the design stage to the planning stage and to production. (Section 15.5)

- Continuing advances in manufacturing operations such as computer-aided process planning, computer simulation of manufacturing processes and systems, group technology, cellular manufacturing, flexible manufacturing systems, and just-in-time manufacturing contribute significantly to improved productivity. (Sections 15.6–15.11)

- Communications networks and their global standardization are essential to CIM strategies. (Section 15.12)

- Artificial intelligence is likely to create new opportunities in all aspects of manufacturing science, engineering, and technology. (Section 15.13)

- A fully automated factory of the future appears to be theoretically possible. However, there are important issues to be considered regarding its impact on employment. (Section 15.14)
- Economic considerations in the design and implementation of computer-integrated manufacturing systems, especially flexible manufacturing systems, are particularly crucial, because of the major capital expenditures involved. (Section 15.15)

# BIBLIOGRAPHY

Amirouche, F. M. L., *Computer-Aided Design and Manufacturing*, Prentice Hall, 1993.

Askin, R. G., and C. R. Standridge, *Modeling and Analysis of Manufacturing Systems*, Wiley, 1993.

Badiru, A. B., *Expert Systems Applications in Engineering and Manufacturing*, Prentice Hall, 1992.

Bedworth, D. D., M. R. Henderson, and P. M. Wolf, *Computer-Integrated Design and Manufacturing*, McGraw-Hill, 1991.

Burbidge, J. L., *Production Flow Analysis for Planning Group Technology*, Oxford, 1997.

Chang, T.-C., R. A. Wysk, and H. P. Wang, *Computer-Aided Manufacturing*, 2d. ed., Prentice Hall, 1997.

Cheng, T. C. E., and S. Podolsky, *Just-in-Time Manufacturing: An Introduction*, 2d. ed., Chapman & Hall, 1996.

Chorafas, D., *Expert Systems in Manufacturing*, Van Nostrand Reinhold, 1992.

Corbett, J., M. Dooner, J. Meleka, and C. Pym, *Design for Manufacture: Strategies, Principles and Techniques*, Addison-Wesley, 1991.

Driankov, D., H. Hellendoorn, and M. Reinfrank, *Introduction to Fuzzy Control*, 2d. ed., Springer, 1996.

Famili, A., D. S. Nau, and S. H. Kim, *Artificial Intelligence Applications in Manufacturing*, MIT Press, 1992.

Fausett, L. V., *Fundamentals of Neural Networks*, Prentice Hall, 1994.

Foston, A. L., C. L. Smith, and T. Au, *Fundamentals of Computer-Integrated Manufacturing*, Prentice Hall, 1991.

Gu, P., and D. H. Norrie, *Intelligent Manufacturing Planning*, Chapman & Hall, 1995.

Hannam, R., *CIM: From Concept to Realisation*, Addison-Wesley, 1998.

Haykin, S. S., *Neural Networks: A Comprehensive Foundation*, 2d. ed., Prentice Hall, 1998.

Higgins, P., L. R. Roy, and L. Tierney, *Manufacturing Planning and Control: Beyond MRP II*, Chapman & Hall, 1996.

Hitomi, K., *Manufacturing Systems Engineering*, 2d. ed., Taylor & Francis, 1996.

Irani, S. (ed.), *Handbook of Cellular Manufacturing Systems*, Wiley, 1999.

Jamshidi, M., *Design and Implementation of Intelligent Manufacturing Systems: From Expert Systems, Neural Networks, to Fuzzy Logic*, Prentice Hall, 1995.

Kasabov, N. K., *Foundations of Neural Networks, Fuzzy Systems, and Knowledge Engineering*, MIT Press, 1996.

Koenig, D. T., *Manufacturing Engineering: Principles for Optimization*, 2d. ed., Taylor and Francis, 1994.

Krishnamoorty, C. S., and S. Rajeev, *Artificial Intelligence and Expert Systems for Engineers*, CRC Press, 1996.

Kusiak, A., *Concurrent Engineering: Automation, Tools, and Techniques*, Wiley, 1993.

Lee, K., *Principles of CAD/CAM Systems*, Addison-Wesley, 1999.

Leondes, C. T. (ed.), *Fuzzy Logic and Expert Systems Applications*, Academic Press, 1998.

Liebowitz, J. (ed.), *The Handbook of Applied Expert Systems*, CRC Press, 1997.

Louis, R. S., *Integrating Kanban with MRP II: Automating a Pull System for Enhanced JIT Inventory Management*, Productivity Press, 1997.

Luggen, W. W., *Flexible Manufacturing Cells and Systems*, Prentice Hall, 1991.

Maus, R., and J. Keyes (eds.), *Handbook of Expert Systems in Manufacturing*, McGraw-Hill, 1991.

McMahon, C., and J. Browne, *CADCAM—Principles, Practice and Manufacturing Management*, Addison-Wesley, 1999.

Mitchell, F. H., Jr., *CIM Systems: An Introduction to Computer-Integrated Manufacturing*, Prentice Hall, 1991.

Monden, Y., *Toyota Production System: An Integrated Approach to Just-in-Time*, 3d. ed., Institute of Industrial Engineers, 1998.

Nee, A. Y. C., K. Whybrew, and A. S. Kumar, *Advanced Fixture Design for FMS*, Springer, 1994.

Popovic, D., and V. Bhatkar, *Methods and Tools for Applied Artificial Intelligence*, Dekker, 1994.

Rehg, J. A., *Introduction to Robotics in CIM Systems*, 4th ed., Prentice Hall, 2000.

Rehg, J. A., and H. W. Kraebber, *Computer-Integrated Manufacturing*, 2d. ed., Prentice Hall, 2001.

Rembold, U., B. O. Nnaji, and A. Storr, *Computer Integrated Manufacturing and Engineering*, Addison-Wesley, 1993.

Sandras, W. W., *Just-in-Time: Making It Happen*, Wiley, 1997.

Singh, N., *Systems Approach to Computer-Integrated Design and Manufacturing*, Wiley, 1995.

Singh, N., and D. Rajamani, *Cellular Manufacturing Systems: Design, Planning and Control*, Chapman & Hall, 1996.

Tempelmeier, H., and H. Kuhn, *Flexible Manufacturing Systems*, Wiley, 1993.

Vajpayee, S. K., *Principles of Computer-Integrated Manufacturing*, Prentice Hall, 1995.

Vollmann, T. E., W. L. Berry, and D. C. Whybark, *Manufacturing Planning and Control Systems*, 4th ed., Irwin, 1997.

Williams, D. J., *Manufacturing Systems*, 2d. ed., Chapman & Hall, 1994.

Wu, J.-K., *Neural Networks and Simulation Methods*, Dekker, 1994.

# QUESTIONS

**15.1**   In what ways have computers had an impact on manufacturing?

**15.2**   What advantages are there in viewing manufacturing as a system? What are the components of a manufacturing system?

**15.3**   Discuss the benefits of computer-integrated manufacturing operations.

**15.4**   What is a database? Why is it necessary in manufacturing? Why should the management of a manufacturing company have access to databases?

**15.5**   Explain how a CAD system operates.

**15.6**   What are the advantages of CAD systems over traditional methods of design? Do CAD systems have any limitations?

**15.7**   Describe the purposes of process planning. How are computers used in such planning?

**15.8**   Explain the features of two types of CAPP systems.

**15.9**   Describe the features of a routing sheet. Why is a routing sheet necessary in manufacturing?

**15.10**   What is group technology? Why was it developed? Explain its advantages.

**15.11**   What is a manufacturing cell? Why was it developed?

**15.12**   Describe the principle of flexible manufacturing systems. Why do they require major capital investment?

**15.13**   Why is a flexible manufacturing system capable of producing a wide range of lot sizes?

**15.14**   What are the benefits of just-in-time production? Why is it called a *pull system*?

**15.15**   Explain the function of a local area network.

**15.16**   What are the advantages of a communications standard?

**15.17**   What is meant by the term *factory of the future*?

**15.18**   What are the differences between ring and star networks? What is the significance of these differences?

**15.19**   What is kanban? Why was it developed?

**15.20**   What is an FMC, and what is an FMS? What are the differences between them?

**15.21**   Describe the elements of artificial intelligence. Why is machine vision a part of it?

**15.22**   Explain why humans will still be needed in the factory of the future.

**15.23**   How would you describe the principle of computer-aided manufacturing to an older worker in a manufacturing facility who is not familiar with computers?

**15.24**   Give examples of primitives of solids other than those shown in Fig. 15.4b.

**15.25**   Explain the logic behind the arrangements shown in Fig. 15.11b.

**15.26**   Describe your observations regarding Fig. 15.17.

**15.27**   What should be the characteristics of an effective guidance system for an automated guided vehicle?

**15.28**   Give examples in manufacturing for which artificial intelligence could be effective.

**15.29**   Describe your opinions concerning the voice-recognition capabilities of future machines and controls.

**15.30**   Would machining centers be suitable for just-in-time production? Explain.

**15.31**   Give an example of a push system and of a pull system, to clarify the fundamental difference between the two methods.

**15.32**   Give a specific example in which the variant system of CAPP is desirable and an example in which the generative system is desirable.

**15.33**   Artificial neural networks are particularly useful where the problems are ill defined and the data are fuzzy. Give examples in manufacturing where this is the case.

**15.34**    Is there a minimum to the number of machines in a manufacturing cell? Explain.

**15.35**    List as many three-letter acronyms used in manufacturing (such as CNC) as you can, and give a brief definition of each, for your future reference.

**15.36**    What are the disadvantages of zero inventory?

**15.37**    Why are robots a major component of an FMC?

**15.38**    Is it possible to exercise JIT in global companies?

**15.39**    A term sometimes used to describe factories of the future is *untended factories*. Can a factory ever be completely untended? Explain your answer.

**15.40**    What are the advantages of hierarchical coding?

**15.41**    Assume that you are asked to rewrite Section 15.14, on the factory of the future. Briefly outline your thoughts regarding this topic.

**15.42**    Assume that you own a manufacturing company and that you are aware that you have not taken full advantage of the technological advances in manufacturing. Now, however, you would like to do so, and you have the necessary capital. Describe how you would go about analyzing your company's needs and how you would plan to implement these technologies. Consider technical as well as human aspects.

**15.43**    With specific examples, describe your thoughts concerning the state of manufacturing in the United States as compared with its state in other industrialized countries.

**15.44**    It has been suggested by some that artificial intelligence systems will ultimately be able to replace the human brain. Do you agree? Explain your response.

# DESIGN

**15.45**    Review various manufactured parts described in this text, and group them in a manner similar to that shown in Fig. 15.10.

**15.46**    Think of a specific product, and make a decision-tree chart similar to that shown in Fig. 15.12.

**15.47**    Describe the trends in product designs and features that have had a major impact on manufacturing.

**15.48**    Think of a specific product, and design a manufacturing cell for making it (see Fig. 15.16), describing the features of the machines and equipment involved. Explain how the cell arrangement would change, if at all, if design changes are made to the product.

**15.49**    Surveys have indicated that 95% of all the different parts made in the United States are produced in lots of 50 or less. Comment on this observation, and describe your thoughts regarding the implementation of the technologies outlined in Chapters 14 and 15.

**15.50**    Think of a simple product, and make a routing sheet for its production, similar to that shown in Fig. 15.8. If the same part is given to another person, what is the likelihood that the routing sheet developed will be the same? Explain.

**15.51**    Review Fig. 15.8, and then suggest a routing sheet for the manufacture of each of the following: (a) an automotive connecting rod, (b) a compressor blade, (c) a glass bottle, (d) an injection-molding die, and (e) a bevel gear.

**15.52**    What types of production machines would not be suitable for a manufacturing cell? What design or production features make them unsuitable? Explain.

# Product Design and Competitive Manufacturing

## 16.1 Introduction

In an increasingly competitive global marketplace, manufacturing high-quality products at the lowest possible cost requires an understanding of the often complex relationships among many factors. We have seen that product design, selection of materials, and selection of manufacturing processes are all interrelated. Furthermore, designs are periodically modified to improve product performance, to take advantage of available new materials or use cheaper materials, to make the products easier and faster to manufacture and assemble, and to strive for zero-based rejection and waste.

Because of the very wide variety of materials and manufacturing processes now available, the task of producing a high-quality product by selecting the best materials and the best processes, while minimizing costs, has become a major challenge, as well as an opportunity. This chapter begins with various considerations in **product design**, and the concept of **robust design**. Product design involves numerous aspects of not only the design itself, but also the ease of manufacturing that product and the time required to produce it. Again, there are many opportunities to simplify designs, reduce the number of components, and reduce the size and certain dimensions of the components to save money on especially costly materials.

**Product quality** and **life expectancy** are then discussed, outlining relevant parameters involved, including the concept of return on quality. Increasingly important are **life-cycle assessment** and **life-cycle engineering** of products, services, and systems, particularly as regards their potential adverse impact on the environment (air, water, and ground). The major emphasis in sustainable manufacturing is to reduce or eliminate any adverse effects of manufacturing on the environment and society in general, while allowing a company to be profitable.

Although the *selection of materials* for products traditionally requires much experience, there are now several databases and expert systems available that facilitate the selection process in order to meet several specific requirements. Also, in reviewing the materials used in existing products, from simple hand tools to automobiles and aircraft, there are numerous opportunities for *substitution of materials* for better performance and, especially, for cost savings.

In the production phase, it is imperative that we properly assess the *capabilities of manufacturing processes* as a basis of and an essential guide to the ultimate selection of an appropriate process or series of processes. As described throughout this text, depending on the product design and the materials specified, there is usually more than one method of manufacturing a product, with its usual number of components and

16.1 Introduction
16.2 Product Design and Robust Design
16.3 Product Quality and Life Expectancy
16.4 Life-Cycle Assessment, Life-Cycle Engineering, and Sustainable Manufacturing
16.5 Selection of Materials for Products
16.6 Substitution of Materials in Products
16.7 Capabilities of Manufacturing Processes
16.8 Selection of Manufacturing Processes
16.9 Manufacturing Costs and Cost Reduction
16.10 Value Analysis

subassemblies. An improper selection can have a major effect not only on product quality, but also on product cost.

Although the *economics* of particular manufacturing processes have been described at the end of individual chapters, this chapter takes a broader view and summarizes the important overall manufacturing cost factors in a competitive marketplace. The **cost** of a product often, but not always, determines its marketability and its customer acceptance. Meeting this challenge requires not only a thorough and up-to-date knowledge of the characteristics of materials, state-of-the-art processes and manufacturing systems, and economics, but also innovative and creative approaches to product design and manufacturing. Finally, we describe the method of **value analysis**, which is a powerful tool that can be used to evaluate the cost of each step in production relative to its contribution to the value of the product.

# 16.2 | Product Design and Robust Design

Although it is beyond the scope of this book to describe the principles and methodology of product design and various optimization techniques, we have highlighted, throughout various chapters, those aspects that are relevant to **design for manufacture and assembly** (DFMA), as well as to competitive manufacturing. The references listed in Table 16.1 give design guidelines for manufacturing processes.

Major advances are being made in design for manufacture and assembly, for which software packages are now available. Although their use requires considerable training and knowledge, they help designers develop products that require fewer components, reduced assembly time, and reduced time for manufacture and thus that have reduced total cost.

## 16.2.1 Product design considerations

In addition to the design guidelines outlined throughout this text, there are general product design considerations. Designers often must ask themselves if they've addressed considerations such as the following:

a) Can the product design be simplified and the number of its components reduced without adversely affecting its intended functions and performance?

b) Have all alternative designs been investigated?

c) Can unnecessary features of the product, or some of its components, be eliminated or combined with other features?

d) Have modular design and building-block concepts been considered for a family of similar products and for servicing and repair, upgrading, and installation options?

e) Can the design be made smaller and lighter?

f) Are the specified dimensional tolerances and surface finish excessively stringent? Can they be relaxed without any adverse effects?

g) Will the product be difficult or excessively time consuming to assemble and disassemble for maintenance, servicing, or recycling?

h) Have subassemblies been considered?

i) Is the use of fasteners, and their quantity and variety, minimized?

j) Are some of the components commercially available?

**TABLE 16.1**

## References to Various Topics in this Text

### PROCESSES: DESIGN CONSIDERATIONS

| | |
|---|---|
| Metal casting | Section 5.12 |
| Bulk deformation | Various sections in Ch. 6 |
| Sheet-metal forming | Section 7.15 |
| Machining | Section 8.13 |
| Abrasives | Section 9.16 |
| Polymers | Section 10.13 |
| Powder metallurgy and ceramics | Sections 11.6 and 11.12 |
| Joining | Section 12.16 |

### MATERIAL PROPERTIES

Tables 2.1, 2.3, and 2.5; Figs. 2.4, 2.6, 2.12, 2.13, and 2.29
Tables 3.2, 3.3, 3.6, 3.8, and 3.9 through 3.13
Tables 5.4, 5.5, and 5.6; Fig. 5.19
Table 7.3
Table 8.6; Fig. 8.31
Table 9.1
Tables 10.1, 10.4, 10.5, and 10.8; Fig. 10.15
Tables 11.3, 11.4, 11.7 and 11.9; Figs. 11.20, 11.21, and 11.22
Table 12.5

### MANUFACTURING CHARACTERISTICS OF MATERIALS

Table 3.7
Tables 5.2 and 5.3
Table 7.2
Section 8.5
Tables 11.5, 11.8, and 11.9
Section 12.5
Table 16.6

### CAPABILITIES OF MANUFACTURING PROCESSES

Fig. 4.20
Tables 5.2, 5.7, and 5.8
Table 6.1
Table 7.1
Table 8.7; Fig. 8.27
Table 9.4; Fig. 9.27
Tables 10.6, 10.7, and 10.9
Tables 11.5 and 11.8
Tables 12.1, 12.2, and 12.5
Table 13.5; Fig. 14.3; Table 15.1; Fig. 16.3

### DIMENSIONAL TOLERANCES AND SURFACE FINISH

Fig. 4.20
Table 5.2
Tables 8.7 and 8.27
Fig. 9.27
Fig. 16.4

### GENERAL COSTS

Table 5.2; Fig. 5.42
Fig. 7.74
Fig. 9.40
Table 10.9
Table 11.9
Tables 12.1 and 12.2
Table 16.7; Fig. 16.5

### MATERIAL COSTS

Table 10.4
Table 11.9
Table 16.3

As some examples of the foregoing considerations, note how (1) the size of products such as radios, electronic equipment, cameras, computers, calculators, and cell phones has been reduced; (2) repair of products is now often done by replacing subassemblies and modules; and (3) there are fewer traditional fasteners and more snap-on type of assemblies used in products.

## 16.2.2 Product design and quantity of materials

With the high production rates and reduced labor permitted by automation, the cost of materials becomes a significant portion of a product's cost. Although the material cost cannot be reduced below the market level, reductions can be made in the *quantity* of materials used in the components to be produced. The overall shape of the product is usually optimized during the design and prototype stages, through techniques such as finite-element analysis, minimum-weight design, design optimization, and computer-aided design and manufacturing. These methods have greatly facilitated design analysis, material selection, material usage, and optimization.

Typically, reductions in the amount of material used can be achieved by decreasing the component's volume. This approach may require the selection of materials that have high strength-to-weight or stiffness-to-weight ratios. (See Section 3.9.1.) Obviously, higher ratios can also be obtained by improving the product's design and by selecting different cross-sections, such as ones that have a high moment of inertia (as in I-beams), or by using tubular or hollow components instead of solid bars.

Implementing such design changes and minimizing the amount of materials used can, however, lead to thin cross-sections and present significant problems in manufacturing. Consider the following examples:

a)  Casting or molding of thin sections can present difficulties in mold filling and in maintaining the specified dimensional accuracy and surface finish.

b)  Forging of thin sections requires high forces, due to effects such as friction and chilling of thin sections.

c)  Impact extrusion of thin-walled parts can become difficult, especially when high dimensional accuracy is required.

d)  Formability in sheet-metal working may be reduced as sheet thickness decreases; furthermore, reducing the thickness of the sheet can lead to wrinkling, due to compressive stresses developed in the plane of the sheet during forming.

e)  Machining and grinding of thin workpieces can present difficulties such as part distortion, poor dimensional accuracy, and chatter; consequently, advanced material-removal processes have to be considered.

f)  Welding of thin sheets or structures can cause distortion, due to thermal gradients developed during the welding cycle.

Conversely, making parts that have thick cross-sections can also have adverse effects. Some examples are as follows:

a)  In processes such as casting and injection molding, the production rate can be slower, because of the increased cycle time required to allow for part cooling in the mold.

b)  Unless controlled, porosity can develop in thicker sections of castings.

c)  The bendability of sheet metals decreases as their thickness increases.

d)  In powder metallurgy, there can be significant variations in density and properties throughout thick parts.

e)  Welding of thick sections can present problems.

f)  In die-cast parts, thinner sections will have higher strength per unit thickness than will thicker sections, because of the smaller grain size of the former.

g)  Processing of polymer parts requires increased cycle times as their thickness or volume increases, because of the longer time required for the parts to cool sufficiently to be removed from the molds.

## 16.2.3  Robustness and robust design

An increasingly important trend in product design is *robustness*. Originally introduced by G. Taguchi (see also Section 4.9.3), robustness can be defined as a design, a process, or a system that continues to function within acceptable parameters despite variabilities in its environment. These variabilities are called *noise*, defined as those factors that are difficult or impossible to control. Examples of noise in manufacturing include (a) ambient temperature and humidity in a production facility, (b) random and unanticipated vibrations of the shop floor, (c) batch-to-batch variations in the dimensions and properties of the incoming raw materials, and (d) the performance of different operators and machines at different times during the day or on different days. Thus, a robust process does not change with changing noise, and in a *robust design*, the part will function sufficiently well even if unanticipated events occur.

As a simple illustration, consider a sheet-metal mounting bracket to be attached to a wall with two bolts (Fig. 16.1a). The positioning of the two mounting holes will, of course, include some error, such as caused by the type of manufacturing process and machine involved (e.g., an error due to punching, drilling, or rimming). This error will, in turn, prevent the top edge of the bracket from being horizontal. In a more robust design (Fig. 16.1b), the mounting holes are twice as far apart from each other. Even though the same manufacturing method is used and the production cost remains the same, the robust design has a variability that is one-half of that of the original design. In an even more robust design, the type of bolts used would be such that they will not loosen in case of vibration or during their use over a period of time.

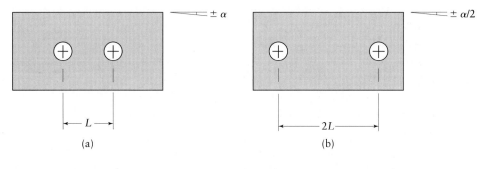

(a)                                     (b)

**FIGURE 16.1**   A simple example of robust design. (a) Location of two mounting holes on a sheet-metal bracket, where the deviation of the top surface of the bracket from being perfectly horizontal is $\pm\, \alpha$. (b) New location of holes, whereby the deviation of the top surface of the bracket from being perfectly horizontal is reduced to $\pm\, \alpha/2$.

---

**EXAMPLE 16.1**   An example of DFMA application

Numerous examples can be given regarding the benefits of applying DFMA principles in the early stages of product concept and development. These principles can also be applied to the modification of existing designs and the selection of appropriate production methods. In this example, we consider the redesign of the pilot's instrument panel for a military helicopter designed and built by McDonnell Douglas.

The components of the panel consisted of sheet metal, extrusions, and rivets. Through the use of DFMA software and analysis of the panel, it was estimated that the redesign would lead to the following changes: The number of parts would be reduced from 74 to 9; the weight of the panel would be reduced from 3.00 kg to 2.74 kg; fabrication time would be reduced from 305 hrs to 20 hrs; assembly time would be reduced from 149 hrs to 8 hours; and total time to produce the part would be reduced from 697 hrs to 181 hrs. It was estimated that, as a result of redesign, the cost savings would be 74%. Based on these results, other components of the instrument panel also were subjected to similar analysis.

# 16.3 | Product Quality and Life Expectancy

Product quality and the techniques involved in quality assurance and control have been described in Section 4.9. The word *quality* is difficult to define precisely, partly because it includes not only well-defined technical considerations, but also human, and hence subjective, opinion. However, a high-quality product generally is considered to have the following characteristics:

1. It satisfies the needs and expectations of the customer.
2. It is compatible with the customer's working environment.
3. It functions reliably over its intended life.
4. Its aesthetics are pleasing.
5. It provides a high level of safety.
6. Installation, maintenance, and future improvements are easy to perform.
7. Maintenance and upgrade costs are low.

Product quality has always been a major consideration in manufacturing. In view of the global economy and competition, a major priority is the concept of *continuous improvement in quality*, as exemplified by the Japanese term **kaizen**, meaning never-ending improvement. Still, the level of quality that a manufacturer chooses for its products depends on the market for which the products are intended. Thus, for example, low-cost, low-quality products have their own market niche (as is easily proved by a visit to a large department store or a home-improvement center), just as there is a market, especially in prosperous times, for high-quality, expensive products, such as a Rolls-Royce automobile, audio equipment, and sporting equipment like tennis rackets, golf clubs, and skis.

We have seen that, through *concurrent engineering* (see Section 1.2), design and manufacturing engineers have the responsibility and latitude to select and specify materials for the particular products to be made. Quality considerations are always a part of such tasks. Let's take a simple case of selecting materials for a screwdriver stem. Based on the functions of a screwdriver, one can specify stem materials that have high yield and tensile strength, torsional stiffness, and resistance to wear and corrosion. A screwdriver made from such materials will perform better and last longer than one made of materials with inferior properties. On the other hand, materials with better properties may generally be more expensive and may even be more difficult to produce than others. Consequently, the manufacturing operations and systems involved must thoroughly be reviewed to keep final product costs low, which can be done based on the numerous technical and economic considerations discussed throughout this text. Unfortunately, design engineers are often faced with this dilemma, which has been succinctly stated as "Good, fast, or cheap; pick any two."

Because of the significant costs that can be incurred in manufacturing activities, the concept of **return on quality** (ROQ) has been advanced. The basic components are: (a) because of its major influence on customer satisfaction, quality should be viewed as an investment; (b) there has to be a certain limit on how much money should be spent on quality improvements; (c) the specific area or issue for which the expenditure should be made toward quality improvement must be properly assessed; and (d) the incremental improvement in quality must be carefully reviewed with regard to the additional costs involved.

As has been discussed previously, the concept of quality is fairly subjective. Clearly, parts or products that do not serve their intended purpose detract from quality, and there is a clearly defined cost that can be calculated for such defects. For example, if a critical part tolerance exceeds a design specification, the part is usually discarded or reworked, and the inspection, material, machining, labor, handling, and other costs can be accounted for and a ROQ quantitatively expressed.

Contrary to a common perception, high-quality products do not necessarily cost more. In some industries, the ROQ is minimized at a value of zero defects, while in others, the cost of eliminating the final few defects is very high. However, indirect factors still may encourage elimination of these final few defects. For example, customer satisfaction is a qualitative factor that is difficult to include in calculations, but satisfaction is increased and customers are more likely to be retained when there are no defects in the product they purchase. Although there can be significant variations, it has been estimated that the relative costs involved in identifying and repairing defects in products grow by orders of magnitude, a concept called the *rule of ten* (see also Section 16.9.2), as follows:

| Stage | Relative cost of repair |
|---|---|
| Fabrication of the part | 1 |
| Subassembly | 10 |
| Final assembly | 100 |
| At product distributor | 1000 |
| At the customer | 10,000 |

**Life expectancy of products.**  Numerous surveys have indicated that the *average life expectancies* of certain products, in years, are as follows (in alphabetical order):

| | | | |
|---|---|---|---|
| Car battery | 4 | Machinery | 30 |
| Central air-conditioning unit | 15 | Motor vehicles | 10 |
| Dishwasher | 10 | Refrigerator | 17 |
| Disposal | 10 | Vacuum cleaner | 10 |
| Dryer, gas | 13 | Washer | 13 |
| Furnace, gas | 18 | Water heater, electric | 14 |
| Hair dryer | 5 | Water heater, gas | 12 |

As described in various sections in this chapter, numerous examples can be given of situations where a choice has to be made between different processes and materials to manufacture a product. Consider, for example, (a) sheet-metal vs. cast frying pans, (b) carbon-steel vs. stainless-steel exhaust systems for automotive vehicles, (c) wood vs. metal handles for hammers, (d) plastic vs. metal outdoor chairs, and (e) wood vs. reinforced-plastic ladders. Depending on the materials and processes employed, and hence on the quality of the product, the life expectancy of products can vary significantly; life expectancy is also determined in part by the frequency of maintenance.

# 16.4 | Life-Cycle Assessment, Life-Cycle Engineering, and Sustainable Manufacturing

Although there are a number of similar definitions, according to the ISO 14000 standard (Section 4.9.4), *life-cycle assessment* (LCA) may be defined as a systematic set of procedures for compiling and examining the inputs and outputs of materials and energy, and the associated environmental impacts or burdens directly attributable to the functioning of a product, process, or service system throughout its entire life cycle. Like LCA, *life-cycle engineering* (LCE) is concerned with environmental factors, but the latter also deals in greater detail with design, optimization, and various technical considerations regarding each component of a product or process life cycle. A major aim of LCE is to give thought to the reuse and recycling of products from the earliest stage of the design process (*green design*, or *green engineering*).

Life cycle, also called *cradle-to-grave*, can be defined as consecutive and interlinked stages of a product or service system, including the (a) extraction of natural resources; (b) processing of raw materials; (c) manufacture of products; (d) transportation and distribution of the product to the customer; (e) use, reuse, and maintenance of the product; and (f) disposal of the product or recovery and recycling of its components.

Note that all these factors are equally applicable to each of the various products listed in foregoing table. Each product has its own share of metallic and nonmetallic materials that have been processed into individual parts and assembled, and each product has its own life cycle. Note also that some products are disposable, while others are reusable. In addition to materials, other important considerations include the use of fluids (such as lubricants, coolants, toxic solutions, and various fluids used in heat treating and plating) that can have serious and adverse environmental effects. Government regulations in many countries are now moving towards the direction of making manufacturers life-cycle accountable.

Although life-cycle analysis and engineering is a comprehensive, powerful, and necessary tool, its implementation can be expensive, challenging, and time consuming. This is largely because of uncertainties in the input data regarding materials, processes, long-term effects, costs, etc., and the time required to gather reliable data to properly assess the often-complex interrelationships among various components of the whole system. A number of software packages are being developed to expedite such analysis in certain industries, particularly the chemical and process industries, because of the higher potential for environmental damage in their operations.

**Sustainable manufacturing.** In recent years, we all have become increasingly aware that, ultimately, our natural resources are limited, which clearly necessitates the need for conservation of materials and energy. The term *sustainable manufacturing* is now being used to emphasize the need for conserving these resources, particularly through maintenance and reuse. This conservation is to be undertaken in order to (a) increase the life cycle of products; (b) eliminate damage to the environment; and (c) increase our collective social well-being, especially of future generations, while (d) maintaining a company's profitability.

---

**EXAMPLE 16.2    Sustainable manufacturing in the production of Nike athletic shoes**

Among various examples, the production of Nike athletic shoes has indicated the benefits of sustainable manufacturing. The athletic shoes are assembled using adhesives (Section 12.13). Up to around 1990, the adhesives used contained petroleum-based solvents; these adhesives pose health hazards to humans and contribute to petrochemical smog. In order to improve this situation, the company worked with chemical suppliers to successfully develop water-based adhesive technology, which is now used for the majority of the assembly operations. As a result, solvent use in all manufacturing processes in Nike's subcontracted facilities in Asia was reduced by 67% since 1995, and in 1997, 834,000 gallons of hazardous solvents were replaced with 1290 tons of water-based adhesives.

As another example, the rubber outsoles of the shoe were made by a process that resulted in significant amount of extra rubber around the periphery of the sole, called *flashing*, similar to the flash shown in Figs. 6.15a and 10.29c. With about 40 factories using thousands of molds and producing over a million outsoles a day, this flashing constitutes the largest source of waste in the manufacturing process for the shoes. In order to reduce this waste, the company developed a technology that grinds the flashing into 500-$\mu$m rubber powder, which is then added to the rubber mixture used to make subsequent outsoles. As a result of this process, waste was reduced by 40%, and furthermore, it was found that the mixed rubber had better abrasion resistance, durability, and overall performance than the highest premium rubber.

# 16.5 | Selection of Materials for Products

A clear understanding of the functional requirements for each of the individual components of a product is essential. Although the general criteria for selecting materials have been described in Section 1.5, this chapter will discuss them in further detail.

## 16.5.1 General properties of materials

As described in Chapter 2, mechanical properties of materials include strength; toughness; ductility; stiffness; hardness; and resistance to fatigue, creep, and impact. Physical properties (see Section 3.9) include density, melting point, specific heat, thermal and electrical conductivity, thermal expansion, and magnetic properties. The chemical properties (see Section 3.9.7) that are of primary concern in manufacturing are susceptibility to oxidation and susceptibility to corrosion. The relevance of these properties to product design and manufacturing has been described in various chapters, and several tables relating to properties of metallic and nonmetallic materials are included therein. Table 16.1 lists the sections in this text that are relevant to material properties.

Selection of materials is now easier and faster, because of the availability of computerized and extensive *databases*, which provide much greater accessibility to information than previously available. However, to facilitate the selection of materials and the determination of the parameters described later in this section, expert-system software (*smart databases*) has been developed. With proper input of product design and functional requirements, these systems are capable of identifying appropriate materials for a particular application, just as an expert or a team of experts would.

The following considerations are important in materials selection for products:

a) Do the materials selected have properties that unnecessarily exceed minimum requirements and specifications?

b) Can some materials be replaced by others that are less expensive?

c) Do the materials selected have the appropriate manufacturing characteristics?

d) Are the raw materials (stock) to be ordered available in standard shapes, dimensions, tolerances, and surface finishes?

e) Is the material supply reliable?

f) Are there likely to be significant price increases or fluctuations for the materials?

g) Can the materials be obtained in the required quantities in the desired time frame?

## 16.5.2 Shapes of commercially available materials

For obvious economic reasons, the shapes and sizes in which raw materials are commercially available are an important consideration (Table 16.2). Materials are generally available in various forms: castings, extrusions, forgings, bar, plate, sheet, foil, rod, wire, and metal powders. Purchasing materials in shapes that require the least additional processing is an additional consideration. However, such characteristics as surface quality, dimensional tolerances, and straightness of the raw materials must also be taken into account. Obviously, the better and the more consistent these characteristics are, the less additional processing effort and time are required.

For example, if we want to produce simple shafts that have good dimensional accuracy, roundness, straightness, and surface finish, we could purchase round bars that are already turned and centerless ground to the dimensions specified. Unless our facilities are capable of producing round bars economically, it is cheaper to purchase them. On the other hand, if we need to make a stepped shaft (a shaft that has different diameters along its length), we could purchase a round bar that has a diameter at least equal to the largest diameter of the final stepped shaft and turn it on a lathe. We may also process it by some other means to reduce the diameter. If the stock has broad dimensional tolerances or is warped or out of round, we must order a larger size to ensure proper dimensions of the final shaft.

As we have seen throughout various chapters, each manufacturing operation produces parts that have specific shapes, dimensional tolerances, and surface finishes. Consider the following examples:

a) Hot-rolled and hot-drawn products have a rougher surface finish and greater dimensional tolerances than cold-rolled and cold-drawn products (Chapter 6).

b) Round bars turned on a lathe have a rougher surface finish than bars that are ground on cylindrical or centerless grinding machines (Chapters 8 and 9).

c) The wall thickness of seamless tubing made by the tube-rolling process is less uniform than that of welded tubing (Chapters 6 and 12).

d) Extrusions have smaller cross-sectional dimensional tolerances than parts made by roll forming (Chapters 6 and 7).

e) Castings generally have less dimensional accuracy and poorer surface finish than parts made by cold extrusion or powder metallurgy (Chapters 5, 6, and 11).

TABLE 16.2

| **Commercially Available Forms of Materials** | |
| --- | --- |
| Material | Available as |
| Aluminum | B, F, I, P, S, T, W, |
| Ceramics | B, p, s, T |
| Copper and brass | B, f, I, P, s, T, W |
| Elastomers | b, P, T |
| Glass | B, P, s, T, W |
| Graphite | B, P, s, T, W |
| Magnesium | B, I, P, S, T, w |
| Plastics | B, f, P, T, w |
| Precious metals | B, F, I, P, t, W |
| Steels and stainless steels | B, I, P, S, T, W |
| Zinc | F, I, P, W |

*Note*: B = bar and rod; F = foil; I = ingots; P = plate and sheet; S = structural shapes; T = tubing; W = wire. Lowercase letters indicate limited availability. Most of the metals are also available in powder form, including prealloyed powders.

### 16.5.3 Manufacturing characteristics of materials

Manufacturing characteristics of materials typically include castability, workability, formability, machinability, grindability, weldability, and hardenability by heat treatment. Because raw materials have to be formed, shaped, machined, ground, fabricated, or heat treated into individual components having specific shapes and dimensions, these properties are crucial to the proper selection of materials. Table 16.1 lists references to general manufacturing characteristics of materials.

Recall also that the quality of a raw material can greatly influence its manufacturing characteristics. The following are some examples:

a) A rod or bar with a longitudinal seam (lap) will develop cracks during simple upsetting and heading operations.

b) Bars with internal defects and inclusions will crack during seamless-tube production.

c) Porous castings will produce poor surface finish when machined.

d) Blanks that are heat treated nonuniformly and bars that are not stress relieved will distort during subsequent operations, such as machining or drilling a hole.

e) Incoming stock that has variations in composition and microstructure cannot be heat treated or machined consistently.

f) Sheet-metal stock that has variations in its cold-worked condition or thickness will exhibit springback during bending and other forming operations, because of differences in yield stress and strains in bending.

g) If prelubricated sheet-metal blanks have nonuniform lubricant distribution and thickness, their formability, surface finish, and overall quality will be adversely affected.

### 16.5.4 Reliability of material supply

It is well known that geopolitical factors can significantly affect the supply of strategic materials. Other factors, such as strikes, shortages, and the reluctance of suppliers to produce materials in a particular shape, quality, or quantity, also affect reliability of supply. Even though the availability of materials may not be a problem throughout a country as a whole, it can cause difficulties for a certain business, because of the location of a particular manufacturing plant.

### 16.5.5 Cost of materials and processing

Because of a raw material's processing history, the unit cost of the raw material depends not only on the material itself, but also on its shape, size, and condition. For example, because more operations are involved in the production of thin wire than in that of round rod, the unit cost of the wire is much higher. Similarly, powder metals generally are more expensive than bulk metals. Furthermore, the cost of materials typically decreases as the quantity purchased increases. As a result, certain segments of industry, such as automotive companies, purchase materials in very large quantities; the larger the quantity, the lower is the cost per unit weight (bulk discount).

The cost of materials may be determined in terms of cost per unit weight or cost per unit volume. Table 16.3 shows the cost per unit volume for wrought metals and plastics relative to that for carbon steel. The benefit of using cost per unit volume

**TABLE 16.3**

| Approximate Cost Per Unit Volume for Wrought Metals and Plastics Relative to the Cost of Carbon Steel | | | |
|---|---|---|---|
| Gold | 60,000 | Magnesium alloys | 2–4 |
| Silver | 600 | Aluminum alloys | 2–3 |
| Molybdenum alloys | 200–250 | High-strength low-alloy steels | 1.4 |
| Nickel | 35 | Gray cast iron | 1.2 |
| Titanium alloys | 20–40 | Carbon steel | 1 |
| Copper alloys | 5–6 | Nylons, acetals, and silicon | 1.1–2 |
| Stainless steels | 2–9 | Rubber* | 0.2–1 |
| | | Other plastics and elastomers* | 0.2–2 |

* As molding compounds.
*Note*: Costs vary significantly with the quantity of puchase, supply and demand, size and shape, and various other factors.

can be seen by the following simple example: In the design of a steel cantilevered rectangular beam that is to support a certain load at its end, a maximum deflection is specified. Using equations developed in the mechanics of solids and assuming that the weight of the beam can be neglected, we can determine an appropriate cross-section of the beam. Since all dimensions are now known, the volume of the beam can be calculated, and if we know the cost of the material per unit volume, we can easily calculate the cost of the beam. If the cost is measured per unit weight, however, we first have to calculate the beam's weight and then determine the cost.

The cost of a particular material is subject to fluctuations caused by factors as simple as supply and demand or as complex as geopolitics. If a product is no longer cost competitive, alternative and less costly materials can be selected. For example, the copper shortage in the 1940s led the U.S. government to mint pennies from zinc-plated steel. Similarly, when the price of copper increased substantially during the 1960s, the electrical wiring being installed in homes was, for a time, made of aluminum. It should be noted, however, that this substitution led to the redesign of switches and outlets to avoid excessive heating at the junctions.

When scrap is produced during manufacturing, as in sheet-metal fabrication, forging, and machining, the value of the scrap is deducted from the material's cost, in order to obtain net material cost. Table 16.4 lists the percentages of scrap produced in selected manufacturing processes. Note that in machining, scrap can be very high, whereas rolling, ring rolling, and powder metallurgy (all of which are net- or near-net-shape processes) produce the least scrap. As expected, the value of the scrap depends on the type of metal and on the demand for the scrap; typically, it is between 10% to 40% of the original cost of the material.

**TABLE 16.4**

| Typical Scrap Produced in Various Manufacturing Processes | | | |
|---|---|---|---|
| Process | Scrap (%) | Process | Scrap (%) |
| Machining | 10–60 | Hot extrusion | 15 |
| Hot closed-die forging | 20–25 | Permanent-mold casting | 10 |
| Sheet-metal forming | 10–25 | Powder metallurgy | < 5 |
| | | Rolling and ring rolling | < 1 |

# 16.6 | Substitution of Materials in Products

Although new products continually appear on the market, the majority of the design and manufacturing effort is concerned with improving existing products. Major product improvements can result from (a) substitution of materials, (b) implementation of new or improved processing techniques, (c) better control of the processing parameters, and (d) increased plant automation.

There is hardly a product on the market today for which substitution of materials has not played a major role in helping companies maintain their competitive positions. Automobile and aircraft manufacturing are particular examples of major industries in which substitution of materials is an important and ongoing activity. A similar trend is evident in the manufacture of sporting goods and in various other consumer products.

There are several reasons for substituting materials in existing products:

1. Reduce the costs of materials and processing;
2. Improve manufacturing and assembly, installation, and conversion to automated assembly;
3. Improve the performance of products—for example, by reducing weight and by improving resistance to wear, fatigue, and corrosion;
4. Increase the stiffness-to-weight and strength-to-weight ratios of structures;
5. Reduce the need for maintenance and repair;
6. Reduce vulnerability to the unreliability of domestic and overseas supply of certain materials;
7. Improve compliance with legislation and regulations prohibiting the use of materials that have adverse environmental impact;
8. Improve the perceived aesthetics of a product;
9. Reduce performance variations or environmental sensitivity in the product, i.e., to improve *robustness* (see Section 16.2.3).

From a brief review of this list, it is evident that there are numerous important technological and economic factors involved in material substitution, and access to much reliable and extensive data is essential before a decision can be made to substitute materials. Note, for instance, that an important factor in substitution is the compatibility of the materials used; a typical example of incompatibility is galvanic corrosion in metal pairs, as described in Section 3.9.7. Another major factor is the influence that the materials selected for substitution may have on the manufacturing processes used. Thus, for example, in spite of some similarities that exist, the production of ceramic or plastic parts and the production of metal parts typically involve different processes, machinery, assembly, production rates, and costs.

a) **Substitution of materials in the automobile industry.** The automobile industry provides a number of good examples of the effective substitution of materials in order to achieve one or more of the foregoing objectives:

1. Several parts of the metal body of automobiles have been replaced with plastic or reinforced-plastic parts.
2. Metal bumpers, gears, pumps, fuel tanks, housings, covers, clamps, and various other components have been replaced with plastic substitutes.
3. Engine components have been replaced with ceramic and reinforced-plastic parts.
4. All-metal drive shafts have been replaced with composite-material drive shafts.
5. Cast-iron engine blocks have been changed to cast-aluminum blocks; forged crankshafts to cast crankshafts; and forged connecting rods to cast, powder-metallurgy, or composite-material connecting rods.

6. Steel structural elements have been replaced with extruded aluminum sections, and rolled sheet-steel body panels have been changed to rolled aluminum sheet (see Fig. 1.5).

Because the automobile industry is a major consumer of both metallic and nonmetallic materials, there is constant and rigorous competition among suppliers, particularly for steel, aluminum, and plastics. Industry engineers and management are continually investigating the relative advantages and limitations of these principal materials in their applications, recycling and other environmental considerations, and, in particular, the materials' relative costs and benefits.

b) **Substitution of materials in the aircraft industry.** In the aircraft and aerospace industries, conventional aluminum alloys (2000 and 7000 series) are being replaced with aluminum–lithium alloys and titanium alloys, because of the latter two alloys' higher strength-to-weight ratios. Forged parts are being replaced with powder-metallurgy parts that are manufactured with better control of impurities and microstructure; the powder-metallurgy parts also require less machining and produce less scrap of expensive materials. Furthermore, advanced composite materials and honeycomb structures are replacing traditional aluminum airframe components (Fig. 16.2), and metal-matrix composites are replacing some of the aluminum and titanium parts used previously in structural components.

---

**EXAMPLE 16.3  Changes made in materials between C-5A and C-5B military cargo aircraft**

Table 16.5 shows the changes made in materials for various components of the C-5A and C-5B military cargo aircraft and the reasons for the changes.
*Source*: H. B. Allison.

---

**FIGURE 16.2** Advanced materials used on the Lockheed C-5A transport aircraft. (FRP stands for *fiber-reinforced plastic*.)

- FRP structure
- FRP–aluminum honeycomb
- Aluminum honeycomb
- Metal-to-metal honeycomb
- Titanium-faced honeycomb

**TABLE 16.5**

| Changes in the Materials From C-5A to C-5B Military Cargo Aircraft | | | |
|---|---|---|---|
| Item | C-5A material | C-5B material | Reason for change |
| Wing panels | 7075-T6511 | 7175-T73511 | Durability |
| Main frame | | | |
|   Forgings | 7075-F | 7049-01 | Stress-corrosion resistance |
|   Machined frames | 7075-T6 | 7049-T73 | |
|   Frame straps | 7075-T6 plate | 7050-T7651 plate | |
| Fuselage skin | 7079-T6 | 7475-T61 | Material availability |
| Fuselage under-floor end fittings | 7075-T6 forging | 7049-T73 forging | Stress-corrosion resistance |
| Wing/pylon attach fitting | 4340 alloy steel | PH13-8Mo | Corrosion prevention |
| Aft ramp's lock hooks | D6-AC | PH13-8Mo | Corrosion prevention |
| Hydraulic lines | AM350 stainless steel | 21-6-9 stainless steel | Improved field repair |
| Fuselage fail-safe straps | Ti-6Al-4V | 7475-T61 aluminum | Titanium strap debonding |

# 16.7 | Capabilities of Manufacturing Processes

We have seen that each manufacturing process has its particular advantages and limitations. Casting and injection molding, for example, can generally produce more complex shapes than can forging and powder metallurgy (because the molten metal or plastic is capable of filling complex mold and die cavities). On the other hand, forgings can generally be made into complex shapes by subsequent machining and finishing operations, and they have a toughness that is generally superior to that of castings and powder-metallurgy products. References to the general characteristics and capabilities of manufacturing processes are listed in Table 16.1.

Recall that the shape of a product may be such that it can best be fabricated from several parts, by joining them with fasteners or with such techniques as brazing, welding, and adhesive bonding. The reverse may be true for another product: Manufacturing it in one piece may be more economical, because of the significant assembly costs otherwise involved. Other factors that must be considered in process selection are the minimum section size and dimensions that can be satisfactorily produced (Fig. 16.3). For example, very thin sections can be produced by cold rolling, but processes such as sand casting and forging prohibit the fabrication of thin sections.

a) **Dimensional tolerances and surface finish.** The *dimensional tolerances* and *surface finish* produced are particularly important in subsequent assembly operations and in the proper operation of machines and instruments. The dimensional tolerances and surface finishes obtained by various manufacturing processes are qualitatively illustrated in Fig. 16.4, and references to the capabilities of manufacturing processes with respect to these features are provided in Table 16.1.

In order to obtain closer dimensional tolerances and better surface finish, additional finishing operations, better control of processing parameters, and the use of higher quality equipment and controls may be required. However, the closer the tolerance and the finer the surface finish specified, the higher is the

**FIGURE 16.3** Minimum part dimensions obtainable by various manufacturing processes. *Source*: J. A. Schey, *Introduction to Manufacturing Processes*, 2d. ed., McGraw-Hill, 1987.

cost of manufacturing, as shown in Figs. 9.40 and 16.5. Also, the finer the surface finish required, the longer is the manufacturing time (Fig. 16.6), the higher is the number of processes involved, and, hence, the higher is the product cost (Fig. 16.5). In the machining of aircraft structural members made of titanium alloys, for example, as much as 60% of the cost of machining the part may be expended in the final machining pass, in order to maintain proper tolerances and surface finish. Unless it is specifically required otherwise by proper technological and economic justification, parts should be made with as rough a surface finish and as wide a tolerance as will be functionally and aesthetically acceptable. In this regard, the importance of continual interaction and communication between the product designer and the manufacturing engineer becomes obvious.

b) **Production volume.** Depending on the type of product, the *production volume*, or *quantity* (*lot size*), can vary widely. For example, paper clips, bolts, washers, spark plugs, bearings, and ballpoint pens are produced in very large quantities. On the other hand, jet engines for large commercial aircraft, diesel engines for locomotives, and propellers for cruise ships are manufactured in limited quantities. Production quantity plays a significant role in process and equipment selection. In fact, an entire manufacturing discipline is devoted to determining mathematically the optimum production quantity, called the *economic order quantity*.

c) **Production rate.** A significant factor in manufacturing process selection is the *production rate*, defined as the number of pieces to be produced per unit of time (such as per hour, per month, or per year). Processes such as powder metallurgy, die casting, deep drawing, and roll forming are high-production-rate operations. By contrast, sand casting, conventional and electrochemical machining, spinning, superplastic forming, adhesive and diffusion bonding, and the processing of reinforced plastics are relatively slow operations. These production rates can, of course, be increased by automation or by using multiple machines. Note, however, that a lower production rate does not necessarily mean that the manufacturing process is inherently uneconomical.

d) **Lead time.** *Lead time* is defined as the length of time between receipt of an order and the delivery of the product to the customer. The selection of a manufacturing process is greatly influenced by the time required to start production.

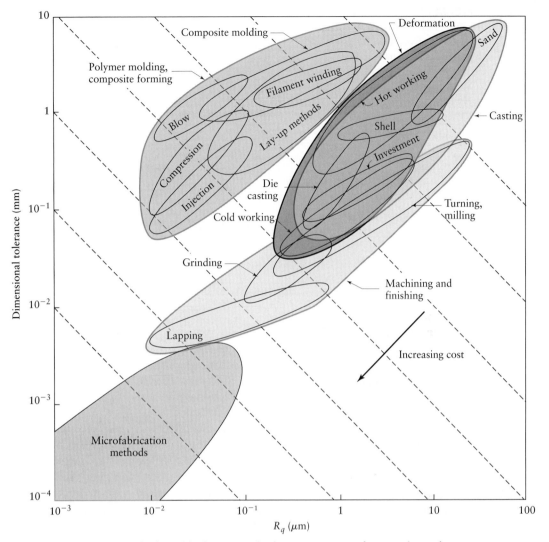

**FIGURE  16.4**    A plot of achievable dimensional tolerance versus surface roughness for assorted manufacturing operations. The dashed lines indicate cost factors; an increase in precision corresponding to the separation of two neighboring lines gives an increase in cost, for a given process, of a factor of two. *Source*: M. F. Ashby, *Materials Selection in Design*, Butterworth-Heineman, 1999.

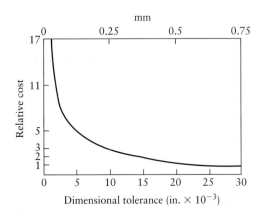

**FIGURE  16.5**
Relationship between relative manufacturing cost and dimensional tolerance.

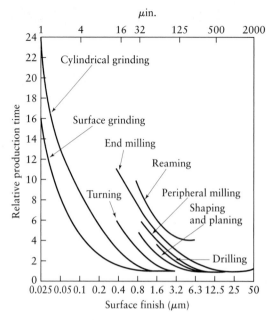

**FIGURE 16.6** Relative production time as a function of surface finish produced by various manufacturing processes. *Source*: *American Machinist*. See also Fig. 9.40.

Such processes as forging, extrusion, die casting, roll forming, and sheet-metal forming typically require dies and tooling that can take a considerable amount of time to produce. Depending on the complexity of the shape of the die, its size, and the material from which the die is to be made (see Table 3.5), lead times can range from weeks to months. In contrast, material-removal processes such as machining and grinding have significant built-in flexibility. As described in Chapters 8 and 9, these processes use tooling and abrasive wheels that can be adapted to most requirements in a very short time. Note also the capabilities of machining centers, flexible manufacturing cells, and flexible manufacturing systems, which are able to respond quickly and effectively to product changes.

## 16.7.1 Robustness in manufacturing processes and machinery

Robustness has been described in Section 16.2.3 in terms of a design, a process, or a system. In order to appreciate its importance in manufacturing processes, let's consider a simple injection-molded plastic gear, and let's assume that there are significant variations in quality as the products are being made. As described in Section 10.10.2, there are several well-understood variables in injection molding of plastics, including the quality of the raw material (pellets), temperature, and time. These are independent variables and hence they can be controlled. However, as noted in Section 16.2.3, there are certain other variables (*noise*) that are largely beyond our control, such as variations in the ambient temperature and humidity in the plant, dust in the air entering the plant from an open door (and thus contaminating the pellets being fed into the hoppers of the injection-molding machine), and the variability in performance of different operators during different shifts. Obviously, these variables are difficult or impossible to control.

In order to obtain sustained good quality, it is first necessary to understand the effects, if any, of each element of this noise on product quality. Why and how does the ambient temperature affect the quality of molded gears? Why and how does the dust coating on a pellet affect its performance in the machine chamber? How different are the performances of different operators during different shifts? After we answer questions such as the foregoing, it will be possible to establish new operating parameters so that variations in, say, ambient temperature do not adversely affect gear quality.

# 16.8 | Selection of Manufacturing Processes

This section describes the importance of proper selection of manufacturing processes and machinery, and how the selection process relates to the characteristics of materials, the dimensional tolerances and surface finish obtained, and manufacturing cost. As described in various chapters, especially Chapters 14 and 15, most manufacturing processes have been automated and are increasingly computer controlled, in order to optimize all aspects of operations. Computerization is also effectively increasing product reliability and product quality and reducing labor costs.

The choice of a manufacturing process is dictated by various considerations (Table 16.6):

1. Characteristics and properties of the workpiece material;
2. Shape, size, and thickness of the part;
3. Dimensional-tolerance and surface-finish requirements;
4. Functional requirements of the part;
5. Production volume (quantity);
6. Level of automation required to meet production volume and production rate;
7. Costs involved in individual and combined aspects of the manufacturing operation.

Some materials can be processed at room temperature, whereas others require elevated temperatures and thus need furnaces and appropriate tooling. Some materials are easy to work, because they are soft and ductile; others, being hard, brittle, and abrasive, require special processing techniques and tool and die materials.

Different materials have different manufacturing characteristics, such as castability, forgeability, workability, machinability, and weldability, and few materials have favorable characteristics in all the relevant categories. For example, a material that is castable or forgeable may later present difficulties in the machining, grinding, or finishing operations that may be required in order to produce a product with acceptable surface finish, dimensional accuracy, and quality.

**Process selection considerations.** A summary of the factors involved in process selection is given as follows, in the form of questions to be asked during this stage:

a) Have all alternative manufacturing processes been investigated?
b) Are the methods chosen economical for the type of material, the shape to be produced, and the required production rate?
c) Can the requirements for dimensional tolerances, surface finish, and product quality be met consistently?
d) Can the part be formed and shaped to final dimensions without the use of additional processes?
e) Are secondary processes such as machining, grinding, and finishing necessary?
f) Is the tooling required available in the plant? Can it be purchased as a standard item?
g) Is scrap produced? If so, is it minimized? What is the value of the scrap?
h) Are all processing parameters optimized?
i) Have all the automation and computer-control possibilities been explored for all phases of the manufacturing cycle?
j) Can group technology be implemented for parts with similar geometric and manufacturing attributes?

TABLE 16.6

## General Applications of Manufacturing Processes for Various Metals and Alloys

| | Carbon Steels | Alloy steels | Stainless steels | Tool and die steels | Aluminum alloys | Magnesium alloys | Copper alloys | Nickel alloys | Titanium alloys | Refractory alloys |
|---|---|---|---|---|---|---|---|---|---|---|
| Casting | | | | | | | | | | |
|   Sand | A | A | A | B | A | A | A | A | B | A |
|   Plaster | - | - | - | - | A | A | A | - | - | - |
|   Ceramic | A | A | A | A | B | B | A | A | B | A |
|   Investment | A | A | A | - | A | B | A | A | A | A |
|   Permanent | B | B | - | - | A | A | A | - | - | - |
|   Die | - | - | - | - | A | A | A | - | - | - |
| Forging, hot | A | A | A | A | A | A | A | A | A | A |
| Extrusion | | | | | | | | | | |
|   Hot | A | A | A | B | A | A | A | A | A | A |
|   Cold | A | B | A | - | A | - | A | B | - | - |
|   Impact | - | - | - | - | A | A | A | - | - | - |
| Rolling | A | A | A | - | A | A | A | A | A | B |
| Powder metals | A | A | A | A | A | A | A | A | A | A |
| Sheet-metal forming | A | A | A | - | A | A | A | A | A | B |
| Machining | A | A | A | A | - | A | A | A | B | A |
|   Chemical | A | B | B | B | A | A | B | B | B | B |
|   ECM | - | A | B | A | - | - | B | A | A | A |
|   EDM | - | B | B | A | B | - | B | B | B | B |
| Grinding | A | A | A | A | A | A | A | A | A | A |
| Welding | A | A | A | - | A | A | A | A | A | A |

*Note:* (A) Generally processed by this method; (B) can be processed by this method, but may present some difficulties; (-) usually not processed by this method. Product quality and productivity depend greatly on the techniques and equipment used, operator skill, and proper control of processing variables.

**FIGURE 16.7** Various methods of making a simple part: (a) casting or powder metallurgy, (b) forging or upsetting, (c) extrusion, (d) machining, and (e) joining two pieces.

**k)** Are inspection techniques and quality control being implemented properly?

**l)** Does each component of the product have to be manufactured in the plant? Are some of its parts commercially available as standard items from external sources?

**An example of process selection in making a simple part.** Assume that you are asked to make the part shown in Fig. 16.7. You should first determine the part's function, the types of load and environment to which it is to be subjected, the dimensional tolerances and surface finish required, and so on. For the sake of discussion, let's assume that the part is round, that it is 125 mm (5 in.) long, and that the large and small diameters are 38 mm and 25 mm (1.5 in. and 1.0 in.), respectively. Let's further assume that, because of functional requirements (such as stiffness, hardness, and resistance to elevated temperatures), this part should be made of metal.

Which manufacturing process would you choose, and how would you organize the production facilities to manufacture a cost-competitive, high-quality product? Recall that, as much as possible, parts should be produced at or near their final shape (net- or near-net-shape manufacturing), an approach that largely eliminates much secondary processing (such as machining, grinding, and other finishing operations) and thus reduces the total manufacturing time and manufacturing cost.

This part is relatively simple and could be suitably manufactured by different methods: (a) casting or powder metallurgy, (b) upsetting, (c) extrusion, (d) machining, and (e) joining two separate pieces together.

For net-shape processing, the two logical methods are casting and powder metallurgy; each process has its own characteristics, need for specific tooling and labor skill, and costs. This part can also be made by cold, warm, or hot forming. One method, for example, is the upsetting of a round bar 25 mm (1 in.) in diameter in a suitable die to form the larger end (heading). Another possibility is the partial extrusion of a bar 38 mm (1.5 in.) in diameter in order to reduce its diameter to 25 mm. Note that each of these processes shapes the material while yielding little or no material waste.

This part can also be made by machining a 38-mm-diameter bar stock to obtain the 25-mm-diameter section. Machining, however, will take a much longer time than forming, and some material will inevitably be wasted as metal chips. On the other hand, machining does not require special tooling (unlike net-shape processes, which generally require special dies), and this operation can easily be carried out on a lathe. Note, finally, that this part could be made in two separate pieces that are later joined by welding, brazing, or adhesive bonding.

Because of the different operations required in producing the raw materials, costs depend not only on the type of material (e.g., ingot, powder, drawn rod, extrusion), but also on its size and shape. Thus, per unit weight, (a) square bars are more expensive than round bars, (b) cold-rolled plate is more expensive than hot-rolled plate or sheet, and (c) hot-rolled bars are much less expensive than powders of the same metal.

In addition to technical requirements, process selection depends on factors such as the required production quantity and production rate, as described in Section 16.6. In sum, it appears that if only a few parts are needed, machining of this part is the most economical method. As the production quantity increases, however, producing this part by a heading operation or by cold extrusion would be the proper choice. Joining would be the appropriate choice if it were required that the top and bottom pieces of this part be made of different metals.

---

### EXAMPLE 16.4 Manufacturing a sheet-metal part by various methods

As shown throughout this text, there is often more than one method of manufacturing a part. Consider, for instance, forming of a simple dish-shaped part from sheet metal (Fig. 16.8). Such a part can be formed by placing a flat piece of sheet metal between a pair of male and female dies and closing the dies by applying a vertical force in a press. Parts can be formed at high production rates by this method, generally known as *stamping* or *pressworking*.

Assume now that the size of the part is very large, say, 2 m (80 in.) in diameter, and that only 50 parts are required. We now have to reconsider the total operation and ask a number of questions: Is it economical to manufacture a set of dies 2 m in diameter (which are very costly) when the production run (quantity) is so low? Are machines available with sufficient capacity to accommodate such large dies? Does the part have to be made in one piece? What are the alternative methods of manufacturing it?

This part can also be made by welding together smaller pieces of metal formed by other methods. Note, for example, that ships and municipal water tanks are made by this method. Would a part manufactured by welding be acceptable for its intended purpose? Will it have the required properties and the correct shape after welding, or will it require additional processing?

This part can also be made by explosive forming, as shown in Fig. 16.8b. It should be noted, however, that deformation of the material in explosive forming takes place at a very high rate. Consequently, several questions have to be asked:

a) Is the material capable of undergoing deformation at high rates without fracture?

b) Does the high rate of deformation have any detrimental effects on the final properties of the formed part?

c) Can the dimensional tolerances be held within acceptable limits?

d) Is the life of the die sufficiently long under the high transient pressures generated in this process?

**FIGURE 16.8** Two methods of making a dish-shaped sheet-metal part: (a) pressworking, using a male and female die, and (b) explosive forming, using one die only.

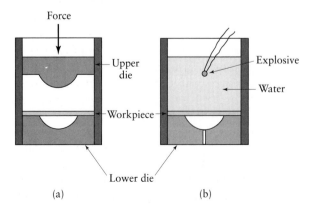

(a)                    (b)

e) Can this operation be performed in a manufacturing plant within a city, or should it be carried out in open land?

f) Although explosive forming requires only one die (an obvious advantage), is the overall operation economical?

From this general discussion, it can be appreciated that for each part or component, a similar approach is required to arrive at a conclusion as to which process is the most suitable and economical.

# 16.9 | Manufacturing Costs and Cost Reduction

In order for a product to be successful, its cost must be competitive with that of similar products, particularly in the global marketplace. The total cost of a product is spread over several categories, such as material costs, tooling costs, fixed costs, variable costs, direct labor costs, and indirect labor costs. References to some cost data are listed in Table 16.1.

Manufacturing organizations use several methods of cost accounting. The methodologies employed by these accounting procedures can be complex (even controversial), and their selection depends on the particular company and its type of operations. Furthermore, because of the technical and operational factors involved, calculating the individual cost factors is difficult, time consuming, and not always accurate or reliable.

Trends in costing systems (*cost justification*) include the following considerations: (a) the intangible benefits of quality improvements and inventory reduction, (b) life-cycle costs; (c) machine usage; (d) the cost of purchasing, compared with that of leasing, machinery; (e) the financial risks involved in implementing automation and the new technologies available. The costs to a manufacturer that are directly attributable to *product liability* (Section 1.9) and defense claims have been a matter of concern and discussion among all parties involved. Every modern product has a built-in added cost to cover possible liability claims. For example, it is estimated that liability suits against car manufacturers in the United States add about $500 to the indirect cost of an automobile and that 20% of the price of a ladder is attributed to potential product liability costs.

a) **Material costs.** *Material costs* are described throughout earlier sections; see also Table 16.1.

b) **Tooling costs.** *Tooling costs* are the costs involved in making the tools, dies, molds, patterns, and special jigs and fixtures necessary for manufacturing a product. Tooling costs are greatly influenced by the production process selected. For example, the tooling cost of die casting is higher than that of sand casting. Similarly, the tooling cost of machining or grinding is much lower than that of powder metallurgy, forging, or extrusion. Still, extrusion dies can sometimes be easily produced by electrical discharge machining, and, depending on the material and part size, it can be far more economical to extrude a complicated cross section than invest in expensive tooling required for roll forming.

In machining operations, carbide tools are more expensive than are high-speed steel tools, but the life of carbide tools is longer. If a part is to be manufactured by spinning, the tooling cost for conventional spinning is much lower than that of shear spinning. Tooling for rubber-forming processes is less expensive than that of the die sets (male and female) used for the deep drawing and stamping of sheet metals. High tooling costs might nonetheless be justified by the high-volume production of a single item. The expected life of tools and dies and their obsolescence because of product changes are also important considerations.

c) **Fixed costs.** *Fixed costs* include the costs of electric power, fuel, taxes on real estate, rent, insurance, and capital (including depreciation and interest). The company must meet these costs, regardless of whether it actually made a particular product. Consequently, fixed costs are not sensitive to production volume.

d) **Capital costs.** *Capital costs* represent investments in buildings, land, machinery, tooling, and equipment; these are major expenses for most manufacturing facilities (Table 16.7). Note the wide range of processes in each category and the fact that some machines cost a million dollars or more. (Coincidentally, the cost per pound of traditional machine tools has historically been approximately equal to that of beefsteak; however, this cost can easily double for machinery that has computer controls and sophisticated auxiliary equipment.)

As a simple example of capital costs, let's assume that a company has decided to manufacture a variety of valves. A new plant has to be built or an old plant has to be remodeled, and all the necessary machinery, support equipment, and facilities must be purchased. In order to cast and machine the valve bodies, the following items must be purchased: melting furnaces, casting equipment, machine tools, quality control equipment, and related equipment and machinery. If valve designs are to be modified often or if product lines are likely to vary greatly, the production machinery and equipment must have sufficient flexibility to accommodate these requirements. Machining centers (Section 8.10) and flexible manufacturing cells and systems (Chapter 15) are especially suitable for this purpose.

**TABLE 16.7**

## Approximate Ranges of Machinery Base Prices

| Type of machinery | Price range ($000) | Type of machinery | Price range ($000) |
|---|---|---|---|
| Broaching | 10–300 | Machining center | 50–1000 |
| Drilling | 10–100 | Mechanical press | 20–250 |
| Electrical discharge | 30–150 | Milling | 10–250 |
| Electromagnetic | 50–150 | Ring rolling | >500 |
| Extruder | 30–80 | Robot | 20–200 |
| Fused-deposition modeling | 60–120 | Roll forming | 5–100 |
| Gear shaping | 100–200 | Rubber forming | 50–500 |
| Grinding | | Stereolithography | 100–500 |
|    Cylindrical | 40–150 | Stretch forming | >1000 |
|    Surface | 20–100 | Transfer line | >1000 |
| Headers | 100–150 | Welding | |
| Injection molding | 20–200 |    Electron beam | 75–1000 |
| Jig boring | 50–150 |    Gas tungsten arc | 1–5 |
| Horizontal boring mill | 100–400 |    Laser beam | 40–1000 |
| Flexible manufacturing system | >1000 |    Resistance, spot | 20–50 |
| Lathe | 10–100 |    Ultrasonic | 50–200 |
|    Automatic | 30–250 | | |
|    Vertical turret | 100–400 | | |

*Note*: Prices vary significantly, depending on size, capacity, options, and level of automation and computer controls.

The equipment and machinery listed are capital cost items, and they require major investment. In view of the generally high equipment costs (see Table 16.7), particularly those involving transfer lines and flexible manufacturing cells and systems, high production rates and quantities are necessary to justify such large expenditures and to maintain product costs at or below the all-important competitive level. Lower unit costs can be achieved by continuous production involving around-the-clock operation, but only as long as demand warrants it. Proper equipment maintenance is essential to ensure high productivity (Section 14.2.7). Any breakdown of machinery leading to downtime can be very expensive, costing typically from a few hundred dollars per hour to thousands of dollars per hour.

e) **Labor costs.** *Labor costs* are generally divided into direct and indirect costs. *Direct labor costs* are for the labor directly involved in manufacturing a part (productive labor). These costs include the cost of all labor, from the time raw materials are first handled to the time the product is finished. This period is generally referred to as *floor-to-floor time*. For example, it is the time during which a machine operator picks up a round bar from a bin, machines it into the shape of a threaded rod, and places it into another bin.

Direct labor costs are calculated by multiplying the labor rate (hourly wage, including benefits) by the time the worker spends producing the part. The time required for producing a particular part depends not only on its size, shape, and dimensional accuracy, but also on the workpiece material. For example, as can be seen in Table 8.8, the highest recommended cutting speeds for high-temperature alloys are lower than those for aluminum, cast iron, or copper alloys. Consequently, the cost of machining aerospace materials is higher than that for machining the more common metals and alloys.

Labor costs in manufacturing and assembly vary greatly from country to country. For the year 2000, the relative hourly compensation for production workers in manufacturing, based on a scale of 100 for the United States, are as follows:

| | | | |
|---|---|---|---|
| Germany | 116 | Italy | 74 |
| Japan and Norway | 111 | Korea | 41 |
| Switzerland | 107 | Taiwan | 30 |
| United States | 100 | Mexico | 12 |
| Canada | 81 | China | 5 |

It is not surprising that many of the products one purchases today, from clothing to high-tech toys, are either made or assembled in countries where labor costs are low. In contrast, firms located in countries with high labor rates tend to emphasize high value-added manufacturing tasks or high automation levels, so that the labor component of the cost is significantly reduced.

*Indirect labor costs* are costs that are generated in the servicing of the total manufacturing operation. The total is composed of such activities as supervision, repair, maintenance quality control, engineering, research, and sales; it also includes the cost of office staff. Because these costs do not contribute directly to the production of finished parts or are not chargeable to a specific product, they are referred to as overhead (burden rate) and charged proportionally to all products. The personnel involved in these activities are categorized as nonproductive labor.

### 16.9.1 Manufacturing costs and production volume

One of the significant factors in manufacturing costs is *production volume*. Obviously, large production volume requires high production rates, which in turn, requires the use of mass-production techniques that involve special machinery (dedicated machinery) and employ proportionally less direct labor. High production rates also require the use of plants that operate on two or three shifts. At the other extreme, small production volume usually means a larger direct labor involvement.

As described in Section 14.2, small batch production is usually done on general-purpose machines, such as lathes, milling machines, and hydraulic presses. The equipment is versatile, and parts with different shapes and sizes can be produced by appropriate changes in the tooling. Direct labor costs are, however, high, because these machines are usually operated by skilled labor.

For larger quantities (medium-batch production), these same general-purpose machines can be equipped with various jigs and fixtures or else be computer controlled. To reduce labor costs further, machining centers and flexible manufacturing systems are important alternatives. Generally, for quantities of 100,000 or more, the machines are designed for specific purposes and they perform a variety of specific operations with very little direct labor.

### 16.9.2 Cost reduction

*Cost reduction* first requires an assessment of how the costs just described are incurred and interrelated, with relative costs depending on numerous factors. Consequently, the unit cost of the product will vary widely, depending on design and manufacturing choices. For example, some parts may be made from expensive materials, but ones that require very little processing, such as minted gold coins. Here, the cost of materials relative to that of direct labor is high. By contrast, some products may require several complex, expensive production steps to process relatively inexpensive materials, such as carbon steels. For instance, an electric motor is made of relatively inexpensive materials, yet many different manufacturing processes are involved in the production of the housing, rotor, bearings, brushes, wire windings, and various other components. Note also that, in such cases, assembly operations can become a significant portion of the overall cost.

An approximate, but typical, breakdown of costs in manufacturing today is as follows:

| | |
|---|---|
| Design | 5% |
| Material | 50% |
| Direct labor | 15% |
| Overhead | 30% |

In the 1960s, labor accounted for as much as 40% of the production cost; today, it can be as low as 5%, depending on the type of product and the level of automation. Note, in the preceding breakdown, the very small contribution of the design phase; yet, in design for manufacture and assembly, including concurrent engineering, the design phase generally has the largest influence on the cost and success of a product in the marketplace. The engineering changes that often are made in the development of products can have a significant influence on costs. In addition to the nature and extent of the changes made, the stage at which they are made is very significant. The cost of engineering changes made (from design stage to the final production) increases by orders of magnitude (*rule of ten*) when changes are made at later stages. (See also Section 16.3.)

Overall cost reductions can be achieved by a thorough analysis of all the costs incurred in each step in manufacturing a product. The methods employed are described in detail in some of the references in the bibliography at the end of the chapter. The various opportunities for cost reduction have been stated throughout this text; among them are the following:

1. Simplifying part design and reducing the number of subassemblies required;

2. Specifying broader dimensional tolerances and allowing rougher surface finish;

3. Using less expensive materials;

4. Investigating alternative methods of manufacturing;

5. Using more efficient machines, equipment, and controls.

The introduction of more automation and of up-to-date technology in a manufacturing facility is an obvious means of reducing some costs. However, this approach must be undertaken with due care and only after a thorough cost–benefit analysis, with reliable data input and a consideration of the technical as well as the human factors involved. Implementing advanced technology, which can be very expensive, should occur only after a complete analysis of the more obvious cost factors, known as **return on investment** (ROI). (See also *return on quality*, Section 16.3.)

It is commonly observed that, over a period of time, the prices of some products (such as calculators, computers, and digital watches) have decreased, while the prices of other products (such as automobiles, aircraft, houses, and books) have gone up. Such differences generally are due to changes in the various costs over time, including changes due to labor, machinery, domestic and international competition, and worldwide economic trends (such as demand, exchange rates, and tariffs), and to the inevitable impact of computers on all aspects of product design, manufacturing, and marketing.

# 16.10 | Value Analysis

There are several areas of activity in manufacturing in which cost reduction is possible. Manufacturing adds *value* to materials as they become discrete products and are then marketed. Because this value is added in individual stages during the creation of the product, the utilization of value analysis (also called **value engineering** or **value management**) is important.

First developed at the General Electric Company in the 1940s, *value analysis* (VA) is a methodology that evaluates in detail each step in design, materials, processes, and operations so as to manufacture a product that performs all its intended functions and does so at the lowest possible cost. In this analysis, a monetary value is established for each of two product attributes:

a) *Use value*, reflecting the functions of the product, and

b) *Esteem*, or *prestige*, *value*, reflecting the attractiveness of the product that makes its ownership desirable.

The *value* of a product is then defined as the following ratio:

$$\text{Value} = \frac{\text{Product function and performance}}{\text{Product cost}} \qquad (16.1)$$

Thus, the goal of value analysis is to obtain maximum performance per unit cost. Value analysis generally consists of the following phases:

1. *Information phase*, to determine the costs of materials and labor and to gather data on specifications, quality, and other relevant topics;

2. *Analysis phase*, to define functions and identify problem areas and opportunities and to prioritize opportunities;

3. *Creativity phase*, to seek new ideas and ways to respond to problems and opportunities, without judging the value of each of these ideas;

4. *Evaluation phase*, to select and refine the best ideas to be developed and evaluated and to identify the costs involved;

5. *Implementation phase*, to present facts, costs, and values to the company management; to develop a plan; and to motivate positive action, all in order to obtain management's commitment to provide the resources necessary to accomplish the task;

6. *Review* the overall value analysis process and identify any adjustments that need to be made.

Value analysis is an important and all-encompassing interdisciplinary activity. It is usually coordinated by a value engineer and is conducted jointly by designers and engineers, by quality control, purchasing, and marketing personnel, and by managers. In order for value analysis to be effective, it must have the full support of a company's top management.

The implementation of value analysis in manufacturing can result in one or more of the following benefits:

a) Significant cost reduction;

b) Reduced lead times;

c) Reduction in product weight and size;

d) Reduction in manufacturing times;

e) Better product quality and product performance.

---

**CASE STUDY** | Concurrent Engineering for Intravenous Solution Containers

Baxter Healthcare Corporation manufactures over 1 million intravenous (IV) solution containers every day in the United States, providing critical therapies within the health care industry. Recognizing that patients' lives depend upon the safe delivery of medical solutions, the introduction of new or improved products is highly regulated by internal company standards, and external government agencies. A well-defined product development process provides the framework to meet consistently the regulated quality, reliability, and manufacturing design requirements on a consistent basis. More importantly, a concurrent engineering environment catalyzes the development process to minimize development cost and time to market.

In the 1990s, Baxter focused development efforts on a new set of materials for flexible IV containers. The container system being developed was expected to be more environmentally friendly, be compatible with a larger variety of new critical-care drugs, exceed all quality requirements, and remain cost effective. Some of the key design issues shown in the table below would allow the product to be safe for patients and maintain economic viability.

| Container product requirements | Container processing requirements |
|---|---|
| • Provide physical and chemical shelf-life of up to several years without compromising the solution. | administration set devices. |
| • Allow for addition and mixing of various drug solutions in the hospital or alternate site pharmacy. | • Implement the new product with existing equipment and plant personnel. |
| • Provide a surface for printing of fade-, flake-, and smear-proof labeling. | • Allow product manufacturing at rates up to one million per day without additional production cells. |
| • Provide a surface for adherence of various pressure-sensitive adhesive labels at room, refrigerated, and elevated temperatures. | • Maintain fabrication throughput at over 60 containers per minute per machine. |
| • Withstand pressurization when an infusion device (cuff) is placed on the container for controlled solution delivery to the patient. | • Maintain printing and filling speeds over 120 per minute. |
| • Maintain compatibility with available | • Withstand steam sterilization temperatures, pressures, and times. |
|  | • Allow packing in cartons that provide fullest density of pallet loading. |
|  | • Withstand air and ground shipping and handling through areas as diverse as Arizona and Alaska. |

**The matrixed team.**  Given the company's extensive requirements, Baxter formed a multifunctional team of over 25 individuals. As the requirements and goals indicated, marketing, manufacturing, and development team members worked side by side with specialists in materials science, regulatory affairs, clinical affairs, toxicology, chemical stability, and sterility assurance. The product team members each recognized their responsibility for the success or failure of the product design effort.

**The active team.**  All team members contributed to the master requirements definition during the product conceptualization phase. All team members led and communicated the test and development activities within their respective fields to the team at large during the development phase. All disciplines offered and accepted peer-review criticism of product or process designs at major development steps. All team members made sure that the product designs, quality assessment methods, and fabrication techniques were transferred efficiently to the designated plants during the implementation phase. The product team avoided costly design iterations and revisions and minimized duplicative efforts by maintaining a matrixed, active team throughout the process.

**The virtual team.**  Many team members were located at manufacturing plants or development centers throughout northern Illinois and the greater United States. By necessity, team meetings leveraged teleconference and videoconference capabilities, which also minimized travel expense. Electronic mail access provided rapid and broad communication of impending issues and resolutions. A companywide intranet site was developed to share documents and design prints, which assured that all locations were utilizing the most recent project advancements. The team customized emerging communication technologies to enhance the concurrent development of new materials, processing, and fabrication technologies.

**Finished goods.** The engineering effort put forth by the matrixed, active, and virtual teams resulted in a product within three years. New materials and components were developed that met all product and process requirements and ultimately satisfied patient needs. The new product was transparent, and patient-care professionals continued to use their existing training and well-practiced techniques. Disposal of the product was streamlined; the new materials can be safely disposed of or recycled.

Reliability and user satisfaction remained high. The new materials were compatible with an extended array of solution format drugs packaged in IV containers. Indeed, team success will be measured by the number of additional products introduced in subsequent years.

*Source*: K. Anderson, Baxter Healthcare Corporation, and Shelly Petronis, formerly Baxter Healthcare Corporation, currently Zimmer Holdings, Inc.

# SUMMARY

- Competitive aspects of production and costs are among the most significant considerations in manufacturing. Regardless of how well a product meets design specifications and quality standards, it must also meet economic criteria in order to be competitive in the domestic and global marketplace. (Section 16.1)

- Various guidelines for designing products for economical production have been established, including a reduction in material volume. Robust design is an important concept. (Section 16.2)

- Product quality and life expectancy are significant concerns because of their impact on customer satisfaction and the marketability of the product. (Section 16.3)

- Life cycle assessment and life cycle engineering are among increasingly important considerations in manufacturing, particularly as regards reducing any adverse impact the product might have on the environment. Sustainable manufacturing is another concept aimed at reducing waste of natural resources such as materials and energy. (Section 16.4)

- Selecting an optimal material from among numerous candidates is one competitive aspect of manufacturing and depends on factors such as properties, commercially available shapes, reliability of supply, and costs of materials and processing. (Section 16.5)

- Substitution of materials, modification of product design, and relaxing of dimensional tolerance and surface finish requirements are important methods of reducing costs. (Section 16.6)

- The capabilities of manufacturing processes vary widely; consequently, their proper selection for a particular product to meet specific design and functional requirements is critical. (Sections 16.7 and 16.8)

- The total cost of a product includes several elements. Material costs can be reduced through careful selection, so that the least expensive material can be identified and used, without compromising design, service requirements, and quality. Although labor costs are continually becoming an increasingly smaller percentage of production costs, they can be reduced further through the use of automated and computer-controlled machinery. (Sections 16.9)

- Value analysis is a methodology that evaluates each step in design, materials and process selection, and production operations to manufacture a product that performs its intended functions at the lowest possible cost. (Section 16.10)

# BIBLIOGRAPHY

Anderson, D. M., *Design for Manufacturability, Optimizing Cost, Quality, and Time-to-Market*, 2d. ed., CIM Press, 2001.

Ashby, M. F., *Materials Selection in Mechanical Design*, 2d. ed., Pergamon, 1999.

*ASM Handbook*, Vol. 20: *Materials Selection and Design*, ASM International, 1997.

Baxter, M., *Product Design: A Practical Guide to Systematic Methods of New Product Development*, Chapman & Hall, 1995.

Billatos, S., and N. Basaly, *Green Technology and Design for the Environment*, Taylor & Francis, 1997.

Boothroyd, G., P. Dewhurst, and W. Knight, *Product Design for Manufacture and Assembly*, 2d. ed., Dekker, 2001.

Bralla, J. G., *Design for Excellence, DFX*, McGraw-Hill, 1996.

_____, *Design for Manufacturability Handbook*, 2d. ed., McGraw-Hill, 1999.

Brown, J., *Value Engineering: A Blueprint*, Industrial Press, 1992.

Cattanach, R. E. (ed.), *The Handbook of Environmentally Conscious Manufacturing*, McGraw-Hill, 1994.

Demaid, A., and J. H. W. DeWit (eds.), *Case Studies in Manufacturing with Advanced Materials*, North-Holland, 1995.

Dettmer, W. H., *Breaking the Constraints to World-Class Performance*, ASQ Quality Press, 1998.

Dieter, G. E., Jr., *Engineering Design: A Materials and Processing Approach*, 3d. ed., McGraw-Hill, 1999.

Dorrf, R. C., and A. Kusiak (eds.), *Handbook of Design, Manufacturing and Automation*, Wiley, 1995.

Farag, M. M., *Materials Selection for Engineering Design*, Prentice Hall, 1997.

Fleischer, M., and J. K. Liker, *Concurrent Engineering Effectiveness: Integrating Product Development Across Organizations*, Hanser Gardner, 1997.

Graedel, T. E., and B. R. Allenby, *Design for the Environment*, Prentice Hall, 1997.

Halevi, G., *Restructuring the Manufacturing Process: Applying the Matrix Method*, St. Lucie Press, 1998.

Harper, C. A. (ed.), *Handbook of Materials for Product Design*, McGraw-Hill, 2001.

Hartley, J. R., and S. Okamoto, *Concurrent Engineering: Shortening Lead Times, Raising Quality, and Lowering Costs*, Productivity Press, 1998.

Helander, M., and M. Nagamachi, *Design for Manufacturability*, Taylor and Francis, 1992.

Kusiak, A. (ed.), *Concurrent Engineering: Automation, Tools and Techniques*, Wiley, 1993.

Lesco, J., *Industrial Design: Guide to Materials and Manufacturing*, Van Nostrand Reinhold, 1998.

Lindbeck, J. R., *Product Design and Manufacturing*, Prentice Hall, 1995.

Magrab, E. B., *Integrated Product and Process Design and Development: The Product Realization Process*, CRC Press, 1997.

Mahoney, R. M., *High-Mix Low-Volume Manufacturing*, Prentice Hall, 1997.

Mangonon, P. C., *The Principles of Materials Selection for Design*, Prentice Hall, 1999.

Mather, H., *Competitive Manufacturing*, 2d. ed., CRC Press, 1998.

Meyers, F. E., *Motion and Time Study for Lean Manufacturing*, Prentice Hall, 1998.

Otto, K. N., and K. L. Wood, *Product Design: Techniques in Reverse Engineering and New Product Development*, Prentice Hall, 2001.

Paashuis, V., *The Organization of Integrated Product Development*, Springer, 1997.

Pahl, G., and W. Beitz, *Engineering Design: A Systematic Approach.*, 2d. ed., Springer, 1996.

Park, R. J., *Value Engineering: A Plan for Invention*, St. Lucie Press, 1999.

Park, S., *Robust Design and Analysis for Quality Engineering*, Chapman & Hall, 1997.

Poli, C., *Design for Manufacturing: A Structured Approach*, Butterworth-Heinemann, 2001.

Prasad, B., *Concurrent Engineering Fundamentals*, 2 vols, Prentice Hall, 1995.

Pugh, S., D. Clausing, and R. Andrade (eds.), *Creating Innovative Products Using Total Design*, Addison-Wesley, 1996.

Pugh, S., *Total Design: Integrated Methods for Successful Product Engineering*, Addison-Wesley, 1991.

Rhyder, R. F., *Manufacturing Process Design and Optimization*, Dekker, 1997.

Roosenburg, N. F. M., and J. Eekels, *Product Design: Fundamentals and Methods*, Wiley, 1995.

Salomone, T. A., *What Every Engineer Should Know about Concurrent Engineering*, Dekker, 1995.

Shina, S. G. (ed.), *Successful Implementation of Concurrent Engineering Products and Processes*, Wiley, 1997.

Sims, E. R., Jr., *Precision Manufacturing Costing*, Dekker, 1995.

Stoll, H. W., *Product Design Methods and Practices*, Dekker, 1999.

Swift, K. G., and J. D. Booker, *Process Selection: From Design to Manufacture*, Wiley, 1997.

*Tool and Manufacturing Engineers Handbook*, 4th ed., Vol. 5: *Manufacturing Engineering Management*, Society of Manufacturing Engineers, 1988.

_____, Vol. 6: *Design for Manufacturability*, 1992.

_____, Vol. 7: *Continuous Improvement*, 1993.

Ullman, D. G., *The Mechanical Design Process*, 2d. ed., McGraw-Hill, 1997.

Ulrich, K. T., and S. D. Eppinger, *Product Design and Development*, 2d. ed., McGraw-Hill, 1999.

Urban, G., and J. Hauser, *Design and Marketing of New Products*, 2d. ed., Prentice Hall, 1993.

Walker, J. M. (ed.), *Handbook of Manufacturing Engineering*, Dekker, 1996.

Wang, B., *Integrated Product, Process, and Enterprise Design*, Chapman & Hall, 1997.

Wenzel, H., M. Hauschild, and L. Alting, *Environmental Assessment of Products*, Vol. 1, Chapman & Hall, 1997.

Wenzel, H., and M. Hauschild, *Environmental Assessment of Products*, Vol. 2, Chapman & Hall, 1997.

Wroblewski, A. J., and S. Vanka, *MaterialTool: A Selection Guide of Materials and Processes for Designers*, Prentice Hall, 1997.

# QUESTIONS

**16.1**  List and describe the major considerations involved in selecting materials for products.

**16.2**  Why is a knowledge of available shapes of materials important? Give five different examples.

**16.3**  Describe what is meant by the *manufacturing characteristics* of materials. Give three examples demonstrating the importance of this information.

**16.4**  Why is material substitution an important aspect of manufacturing engineering? Give five examples from your own experience or observations.

**16.5**  Why has material substitution been particularly critical in the automotive and aerospace industries?

**16.6**  What factors are involved in the selection of manufacturing processes? Explain why these factors are important.

**16.7**  What is meant by *process capabilities*? Select four different, specific manufacturing processes, and describe their capabilities.

**16.8**  Is production volume significant in process selection? Explain your answer.

**16.9**  Discuss the advantages of long lead times, if any, in production.

**16.10**  What is meant by an *economic order quantity*?

**16.11**  Describe the costs involved in manufacturing. Explain how you could reduce each of these costs.

**16.12**  What is value analysis? What are its benefits?

**16.13**  What is meant by *trade-off*? Why is it important in manufacturing?

**16.14**  Explain the difference between direct labor cost and indirect labor cost.

**16.15**  Explain why the larger the quantity per package of food products, the lower is the cost per unit weight.

**16.16**  Explain why the value of the scrap produced in a manufacturing process depends on the type of material.

**16.17**  Comment on the magnitude and range of scrap shown in Table 16.4.

**16.18**  Describe your observations concerning the information given in Table 16.6.

**16.19**  Other than the size of the machine, what factors are involved in the range of prices in each machine category shown in Table 16.7?

**16.20**  Explain how the high cost of some of the machinery listed in Table 16.7 can be justified.

**16.21**  Explain the reasons for the relative positions of the curves shown in Fig. 16.3.

**16.22**  What factors are involved in the shape of the curve shown in Fig. 16.5?

**16.23**  Make suggestions as to how to reduce the dependence of production time on surface finish (shown in Fig. 16.6).

**16.24**  Is it always desirable to purchase stock that is close to the final dimensions of a part to be manufactured? Explain your answer, and give some examples.

**16.25**  What course of action would you take if the supply of a raw material selected for a product line became unreliable?

**16.26**  Describe the potential problems involved in reducing the quantity of materials in products.

**16.27**  Present your thoughts concerning the replacement of aluminum beverage cans with steel ones.

**16.28**  There is a period, between the time that an employee is hired and the time that the employee finishes with training, during which the employee is paid and receives benefits, but produces nothing. Where should such costs be placed among the categories given in this chapter?

**16.29**  Why is there a strong desire in industry to practice near-net-shape manufacturing? Give several examples.

**16.30**  Estimate the position of the following processes in Fig. 16.4: (a) centerless grinding, (b) electrochemical machining, (c) chemical milling, and (d) extrusion.

**16.31** From your own experience and observations, comment on the size, shape, and weight of specific products as they have changed over the years.

**16.32** In Section 16.9.2, a breakdown of costs in today's manufacturing environment suggests that design costs contribute only 5% to total costs. Explain why this suggestion is reasonable.

**16.33** Make a list of several (a) disposable and (b) reusable products. Discuss your observations, and explain how you would go about making more products that are reusable.

**16.34** Describe your own concerns regarding life-cycle assessment of products.

# DESIGN

**16.35** As you can see, Table 16.6 on manufacturing processes includes only metals and their alloys. On the basis of the information given in this book and other sources, prepare a similar table for nonmetallic materials, including ceramics, plastics, reinforced plastics, and metal–matrix and ceramic–matrix composite materials.

**16.36** Review Fig. 1.2, and present your thoughts concerning the two flowcharts. Would you want to make any modifications to them and if so, what would the modifications be and why?

**16.37** Over the years, numerous consumer products have become obsolete, or nearly so, such as rotary-dial telephones, analog radio tuners, turntables, and vacuum tubes. By contrast, many new products have entered the market. Make a comprehensive list of obsolete products and one of new products. Comment on the possible reasons for the changes you have observed. Discuss how different manufacturing methods and systems have evolved in order to make the new products.

**16.38** Select three different household products, and make a survey of the changes in their prices over the past 10 years. Discuss the reasons for the changes.

**16.39** Figure 2.2a shows the shape of a typical tension-test specimen having a round cross section. Assuming that the starting material (stock) is a round rod and that only one specimen is needed, discuss the processes and the machinery by which the specimen can be made, including their relative advantages and limitations. Describe how the process you selected can be changed for economical production as the number of specimens required increases.

**16.40** Table 16.2 lists several materials and their commercially available shapes. By contacting suppliers of materials, extend the list to include (a) titanium, (b) superalloys, (c) lead, (d) tungsten, and (e) amorphous metals.

**16.41** Select three different products commonly found in homes. State your opinions about (a) what materials were used in each product, and why, and (b) how the products were made, and why those particular manufacturing processes were used.

**16.42** Inspect the components under the hood of your automobile. Identify several parts that have been produced to net-shape or near-net-shape condition. Comment on the design and production aspects of these parts and on how the manufacturer achieved the near-net-shape condition for the parts.

**16.43** Comment on the differences, if any, between the designs, the materials, and the processing and assembly methods used to make such products as hand tools, ladders for professional use, and ladders for consumer use.

**16.44** Other than powder metallurgy, which processes could be used (singly or in combination) in the making of the parts shown in Fig. 11.1? Would they be economical?

**16.45** Discuss production and assembly methods that can be employed to build the presses used in sheet-metal forming operations described in Chapter 7.

**16.46** The shape capabilities of some machining processes are shown in Fig. 8.40. Inspect the various shapes produced, and suggest alternative processes for producing them. Comment on the properties of workpiece materials that would influence your suggestions.

**16.47** The following figure shows a steel bracket that can be made either by casting or by the stamping of sheet metal:

(a) Casting                    (b) Stamping

Describe the manufacturing processes and the sequences involved in each method. Assume that the approximate dimensions of the bracket are 100 mm by 100 mm by 25 mm wide (4 in. by 4 in. by 1 in.).

**16.48**    If the dimensions of the part in Question 16.47 were (a) 10 times larger and (b) 20 times larger, how different would your answer be? Explain.

**16.49**    A new model of an internal combustion engine is shown in Fig. 1.1. On the basis of the topics covered in this text, select any three individual components of such an engine, and describe the materials and processes that you would use in making those components. Remember that the parts must be manufactured at very large numbers and at minimum cost, yet maintain their quality, integrity, and reliability during service.

**16.50**    Discuss the trade-offs involved in selecting between the two materials for each of the applications listed:

   a)  Sheet metal vs. reinforced plastic chairs;

   b)  Forged vs. cast crankshafts;

   c)  Forged vs. powder-metallurgy connecting rods;

   d)  Plastic vs. sheet-metal light-switch plates;

   e)  Glass vs. metal water pitchers;

   f)  Sheet-metal vs. cast hubcaps;

   g)  Steel vs. copper nails;

   h)  Wood vs. metal handles for hammers.

Also, discuss the typical conditions to which these products are subjected in their normal use.

**16.51**    Discuss the manufacturing process or processes suitable for making the products listed in Question 16.50. Explain whether the products would require additional operations (such as coating, plating, heat treating, and finishing). If so, make recommendations and give the reasons for them.

**16.52**    Discuss the factors that influence the choice between the following pairs of processes to make the products indicated:

   a)  Sand casting vs. die casting of a fractional electric-motor housing;

   b)  Machining vs. forming of a large-diameter bevel gear;

   c)  Forging vs. powder-metallurgy production of a cam;

   d)  Casting vs. stamping a sheet-metal frying pan;

   e)  Making outdoor summer furniture from aluminum tubing vs. cast iron;

   f)  Welding vs. casting of machine-tool structures;

   g)  Thread rolling vs. machining of a bolt for high-strength application;

   h)  Thermoforming a plastic vs. molding a thermoset to make the blade for an inexpensive household fan.

**16.53**    The following figure shows a sheet-metal part made of steel:

Discuss how this part could be made and how your selection of a manufacturing process may change (a) as the number of parts required increases from 10 to thousands and (b) as the length of the part increases from 2 m to 20 m.

**16.54**    The part shown in the following figure is a carbon-steel segment (partial) gear:

The smaller hole at the bottom is for clamping the part onto a round shaft, using a screw and a nut. Suggest a sequence of manufacturing processes to make this part. Consider such factors as the influence of the number of parts required, dimensional tolerances, and surface finish. Discuss such processes as machining from a bar stock, extrusion, forging, and powder metallurgy.

**16.55**    Several methods can be used to make the part shown in the following figure:

List these methods, and for each one, explain such factors as the machinery and equipment required, the amount of scrap produced, the strength of the part, etc. What would be the influence of the sharpness of the corners at the two diameters on the choice of processes or the need for secondary processing? Also, comment on how different your answers would be if the material were (a) ferrous, (b) nonferrous, (c) thermoplastic, (d) thermoset, and (e) ceramic.

**16.56** The following illustration shows various components of a modular artificial knee, which allows surgeons to choose the particular components that best fit an individual patient's needs:

Femoral stem extension

Patella

Tibial base plate

Femoral implant

Tibial articular surface

Tibial stem extension

Choose any three components, and describe suitable materials and manufacturing processes for producing them. (*Source for illustration*: Zimmer, Inc.)

**16.57** Inspect several small and large products that you are familiar with. Present a plan on how you would disassemble them and recycle their individual components. Discuss any difficulties you may encounter in the process of doing so. Are there any components that you don't think could be recycled, and if so, why?

# Index

Ablation, 579
Abrasive
    belts, 529
    friction, 137
    grains, 506
Abrasive machining, 504
    economics, 549
Abrasive water-jet machining, 548
Abrasive wear, 142
Abrasive-flow machining, 532
Abrasive-jet
    cutting, 347
    machining, 548
Abrasives, 504
    bonded, 506
    coated, 529
Accelerated
    life testing, 782
    strain aging, 99
Accelerometer manufacture, 800
Acceptance, angle of, 291
Accuracy. *See* Tolerances
Acetals, 577
Acoustic
    emission, 168, 433
    impact, 168
    holography, 169
    impedance, 369
Acrylic adhesives, 731, 744
Acrylic concrete, 489
Acrylics, 577
Acrylonitrile-butadiene-styrene, 577
Adaptive control, 18, 822
    constraint, 823
    optimization, 823
Adaptive-resonance artificial neural
        network, 883
Added value, 2
Addition polymerization, 562
Additive manufacturing, 610
Additives
    extreme pressure, 148
    in ceramics, 664
    in free-machining steels, 438
    in oils, 148
    in polymers, 567
    lubricants, 148
Adhesion theory of friction, 134

Adhesive
    systems, 731
    wear, 141
Adhesives, 730
Adhesive bonding, 135, 689, 730, 738
    characteristics, 733
Adiabatic, 72
Adjustable-force clamping, 838
Adsorption, 130, 656
Advanced
    composites, 589
    machining processes,
        characteristics, 534
Age hardening, 239
Aggregation, 159
Agile manufacturing, 23
Aging, 239
Air
    bending, 356, 357
    gages, 164
Alclad, 152
Alkali soaps, 149
Alkyds, 578
Alligatoring, 293
Allotropic transformation, 191
Allotropism, 83
Allowable wear land, 430
Allowances, in casting, 247
Alloy steels, 108, 109
Alloying, 83
    elements in iron, 193
    mechanical, 633
Alloys, 187
    solidification, 196
Alpha ferrite, 191
Alumina, 656
Alumina-base ceramics, 449
Aluminizing, 157
Aluminum
    alloys properties, 114
    and alloys, 113
Aluminum oxide, 449, 506
    abrasives, 506
    ceramic tools, 449
    coatings, 447
Aminos, 578
Amorphous, 197
    alloys, 123, 235
    layer, 129

polymers, 565
Anaerobic adhesive system, 731
Analog sensors, 833
Analog-to-digital converters, 833
Angle
    gage blocks, 161
    measuring instruments, 161
    milling cutter, 474
    of acceptance, 291
    of twist, 46
Anisotropic etching, 768
Anisotropy, 92, 340, 352
    in crystals, 85
    in polymers, 573
    normal, 340, 378
    of thermal expansion, 106, 660
    planar, 340, 378
    ratio (etching), 769
    vs. limiting drawing ratio, 380
Annealing, 240
    of glass, 674
    stress relief, 58
Anode, 155
Anodic oxidation, 156
Anodizing, 156
Anthropomorphic, 835
Antioch process, 222
Apron, 461
Aramids, 577, 584
Arbor cutters, 475
Arc
    cutting, 547
    evaporation, 154
    of contact, 289
    strikes, 707
Arcwelding processes, 690
    consumable electrode, 690
    nonconsumable electrode, 697
Archard, J. F., 141
Arithmetic
    average, 131
    mean, 131, 172
Arm and wrist (robot), 828
Arrays (sensors), 834
Arrowhead fracture, 309
Artificial
    aging, 239
    intelligence, 19, 880
    neural network, 19, 882

Ashing, 763, 771
Aspect ratio, 634
Asperities, 134
Aspiration, 200
Assembly, 9, 839
    automated, 839, 843
    design for, 10, 842, 843
    manual, 843
    robotic, 844
    systems, 839
    verification, 835
Assignable variations, 171
Associative memories, 883
ASTM
    designations, 109
    grain size, 90
Astroloy, 118
Atomic force microscope, 132
Atomic structure (metals), 82
Atomic-hydrogen welding, 698
Atomization, 631
Attributes, 866
Attritious wear, 517
Ausforming, 243
Austempering, 243
Austenite, 191
    former, 193
    retained, 237
Austenitic stainless steels, 110
Austenitizing temperature, 243
Autoclave, 225, 607
Autocollimator, 162
Autogenous weld, 702
Automated
    assembly, 19, 843
    guided vehicle, 20, 826
    inspection, 165, 170
    material handling, 19
    measurement, 165
    storage/retrieval systems, 826
Automatic
    bar machine, 465
    lathe, 461
    pallet changer, 483
    screw machine, 464
    tool changers, 483
Automatically programmed tools, 821
Automation, 17, 809
    goals, 811
    hard, 813
    history, 810
    of manufacturing processes, 808
    soft, 815
Automotive
    air bag sensor manufacture, 800
    bearing cap manufacture, 648
    CAD in components, 886

engine, 1
    piston casting, 250
Autonomous guidance, 827
Availability of materials, 13
Average of averages, 174

Babbitts, 122
Back
    etching, 770
    rake angle, 457
    tension, 288
Backlash, 818, 820
Backward
    extrusion, 300
    spinning, 366
Baekeland, L. H., 559
Bainite, 237
Bainitic steel, 237
Bakelite, 559
Ball
    bearing manufacturing, 276
    burnishing, 152
    grid array, 781
    mill, 632
Ballistic-particle manufacturing,
        616
Ballizing, 152
Bamboo defect, 309
Band
    files, 480
    saw, 347, 479
Bar codes, 827
Bar drawing. *See* Rod drawing
Barrel finishing, 532
Barreling, 43, 262, 278
Barrier antireflective coating, 763
Barrier antireflective layer,
Basal plane, 83
Base metal (welding), 702
Basic
    hole system, 166
    shaft system, 166
    size, 166
Bat, 666
Batch production, 812
Bauschinger effect, 45, 384
Beach marks, 101
Beading, 357
Beads
    draw, 376, 383
    in foam molding, 603
Bed, 460
Bed-of-nails device, 838
Bed-type machines, 475
Beilby layer, 129
Bellows manufacturing, 362
Belt (abrasive), 529

Bench
    grinder, 524
    lathe, 461
Benchmark, 22
Bend allowance, 349
    angle, 349
    radius, 349
    tests (flexure), 48
    tests welded joints, 710
Bendability, 351
Bending, 48
    force, 355
    sheet and plate, 349
    tubes and sections, 359
Bernoulli's theorem, 199
Beryllium, 121
    bronze, 116
Bevel
    angle, 349
    protractor, 161
Beveling of punches, 346
Bezier curves, 858
Biaxial tension, 68
Bilateral tolerance, 166
Bill of materials, 863
Billet, 283, 295
Bimetallic strip welding, 714
Binary phase diagram, 189
Binders (for sand), 218
Bioceramics, 662
Biodegradable plastics, 580
Biological effects (metalworking
        fluids), 452
Bipolar junction transistor, 755, 782
Black ceramics, 449, 658
Blackening, 156
Blanchard-type grinder, 522
Blank
    design, 392
    tailor welded, 347
Blankholder, 375
    force, 383
Blanking, 344
    chemical, 535
    fine, 344
Blasting, shot, 532
Bleeder cloth, 607
Blending (metal powders), 635
Blends (polymer), 567
Blind riser, 199
Blocking, 279
Bloom, 283, 295
Blow molding, 598
Blowholes, 215
Blowing
    glass, 672
    agent, 597

Blown film, 594
Blue brittleness, 99
Bluetooth, 879
Body-centered cubic, 82
Bond
    fracture, 518
    types for abrasives, 508
Bonded abrasives, 506
    designations, 509
Bonding
    adhesive, 689, 730
    clad, 152
    materials for abrasives, 506
    of abrasives, 510
    of polymers, 562
Boring, 455, 465
    bar, 465
    machines, 465
    mills, 465
Boron
    etch stop, 770
    fibers, 585
Boronizing, 240
Borosilicate glass, 669
Boss, 206
Bottle manufacturing, 599, 672
Bottoming
    in bending, 355
    taps, 468
Boundary
    lubricants, 146, 147
    representation, 858
Branched polymers, 563
Brass, 116
Braze
    metals, 724
    welding, 723, 726
Brazing, 689, 723
    filler metals, 724
    of cutting tools, 444
Breaking down
    lubricants, 147
    of ingot, 294
Breaking stress, 29
Bridge die, 311
Bridgman, P. W., 64
Brinell hardness, 50
Brinelling, 50
Brittle
    fracture, 99
    materials testing, 45, 48
Broach, 477
Broaching, 477
Bronze, 116
B-splines, 858
Buckling, 96, 340
Buckyballs, 149, 675

Buffer, 813, 841
Buffered
    hydrofluoric acid, 771
    oxide etch, 771
Buffing, 531
Building-block principle, 813
Built-up edge, 410
Bulge test, 386
Bulging (tubes), 362
Bulk
    micromachining, 785
    modulus, 66
Bulk-deformation processes, 260
    general characteristics, 261
Bulk-molding compound, 607
Bull block, 321
Bundle drawing, 320
Bur, 480
Burden rate, 917
Burning (in grinding), 516
Burnishing, 152, 408
Burr, 343, 480, 532
Bus topology, 878
Business
    execution, 854
    planning, 854
Butt joint, 689

CAD
    models for automotive
        components, 886
    systems, 856
CAD/CAM systems, 861
Calcium-deoxidized steels, 438
Calendering, 604
Caliper gages, 160
Calipers, 160
Cam plastometer, 43
Camber, 292
Can manufacturing, 394
Canning (extrusion), 311
Capacitor-discharge stud welding, 720
Capacity planning, 863
Capital costs, 916
Capstan, 321
Carbide, 192
Carbides, 444, 657
    coated, 447
Carbon. *See also* Graphite
    black, 567
    fibers, 584
    foam, 676
    in stainless steels, 110
    in steels, 108
    steels, 108
Carbon-arc welding, 697
Carbon-dioxide process, 221

Carbon-fiber reinforced plastic, 584
Carbonitriding, 152, 240
Carbonyls, 632
Carburizing, 152, 240
Carpal tunnel syndrome, 840
Carriage, 460
Carrier sense multiple access with
        collision detection, 878
Case hardening, 152, 240
Cast
    irons, 193, 211
    stainless steels, 214
    steels, 214
    structures, 195
    structures, in welding, 702
Castability, 202
Cast-cobalt alloy tools, 443
Casting, 186
    alloys, 209
    continuous, 216
    defects, 204
    design for expendable-mold
        processes, 246
    design for permanent-mold casting,
        248
    economics, 249
    expendable mold, 218, 224
    expendable pattern, 224
    ferrous alloys, 211
    nonferrous alloys, 212, 214
    of ceramics, 664
    of ingots, 215
    of pistons, 250
    of plastics, 603
    permanent mold, 227
    permanent patterns, 218
    sand, 220
Casting processes, 218, 249
    general characteristics, 210
Catastrophic failure, 100
Cathode, 155
Cavities. *See* Porosity
Cellular manufacturing, 19, 872
Cellulosics, 577
Cemented carbides. *See* Carbides
Cementite, 192
Center cracking, 309, 318
Center drill, 466
Center-line average, 131
Center-type grinding, 522
Centerburst defect, 309
Centerless grinding, 523
Centrifugal casting
    glass, 673
    metals, 230, 232
    polymers, 603
Centrifugal spraying (glass), 672

Centrifuge casting, 232
Ceramic
    bond, 508
    coatings, 157, 447
    grinding wheel (*see* Vitrified bond)
    matrix composites, 678
    mold casting, 222
    molds, 643
    packages (chip), 779
Ceramics, 655
    applications, 661
Coatings, 153
    cutting tools, 449
    friction of,
    general characteristics, 655
    joining, 739
    processing, 663, 664
    properties, 659
    structure, 656
    wear of, 144
Ceramic-shell investment casting, 226
Cermets, 449, 658
Certification, quality, 179
Chain-reaction polymerization, 562
Chain-type code, 870
Chamfer, cutting tools, 446
Chance variations, 171
Channel furnace, 207
Chaplets, 219
Charpy test, 56
Chatter, 486, 487
    fifth-octave, 294
    in grinding, 526
    in machining, 486
    in rolling, 294
    marks, 526
    third-octave, 294
    torsional, 294
Checking. *See* Heat checking
Chemical
    blanking, 535
    cleaning, 159
    guide path, 827
    machining, 533
    mechanical planarization (*see*
        Chemical mechanical polishing)
    mechanical polishing, 531, 776
    milling, 533
    properties, 13
    reaction priming, 156
    synthesis processes, 678
    tempering, 674
    wear, 143
Chemical vapor deposition, 154, 446
    Low pressure, 760
    medium temperature, 447
    plasma-enhanced, 760

vapor infiltration, 678
Chemically
    assisted ion-beam etching, 774
    reactive adhesive, 730
Chevron cracking, 309
Chill tempering, 674
Chills, 206
Chip
    breakers, 413, 446
    compression ratio, 407
    curl, 412
    depth of cut, 470
    flow angle, 413
    morphology, 408, 515
    types, 409
    velocity, 407
Chip formation
    abrasive wear, 137
    cutting, 405
    grinding, 512
Chip (semiconductor), 755
Chipping (cutting tools), 432
Chips (metal), 405
Chisel-edge angle, 466
Chisel-point drill, 466
Choke core, 200
Choking, 200
Chromium
    in stainless steels, 109
    plating, 155
Chuck, 461
Chuckers, 461
Chucking machines, 461
Chvorinov's rule, 203
Circuit board, 782
Circular
    cutters, 474
    interpolation, 820
    saws, 479
    tracing, 162
Circumscribing-circle diameter, 301
CL process, 223
CLA process, 223
Clad bonding, 152
Cladding, 152, 301, 712
Clamps, 837
Classification and coding systems, 868
Clay, 656
Cleaning
    fluids, 159
    of castings, 244
    processes, 158
Clearance, 166, 342
    angle, 406
    in deep drawing, 383
Cleavage plane, 99
Climb milling, 470

Closed-die forging, 273
Closed-loop control, 818, 829
Cluster mill, 296
CLV process, 223
Coalescence of voids, 97
Coarse pearlite, 236
Coated
    abrasives, 529
    sheet metal, 341
    tools, 446
    tools, multiphase, 447
Coatings
    ceramic, 153, 157, 447
    conversion, 156, 321
    for cutting tools, 446
    for electrodes, 696
    for rod and wire drawing, 321
    for sheet metal, 341
Coaxial extrusion, 301
Cobalt (in tungsten carbide), 444
Cobalt-base superalloys, 118
Cocuring, 738
Coding systems, 869
Coefficient of friction, 135
    in cutting, 416
    in metalworking processes, 137
    in ring compression, 138
    measurement, 137
Cogging, 276
Coining, 274, 355, 649
Co-injection molding, 596
Cold
    cracking (welds), 706
    extrusion, 306
    forming of plastics, 604
    forming (*see individual processes*)
    isostatic pressing, 641
    pressure welding, 136
    roll forming, 358
    saws, 479
    shut, 205, 277
    welding, 712
    working, 92, 95, 261
Cold-box mold, 219
Cold-chamber process, 229
Cold-pressed ceramics, 449
Cold-runner molds, 596
Cold-setting processes, 219
Cold-work steels, 112
Collapsibility, 218
Collapsible taps, 468
Collapsing, 340
Collet, 461
Colloidal, 149
Colorants, 568
Coloring, 156
Columbium. *See* Niobium

Column-and-knee milling machine, 475
Columnar
   dendrites, 196
   grains, 195
Combination square, 161
Commercially available forms of materials, 902
Comminution, 632, 663
Communications
   network, 878
   standards, 879
Compacted-graphite iron, 213
Compacting pressure various metal powders, 640
Compaction of metal powders, 636
Compacts, 450
   diamond, 450
Comparative length measurement, 161
Comparator, 162
Competitive aspects of manufacturing, 893
Compilers, 821
Compliant end effector, 828, 834
Composite
   coatings, 448
   molds, 221
Composite materials, 582, 676
   ceramic-matrix, 678
   metal-matrix, 676
   processing, 676
Compound
   dies, 346
   rest, 461
Compounded oils, 148
Compressibility metal powders, 637
Compression
   molding, 601, 607
   test, 43
Compressive strength various materials, 659
Computer
   integrated manufacturing, 854
   numerical control, 17, 461, 816
   simulation, 8, 864
   vision, 831, 835
Computer-aided
   design, 8, 856
   engineering, 8, 856
   manufacturing, 8, 861
   part programming, 821
   process planning, 19, 861
Computer-controlled lathes, 461
Computer-integrated manufacturing, 17, 852
Concentration gradient, 197

Concurrent engineering, 6
Condensation polymerization, 561
Conditioning of rolls, 295
Conductive graphite fibers, 585
Conductivity
   electrical, 106
   thermal, 105
Conductor, 106
Connecting rod forging, 279
Consistency index, 571
Constitutional diagram, 189
Constructive solid geometry, 858
Consumable electrode, 690
Contact
   area (apparent), 134
   molding, 608
Contacts, 776
Contaminants in welds, 704
Continuity, 200
Continuous
   casting, 215
   chip, 408
   fibers, 586
   laser, 700
   path system, 819
   products, 2
   systems (assembly), 841
Contour roll forming, 358
Contouring
   lathe, 461
   system, 819
Contraction (solidification), 204
Control
   adaptive, 822
   charts, 21, 174
   limits, 175
   numerical, 17, 815
   systems (robots), 829
Controllers, 815, 829
Conventional
   abrasives, 505
   milling, 469
   spinning, 364
Conversion coatings, 156, 321
Convex shear die, 346
Conveyance card, 877
Coolants, 451
Cooling rate (castings), 197
Cooling. *See* Cutting fluids *and individual processes*
Coordinate measuring machine, 163
Cope, 199, 220
Cope-and-drag investment casting, 222
Copolymers, 564
Copper and alloys, 115
Copy milling machines, 476

Core
   blowers, 219
   boxes, 219
   drill, 466
   furnace, 207
   prints, 219
Cored dendrites, 197
Coreless induction furnace, 207
Cores, 199, 219
Corner
   joint, 689
   radii, in deep drawing, 383
Correction for necking, 63
Corrosion, 107
   tests (welds), 711
Corrosive wear, 143
Corrugation process (honeycomb), 374
Corundum, 656
Cost
   accounting, 915
   justification, 915
   materials and processing, 903
   reduction, 13, 16, 22, 915, 918
   machinery, 915[
Cost-benefit analysis, 919
Costs. *See* Economics of manufacture *and individual processes*
Counterblow hammers, 282
Counterboring, 466
Counter-gravity low-pressure process, 223
Countersinking, 466
Covalent bond, 562, 656
Crack velocity, 100
Cracks. *See also* Defects *and individual processes*
   in welds, 705
Crank press, 281
Crankshaft drill, 466
Crater wear, 431
Craters, 130
Crazing, 573
Creep, 55, 91
   in polymers, 573
   rupture, 56
   tests (welds), 711
Creep-feed grinding, 525
Crevice corrosion, 107
Crimping, 736
Critical shear stress, 84
Cross-feeding, 521
Cross-linked polymers, 564, 575
Cross-rolling, 276
Cross-tension test (spot welds), 711
Crow's foot package, 780
Crown, 291
Crucible furnace, 208

Crush
  dressing, 519
  forming, 519
Crushing, 663
Cryogenic
  dry etching, 775
  grinding, 526
  machining, 453
  tempering, 243
  treatment, cutting tools, 451
Crystal
  growing, 234, 757
  pulling, 234
Crystalline structure, 82
Crystallinity
  effects of, 565
  polymers, 565
Crystallites, 565
Crystallographic anisotropy, 93, 340
Crystals, 82
Cubic
  boron nitride, 449, 506, 657
  interpolation, 820
Cumulative distribution, 634
Cumulative trauma disorders, 840
Cup (casting), 199
Cup-and-cone fracture, 97
Cupola, 208
Cupping tests, 385
Cure depth (stereolithography), 612
Curing (thermosets), 575
Curl (chip), 412
Cutoff (roughness), 132
Cutters. *See individual processes*
Cutting. *See also* Machining *and*
       *individual processes*
  dry, 452
  forces, 415, 458
  processes, 454
  processes, general characteristics, 455
  ratio, 407
  temperature, 424
Cutting fluids, 451, 458
  applications, 452
  effects of, 452
Cutting off, 455
  abrasive, 525
  turning, 455
Cutting speeds
  drilling, 467
  milling, 475
  turning, 459
Cutting tools
  geometry, 457
  materials, 440
  reconditioning, 451
  toollife curves, 428

wear, 426
Cutting-edge angle, 457
Cyaniding, 152, 240
Cyanoacrylate adhesive systems, 731
Cylindrical grinding, 522
Czochralski process, 234, 758

Dacron, 577
Dampers, 294
Damping
  external, 488
  in cast iron, 487
  machine tools, 487
Data
  acquisition system, 855
  fusion, 882
Database, 855
  smart, 901
Datum, 132
Dead-metal zone, 262
  in extrusion, 302
Deburring, 532
Decarburization, 240
Decision-tree coding, 870
Dedicated
  fixtures, 837
  machines, 812
Deep drawing, 374
  drawability, 378
  lubrication, 385
Deep hole drilling, 466
Deep reactive ion etching, 774, 792
Deep ultraviolet lithography, 764
Defect prevention, 178
Defects, 87, 100, 130, 309. *See also*
       *individual processes*
  casting, 204, 205
  drawing, 318
  forging, 277
  surface, 130
Deflection (rolls), 291
Deflocculent, 664
Deformation
  plastic, 26
  processing, 26[
  rate, 38
  single crystals, 83
  speed in metalworking, 39
  work, 68
Deformation zone
  extrusion, 310
  geometry, 270
Degradation (polymers), 107, 564, 568
Degree
  of crystallinity, 565
  of polymerization, 562
Delamination, reinforced plastics, 586

Delayed tack adhesives, 731
Delivery reel, 288
Delrin, 577
Delta ferrite, 191
Demand scheduling, 877
Deming, W. E., 178
Dendrite multiplication, 198
Dendrites, 196
  arm spacing, 197
Density, 103, 104
  various materials, 104, 659
Dent resistance, 390
Deoxidation, 206
Deposition
  chemical vapor, 154, 446
  physical vapor, 153, 446
Depth-of-cut, 405, 426
  line, 433
Derivative CAPP system, 862
Design
  attributes, 869
Design considerations
  abrasive machining, 549
  adhesive bonding, 741
  advanced machining, 549
  assembly, 10, 842, 843
  bending, 392
  blanking, 392
  brazing and joining, 741
  broaching, 490
  casting, 245
  disassembly, 10, 844
  drilling, 490
  fixturing, 842
  gear manufacture, 490
  glass products, 675
  heat treating, 244
  mechanical fastening, 742
  milling, 490
  polymers, 619
  powder metallurgy, 651
  printed circuit boards, 784
  reaming, 490
  reinforced plastics, 619
  sheet-metal forming, 391
  tapping, 490
  thread cutting, 490
  turning, 489
  welding, 740
  *See also individual processes*
Design for the environment, 11
Design for manufacture, 9
  manufacturing cell, 873
  of experiments, 21
  process, 6
Design for recycling, 11
Design for service, 10, 844

Designations. *See individual materials and processes*
Desk-top manufacturing, 610
Desorption, 147
Destructive testing, 169, 710
  welded joints, 710
Detonation
  gun, 153
  spraying, 153
Deviation-type measuring instruments, 161
Devitrification, 670
Dial indicator, 161
Diamond, 676
  abrasives, 506
  coatings, 158, 448
  cutting tools, 450
  film, 158
  pyramid hardness(*see* Vickers hardness)
  saws, 479
  turning, 464
  wire saw, 479
Diamond-like carbon, 158, 676
Diamond-type lattice , 757
Die
  characteristic (extruder), 593
  cutting, 344
  design in forging, 279
  distortion, 279
  failures, 324, 325
  geometry (*see individual processes*)
  inserts, 320, 324
  life (*see individual processes*)
  manufacturing methods, 322
  materials, 112, 280
  pressure (*see individual processes*)
  quenching, 238
  sinking, 541
  steels, 111
  swell, 593
Die casting, 228
  alloys, 231
  cold chamber, 229
  hot chamber, 228
Die(chip), 777
Dielectric
  fluid, 540
  strength, 106
Dielectrics, 106
Dies, 323. *See also individual processes*
Diffraction gratings, 160
Diffuse necking, 40, 339
Diffusion, 94, 765
  adhesives, 731
  bonding, 372, 722

brazing, 726
coating, 155
in crater wear, 432
sintering, 645
welding, 722
Digital
  micromirror, 785
  pixel technology device, 785
  readout, 160
  sensors, 833
Dilatation, 66
Dimensional tolerances, 15, 165, 907. *See also* Tolerances.
Dimples (ductile fracture), 97
Dimpling, 358
Dip brazing, 725
Direct
  emulsion, 149
  engineering, 7
  extrusion, 301
  labor costs, 917
  manufacturing, 617
  numerical control, 817
Directional solidification, 233
Directionality, 130, 340. *See also* Anisotropy
Disassembly, 10, 844
Discontinuities
  in castings, 205
  in welds, 704
Discontinuous
  chips, 410
  fibers, 586
Discrete products, 2
Discrete-event simulation, 864
Disk test, 45
Dislocations, 87
  density, 88
Dispersion, 159, 172, 173
Distortion, 279. *See also* Warping
Distortion-energy criterion, 62
Distributed numerical control, 817
Distribution (statistical), 172
Dividers, 161
Doctor-blade process, 664
Domain, 881
Dopants, 106, 235, 756
Doping, 580
Double
  action press, 385
  barreling, 262
  bevel, 346
Down milling, 470
Draft
  angles in forging, 280
  in castings, 247
  in rolling, 289

Drag, 199, 220
  flow, 591
Drain casting, 664
Draw
  beads, 376, 383
  bench, 321
  cut, 477
Drawability, *See* Deep drawability
Drawing
  deep, 374
  dies, 320
  exchange format, 856
  glass, 671
  out, 276 (*see also* Cogging)
  stress, 312
  strip, 317
  tubes, 318
  without a blankholder, 384
Dressing, 518
  grinding wheels, 519
Drift pin, 735
Drill press, 467
Drilling, 455, 465
  machines, 467
Drills, 465
  geometry, 466
  life, 467
Drive-in diffusion, 765
Dross, 201
  traps, 201
Dry
  chemical etching, 774
  drawing, 320
  etching, 773
  machining, 452, 493
  oxidation, 761
  pressing (ceramics), 665
  spinning, 584
Drying (ceramics), 667
Dual
  in-line package, 780
  ion-beam assisted depositon, 154
  phase steels, 109
Ductile
  fracture, 96
  iron, 213
Ductile regime
  cutting, 464
  grinding, 521
Ductility, 30
Dummy
  bar, 216
  block, 301
Duplex structure, steels, 111
Duplicating
  lathe, 461
  milling machine, 476

Duranickel, 117
Durometer, 52
Dynamic
    loading, 56
    scheduling, 876
    stiffness, 487
Dynamometer, 419

Earing, 380
E-beam lithography. *See* Electron-
    beam lithography
Eccentric press, 281
Ecological considerations, 150
Economic
    order quantity, 908
    production quantity (plastics), 621
Economics
    abrasive machining, 549
    assembly, 844
    casting, 249
    computer integrated
        manufacturing, 885
    joining, 743
    machining, 490
    plastics processing, 621
    powder metallurgy, 653
    sheet metal forming, 394
    *See also* Costs *and individual*
        *processes*
Eddy-current inspection, 169
Edge
    condition, 352
    cracks, 293
    dislocation, 88
    joint, 689
Edger mill, 292
Edging, 276
    roll, 298
Effective
    strain, 67
    stress, 67
Efficiency (in deformation processing),
    70
Ejectors, 144, 229, 596
Elastic
    deformation, 83
    limit (*see* Proportional limit)
    modulus (*see* Modulus of elasticity)
    recovery, 29, 352
Elastomers, 581
    processing, 604
Electrical
    conductivity, 106
    properties (polymers), 574
    resistivity, 106
Electrical-discharge
    grinding, 543

machining, 540
    wire cutting (*see* Wire EDM)
Electrically conducting
    adhesives, 731, 734
    polymers, 580
Electric-arc furnace, 207
Electric-discharge forming, 370
Electrochemical
    discharge grinding, 543
    fabrication, 799
    grinding, 539
    honing, 530
    machining, 537
Electrode
    coatings, 696
    wear (EDM), 542
Electrodes
    EDM, 540, 541
    welding, 695
Electrodischarge machining, 540
Electroforming, 156
    and LIGA, 795
Electrogas welding, 694
Electrohydraulic forming, 370
Electroless plating, 155
Electrolytic
    cleaning, 159
    deposition, 632
    grinding (*see* Electrochemical
        grinding)
Electromagnetic bonding, 738
Electromigration, 776
Electron-beam
    lithography, 765
    machining, 546
    welding, 699
Electronic gage, 161
Electronic-grade silicon, 758
Electroplating, 155
Electropolishing, 531
Electrorheological fluid, 839
Electroslag welding, 695
Elements (alloying) in iron, 193. *See*
    *also individual elements*
Elongation, 30, 337
    various materials, 28
Embossing, 362, 393
Embrittlement, 91
    hydrogen, 102
Emery, 656
    Cloth, 506, 529
Emulsification, 159
Emulsions, 148, 452
Enameling, 157
Encapsulation, 603
Encoder, 818
End

cutting edge angle, 414, 456
    effector, 828
    grains, 277
    milling, 474
    relief angle, 414, 456
End-feed grinding, 524
End-of-arm tooling, 828
Endurance limit, 54. *See also* Fatigue
    limit vs. strength, 55
Energy
    ideal, 69
    limited, 281
    per unit volume, 68
    requirements (*see* Specific energy)
Engine lathe, 460
Engineered materials, 12, 582
Engineering
    metrology, 160
    strain, 26
    strain rate, 38, 43
    stress, 27
Enterprise resource planning, 864
Entrainment (entrapment) of
    lubricants, 148
Entry zone rolling, 286, 288
Environment, design for, 11
Environmental
    considerations, 10
    effects, 452
    management systems, 180
    stress cracking, 574
Environmentally conscious design and
    manufacturing, 10
Environments (expert systems), 881
Epitaxial layers, 759, 761
EPOCH hip manufacture, 622
Epoxies, 578
Epoxy-based adhesive systems, 731
Equiaxed grains, 195
Equilibrium diagram, 189
Equipment. *See individual processes*
Equivalent
    strain (*see* Effective strain)
    stress (*see* Effective stress)
Ergonomics, 22
Erichsen test, 385
Erosion, 143
    in forging dies, 144
Error of form, 131
Esteem value, 919
Etch
    rates, 767
    stop, 766, 769
Etchant, 533
Etching, 766, 767, 768
    anisotropic, 768, 769
    chemically assisted ion-beam, 774

Etching, (*cont.*)
  cryogenic, 775
  deep reactive ion, 774
  dry, 773, 774
  isotropic, 766
  of silicon dioxide, 770
  of silicon nitride, 771
  physical-chemical, 774
  reactive ion beam, 774
  reactive plasma, 774
  sputter, 773
  wet, 766
Ethernet, 878
Eutectic, 197
Eutectoid
  cementite, 193
  ferrite, 193
  reaction, 193
Evaporation (film deposition), 759
Evaporative
  adhesives, 731
  pattern casting, 224
Execution functions, 853
Exit zone, rolling, 286, 288
Exothermic reaction (grinding
    sparks), 516
Expanded polystyrene process, 224
Expanding. *See* Bulging
Expansion
  process (honeycomb), 373
  thermal, 105
Expendable
  mold, 218, 224
  pattern, 224
Experimental design, 179
Expert systems, 9, 19, 880
Explosion welding, 721
Explosive
  forming, 368
  hardening, 152
External
  damping, 488
  grinding (*see* Cylindrical grinding)
  spinning, 367
Extreme ultraviolet lithography, 764
Extreme-pressure additives, 148
Extruder, 590
  characteristic, 592
  flow rate, 592
Extrusion, 300
  blow molding, 598
  ceramics, 665
  cold, 306
  constant, 305
  defects, 302, 309
  hot, 310
  hydrostatic, 307

  impact, 307
  plastics, 590
  powders, 643
  ratio, 302
  various materials, 306

Face milling, 471
Face-centered cubic, 82
Facing, 454
Factory of the future, 20, 883
Failure, 96. *See also* Fracture
  in time, 781
Families of parts, 866
Faraday's constant, 539
Fasteners, 734
Fastening, 689
Fatigue, 54
  fracture, 54, 101
  fracture surfaces, 101
  limit, 54
  static, 660
  strength, effect of manufacturing
      process, 101
  strength, improvement, 101
  thermal, 106, 143
  wear, 143
Faying surfaces, 713, 716
Feed
  force, 458
  marks, cutting, 436
  marks, milling, 471
  rod, 461
  *See also individual processes*
Feedback, 818
  devices, 829
Feeders, 199
Feldspar, 656
Ferrimagnetism, 107
Ferrite, 191
  former, 193
Ferritic
  gray iron, 213
  stainless steels, 110
Ferromagnetism, 107
Ferrous alloys, 108, 211
  ingots, 215
Ferrule, 720
Fiber pullout, 144, 586
Fiberglass, 670
Fibering (mechanical), 93, 98, 340
Fiber-reinforced epoxy, 579
Fibers (reinforcing), 582, 584
Fibrous, 97
Field effect transistor, 745
Fifth-octave chatter, 294
Filament winding, 609
Files, 480

Filing, 480
Filled gold, 722
Filler
  in polymers, 567
  metals, 724
  wire, 697
Film
  blown, 594
  deposition, 759
  diamond, 158
  extrusion, 593
Fine
  blanking, 344
  ceramics, 656
  grain graphite, 675
  pearlite, 236
Finish. *See* Surface finish
Finishing cuts, 455, 458
Finishing operations, 529
  casting, 244
  ceramics, 667
  gear, 481
  powder metallurgy, 649
Finite-element method, 269
Fire polishing, 671, 674
Firing, 667
Fir-tree cracking, 309
Fishmouthing.
      *See* Alligatoring
Fishtailing, 309
Fixed
  costs, 916
  gages, 164
  position automation, 813
  sequence robot, 830
Fixtures, 837
  flexible, 837
Flame
  cutting, 347
  hardening, 152, 240
  retardants, 568
  wire spraying, 153
Flammability (polymers), 568
Flanging, 357
Flank wear, 428
Flaring, 358
Flash, 272
  in casting, 230, 247
  in compression molding, 602
  in forging, 280
  in friction welding, 715
  welding, 719
Flashless forging, 273
Flask, 199
Flaskless molding, 220
Flat, wear, 514
Flat rolling, 282, 294

Flatness, 162
Flattening
    by bending, 358
    by roller leveling, 295
    of rolls, 291
Flaws. *See* Defects
Flexible
    assembly, 841
    automation, 815
    fixturing, 837
    manufacturing cell, 873
    manufacturing system, 19, 874
    PVC, 578
Flexure. *See* Bending
Flights, 591
Flint, 656
Flip-chip, 781
Float glass, 671
Floating-zone method, 235
Flood cooling, 452
Floor-to-floor time, 917
Flow
    forming, 327
    lines, 814
    rules, 65
    stress, 34, 37, 69, 264
    turning, 365
Fluid flow, in casting, 199, 248
Fluid-forming process, 363
Fluid-to-fluid extrusion, 308
Fluidity, 201, 202
Fluidized bed, 226
Fluorocarbons, 577
Flush fine machining, 494
Flute, 466
Flux, 207, 724, 728
Flux-cored arc welding, 693
Fly cutting, 475
Flyer plate, 721
Foam
    carbon, 676
    casting, 603
    metal, 124
    molding, 603
Foil, 282, 295
Folding, 340
Folds, 130
Force
    bending, 355
    cutting, 458
    forging, 263, 266
    grinding, 513
    hubbing, 275
    per unit width, 265
    shearing, 343
    spinning, 365
    *See also individual processes*

Forced vibration, 486
Forgeability, 47, 278
Forging, 260, 261
    closed-die, 273
    dies, 279
    equipment selection, 282
    force, 263, 266, 272
    hot die, 273
    impression-die, 272
    incremental, 274
    isothermal, 273
    lubrication, 280
    open-die, 261
    orbital, 274
    powder metals, 649
    precision, 273
    pressure, 265,266
    rocking die, 274
    rotary, 274
    swing, 274
    temperature, 281
Form
    block, 359
    cutting, 480
    grinding, 522
    milling, 474
    tools, cutting, 454
Formability. *See individual processes*
Forming. *See individual processes*
Forming-limit diagram, 386
    various materials, 387
Forward
    extrusion, 301
    slip, 284
    spinning, 366
Foundations (machine), 489
Foundries, 208
Foundry automation, 208
Four-high mill, 296
Four-point bending, 48
Four-slide machine, 357
Fracture, 96
    abrasive bonds, 518
    abrasive grains, 518
    brittle, 99
    ductile, 97
    modes, 101
    stress, 29
    surfaces, 97, 100
    toughness, 34, 710
    *See also* Cracking *and* Defects
Framework systems, 881
Free
    bending (*see* Air bending)
    cutting wheels, 517
    space, 573
    volume, 573

Free-form fabrication, 610
Free-machining steels, 437
Free-standing diamond films, 158
Freezing range, 197
Frequency distribution, 172, 173, 634
Fretting corrosion, 144
Friability, 505, 518
Friction, 134
    and forming limit diagram, 389
    angle, 416
    factor, 136
    hill, forging, 265
    in ceramics, 139
    in metals, 134
    in metalworking (*see individual processes*)
    in polymers, 139
    in ring compression, 138
    in rolling, 291
    measurement, 137
    reduction, 139
    work, 303
Frictionsawing, 347, 479
Frictionstir welding, 715
Frictional stress, 264, 267
Friction welding, 714
    inertia, 715
    linear, 715
    polymers, 738
    stir, 715
Front tension, 288
Fukui test, 385
Full
    annealing, 242
    indicator movement, 162
Fuller, B., 675
Fullerenes, 149, 676
Fullering, 276
Full-mold process, 224
Fully gray iron, 213
Functional layout, 867
Furnace
    brazing, 725
    selection, 208
Furnaces (melting), 207
Fused
    aluminum oxide, 506
    deposition modeling, 614
    silica, 669
Fusion
    bonding silicon, 792
    weld joints, 701
    welding, 688
    zone, 702
Fuzzy
    logic, 883
    models, 883

Gage
   blocks, 161, 164
   length, 27
   numbers sheet, 295
Gages, 163
Gain, 132, 822
   scheduling, 822
Galling, 141
Gallium arsenide, 757
Galvanic corrosion, 107
Galvanized steel, 341
Galvanizing, 122, 157
Gamma iron, 191
Gantry robot, 830
Gap bed lathe, 461
Garnet, 505
Gas
   metal arc welding, 692
   mixtures (forming), 373
   tungsten arc welding, 697
Gas-assist molding, 597
Gases
   in casting, 206
   shielding (*see individual processes*)
Gating system, 199
Gaussian, 172
G-code, 822
Gear manufacturing
   broaching, 480
   finishing, 481
   form cutting, 480
   generating, 480
   rolling, 298
   shapers, 480
Gearteeth measurement, 162
Generalized Hooke's law, 61
Generative CAPP system, 863
Gentle grinding, 517
Geometric
   effects (lubrication), 147
   modeling, 857
   softening, 36
Germanium, 756
Giant molecules, 561
Glass, 668
   ceramics, 670
   fibers, 584
   former, 669
   lubrication, 150, 311
   point, 566
   processing, 671
   temperature, 566
   tubing, 671
   wool, 672
Glasses
   as bonding agents, 508
   joining, 739

properties, 669
strengthening, 674
tempering, 673
types, 669
Glass-transition temperature, 566
   various polymers, 566
Glassy behavior, 566
Glazing, 157, 667
   of grinding wheels, 519
Global competitiveness, 22
Globular transfer, 693
GO gage, 164
Gold, 122
Goodyear C., 581
Grade (bonded abrasives), 509, 510
Graduated instruments, 160
Grain, 89
   abrasive, 506
   boundary, 87, 89
   boundary embrittlement, 91
   boundary sliding, 55, 91
   depth of cut, 512
   flow patterns, forging, 277
   force, 513
   fracture, 518
   growth, 95
   size, 90, 340
   structure in casting, 196
Grand average, 174
Granite-epoxy composite, 489
Graphite, 149, 675. *See also* Carbon
   fibers, 584
   lubricant, 149
   molds, 221, 228
Graphitization, 194
Gravity segregation, 198
Gray (cast) iron, 212
Greases, 149
Green
   ceramics, 664
   compact, 636
   density, 637
   design, 11, 900
   engineering, 900
   molding sand, 218
   strength, 644
Grindability, 520
Grinding
   cryogenic, 526
   electrical-discharge, 543
   electrochemical, 539
   fluids, 526
   mechanics, 510
   operations, 521
   ratio, 519
   specific energy, 513

Grinding wheels, 506. *See also*
   Bonded abrasives
   dressing, 519
   glazing, 519
   loading, 519
   selection, 520
   types, 507
   wear, 517
Grit
   blasting, 532
   number (abrasive), 506
Grooving, 455
Group
   layout, 867
   machine cell, 872
   technology, 19, 866
Gull wing package, 780
Gun drilling, 466
Gutter, 280

Hacksaws, 479
Hafnium nitride, 448
Hall-Petch equation, 90
Hammers (forging), 282
Hand lay-up, 608
Hard
   acting wheel, 520
   automation, 813
   chromium plating, 155
   facing, 153
   glass, 669
   turning, 464, 527
   wired, 815, 819
Hardenability, 238
Hardeners, 207
Hardening
   case, 152, 240
   explosive, 152
   spark, 153
Hardness, 48, 49
   bonded abrasives, 510
   hot, 440
   scales, 49
   tests, 49
   various abrasives, 505
   various materials, 506, 659
   vs. strength, 52
Hastelloy, 116
Hazards. *See individual processes*
Headers, 275
Heading, 274
Headstock, 461
Heat, 71
   checking, 143, 230, 325, 516
   flow, 202
   in resistance welding, 716
   transfer, 202

transfer in casting, 248
Heat treatment, 235, 241
    ferrous alloys, 235
    nonferrous alloys, 238
    welds, 708
Heat-affected zone, 130, 702, 703
    friction sawing, 479
Heat-resistant alloys, 117.
    *See also* Superalloys
Helix angle, 466
Hemming, 356, 358
Heterogeneous nucleation, 198
Heuristic, 881
Hexagonal close-packed, 83
Hexapod, 489
Hierarchical coding, 869
High speed
    automated assembly, 840, 843
    machining, 463, 493
    milling, 474
High-carbon steel, 109
High-efficiency machining range, 492
High-energy
    beam machining, 544
    rate forming, 368
High-frequency resistance welding,
    719
High-pressure refrigerated coolant,
    452
High-removal rate machining, 462
High-speed steels, 111, 443
High-strength low-alloy steels, 109
High-temperature
    alloys (*see* Superalloys)
    polymers, 579
    superconductors, 679
High-velocity oxyfuel gas spraying,
    153
Hinge (MEMS) manufacture, 790
History of
    automation, 17, 810
    manufacturing, 2
HNA etching system, 771
Hob, 480
Hold-down ring (*see* Blankholder)
Hole preparation (fastening), 735
Holemaking processes. *See also*
        Advanced machining processes
    drilling, 466
    punching, 342
Hollow end mill, 474
Holographic interferometry, 169
Holography, 169
Homogeneous deformation, 261
Homogenization, 242
Homologous temperature, 96
Homopolymers, 564

Honeycomb structures
    (manufacturing), 373
Honing, 529
Hooker process, 308
Hooke's law, 28, 61
Horizontal upsetters. *See* Headers
Host atoms, 187
Hot
    cracking (welds), 706
    dipping, 157
    extrusion, 310
    forging, 261
    hardness, 53, 440
    isostatic pressing, 206, 642, 666
    machining, 440
    pressing (ceramics), 666
    pressure welding, 722
    roll bonding, 712
    rolling, 291
    shortness, 91, 305
    spots (casting), 246
    tearing, 197, 706
    working, 95
Hot-chamber process, 228
Hot-die forging, 273
Hot-melt adhesives, 731
Hot-plate welding, 738
Hot-pressed ceramics, 449, 658
Hot-rolled wire rods, 295
Hot-runner mold, 596
Hot-tool welding, 738
Hot-twist test, 279
Hot-work steels, 112
Hubbing, 275
Human factors engineering, 22
Hybrid, 583
    code, 870
Hydraulic press, 281, 312
Hydrodynamic
    fluid film, 146
    lubrication, 291
    machining, 547
Hydroform process, 363
Hydrogen
    bonds, 563
    embrittlement, 102
    in aluminum, 206
Hydroplastic forming, 665
Hydroplasticity, 656
Hydrospinning, 365
Hydrostatic
    extrusion, 301, 307
    pressure, effects of, 42,
        279, 351
    tensile stress, 309
Hygroscopy, 574
Hysteresis, 581

Ice-cold molding, 597
Ideal
    deformation, 302, 312
    energy, 69
    work, 303
Idler roll, 298
If-then rules, 881
Image recognition, 835
Impact, 56
    extrusion, 301, 307
    molding, 220
    plating, 152
    toughness, 56
    wear, 144
Imperfections (crystals), 87
Impingement, 344
Impregnation, 650
Impression-die forging, 272
Impurities, 91, 97, 106
Impurity atom, 87
In good statistical control, 176
Incipient melting, 72
Inclination angle, 413
Inclusions, 87, 130, 202, 205, 352
    effects of, 97
Incoloy, 118
Incomplete
    fusion, 704
    penetration, 704
Inconel, 116
Incremental forging, 274
Indenters, 49
Indexing
    assembly, 840
    inserts, 445
Indirect
    emulsion, 149
    extrusion, 301
    labor costs, 917
    reading instruments, 160
Induction
    brazing, 725
    furnace, 207
    hardening, 152, 240
    welding, 738
Industrial
    ceramics, 656
    diamond, 676
    engineers, 24
    graphite, 675
    Revolution, 4
    robots, 18, 827
Inertia friction welding, 715
Infeed grinding, 524
Inference, 880
    engine, 881
Infiltration, 650

Infrared
    brazing, 726
    radiation, 426
Ingot casting, 215
Inhomogeneity factor, 314
Inhomogeneous deformation, 262
    drawing, 314
Initial graphics exchange specification, 857
Injection
    blow molding, 598
    foam molding, 597
Injection molding
    and LIGA, 795
    ceramics, 666
    foam, 598
    polymers, 595
    powder metals, 642
    reaction, 597
Ink-jet printer manufacture, 792
Inlet angle (lubrication), 147
In-line indexing, 841
Inoculant, 198
Inorganic adhesives, 730, 731
In-process
    annealing, 242
    measurement, 165
Insert molding, 230, 596
Inserts (cutting tool), 444
Inspection, 160
    automated, 170
    castings, 244
Instability, 68, 338
    in tension, 35
    in torsion, 47
Instant masking, 799
Insulators, 106
Integral transfer devices, 826
Integrated
    circuit, 753
    communications systems, 854
    mills, 297
    snap fasteners, 738
Intelligent
    fixturing systems, 837
    robot, 831
Interactive computer graphics, 856
Interchangeable manufacturing, 171
Interconnections, 776
Interferometry, 162
Intergranular, 100, 130
    corrosion, 107
    fracture, 100
Intermediate
    annealing, 242
    shape, 279
Intermediates (glass), 669

Intermetallic compounds, 188
    in welding, 712
Internal
    centerless grinding, 524
    grinding, 523
    spinning, 367
Internet, 880
Interpolation, 819
Interstitial
    atom, 87
    solid solutions, 188
Intranet, 880
Invar, 106, 116
Inventory
    control, 863
    zero, 876
Inverse
    lever rule, 190
    segregation, 198
Inverted extrusion, 301
Investment casting, 225
    and LIGA, 795
Ion
    beam lithography, 765
    implantation, 154, 448, 765
    plating, 154
Ionic bond, 563, 656
Iron, 191
    effects of various elements, 193
Iron-base superalloys, 118
Iron-carbon system, 191
Iron-iron carbide phase diagram, 191, 194
Ironing, 318, 377
Islands of automation, 879
ISO 9000 standard, 21, 179
ISO 14000 standard, 179, 180
ISO/OSI reference model, 879
Isostatic pressing, 641
    ceramics, 666
Isothermal
    forging, 273
    transformation diagram, 236
Isotherms, 252
Isotropic, 62
    etchants, 768
    etching, 766
Izod test, 56

Jacketing, 311
Jaws
    soft,
    machinable, 837
Jet engine, 113
Jig borer,
Jiggering, 666
Jigs, 837

Job shops, 811
Joggling, 356
Joining
    ceramics, 739
    glasses, 739
    plastics, 689
    processes, 688, 689
    processes, general characteristics, 690
    thermoplastics, 737
    thermosets, 739
Joint design, 740. *See also individual joining processes*
Jolting, 220
Jominy test, 238
Junction growth, 135
Junctions, 135
Juran J. M., 178
Just-in-time production, 19, 876

Kaizen, 898
Kanban, 877
Kaolinite, 656
Kelvin model, 570
Kerf, 478, 544, 547
Kevlar, 584
Keyhole technique (welding), 699
Key-seat cutter, 474
Kickout. *See* Ejectors
Killed steel, 215
KK-3 coding system, 870
Knee, 476
Knockout. *See* Ejectors
Knoop hardness, 51
Knowledge
    based system, 880
    engineering, 880
Knuckle-joint press, 281
Knurling, 455
Kodel, 577
Kovar, 106, 779

Labor costs, 917
Labor intensive, 812
Lactic-based polymers, 580
Ladle, 215, 229
Lamellae (pearlite), 193
Lamellar tears, 707
Laminar flow, 201
Laminate strengthening, 674
Laminated glass, 674
Laminated-object manufacturing, 616
Lampblack, 675
Lancing, 344
Land, 272
    drawing die, 323
    forging die, 280

negative, 446
wear, 430
Lap joint, 689
Lapping, 530
Laps, 130, 277. *See also* Defects
Laser
    beams, 162
    forming, 373
    peening, 151
    types, 545
Laser-assisted forming, 373
Laser-beam
    cutting, 347
    machining, 544
    torch, 546
    welding, 700
Latex, 581
Lathes, 460
Lattice structure, 82
Lay, 130, 131
Layered structure, 675
Layout, 346
    functional, 867
    group, 867
    machines, 163
Lay-up, 608
Leaching, 107, 325
Lead, 121
    alkali glass, 669
    angle, 471
    glass, 461
    screw, 461
    time, 853, 908
Leaded steels, 437
Lead-free
    solders, 727
    steels, 437
Lean production (lean
        manufacturing), 22
Leathery behavior, 566
Lee, E. H., 420
Lehr, 671
Leveling rolls, 295
Lever rule, 190
Levitation melting, 208
Lévy-Mises equations, 65
Lexan, 577
Liability (product), 21
Lifecycle, 6
    cycle assessment, 899
    cycle engineering, 900
Life expectancy, 899
    testing, 782
LIGA, 794
Light section microscope, 164
Light-curing, acrylic adhesives,
    744

Light-metal castings, 214
Limiting
    dome height test, 389
    drawing ratio, 378
    drawing ratio and anisotropy, 380
Line
    representation, 857
    width (stereolithography), 613
Linear
    array, 835
    defects, 87
    elastic behavior, 28
    friction welding, 715
    interpolation, 819
    polymers, 563
    variable differential transformer,
        161
Line-graduated instruments, 160
Linewidth, 764
Lip-relief angle, 466
Liquid
    metal embrittlement, 91
    nitrogen, 453
    penetrants, 167
Liquid phase
    epitaxy, 761
    processing, 677
    sintering, 646
Liquid surface acoustical holography,
    169
Liquidus, 187
Liquified gases (for forming), 373
Lithium aluminum silicate, 658
Lithography, 762
    deep ultraviolet, 764
    electron beam, 765
    extreme ultraviolet, 764
    ion beam, 765
    X ray, 764
Load limited, 281
Loading (grinding wheels), 519
Load-stroke curve
    forging, 273
    shearing, 343
Local area network, 878
Localized necking, 338
Lock seams, 736
Logarithmic strain. *See* True strain
Long
    chain molecules, 561
    fibers, 586
Loss
    function, 179
    of quality, 178
Lost
    foam process, 224
    pattern casting, 224

wax process, 225
Lot size, 172, 811, 908
Low carbon
    stainless steels, 110
    steel, 108
Lower
    control limit, 175
    yield point, 339
Low-expansion alloys, 106
Low-melting point metals, 121
Low-pressure
    casting, 228
    chemical vapor deposition, 760
Low-stress grinding, 517
Low-temperature superconductors,
    679
Lubricant
    break down, 147
    polymer, 568
Lubricants, 28,451. *See also*
        Metalworking fluids *and*
        *individual processes*
    boundary, 147
    selection, 150
Lubrication, 145, 146
    boundary, 146
    deep drawing, 385
    drawing, 320
    extrusion, 311
    forging, 280
    geometric effects, 147
    mixed, 146
    regimes, 146
    rolling, 295
    thick film, 146
    thin film, 146
    *See also individual processes*
Lucite, 577
Lueder's bands, 295, 340

Machinability, 437
    metals, 437
    ratings, 437
    stainless steels, 439
    steels, 437
    various materials, 439, 440
Machinable
    ceramics, 660
    jaws, 837
Machine tool structures, 487
Machine tools. *See individual*
        *processes*
Machine vision, 835, 882
Machining, 404. *See also individual*
        *processes*
    allowance, 247
    centers, 482, 483, 811

Machining (*cont.*)
   dry, 452
   economics, 490
   high speed, 463
Machinist's rule, 160
Macromolecules, 561
Macroporosity, 207
Macrosegregation, 197
Magazine (tool), 483
Magnesium and alloys, 115
Magnetic properties, 107 (*see also*
    *specific materials*)
   strips, 827
Magnetic-field assisted polishing, 532
Magnetic-float polishing, 531
Magnetic-particle inspection, 168
Magnetic-pulse forming, 370
Magnetorheological fluid, 839
Magnetostriction, 107
Magnetron sputtering, 760
Main
   bearing caps manufacture, 648
   turret, 461
Major strain, 388
Malleable iron, 213
Mandrel drawing, 322
Mandrels, 322. *See also individual*
    *processes*
Manganese sulfide inclusions, 438
Manipulators, 826
   robots, 828
Mannesmann process, 300
Manual
   assembly, 840, 843
   part programming, 821
Manufacturing, 2
   attributes, 869
   automation protocol, 879
   cell, 872
   cell design, 873
   cells, 19
   characteristics of materials, 903
   communications networks, 878
   costs, 16, 915 (*see also* Costs)
   design for, 9
   engineers, 24
   history, 2
   methods, 9
   process selection, 14, 907
   properties, 13
   resource planning, 863
   systems, 853
   *See also individual processes*
Maraging, 239
Margin (drill), 466
Marking, 546
Marquenching, 243

Martempering, 243
Martensite, 237
Martensitic
   gray iron, 213
   stainless steels, 111
Mash-seam welding, 719
Mask, 762
Maskant, 533
Masking, 533, 759
Masonry drills, 726
Mass
   continuity, 200
   production, 812
Master alloys, 207
Masticated, 605
Match-plate pattern, 219
Material
   costs, 13
   handling and movement, 825
   handling system, 872
   manufacturing characteristics, 903
   properties, 13
   removal processes, cutting, 404
   selection, 11, 901
   substitution, 909
Material-requirements planning, 863
Materials. *See specific materials*
Matrix
   array, 835
   composite materials, 582
   materials, 586
Mats, 586
Maturation period, 606
Maximum,
   reduction (*see individual processes*)
   roughness height, 132
Maximum-shear-stress criterion, 61
Maxwell model, 570
Mean stress, 66
Measurement of roughness, 132
Mechanical
   alloying, 633
   assembly, 734
   behavior of materials, 26
   cleaning, 159
   fastening, 689, 734
   fibering, 93, 98, 340, 352
   plating, 152
   press, 281
   properties, 12 (*see also specific*
    materials*)
   shock, 433
Mechanization, 808
Medium temperature chemical vapor
    deposition, 447
Medium-carbon steel, 109
Melamine, 578

Melt spinning, 123, 235, 584
Melting
   furnaces, 207
   point, 104
   point,various materials, 104, 566
Melting practice, 207Mer, 561
Merchant, M. E., 419
Mesh, 269
Metal
   bonds (grinding wheels), 510
   flow pattern extrusion, 301
   foam, 124
   inert-gas welding, 692
   injection molding, 642
   powders, 631
   removal rate turning, 457
Metallic
   coatings, 149
   glasses, 123, 235
Metallization, 776
Metallizing, 153
Metallurgical
   burn, 516
   transformation, 130
Metal-matrix composites, 676
   brake rotor, 677
Metal-oxide semiconductor, 754
Metals. *See individual metals*
Metalworking fluids, 148, 281. *See*
    also individual processes*
Metalworking. *See individual*
    processes*
Metastable, 194
Method of
   attributes, 172
   variables, 172
Metrology, 160
MICLASS, 870
Microabrasive blasting, 532
Microcellular carbon foam, 676
Microchipping, 528
Microelectromechanical
   devices, 785
   systems, 15, 756, 785
Microelectronic device fabrication,
    753. *See also individual*
    processes*
Microencapsulated powders, 633
Microforming, 373
Micrograin
   carbides, 444
   graphite, 675
Microhardness, 51
Micromachining, 785
Micrometer, 160
   depth gage, 160
Micromirror, 785

Microporosity, 206. *See also* Porosity
Microreplication, 529
Microscopes, 164
Microsegregation, 197
Microstereolithography, 799
Microstructure. *See Structure and also specific materials*
Microwelds, 135
Mild
    steel, 108
    wear, 141
Mill (rolling), 296
Miller indices, 756
Milling, 468
    cutters, 468, 474
    machines, 475
Mineral oils, 148
Minimills, 297
Minimum
    bend radius, 350
    bend radius, various metals, 350
    cost (*see individual processes*)
    dimensions, various processes, 908
Minor strain, 388
Minting coins, 274
Miscible blends, 567
Mist cooling, 452
Mixed lubrication, 146
Mizuno, M., 420
Mobile robot, 830
Mode size (powder metals), 634
Modeling, 853, 857
    casting, 248
    sheet forming, 391
Modified austempering, 243
Modifiers (glass), 669
Modular, 813
    fixture design, 844
    fixtures, 837
Module, 484
Modulus, bulk, 66
Modulus of elasticity, 28
    various fibers, 582
    various materials, 28, 560, 659
Modulus of resilience, 28
Modulus of rigidity, 46
Modulus of rupture, 48
Mohr's circles, 61
Mohs hardness, 51
Moisture effects
    ceramics, 667
    polymers, 574
    sand, 218
Mold wash, 201
Molding. *See individual processes*
Molding machines, 220
Molds. *See individual processes*

Molecular weight (polymers), 562
Molecular-beam epitaxy, 761
Molybdenum
    and alloys, 120
    disulfide, 149
Monel, 116, 117
Monocode, 869
Monomer, 561
Mounted wheels (grinding), 507
Move card, 877
M-series highspeed steel, 443
Mulling machines, 218
Mullite, 657
MultiClass coding system, 870
Multi-component injection molding, 596
Multilayer
    blow molding, 598
    X ray lithography, 798
Multiphase coatings (tools), 447
Mushy zone, 196
Mylar, 577

Nanocoatings, 448
Nanocrystalline tool
    materials, 451
Nanoelectromechanical systems, 15
Nanofabrication, 15
Nanogrinding, 796
Nanomaterials, 123
Nanophase ceramics, 658, 667
Nanopowders, 633
Nanotechnology, 15
Natural
    adhesives, 730, 731
    aging, 239
    language processing, 882
    rubber, 581
    strain (*see* True strain)
Near-net shape manufacturing, 17
Necking, 29
    correction for, 63
    diffuse, 40
    localized, 338
    strain, 36
Negative
    dopants, 756
    springback, 354
Nesting, 346, 392
Net-shape, 17
    forging, 273
    forming, 630
Network polymers, 564
Neural networks, 882
Neurons, 882
Neutral point, 287, 293
Newtonian, 42, 571

Nextel, 584
Nib (in dies), 320
Nibbling, 345
Nichrome, 116
Nickel and alloys, 116
Nickel-base superalloys, 118
Nimonic, 118
Niobium (columbium), 120
Nitrides, 657
Nitriding, 152, 240
Nitrogen-ion implantation, 448
No-bake mold, 219
Noble metals, 122
Nodular iron, 213
Noise, in robust design, 897
Nominal
    size, 165
    strain, 26
    stress, 27
Nonconsumable electrode, 690
Nondestructive testing, 167
    welded joints, 711
Nonferrous alloys, 112, 212, 214
Nonhomogeneous chips, 410
Nonproductive labor, 917
Nonsynchronous system
    (assembly), 841
Nontraditional machining
    processes, 534
Nontransferred arc, 698
Nonuniform
    deformation, 262
    rational B-splines, 858
Normal
    force, cutting, 415
    rake angle, 413
Normal anisotropy, 340, 378
    various metals, 379
    vs. elastic modulus, 380
Normal distribution curve, 172
Normal segregation, 198
Normalizing, 242
Nose radius, 457
No-slip point (rolling), 283, 288
NOT GO gage, 164
Notch sensitivity, 57
Notching, 344
    cutting tools, 433
No-wear EDM, 541
*n*-type dopants, 756
Nucleating agent, 198
Nucleation, 89, 198
Nugget (weld), 717
Numerical control, 815
    programming for, 821
    robot, 830
Nylons, 577

Oblique cutting, 413
Octree, 859
Offset, 29
   forming, 356
Oils, 148, 452
Olsen test, 385
On-line inspection, 170
One-dimensional defects, 87
One-piece pattern, 219
Open riser, 199
Open-die forging, 261
Open-loop control, 818, 829
Operation (routing) sheet, 863
Opitz coding system, 870
Optical
   comparators, 162
   flat, 162
   projectors, 162
Optical properties
   ceramics, 660
   glasses, 670
   polymers, 565
Optimization (adaptive control), 823.
      See also individual processes
Optimum
   cutting speed, 430, 492
   die angle, 303, 316
   tool life, 492
Orange peel, 95, 340
Orbital
   forging, 274
   welding, 738
Organic
   acid polymers, 580
   coatings, 157
   grinding wheels, 508
   materials, 559
Orientation (polymers), 573
Orientation-dependent etching, 769
Orthogonal cutting, 406
Osprey process, 643
Out of control, 176
Out of roundness, 162
Overaging, 239
Overarm, 476
Overbending, 354
Overcontrol, 176
Overcut, 541
Overhead, 917
Overlap, 705
Overmolding, 597
Oxalate coatings, 156
Oxidation, 108, 761
   wear, 143
Oxide
   ceramics, 656
   films, 129

layers and wear, 142
   powder in tube process, 679
Oxides. See specific materials
Oxychloride bond, 510
Oxyfuel gas
   cutting, 347, 547
   spraying, 153
Oxygen embrittlement, 102

Pack rolling, 295
Packaging (chip) 778
Painting, 158
Pallet, 483, 876
Paperless design, 861
Parabolic interpolation, 820
Paraffin, 149
Parallel
   kinematic linked machines, 489
   processing, 882
Parametric model, 859
Parison, 599
Part
   checking station, 483
   families, 869
   programming, 821
Partially,
   deoxidized steel, 216
   stabilized zirconia, 657
Particle size distribution, 633
Parting, 344
   agent, 219, 281
Parting line
   casting, 247
   forging, 280
Passivation, 108, 109, 130
   layer, 777
Paste soldering, 728, 784
Patenting, 243
Pattern recognition, 831
Patternmaker's shrinkage allowance,
      247
Patterns, 219
Payload (robots), 828
Payoff reel, 288
Pearlite, 193, 236
Pearlitic gray iron, 213
Pedestal grinder, 524
Peel test (spot welds), 711
Peeling (adhesives), 732
Peen
   forming, 372
   plating, 152
Peening, 151
   laser, 151
   shot, 58
   water jet, 151
Pellets, 593

Pencil sources, 765
Penetrants (liquid), 167
Percussion welding, 721
Perforating, 344
Peripheral milling, 469
Permanent
   deformation, 83
   patterns (casting), 218
Permanent-mold casting, 227
Permeability (sand), 218
Personal area networks, 879
Pewter, 122
Phase
   boundaries, 87
   diagrams, 189
   transformation, 236
Phase-change materials, 838
Phenolics, 579
Phosphate coating, 156
Phosphor bronze, 116
Photochemical blanking, 535
Photoetching, 535
Photolithography, 762
Photomask, 762
Photopolymer, 610
Photoresist, 763, 771
Physical
   tempering, 674
   vapor deposition, 153, 446
Physical properties, 13, 103.
      See also specific materials
Physical-chemical etching, 774
Pick-and-place robot, 830
Pickling, 321
Piece-part production, 812
Piercing, 275, 358
Piezoelectric effect, 107
Pilger mill, 300
Pinch rolls, 216
Pin-grid array, 781
Pipe
   defect, 302, 309
   in ingots, 215
Pipe manufacturing, 283, 300
Piranha solutions, 771
Pitting, 107, 143
Planar
   anisotropy, 340, 378, 379
   imperfections, 87
Planarization, 776, 796
Plane strain, 44, 64, 264, 338
   compression, 44, 64
   drawing, 317
   metalworking processes, 271
   rolling, 284
Plane stress, 64, 266
Planers, 476

Planer-type milling machine, 476
Planing, 476
Planning functions, 854
Plant
    engineering, 24
    layout, 825
Plant-scale simulation, 865
Plasma, 153, 698
    spraying, 153
Plasma-arc
    cutting, 547
    welding, 698
Plasma-enhanced chemical vapor
        deposition, 760
Plaster of paris, 222
Plaster-mold casting, 222
Plastic
    anisotropy, 340, 378
    deformation, 26, 29, 65, 83, 92, 130
    forming (ceramics), 665
    modulus, 37
Plasticizer, 567, 664
Plastics, 559
    for coatings, 149
    friction, 139
    general recommendations, 576
    joining, 737
    machining, 440
    processing, 590
    processing characteristics, 591
    reinforced, 582
    structure, 560
    wear, 144
Plastisols, 600
Plastometer, 43
Plate, 282, 336
    glass, 671
Plating
    electroless, 155
    electroplating, 155
    mechanical, 152
    peen, 152
Platinum, 122
Playback robot, 830
Plexiglas, 577
Plowing in
    friction, 137
    grinding, 513
Plug gages, 164
Plugs, 164
    tube bending, 359
    tube bulging, 362
Plunge grinding, 521, 522, 524
Pneumatic gages, 164
Point
    angle, 466
    defects, 87

Point-to-point control, 819
Pointing, 320
Poisson's effect, 337
Poisson's ratio, 28
    various materials, 28, 560, 659
Polarity (arc welding), 691
Polishing, 531
Polyamides, 577
Polyblends, 567
Polycarbonates, 577
Polycodes, 870
Polycrystalline diamond, 450
Polycrystals, 89
Polyesters, 577, 579
Polyethylenes, 578
Polygonization, 93
Polyimide bond, 510
Polyimides, 578, 579
Polymer, 559, 561
    precursor, 678
Polymeric coatings (lubrication), 149
Polymerization, 561
    degree of, 562
Polymers. *See also* Plastics
    structure, 560
Polymethylmethacrylate. *See* Acrylics
Polymorphic transformation. *See*
        Allotropic transformation
Polymorphism, 83
Polypropylenes, 578
Polystyrenes, 578
    beads, 603
Polysulfones, 578
Polytetrafluoroethylene. *See*
        Fluorocarbons
Polyurethane, 581
    foam processing, 604
Polyvinyl chloride, 578
Population, 172
    mean, 174
Porcelain, 661
    enameling, 157
Porosity, 203
    bonded abrasives, 510
    casting, 206
    ceramics, 659
    ingots, 215
    powder metals, 646
    welds, 704
Portable grinders, 525
Porthole die, 311
Position measurement, 818
Positioning, 819
    accuracy, 820
Positive
    dopants, 756
    mold, 602

Postbaking, 763
Postprocess inspection, 170
Postuniform elongation, 41, 338
Potting, 603
Pouring basin, 199
Powder
    forging, 649
    injection molding, 642
    metal bearing caps, 647
    metallurgy, 630
    rolling, 643
Powders (metal), 631
Power
    hammer, 282
    in rolling, 290
    requirements (*see* Specific energy)
    spinning, 365
Powerhead production units, 813
Power-law index (polymers), 571
Prebaking, 763
Precious metals, 122
Precipitates, 238
Precipitation
    hardening, 238, 239
    hardening steels, 111
    metal powders, 633
Precision
    casting, 222
    forging, 273
    machining, 15, 464
Precoated sheet steel, 157, 341
Precursor, 584, 678
Preexpanded beads, 603
Preferred orientation, 93
Preforming. *See individual processes*
Prepregs, 606
Press
    brake, 356
    fit, 737
    working, 336
Presses
    hydraulic, 281, 312
    mechanical, 281
    screw, 281
    selection (sheet forming), 391
Pressing
    ceramics, 665
    glass, 673
    powders, 642, 649
    stem, 301
Pressure
    casting, 228
    die casting, 228
    die (*see individual processes*)
    distribution (rolling), 287
    in explosive forming, 368
    in forging, 265, 266

Pressure (*cont.*)
  pouring, 228
  sensitive adhesives, 731
  sintering, 666
Pressureless compaction, 643
Pressure-multiplying factor, forging, 272
Pressworking. *See* Sheet-metal forming processes
Prestige value, 919
Prestressed ceramics, 660
Primary
  bonds, 562
  metals, 207
  shear zone, 408
  working, 260
Primitives of solids, 858
Principal stresses, 61, 265
Printed
  circuit board, 782
  wiring board, 783
Prism-shaped machines, 489
Probes, 483
Process
  annealing, 242
  capabilities, 178, 907 (*see also individual processes*)
  optimization, 864
  planning, 861
  selection, 14, 911
  simulation, 864
  viability, 864
Producibility, 9
Product
  data exchange specification, 857
  design, 6, 894, 896
  integrity, 21
  liability, 21, 915
  life, 6
  quality, 20, 167, 898
  safety, 22
Production, 2
  card, 877
  flow analysis, 869
  methods, 9
  quantity (volume), 908
  rate, 811, 908 (*see also individual processes*)
  volume, 918
Productive labor, 917
Productivity, 808
Proeutectiod
  cementite, 193
  ferrite, 193
Profile
  measurement, 162
  milling machine, 476

Profilometer, 132
Programmable
  automation, 815
  logic controllers, 815
  numerical control, 821
Programming languages, 821
Progressive dies, 347
Projection welding, 719
Proof stressing, 708
Properties of materials. *See specific materials*
Proportional limit, 28
Prototype, 8
Protractor, 161
Pseudoplastic, 571
*p*-type dopants, 756
Puckering, 383
Pulforming, 610
Pull
  broach, 478
  system, 876
Pulsed
  arc, 693
  electrochemical machining, 538
  laser, 700
Pultrusion, 609
Pulverization, 632
Punch. *See individual processes*
Punch force
  bending, 355
  deep drawing, 381
Punching, 344
Pure
  drawing, 376
  metals (solidification), 195
  shear, 46
  stretching, 376
Push
  broach, 478
  cut, 477
  system, 876
Pyroceram, 670
Pyrolysis, 584
Pyrophoric, 116

QS 9000, 21
  standard, 180
Quality
  assurance, 20, 170
  assurance standard, 179
  circle, 178
  control, 171, 178
  engineering, 178
Quality Management and Quality Assurance Standards, 179
Quality-process certification, 180

Quantity
  of materials, 896
  of production, 908
Quartz, 658
Quench cracking, 237
Quenching, 238
  media, 238
  severity, 238

Rack shaper, 480
Radial
  drill, force (drilling), 458
  forging, 321
Radiation effects, 42
Radio frequency
  sputtering, 154
  tags, 827
  welding, 738
Radiography, 145, 169
Rail transfer, 814
Rake angle, 406, 456
Ram, 281, 300
Rammed graphite molding, 221
Random
  assembly, 840
  sampling, 172
Range (statistics), 173
Rapid prototyping, 8, 208, 610
  materials, 612
  processes, 612
Rapid solidification, 235
Rapid thermal
  annealing, 761
  oxidation, 761
  processing, 761
Rapid tooling, 617
Rare-earth magnet manufacture, 797
Razor blades, welding, 700
Reaction
  bonding, 678
  sintering, 678
Reaction-injection molding, 597
Reactive
  hot melt adhesives, 731
  ion-beam etching, 774
  plasma etching, 774
  sputtering, 154
Reagents, 533
Real area of contact, 134
Reamers, 467
Reaming, 467
Reciprocating screw extruder, 595
Reconfigurable machines, 23, 484
Recovery, 93
Recrystallization
  glass ceramics, 670

metals, 94
temperature, metals,93
Recyclability of polymers, 580
Recycling, 14
design for, 11
Red-crescent machining, 493
Red hardness. *See* Hot hardness
Redrawing, 384
Reduction
in height, 261
lithography, 763
of area, 30
powder metals, 631
Redundant work, 69, 303, 314
Refrigerated coolant, 452
Reflow soldering, 728, 784
Reflux system, 771
Refractory metals and alloys, 120
Regenerative chatter, 487
Registration, 763
Regulating wheel, 523
Reinforced plastics, 582
applications, 589
elastic modulus, 588
processing, 605
properties, 586
strength, 588
Reinforced resinoid wheels, 508
wheels (grinding), 508
Relative grain force, 513
Relaxation
stress, 573
time, 573
Release
agent (*see* Parting agent)
cloth, 607
Reliability
chip, 781
of material supply, 903
Relief
angle, 406, 457
notch in bending, 392
René, 118
Repeatability, 820
Rephosporized and resulfurized
carbon steels, 109
Replicast CS process, 224
Representative
strain, 67
stress, 67
Repressing (powder metal), 642, 649
Residual stresses, 57, 130, 340
bending, 57
glasses, 674
grinding, 517
reduction of, 58
rod drawing, 319

rolling, 293
welding, 707
Resilience, 28
modulus of, 28
Resin transfer molding, 608
Resinoid bond, 508
Resins. *See* Polymers
Resist thickness, 763
Resistance brazing, 725
Resistance welding, 716
high-frequency, 719
projection, 719
seam, 718
spot, 717
Resistive-implant welding, 738
Resistivity (electrical), 106
Resolution, 820
Resolver, 818
Response time, 823
Resulfurized
carbon steel, 109
steels, 438
Resultant force (cutting), 415
Retained austenite, 237
Reticle, 762
Retrieval system, 826
Return on
investment, 919
quality, 898
Reverse
extrusion, 301
polarity, 691
redrawing, 384
Reynolds number, 201
RF sputtering, 760
Rheocasting, 198, 233, 677
Rib, 280
Rifling, 322
Rigid PVC, 578
Rimmed steel, 216
Ring
compression test, 137
gages, 164
rolling, 297, 298
topology, 878
Risers, 199, 246
Rivets, 735
R.M.S. Titanic, 103
Robots, 18
applications, 831, 832
assembly, 840
selection, 832
Robotic
assembly, 19, 844
deburring, 533
Robust design, 897
Robustness, 897, 910

Rocking-die forging, 274
Rockwell hardness, 50
superficial, 50
Rod drawing, 312
Roll
bending, 357
bonding, 712
compaction, 643
deflections, 291
flattening, 291, 292
force, 289
forging, 276
forming, 358
grinder, 522
spot welding, 719
stands, 296
torque, 289
welding, 712
Roller
burnishing, 152
leveling, 295
Rolling
ceramic (slip), 664
chatter, 294
glass, 671
lubrication, 295
metal powders, 643
metals, 282
mills, 296
power, 290
strain rate, 291
Roll-pass design, 297
Rolls, 282
Root-mean-square average, 132
Rotary
bending, 354
encoder, 818
file, 480
forging, 274
indexing, 841
knives, 345
swaging, 321
table grinder, 522
table milling machine, 476
transfer, 814
tube piercing, 300
ultrasonic machining, 528
Rotational molding, 598
Roughing cuts, 455, 458
Roughness. *See* Surface finish *and*
Surface roughness
Rounding roll, 298
Roundness, 162
Routing sheet, 862
Roving, 586
Rubber
bond, 510

Rubber (*cont.*)
   forming, 363
   modified polymers, 567
Rubbers. *See* Elastomers
Rubbery behavior, 566
Rule of ten, 918
Runner choke, 200
Runnerless mold, 596
Runners, 199
Running-in, 140
Rupture, 56
   modulus of, 48

Saddle, 476
Safety. *See also individual processes*
   glass (*see* Laminated glass)
   in machining magnesium, 439
   in powder metallurgy, 636
Sag point, 600
Sagging, 673
Sample size, 172
Sampling, 172
Sand
   casting, 218, 220
   molding machines, 220
   molds, 218
   slingers, 220
Sandpaper, 529
Sands (casting), 218
Sandwich molding, 596
Saponification, 159
Saran, 578
Sata, T., 420
Sawing, 478
   EDM, 543
Saws, 478
Scanning,
   acoustical holography, 169
   electron microscope, 164
Scarfing, 293
Scheduling, 876
Scleroscope, 51
Scoring, 393
Scrap, 904
   shearing, 346
Screening, 633
   in soldering, 728
Screw,
   dislocation, 88
   extruder, 590
   machines, 464
   press, 281
   threads, cutting, 464
Scribe line, 778
Scuffing, 141
Seam
   defect, 130, 278, 318

welding, 718
Seaming, 358, 736
Seamless tube production, 300
Search units (probes), 168
Season cracking, 102
Secondary
   bonds, 563
   shear zone, 408
   tensile stresses, 278, 309
   working, 260
Second-phase particles, 188
Sedimentation, 633
Seeded gel abrasives, 506
Segmented
   chips, 410
   dies, 362
Segregation, 197
Seizure, 141, 307
Sejournet process, 311
Selection criteria,
   manufacturing processes, 14, 907,
     911
   materials, 12, 901
   machine tools, 485
   *See also specific materials and
     processes*
Selective
   assembly, 840
   automation, 813
   laser sintering, 615
   leaching, 107
   oxidation, 762
Selectivity, 766, 769
Self-excited vibration, 294, 486, 487
Self-guided vehicle, 826
Self-interstitial, 87
Self-lubricating materials, 139
Self-organizing artificial neural
   network, 883
Self-tapping, threaded fasteners, 735
Semicentrifugal casting, 232
Semiconductors, 106, 756
Semi-killed steel, 216
Semipermanent
   composite mold, 221
   mold, 227
Semipositive mold, 602
Semisolid metalworking (forming),
   233
Semisynthetic
   cutting fluids, 452
   solutions, 149
Sendzimir mill, 296
Sensor
   fusion, 836
   selection, 836
   validation, 836

Sensors, 170, 833
Sensory robot, 831
Serrated chips, 410
Service, 10
   design for, 844
   life, 14
Servomotor, 817
Setback, 349
Severe wear, 142
Severity of quench, 238
Shaffer, B. W., 420
Shank cutter, 475
Shape factor
   extrusion, 301, 311
   powder metals, 634
Shapeindex (powder metals), 634
Shapememory alloys, 123
Shaperolling, 297
Shapers, 477
Shaping (grinding wheels), 519
Shared manufacturing, 19
Shaving, 346, 414
Shaw process, 222
Shear
   angle, 406
   angle, relationships, 419
   die, 311
   factor, 136
   forming, 327
   modulus, 46
   plane, 406
   spinning, 365
Shear strain, 27, 46
   cutting, 407
   rate, 408
Shear strength
   adhesives, 732
   single crystals, 85
   stress, 46
   stress critical, 84
Shear yield stress, 61
Shearing, 341
Sheet, 282, 336
   extrusion (polymers), 593
   glass, 671
   plastic, 593
Sheet metal, 336
   characteristics, 336
   formability, 385
   forming, 336, 337 (*see also
     individual processes*)
Sheet-molding compound, 606
Shell
   in casting, 195
   mills, 474
   mold-casting, 221
Shellac bond, 510

Shells (expert system), 881
Shielded metal arc welding, 690
Shielding gas. *See individual processes*
Shock-resisting steels, 112
Short fibers, 586
Short-circuiting arc, 693
Shot
    blasting, 532
    chamber, 229
    peening, 58, 151, 372
Shrink
    fit, 737
    flanging, 357
Shrinkage
    allowance, 247
    cavity, 207
    in casting, 204, 247
    in ceramics, 667
    in sintering, 646
    various metals, 205
Sialon, 450, 658
Side
    cutting edge angle, 414, 456
    rake angle, 414, 457
    relief angle, 414, 456
Silane, 586
Silica, 658
Silicate bond, 510
Silicate-bonded sand process, 221
Silicates, 658
Silicon
    carbide, 506, 657
    dioxide, 757
    dioxide etching, 770
    etching, 771
    fusion bonding and deep reactive
        ion etching, 792
    micromachining by single step
        plasma etching, 791
    nitride, 658
    nitride etching, 771
    nitride-base ceramics, 450
Silicone adhesive systems, 731
Silicones, 579, 581
Silver, 122
Solder, 726
Simple shear, 46
Simulation, computer, 853, 864
Simultaneous engineering. *See*
        Concurrent engineering
Sine bar, 161
Single
    crystal silicon reactive etching and
        metallization, 791
    machine cell, 872
    minute exchange of dies, 391

Single-crystal
    casting, 233
    growing, 234
    turbine blades, 234
Sinter forging, 649
Sintered carbides. *See* Carbides
Sintering
    ceramics, 667
    powder metals, 644
    spark, 646
    temperature various powders, 644
Six sigma, 174
Size effect, 102
    grinding, 514
Sizing, 274, 649
    pass, 319
Skeleton, 860
Skelp, 283
Skew rolling, 276
Skill required. *See individual processes*
Skin
    in casting, 195
    pass, 295
    rolling, 340
Skin-dried mold, 219
Skiving, 414, 594
Slab, 283, 295
    method of analysis, 264, 284
    milling, 469
Slag, 201, 207, 691
    inclusions, 704
Sliding wear, 141
Slip, 84
    bands, 85
    casting, 664
    ceramics, 664
    line analysis, 269
    planes, 84
    system, 85
Slip-line field, 270
Slitting, 344, 345
    saws, 474
Slotting, 474
Slotters, 477
Slurry
    abrasive, 527, 528
    in casting, 222, 224, 226
    infiltration, 678
Slush
    casting, 203, 228
    molding, 600
Small batch production, 812
Smart
    databases, 901
    end effector, 836
    materials, 23
    sensors, 836

Smearing, 141
S–N curves, 54
Snagging, 525
    in sawing, 479
Snap
    fit, 737
    gages, 164
Snap-in fasteners, 737
Soaking pits, 215
Soaps, 149
Soda-lime glass, 669
Sodium silicate process, 221
Soft
    automation, 815
    forming (ceramics), 665
    glass, 669
Soft-acting wheel, 520
Softening, strain, 45, 384
Soils, 159
Solderability, 728
Soldering, 689, 727
    methods, 728
Solders, 727
Solgel, 678
Solid
    freeform fabrication, 799
    lubricants, 149
    model, 857, 858
    rocket casing manufacture, 325
    solutions, 187, 190
Solid-base curing, 616
Solid-ground curing, 616,
Solid-metal embrittlement, 91
Solid-phase
    forming (plastics), 604
    processing, 677
Solid-state
    bonding, 645
    welding, 688
Solidification, 187
    alloys, 188
    contraction (various metals), 205
    front, 197
    pattern, 196, 201
    pure metals, 187
    time, 203
    weld metal, 702
Solidus, 187
Solubility (gases), 206
Soluble oils. *See* Emulsions
Solute, 187
    atoms, 187
Solution treatment, 239
Solvent, 187
    bonding, 738
    cleaning, 159
    crazing, 574

Sonotrode, 527, 713
Soot, 567
Space shuttle, 325
Spade drill, 466
Spalling, 660
Spark
    hardening, 153
    sintering, 646
Spark erosion machining, 540
Sparks (grinding), 516
Specific
    gravity, 103
    heat, 105
    stiffness, 103
    strength, 103
    various materials, 104
    volume, 566
Specific energy, 29, 68, 513
    cutting, 420, 421
    grinding, 514
Spectra fiber, 584
Spectroscopy, 145
Speed cracking, 305, 309
Speeds. *See individual processes*
Spheroidite, 237
Spider die, 311
Spin
    forging, 365
    welding, 738
Spinel, 657
Spinnability, 365
Spinnerette, 584
Spinning, 364, 763
    conventional, 364
    glass, 673
    power, 365
    shear, 365
    tube, 366
Splat cooling, 123, 235
Splatter, 130
Split female mold, 598
Split pattern, 219
Spot weld (tests), 711
Spray
    atomization and deposition, 677
    deposition, 643
    thermal, 153
    transfer, 692
Spray-up, 608
Spread, 172
Spreading (rolling), 292
Springback, 340, 352
    compensation for, 354
    design, 200
    factor, 353
    negative, 354
Sprue, 199

Sputter etching, 773
Sputtering, 154, 760
Square
    die, 311
    turret, 461
Squeeze casting, 232
Staining, by lubricants, 147, 150
Stainless steels, 109
Stamping, 393. *See also* Sheet-metal
        forming
Stand grinder, 524
Stand-alone machines, 811
Standard
    deviation, 173
    for the exchange of product model
        data, 857
Standoff, 369
Stands (rolling), 296
Stapling, 736
Star topology, 878
Starch-based polymers, 580
Starter bar, 216
State of stress, 64
Static fatigue, 660
Statistical
    process control, 21, 174
    quality control, 172
Statistics, 171
Steckel rolling, 296
Steel-rule die, 345
Steels, 108, 109. *See also specific steels*
    properties, 110
Stellite, 443
Stenciling, 728
Stents manufacturing, 551
Step drill, 466
Step-growth polymerization, 562
Step-reaction polymerization, 562
Step-type gage, 164
Stepper, 763
Stereolithography, 610
    cure depth, 612
    file format, 857
    line width, 613
Sterling silver, 122
Stewart platform, 489
Stick welding, 690
Sticking, 136
    condition, 267
Stiction in MEMS, 789
Stiffness (machine tools), 487
Stiffness-to-weight ratio, 103
Stinger, 344
Stitching, 736
Stockless production, 877
Stones (honing), 529
Stop-off, 372, 712

Storage and retrieval systems, 826
Stored energy, 71, 94
Straddle milling, 474
Straight polarity, 691
Straightening
    rods, 319
    sheet, 295
Straightness, 162
Strain
    aging, 99
    aging, accelerated, 99
    effective, 67
    engineering, 26
    logarithmic, 31
    natural, 31
    nominal, 26
    ratio, 378
    shear, 27, 46
    true, 31
    volume, 66
Strain hardening, 32, 88
    exponent, 32
    various materials, 33
Strain rate, 38
    drawing, 316
    extrusion, 305
    metalworking, 41
    rolling, 291
Strain softening, 45, 384
Strain-rate sensitivity, 40
    exponent, 40
    exponent, various metals, 41
    polymers, 570
Strand casting, 216
Strength. *See also specific materials*
    coefficient, 33, 40
    coefficient, various metals, 33, 41
    reinforced plastics, 588
    single crystals, 85
Strength-to-weight ratio, 103
Strengthening (glass), 674
Stress
    cracking, 58, 102
    effective, 67
    engineering, 27
    flow, 34, 37, 69
    mean, 66
    relaxation, 56, 58, 573
    relief annealing, 242
    relieving, 242
    relieving welds, 708
    shear, 46
    true, 31
    whitening, 574
Stress-corrosion cracking, 58, 102,
        107, 340

Stress-relief annealing, 58
    welds, 708
Stress-strain curves, 32
    construction, 34
    types, 37
    various metals, 33
    various polymers, 569, 570
Stresses, thermal, 105
Stretch
    bending, 355
    flanging, 357
    forming, 360
Stretcher strains, 295, 340
Striations, 101
Stringers, 352
Strip
    casting, 217
    drawing, 318
    rolling, 284
Stripper, 375
Stroke limited, 281
Structural
    adhesives, 730
    defects (rolling), 293
    foam molding, 597
    grade steels, 109
Structure. See also specific materials
    bonded abrasives, 510
    ceramics, 656
    glass ceramics, 670
    glasses, 668
    insensitive, 88
    metals, 82
    polymers, 560
    sensitive, 88
Structure-property relationships
        (casting), 197
Stud
    arc welding, 720
    welding, 720
Stuffing box, 321
Stylus, 132
Styrofoam, 603
Subassemblies, 839
Subcritical annealing, 242
Submerged arc welding, 692
Substitution of materials, 909
Substitutional
    atom, 87
    solid solutions, 188
Substrate, 129
Superabrasives, 505
Superalloys, 117, 118, 215
Superconductivity, 106
Superconductors, processing, 679
Supercooling, 668
Superficial hardness, 50

Superfinishing, 530
Superheat, 202
Superplastic forming, 371, 722
Superplasticity, 31, 40
Surface
    cleaning, 158
    cracking, 277, 305, 309
    damage (welding), 707
    defects, 130, 293 (see also
        individual processes)
    energy, 87
    fatigue wear, 143
    grinder, 521
    grinding, 511,521
    hardening, 240
    integrity, 130, 434
    measurement, 132
    micromachining, 787
    model, 858
    mount, 780
    preparation (adhesive bonding),
        732
    profilometer, 132
    rolling, 58, 152
    shear stress, 267
    structure, 129
    tension, 201
    texture, 130
    texturing, 153
    treatments, 151
Surface finish, 907 (see also Surface
        roughness)
    and cost, 550
    cutting, 434
Surface roughness, 131
    and cost, 550
    and lubrication, 147
    in machining, 435, 436
    in practice, 133
    various machining processes, 536
Surfaces, 128
Sustainable manufacturing, 900
Swaging, 321
Swept volumes, 858
Swift test, 385
Swing
    lathe, 461
    forging, 274
Swing-frame grinder, 525
Synchronous system (assembly),
        841
Synthetic
    cutting fluids, 452
    diamond, 676
    rubbers, 581
    solutions, 149

Synthetic organic
    adhesives, 730, 731
    polymers, 560
Synthetics, 526

T joint, 689
T slot, 474
    cutters, 474
T-series high-speed steel, 443
Tactile sensing, 834
Taguchi, G., 178
Taguchi loss function, 179
Tailor-welded blanks, 347
Tailpipe (defect), 309
Tailstock, 461
Tandem
    drawing, 320
    rolling, 296
Tantalum and alloys, 121
Tap, 468
Tapping, 468
Taylor F. W., 428
Teach pendant, 830
Tearing
    casting, 197
    deep drawing, 381
    welding, 707
Technical and office protocol, 879
Teemed, 215
Teflon, 145,577
Telescoping gages, 161
Temper
    designations, 114
    embrittlement, 92, 243
    rolling, 295, 340
Temperature
    cutting, 424
    effects, 38
    forging, 281
    gradients, 58
    See also individual processes
Temperature rise, 71
    friction, 138
    grinding, 515
    plastic deformation, 71
Tempered
    glass, 674
    martensite, 237
Tempering
    in grinding, 516
    of glass, 674
    of metals, 237, 243
Tensile strength. See Ultimate tensile
        strength
Tension test, 27
    welds, 710
Tension-shear test (welds), 710

Tensions, in rolling, 288
Terpolymers, 564
Terne, 157
Testing. *See also individual methods*
    destructive, 169
    integrated circuits, 778
    nondestructive, 167
Texture, 92
    surface, 130
Theoretical
    tensile strength, 86
    shear strength, 84
Thermal
    aging, 568
    barrier, 281
    camber, 292
    conductivity, 105
    conductivity, various materials, 104
    cracks, 660
    energy deburring, 533
    expansion, 105
    expansion, anisotropy, 106
    expansion, various materials, 104
    fatigue, 106, 143, 433
    ink-jet printer manufacture, 792
    inspection, 169
    oxidation, 761
    processing, 761
    properties (*see specific materials*)
    shock, 106
    spraying, 153
    stresses, 105
    tempering, 674
Thermally
    assisted machining, 440
    conducting adhesives, 731
Thermocompression, 778
Thermoforming, 600
Thermographic inspection, 169
Thermomechanical processing, 243
Thermoplastics, 568
    applications, 575
Thermosets, 575, 564
    applications, 578
    joining, 739
Thermosetting plastics. *See*
    Thermosets
Thermosonic stitching, 778, 779
Thick-film lubrication, 146
Thick-molding compound, 607
Thin-film lubrication, 146
Third-octave chatter, 294
Thixoforming (thixocasting), 233
Thixotropic behavior, 233
Thread
    cutting, 464

gages, 164
grinding, 523
measurement, 162
rolling, 298, 464
Threaded fasteners, 735
Threading, 455, 464
Three-body wear, 143
Three-high mill, 296
Three-plate mold, 596
Three-point bending, 48, 710
Through hardening, 240
Through-feed grinding, 521, 523
Thrust force, 415, 458
    cutting, 415, 415, 458
    drilling, 467
Tie line, 190
Time-temperature-transformation
    diagram, 236
Tin, 122
    cry, 85
    plate, 122
Titanic. *See* R.M.S. Titanic
Titanium
    aluminide intermetallics, 120
    aluminum nitride, 448
    and alloys, 118
    carbide, 444, 447, 657
    carbonitride, 447
    nitride, 447, 658
Token
    bus, 878
    ring, 878
Tolerance and surface roughness, 166,
    909
Tolerances, 165, 907. *See also*
    Dimensional tolerances *and*
    individual processes*
    in casting, 247
    various machining processes, 536
Tool
    checking station, 483
    condition monitoring, 433
    failure, 426
    geometry, 456
    holders, milling, 475
    materials, 440
    post, 461
    steels, 111, 112
    storage, 483
Tool wear, 426
    measurement, 433
Tool-exchange arm, 483
Tool-life curves, 428
Tool-post grinder, 52
Tooling costs, 915
Toolmaker's microscope, 164, 433

Toolroom lathe, 461
Tools. *See specific materials and*
    *individual processes*
Torlon, 578
Torch brazing, 725
Torque in rolling, 289
Torsion test, 46
Torsional chatter, 294
Total
    elongation, 30
    indicator reading, 162
    production quantity, 811
    productive mainentance, 815
Total quality
    control, 178
    management, 20, 21, 171
Touch probes, 483
Toughness, 34
Tracer lathe, 461
Tractrix, 384
Transducers, 434, 713, 833
Transfer
    dies, 347
    lines, 814
    machines, 813
    mechanisms, 814, 840
    molding, 602
    systems (assembly), 841
Transfer/injection molding, 608
Transferred arc, 698
Transformation
    diagrams, 236
    phase, 236
    toughened zirconia, 657
Transgranular fracture, 100
Transistor, 754, 755
Transition temperature, 98
Transits, 162
Transverse rupture strength, 48
Traverse grinding, 521
Tree (investment casting), 226
Trepanning, 466
Tresca
    criterion, 61
    friction, 136
Triaxial stresses, 60
Tribology, 128, 134
Trim loss, 346
Trimming, 279
True
    centrifugal casting, 230
    strain, 31
    strain in metalworking, 39
    strain rate, 38, 43
    stress, 31
    stress-true strain curves, 32, 33

Truing, 519
Tube
    bending, 359
    drawing, 318
    glass, 671
    hydroforming, 363
    production, 300
    rolling, 300
    sinking, 318
    spinning, 366
    swaging, 322
Tundish, 216
Tungsten
    and alloys, 120
    carbide, 444, 657
    carbid production, 650
    electrode, 697
    inert-gas welding, 697
Turbine blades, 81, 233
    casting, 233
    single crystal, 234
Turbulence (in casting), 201
Turk's head, 320
Turn broaching, 478
Turning, 454
    centers, 482
Turret lathe, 461
Twinning, 85
    plane, 85
Twist
    drill, 466
    test (spot welds), 711
Two-body wear, 143
Two-dimensional imperfections, 87
Two-high mill, 296
Two-phase
    processing, 677
    systems, 188
Two-plate mold, 596
Tygon, 578
Type metals, 121
Udimet, 118

Ultimate tensile strength, 29
    ideal, 87
    various materials, 28, 560
    vs. fatigue strength, 55
    vs. hardness, 53
Ultrahigh speed machining, 463
Ultraprecision
    machining, 464
    manufacturing, 13
Ultrasonic
    cleaning, 159
    inspection, 168
    machining, 527
    rotary, 528

soldering, 728
stitching, 778
welding, 713
Ultrasonics vibration
    friction, 139
    drawing, 321
Ultralarge-scale integration, 756
Unconventional machining processes, 534
Undeformed chip
    length, 512
    thickness, 470, 471, 512
Undercutting, 705
Underfilling, 705
Underwater spark forming, 370
Unfused alumina, 506
Uniform
    deformation, 262
    elongation, 30, 337
    strain, 30
Unit
    cell, 82
    costs, 249
Universal
    grinder, 522
    machining center, 484
    tool and cutter grinder, 524
Universe (statistics), 172
Untended factories, 883
Up
    grinding, 511
    milling, 469
Upper
    bound technique, 270
    control limit, 175
    yield point, 339
Upset welding, 719
Upsetters. See Headers
Upsetting, 261
    test, 278
Urea, 578
Urethane adhesive systems, 731
Use value, 919

V dies, 356
V process, 220
Vacancy, 87
Vacuum
    casting, 223
    evaporation, 154
    forming, 600
    molding, 220
Vacuum-bag molding, 607
Value, 2, 919
    engineering, 919
    management, 919
    van der Waals bonds, 563

Vapor
    blanket, 238
    condensation, 633
    degreasing, 159
Vapordeposition, 153
    chemical, 154
    physical, 153
Vaporphase
    epitaxy, 761
    transport, 646
Variability, 171
Variable
    blankholder force, 383
    sequence robot, 830
Variant CAPP system, 862
Vent, 199
Vernier calipers, 160
Vertical
    etching, 769
    flaskless molding, 220
Very high speed machining, 463
Very large-scale integration, 756
Via, 776
Vibration, 487. See also Chatter
    in rolling, 294
    machining, 486
Vibratory
    finishing, 532
    welding, 738
Vickers hardness, 51
Vinyls. See Polyvinyl chloride
Viscoelastic, 570
    modulus, 570, 572
Viscosity
    in casting, 201
    index (in casting), 201
    vs. strain rate, various polymers, 571
    vs. temperature, various polymers, 571
Viscous behavior, 566
Visual sensing, 835
Vitreous. See Glasses
Vitrified bond, 508
Voice recognition, 882
Voids, 87. See also Porosity
Voigt model, 570
Volume strain, 66
    von Mises criterion, 62
Voxels, 860
Vulcanization, 564, 581

Wafer cleaning, 771
Wafers, 758
Warm working, 95
Warping, 58

Waspaloy, 118
Water absorption (polymers), 574
Waterjet
    abrasive,548
    cutting, 347, 547
    peening, 151
Water-base emulsions, 526
Water-soluble oils. *See* Emulsions
Waterbreak test, 159
Waterglass, 221
Wave soldering, 729, 784
Waviness, 130
    height, 131
    width, 131
Wavy edges rolling, 293
Wax patterns, 225
Waxes, 149
Ways, 461
Wear, 140
    abrasive, 142
    adhesive, 141
    attritious, 517
    ceramics, 144
    chemical, 143
    coefficient, 141
    corrosive, 143
    cutting tools, 426
    electrodes EDM, 542
    erosive, 143
    fatigue, 143
    flat, 514
    flat, grinding, 515
    fragment, 141
    grinding wheels, 517
    impact, 144
    land (allowable), 430
    measurement, 145
    parts, 140
    plastics, 144
    plates, 140
Web, 280
Weld
    contaminants, 704
    metal, 702
    nugget, 717

overlay, 153
porosity, 704
profile, 704
quality, 704
Weldability, 708
    various metals, 709
Welded joint, 704
    testing, 709
Welding
    electrodes, 695
    gun, 692
    process selection, 711
    processes, 689
    razor blades, 700
Welding-chamber method (extrusion
    dies), 311
Well (casting), 199
Wet
    drawing, 321
    etching, 766
    forming, 665
    oxidation, 762
    pressing (ceramics), 666
Wetting agent, 664
Wheel depth of cut, 511
Whisker-reinforced tool materials, 451
Whiskers, 102, 451, 585
White
    cast iron, 213
    ceramics, 449, 657
    metals, 122
Whitney, E., 171
Wide area network, 878
Wiping die, 356
Wire
    arc, 153
    bonding, 778
    brushing, 529
    drawing, 312
    EDM, 543
    rod, 295
    saw, 479
Wireframe, 857
Wireless local area networks, 879
Wiring board, 783

Work
    envelope, 483, 830
    hardening, 88 (*see also* Strain
        hardening) of deformation, 68, 63
    roll, 296
Workability. *See individual processes*
Work-hardened layer, 129
Workholding devices. *See individual
    processes*
Workstations, 856
Worktable. *See individual processes*
World-class manufacturing, 22
Worms, 340
Wrap-around bend test (welds), 710
Wrinkling, 340, 377
Wrought structure, 282, 295

X ray lithography, 764
X rays, 169
Xenon-ion implantation, 448

Yarn, 586, 605
Yield
    criteria, 60
    criteria, experimental verification,
        66
    point, 28
    stress, 28
    various materials, 28
Yield, chip, 781
Yield-point elongation, 295, 339
Young's modulus. *See* Modulus of
    elasticity

Z mill, 296
Zero inventory, 877
Zero-base waste, 22
Zinc, 121
    casting alloys, 214
    phosphate coating, 156
Zipper cracks, 293
Zirconia, 657
Zirconium, 121
    nitride, 448
    oxide, 451

## CONVERSION FACTORS FOR SI UNITS

| PROPERTY | TO CONVERT FROM | TO | MULTIPLY BY |
|---|---|---|---|
| Acceleration | $ft/s^2$ | m/s | $23.048 \times 10^{-1}$ |
| Angle | degrees | rad | $1.745 \times 10^{-2}$ |
| | minute | rad | $2.909 \times 10^{-4}$ |
| | second | rad | $4.848 \times 10^{-6}$ |
| Area | $in^2$ | $m^2$ | $6.452 \times 10^{-3}$ |
| | $ft^2$ | $m^2$ | $9.290 \times 10^{-2}$ |
| | $in^2$ | $mm^2$ | $6.452 \times 10^2$ |
| | $ft^2$ | $mm^2$ | $9.920 \times 10^4$ |
| Density | $lb/in^3$ | $kg/m^3$ | $2.768 \times 10^4$ |
| Energy | $ft \cdot b$ | J | 1.356 |
| | Btu | J | $1.054 \times 10^3$ |
| | calorie | J | 4.184 |
| | $watt \cdot h$ | J | $3.600 \times 10^3$ |
| Force | kgf | N | 9.807 |
| | lb | N | 4.448 |
| Fracture toughness | $ksi \cdot in^{1/2}$ | $Mn \cdot m^{-3/2}$ | 1.099 |
| Length | in | m | $2.540 \times 10^{-2}$ |
| | ft | m | $3.048 \times 10^{-1}$ |
| Mass | lb | kg | $4.536 \times 10^{-1}$ |
| | tonne (metric) | kg | $1.000 \times 10^{-3}$ |
| | ton (short) | kg | $9.072 \times 10^2$ |
| Power | hp | W | $7.457 \times 10^2$ |
| | Btu/min | W | $1.757 \times 10$ |
| | $ft \cdot lb/min$ | W | $2.260 \times 10^{-2}$ |
| Pressure, stress | $lb/in^2$ | Pa | $6.895 \times 10^3$ |
| | bar | Pa | $1.000 \times 10^5$ |
| | atmosphere | Pa | $1.013 \times 10^5$ |
| Thermal | $Btu/h \cdot ft \cdot °F$ | $W/m \cdot K$ | 1.730 |
| | $cal/s \cdot fcm \cdot °C$ | $W/m \cdot K$ | $4.184 \times 10^2$ |
| | $Btu/lb \cdot °F$ | $J/kg \cdot K$ | $4.184 \times 10^3$ |
| Torque | in | $N \cdot m$ | $1.130 \times 10^{-1}$ |
| | $ft \cdot lb$ | $N \cdot m$ | 1.356 |
| Velocity | ft/min | m/s | $5.080 \times 10^{-3}$ |
| | rpm | rad/s | $1.047 \times 10^{-1}$ |
| Volume | $in^3$ | $m^3$ | $1.639 \times 10^{-5}$ |
| | $ft^3$ | $m^3$ | $2.832 \times 10^{-2}$ |
| | $in^3$ | $mm^3$ | $1.639 \times 10^4$ |
| | $ft^3$ | $mm^3$ | $2.832 \times 10^7$ |
| | gallon (U.S.) | liter | 3.785 |

## CONVERSION FACTORS FOR SI UNITS (continued)

| | TO CONVERT FROM | TO | MULTIPLY BY |
|---|---|---|---|
| OTHER CONVERSION | lb/in$^2$ | kgf/mm$^2$ | $7.030 \times 10^{-4}$ |
| | lb/in$^2$ | g/cm$^3$ | $2.768 \times 10$ |
| | kg/mm$^2$ | MPa | 9.807 |
| | Btu/h · ft · °F | cal/s · cm · °C | $4.134 \times 10^{-4}$ |
| | ounce | g | $2.835 \times 10$ |
| | torr | Pa | $1.33 \times 10^2$ |
| | poise | Pa · s | $10^{-1}$ |
| | microinch | micron | $3.937 \times 10$ |
| | erg | J | $10^{-7}$ |
| | dyne | N | $10^{-5}$ |
| | micron | m | $10^{-6}$ |
| | angstrom | m | $10^{-10}$ |
| | ton (short) | lb | $2.000 \times 10^3$ |
| | ton (long) | lb | $2.240 \times 10^3$ |

TEMPERATURE CONVERSIONS

°C = 5/9 (°F − 32)

°F = 9/5°C + 32

K = °C − 273

°R = °F + 460

## MULTIPLICATION FACTORS

| PREFIX | SYMBOL | | PREFIX | SYMBOL | |
|---|---|---|---|---|---|
| tera | T | $10^{12}$ | centi | c | $10^{-2}$ |
| giga | G | $10^9$ | milli | m | $10^{-3}$ |
| mega | M | $10^6$ | micro | $\mu$ | $10^{-6}$ |
| kilo | k | $10^3$ | nano | n | $10^{-9}$ |
| deci | d | $10^{-1}$ | pico | p | $10^{-12}$ |